DIMENSION	METRIC	METRIC/ENGLISH
Power, heat transfer rate	1 W = 1 J/s 1 kW = 1000 W = 1.341 hp 1 hp‡ = 745.7 W	1 kW = 3412.14 Btu/h = 737.56 lbf · ft/s 1 hp = 550 lbf · ft/s = 0.7068 Btu/s = 42.41 Btu/min = 2544.5 Btu/h = 0.74570 kW 1 Btu/h = 1.055056 kJ/h 1 ton of refrigeration = 200 Btu/min
Pressure	1 Pa = 1 N/m^2 1 kPa = 10^3 Pa = 10^{-3} MPa 1 atm = 101.325 kPa = 1.01325 bars = 760 mmHg at 0°C = 1.03323 kgf/cm^2 1mmHg = 0.1333 kPa	1 Pa = 1.4504 × 10^{-4} psia = 0.020886 lbf/ft^2 1 psia = 144 lbf/ft^2 = 6.894757 kPa 1 atm = 14.696 psia = 29.92 inHg at 30°F 1 inHg = 3.387 kPa
Specific heat	1 kJ/kg · °C = 1 kJ/kg · K = 1 J/g · °C	1 Btu/lbm · °F = 4.1868 kJ/kg · °C 1 Btu/lbmol · R = 4.1868 kJ/kmol · K 1 kJ/kg · °C = 0.23885 Btu/lbm · °F = 0.23885 Btu/lbm · R
Specific volume	1 m^3/kg = 1000 L/kg = 1000 cm^3/g	1 m^3/kg = 16.02 ft^3/lbm 1 ft^3/lbm = 0.062428 m^3/kg
Temperature	$T(\mathrm{K}) = T(^\circ\mathrm{C}) + 273.15$ $\Delta T(\mathrm{K}) = \Delta T(^\circ\mathrm{C})$	$T(\mathrm{R}) = T(^\circ\mathrm{F}) + 459.67 = 1.8T(\mathrm{K})$ $T(^\circ\mathrm{F}) = 1.8\,T(^\circ\mathrm{C}) + 32$ $\Delta T(^\circ\mathrm{F}) = \Delta T(\mathrm{R}) = 1.8^* \Delta T(\mathrm{K})$
Thermal conductivity	1 W/m · °C = 1 W/m · K	1 W/m · °C = 0.57782 Btu/h · ft · °F
Thermal resistance	1°C/W = 1 K/W	1 K/W = 0.52750°F/(h · Btu)
Velocity	1 m/s = 3.60 km/h	1 m/s = 3.2808 ft/s = 2.237 mi/h 1 mi/h = 1.609 km/h
Volume	1 m^3 = 1000 L = 10^6 cm^3 (cc)	1 m^3 = 6.1024 × 10^4 in^3 = 35.315 ft^3 = 264.17 gal (U.S.) 1 U.S. gallon = 231 in^3 = 3.7854 L

*Exact conversion factor between metric and English units.

‡Mechanical horsepower. The electrical horsepower is taken to be exactly 746 W.

Thermodynamics: An Engineering Approach

McGRAW-HILL SERIES IN MECHANICAL ENGINEERING

CONSULTING EDITORS
Jack P. Holman, *Southern Methodist University*
John R. Lloyd, *Michigan State University*

Anderson:	*Computational Fluid Dynamics: The Basics with Applications*
Anderson:	*Modern Compressible Flow: With Historical Perspective*
Arora:	*Introduction to Optimum Design*
Borman/Ragland:	*Combustion Engineering*
Bray and Stanley:	*Nondestructive Evaluation: A Tool for Design, Manufacturing, and Service*
Burton:	*Introduction to Dynamic Systems Analysis*
Culp:	*Principles of Energy Conversion*
Dally:	*Packaging of Electronic Systems: A Mechanical Engineering Approach*
Dieter:	*Engineering Design: A Materials & Processing Approach*
Doebelin:	*Engineering Experimentation: Planning, Execution, Reporting*
Driels:	*Linear Control Systems Engineering*
Eckert and Drake:	*Analysis of Heat and Mass Transfer*
Edwards and McKee:	*Fundamentals of Mechanical Component Design*
Gebhart:	*Heat Conduction and Mass Diffusion*
Gibson:	*Principles of Composite Material Mechanics*
Hamrock:	*Fundamentals of Fluid Film Lubrication*
Heywood:	*Internal Combustion Engine Fundamentals*
Hinze:	*Turbulence*
Holman:	*Experimental Methods for Engineers*
Howell and Buckius:	*Fundamentals of Engineering Thermodynamics*
Hutton:	*Applied Mechanical Vibrations*
Jaluria:	*Design and Optimization of Thermal Systems*
Juvinall:	*Engineering Considerations of Stress, Strain, and Strength*
Kane and Levinson:	*Dynamics: Theory and Applications*
Kays and Crawford:	*Convective Heat and Mass Transfer*
Kelly:	*Fundamentals of Mechanical Vibrations*
Kimbrell:	*Kinematics Analysis and Synthesis*
Kreider and Rabl:	*Heating and Cooling of Buildings*
Martin:	*Kinematics and Dynamics of Machines*
Mattingly:	*Elements of Gas Turbine Propulsion*
Modest:	*Radiative Heat Transfer*
Norton:	*Design of Machinery*
Oosthuizen and Carscallen:	*Compressible Fluid Flow*
Phelan:	*Fundamentals of Mechanical Design*
Raven:	*Automatic Control Engineering*
Reddy:	*An Introduction to the Finite Element Method*
Rosenberg and Karnopp:	*Introduction to Physical Systems Dynamics*
Schlichting:	*Boundary-Layer Theory*
Shames:	*Mechanics of Fluids*
Sherman:	*Viscous Flow*
Shigley:	*Kinematic Analysis of Mechanisms*
Shigley and Mischke:	*Mechanical Engineering Design*
Shigley and Uicker:	*Theory of Machines and Mechanisms*

Stiffler:	*Design with Microprocessors for Mechanical Engineers*
Stoecker and Jones:	*Refrigeration and Air Conditioning*
Turns:	*An Introduction to Combustion: Concepts and Applications*
Ullman:	*The Mechanical Design Process*
Vanderplaats:	*Numerical Optimization: Techniques for Engineering Design, with Applications*
Wark:	*Advanced Thermodynamics for Engineers*
White:	*Viscous Fluid Flow*
Zeid:	*CAD/CAM Theory and Practice*

THERMODYNAMICS

An Engineering Approach

THIRD EDITION

Dr. Yunus A. Çengel
University of Nevada, Reno

Dr. Michael A. Boles
North Carolina State University

Boston Burr Ridge, IL Dubuque, IA Madison, WI New York San Francisco St. Louis
Bangkok Bogotá Caracas Lisbon London Madrid Mexico City
Milan New Delhi Seoul Singapore Sydney Taipei Toronto

WCB/McGraw-Hill

A Division of The **McGraw-Hill** *Companies*

Thermodynamics:
An Engineering Approach

This book is printed on acid-free paper.

3 4 5 6 7 8 9 0 DOC/DOC 9 0 9 8

ISBN 0-07-011927-9

Vice president and editorial director: *Kevin T. Kane*
Publisher: *Tom Casson*
Senior sponsoring editor: *Debra Riegert*
Marketing manager: *John T. Wannemacher*
Senior project manager: *Gladys True*
Production supervisor: *Heather D. Burbridge*
Designer: *Kiera Cunningham*
Compositor: *Shepard Poorman Communications*
Typeface: *10/12 Times Roman*
Printer: *R. R. Donnelley & Sons Company*

Library of Congress Cataloging-in-Publication Data

Çengel, Yunus A.
Thermodynamics : an engineering approach / Yunus A. Çengel, Michael A. Boles. — 3rd ed.
p. cm.
Includes index.
ISBN 0-07-011927-9
1. Thermodynamics. I. Boles, Michael A. II. Title.
TJ265.C43 1998
621.402′1—dc21 97-28300

When ordering the title, use ISBN 0-07-115247-4.

If you are interested in ordering the text with the EES Problems Disk, please refer to ISBN 0-07-913238-3.

http://www.mhhe.com

About the Authors

Yunus A. Çengel received his Ph.D. in mechanical engineering from North Carolina State University and joined the faculty of mechanical engineering at the University of Nevada, Reno, where he has been teaching undergraduate and graduate courses in thermodynamics and heat transfer while conducting research. He has published primarily in the areas of thermodynamics, radiation heat transfer, natural convection, solar energy, geothermal energy, energy conservation, and engineering education. He has also authored the books *Introduction to Thermodynamics and Heat Transfer* and *Heat Transfer: A Practical Approach*, both published by McGraw-Hill. He has led teams of engineering students to numerous manufacturing facilities in Northern Nevada and California to conduct energy audits, and has prepared energy conservation reports for them. Dr. Çengel has been voted outstanding teacher by the ASME student sections in both North Carolina State University and the University of Nevada, Reno. He is a member of the American Society of Mechanical Engineers (ASME) and the American Society for Engineering Education (ASEE).

Michael A. Boles is Associate Professor of Mechanical and Aerospace Engineering at North Carolina State University where he earned his Ph.D. in mechanical engineering and is an Alumni Distinguished Professor. Dr. Boles has received numerous awards and citations for excellence as an engineering educator. He is a past recipient of the SAE Ralph R. Teetor Education Award and has been twice elected to the Academy of Outstanding Teachers. On several occasions the ASME student section has recognized him as outstanding teacher and the faculty member having the most impact on mechanical engineering students. Dr. Boles specializes in heat transfer and has been involved in the analytical and numerical solution of phase change and drying of porous media. He is a member of the American Society of Mechanical Engineers, the American Society for Engineering Education, and Sigma Xi.

Professors Çengel and Boles received the ASEE Meriam/Wiley Distinguished Author Award in 1992 in recognition of their excellence in the authorship of the first edition of this text.

Contents

Table of Examples

Chapter 1 ■ BASIC CONCEPTS OF THERMODYNAMICS

Chapter 2 ■ PROPERTIES OF PURE SUBSTANCES

Chapter 3 ■ THE FIRST LAW OF THERMODYNAMICS: CLOSED SYSTEMS

Chapter 11 ■ THERMODYNAMIC PROPERTY RELATIONS

Chapter 12 ■ GAS MIXTURES

Chapter 13 ■ GAS-VAPOR MIXTURES AND AIR-CONDITIONING

Chapter 14 ■ CHEMICAL REACTIONS

Chapter 15 ■ CHEMICAL AND PHASE EQUILIBRIUM

Chapter 16 ■ THERMODYNAMICS OF HIGH-SPEED GAS FLOW

Preface

Thermodynamics is a basic science that deals with energy and has long been an essential part of engineering curricula all over the world. This introductory text is intended for use in *undergraduate* engineering courses and contains sufficient material for two sequential courses in thermodynamics.

The traditional *classical,* or macroscopic, approach is used throughout the text, with microscopic arguments serving in a supporting role where appropriate. This approach is more in line with the students' intuition and makes learning the subject matter much easier.

Our philosophy that contributed to the popularity of the first two editions remains unchanged: talk directly to the minds of tomorrow's engineers in a simple, yet precise manner and encourage creative thinking and development of a deeper understanding of the subject matter. Our goal in this undertaking was to offer an engineering textbook that is *read by the students* with interest and enthusiasm instead of one that is used as a reference book to solve problems. We wanted to touch the curious minds and take them to a pleasant journey in the wonderful world of thermodynamics where they could explore the wonders of this exciting science. The enthusiastic response we received from the users of the earlier editions from small colleges to large universities indicates that our objectives have been achieved.

Thermodynamics is often perceived as a difficult subject, and the majority of students dread the experience. The authors believe, on the contrary, that thermodynamics is a *simple* subject, and an observant mind should have no difficulty understanding it. After all, the principles of thermodynamics are based on our *everyday experiences* and *experimental observations.*

Thermodynamics is a mature basic science, and the topics covered in introductory texts are well established. Primarily, the texts differ only in the approach used. In this text, a *more physical, intuitive* approach is used throughout. Frequently, *parallels* are drawn between the subject matter and students' everyday experiences so that they can relate the subject matter to what they already know.

Yesterday's engineer spent a major portion of his or her time substituting values into formulas and obtaining numerical results. But today all the formula manipulations and number crunching are being left to the computers. Tomorrow's engineer will have to have a clear understanding and firm grasp of the basic principles in order to understand, formulate, and interpret the results of even the most complex problems. A conscious effort is made in this text to lead students in this direction.

The material is introduced at a level that average students can follow comfortably. It speaks *to* the students, not over them. In fact, the material is *self-instructive,* thus freeing the instructor to use class time more productively.

The order of coverage is from simple to general. That is, it starts with the simplest case and adds complexities one at a time. In this way, the basic principles are repeatedly applied to different systems, and students master how to apply the principles instead of how to simplify a general formula. Since thermodynamic principles are based on experimental observations, all derivations in this text are based on *physical grounds;* thus, they are easy to follow and understand.

The subject material is covered in a logical order. First, various concepts are reviewed and some new ones are defined in order to establish a firm basis for the development of thermodynamic principles. Then the properties of pure substances are discussed and the use of property tables is illustrated. At this point the ideal-gas approximation is introduced, together with other equations of state, and the deviation from ideal-gas behavior is examined through the use of compressibility charts. After the introduction of heat and work, the conservation of energy principle is developed for a closed system. Following a discussion of flow energy, the conservation of energy principle for control volumes is developed, first for steady-flow systems and then for general unsteady-flow systems. The development of the second-law relations follows the same order, with special emphasis given to entropy generation. The concepts of exergy, reversible work, and exergy destruction are developed using familiar examples before they are applied to more complex engineering systems. The principles of thermodynamics are then applied to various areas of engineering.

NEW IN THE THIRD EDITION

Pedagogical Changes

All the popular features of the previous editions are retained while new ones are added. With the exception of major revisions of Chapters 6 and 7 and the addition of some new sections in other chapters, the main body of the text remains largely unchanged. The most significant changes in this edition are highlighted below.

- **The use of a formal sign convention for heat and work is de-emphasized.** Instead, an intuitive and unified approach is adopted for interactions. The mass, energy, entropy, and exergy balances for *any system* undergoing *any process* are expressed as

  Mass balance: $$m_{in} - m_{out} = \Delta m_{system}$$

  Energy balance: $$\underbrace{E_{in} - E_{out}}_{\substack{\text{Net energy transfer} \\ \text{by heat, work, and mass}}} = \underbrace{\Delta E_{system}}_{\substack{\text{Change in internal, kinetic,} \\ \text{potential, etc., energies}}}$$

  Entropy balance: $$\underbrace{S_{in} - S_{out}}_{\substack{\text{Net entropy transfer} \\ \text{by heat and mass}}} + \underbrace{S_{gen}}_{\substack{\text{Entropy} \\ \text{generation}}} = \underbrace{\Delta S_{system}}_{\substack{\text{Change} \\ \text{in entropy}}}$$

  Exergy balance: $$\underbrace{X_{in} - X_{out}}_{\substack{\text{Net exergy transfer} \\ \text{by heat, work, and mass}}} - \underbrace{X_{destroyed}}_{\substack{\text{Exergy} \\ \text{destruction}}} = \underbrace{\Delta X_{system}}_{\substack{\text{Change} \\ \text{in exergy}}}$$

  The relations above reinforce that during an actual process, mass and energy are conserved, entropy is generated, and exergy is destroyed. Students are encouraged to use these forms of balances in early chapters after they specify the system and to simplify them for the particular problem. A more relaxed approach is used in later chapters.
- **An all new software based on the popular EES (Engineering Equation Solver) program allows students to solve design problems and to ask "what if" questions.** The software, described in detail later in the Preface and in Appendix 3, now incorporates a user-friendly equation solver and enables the user to perform parametric and optimization studies on thermodynamic systems and displays the results in tabular and graphical forms.
- **A more structured approach is used in the example problems while maintaining the informal conversational style.** Starting with Chap. 3, in problems that require the application of a balance, the *system* is clearly identified, any *observations* about the specifics of the problem are stated, and any *assumptions* made and their justifications are listed.
- **The problems repeated in different unit systems are deleted to save space.** However, the problems in English units are still designated by "E" following the problem number for easy recognition; the *analysis* is presented, and the results are *discussed*.
- **Some new examples and numerous new problems are added, including some extensive design problems.** Many real-world problems dealing with *thermoeconomics* and the dollar value of various energy conservation measures are incorporated into various chapters to help students develop an understanding of the monetary value of energy. Some new problems deal with the *environmental impact* of energy conversion and energy conservation, and the emission of pollutants and greenhouse gases into the atmosphere.

- In response to the international treaties banning the use of the *refrigerant-12* because of its destructive effect on the protective ozone layer of the earth, **the refrigerant-12 tables in the appendix are deleted** (they are kept in the software for this edition), and all the examples and problems dealing with refrigerant-12 are replaced by those dealing with refrigerant-134a.
- **Recent information dealing with new developments in thermodynamic systems,** such as the new generation of gas turbines with remarkable efficiencies, is incorporated throughout the text.

Content Changes

With the exception of some fine-tuning, the main body of the text remains largely unchanged. The noteworthy changes in various chapters are summarized below for those who are familiar with the previous edition.

- In Chap. 1, the discussions of internal energy are revised and expanded and more discussions are added on pressure measurement devices. Also, former Sec. 3-10, "Thermodynamic Aspects of Biological Systems," is moved to this chapter to expose students to this exciting topic early in the course. More discussions on physical fitness are added, including a sensible diet proposed by the authors.
- In Chap. 2, a new section, "Vapor Pressure and Phase Equilibrium" is added to expose students to the dynamics of phase-change processes and mass transfer. Also, Sec. 2-3 is expanded with discussions on some interesting consequences of the saturation temperature–pressure dependence.
- In Chap. 3, a new section "Refrigeration and Freezing of Foods" is added to show the relevance of thermodynamics to everyday life. Also, the first-law relation is redeveloped using an intuitive approach.
- In Chap. 4, the development of the first-law relations is greatly revised.
- In Chap. 5, two new sections, "Energy Conversion Efficiencies" and "Household Refrigerators," are added. In the former section, combustion efficiency, motor efficiency, lighting efficiency, overall efficiency, and the heating values of fuels are presented. In the latter section, several ways of conserving energy while using a refrigerator are discussed.
- In Chap. 6, a new section "Reducing the Cost of Compressed Air" is added to expose future engineers to several practical ways of minimizing the power consumption of compressors, which is a significant fraction of the total energy consumption of most manufacturing facilities. Also, the entropy balance relation is redeveloped and moved to the end of the chapter.
- In Chap. 7, the term "availability" is replaced by "exergy" because of the universal use of the latter. Also, "irreversibility" is mostly replaced by "exergy destruction." This chapter is extensively revised to incorporate the exergy balance concept.

- In Chap. 8, the discussions on gas turbine efficiency are expanded to incorporate recent developments.
- In Chap. 14, the energy and entropy balance relations are redeveloped.
- In Chap. 15, the section "Phase Equilibrium" is greatly expanded to include Henry's law and solid–gas phase equilibrium.

Pedagogical Tools

Figures are important learning tools that help the students get the "picture." They attract attention and stimulate curiosity and interest. The text makes effective use of graphics and probably contains more figures and illustrations than any other thermodynamics book. Some of the figures do not function in the traditional sense. Rather, they serve as a means of emphasizing some key statements that would otherwise go unnoticed as paragraph summaries. The popular cartoon feature "Blondie" is used to make some important points in a humorous way and also to break the ice and ease the nerves. Who says studying thermodynamics can't be fun?

Each chapter contains **numerous worked-out examples** that clarify the material and illustrate the use of the basic principles. A consistent and systematic approach is used in the solution of the example problems, with particular attention to the proper use of units. A **sketch** and a **process diagram** are included for most examples to clearly illustrate the geometry and the type of process involved.

A **short summary** is included at the end of each chapter for a quick overview of basic concepts and important relations. This is followed by a list of references that are appropriate for the level of students studying thermodynamics for the first time. Summaries can also be used as formula sheets during exams by instructors who prefer closed-book tests and yet want to make the equations available to the students.

The **end-of-chapter problems** are grouped under specific topics in the order they are covered to make problem selection easier for both instructors and students. The problems within each group start with concept questions, indicated by "C," to check the student's level of understanding of basic concepts. The problems involving numerical calculations are arranged in increasing complexity, with the later ones requiring more comprehensive analysis. The problems grouped under **Review Problems** are more comprehensive in nature, and are not directly tied to any specific section of a chapter. The problems under **Computer, Design, and Essay Problems** are intended to encourage students to use computers in problem solving, to make engineering judgments, to conduct independent searches on topics of interest, and to communicate their findings in a professional manner. Some safety-related problems are incorporated throughout to enhance safety awareness among engineering students. Answers to selected problems are presented immediately following the problems for convenience to the students.

EES programs, discussed below, have been developed to solve many of the example problems in the book and they are included on the accompanying Problems Disk. The EES problems developed for this book are denoted in the text with a disk symbol. The Load Textbook command in EES will generate a separate menu for these programs so that they can easily be accessed. Each program provides detailed comments and on-line help. These programs should help the student master the important concepts without the calculational burden that has been previously required. The text is also available without the disk for those who do not wish to incorporate the software in the course. The menu-driven software that accompanied the second edition is available to adopters who prefer to continue to use it.

In recognition of the fact that English units are still widely used in some industries, **both SI and English units are used in this text,** with an emphasis on SI. The material in this text can be covered using combined SI/English units or SI units alone, depending on the preference of the instructor. The property tables and charts in the appendix are presented in both units, except the ones that involve dimensionless quantities. Problems, tables, and charts in English units are designated by "E" after the number for easy recognition. Frequently used conversion factors and physical constants are listed on the inside cover pages of the text for easy reference.

SUPPLEMENTS

EES (Engineering Equation Solver) is a general program that solves algebraic and initial-value differential equations. EES can also do optimization, parametric analysis, and linear and nonlinear regression and provide publication quality plotting capability. EES has an intuitive interface that is very easy to master. Equations can be entered in any form and in any order since EES automatically rearranges the equations to solve them in the most efficient manner. The EES engine is available to adopters of the text with the problems disk.

EES is particularly useful in thermodynamics problems since most property data needed for solving problems in these areas are provided by the program. For example, the steam tables are implemented such that any thermodynamic property can be obtained from a built-in function call in terms of any other properties. Similar capability is provided for all substances. EES also allows the user to enter property data or functional relationships with lookup tables, with internal functions written with EES, or with externally compiled functions written in Pascal, C, C++, or Fortran. Interesting practical problems that may have implicit solutions are often not assigned because of the mathematical complexity involved. EES allows the user to concentrate on concepts by freeing him or her from mundane chores.

The **Instructor's Manual** is prepared with a scientific word processor, and it provides complete and detailed solutions to end-of-chapter problems. The solutions may be photocopied for posting or preparing transparencies for classroom discussions.

The **Instructor's Resource CD** that is available to adopters contains PowerPoint presentation of key figures in the text and lecture outlines as well as the complete solutions manual, the EES problems disk, and the tables and charts from the appendix.

ACKNOWLEDGMENTS

We would like to take this opportunity to thank the many users of the first two editions at more than 150 colleges and universities in the United States and Canada and many others in several countries overseas for their comments, praises, and constructive criticism. Special thanks are due to faculty who reviewed the third edition with a critical eye and made valuable suggestions for improvements:

Ralph C. Aldredge III, *University of California, Davis*

Major Michael Fabian, *United States Air Force Academy*

Bakhtiar Farouk, *Drexel University*

Costas P. Grigoropoulos, *University of California, Berkeley*

Charles M. Harman, *Duke University*

Karim J. Nasr, *GMI Engineering and Management Institute*

G. P. Peterson, *The Texas A&M University*

Ramendra P. Roy, *Arizona State University*

Elaine P. Scott, *Virginia Polytechnic Institute and State University*

J. Edwards Sunderland, *University of Massachusetts*

We would also like to thank those who reviewed the second edition, including Joseph Augostos, *Manhattan College*; Daisie Boettner, *United States Military Academy*; Frank J. DeLuise, *University of Rhode Island*; Jerry Drummond, *The University of Akron*; Jerry Dunn, *Texas Tech University*; Jeffrey W. Hodgson, *University of Tennessee*; Vincent J. Lopardo, *United States Naval Academy*; A. K. MacPherson, *Lehigh University*; M. Pinar Menguc, *University of Kentucky*; Alan Parkinson, *Brigham Young University*; Larry Simmons, *University of Portland*; and James Strozier, *University of Utah*. We are grateful to them for their expert advice and suggestions for improvements. We are also grateful to our colleagues in many other countries who undertook the relentless task of translating the second edition of this book into several languages including Chinese, Greek, Japanese, Korean, Spanish, and Turkish.

We are gratified by the enthusiastic response this book has received, and we hope that the improvements made in the new edition will make it even more appealing. Your comments and criticisms are always welcome and will be greatly appreciated.

Finally, we would like to express our appreciation to our wives Zehra and Sylvia and our children for their continued patience, understanding, encouragement, and support throughout this project.

Yunus A. Çengel
Michael A. Boles

Nomenclature

a	Acceleration, m/s^2
a	Specific Helmholtz function, $u - Ts$, kJ/kg
A	Area, m^2
A	Helmholtz function, $U - TS$, kJ
AF	Air–fuel ratio
C	Speed of sound, m/s
C	Specific heat, kJ/(kg · K)
C_p	Constant pressure specific heat, kJ/(kg · K)
C_v	Constant volume specific heat, kJ/(kg · K)
COP	Coefficient of performance
COP_{HP}	Coefficient of performance of a heat pump
COP_{R}	Coefficient of performance of a refrigerator
d, D	Diameter, m
e	Specific total energy, kJ/kg
E	Total energy, kJ
EER	Energy efficiency rating
F	Force, N
FA	Fuel–air ratio
g	Gravitational acceleration, m/s^2

g	Specific Gibbs function, $h - Ts$, kJ/kg
G	Total Gibbs function, $H - TS$, kJ
h	Convection heat transfer coefficient, W/(m$^2 \cdot$°C)
h	Specific enthalpy, $u + Pv$, kJ/kg
H	Total enthalpy, $U + PV$, kJ
$\overline{h}_c$	Enthalpy of combustion, kJ/kmol fuel
$\overline{h}_f$	Enthalpy of formation, kJ/kmol
$\overline{h}_R$	Enthalpy of reaction, kJ/kmol
HHV	Higher heating value, kJ/kmol fuel
i	Specific irreversibility, kJ/kg
I	Electric current, A
I	Total irreversibility, kJ
k	Specific heat ratio, C_p/C_v
k_s	Spring constant
k_t	Thermal conductivity
K_p	Equilibrium constant
ke	Specific kinetic energy, $V^2/2$, kJ/kg
KE	Total kinetic energy, $mV^2/2$, kJ
LHV	Lower heating value, kJ/kmol fuel
m	Mass, kg
$\dot{m}$	Mass flow rate, kg/s
M	Molar mass, kg/kmol
MEP	Mean effective pressure, kPa
mf	Mass fraction
n	Polytropic exponent
N	Number of moles, kmol
P	Pressure, kPa
P_{cr}	Critical pressure, kPa
P_i	Partial pressure, kPa
P_m	Mixture pressure, kPa
P_r	Relative pressure
P_R	Reduced pressure
P_v	Vapor pressure, kPa
P_0	Surroundings pressure, kPa
pe	Specific potential energy, gz, kJ/kg
PE	Total potential energy, mgz, kJ
q	Heat transfer per unit mass, kJ/kg
Q	Total heat transfer, kJ
$\dot{Q}$	Heat transfer rate, kW
Q_H	Heat transfer with high-temperature body, kJ

Q_L	Heat transfer with low-temperature body, kJ
r	Compression ratio
R	Gas constant, kJ/(kg · K)
r_c	Cutoff ratio
r_p	Pressure ratio
R_u	Universal gas constant, kJ/(kmol · K)
s	Specific entropy, kJ/(kg · K)
S	Total entropy, kJ/K
s_{gen}	Specific entropy generation, kJ/(kg · K)
S_{gen}	Total entropy generation, kJ/K
t	Time, s
T	Temperature, °C or K
T_{cr}	Critical temperature, K
T_{db}	Dry-bulb temperature, °C
T_{dp}	Dew-point temperature, °C
T_f	Bulk fluid temperature, °C
T_H	Temperature of high-temperature body, K
T_L	Temperature of low-temperature body, K
T_R	Reduced temperature
T_{wb}	Wet-bulb temperature, °C
T_0	Surroundings temperature, °C or K
u	Specific internal energy, kJ/kg
U	Total internal energy, kJ
v	Specific volume, m^3/kg
v_{cr}	Critical specific volume, m^3/kg
v_r	Relative specific volume
v_R	Pseudoreduced specific volume
V	Total volume, m^3
$\mathcal{V}$	Velocity, m/s
w	Work per unit mass, kJ/kg
W	Total work, kJ
$\dot{W}$	Power, kW
W_{in}	Work input, kJ
W_{out}	Work output, kJ
W_{rev}	Reversible work, kJ
x	Quality
x	Specific exergy, kJ/kg
X	Total exergy, kJ
y	Mole fraction
z	Elevation, m

Z	Compressibility factor
Z_h	Enthalpy departure factor
Z_s	Entropy departure factor

Greek Letters

α	Absorptivity
α	Isothermal compressibility, 1/kPa
β	Volume expansivity, 1/K
Δ	Finite change in quantity
ϵ	Emissivity
η_{th}	Thermal efficiency
η_{II}	Second-law efficiency
θ	Total energy of a flowing fluid, kJ/kg
μ_{JT}	Joule-Thomson coefficient, K/kPa
μ	Chemical potential, kJ/kg
ν	Stoichiometric coefficient
ρ	Density, kg/m^3
ρ_s	Specific weight or relative density
σ	Stefan–Boltzmann constant
σ_n	Normal stress, N/m^2
σ_s	Surface tension, N/m
τ	Torque, Nm
ϕ	Relative humidity
ϕ	Specific closed system availability, kJ/kg
Φ	Total closed system availability, kJ
ψ	Stream availability, kJ/kg
ω	Specific or absolute humidity, kg H_2O/kg dry air

Subscripts

a	Air
abs	Absolute
act	Actual
atm	Atmospheric
av	Average
c	Combustion
cr	Critical point
cv	Control volume
e	Exit conditions
f	Saturated liquid

fg	Difference in property between saturated liquid and saturated vapor
g	Saturated vapor
gen	Generation
H	High temperature (as in T_H and Q_H)
i	Inlet conditions
i	ith component
L	Low temperature (as in T_L and Q_L)
m	Mixture
r	Relative
R	Reduced
rev	Reversible
s	Isentropic
sat	Saturated
surr	Surroundings
sys	System
v	Water vapor
0	Dead state
1	Initial or inlet state
2	Final or exit state

Superscripts

$\dot{}$ (dot)	Quantity per unit time
$\bar{}$ (bar)	Quantity per unit mole
° (circle)	Standard reference state

Basic Concepts of Thermodynamics

CHAPTER 1

Every science has a unique vocabulary associated with it, and thermodynamics is no exception. Precise definition of basic concepts forms a sound foundation for the development of a science and prevents possible misunderstandings. In this chapter, the unit systems that will be used are reviewed, and the basic concepts of thermodynamics such as system, energy, property, state, process, cycle, pressure, and temperature are explained. Careful study of these concepts is essential for a good understanding of the topics in the following chapters. Finally, the thermodynamic aspects of biological systems, including weight control and exercise are considered.

1-1 ■ THERMODYNAMICS AND ENERGY

Thermodynamics can be defined as the science of energy. Although everybody has a feeling of what energy is, it is difficult to give a precise definition for it. Energy can be viewed as the ability to cause changes.

The name *thermodynamics* stems from the Greek words *therme* (heat) and *dynamis* (power), which is most descriptive of the early efforts to convert heat into power. Today the same name is broadly interpreted to include all aspects of energy and energy transformations, including power production, refrigeration, and relationships among the properties of matter.

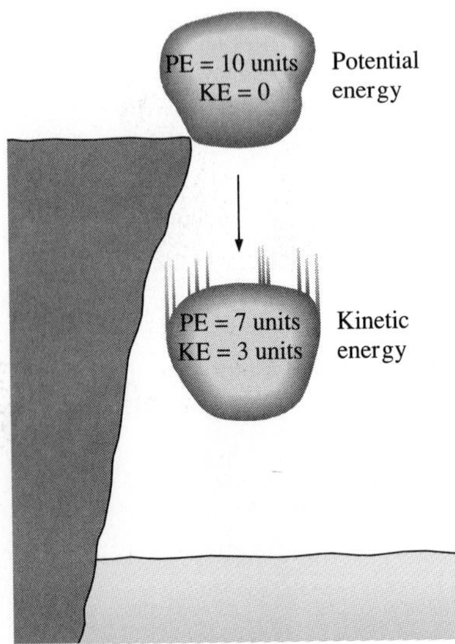

FIGURE 1-1
Energy cannot be created or destroyed; it can only change forms (the first law).

One of the most fundamental laws of nature is the **conservation of energy principle.** It simply states that during an interaction, energy can change from one form to another but the total amount of energy remains constant. That is, energy cannot be created or destroyed. A rock falling off a cliff, for example, picks up speed as a result of its potential energy being converted to kinetic energy (Fig. 1-1). The conservation of energy principle also forms the backbone of the diet industry: a person who has a greater energy input (food) than energy output (exercise) will gain weight (store energy in the form of fat), and a person who has a smaller energy input than output will lose weight (Fig. 1-2).

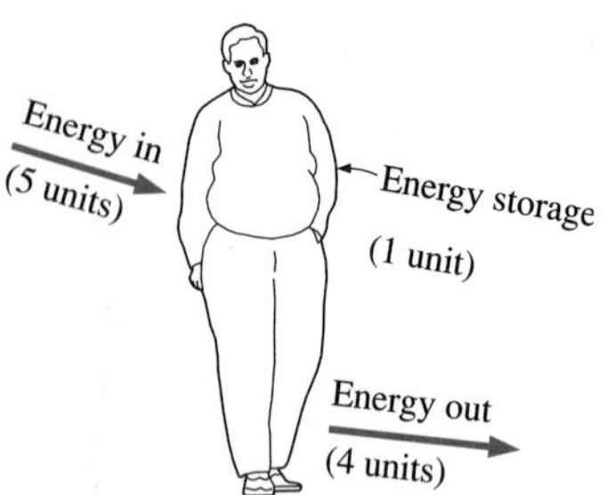

FIGURE 1-2
Conservation of energy principle for the human body.

The **first law of thermodynamics** is simply an expression of the conservation of energy principle, and it asserts that *energy* is a thermodynamic property. The **second law of thermodynamics** asserts that energy has *quality* as well as *quantity,* and actual processes occur in the direction of decreasing quality of energy. For example, a cup of hot coffee left on a table eventually cools, but a cup of cool coffee on the same table never gets hot by itself (Fig. 1-3). The high-temperature energy of the coffee is degraded (transformed into a less useful form at a lower temperature) once it is transferred to the surrounding air.

Although the principles of thermodynamics have been in existence since the creation of the universe, thermodynamics did not emerge as a science until the construction of the first successful atmospheric steam engines in England by Thomas Savery in 1697 and Thomas Newcomen in 1712. These engines were very slow and inefficient, but they opened the way for the development of a new science.

The first and second laws of thermodynamics emerged simultaneously in the 1850s, primarily out of the works of William Rankine, Rudolph Clausius, and Lord Kelvin (formerly William Thomson). The term *thermodynamics* was first used in a publication by Lord Kelvin in 1849. The first thermodynamic textbook was written in 1859 by William Rankine, a professor at the University of Glasgow.

It is well known that a substance consists of a large number of particles called *molecules.* The properties of the substance naturally depend on the behavior of these particles. For example, the pressure of a gas in a container is the result of momentum transfer between the molecules and the walls of the container. But one does not need to know the behavior of the gas particles to determine the pressure in the container. It would be sufficient to attach a pressure gage to the container. This macroscopic approach to the study of thermodynamics that does not require a knowledge of the behavior of individual particles is called **classical thermodynamics.** It provides a direct and easy way to the solution of engineering problems. A more elaborate approach, based on the average behavior of large groups of individual particles, is called **statistical thermodynamics.** This microscopic approach is rather involved and is used in this text only in the supporting role.

FIGURE 1-3
Heat can flow only from hot to cold bodies (the second law).

Application Areas of Thermodynamics

Every engineering activity involves an interaction between energy and matter; thus it is hard to imagine an area that does not relate to thermodynamics in some respect. Therefore, developing a good understanding of thermodynamic principles has long been an essential part of engineering education.

One does not need to go very far to see some application areas of thermodynamics. In fact, one does not need to go anywhere. These areas are right where one lives. An ordinary house is, in some respects, an exhibition hall filled with thermodynamic wonders. Many ordinary household utensils and appliances are designed, in whole or in part, by using the principles of thermodynamics. Some examples include the electric or gas range, the heating and air-conditioning systems, the refrigerator, the humidifier, the pressure cooker, the water heater, the shower, the iron, and even the computer, the TV, and the VCR set. On a larger scale, thermodynamics plays a major part in the design and analysis of automotive engines, rockets, jet engines, and conventional or nuclear power plants (Fig. 1-4). We should also mention the human body as an interesting application area of thermodynamics.

1-2 ■ A NOTE ON DIMENSIONS AND UNITS

Any physical quantity can be characterized by **dimensions.** The arbitrary magnitudes assigned to the dimensions are called **units.** Some basic dimensions such as mass m, length L, time t, and temperature T are selected as **primary** or **fundamental dimensions,** while others such as velocity $\mathcal{V}$, energy E, and volume V are expressed in terms of the primary dimensions and are called **secondary dimensions,** or **derived dimensions.**

A number of unit systems have been developed over the years. Despite strong efforts in the scientific and engineering community to unify the world with a single unit system, two sets of units are still in common use today: the **English system,** which is also known as the *United States Customary System* (USCS), and the metric **SI** (from *Le Système International d'Unités*), which is also known as the *International System*. The SI is a simple and logical system based on a decimal relationship between the various units, and it is being used for scientific and engineering work in most of the industrialized nations, including England. The English system, however, has no numerical base, and various units in this system are related to each other rather arbitrarily (12 in. in 1 ft, 16 oz in 1 lb, 4 qt in 1 gal, etc.) which makes it confusing and difficult to learn. The United States is the only industrialized country that has not yet fully converted to the metric system.

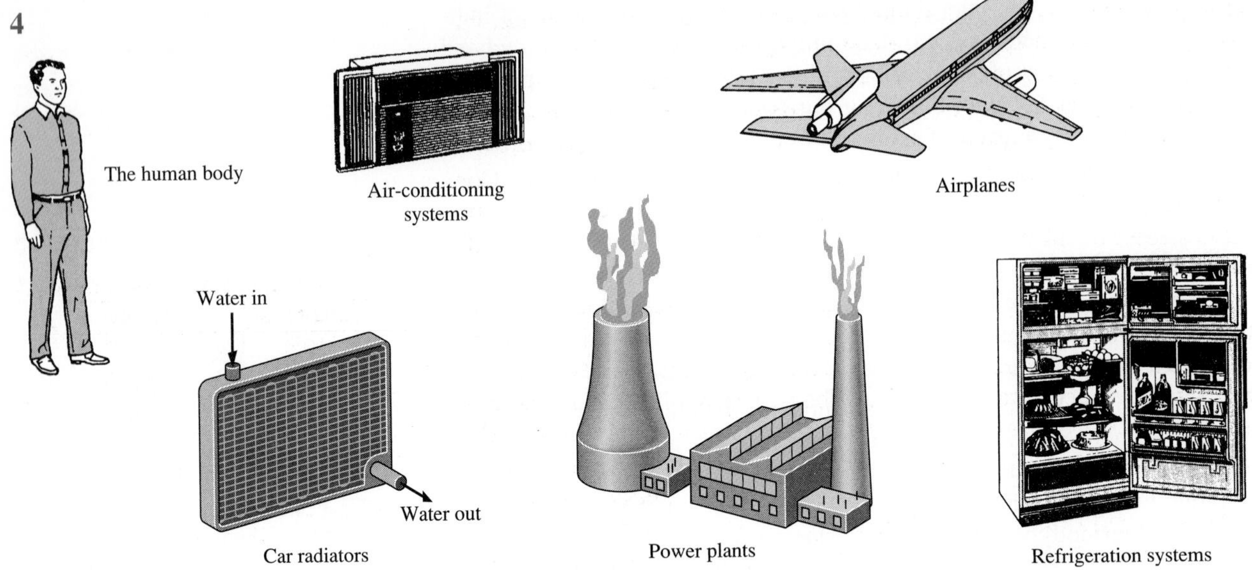

FIGURE 1-4
Some application areas of thermodynamics.

The systematic efforts to develop a universally acceptable system of units dates back to 1790 when the French National Assembly charged the French Academy of Sciences to come up with such a unit system. An early version of the metric system was soon developed in France, but it did not find much universal acceptance until 1875 when *The Metric Convection Treaty* was prepared and signed by 17 nations, including the United States. In this international treaty, meter and gram were established as the metric units for length and mass, respectively, and a *General Conference of Weights and Measures* (CGPM) was established that was to meet every six years. In 1960, the CGPM produced the SI, which was based on six fundamental quantities and their units adopted in 1954 at the Tenth General Conference of Weights and Measures: *meter* (m) for length, *kilogram* (kg) for mass, *second* (s) for time, *ampere* (A) for electrical current, degree *Kelvin* (°K) for temperature, and *candela* (cd) for luminous intensity (amount of light). In 1971, the CGPM added a seventh fundamental quantity and unit: *mole* (mol) for the amount of matter.

TABLE 1-1

The seven fundamental dimensions and their units in SI

Dimension	Unit
Length	meter (m)
Mass	kilogram (kg)
Time	second (s)
Temperature	kelvin (K)
Electric current	ampere (A)
Amount of light	candela (c)
Amount of matter	mole (mol)

Based on the notational scheme introduced in 1967, the degree symbol was officially dropped from the absolute temperature unit, and all unit names were to be written without capitalization even if they were derived from proper names (Table 1-1). However, the abbreviation of a unit was to be capitalized if the unit was derived from a proper name. For example, the SI unit of force, which is named after Sir Isaac Newton (1647–1723), is *newton* (not Newton), and it is abbreviated as *N*. Also, the full name of a unit may be pluralized, but its abbreviation cannot. For example, the length of an object can be 5 m or 5 meters, *not* 5 ms or 5 meter. Finally, no period is to be used in unit abbreviations unless they appear at the end of a sentence. For example, the proper abbreviation of meter is m (not m.).

The recent move toward the metric system in the United States seems to have started in 1968 when Congress, in response to what was happening in the rest of the world, passed a Metric Study Act. Congress continued to promote a voluntary switch to the metric system by passing the Metric Conversion Act in 1975. A trade bill passed by Congress in 1988 set a September 1992 deadline for all federal agencies to convert to the metric system. But the deadlines were relaxed later with no clear plans for the future.

The industries that are heavily involved in international trade (such as the automotive, soft drink, and liquor industries) have been quick in converting to the metric system for economic reasons (having a single worldwide design, fewer sizes, smaller inventories, etc.). Today, nearly all the cars manufactured in the United States are metric. Most car owners probably do not realize this until they try an inch socket wrench on a metric bolt. Most industries, however, resisted the change, thus slowing down the conversion process.

Presently the United States is a dual-system society, and it will stay that way until the transition to the metric system is completed. This puts an extra burden on today's engineering students, since they are expected to retain their understanding of the English system while learning, thinking, and working in terms of the SI. Given the position of the engineers in the transition period, both unit systems are used in this text, with particular emphasis on SI units.

As pointed out earlier, the SI is based on a decimal relationship between units. The prefixes used to express the multiples of the various units are listed in Table 1-2. They are standard for all units, and the student is encouraged to memorize them because of their widespread use (Fig. 1-5).

TABLE 1-2

Standard prefixes in SI units

Multiple	Prefix
10^{12}	tetra, T
10^{9}	giga, G
10^{6}	mega, M
10^{3}	kilo, k
10^{2}	hecto, h
10^{1}	deka, da
10^{-1}	deci, d
10^{-2}	centi, c
10^{-3}	milli, m
10^{-6}	micro, μ
10^{-9}	nano, n
10^{-12}	pico, p

Some SI and English Units

In SI, the units of mass, length, and time are the kilogram (kg), meter (m), and second (s), respectively. The respective units in the English system are the pound-mass (lbm), foot (ft), and second (s). The pound symbol *lb* is actually the abbreviation of *libra*, which was the ancient Roman unit of weight. The English retained this symbol even after the end of the Roman occupation of Britain in 410. The mass and length units in the two systems are related to each other by

$$1 \text{ lbm} = 0.45359 \text{ kg}$$
$$1 \text{ ft} = 0.3048 \text{ m}$$

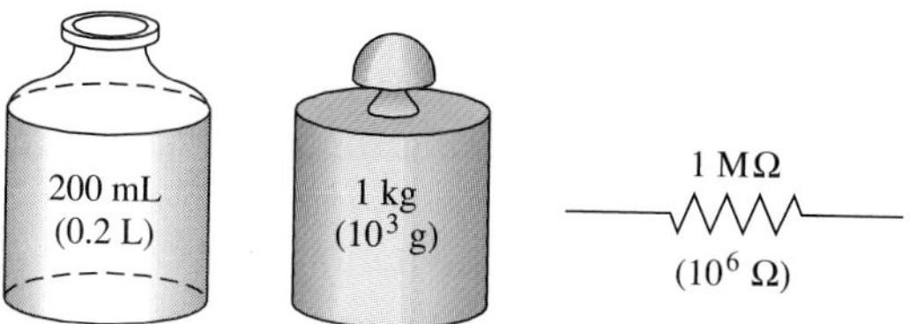

FIGURE 1-5
The SI unit prefixes are used in all branches of engineering.

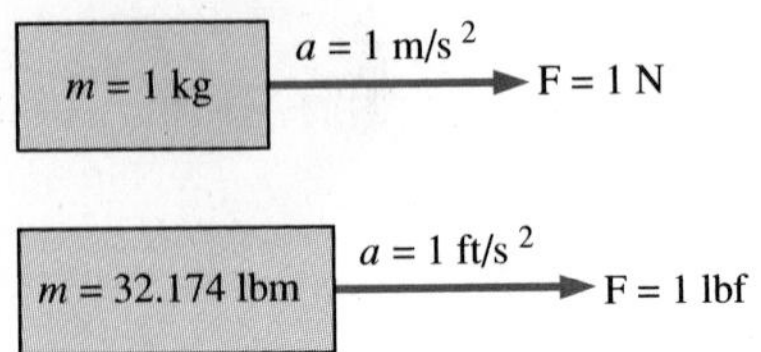

FIGURE 1-6

The definition of the force units.

In the English system, force is usually considered to be one of the primary dimensions and is assigned a nonderived unit. This is a source of confusion and error that necessitates the use of a conversion factor (g_c) in many formulas. To avoid this nuisance, we consider force to be a secondary dimension whose unit is derived from Newton's second law, i.e.,

$$\text{Force} = (\text{Mass})(\text{Acceleration})$$

or

$$F = ma \tag{1-1}$$

In SI, the force unit is the newton (N), and it is defined as the *force required to accelerate a mass of 1 kg at a rate of 1 m/s²*. In the English system, the force unit is the **pound-force** (lbf) and is defined as the *force required to accelerate a mass of 32.174 lbm (1 slug) at a rate of 1 ft/s²* (Fig. 1-6). That is,

$$1 \text{ N} = 1 \text{ kg} \cdot \text{m/s}^2$$
$$1 \text{ lbf} = 32.174 \text{ lbm} \cdot \text{ft/s}^2$$

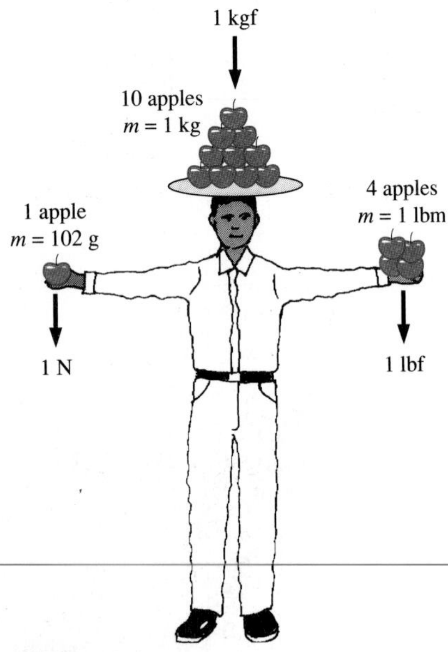

FIGURE 1-7

The relative magnitudes of the force units newton (N), kilogram-force (kgf), and pound-force (lbf).

A force of 1 newton is roughly equivalent to the weight of a small apple (m = 102 g), whereas a force of 1 pound-force is roughly equivalent to the weight of 4 medium apples (m_{total} = 454 g), as shown in Fig. 1-7. Another force unit in common use in many European countries is the *kilogram-force* (kgf), which is the weight of 1 kg mass at sea level (1 kgf = 9.807 N).

The term **weight** is often incorrectly used to express mass, particularly by the "weight watchers." Unlike mass, weight W is a *force*. It is the gravitational force applied to a body, and its magnitude is determined from Newton's second law,

$$W = mg \qquad \text{(N)} \tag{1-2}$$

where m is the mass of the body and g is the local gravitational acceleration (g is 9.807 m/s^2 or 32.174 ft/s^2 at sea level and 45° latitude). The ordinary bathroom scale measures the gravitational force acting on a body. The weight of a unit volume of a substance is called the **specific weight** w and is determined from $w = \rho g$, where ρ is density.

The mass of a body will remain the same regardless of its location in the universe. Its weight, however, will change with a change in gravitational acceleration. A body will weigh less on top of a mountain since g decreases with altitude. On the surface of the moon, an astronaut will weigh about one-sixth of what she or he normally weighs on earth (Fig. 1-8).

FIGURE 1-8

A body weighing 150 pounds on earth will weigh only 25 pounds on the moon.

At sea level a mass of 1 kg will weigh 9.807 N, as illustrated in Fig. 1-9. A mass of 1 lbm, however, will weigh 1 lbf, which misleads people to believe that pound-mass and pound-force can be used interchangeably as pound (lb), which is a major source of error in the English system.

It should be noted that the *gravity force* acting on a mass is due to the *attraction* between the masses, and thus it is proportional to the magnitudes of the masses and inversely proportional to the square of the distance between them. Therefore, the gravitational acceleration g at a location depends on the *local density* of the earth's crust, the *distance* to the center of the earth, and to a lesser extent, the *positions* of the moon and the sun. The value of g varies with location from 9.8295 m/s^2 at 4500 m below sea level to 7.3218 m/s^2 at 100,000 m above sea level. However, at altitudes up to 30,000 m, the variation of g from the sea level value of 9.807 m/s^2 is less than 1 percent. Therefore, for most practical purposes, the gravitational acceleration can be assumed to be *constant* at 9.8 m/s^2. It is interesting to note that at locations below sea level, the value of g increases with distance from the sea level, reaches a maximum at about 4500 m, and then starts decreasing. (What do you think the value of g will be at the center of the earth?)

The primary cause of confusion between mass and weight is that mass is usually measured *indirectly* by measuring the *gravity force* it exerts. This approach also assumes that the forces exerted by other effects such as air buoyancy and fluid motion are negligible. This is like a car odometer that measures the velocity of a car by measuring the number of revolutions of a wheel and multiplying it by the wheel perimeter. The correct way of measuring mass is to compare it to a known mass. But this is cumbersome, and it is mostly used for calibration and measuring precious metals.

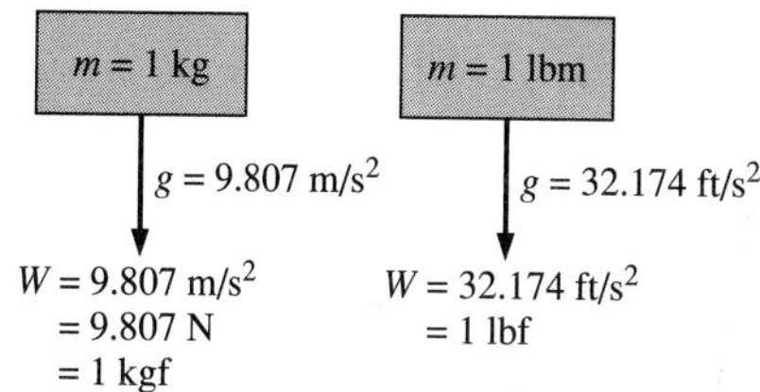

FIGURE 1-9
The weight of a unit mass at sea level.

Work, which is a form of energy, can simply be defined as force times distance; therefore, it has the unit "newton-meter (N · m)," which is called a **joule** (J). That is,

$$1 \text{ J} = 1 \text{ N} \cdot \text{m}$$

A more common unit for energy in SI is the kilojoule (1 kJ = 10^3 J). In the English system, the energy unit is the **Btu** (British thermal unit), which is defined as the energy required to raise the temperature of 1 lbm of water at 68°F by 1°F. In the metric system, the amount of energy needed to raise the temperature of 1 g of water at 15°C by 1°C is defined as 1 **calorie** (cal), and 1 cal = 4.1868 J. The magnitudes of the kilojoule and Btu are almost identical (1 Btu = 1.055 kJ).

FIGURE 1-10*
To be dimensionally homogeneous, all the terms in an equation must have the same unit.

Dimensional Homogeneity

We all know from grade school that apples and oranges do not add. But we somehow manage to do it (by mistake, of course). In engineering, all equations must be *dimensionally homogeneous*. That is, every term in an equation must have the same unit (Fig. 1-10). If, at some stage of an analysis, we find ourselves in a position to add two quantities that have different units, it is a clear indication that we have made an error at an earlier stage. So checking units can serve as a valuable tool to spot errors.

EXAMPLE 1-1 Spotting Errors from Unit Inconsistencies

While solving a problem, a person ended up with the following equation at some stage:

$$E = 25 \text{ kJ} + 7 \text{ kJ/kg}$$

where E is the total energy and has the unit of kilojoules. Determine the error that may have caused it.

---

*BLONDIE cartoons are reprinted with special permission of King Features Syndicate, Inc.

Solution The two terms on the right-hand side do not have the same units, and therefore they cannot be added to obtain the total energy. Multiplying the last term by mass will eliminate the kilograms in the denominator, and the whole equation will become dimensionally homogeneous, that is, every term in the equation will have the same unit. Obviously this error was caused by forgetting to multiply the last term by mass at an earlier stage.

We all know from experience that units can give terrible headaches if they are not used carefully in solving a problem. But with some attention and skill, units can be used to our advantage. They can be used to check formulas; they can even be used to derive formulas, as explained in the following example.

EXAMPLE 1-2 Obtaining Formulas from Unit Considerations

A tank is filled with oil whose density is $\rho = 850$ kg/m^3. If the volume of the tank is $V = 2$ m^3, determine the amount of mass m in the tank.

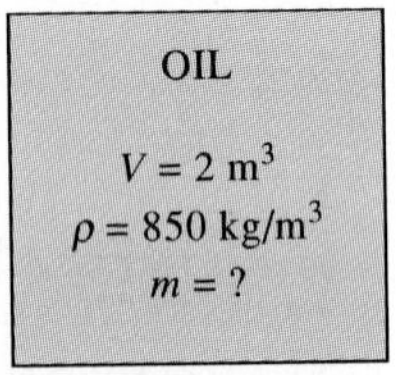

OIL
$V = 2$ m^3
$\rho = 850$ kg/m^3
$m = ?$

FIGURE 1-11
Sketch for Example 1-2.

Solution A sketch of the system described above is given in Fig. 1-11. Suppose we forgot the formula that relates mass to density and volume. But we know that mass has the unit of kilograms. That is, whatever calculations we do, we should end up with the unit of kilograms. Putting the given information into perspective, we have

$$\rho = 850 \text{ kg/m}^3 \qquad \text{and} \qquad V = 2 \text{ m}^3$$

It is obvious that we can eliminate m^3 and end up with kg by multiplying these two quantities. Therefore, the formula we are looking for is

$$m = \rho V$$

Thus,

$$m = (850 \text{ kg/m}^3)(2 \text{ m}^3) = \mathbf{1700\ kg}$$

The student should keep in mind that a formula that is not dimensionally homogeneous is definitely wrong, but a dimensionally homogeneous formula is not necessarily right.

1-3 ■ CLOSED AND OPEN SYSTEMS

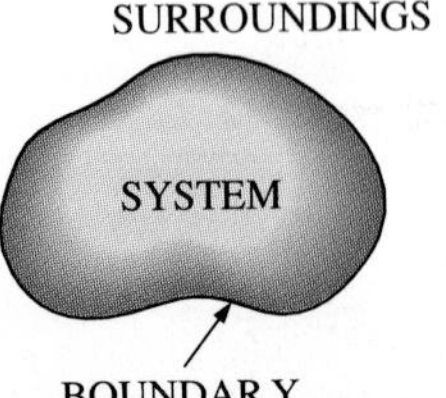

FIGURE 1-12
System, surroundings, and boundary.

A **thermodynamic system,** or simply a **system,** is defined as a *quantity of matter or a region in space chosen for study*. The mass or region outside the system is called the **surroundings.** The real or imaginary surface that separates the system from its surroundings is called the **boundary.** These terms are illustrated in Fig. 1-12. The boundary of a system can be *fixed* or *movable*. Note that the boundary is the contact surface shared by both the system and the surroundings. Mathematically speaking, the boundary has zero thickness, and thus it can neither contain any mass nor occupy any volume in space.

Systems may be considered to be *closed* or *open*, depending on whether a fixed mass or a fixed volume in space is chosen for study. A **closed system** (also known as a **control mass**) consists of a fixed amount of mass, and no mass can cross its boundary. That is, no mass can enter or leave a closed system, as shown in Fig. 1-13. But energy, in the form of heat or work, can cross the boundary, and the volume of a closed system does not have to be fixed. If, as a special case, even energy is not allowed to cross the boundary, that system is called an **isolated system.**

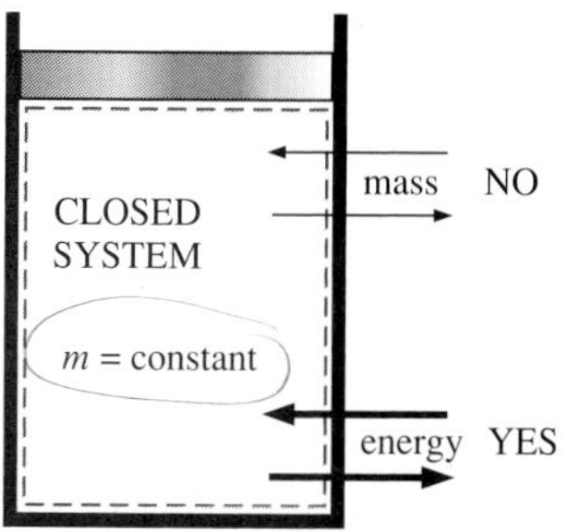

FIGURE 1-13
Mass cannot cross the boundaries of a closed system, but energy can.

Consider the piston-cylinder device shown in Fig. 1-14. Let us say that we would like to find out what happens to the enclosed gas when it is heated. Since we are focusing our attention on the gas, it is our system. The inner surfaces of the piston and the cylinder form the boundary, and since no mass is crossing this boundary, it is a closed system. Notice that energy may cross the boundary, and part of the boundary (the inner surface of the piston, in this case) may move. Everything outside the gas, including the piston and the cylinder, is the surroundings.

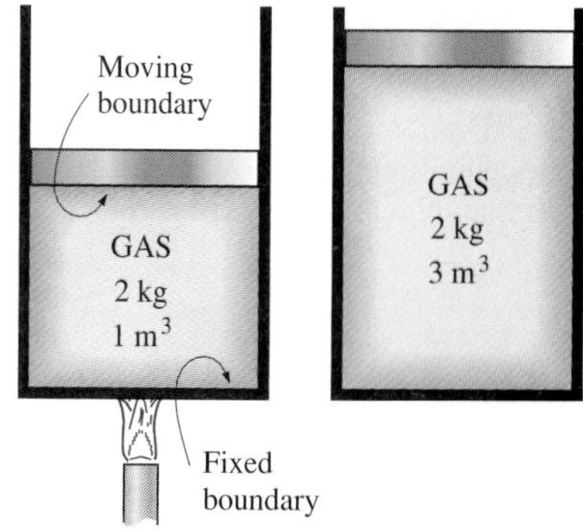

FIGURE 1-14
A closed system with a moving boundary.

An **open system,** or a **control volume,** as it is often called, is a properly selected region in space. It usually encloses a device that involves mass flow such as a compressor, turbine, or nozzle. Flow through these devices is best studied by selecting the region within the device as the control volume. Both mass and energy can cross the boundary of a control volume, which is called a **control surface.** This is illustrated in Fig. 1-15.

As an example of an open system, consider the water heater shown in Fig. 1-16. Let us say that we would like to determine how much heat we must transfer to the water in the tank in order to supply a steady stream of hot water. Since hot water will leave the tank and be replaced by cold water, it is not convenient to choose a fixed mass as our system for the analysis. Instead, we can concentrate our attention on the volume formed by the interior surfaces of the tank and consider the hot and cold water streams as mass leaving and entering the control volume. The interior surfaces of the tank form the control surface for this case, and mass is crossing the control surface at two locations.

The thermodynamic relations that are applicable to closed and open systems are different. Therefore, it is extremely important that we recognize the type of system we have before we start analyzing it.

In all thermodynamic analyses, the system under study *must* be defined carefully. In most cases, the system investigated is quite simple and obvious, and defining the system may seem like a tedious and unnecessary task. In other cases, however, the system under study may be rather involved, and a proper choice of the system may greatly simplify the analysis.

FIGURE 1-15
Both mass and energy can cross the boundaries of a control volume.

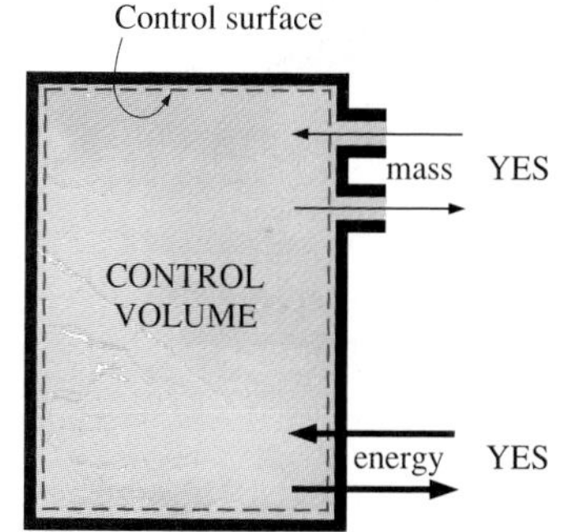

1-4 ■ FORMS OF ENERGY

Energy can exist in numerous forms such as thermal, mechanical, kinetic, potential, electric, magnetic, chemical, and nuclear, and their sum constitutes the **total energy** E of a system. The total energy of a system on a *unit mass* basis is denoted by e and is defined as

$$e = \frac{E}{m} \quad \text{(kJ/kg)} \tag{1-3}$$

Thermodynamics provides no information about the absolute value of the total energy. It only deals with the *change* of the total energy, which is what matters in engineering problems. Thus the total energy of a system can be assigned a value of zero ($E = 0$) at some convenient reference point. The change in total energy of a system is independent of the reference point selected. The decrease in the potential energy of a falling rock, for example, depends on only the elevation difference and not the reference level selected.

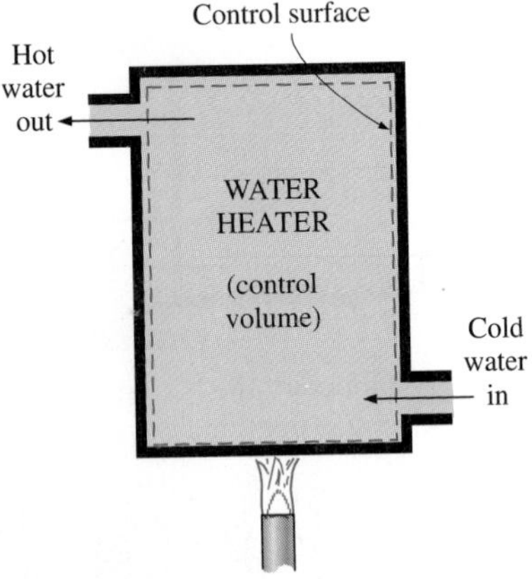

FIGURE 1-16
An open system (a control volume) with one inlet and one exit.

In thermodynamic analysis, it is often helpful to consider the various forms of energy that make up the total energy of a system in two groups: *macroscopic* and *microscopic*. The **macroscopic** forms of energy, on one hand, are those a system possesses as a whole with respect to some outside reference frame, such as kinetic and potential energies (Fig. 1-17). The **microscopic** forms of energy, on the other hand, are those related to the molecular structure of a system and the degree of the molecular activity, and they are independent of outside reference frames. The sum of all the microscopic forms of energy is called the **internal energy** of a system and is denoted by *U*. The term *energy* was coined in 1807 by Thomas Young, and its use in thermodynamics was proposed in 1852 by Lord Kelvin. The term *internal energy* and its symbol *U* first appeared in the works of Rudolph Clausius and William Rankine in the second half of the nineteenth century, and it eventually replaced the alternative terms *inner work, internal work,* and *intrinsic energy* commonly used at the time.

The macroscopic energy of a system is related to motion and the influence of some external effects such as gravity, magnetism, electricity, and surface tension. The energy that a system possesses as a result of its motion relative to some reference frame is called **kinetic energy** KE. When all parts of a system move with the same velocity, the kinetic energy is expressed as

FIGURE 1-17
The macroscopic energy of an object changes with velocity and elevation.

$$\mathrm{KE} = \frac{m\mathcal{V}^2}{2} \quad \text{(kJ)} \tag{1-4}$$

or, on a unit mass basis,

$$\mathrm{ke} = \frac{\mathcal{V}^2}{2} \quad \text{(kJ/kg)} \tag{1-5}$$

where the script $\mathcal{V}$ denotes the velocity of the system relative to some fixed reference frame. The kinetic energy of a rotating body is given by $\frac{1}{2}I\omega^2$ where I is the moment of inertia of the body and ω is the angular velocity.

The energy that a system possesses as a result of its elevation in a gravitational field is called **potential energy** PE and is expressed as

$$\mathrm{PE} = mgz \quad \text{(kJ)} \tag{1-6}$$

or, on a unit mass basis,

$$\mathrm{pe} = gz \qquad \text{(kJ/kg)} \tag{1-7}$$

where g is the gravitational acceleration and z is the elevation of the center of gravity of a system relative to some arbitrarily selected reference plane.

The magnetic, electric, and surface tension effects are significant in some specialized cases only and are not considered in this text. In the absence of these effects, the total energy of a system consists of the kinetic, potential, and internal energies and is expressed as

$$E = U + \mathrm{KE} + \mathrm{PE} = U + \frac{m\mathcal{V}^2}{2} + mgz \qquad \text{(kJ)} \tag{1-8}$$

or, on a unit mass basis,

$$e = u + \mathrm{ke} + \mathrm{pe} = u + \frac{\mathcal{V}^2}{2} + gz \qquad \text{(kJ/kg)} \tag{1-9}$$

Most closed systems remain stationary during a process and thus experience no change in their kinetic and potential energies. Closed systems whose velocity and elevation of the center of gravity remain constant during a process are frequently referred to as **stationary systems.** The change in the total energy ΔE of a stationary system is identical to the change in its internal energy ΔU. In this text, a closed system is assumed to be stationary unless it is specifically stated otherwise.

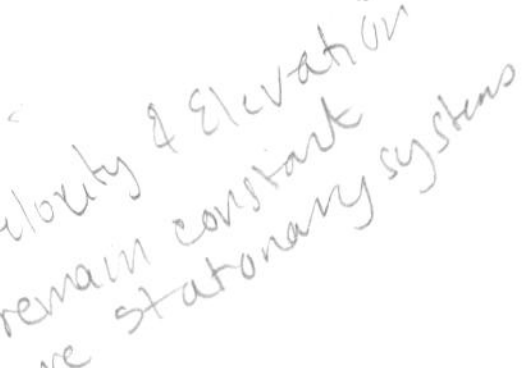

Some Physical Insight to Internal Energy

Internal energy is defined above as the sum of all the *microscopic* forms of energy of a system. It is related to the *molecular structure* and the degree of *molecular activity* and may be viewed as the sum of the *kinetic* and *potential* energies of the molecules.

To have a better understanding of internal energy, let us examine a system at the molecular level. The molecules of a gas move through space with some velocity, and thus possess some kinetic energy. This is known as the *translational energy*. The atoms of polyatomic molecules rotate about an axis, and the energy associated with it is the *rotational kinetic energy*. The atoms of a polyatomic molecule may also vibrate about their common center of mass, and the energy associated with this back-and-forth motion is the *vibrational kinetic energy*. For gases, the kinetic energy is mostly due to translational and rotational motions, with vibrational motion becoming significant at higher temperatures. The electrons in an atom rotate about the nucleus, and thus possess *rotational kinetic energy*. Electrons at outer orbits have larger kinetic energies. Electrons also spin about their axes, and the energy associated with this motion is the *spin energy*. Other particles in the nucleus of an atom also possess spin energy. The portion of the internal energy of a system associated with the kinetic energies of the molecules is called the **sensible energy** (Fig. 1-18). The average velocity and the degree of activity of the

FIGURE 1-18

The various forms of microscopic energies that make up *sensible* energy.

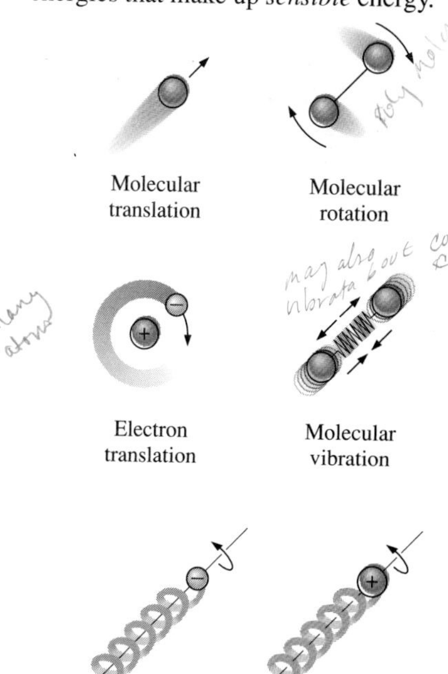

molecules are proportional to the temperature of the gas. Therefore, at higher temperatures, the molecules will possess higher kinetic energies, and as a result the system will have a higher internal energy.

The internal energy is also associated with various *binding forces* between the molecules of a substance, between the atoms within a molecule, and between the particles within an atom and its nucleus. The forces that bind the *molecules* to each other are, as one would expect, strongest in solids and weakest in gases. If sufficient energy is added to the molecules of a solid or liquid, they will overcome these molecular forces and break away, turning the substance into a gas. This is a phase-change process. Because of this added energy, a system in the gas phase is at a higher internal energy level than it is in the solid or the liquid phase. The internal energy associated with the phase of a system is called the **latent energy.** The phase-change process can occur without a change in the chemical composition of a system. Most thermodynamic problems fall into this category, and one does not need to pay any attention to the forces binding the atoms in a molecule to each other.

An atom consists of positively charged protons and neutrons bound together by very strong nuclear forces in the nucleus, and electrons orbiting around it. The internal energy associated with the atomic bonds in a molecule is called **chemical energy.** During a chemical reaction, such as a combustion process, some chemical bonds are destroyed while others are formed. As a result, the internal energy changes. The nuclear forces are much larger than the forces that bind the electrons to the nucleus. The tremendous amount of energy associated with the strong bonds within the nucleus of the atom itself is called **nuclear energy** (Fig. 1-19). Obviously, we need not be concerned with nuclear energy in thermodynamics unless, of course, we have a fusion or fission reaction on our hands. A chemical reaction involves changes in the structure of the electrons of the atoms, but a nuclear reaction involves changes in the core or nucleus. Therefore, an atom preserves its identity during a chemical reaction but loses it during a nuclear reaction. Atoms may also possess *electric* and *magnetic dipole-moment energies* when subjected to external electric and magnetic fields due to the twisting of the magnetic dipoles produced by the small electric currents associated with the orbiting electrons.

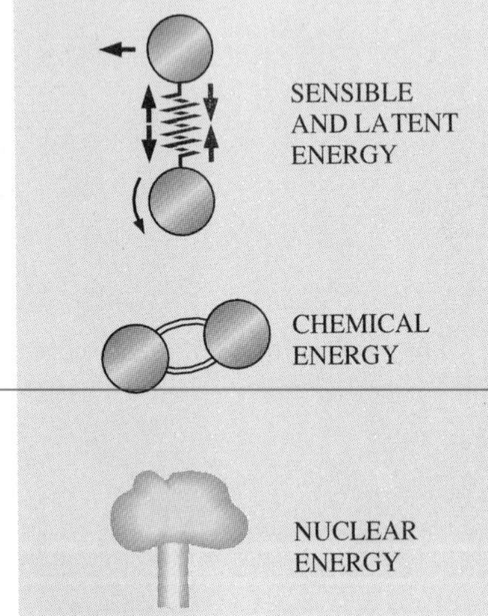

FIGURE 1-19
The internal energy of a system is the sum of all forms of the microscopic energies.

The forms of energy discussed above, which constitute the total energy of a system, can be *contained* or *stored* in a system, and thus can be viewed as the *static* forms of energy. The forms of energy not stored in a system can be viewed as the *dynamic* forms of energy or as *energy interactions.* The dynamic forms of energy are recognized at the system boundary as they cross it, and they represent the energy gained or lost by a system during a process. The only two forms of energy interactions associated with a closed system are **heat transfer** and **work.** An energy interaction is heat transfer if its driving force is a temperature difference. Otherwise it is work, as explained in Chap. 3. A control volume can also exchange energy via mass transfer since any time mass is transferred into or out of a system, the energy contained in the mass is also transferred with it.

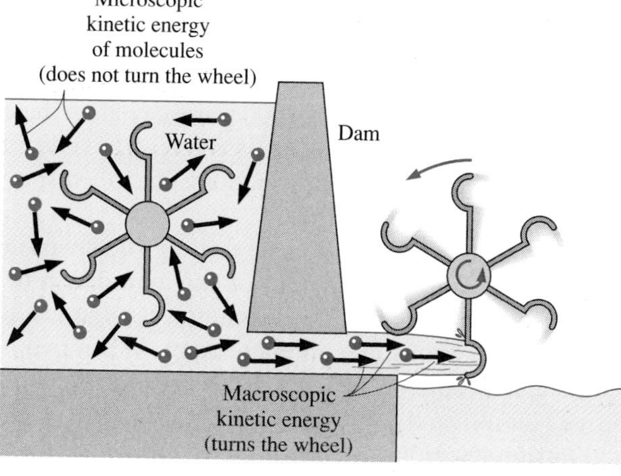

FIGURE 1-20

The *macroscopic* kinetic energy is an organized form of energy and is much more useful than the disorganized *microscopic* kinetic energies of the molecules.

In daily life, we frequently refer to the sensible and latent forms of internal energy as *heat,* and we talk about heat content of bodies. In thermodynamics, however, we usually refer to those forms of energy as **thermal energy** to prevent any confusion with *heat transfer.*

Distinction should be made between the macroscopic kinetic energy of an object as a whole and the microscopic kinetic energies of its molecules that constitute the sensible internal energy of the object (Fig. 1-20). The kinetic energy of an object is an *organized* form of energy associated with the orderly motion of all molecules in one direction in a straight path or around an axis. In contrast, the kinetic energies of the molecules are completely *random* and highly *disorganized.* As you will see in later chapters, the organized energy is much more valuable than the disorganized energy, and a major application area of thermodynamics is the conversion of disorganized energy (heat) into organized energy (work). You will also see that the organized energy can be converted to disorganized energy completely, but only a fraction of disorganized energy can be converted to organized energy by specially built devices called *heat engines* (like car engines and power plants). A similar argument can be given for the macroscopic potential energy of an object as a whole and the microscopic potential energies of the molecules.

More on Nuclear Energy

The best known fission reaction involves the split of the uranium atom (the U-235 isotope) into other elements, and is commonly used to generate electricity in nuclear power plants (429 of them in 1990, generating 311,000 MW worldwide), to power nuclear submarines and aircraft carriers, and even to power spacecraft as well as building nuclear bombs. The first nuclear chain reaction was achieved by Enrico Fermi in 1942, and the first large-scale nuclear reactors were built in 1944 for the purpose of producing material for nuclear weapons. When a uranium-235 atom absorbs a neutron and splits during a fission process, it produces a cesium-140 atom, a rubidium-93 atom, 3 neutrons, and 3.2×10^{-11} J of energy. In practical terms, the complete fission of 1 kg of uranium-235 releases 6.73×10^{10} kJ of heat, which is more than the heat released when 3000 tons of coal are burned. Therefore, for the same amount of fuel, a nuclear fission reaction releases several million times more energy than a chemical reaction. The safe disposal of used nuclear fuel, however, remains a concern.

Nuclear energy by fusion is released when two small nuclei combine into a larger one. The huge amount of energy radiated by the sun and the other stars originates from such a fusion process that involves the combination of two hydrogen atoms into a helium atom. When two heavy hydrogen (deuterium) nuclei combine during a fusion process, they produce a helium-3 atom, a free neutron, and 5.1×10^{-13} J of energy (Fig. 1-21).

Fusion reactions are much more difficult to achieve in practice because of the strong repulsion between the positively charged nuclei, called the *Coulomb repulsion*. To overcome this repulsive force and to enable the two nuclei to fuse together, the energy level of the nuclei must be raised by heating them to about 100 million °C. But such high temperatures are found only in the stars or in exploding atomic bombs (the A-bomb). In fact, the uncontrolled fusion reaction in a hydrogen bomb (the H-bomb) is initiated by a small atomic bomb. The uncontrolled fusion reaction was achieved in early 1930s, but all the efforts to achieve controlled fusion by massive lasers, powerful magnetic fields, and electric currents to generate power since then have failed.

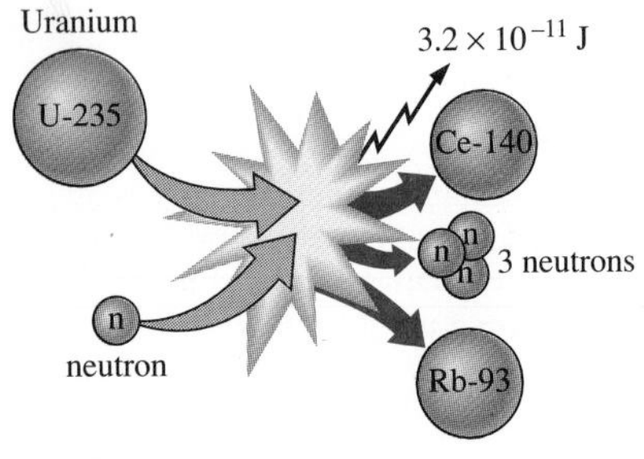

(a) Fission of uranium

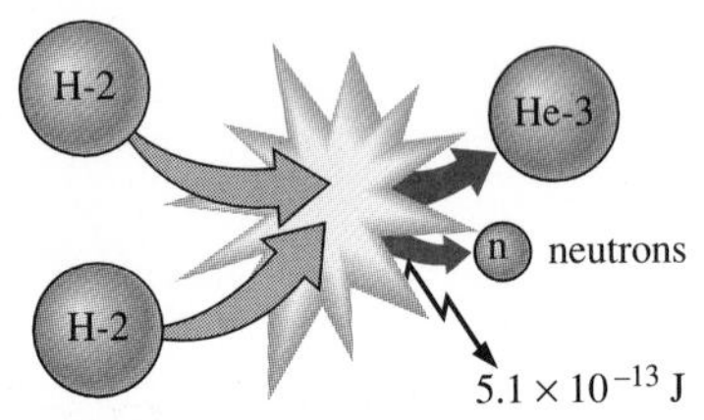

(a) Fusion of hydrogen

FIGURE 1-21
The fission of uranium and the fusion of hydrogen during nuclear reactions, and the release of nuclear energy.

EXAMPLE 1-3 A Car Powered by Nuclear Fuel

An average car consumes about 5 L of gasoline a day, and the capacity of the fuel tank of a car is about 50 L. Therefore, a car needs to be refueled once every 10 days. Also, the density of gasoline ranges from 0.72 to 0.78 kg/L, and its lower heating value is about 44,000 kJ/kg (that is, 44,000 kJ of heat is released when 1 kg of gasoline is completely burned). Suppose all the problems associated with the radioactivity and waste disposal of nuclear fuels are resolved, and a car is to be powered by U-235. If a new car comes equipped with 0.1-kg of the nuclear fuel U-235, determine if this car will ever need refueling under average driving conditions (Fig. 1-22).

FIGURE 1-22
Schematic for Example 1-3.

Solution Taking the average density of the gasoline to be 0.75 kg/L, the mass of gasoline used per day by the car is determined to be

$$m_{\text{gasoline}} = (\rho V)_{\text{gasoline}} = (0.75 \text{ kg/L})(5 \text{ L/day}) = 3.75 \text{ kg/day}$$

Noting that the heating value of gasoline is 44,000 kJ/kg, the energy supplied to the car per day is

$$E_{\text{day}} = (m_{\text{gasoline}})(\text{Heating value}) = (3.75 \text{ kg/day})(44{,}00 \text{ kJ/kg}) = 165{,}000 \text{ kJ/day}$$

The complete fission of 0.1 kg of uranium-235 releases

$$(6.73 \times 10^{10} \text{ kJ/kg})(0.1 \text{ kg}) = 6.73 \times 10^{9} \text{ kJ}$$

of heat, which is sufficient to meet the energy needs of the car for

$$\text{No. of days} = \frac{\text{Energy content of fuel}}{\text{Daily energy use}} = \frac{6.73 \times 10^9 \text{ kJ}}{165{,}000 \text{ kJ/day}} = \mathbf{40{,}790 \text{ days}}$$

which is equivalent to about 112 years. Considering that no car will last more than 100 years, this car will never need refueling. It appears that nuclear fuel no larger than a cherry is sufficient to power a car during its lifetime.

1-5 ■ PROPERTIES OF A SYSTEM

Any characteristic of a system is called a **property.** Some familiar properties are pressure P, temperature T, volume V, and mass m. The list can be extended to include less familiar ones such as viscosity, thermal conductivity, modulus of elasticity, thermal expansion coefficient, electric resistivity, and even velocity and elevation.

Not all properties are independent, however. Some are defined in terms of other ones. For example, **density** is defined as *mass per unit volume,*

$$\rho = \frac{m}{V} \qquad (\text{kg/m}^3) \tag{1-10}$$

Sometimes the density of a substance is given relative to the density of a well-known substance. Then it is called **specific gravity,** or **relative density,** and is defined as *the ratio of the density of a substance to the density of some standard substance at a specified temperature* (usually water at 4°C, for which $\rho_{H_2O} = 1000$ kg/m^3). That is,

$$\rho_s = \frac{\rho}{\rho_{H_2O}} \tag{1-11}$$

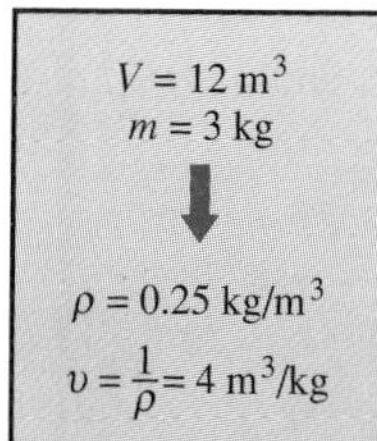

FIGURE 1-23
Density is mass per unit volume; specific volume is volume per unit mass.

Note that specific gravity is a dimensionless quantity.

A more frequently used property in thermodynamics is the **specific volume.** It is the reciprocal of density (Fig. 1-23) and is defined as *the volume per unit mass*:

$$v = \frac{V}{m} = \frac{1}{\rho} \qquad (\text{m}^3/\text{kg}) \tag{1-12}$$

Note that in classical thermodynamics, the atomic structure of a substance (thus, the spaces between and within the molecules) is disregarded, and the substance is viewed to be a continuous, homogeneous matter with no microscopic holes, that is, a **continuum.** This idealization is valid as long as we work with volumes, areas, and lengths that are large relative to the intermolecular spacings.

FIGURE 1-24
Intensive properties are independent of the size of the system.

ρ, kg/m^3
υ, m^3/kg
e, kJ/kg
u, kJ/kg

Properties are considered to be either *intensive* or *extensive.* **Intensive properties** are those that are independent of the size of a system, such as temperature, pressure, and density (Fig. 1-24). **Extensive properties** are

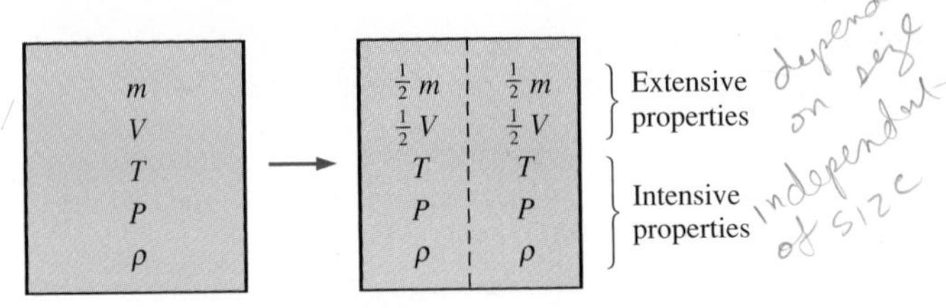

FIGURE 1-25

Criteria to differentiate intensive and extensive properties.

those whose values depend on the size—or extent—of the system. Mass m, volume V, and total energy E are some examples of extensive properties. An easy way to determine whether a property is intensive or extensive is to divide the system into two equal parts with a partition, as shown in Fig. 1-25. Each part will have the same value of intensive properties as the original system, but half the value of the extensive properties.

Generally, uppercase letters are used to denote extensive properties (with mass m being a major exception), and lowercase letters are used for intensive properties (with pressure P and temperature T being the obvious exceptions).

Extensive properties per unit mass are called **specific properties.** Some examples of specific properties are specific volume ($v = V/m$), specific total energy ($e = E/m$), and specific internal energy ($u = U/m$).

1-6 ■ STATE AND EQUILIBRIUM

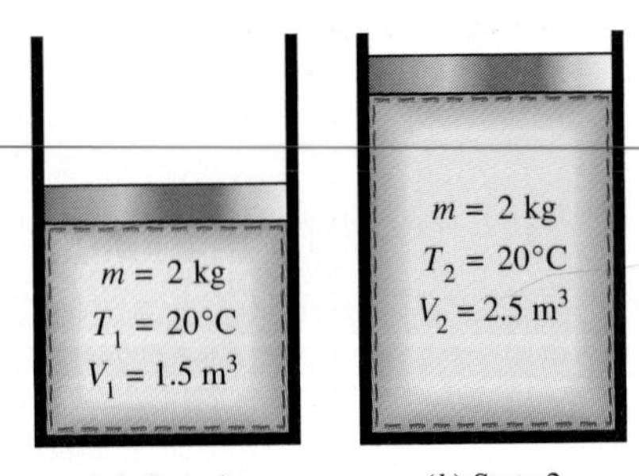

(*a*) State 1 (*b*) State 2

FIGURE 1-26

A system at two different states.

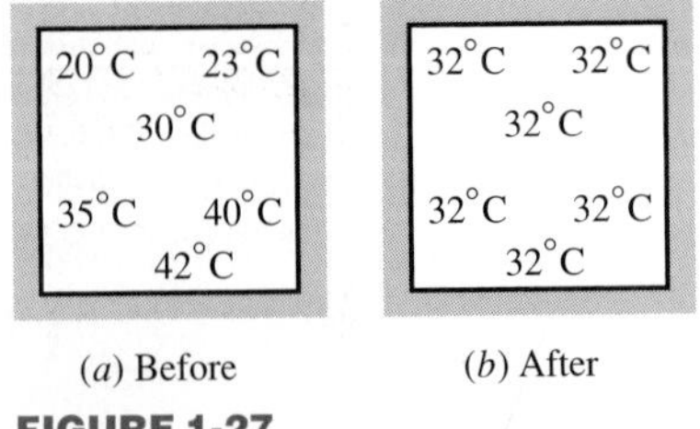

(*a*) Before (*b*) After

FIGURE 1-27

A closed system reaching thermal equilibrium.

Consider a system not undergoing any change. At this point, all the properties can be measured or calculated throughout the entire system, which gives us a set of properties that completely describe the condition, or the **state,** of the system. At a given state, all the properties of a system have fixed values. If the value of even one property changes, the state will change to a different one. In Fig. 1-26 a system is shown at two different states.

Thermodynamics deals with **equilibrium** states. The word *equilibrium* implies a state of balance. In an equilibrium state there are no unbalanced potentials (or driving forces) within the system. A system in equilibrium experiences no changes when it is isolated from its surroundings.

There are many types of equilibrium, and a system is not in thermodynamic equilibrium unless the conditions of all the relevant types of equilibrium are satisfied. For example, a system is in **thermal equilibrium** if the temperature is the same throughout the entire system, as shown in Fig. 1-27*b*. That is, the system involves no temperature differential, which is the driving force for heat flow. **Mechanical equilibrium** is related to pressure, and a system is in mechanical equilibrium if there is no change in pressure at any point of the system with time. However, the pressure may vary within the system with elevation as a result of gravitational effects. But the higher pressure at a bottom layer is balanced by the extra weight it must carry, and, therefore, there is no imbalance of forces. The variation of pressure as a result of gravity in most thermodynamic systems is relatively small and usually disregarded. If a system involves two phases, it is in **phase equilibrium** when the mass of each phase reaches an equilibrium level and stays there. Finally, a system is in **chemical equilibrium** if its chemical composition does not change with time, that is, no chemical reactions occur. A system will not be in equilibrium unless all the relevant equilibrium criteria are satisfied.

1-7 ■ PROCESSES AND CYCLES

Any change that a system undergoes from one equilibrium state to another is called a **process,** and the series of states through which a system passes during a process is called the **path** of the process (Fig. 1-28). To describe a process completely, one should specify the initial and final states of the process, as well as the path it follows, and the interactions with the surroundings.

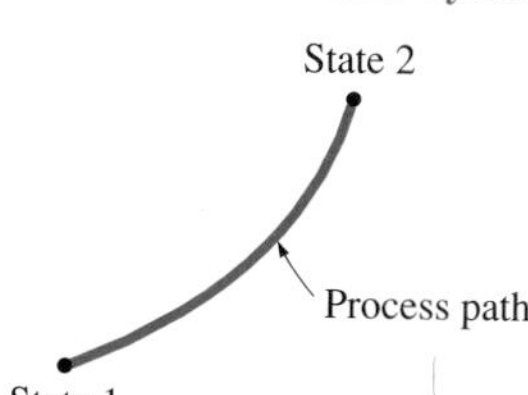

FIGURE 1-28
A process between states 1 and 2 and the process path.

When a process proceeds in such a manner that the system remains infinitesimally close to an equilibrium state at all times, it is called a **quasi-static,** or **quasi-equilibrium, process.** A quasi-equilibrium process can be viewed as a sufficiently slow process that allows the system to adjust itself internally so that properties in one part of the system do not change any faster than those at other parts.

This is illustrated in Fig. 1-29. When a gas in a piston-cylinder device is compressed suddenly, the molecules near the face of the piston will not have enough time to escape and they will have to pile up in a small region in front of the piston, thus creating a high-pressure region there. Because of this pressure difference, the system can no longer be said to be in equilibrium, and this makes the entire process non-quasi-equilibrium. However, if the piston is moved slowly, the molecules will have sufficient time to redistribute and there will not be a molecule pileup in front of the piston. As a result, the pressure inside the cylinder will always be uniform and will rise at the same rate at all locations. Since equilibrium is maintained at all times, this is a quasi-equilibrium process.

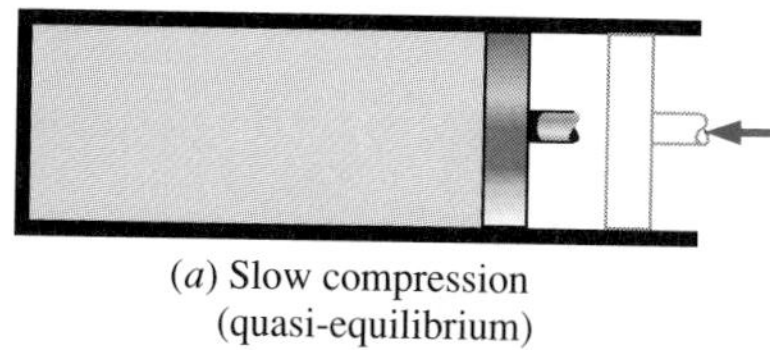

(*a*) Slow compression (quasi-equilibrium)

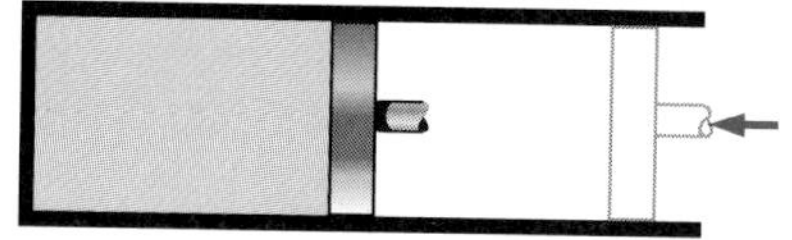

(*b*) Very fast compression (non-quasi-equilibrium)

FIGURE 1-29
Quasi-equilibrium and non-quasi-equilibrium compression processes.

It should be pointed out that a quasi-equilibrium process is an idealized process and is not a true representation of an actual process. But many actual processes closely approximate it, and they can be modeled as quasi-equilibrium with negligible error. Engineers are interested in quasi-equilibrium processes for two reasons. First, they are easy to analyze; second, work-producing devices deliver the most work when they operate on quasi-equilibrium processes (Fig. 1-30). Therefore, quasi-equilibrium processes serve as standards to which actual processes can be compared.

FIGURE 1-30
Work-producing devices operating in a quasi-equilibrium manner deliver the most work.

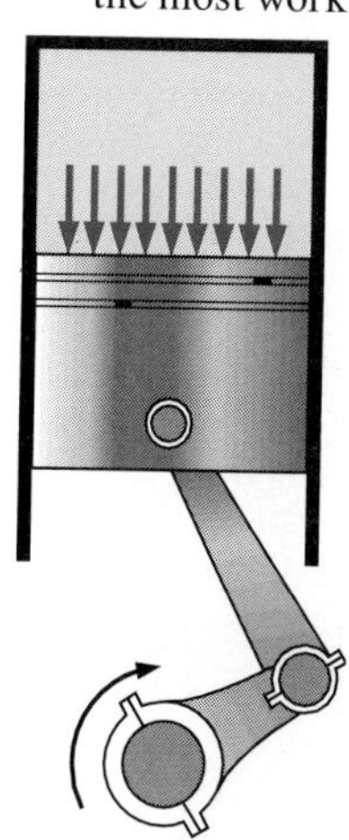

Process diagrams plotted by employing thermodynamic properties as coordinates are very useful in visualizing the processes. Some common properties that are used as coordinates are temperature T, pressure P, and volume V (or specific volume v). Figure 1-31 shows the P-V diagram of a compression process of a gas.

Note that the process path indicates a series of equilibrium states through which the system passes during a process and has significance for quasi-equilibrium processes only. For non-quasi-equilibrium processes, we are not able to specify the states through which the system passes during the process and so we cannot speak of a process path. A non-quasi-equilibrium process is denoted by a dashed line between the initial and final states instead of a solid line.

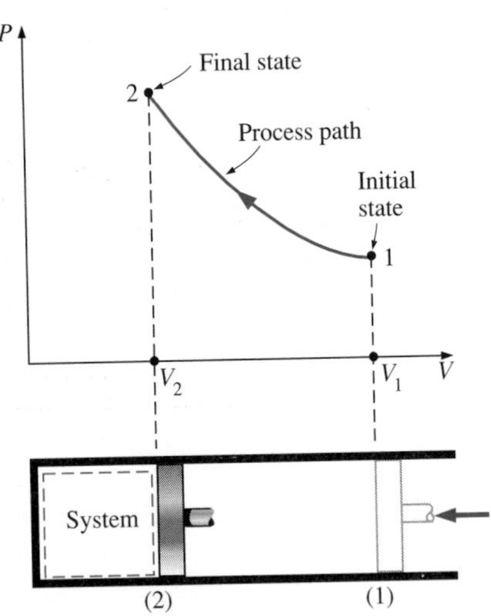

FIGURE 1-31
The *P-V* diagram of a compression process.

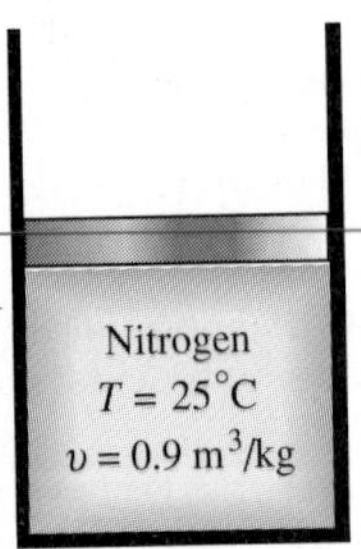

FIGURE 1-32
The state of nitrogen is fixed by two independent, intensive properties.

FIGURE 1-33
The pressure of a fluid at rest increases with depth (as a result of added weight).

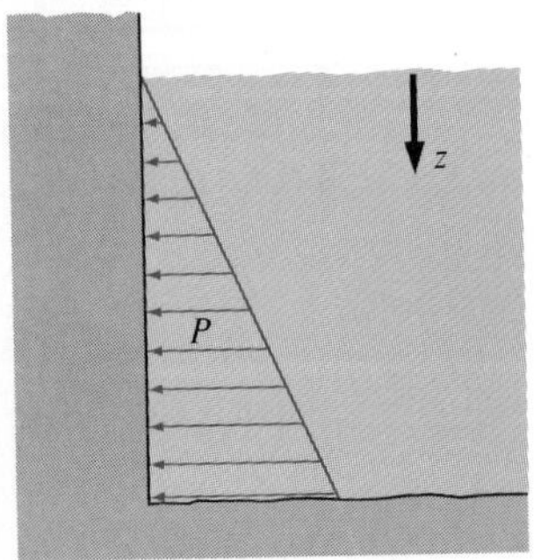

The prefix *iso-* is often used to designate a process for which a particular property remains constant. An **isothermal process,** for example, is a process during which the temperature T remains constant; an **isobaric process** is a process during which the pressure P remains constant; and an **isochoric** (or **isometric**) **process** is a process during which the specific volume v remains constant.

A system is said to have undergone a **cycle** if it returns to its initial state at the end of the process. That is, for a cycle the initial and final states are identical.

1-8 ■ THE STATE POSTULATE

As noted earlier, the state of a system is described by its properties. But we know from experience that we do not need to specify all the properties in order to fix a state. Once a sufficient number of properties are specified, the rest of the properties assume certain values automatically. That is, specifying a certain number of properties is sufficient to fix a state. The number of properties required to fix the state of a system is given by the **state postulate**:

The state of a simple compressible system is completely specified by two independent, intensive properties.

A system is called a **simple compressible system** in the absence of electrical, magnetic, gravitational, motion, and surface tension effects. These effects are due to external force fields and are negligible for most engineering problems. Otherwise, an additional property needs to be specified for each effect that is significant. If the gravitational effects are to be considered, for example, the elevation z needs to be specified in addition to the two properties necessary to fix the state.

The state postulate requires that the two properties specified be **independent** to fix the state. Two properties are independent if one property can be varied while the other one is held constant. Temperature and specific volume, for example, are always independent properties, and together they can fix the state of a simple compressible system (Fig. 1-32). Temperature and pressure, however, are independent properties for single-phase systems, but are dependent properties for multiphase systems. At sea level (P = 1 atm), water boils at 100°C, but on a mountaintop where the pressure is lower, water boils at a lower temperature. That is, $T = f(P)$ during a phase-change process; thus, temperature and pressure are not sufficient to fix the state of a two-phase system. Phase-change processes are discussed in detail in the next chapter.

1-9 ■ PRESSURE

Pressure is *the force exerted by a fluid per unit area*. We speak of pressure only when we deal with a gas or a liquid. The counterpart of pressure in solids is *stress*. For a fluid at rest, the pressure at a given point is the same in all directions. The pressure in a fluid increases with depth as a result of the weight of the fluid, as shown in Fig. 1-33. This is due to the fluid at lower levels carrying more weight than the fluid at upper levels. The pressure varies in the vertical direction as a result of gravitational effects, but there is no variation in the horizontal direction. The pressure in a tank containing a gas may be considered to be uniform since the weight of the gas is too small to make a significant difference (Fig. 1-34).

Since pressure is defined as force per unit area, it has the unit of newtons per square meter (N/m^2), which is called a **pascal** (Pa). That is,

$$1 \text{ Pa} = 1 \text{ N/m}^2$$

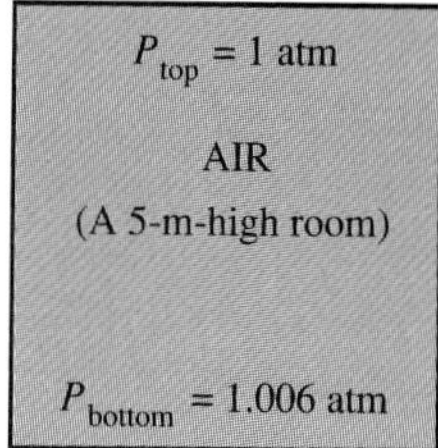

FIGURE 1-34

In a container filled with a gas, the variation of pressure with height is negligible.

The pressure unit pascal is too small for pressures encountered in practice. Therefore, its multiples *kilopascal* (1 kPa = 10^3 Pa) and *megapascal* (1 MPa = 10^6 Pa) are commonly used. Two other common pressure units are the *bar* and *standard atmosphere*:

$$1 \text{ bar} = 10^5 \text{ Pa} = 0.1 \text{ MPa} = 100 \text{ kPa}$$
$$1 \text{ atm} = 101{,}325 \text{ Pa} = 101.325 \text{ kPa} = 1.01325 \text{ bars}$$

In the English system, the pressure unit is *pound-force per square inch* (lbf/in^2, or psi), and 1 atm = 14.696 psi.

The actual pressure at a given position is called the **absolute pressure,** and it is measured relative to absolute vacuum, that is, absolute zero pressure. Most pressure-measuring devices, however, are calibrated to read zero in the atmosphere (Fig. 1-35), and so they indicate the difference between the absolute pressure and the local atmospheric pressure. This difference is called the **gage pressure.** Pressures below atmospheric pressure are called **vacuum pressures** and are measured by vacuum gages that indicate the difference between the atmospheric pressure and the absolute pressure. Absolute, gage, and vacuum pressures are all positive quantities and are related to each other by

FIGURE 1-35

A pressure gage open to the atmosphere reads zero.

$$P_{gage} = P_{abs} - P_{atm} \quad \text{(for pressures above } P_{atm}\text{)} \qquad (1\text{-}13)$$

$$P_{vac} = P_{atm} - P_{abs} \quad \text{(for pressures below } P_{atm}\text{)} \qquad (1\text{-}14)$$

This is illustrated in Fig. 1-36.

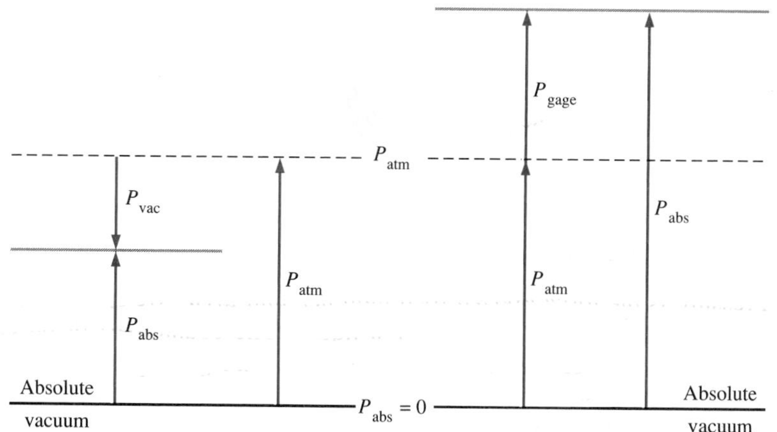

FIGURE 1-36

Absolute, gage, and vacuum pressures.

EXAMPLE 1-4 Absolute Pressure of a Vacuum Chamber

A vacuum gage connected to a chamber reads 5.8 psi at a location where the atmospheric pressure is 14.5 psi. Determine the absolute pressure in the chamber.

Solution The absolute pressure is easily determined from Eq. 1-14:

$$P_{abs} = P_{atm} - P_{vac} = 14.5 - 5.8 = \mathbf{8.7\ psi}$$

In thermodynamic relations and tables, absolute pressure is almost always used. Throughout this text, the pressure P will denote *absolute pressure* unless it is otherwise specified. Often the letters "a" (for absolute pressure) and "g" (for gage pressure) are added to pressure units (such as psia and psig) in order to clarify what is meant.

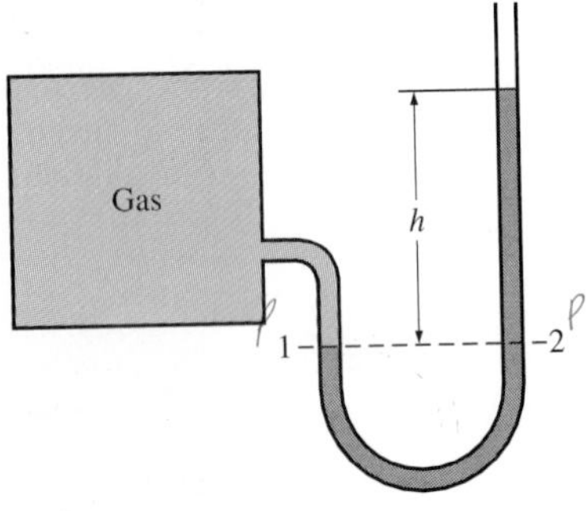

FIGURE 1-37
The basic manometer.

Manometer

Small and moderate pressure differences are often measured by using a device known as a **manometer,** which mainly consists of a glass or plastic U-tube containing a fluid such as mercury, water, alcohol, or oil. To keep the size of the manometer at a manageable level, heavy fluids such as mercury are used if large pressure differences are anticipated.

Consider the manometer shown in Fig. 1-37 that is used to measure the pressure in the tank. Since the gravitational effects of gases are negligible, the pressure anywhere in the tank and at position 1 has the same value. Furthermore, since pressure in a fluid does not vary in the horizontal direction within a fluid, the pressure at 2 is the same as the pressure at 1, or $P_2 = P_1$.

The differential fluid column of height h is in static equilibrium, and its free-body diagram is shown in Fig. 1-38. A force balance in the vertical direction gives

$$AP_1 = AP_{atm} + W$$

where $$W = mg = \rho Vg = \rho Ahg$$

Thus, $$P_1 = P_{atm} + \rho gh \qquad \text{(kPa)} \qquad (1\text{-}15)$$

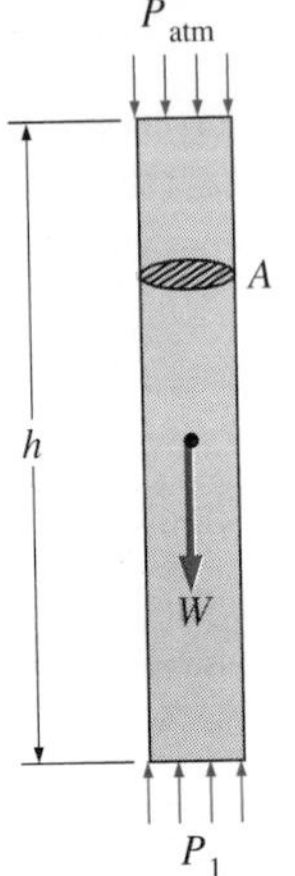

FIGURE 1-38
The free-body diagram of a fluid column of height h.

In the above relations, W is the weight of the fluid column, ρ is the density of the fluid and is assumed to be constant, g is the local gravitational acceleration, A is the cross-sectional area of the tube, and P_{atm} is the atmospheric pressure. The pressure difference can be expressed as

$$\Delta P = P_1 - P_{atm} = \rho gh \qquad \text{(kPa)} \qquad (1\text{-}16)$$

Note that the cross-sectional area of the tube has no effect on the height differential h, and thus the pressure exerted by the fluid.

EXAMPLE 1-5 Measuring Pressure with a Manometer

A manometer is used to measure the pressure in a tank. The fluid used has a specific gravity of 0.85, and the manometer column height is 55 cm, as shown in

Fig. 1-39. If the local atmospheric pressure is 96 kPa, determine the absolute pressure within the tank.

Solution The gravitational acceleration is not specified, so we assume the standard value of 9.807 m/s². The density of the fluid is obtained by multiplying its specific gravity by the density of water, which is taken to be 1000 kg/m³:

$$\rho = (\rho_s)(\rho_{H_2O}) = (0.85)(1000 \text{ kg/m}^3) = 850 \text{ kg/m}^3$$

From Eq. 1-15,

$$\begin{aligned} P &= P_{atm} + \rho g h \\ &= 96 \text{ kPa} + (850 \text{ kg/m}^3)(9.807 \text{ m/s}^2)(0.55 \text{ m})\left(\frac{1 \text{ kPa}}{1000 \text{ N/m}^2}\right) \\ &= \mathbf{100.6 \text{ kPa}} \end{aligned}$$

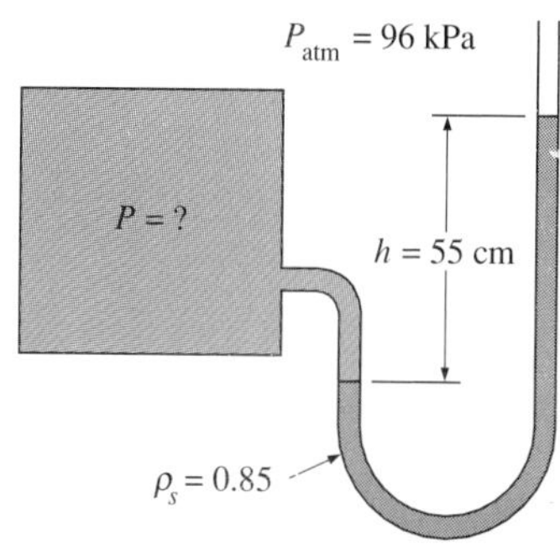

FIGURE 1-39
Sketch for Example 1-5.

Another type of commonly used mechanical pressure measurement device is the **Bourdon tube,** named after the French inventor Eugene Bourdon, which consists of a hollow metal tube bent like a hook whose end is closed and connected to a dial indicator needle (Fig. 1-40). When the fluid inside the tube is pressurized, the tube stretches and moves the needle on the dial in proportion to the pressure applied.

Electronics have made their way into every aspect of life, including pressure measurement devices. Modern pressure sensors, called **pressure transducers,** are made of semiconductor materials such as silicon and convert the pressure effect to an electrical effect such as a change in voltage, resistance, or capacitance. Pressure transducers are smaller and faster, and they are more sensitive, reliable, and precise than their mechanical counterparts. They can measure pressures from less than a millionth of atm to several thousands of atm.

A wide variety of pressure transducers are available to measure gage, absolute, and differential pressures in a wide range of applications. Gage pressure transducers use the atmospheric pressure as a reference by venting the back side of the pressure-sensing diaphragm to the atmosphere, and they give a zero signal output at atmospheric pressure regardless of altitude. The absolute pressure transducers are calibrated to have a zero signal output at full vacuum. Differential pressure transducers measure the pressure between two locations directly instead of using two pressure transducers and taking their difference.

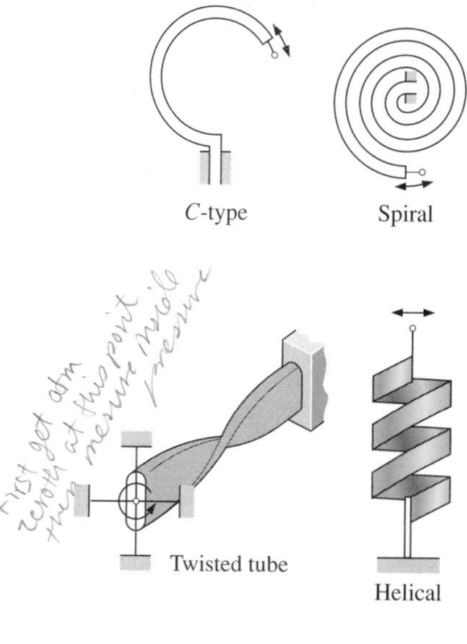

FIGURE 1-40
Various types of Bourdon tubes used to measure pressure.

The emergence of an electric potential in a crystalline substance when subjected to mechanical pressure is called the **piezoelectric** (or press-electric) **effect.** This phenomenon, first discovered by brothers Pierre and Jacques Curie in 1880, forms the basis for the widely used **strain-gage** pressure transducers. The sensors of such transducers are made of thin metal wires or foil whose electrical resistance changes when strained under the influence of fluid pressure. The change in the resistance is determined by supplying electric current to the sensor and measuring the corresponding change in voltage drop that is proportional to the applied pressure.

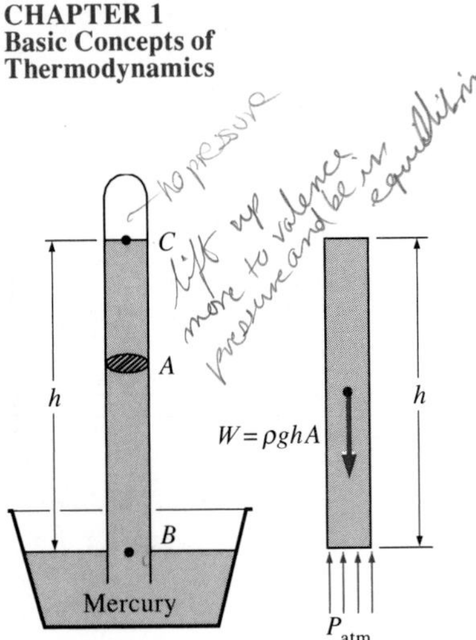

FIGURE 1-41
The basic barometer.

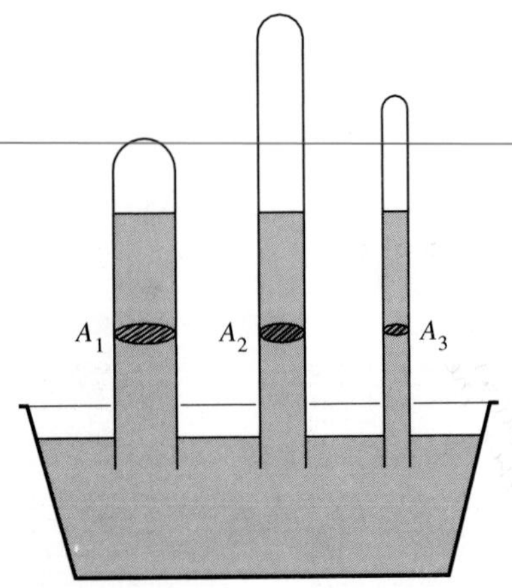

FIGURE 1-42
The length or the cross-sectional area of the tube has no effect on the height of the fluid column of a barometer.

Barometer

The atmospheric pressure is measured by a device called a **barometer**; thus the atmospheric pressure is often called the *barometric pressure*.

As Torricelli (1608–1647) discovered some centuries ago, the atmospheric pressure can be measured by inverting a mercury-filled tube into a mercury container that is open to the atmosphere, as shown in Fig. 1-41. The pressure at point B is equal to the atmospheric pressure, and the pressure at C can be taken to be zero since there is only mercury vapor above point C and the pressure it exerts is negligible. Writing a force balance in the vertical direction gives

$$P_{atm} = \rho g h \qquad \text{(kPa)} \qquad \text{(1-17)}$$

where ρ is the density of mercury, g is the local gravitational acceleration, and h is the height of the mercury column above the free surface. Note that the length and the cross-sectional area of the tube have no effect on the height of the fluid column of a barometer (Fig. 1-42).

A frequently used pressure unit is the *standard atmosphere,* which is defined as the pressure produced by a column of mercury 760 mm in height at 0°C (ρ_{Hg} = 13,595 kg/m^3) under standard gravitational acceleration (g = 9.807 m/s^2). If water instead of mercury were used to measure the standard atmospheric pressure, a water column of about 10.3 m would be needed. Pressure is sometimes expressed (especially by weather forecasters) in terms of the height of the mercury column. The standard atmospheric pressure, for example, is 760 mmHg (29.92 inHg) at 0°C.

The average atmospheric pressure P_{atm} changes from 101.325 kPa at sea level to 89.88, 79.50, 54.05, 26.5, and 5.53 kPa at altitudes of 1000, 2000, 5000, 10,000, and 20,000 meters, respectively. The average atmospheric pressure in Denver (elevation = 1610 m), for example, is 83.4 kPa.

Remember that the atmospheric pressure at a location is simply the weight of the air above that location per unit surface area. Therefore, it changes not only with elevation but also with weather conditions.

EXAMPLE 1-6 Measuring Atmospheric Pressure with a Barometer

Determine the atmospheric pressure at a location where the barometric reading is 740 mmHg and the gravitational acceleration is g = 9.7 m/s^2. Assume the temperature of mercury to be 10°C, at which its density is 13,570 kg/m^3.

Solution From Eq. 1-17, the atmospheric pressure is determined to be

$$\begin{aligned} P_{atm} &= \rho g h \\ &= (13{,}570\ \text{kg/m}^3)(9.7\ \text{m/s}^2)(0.74\ \text{m})\left(\frac{1\ \text{N}}{1\ \text{kg}\cdot\text{m/s}^2}\right)\left(\frac{1\ \text{kPa}}{1000\ \text{N/m}^2}\right) \\ &= \mathbf{97.41\ kPa} \end{aligned}$$

EXAMPLE 1-7 Effect of Piston Weight on Pressure in a Cylinder

The piston of a piston-cylinder device containing a gas has a mass of 60 kg and a cross-sectional area of 0.04 m^2, as shown in Fig. 1-43. The local atmospheric pressure is 0.97 bar, and the gravitational acceleration is 9.8 m/s^2.

(*a*) Determine the pressure inside the cylinder. (*b*) If some heat is transferred to the gas and its volume doubles, do you expect the pressure inside the cylinder to change?

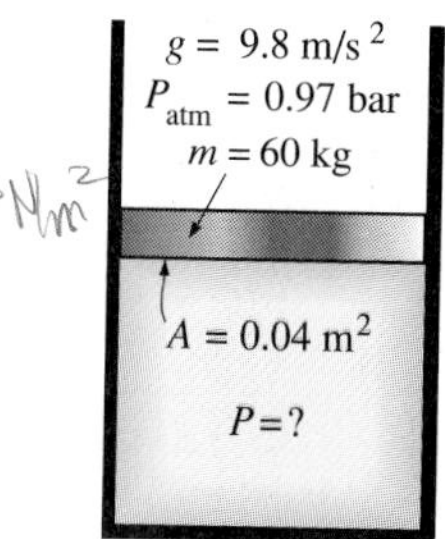

FIGURE 1-43

Sketch for Example 1-7.

Solution (*a*) The gas pressure in the piston-cylinder device depends on the atmospheric pressure and the weight of the piston. Drawing the free-body diagram of the piston (Fig. 1-44) and balancing the vertical forces yield

$$PA = P_{atm}A + W$$

$$P = P_{atm} + \frac{mg}{A}$$

$$= 0.97 \text{ bar} + \frac{(60 \text{ kg})(9.8 \text{ m/s}^2)}{0.04 \text{ m}^2}\left(\frac{1 \text{ N}}{1 \text{ kg}\cdot\text{m/s}^2}\right)\left(\frac{1 \text{ bar}}{10^5 \text{ N/m}^2}\right)$$

$$= \mathbf{1.117 \text{ bars}}$$

(*b*) The volume change will have no effect on the free-body diagram drawn in part (*a*), and therefore the pressure inside the cylinder will remain the same.

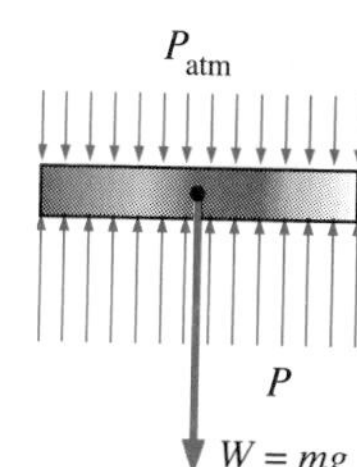

FIGURE 1-44

Free-body diagram of the piston.

1-10 ■ TEMPERATURE AND THE ZEROTH LAW OF THERMODYNAMICS

Although we are familiar with **temperature** as a measure of "hotness" or "coldness," it is not easy to give an exact definition for it. Based on our physiological sensations, we express the level of temperature qualitatively with words like *freezing cold, cold, warm, hot,* and *red-hot*. However, we cannot assign numerical values to temperatures based on our sensations alone. Furthermore, our senses may be misleading. A metal chair, for example, will feel much colder than a wooden one even when both are at the same temperature.

Fortunately, several properties of materials change with temperature in a *repeatable* and *predictable* way, and this forms the basis for accurate temperature measurement. The commonly used mercury-in-glass thermometer, for example, is based on the expansion of mercury with temperature. Temperature is also measured by using several other temperature-dependent properties.

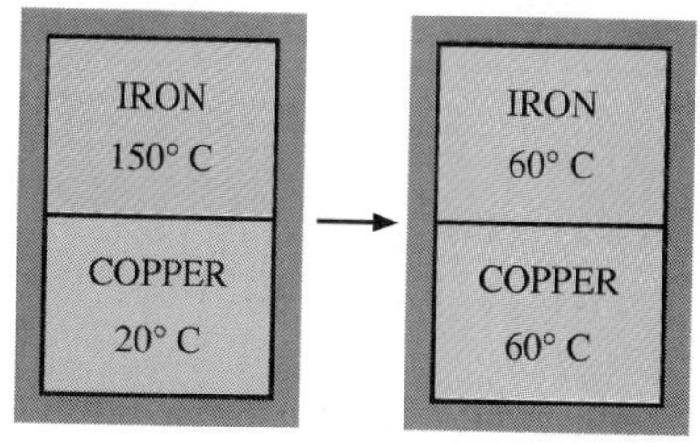

FIGURE 1-45

Two bodies reaching thermal equilibrium after being brought into contact in an isolated enclosure.

It is a common experience that a cup of hot coffee left on the table eventually cools off and a cold drink eventually warms up. That is, when a body is brought into contact with another body that is at a different temperature, heat is transferred from the body at higher temperature to the one at lower temperature until both bodies attain the same temperature (Fig. 1-45). At that point, the heat transfer stops, and the two bodies are said to have reached **thermal equilibrium.** The equality of temperature is the only requirement for thermal equilibrium.

The **zeroth law of thermodynamics** states that if two bodies are in thermal equilibrium with a third body, they are also in thermal equilibrium with each other. It may seem silly that such an obvious fact is called one of the basic laws of thermodynamics. However, it cannot be concluded from the other laws of thermodynamics, and it serves as a basis for the validity of temperature measurement. By replacing the third body with a thermometer, the zeroth law can be restated as *two bodies are in thermal equilibrium if both have the same temperature reading even if they are not in contact.*

The zeroth law was first formulated and labeled by R. H. Fowler in 1931. As the name suggests, its value as a fundamental physical principle was recognized more than half a century after the formulation of the first and the second laws of thermodynamics. It was named the zeroth law since it should have preceded the first and the second laws of thermodynamics.

Temperature Scales

Temperature scales enable scientists to use a common basis for temperature measurements, and several have been introduced throughout history. All temperature scales are based on some easily reproducible states such as the freezing and boiling points of water, which are also called the *ice point* and the *steam point,* respectively. A mixture of ice and water that is in equilibrium with air saturated with vapor at 1 atm pressure is said to be at the ice point, and a mixture of liquid water and water vapor (with no air) in equilibrium at 1 atm pressure is said to be at the steam point.

The temperature scales used in the SI and in the English system today are the **Celsius scale** (formerly called the *centigrade scale*; in 1948 it was renamed after the Swedish astronomer A. Celsius, 1701–1744, who devised it) and the **Fahrenheit scale** (named after the German instrument maker G. Fahrenheit, 1686–1736), respectively. On the Celsius scale, the ice and steam points are assigned the values of 0 and 100°C, respectively. The corresponding values on the Fahrenheit scale are 32 and 212°F. These are often referred to as *two-point scales* since temperature values are assigned at two different points.

In thermodynamics, it is very desirable to have a temperature scale that is independent of the properties of any substance or substances. Such a temperature scale is called a **thermodynamic temperature scale,** which is developed in Chap. 5 in conjunction with the second law of thermodynamics. The thermodynamic temperature scale in the SI is the **Kelvin scale,** named after Lord Kelvin (1824–1907). The temperature unit on this scale is the **kelvin,** which is designated by K (not °K; the degree symbol was officially dropped from kelvin in 1967). The lowest temperature on the Kelvin scale is 0 K. Using nonconventional refrigeration techniques, scientists have approached absolute zero kelvin (they achieved 0.000000002 K in 1989).

The thermodynamic temperature scale in the English system is the **Rankine scale,** named after William Rankine (1820–1872). The temperature unit on this scale is the **rankine,** which is designated by R.

A temperature scale that turns out to be identical to the Kelvin scale is the **ideal gas temperature scale.** The temperatures on this scale are measured using a **constant-volume gas thermometer,** which is basically a rigid vessel filled with a gas, usually hydrogen or helium, at low pressure. This thermometer is based on the principle that *at low pressures, the temperature of a gas is proportional to its pressure at constant volume.* That is, the temperature of a gas of fixed volume varies *linearly* with pressure at sufficiently low pressures. Then the relationship between the temperature and the pressure of the gas in the vessel can be expressed as

$$T = a + bP \tag{1-18}$$

where the values of the constants a and b for a gas thermometer are determined experimentally. Once a and b are known, the temperature of a medium can be calculated from the relation above by immersing the rigid vessel of the gas thermometer into the medium and measuring the gas pressure when thermal equilibrium is established between the medium and the gas in the vessel whose volume is held constant.

An ideal gas temperature scale can be developed by measuring the pressures of the gas in the vessel at two reproducible points (such as the ice and the steam points) and assigning suitable values to temperatures at those two points. Considering that only one straight line passes through two fixed points on a plane, these two measurements are sufficient to determine the constants a and b in Eq. 1-18. Then the unknown temperature T of a medium corresponding to a pressure reading P can be determined from that equation by a simple calculation. The values of the constants will be different for each thermometer, depending on the type and the amount of the gas in the vessel, and the temperature values assigned at the two reference points. If the ice and steam points are assigned the values 0 and 100, respectively, then the gas temperature scale will be identical to the Celsius scale. In this case the value of the constant a (which corresponds to an absolute pressure of zero) is determined to be −273.15°C regardless of the type and the amount of the gas in the vessel of the gas thermometer. That is, on a P-T diagram, all the straight lines passing through the data points in this case will intersect the temperature axis at −273.15°C when extrapolated, as shown in Fig. 1-46. This is the lowest temperature that can be obtained by a gas thermometer, and thus we can obtain an *absolute gas temperature scale* by assigning a value of zero to the constant a in Eq. 1-18. In that case Eq. 1-18 reduces to $T = bP$, and thus we need to specify the temperature at only *one* point to define an absolute gas temperature scale.

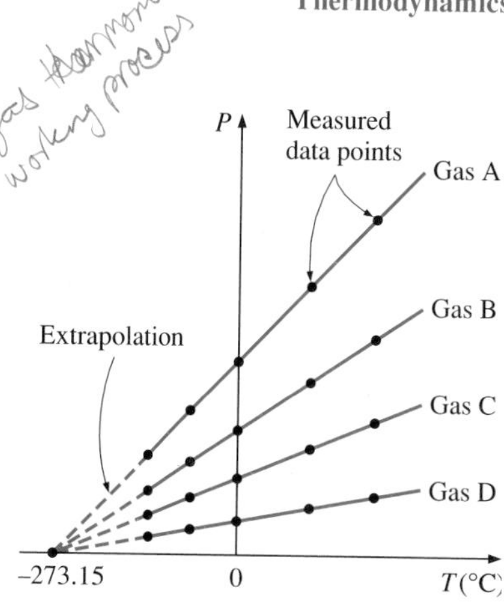

FIGURE 1-46

P versus T plots of the experimental data obtained from a constant-volume gas thermometer using four different gases at different (but low) pressures.

It should be noted that the absolute gas temperature scale is not a thermodynamic temperature scale, since it cannot be used at very low temperatures (due to condensation) and at very high temperatures (due to dissociation and ionization). However, absolute gas temperature is identical to the thermodynamic temperature in the temperature range in which the gas thermometer can be used, and thus we can view the thermodynamic temperature scale at this point as an absolute gas temperature scale that utilizes an "ideal" or "imaginary" gas that always acts as a low-pressure gas regardless of the temperature. If such a gas thermometer existed, it would read zero kelvin at absolute zero pressure, which corresponds to −273.15°C on the Celsius scale (Fig. 1-47).

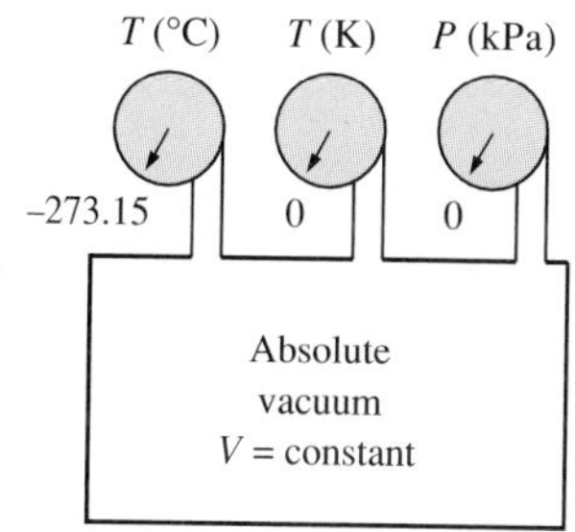

FIGURE 1-47

A constant-volume gas thermometer would read −273.15°C at absolute zero pressure.

The Kelvin scale is related to the Celsius scale by

$$T(\text{K}) = T(^\circ\text{C}) + 273.15 \tag{1-19}$$

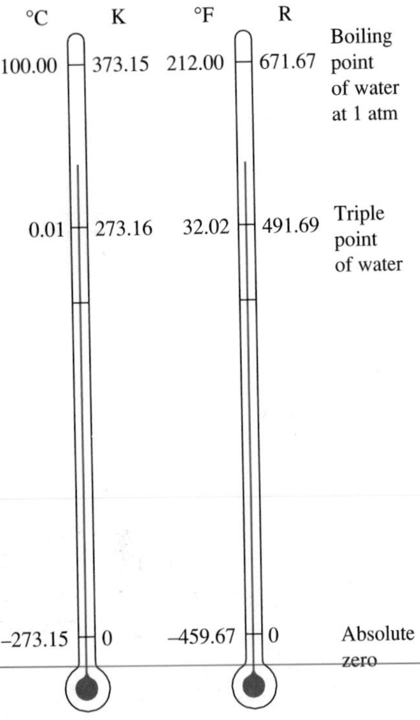

FIGURE 1-48
Comparison of temperature scales.

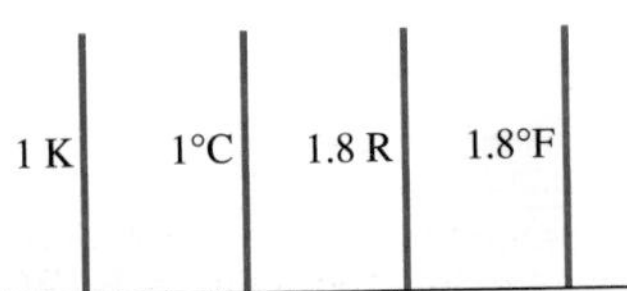

FIGURE 1-49
Comparison of magnitudes of various temperature units.

The Rankine scale is related to the Fahrenheit scale by

$$T(\text{R}) = T(°\text{F}) + 459.67 \tag{1-20}$$

It is common practice to round the constant in Eq. 1-19 to 273 and that in Eq. 1-20 to 460.

The temperature scales in the two unit systems are related by

$$T(\text{R}) = 1.8\,T(\text{K}) \tag{1-21}$$

$$T(°\text{F}) = 1.8\,T(°\text{C}) + 32 \tag{1-22}$$

A comparison of various temperature scales is given in Fig. 1-48.

At the Tenth Conference on Weights and Measures in 1954, the Celsius scale was redefined in terms of a single fixed point and the absolute temperature scale. The selected single point is the *triple point* of water (the state at which all three phases of water coexist in equilibrium), which is assigned the value 0.01°C. The magnitude of the degree is defined from the absolute temperature scale. As before, the boiling point of water at 1 atm pressure is 100.00°C. Thus the new Celsius scale is essentially the same as the old one.

On the Kelvin scale, the size of the temperature unit *kelvin* is defined as "the fraction 1/273.16 of the thermodynamic temperature of the triple point of water, which is assigned the value of 273.16 K." The ice point on the Celsius and Kelvin scales are 0°C and 273.15 K, respectively.

Note that the magnitudes of each division of 1 K and 1°C are identical (Fig. 1-49). Therefore, when we are dealing with temperature differences ΔT, the temperature interval on both scales is the same. Raising the temperature of a substance by 10°C is the same as raising it by 10 K. That is,

$$\Delta T(\text{K}) = \Delta T(°\text{C}) \tag{1-23}$$

$$\Delta T(\text{R}) = \Delta T(°\text{F}) \tag{1-24}$$

Some thermodynamic relations involve the temperature T and often the question arises of whether it is in K or °C. If the relation involves temperature differences (such as $a = b\,\Delta T$), it makes no difference and either can be used. But if the relation involves temperatures only instead of temperature differences (such as $a = bT$) then K must be used. When in doubt, it is always safe to use K because there are virtually no situations in which the use of K is incorrect, but there are many thermodynamic relations that will yield an erroneous result if °C is used.

EXAMPLE 1-8 Expressing Temperature Rise in Different Units

During a heating process, the temperature of a system rises by 10°C. Express this rise in temperature in K, °F, and R.

Solution This problem deals with temperature changes, which are identical in Kelvin and Celsius scales. Then from Eq. 1-23,

$$\Delta T(\text{K}) = \Delta T(°\text{C}) = \mathbf{10\ K}$$

The temperature changes in Fahrenheit and Rankine scales are also identical and are related to the changes in Celsius and Kelvin scales through Eqs. 1-21 and 1-24:

$$\Delta T(\text{R}) = 1.8\,\Delta T(\text{K}) = (1.8)(10) = \mathbf{18\ R}$$

and

$$\Delta T(°\text{F}) = \Delta T(\text{R}) = \mathbf{18°F}$$

1-11 ■ THERMODYNAMIC ASPECTS OF BIOLOGICAL SYSTEMS

An important and exciting application area of thermodynamics is biological systems, which are the sites of rather complex and intriguing energy transfer and transformation processes. Biological systems are not in thermodynamic equilibrium, and thus they are not easy to analyze. Despite their complexity, biological systems are primarily made up of four simple elements: hydrogen, oxygen, carbon, and nitrogen. In the human body, hydrogen accounts for 63 percent, oxygen 25.5 percent, carbon 9.5 percent, and nitrogen 1.4 percent of all the atoms. The remaining 0.6 percent of the atoms comes from 20 other elements essential for life. By mass, about 72 percent of the human body is water.

The building blocks of living organisms are *cells,* which resemble miniature factories performing functions that are vital for the survival of organisms. A biological system can be as simple as a single cell. The human body contains about 100 trillion cells with an average diameter of 0.01 mm. The membrane of the cell is a semipermeable wall that allows some substances to pass through it while excluding others.

FIGURE 1-50
An average person dissipates energy to the surroundings at a rate of 84 W when resting.

In a typical cell, thousands of chemical reactions occur every second during which some molecules are broken down and energy is released and some new molecules are formed. This high level of chemical activity in the cells, which maintains the human body at a temperature of 37°C while performing the necessary bodily tasks, is called **metabolism.** In simple terms, metabolism refers to the burning of foods such as carbohydrates, fat, and protein. The rate of metabolism in the resting state is called the *basal metabolic rate,* which is the rate of metabolism required to keep a body performing the necessary functions (such as breathing and blood circulation) at zero external activity level. The metabolic rate can also be interpreted as the energy consumption rate for a body. For an average male (30 years old, 70 kg, 1.8-m^2 body surface area), the basal metabolic rate is 84 W. That is, the body dissipates energy to the environment at a rate of 84 W (joules per second), which means that the body is converting chemical energy of the food (or of the body fat if the person has not eaten) into thermal energy at a rate of 84 W (Fig. 1-50). The metabolic rate increases with the level of activity, and it may exceed 10 times the basal metabolic rate when a body is doing strenuous exercise. That is, two people doing heavy exercising in a room may be supplying more energy to the room than a 1-kW electrical resistance heater (Fig. 1-51). The fraction of sensible heat varies from about 40 percent in the

FIGURE 1-51
Two fast-dancing people supply more energy to a room than a 1-kW electric resistance heater.

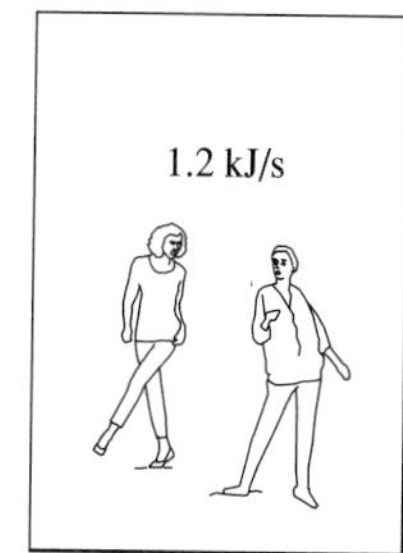

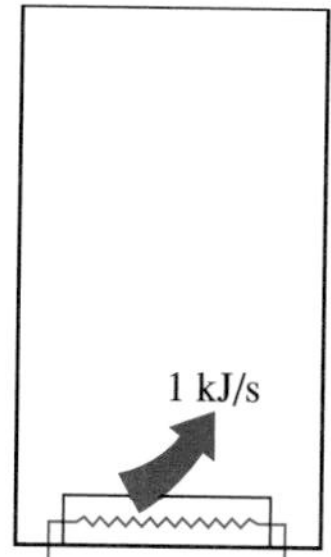

case of heavy work to about 70 percent in the case of light work. The rest of the energy is rejected from the body by perspiration in the form of latent heat.

The basal metabolic rate varies with sex, body size, general health conditions, and so forth, and decreases considerably with age. This is one of the reasons people tend to put on weight in their late twenties and thirties even though they do not increase their food intake. The brain and the liver are the major sites of metabolic activity. These two organs are responsible for almost 50 percent of the basal metabolic rate of an adult human body although they constitute only about 4 percent of the body mass. In small children, it is remarkable that about half of the basal metabolic activity occurs in the brain alone.

The metabolic rate of an animal can be measured directly (*direct calorimetry*) or indirectly (*indirect calorimetry*). In direct calorimetry, the animal is placed in a well-insulated closed box equipped with a water circulating system through all sides of the box. The metabolic energy released by the animal is eventually transferred to the water, and the metabolic rate is determined by measuring the temperature rise in water during the period of observation. Although simple in concept, direct calorimetry is difficult to carry out in practice. Therefore, practically all metabolic measurements today are done with indirect calorimetry, which is much simpler and just as accurate as direct calorimetry.

In indirect calorimetry, the metabolism rate is determined from the measurements of the rates of O_2 consumption and CO_2 production of the body. The ratio of the number of moles of CO_2 produced to the number of moles of O_2 consumed is called the *respiratory quotient* (RQ), whose value depends on the type of food consumed. For example, RQ = 1.0 for glucose ($C_6H_{12}O_6$) since equal number of moles of O_2 and CO_2 are produced when glucose is oxidized (burned). The RQ is 0.84 for protein and 0.707 for fat. In practice, the protein in the diet is ignored in the calculation of the metabolic rate. The error in ignoring the protein is negligible since the protein comprises a small fraction of the diet, and it has an RQ between those of carbohydrate and fat. Under basal conditions, the RQ of an average adult male is 0.80, which corresponds to a metabolic rate of 20.1 kJ/L of O_2 consumed. Thus a good estimate of the average basal metabolic rate of a person is obtained by measuring the number of liters of O_2 the person consumes per unit time, and multiplying this value by 20.1 kJ/L O_2. For example, an average resting adult male consumes O_2 at a rate of 0.250 L/min, which corresponds to a basal metabolic rate of 84 W. In the absence of any food intake, the starving person consumes his or her own body fat and protein. The average basal metabolic rate in this case is 21.3 kJ/L O_2.

The biological reactions in cells occur essentially at constant temperature, pressure, and volume. The temperature of the cell tends to rise when some chemical energy is converted to heat, but this energy is quickly transferred to the circulatory system, which transports it to outer parts of the body and eventually to the environment through the skin.

The muscle cells function very much like an engine, converting the chemical energy into mechanical energy (work) with a conversion efficiency of close to 20 percent. When the body does no net work on the environment (such as moving some furniture upstairs), the entire work is also converted to heat. In that case, the entire chemical energy in the food released during metabolism in the body is eventually transferred to the environment. A TV set that consumes electricity at a rate of 300 W must reject heat to its environment at a rate of 300 W in steady operation regardless of what goes on in the set. That is, turning on a 300-W TV set or three 100-W light bulbs will produce the same heating effect in a room as a 300-W resistance heater (Fig. 1-52). This is a consequence of the conservation of energy principle, which requires that the energy input into a system must equal the energy output when the total energy content of a system remains constant during a process.

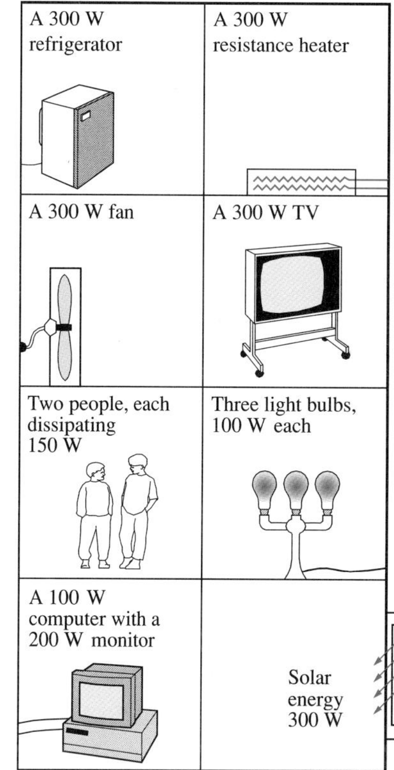

FIGURE 1-52
Some arrangements that supply a room the same amount of energy as a 300-W electric resistance heater.

Food and Exercise

The energy requirements of a body are met by the food we eat. The nutrients in the food are considered in three major groups: carbohydrates, proteins, and fats. *Carbohydrates* are characterized by having hydrogen and oxygen atoms in a 2:1 ratio in their molecules. The molecules of carbohydrates range from very simple (as in plain sugar) to very complex or large (as in starch). Bread and plain sugar are the major sources of carbohydrates. *Proteins* are very large molecules that contain carbon, hydrogen, oxygen, and nitrogen, and they are essential for the building and repairing of the body tissues. Proteins are made up of smaller building blocks called *amino acids*. Complete proteins such as meat, milk, and eggs have all the amino acids needed to build body tissues. Plant source proteins such as those in fruits, vegetables, and grains lack one or more amino acids, and are called incomplete proteins. *Fats* are relatively small molecules that consist of carbon, hydrogen, and oxygen. Vegetable oils and animal fats are major sources of fats. Most foods we eat contain all three nutrition groups at varying amounts. The typical average American diet consists of 45 percent carbohydrate, 40 percent fat, and 15 percent protein, although it is recommended that in a healthy diet less than 30 percent of the calories should come from fat.

The energy content of a given food is determined by burning a small sample of the food in a device called a *bomb calorimeter,* which is basically a well-insulated rigid tank (Fig. 1-53). The tank contains a small combustion chamber surrounded by water. The food is ignited and burned in the combustion chamber in the presence of excess oxygen, and the energy released is transferred to the surrounding water. The energy content of the food is calculated on the basis of the conservation of energy principle by measuring the temperature rise of the water. The carbon in the food is converted into CO_2 and hydrogen into H_2O as the food burns. The same chemical reactions occur in the body, and thus the same amount of energy is released.

Using dry (free of water) samples, the average energy contents of the three basic food groups are determined by bomb calorimeter measurements to be 18.0 MJ/kg for carbohydrates, 22.2 MJ/kg for proteins, and 39.8 MJ/kg for fats. These food groups are not entirely metabolized in the human body, however. The fraction of metabolizable energy contents are 95.5 percent for carbohydrates, 77.5 percent for proteins, and 97.7 percent for fats. That

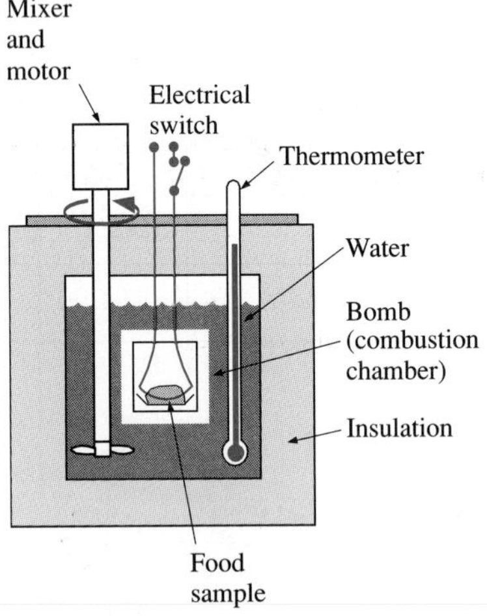

FIGURE 1-53
Schematic of a bomb calorimeter used to determine the energy content of food samples.

Fat: (8g)(9.3 Cal/g) = 74.4 Cal
Protein: (2g)(4.1 Cal/g) = 8.2 Cal
Carbohydrates: (21g)(4.1 Cal/g) = 86.1 Cal
Other: (1g) (0 Cal/g) = 0

TOTAL (for 32 g): 169 Cal

FIGURE 1-54
Evaluating the calorie content of one serving of chocolate chip cookies (values are for Chips Ahoy cookies made by Nabisco).

is, the fats we eat are almost entirely metabolized in the body, but close to one quarter of the protein we eat is discarded from the body unburned. This corresponds to 4.1 Calories/g for proteins and carbohydrates and 9.3 Calories/g for fats (Fig. 1-54) commonly seen in nutrition books and on food labels. The energy contents of the foods we normally eat are much lower than the values above because of the large water content (water adds bulk to the food but it cannot be metabolized or burned, and thus it has no energy value). Most vegetables, fruits, and meats, for example, are mostly water. The average metabolizable energy contents of the three basic food groups are 4.2 MJ/kg for carbohydrates, 8.4 MJ/kg for proteins, and 33.1 MJ/kg for fats. Note that 1 kg of natural fat contains almost 8 times the metabolizable energy of 1 kg of natural carbohydrates. Thus, a person who fills his stomach with fatty foods is consuming much more energy than a person who fills his stomach with carbohydrates such as bread or rice.

The metabolizable energy content of foods is usually expressed by nutritionists in terms of the capitalized *Calories*. One Calorie is equivalent to one *kilocalorie* (1000 calories), which is equivalent to 4.1868 kJ. That is,

$$1 \text{ Cal (Calorie)} = 1000 \text{ calories} = 1 \text{ kcal (kilocalorie)} = 4.1868 \text{ kJ}$$

The calorie notation often causes confusion since it is not always followed in the tables or articles on nutrition. When the topic is food or fitness, a calorie normally means a kilocalorie whether it is capitalized or not.

The **daily calorie needs** of people vary greatly with age, gender, the state of health, the activity level, the body weight, and the composition of the body as well as other factors. A small person needs fewer calories than a larger person of the same sex and age. An average man needs about 2400 to 2700 Calories a day. The daily need of an average woman varies from 1800 to 2200 Calories. The daily calorie needs are about 1600 for sedentary women and some older adults; 2000 for sedentary men and most older adults; 2200 for most children, teenage girls, and active women; 2800 for teenage boys, active men, and some very active women; and above 3000 for very active men. The *average* value of calorie intake is usually taken to be 2000 Calories per day. The daily calorie needs of a person can be determined by multiplying the body weight in pounds (which is 2.205 times the body weight in kg) by 11 for a sedentary person, 13 for a moderately active person, 15 for a moderate exerciser or physical laborer, and 18 for an extremely active exerciser or physical laborer. The extra calories a body consumes are usually stored as fat, which serves as the spare energy of the body for use when the energy intake of the body is less than the needed amount.

Like other natural fat, 1 kg of human body fat contains about 33.1 MJ of metabolizable energy. Therefore, a starving person (zero energy intake) who uses up 2200 Calories (9211 kJ) a day can meet his daily energy intake requirements by burning only 9211/33,100 = 0.28 kg of body fat. So it is no surprise that people are known to survive over 100 days without eating. (They still need to drink water, however, to replenish the water lost through the lungs and the skin to avoid the dehydration that may occur in just a few days.) Although the desire to get rid of the excess fat in a thin world may be overwhelming at times, starvation diets are not recommended because the body soon starts to consume its own muscle tissue in addition to fat. A healthy diet should involve regular exercise while allowing a reasonable amount of calorie intake.

The average metabolizable energy contents of various foods and the energy consumption during various activities are given in Tables 1-3 and 1-4. Considering that no two hamburgers are alike, and that no two people walk exactly the same way, there is some uncertainty in these values, as you would expect. Therefore, you may encounter somewhat different values in other books or magazines for the same items.

The rates of energy consumption listed in Table 1-4 during some activities are for a 68-kg adult. The energy consumed for smaller or larger adults can be determined using the proportionality of the metabolism rate and the body size. For example, the rate of energy consumption by a 68-kg bicyclist is listed in Table 1-4 to be 639 Calories/h. Then the rate of energy consumption by a 50-kg bicyclist is

$$(50\text{ kg})\frac{639\text{ Cal/h}}{68\text{ kg}} = 470\text{ Cal/h}$$

For a 100-kg person, it would be 960 Calories/h.

The thermodynamic analysis of the human body is rather complicated since it involves mass transfer (during breathing, perspiring, etc.) as well as energy transfer. As such, it should be treated as an open system. However, the energy transfer with mass is difficult to quantify. Therefore, the human body is often modeled as a closed system for simplicity by treating energy transported with mass as just energy transfer. For example, eating is modeled as the transfer of energy into the human body in the amount of the metabolizable energy content of the food.

TABLE 1-3

Approximate metabolizable energy content of some common foods (1 Calorie = 4.1868 kJ = 3.968 Btu)

Food	Calories
Apple, (one, medium)	70
Baked potato (plain)	250
Baked potato with cheese	550
Bread (white, one slice)	70
Butter (one teaspoon)	35
Cheeseburger	325
Chocolate candy bar (20 g)	105
Cola (200 ml)	87
Egg (one)	80
Fish sandwich	450
French fries (regular)	250
Hamburger	275
Hot dog	300
Ice cream (100 ml, 10% fat)	110
Lettuce salad with French dressing	150
Milk (skim, 200 ml)	76
Milk (whole, 200 ml)	136
Peach (one, medium)	65
Pie (one $\frac{1}{8}$ slice, 23 cm diameter)	300
Pizza (large, cheese, one $\frac{1}{8}$ slice)	350

EXAMPLE 1-9 Burning Off Lunch Calories

A 90-kg man had two hamburgers, a regular serving of french fries, and a 200-ml Coke for lunch (Fig. 1-55). Determine how long it will take for him to burn the lunch calories off (*a*) by watching TV and (*b*) by fast swimming. What would your answers be for a 45-kg man?

Solution (*a*) We take the human body as our system and treat it as a closed system whose energy content remains unchanged during the process. Then the conservation of energy principle requires that the energy input into the body must be equal to the energy output. The net energy input in this case is the metabolizable energy content of the food eaten. It is determined from Table 1-3 to be

$$\begin{aligned} E_{\text{in}} &= 2 \times E_{\text{hamburger}} + E_{\text{fries}} + E_{\text{cola}} \\ &= 2 \times 275 + 250 + 87 \\ &= 887\text{ Cal} \end{aligned}$$

Then $E_{\text{out}} = E_{\text{in}} = 887$ Calories. The rate of energy output for a 68-kg man watching TV is given in Table 1-4 to be 72 Calories/h. For a 90-kg man it becomes

$$\dot{E}_{\text{out}} = (90\text{ kg})\frac{72\text{ Cal/h}}{68\text{ kg}} = 95.3\text{ Cal/h}$$

Therefore, it will take

$$\Delta t = \frac{887 \text{ Cal}}{95.3 \text{ Cal/h}} = \mathbf{9.3\ h}$$

to burn the lunch calories off by watching TV.

(*b*) It can be shown in a similar manner that it takes only **47 min** to burn the lunch calories off by fast swimming.

The 45-kg man is half as large as the 90-kg man. Therefore, expending the same amount of energy will take twice as long in each case: **18.6 h** by watching TV and **94 min** by fast swimming.

TABLE 1-4

Approximate energy consumption of a 68-kg adult during some activities (1 Calorie = 4.1868 kJ = 3.968 Btu)

Activity	Calories/h
Basal metabolism	72
Basketball	550
Bicycling (21 km/h)	639
Cross-country skiing (13 km/h)	936
Driving a car	180
Eating	99
Fast dancing	600
Fast running (13 km/h)	936
Jogging (8 km/h)	540
Swimming (fast)	860
Swimming (slow)	288
Tennis (advanced)	480
Tennis (beginner)	288
Walking (7.2 km/h)	432
Watching TV	72

FIGURE 1-55

A typical lunch discussed in Example 1-9.

Most diets are based on *calorie counting*; that is, the conservation of energy principle: a person who consumes more calories than his body burns will gain weight whereas a person who consumes less calories than his body burns will lose weight. Yet, people who eat whatever they want whenever they want without gaining any weight are living proof that the calorie-counting technique alone does not work in dieting. Obviously there is more to dieting than keeping track of calories. It should be noted that the phrases *weight gain* and *weight loss* are misnomers. The correct phrases should be *mass gain* and *mass loss*. A man who goes to space loses practically all of his weight but none of his mass. When the topic is food and fitness, *weight* is understood to mean *mass*, and weight is expressed in mass units.

Researchers on nutrition proposed several theories on dieting. One theory suggests that some people have very "food efficient" bodies. These people need fewer calories than other people do for the same activity, just like a fuel-efficient car needing less fuel for traveling a given distance. It is interesting that we want our cars to be fuel efficient but we do not want the same high efficiency for our bodies. One thing that frustrates the dieters is that the body interprets dieting as *starvation* and starts using the energy reserves of the body more stringently. Shifting from a normal 2000-Calorie daily diet to an 800-Calorie diet without exercise is observed to lower the basal metabolic rate by 10 to 20 percent. Although the metabolic rate returns to normal once the dieting stops, extended periods of low-calorie dieting without adequate exercise may result in the loss of considerable muscle tissue together with fat. With less muscle tissue to burn calories, the metabolic rate of the body declines and stays below normal even after a person starts eating normally. As a result, the person regains the weight he or she has lost in the form of fat, plus more. The basal metabolic rate remains about the same in people who exercise while dieting.

Regular moderate exercise is part of any healthy dieting program for good reason: it builds or preserves muscle tissue that burns calories much faster than the fat tissue does. It is interesting that aerobic exercise continues burning calories for several hours after the workout, raising the overall metabolic rate considerably.

Another theory suggests that people with *too many fat cells* developed during childhood or adolescence are much more likely to gain weight. Some people believe that the fat content of the bodies is controlled by the setting of a "fat control" mechanism, much like the temperature of a house is controlled by the thermostat setting.

Some people put the blame for weight problems simply on the *genes*. Considering that 80 percent of the children of overweight parents are also overweight, heredity may indeed play an important role in the way a body stores fat. Researchers from the University of Washington and the Rockefeller University have identified a gene, called the RIIbeta, that seems to control the rate of metabolism. The body tries to keep the body fat at a particular level, called the **set point,** that differs from person to person (Fig. 1-56). This is done by *speeding up* the metabolism and thus burning extra calories much faster when a person tends to gain weight and by *slowing down* the metabolism and thus burning calories at a slower rate when a person tends to lose weight. Therefore, a person who just became slim burns fewer calories than does a person of the same size who has always been slim. Even exercise does not seem to change that. Then to keep the weight off, the newly slim person should consume no more calories than he or she can burn. Note that in people with high metabolic rates, the body dissipates the extra calories as body heat instead of storing them as fat, and thus there is no violation of the conservation of energy principle.

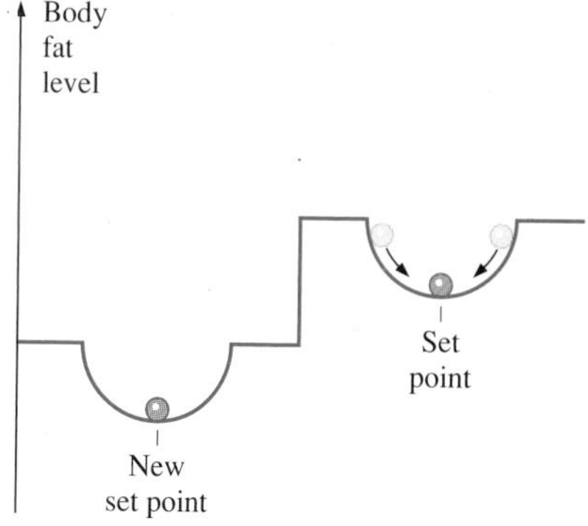

FIGURE 1-56

The body tends to keep the body fat level at a *set point* by speeding up metabolism when a person splurges and by slowing it down when the person starves.

In some people, a *genetic flaw* is believed to be responsible for the extremely low rates of metabolism. Several studies concluded that losing weight for such people is nearly impossible. That is, obesity is a biological phenomenon. But even such people will not gain weight unless they eat more than their body can burn. They just must learn to be content with little food to remain slim, and forget about ever having a normal "eating" life. For most people, genetics determine the range of normal weights. A person may end up at the high or low end of that range, depending on eating and exercise habits. This also explains why some genetically identical twins are not so identical when it comes to body weight. *Hormone imbalance* is also believed to cause excessive weight gain or loss.

Based on his experience, the first author of this book has also developed a diet called the "*sensible diet.*" It consists of two simple rules: eat *whatever* you want *whenever* you want *as much as* you want provided that (1) you do not eat unless you are hungry and (2) you stop eating before you get stuffed. In other words, *listen to your body and don't impose on it.* Don't expect to see this unscientific diet advertised anywhere since there is nothing to be sold and thus no money to be made. Also, it is not as easy as it sounds since food is at the center stage of most leisure activities in social life, and eating and drinking have become synonymous with having a good time. However, it is comforting to know that the human body is quite forgiving of occasional impositions.

The discussions above suggest that the first law of thermodynamics does not give us the complete picture in energy conversion processes, and there is a need for another fundamental principle to complement it. That principle is the second law, which we will study in later chapters.

Being *overweight* is associated with a long list of health risks from high blood pressure to some forms of cancer, especially for people who have a weight-related medical condition such as diabetes, hypertension, and heart

TABLE 1-5

The range of healthy weight for adults of various heights (Source: National Institute of Health)

English units		SI units	
Height, in.	Healthy weight, lbm*	Height, m	Healthy weight, kg*
58	91–119	1.45	40–53
60	97–127	1.50	43–56
62	103–136	1.55	46–60
64	111–146	1.60	49–64
66	118–156	1.65	52–68
68	125–165	1.70	55–72
70	133–175	1.75	58–77
72	140–185	1.80	62–81
74	148–195	1.85	65–86
76	156–205	1.90	69–90

*The upper and lower limits of healthy range correspond to mass body indexes of 19 and 25, respectively.

disease. Therefore, people often wonder if their weight is in the proper range. Well, the answer to this question is not written in stone, but if you cannot see your toes or you can pinch your love handles more than an inch, you don't need an expert to tell you that you went over your range. On the other hand, some people who are obsessed with the weight issue try to lose more weight even though they are actually underweight. Therefore, it is useful to have a scientific criterion to determine physical fitness. The range of healthy weight for adults is usually expressed in terms of the **body mass index** (BMI), defined, in SI units, as

$$\text{BMI} = \frac{W(\text{kg})}{H^2(\text{m}^2)} \quad \text{with} \quad 19 \le \text{BMI} \begin{cases} < 19 & \text{underweight} \\ \le 25 & \text{healthy weight} \\ > 25 & \text{overweight} \end{cases} \tag{1-25}$$

where W is the weight (actually, the mass) of the person in kg and H is the height in m. Therefore, a BMI of 25 is the upper limit for the healthy weight and a person with a BMI of 27 is 8 percent overweight. It can be shown that the formula above is equivalent in English units to BMI = 705 W/H^2 where W is in pounds and H is in inches. The proper range of weight for adults of various heights is given in Table 1-5 in both SI and English units.

EXAMPLE 1-10 Losing Weight by Switching to Fat-Free Chips

The fake fat olestra passes through the body undigested, and thus adds zero calorie to the diet. Although foods cooked with olestra taste pretty good, they may cause abdominal discomfort and the long-term effects are unknown. A 1-oz (28.3-g) serving of regular potato chips has 10 g of fat and 150 Calories, whereas 1 oz of the so-called fat-free chips fried in olestra has only 75 Calories. Consider a person who eats 1 oz of regular potato chips every day at lunch without gaining or losing any weight. Determine how much weight this person will lose in one year if he or she switches to fat-free chips (Fig. 1-57).

FIGURE 1-57
Schematic for Example 1-10.

Solution The person who switches to the fat-free chips will consume 75 fewer Calories a day. Then the annual reduction in the calories consumed becomes

$$E_{\text{reduced}} = (75 \text{ Cal/day})(365 \text{ days/year}) = 27{,}375 \text{ Cal/year}$$

The metabolizable energy content of 1 kg of body fat is 33,100 kJ. Therefore, assuming the deficit in the calorie intake is made up by burning body fat, the person who switches to fat-free chips will lose

$$m_{\text{fat}} = \frac{E_{\text{reduced}}}{\text{Energy content of fat}} = \frac{27{,}375 \text{ Cal}}{33{,}100 \text{ kJ/kg}}\left(\frac{4.1868 \text{ kJ}}{1 \text{ Cal}}\right) = \mathbf{3.46\ kg}$$

(about 7.6 pounds) of body fat that year.

1-12 ■ SUMMARY

In this chapter, the basic concepts of thermodynamics are introduced and discussed. *Thermodynamics* is the science that primarily deals with energy. The *first law of thermodynamics* is simply an expression of the conservation of energy principle, and it asserts that *energy* is a thermodynamic property. The *second law of thermodynamics* asserts that energy has *quality* as well as *quantity,* and actual processes occur in the direction of decreasing quality of energy.

A system of fixed mass is called a *closed system,* or *control mass,* and a system that involves mass transfer across its boundaries is called an *open system,* or *control volume.* The mass-dependent properties of a system are called *extensive properties* and the others, *intensive properties. Density* is mass per unit volume, and *specific volume* is volume per unit mass.

The sum of all forms of energy of a system is called *total energy,* which is considered to consist of internal, kinetic, and potential energies. *Internal energy* represents the molecular energy of a system and may exist in sensible, latent, chemical, and nuclear forms.

A system is said to be in *thermodynamic equilibrium* if it maintains thermal, mechanical, phase, and chemical equilibrium. Any change from one state to another is called a *process.* A process with identical end states is called a *cycle.* During a *quasi-static* or *quasi-equilibrium process,* the system remains practically in equilibrium at all times. The state of a simple, compressible system is completely specified by two independent, intensive properties.

Force per unit area is called *pressure,* and its unit is the *pascal.* The absolute, gage, and vacuum pressures are related by

$$P_{gage} = P_{abs} - P_{atm} \qquad \text{(kPa)}$$
$$P_{vac} = P_{atm} - P_{abs} \qquad \text{(kPa)}$$

Small to moderate pressure differences are measured by a *manometer,* and a differential fluid column of height h corresponds to a pressure difference of

$$\Delta P = \rho g h \qquad \text{(kPa)}$$

where ρ is the fluid density and g is the local gravitational acceleration. The atmospheric pressure is measured by a *barometer* and is determined from

$$P_{atm} = \rho g h \qquad \text{(kPa)}$$

where h is the height of the liquid column above the free surface.

The *zeroth law of thermodynamics* states that two bodies are in thermal equilibrium if both have the same temperature reading even if they are not in contact.

The temperature scales used in the SI and the English system today are the *Celsius scale* and the *Fahrenheit scale,* respectively. The absolute temperature scale in the SI is the *Kelvin scale,* which is related to the Celsius scale by

$$T(\text{K}) = T(^\circ\text{C}) + 273.15$$

In the English system, the absolute temperature scale is the *Rankine scale,* which is related to the Fahrenheit scale by

$$T(\text{R}) = T(^\circ\text{F}) + 459.67$$

The magnitudes of each division of 1 K and 1°C are identical, and so are the magnitudes of each division of 1 R and 1°F. Therefore,

$$\Delta T(\text{K}) = \Delta T(°\text{C})$$

and

$$\Delta T(\text{R}) = \Delta T(°\text{F})$$

An important application area of thermodynamics is the biological system. Most diets are based on the simple energy balance: the net energy gained by a person in the form of fat is equal to the difference between the energy intake from food and the energy expended by exercise.

REFERENCES AND SUGGESTED READING

1. American Society for Testing and Materials. *Standards for Metric Practice.* ASTM E 380-79, January 1980.

2. R. T. Balmer. *Thermodynamics.* St. Paul, MN: West Publishing, 1990.

3. A. Bejan. *Advanced Engineering Thermodynamics.* New York: John Wiley & Sons, 1988.

4. W. Z. Black and J. G. Hartley. *Thermodynamics.* New York: Harper and Row, 1985.

5. M. Snowman. *Food and Fitness.* Syracuse, NY: New Readers Press, 1986.

6. G. J. Van Wylen and R. E. Sonntag. *Fundamentals of Classical Thermodynamics.* 3d ed. New York: John Wiley & Sons, 1985.

7. K. Wark. *Thermodynamics.* 5th ed. New York: McGraw-Hill, 1988.

PROBLEMS*

Thermodynamics and Energy

1-1C What is the difference between the classical and the statistical approaches to thermodynamics?

1-2C Why does a bicyclist pick up speed on a downhill road even when he is not pedaling? Does this violate the conservation of energy principle?

1-3C An office worker claims that a cup of cold coffee on his table warmed up to 80°C by picking up energy from the surrounding air, which is at 25°C. Is there any truth to his claim? Does this process violate any thermodynamic laws?

1-4C A person claims that even drinking water causes him to gain weight. Is there any truth to this claim?

*Students are encouraged to answer *all* the concept "C" questions.

Mass, Force, and Acceleration

1-5C What is the difference between pound-mass and pound-force?

1-6C What is the net force acting on a car cruising at a constant velocity of 70 km/h (*a*) on a level road and (*b*) on an uphill road?

1-7 A 3-kg plastic tank that has a volume of 0.2 m^3 is filled with liquid water. Assuming the density of water is 1000 kg/m^3, determine the weight of the combined system.

1-8 Determine the mass and the weight of the air contained in a room whose dimensions are 6 m $\times$ 6 m $\times$ 8 m. Assume the density of the air is 1.16 kg/m^3. *Answers:* 334.1 kg, 3277 N

1-9 At 45° latitude, the gravitational acceleration as a function of elevation z above sea level is given by $g = a - bz$, where $a = 9.807$ m/s^2 and $b = 3.32 \times 10^{-6}$ s^{-2}. Determine the height above sea level where the weight of an object will decrease by 1 percent. *Answer:* 29,539 m

1-10E A 150-lbm astronaut took his bathroom scale (a spring scale) and a beam scale (compares masses) to the moon where the local gravity is $g = 5.48$ ft/s^2. Determine how much he will weigh (*a*) on the spring scale and (*b*) on the beam scale. *Answers:* (*a*) 25.5 lbf; (*b*) 150 lbf

1-11 The acceleration of high-speed aircraft is sometimes expressed in g's (in multiples of the standard acceleration of gravity). Determine the net upward force, in N, that a 90-kg man would experience in an aircraft whose acceleration is 6 g's.

1-12 A 5-kg rock is thrown upward with a force of 150 N at a location where the local gravitational acceleration is 9.79 m/s^2. Determine the acceleration of the rock, in m/s^2.

1-13 The value of the gravitational acceleration g decreases with elevation from 9.807 m/s^2 at sea level to 9.4175 m/s^2 at an altitude of 13,000 m, where large passenger planes cruise. Determine the percent reduction in the weight of an airplane cruising at 13,000 m relative to its weight at sea level.

Forms of Energy, Systems, State, Properties

FIGURE P1-14C

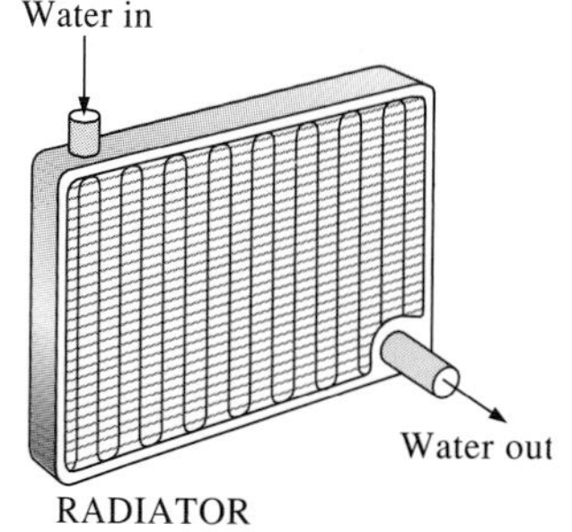

1-14C Most of the energy generated in the engine of a car is rejected to the air by the radiator through the circulating water. Should the radiator be analyzed as a closed system or as an open system? Explain.

1-15C A can of soft drink at room temperature is put into the refrigerator so that it will cool. Would you model the can of soft drink as a closed system or as an open system? Explain.

1-16C Portable electric heaters are commonly used to heat small rooms. Explain the energy transformation involved during this heating process.

1-17C Consider the process of heating water on top of an electric range. What are the forms of energy involved during this process? What are the energy transformations that take place?

1-18C What is the difference between the macroscopic and microscopic forms of energy?

1-19C What is total energy? Identify the different forms of energy that constitute the total energy.

1-20C List the forms of energy that contribute to the internal energy of a system.

1-21C How are heat, internal energy, and thermal energy related to each other?

1-22C What is the difference between intensive and extensive properties?

1-23C For a system to be in thermodynamic equilibrium, do the temperature and the pressure have to be the same everywhere?

1-24C What is a quasi-equilibrium process? What is its importance in engineering?

1-25C Define the isothermal, isobaric, and isochoric processes.

1-26C What is the state postulate?

1-27C Is the state of the air in an isolated room completely specified by the temperature and the pressure? Explain.

1-28 Consider a nuclear power plant that produces 1000 MW of power and has a conversion efficiency of 30 percent (that is, for each unit of fuel energy used, the plant produces 0.3 unit of electrical energy). Assuming continuous operation, determine the amount of nuclear fuel consumed by this plant per year.

1-29 Repeat Prob. 1-28 for a coal power plant that burns coal whose heating value is 28,000 kJ/kg.

1-30 When a hydrocarbon fuel is burned, almost all of the carbon in the fuel burns completely to form CO_2 (carbon dioxide), which is the principal gas causing the greenhouse effect and thus global climate change. On average, 0.59 kg of CO_2 is produced for each kWh of electricity generated from a power plant that burns natural gas. A typical new household refrigerator uses about 700 kWh of electricity per year. Determine the amount of CO_2 production that is due to the refrigerators in a city with 200,000 households.

1-31 Repeat Prob. 1-30 assuming the electricity is produced by a power plant that burns coal. The average production of CO_2 in this case is 1.1 kg per kWh.

1-32E Consider a household that uses 8000 kWh of electricity per year and 1500 gallons of fuel oil during a heating season. The average amount of CO_2 produced is 26.4 lbm/gallon of fuel oil and 1.54 lbm/kWh of electricity. If this household reduces its oil and electricity usage by 20 percent as a result of implementing some energy conservation measures, determine the reduction in the amount of CO_2 emissions by that household per year.

1-33 A typical car driven 12,000 miles a year emits to the atmosphere about 11 kg per year of NO_x (nitrogen oxides), which causes smog in major population areas. Natural gas burned in the furnace emits about 4.3 g of NO_x per therm, and the electric power plants emit about 7.1 g of NO_x per kWh of electricity produced. Consider a household that has 2 cars and consumes 9000 kWh of electricity and 1200 therms of natural gas. Determine the amount of NO_x emission to the atmosphere per year for which this household is responsible.

Pressure

1-34C What is the difference between gage pressure and absolute pressure?

1-35C Explain why some people experience nose bleeding and some others experience shortness of breath at high elevations.

1-36 A vacuum gage connected to a tank reads 30 kPa at a location where the barometric reading is 755 mmHg. Determine the absolute pressure in the tank. Take $\rho_{Hg} = 13{,}590 \text{ kg/m}^3$. *Answer:* 70.6 kPa

1-37E A pressure gage connected to a tank reads 50 psi at a location where the barometric reading is 29.1 inHg. Determine the absolute pressure in the tank. Take $\rho_{Hg} = 848.4 \text{ lbm/ft}^3$. *Answer:* 64.29 psia

1-38 A pressure gage connected to a tank reads 500 kPa at a location where the atmospheric pressure is 94 kPa. Determine the absolute pressure in the tank.

1-39 The barometer of a mountain hiker reads 930 mbars at the beginning of a hiking trip and 780 mbars at the end. Neglecting the effect of altitude on local gravitational acceleration, determine the vertical distance climbed. Assume an average air density of 1.20 kg/m^3 and take $g = 9.7 \text{ m/s}^2$.
Answer: 1289 m

1-40 The basic barometer can be used to measure the height of a building. If the barometric readings at the top and at the bottom of a building are 730 and 755 mmHg, respectively, determine the height of the building. Assume an average air density of 1.18 kg/m^3.

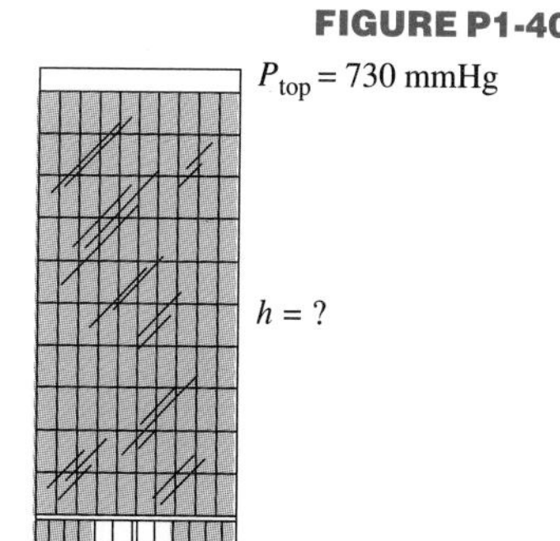

FIGURE P1-40

1-41 Determine the pressure exerted on a diver at 30 m below the free surface of the sea. Assume a barometric pressure of 101 kPa and a specific gravity of 1.03 for seawater. *Answer:* 404.0 KPa

1-42E Determine the pressure exerted on the surface of a submarine cruising 300 ft below the free surface of the sea. Assume that the barometric pressure is 14.7 psia and the specific gravity of seawater is 1.03.

1-43 A gas is contained in a vertical, frictionless piston-cylinder device. The piston has a mass of 4 kg and cross-sectional area of 35 cm^2.

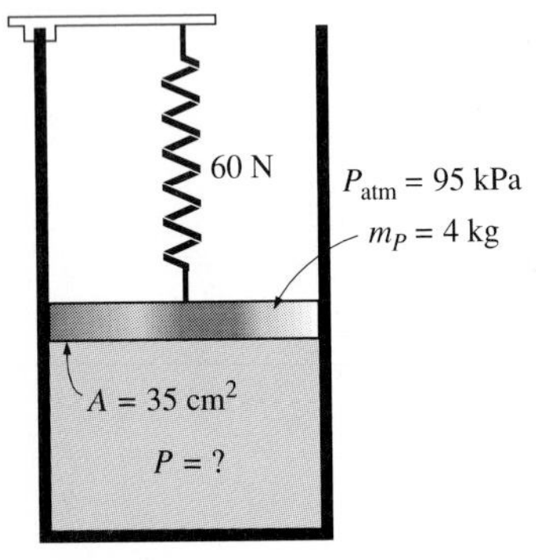

FIGURE P1-43

A compressed spring above the piston exerts a force of 60 N on the piston. If the atmospheric pressure is 95 kPa, determine the pressure inside the cylinder. *Answer:* 123.4 kPa

1-44 Both a gage and a manometer are attached to a gas tank to measure its pressure. If the reading on the pressure gage is 80 kPa, determine the distance between the two fluid levels of the manometer if the fluid is (*a*) mercury (ρ = 13,600 kg/m^3) or (*b*) water (ρ = 1000 kg/m^3).

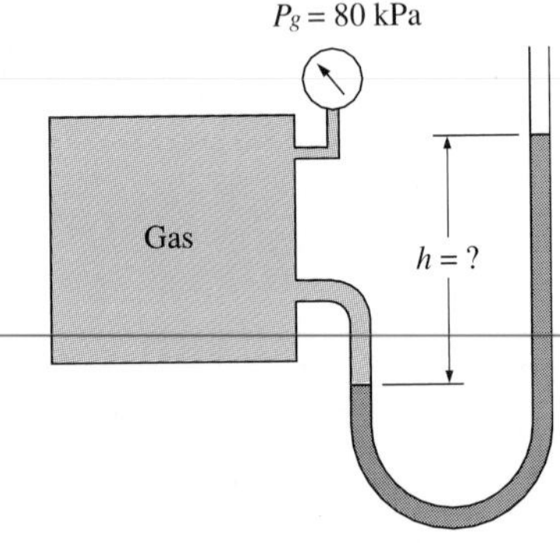

FIGURE P1-44

1-45 A manometer containing oil (ρ = 850 kg/m^3) is attached to a tank filled with air. If the oil-level difference between the two columns is 45 cm and the atmospheric pressure is 98 kPa, determine the absolute pressure of the air in the tank. *Answer:* 101.75 kPa

1-46 A mercury manometer (ρ = 13,600 kg/m^3) is connected to an air duct to measure the pressure inside. The difference in the manometer levels is 15 mm, and the atmospheric pressure is 100 kPa.

(*a*) Judging from Fig. P1-46, determine if the pressure in the duct is above or below the atmospheric pressure.

(*b*) Determine the absolute pressure in the duct.

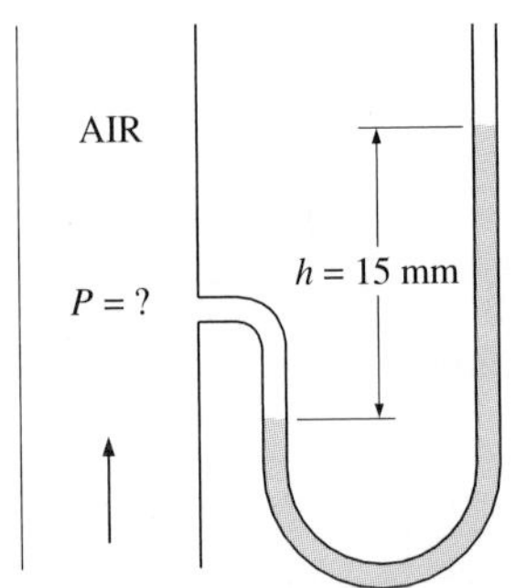

FIGURE P1-46

Temperature

1-47C What is the zeroth law of thermodynamics?

1-48C What are the ordinary and absolute temperature scales in the SI and the English system?

1-49C Consider an alcohol and a mercury thermometer that read exactly 0°C at the ice point and 100°C at the steam point. The distance between the two points is divided into 100 equal parts in both thermometers. Do you think these thermometers will give exactly the same reading at a temperature of, say, 60°C? Explain.

1-50C The deep body temperature of a healthy person is 37°C. What is it in kelvins? *Answer:* 310 K

1-51E Consider a system whose temperature is 18°C. Express this temperature in R, K, and °F.

1-52 The temperature of a system rises by 30°C during a heating process. Express this rise in temperature in kelvins. *Answer:* 30 K

1-53E The temperature of a system drops by 27°F during a cooling process. Express this drop in temperature in K, R, and °C.

1-54 Consider two closed systems A and B. System A contains 2000 kJ of thermal energy at 20°C whereas system B contains 200 kJ of thermal energy at 50°C. Now the systems are brought into contact with each other. Determine the direction of any heat transfer between the two systems.

Biological Systems

1-55C What is metabolism? What is basal metabolic rate? What is the value of basal metabolic rate for an average man?

1-56C For what is the energy released during metabolism in humans used?

1-57C Is the metabolizable energy content of a food the same as the energy released when it is burned in a bomb calorimeter? If not, how does it differ?

1-58C Is the number of prospective occupants an important consideration in the design of heating and cooling systems of classrooms? Explain.

1-59C What do you think of a diet program that allows for generous amounts of bread and rice provided that no butter or margarine is added?

1-60C Consider two identical rooms, one with a 2-kW electric resistance heater and the other with three couples fast dancing. In which room will the air temperature rise faster?

1-61 Consider two identical 80-kg men who are eating identical meals and doing identical things except that one of them jogs for 30 min every day while the other watches TV. Determine the weight difference between the two in a month. *Answer:* 1.045 kg

1-62 Consider a classroom that is losing heat to the outdoors at a rate of 20,000 kJ/h. If there are 30 students in class, each dissipating sensible heat at a rate of 100 W, determine if it is necessary to turn the heater in the classroom on to prevent the room temperature from dropping.

1-63 A 68-kg woman is planning to bicycle for an hour. If she is to meet her entire energy needs while bicycling by eating 30-g chocolate candy bars, determine how many candy bars she needs to take with her.

1-64 A 55-kg man gives in to temptation and eats an entire 1-L box of ice cream. How long does this man need to jog to burn off the calories he consumed from the ice cream? *Answer:* 2.52 h

1-65 Consider a man who has 20 kg of body fat when he goes on a hunger strike. Determine how long he can survive on his body fat alone.

1-66 Consider two identical 50-kg women, Candy and Wendy, who are doing identical things and eating identical food except that Candy eats her baked potato with four teaspoons of butter while Wendy eats hers plain every evening. Determine the difference in the weights of Candy and Wendy after one year. *Answer:* 6.5 kg

1-67 A woman who used to drink about one liter of regular cola every day switches to diet cola (zero calorie) and starts eating two slices of apple pie every day. Is she now consuming fewer or more calories?

1-68 A 60-kg man used to have an apple every day after dinner without losing or gaining any weight. He now eats a 200-ml serving of ice cream instead of an apple and walks 20 min every day. On this new diet, how much weight will he lose or gain per month? *Answer:* 0.087-kg gain

1-69 The average specific heat of the human body is 3.6 kJ/(kg · °C). If the body temperature of a 80-kg man rises from 37°C to 39°C during strenuous exercise, determine the increase in the thermal energy of the body as a result of this rise in body temperature.

1-70 Alcohol provides 7 Calories per gram, but it provides no essential nutrients. A 1.5 ounce serving of 80-proof liquor contains 100 Calories in alcohol alone. Sweet wines and beer provide additional calories since they also contain carbohydrates. About 75 percent of American adults drink some sort of alcoholic beverage, which adds an average of 210 Calories a day to their diet. Determine how many pounds less an average American adult will weigh per year if he or she quit drinking alcoholic beverages and started drinking diet soda.

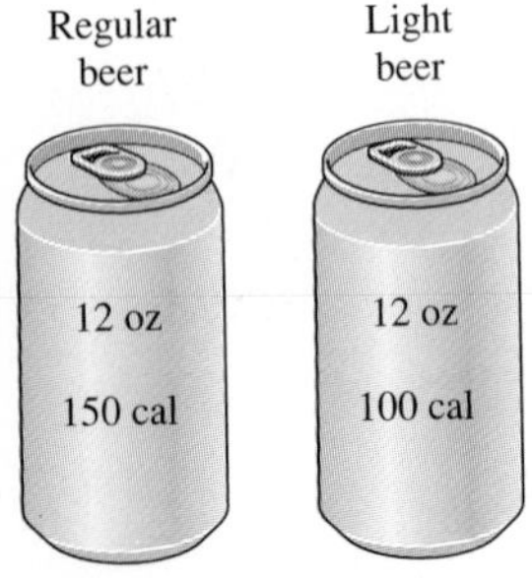

FIGURE P1-71

1-71 A 12-oz serving of a regular beer contains 13 g of alcohol and 13 g of carbohydrates, and thus 150 Calories. A 12-oz serving of a light beer contains 11 g of alcohol and 5 g of carbohydrates, and thus 100 Calories. An average person burns 700 Calories per hour while exercising on a treadmill. Determine how long it will take to burn the calories from a 12-oz can of (*a*) regular beer and (*b*) light beer on a treadmill.

1-72 A 5-oz serving of a Bloody Mary contains 14 g of alcohol and 5 g of carbohydrates, and thus 116 Calories. A 2.5-oz serving of a martini contains 22 g of alcohol and a negligible amount of carbohydrates, and thus 156 Calories. An average person burns 600 Calories per hour while exercising on a cross-country ski machine. Determine how long it will take to burn the calories from one serving of (*a*) a Bloody Mary and (*b*) a martini on this cross-country ski machine.

1-73 A 176-pound man and a 132-pound woman went to Burger King for lunch. The man had a BK Big Fish sandwich (720 Cal), medium french fries (400 Cal), and a large Coke (225 Cal). The woman had a basic hamburger (330 Cal), medium french fries (400 Cal), and a diet Coke (0 Cal). After lunch, they start shoveling snow and burn calories at a rate of 360 Cal/h for the woman and 480 Cal/h for the man. Determine how long each one of them needs to shovel snow to burn off the lunch calories.

1-74 Consider two friends who go to Burger King every day for lunch. One of them orders a Double Whopper sandwich, large fries, and a large Coke (total Calories = 1600) while the other orders a Whopper Junior, small fries, and a small Coke (total Calories = 800) every day. If these two friends are very much alike otherwise and they have the same metabolic rate, determine the weight difference between these two friends in a year.

1-75E A 150-pound person goes to Hardee's for dinner and orders a regular roast beef (270 Cal) and a big roast beef (410 Cal) sandwich together with a 12-oz can of Pepsi (150 Cal). A 150-pound person burns 400 Calories per hour while climbing stairs. Determine how long this person needs to climb stairs to burn off the dinner calories.

1-76 A person eats a McDonald's Big Mac sandwich (530 Cal), a second person eats a Burger King Whopper sandwich (640 Cal), and a third person eats 50 olives with regular french fries (350 Cal) for lunch. Determine who consumes the most calories. An olive contains about 5 Calories.

Review Problems

1-77 Balloons are often filled with helium gas because it weighs only about one-seventh of what air weighs under identical conditions. The buoyancy force which can be expressed as $F_b = \rho_{air} g V_{balloon}$, will push the balloon upward. If the balloon has a diameter of 10 m and carries two people, 70 kg each, determine the acceleration of the balloon when it is first released. Assume the density of air is $\rho = 1.16$ kg/m^3, and neglect the weight of the ropes and the cage. *Answer:* 16.5 m/s^2

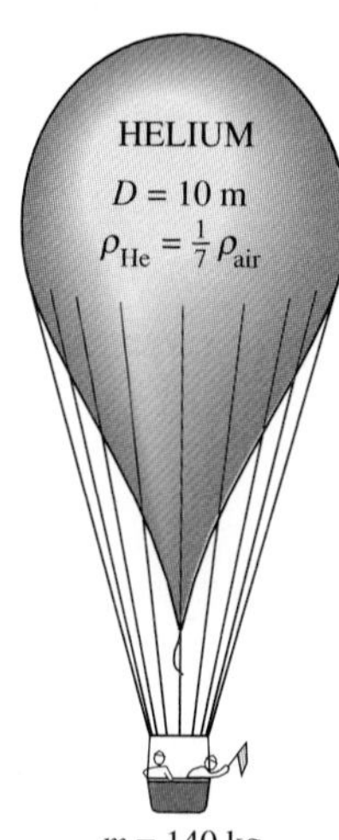

FIGURE P1-77

1-78 Determine the maximum amount of load, in kg, the balloon described in Prob. 1-77 can carry. *Answer:* 520.6 kg

1-79 The basic barometer can be used as an altitude-measuring device in airplanes. The ground control reports a barometric reading of 753 mmHg while the pilot's reading is 690 mmHg. Estimate the altitude of the plane from ground level if the average air density is 1.20 kg/m^3 and $g = 9.8$ m/s^2.
Answer: 714 m

1-80 The lower half of a 10-m-high cylindrical container is filled with water ($\rho = 1000$ kg/m^3) and the upper half with oil that has a specific gravity of 0.85. Determine the pressure difference between the top and bottom of the cylinder. *Answer:* 90.7 kPa

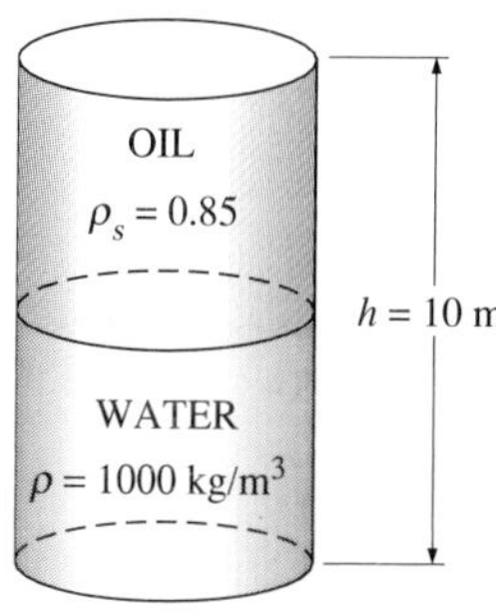

FIGURE P1-80

1-81 A vertical, frictionless piston-cylinder device contains a gas at 500 kPa. The atmospheric pressure outside is 100 kPa, and the piston area is 30 cm^2. Determine the mass of the piston. Assume standard gravitational acceleration.

1-82 A pressure cooker cooks a lot faster than an ordinary pan by maintaining a higher pressure and temperature inside. The lid of a pressure cooker is well sealed, and steam can escape only through an opening in the middle of the lid. A separate piece of certain mass, the petcock, sits on top of this opening and prevents steam from escaping until the pressure force overcomes the weight of the petcock. The periodic escape of the steam in this manner prevents any potentially dangerous pressure buildup and keeps the pressure inside at a constant value.

Determine the mass of the petcock of a pressure cooker whose operation pressure is 100 kPa gage and has an opening cross-sectional area of 4 mm^2. Assume an atmospheric pressure of 101 kPa, and draw the freebody diagram of the petcock. *Answer:* 40.8 g

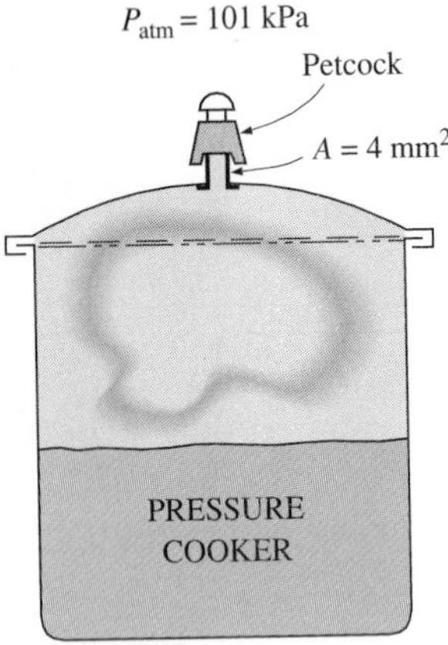

FIGURE P1-82

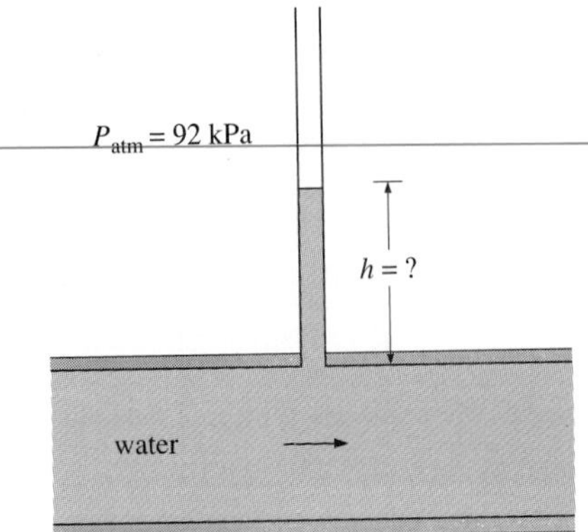

FIGURE P1-83

1-83 A glass tube is attached to a water pipe as shown in Fig. P1-83. If the water pressure at the bottom of the tube is 115 kPa and the local atmospheric pressure is 92 kPa, determine how high the water will rise in the tube, in m. Assume $g = 9.8 \text{ m/s}^2$ at that location and take the density of water to be 1000 kg/m³.

1-84 The average atmospheric pressure on earth is approximated as a function of altitude by the relation

$$P_{atm} = 101.325(1 - 0.02256z)^{5.256}$$

where P_{atm} is the atmospheric pressure in kPa and z is the altitude in km (1 km = 1000 m) with $z = 0$ at sea level. Determine the approximate atmospheric pressures at Atlanta (z = 306 m), Denver (z = 1610 m), Mexico City (z = 2309 m), and the top of Mount Everest (z = 8848 m).

1-85 The weight of bodies may change somewhat from one location to another as a result of the variation of the gravitational acceleration g with elevation. Accounting for this variation using the relation in Prob. 1-9, determine the weight of an 80-kg person at sea level (z = 0), in Denver (z = 1610 m), and on the top of Mount Everest (z = 8848 m).

1-86E The efficiency of a refrigerator increases by 3 percent for each °C rise in the minimum temperature in the device. What is the increase in the efficiency for each (*a*) K, (*b*) °F, and (*c*) R rise in temperature?

1-87E The boiling temperature of water decreases by about 3°C for each 1000 m rise in altitude. What is the decrease in the boiling temperature in (*a*) K, (*b*) °F, and (*c*) R for each 1000 m rise in altitude?

1-88E The average body temperature of a person rises by about 2°C during strenuous exercise. What is the rise in the body temperature in (*a*) K, (*b*) °F, and (*c*) R during strenuous exercise?

1-89E Hyperthermia of 5°C (i.e., 5°C rise above the normal body temperature) is considered fatal. Express this fatal level of hyperthermia in (*a*) K, (*b*) °F, and (*c*) R.

1-90E A house is losing heat at a rate of 3000 kJ/h per °C temperature difference between the indoor and the outdoor temperatures. Express the rate of heat loss from this house per (*a*) K, (*b*) °F, and (*c*) R difference between the indoor and the outdoor temperature.

1-91 The average temperature of the atmosphere in the world is approximated as a function of altitude by the relation

$$T_{atm} = 288.15 - 6.5z$$

where T_{atm} is the temperature of the atmosphere in K and z is the altitude in km with $z = 0$ at sea level. Determine the average temperature of the atmosphere outside an airplane that is cruising at an altitude of 12,000 m.

1-92 Joe Smith, an old-fashioned engineering student, believes that the boiling point of water is best suited for use as the reference point on temperature scales. Unhappy that the boiling point corresponds to some odd number in the current absolute temperature scales, he has proposed a new absolute temperature scale that he calls the Smith scale. The temperature unit on this scale is *smith*, denoted by S, and the boiling point of water on this scale is assigned to be 1000 S. From a thermodynamic point of view, discuss if it is an acceptable temperature scale. Also, determine the ice point of water on the Smith scale and obtain a relation between the Smith and Celsius scales.

1-93 A man goes to a traditional market to buy a steak for dinner. He finds a 12-ounce steak (1 lbm = 16 ounces) for $3.15. He then goes to the adjacent international market and finds a 320-gram steak of identical quality for $2.80. Which steak is a better buy?

1-94 Milk is to be transported from Texas to California for a distance of 2100 km in a 7-m-long, 2-m-external-diameter cylindrical tank. The walls of the tank are constructed of 5-cm-thick urethane insulation sandwiched between two metal sheets of negligible thickness. Determine the amount of milk in the tank in kg and in gallons (1 gal = 3.78 L).

1-95 An engineer who is working on the heat transfer analysis of a brick building in English units needs the thermal conductivity of brick. But the only value he can find from his handbooks is 0.72 W/(m · °C), which is in SI units. To make matters worse, the engineer does not have a direct conversion factor between the two unit systems for thermal conductivity (he should have kept his heat transfer textbook instead of selling it back to the bookstore). Can you help him out?

1-96 It is well known that cold air feels much colder in windy weather than what the thermometer reading indicates because of the "chilling effect" of the wind. This effect is due to the increase in the convection heat transfer coefficient with increasing air velocities. The *equivalent wind chill temperature* in °F is given by [ASHRAE, *Handbook of Fundamentals* (Atlanta, GA, 1993), p. 8.15]

$$T_{\text{equiv}} = 91.4 - (91.4 - T_{\text{ambient}})(0.475 - 0.0203\mathcal{V} + 0.304\sqrt{\mathcal{V}})$$

where $\mathcal{V}$ is the wind velocity in mph and T_{ambient} is the ambient air temperature in °F in calm air, which is taken to be air with light winds at speeds up to 4 mph. The constant 91.4°F in the above equation is the mean skin temperature of a resting person in a comfortable environment. Windy air at temperature T_{ambient} and velocity $\mathcal{V}$ will feel as cold as the calm air at temperature T_{equiv}. Using proper conversion factors, obtain an equivalent relation in SI units where $\mathcal{V}$ is the wind velocity in km/h and T_{ambient} is the ambient air temperature in °C.

Answer: $T_{\text{equiv}} = 33.0 - (33.0 - T_{\text{ambient}})(0.475 - 0.0126\mathcal{V} + 0.240\sqrt{\mathcal{V}})$

1-97 The reactive force developed by a jet engine to push an airplane forward is called thrust, and the thrust developed by the engine of a Boeing 777 is about 85,000 pounds. Express this thrust in N and kgf.

1-98 A 100-kg man decides to lose 5 kg without cutting down his intake of 3000 Calories a day. Instead, he starts fast swimming, fast dancing, jogging, and biking each for an hour every day. He sleeps or relaxes the rest of the day. Determine how long it will take him to lose 5 kg.

1-99 The range of healthy weight for adults is usually expressed in terms of the *body mass index* (BMI), defined, in SI units, as

$$\text{BMI} = \frac{W(\text{kg})}{H^2(\text{m}^2)}$$

where W is the weight (actually, the mass) of the person in kg and H is the height in m, and the range of healthy weight is $19 \leq \text{BMI} \leq 25$. Convert the formula above to English units such that the weight is in pounds and the height in inches. Also, calculate your own BMI, and if it is not in the healthy range, determine how many pounds (or kg) you need to gain or to lose to be fit. *Answer:* BMI = $705\ W/H^2$

Computer, Design, and Essay Problems

1-100 Write an interactive computer program to express a given temperature in °C, °F, K, and R in terms of the other three units.

1-101 Write an interactive computer program to express a pressure given in SI units in terms of the height of water and mercury columns.

1-102 Write an essay on different temperature measurement devices. Explain the operational principle of each device, its advantages and disadvantages, its cost, and its range of applicability. Which device would you recommend for use in the following cases: taking the temperatures of patients in a doctor's office, monitoring the variations of temperature of a car engine block at several locations, and monitoring the temperatures in the furnace of a power plant.

1-103 Write an essay on different pressure-measurement devices. Explain the operational principle of each device, its advantages and disadvantages, its cost, and its range of applicability. Give examples of application areas for which each device is best suited.

1-104 Write an essay on the various mass- and volume-measurement devices used throughout history. Also, explain the development of the modern units for mass and volume.

1-105 Prepare a nutritional analysis of everything you have eaten and drunk during the last 24 h. Determine your total calorie intake and compare it to your calorie needs for that day. Also, determine the percentage of calories that came from fats. Do you consider your diet during the last 24 h to be a healthy one? What changes would make it a healthier diet?

1-106 Prepare a tasty, healthy, and economical menu for yourself, for a one-week period, complete with a nutritional analysis.

Properties of Pure Substances

CHAPTER 2

In this chapter, the concept of a pure substance is introduced, and the physics of phase-change processes are discussed. Various property diagrams and *P-v-T* surfaces of pure substances are illustrated, and the hypothetical substance "ideal gas" and the ideal-gas equation of state are discussed. The compressibility factor, which accounts for the deviation of real gases from ideal-gas behavior, is introduced, and its use is illustrated. Finally, some of the best-known equations of state are presented.

2-1 ■ PURE SUBSTANCE

A substance that has a fixed chemical composition throughout is called a **pure substance.** Water, nitrogen, helium, and carbon dioxide, for example, are all pure substances.

A pure substance does not have to be of a single chemical element or compound, however. A mixture of various chemical elements or compounds also qualifies as a pure substance as long as the mixture is homogeneous. Air, for example, is a mixture of several gases, but it is often considered to be a pure substance because it has a uniform chemical composition (Fig. 2-1). However, a mixture of oil and water is not a pure substance. Since oil is not soluble in water, it will collect on top of the water, forming two chemically dissimilar regions.

FIGURE 2-1
Nitrogen and gaseous air are pure substances.

A mixture of two or more phases of a pure substance is still a pure substance as long as the chemical composition of all phases is the same (Fig. 2-2). A mixture of ice and liquid water, for example, is a pure substance because both phases have the same chemical composition. A mixture of liquid air and gaseous air, however, is not a pure substance since the composition of liquid air is different from the composition of gaseous air, and thus the mixture is no longer chemically homogeneous. This is due to different components in air having different condensation temperatures at a specified pressure.

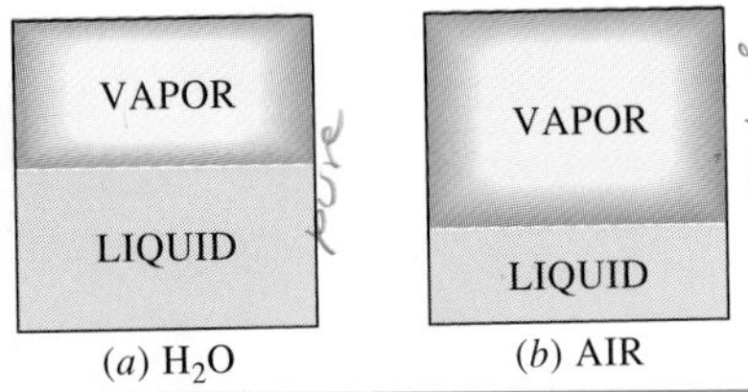

FIGURE 2-2
A mixture of liquid and gaseous water is a pure substance, but a mixture of liquid and gaseous air is not.

2-2 ■ PHASES OF A PURE SUBSTANCE

We all know from experience that substances exist in different phases. At room temperature and pressure, copper is a solid, mercury is a liquid, and nitrogen is a gas. Under different conditions, each may appear in a different phase. Even though there are three principal phases—solid, liquid, and gas—a substance may have several phases within a principal phase, each with a different molecular structure. Carbon, for example, may exist as graphite or diamond in the solid phase. Helium has two liquid phases; iron has three solid phases. Ice may exist at seven different phases at high pressures. A phase is identified as having a distinct molecular arrangement that is homogeneous throughout and separated from the others by easily identifiable boundary surfaces. The two phases of H_2O in iced water represent a good example of this.

When studying phases or phase changes in thermodynamics, one does not need to be concerned with the molecular structure and behavior of different phases. However, it is very helpful to have some understanding of the molecular phenomena involved in each phase, and a brief discussion of phase transformations is given below.

Molecular bonds are strongest in solids and weakest in gases. One reason is that molecules in solids are closely packed together, whereas in gases they are separated by great distances.

The molecules in a **solid** are arranged in a three-dimensional pattern (lattice) that is repeated throughout (Fig. 2-3). Because of the small distances between molecules in a solid, the attractive forces of molecules on each other

FIGURE 2-3
The molecules in a solid are kept at their positions by the large springlike intermolecular forces.

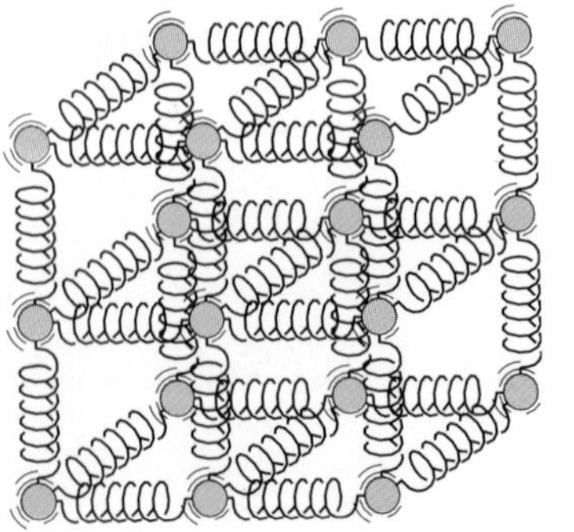

are large and keep the molecules at fixed positions (Fig. 2-4). Note that the attractive forces between molecules turn to repulsive forces as the distance between the molecules approaches zero, thus preventing the molecules from piling up on top of each other. Even though the molecules in a solid cannot move relative to each other, they continually oscillate about their equilibrium positions. The velocity of the molecules during these oscillations depends on the temperature. At sufficiently high temperatures, the velocity (and thus the momentum) of the molecules may reach a point where the intermolecular forces are partially overcome and groups of molecules break away (Fig. 2-5). This is the beginning of the melting process.

The molecular spacing in the **liquid** phase is not much different from that of the solid phase, except the molecules are no longer at fixed positions relative to each other. In a liquid, chunks of molecules float about each other; however, the molecules maintain an orderly structure within each chunk and retain their original positions with respect to one another. The distances between molecules generally experience a slight increase as a solid turns liquid, with water being a rare exception.

In the **gas** phase, the molecules are far apart from each other, and a molecular order is nonexistent. Gas molecules move about at random, continually colliding with each other and the walls of the container they are in. Particularly at low densities, the intermolecular forces are very small, and collisions are the only mode of interaction between the molecules. Molecules in the gas phase are at a considerably higher energy level than they are in the liquid or solid phases. Therefore, the gas must release a large amount of its energy before it can condense or freeze.

FIGURE 2-4
In a solid, the attractive and repulsive forces between the molecules tend to maintain them at relatively constant distances from each other.

2-3 ■ PHASE-CHANGE PROCESSES OF PURE SUBSTANCES

There are many practical situations where two phases of a pure substance coexist in equilibrium. Water exists as a mixture of liquid and vapor in the boiler and the condenser of a steam power plant. The refrigerant turns from liquid to vapor in the freezer of a refrigerator. Even though many

FIGURE 2-5
The arrangement of atoms in different phases: (*a*) molecules are at relatively fixed positions in a solid, (*b*) chunks of molecules float about each other in the liquid phase, and (*c*) molecules move about at random in the gas phase.

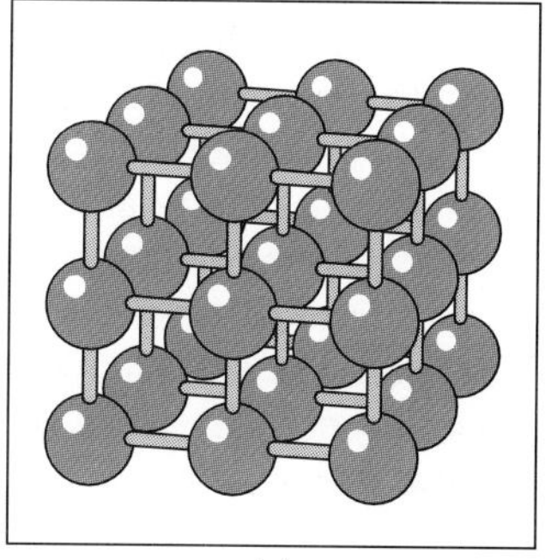

(*a*)

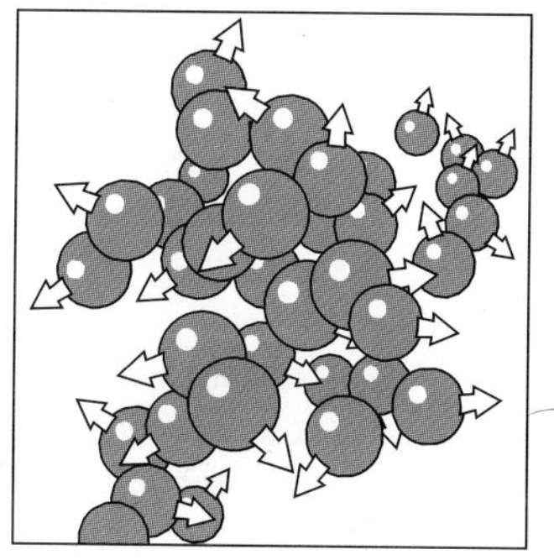

(*b*)

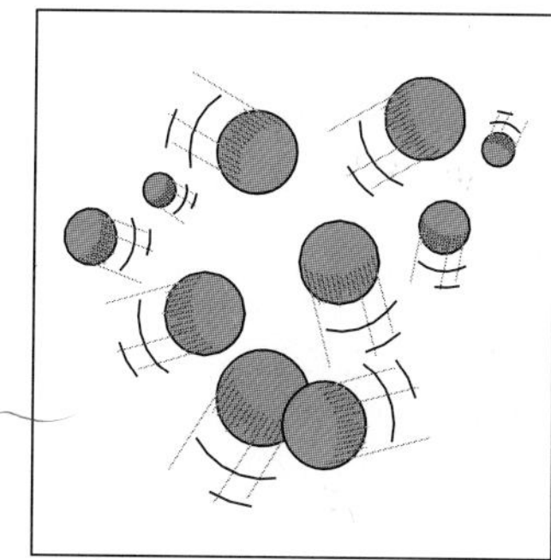

(*c*)

homeowners consider the freezing of water in underground pipes as the most important phase-change process, attention in this section is focused on the liquid and vapor phases and the mixture of these two. As a familiar substance, water will be used to demonstrate the basic principles involved. Remember, however, that all pure substances exhibit the same general behavior.

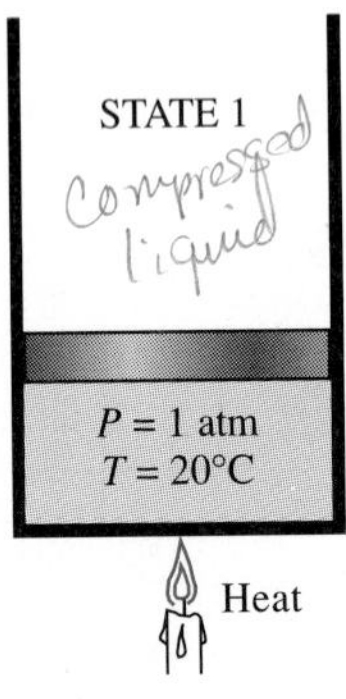

FIGURE 2-6
At 1 atm and 20°C, water exists in the liquid phase (*compressed liquid*).

Compressed Liquid and Saturated Liquid

Consider a piston-cylinder device containing liquid water at 20°C and 1 atm pressure (state 1, Fig. 2-6). Under these conditions, water exists in the liquid phase, and it is called a **compressed liquid,** or a **subcooled liquid,** meaning that it is *not about to vaporize*. Heat is now transferred to the water until its temperature rises to, say, 40°C. As the temperature rises, the liquid water expands slightly, and so its specific volume increases. To accommodate this expansion, the piston will move up slightly. The pressure in the cylinder remains constant at 1 atm during this process since it depends on the outside barometric pressure and the weight of the piston, both of which are constant. Water is still a compressed liquid at this state since it has not started to vaporize.

As more heat is transferred, the temperature will keep rising until it reaches 100°C (state 2, Fig. 2-7). At this point water is still a liquid, but any heat addition will cause some of the liquid to vaporize. That is, a phase-change process from liquid to vapor is about to take place. A liquid that is *about to vaporize* is called a **saturated liquid.** Therefore, state 2 is a saturated liquid state.

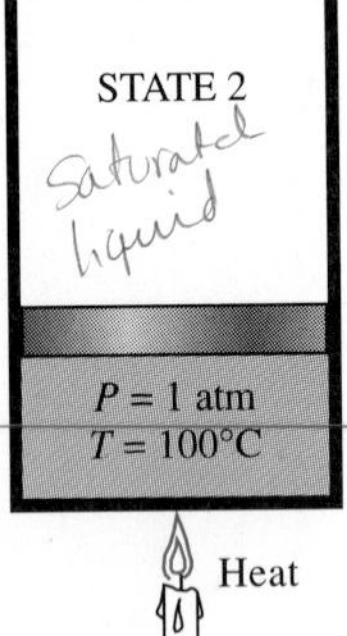

FIGURE 2-7
At 1 atm pressure and 100°C, water exists as a liquid that is ready to vaporize (*saturated liquid*).

Saturated Vapor and Superheated Vapor

Once boiling starts, the temperature will stop rising until the liquid is completely vaporized. That is, the temperature will remain constant during the entire phase-change process if the pressure is held constant. This can easily be verified by placing a thermometer into boiling water on top of a stove. At sea level (P = 1 atm), the thermometer will always read 100°C if the pan is uncovered or covered with a light lid. During a boiling process, the only change we will observe is a large increase in the volume and a steady decline in the liquid level as a result of more liquid turning to vapor.

FIGURE 2-8
As more heat is transferred, part of the saturated liquid vaporizes (*saturated liquid–vapor mixture*).

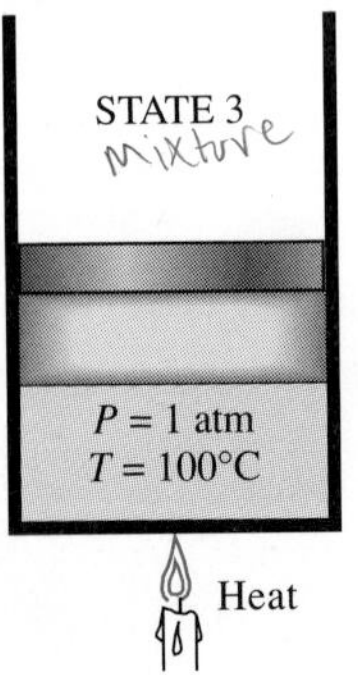

Midway about the vaporization line (state 3, Fig. 2-8), the cylinder contains equal amounts of liquid and vapor. As we continue transferring heat, the vaporization process will continue until the last drop of liquid is vaporized (state 4, Fig. 2-9). At this point, the entire cylinder is filled with vapor that is on the borderline of the liquid phase. Any heat loss from this vapor will cause some of the vapor to condense (phase change from vapor to liquid). A vapor that is *about to condense* is called a **saturated vapor.** Therefore, state 4 is a saturated vapor state. A substance at states between 2 and 4 is often referred to as a **saturated liquid–vapor mixture** since the *liquid and vapor phases coexist in equilibrium* at these states.

Once the phase-change process is completed, we are back to a single-phase region again (this time vapor), and further transfer of heat will result

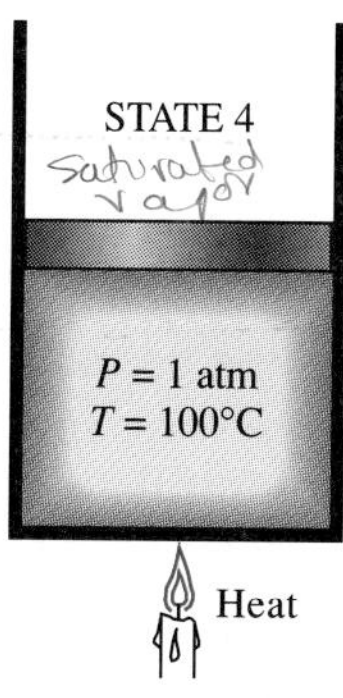

FIGURE 2-9
At 1 atm pressure, the temperature remains constant at 100°C until the last drop of liquid is vaporized (*saturated vapor*).

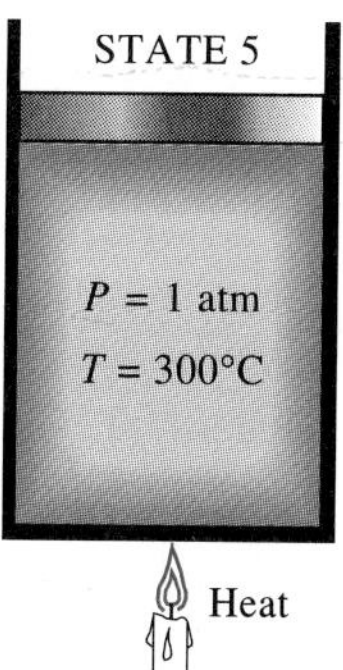

FIGURE 2-10
As more heat is transferred, the temperature of the vapor starts to rise (*superheated vapor*).

in an increase in both the temperature and the specific volume (Fig. 2-10). At state 5, the temperature of the vapor is, let us say, 300°C; and if we transfer some heat from the vapor, the temperature may drop somewhat but no condensation will take place as long as the temperature remains above 100°C (for $P = 1$ atm). A vapor that is *not about to condense* (i.e., not a saturated vapor) is called a **superheated vapor.** Therefore, water at state 5 is a superheated vapor. The constant-pressure phase-change process described above is illustrated on a *T-v* diagram in Fig. 2-11.

If the entire process described above is reversed by cooling the water while maintaining the pressure at the same value, the water will go back to state 1, retracing the same path, and in so doing, the amount of heat released will exactly match the amount of heat added during the heating process.

In our daily life, water implies liquid water and steam implies water vapor. In thermodynamics, however, both water and steam usually mean only one thing: H_2O.

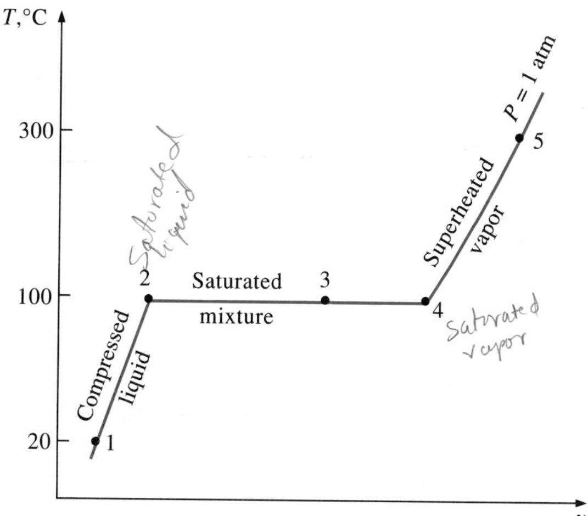

FIGURE 2-11
T-v diagram for the heating process of water at constant pressure.

Saturation Temperature and Saturation Pressure

It probably came as no surprise to you that the water started "boiling" at 100°C. Strictly speaking, the statement "water boils at 100°C" is incorrect. The correct statement is "water boils at 100°C at 1 atm pressure." The only reason the water started boiling at 100°C was because we held the pressure constant at 1 atm (101.325 kPa). If the pressure inside the cylinder were raised to 500 kPa by adding weights on top of the piston, the water would start boiling at 151.9°C. That is, *the temperature at which water starts boiling depends on the pressure; therefore, if the pressure is fixed, so is the boiling temperature.*

At a given pressure, the temperature at which a pure substance changes phase is called the **saturation temperature** T_{sat}. Likewise, at a given temperature, the pressure at which a pure substance changes phase is called the **saturation pressure** P_{sat}. At a pressure of 101.325 kPa, T_{sat} is 100°C. Conversely, at a temperature of 100°C, P_{sat} is 101.325 kPa.

Saturation tables that list the saturation pressure against the temperature (or the saturation temperature against the pressure) are available for practically all substances. A partial listing of such a table is given in Table 2-1 for water. This table indicates that the pressure of water changing phase (boiling or condensing) at 25°C must be 3.17 kPa, and the pressure of water must be maintained at 3973 kPa (about 40 atm) to have it boil at 250°C. Also, water can be frozen by dropping its pressure below 0.61 kPa.

TABLE 2-1

Saturation (boiling) pressure of water at various temperatures

Temperature, T °C	Saturation pressure, P_{sat} kPa
−10	0.26
−5	0.40
0	0.61
5	0.87
10	1.23
15	1.71
20	2.34
25	3.17
30	4.25
40	7.38
50	12.35
100	101.3 (1 atm)
150	475.8
200	1554
250	3973
300	8581

It takes a large amount of energy to melt a solid or vaporize a liquid. The amount of energy absorbed or released during a phase-change process is called the **latent heat.** More specifically, the amount of energy absorbed during melting is called the **latent heat of fusion** and is equivalent to the amount of energy released during freezing. Similarly, the amount of energy absorbed during vaporization is called the **latent heat of vaporization** and is equivalent to the energy released during condensation. The magnitudes of the latent heats depend on the temperature or pressure at which the phase change is occurring. At 1 atm pressure, the latent heat of fusion of water is 333.7 kJ/kg and the latent heat of vaporization is 2257.1 kJ/kg.

During a phase-change process, pressure and temperature are obviously dependent properties, and there is a definite relation between them, that is, $T_{sat} = f(P_{sat})$. A plot of T_{sat} versus P_{sat}, such as the one given for water in Fig. 2-12, is called a **liquid–vapor saturation curve.** A curve of this kind is characteristic of all pure substances.

It is clear from Fig. 2-12 that T_{sat} increases with P_{sat}. Thus, a substance at higher pressures will boil at higher temperatures. In the kitchen, higher boiling temperatures mean shorter cooking times and energy savings. A beef stew, for example, may take 1 to 2 h to cook in a regular pan that operates at 1 atm pressure, but only 20 to 30 min in a pressure cooker operating at 2 atm absolute pressure (corresponding boiling temperature: 120°C).

The atmospheric pressure, and thus the boiling temperature of water, decreases with elevation. Therefore, it takes longer to cook at higher altitudes than it does at sea level (unless a pressure cooker is used). For example, the

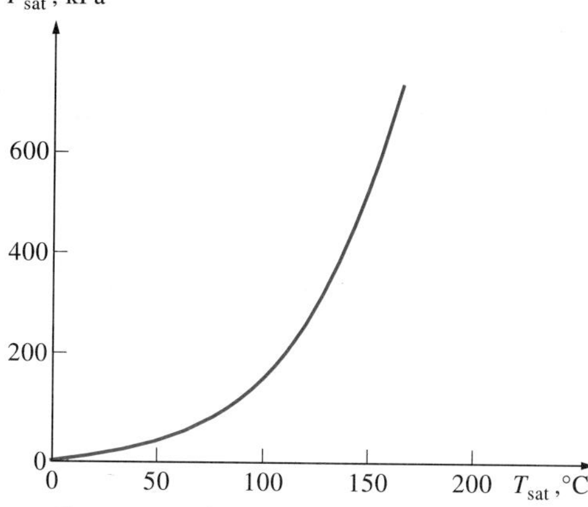

FIGURE 2-12

The liquid–vapor saturation curve of a pure substance (numerical values are for water).

standard atmospheric pressure at an elevation of 2000 m is 79.50 kPa, which corresponds to a boiling temperature of 93.2°C as opposed to 100°C at sea level (zero elevation). The variation of the boiling temperature of water with altitude at standard atmospheric conditions is given in Table 2-2. For each 1000 m increase in elevation, the boiling temperature drops by a little over 3°C. Note that the atmospheric pressure at a location, and thus the boiling temperature, changes slightly with the weather conditions. But the corresponding change in the boiling temperature is no more than about 1°C.

TABLE 2-2

Variation of the standard atmospheric pressure and the boiling (saturation) temperature of water with altitude

Elevation (m)	Atmospheric pressure (kPa)	Boiling temperature (°C)
0	101.33	100.0
1000	89.55	96.3
2000	79.50	93.2
5000	54.05	83.0
10,000	26.50	66.2
20,000	5.53	34.5

Some Consequences of T_{sat} and P_{sat} Dependence

We mentioned earlier that a substance at a specified pressure will boil at the saturation temperature corresponding to that pressure. This phenomenon allows us to control the boiling temperature of a substance by simply controlling the pressure, and it has numerous applications in practice. Below we give some examples. In most cases, the natural drive to achieve phase equilibrium by allowing some liquid to evaporate is at work behind the scenes.

Consider a sealed can of *liquid refrigerant-134a* in a room at 25°C. If the can has been in the room long enough, the temperature of the refrigerant in the can will also be 25°C. Now, if the lid is opened slowly and some refrigerant is allowed to escape, the pressure in the can will start dropping until it reaches the atmospheric pressure. If you are holding the can, you will notice its temperature dropping rapidly, and even ice forming outside the can if the air is humid. A thermometer inserted in the can will register −26°C when the pressure drops to 1 atm (Table A-3*a*), which is the saturation temperature of refrigerant-134a at that pressure. The temperature of the liquid refrigerant will remain at −26°C until the last drop of it vaporizes.

Another aspect of this interesting physical phenomenon is that a liquid cannot vaporize unless it absorbs energy in the amount of the latent heat of vaporization, which is 217 kJ/kg for refrigerant-134a at 1 atm. Therefore, the rate of vaporization of the refrigerant depends on the rate of heat transfer to the can: the larger the rate of heat transfer, the higher the rate of vaporization.

The rate of heat transfer to the can and thus the rate of vaporization of the refrigerant can be minimized by insulating the can heavily. In the limiting case of no heat transfer, the refrigerant will remain in the can as a liquid at −26°C indefinitely.

The boiling temperature of *nitrogen* at atmospheric pressure is −196°C (see Table A-3*a*). This means the temperature of liquid nitrogen exposed to the atmosphere must be −196°C since some nitrogen will be evaporating. The temperature of liquid nitrogen will remain constant at −196°C until it is depleted. For this reason, nitrogen is commonly used in low-temperature scientific studies (such as superconductivity) and cryogenic applications to maintain a test chamber at a constant temperature of −196°C. This is done by placing the test chamber into a liquid nitrogen bath that is open to the atmosphere. Any heat transfer from the environment to the test section is absorbed by the nitrogen, which evaporates isothermally and keeps the test chamber temperature constant at −196°C (Fig. 2-13). The entire test section must be insulated heavily to minimize heat transfer and thus liquid nitrogen consumption. Liquid nitrogen is also used for medical purposes to burn off unsightly spots on the skin. This is done by soaking a cotton swap in liquid nitrogen and wetting the desired area with it. As the nitrogen evaporates, it freezes the affected skin by rapidly absorbing heat from it.

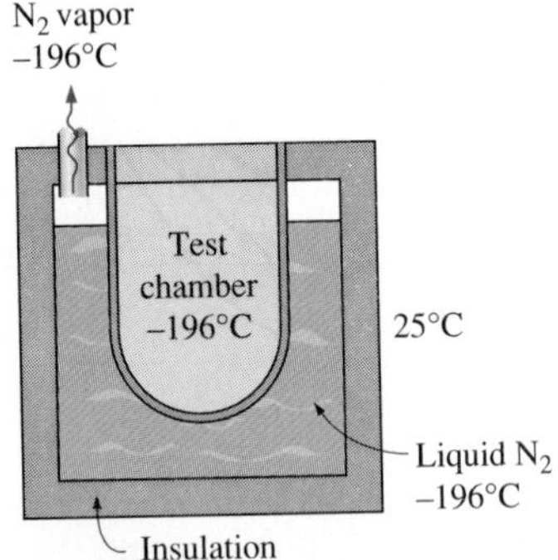

FIGURE 2-13
The temperature of liquid nitrogen exposed to the atmosphere remains constant at −196°C, and thus it maintains the test chamber at −196°C.

A practical way of cooling leafy vegetables is **vacuum cooling,** which is based on *reducing the pressure* of the sealed cooling chamber to the saturation pressure at the desired low temperature and evaporating some water from the products to be cooled. The heat of vaporization during evaporation is absorbed from the products, which lowers the product temperature. The saturation pressure of water at 0°C is 0.61 kPa, and the products can be cooled to 0°C by lowering the pressure to this level. The cooling rate can be increased by lowering the pressure below 0.61 kPa, but this is not desirable because of the danger of freezing and the added cost.

In vacuum cooling, there are two distinct stages. In the first stage, the products at ambient temperature, say at 25°C, are loaded into the flash chamber, and the operation begins. The temperature in the chamber remains constant until the *saturation pressure* is reached, which is 3.17 kPa at 25°C. In the second stage that follows, saturation conditions are maintained inside at progressively *lower pressures* and the corresponding *lower temperatures* until the desired temperature, usually slightly above 0°C, is reached (Fig. 2-14).

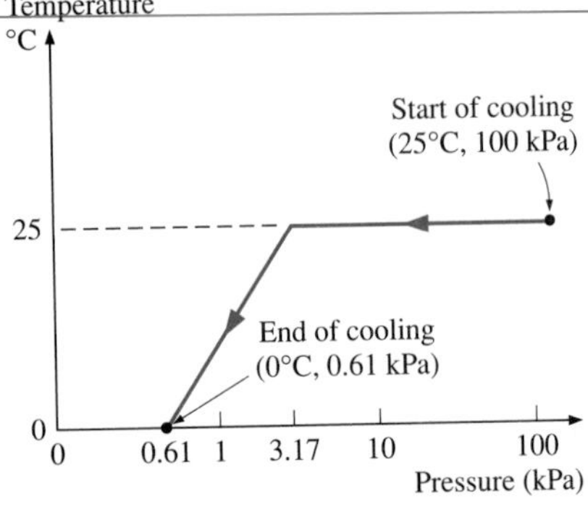

FIGURE 2-14
The variation of the temperature of fruits and vegetables with pressure during vacuum cooling from 25°C to 0°C.

Vacuum cooling is usually more expensive than the conventional refrigerated cooling, and its use is limited to applications that result in much faster cooling. Products with large surface area per unit mass and a high tendency to release moisture such as *lettuce* and *spinach* are well-suited for vacuum cooling. Products with low surface area to mass ratio are not suitable, especially those that have relatively impervious peels such as tomatoes and cucumbers. Some products such as mushrooms and green peas can be vacuum cooled successfully by wetting them first.

The vacuum cooling described above becomes **vacuum freezing** if the pressure (actually, the vapor pressure) in the vacuum chamber is dropped below 0.6 kPa, the saturation pressure of water at 0°C. The idea of making ice by using a vacuum pump is nothing new. Dr. William Cullen actually made ice in Scotland in 1775 by evacuating the air in a water tank (Fig. 2-15). **Package icing** is commonly used in small-scale cooling applications to remove heat and keep the products cool during transit by taking advantage of the large latent heat of fusion of water, but its use is limited to products that are not harmed by contact with ice. Also, ice provides *moisture* as well as *refrigeration*.

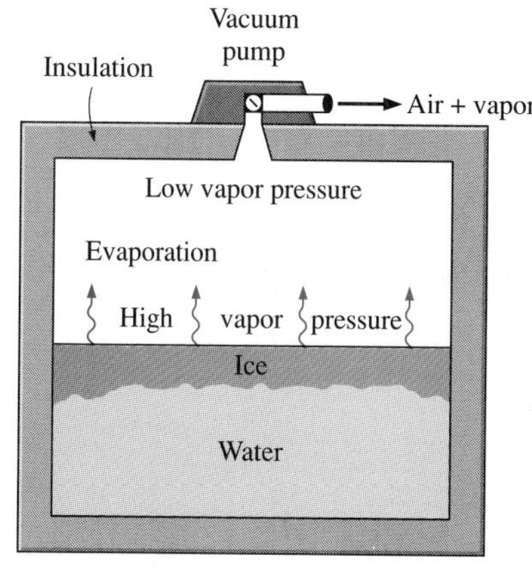

FIGURE 2-15
In 1775, ice was made by evacuating the air in a water tank.

2-4 ■ PROPERTY DIAGRAMS FOR PHASE-CHANGE PROCESSES

The variations of properties during phase-change processes are best studied and understood with the help of property diagrams. Below we develop and discuss the *T-v*, *P-v*, and *P-T* diagrams for pure substances.

1 The *T-v* Diagram

The phase-change process of water at 1 atm pressure was described in detail in the last section and plotted on a *T-v* diagram in Fig. 2-11. Now we repeat this process at different pressures to develop the *T-v* diagram for water.

Let us add weights on top of the piston until the pressure inside the cylinder reaches 1 MPa. At this pressure, water will have a somewhat smaller specific volume than it did at 1 atm pressure. As heat is transferred to the water at this new pressure, the process will follow a path that looks very much like the process path at 1 atm pressure, as shown in Fig. 2-16, but there are some noticeable differences. First, water will start boiling at a much higher temperature (179.9°C) at this pressure. Second, the specific volume of the saturated liquid is larger and the specific volume of the saturated vapor is smaller than the corresponding values at 1 atm pressure. That is, the horizontal line that connects the saturated liquid and saturated vapor states is much shorter.

As the pressure is increased further, this saturation line will continue to get shorter, as shown in Fig. 2-16, and it will become a point when the pressure reaches 22.09 MPa for the case of water. This point is called the **critical point,** and it may be defined as *the point at which the saturated liquid and saturated vapor states are identical.*

The temperature, pressure, and specific volume of a substance at the critical point are called, respectively, the *critical temperature* T_{cr}, *critical pressure* P_{cr}, and *critical specific volume* v_{cr}. The critical-point properties of water are $P_{cr} = 22.09$ MPa, $T_{cr} = 374.14$°C, and $v_{cr} = 0.003155$ m^3/kg. For helium, they are 0.23 MPa, −267.85°C, and 0.01444 m^3/kg. The critical properties for various substances are given in Table A-1 in the appendix.

At pressures above the critical pressure, there will not be a distinct phase-change process (Fig. 2-17). Instead, the specific volume of the substance will continually increase, and at all times there will be only one phase present. Eventually, it will resemble a vapor, but we can never tell when the change has occurred. Above the critical state, there is no line that separates the compressed liquid region and the superheated vapor region. However, it is

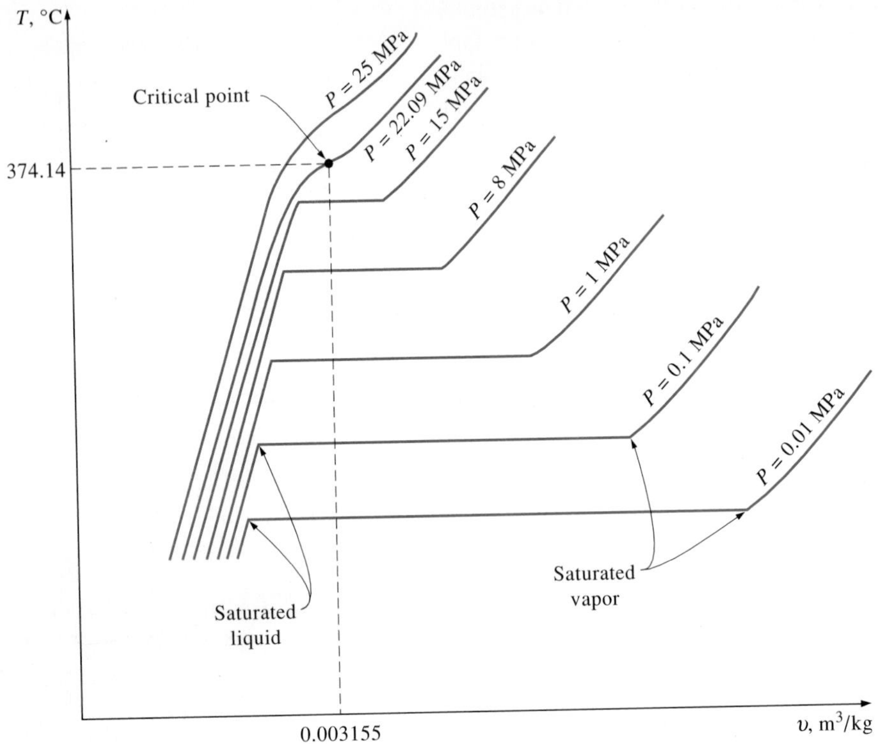

FIGURE 2-16

T-v diagram of constant-pressure phase-change processes of a pure substance at various pressures (numerical values are for water).

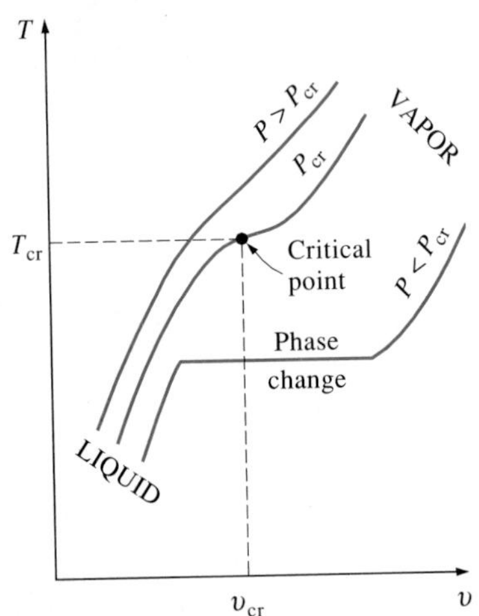

FIGURE 2-17

At supercritical pressures ($P > P_{cr}$), there is no distinct phase-change (boiling) process.

customary to refer to the substance as superheated vapor at temperatures above the critical temperature and as compressed liquid at temperatures below the critical temperature.

The saturated liquid states in Fig. 2-16 can be connected by a line called the **saturated liquid line,** and saturated vapor states in the same figure can be connected by another line, called the **saturated vapor line.** These two lines meet at the critical point, forming a dome as shown in Fig. 2-18. All the compressed liquid states are located in the region to the left of the saturated liquid line, called the **compressed liquid region.** All the superheated vapor states are located to the right of the saturated vapor line, called the **superheated vapor region.** In these two regions, the substance exists in a single phase, a liquid or a vapor. All the states that involve both phases in equilibrium are located under the dome, called the **saturated liquid–vapor mixture region,** or the **wet region.**

2 The *P-v* Diagram

The general shape of the *P-v* diagram of a pure substance is very much like the *T-v* diagram, but the $T =$ constant lines on this diagram have a downward trend, as shown in Fig. 2-19.

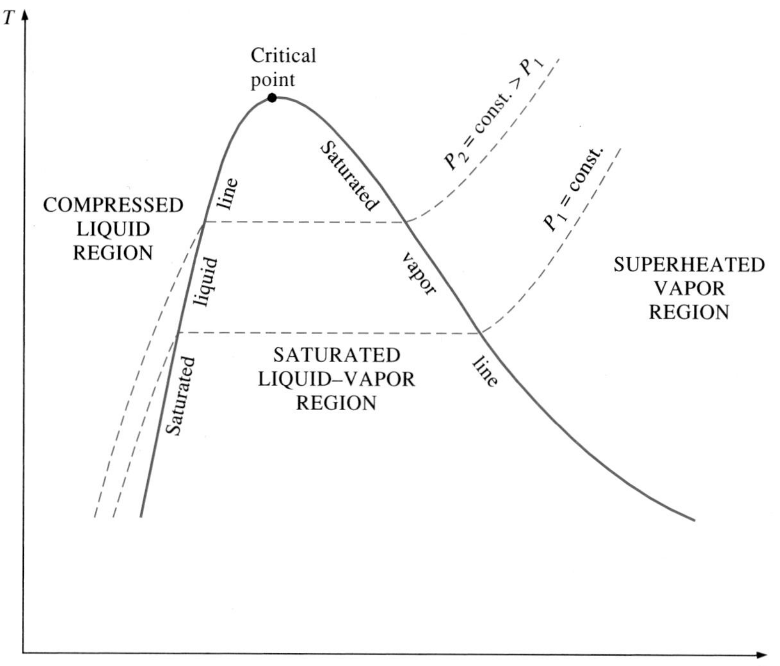

FIGURE 2-18
T-v diagram of a pure substance.

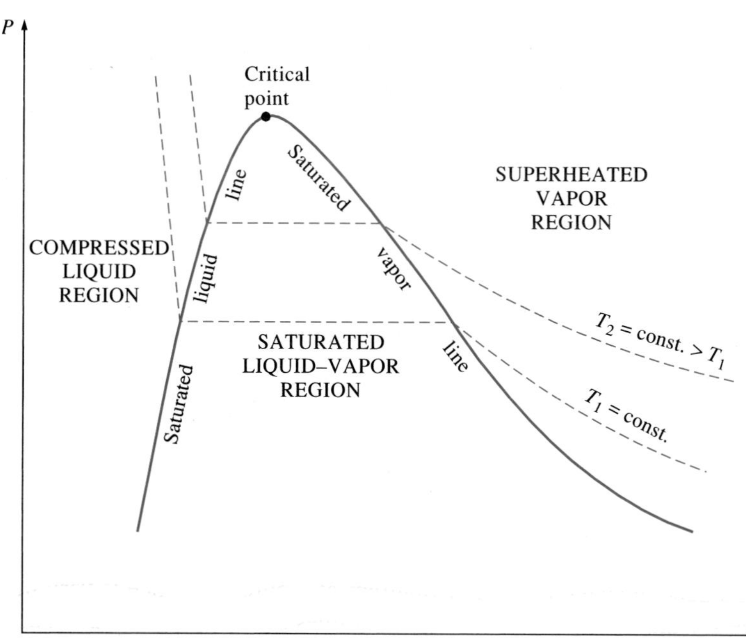

FIGURE 2-19
P-v diagram of a pure substance.

FIGURE 2-20
The pressure in a piston-cylinder device can be reduced by reducing the weight of the piston.

Consider again a piston-cylinder device that contains liquid water at 1 MPa and 150°C. Water at this state exists as a compressed liquid. Now the weights on top of the piston are removed one by one so that the pressure inside the cylinder decreases gradually (Fig. 2-20). The water is allowed to exchange heat with the surroundings so its temperature remains constant. As the pressure decreases, the volume of the water will increase slightly. When the pressure reaches the saturation-pressure value at the specified temperature (0.4758 MPa), the water will start to boil. During this vaporization process, both the temperature and the pressure remain constant, but the specific volume increases. Once the last drop of liquid is vaporized, further reduction in pressure results in a further increase in specific volume. Notice that during the phase-change process, we did not remove any weights. Doing so would cause the pressure and therefore the temperature to drop [since $T_{sat} = f(P_{sat})$], and the process would no longer be isothermal.

If the process is repeated for other temperatures, similar paths will be obtained for the phase-change processes. Connecting the saturated liquid and the saturated vapor states by a curve, we obtain the *P-v* diagram of a pure substance, as shown in Fig. 2-19.

Extending the Diagrams to Include the Solid Phase

The two equilibrium diagrams developed so far represent the equilibrium states involving the liquid and the vapor phases only. But these diagrams can easily be extended to include the solid phase as well as the solid–liquid and the solid–vapor saturation regions. The basic principles discussed in conjunction with the liquid–vapor phase-change process apply equally to the solid–liquid and solid–vapor phase-change processes. Most substances contract during a solidification (i.e., freezing) process. Others, like water, expand as they freeze. The *P-v* diagrams for both groups of substances are given in Figs. 2-21 and 2-22. These two diagrams differ only in the solid–liquid saturation region. The *T-v* diagrams look very much like the *P-v* diagrams, especially for substances that contract on freezing.

The fact that water expands upon freezing has vital consequences in nature. If water contracted on freezing as most other substances do, the ice formed would be heavier than the liquid water, and it would settle to the bottom of rivers, lakes, or oceans instead of floating at the top. The sun's rays would never reach these ice layers, and the bottoms of many rivers, lakes, or oceans would be covered with ice year round, seriously disrupting marine life.

We are all familiar with two phases being in equilibrium, but under some conditions all three phases of a pure substance coexist in equilibrium (Fig. 2-23). On *P-v* or *T-v* diagrams, these triple-phase states form a line called the **triple line.** The states on the triple line of a substance have the same pressure and temperature but different specific volumes. The triple line appears as a point on the *P-T* diagrams and, therefore, is often called the **triple point.** The triple-point temperatures and pressures of various sub-

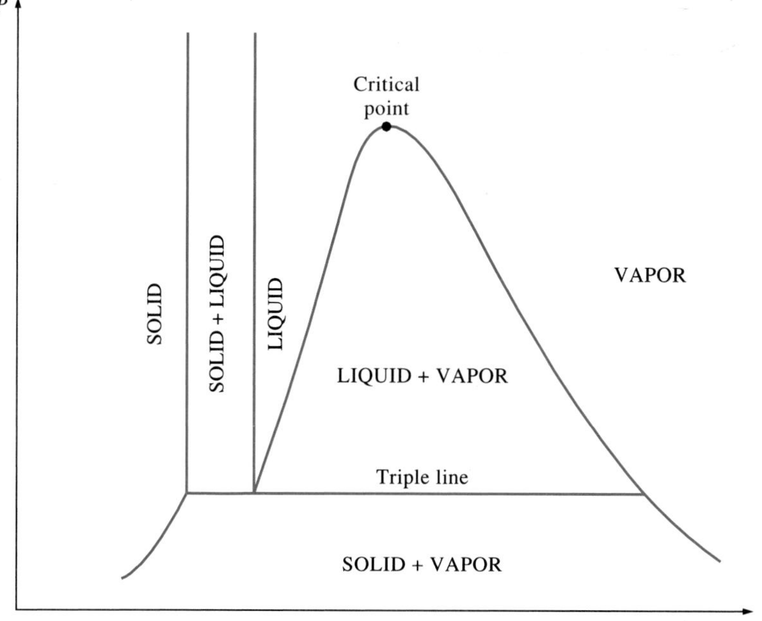

FIGURE 2-21

P-v diagram of a substance that contracts on freezing.

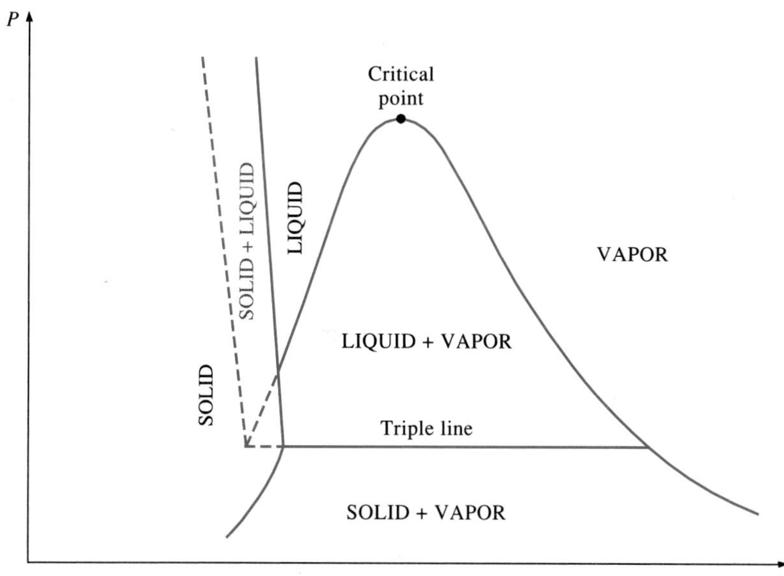

FIGURE 2-22

P-v diagram of a substance that expands on freezing (such as water).

stances are given in Table 2-3. For water, the triple-point temperature and pressure are 0.01°C and 0.6113 kPa, respectively. That is, all three phases of water will exist in equilibrium only if the temperature and pressure have precisely these values. No substance can exist in the liquid phase in stable equilibrium at pressures below the triple-point pressure. The same can be said for temperature for substances that contract on freezing. However, substances at high pressures can exist in the liquid phase at temperatures below the triple-point temperature. For example, water cannot exist in liquid form in equilibrium at atmospheric conditions at temperatures below 0°C, but it can exist as a liquid at −20°C at 200 MPa pressure. Also, ice exists at seven different solid phases at pressures above 100 MPa.

There are two ways a substance can pass from the solid to vapor phase: either it melts first into a liquid and subsequently evaporates, or it evaporates

FIGURE 2-23

At triple-point pressure and temperature, a substance exists in three phases in equilibrium.

TABLE 2-3

Triple-point temperatures and pressures of various substances

Substance	Formula	T_{tp} K	P_{tp} kPa
Acetylene	C_2H_2	192.4	120
Ammonia	NH_3	195.40	6.076
Argon	A	83.81	68.9
Carbon (graphite)	C	3900	10,100
Carbon dioxide	CO_2	216.55	517
Carbon monoxide	CO	68.10	15.37
Deuterium	D_2	18.63	17.1
Ethane	C_2H_6	89.89	8×10^{-4}
Ethylene	C_2H_4	104.0	0.12
Helium 4 (λ point)	He	2.19	5.1
Hydrogen	H_2	13.84	7.04
Hydrogen chloride	HCl	158.96	13.9
Mercury	Hg	234.2	1.65×10^{-7}
Methane	CH_4	90.68	11.7
Neon	Ne	24.57	43.2
Nitric oxide	NO	109.50	21.92
Nitrogen	N_2	63.18	12.6
Nitrous oxide	N_2O	182.34	87.85
Oxygen	O_2	54.36	0.152
Palladium	Pd	1825	3.5×10^{-3}
Platinum	Pt	2045	2.0×10^{-4}
Sulfur dioxide	SO_2	197.69	1.67
Titatium	Ti	1941	5.3×10^{-3}
Uranium hexafluoride	UF_6	337.17	151.7
Water	H_2O	273.16	0.61
Xenon	Xe	161.3	81.5
Zinc	Zn	692.65	0.065

Source: Data from National Bureau of Standards (U.S.) Circ., 500 (1952).

directly without melting first. The latter occurs at pressures below the triple-point value, since a pure substance cannot exist in the liquid phase at those pressures (Fig. 2-24). Passing from the solid phase directly into the vapor phase is called **sublimation.** For substances that have a triple-point pressure above the atmospheric pressure such as solid CO_2 (dry ice), sublimation is the only way to change from the solid to vapor phase at atmospheric conditions.

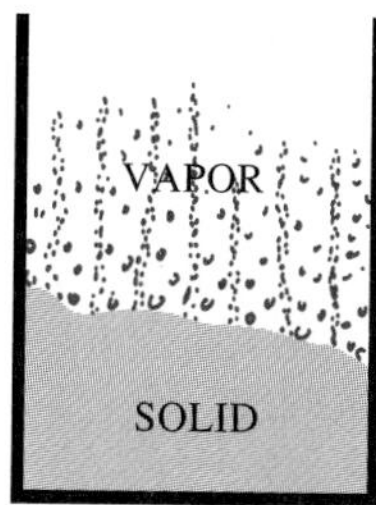

FIGURE 2-24
At low pressures (below the triple-point value), solids evaporate without melting first (*sublimation*).

3 The *P-T* Diagram

Figure 2-25 shows the *P-T* diagram of a pure substance. This diagram is often called the **phase diagram** since all three phases are separated from each other by three lines. The sublimation line separates the solid and vapor regions, the vaporization line separates the liquid and vapor regions, and the melting (or fusion) line separates the solid and liquid regions. These three lines meet at the triple point, where all three phases coexist in equilibrium. The vaporization line ends at the critical point because no distinction can be made between liquid and vapor phases above the critical point. Substances that expand and contract on freezing differ only in the melting line on the *P-T* diagram.

The *P-v-T* Surface

In Chap. 1, we indicated that the state of a simple compressible substance is fixed by any two independent, intensive properties. Once the two appropriate properties are fixed, all the other properties become dependent properties.

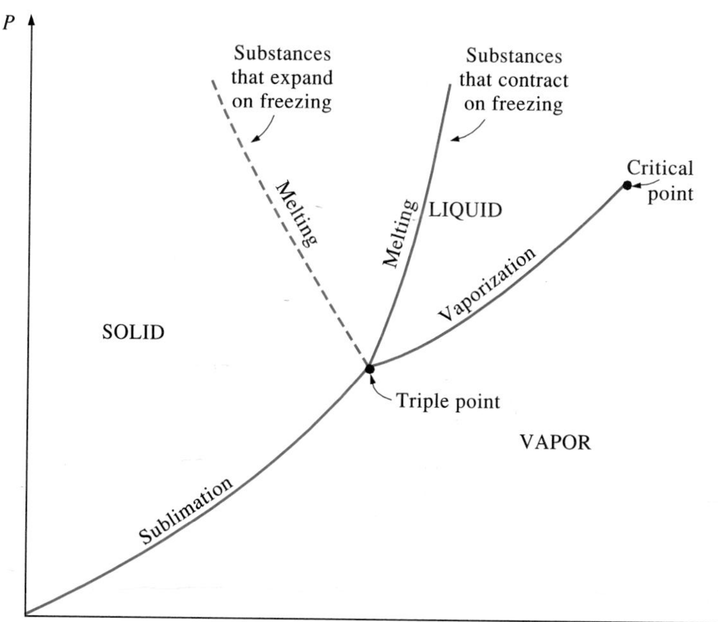

FIGURE 2-25
P-T diagram of pure substances.

Remembering that any equation with two independent variables in the form $z = z(x, y)$ represents a surface in space, we can represent the *P-v-T* behavior of a substance as a surface in space, as shown in Figs. 2-26 and 2-27. Here *T* and *v* may be viewed as the independent variables (the base) and *P* as the dependent variable (the height).

All the points on the surface represent equilibrium states. All states along the path of a quasi-equilibrium process lie on the *P-v-T* surface since such a process must pass through equilibrium states. The single-phase regions appear as curved surfaces on the *P-v-T* surface, and the two-phase regions as surfaces perpendicular to the *P-T* plane. This is expected since the projections of two-phase regions on the *P-T* plane are lines.

All the two-dimensional diagrams we have discussed so far are merely projections of this three-dimensional surface onto the appropriate planes. A *P-v* diagram is just a projection of the *P-v-T* surface on the *P-v* plane, and a *T-v* diagram is nothing more than the bird's-eye view of this surface. The *P-v-T* surfaces present a great deal of information at once, but in a thermodynamic analysis it is more convenient to work with two-dimensional diagrams, such as the *P-v* and *T-v* diagrams.

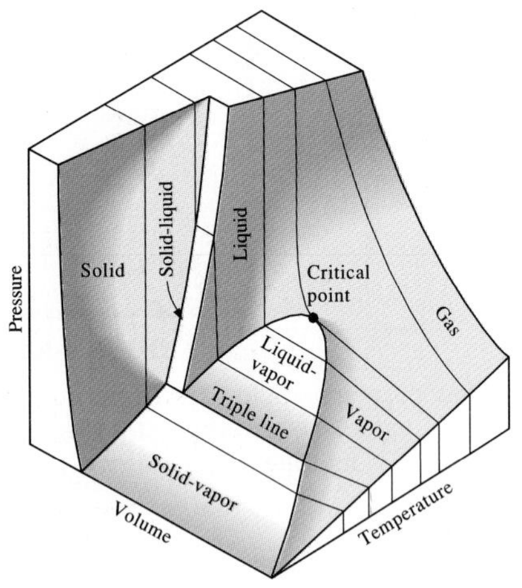

FIGURE 2-26

P-v-T surface of a substance that *contracts* on freezing.

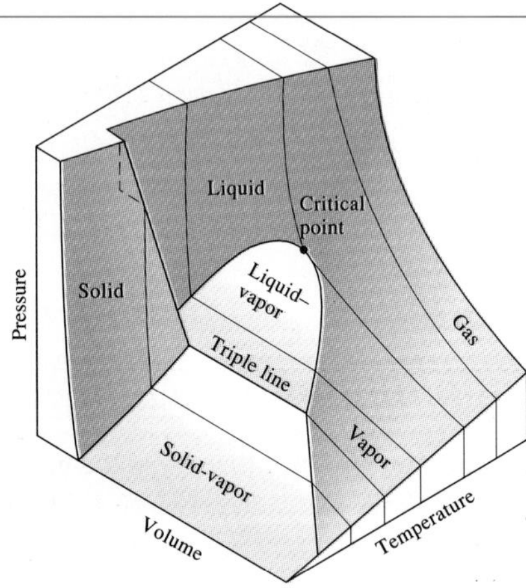

FIGURE 2-27

P-v-T surface of a substance that *expands* on freezing (like water).

2-5 ■ VAPOR PRESSURE AND PHASE EQUILIBRIUM

The pressure in a gas container is due to the individual molecules striking the wall of the container and exerting a force on it. This force is proportional to the average velocity of the molecules and the number of molecules per unit volume of the container (i.e., molar density). Therefore, the pressure exerted by a gas is a strong function of the density and the temperature of the gas. For a gas mixture, the pressure measured by a sensor such as a transducer is the sum of the pressures exerted by the individual gas species, called the *partial pressure*. It can be shown (see Chap. 12) that the partial pressure of a gas in a mixture is proportional to the number of moles (or the mole fraction) of that gas.

Atmospheric air can be viewed as a mixture of dry air (air with zero moisture content) and water vapor (also referred to as moisture), and the atmospheric pressure is the sum of the pressure of dry air P_a and the pressure of water vapor, called the **vapor pressure** P_v (Fig. 2-28). That is,

$$P_{atm} = P_a + P_v \tag{2-1}$$

The vapor pressure constitutes a small fraction (usually under 3 percent) of the atmospheric pressure since air is mostly nitrogen and oxygen, and the water molecules constitute a small fraction (usually under 3 percent) of the total molecules in the air. However, the amount of water vapor in the air has a major impact on thermal comfort and many processes such as drying.

Air can hold a certain amount of moisture only, and the ratio of the actual amount of moisture in the air at a given temperature to the maximum amount air can hold at that temperature is called the **relative humidity** ϕ. The relative humidity ranges from 0 for dry air to 100 percent for **saturated air** (air that cannot hold any more moisture). The vapor pressure of saturated air at a

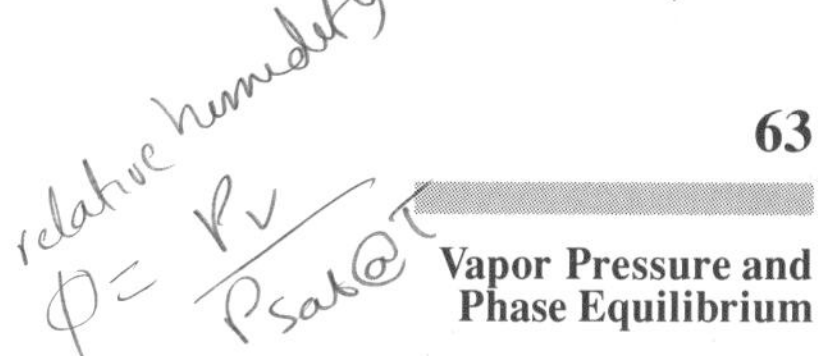

given temperature is equal to the saturation pressure of water at that temperature. For example, the vapor pressure of saturated air at 25°C is 3.17 kPa.

The amount of moisture in the air is completely specified by the temperature and the relative humidity, and the vapor pressure is related to relative humidity ϕ by

$$P_v = \phi P_{\text{sat @ }T} \tag{2-2}$$

where $P_{\text{sat @ }T}$ is the saturation pressure of water at the specified temperature. For example, the vapor pressure of air at 25°C and 60 percent relative humidity is

$$P_v = \phi P_{\text{sat @ 25°C}} = 0.6 \times (3.17 \text{ kPa}) = 1.90 \text{ kPa}$$

The desirable range of relative humidity for thermal comfort is 40 to 60 percent.

Note that the amount of moisture air can hold is proportional to the saturation pressure, which increases with temperature. Therefore, air can hold more moisture at higher temperatures. Dropping the temperature of moist air reduces its moisture capacity and may result in the condensation of some of the moisture in the air as suspended water droplets (fog) or as a liquid film on cold surfaces (dew). So it is no surprise that fog and dew are common occurrences at humid locations especially in the early morning hours when the temperatures are the lowest. Both fog and dew disappear (evaporate) as the air temperature rises shortly after sunrise. You also may have noticed that electronic devices such as camcorders come with warnings against bringing them into moist indoors when the devices are cold to avoid moisture condensation on the sensitive electronics of the devices.

It is a common observation that whenever there is an imbalance of a commodity in a medium, nature tends to redistribute it until a "balance" or "equality" is established. This tendency is often referred to as the *driving force,* which is the mechanism behind many naturally occurring transport phenomena such as heat transfer, fluid flow, electric current, and mass transfer. If we define the amount of a commodity per unit volume as the *concentration* of that commodity, we can say that the flow of a commodity is always in the direction of decreasing concentration, that is, from the region of high concentration to the region of low concentration (Fig. 2-29). The commodity simply creeps away during redistribution, and thus the flow is a *diffusion process.*

We know from experience that a wet T-shirt hanging in an open area eventually dries, a small amount of water left in a glass evaporates, and the aftershave in an open bottle quickly disappears. These and many other similar examples suggest that there is a driving force between the two phases of a substance that forces the mass to transform from one phase to another. The magnitude of this force depends on the relative concentrations of the two phases. A wet T-shirt will dry much faster in dry air than it would in humid air. In fact, it will not dry at all if the relative humidity of the environment is 100 percent and thus the air is saturated. In this case, there will be no transformation from the liquid phase to the vapor phase, and the two phases will

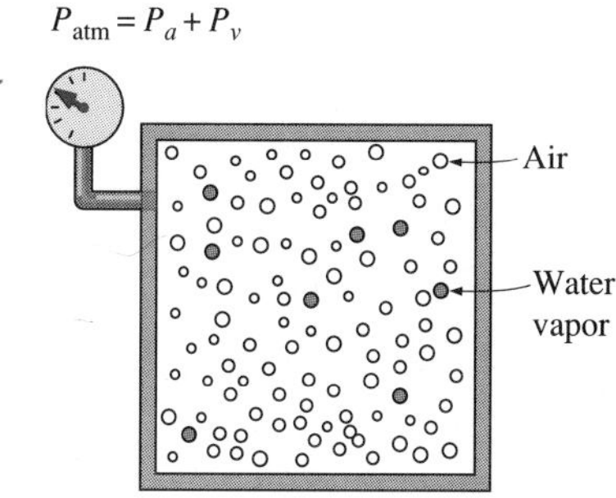

FIGURE 2-28

Atmospheric pressure is the sum the dry air pressure P_a and the vapor pressure P_v.

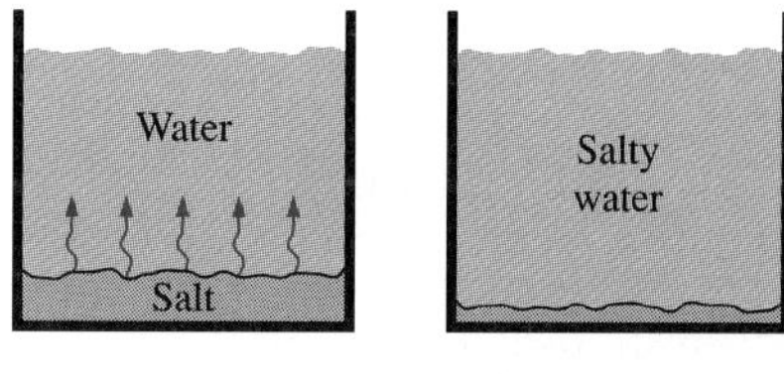

FIGURE 2-29

Whenever there is a concentration difference of a physical quantity in a medium, nature tends to equalize things by forcing a flow from the high to the low concentration region.

be in **phase equilibrium.** For liquid water that is open to the atmosphere, the criterion for phase equilibrium can be expressed as follows: *The vapor pressure in the air must be equal to the saturation pressure of water at the water temperature.* That is (Fig. 2-30),

$$\text{Phase equilibrium criterion for water exposed to air: } P_v = P_{\text{sat, water @ } T} \quad (2\text{-}3)$$

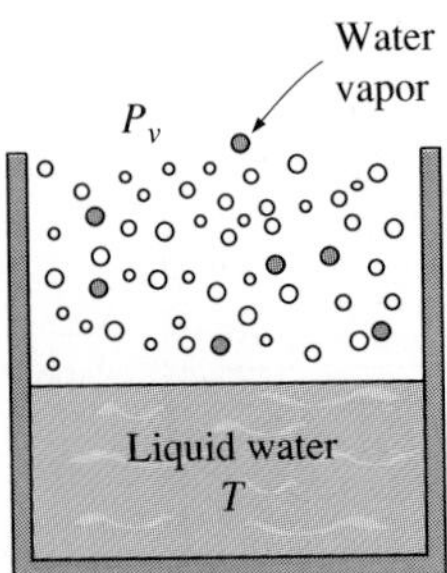

FIGURE 2-30
When open to the atmosphere, water is in phase equilibrium with the vapor in the air if the vapor pressure is equal to the saturation temperature of water.

Therefore, if the vapor pressure in the air is less than the saturation pressure of water at the water temperature, some liquid will evaporate. The larger the difference between the vapor and saturation pressures, the higher the rate of evaporation. The evaporation will have a cooling effect on water, and thus reduce its temperature. This, in turn, will reduce the saturation pressure of water and thus the rate of evaporation until some kind of quasi-steady operation is reached. This explains why water is usually at a considerably lower temperature than the surrounding air, especially in dry climates. It also suggests that the rate of evaporation of water can be increased by increasing the water temperature and thus the saturation pressure of water.

Note that the air at the water surface will always be saturated because of the direct contact with water, and thus the vapor pressure. Therefore, the vapor pressure at the lake surface will simply be the saturation pressure of water at the temperature of the water at the surface. If the air is not saturated, then the vapor pressure will decrease to the value in the air at some distance from the water surface, and the difference between these two vapor pressures is the driving force for the evaporation of water.

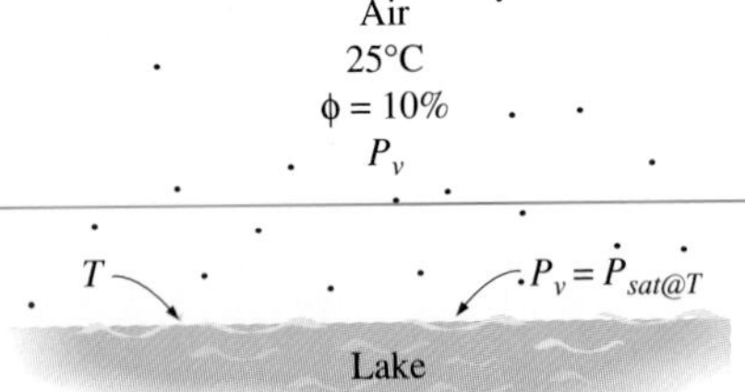

FIGURE 2-31
Schematic for Example 2-1.

EXAMPLE 2-1 Temperature Drop of a Lake Due to Evaporation

On a summer day, the air temperature over a lake is measured to be 25°C. Determine water temperature of the lake when phase equilibrium conditions are established between the water in the lake and the vapor in the air for relative humidities of 10, 80, and 100 percent for the air (Fig. 2-31).

Solution The saturation pressure of water at 25°C, from Table 2-1, is 3.17 kPa. Then the vapor pressures at relative humidities of 10, 80, and 100 percent are determined from Eq. 2-2 to be

Relative humidity = 10%: $P_{v1} = \phi_1 P_{\text{sat @ 25°C}} = 0.1 \times (3.17 \text{ kPa}) = 0.317 \text{ kPa}$

Relative humidity = 80%: $P_{v2} = \phi_2 P_{\text{sat @ 25°C}} = 0.8 \times (3.17 \text{ kPa}) = 2.536 \text{ kPa}$

Relative humidity = 100%: $P_{v3} = \phi_3 P_{\text{sat @ 25°C}} = 1.0 \times (3.17 \text{ kPa}) = 3.17 \text{ kPa}$

The saturation temperatures corresponding to these pressures are determined from Table 2-1 by interpolation to be

$$T_1 = \mathbf{-8.0°C} \qquad T_2 = \mathbf{21.2°C} \qquad \text{and} \qquad T_3 = \mathbf{25°C}$$

Therefore, water will freeze in the first case even though the surrounding air is hot. In the last case the water temperature will be the same as the surrounding air temperature.

You are probably skeptical about the lake freezing when the air is at 25°C, and you are right. The water temperature will drop to −8°C in the limiting case of no heat transfer to the water surface. In practice the water temperature will drop below the air temperature, but it will not drop to −8°C because (1) it is very unlikely for the air over the lake to be so dry (a relative humidity of just 10 percent) and (2) as the water temperature near the surface drops, heat transfer from the air and the lower parts of the water body will tend to make up for this heat loss and prevent the water temperature from dropping too much. The water temperature will stabilize when the heat gain from the surrounding air and the water body equals the heat loss by evaporation, that is, when a *dynamic balance* is established between heat and mass transfer instead of phase equilibrium. If you try this experiment at home using a shallow layer of water in a well-insulated pan, you can actually freeze the water if the air is really dry and relatively cool.

The natural tendency of water to evaporate in order to achieve phase equilibrium with the water vapor in the surrounding air forms the basis for the operation of the **evaporative coolers** (also called the *swamp coolers*). In such coolers, hot and dry outdoor air is forced to flow through a wet cloth before entering a building. Some of the water evaporates by absorbing heat from the air, and thus cooling it. Evaporative coolers are commonly used in dry climates and provide effective cooling. They are much cheaper to run than air conditioners since they are inexpensive to buy, and the fan of an evaporative cooler consumes much less power than the compressor of an air conditioner.

Boiling and evaporation are often used interchangeably to indicate *phase change from liquid to vapor*. Although they refer to the same physical process, they differ in some aspects. **Evaporation** occurs at the *liquid–vapor interface* when the vapor pressure is less than the saturation pressure of the liquid at a given temperature. Water in a lake at 20°C, for example, will evaporate to air at 20°C and 60 percent relative humidity since the saturation pressure of water at 20°C is 2.34 kPa, and the vapor pressure of air at 20°C and 60 percent relative humidity is 1.4 kPa. Other examples of evaporation are the drying of clothes, fruits, and vegetables; the evaporation of sweat to cool the human body; and the rejection of waste heat in wet cooling towers. Note that evaporation involves no bubble formation or bubble motion (Fig. 2-32).

Boiling, on the other hand, occurs at the *solid–liquid interface* when a liquid is brought into contact with a surface maintained at a temperature T_s sufficiently above the saturation temperature T_{sat} of the liquid. At 1 atm, for

FIGURE 2-32

A liquid-to-vapor phase change process is called evaporation if it occurs at a liquid–vapor interface, and boiling if it occurs at a solid–liquid interface.

example, liquid water in contact with a solid surface at 110°C will boil since the saturation temperature of water at 1 atm is 100°C. The boiling process is characterized by the rapid motion of *vapor bubbles* that form at the solid–liquid interface, detach from the surface when they reach a certain size, and attempt to rise to the free surface of the liquid. When cooking, we do not say water is boiling unless we see the bubbles rising to the top.

2-6 ■ PROPERTY TABLES

For most substances, the relationships among thermodynamic properties are too complex to be expressed by simple equations. Therefore, properties are frequently presented in the form of tables. Some thermodynamic properties can be measured easily, but others cannot and are calculated by using the relations between them and measurable properties. The results of these measurements and calculations are presented in tables in a convenient format. In the following discussion, the steam tables will be used to demonstrate the use of thermodynamic property tables. Property tables of other substances are used in the same manner.

For each substance, the thermodynamic properties are listed in more than one table. In fact, a separate table is prepared for each region of interest such as the superheated vapor, compressed liquid, and saturated (mixture) regions. Property tables are given in the appendix in both SI and English units. The tables in English units carry the same number as the corresponding tables in SI, followed by an identifier E. Tables A-6 and A-6E, for example, list properties of superheated water vapor, the former in SI and the latter in English units. Before we get into the discussion of property tables, we will define a new property called *enthalpy*.

Enthalpy—A Combination Property

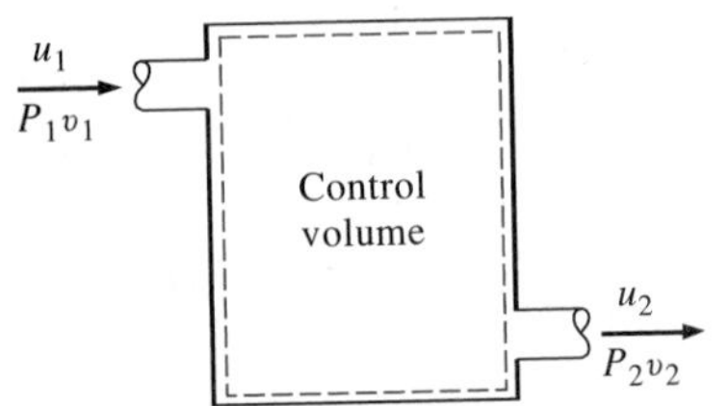

FIGURE 2-33
The combination $u + Pv$ is frequently encountered in the analysis of control volumes.

A person looking at the tables carefully will notice two new properties: enthalpy h and entropy s. Entropy is a property associated with the second law of thermodynamics, and we will not use it until it is properly defined in Chap. 6. However, it is appropriate to introduce enthalpy at this point.

In the analysis of certain types of processes, particularly in power generation and refrigeration (Fig. 2-33), we frequently encounter the combination of properties $U + PV$. For the sake of simplicity and convenience, this combination is defined as a new property, **enthalpy,** and given the symbol H:

$$H = U + PV \qquad \text{(kJ)}$$

or, per unit mass,

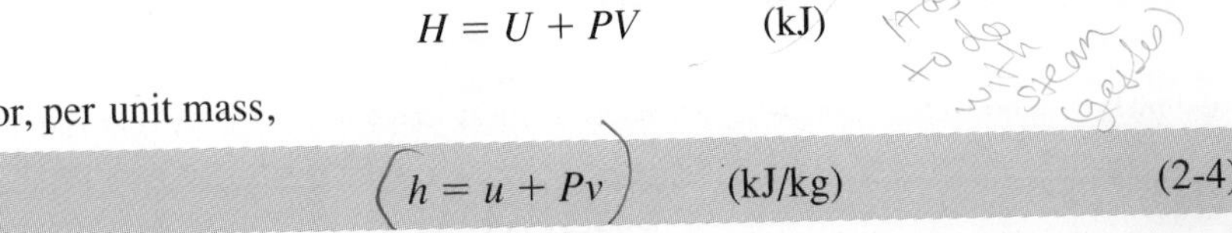

$$h = u + Pv \qquad \text{(kJ/kg)} \qquad \text{(2-4)}$$

Both the total enthalpy H and specific enthalpy h are simply referred to as enthalpy since the context will clarify which one is meant. Notice that the equations given above are dimensionally homogeneous. That is, the unit of the pressure–volume produce may differ from the unit of the internal energy by only a factor (Fig. 2-34). For example, it can be easily shown that $1 \text{ kPa} \cdot \text{m}^3 = 1 \text{ kJ}$. In some tables encountered in practice, the internal energy u is frequently not listed, but it can always be determined from $u = h - Pv$.

The widespread use of the property enthalpy is due to Professor Richard Mollier, who recognized the importance of the group $u + Pv$ in the analysis of steam turbines and in the representation of the properties of steam in tabular and graphical form (as in the famous Mollier chart). Mollier referred to the group $u + Pv$ as *heat contents* and *total heat*. These terms were not quite consistent with the modern thermodynamic terminology and were replaced in 1930s by the term *enthalpy* (from the Greek word *enthalpien*, which means *to heat*).

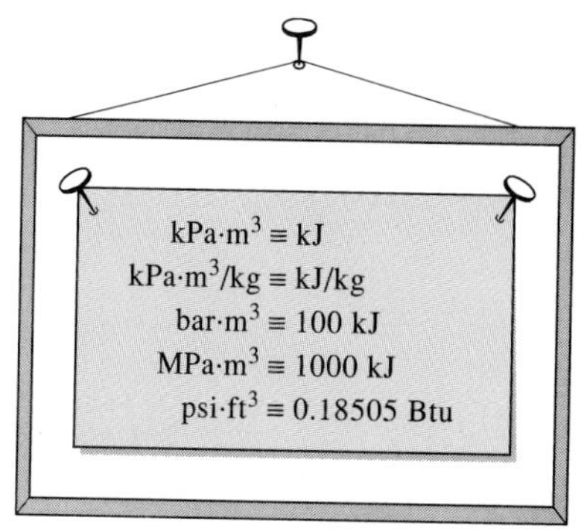

FIGURE 2-34

The product *pressure* × *volume* has energy units.

1a Saturated Liquid and Saturated Vapor States

The properties of saturated liquid and saturated vapor for water are listed in Tables A-4 and A-5. Both tables give the same information. The only difference is that in Table A-4 properties are listed under temperature and in Table A-5 under pressure. Therefore, it is more convenient to use Table A-4 when *temperature* is given and Table A-5 when *pressure* is given. The use of Table A-4 is illustrated in Fig. 2-35.

The subscript f is used to denote properties of a saturated liquid, and the subscript g to denote the properties of saturated vapor. These symbols are commonly used in thermodynamics and originated from German. Another subscript commonly used is fg, which denotes the difference between the saturated vapor and saturated liquid values of the same property. For example,

$$v_f = \text{specific volume of saturated liquid}$$
$$v_g = \text{specific volume of saturated vapor}$$
$$v_{fg} = \text{difference between } v_g \text{ and } v_f \text{ (that is, } v_{fg} = v_g - v_f)$$

The quantity h_{fg} is called the **enthalpy of vaporization** (or latent heat of vaporization). It represents the amount of energy needed to vaporize a unit mass of saturated liquid at a given temperature or pressure. It decreases as the temperature or pressure increases, and becomes zero at the critical point.

Temp. °C T	Sat. press kPa P_{sat}	Specific volume m^3/kg: Sat. liquid v_f	Sat. vapor v_g
85	57.83	0.001 033	2.828
90	70.14	0.001 036	2.361
95	84.55	0.001 040	1.982

Specific temperature

Corresponding saturation pressure

Specific volume of saturated liquid

Specific volume of saturated vapor

FIGURE 2-35

A partial list of Table A-4.

EXAMPLE 2-2 Finding the Pressure of Saturated Liquid

A rigid tank contains 50 kg of saturated liquid water at 90°C. Determine the pressure in the tank and the volume of the tank.

Solution The state of the saturated liquid water is shown on a *T-v* diagram in Fig. 2-36. Since saturation conditions exist in the tank, the pressure must be the saturation pressure at 90°C:

$$P = P_{sat\,@\,90°C} = \mathbf{70.14\ kPa} \qquad \text{(Table A-4)}$$

The specific volume of the saturated liquid at 90°C is

$$v = v_{f\,@\,90°C} = 0.001036 \text{ m}^3/\text{kg} \qquad \text{(Table A-4)}$$

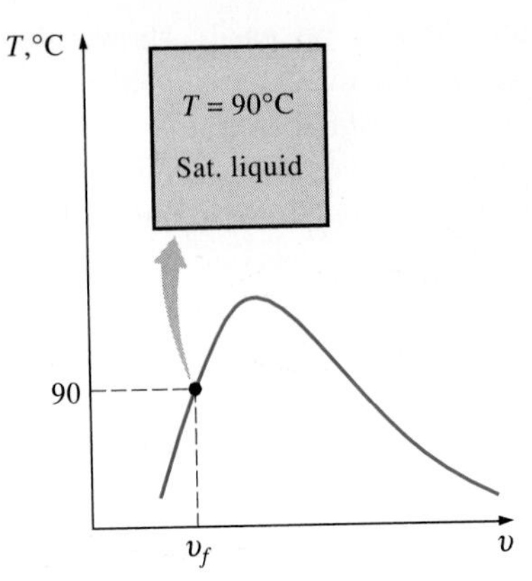

FIGURE 2-36
Schematic and *T-v* diagram for Example 2-2.

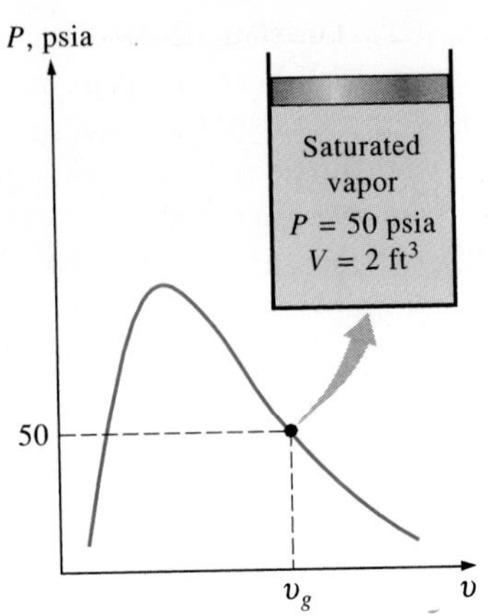

FIGURE 2-37
Schematic and *P-v* diagram for Example 2-3.

Then the total volume of the tank is determined to be

$$V = mv = (50 \text{ kg})(0.001036 \text{ m}^3/\text{kg}) = \mathbf{0.0518 \text{ m}^3}$$

EXAMPLE 2-3 Finding the Temperature of Saturated Vapor

A piston-cylinder device contains 2 ft³ of saturated water vapor at 50-psia pressure. Determine the temperature of the vapor and the mass of the vapor inside the cylinder.

Solution The state of the saturated water vapor is shown on a *P-v* diagram in Fig. 2-37. Since the cylinder contains saturated vapor at 50 psia, the temperature inside must be the saturation temperature at this pressure:

$$T = T_{\text{sat @ 50 psia}} = \mathbf{281.03°F} \qquad \text{(Table A-5E)}$$

The specific volume of the saturated vapor at 50 psia is

$$v = v_{g\text{ @ 50 psia}} = 8.518 \text{ ft}^3/\text{lbm} \qquad \text{(Table A-5E)}$$

Then the mass of water vapor inside the cylinder becomes

$$m = \frac{V}{v} = \frac{2 \text{ ft}^3}{8.518 \text{ ft}^3/\text{lbm}} = \mathbf{0.235 \text{ lbm}}$$

FIGURE 2-38
Schematic and *P-v* diagram for Example 2-4.

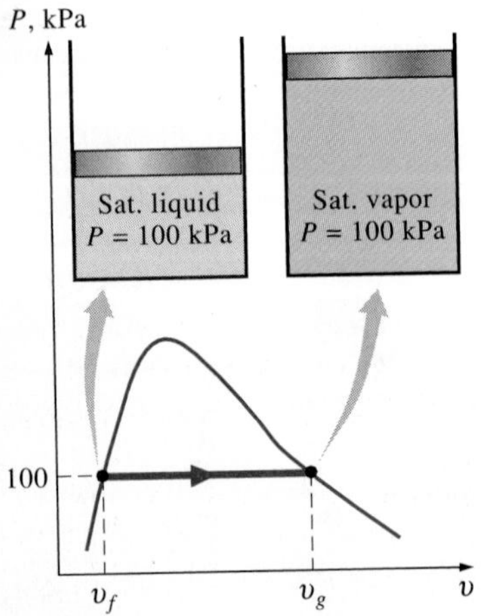

EXAMPLE 2-4 The Volume and Energy Change during Evaporation

A mass of 200 g of saturated liquid water is completely vaporized at a constant pressure of 100 kPa. Determine (*a*) the volume change and (*b*) the amount of energy added to the water.

Solution (*a*) The process described is illustrated on a *P-v* diagram in Fig. 2-38. The volume change per unit mass during a vaporization process is v_{fg}, which is the difference between v_g and v_f. Reading these values from Table A-5 at 100 kPa and substituting yield

$$v_{fg} = v_g - v_f = (1.6940 - 0.001043)\ \text{m}^3/\text{kg} = 1.6930\ \text{m}^3/\text{kg}$$

Thus, $\Delta V = mv_{fg} = (0.2\ \text{kg})(1.6930\ \text{m}^3/\text{kg}) = \mathbf{0.3386\ m^3}$

Note that we have considered the first four decimal digits of v_f and disregarded the rest. This is because v_g has significant numbers to the first four decimal places only, and we do not know the numbers in the other decimal places. Taking v_f as it is would mean that we are assuming v_g = 1.694000, which is not necessarily the case. It could very well be that v_g = 1.694038 since this number, too, would truncate to 1.6940. All the digits in our result (1.6930) are significant. But if we used v_f as it is, we would obtain v_{fg} = 1.692957, which falsely implies that our result is accurate to the sixth decimal place.

(*b*) The amount of energy needed to vaporize the unit mass of a substance at a given pressure is the enthalpy of vaporization at that pressure, which, at 100 kPa, is h_{fg} = 2258.0 kJ/kg. Thus, the amount of energy added is

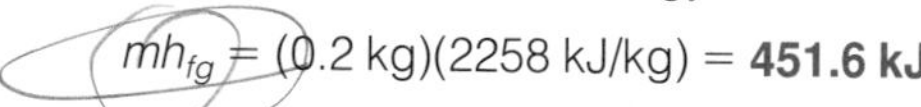

$$mh_{fg} = (0.2\ \text{kg})(2258\ \text{kJ/kg}) = \mathbf{451.6\ kJ}$$

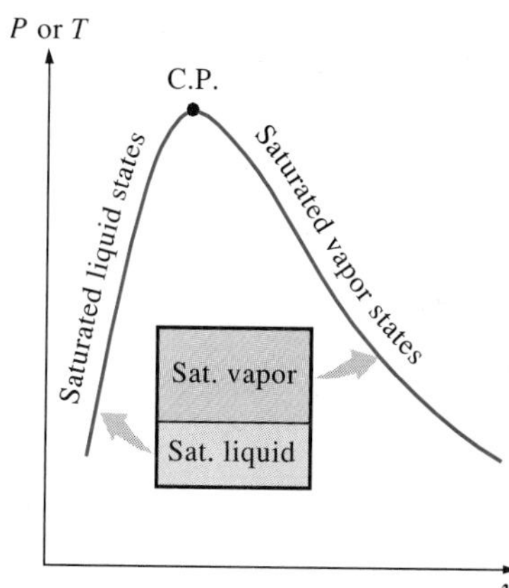

FIGURE 2-39
The relative amounts of liquid and vapor phases in a saturated mixture are specified by the *quality x*.

1b Saturated Liquid–Vapor Mixture

During a vaporization process, a substance exists as part liquid and part vapor. That is, it is a mixture of saturated liquid and saturated vapor (Fig. 2-39). To analyze this mixture properly, we need to know the proportions of the liquid and vapor phases in the mixture. This is done by defining a new property called the **quality** x as the ratio of the mass of vapor to the total mass of the mixture:

$$x = \frac{m_{\text{vapor}}}{m_{\text{total}}} \tag{2-5}$$

where $m_{\text{total}} = m_{\text{liquid}} + m_{\text{vapor}} = m_f + m_g$

Quality has significance for *saturated mixtures* only. It has no meaning in the compressed liquid or superheated vapor regions. Its value is always between 0 and 1. The quality of a system that consists of *saturated liquid* is 0 (or 0 percent), and the quality of a system consisting of *saturated vapor* is 1 (or 100 percent). In saturated mixtures, quality can serve as one of the two independent intensive properties needed to describe a state. Note that *the properties of the saturated liquid are the same whether it exists alone or in a mixture with saturated vapor.* During the vaporization process, only the amount of saturated liquid changes, not its properties. The same can be said about a saturated vapor.

A saturated mixture can be treated as a combination of two subsystems: the saturated liquid and the saturated vapor. However, the amount of mass for each phase is usually not known. Therefore, it is often more convenient to imagine that the two phases are mixed very well, forming a homogeneous appearance (Fig. 2-40). Then the properties of this "mixture" will simply be the average properties of the saturated liquid–vapor mixture under consideration. Here is how it is done:

FIGURE 2-40
A two-phase system can be treated as a homogeneous mixture for convenience.

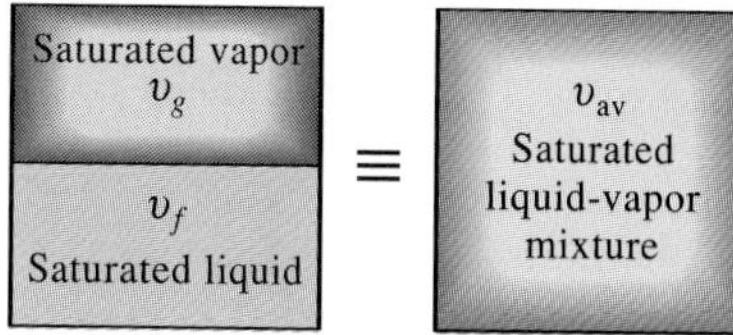

Consider a tank that contains a saturated liquid–vapor mixture. The volume occupied by saturated liquid is V_f, and the volume occupied by saturated vapor is V_g. The total volume V is the sum of these two:

$$V = V_f + V_g$$

$$V = mv \longrightarrow m_t v_{av} = m_f v_f + m_g v_g$$

$$m_f = m_t - m_g \longrightarrow m_t v_{av} = (m_t - m_g)v_f + m_g v_g$$

Dividing by m_t yields

$$v_{av} = (1 - x)v_f + xv_g$$

since $x = m_g/m_t$. This relation can also be expressed as

$$v_{av} = v_f + xv_{fg} \qquad (\text{m}^3/\text{kg}) \qquad (2\text{-}6)$$

where $v_{fg} = v_g - v_f$. Solving for quality, we obtain

$$x = \frac{v_{av} - v_f}{v_{fg}} \qquad (2\text{-}7)$$

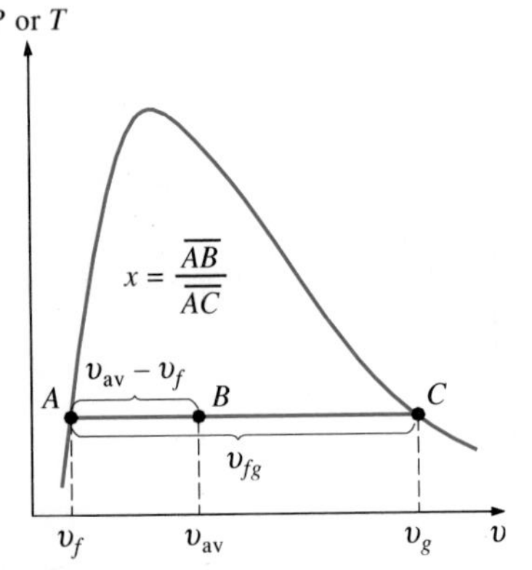

FIGURE 2-41
Quality is related to the horizontal distances on P-v and T-v diagrams.

Based on this equation, quality can be related to the horizontal distances on a P-v or T-v diagram (Fig. 2-41). At a given temperature or pressure, the numerator of Eq. 2-7 is the distance between the actual state and the saturated liquid state, and the denominator is the length of the entire horizontal line that connects the saturated liquid and saturated vapor states. A state of 50 percent quality will lie in the middle of this horizontal line.

The analysis given above can be repeated for internal energy and enthalpy with the following results:

$$u_{av} = u_f + xu_{fg} \qquad (\text{kJ/kg})$$

$$h_{av} = h_f + xh_{fg} \qquad (\text{kJ/kg})$$

All the results are of the same format, and they can be summarized in a single equation as

$$y_{av} = y_f + xy_{fg} \qquad (2\text{-}8)$$

where y is v, u, or h. The subscript "av" (for "average") is usually dropped for simplicity. The values of the average properties of the mixtures are always *between* the values of the saturated liquid and the saturated vapor properties (Fig. 2-42). That is,

$$y_f \leq y_{av} \leq y_g$$

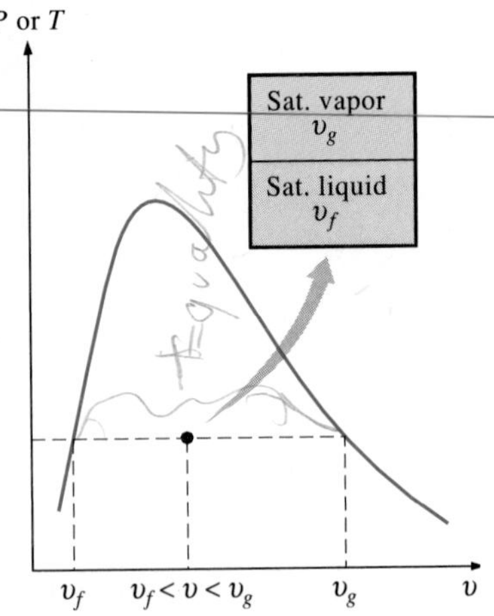

FIGURE 2-42
The v value of a saturated liquid–vapor mixture lies between the v_f and v_g values at the specified T or P.

Finally, all the saturated-mixture states are located under the saturation curve, and to analyze saturated mixtures, all we need are saturated liquid and saturated vapor data (Tables A-4 and A-5 in the case of water).

EXAMPLE 2-5 The Pressure and Volume of a Saturated Mixture

A rigid tank contains 10 kg of water at 90°C. If 8 kg of the water is in the liquid form and the rest is in the vapor form, determine (*a*) the pressure in the tank and (*b*) the volume of the tank.

Solution (*a*) The state of the saturated liquid–vapor mixture is shown in Fig. 2-43. Since the two phases coexist in equilibrium, we have a saturated mixture and the pressure must be the saturation pressure at the given temperature:

$$P = P_{\text{sat @ 90°C}} = \mathbf{70.14\ kPa} \qquad \text{(Table A-4)}$$

(*b*) At 90°C, v_f and v_g values are $v_f = 0.001036\ \text{m}^3/\text{kg}$ and $v_g = 2.361\ \text{m}^3/\text{kg}$ (Table A-4).

One way of finding the volume of the tank is to determine the volume occupied by each phase and then add them:

$$\begin{aligned} V = V_f + V_g &= m_f v_f + m_g v_g \\ &= (8\ \text{kg})(0.001\ \text{m}^3/\text{kg}) + (2\ \text{kg})(2.361\ \text{m}^3/\text{kg}) \\ &= \mathbf{4.73\ m^3} \end{aligned}$$

Another way is to first determine the quality x, then the average specific volume v, and finally the total volume:

$$x = \frac{m_g}{m_t} = \frac{2\ \text{kg}}{10\ \text{kg}} = 0.2$$

$$\begin{aligned} v &= v_f + x v_{fg} \\ &= 0.001\ \text{m}^3/\text{kg} + (0.2)[(2.361 - 0.001)\ \text{m}^3/\text{kg}] \\ &= 0.473\ \text{m}^3/\text{kg} \end{aligned}$$

and $\qquad V = mv = (10\ \text{kg})(0.473\ \text{m}^3/\text{kg}) = 4.73\ \text{m}^3$

The first method appears to be easier in this case since the masses of each phase are given. But in most cases, the masses of each phase are not available, and the second method becomes more convenient.

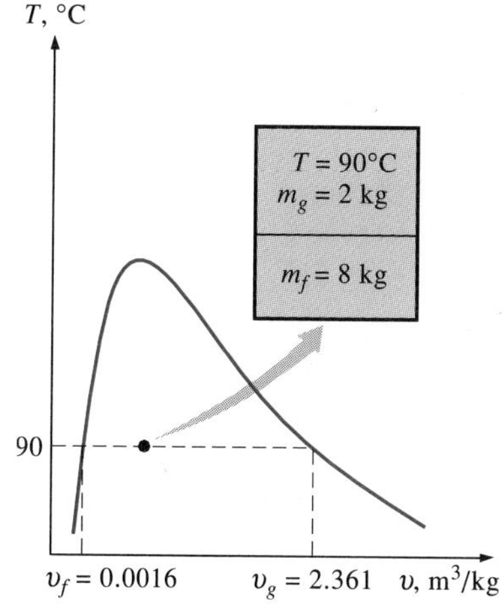

FIGURE 2-43

Schematic and *T*-*v* diagram for Example 2-5.

EXAMPLE 2-6 The Properties of Saturated Liquid–Vapor Mixture

An 80-L vessel contains 4 kg of refrigerant-134a at a pressure of 160 kPa. Determine (*a*) the temperature of the refrigerant, (*b*) the quality, (*c*) the enthalpy of the refrigerant, and (*d*) the volume occupied by the vapor phase.

Solution (*a*) The state of the saturated liquid–vapor mixture is shown in Fig. 2-44. At this point we do not know whether the refrigerant is in the compressed liquid, superheated vapor, or saturated mixture region. This can be determined by comparing a suitable property to the saturated liquid and saturated vapor data. From the information given, we can determine the specific volume:

$$v = \frac{V}{m} = \frac{0.080\ \text{m}^3}{4\ \text{kg}} = 0.02\ \text{m}^3/\text{kg}$$

At 160 kPa, we read

$$\begin{aligned} v_f &= 0.0007435\ \text{m}^3/\text{kg} \\ v_g &= 0.1229\ \text{m}^3/\text{kg} \end{aligned} \qquad \text{(Table A-12)}$$

Obviously, $v_f < v < v_g$, and, therefore, the refrigerant is in the saturated mixture region. Thus, the temperature must be the saturation temperature at the specified pressure:

$$T = T_{\text{sat @ 160kPa}} = \mathbf{-15.62°C}$$

Table A-12

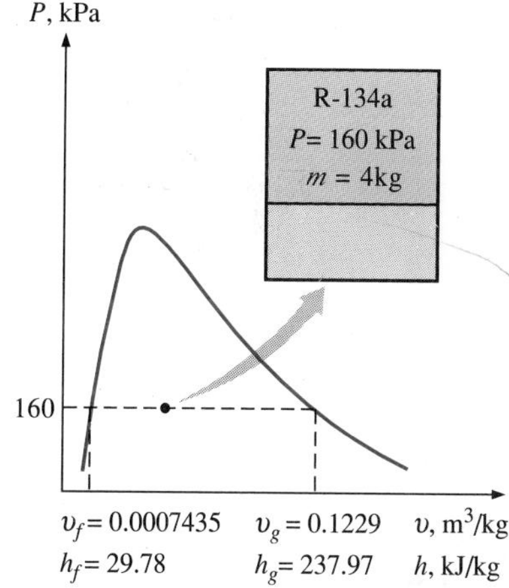

FIGURE 2-44

Schematic and *P*-*v* diagram for Example 2-6.

(*b*) Quality can be determined from Eq. 2-7:

$$x = \frac{v - v_f}{v_{fg}} = \frac{0.02 - 0.0007}{0.1229 - 0.0007} = \mathbf{0.158}$$

(*c*) At 160 kPa, we also read from Table A-12 that h_f = 29.78 kJ/kg and h_{fg} = 208.18 kJ/kg. Then,

$$\begin{aligned} h &= h_f + xh_{fg} \\ &= 29.78 \text{ kJ/kg} + (0.158)(208.18 \text{ kJ/kg}) \\ &= \mathbf{62.7\ kJ/kg} \end{aligned}$$

(*d*) The mass of the vapor can be determined from

$$m_g = xm_t = (0.158)(4 \text{ kg}) = 0.632 \text{ kg}$$

and the volume occupied by the vapor phase is

$$V_g = m_g v_g = (0.632 \text{ kg})(0.1229 \text{ m}^3/\text{kg}) = \mathbf{0.0777\ m^3} \text{ (or 77.7 L)}$$

The rest of the volume (2.3 L) is occupied by the liquid.

Property tables are also available for saturated solid–vapor mixtures. Properties of saturated ice–water vapor mixtures, for example, are listed in Table A-8. Saturated solid–vapor mixtures can be handled just as saturated liquid–vapor mixtures.

2 Superheated Vapor

In the region to the right of the saturated vapor line, a substance exists as superheated vapor. Since the superheated region is a single-phase region (vapor phase only), temperature and pressure are no longer dependent properties and they can conveniently be used as the two independent properties in the tables. The format of the superheated vapor tables is illustrated in Fig. 2-45.

In these tables, the properties are listed versus temperature for selected pressures starting with the saturated vapor data. The saturation temperature is given in parentheses following the pressure value.

Superheated vapor is characterized by

Lower pressures ($P < P_{sat}$ at a given T)

Higher temperatures ($T > T_{sat}$ at a given P)

Higher specific volumes ($v > v_g$ at a given P or T)

Higher internal energies ($u > u_g$ at a given P or T)

Higher enthalpies ($h > h_g$ at a given P or T)

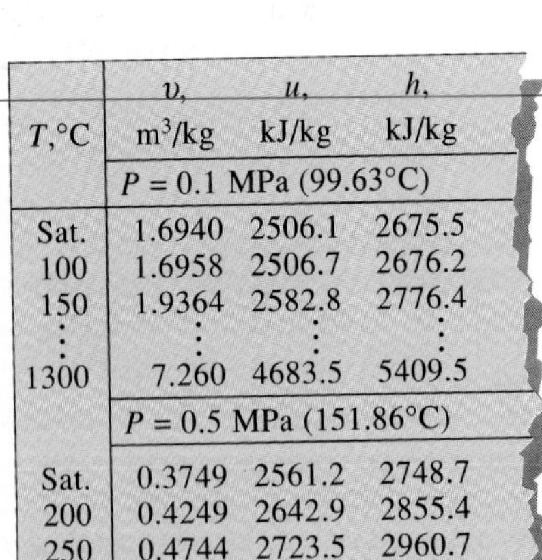

T,°C	v, m³/kg	u, kJ/kg	h, kJ/kg
	P = 0.1 MPa (99.63°C)		
Sat.	1.6940	2506.1	2675.5
100	1.6958	2506.7	2676.2
150	1.9364	2582.8	2776.4
⋮	⋮	⋮	⋮
1300	7.260	4683.5	5409.5
	P = 0.5 MPa (151.86°C)		
Sat.	0.3749	2561.2	2748.7
200	0.4249	2642.9	2855.4
250	0.4744	2723.5	2960.7

FIGURE 2-45
A partial listing of Table A-6.

EXAMPLE 2-7 Finding the Internal Energy of Superheated Vapor

Determine the internal energy of water at 20 psia and 400°F.

Solution At 20 psia, the saturation temperature is 227.96°F. Since $T > T_{sat}$, the water is in the superheated vapor region. Then the internal energy is determined from the superheated vapor table (Table A-6E) to be

$$u = \mathbf{1145.1\ Btu/lbm}$$

at the given temperature and pressure.

EXAMPLE 2-8 Finding the Temperature of Superheated Vapor

Determine the temperature of water at a state of P = 0.5 MPa and h = 2890 kJ/kg.

Solution At 0.5 MPa, the enthalpy of saturated water vapor is h_g = 2748.7 kJ/kg. Since $h > h_g$, as shown in Fig. 2-46, we again have superheated vapor. Under 0.5 MPa in Table A-6 we read

T °C	h kJ/kg
200	2855.4
250	2960.7

Obviously, the temperature is between 200 and 250°C. By linear interpolation it is determined to be

$$T = \mathbf{216.4°C}$$

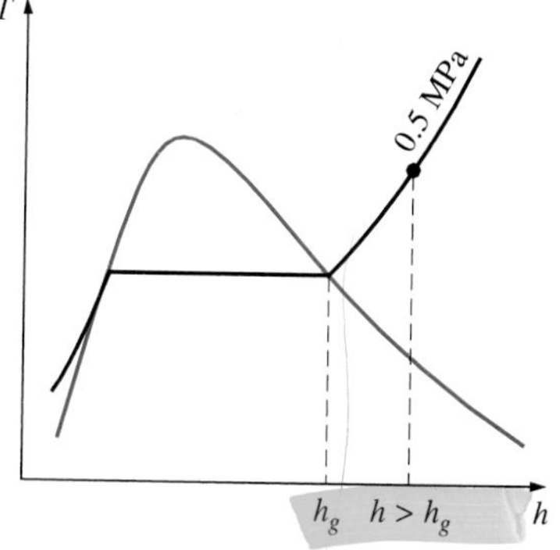

FIGURE 2-46

At a specified P, superheated vapor exists at a higher h than the saturated vapor (Example 2-8).

3 Compressed Liquid

There are not many data for compressed liquid in the literature, and Table A-7 is the only compressed liquid table in this text. The format of Table A-7 is very much like the format of the superheated vapor tables. One reason for the lack of compressed liquid data is the relative independence of compressed liquid properties from pressure. Variation of properties of compressed liquid with pressure is very mild. Increasing the pressure 100 times often causes properties to change less than 1 percent. The property most affected by pressure is enthalpy.

In the absence of compressed liquid data, a general approximation is *to treat compressed liquid as saturated liquid at the given temperature* (Fig. 2-47). This is because the compressed liquid properties depend on temperature more strongly than they do on pressure. Thus,

$$y \cong y_{f @ T}$$

for compressed liquids, where y is v, u, or h. Of these three properties, the property whose value is most sensitive to variations in the pressure is the enthalpy h. Although the above approximation results in negligible error in v and u, the error in h may reach undesirable levels. However, the error in h at very high pressures can be reduced significantly by evaluating it from

$$h \cong h_{f @ T} + v_f(P - P_{sat})$$

Given: P and T

$v \cong v_{f @ T}$
$u \cong u_{f @ T}$
$h \cong h_{f @ T}$

FIGURE 2-47

A compressed liquid may be approximated as a saturated liquid at the same temperature.

instead of taking it to be just h_f. Here P_{sat} is the saturation pressure at the given temperature.

In general, a compressed liquid is characterized by

Higher pressures ($P > P_{sat}$ at a given T)

Lower temperatures ($T < T_{sat}$ at a given P)

Lower specific volumes ($v < v_f$ at a given P or T)

Lower internal energies ($u < u_f$ at a given P or T)

Lower enthalpies ($h < h_f$ at a given P or T)

But these effects are not as pronounced as they are for the superheated vapor.

EXAMPLE 2-9 Approximating Compressed Liquid as Saturated Liquid

Determine the internal energy of compressed liquid water at 80°C and 5 MPa, using (*a*) data from the compressed liquid table and (*b*) saturated liquid data. What is the error involved in the second case?

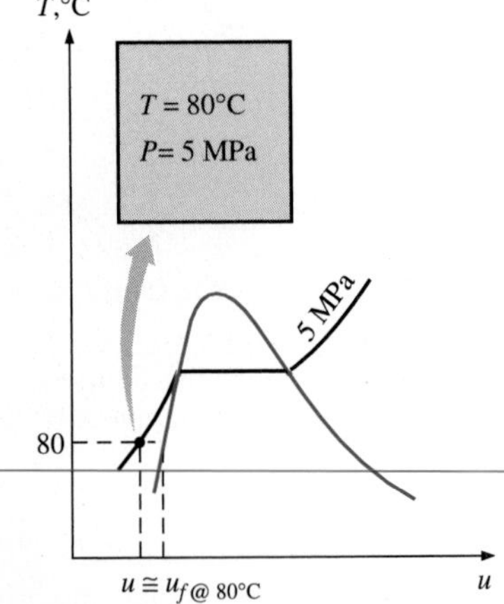

FIGURE 2-48
Schematic and T-v diagram for Example 2-9.

Solution At 80°C, the saturation pressure of water is 47.39 kPa, and since 5 MPa > P_{sat}, we obviously have compressed liquid, as shown in Fig. 2-48.

(*a*) From the compressed liquid table (Table A-7)

$$\left.\begin{aligned} P &= 5\text{MPa} \\ T &= 80°\text{C} \end{aligned}\right\} \quad u = \mathbf{333.72\ kJ/kg}$$

(*b*) From the saturation table (Table A-4), we read

$$u \cong u_{f@\,80°\text{C}} = \mathbf{334.86\ kJ/kg}$$

The error involved is

$$\frac{334.86 - 333.72}{333.72} \times 100 = \mathbf{0.34\%}$$

which is less than 1 percent.

Reference State and Reference Values

The values of u, h, and s cannot be measured directly, and they are calculated from measurable properties using the relations between thermodynamic properties developed in Chap. 11. However, those relations give the *changes* in properties, not the values of properties at specified states. Therefore, we need to choose a convenient *reference state* and assign a value of *zero* for a convenient property or properties at that state. For water, the state of saturated liquid at 0.01°C is taken as the reference state, and the internal energy and entropy are assigned zero values at that state. For refrigerant-134a, the state of saturated liquid at −40°C is taken as the reference state, and the enthalpy and entropy are assigned zero values at that state. Note that some properties may have negative values as a result of the reference state chosen.

It should be mentioned that sometimes different tables list different values for some properties at the same state as a result of using a different reference state. However, in thermodynamics we are concerned with the *changes* in properties, and the reference state chosen is of no consequence in calculations as long as we use values from a single consistent set of tables or charts.

EXAMPLE 2-10 The Use of Steam Tables to Determine Properties
Determine the missing properties and the phase descriptions in the following table for water:

	T °C	P kPa	u kJ/kg	x	Phase description
(*a*)		200		0.6	
(*b*)	125		1600		
(*c*)		1000	2950		
(*d*)	75	500			
(*e*)		850		0.0	

Solution (*a*) The quality is given to be $x = 0.6$, which implies that 60 percent of the mass is in the vapor phase and the remaining 40 percent is in the liquid phase. Therefore, we have saturated liquid–vapor mixture at a pressure of 200 kPa. Then the temperature must be the saturation temperature at the given pressure:

$$T = T_{\text{sat @ 200kPa}} = \mathbf{120.23°C} \qquad \text{(Table A-5)}$$

At 200 kPa, we also read from Table A-5 that u_f = 504.49 kJ/kg and u_{fg} = 2025.0 kJ/kg. Then the average internal energy of the mixture is determined from Eq. 2-8 to be

$$\begin{aligned} u &= u_f + xu_{fg} \\ &= 504.49 \text{ kJ/kg} + (0.6)(2025.0 \text{ kJ/kg}) \\ &= \mathbf{1719.49 \text{ kJ/kg}} \end{aligned}$$

(*b*) This time the temperature and the internal energy are given, but we do not know which table to use to determine the missing properties because we have no clue as to whether we have saturated mixture, compressed liquid, or superheated vapor. To determine the region we are in, we first go to the saturation table (Table A-4) and determine the u_f and u_g values at the given temperature. At 125°C, we read u_f = 524.74 kJ/kg and u_g = 2534.6 kJ/kg. Next we compare the given u value to these u_f and u_g values, keeping in mind that

if	$u < u_f$	we have *compressed liquid*
if	$u_f \leq u \leq u_g$	we have *saturated mixture*
if	$u > u_g$	we have *superheated vapor*

In our case the given u value is 1600, which falls between the u_f and u_g values at 125°C. Therefore, we have saturated liquid–vapor mixture. Then the pressure must be the saturation pressure at the given temperature:

$$P = P_{\text{sat @ 125°C}} = \mathbf{232.1 \text{ kPa}} \qquad \text{(Table A-4)}$$

The quality is determined from

$$x = \frac{u - u_f}{u_{fg}} = \frac{1600 - 524.74}{2009.9} = \mathbf{0.535}$$

The criteria above for determining whether we have compressed liquid, saturated mixture, or superheated vapor can also be used when enthalpy *h* or specific volume *v* is given instead of internal energy *u*, or when pressure is given instead of temperature.

(*c*) This is similar to case (*b*), except pressure is given instead of temperature. Following the argument given above, we read the u_f and u_g values at the specified pressure. At 1 MPa, we have u_f = 761.68 kJ/kg and u_g = 2583.6 kJ/kg. The specified *u* value is 2950 kJ/kg, which is greater than the u_g value at 1 MPa. Therefore, we have superheated vapor, and the temperature at this state is determined from the superheated vapor table by interpolation to be

$$T = \mathbf{395.6°C} \qquad \text{(Table A-6)}$$

We would leave the quality column blank in this case since quality has no meaning for a superheated vapor.

(*d*) In this case the temperature and pressure are given, but again we cannot tell which table to use to determine the missing properties because we do not know whether we have saturated mixture, compressed liquid, or superheated vapor. To determine the region we are in, we go to the saturation table (Table A-5) and determine the saturation temperature value at the given pressure. At 500 kPa, we have T_{sat} = 151.86°C. We then compare the given *T* value to this T_{sat} value, keeping in mind that

if	$T < T_{sat\,@\,given\,P}$	we have *compressed liquid*
if	$T = T_{sat\,@\,given\,P}$	we have *saturated mixture*
if	$T > T_{sat\,@\,given\,P}$	we have *superheated vapor*

In our case, the given *T* value is 75°C, which is less than the T_{sat} value at the specified pressure. Therefore, we have compressed liquid (Fig. 2-49), and normally we would determine the internal energy value from the compressed liquid table. But in this case the given pressure is much lower than the lowest pressure value in the compressed liquid table (which is 5 MPa), and therefore we are justified to treat the compressed liquid as saturated liquid at the given temperature (*not* pressure):

$$u \cong u_{f\,@\,75°C} = \mathbf{313.90\ kJ/kg} \qquad \text{(Table A-4)}$$

We would leave the quality column blank in this case since quality has no meaning in the compressed liquid region.

FIGURE 2-49
At a given *P* and *T*, a pure substance will exist as a compressed liquid if $T < T_{sat\,@\,P}$.

(*e*) The quality is given to be x = 0.0, and thus we have saturated liquid at the specified pressure of 850 kPa. Then the temperature must be the saturation temperature at the given pressure, and the internal energy must have the saturated liquid value:

$$T = T_{sat\,@\,850\,kPa} = \mathbf{172.96°C} \qquad \text{(Table A-5)}$$

$$u = u_{f\,@\,850\,kPa} = \mathbf{731.27\ kJ/kg} \qquad \text{(Table A-5)}$$

2-7 ■ THE IDEAL-GAS EQUATION OF STATE

The Ideal-Gas Equation of State

One way of reporting property data for pure substances is to list values of properties at various states. The property tables provide very accurate information about the properties, but they are very bulky and vulnerable to typographical errors. A more practical and desirable approach would be to have some simple relations among the properties that are sufficiently general and accurate.

Any equation that relates the pressure, temperature, and specific volume of a substance is called an **equation of state.** Property relations that involve other properties of a substance at equilibrium states are also referred to as equations of state. There are several equations of state, some simple and others very complex. The simplest and best-known equation of state for substances in the gas phase is the ideal-gas equation of state. This equation predicts the *P-v-T* behavior of a gas quite accurately within some properly selected region.

Gas and *vapor* are often used as synonymous words. The vapor phase of a substance is customarily called a *gas* when it is above the critical temperature. *Vapor* usually implies a gas that is not far from a state of condensation.

In 1662, Robert Boyle, an Englishman, observed during his experiments with a vacuum chamber that the pressure of gases is inversely proportional to their volume. In 1802, J. Charles and J. Gay-Lussac, Frenchmen, experimentally determined that at low pressures the volume of a gas is proportional to its temperature. That is,

$$P = R\left(\frac{T}{v}\right)$$

or

$$Pv = RT \qquad (2\text{-}9)$$

where the constant of proportionality R is called the **gas constant.** Equation 2-9 is called the **ideal-gas equation of state,** or simply the **ideal-gas relation,** and a gas that obeys this relation is called an **ideal gas.** In this equation, P is the absolute pressure, T is the absolute temperature, and v is the specific volume.

Substance	R, kJ/(kg·K)
Air	0.2870
Helium	2.0769
Argon	0.2081
Nitrogen	0.2968

FIGURE 2-50
Different substances have different gas constants.

The gas constant R is different for each gas (Fig. 2-50) and is determined from

$$R = \frac{R_u}{M} \qquad [\text{kJ/(kg}\cdot\text{K) or kPa}\cdot\text{m}^3\text{/(kg}\cdot\text{K)}] \qquad (2\text{-}10)$$

where R_u is the **universal gas constant** and M is the molar mass (also called *molecular weight*) of the gas. The constant R_u is the same for all substances, and its value is

$$R_u = \begin{cases} 8.314 \text{ kJ/(kmol} \cdot \text{K)} \\ 8.314 \text{ kPa} \cdot \text{m}^3\text{/(kmol} \cdot \text{K)} \\ 0.08314 \text{ bar} \cdot \text{m}^3\text{/(kmol} \cdot \text{K)} \\ 1.986 \text{ Btu/(lbmol} \cdot \text{R)} \\ 10.73 \text{ psia} \cdot \text{ft}^3\text{/(lbmol} \cdot \text{R)} \\ 1545 \text{ ft lbf/(lbmol} \cdot \text{R)} \end{cases} \tag{2-11}$$

The **molar mass** M can simply be defined as *the mass of one mole* (also called a *gram-mole,* abbreviated gmol) *of a substance in grams,* or *the mass of one kmol* (also called a *kilogram-mole,* abbreviated kgmol) *in kilograms.* In English units, it is the mass of 1 lbmol (1 pound-mole = 0.4536 kmol) in lbm (1 pound-mass = 0.4536 kg). Notice that the molar mass of a substance has the same numerical value in both unit systems because of the way it is defined. When we say the molar mass of nitrogen is 28, it simply means the mass of 1 kmol of nitrogen is 28 kg, or the mass of 1 lbmol of nitrogen is 28 lbm. That is, M = 28 kg/kmol = 28 lbm/lbmol. The mass of a system is equal to the product of its molar mass M and the mole number N:

$$m = MN \quad \text{(kg)} \tag{2-12}$$

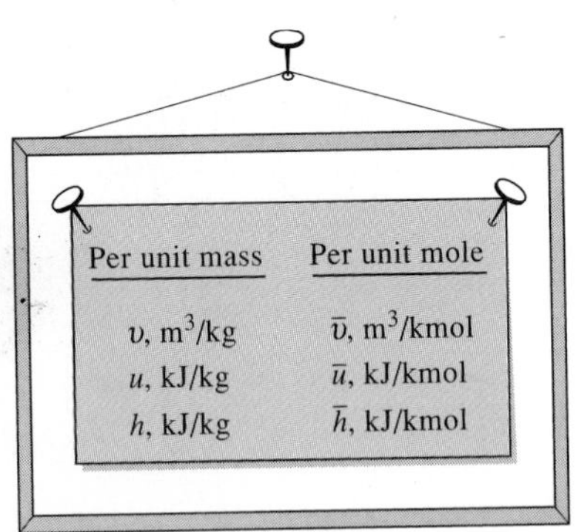

FIGURE 2-51
Properties per unit mole are denoted with a bar on the top.

The values of R and M for several substances are given in Table A-1.

The ideal-gas equation of state can be written in several different forms:

$$V = mv \longrightarrow PV = mRT \tag{2-13}$$

$$mR = (MN)R = NR_u \longrightarrow PV = NR_uT \tag{2-14}$$

$$V = N\bar{v} \longrightarrow P\bar{v} = R_uT \tag{2-15}$$

where $\bar{v}$ is the molar specific volume, that is, the volume per unit mole (in m^3/kmol or ft^3/lbmol). A *bar* above a property will denote values on a *unit-mole basis* throughout this text (Fig. 2-51).

By writing Eq. 2-13 twice for a fixed mass and simplifying, the properties of an ideal gas at two different states are related to each other by

$$\frac{P_1V_1}{T_1} = \frac{P_2V_2}{T_2} \tag{2-16}$$

FIGURE 2-52
The ideal-gas relation often is not applicable to real gases; thus, care should be exercised when using it.

An ideal gas is an *imaginary* substance that obeys the relation $Pv = RT$ (Fig. 2-52). It has been experimentally observed that the ideal-gas relation given above closely approximates the P-v-T behavior of real gases at low densities. At low pressures and high temperatures, the density of a gas decreases, and the gas behaves as an ideal gas under these conditions. What constitutes low pressure and high temperature is explained in the next section.

In the range of practical interest, many familiar gases such as air, nitrogen, oxygen, hydrogen, helium, argon, neon, krypton, and even heavier gases such as carbon dioxide can be treated as ideal gases with negligible error (often less than 1 percent). Dense gases such as water vapor in steam power plants and refrigerant vapor in refrigerators, however, should not be treated as ideal gases. Instead, the property tables should be used for these substances.

EXAMPLE 2-11 Finding the Mass of an Ideal Gas

Determine the mass of the air in a room whose dimensions are 4 m × 5 m × 6 m at 100 kPa and 25°C.

Solution A sketch of the room is given in Fig. 2-53. Air at specified conditions can be treated as an ideal gas. From Table A-1, the gas constant of air is $R = 0.287$ kPa · m³/(kg · K), and the absolute temperature is $T = 25°C + 273 = 298$ K. The volume of the room is

$$V = (4 \text{ m})(5 \text{ m})(6 \text{ m}) = 120 \text{ m}^3$$

By substituting these values into Eq. 2-13, the mass of air in the room is determined to be

$$m = \frac{PV}{RT} = \frac{(100 \text{ kPa})(120 \text{ m}^3)}{[0.287 \text{ kPa} \cdot \text{m}^3/(\text{kg} \cdot \text{K})](298 \text{ K})} = \mathbf{140.3 \text{ kg}}$$

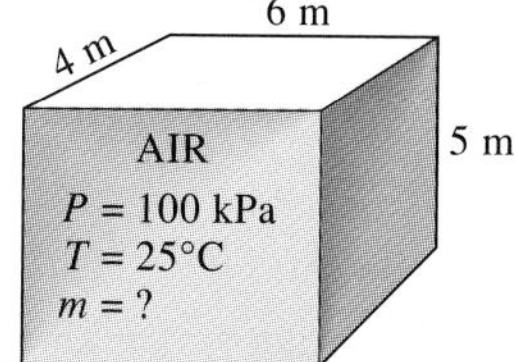

FIGURE 2-53
Schematic for Example 2-11.

Is Water Vapor an Ideal Gas?

This question cannot be answered with a simple yes or no. The error involved in treating water vapor as an ideal gas is calculated and plotted in Fig. 2-54. It is clear from this figure that at pressures below 10 kPa, water vapor can be treated as an ideal gas, regardless of its temperature, with negligible error (less than 0.1 percent). But at higher pressures, the ideal-gas assumption yields unacceptable errors, particularly in the vicinity of the critical point and the saturated vapor line (over 100 percent). Therefore, in air-conditioning applications, the water vapor in the air can be treated as an ideal gas with essentially no error since the pressure of the water vapor is very low. In steam power plant applications, however, the pressures involved are usually very high; therefore, ideal-gas relations should not be used.

For water vapor with P ≤ 10kPa can be treated as ideal gas law

2-8 ■ COMPRESSIBILITY FACTOR—A Measure of Deviation from Ideal-Gas Behavior

The ideal-gas equation is very simple and thus very convenient to use. But, as illustrated in Fig. 2-54, gases deviate from ideal-gas behavior significantly at states near the saturation region and the critical point. This deviation from ideal-gas behavior at a given temperature and pressure can accurately be accounted for by the introduction of a correction factor called the **compressibility factor** Z. It is defined as

$$Z = \frac{Pv}{RT} \tag{2-17}$$

or

$$Pv = ZRT \tag{2-18}$$

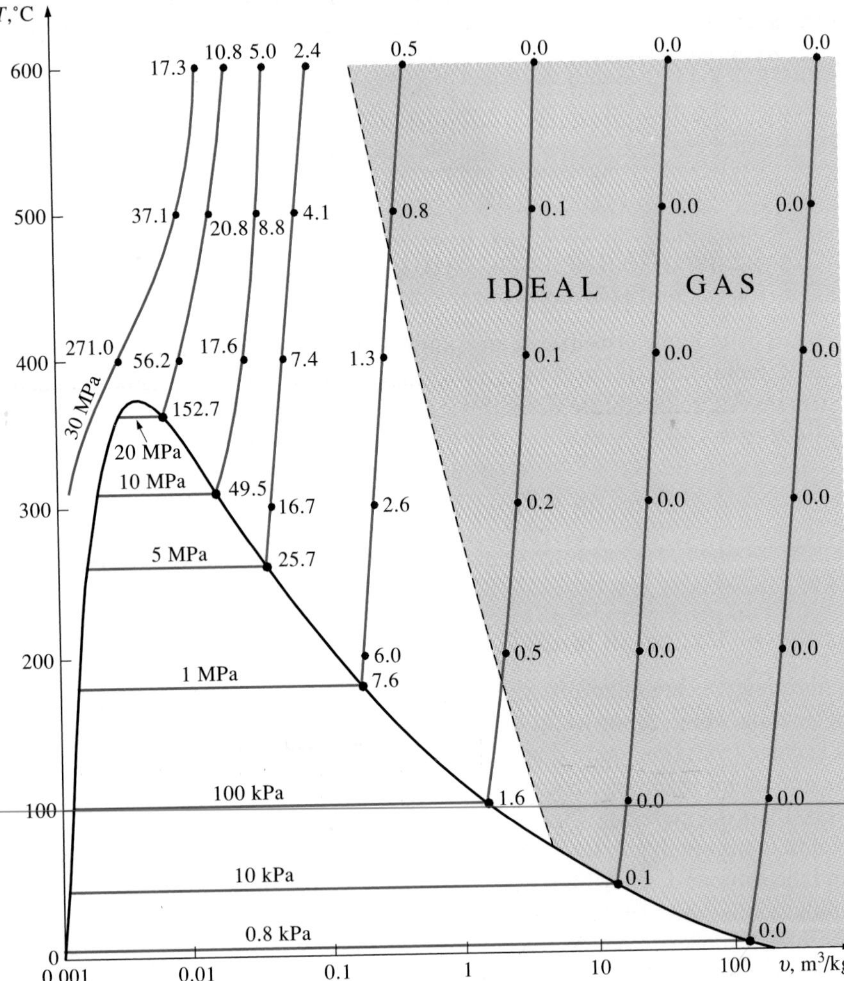

FIGURE 2-54

Percentage of error ($[|v_{table} - v_{ideal}|/v_{table}] \times 100$) involved in assuming steam to be an ideal gas, and the region where steam can be treated as an ideal gas with less than 1 percent error.

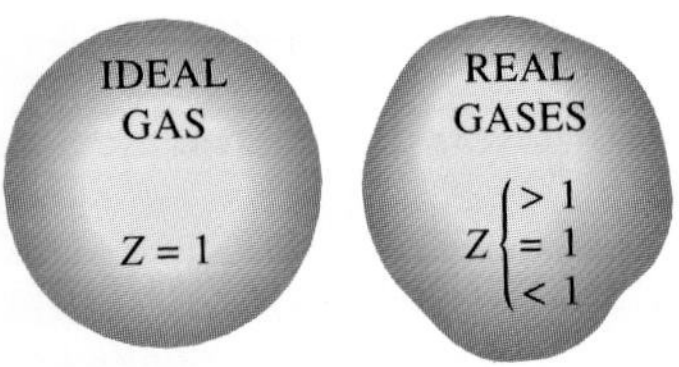

FIGURE 2-55

The compressibility factor is unity for ideal gases.

It can also be expressed as

$$Z = \frac{v_{actual}}{v_{ideal}} \qquad (2\text{-}19)$$

where $v_{ideal} = RT/P$. Obviously, $Z = 1$ for ideal gases. For real gases Z can be greater than or less than unity (Fig. 2-55). The farther away Z is from unity, the more the gas deviates from ideal-gas behavior.

We have repeatedly said that gases follow the ideal-gas equation closely at low pressures and high temperatures. But what exactly constitutes low pressure or high temperature? Is $-100°C$ a low temperature? It definitely is for most substances, but not for air. Air (or nitrogen) can be treated as an ideal gas at this temperature and atmospheric pressure with an error under 1 percent. This is because nitrogen is well over its critical temperature ($-147°C$) and away from the saturation region. But at this temperature and pressure, most substances would exist in the solid phase. Therefore, the pressure or temperature of a substance is high or low relative to its critical temperature or pressure.

Gases behave differently at a given temperature and pressure, but they behave very much the same at temperatures and pressures normalized with respect to their critical temperatures and pressures. The normalization is done as

$$P_R = \frac{P}{P_{cr}} \quad \text{and} \quad T_R = \frac{T}{T_{cr}} \tag{2-20}$$

Here P_R is called the **reduced pressure** and T_R the **reduced temperature.** The Z factor for all gases is approximately the same at the same reduced pressure and temperature (Fig. 2-56). This is called the **principle of corresponding states.** In Fig. 2-57, the experimentally determined Z values are

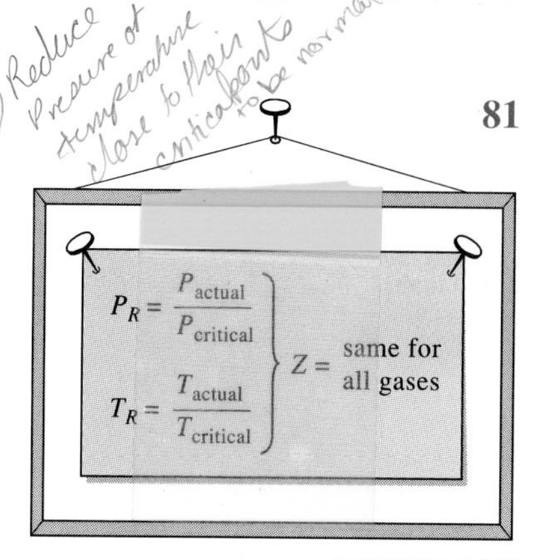

FIGURE 2-56
The compressibility factor is the same for all gases at the same reduced pressure and temperature (*principle of corresponding states*).

FIGURE 2-57
Comparison of Z factors for various gases. [*Source:* Gour-Jen Su, "Modified Law of Corresponding States," *Ind. Eng. Chem.* (international ed.) 38 (1946), p. 803.]

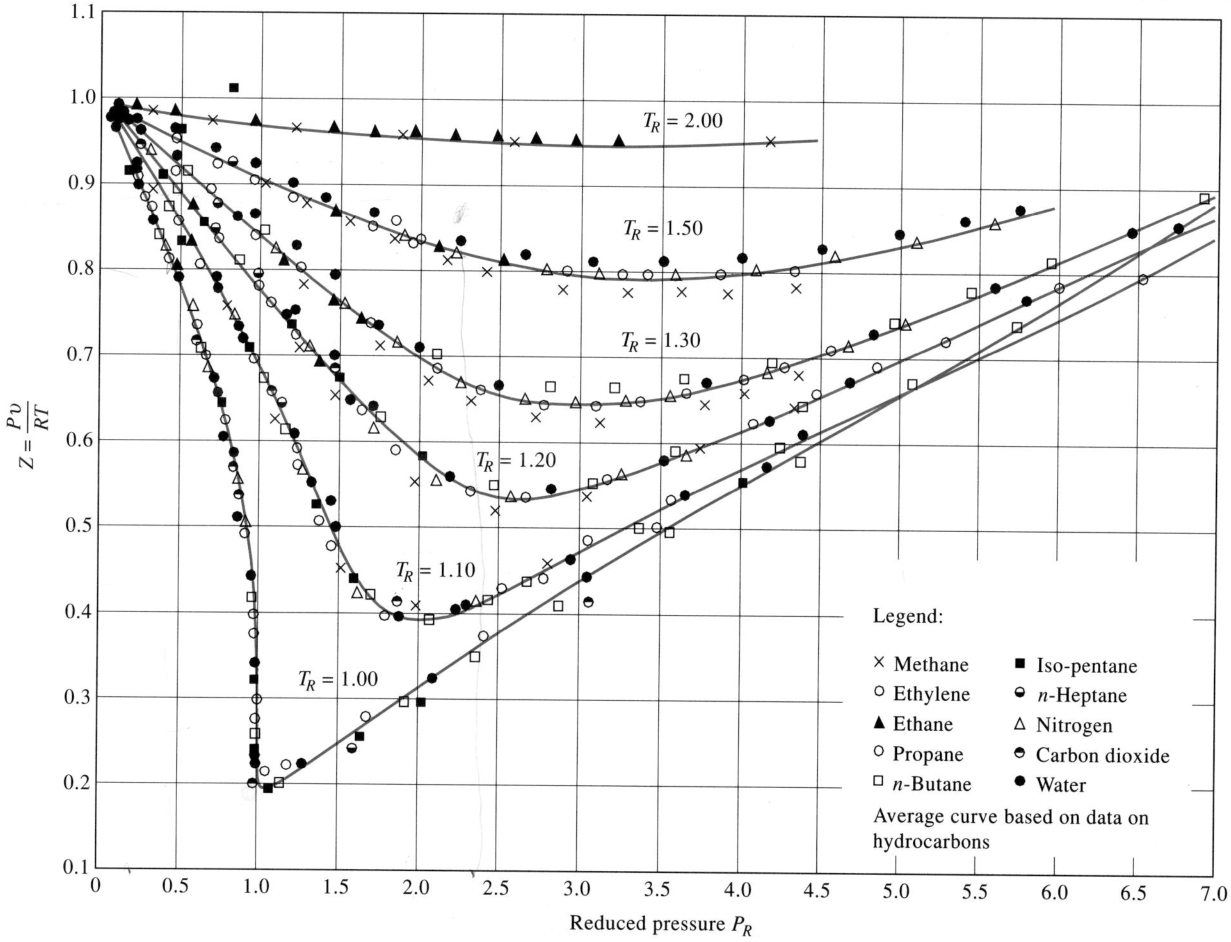

plotted against P_R and T_R for several gases. The gases seem to obey the principle of corresponding states reasonably well. By curve-fitting all the data, we obtain the **generalized compressibility chart** that can be used for all gases. This chart is given in the appendix in three separate parts (Figs. A-29*a*, *b*, and *c*), each for a different range of reduced pressures for more accurate reading. The use of a compressibility chart requires a knowledge of critical-point data, and the results obtained are accurate to within a few percent.

The following observations can be made from the generalized compressibility chart:

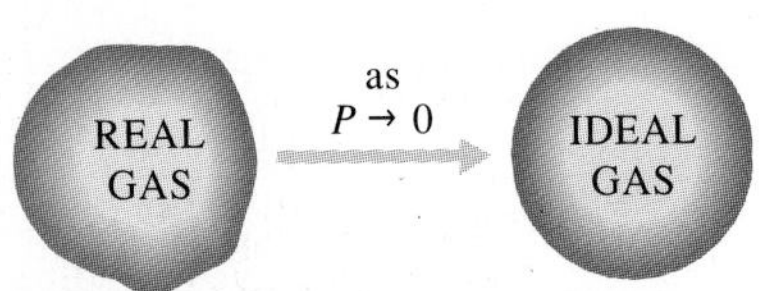

FIGURE 2-58
At very low pressures, all gases approach ideal-gas behavior (regardless of their temperature).

1. At very low pressures ($P_R \ll 1$), the gases behave as an ideal gas regardless of temperature (Fig. 2-58),

2. At high temperatures ($T_R > 2$), ideal-gas behavior can be assumed with good accuracy regardless of pressure (except when $P_R \gg 1$).

3. The deviation of a gas from ideal-gas behavior is greatest in the vicinity of the critical point (Fig. 2-59).

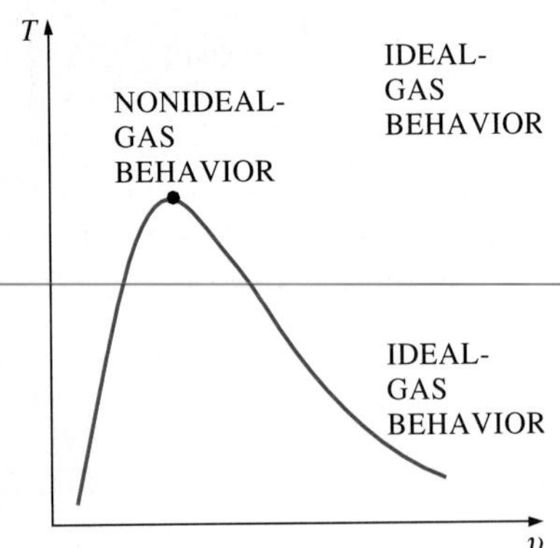

FIGURE 2-59
Gases deviate from the ideal-gas behavior most in the neighborhood of the critical point.

EXAMPLE 2-12 Using Generalized Charts to Determine Specific Volume

Determine the specific volume of refrigerant-134a at 1 MPa and 50°C, using (*a*) the ideal-gas equation of state and (*b*) the generalized compressibility chart. Compare the values obtained to the actual value of 0.02171 m³/kg and determine the error involved in each case.

Solution The gas constant, the critical pressure, and the critical temperature of refrigerant-134a are determined from Table A-1 to be

$$R = 0.0815\ \text{kPa}\cdot\text{m}^3/(\text{kg}\cdot\text{K})$$
$$P_{cr} = 4.067\ \text{MPa}$$
$$T_{cr} = 374.3\ \text{K}$$

(*a*) The specific volume of refrigerant-134a under the ideal-gas assumption is determined from the ideal-gas relation to be

$$v = \frac{RT}{P} = \frac{[0.0815\ \text{kPa}\cdot\text{m}^3/(\text{kg}\cdot\text{K})](323\ \text{K})}{1000\ \text{kPa}} = \mathbf{0.02632\ m^3/kg}$$

Therefore, treating the refrigerant-134a vapor as an ideal gas would result in an error of (0.02632 − 0.02171)/0.02171 = **0.212**, or 21.2 percent in this case.

(*b*) To determine the correction factor *Z* from the compressibility chart, we first need to calculate the reduced pressure and temperature:

$$\left.\begin{aligned} P_R &= \frac{P}{P_{cr}} = \frac{1\ \text{MPa}}{4.067\ \text{MPa}} = 0.246 \\ T_R &= \frac{T}{T_{cr}} = \frac{323\ \text{K}}{374.3\ \text{K}} = 0.863 \end{aligned}\right\} Z = 0.84$$

Thus $\quad v = Zv_{\text{ideal}} = (0.84)(0.02632\ \text{m}^3/\text{kg}) = \mathbf{0.02211\ m^3/kg}$

The error in this result is less than **2 percent**. Therefore, in the absence of exact tabulated data, the generalized compressibility chart can be used with confidence.

When P and v, or T and v, are given instead of P and T, the generalized compressibility chart can still be used to determine the third property, but it would involve tedious trial and error. Therefore, it is necessary to define one more reduced property called the **pseudo-reduced specific volume** v_R as

$$v_R = \frac{v_{\text{actual}}}{RT_{\text{cr}}/P_{\text{cr}}} \qquad \text{(2-21)}$$

Note that v_R is defined differently from P_R and T_R. It is related to T_{cr} and P_{cr} instead of v_{cr}. Lines of constant v_R are also added to the compressibility charts, and this enables one to determine T or P without having to resort to time-consuming iterations (Fig. 2-60).

$$\left.\begin{aligned} P_R &= \frac{P}{P_{\text{cr}}} \\ v_R &= \frac{v}{RT_{\text{cr}}/P_{\text{cr}}} \end{aligned}\right\} \begin{aligned} &Z = \ldots \\ &\text{(Fig. A-30)} \end{aligned}$$

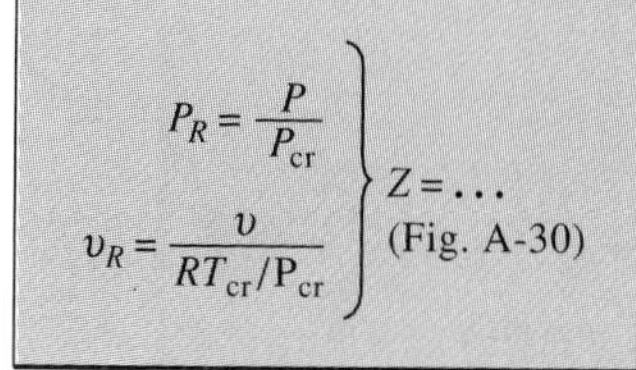

FIGURE 2-60
The compressibility factor can also be determined from a knowledge of P_R and v_R.

EXAMPLE 2-13 Using Generalized Charts to Determine Pressure

Determine the pressure of water vapor at 600°F and 0.514 ft³/lbm, using (*a*) the steam tables, (*b*) the ideal-gas equation, and (*c*) the generalized compressibility chart.

Solution A sketch of the system is given in Fig. 2-61. The gas constant, the critical pressure, and the critical temperature of steam are determined from Table A-1E to be

$$R = 0.5956 \text{ psia} \cdot \text{ft}^3/(\text{lbm} \cdot \text{R})$$
$$P_{cr} = 3204 \text{ psia}$$
$$T_{cr} = 1165.3 \text{ R}$$

H_2O

$T = 600°\text{F}$
$v = 0.514 \text{ ft}^3/\text{lbm}$
$P = ?$

FIGURE 2-61
Schematic for Example 2-13.

(*a*) The pressure of steam at the specified state is determined from Table A-6E to be

$$\left.\begin{aligned} v &= 0.514 \text{ ft}^3/\text{lbm} \\ T &= 600°\text{F} \end{aligned}\right\} \quad P = \mathbf{1000\ psia}$$

This is the experimentally determined value, and thus it is the most accurate.

(*b*) The pressure of steam under the ideal-gas assumption is determined from the ideal-gas relation to be

$$P = \frac{RT}{v} = \frac{[0.5956 \text{ psia} \cdot \text{ft}^3/(\text{lbm} \cdot \text{R})](1060 \text{ R})}{0.514 \text{ ft}^3/\text{lbm}} = \mathbf{1228.3\ psia}$$

Therefore, treating the steam as an ideal gas would result in an error of $(1228.3 - 1000)/1000 = 0.228$, or 22.8 percent in this case.

(*c*) To determine the correction factor Z from the compressibility chart (Fig. A-29), we first need to calculate the pseudo-reduced specific volume and the reduced temperature:

$$\left.\begin{aligned} v_R &= \frac{v_{\text{actual}}}{RT_{cr}/P_{cr}} = \frac{(0.514 \text{ ft}^3/\text{lbm})(3204 \text{ psia})}{[0.5956 \text{ psia} \cdot \text{ft}^3/(\text{lbm} \cdot \text{R})](1165.3 \text{ R})} \\ &= 2.373 \\ T_R &= \frac{T}{T_{cr}} = \frac{1060 \text{ R}}{1165.3 \text{ R}} = 0.91 \end{aligned}\right\} \quad P_R = 0.33$$

Thus, $$P = P_R P_{cr} = (0.33)(3204 \text{ psia}) = \mathbf{1057.3\ psia}$$

Using the compressibility chart reduced the error from 22.8 to 5.7 percent, which is acceptable for most engineering purposes (Fig. 2-62). A bigger chart, of course, would give better resolution and reduce the reading errors. Notice that we did not have to determine Z in this problem since we could read P_R directly from the chart.

	P, psia
Exact	1000.0
Z chart	1057.3
Ideal gas	1228.3

(from Example 2-12)

FIGURE 2-62

Results obtained by using the compressibility chart are usually within a few percent of the experimentally determined values.

2-9 ■ OTHER EQUATIONS OF STATE

The ideal-gas equation of state is very simple, but its range of applicability is limited. It is desirable to have equations of state that represent the *P-v-T* behavior of substances accurately over a larger region with no limitations. Such equations are naturally more complicated. Several equations have been proposed for this purpose (Fig. 2-63), but we shall discuss only three: the *van der Waals* equation because it is one of the earliest, the *Beattie-Bridgeman* equation of state because it is one of the best known and is reasonably accurate, and the *Benedict-Webb-Rubin* equation because it is one of the more recent and is very accurate.

van der Waals
Berthelet
Redlich-Kwang
Beattie-Bridgeman
Benedict-Webb-Rubin
Strobridge
Virial

FIGURE 2-63

Several equations of state have been proposed throughout history.

Van der Waals Equation of State

The van der Waals equation of state was proposed in 1873, and it has two constants that are determined from the behavior of a substance at the critical point. The van der Waals equation of state is given by

$$\left(P + \frac{a}{v^2}\right)(v - b) = RT \tag{2-22}$$

Van der Waals intended to improve the ideal-gas equation of state by including two of the effects not considered in the ideal-gas model: the *intermolecular attraction forces* and the *volume occupied by the molecules themselves.* The term a/v^2 accounts for the intermolecular attraction forces, and b accounts for the volume occupied by the gas molecules. In a room at atmospheric pressure and temperature, the volume actually occupied by molecules is only about one-thousandth of the volume of the room. As the pressure increases, the volume occupied by the molecules becomes an increasingly significant part of the total volume. Van der Waals proposed to correct this by replacing v in the ideal-gas relation with the quantity $v - b$, where b represents the volume occupied by the gas molecules per unit mass.

The determination of the two constants appearing in this equation is based on the observation that the critical isotherm on a *P-v* diagram has a horizontal inflection point at the critical point (Fig. 2-64). Thus, the first and the second derivatives of P with respect to v at the critical point must be zero. That is,

$$\left(\frac{\partial P}{\partial v}\right)_{T=T_{cr}=\text{const}} = 0 \quad \text{and} \quad \left(\frac{\partial^2 P}{\partial v^2}\right)_{T=T_{cr}=\text{const}} = 0 \tag{2-23}$$

FIGURE 2-64

Critical isotherm of a pure substance has an inflection point at the critical state.

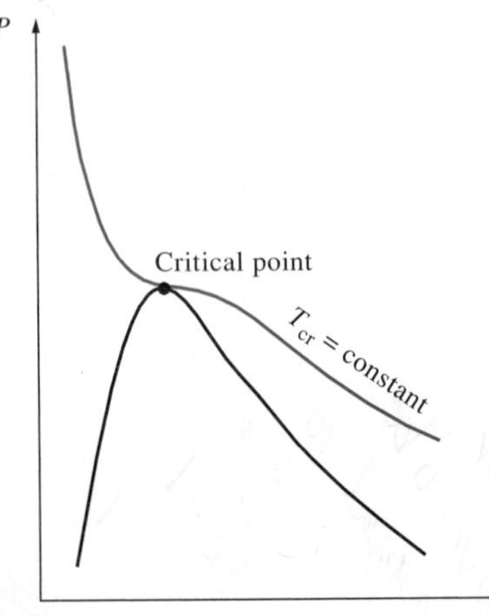

By performing the differentiations and eliminating v_{cr}, the constants a and b are determined to be

$$a = \frac{27R^2T_{cr}^2}{64P_{cr}} \quad \text{and} \quad b = \frac{RT_{cr}}{8P_{cr}} \tag{2-24}$$

The constants a and b can be determined for any substance from the critical-point data alone (Table A-1).

The accuracy of the van der Waals equation of state is often inadequate, but it can be improved by using the values of a and b that are based on the actual behavior of the gas over a wider range instead of a single point. Despite its limitations, the van der Waals equation of state has a historical value in that it was one of the first attempts to model the behavior of real gases. The van der Waals equation of state can also be expressed on a unit-mole basis by replacing the v in Eq. 2-22 by $\bar{v}$ and the R in Eqs. 2-22 and 2-24 by R_u.

Beattie-Bridgeman Equation of State

The Beattie-Bridgeman equation, proposed in 1928, is an equation of state based on five experimentally determined constants. It was proposed in the form of

$$P = \frac{R_uT}{\bar{v}^2}\left(1 - \frac{c}{\bar{v}T^3}\right)(\bar{v} + B) - \frac{A}{\bar{v}^2} \tag{2-25}$$

where

$$A = A_0\left(1 - \frac{a}{\bar{v}}\right) \quad \text{and} \quad B = B_0\left(1 - \frac{b}{\bar{v}}\right) \tag{2-26}$$

The constants appearing in the above equation are given in Table A-29*a* for various substances. The Beattie-Bridgeman equation is known to be reasonably accurate for densities up to about $0.8\rho_{cr}$, where ρ_{cr} is the density of the substance at the critical point.

Benedict-Webb-Rubin Equation of State

Benedict, Webb, and Rubin extended the Beattie-Bridgeman equation in 1940 by raising the number of constants to eight. It is expressed as

$$P = \frac{R_uT}{\bar{v}} + \left(B_0R_uT - A_0 - \frac{C_0}{T^2}\right)\frac{1}{\bar{v}^2} + \frac{bR_uT - a}{\bar{v}^3} + \frac{a\alpha}{\bar{v}^6} + \frac{c}{\bar{v}^3T^2}\left(1 + \frac{\gamma}{\bar{v}^2}\right)e^{-\gamma/\bar{v}^2} \tag{2-27}$$

The values of the constants appearing in this equation are given in Table A-29*b*. This equation can handle substances at densities up to about $2.5\rho_{cr}$. In 1962, Strobridge further extended this equation by raising the number of constants to 16 (Fig. 2-65).

FIGURE 2-65

Complex equations of state represent the *P-v-T* behavior of gases more accurately over a wide range.

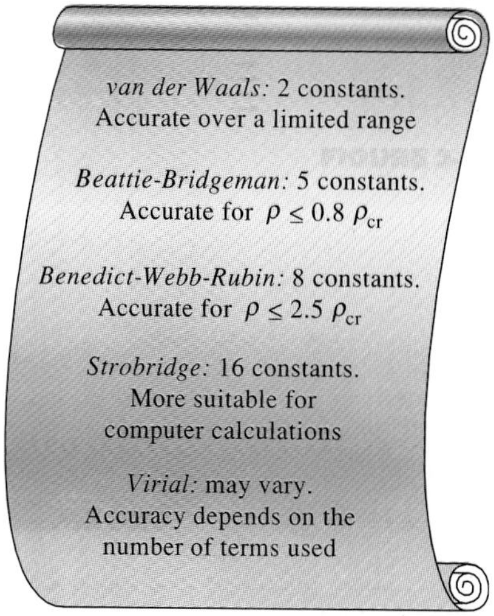

Virial Equation of State

The equation of state of a substance can also be expressed in a series form as

$$P = \frac{RT}{v} + \frac{a(T)}{v^2} + \frac{b(T)}{v^3} + \frac{c(T)}{v^4} + \frac{d(T)}{v^5} + \cdots \tag{2-28}$$

This and similar equations are called the *virial equations of state,* and the coefficients $a(T)$, $b(T)$, $c(T)$, and so on, that are functions of temperature alone are called *virial coefficients.* These coefficients can be determined experimentally or theoretically from statistical mechanics. Obviously, as the pressure approaches zero, all the virial coefficients will vanish and the equation will reduce to the ideal-gas equation of state. The P-v-T behavior of a substance can be represented accurately with the virial equation of state over a wider range by including a sufficient number of terms. All equations of state discussed above are applicable to the gas phase of the substances only, and thus should not be used for liquids or liquid–vapor mixtures.

Complex equations represent the P-v-T behavior of substances reasonably well and are very suitable for digital computer applications. For hand calculations, however, it is suggested that the reader use the property tables or the simpler equations of state for convenience. This is particularly true for specific-volume calculations since all the equations above are implicit in v and will require a trial-and-error approach. The accuracy of the van der Waals, Beattie-Bridgeman, and Benedict-Webb-Rubin equations of state is illustrated in Fig. 2-66. It is obvious from this figure that the Benedict-Webb-Rubin equation of state is the most accurate.

EXAMPLE 2-14 Different Methods of Evaluating Gas Pressure

Predict the pressure of nitrogen gas at $T = 175$ K and $v = 0.00375$ m³/kg on the basis of (*a*) the ideal-gas equation of state, (*b*) the van der Waals equation of state, (*c*) the Beattie-Bridgeman equation of state, and (*d*) the Benedict-Webb-Rubin equation of state. Compare the values obtained to the experimentally determined value of 10,000 kPa.

Solution (*a*) By using the ideal-gas equation of state, the pressure is found to be

$$P = \frac{RT}{v} = \frac{[0.279\ \text{kPa}\cdot\text{m}^3/(\text{kg}\cdot\text{K})](175\ \text{K})}{0.00375\ \text{m}^3/\text{kg}} = \mathbf{13{,}860\ kPa}$$

which is in error by 38.6 percent.

(*b*) The van der Waals constants for nitrogen are determined from Eq. 2-24 to be

$$a = 0.175\ \text{m}^6\cdot\text{kPa/kg}^2$$
$$b = 0.00138\ \text{m}^3/\text{kg}$$

From Eq. 2-22,

$$P = \frac{RT}{v-b} - \frac{a}{v^2} = \mathbf{9465\ kPa}$$

which is in error by 5.4 percent.

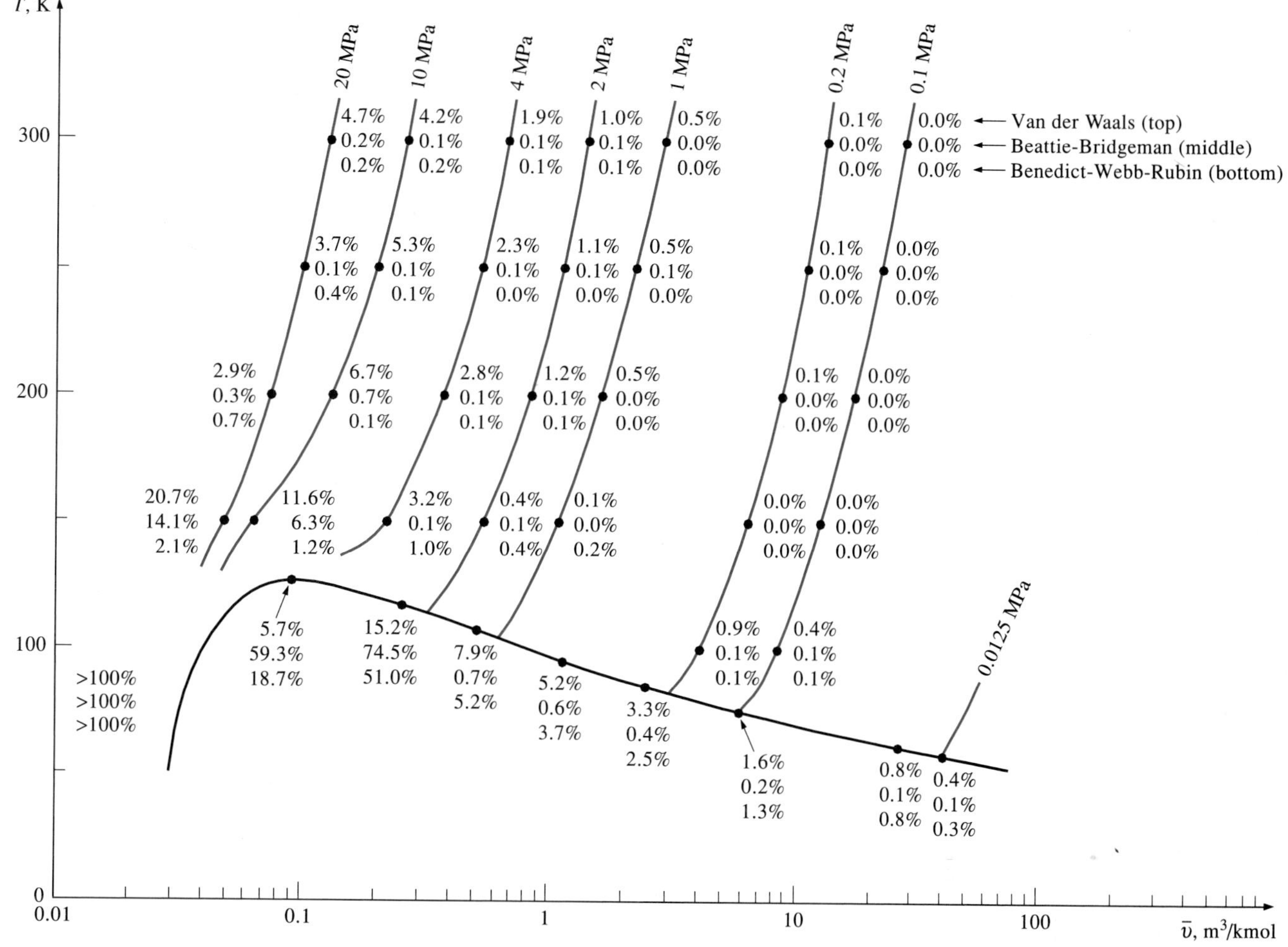

FIGURE 2-66

Percentage of error involved in various equations of state for nitrogen (% error $= [(|v_{\text{table}} - v_{\text{equation}}|)/v_{\text{table}}] \times 100$).

(*c*) The constants in the Beattie-Bridgeman equation are determined from Table A-29*a* to be

$$A = 102.29$$
$$B = 0.05378$$
$$c = 4.2 \times 10^4$$

Also, $\bar{v} = Mv = (28.013 \text{ kg/mol})(0.00375 \text{ m}^3/\text{kg}) = 0.10505 \text{ m}^3/\text{kmol}$. Substituting these values into Eq. 2-25, we obtain

$$P = \frac{R_u T}{\bar{v}^2}\left(1 - \frac{c}{\bar{v}T^3}\right)(\bar{v} + B) - \frac{A}{\bar{v}^2} = \mathbf{10{,}110\ kPa}$$

which is in error by 1.1 percent.

(*d*) The constants in the Benedict-Webb-Rubin equation are determined from Table A-29*b* to be

$$a = 2.54 \qquad A_0 = 106.73$$
$$b = 0.002328 \qquad B_0 = 0.04074$$
$$c = 7.379 \times 10^4 \qquad C_0 = 8.164 \times 10^5$$
$$\alpha = 1.272 \times 10^{-4} \qquad \gamma = 0.0053$$

Substituting these values into Eq. 2-27, we obtain

$$P = \frac{R_u T}{\bar{v}} + \left(B_0 R_u T - A_0 - \frac{C_0}{T^2}\right)\frac{1}{\bar{v}^2} + \frac{bR_u T - a}{\bar{v}^3} + \frac{a\alpha}{\bar{v}^6} + \frac{c}{\bar{v}^3 T^2}\left(1 + \frac{\gamma}{\bar{v}^2}\right)e^{-\gamma/\bar{v}^2}$$

$$= \mathbf{10{,}009\ kPa}$$

which is in error by only 0.09 percent. Thus, the accuracy of the Benedict-Webb-Rubin equation of state is rather impressive in this case.

2-10 ■ SUMMARY

A substance that has a fixed chemical composition throughout is called a *pure substance*. A pure substance exists in different phases depending on its energy level. In the liquid phase, a substance that is not about to vaporize is called a *compressed* or *subcooled liquid*. In the gas phase, a substance that is not about to condense is called a *superheated vapor*. During a phase-change process, the temperature and pressure of a pure substance are dependent properties. At a given pressure, a substance changes phase at a fixed temperature, called the *saturation temperature*. Likewise, at a given temperature, the pressure at which a substance changes phase is called the *saturation pressure*. During a boiling process, both the liquid and the vapor phases coexist in equilibrium, and under this condition the liquid is called *saturated liquid* and the vapor *saturated vapor*.

In a saturated liquid–vapor mixture, the mass fraction of the vapor phase is called the *quality* and is defined as

$$x = \frac{m_{\text{vapor}}}{m_{\text{total}}}$$

The quality may have values between 0 (saturated liquid) and 1 (saturated vapor). It has no meaning in the compressed liquid or superheated vapor regions. In the saturated mixture region, the average value of any intensive property y is determined from

$$y = y_f + xy_{fg}$$

where f stands for saturated liquid and g for saturated vapor.

In the absence of compressed liquid data, a general approximation is to treat a compressed liquid as a saturated liquid at the given *temperature*, that is,

$$y \cong y_{f@T}$$

where y stands for v, u, or h.

The state beyond which there is no distinct vaporization process is called the *critical point*. At supercritical pressures, a substance gradually and uniformly expands from the liquid to vapor phase. All three phases of a substance coexist in equilibrium at states along the *triple line* characterized by triple-line temperature and pressure. Various properties of some pure substances are listed in the appendix. As can be noticed from these tables, the compressed liquid has lower v, u, and h values than the saturated liquid at the same T or P. Likewise, superheated vapor has higher v, u, and h values than the saturated vapor at the same T or P.

Any relation among the pressure, temperature, and specific volume of a substance is called an *equation of state*. The simplest and best-known equation of state is the *ideal-gas equation of state*, given as

$$Pv = RT$$

where R is the gas constant. Caution should be exercised in using this relation since an ideal gas is a fictitious substance. Real gases exhibit ideal-gas behavior at relatively low pressures and high temperatures.

The deviation from ideal-gas behavior can be properly accounted for by using the *compressibility factor* Z, defined as

$$Z = \frac{Pv}{RT} \qquad \text{or} \qquad Z = \frac{v_{\text{actual}}}{v_{\text{ideal}}}$$

The Z factor is approximately the same for all gases at the same *reduced temperature* and *reduced pressure*, which are defined as

$$T_R = \frac{T}{T_{\text{cr}}} \qquad \text{and} \qquad P_R = \frac{P}{P_{\text{cr}}}$$

where P_{cr} and T_{cr} are the critical pressure and temperature, respectively. This is known as the *principle of corresponding states*. When either P or T is unknown, it can be determined from the compressibility chart with the help of the *pseudo-reduced specific volume*, defined as

$$v_R = \frac{v_{\text{actual}}}{RT_{\text{cr}}/P_{\text{cr}}}$$

The P-v-T behavior of substances can be represented more accurately by the more complex equations of state. Three of the best known are

van der Waals:

$$\left(P + \frac{a}{v^2}\right)(v - b) = RT$$

where

$$a = \frac{27R^2T_{\text{cr}}^2}{64P_{\text{cr}}} \qquad \text{and} \qquad b = \frac{RT_{\text{cr}}}{8P_{\text{cr}}}$$

Beattie-Bridgeman:

$$P = \frac{R_uT}{\bar{v}^2}\left(1 - \frac{c}{\bar{v}T^3}\right)(\bar{v} + B) - \frac{A}{\bar{v}^2}$$

where

$$A = A_0\left(1 - \frac{a}{\bar{v}}\right) \qquad \text{and} \qquad B = B_0\left(1 - \frac{b}{\bar{v}}\right)$$

Benedict-Webb-Rubin:

$$P = \frac{R_u T}{\bar{v}} + \left(B_0 R_u T - A_0 - \frac{C_0}{T^2}\right)\frac{1}{\bar{v}^2} + \frac{bR_u T - a}{\bar{v}^3} + \frac{a\alpha}{\bar{v}^6} + \frac{c}{\bar{v}^3 T^2}\left(1 + \frac{\gamma}{\bar{v}^2}\right)e^{-\gamma/\bar{v}^2}$$

The constants appearing in the Beattie-Bridgeman and Benedict-Webb-Rubin equations are given in Table A-29 for various substances.

REFERENCES AND SUGGESTED READING

1. A. Bejan. *Advanced Engineering Thermodynamics.* New York: John Wiley & Sons, 1988.

2. M. D. Burghardt. *Engineering Thermodynamics with Applications.* New York: Harper and Row, 1986.

3. J. R. Howell and R. O. Buckius. *Fundamentals of Engineering Thermodynamics.* New York: McGraw-Hill, 1987.

4. J. B. Jones and G. A. Hawkins. *Engineering Thermodynamics.* 2nd ed. New York: John Wiley & Sons, 1986.

5. G. J. Van Wylen and R. E. Sonntag. *Fundamentals of Classical Thermodynamics.* 3rd ed. New York: John Wiley & Sons, 1985.

6. K. Wark. *Thermodynamics.* 5th ed. New York: McGraw-Hill, 1988.

PROBLEMS*

Pure Substances, Phase-Change Processes, Phase Diagrams

2-1C Is iced water a pure substance? Why?

2-2C What is the difference between saturated liquid and compressed liquid?

2-3C What is the difference between saturated vapor and superheated vapor?

2-4C Is there any difference between the properties of saturated vapor at a given temperature and the vapor of a saturated mixture at the same temperature?

2-5C Is there any difference between the properties of saturated liquid at a given temperature and the liquid of a saturated mixture at the same temperature?

*Students are encouraged to answer *all* the concept "C" questions.

2-6C Is it true that water boils at higher temperatures at higher pressures? Explain.

2-7C If the pressure of a substance is increased during a boiling process, will the temperature also increase or will it remain constant? Why?

2-8C Why are the temperature and pressure dependent properties in the saturated mixture region?

2-9C What is the difference between the critical point and the triple point?

2-10C Is it possible to have water vapor at −10°C?

2-11C A househusband is cooking beef stew for his family in a pan that is (*a*) uncovered, (*b*) covered with a light lid, and (*c*) covered with a heavy lid. For which case will the cooking time be the shortest? Why?

2-12C How does the boiling process at supercritical pressures differ from the boiling process at subcritical pressures?

Property Tables

2-13C In what kind of pot will a given volume of water boil at a higher temperature: a tall and narrow one or a short and wide one? Explain.

2-14C A perfectly fitting pot and its lid often stick after cooking, and it becomes very difficult to open the lid when the pot cools down. Explain why this happens and what you would do to open the lid.

2-15C It is well known that warm air in a cooler environment rises. Now consider a warm mixture of air and gasoline on top of an open gasoline can. Do you think this gas mixture will rise in a cooler environment?

2-16C In 1775, Dr. William Cullen made ice in Scotland by evacuating the air in a water tank. Explain how that device works, and discuss how the process can be made more efficient.

2-17C Does the amount of heat absorbed as 1 kg of saturated liquid water boils at 100°C have to be equal to the amount of heat released as 1 kg of saturated water vapor condenses at 100°C?

2-18C Does the reference point selected for the properties of a substance have any effect in thermodynamic analysis? Why?

2-19C What is the physical significance of h_{fg}? Can it be obtained from a knowledge of h_f and h_g? How?

2-20C Is it true that it takes more energy to vaporize 1 kg of saturated liquid water at 100°C than it would to vaporize 1 kg of saturated liquid water at 120°C?

2-21C What is quality? Does it have any meaning in the superheated vapor region?

2-22C Which process requires more energy: completely vaporizing 1 kg of saturated liquid water at 1 atm pressure or completely vaporizing 1 kg of saturated liquid water at 8 atm pressure?

2-23C Does h_{fg} change with pressure? How?

2-24C Can quality be expressed as the ratio of the volume occupied by the vapor phase to the total volume?

2-25C In the absence of compressed liquid tables, how is the specific volume of a compressed liquid at a given P and T determined?

2-26 Complete the following table for H_2O:

T °C	P kPa	v m³/kg	Phase description
50		4.16	
	200		Saturated vapor
250	400		
110	600		

2-27E Complete the following table for H_2O:

T °F	P psia	u Btu/lbm	Phase description
250		851	
	20		Saturated liquid
500	120		
400	400		

2-28 Complete the following table for H_2O:

T °C	P kPa	h kJ/kg	x	Phase description
	325		0.4	
160		1682		
	950		0.0	
80	500			
	800	3161.7		

2-29 Complete the following table for refrigerant-134a:

T °C	P kPa	v m³/kg	Phase description
−12	600		
20		0.022	
	320		Saturated vapor
100	600		

2-30 Complete the following table for refrigerant-134a:

T °C	P kPa	u kJ/kg	Phase description
30		120	
−8			Saturated liquid
	400	300	
8	600		

2-31E Complete the following table for refrigerant-134a:

T °F	P psia	h Btu/lbm	x	Phase description
	70	64		
20			0.7	
10	70			
	180	128.77		
110			1.0	

2-32 Complete the following table for H_2O:

T °C	P kPa	v m³/kg	Phase description
125		0.53	
	1000		Saturated liquid
25	750		
500		0.130	

2-33 Complete the following table for H_2O:

T °C	P kPa	u kJ/kg	Phase description
	325	2452	
170			Saturated vapor
190	2000		
	4000	3040	

2-34E The temperature in a pressure cooker during cooking at sea level is measured to be 250°F. Determine the absolute pressure inside the cooker in psia and in atm. Would you modify your answer if the place were at a higher elevation?

FIGURE P2-34E

2-35E The atmospheric pressure at a location is usually specified at standard conditions, but it changes with the weather conditions. As the weather forecasters frequently state, the atmospheric pressure drops during stormy weather and it rises during clear and sunny days. If the pressure difference between the two extreme conditions is given to be 0.3 in. of mercury, determine how much the boiling temperatures of water will vary as the weather changes from one extreme to the other.

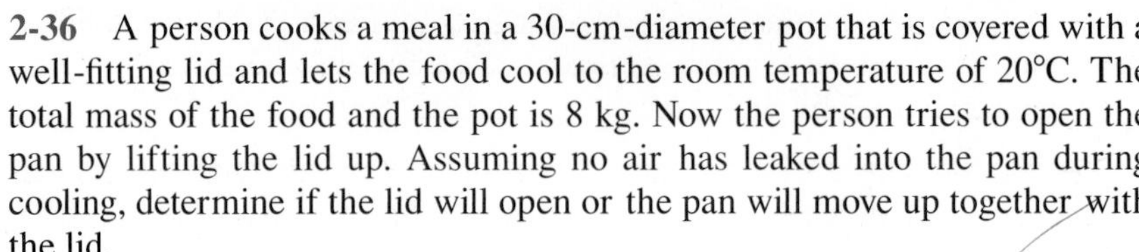

2-36 A person cooks a meal in a 30-cm-diameter pot that is covered with a well-fitting lid and lets the food cool to the room temperature of 20°C. The total mass of the food and the pot is 8 kg. Now the person tries to open the pan by lifting the lid up. Assuming no air has leaked into the pan during cooling, determine if the lid will open or the pan will move up together with the lid.

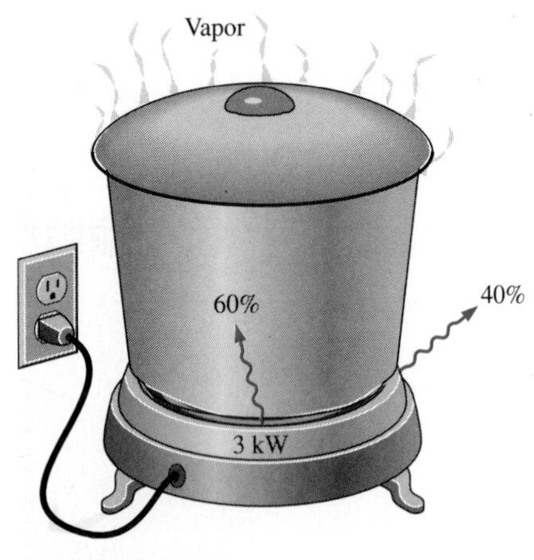

FIGURE P2-37

2-37 Water is to be boiled at seal level in a 30-cm-diameter stainless steel pan placed on top of a 3-kW electric burner. If 60 percent of the heat generated by the burner is transferred to the water during boiling, determine the rate of evaporation of water.

2-38 Repeat Prob. 2-37 for a location at an elevation of 1500 m where the atmospheric pressure is 84.5 kPa and thus the boiling temperature of water is 95°C.

2-39 Water is boiled at 1 atm pressure in a 20-cm-internal-diameter stainless steel pan on an electric range. If it is observed that the water level in the pan drops by 10 cm in 30 min, determine the rate of heat transfer to the pan.

2-40 Repeat Prob. 2-39 for a location at 2000-m elevation where the standard atmospheric pressure is 79.5 kPa.

2-41 Saturated steam coming off the turbine of a steam power plant at 30°C condenses on the outside of a 4-cm-outer-diameter, 20-m-long tube at a rate of 45 kg/h. Determine the rate of heat transfer from the steam to the cooling water flowing through the pipe.

2-42 The average atmospheric pressure in Denver (elevation = 1610 m) is 83.4 kPa. Determine the temperature at which water in an uncovered pan will boil in Denver. *Answer:* 94.4°C.

2-43 Water in a 5-cm-deep pan is observed to boil at 98°C. At what temperature will the water in a 40-cm-deep pan boil? Assume both pans are full of water.

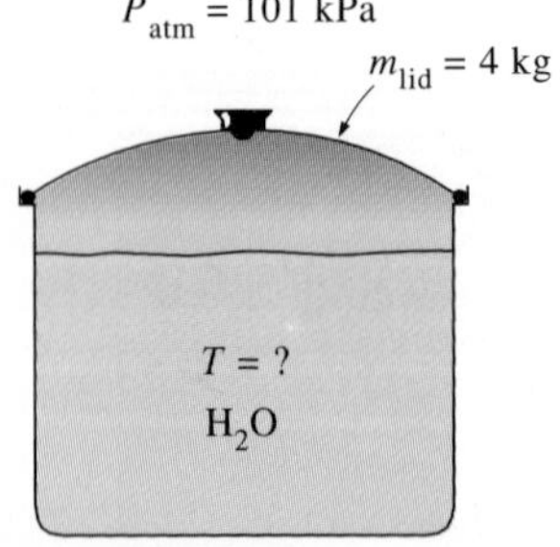

FIGURE P2-44E

2-44 A cooking pan whose inner diameter is 20 cm is filled with water and covered with a 4-kg lid. If the local atmospheric pressure is 101 kPa, determine the temperature at which the water will start boiling when it is heated.
Answer: 100.2°C

2-45 Water is being heated in a vertical piston-cylinder device. The piston has a mass of 20 kg and a cross-sectional area of 100 cm^2. If the local atmospheric pressure is 100 kPa, determine the temperature at which the water will start boiling.

2-46 A rigid tank with a volume of 2.5 m^3 contains 5 kg of saturated liquid–vapor mixture of water at 75°C. Now the water is slowly heated. Determine the temperature at which the liquid in the tank is completely vaporized. Also, show the process on a T-v diagram with respect to saturation lines.
Answer: 140.7°C

2-47 A rigid vessel contains 2 kg of refrigerant-134a at 900 kPa and 80°C. Determine the volume of the vessel and the total internal energy.
Answers: 0.0572 m^3, 577.7 kJ

2-48E A 5-ft^3 rigid tank contains 5 lbm of water at 20 psia. Determine (*a*) the temperature, (*b*) the total enthalpy, and (*c*) the mass of each phase of water.

2-49 A 0.5-m^3 vessel contains 10 kg of refrigerant-134a at −20°C. Determine (*a*) the pressure, (*b*) the total internal energy, and (*c*) the volume occupied by the liquid phase.
Answers: (*a*) 132.99 kPa, (*b*) 889.5 kJ, (*c*) 0.00487 m^3

2-50 A piston-cylinder device contains 0.1 m^3 of liquid water and 0.9 m^3 of water vapor in equilibrium at 800 kPa. Heat is transferred at constant pressure until the temperature reaches 350°C.

(*a*) What is the initial temperature of the water?
(*b*) Determine the total mass of the water.
(*c*) Calculate the final volume.
(*d*) Show the process on a *P*-*v* diagram with respect to saturation lines.

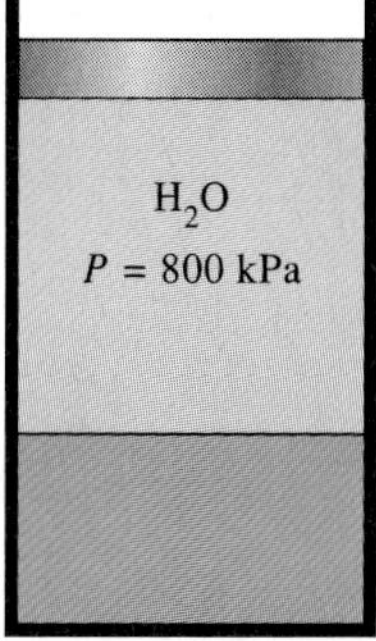

FIGURE P2-50

2-51E Superheated water vapor at 180 psia and 500°F is allowed to cool at constant volume until the temperature drops to 250°F. At the final state, determine (*a*) the pressure, (*b*) the quality, and (*c*) the enthalpy. Also, show the process on a *T*-*v* diagram with respect to saturation lines.
Answers: (*a*) 29.82 psia, (*b*) 0.219, (*c*) 425.7 Btu/lbm

2-52 A piston-cylinder device initially contains 50 L of liquid water at 25°C and 300 kPa. Heat is added to the water at constant pressure until the entire liquid is vaporized.

(*a*) What is the mass of the water?
(*b*) What is the final temperature?
(*c*) Determine the total enthalpy change.
(*d*) Show the process on a *T*-*v* diagram with respect to saturation lines.

Answers: (*a*) 49.85 kg, (*b*) 133.55°C, (*c*) 130,627 kJ

2-53 A 0.5-m^3 rigid vessel initially contains saturated liquid–vapor mixture of water at 100°C. The water is now heated until it reaches the critical state. Determine the mass of the liquid water and the volume occupied by the liquid at the initial state. *Answers:* 158.28 kg, 0.165 m^3

2-54 Determine the specific volume, internal energy, and enthalpy of compressed liquid water at 100°C and 15 MPa using the saturated liquid approximation. Compare these values to the ones obtained from the compressed liquid tables.

2-55E A 15-ft^3 rigid tank contains saturated mixture of refrigerant-134a at 30 psia. If the saturated liquid occupies 10 percent of the volume, determine the quality and the total mass of the refrigerant in the tank.

2-56 A piston-cylinder device contains 0.8 kg of steam at 300°C and 1 MPa. Steam is cooled at constant pressure until one-half of the mass condenses.

(*a*) Show the process on a *T*-*v* diagram.
(*b*) Find the final temperature.
(*c*) Determine the volume change.

2-57 A rigid tank contains water vapor at 300°C and an unknown pressure. When the tank is cooled to 180°C, the vapor starts condensing. Estimate the initial pressure in the tank. *Answer:* 1.325 MPa

Vapor Pressure and Phase Equilibrium

2-58 Consider a glass of water in a room that is at 20°C and 60 percent relative humidity. If the water temperature is 15°C, determine the vapor pressure (*a*) at the free surface of the water and (*b*) at a location in the room far from the glass.

2-59 During a hot summer day at the beach when the air temperature is 30°C, someone claims the vapor pressure in the air to be 5.2 kPa. Is this claim reasonable?

2-60 On a certain day, the temperature and relative humidity of air over a large swimming pool are measured to be 20°C and 40 percent, respectively. Determine the water temperature of the pool when phase equilibrium conditions are established between the water in the pool and the vapor in the air.

2-61 Consider two rooms that are identical except that one is maintained at 30°C and 40 percent relative humidity while the other is maintained at 20°C and 70 percent relative humidity. Noting that the amount of moisture is proportional to the vapor pressure, determine which room contains more moisture.

2-62E A thermos bottle is half-filled with water and is left open to the atmospheric air at 70°F and 35 percent relative humidity. If heat transfer to the water through the thermos walls and the free surface is negligible, determine the temperature of water when phase equilibrium is established.

2-63 During a hot summer day when the air temperature is 35°C and the relative humidity is 70 percent, you buy a supposedly "cold" canned drink from a store. The store owner claims that the temperature of the drink is below 10°C. Yet the drink does not feel so cold and you are skeptical since you notice no condensation forming outside the can. Can the store owner be telling the truth?

Ideal Gas

2-64 Propane and methane are commonly used for heating in winter, and the leakage of these fuels, even for short periods, poses a fire danger for homes. Which gas leakage do you think poses a greater risk for fire? Explain.

2-65C Under what conditions is the ideal-gas assumption suitable for real gases?

2-66C What is the difference between R and R_u? How are these two related?

2-67C What is the difference between mass and molar mass? How are these two related?

2-68 A spherical balloon with a diameter of 6 m is filled with helium at 20°C and 200 kPa. Determine the mole number and the mass of the helium in the balloon. *Answers:* 9.28 kmol, 37.15 kg

2-69 The pressure in an automobile tire depends on the temperature of the air in the tire. When the air temperature is 25°C, the pressure gage reads 210 kPa. If the volume of the tire is 0.025 m³, determine the pressure rise in the tire when the air temperature in the tire rises to 50°C. Also, determine the amount of air that must be bled off to restore pressure to its original value at this temperature. Assume the atmospheric pressure to be 100 kPa.

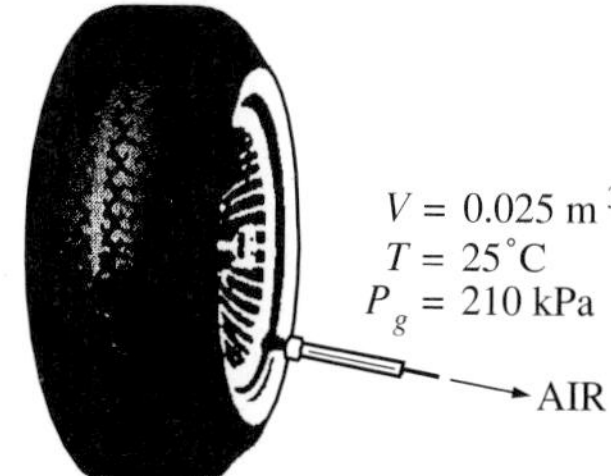

FIGURE P2-69

2-70E The air in an automobile tire with a volume of 0.53 ft³ is at 90°F and 20 psig. Determine the amount of air that must be added to raise the pressure to the recommended value of 30 psig. Assume the atmospheric pressure to be 14.6 psia and the temperature and the volume to remain constant. *Answer:* 0.0260 lbm

2-71 The pressure gage on a 1.2-m³ oxygen tank reads 500 kPa. Determine the amount of oxygen in the tank if the temperature is 24°C and the atmospheric pressure is 97 kPa.

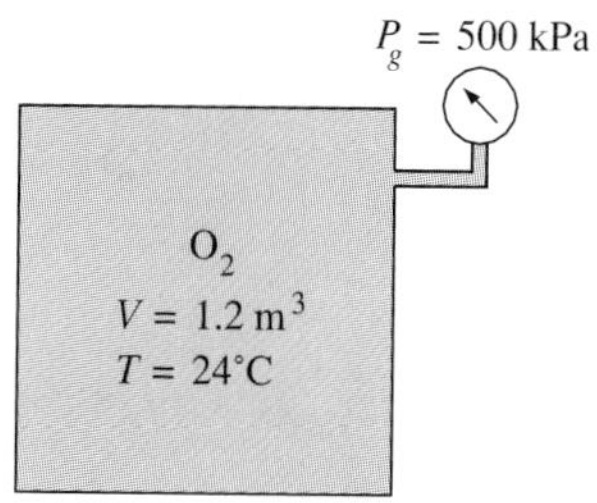

FIGURE P2-71

2-72E A rigid tank contains 20 lbm of air at 20 psia and 70°F. More air is added to the tank until the pressure and temperature rise to 35 psia and 90°F, respectively. Determine the amount of air added to the tank. *Answer:* 13.73 lbm

2-73 A 800-L rigid tank contains 10 kg of air at 25°C. Determine the reading on the pressure gage if the atmospheric pressure is 97 kPa.

2-74 A 1-m³ tank containing air at 25°C and 500 kPa is connected through a valve to another tank containing 5 kg of air at 35°C and 200 kPa. Now the valve is opened, and the entire system is allowed to reach thermal equilibrium with the surroundings, which are at 20°C. Determine the volume of the second tank and the final equilibrium pressure of air. *Answer:* 284.1 kPa

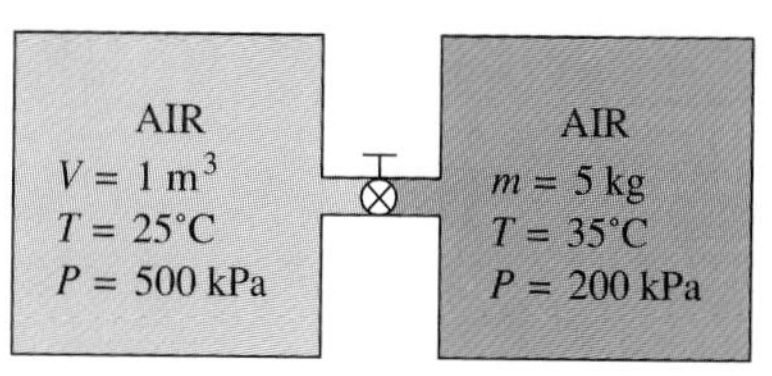

FIGURE P2-74

Compressibility Factor

2-75C What is the physical significance of the compressibility factor Z?

2-76C What is the principle of corresponding states?

2-77C How are the reduced pressure and reduced temperature defined?

2-78 Determine the specific volume of superheated water vapor at 10 MPa and 400°C, using (*a*) the ideal-gas equation, (*b*) the generalized compressibility chart, and (*c*) the steam tables. Also determine the error involved in the first two cases.

Answers: (*a*) 0.03106 m^3/kg, 17.6 percent; (*b*) 0.02609 m^3/kg, 1.2 percent; (*c*) 0.02641 m^3/kg

2-79 Determine the specific volume of refrigerant-134a vapor at 1.4 MPa and 140°C based on (*a*) the ideal gas equation, (*b*) the generalized compressibility chart, and (*c*) the experimental data from tables. Also, determine the error involved in the first two cases.

2-80 Determine the specific volume of nitrogen gas at 10 MPa and 150 K based on (*a*) the ideal-gas equation and (*b*) the generalized compressibility chart. Compare these results with the experimental value of 0.002388 m^3/kg, and determine the error involved in each case.

Answers: (*a*) 0.004452 m^3/kg, 86.4 percent; (*b*) 0.002404 m^3/kg, 0.7 percent

2-81 Determine the specific volume of superheated water vapor at 1.6 MPa and 225°C based on (*a*) the ideal-gas equation, (*b*) the generalized compressibility chart, and (*c*) the steam tables. Determine the error involved in the first two cases.

2-82E Refrigerant-134a at 400 psia has a specific volume of 0.1386 ft^3/lbm. Determine the temperature of the refrigerant based on (*a*) the ideal gas equation, (*b*) the generalized compressibility chart, and (*c*) the refrigerant tables.

2-83 A 0.01677-m^3 tank contains 1 kg of refrigerant-134a at 110°C. Determine the pressure of the refrigerant, using (*a*) the ideal gas equation, (*b*) the generalized compressibility chart, and (*c*) the refrigerant tables.

Answers: (*a*) 1.861 MPa, (*b*) 1.586 MPa, (*c*) 1.6 MPa

2-84 Somebody claims that oxygen gas at 160 K and 3 MPa can be treated as an ideal gas with an error of less than 10 percent. Is this claim valid?

2-85 What is the percentage of error involved in treating carbon dioxide at 3 MPa and 10°C as an ideal gas? *Answer:* 25 percent

2-86 What is the percentage of error involved in treating carbon dioxide at 5 MPa and 350 K as an ideal gas?

Other Equations of State

2-87C What is the physical significance of the two constants that appear in the van der Waals equation of state? On what basis are they determined?

2-88 A 3.27-m^3 tank contains 100 kg of nitrogen at 225 K. Determine the pressure in the tank, using (*a*) the ideal-gas equation, (*b*) the van der Waals equation, and (*c*) the Beattie-Bridgeman equation. Compare your results with the actual value of 2000 kPa.

2-89 A 1-m^3 tank contains 2.841 kg of steam at 0.6 MPa. Determine the temperature of the steam, using (*a*) the ideal gas equation, (*b*) the van der Waals equation, and (*c*) the steam tables.

Answers: (*a*) 457.6 K, (*b*) 465.9 K, (*c*) 473 K

2-90E Refrigerant-134a at 100 psia has a specific volume of 0.5388 ft^3/lbm. Determine the temperature of the refrigerant based on (*a*) the ideal-gas equation, (*b*) the van der Waals equation, and (*c*) the refrigerant tables.

2-91 Nitrogen at 150 K has a specific volume of 0.041884 m^3/kg. Determine the pressure of the nitrogen, using (*a*) the ideal-gas equation and (*b*) the Beattie-Bridgeman equation. Compare your results to the experimental value of 1000 kPa. *Answers:* (*a*) 1063 kPa, (*b*) 1000.4 kPa

Review Problems

2-92 A smoking lounge is to accommodate 15 heavy smokers. The minimum fresh air requirements for smoking lounges are specified to be 30 L/s per person (ASHRAE, *Standard 62,* 1989). Determine the minimum required flow rate of fresh air that needs to be supplied to the lounge, and the diameter of the duct if the air velocity is not to exceed 8 m/s.

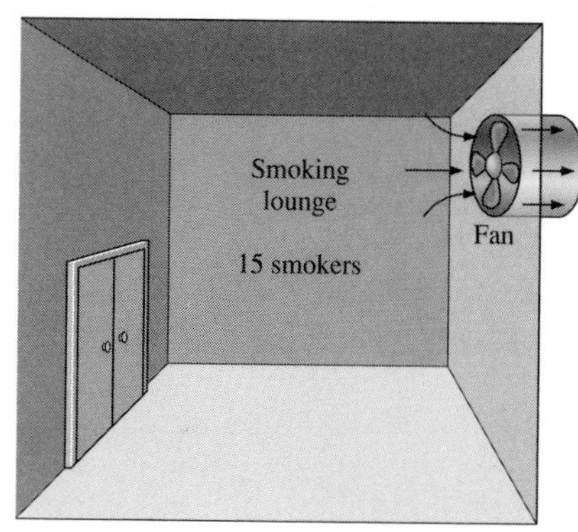

FIGURE P2-92

2-93 The minimum fresh air requirements of a residential building are specified to be 0.35 air change per hour (ASHRAE, *Standard 62,* 1989). That is, 35 percent of the entire air contained in a residence should be replaced by fresh outdoors air every hour. If the ventilation requirements of a 2.7-m-high, 200-m^2 residence is to be met entirely by a fan, determine the size of the fan, in L/min, that needs to be installed. Also determine the diameter of the duct if the air velocity is not to exceed 6 m/s.

2-94 The gage pressure of an automobile tire is measured to be 200 kPa before a trip and 220 kPa after the trip at a location where the atmospheric pressure is 90 kPa. Assuming the volume of the tire remains constant at 0.022 m^3, determine the percent increase in the absolute temperature of the air in the tire.

2-95 Although balloons have been around since 1783 when the first balloon took to the skies in France, a real breakthrough in ballooning occurred in 1960 with the design of the modern hot-air balloon fueled by inexpensive propane and constructed of lightweight nylon fabric. Over the years, ballooning has become a sport and a hobby for many people around the world. Unlike balloons filled with the light helium gas, hot-air balloons are open to the atmosphere. Therefore, the pressure in the balloon is always the same as the local atmospheric pressure, and the balloon is never in danger of exploding.

Hot-air balloons range from about 15 to 25 m in diameter. The air in the balloon cavity is heated by a propane burner located at the top of the passenger cage. The flames from the burner that shoot into the balloon heat the air in the balloon cavity, raising the air temperature at top of the balloon from 65°C to over 120°C. The air temperature is maintained at the desired levels by periodically firing the propane burner. The buoyancy force that pushes the balloon upward is proportional to the density of the cooler air outside the balloon and the volume of the balloon, and can be expressed as

FIGURE P2-95
A hot-air balloon.

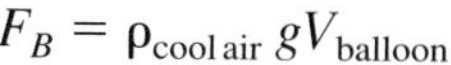

$$F_B = \rho_{\text{cool air}}\, g V_{\text{balloon}}$$

where g is the gravitational acceleration. When air resistance is negligible, the buoyancy force is opposed by (1) the weight of the hot air in the balloon, (2) the weight of the cage, the ropes, and the balloon material, and (3) the weight of the people and other load in the cage. The operator of the balloon can control the height and the vertical motion of the balloon by firing the burner or by letting some hot air in the balloon escape, to be replaced by cooler air. The forward motion of the balloon is provided by the winds.

Consider a 20-m-diameter hot-air balloon that, together with its cage, has a mass of 80 kg when empty. This balloon is hanging still in the air at a location where the atmospheric pressure and temperature are 90 kPa and 15°C, respectively, while carrying three 65-kg people. Determine the average temperature of the air in the balloon. What would your response be if the atmospheric air temperature were 30°C?

2-96 Consider an 18-m-diameter hot-air balloon that, together with its cage, has a mass of 120 kg when empty. The air in the balloon, which is now carrying two 70-kg people, is heated by propane burners at a location where the atmospheric pressure and temperature are 93 kPa and 12°C, respectively. Determine the average temperature of the air in the balloon when the balloon first starts rising. What would your response be if the atmospheric air temperature were 25°C?

2-97E Water in a pressure cooker is observed to boil at 250°F. What is the absolute pressure in the pressure cooker, in psia?

2-98 A rigid tank with a volume of 0.07 m^3 contains 1 kg of refrigerant-134a vapor at 400 kPa. The refrigerant is now allowed to cool. Determine the pressure when the refrigerant first starts condensing. Also, show the process on a P-v diagram with respect to saturation lines.

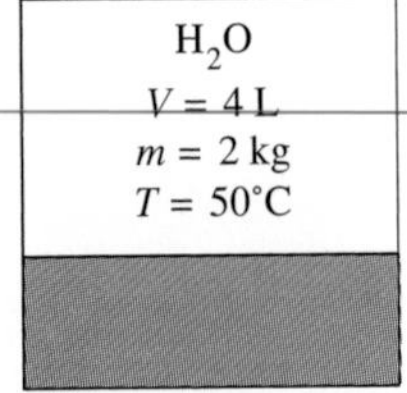

FIGURE P2-99

2-99 A 4-L rigid tank contains 2 kg of saturated liquid–vapor mixture of water at 50°C. The water is now slowly heated until it exists in a single phase. At the final state, will the water be in the liquid phase or the vapor phase? What would your answer be if the volume of the tank were 400 L instead of 4 L?

2-100 A 10-kg mass of superheated refrigerant-134a at 0.8 MPa and 40°C is cooled at constant pressure until it exists as a compressed liquid at 20°C.

(*a*) Show the process on a T-v diagram with respect to saturation lines.
(*b*) Determine the change in volume.
(*c*) Find the change in total internal energy.

Answers: (*b*) −0.261 m^3, (*c*) −1753 kJ

2-101 A 0.5-m^3 rigid tank containing hydrogen at 20°C and 600 kPa is connected by a valve to another 0.5-m^3 rigid tank that holds hydrogen at 30°C and 150 kPa. Now the valve is opened and the system is allowed to reach thermal equilibrium with the surroundings, which are at 15°C. Determine the final pressure in the tank.

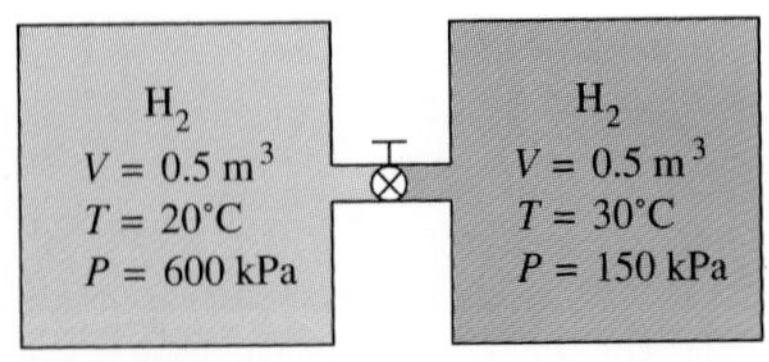

FIGURE P2-101

2-102 A 20-m^3 tank contains nitrogen at 25°C and 800 kPa. Some nitrogen is allowed to escape until the pressure in the tank drops to 600 kPa. If the temperature at this point is 20°C, determine the amount of nitrogen that has escaped. *Answer:* 42.9 kg

2-103 Steam at 400°C has a specific volume of 0.02 m^3/kg. Determine the pressure of the steam based on (*a*) the ideal-gas equation, (*b*) the generalized compressibility chart, and (*c*) the steam tables.
Answers: (*a*) 15,529 kPa, (*b*) 12,591 kPa, (*c*) 12,500 kPa

2-104 A tank whose volume is unknown is divided into two parts by a partition. One side of the tank contains 0.01 m^3 of refrigerant-134a that is a saturated liquid at 0.8 MPa, while the other side is evacuated. The partition is now removed, and the refrigerant fills the entire tank. If the final state of the refrigerant is 25°C and 200 kPa, determine the volume of the tank.

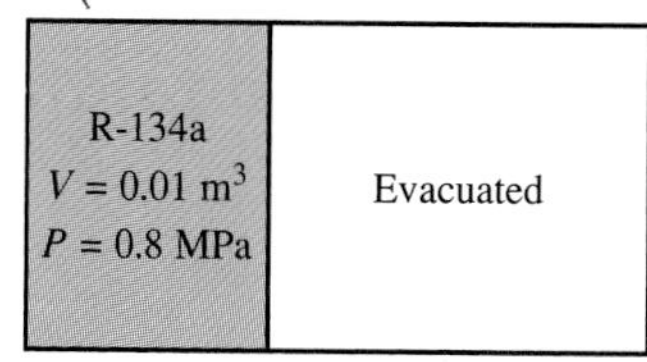

FIGURE P2-104

2-105 Liquid propane is commonly used as a fuel for heating homes, powering vehicles such as forklifts, and filling portable picnic tanks. Consider a propane tank that initially contains 5 L of liquid propane at the environment temperature of 20°C. If a hole develops in the connecting tube of a propane tank and the propane starts to leak out, determine the temperature of propane when the pressure in the tank drops to 1 atm. Also, determine the total amount of heat transfer from the environment to the tank to vaporize the entire propane in the tank.

2-106 Repeat Prob. 2-105 for isobutane.

Computer, Design, and Essay Problems

2-107 Write a computer program to express $T_{sat} = f(P_{sat})$ for steam as a fifth-degree polynomial where the pressure is in kPa and the temperature is in °C. Use tabulated data from Table A-4. What is the accuracy of this equation?

2-108 It is claimed that fruits and vegetables are cooled by 6°C for each percentage point of weight loss as moisture during vacuum cooling. Using calculations, demonstrate if this claim is reasonable.

2-109 Write a computer program to determine the specific volume of a substance at a given temperature and pressure, using the Beattie-Bridgeman equation. Check your program by evaluating the specific volume of refrigerant-134a at 10 different states and comparing them to the tabulated values.

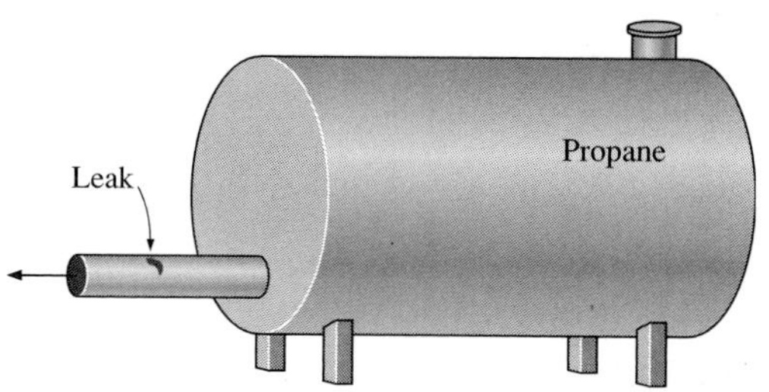

FIGURE P2-105

2-110 A solid normally absorbs heat as it melts, but there is a known exception at temperatures close to absolute zero. Find out which solid it is and give a physical explanation for it.

2-111 Numerous equations of state have been proposed throughout history. Write an essay on two equations of state not discussed in this chapter, describe them in sufficient detail, and discuss the accuracy and range of applicability of each equation.

2-112 It is well known that water freezes at 0°C at atmospheric pressure. The mixture of liquid water and ice at 0°C is said to be at stable equilibrium since it cannot undergo any changes when it is isolated from its surrounding. However, when water is free of impurities and the inner surfaces of the container are smooth, the temperature of water can be lowered to −2°C or even lower without any formation of ice at atmospheric pressure. But at that state even a small disturbance can initiate the formation of ice abruptly, and the water temperature stabilizes at 0°C following this sudden change. The water at −2°C is said to be in a *metastable state*. Write an essay on metastable states and discuss how they differ from stable equilibrium states.

2-113 Using a thermometer, measure the boiling temperature of water and calculate the corresponding saturation pressure. From this information, estimate the altitude of your town and compare it with the actual altitude value.

Software Problems

2-114 Repeat Prob. 2-26 using the enclosed software.

2-115E Repeat Prob. 2-27E using the enclosed software.

2-116 Repeat Prob. 2-28 using the enclosed software.

2-117 Repeat Prob. 2-29 using the enclosed software.

2-118 Repeat Prob. 2-30 using the enclosed software.

2-119E Repeat Prob. 2-31E using the enclosed software.

2-120 Repeat Prob. 2-32 using the enclosed software.

2-121 Repeat Prob. 2-33 using the enclosed software.

The First Law of Thermodynamics: Closed Systems

CHAPTER

3

The first law of thermodynamics is simply a statement of the conservation of energy principle, and it asserts that total energy is a thermodynamic property. In this chapter, energy transfer with heat and work is introduced, and the mechanisms of heat transfer as well as various work modes are discussed. The first-law relation for closed systems is developed in a step-by-step manner using an intuitive approach. Specific heats are defined, and relations are obtained for the internal energy and enthalpy of ideal gases in terms of specific heats and temperature. This approach is also applied to solids and liquids, which are approximated as incompressible substances. Finally, the refrigeration and freezing of foods are discussed.

3-1 ■ INTRODUCTION

In Chap. 1, it was pointed out that energy can be neither created nor destroyed; it can only change forms. This principle is based on experimental observations and is known as the *first law of thermodynamics,* or the *conservation of energy principle.* The first law can simply be stated as follows: During an interaction between a system and its surroundings, the amount of energy gained by the system must be exactly equal to the amount of energy lost by the surroundings.

Energy can cross the boundary of a closed system in two distinct forms: *heat* and *work* (Fig. 3-1). It is important to distinguish between these two forms of energy. Therefore, they will be discussed first, to form a sound basis for the development of the first law of thermodynamics.

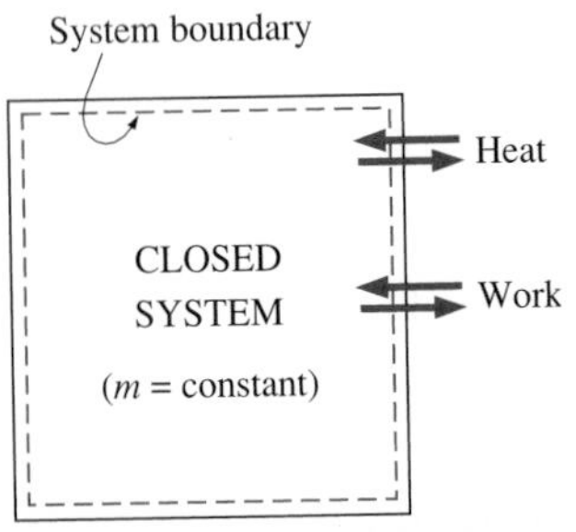

FIGURE 3-1

Energy can cross the boundaries of a closed system in the form of heat and work.

3-2 ■ HEAT TRANSFER

We know from experience that a can of cold soda left on a table eventually warms up and that a hot baked potato on the same table cools down (Fig. 3-2). When a body is left in a medium that is at a different temperature, energy transfer takes place between the body and the surrounding medium until thermal equilibrium is established, that is, the body and the medium reach the same temperature. The direction of energy transfer is always from the higher-temperature body to the lower-temperature one. Once the temperature equality is established, energy transfer stops. In the processes described above, energy is said to be transferred in the form of heat.

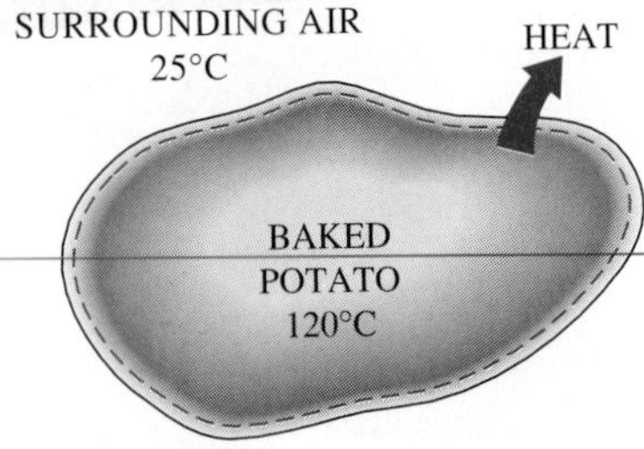

FIGURE 3-2

Heat is transferred from hot bodies to colder ones by virtue of a temperature difference.

Heat is defined as *the form of energy that is transferred between two systems (or a system and its surroundings) by virtue of a temperature difference.* That is, an energy interaction is heat only if it takes place because of a temperature difference. Then it follows that there cannot be any heat transfer between two systems that are at the same temperature.

In daily life, we frequently refer to the sensible and latent forms of internal energy as *heat,* and we talk about the heat content of bodies. In thermodynamics, however, we usually refer to those forms of energy as *thermal energy* to prevent any confusion with *heat transfer.*

Several phrases in common use today—such as *heat flow, heat addition, heat rejection, heat absorption, heat removal, heat gain, heat loss, heat storage, heat generation, electrical heating, resistance heating, frictional heating, gas heating, heat of reaction, liberation of heat, specific heat, sensible heat, latent heat, waste heat, body heat, process heat, heat sink,* and *heat source*—are not consistent with the strict thermodynamic meaning of the term *heat,* which limits its use to the *transfer* of thermal energy during a process. However, these phrases are deeply rooted in our vocabulary, and they are used by both ordinary people and scientists without causing any misunderstanding since they are usually interpreted properly instead of being taken literally. (Besides, no acceptable alternatives exist for some of these phrases.) For example, the phrase *body heat* is understood to mean *the thermal energy content* of a body. Likewise, *heat flow* is understood to mean *the*

— First law of termodynamics or
conservation of energy principle.

ergy can cross the boundary of
sed system by heat or work

body is left in a different temperature
the body will transfere and
heat until thermal equilibrium is

transfer of thermal energy, not the flow of a fluidlike substance called heat, although the latter incorrect interpretation, which is based on the caloric theory, is the origin of this phrase. Also, the transfer of heat into a system is frequently referred to as *heat addition* and the transfer of heat out of a system as *heat rejection.* Perhaps there are thermodynamic reasons for being so reluctant to replace *heat* by *thermal energy*: It takes less time and energy to say, write, and comprehend *heat* than it does *thermal energy.*

Heat is energy in transition. It is recognized only as it crosses the boundary of a system. Consider the hot baked potato one more time. The potato contains energy, but this energy is heat transfer only as it passes through the skin of the potato (the system boundary) to reach the air, as shown in Fig. 3-3. Once in the surroundings, the transferred heat becomes part of the internal energy of the surroundings. Thus, in thermodynamics, the term *heat* simply means *heat transfer.*

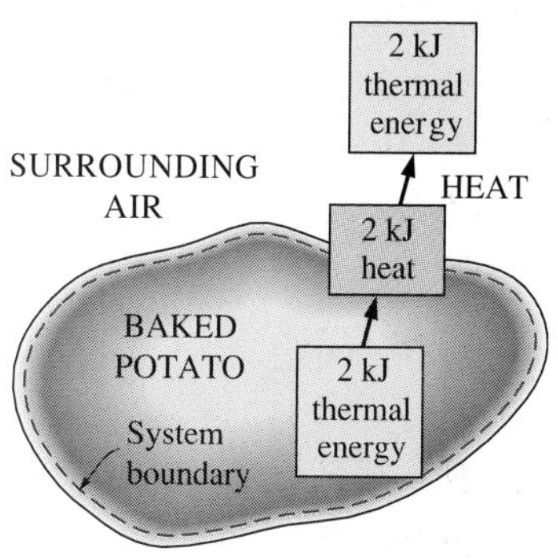

FIGURE 3-3
Energy is recognized as heat only as it crosses the system boundary.

A process during which there is no heat transfer is called an **adiabatic process** (Fig. 3-4). The word *adiabatic* comes from the Greek word *adiabatos,* which means *not to be passed.* There are two ways a process can be adiabatic: Either the system is well insulated so that only a negligible amount of heat can pass through the boundary, or both the system and the surroundings are at the same temperature and therefore there is no driving force (tem-
at transfer. An adiabatic process should not be
l process. Even though there is no heat transfer
the energy content and thus the temperature of a
by other means such as work.

eat has energy units, kJ (or Btu) being the most
heat transferred during the process between two
ted by Q_{12}, or just Q. Heat transfer *per unit mass*
is determined from

$$= \frac{Q}{m} \qquad \text{(kJ/kg)} \qquad (3\text{-}1)$$

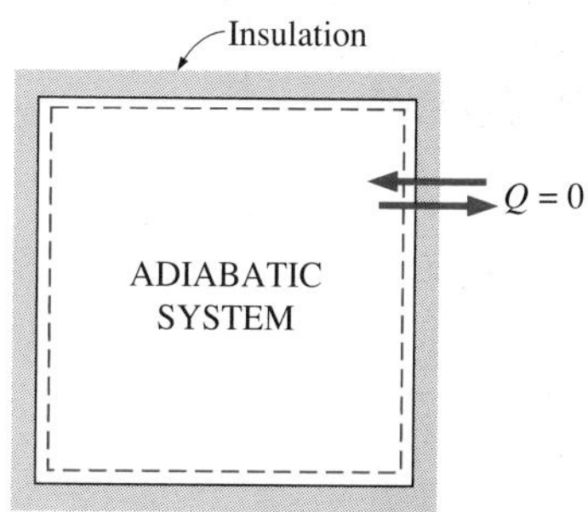

FIGURE 3-4
During an adiabatic process, a system exchanges no heat with its surroundings.

to know the *rate of heat transfer* (the amount
time) instead of the total heat transferred over
The heat transfer rate is denoted $\dot{Q}$, where the
rivative, or "per unit time." The heat transfer
is equivalent to kW. When $\dot{Q}$ varies with time,
uring a process is determined by integrating $\dot{Q}$
rocess:

$$\int_{t_1}^{t_2} \dot{Q}\, dt \qquad \text{(kJ)} \qquad (3\text{-}2)$$

ng a process, the relation above reduces to

$$Q = \dot{Q}\, \Delta t \qquad \text{(kJ)} \qquad (3\text{-}3)$$

where $\Delta t = t_2 - t_1$ is the time interval during which the process occurs.

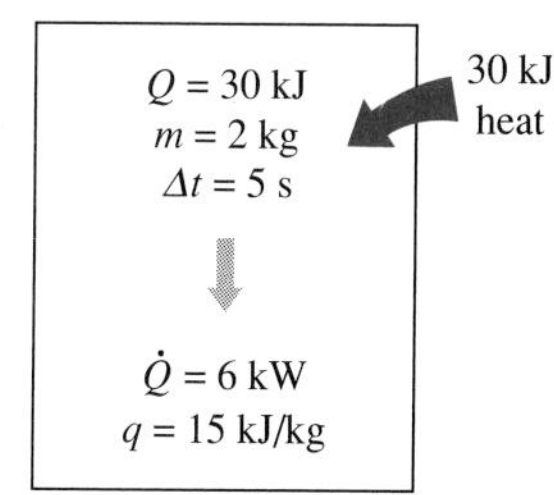

FIGURE 3-5
The relationships among q, Q, and $\dot{Q}$.

Historical Background

Heat was always perceived to be something that produces in us a sensation of warmth, and one would think that the nature of heat is one of the first things understood by man. But it was only in the middle of the nineteenth century that we had a true physical understanding of the nature of heat, thanks to the development at that time of the **kinetic theory,** which treats the molecules as tiny balls that are in motion and thus possess kinetic energy. Heat is then defined as the energy associated with the random motion of atoms and molecules.

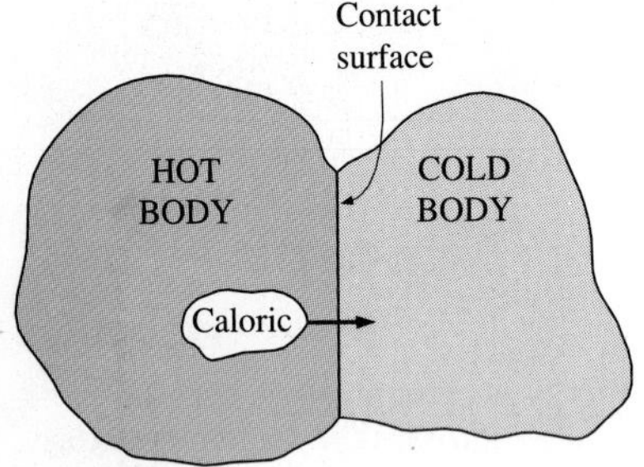

FIGURE 3-6
In the early nineteenth century, heat was thought to be an invisible fluid called the *caloric* that flowed from warmer bodies to the cooler ones.

Although it was suggested in the eighteenth and early nineteenth centuries that heat is the manifestation of motion at the molecular level (called the *live force*), the prevailing view of heat until the middle of the nineteenth century was based on the **caloric theory,** which was proposed by the French chemist Antoine Lavoisier (1743–1794) in 1789. The caloric theory asserts that heat is a fluidlike substance called the **caloric** that is a massless, colorless, odorless, and tasteless substance that can be poured from one body into another (Fig. 3-6). When caloric was added to a body, its temperature increased; and when caloric was removed from a body, its temperature decreased. When a body could not contain any more caloric, in much the same way as when a glass of water could not dissolve any more salt or sugar, the body was said to be *saturated* with caloric. This interpretation gave rise to the terms *saturated liquid* and *saturated vapor* still in use today.

Caloric theory came under attack soon after its introduction. The caloric theory maintained that heat is a substance, and it cannot be created or destroyed. Yet it was known that heat can be generated indefinitely by rubbing one's hands together or rubbing two pieces of wood together. In 1798, Benjamin Thompson (Count Rumford) (1753–1814) showed in his papers that heat can be generated continuously through friction. The validity of the caloric theory was also challenged by several others. But it was the careful experiments of the Englishman James P. Joule (1818–1889) published in 1843 that finally convinced the skeptics that heat was not a substance after all, and put the caloric theory to rest. Although the caloric theory was totally abandoned in the middle of the nineteenth century, it contributed greatly to the development of thermodynamics.

Modes of Heat Transfer

Heat can be transferred in three different ways: *conduction, convection,* and *radiation*. A detailed study of these heat transfer modes is given later in this text. Below we will give a brief description of each mode to familiarize the reader with the basic mechanisms of heat transfer. All modes of heat transfer require the existence of a temperature difference, and all modes of heat transfer are from the high-temperature medium to a lower-temperature one.

Conduction is the transfer of energy from the more energetic particles of a substance to the adjacent less energetic ones as a result of interactions between the particles. Conduction can take place in solids, liquids, or gases. In gases and liquids, conduction is due to the collisions of the molecules during their random motion. In solids, it is due to the combination of vibrations of the molecules in a lattice and the energy transport by free electrons. A cold canned drink in a warm room, for example, eventually warms up to the room temperature as a result of heat transfer from the room to the drink through the aluminum can by conduction (Fig. 3-7).

It is observed that the rate of heat conduction $\dot{Q}_{\text{cond}}$ through a layer of constant thickness Δx is proportional to the temperature difference ΔT across the layer and the area A normal to the direction of heat transfer, and is inversely proportional to the thickness of the layer. Therefore,

$$\dot{Q}_{\text{cond}} = k_t A \frac{\Delta T}{\Delta x} \qquad \text{(W)} \qquad (3\text{-}4)$$

where the constant of proportionality k_t is the **thermal conductivity** of the material, which is a measure of the ability of a material to conduct heat (Table 3-1). Materials such as copper and silver, which are good electric conductors, are also good heat conductors, and therefore have high k_t values. Materials such as rubber, wood, and styrofoam are poor conductors of heat, and therefore have low k_t values.

In the limiting case of $\Delta x \to 0$, the equation above reduces to the differential form

$$\dot{Q}_{\text{cond}} = -k_t A \frac{dT}{dx} \qquad \text{(W)} \qquad (3\text{-}5)$$

which is known as **Fourier's law** of heat conduction. It indicates that the rate of heat conduction in a direction is proportional to the *temperature gradient* in that direction. Heat is conducted in the direction of decreasing temperature, and the temperature gradient becomes negative when temperature decreases with increasing x. Therefore, a negative sign is added in Eq. 3-5 to make heat transfer in the positive x direction a positive quantity.

Temperature is a measure of the kinetic energies of the molecules. In a liquid or gas, the kinetic energy of the molecules is due to the random motion of the molecules as well as the vibrational and rotational motions. When two molecules possessing different kinetic energies collide, part of the kinetic energy of the more energetic (higher-temperature) molecule is transferred to the less energetic (lower-temperature) particle, in much the same way as when two elastic balls of the same mass at different velocities collide, part of the kinetic energy of the faster ball is transferred to the slower one.

In solids, heat conduction is due to two effects: the lattice vibrational waves induced by the vibrational motions of the molecules positioned at relatively fixed positions in a periodic manner called a *lattice*, and the energy transported via the free flow of electrons in the solid. The thermal conductivity of a solid is obtained by adding the lattice and the electronic components. The thermal conductivity of pure metals is primarily due to the electronic component, whereas the thermal conductivity of nonmetals is primarily due to the lattice component. The lattice component of thermal conductivity strongly depends on the way the molecules are arranged. For example, the thermal conductivity of diamond, which is a highly ordered crystalline solid,

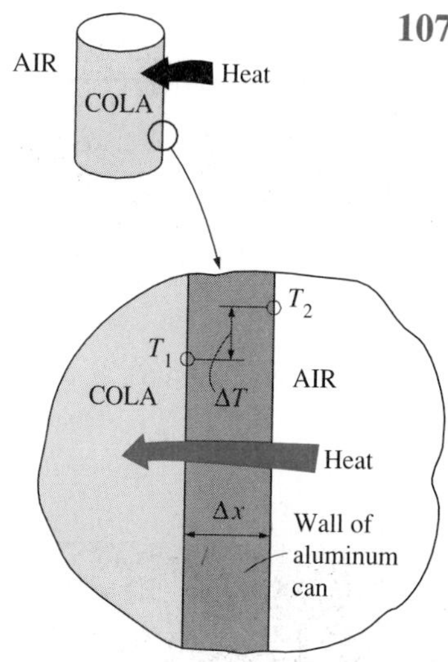

FIGURE 3-7

Heat conduction from warm air to a cold canned drink through the wall of the aluminum can.

TABLE 3-1

Thermal conductivities of some materials at room conditions

Material	Thermal conductivity, W/(m · K)
Diamond	2300
Silver	429
Copper	401
Gold	317
Aluminum	237
Iron	80.2
Mercury (ℓ)	8.54
Glass	1.4
Brick	0.72
Water (ℓ)	0.613
Human skin	0.37
Wood (oak)	0.17
Helium (*g*)	0.152
Soft rubber	0.13
Glass fiber	0.043
Air (*g*)	0.026
Urethane, rigid foam	0.026

is much higher than the thermal conductivities of pure metals, as can be seen from Table 3-1.

Convection is the mode of energy transfer between a solid surface and the adjacent liquid or gas that is in motion, and it involves the combined effects of *conduction* and *fluid motion*. The faster the fluid motion, the greater the convection heat transfer. In the absence of any bulk fluid motion, heat transfer between a solid surface and the adjacent fluid is by pure conduction. The presence of bulk motion of the fluid enhances the heat transfer between the solid surface and the fluid, but it also complicates the determination of heat transfer rates.

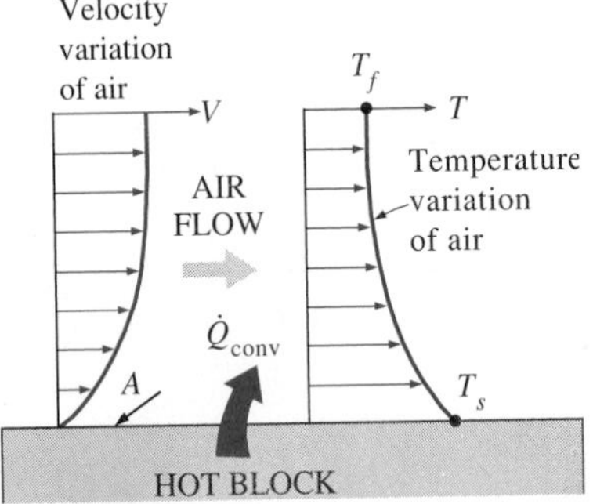

FIGURE 3-8
Heat transfer from a hot surface to air by convection.

Consider the cooling of a hot block by blowing of cool air over its top surface (Fig. 3-8). Energy is first transferred to the air layer adjacent to the surface of the block by conduction. This energy is then carried away from the surface by convection; that is, by the combined effects of conduction within the air, which is due to random motion of air molecules, and the bulk or macroscopic motion of the air, which removes the heated air near the surface and replaces it by the cooler air.

Convection is called **forced convection** if the fluid is *forced* to flow in a tube or over a surface by external means such as a fan, pump, or the wind. In contrast, convection is called **free** (or **natural**) **convection** if the fluid motion is caused by buoyancy forces induced by density differences due to the variation of temperature in the fluid (Fig. 3-9). For example, in the absence of a fan, heat transfer from the surface of the hot block in Fig. 3-8 will be by natural convection since any motion in the air in this case will be due to the rise of the warmer (and thus lighter) air near the surface and the fall of the cooler (and thus heavier) air to fill its place. Heat transfer between the block and the surrounding air will be by conduction if the temperature difference between the air and the block is not large enough to overcome the resistance of air to move and thus to initiate natural convection currents.

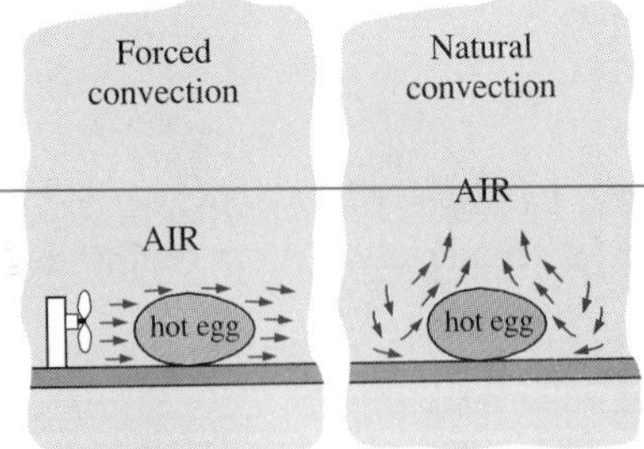

FIGURE 3-9
The cooling of a boiled egg by forced and natural convection.

Heat transfer processes that involve *change of phase* of a fluid are also considered to be convection because of the fluid motion induced during the process such as the rise of the vapor bubbles during *boiling* or the fall of the liquid droplets during *condensation*.

The rate of heat transfer by convection $\dot{Q}_{conv}$ is determined from **Newton's law of cooling**, expressed as

$$\dot{Q}_{conv} = hA(T_s - T_f) \qquad \text{(W)} \qquad (3\text{-}6)$$

where h is the **convection heat transfer coefficient,** A is the surface area through which heat transfer takes place, T_s is the surface temperature, and T_f is bulk fluid temperature away from the surface. (At the surface, the fluid temperature equals the surface temperature of the solid.)

The convection heat transfer coefficient h is not a property of the fluid. It is an experimentally determined parameter whose value depends on all the variables that influence convection such as the surface geometry, the nature of fluid motion, the properties of the fluid, and the bulk fluid velocity. Typical values of h, in W/(m$^2\cdot$K), are 2–25 for the free convection of gases, 50–1000 for the free convection of liquids, 25–250 for the forced convection of gases,

50–20,000 for the forced convection of liquids, and 2500–100,000 for convection in boiling and condensation processes.

Radiation is the energy emitted by matter in the form of electromagnetic waves (or photons) as a result of the changes in the electronic configurations of the atoms or molecules. Unlike conduction and convection, the transfer of energy by radiation does not require the presence of an intervening medium (Fig. 3-10). In fact, energy transfer by radiation is fastest (at the speed of light) and it suffers no attenuation in a vacuum. This is exactly how the energy of the sun reaches the earth.

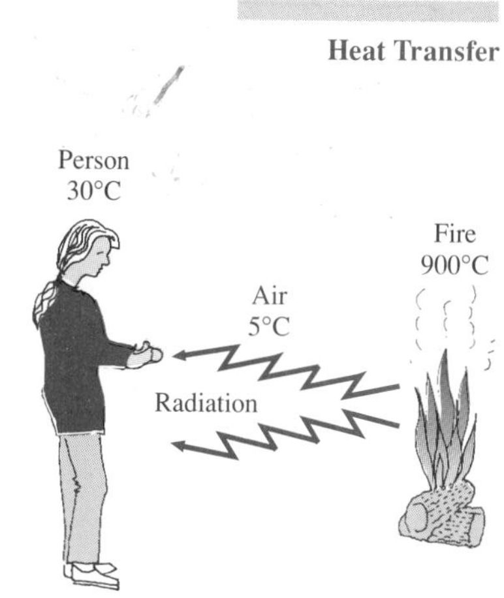

FIGURE 3-10

Unlike conduction and convection, heat transfer by radiation can occur between two bodies, even when they are separated by a medium colder than both of them.

In heat transfer studies, we are interested in *thermal radiation,* which is the form of radiation emitted by bodies because of their temperature. It differs from other forms of electromagnetic radiation such as X-rays, gamma rays, microwaves, radio waves, and television waves that are not related to temperature. All bodies at a temperature above absolute zero emit thermal radiation.

Radiation is a *volumetric phenomenon,* and all solids, liquids, and gases emit, absorb, or transmit radiation to varying degrees. However, radiation is usually considered to be a *surface phenomenon* for solids that are opaque to thermal radiation such as metals, wood, and rocks since the radiation emitted by the interior regions of such material can never reach the surface, and the radiation incident on such bodies is usually absorbed within a few microns from the surface.

The maximum rate of radiation that can be emitted from a surface at an *absolute* temperature T_s is given by the *Stefan-Boltzmann law* as

$$\dot{Q}_{\text{emit, max}} = \sigma A T_s^4 \qquad \text{(W)} \qquad (3\text{-}7)$$

where A is the surface area and $\sigma = 5.67 \times 10^{-8}$ W/(m$^2 \cdot$ K^4) is the **Stefan-Boltzmann constant.** The idealized surface that emits radiation at this maximum rate is called a **blackbody,** and the radiation emitted by a blackbody is called **blackbody radiation.** The radiation emitted by all *real* surfaces is less than the radiation emitted by a blackbody at the same temperatures and is expressed as

$$\dot{Q}_{\text{emit}} = \varepsilon \sigma A T_s^4 \qquad \text{(W)} \qquad (3\text{-}8)$$

where ε is the **emissivity** of the surface. The property emissivity, whose value is in the range $0 \leq \varepsilon \leq 1$, is a measure of how closely a surface approximates a blackbody for which $\varepsilon = 1$. The emissivities of some surfaces are given in Table 3-2.

TABLE 3-2

Emissivity of some materials at 300 K

Material	Emissivity
Aluminum foil	0.07
Anodized aluminum	0.82
Polished copper	0.03
Polished gold	0.03
Polished silver	0.02
Polished stainless steel	0.17
Black paint	0.98
White paint	0.90
White paper	0.92–0.97
Asphalt pavement	0.85–0.93
Red brick	0.93–0.96
Human skin	0.95
Wood	0.82–0.92
Soil	0.93–0.96
Water	0.96
Vegetation	0.92–0.96

Another important radiation property of a surface is its **absorptivity,** α, which is the fraction of the radiation energy incident on a surface that is absorbed by the surface. Like emissivity, its value is in the range $0 \leq \alpha \leq 1$. A blackbody absorbs the entire radiation incident on it. That is, a blackbody is a perfect absorber ($\alpha = 1$) as well as a perfect emitter.

In general, both ε and α of a surface depend on the temperature and the wavelength of the radiation. **Kirchhoff's law** of radiation states that the emissivity and the absorptivity of a surface are equal at the same temperature and wavelength. In most practical applications, the dependence of ε and α on the temperature and wavelength is ignored, and the average absorptivity of a surface is taken to be equal to its average emissivity. The rate at which a surface absorbs radiation is determined from (Fig. 3-11)

$$\dot{Q}_{abs} = \alpha \dot{Q}_{inc} \qquad \text{(W)} \qquad (3\text{-}9)$$

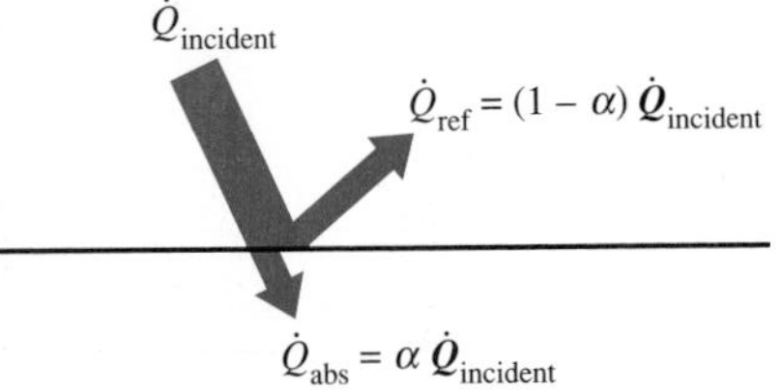

FIGURE 3-11

The absorption of radiation incident on an opaque surface of absorptivity α.

where $\dot{Q}_{inc}$ is the rate at which radiation is incident on the surface and α is the absorptivity of the surface. For opaque (nontransparent) surfaces, the portion of incident radiation that is not absorbed by the surface is reflected back.

The difference between the rates of radiation emitted by the surface and the radiation absorbed is the *net* radiation heat transfer. If the rate of radiation absorption is greater than the rate of radiation emission, the surface is said to be *gaining* energy by radiation. Otherwise, the surface is said to be *losing* energy by radiation. In general, the determination of the net rate of heat transfer by radiation between two surfaces is a complicated matter since it depends on the properties of the surfaces, their orientation relative to each other, and the interaction of the medium between the surfaces with radiation. However, in the special case of a relatively small surface of emissivity ε and surface area A at *absolute* temperature T_s that is completely enclosed by a much larger surface at *absolute* temperature T_{surr} separated by a gas (such as air) that does not intervene with radiation (i.e., the amount of radiation emitted, absorbed, or scattered by the medium is negligible), the net rate of radiation heat transfer between these two surfaces is determined from (Fig. 3-12)

$$\dot{Q}_{rad} = \varepsilon\sigma A(T_s^4 - T_{surr}^4) \qquad \text{(W)} \qquad (3\text{-}10)$$

In this special case, the emissivity and the surface area of the surrounding surface do not have any effect on the net radiation heat transfer.

EXAMPLE 3-1 Heat Transfer from a Person

Consider a person standing in a breezy room at 20°C. Determine the total rate of heat transfer from this person if the exposed surface area and the average outer surface temperature of the person are 1.6 m^2 and 29°C, respectively, and the convection heat transfer coefficient is 6 W/(m^2 · °C) (Fig. 3-13).

Solution A person is standing in a breezy room. The total rate of heat loss from the person is to be determined.

LARGE ENCLOSURE

ϵ, A, T_s

SMALL BODY

$\dot{Q}_{rad}$

T_{surr}

FIGURE 3-12

Radiation heat transfer between a body and the inner surfaces of a much larger enclosure that completely surrounds it.

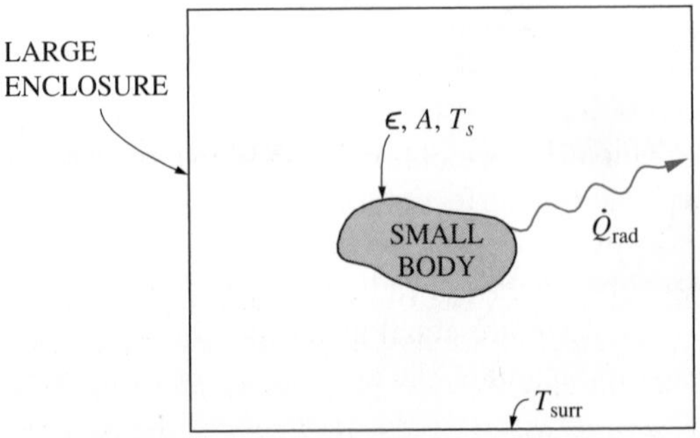

Heat can be transfer in three way ; <u>conduction</u>, <u>convection</u> and <u>radiation</u>

Assumptions **1** The emissivity and heat transfer coefficient are constant and uniform. **2** Heat conduction through the feet is negligible. **3** Heat loss by evaporation is disregarded.

Analysis The heat transfer between the person and the air in the room will be by convection (instead of conduction) since it is conceivable that the air in the vicinity of the skin or clothing will warm up and rise as a result of heat transfer from the body, initiating natural convection currents. It appears that the experimentally determined value for the rate of convection heat transfer in this case is 6 W per unit surface area (m^2) per unit temperature difference (in K or °C) between the person and the air away from the person. Thus, the rate of convection heat transfer from the person to the air in the room is, from Eq. 3-6,

$$\begin{aligned}\dot{Q}_{\text{conv}} &= hA(T_s - T_f)\\ &= [6\ \text{W/(m}^2\cdot{}^\circ\text{C)}](1.6\ \text{m}^2)(29-20)^\circ\text{C}\\ &= 86.4\ \text{W}\end{aligned}$$

The person will also lose heat by radiation to the surrounding wall surfaces. We take the temperature of the surfaces of the walls, ceiling, and the floor to be equal to the air temperature in this case for simplicity, but we recognize that this does not need to be the case. These surfaces may be at a higher or lower temperature than the average temperature of the room air, depending on the outdoor conditions and the structure of the walls. Considering that air does not intervene with radiation and the person is completely enclosed by the surrounding surfaces, the net rate of radiation heat transfer from the person to the surrounding walls, ceiling, and the floor is, from Eq. 3-10,

$$\begin{aligned}\dot{Q}_{\text{rad}} &= \varepsilon\sigma A(T_s^4 - T_{\text{surr}}^4)\\ &= (0.95)[5.67\times10^{-8}\ \text{W/(m}^2\cdot\text{K}^4)](1.6\ \text{m}^2)[(29+273)^4-(20+273)^4]\text{K}^4\\ &= 81.7\ \text{W}\end{aligned}$$

Note that we must use *absolute* temperatures in radiation calculations. Also note that we used the emissivity value for the skin and clothing at room temperature since the emissivity is not expected to change significantly at a slightly higher temperature.

Then the rate of total heat transfer from the body is determined by adding these two quantities to be

$$\dot{Q}_{\text{total}} = \dot{Q}_{\text{conv}} + \dot{Q}_{\text{rad}} = 86.4 + 81.7 = \mathbf{168.1\ W}$$

The heat transfer would be much higher if the person were not dressed since the exposed surface temperature would be higher. Thus, an important function of the clothes is to serve as a barrier against heat transfer.

Discussion In the above calculations, heat transfer through the feet to the floor by conduction, which is usually very small, is neglected. Heat transfer from the skin by perspiration, which is the dominant mode of heat transfer in hot environments, is not considered here.

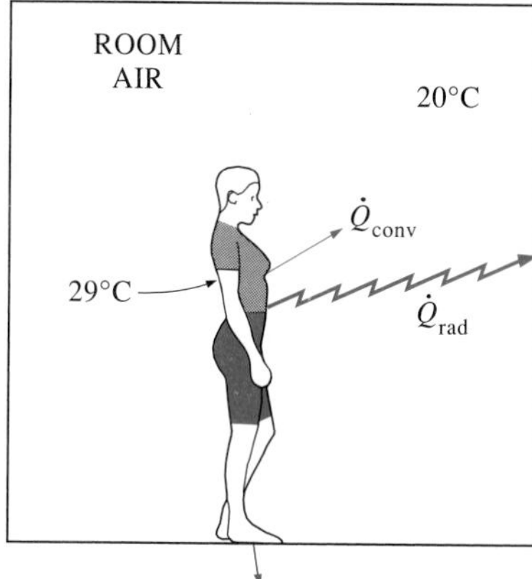

FIGURE 3-13
Heat transfer from the person described in Example 3-1.

3-3 ■ WORK

Work, like heat, is an energy interaction between a system and its surroundings. As mentioned earlier, energy can cross the boundary of a closed system in the form of heat or work. Therefore, *if the energy crossing the boundary of a closed system is not heat, it must be work*. Heat is easy to recognize:

Its driving force is a temperature difference between the system and its surroundings. Then we can simply say that an energy interaction that is not caused by a temperature difference between a system and its surroundings is work. More specifically, *work is the energy transfer associated with a force acting through a distance.* A rising piston, a rotating shaft, and an electric wire crossing the system boundaries are all associated with work interactions.

Work is also a form of energy transferred like heat and, therefore, has energy units such as kJ. The work done during a process between states 1 and 2 is denoted by W_{12}, or simply W. The work done *per unit mass* of a system is denoted w and is defined as

$$w = \frac{W}{m} \qquad \text{(kJ/kg)} \tag{3-11}$$

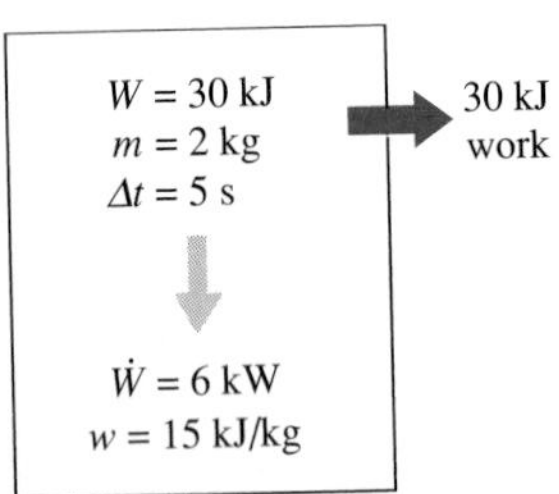

FIGURE 3-14

The relationships among w, W, and $\dot{W}$.

The work done *per unit time* is called **power** and is denoted $\dot{W}$ (Fig. 3-14). The unit of power is kJ/s, or kW.

Heat and work are *directional quantities,* and thus the complete description of a heat or work interaction requires the specification of both the *magnitude* and *direction*. One way of doing that is to adopt a sign convention. The generally accepted **formal sign convection** for heat and work interactions is as follows: *heat transfer to a system and work done by a system are positive; heat transfer from a system and work done on a system are negative*. Another way is to use the subscripts *in* and *out* to indicate direction (Fig. 3-15). For example, a work input of 5 kJ can be expressed as W_{in} = 5 kJ, while a heat loss of 3 kJ can be expressed as Q_{out} = 3 kJ. When the direction of a heat or work interaction is not known, we can simply *assume* a direction for the interaction (using the subscript *in* or *out*) and solve for it. A positive result indicates the assumed direction is right. A negative result, on the other hand, indicates that the direction of the interaction is the opposite of the assumed direction. This is just like assuming a direction for an unknown force when solving a statics problem, and reversing the direction when a negative result is obtained for the force. We will use this *intuitive approach* in this book as it eliminates the need to adopt a formal sign convention and the need to carefully assign negative values to some interactions.

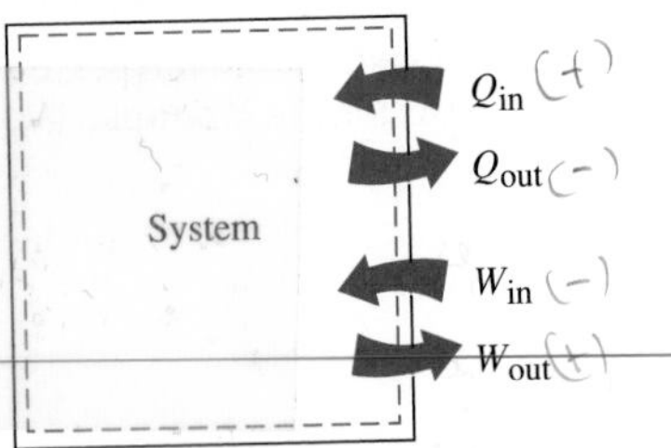

FIGURE 3-15

Specifying the directions of heat and work.

Note that a quantity that is transferred to or from a system during an interaction is not a property since the amount of such a quantity depends on more than just the state of the system. Heat and work are *energy transfer mechanisms* between a system and its surroundings, and there are many similarities between them:

1. Both are recognized at the boundaries of a system as they cross them. That is, both heat and work are *boundary* phenomena.
2. Systems possess energy, but not heat or work.
3. Both are associated with a *process,* not a state. Unlike properties, heat or work has no meaning at a state.

4. Both are *path functions* (i.e., their magnitudes depend on the path followed during a process as well as the end states).

Path functions have **inexact differentials** designated by the symbol δ. Therefore, a differential amount of heat or work is represented by δQ or δW, respectively, instead of dQ or dW. Properties, however, are **point functions** (i.e., they depend on the state only, and not on how a system reaches that state), and they have **exact differentials** designated by the symbol d. A small change in volume, for example, is represented by dV, and the total volume change during a process between states 1 and 2 is

$$\int_1^2 dV = V_2 - V_1 = \Delta V$$

That is, the volume change during process 1-2 is always the volume at state 2 minus the volume at state 1, regardless of the path followed (Fig. 3-16). The total work done during process 1-2, however, is

$$\int_1^2 \delta W = W_{12} \qquad (not\ \Delta W)$$

That is, the total work is obtained by following the process path and adding the differential amounts of work (δW) done along the way. The integral of δW *is not* $W_2 - W_1$ (i.e., the work at state 2 minus work at state 1), which is meaningless since work is not a property and systems do not possess work at a state.

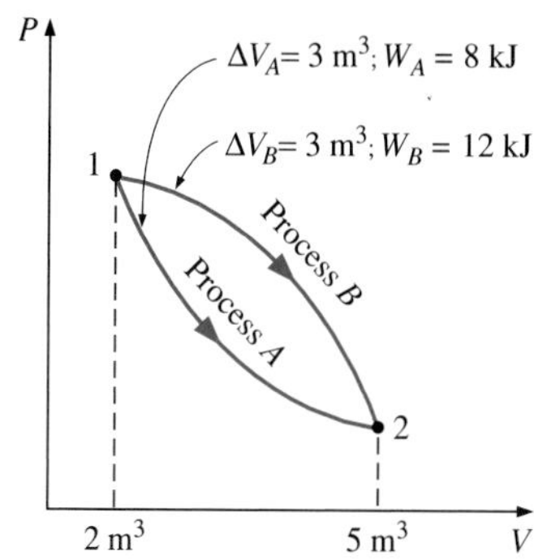

FIGURE 3-16
Properties are point functions; but heat and work are path functions (their magnitudes depend on the path followed).

EXAMPLE 3-2 Burning of a Candle in an Insulated Room

A candle is burning in a well-insulated room. Taking the room (the air plus the candle) as the system, determine (*a*) if there is any heat transfer during this burning process and (*b*) if there is any change in the internal energy of the system.

(Insulation)
ROOM

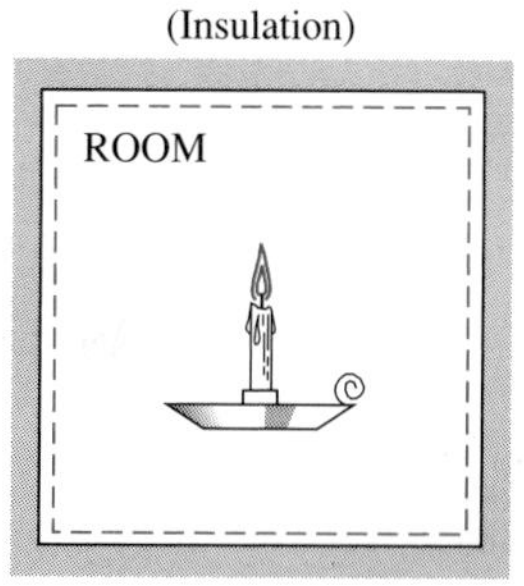

FIGURE 3-17
Schematic for Example 3-2.

Solution (*a*) The interior surfaces of the room form the system boundary, as indicated by the dashed lines in Fig. 3-17. As pointed out earlier, heat is recognized as it crosses the boundaries. Since the room is well insulated, we have an adiabatic system and no heat will pass through the boundaries. Therefore, $Q = 0$ for this process.

(*b*) As discussed in Chap. 1, the internal energy involves energies that exist in various forms (sensible, latent, chemical, nuclear). During the process described above, part of the chemical energy is converted to sensible energy. Since there is no increase or decrease in the total internal energy of the system, $\Delta U = 0$ for this process.

EXAMPLE 3-3 Heating of a Potato in an Oven

A potato initially at room temperature (25°C) is being baked in an oven that is maintained at 200°C, as shown in Fig. 3-18. Is there any heat transfer during this baking process?

FIGURE 3-18
Schematic for Example 3-3.

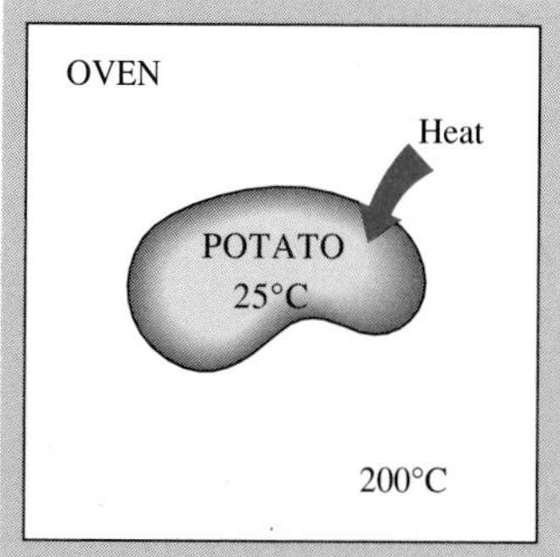

Solution This is not a well-defined problem since the system is not specified. Let us assume that we are observing the potato, which will be our system. Then the skin of the potato may be viewed as the system boundary. Part of the energy in the oven will pass through the skin to the potato. Since the driving force for this energy transfer is a temperature difference, this is a heat transfer process.

EXAMPLE 3-4 Heating of an Oven by Work Transfer

A well-insulated electric oven is being heated through its heating element. If the entire oven, including the heating element, is taken to be the system, determine whether this is a heat or work interaction.

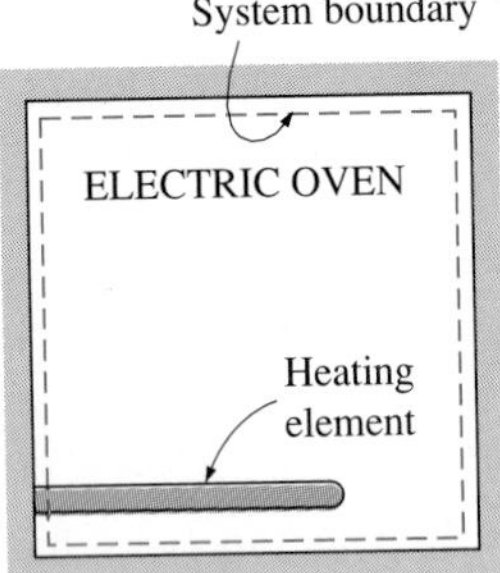

FIGURE 3-19
Schematic for Example 3-4.

Solution For this problem, the interior surfaces of the oven form the system boundary, as shown in Fig. 3-19. The energy content of the oven obviously increases during this process, as evidenced by a rise in temperature. This energy transfer to the oven is not caused by a temperature difference between the oven and the surrounding air. Instead, it is caused by *electrons* crossing the system boundary and thus doing work. Therefore, this is a work interaction.

EXAMPLE 3-5 Heating of an Oven by Heat Transfer

Answer the question in Example 3-4 if the system is taken as only the air in the oven without the heating element.

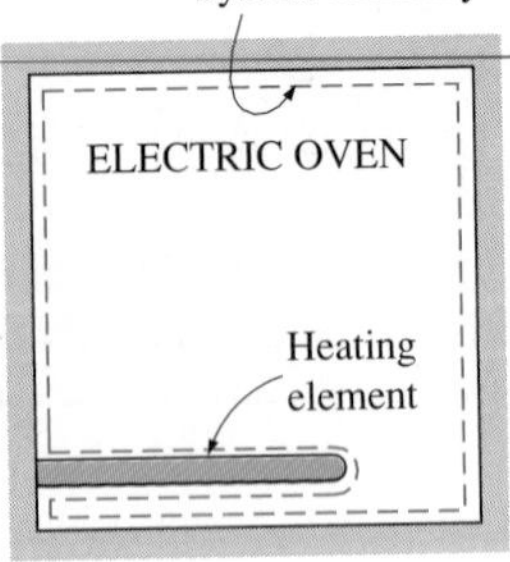

FIGURE 3-20
Schematic for Example 3-5.

Solution This time, the system boundary will include the outer surface of the heating element and will not cut through it, as shown in Fig. 3-20. Therefore, no electrons will be crossing the system boundary at any point. Instead, the energy generated in the interior of the heating element will be transferred to the air around it as a result of the temperature difference between the heating element and the air in the oven. Therefore, this is a heat transfer process.

Discussion For both cases, the amount of energy transfer to the air is the same. These two examples show that the same interaction can be heat or work depending on how the system is selected.

Electrical Work

It was pointed out in Example 3-4 that electrons crossing the system boundary do electrical work on the system. In an electric field, electrons in a wire move under the effect of electromotive forces, doing work. When N coulombs of electrons move through a potential difference V, the electrical work done is

$$W_e = VN$$

which can also be expressed in the rate form as

$$\dot{W}_e = VI \qquad \text{(W)} \tag{3-12}$$

where $\dot{W}_e$ is the **electrical power** and I is the number of electrons flowing per unit time, that is, the *current* (Fig. 3-21). In general, both V and I vary with time, and the electrical work done during a time interval Δt is expressed as

$$W_e = \int_1^2 VI\,dt \qquad \text{(kJ)} \tag{3-13}$$

When both V and I remain constant during the time interval Δt, it reduces to

$$W_e = VI\,\Delta t \qquad \text{(kJ)} \tag{3-14}$$

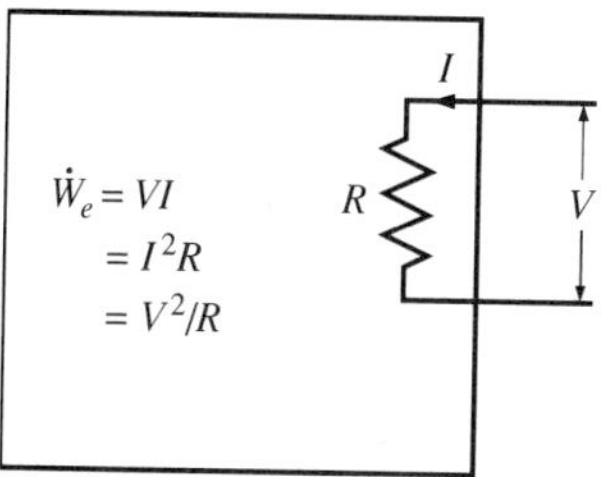

FIGURE 3-21
Electrical power in terms of resistance R, current I, and potential difference V.

EXAMPLE 3-6 Identifying Heat and Work Interactions

A small tank containing iced water at 0°C is placed in the middle of a large, well-insulated tank filled with oil, as shown in Fig. 3-22. The entire system is initially in thermal equilibrium at 0°C. The electric heater in the oil is now turned on, and 10 kJ of electrical work is done on the oil. After a while, it is noticed that the entire system is again at 0°C, but some ice in the small tank has melted. Considering the oil to be system A and the iced water to be system B, discuss the heat and work interactions for system A, system B, and the combined system (oil and iced water).

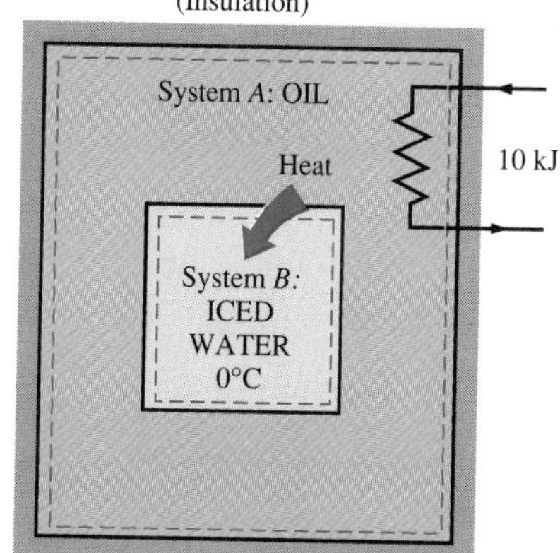

FIGURE 3-22
Schematic for Example 3-6.

Solution The boundaries of each system are indicated by dashed lines in the figure. Notice that the boundary of system B also forms the inner part of the boundary of system A.

System A: When the heater is turned on, electrons cross the outer boundary of system A, doing electrical work. This work is done on the system, and therefore $W_{A,\text{in}} = 10$ kJ. Because of this added energy, the temperature of the oil will rise, creating a temperature gradient that results in a heat flow process from the oil to the iced water through their common boundary. Since the oil is restored to its initial temperature of 0°C, the energy lost as heat must equal the energy gained as work. Therefore, $Q_{A,\text{out}} = 10$ kJ.

System B: The only energy interaction at the boundaries of system B is the heat flow from system A. All the heat lost by the oil is gained by the iced water. Thus, $W_B = 0$ and $Q_{B,\text{in}} = 10$ kJ.

Combined system: The outer boundary of system A forms the entire boundary of the combined system. The only energy interaction at this boundary is the electrical work. Since the tank is well insulated, no heat will cross this boundary. Therefore, $W_{\text{comb, in}} = 10$ kJ and $Q_{\text{comb}} = 0$. Notice that the heat flow from the oil to the iced water is an internal process for the combined system and, therefore, is not recognized as heat. It is simply the redistribution of the internal energy.

3-4 ■ MECHANICAL FORMS OF WORK

There are several different ways of doing work, each in some way related to a force acting through a distance (Fig. 3-23). In elementary mechanics, the work done by a constant force F on a body displaced a distance s in the direction of the force is given by

$$W = Fs \qquad \text{(kJ)} \tag{3-15}$$

FIGURE 3-23
The work done is proportional to the force applied (F) and the distance traveled (s).

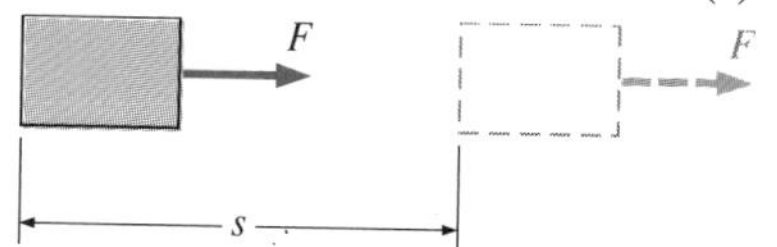

If the force F is not constant, the work done is obtained by adding (i.e., integrating) the differential amounts of work,

$$W = \int_1^2 F\,ds \qquad \text{(kJ)} \tag{3-16}$$

Obviously one needs to know how the force varies with displacement to perform this integration. Equations 3-15 and 3-16 give only the magnitude of the work. The sign is easily determined from physical considerations: The work done on a system by an external force acting in the direction of motion is negative, and work done by a system against an external force acting in the opposite direction to motion is positive.

There are two requirements for a work interaction between a system and its surroundings to exist: (1) there must be a *force* acting on the boundary, and (2) the boundary must *move*. Therefore, the presence of forces on the boundary without any displacement of the boundary does not constitute a work interaction. Likewise, the displacement of the boundary without any force to oppose or drive this motion (such as the expansion of a gas into an evacuated space) is not a work interaction since no energy is transferred.

In many thermodynamic problems, mechanical work is the only form of work involved. It is associated with the movement of the boundary of a system or with the movement of the entire system as a whole (Fig. 3-24). Some common forms of mechanical work are discussed below.

FIGURE 3-24
If there is no movement, no work is done.

1 Moving Boundary Work

One form of mechanical work frequently encountered in practice is associated with the expansion or compression of a gas in a piston-cylinder device. During this process, part of the boundary (the inner face of the piston) moves back and forth. Therefore, the expansion and compression work is often called **moving boundary work,** or simply **boundary work** (Fig. 3-25). Some call it the $P\ dV$ work for reasons explained below. Moving boundary work is the primary form of work involved in *automobile engines.* During their expansion, the combustion gases force the piston to move, which in turn forces the crankshaft to rotate.

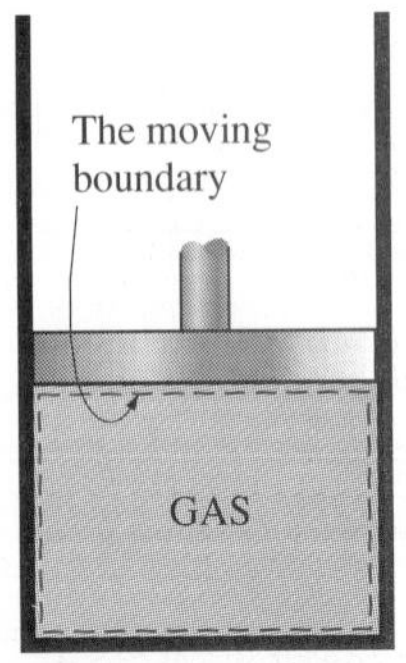

FIGURE 3-25
The work associated with a moving boundary is called *boundary work*.

The moving boundary work associated with real engines or compressors cannot be determined exactly from a thermodynamic analysis alone because the piston usually moves at very high speeds, making it difficult for the gas inside to maintain equilibrium. Then the states through which the system passes during the process cannot be specified, and no process path can be drawn. Work, being a path function, cannot be determined analytically without a knowledge of the path. Therefore, the boundary work in real engines or compressors is determined by direct measurements.

In this section, we analyze the moving boundary work for a *quasi-equilibrium process,* a process during which the system remains in equilibrium at all times. A quasi-equilibrium process, also called a *quasi-static process,* is closely approximated by real engines, especially when the piston moves at low velocities. Under identical conditions, the work output of the engines is found to be a maximum, and the work input to the compressors to be a minimum when quasi-equilibrium processes are used in place of non-quasi-equilibrium processes. Below, the work associated with a moving boundary is evaluated for a quasi-equilibrium process.

Consider the gas enclosed in the piston-cylinder device shown in Fig. 3-26. The initial pressure of the gas is P, the total volume is V, and the cross-sectional area of the piston is A. If the piston is allowed to move a distance ds in a quasi-equilibrium manner, the differential work done during this process is

$$\delta W_b = F\,ds = PA\,ds = P\,dV \tag{3-17}$$

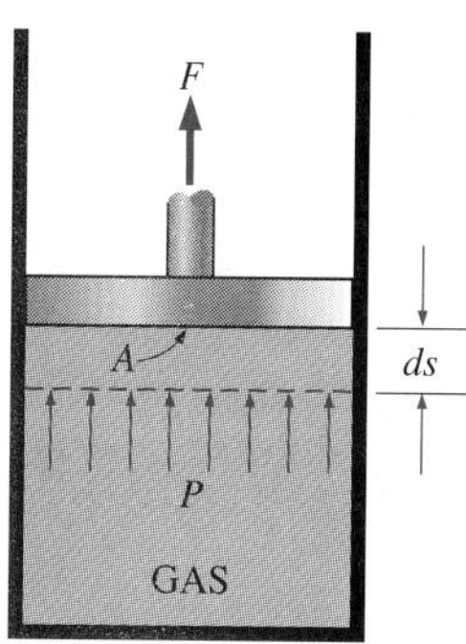

FIGURE 3-26
A gas does a differential amount of work δW_b as it forces the piston to move by a differential amount ds.

That is, the boundary work in the differential form is equal to the product of the absolute pressure P and the differential change in the volume dV of the system. This expression also explains why the moving boundary work is sometimes called the $P\,dV$ work.

Note in Eq. 3-17 that P is the absolute pressure, which is always positive. However, the volume change dV is positive during an expansion process (volume increasing) and negative during a compression process (volume decreasing). Thus, the boundary work is positive during an expansion process and negative during a compression process. Therefore, Eq. 3-17 can be viewed as an expression for boundary work output, $W_{b,\text{out}}$. A negative result indicates boundary work input (compression).

The total boundary work done during the entire process as the piston moves is obtained by adding all the differential works from the initial state to the final state:

$$W_b = \int_1^2 P\,dV \qquad \text{(kJ)} \tag{3-18}$$

This integral can be evaluated only if we know the functional relationship between P and V during the process. That is, $P = f(V)$ should be available. Note that $P = f(V)$ is simply the equation of the process path on a P-V diagram.

The quasi-equilibrium expansion process described above is shown on a P-V diagram in Fig. 3-27. On this diagram, the differential area dA is equal to $P\,dV$, which is the differential work. The total area A under the process curve 1-2 is obtained by adding these differential areas:

$$\text{Area} = A = \int_1^2 dA = \int_1^2 P\,dV$$

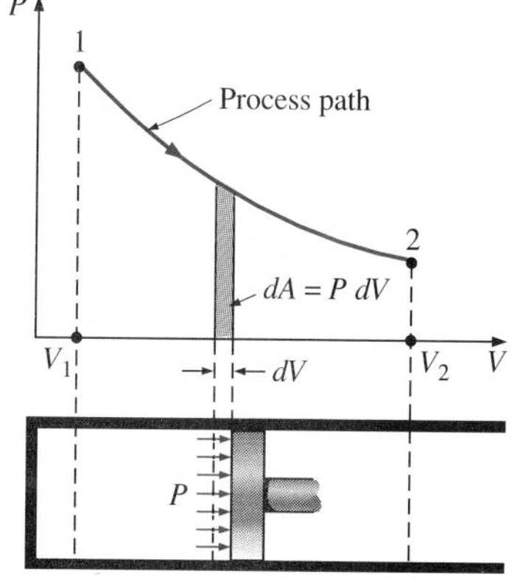

FIGURE 3-27
The area under the process curve on a P-V diagram represents the boundary work.

A comparison of this equation with Eq. 3-18 reveals that *the area under the process curve on a P-V diagram is equal, in magnitude, to the work done during a quasi-equilibrium expansion or compression process of a closed system*. (On the P-v diagram, it represents the boundary work done per unit mass.)

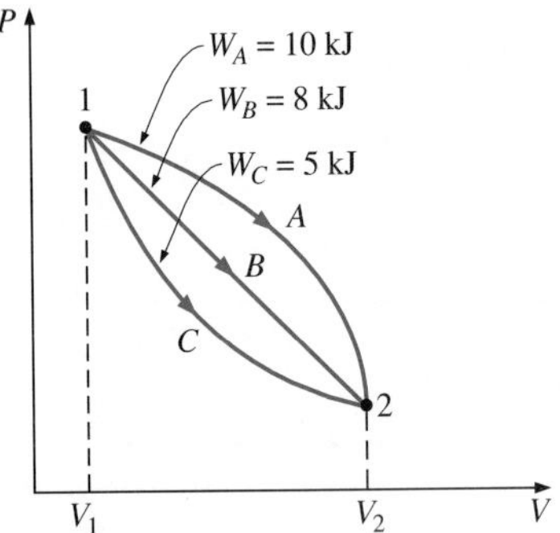

FIGURE 3-28
The boundary work done during a process depends on the path followed as well as the end states.

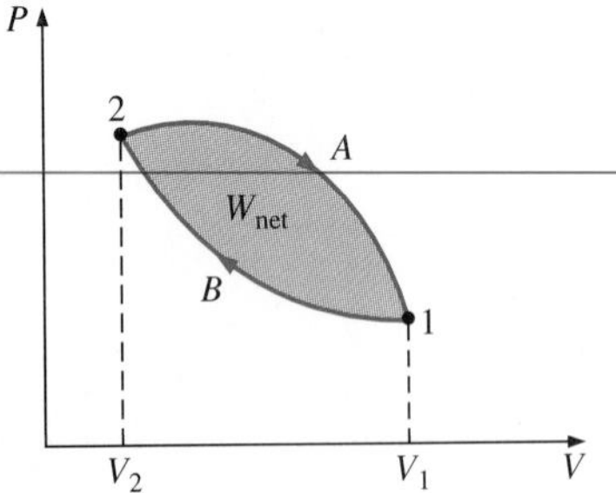

FIGURE 3-29
The net work done during a cycle is the difference between the work done by the system and the work done on the system.

A gas can follow several different paths as it expands from state 1 to state 2. In general, each path will have a different area underneath it, and since this area represents the magnitude of the work, the work done will be different for each process (Fig. 3-28). This is expected, since work is a path function (i.e., it depends on the path followed as well as the end states). If work were not a path function, no cyclic devices (car engines, power plants) could operate as work-producing devices. The work produced by these devices during one part of the cycle would have to be consumed during another part, and there would be no net work output. The cycle shown in Fig. 3-29 produces a net work output because the work done by the system during the expansion process (area under path *A*) is greater than the work done on the system during the compression part of the cycle (area under path *B*), and the difference between these two is the net work done during the cycle (the colored area).

If the relationship between *P* and *V* during an expansion or a compression process is given in terms of experimental data instead of in a functional form, obviously we cannot perform the integration analytically. But we can always plot the *P-V* diagram of the process, using these data points, and calculate the area underneath graphically to determine the work done.

Strictly speaking, the pressure *P* in Eq. 3-18 is the pressure at the inner surface of the piston. It becomes equal to the pressure of the gas in the cylinder only if the process is quasi-equilibrium and thus the entire gas in the cylinder is at the same pressure at any given time. Equation 3-18 can also be used for non-quasi-equilibrium processes provided that the pressure *at the inner face of the piston* is used for *P*. (Besides, we cannot speak of the pressure of a *system* during a non-quasi-equilibrium process since properties are defined for equilibrium states.) Therefore, we can generalize the boundary work relation by expressing it as

$$W_b = \int_1^2 P_i dV$$

where P_i is the pressure at the inner face of the piston.

Note that work is a mechanism for energy interaction between a system and its surroundings, and W_b represents the amount of energy transferred from the system during an expansion process (or to the system during a compression process). Therefore, it has to appear somewhere else and we must be able to account for it since energy is conserved. In a car engine, for example, the boundary work done by the expanding hot gases is used to overcome friction between the piston and the cylinder, to push atmospheric air out of the way, and to rotate the crankshaft. Therefore,

$$W_b = W_{\text{friction}} + W_{\text{atm}} + W_{\text{crank}} = \int_1^2 (F_{\text{friction}} + P_{\text{atm}}A + F_{\text{crank}})dx$$

Of course the work used to overcome friction will appear as frictional heat and the energy transmitted through the crankshaft will be transmitted to other components (such as the wheels) to perform certain functions. But note that the energy transferred by the system as work must equal the energy

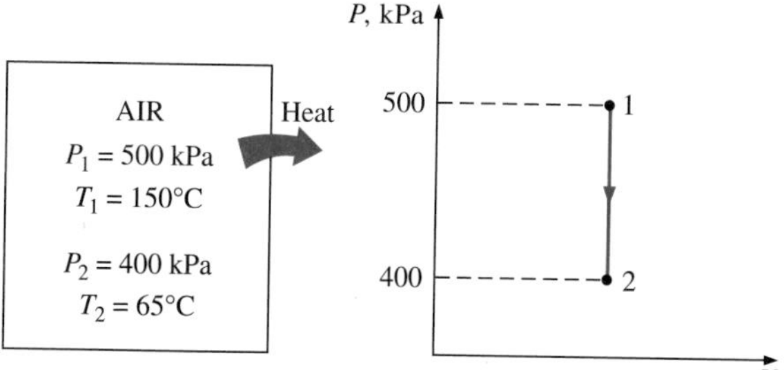

FIGURE 3-30

Schematic and *P*-*V* diagram for Example 3-7.

received by the crankshaft, the atmosphere, and the energy used to overcome friction.

The use of the boundary work relation (Eq. 3-18) is not limited to the quasi-equilibrium processes of gases only. It can also be used for solids and liquids.

EXAMPLE 3-7 Boundary Work during a Constant Volume Process

A rigid tank contains air at 500 kPa and 150°C. As a result of heat transfer to the surroundings, the temperature and pressure inside the tank drop to 65°C and 400 kPa, respectively. Determine the boundary work done during this process.

Solution A sketch of the system and the *P*-*V* diagram of the process are shown in Fig. 3-30.

Analysis The boundary work can be determined from Eq. 3-18 to be

$$W_b = \int_1^2 P \overset{0}{\cancel{dV}} = 0$$

This is expected since a rigid tank has a constant volume and $dV = 0$ in the above equation. Therefore, there is no boundary work done during this process. That is, the boundary work done during a constant-volume process is always zero. This is also evident from the *P*-*V* diagram of the process (the area under the process curve is zero).

EXAMPLE 3-8 Boundary Work during a Constant Pressure Process

A frictionless piston-cylinder device contains 10 lbm of water vapor at 60 psia and 320°F. Heat is now transferred to the steam until the temperature reaches 400°F. If the piston is not attached to a shaft and its mass is constant, determine the work done by the steam during this process.

Solution A sketch of the system and the *P*-*v* diagram of the process are shown in Fig. 3-31.

Assumption The expansion process is quasi-equilibrium.

Analysis Even though it is not explicitly stated, the pressure of the steam within the cylinder remains constant during this process since both the atmospheric pressure and the weight of the piston remain constant. Therefore, this is a constant-pressure process, and, from Eq. 3-18

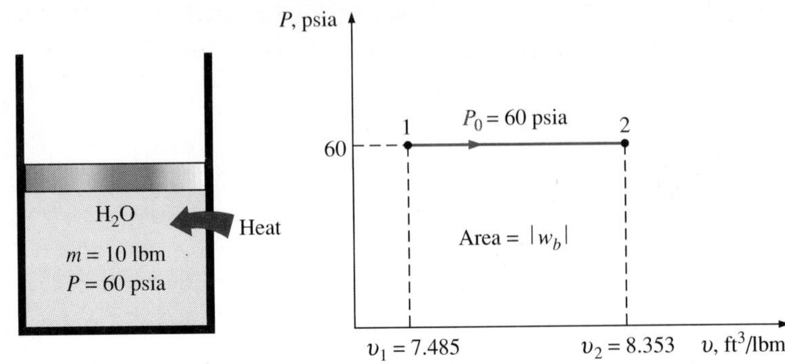

FIGURE 3-31
Schematic and *P-v* diagram for Example 3-8.

$$W_b = \int_1^2 P\,dV = P_0 \int_1^2 dV = P_0(V_2 - V_1)$$

or

$$W_b = mP_0(v_2 - v_1) \tag{3-19}$$

since $V = mv$. From the superheated vapor table (Table A-6E), the specific volumes are determined to be $v_1 = 7.485$ ft^3/lbm at state 1 (60 psia, 320°F) and $v_2 = 8.353$ ft^3/lbm at state 2 (60 psia, 400°F). Substituting these values yields

$$W_b = (10\text{ lbm})(60\text{ psia})[(8.353 - 7.485)\text{ ft}^3\text{/lbm}]\left(\frac{1\text{ Btu}}{5.404\text{ psia}\cdot\text{ft}^3}\right)$$
$$= \mathbf{96.4\ Btu}$$

Discussion The positive sign indicates that the work is done by the system. That is, the steam used 96.4 Btu of its energy to do this work. The magnitude of this work could also be determined by calculating the area under the process curve on the *P-V* diagram, which is $P_0\,\Delta V$ for this case.

EXAMPLE 3-9 Boundary Work during an Isothermal Process

A piston-cylinder device initially contains 0.4 m^3 of air at 100 kPa and 80°C. The air is now compressed to 0.1 m^3 in such a way that the temperature inside the cylinder remains constant. Determine the work done during this process.

Solution A sketch of the system and the *P-V* diagram of the process are shown in Fig. 3-32.

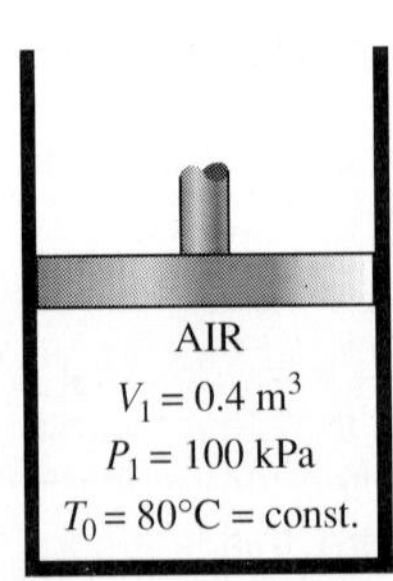

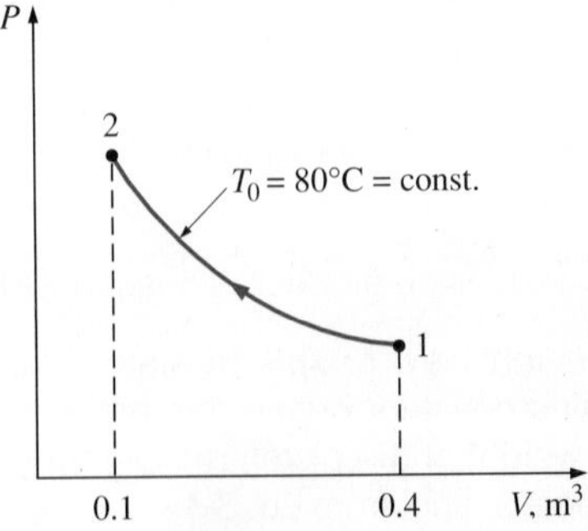

FIGURE 3-32
Schematic and *P-V* diagram for Example 3-9.

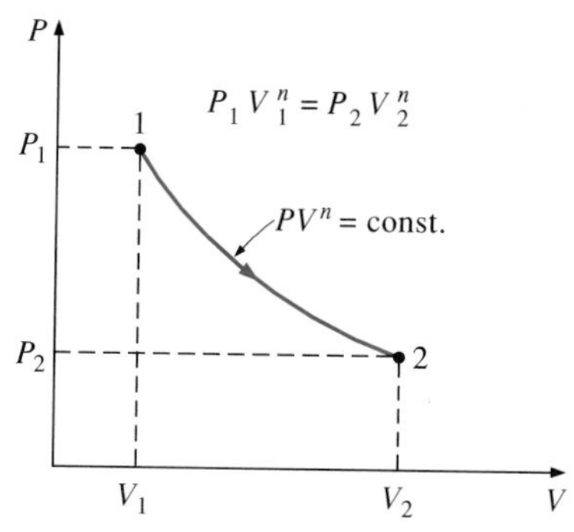

FIGURE 3-33
Schematic and *P-V* diagram for a polytropic process.

Assumptions **1** The compression process is quasi-equilibrium. **2** At the specified conditions, air can be considered to be an ideal gas since it is at a high temperature and low pressure relative to its critical-point values.

Analysis For an ideal gas at constant temperature T_0,

$$PV = mRT_0 = C \quad \text{or} \quad P = \frac{C}{V}$$

where C is a constant. Substituting this into Eq. 3-18, we have

$$W_b = \int_1^2 P\,dV = \int_1^2 \frac{C}{V}\,dV = C\int_1^2 \frac{dV}{V} = C\ln\frac{V_2}{V_1} = P_1V_1\ln\frac{V_2}{V_1} \qquad (3\text{-}20)$$

In the above equation, P_1V_1 can be replaced by P_2V_2 or mRT_0. Also, V_2/V_1 can be replaced by P_1/P_2 for this case since $P_1V_1 = P_2V_2$.

Substituting the numerical values into the above equation yields

$$W_b = (100\text{ kPa})(0.4\text{ m}^3)\left(\ln\frac{0.1}{0.4}\right)\left(\frac{1\text{ kJ}}{1\text{ kPa}\cdot\text{m}^3}\right)$$

$$= \mathbf{-55.45\ kJ}$$

Discussion The negative sign indicates that this work is done on the system, which is always the case for compression processes.

Polytropic Process

During expansion and compression processes of real gases, pressure and volume are often related by $PV^n = C$, where n and C are constants. A process of this kind is called a polytropic process (Fig. 3-33). Below we develop a general expression for the work done during a polytropic process. The pressure for a polytropic process can be expressed as

$$P = CV^{-n} \qquad (3\text{-}21)$$

Substituting this relation into Eq. 3-18, we obtain

$$W_b = \int_1^2 P\,dV = \int_1^2 CV^{-n}\,dV = C\frac{V_2^{-n+1} - V_1^{-n+1}}{-n+1} = \frac{P_2V_2 - P_1V_1}{1-n} \qquad (3\text{-}22)$$

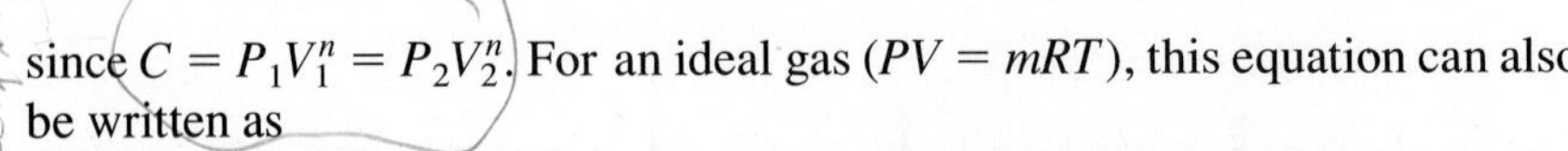

since $C = P_1V_1^n = P_2V_2^n$. For an ideal gas ($PV = mRT$), this equation can also be written as

$$W_b = \frac{mR(T_2 - T_1)}{1 - n}, n \neq 1 \qquad \text{(kJ)} \qquad (3\text{-}23)$$

The special case of $n = 1$ is equivalent to the isothermal process discussed in the previous example.

2 Gravitational Work

Gravitational work can be defined as the work done by or against a gravitational force field. In a gravitational field, the force acting on a body is

$$F = mg$$

where m is the mass of the body and g is the acceleration of gravity, which is assumed to be constant. Then the work required to raise this body from level z_1 to level z_2 is

$$W_g = \int_1^2 F\,dz = mg\int_1^2 dz = mg(z_2 - z_1) \qquad \text{(kJ)} \qquad (3\text{-}24)$$

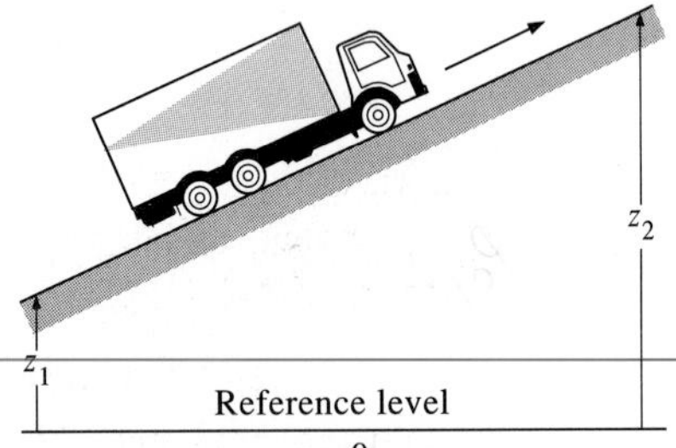

FIGURE 3-34
Vehicles require more power (gravitational work per unit time) as they climb a hill.

where $z_2 - z_1$ is the vertical distance traveled (Fig. 3-34). This expression is easily recognized as the *change in potential energy*. We conclude from Eq. 3-24 that the gravitational work depends on the end states only and is independent of the path followed. Also, the work done is equal, in magnitude, to the change in the potential energy of the system.

EXAMPLE 3-10 Power Needs of a Car to Climb a Hill

Consider a 1200-kg car cruising steadily on a level road at 90 km/h. Now the car starts climbing a hill that is sloped 30° from the horizontal (Fig. 3-35). If the velocity of the car is not to drop during climbing, determine the additional power that must be delivered by the engine.

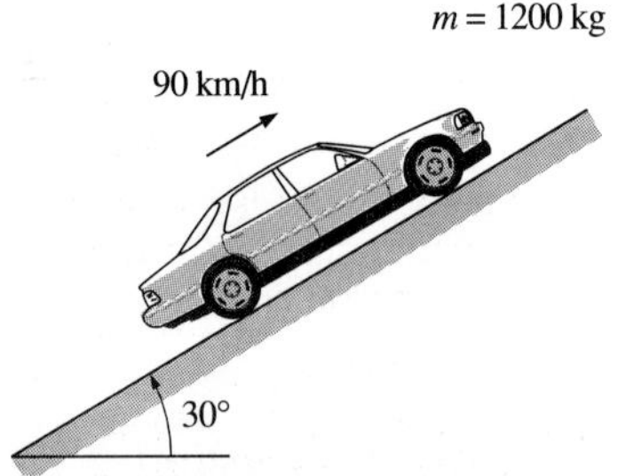

FIGURE 3-35
Schematic for Example 3-10.

Solution The additional power required is simply the gravitational work that needs to be done per unit time to raise the elevation of the car, which is equal to the change in the potential energy of the car per unit time:

$$\begin{aligned}\dot{W}_g &= mg\Delta z/\Delta t = mg\mathcal{V}_{\text{vertical}} \\ &= (1200\text{ kg})(9.8\text{ m/s}^2)(90\text{ km/h})(\sin 30°)\left(\frac{1\text{ m/s}}{3.6\text{ km/h}}\right)\left(\frac{1\text{ kJ/kg}}{1000\text{ m}^2/\text{s}^2}\right) \\ &= 147\text{ kJ/s} = \mathbf{147\text{ kW}} \qquad \text{(or 197 hp)}\end{aligned}$$

Therefore, the car engine will have to produce almost 200 hp of additional power while climbing the hill if the car is to maintain its velocity.

The work associated with a change in velocity of a system is called **accelerational work.** The accelerational work required to accelerate a body of mass m from an initial velocity of $\mathcal{V}_1$ to a final velocity of $\mathcal{V}_2$ (Fig. 3-36) is determined from the definition of acceleration and Newton's second law:

$$\left.\begin{aligned} F &= ma \\ a &= \frac{d\mathcal{V}}{dt} \end{aligned}\right\} \qquad F = m\frac{d\mathcal{V}}{dt}$$

The differential displacement ds is related to velocity $\mathcal{V}$ by

$$\mathcal{V} = \frac{ds}{dt} \longrightarrow ds = \mathcal{V}\,dt$$

Substituting the F and ds relations into the work expression (Eq. 3-16), we obtain

$$W_a = \int_1^2 F\,ds = \int_1^2 \left(m\frac{d\mathcal{V}}{dt}\right)(\mathcal{V}\,dt) = m\int_1^2 \mathcal{V}\,d\mathcal{V} = \tfrac{1}{2}m(\mathcal{V}_2^2 - \mathcal{V}_1^2) \qquad (3\text{-}25)$$

The work done to accelerate a body is independent of the path followed and is equivalent to the *change in the kinetic energy* of the body.

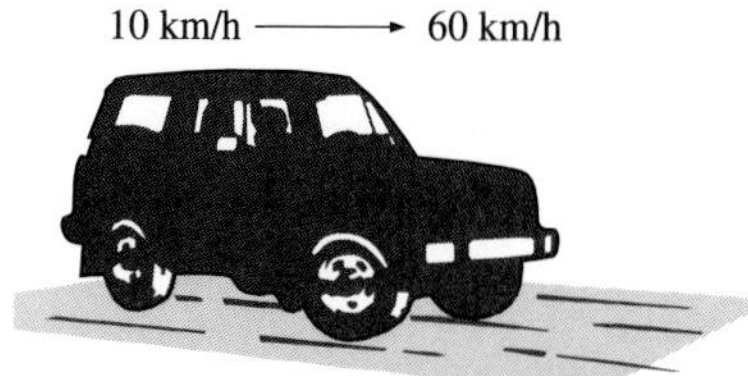

FIGURE 3-36
Vehicles require more power (accelerational work per unit time) as they accelerate.

EXAMPLE 3-11 Power Needs of a Car to Accelerate

Determine the power required to accelerate a 900-kg car shown in Fig. 3-37 from rest to a velocity of 80 km/h in 20 s on a level road.

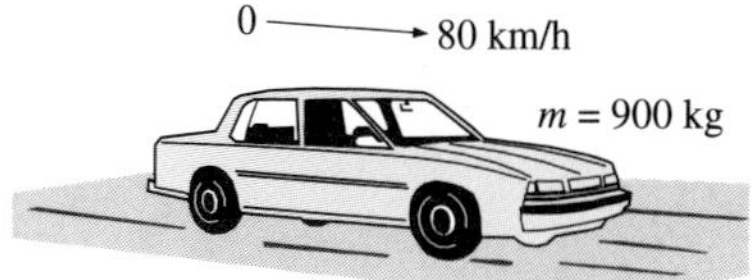

FIGURE 3-37
Schematic for Example 3-11.

Solution The accelerational work is determined from Eq. 3-25 to be

$$W_a = \tfrac{1}{2}m(\mathcal{V}_2^2 - \mathcal{V}_1^2) = \tfrac{1}{2}(900\text{ kg})\left[\left(\frac{80{,}000\text{ m}}{3600\text{ s}}\right)^2 - 0^2\right]\left(\frac{1\text{ kJ/kg}}{1000\text{ m}^2/\text{s}^2}\right)$$
$$= 222.2\text{ kJ}$$

The average power is determined from

$$\dot{W}_a = \frac{W_a}{\Delta t} = \frac{222.2\text{ kJ}}{20\text{ s}} = \mathbf{11.1\ kW}\text{ (or 14.9 hp)}$$

This is in addition to the power required to combat friction, rolling resistance, and other imperfections.

4 Shaft Work

Energy transmission with a rotating shaft is very common in engineering practice (Fig. 3-38). Often the torque τ applied to the shaft is constant, which means that the force F applied is also constant. For a specified constant torque, the work done during n revolutions is determined as follows: A force

FIGURE 3-38
Energy transmission through rotating shafts is commonly encountered in practice.

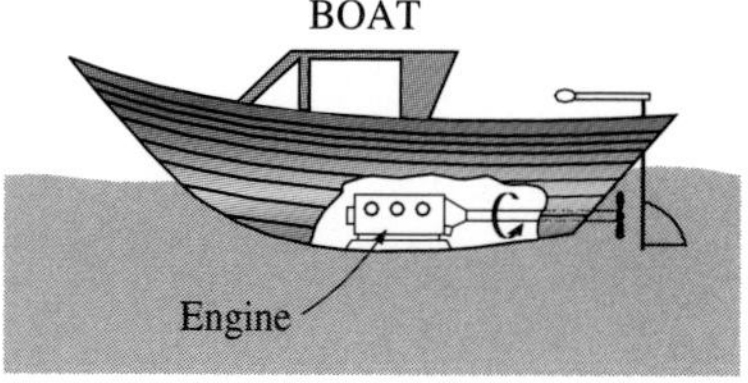

F acting through a moment arm r generates a torque τ of (Fig. 3-39)

$$\tau = Fr \longrightarrow F = \frac{\tau}{r}$$

This force acts through a distance s, which is related to the radius r by

$$s = (2\pi r)n$$

Then the shaft work is determined from Eq. 3-15:

$$W_{sh} = Fs = \left(\frac{\tau}{r}\right)(2\pi rn) = 2\pi n\tau \qquad \text{(kJ)} \qquad (3\text{-}26)$$

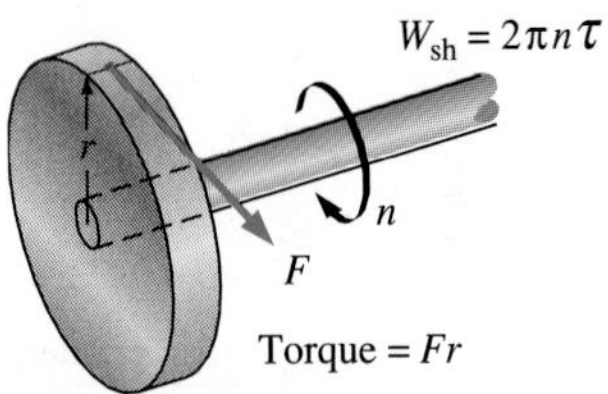

FIGURE 3-39

Shaft work is proportional to the torque applied and the number of revolutions of the shaft.

The power transmitted through the shaft is the shaft work done per unit time, which can be expressed as

$$\dot{W}_{sh} = 2\pi\dot{n}\tau \qquad \text{(kW)} \qquad (3\text{-}27)$$

where $\dot{n}$ is the number of revolutions per unit time.

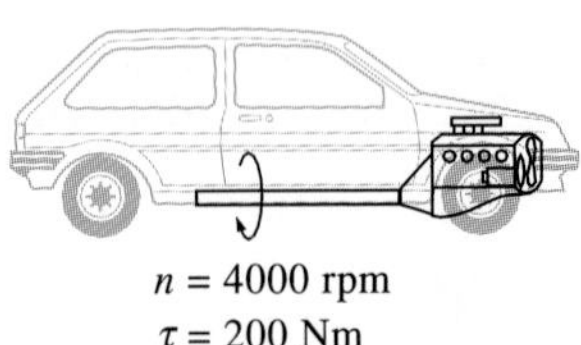

FIGURE 3-40

Schematic for Example 3-12.

EXAMPLE 3-12 Power Transmission by the Shaft of a Car

Determine the power transmitted through the shaft of a car when the torque applied is 200 N · m and the shaft rotates at a rate of 4000 revolutions per minute (rpm).

Solution A sketch of the car is given in Fig. 3-40. The shaft power is determined from Eq. 3-27:

$$\dot{W}_{sh} = 2\pi\dot{n}\tau = (2\pi)\left(4000\,\frac{1}{\text{min}}\right)(200\text{ N}\cdot\text{m})\left(\frac{1\text{ min}}{60\text{ s}}\right)\left(\frac{1\text{ kJ}}{1000\text{ N}\cdot\text{m}}\right)$$
$$= \mathbf{83.7\ kW} \text{ (or 112.2 hp)}$$

5 Spring Work

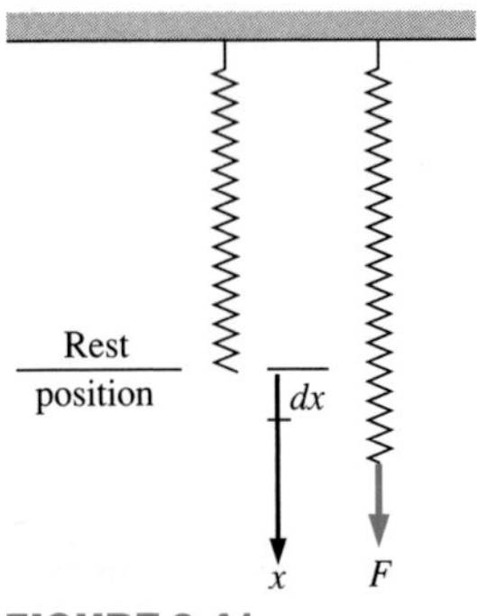

FIGURE 3-41

Elongation of a spring under the influence of a force.

It is common knowledge that when a force is applied on a spring, the length of the spring changes (Fig. 3-41). When the length of the spring changes by a differential amount dx under the influence of a force F, the work done is

$$\delta W_{spring} = F\,dx \qquad (3\text{-}28)$$

To determine the total spring work, we need to know a functional relationship between F and x. For linear elastic springs, the displacement x is proportional to the force applied (Fig. 3-42). That is,

$$F = kx \qquad \text{(kN)} \qquad (3\text{-}29)$$

where k is the spring constant and has the unit kN/m. The displacement x is measured from the undisturbed position of the spring (that is, $x = 0$ when $F = 0$). Substituting Eq. 3-29 into Eq. 3-28 and integrating yield

$$W_{spring} = \tfrac{1}{2}k(x_2^2 - x_1^2) \qquad \text{(kJ)} \qquad (3\text{-}30)$$

where x_1 and x_2 are the initial and the final displacements of the spring, respectively, measured from the undisturbed position of the spring.

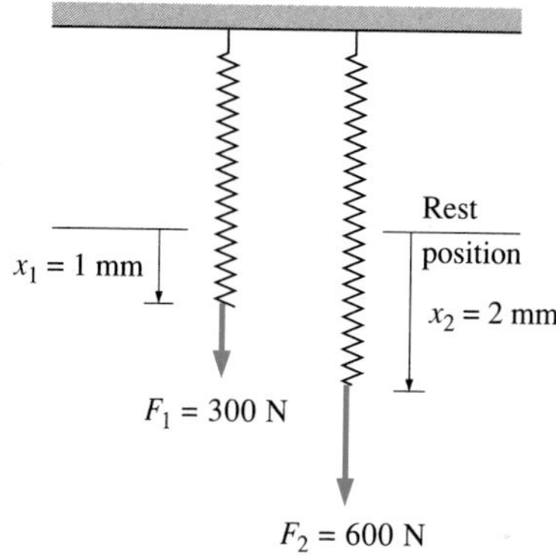

FIGURE 3-42

The displacement of a linear spring doubles when the force is doubled.

EXAMPLE 3-13 Expansion of a Gas against a Spring

A piston-cylinder device contains 0.05 m^3 of a gas initially at 200 kPa. At this state, a linear spring that has a spring constant of 150 kN/m is touching the piston but exerting no force on it. Now heat is transferred to the gas, causing the piston to rise and to compress the spring until the volume inside the cylinder doubles. If the cross-sectional area of the piston is 0.25 m^2, determine (*a*) the final pressure inside the cylinder, (*b*) the total work done by the gas, and (*c*) the fraction of this work done against the spring to compress it.

Solution A sketch of the system and the *P-V* diagram of the process are shown in Fig. 3-43.

Assumptions **1** The expansion process is quasi-equilibrium. **2** The spring is linear in the range of interest.

Analysis (*a*) The enclosed volume at the final state is

$$V_2 = 2V_1 = (2)(0.05 \text{ m}^3) = 0.1 \text{ m}^3$$

Then the displacement of the piston (and of the spring) becomes

$$x = \frac{\Delta V}{A} = \frac{(0.1 - 0.05) \text{ m}^3}{0.25 \text{ m}^2} = 0.2 \text{ m}$$

The force applied by the linear spring at the final state is determined from Eq. 3-29 to be

$$F = kx = (150 \text{ kN/m})(0.2 \text{ m}) = 30 \text{ kN}$$

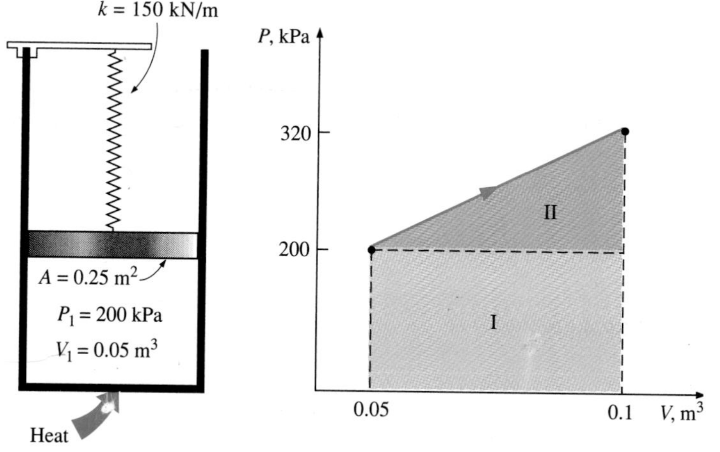

FIGURE 3-43

Schematic and *P-V* diagram for Example 3-13.

The additional pressure applied by the spring on the gas at this state is

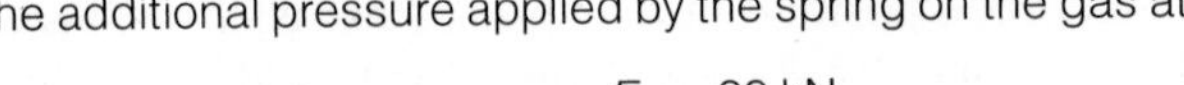

$$P = \frac{F}{A} = \frac{30 \text{ kN}}{0.25 \text{ m}^2} = 120 \text{ kPa}$$

Without the spring, the pressure of the gas would remain constant at 200 kPa while the piston is rising. But under the effect of the spring, the pressure rises linearly from 200 kPa to

$$200 + 120 = \mathbf{320 \text{ kPa}}$$

at the final state.

(*b*) An easy way of finding the work done is to plot the process on a *P-V* diagram and find the area under the process curve. From Fig. 3-43 the area under the process curve (a trapezoid) is determined to be

$$W = \text{area} = \frac{(200 + 320) \text{ kPa}}{2}[(0.1 - 0.05) \text{ m}^3]\left(\frac{1 \text{ kJ}}{1 \text{ kPa} \cdot \text{m}^3}\right) = \mathbf{13 \text{ kJ}}$$

Note that the work is done by the system.

(*c*) The work represented by the rectangular area (region I) is done against the piston and the atmosphere, and the work represented by the triangular area (region II) is done against the spring. Thus,

$$W_{\text{spring}} = \tfrac{1}{2}[(320 - 200) \text{ kPa}](0.05 \text{ m}^3)\left(\frac{1 \text{ kJ}}{1 \text{ kPa} \cdot \text{m}^3}\right) = \mathbf{3 \text{ kJ}}$$

This result could also be obtained from Eq. 3-30:

$$W_{\text{spring}} = \tfrac{1}{2}k(x_2^2 - x_1^2) = \tfrac{1}{2}(150 \text{ kN/m})[(0.2 \text{ m})^2 - 0^2]\left(\frac{1 \text{ kJ}}{1 \text{ kN} \cdot \text{m}}\right) = 3 \text{ kJ}$$

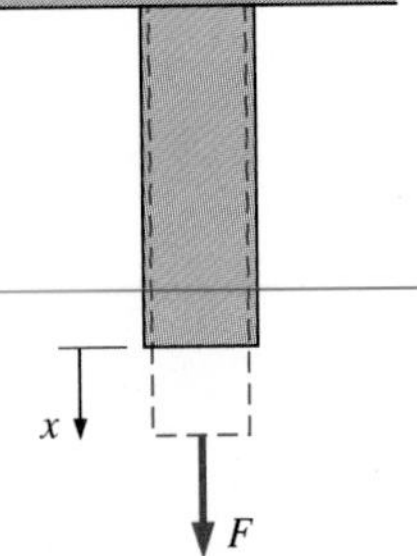

FIGURE 3-44
Solid bars behave as springs under the influence of a force.

Work Done on Elastic Solid Bars

Solids are often modeled as linear springs because under the action of a force they contract or elongate, as shown in Fig. 3-44, and when the force is lifted, they return to their original lengths, like a spring. This is true as long as the force is in the elastic range, that is, not large enough to cause permanent (plastic) deformations. Therefore, the equations given for a linear spring can also be used for elastic solid bars. Alternately, we can determine the work associated with the expansion or contraction of an elastic solid bar by replacing pressure *P* by its counterpart in solids, *normal stress* $\sigma_n = F/A$, in the boundary work expression:

$$W_{\text{elastic}} \int_1^2 \sigma_n dV = \int_1^2 \sigma_n A \, dx \qquad \text{(kJ)} \qquad (3\text{-}31)$$

where *A* is the cross-sectional area of the bar. Note that the normal stress has pressure units.

FIGURE 3-45
Stretching a liquid film with a movable wire.

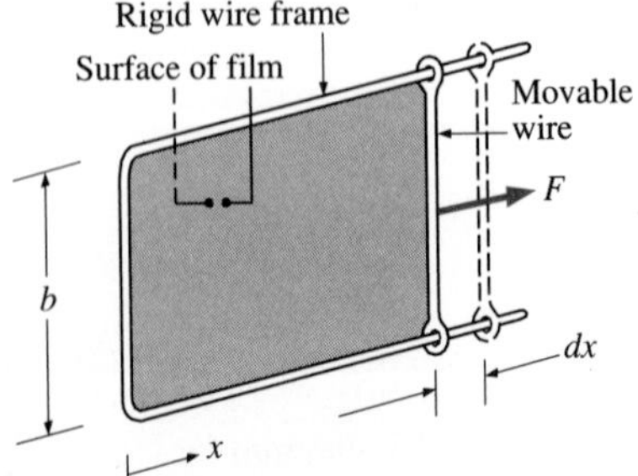

Work Associated with the Stretching of a Liquid Film

Consider a liquid film such as soap film suspended on a wire frame (Fig. 3-45). We know from experience that it will take some force to stretch this film by the movable portion of the wire frame. This force is used to overcome the microscopic forces between molecules at the liquid-air interfaces. These microscopic forces are perpendicular to any line in the surface, and the force generated by these forces per unit length is called the **surface tension** σ_s, whose unit is N/m. Therefore, the work associated with the stretching of a film is also called *surface tension work*. It is determined from

$$W_{\text{surface}} = \int_1^2 \sigma_s dA \qquad \text{(kJ)} \qquad (3\text{-}32)$$

where $dA = 2b\,dx$ is the change in the surface area of the film. The factor 2 is due to the fact that the film has two surfaces in contact with air. The force acting on the movable wire as a result of surface tension effects is $F = 2b\sigma_s$ where σ_s is the surface tension force per unit length.

6 Nonmechanical Forms of Work

The treatment above represents a fairly comprehensive coverage of mechanical forms of work. But some work modes encountered in practice are not mechanical in nature. However, these nonmechanical work modes can be treated in a similar manner by identifying a *generalized force F* acting in the direction of a *generalized displacement x*. Then the work associated with the differential displacement under the influence of this force is determined from $\delta W = F \cdot dx$.

Some examples of nonmechanical work modes are **electrical work,** where the generalized force is the *voltage* (the electrical potential) and the generalized displacement is the *electrical charge* as discussed in the last section; **magnetic work,** where the generalized force is the *magnetic field strength* and the generalized displacement is the total *magnetic dipole moment*; and **electrical polarization work,** where the generalized force is the *electric field strength* and the generalized displacement is the *polarization of the medium* (the sum of the electric dipole rotation moments of the molecules). Detailed consideration of these and other nonmechanical work modes can be found in specialized books on these topics.

3-5 ■ THE FIRST LAW OF THERMODYNAMICS

So far, we have considered various forms of energy such as heat Q, work W, and total energy E individually, and no attempt has been made to relate them to each other during a process. The *first law of thermodynamics,* also known as *the conservation of energy principle,* provides a sound basis for studying the relationships among the various forms of energy and energy interactions. Based on experimental observations, the first law of thermodynamics states that *energy can be neither created nor destroyed; it can only change forms.* Therefore, every bit of energy should be accounted for during a process. The first law cannot be proved mathematically, but no process in nature is known to have violated the first law, and this should be taken as sufficient proof.

We all know that a rock at some elevation possesses some potential energy, and part of this potential energy is converted to kinetic energy as the rock falls (Fig. 3-46). Experimental data show that the decrease in potential energy ($mg\ \Delta z$) exactly equals the increase in kinetic energy $[m(\mathcal{V}_2^2 - \mathcal{V}_1^2)/2]$ when the air resistance is negligible, thus confirming the conservation of energy principle.

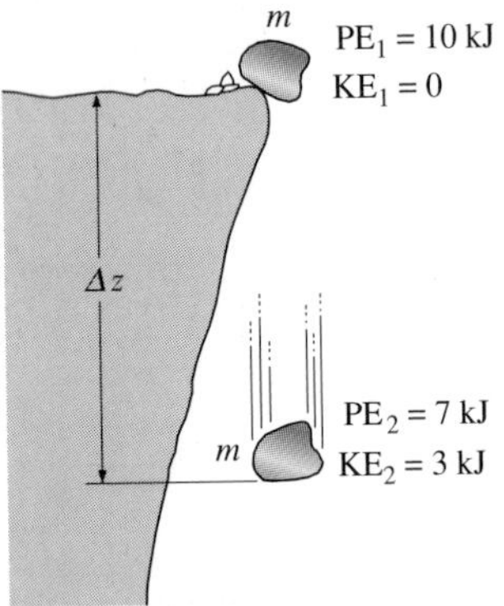

FIGURE 3-46
Energy cannot be created or destroyed; it can only change forms.

Consider a system undergoing a series of *adiabatic* processes from a specified state 1 to another specified state 2. Being adiabatic, these processes obviously cannot involve any heat transfer, but they may involve several kinds of work interactions. Careful measurements during these experiments indicate the following: *For all adiabatic processes between two specified states of a closed system, the net work done is the same regardless of the nature of the closed system and the details of the process.* Considering that there are an infinite number of ways to perform work interactions under adiabatic conditions, the statement above appears to be very powerful, with a potential for far-reaching implications. This statement, which is largely based on the experiments of Joule in the first half of the nineteenth century, cannot be drawn from any other known physical principle and is recognized as a fundamental principle. This principle is called the **first law of thermodynamics** or just the **first law.**

A major consequence of the first law is the existence and the definition of the property *total energy E*. Considering that the net work is the same for all adiabatic processes of a closed system between two specified states, the value of the net work must depend on the end states of the system only, and thus it must correspond to a change in a property of the system. This property is the *total energy*. Note that the first law makes no reference to the value of the total energy of a closed system at a state. It simply states that the *change* in the total energy during an adiabatic process must be equal to the net work done. Therefore, any convenient arbitrary value can be assigned to total energy at a specified state to serve as a reference point.

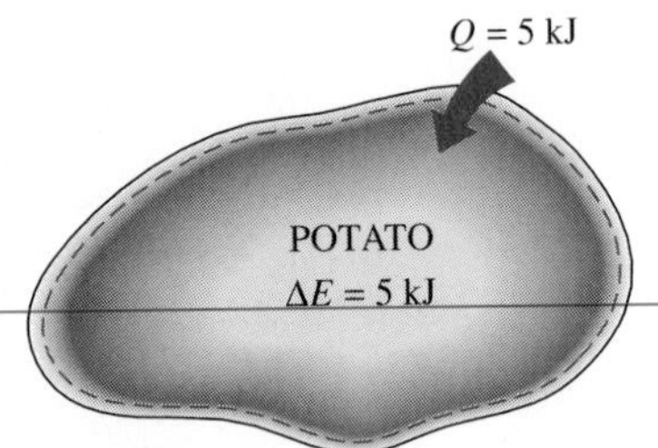

FIGURE 3-47
The increase in the energy of a potato in an oven is equal to the amount of heat transferred to it.

Implicit in the first law statement is the conservation of energy. Although the essence of the first law is the existence of the property *total energy*, the first law is often viewed as a statement of the *conservation of energy* principle. Below we develop the first law or the conservation of energy relation for closed systems with the help of some familiar examples using intuitive arguments.

First, we consider some processes that involve heat transfer but no work interactions. The potato baked in the oven is a good example for this case (Fig. 3-47). As a result of heat transfer to the potato, the energy of the potato will increase. If we disregard any mass transfer (moisture loss from the potato), the increase in the total energy of the potato becomes equal to the amount of heat transfer. That is, if 5 kJ of heat is transferred to the potato, the energy increase of the potato will also be 5 kJ.

FIGURE 3-48
In the absence of any work interactions, energy change of a system is equal to the net heat transfer.

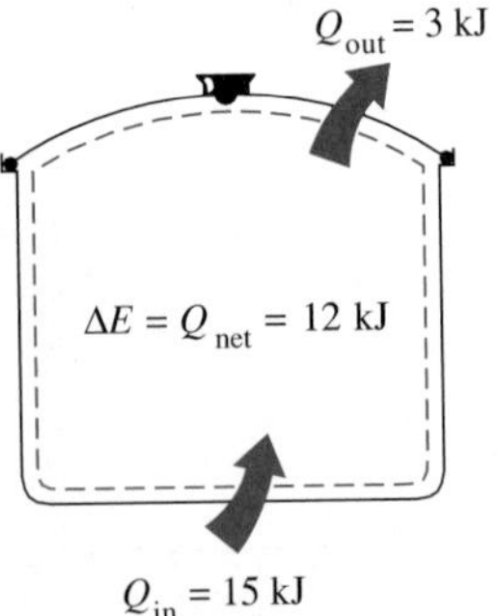

As another example, consider the heating of water in a pan on top of a range (Fig. 3-48). If 15 kJ of heat is transferred to the water from the heating element and 3 kJ of it is lost from the water to the surrounding air, the increase in energy of the water will be equal to the net heat transfer to water, which is 12 kJ.

Now consider a well-insulated (i.e., adiabatic) room heated by an electric heater as our system (Fig. 3-49). As a result of electrical work done, the energy of the system will increase. Since the system is adiabatic and cannot have any heat transfer to or from the surroundings ($Q = 0$), the conservation of energy principle dictates that the electrical work done on the system must equal the increase in energy of the system.

Next, let us replace the electric heater with a paddle wheel (Fig. 3-50). As a result of the stirring process, the energy of the system will increase. Again, since there is no heat interaction between the system and its surroundings ($Q = 0$), the paddle-wheel work done on the system must show up as an increase in the energy of the system.

Many of you have probably noticed that the temperature of air rises when it is compressed (Fig. 3-51). This is because energy is transferred to the air in the form of boundary work. In the absence of any heat transfer ($Q = 0$), the entire boundary work will be stored in the air as part of its total energy. The conservation of energy principle again requires that the increase in the energy of the system be equal to the boundary work done on the system.

We can extend the discussions above to systems that involve various heat and work interactions simultaneously. For example, if a system gains 12 kJ of heat during a process while 6 kJ of work is done on it, the increase in the energy of the system during that process is 18 kJ (Fig. 3-52). That is, the change in the energy of a system during a process is simply equal to the net energy transfer to (or from) the system.

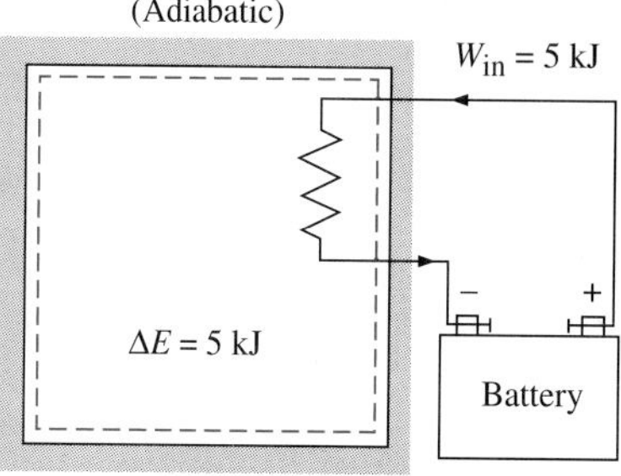

FIGURE 3-49
The work (electrical) done on an adiabatic system is equal to the increase in the energy of the system.

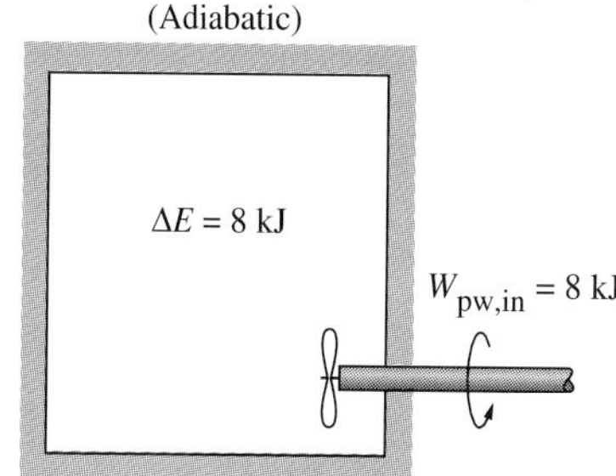

FIGURE 3-50
The work (shaft) done on an adiabatic system is equal to the increase in the energy of the system.

Energy Balance

In the light of the discussions above, the conservation of energy principle may be expressed as follows: *The net change (increase or decrease) in the total energy of the system during a process is equal to the difference between the total energy entering and the total energy leaving the system during that process.* That is, during a process,

$$\begin{pmatrix}\text{Total energy}\\\text{entering the system}\end{pmatrix} - \begin{pmatrix}\text{Total energy}\\\text{leaving the system}\end{pmatrix} = \begin{pmatrix}\text{Change in the total}\\\text{energy of the system}\end{pmatrix}$$

or

$$E_{in} - E_{out} = \Delta E_{system}$$

This relation is often referred to as the **energy balance** and is applicable to any kind of system undergoing any kind of process. The successful use of this relation to solve engineering problems depends on understanding the various forms of energy and recognizing the forms of energy transfer.

FIGURE 3-51
The work (boundary) done on an adiabatic system is equal to the increase in the energy of the system.

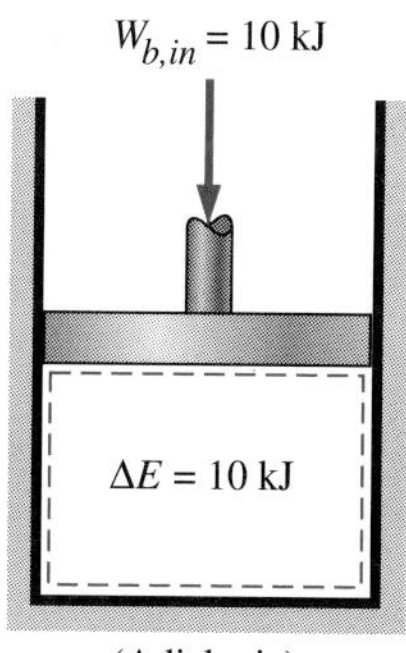

Energy Change of a System, ΔE_{system}

The determination of the energy change of a system during a process involves the evaluation of the energy of the system at the beginning and at the end of the process, and taking their difference. That is,

$$\text{Energy change} = \text{Energy at final state} - \text{Energy at initial state}$$

or

$$\Delta E_{\text{system}} = E_{\text{final}} - E_{\text{initial}} = E_2 - E_1 \quad (3\text{-}33)$$

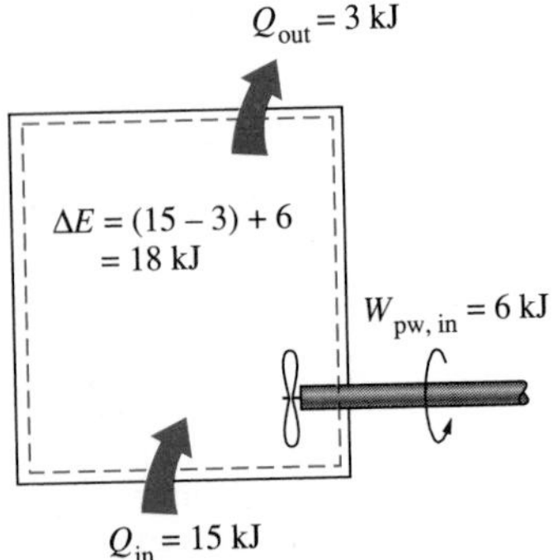

FIGURE 3-52

The energy change of a system during a process is equal to the *net* work and heat transfer between the system and its surroundings.

Note that energy is a property, and the value of a property does not change unless the state of the system changes. Therefore, the energy change of a system is zero if the state of the system does not change during the process. Also, energy can exist in numerous forms such as internal (sensible, latent, chemical, and nuclear), kinetic, potential, electrical, and magnetic, and their sum constitutes the *total energy E* of a system. In the absence of electric, magnetic, and surface tension effects (i.e., for simple compressible systems), the change in the total energy of a system during a process is the sum of the changes in its internal, kinetic, and potential energies and can be expressed as

$$\Delta E = \Delta U + \Delta \mathrm{KE} + \Delta \mathrm{PE} \quad (3\text{-}34)$$

where

$$\Delta U = m(u_2 - u_1)$$

$$\Delta \mathrm{KE} = \tfrac{1}{2}m(\mathcal{V}_2^2 - \mathcal{V}_1^2)$$

$$\Delta \mathrm{PE} = mg(z_2 - z_1)$$

When the initial and final states are specified, the values of the specific internal energies u_1 and u_2 can be determined directly from the property tables or thermodynamic property relations.

Most systems encountered in practice are stationary, that is, they do not involve any changes in their velocity or elevation during a process (Fig. 3-53). Thus, for **stationary systems,** the changes in kinetic and potential energies are zero (that is, $\Delta \mathrm{KE} = \Delta \mathrm{PE} = 0$), and the total energy change relation above reduces to $\Delta E = \Delta U$ for such systems. Also, the energy of a system during a process will change even if only one form of its energy changes while the other forms of energy remain unchanged.

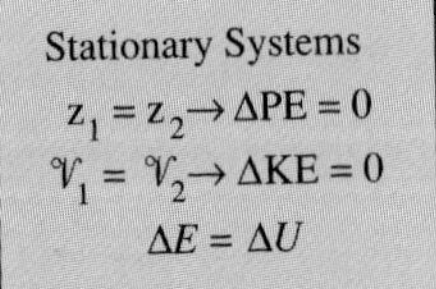

FIGURE 3-53

For stationary systems. $\Delta \mathrm{KE} = \Delta \mathrm{PE} = 0$; thus $\Delta E = \Delta U$.

Mechanisms of Energy Transfer, E_{in} and E_{out}

Energy can be transferred to or from a system in three forms: *heat, work,* and *mass flow*. Energy interactions are recognized at the system boundary as they cross it, and they represent the energy gained or lost by a system during a process. The only two forms of energy interactions associated with a fixed mass or closed system are *heat transfer* and *work*.

1. **Heat Transfer,** Q Heat transfer to a system (heat gain) increases the energy of the molecules and thus the internal energy of the system, and heat transfer from a system (heat loss) decreases it since the energy transferred out as heat comes from the energy of the molecules of the system.

2. **Work,** W An energy interaction that is not caused by a temperature difference between a system and its surroundings is work. A rising piston, a rotating shaft, and an electrical wire crossing the system boundaries are all associated with work interactions. Work transfer to a system (i.e., work done

on a system) increases the energy of the system, and work transfer from a system (i.e., work done by the system) decreases it since the energy transferred out as work comes from the energy contained in the system. Car engines, hydraulic, steam, or gas turbines produce work while compressors, pumps, and mixers consume work.

3. **Mass Flow, m** Mass flow in and out of the system serves as an additional mechanism of energy transfer. When mass enters a system, the energy of the system increases because mass carries energy with it (in fact, mass is energy). Likewise, when some mass leaves the system, the energy contained within the system decreases because the leaving mass takes out some energy with it. Systems that involve energy balance with mass flow are considered in detail in Chap. 4.

Noting that energy can be transferred in the forms of heat, work, and mass, and that the net transfer of a quantity is equal to the difference between the amounts transferred in and out, the energy balance can be written more explicitly as

$$E_{in} - E_{out} = (Q_{in} - Q_{out}) + (W_{in} - W_{out}) + (E_{mass,\,in} - E_{mass,\,out}) = \Delta E_{system} \tag{3-35}$$

where the subscripts "in" and "out" denote quantities that enter and leave the system, respectively. All six quantities on the right side of the equation represent "amounts," and thus they are *positive* quantities. The direction of any energy transfer is described by the subscripts "in" and "out." Therefore, we do not need to adopt a formal sign convention for heat and work interactions. When heat or work is to be determined and their direction is unknown, we can assume any direction (in or out) for heat or work and solve the problem. A negative result in that case will indicate that the assumed direction is wrong, and it is corrected by reversing the assumed direction. This is just like assuming a direction for an unknown force when solving a problem in statics and reversing the assumed direction when a negative quantity is obtained.

The heat transfer Q is zero for adiabatic systems, the work W is zero for systems that involve no work interactions, and the energy transport with mass E_{mass} is zero for systems that involve no mass flow across their boundaries (i.e., closed systems).

Energy balance for any system undergoing any kind of process can be expressed more compactly as

$$\underbrace{E_{in} - E_{out}}_{\substack{\text{Net energy transfer} \\ \text{by heat, work, and mass}}} = \underbrace{\Delta E_{system}}_{\substack{\text{Change in internal, kinetic,} \\ \text{potential, etc., energies}}} \quad \text{(kJ)} \tag{3-36}$$

or, in the **rate form**, as

$$\underbrace{\dot{E}_{in} - \dot{E}_{out}}_{\substack{\text{Rate of net energy transfer} \\ \text{by heat, work, and mass}}} = \underbrace{\Delta \dot{E}_{system}}_{\substack{\text{Rate of change in internal,} \\ \text{kinetic, potential, etc., energies}}} \quad \text{(kW)} \tag{3-37}$$

For constant rates, the total quantities during a time interval Δt are related to the quantities per unit time as

$$Q = \dot{Q}\Delta t, \qquad W = \dot{W}\Delta t, \qquad \text{and} \qquad \Delta E = \Delta\dot{E}\Delta t \qquad \text{(kJ)} \quad (3\text{-}38)$$

The energy balance can also be expressed on a **per unit mass** basis as

$$e_{in} - e_{out} = \Delta e_{system} \qquad \text{(kJ/kg)} \qquad (3\text{-}39)$$

which is obtained by dividing all the quantities in Eq. 3-36 by the mass m of the system. Energy balance can also be expressed in the differential form as

$$\delta E_{in} - \delta E_{out} = dE_{system} \qquad \text{or} \qquad \delta e_{in} - \delta e_{out} = de_{system} \qquad (3\text{-}40)$$

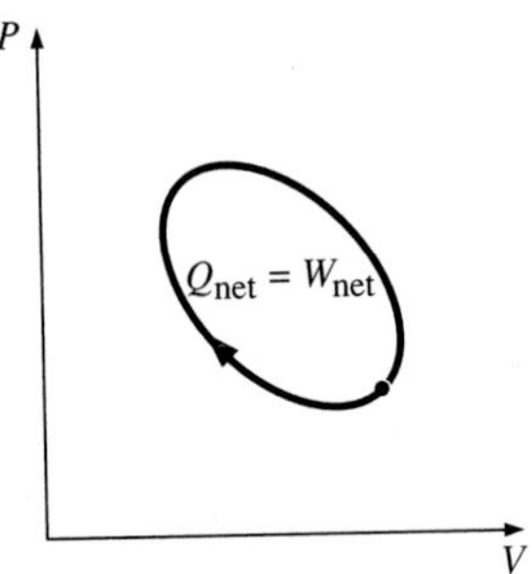

FIGURE 3-54
For a cycle $\Delta E = 0$, thus $Q = W$.

For a closed system undergoing a **cycle,** the initial and final states are identical, and thus $\Delta E_{system} = E_2 - E_1 = 0$. Then the energy balance for a cycle simplifies to $E_{in} - E_{out} = 0$ or $E_{in} = E_{out}$. Noting that a closed system does not involve any mass flow across its boundaries, the energy balance for a cycle can be expressed in terms of heat and work interactions as

$$W_{net,out} = Q_{net,in} \qquad \text{or} \qquad \dot{W}_{net,out} = \dot{Q}_{net,in} \qquad \text{(for a cycle)} \quad (3\text{-}41)$$

That is, the net work output during a cycle is equal to net heat input (Fig. 3-54).

The energy balance (or the first law) relations given above are intuitive in nature and are easy to use when the magnitudes and directions of heat and work transfers are known. But when performing a general analytical study or solving a problem that involves an unknown heat or work interaction, we need to assume a direction for the heat or work interactions. In such cases, it is common practice to assume heat to be transferred *into the system* (heat input) in the amount of Q and work to be done *by the system* (work output) in the amount of W and then to solve the problem. The first law or energy balance relation in that case for a closed system becomes

$$Q - W = \Delta E \qquad (3\text{-}42)$$

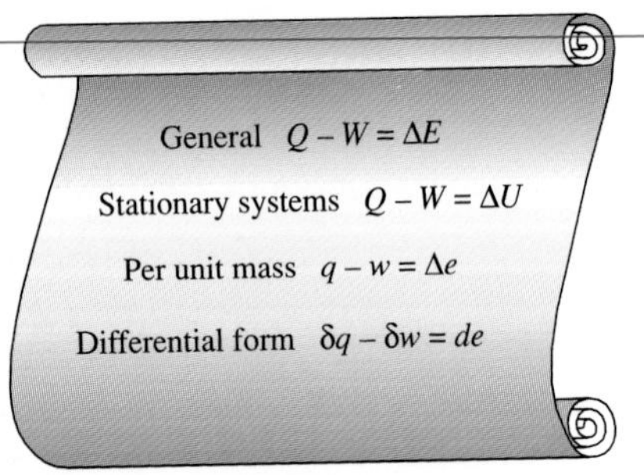

FIGURE 3-55
Various forms of the first-law relation for closed systems.

That is, heat input to a system minus work output by the system is equal to the change in the energy of the system. Obtaining a negative quantity for Q or W simply means that the assumed direction for that quantity is wrong and should be reversed. Various forms of this "traditional" first law relation for closed systems are given in Fig. 3-55.

The first law cannot be proven mathematically, but no process in nature is known to have violated the first law, and this should be taken as sufficient proof. Note that if it were possible to prove the first law on the basis of other physical principles, the first law then would be a consequence of those principles instead of being a fundamental physical law itself.

As energy quantities, heat and work are not that different, and you probably wonder why we keep distinguishing them. After all, the change in the energy content of a system is equal to the amount of energy that crosses the system boundaries, and it makes no difference whether the energy crosses the boundary as heat or work. It seems as if the first-law relations would be much simpler if we had just one quantity that we could call *energy interaction* to represent both heat and work. Well, from the first-law point of view, heat and work are not different at all. But from the second-law point of view, heat and work are very different, as is discussed in later chapters.

EXAMPLE 3-14 Cooling of a Hot Fluid in a Tank

A rigid tank contains a hot fluid that is cooled while being stirred by a paddle wheel. Initially, the internal energy of the fluid is 800 kJ. During the cooling process, the fluid loses 500 kJ of heat, and the paddle wheel does 100 kJ of work on the fluid. Determine the final internal energy of the fluid. Neglect the energy stored in the paddle wheel.

Solution We take the contents of the tank as the *system* (Fig. 3-56). This is a *closed system* since no mass crosses the boundary during the process. We observe that the volume of a rigid tank is constant, and thus there is no boundary work and $v_2 = v_1$. Also, heat is lost from the system and shaft work is done on the system.

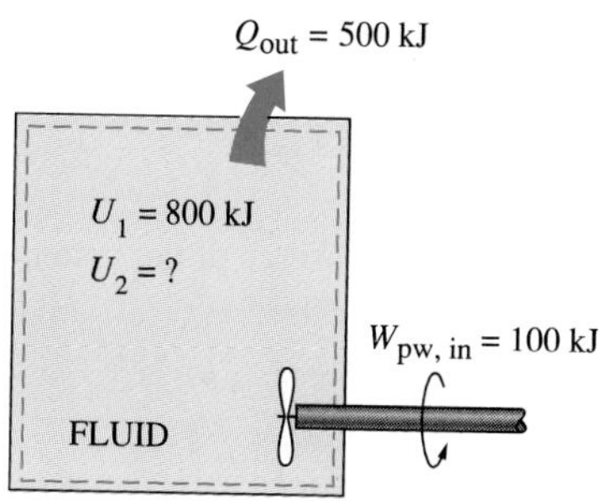

FIGURE 3-56
Schematic for Example 3-14.

Assumptions The tank is stationary and thus the kinetic and potential energy changes are zero, $\Delta KE = \Delta PE = 0$. Therefore, $\Delta E = \Delta U$ and internal energy is the only form of the system's energy that may change during this process.

Analysis Applying the energy balance on the system gives

$$\underbrace{E_{in} - E_{out}}_{\substack{\text{Net energy transfer}\\ \text{by heat, work, and mass}}} = \underbrace{\Delta E_{system}}_{\substack{\text{Change in internal, kinetic,}\\ \text{potential, etc., energies}}}$$

$$W_{pw,\,in} - Q_{out} = \Delta U = U_2 - U_1$$

$$100 \text{ kJ} - 500 \text{ kJ} = U_2 - 800 \text{ kJ}$$

$$U_2 = \mathbf{400\ kJ}$$

Therefore, the final internal energy of the system is 400 kJ.

3-6 ■ A SYSTEMATIC APPROACH TO PROBLEM SOLVING

To this point, we have concentrated our efforts on understanding the basics of thermodynamics. Armed with this knowledge, we are now in a position to tackle significant engineering problems. Knowledge is certainly an essential part of problem solving. But thermodynamic problems, particularly the complicated ones, also require a systematic approach. By using a step-by-step approach, an engineer can solve a series of simple problems instead of one large, formidable problem (Fig. 3-57).

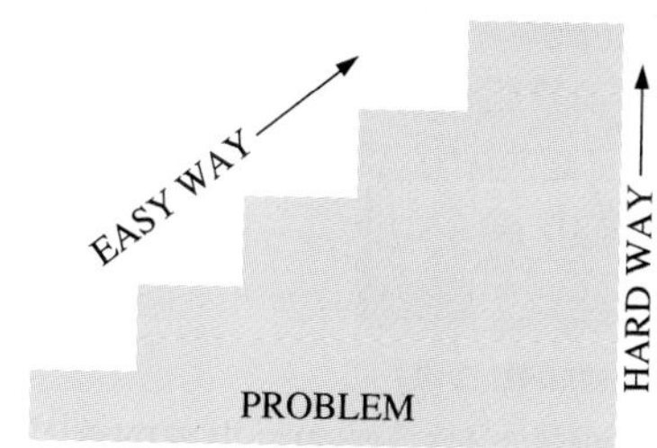

FIGURE 3-57
A step-by-step approach can greatly simplify problem solving.

The proper approach to solving thermodynamic problems is illustrated below with the help of a sample problem. Readers are urged to master this approach and use it zealously, since this will help them avoid some of the common pitfalls of problem solving.

SAMPLE PROBLEM

A 0.1-m^3 rigid tank contains steam initially at 500 kPa and 200°C. The steam is now allowed to cool until its temperature drops to 50°C. Determine the amount of heat transfer during this process and the final pressure in the tank.

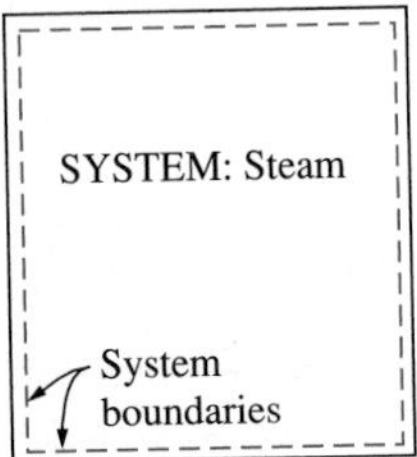

FIGURE 3-58
Step 1: Draw a sketch of the system and system boundaries.

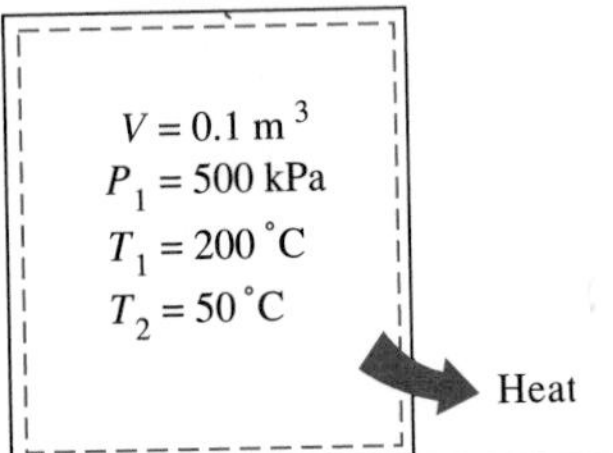

FIGURE 3-59
Step 2: List the given information on the sketch.

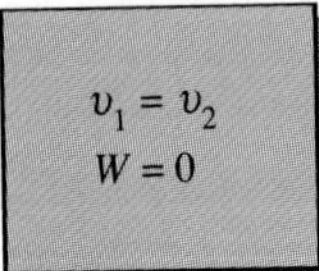

FIGURE 3-60
Step 3: Look for simplifications.

FIGURE 3-61
Step 4: Make realistic assumptions, if necessary.

Assume:

(1) $\Delta PE = 0$ (since there is no mention of elevation change).

(2) $\Delta KE = 0$ (since there is no mention of velocity change).

Step 1: Identify the System and Draw a Sketch of It

It is always a good practice to start solving a problem by drawing a sketch of the physical system. It does not have to be something elaborate, but it should resemble the physical system on hand. The system that is about to be analyzed should be identified on the sketch by drawing its boundaries using dashed lines (Fig. 3-58). In this way the region or system to which the conservation equations are applied is clearly specified.

For simple problems, the choice of the system is quite obvious (the steam in the tank, in this case), but large problems may involve several devices and even several different substances, and may require separate analyses for various parts of the system. For such cases, it is essential that the system be identified before each analysis.

Step 2: List the Given Information on the Sketch

In a typical problem, the information is scattered, and listing the given data with the proper symbols on the sketch enables one to see the entire problem at once (Fig. 3-59). Heat and work interactions, if any, should also be indicated on the sketch with the proper directions.

Step 3: Check for Special Processes

During a process not all the properties change. Also, not all processes involve heat transfer and various work interactions simultaneously. Often a key property, such as temperature or pressure, remains constant during the process, and this greatly simplifies the analysis. In our case, both the temperature and the pressure vary, but the specific volume remains constant since rigid tanks have a fixed volume (V = constant), and the mass of our system is fixed (m = constant) (Fig. 3-60). Then it follows that

$$v = \frac{V}{m} = \text{constant} \longrightarrow v_2 = v_1$$

If the process were isothermal, we would have $T_2 = T_1$, and if it were adiabatic, we would have $Q = 0$. Our system involves no moving boundaries or any other forms of work; thus, the work term is zero, $W = 0$.

Step 4: State Any Assumptions

The simplifying assumptions that are made to solve a problem should be stated and fully justified. Assumptions whose validity is questionable should be avoided. Some commonly made assumptions in thermodynamics are assuming the process to be quasi-equilibrium, neglecting the changes in kinetic and potential energies of a system, treating a gas as an ideal gas, and neglecting the heat transfer to or from insulated systems.

The system in our case can be assumed to be stationary since there is no indication to the contrary. Thus, the changes in kinetic and potential energies can be neglected (Fig. 3-61).

Step 5: Apply the Conservation Equations

Now we are ready to apply the conservation equations such as the conservation of mass and conservation of energy. We should start with the most general forms of the equations and simplify them, using the applicable assumptions (Fig. 3-62). The numerical values should not be introduced into the equations before they are reduced to their simplest forms.

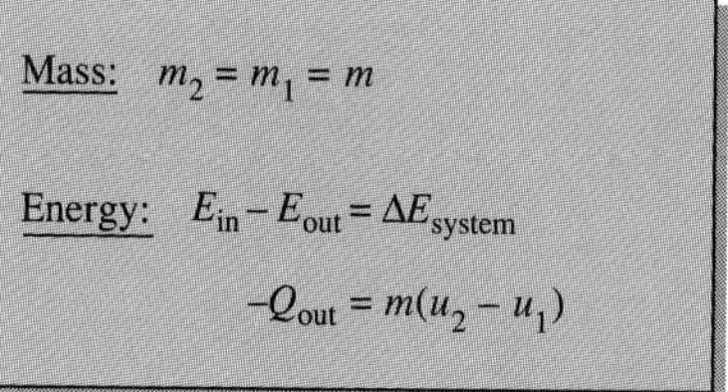

FIGURE 3-62
Step 5: Apply relevant conservation equations and simplify them.

Step 6: Draw a Process Diagram

Process diagrams, such as the *P*-*v* or *T*-*v* diagrams, are extremely helpful in visualizing the initial and final states of a system and the path of the process. If a property remains constant during a process, it should be apparent on the diagram. For our problem, the volume remains constant during the process, and so it will appear as a vertical-line segment on a *T*-*v* diagram (Fig. 3-63). For pure substances, the process diagrams should be plotted relative to the saturation lines. That way, the region where a substance is found at each state will be apparent. From this diagram it is clear that steam is a superheated vapor at the initial state and a saturated mixture at the final state.

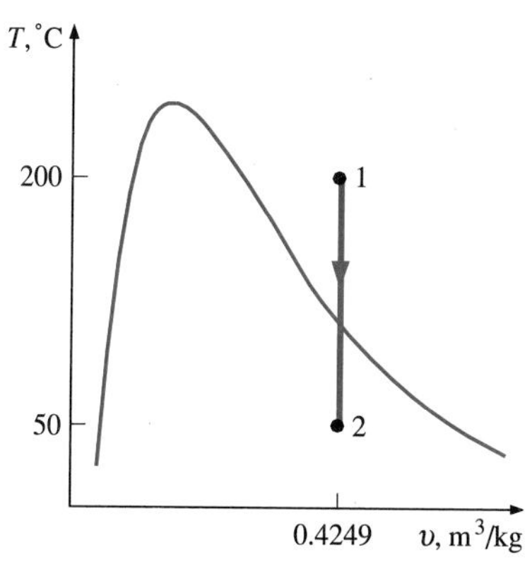

FIGURE 3-63
Step 6: Show the process on a property diagram.

Step 7: Determine the Required Properties and Unknowns

The unknown properties at any state can be determined with the help of thermodynamic relations or tables. Thermodynamic relations are usually valid over some limited range, and therefore, to prevent any errors, their validity should be checked before they are used. The thermodynamic relation that is used incorrectly most often is probably the ideal-gas relation. Even though its use is limited to gases at low pressures (relative to the critical-point value), some use it carelessly for substances that are not even in the gas phase.

When reading properties from the steam or refrigerant tables, we first need to know whether the substance exists as a superheated vapor, a compressed liquid, or a saturated mixture. This can easily be done by comparing the given property values to the corresponding saturation values.

In the sample problem, the temperature and pressure at the initial state are given (Fig. 3-64). Steam exists as a superheated vapor at this state since the temperature is greater than the saturation temperature at the given pressure (that is, $200°C > T_{sat@\,500\,kPa} = 151.9°C$). Then the initial values of the specific volume and internal energy are readily determined from the superheated vapor tables. At the final state, steam exists as a saturated mixture since the specific volume v_2 is greater than v_f but less than v_g at the final temperature, that is, $v_f < v_2 < v_g$.

Particular attention should be paid to the units of various quantities when the numerical values are substituted into the equations. The majority of the errors at this stage are due to using inconsistent units. Finally, any unreasonable results should be interpreted as an indication of possible errors, and the analysis should be checked.

The approach described above is used in the example problems that follow without explicitly stating each step. For some problems, some of the steps may not be applicable or necessary, and they may be skipped. However, we cannot overemphasize the importance of a logical and orderly approach to problem solving. The majority of the difficulties encountered while solving

State 1: $P_1 = 500$ kPa, $T_1 = 200\,°C$ } $v_1 = 0.4249$ m³/kg, $u_1 = 2642.9$ kJ/kg

State 2: $v_2 = v_1 = 0.4269\text{ m}^3\text{/kg}$
$T_2 = 50°C \rightarrow v_f = 0.001\text{ m}^3\text{/kg}$
$v_g = 12.03\text{ m}^3\text{/kg}$
$u_f = 209.32$ kJ/kg
$u_g = 2443.5$ kJ/kg
(Table A-4)

$P_2 = P_{\text{sat @ 50°C}} = 12.349$ kPa
$v_2 = v_f + x_2 v_{fg}$
$0.4249 = 0.001 + x_2(12.03 - 0.001)$

$x_2 = 0.0352$
$u_2 = u_f + x_2 u_{fg}$
$= 209.32 + (0.0352)(2443.5 - 209.32)$
$= 288.0$ kJ/kg

$$m = \frac{V}{v} = \frac{0.1\text{ m}^3}{0.4249\text{ m}^3\text{/kg}} = 0.235\text{ kg}$$

$$\underbrace{E_{in} - E_{out}}_{\substack{\text{Net energy transfer}\\ \text{by heat, work, and mass}}} = \underbrace{\Delta E_{\text{system}}}_{\substack{\text{Change in internal, kinetic,}\\ \text{potential, etc., energies}}}$$

$$-Q_{out} = \Delta U = m(u_2 - u_1)$$
$$Q_{out} = m(u_1 - u_2)$$
$$= (0.235\text{ kg})(2642.9 - 288)$$
$$= 553.4\text{ kJ}$$

FIGURE 3-64
Step 7: Determine the required properties, and solve the problem.

a problem are not due to a lack of knowledge; they are due to a lack of coordination. The individual studying thermodynamics is strongly encouraged to follow these steps in problem solving until he or she develops a personal systematic approach that works best.

EXAMPLE 3-15 Electric Heating of a Gas at Constant Pressure

A piston-cylinder device contains 25 g of saturated water vapor that is maintained at a constant pressure of 300 kPa. A resistance heater within the cylinder is turned on and passes a current of 0.2 A for 5 min from a 120-V source. At the same time, a heat loss of 3.7 kJ occurs. (*a*) Show that for a closed system the boundary work W_b and the change in internal energy ΔU in the first-law relation can be combined into one term, ΔH, for a constant-pressure process. (*b*) Determine the final temperature of the steam.

Solution We take the contents of the cylinder, including the resistance wires, as the *system* (Fig. 3-65). This is a *closed system* since no mass crosses the system boundary during the process. We observe that a piston-cylinder device typically involves a moving boundary and thus boundary work, W_b. The pressure remains constant during the process and thus $P_2 = P_1$. Also, heat is lost from the system and electrical work W_e is done on the system.

Assumptions **1** The tank is stationary and thus the kinetic and potential energy changes are zero, $\Delta KE = \Delta PE = 0$. Therefore, $\Delta E = \Delta U$ and internal energy is the only form of energy of the system that may change during this process. **2** Electrical wires constitute a very small part of the system, and thus the energy change of the wires can be neglected.

Analysis (*a*) This part of the solution involves a general analysis for a closed system undergoing a quasi-equilibrium constant-pressure process, and thus we consider a general closed system. We take the direction of heat transfer Q to be to the system and the work W to be done by the system. We also express the

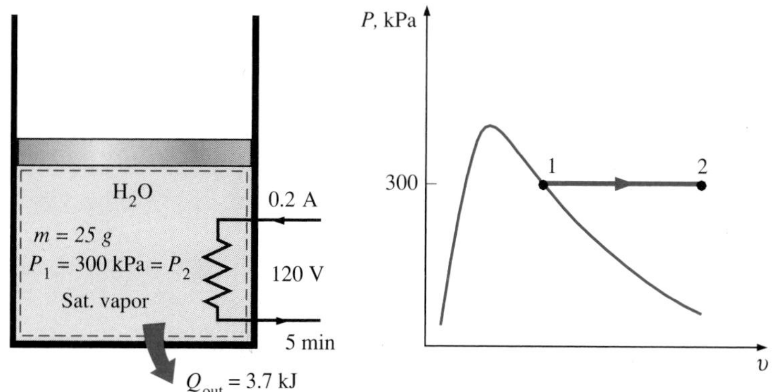

FIGURE 3-65
Schematic and P-V diagram for Example 3-15.

work as the sum of boundary and other forms of work (such as electrical and shaft). Then the energy balance can be expressed as

$$\underbrace{E_{in} - E_{out}}_{\substack{\text{Net energy transfer}\\ \text{by heat, work, and mass}}} = \underbrace{\Delta E_{system}}_{\substack{\text{Change in internal, kinetic,}\\ \text{potential, etc., energies}}}$$

$$Q - W = \Delta U + \Delta KE^{\nearrow 0} + \Delta PE^{\nearrow 0}$$

$$Q - W_{other} - W_b = U_2 - U_1$$

For a constant-pressure process, the boundary work is given by Eq. 3-19 as $W_b = P_0(V_2 - V_1)$. Substituting this into the above relation gives

$$Q - W_{other} - P_0(V_2 - V_1) = U_2 - U_1$$

But $\quad P_0 = P_2 = P_1 \longrightarrow Q - W_{other} = (U_2 + P_2V_2) - (U_1 + P_1V_1)$

Also $H = U + PV$, and thus

$$Q - W_{other} = H_2 - H_1 \qquad \text{(kJ)} \qquad (3\text{-}43)$$

which is the desired relation (Fig. 3-66). *This equation is very convenient to use in the analysis of closed systems undergoing a constant-pressure quasi-equilibrium process since the boundary work is automatically taken care of by the enthalpy terms, and one no longer needs to determine it separately.*

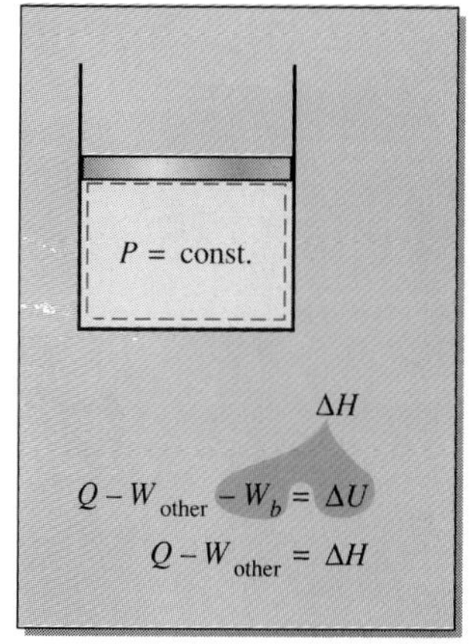

FIGURE 3-66
For a closed system undergoing a quasi-equilibrium, P = constant process, $\Delta U + W_b = \Delta H$.

(*b*) For our case, the only other form of work is the electrical work, which can be determined from Eq. 3-14:

$$W_e = VI\Delta t = (120 \text{ V})(0.2 \text{ A})(300 \text{ s})\left(\frac{1 \text{ kJ/s}}{1000 \text{ VA}}\right) = 7.2 \text{ kJ}$$

State 1: $\left.\begin{array}{l} P_1 = 100 \text{ kPa} \\ \text{sat. vapor} \end{array}\right\} \quad h_1 = h_{g@300 \text{ kPa}} = 2725.3 \text{ kJ/kg} \qquad \text{(Table A-5)}$

The enthalpy at the final state can be determined directly from Eq. 3-43 by expressing heat transfer from the system and work done on the system as negative quantities (since their directions are opposite to the assumed directions). Alternately, we can use the general energy balance relation with the simplification that the boundary work is considered automatically by replacing ΔU by ΔH for a constant-pressure expansion or compression process:

$$\underbrace{E_{in} - E_{out}}_{\substack{\text{Net energy transfer}\\ \text{by heat, work, and mass}}} = \underbrace{\Delta E_{system}}_{\substack{\text{Change in internal, kinetic,}\\ \text{potential, etc., energies}}}$$

$$W_{e,\text{in}} - Q_{out} - W_b = \Delta U$$

$$W_{e,\text{in}} - Q_{out} = \Delta H = m(h_2 - h_1) \qquad \text{(since } P = \text{constant)}$$

$$7.2\text{ kJ} - 3.7\text{ kJ} = (0.025\text{ kg})(h_2 - 2725.3)\text{ kJ/kg}$$

$$h_2 = 2865.3$$

Now the final state is completely specified since we know both the pressure and the enthalpy. The temperature at this state is

$$\textit{State 2:} \qquad \left.\begin{aligned} P_2 &= 300\text{ kPa} \\ h_2 &= 2865.3\text{ kJ/kg} \end{aligned}\right\} \quad T_2 = \mathbf{200°C} \qquad \text{(Table A-6)}$$

Therefore, the steam will be at 200°C at the end of this process.

Discussion Strictly speaking, the potential energy change of the steam is not zero for this process since the center of gravity of the steam rose somewhat. Assuming an elevation change of 1 m (which is rather unlikely), the change in the potential energy of the steam would be (from Eq. 3-24) 0.0002 kJ, which is very small compared to the other terms in the first-law relation. Therefore, in problems of this kind, the potential energy term is always neglected.

EXAMPLE 3-16 Unrestrained Expansion of Water into an Evacuated Tank

A rigid tank is divided into two equal parts by a partition. Initially, one side of the tank contains 5 kg of water at 200 kPa and 25°C, and the other side is evacuated. The partition is then removed, and the water expands into the entire tank. The water is allowed to exchange heat with its surroundings until the temperature in the tank returns to the initial value of 25°C. Determine (*a*) the volume of the tank, (*b*) the final pressure, and (*c*) the heat transfer for this process.

Solution We take the contents of the tank, including the evacuated space, as the *system* (Fig. 3-67). This is a *closed system* since no mass crosses the system boundary during the process. We observe that the water fills the entire tank when the partition is removed (possibly as a liquid–vapor mixture).

Assumptions **1** The system is stationary and thus the kinetic and potential energy changes are zero, $\Delta KE = \Delta PE = 0$ and $\Delta E = \Delta U$. **2** The direction of heat transfer is to the system (heat gain, Q_{in}). A negative result for Q_{in} will indicate the assumed direction is wrong and thus it is heat loss. **3** The volume of the rigid tank is constant, and thus there is no energy transfer as boundary work. **4** The water temperature remains constant during the process. **5** There is no electrical, shaft, or any other kind of work involved.

Analysis (*a*) Initially the water in the tank exists as a compressed liquid since its pressure (200 kPa) is greater than the saturation pressure at 25°C (3.169 kPa). Approximating the compressed liquid as a saturated liquid at the given temperature, we find

$$v_1 \cong v_{f@25°C} = 0.001003\text{ m}^3\text{/kg} \cong 0.001\text{ m}^3\text{/kg} \qquad \text{(Table A-4)}$$

Then the initial volume of the water is

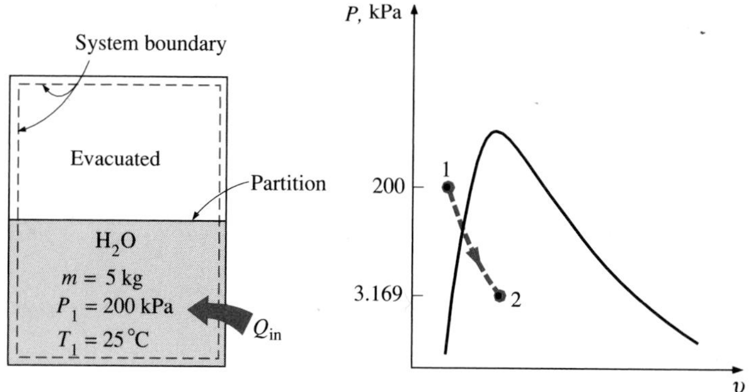

FIGURE 3-67

Schematic and *P*-*v* diagram for Example 3-16.

$$V_1 = mv_1 = (5 \text{ kg})(0.001 \text{ m}^3/\text{kg}) = 0.005 \text{ m}^3$$

The total volume of the tank is twice this amount:

$$V_{\text{tank}} = (2)(0.005 \text{ m}^3) = \mathbf{0.01 \text{ m}^3}$$

(*b*) At the final state, the specific volume of the water is

$$v_2 = \frac{V_2}{m} = \frac{0.01 \text{ m}^3}{5 \text{ kg}} = 0.002 \text{ m}^3/\text{kg}$$

which is twice the initial value of the specific volume. This result is expected since the volume doubles while the amount of mass remains constant.

At 25°C: $v_f = 0.001003 \text{ m}^3/\text{kg}$ and $v_g = 43.36 \text{ m}^3/\text{kg}$ (Table A-4)

Since $v_f < v_2 < v_g$, the water is a saturated liquid–vapor mixture at the final state, and thus the pressure is the saturation pressure at 25°C:

$$P_2 = P_{\text{sat @ 25°C}} = \mathbf{3.169 \text{ kPa}} \qquad \text{(Table A-4)}$$

(*c*) Under stated assumptions and observations, the energy balance on the system can be expressed as

$$\underbrace{E_{\text{in}} - E_{\text{out}}}_{\substack{\text{Net energy transfer} \\ \text{by heat, work, and mass}}} = \underbrace{\Delta E_{\text{system}}}_{\substack{\text{Change in internal, kinetic,} \\ \text{potential, etc., energies}}}$$

$$Q_{\text{in}} = \Delta U = m(u_2 - u_1)$$

Notice that even though the water is expanding during this process, the system chosen involves fixed boundaries only (the dashed lines) and therefore the moving boundary work is zero (Fig. 3-68). Then $W = 0$ since the system does not involve any other forms of work. (Can you reach the same conclusion by choosing the water as our system?) Initially,

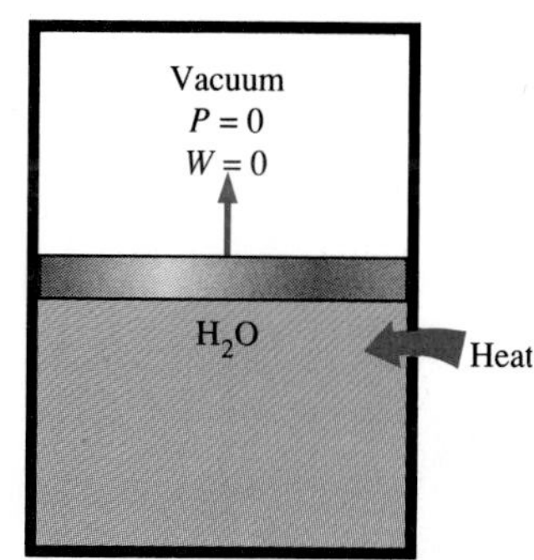

FIGURE 3-68

Expansion against a vacuum involves no work and thus no energy transfer.

$$u_1 \cong u_{f @ 25°C} = 104.88 \text{ kJ/kg}$$

The quality at the final state is determined from the specific-volume information:

$$x_2 = \frac{v_2 - v_f}{v_{fg}} = \frac{0.002 - 0.001}{43.36 - 0.001} = 2.3 \times 10^{-5}$$

Then $u_2 = u_f + x_2 u_{fg}$

$$= 104.88 \text{ kJ/kg} + (2.3 \times 10^{-5})(2304.9 \text{ kJ/kg})$$
$$= 104.93 \text{ kJ/kg}$$

Substituting yields

$$Q_{in} = (5 \text{ kg})[(104.93 - 104.88) \text{ kJ/kg}] = \mathbf{0.25\ kJ}$$

Discussion The positive sign indicates that the assumed direction is correct, and heat is transferred to the water.

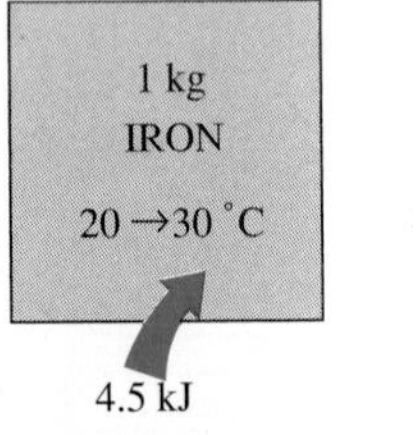

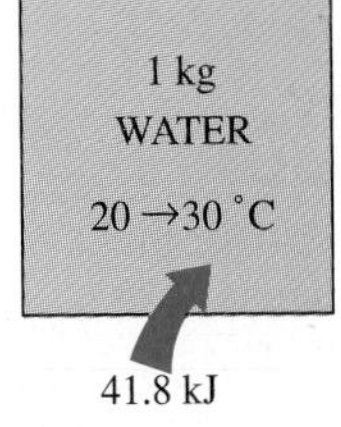

FIGURE 3-69
It takes different amounts of energy to raise the temperature of different substances by the same amount.

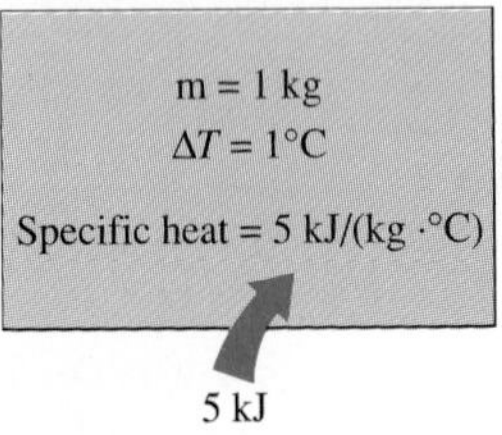

FIGURE 3-70
Specific heat is the energy required to raise the temperature of a unit mass of a substance by one degree in a specified way.

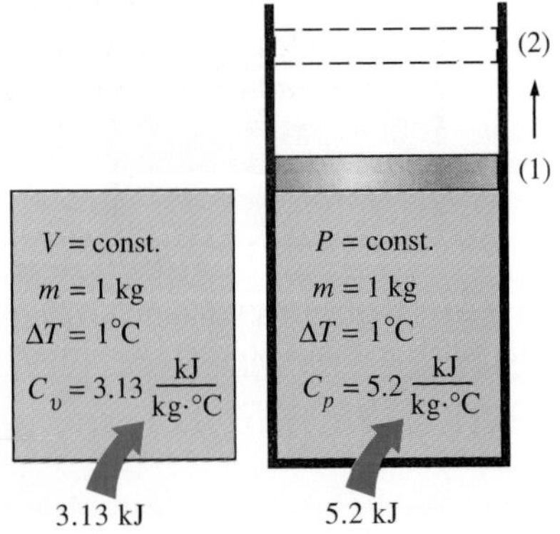

FIGURE 3-71
Constant-volume and constant-pressure specific heats C_v and C_p (values given are for helium gas).

3-7 ■ SPECIFIC HEATS

We know from experience that it takes different amounts of energy to raise the temperature of identical masses of different substances by one degree. For example, we need about 4.5 kJ of energy to raise the temperature of 1 kg of iron from 20 to 30°C, whereas it takes about 9 times this energy (41.8 kJ to be exact) to raise the temperature of 1 kg of liquid water by the same amount (Fig. 3-69). Therefore, it is desirable to have a property that will enable us to compare the energy storage capabilities of various substances. This property is the specific heat.

The **specific heat** is defined as *the energy required to raise the temperature of a unit mass of a substance by one degree* (Fig. 3-70). In general, this energy will depend on how the process is executed. In thermodynamics, we are interested in two kinds of specific heats: **specific heat at constant volume** C_v and **specific heat at constant pressure** C_p.

Physically, the specific heat at constant volume C_v can be viewed as *the energy required to raise the temperature of the unit mass of a substance by one degree as the volume is maintained constant.* The energy required to do the same as the pressure is maintained constant is the specific heat at constant pressure C_p. This is illustrated in Fig. 3-71. The specific heat at constant pressure C_p is always greater than C_v because at constant pressure the system is allowed to expand and the energy for this expansion work must also be supplied to the system.

Now we will attempt to express the specific heats in terms of other thermodynamic properties. First, consider a stationary closed system undergoing a constant-volume process ($w_b = 0$). The first-law relation for this process can be expressed in the differential form as

$$\delta q - \delta w_{\text{other}} = du$$

The left-hand side of this equation ($\delta q - \delta w_{\text{other}}$) represents the amount of energy transferred to the system in the form of heat and/or work. From the definition of C_v, this energy must be equal to $C_v\, dT$, where dT is the differential change in temperature. Thus,

$$C_v\, dT = du \qquad \text{at constant volume}$$

or

$$C_v = \left(\frac{\partial u}{\partial T}\right)_v \tag{3-44}$$

Similarly, an expression for the specific heat at constant pressure C_p can be obtained by considering a constant-pressure process ($w_b + \Delta u = \Delta h$). It yields

$$C_p = \left(\frac{\partial h}{\partial T}\right)_p \tag{3-45}$$

Equations 3-44 and 3-45 are the defining equations for C_v and C_p, and their interpretation is given in Fig. 3-72.

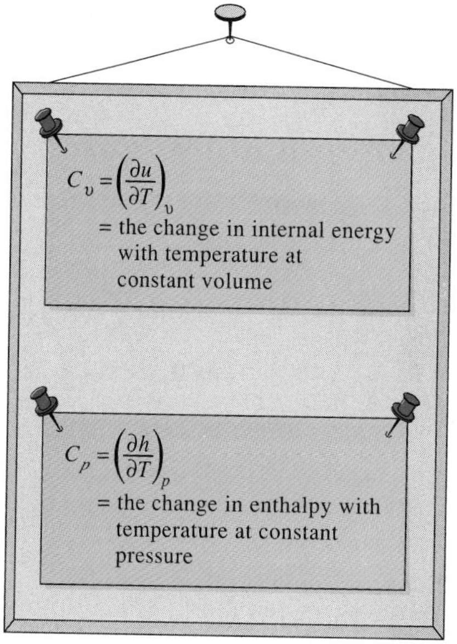

Note that C_v and C_p are expressed in terms of other properties; thus, they must be properties themselves. Like any other property, the specific heats of a substance depend on the state that, in general, is specified by two independent, intensive properties. That is, the energy required to raise the temperature of a substance by one degree will be different at different temperatures and pressures (Fig. 3-73). But this difference is usually not very large.

A few observations can be made from Eqs. 3-44 and 3-45. First, these equations are *property relations* and as such *are independent of the type of processes.* They are valid for *any* substance undergoing *any* process. The only relevance C_v has to a constant-volume process is that C_v happens to be the energy transferred to a system during a constant-volume process per unit mass per unit degree rise in temperature. This is how the values of C_v are determined. This is also how the name *specific heat at constant volume* originated. Likewise, the energy transferred to a system per unit mass per unit temperature rise during a constant-pressure process happens to be equal to C_p. This is how the values of C_p can be determined and also explains the origin of the name *specific heat at constant pressure.*

Another observation that can be made from Eqs. 3-44 and 3-45 is that C_v is related to the changes in *internal energy* and C_p to the changes in *enthalpy*. In fact, it would be more proper to define C_v as *the change in the internal energy of a substance per unit change in temperature at constant volume.* Likewise, C_p can be defined as *the change in the enthalpy of a substance per unit change in temperature at constant pressure.* In other words, C_v is a measure of the variation of internal energy of a substance with temperature, and C_p is a measure of the variation of enthalpy of a substance with temperature.

Both the internal energy and enthalpy of a substance can be changed by the transfer of *energy* in any form, with heat being only one of them. Therefore, the term *specific energy* is probably more appropriate than the term *specific heat,* which implies that energy is transferred (and stored) in the form of heat.

A common unit for specific heats is kJ/(kg · °C) or kJ/(kg · K). Notice that these two units are *identical* since $\Delta T(°\text{C}) = \Delta T(\text{K})$, and 1°C change in

$C_v = \left(\frac{\partial u}{\partial T}\right)_v$
= the change in internal energy with temperature at constant volume

$C_p = \left(\frac{\partial h}{\partial T}\right)_p$
= the change in enthalpy with temperature at constant pressure

FIGURE 3-72
Formal definitions of C_v and C_p.

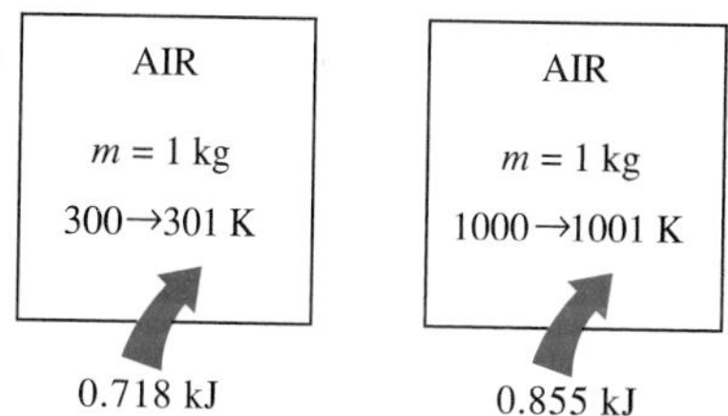

FIGURE 3-73
The specific heat of a substance changes with temperature.

temperature is equivalent to a change of 1 K (see Sec. 1-10). The specific heats are sometimes given on a *molar basis*. They are then denoted by $\overline{C}_v$ and $\overline{C}_p$ and have the unit kJ/(kmol · °C) or kJ/(kmol · K).

3-8 ■ INTERNAL ENERGY, ENTHALPY, AND SPECIFIC HEATS OF IDEAL GASES

In Chap. 2, we defined an ideal gas as a gas whose temperature, pressure, and specific volume are related by

$$Pv = RT$$

It has been demonstrated mathematically (Chap. 11) and experimentally (Joule, 1843) that for an ideal gas the internal energy is a function of the temperature only. That is,

$$u = u(T) \tag{3-46}$$

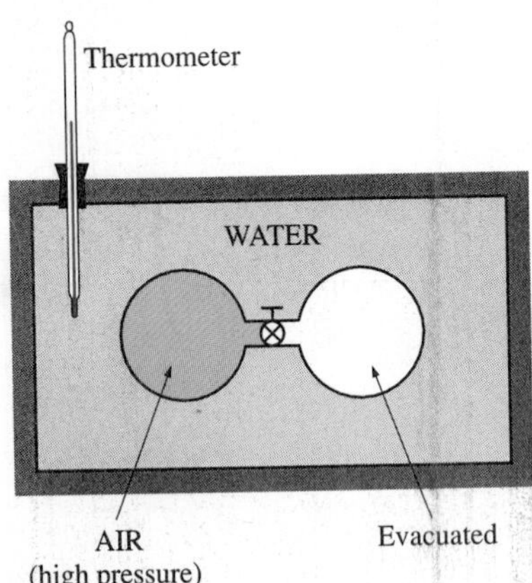

FIGURE 3-74
Schematic of the experimental apparatus used by Joule.

In his classical experiment, Joule submerged two tanks connected with a pipe and a valve in a water bath, as shown in Fig. 3-74. Initially, one tank contained air at a high pressure and the other tank was evacuated. When thermal equilibrium was attained, he opened the valve to let air pass from one tank to the other until the pressures equalized. Joule observed no change in the temperature of the water bath and assumed that no heat was transferred to or from the air. Since there was also no work done, he concluded that the internal energy of the air did not change even though the volume and the pressure changed. Therefore, he reasoned, the internal energy is a function of temperature only and not a function of pressure or specific volume. (Joule later showed that for gases that deviate significantly from ideal-gas behavior, the internal energy is not a function of temperature alone.)

Using the definition of enthalpy and the equation of state of an ideal gas, we have

$$\left.\begin{aligned} h &= u + Pv \\ Pv &= RT \end{aligned}\right\} \qquad h = u + RT$$

Since R is constant and $u = u(T)$, it follows that the enthalpy of an ideal gas is also a function of temperature only:

$$h = h(T) \tag{3-47}$$

$$u = u(T)$$
$$h = h(T)$$
$$C_v = C_v(T)$$
$$C_p = C_p(T)$$

FIGURE 3-75
For ideal gases, u, h, C_v, and C_p vary with temperature only.

Since u and h depend only on temperature for an ideal gas, the specific heats C_v and C_p also depend, at most, on temperature only. Therefore, *at a given temperature, u, h, C_v, and C_p of an ideal gas will have fixed values regardless of the specific volume or pressure* (Fig. 3-75). Thus, for ideal gases, the partial derivatives in Eqs. 3-44 and 3-45 can be replaced by ordinary derivatives. Then the differential changes in the internal energy and enthalpy of an ideal gas can be expressed as

$$du = C_v(T)\, dT \tag{3-48}$$

and

$$dh = C_p(T)\, dT \tag{3-49}$$

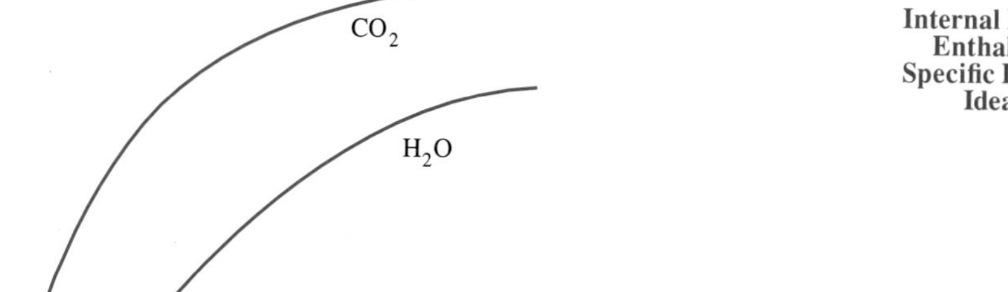

FIGURE 3-76

Ideal-gas constant-pressure specific heats for some gases (see Table A-2*c* for $\bar{C}_p$ equations).

The change in internal energy or enthalpy for an ideal gas during a process from state 1 to state 2 is determined by integrating these equations:

$$\Delta u = u_2 - u_1 = \int_1^2 C_v(T)\, dT \qquad \text{(kJ/kg)} \tag{3-50}$$

and

$$\Delta h = h_2 - h_1 = \int_1^2 C_p(T)\, dT \qquad \text{(kJ/kg)} \tag{3-51}$$

To carry out these integrations, we need to have relations for C_v and C_p as functions of temperature.

At low pressures, all real gases approach ideal-gas behavior, and therefore their specific heats depend on temperature only. The specific heats of real gases at low pressures are called *ideal-gas specific heats,* or *zero-pressure specific heats,* and are often denoted C_{p0} and C_{v0}. Accurate analytical expressions for ideal-gas specific heats, based on direct measurements or calculations from statistical behavior of molecules, are available and are given as third-degree polynomials in the appendix (Table A-2*c*) for several gases. A plot of $\bar{C}_{p0}(T)$ data for some common gases is given in Fig. 3-76.

The use of ideal-gas specific heat data is limited to low pressures, but these data can also be used at moderately high pressures with reasonable accuracy as long as the gas does not deviate from ideal-gas behavior significantly.

AIR

T, K	u, kJ/kg	h, kJ/kg
0	0	0
.	.	.
.	.	.
300	214.17	300.19
310	221.25	310.24
.	.	.
.	.	.

FIGURE 3-77

In the preparation of ideal-gas tables, 0 K is chosen as the reference temperature.

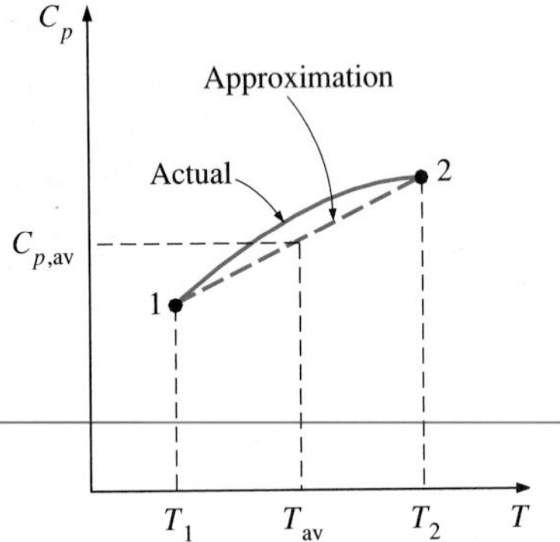

FIGURE 3-78

For small temperature intervals, the specific heats may be assumed to vary linearly with temperature.

FIGURE 3-79

The relation $\Delta u = C_v\,\Delta T$ is valid for *any* kind of process, constant-volume or not.

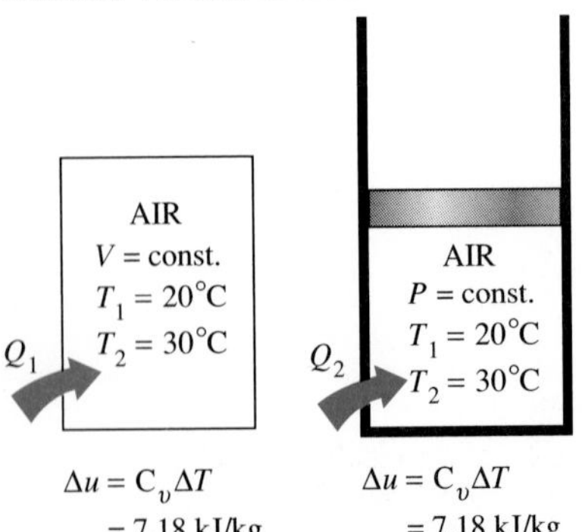

The integrations in Eqs. 3-50 and 3-51 are straightforward but rather time-consuming and thus impractical. To avoid these laborious calculations, u and h data for a number of gases have been tabulated over small temperature intervals. These tables are obtained by choosing an arbitrary reference point and performing the integrations in Eqs. 3-50 and 3-51 by treating state 1 as the reference state. In the ideal-gas tables given in the appendix, zero kelvin is chosen as the reference state, and both the enthalpy and the internal energy are assigned zero values at that state (Fig. 3-77). The choice of the reference state has no effect on Δu or Δh calculations. The u and h data are given in kJ/kg for air (Table A-17) and in kJ/kmol for other gases (N_2, O_2, CO_2, H_2O, and H_2). The unit kJ/kmol is very convenient in the thermodynamic analysis of chemical reactions.

Some observations can be made from Fig. 3-76. First, the specific heats of gases with complex molecules (molecules with two or more atoms) are higher and increase with temperature. Also, the variation of specific heats with temperature is smooth and may be approximated as linear over small temperature intervals (a few hundred degrees or less). Then the specific heat functions in Eqs. 3-50 and 3-51 can be replaced by the constant average specific heat values. Now the integrations in these equations can be performed, yielding

$$u_2 - u_1 = C_{v,\text{av}}(T_2 - T_1) \qquad \text{(kJ/kg)} \qquad \text{(3-52)}$$

and

$$h_2 - h_1 = C_{p,\text{av}}(T_2 - T_1) \qquad \text{(kJ/kg)} \qquad \text{(3-53)}$$

The specific heat values for some common gases are listed as a function of temperature in Table A-2*b*. The average specific heats $C_{p,\text{av}}$ and $C_{v,\text{av}}$ are evaluated from this table at the average temperature $(T_1 + T_2)/2$, as shown in Fig. 3-78. If the final temperature T_2 is not known, the specific heats may be evaluated at T_1 or at anticipated average temperature. Then T_2 can be determined by using these specific heat values. The value of T_2 can be refined, if necessary, by evaluating the specific heats at the new average temperature.

Another way of determining the average specific heats is to evaluate them at T_1 and T_2 and then take their average. Usually both methods give reasonably good results, and one is not necessarily better than the other.

Another observation that can be made from Fig. 3-76 is that the ideal-gas specific heats of *monatomic gases* such as argon, neon, and helium remain constant over the entire temperature range. Thus, Δu and Δh of monatomic gases can easily be evaluated from Eqs. 3-52 and 3-53.

Note that the Δu and Δh relations given above are not restricted to any kind of process. They are valid for all processes. The presence of the constant-volume specific heat C_v in an equation should not lead one to believe that this equation is valid for a constant-volume process only. On the contrary, the relation $\Delta u = C_{v,\text{av}}\,\Delta T$ is valid for *any* ideal gas undergoing *any* process (Fig. 3-79). A similar argument can be given for C_p and Δh.

To summarize, there are three ways to determine the internal energy and enthalpy changes of ideal gases (Fig. 3-80):

1. By using the tabulated u and h data. This is the easiest and most accurate way when tables are readily available.

2. By using the C_v or C_p relations as a function of temperature and performing the integrations. This is very inconvenient for hand calculations but quite desirable for computerized calculations. The results obtained are very accurate.

3. By using average specific heats. This is very simple and certainly very convenient when property tables are not available. The results obtained are reasonably accurate if the temperature interval is not very large.

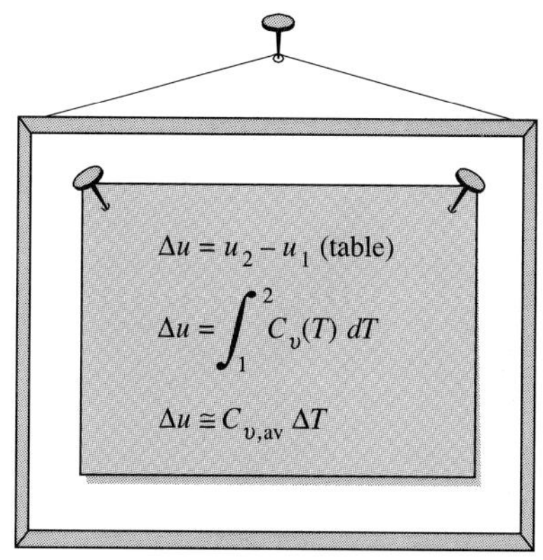

FIGURE 3-80

Three ways of calculating Δu.

Specific-Heat Relations of Ideal Gases

A special relationship between C_p and C_v for ideal gases can be obtained by differentiating the relation $h = u + RT$, which yields

$$dh = du + R\, dT$$

Replacing dh by $C_p\, dT$ and du by $C_v\, dT$ and dividing the resulting expression by dT, we obtain

$$C_p = C_v + R \qquad [\text{kJ/(kg} \cdot \text{K)}] \qquad (3\text{-}54)$$

This is an important relationship for ideal gases since it enables us to determine C_v from a knowledge of C_p and the gas constant R.

When the specific heats are given on a molar basis, R in the above equation should be replaced by the universal gas constant R_u (Fig. 3-81). That is,

$$\overline{C}_p = \overline{C}_v + R_u \qquad [\text{kJ/(kmol} \cdot \text{K)}] \qquad (3\text{-}55)$$

At this point, we introduce another ideal-gas property called the **specific heat ratio** k, defined as

$$k = \frac{C_p}{C_v} \qquad (3\text{-}56)$$

The specific heat ratio also varies with temperature, but this variation is very mild. For monatomic gases, its value is essentially constant at 1.667. Many diatomic gases, including air, have a specific heat ratio of about 1.4 at room temperature.

AIR at 300 K

$C_v = 0.718$ kJ/(kg · K), $R = 0.287$ kJ/(kg · K) } $C_p = 1.005$ kJ/(kg · K)

or,

$\overline{C}_v = 20.80$ kJ/(kmol · K), $R_u = 8.314$ kJ/(kmol · K) } $\overline{C}_p = 29.114$ kJ/(kmol · K)

FIGURE 3-81

The C_p of an ideal gas can be determined from a knowledge of C_v and R.

EXAMPLE 3-17 Evaluation of the ΔU of an Ideal Gas

Air at 300 K and 200 kPa is heated at constant pressure to 600 K. Determine the change in internal energy of air per unit mass, using (*a*) data from the air table (Table A-17), (*b*) the functional form of the specific heat (Table A-2*c*), and (*c*) the average specific heat value (Table A-2*b*).

Solution At specified conditions, air can be considered to be an ideal gas since it is at a high temperature and low pressure relative to its critical-point values. The internal energy change Δu of ideal gases depends on the initial and final temperatures only, and not on the type of process. Thus, the solution given below is valid for any kind of process.

Analysis (*a*) One way of determining the change in internal energy of air is to read the u values at T_1 and T_2 from Table A-17 and take the difference:

$$u_1 = u_{@\,300\,\text{K}} = 214.07 \text{ kJ/kg}$$

$$u_2 = u_{@\,600\,\text{K}} = 434.78 \text{ kJ/kg}$$

Thus, $$\Delta u = u_2 - u_1 = (434.78 - 214.07) \text{ kJ/kg} = \mathbf{220.71 \text{ kJ/kg}}$$

(*b*) The $\bar{C}_p(T)$ of air is given in Table A-2*c* in the form of a third-degree polynomial expressed as

$$\bar{C}_p(T) = a + bT + cT^2 + dT^3$$

where $a = 28.11$, $b = 0.1967 \times 10^{-2}$, $c = 0.4802 \times 10^{-5}$, and $d = -1.966 \times 10^{-9}$. From Eq. 3-55,

$$\bar{C}_v(T) = \bar{C}_p - R_u = (a - R_u) + bT + cT^2 + dT^3$$

From Eq. 3-50,

$$\Delta \bar{u} = \int_1^2 \bar{C}_v(T)\, dT = \int_{T_1}^{T_2} [(a - R_u) + bT + cT^2 + dT^3]\, dT$$

Performing the integration and substituting the values, we obtain

$$\Delta \bar{u} = 6447.15 \text{ kJ/kmol}$$

The change in the internal energy on a unit-mass basis is determined by dividing this value by the molar mass of air (Table A-1):

$$\Delta u = \frac{\Delta \bar{u}}{M} = \frac{6447.15 \text{ kJ/kmol}}{28.97 \text{ kg/kmol}} = \mathbf{222.55 \text{ kJ/kg}}$$

which differs form the exact result by 0.8 percent.

(*c*) The average value of the constant-volume specific heat $C_{v,\text{av}}$ is determined from Table A-2*b* at the average temperature $(T_1 + T_2)/2 = 450$ K to be

$$C_{v,\text{av}} = C_{v\,@\,450\,\text{K}} = 0.733 \text{ kJ/(kg·K)}$$

Thus, $$\Delta u = C_{v,\text{av}}(T_2 - T_1) = [0.733 \text{ kJ/(kg·K)}][(600 - 300) \text{ K}]$$
$$= \mathbf{219.9 \text{ kJ/kg}}$$

This answer differs from the exact result (220.71 kJ/kg) by only 0.4 percent. This close agreement is not surprising since the assumption that C_v varies linearly with

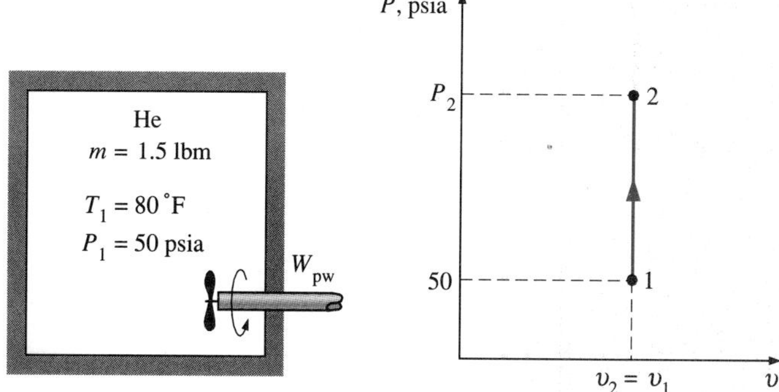

FIGURE 3-82
Schematic and *P*-*v* diagram for Example 3-18.

temperature is a reasonable one at temperature intervals of only a few hundred degrees. If we had used the C_v value at $T_1 = 300$ K instead of at T_{av}, the result would be 215.4 kJ/kg, which is in error by about 2 percent. Errors of this magnitude are acceptable for most engineering purposes.

EXAMPLE 3-18 Heating of a Gas in a Tank by Stirring

An insulated rigid tank initially contains 1.5 lbm of helium at 80°F and 50 psia. A paddle wheel with a power rating of 0.02 hp is operated within the tank for 30 min. Determine (*a*) the final temperature and (*b*) the final pressure of the helium gas.

Solution We take the contents of the tank as the *system* (Fig. 3-82). This is a *closed system* since no mass crosses the system boundary during the process. We observe that there is paddle work done on the system.

Assumptions **1** Helium is an ideal gas since it is at a very high temperature relative to its critical point value of −451°F. **2** Constant specific heats can be used for helium. **3** The system is stationary and thus the kinetic and potential energy changes are zero, $\Delta KE = \Delta PE = 0$ and $\Delta E = \Delta U$. **4** The volume of the tank is constant, and thus there is no boundary work and $V_2 = V_1$. **5** The system is adiabatic and thus there is no heat transfer.

Analysis (*a*) The amount of paddle-wheel work done on the system is

$$W_{pw} = \dot{W}_{pw}\,\Delta t = (0.02 \text{ hp})(0.5 \text{ h})\left(\frac{2545 \text{ Btu/h}}{1 \text{ hp}}\right) = 25.45 \text{ Btu}$$

Under stated assumptions and observations, the energy balance on the system can be expressed as

$$\underbrace{E_{in} - E_{out}}_{\substack{\text{Net energy transfer}\\ \text{by heat, work, and mass}}} = \underbrace{\Delta E_{system}}_{\substack{\text{Change in internal, kinetic,}\\ \text{potential, etc., energies}}}$$

$$W_{pw,\,in} = \Delta U = m(u_2 - u_1) = mC_{v,\,av}(T_2 - T_1)$$

As we pointed out earlier, the ideal-gas specific heats of monatomic gases (helium being one of them) are constant. The C_v value of helium is determined from Table A-2E*a* to be C_v = 0.753 Btu/(lbm · °F). Substituting this and other known quantities into the above equation, we obtain

$$25.45 \text{ Btu} = (1.5 \text{ lbm})[0.753 \text{ Btu/(lbm} \cdot {}^\circ\text{F)}](T_2 - 80^\circ\text{F})$$

$$T_2 = \mathbf{102.5^\circ F}$$

(*b*) The final pressure is determined from the ideal-gas relation

$$\frac{P_1V_1}{T_1} = \frac{P_2V_2}{T_2}$$

where V_1 and V_2 are identical and cancel. Then the final pressure becomes

$$\frac{50 \text{ psia}}{(80 + 460) \text{ R}} = \frac{P_2}{(102.5 + 460) \text{ R}}$$

$$P_2 = \mathbf{53.1 \text{ psia}}$$

EXAMPLE 3-19 Heating of a Gas by a Resistance Heater

A piston-cylinder device initially contains 0.5 m^3 of nitrogen gas at 400 kPa and 27°C. An electric heater within the device is turned on and is allowed to pass a current of 2 A for 5 min from a 120-V source. Nitrogen expands at constant pressure, and a heat loss of 2800 J occurs during the process. Determine the final temperature of the nitrogen, using data from the nitrogen table (Table A-18).

Solution We take the contents of the cylinder as the *system* (Fig. 3-83). This is a *closed system* since no mass crosses the system boundary during the process. We observe that a piston-cylinder device typically involves a moving boundary and thus boundary work, W_b. Also, heat is lost from the system and electrical work W_e is done on the system.

Assumptions **1** Nitrogen is an ideal gas since it is at a high temperature and low pressure relative to its critical point values of −147°C, and 3.39 MPa. **2** The system is stationary and thus the kinetic and potential energy changes are zero, $\Delta KE = \Delta PE = 0$ and $\Delta E = \Delta U$. **3** The pressure remains constant during the process and thus $P_2 = P_1$.

Analysis First, let us determine the electrical work done on the nitrogen:

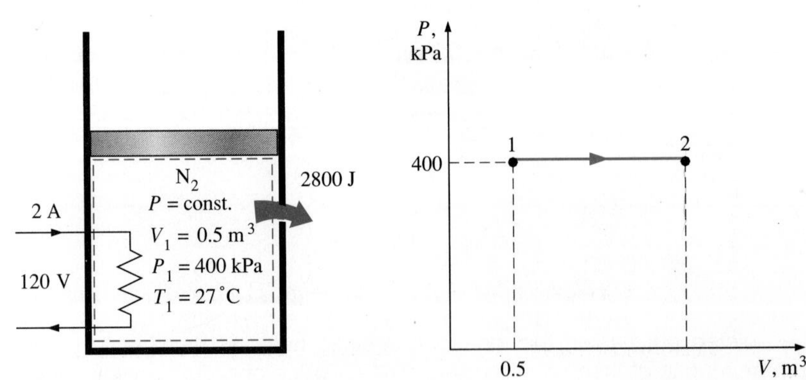

FIGURE 3-83
Schematic and *P-V* diagram for Example 3-19.

$$W_e = VI\,\Delta t = (120\text{ V})(2\text{ A})(5 \times 60\text{ s})\left(\frac{1\text{ kJ/s}}{1000\text{ VA}}\right) = 72\text{ kJ}$$

The number of moles of nitrogen is determined from the ideal-gas relation:

$$N = \frac{P_1 V_1}{R_u T_1} = \frac{(400\text{ kPa})(0.5\text{ m}^3)}{[8.314\text{ kJ/(kmol}\cdot\text{K)}](300\text{ K})} = 0.080\text{ kmol}$$

When gases other than air are involved, it is more convenient to work with mole numbers instead of masses since all the u and h data are given on a mole basis (Fig. 3-84). Under stated assumptions and observations, the energy balance on the system can be expressed as

$$\underbrace{E_{in} - E_{out}}_{\substack{\text{Net energy transfer}\\ \text{by heat, work, and mass}}} = \underbrace{\Delta E_{system}}_{\substack{\text{Change in internal, kinetic,}\\ \text{potential, etc., energies}}}$$

$$W_{e,\text{ in}} - Q_{out} - W_b = \Delta U$$

$$W_{e,\text{ in}} - Q_{out} = \Delta H = m(h_2 - h_1) = N(\bar{h}_2 - \bar{h}_1)$$

since $\Delta U + W_b \equiv \Delta H$ for a closed system undergoing a quasi-equilibrium expansion or compression process at constant pressure. From the nitrogen table, $\bar{h}_1 = \bar{h}_{@\,300\text{ K}} = 8723$ kJ/kmol. The only unknown quantity in the above equation is $\bar{h}_2$, and it is found to be

$$72\text{ kJ} - 2.8\text{ kJ} = (0.08\text{ kmol})(\bar{h}_2 - 8723\text{ kJ/kmol})$$

$$\bar{h}_2 = 9588\text{ kJ/kmol}$$

The temperature corresponding to this enthalpy value is

$$T_2 = \mathbf{56.7°C}$$

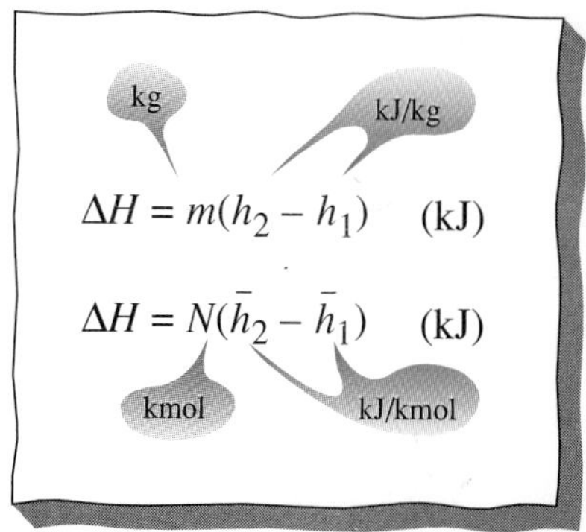

FIGURE 3-84

Two equivalent ways of determining the total enthalpy change ΔH.

EXAMPLE 3-20 Heating of a Gas at Constant Pressure

A piston-cylinder device initially contains air at 150 kPa and 27°C. At this state, the piston is resting on a pair of stops, as shown in Fig. 3-85, and the enclosed volume is 400 L. The mass of the piston is such that a 350-kPa pressure is required to move it. The air is now heated until its volume has doubled. Determine (*a*) the final temperature, (*b*) the work done by the air, and (*c*) the total heat transferred to the air.

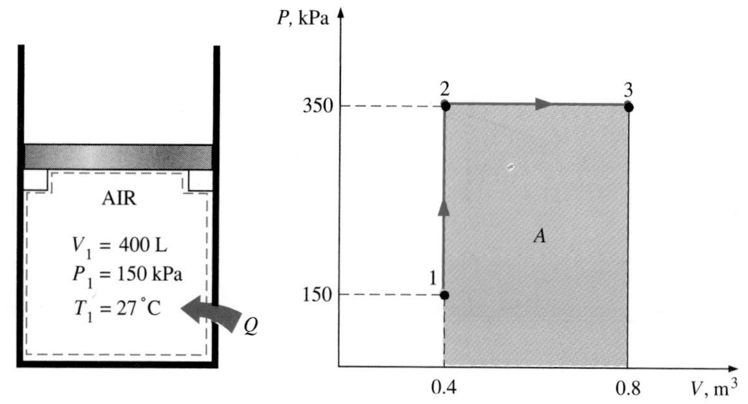

FIGURE 3-85

Schematic and P-V diagram for Example 3-20.

Solution We take the contents of the cylinder as the *system* (Fig. 3-85). This is a *closed system* since no mass crosses the system boundary during the process. We observe that a piston-cylinder device typically involves a moving boundary and thus boundary work, W_b. Also, the boundary work is done by the system, and heat is transferred to the system.

Assumptions **1** Air is an ideal gas since it is at a high temperature and low pressure relative to its critical point values of −141°C and 3.77 MPa. **2** The system is stationary and thus the kinetic and potential energy changes are zero, $\Delta KE = \Delta PE = 0$ and $\Delta E = \Delta U$. **3** The volume remains constant until the piston starts moving, and the pressure remains constant afterwards. **4** There are no electrical, shaft, or other forms of work involved.

Analysis (*a*) The final temperature can be determined easily by using the ideal-gas relation between states 1 and 3 in the following form:

$$\frac{P_1V_1}{T_1} = \frac{P_3V_3}{T_3} \longrightarrow \frac{(150 \text{ kPa})(V_1)}{300 \text{ K}} = \frac{(350 \text{ kPa})(2V_1)}{T_3}$$

$$T_3 = \mathbf{1400\ K}$$

(*b*) The work done could be determined from Eq. 3-18 by integration, but for this case it is much easier to find it from the area under the process curve on a *P-V* diagram, shown in Fig. 3-85:

$$A = (V_2 - V_1)(P_2) = (0.4 \text{ m}^3)(350 \text{ kPa}) = 140 \text{ m}^3 \cdot \text{kPa}$$

Therefore,

$$W_{13} = \mathbf{140\ kJ}$$

The work is done by the system (to raise the piston and to push the atmospheric air out of the way), thus it is work output.

(*c*) Under stated assumptions and observations, the energy balance on the system between the initial and final states (process 1-3) can be expressed as

$$\underbrace{E_{in} - E_{out}}_{\substack{\text{Net energy transfer}\\ \text{by heat, work, and mass}}} = \underbrace{\Delta E_{system}}_{\substack{\text{Change in internal, kinetic,}\\ \text{potential, etc., energies}}}$$

$$Q_{in} - W_{b,\text{out}} = \Delta U = m(u_3 - u_1)$$

The mass of the system can be determined from the ideal-gas equation of state:

$$m = \frac{P_1V_1}{RT_1} = \frac{(150 \text{ kPa})(0.4 \text{ m}^3)}{[0.287 \text{ kPa}\cdot\text{m}^3/(\text{kg}\cdot\text{K})](300 \text{ K})} = 0.697 \text{ kg}$$

The internal energies are determined from the air table (Table A-17) to be

$$u_1 = u_{@\,300\text{ K}} = 214.07 \text{ kJ/kg}$$

$$u_3 = u_{@\,1400\text{ K}} = 1113.52 \text{ kJ/kg}$$

Thus,

$$Q_{in} - 140 \text{ kJ} = (0.697 \text{ kg})[(1113.52 - 214.07) \text{ kJ/kg}]$$

$$Q_{in} = \mathbf{766.9\ kJ}$$

The positive sign verifies that heat is transferred to the system.

3-9 ■ INTERNAL ENERGY, ENTHALPY, AND SPECIFIC HEATS OF SOLIDS AND LIQUIDS

A substance whose specific volume (or density) is constant is called an **incompressible substance.** The specific volumes of solids and liquids essentially remain constant during a process (Fig. 3-86). Therefore, liquids and solids can be approximated as incompressible substances without sacrificing much in accuracy. The constant-volume assumption should be taken to imply that the energy associated with the volume change, such as the boundary work, is negligible compared with other forms of energy. Otherwise, this assumption would be ridiculous for studying the thermal stresses in solids (caused by volume change with temperature) or analyzing liquid-in-glass thermometers.

It can be mathematically shown (Chap. 11) that the constant-volume and constant-pressure specific heats are identical for incompressible substances (Fig. 3-87). Therefore, for solids and liquids, the subscripts on C_p and C_v can be dropped, and both specific heats can be represented by a single symbol C. That is,

$$C_p = C_v = C \qquad (3\text{-}57)$$

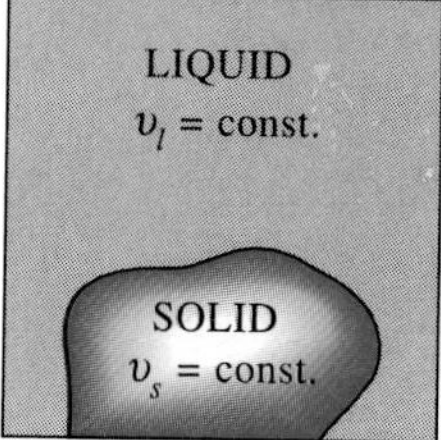

FIGURE 3-86
The specific volumes of incompressible substances remain constant during a process.

This result could also be deduced from the physical definitions of constant-volume and constant-pressure specific heats. Specific heat values for several common liquids and solids are given in Table A-3.

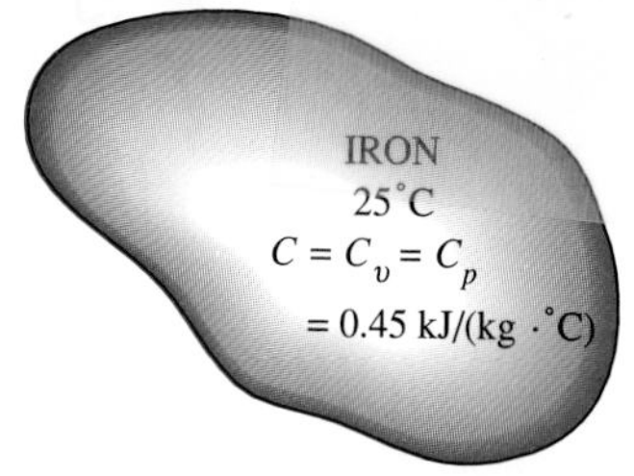

FIGURE 3-87
The C_v and C_p values of incompressible substances are identical and are denoted by C.

Internal Energy Changes

Like those of ideal gases, the specific heats of incompressible substances depend on temperature only. Thus, the partial differentials in the defining equation of C_v (Eq. 3-44) can be replaced by ordinary differentials, which yield

$$du = C_v\,dT = C(T)\,dT \qquad (3\text{-}58)$$

The change in internal energy between states 1 and 2 is then obtained by integration:

$$\Delta u = u_2 - u_1 = \int_1^2 C(T)\,dT \qquad \text{(kJ/kg)} \qquad (3\text{-}59)$$

The variation of specific heat C with temperature should be known before this integration can be carried out. For small temperature intervals, a C value at the average temperature can be used and treated as a constant, yielding

$$\Delta u \cong C_{av}(T_2 - T_1) \qquad \text{(kJ/kg)} \qquad (3\text{-}60)$$

Enthalpy Changes

Using the definition of enthalpy $h = u + Pv$ and noting that v = constant, the differential form of the enthalpy change of incompressible substances can be determined by differentiation to be

$$dh = du + v\,dP + P\,dv^{0} = du + v\,dP \tag{3-61}$$

Integrating,

$$\Delta h = \Delta u + v\Delta P \cong C_{av}\Delta T + v\Delta P \qquad \text{(kJ)} \tag{3-62}$$

For *solids,* the term $v\Delta P$ is insignificant and thus $\Delta h = \Delta u \cong C_{av}\Delta T$. For *liquids,* two special cases are commonly encountered:

1. *Constant pressure processes,* as in heaters ($\Delta P = 0$): $\Delta h = \Delta u \cong C_{av}\Delta T$
2. *Constant temperature processes,* as in pumps ($\Delta T = 0$): $\Delta h = v\Delta P$

For a process between states 1 and 2, the last relation can be expressed as $h_2 - h_1 = v(P_2 - P_1)$. By taking state 2 to be the compressed liquid state at a given T and P and state 1 to be the saturated liquid state at the same temperature, the enthalpy of the compressed liquid can he expressed as

$$h_{@\,P,T} \cong h_{f@\,T} + v_{f@\,T}\,(P - P_{sat}) \tag{3-63}$$

where P_{sat} is the saturation pressure at the given temperature. This is an improvement over the assumption that the enthalpy of the compressed liquid could be taken as h_f at the given temperature (that is, $h_{@\,P,T} \cong h_{f@\,T}$). However, the contribution of the last term is often very small, so it is usually neglected.

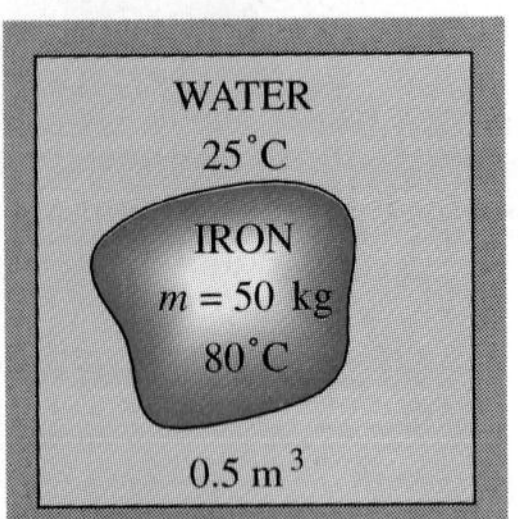

FIGURE 3-88
Schematic for Example 3-21.

EXAMPLE 3-21 Cooling of an Iron Block by Water

A 50-kg iron block at 80°C is dropped into an insulated tank that contains 0.5 m^3 of liquid water at 25°C. Determine the temperature when thermal equilibrium is reached.

Solution We take the entire contents of the tank, water + iron block, as the *system* (Fig. 3-88). This is a *closed system* since no mass crosses the system boundary during the process. We observe that the volume of a rigid tank is constant, and thus there is no boundary work.

Assumptions **1** Both water and the iron block are incompressible substances. **2** Constant specific heats at room temperature can be used for water and the iron. **3** The system is stationary and thus the kinetic and potential energy changes are zero, $\Delta KE = \Delta PE = 0$ and $\Delta E = \Delta U$. **4** There are no electrical, shaft, or other forms of work involved. **5** The system is well-insulated and thus there is no heat transfer.

Analysis The energy balance on the system can be expressed as

$$\underbrace{E_{in} - E_{out}}_{\substack{\text{Net energy transfer}\\ \text{by heat, work, and mass}}} = \underbrace{\Delta E_{system}}_{\substack{\text{Change in internal, kinetic,}\\ \text{potential, etc., energies}}}$$

$$0 = \Delta U$$

The total internal energy U is an extensive property, and therefore it can be expressed as the sum of the internal energies of the parts of the system. Then the total internal energy change of the system becomes

$$\Delta U_{sys} = \Delta U_{iron} + \Delta U_{water} = 0$$
$$[mC(T_2 - T_1)]_{iron} + [mC(T_2 - T_1)]_{water} = 0$$

The specific volume of liquid water at or about room temperature can be taken to be 0.001 m^3/kg. Then the mass of the water is

$$m_{water} = \frac{V}{v} = \frac{0.5 \text{ m}^3}{0.001 \text{ m}^3/\text{kg}} = 500 \text{ kg}$$

The specific heats of iron and liquid water are determined from Table A-3 to be $C_{iron} = 0.45$ kJ/(kg · °C) and $C_{water} = 4.184$ kJ/(kg · °C). Substituting these values into the energy equation, we obtain

$$(50 \text{ kg})[0.45 \text{ kJ/(kg} \cdot \text{°C)}](T_2 - 80\text{°C}) + (500 \text{ kg})[4.184 \text{ kJ/(kg} \cdot \text{°C)}](T_2 - 25\text{°C}) = 0$$
$$T_2 = \mathbf{25.6°C}$$

Therefore, when thermal equilibrium is established, both the water and iron will be at 25.6°C. The small rise in water temperature is due to its large mass and large specific heat.

EXAMPLE 3-22 Enthalpy of Compressed Liquid

Determine the enthalpy of liquid water at 100°C and 15 MPa (*a*) by using compressed liquid tables, (*b*) by approximating it as a saturated liquid, and (*c*) by using the correction given by Eq. 3-63.

Solution At 100°C, the saturation pressure of water is 101.35 kPa, and since $P > P_{sat}$, the water exists as a compressed liquid at the specified state.

Analysis (*a*) From compressed liquid tables, we read

$$\left.\begin{array}{l} P = 15 \text{ MPa} \\ T = 100\text{°C} \end{array}\right\} \quad h = \mathbf{430.28 \text{ kJ/kg}} \qquad \text{(Table A-7)}$$

This is the exact value.

(*b*) Approximating the compressed liquid as a saturated liquid at 100°C, as is commonly done, we obtain

$$h \cong h_{f @ 100°C} = \mathbf{419.04 \text{ kJ/kg}}$$

This value is in error by about 2.6 percent.

(*c*) From Eq. 3-63,

$$\begin{aligned} h_{@ P,T} &= h_{f @ T} + v_f(P - P_{sat}) \\ &= (419.04 \text{ kJ/kg}) + (0.001 \text{ m}^3/\text{kg})[(15{,}000 - 101.35) \text{ kPa}]\left(\frac{1 \text{ kJ}}{1 \text{ kPa} \cdot \text{m}^3}\right) \\ &= \mathbf{434.60 \text{ kJ/kg}} \end{aligned}$$

The correction term reduced the error from 2.6 to about 1 percent. But this improvement in accuracy is often not worth the extra effort involved.

EXAMPLE 3-23 Temperature Rise due to Slapping

If you ever slapped someone or got slapped yourself, you probably remember the burning sensation on your hand or your face. Imagine you had the unfortunate occasion of being slapped by an angry person, which caused the temperature of the affected area of your face to rise by 1.8°C (ouch!). Assuming the slapping hand has a mass of 1.2 kg and about 0.150 kg of the tissue on the face and the hand is affected by the incident, estimate the velocity of the hand just before impact. Take the specific heat of the tissue to be 3.8 kJ/(kg · °C).

FIGURE 3-89
Schematic for Example 3-23.

Solution We will analyze this incident in a professional manner without involving any emotions. First, we identify the system, draw a sketch of it, state our observations about the specifics of the problem, and make appropriate assumptions.

We take the hand and the affected portion of the face as the system (Fig. 3-89). This is a *closed system* since it involves a fixed amount of mass (no mass transfer). We observe that the kinetic energy of the hand decreases during the process, as evidenced by a decrease in velocity from initial value to zero, while the internal energy of the affected area increases, as evidenced by an increase in the temperature. There seems to be no significant energy transfer between the system and its surroundings during this process.

Assumptions **1** The hand is brought to a complete stop after the impact. **2** The face takes the blow well without significant movement. **3** No heat is transferred from the affected area to the surroundings, and thus the process is adiabatic. **4** No work is done on or by the system. **5** The potential energy change is zero, $\Delta PE = 0$ and $\Delta E = \Delta U + \Delta KE$.

Analysis Under the stated assumptions and observations, the energy balance on the system can be expressed as

$$\underbrace{E_{in} - E_{out}}_{\substack{\text{Net energy transfer}\\ \text{by heat, work, and mass}}} = \underbrace{\Delta E_{system}}_{\substack{\text{Change in internal, kinetic,}\\ \text{potential, etc., energies}}}$$

$$0 = \Delta U_{\text{affected tissue}} + \Delta KE_{\text{hand}}$$

$$0 = (mC\Delta T)_{\text{affected tissue}} + [m(0 - \mathcal{V}^2)/2]_{\text{hand}}$$

That is, the decrease in the kinetic energy of the hand must be equal to the increase in the internal energy of the affected area. Solving for the velocity and substituting the given quantities, the impact velocity of the hand is determined to be

$$\mathcal{V}_{\text{hand}} = \sqrt{\frac{2(mC\Delta T)_{\text{affected tissue}}}{m_{\text{hand}}}} = \sqrt{\frac{2(0.15\text{ kg})[3.8\text{ kJ/(kg}\cdot°\text{C)}](1.8°\text{C})}{1.2\text{ kg}}\left(\frac{1000\text{ m}^2/\text{s}^2}{1\text{ kJ/kg}}\right)}$$

$$= \mathbf{41.3\ m/s}\text{ (or 149 km/h)}$$

Refrigeration and freezing of perishable food products is an important application area of thermodynamics. There are many considerations in the cooling and freezing of foods. For example, fruits and vegetables continue to *respire* and *generate heat* during storage; most foods freeze over a *range of temperatures* instead of at a single temperature; the quality of frozen foods is greatly affected by the *rate of freezing*; the *velocity* of refrigerated air affects the rate of moisture loss from the products in addition to the rate of heat transfer; and so forth.

Microorganisms such as *bacteria, yeasts, molds,* and *viruses* are widely encountered in air, water, soil, living organisms, and unprocessed food items and cause *off-flavors* and *odors, slime production, changes* in the texture and appearance, and the eventual *spoilage* of foods. Holding perishable foods at warm temperatures is the primary cause of spoilage, and the prevention of food spoilage and the premature degradation of quality due to microorganisms is the largest application area of refrigeration.

Microorganisms grow best at "warm" temperatures, usually between 20°C and 60°C. The growth rate *declines* at high temperatures, and *death* occurs at still higher temperatures, usually above 70°C for most microorganisms. *Cooling* is an effective and practical way of reducing the growth rate of microorganisms and thus extending the *shelf life* of perishable foods. A temperature of 4°C or lower is considered to be a safe refrigeration temperature. Sometimes, a small increase in refrigeration temperature may cause a large increase in the growth rate, and thus a considerable decrease in shelf life of the food (Fig. 3-90). The growth rate of some microorganisms, for example, doubles for each 3°C rise in temperature.

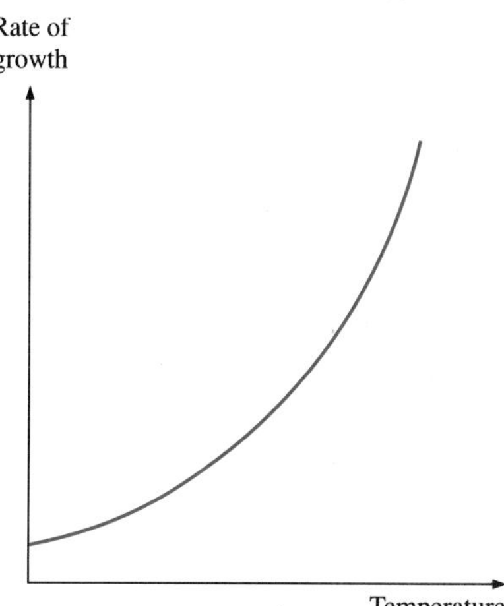

FIGURE 3-90
The rate of growth of microorganisms in a food product increases exponentially with increasing environmental temperature.

The *storage life* of fresh perishable foods such as meats, fish, vegetables, and fruits can be extended by several days by storing them at temperatures just above freezing, usually between 1°C and 4°C. The storage life of foods can be extended by several months by freezing and storing them at subfreezing temperatures, usually between −18°C and −35°C, depending on the particular food.

Refrigeration *slows down* the chemical and biological processes in foods, and the accompanying deterioration and loss of quality and nutrients. Sweet corn, for example, may lose half of its initial sugar content in one day at 21°C, but only 5 percent of it at 0°C. Fresh asparagus may lose 50 percent of its vitamin C content in one day at 20°C, but in 12 days at 0°C. Refrigeration also extends the shelf life of products. The first appearance of unsightly yellowing of broccoli, for example, may be delayed by three or more days by refrigeration.

Early attempts to freeze food items resulted in poor quality products because of the large ice crystals that formed, until it was discovered that the *rate of freezing* has a major effect on the size of ice crystals and the quality, texture, and nutritional and sensory properties of many foods. During *slow freezing,* ice crystals can grow to a large size, whereas during *fast freezing* a large number of ice crystals start forming at once and the size of the ice crystals is much smaller in this case. Large ice crystals are not desirable since they can *puncture* the walls of the cells, causing a degradation of texture and a loss of natural juices during thawing. A *crust* forms on the outer layer of the product rapidly during fast freezing, which minimizes *dehydration* of the food product and seals in the juices, aromatics, and flavoring agents. The product quality is also affected adversely by the temperature fluctuations of the storage room.

The ordinary refrigeration of foods involves *cooling* only without any phase change. The *freezing* of foods, on the other hand, involves three stages: *cooling* to the freezing point (removing the sensible heat), *freezing* (removing the latent heat), and *further cooling* to desired subfreezing temperature (removing the sensible heat of frozen food), as shown in Fig. 3-91.

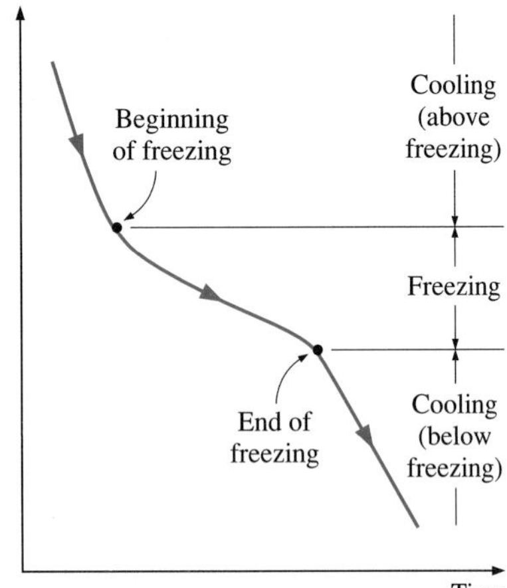

FIGURE 3-91
Typical freezing curve of a food item.

Fresh fruits and vegetables are *live products,* and thus they continue giving off heat, which adds to the refrigeration load of the cold storage room. The storage life of fruits and vegetables can be extended greatly by removing the field heat and cooling as soon after harvesting as possible. The optimum storage temperature of most fruits and vegetables is about 0.5°C to 1°C above their freezing point. But this is not the case for some fruits and vegetables such as bananas and cucumbers, which experience undesirable *physiological changes* when exposed to low (but still above freezing) temperatures, usually between 0 and 10°C. The resulting tissue damage is called the **chilling injury** and is characterized by internal discoloration, soft scald, skin blemishes, soggy breakdown, and failure to ripen. The severity of chilling injury depends on both the temperature and the length of storage at that temperature. The lower the temperature, the greater the damage in a given time. Therefore, products susceptible to chilling injury must be stored at higher temperatures. A list of vegetables susceptible to chilling injury and the lowest safe storage temperature are given in Table 3-3.

TABLE 3-3

Some vegetables susceptible to chilling injury and the lowest safe storage temperature (from ASHRAE *Handbook of Refrigeration*, Ref. 2, Chap. 17, Table 3)

Vegetable	Lowest safe temperature, °C
Cucumbers	10
Eggplants	7
Casaba melons	7 to 10
Watermelons	4
Okra	7
Sweet peppers	7
Potatoes	3 to 4
Pumpkins	10
Hard-shell squash	10
Sweet potatoes	13
Ripe tomatoes	7 to 10
Mature green tomatoes	13

Chilling injury differs from **freezing injury,** which is caused by prolonged exposure of the fruits and vegetables to subfreezing temperatures and thus the actual *freezing* at the affected areas. Freezing injury is characterized by rubbery texture, browning, bruising, and drying due to rapid moisture loss. The freezing points of fruits and vegetables do not differ by much, but their susceptibility to freezing injury differs greatly. Some vegetables are frozen and thawed several times with no significant damage, but others such as tomatoes suffer severe tissue injury and are ruined after one freezing. Products near the refrigerator coils or at the bottom layers of refrigerator cars and trucks are most susceptible to freezing injury. To avoid freezing injury, the rail cars or trucks should be *heated* during transportation in subfreezing weather, and adequate air circulation must be provided in cold storage rooms. Damage also occurs during *thawing* if it is done too fast. It is recommended that thawing be done at 4°C.

Dehydration or *moisture loss* causes a product to shrivel or wrinkle and lose quality. Therefore, proper measures must be taken during cold storage of food items to minimize moisture loss, which also represents a direct loss of the salable amount. A fruit or vegetable that loses 5 percent moisture, for example, will weight 5 percent less and will probably be sold at a lower unit price because of loss of quality. Moisture loss can be minimized by (1) keeping the *storage temperature* of food as low as possible, (2) keeping the *relative humidity* of the storage room as high as possible, and (3) avoiding *high air velocities* in the storage room.

Thermal Properties of Foods

Refrigeration of foods offers considerable challenges to engineers since the structure and composition of foods and their thermal and physical properties vary considerably. Furthermore, the properties of foods also change with *time* and *temperature.* Fruits and vegetables offer an additional challenge since they *generate heat* during storage as they consume oxygen and give off carbon dioxide, water vapor, and other gases.

The thermal properties of foods are dominated by their *water content.* In fact, the specific heat and the latent heat of foods are calculated with reasonable accuracy on the basis of their water content alone. The *specific heats* of foods can be expressed by **Siebel's formula** as

$$C_{p,\text{fresh}} = 3.35a + 0.84 \qquad [\text{kJ/(kg}\cdot{}^\circ\text{C)}] \qquad (3\text{-}64)$$

$$C_{p,\text{frozen}} = 1.26a + 0.84 \qquad [\text{kJ/(kg}\cdot{}^\circ\text{C)}] \qquad (3\text{-}65)$$

where $C_{p,\text{fresh}}$ and $C_{p,\text{frozen}}$ are the specific heats of the food before and after freezing, respectively; a is the fraction of water content of the food ($a = 0.65$ if the water content is 65 percent); and the constant 0.84 kJ/(kg · °C) represents the specific heat of the solid (nonwater) portion of the food. For example, the specific heats of fresh and frozen *chicken* whose water content is 74 percent are

$$C_{p,\text{fresh}} = 3.35a + 0.84 = 3.35 \times 0.74 + 0.84 = 3.32 \text{ kJ/(kg}\cdot{}^\circ\text{C)}$$

$$C_{p,\text{frozen}} = 1.26a + 0.84 = 1.26 \times 0.74 + 0.84 = 1.77 \text{ kJ/(kg}\cdot{}^\circ\text{C)}$$

Siebel's formulas are based on the specific heats of *water* and *ice* at 0°C of 4.19 and 2.10 kJ/(kg · °C), respectively, and thus they result in the specific heat values of water and ice at 0°C for $a = 100\%$ (i.e., pure water). Therefore, Siebel's formulas give the specific heat values at 0°C. However, they can be used over a wide temperature range with reasonable accuracy.

The *latent heat* of a food product during freezing or thawing (the heat of fusion) also depends on its water content and is determined from

$$h_{\text{latent}} = 334a \qquad (\text{kJ/kg}) \qquad (3\text{-}66)$$

where a is again the fraction of water content and 334 kJ/kg is the *latent heat of water* during freezing at 0°C at atmospheric pressure (Fig. 3-92). For example, the latent heat of chicken whose water content is 74 percent is

$$h_{\text{latent, chicken}} = 334a = 334 \times 0.74 = 247 \text{ kJ/kg}$$

Perishable foods are mostly water in content, which turns to ice during freezing. Therefore, we may expect the food items to freeze at 0°C, which is the freezing point of pure water at atmospheric pressure. But the water in

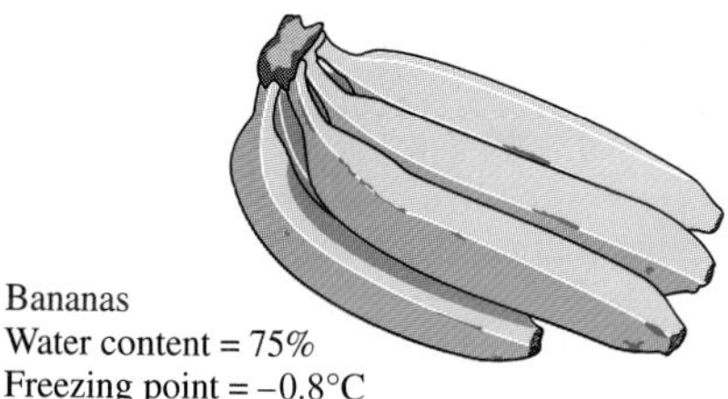

FIGURE 3-92
The specific and latent heats of foods depend on their water content alone.

TABLE 3-4

The drop of freezing point of typical ice cream whose composition is 12.5 percent milkfat, 10.5 percent serum solids, 15 percent sugar, 0.3 percent stabilizer, and 61.7 percent water (from ASHRAE *Handbook of Refrigeration*, Ref. 2, Chap. 14, Table 9)

Percent of water unfrozen	Freezing point of unfrozen part
0	−2.47°C
10	−2.75°C
20	−3.11°C
30	−3.50°C
40	−4.22°C
50	−5.21°C
60	−6.78°C
70	−9.45°C
80	−14.92°C
90	−30.36°C
100	−55.0°C

foods is far from being pure, and thus the freezing temperature of foods will be somewhat below 0°C, depending on the composition of a particular food. In general, food products freeze over a *range of temperatures* instead of a single temperature since the composition of the liquid in the food changes (becomes more concentrated in sugar) and its freezing point drops when some of the liquid water freezes (Table 3-4). Therefore, we often speak of the *average* freezing temperature or, for foods like lettuce that are damaged by freezing, the temperature at which *freezing begins*. The freezing temperature of most foods is between −0.3°C and −2.8°C. In the absence of the exact data, the freezing temperature can be assumed to be −2.0°C for meats and −1.0°C for vegetables and fruits. The freezing temperature and specific and latent heats of common food products are given in Table A-3. The *freezing temperature* in this table represents the temperature at which freezing starts for fruits and vegetables, and the average freezing temperature for other foods. The *water content* values are typical values for mature products and may exhibit some variation. The specific and latent heat values given are valid only for the *listed water content*, but they can be reevaluated easily for different water contents using the formulas above.

Refrigeration of Fruits and Vegetables

Fruits and vegetables are frequently *cooled* to preserve preharvest freshness and flavor, and to extend storage and shelf life. Cooling at the field before the product is shipped to the market or storage warehouse is referred to as **pre-cooling.** The cooling requirements of fruits and vegetables vary greatly, as do the cooling methods. Highly perishable products such as broccoli, ripe tomatoes, carrots, leafy vegetables, apricots, strawberries, peaches, and plums must be cooled as soon as possible after harvesting. Cooling is not necessary or as important for long-lasting fruits and vegetables such as potatoes, pumpkins, green tomatoes, and apples.

Fruits and vegetables are mostly water, and thus their properties are close in value to those of water. Initially, all of the heat removed from the product comes from the exterior of the products, causing a large temperature gradient within the product during fast cooling. But the *mass-average temperature*, which is the equivalent average temperature of the product at a given time, is used in calculations for simplicity.

The heat removed from the products accounts for the majority of the refrigeration load and is determined from

$$\dot{Q}_{\text{product}} = mC_p(T_{\text{initial}} - T_{\text{final}})/\Delta t \qquad \text{(W)} \qquad (3\text{-}67)$$

where $\dot{Q}_{\text{product}}$ is the *average rate of heat removal* from the fruits and vegetables; m is the total mass; C_p is the average specific heat; T_{initial} and T_{final} are the mass-average temperatures of the products before and after cooling, respectively; and Δt is the cooling time (Fig. 3-93). The heat of respiration is negligible when the cooling time is less than a few hours.

Fresh fruits and vegetables are *live products*, and they continue to *respire* for days and even weeks after harvesting at varying rates. During respiration,

FIGURE 3-93
Refrigeration load of an apple cooling facility that cools apples from 25°C to 2°C in 3 h.

$\dot{Q} = mC_p(T_{\text{initial}} - T_{\text{final}})/\Delta t$
$= (0.14\text{ kg})[3600\text{ J/(kg·°C)}](23°\text{C})/(3 \times 3600\text{ s})$
$= 1.07$ W (per apple)

a sugar like glucose combines with O_2 to produce CO_2 and H_2O. **Heat of respiration** is released during this exothermic reaction, which adds to the refrigeration load during cooling of fruits and vegetables. The rate of respiration varies strongly with temperature. For example, the average heats of respiration of strawberries at 0, 10, and 20°C are 44, 213, and 442 mW/kg, respectively. The heat of respiration also varies greatly from one product to another. For example, at 20°C, it is 34 mW/kg for mature potatoes, 77 mW/kg for apples, 167 mW/kg for cabbage, and 913 mW/kg for broccoli. The heat of respiration of most vegetables decreases with time. The opposite is true for fruits that ripen in storage such as apples and peaches. For plums, for example, the heat of respiration increases from 12 mW/kg shortly after harvest to 21 after 6 days and to 27 after 18 days when stored at 5°C. Being infected with decay organisms also increases the respiration rate.

The heat of respiration for some fruits and vegetables is given in Table 3-5. We should use the *initial rates* of respiration when calculating the cooling load of fruits and vegetables for the first day or two, and the *long-term equilibrium rates* when calculating the heat load for long-term cold storage. The *refrigeration load* due to *respiration* is determined from

$$\dot{Q}_{\text{respiration}} = \sum m_i \dot{q}_{\text{respiration},i} \quad \text{(W)} \qquad (3\text{-}68)$$

which is the sum of the mass times the heat of respiration for all the food products stored in the refrigerated space. Fresh fruits and vegetables with the highest rates of respiration are the most perishable, and refrigeration is the most effective way to slow down respiration and decay.

TABLE 3-5

Heat of respiration of some fresh fruits and vegetables at various temperatures (from ASHRAE *Handbook of Fundamentals*, Ref. 1, Chap. 30, Table 2)*

	Heat of respiration, mW/kg	
Product	**5°C**	**20°C**
Apples	13–36	44–167
Strawberries	48–98	303–581
Broccoli	102–475	825–1011
Cabbage	22–87	121–437
Carrots	20–58	64–117
Cherries	28–42	83–95
Lettuce	39–87	169–298
Watermelon	*	51–74
Mushrooms	211	782–939
Onions	10–20	50
Peaches	19–27	176–304
Plums	12–27	53–77
Potatoes	11–35	13–92
Tomatoes	*	71–120

*An asterisk indicates a chilling temperature

EXAMPLE 3-24 Cooling of Bananas in a Cold Storage Room

A typical one-half carlot capacity banana room contains 18 pallets of bananas. Each pallet consists of 24 boxes, and thus the room stores 432 boxes of bananas. A box holds an average of 19 kg of bananas and is made of 2.3 kg of fiberboard. The specific heats of banana and the fiberboard are 3.55 kJ/(kg · °C) and 1.7 kJ/(kg · °C), respectively. The peak heat of respiration of bananas is 0.3 W/kg. The bananas are cooled at a rate of 0.2°C/h. Disregarding any heat gain through the walls or other surfaces, determine the required rate of heat removal from the banana room.

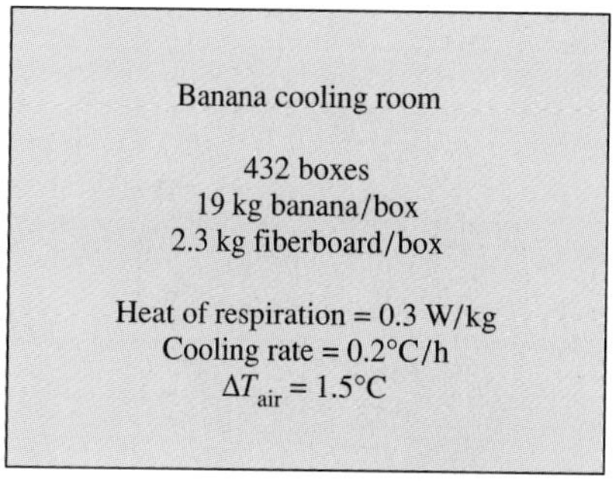

Banana cooling room

432 boxes
19 kg banana/box
2.3 kg fiberboard/box

Heat of respiration = 0.3 W/kg
Cooling rate = 0.2°C/h
ΔT_{air} = 1.5°C

FIGURE 3-94
Schematic for Example 3-24.

Solution We take the contents of the banana room as the *system* (Fig. 3-94). This is a *closed system* since it involves a fixed mass. We observe that the heat of respiration can be treated as an energy input to the system.

Assumptions **1** There is no heat gain through the walls and other surfaces. **2** The energy change of the air in the banana room is negligible. **3** Thermal properties of air, bananas, and boxes are constant. **4** The system is stationary and involves changes in its internal energy only (due to temperature change), and thus $\Delta E = \Delta U$. **5** There is no electrical, shaft, boundary, or any other kind of work involved.

Analysis Under the stated assumptions and observations, the rate form energy balance on the system reduces to

$$\underbrace{\dot{E}_{in} - \dot{E}_{out}}_{\substack{\text{Rate of net energy transfer}\\ \text{by heat, work, and mass}}} = \underbrace{\Delta \dot{E}_{system}}_{\substack{\text{Rate of change in internal, kinetic,}\\ \text{potential, etc., energies}}}$$

$$\dot{E}_{respiration} - \dot{Q}_{out} = \Delta \dot{U}_{banana} + \Delta \dot{U}_{box}$$

Noting that the banana room holds 432 boxes, the total mass of bananas and the boxes is determined to be

$$m_{banana} = \text{(Mass per box)(Number of boxes)} = \text{(19 kg/box)(432 boxes)} = 8208 \text{ kg}$$
$$m_{box} = \text{(Mass per box)(Number of boxes)} = \text{(2.3 kg/box)(432 boxes)} = 993.6 \text{ kg}$$

Then,

$$\dot{Q}_{respiration} = m_{banana}\dot{q}_{respiration} = (8208 \text{ kg})(0.3 \text{ W/kg}) = 2462 \text{ W}$$

$$\Delta \dot{U}_{banana} = (mC\Delta T/\Delta t)_{banana} = (8208 \text{ kg})[3.55 \text{ kJ/(kg} \cdot \text{°C)}](-0.2\text{°C/h})$$
$$= -5828 \text{ kJ/h} = -1619 \text{ W} \qquad \text{(since 1 W} = 3.6 \text{ kJ/h)}$$

$$\Delta \dot{U}_{box} = (mC_p\Delta T/\Delta t)_{box} = (993.6 \text{ kg})[1.7 \text{ kJ/(kg} \cdot \text{°C)}](-0.2\text{°C/h}) = -338 \text{ kJ/h}$$
$$= -94 \text{ W} \qquad \text{(since 1 W} = 3.6 \text{ kJ/h)}$$

since the quantity $\Delta T/\Delta t$ is the rate of change in temperature of the products and is given to be −0.2°C/h (a temperature drop). Substituting,

$$\dot{Q}_{total} = \dot{E}_{respiration} - \Delta \dot{E}_{banana} - \Delta \dot{E}_{box} = 2462 + 1619 + 94 = \mathbf{4175\ W}$$

Therefore, the refrigeration system must remove heat at a rate of 4175 W from the banana room to achieve the desired cooling rate.

Refrigeration of Meats

Meat carcasses in slaughterhouses should be cooled *as fast as possible* to a uniform temperature of about 1.7°C to reduce the growth rate of microorganisms that may be present on carcass surfaces, and thus minimize spoilage. The right level of *temperature, humidity,* and *air motion* should be selected to prevent excessive shrinkage, toughening, and discoloration.

The deep-body temperature of an animal is about 39°C, but this temperature tends to rise a couple of degrees in the midsections after slaughter as a result of the *heat generated* during the biological reactions that occur in the cells. The temperature of the exposed surfaces, on the other hand, tends to drop as a result of heat losses. The thickest part of the carcass is the *round,* and the center of the round is the last place to cool during chilling. Therefore, the cooling of the carcass can best be monitored by inserting a thermometer deep into the central part of the round.

About 70 percent of the beef carcass is water, and the carcass is cooled mostly by *evaporative cooling* as a result of moisture migration toward the surface where evaporation occurs. But this shrinking translates into a loss of salable mass that can amount to 2 percent of the total mass during an overnight chilling. To prevent *excessive* loss of mass, carcasses are usually washed

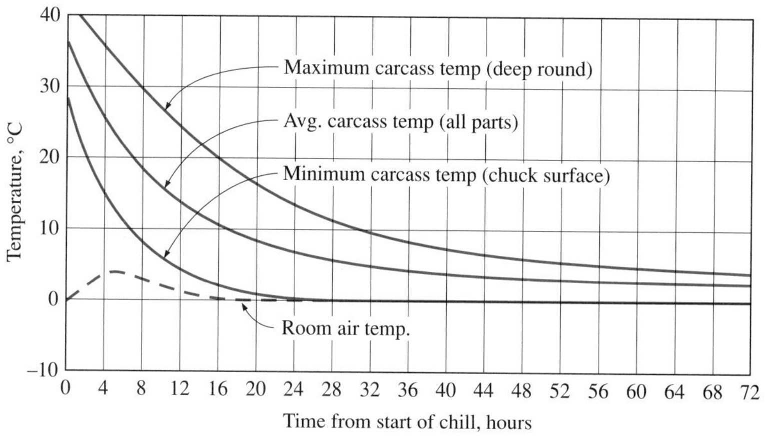

FIGURE 3-95

Typical cooling curve of a beef carcass in the chilling and holding rooms at an average temperature of 0°C (from ASHRAE *Handbook of Refrigeration*, Ref. 2, Chap. 11, Fig. 2).

or sprayed with water prior to cooling. With adequate care, spray chilling can eliminate carcass cooling shrinkage almost entirely.

The average total mass of dressed beef, which is normally split into two sides, is about 300 kg, and the average specific heat of the carcass is about 3.14 kJ/(kg · °C) (Table 3-6). The *chilling room* must have a capacity equal to the daily kill of the slaughterhouse, which may be several hundred. A beef carcass is washed before it enters the chilling room and absorbs a large amount of water (about 3.6 kg) at its surface during the washing process. This does not represent a net mass gain, however, since it is lost by dripping or evaporation in the chilling room during cooling. Ideally, the carcass does not lose or gain any net weight as it is cooled in the chilling room. However, it does lose about 0.5 percent of the total mass in the *holding room* as it continues to cool. The actual product loss is determined by first weighing the dry carcass before washing and then weighing it again after it is cooled.

The variation of temperature of the beef carcass during cooling is given in Fig. 3-95. Note that the average temperature of the carcass is reduced by about 28°C (from 36 to 8°C) in 20 h. The cooling rate of the carcass could be increased by *lowering* the refrigerated air temperature and *increasing* the air velocity, but such measures also increase the risk of *surface freezing*.

Although the average freezing point of lean meat can be taken to be −2°C with a latent heat of 249 kJ/kg, it should be remembered that freezing occurs over a *temperature range*, with most freezing occurring between −1 and −4°C. Therefore, cooling the meat through this temperature range and removing the latent heat takes the most time during freezing.

Meat can be kept at an internal temperature of −2 to −1°C for local use and storage *under a week*. Meat must be frozen and stored at much lower temperatures for *long-term storage*. The lower the storage temperature, the longer the storage life of meat products, as shown in Table 3-7.

The *internal temperature* of carcasses entering the cooling sections varies from 38 to 41°C for hogs and from 37 to 39°C for lambs and calves. It takes about 15 h to cool the hogs and calves to the recommended temperature of 3 to 4°C. The cooling room temperature is maintained at −1 to 0°C, and

TABLE 3-6

Thermal properties of beef

Quantity	Typical value
Average density	1070 kg/m^3
Specific heat	
Above freezing	3.14 kJ/(kg · °C)
Below freezing	1.70 kJ/(kg · °C)
Freezing point	−2.7°C
Latent heat of fusion	249 kJ/kg
Thermal conductivity	0.41 W/(m · °C) (at 6°C)

TABLE 3-7

Storage life of frozen meat products at different storage temperatures (from ASHRAE *Handbook of Refrigeration*, Ref. 2, Chap. 10, Table 7)

	Storage life, months		
	Temperature		
Product	−12°C	−18°C	−23°C
Beef	4–12	6–18	12–24
Lamb	3–8	6–16	12–18
Veal	3–4	4–14	8
Pork	2–6	4–12	8–15
Chopped beef	3–4	4–6	8
Cooked foods	2–3	2–4	

the temperature difference between the refrigerant and the cooling air is kept at about 6°C. *Lamb carcasses* are cooled to an internal temperature of 1 to 2°C, which takes about 12 to 14 h, and are held at that temperature with 85 to 90 percent relative humidity until shipped or processed.

Freezing does not seem to affect the *flavor* of meat much, but it affects the *quality* in several ways. The *rate* and *temperature* of freezing may influence color, tenderness, and drip. Rapid freezing increases tenderness and reduces the tissue damage and the amount of drip after thawing. Storage at low freezing temperatures causes significant changes in *animal fat*. Frozen pork experiences more undesirable changes during storage because of its fat structure, and thus its acceptable storage period is shorter than that of beef, veal, or lamb.

Poultry Products

Poultry products can be preserved by *ice chilling* to 1 to 2°C or *deep chilling* to about −2°C for short-term storage, or by *freezing* them to −18°C or below for long-term storage. Poultry processing plants are completely *automated,* and the small size of the birds makes continuous conveyor-line operation feasible.

The birds are first electrically stunned before cutting to prevent struggling, and to make the killing process more humane (!). Following a 90- to 120-s bleeding time, the birds are *scalded* by immersing them into a tank of warm water, usually at 51 to 55°C, for up to 120 s to loosen the feathers. Then the feathers are removed by feather-picking machines, and the eviscerated carcass is *washed* thoroughly before chilling. The internal temperature of the birds ranges from 24 to 35°C after washing, depending on the temperatures of the ambient air and the washing water as well as the extent of washing.

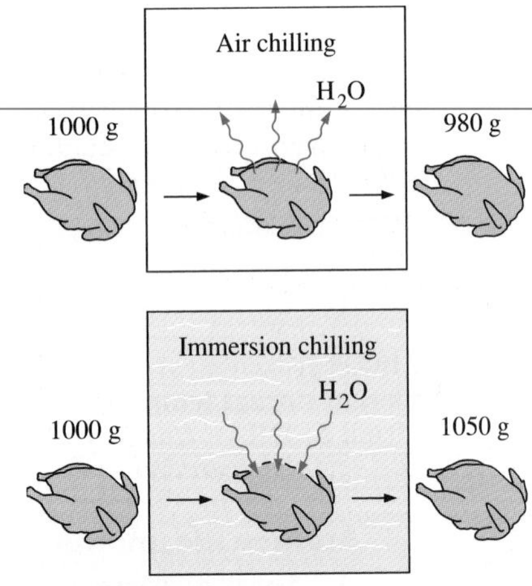

FIGURE 3-96
Air chilling causes dehydration and thus weight loss for poultry, whereas immersion chilling causes a weight gain as a result of water absorption.

To control the microbial growth, the USDA regulations require that poultry be chilled to 4°C or below in less than 4 h for carcasses of less than 1.8 kg, in less than 6 h for carcasses of 1.8 to 3.6 kg, and in less than 8 h for carcasses more than 3.6 kg. Meeting these requirements today is not difficult since the slow *air chilling* is largely replaced by the rapid *immersion chilling* in tanks of slush ice. Immersion chilling has the added benefit that it not only prevents dehydration, but it causes a *net absorption of water* and thus increases the mass of salable product. Cool air chilling of unpacked poultry can cause a moisture loss of 1 to 2 percent while water immersion chilling can cause a moisture absorption of 4 to 15 percent (Fig. 3-96). Water spray chilling can cause a moisture absorption of up to 4 percent. Most water absorbed is held between the flesh and the skin and the connective tissues in the skin. In immersion chilling, some soluble solids are lost from the carcass to the water, but the loss has no significant effect on flavor.

Many slush ice tank chillers today are replaced by *continuous* flow-type immersion slush ice chillers. Continuous slush ice chillers can reduce the internal temperature of poultry from 32 to 4°C in about 30 minutes at a rate up to 10,000 birds per hour. Ice requirements depend on the inlet and exit temperatures of the carcass and the water, but 0.25 kg of ice per kg of carcass is usually adequate. However, *bacterial contamination* such as salmonella remains a concern with this method, and it may be necessary to chloride the water to control contamination.

Tenderness is an important consideration for poultry products as it is for red meat, and preserving tenderness is an important consideration in the cooling and freezing of poultry. Birds cooked or frozen before passing through rigor mortis remain very tough. Natural tenderization begins soon after slaughter and is completed within 24 h when birds are held at 4°C. Tenderization is rapid during the first 3 hours and slows down thereafter. Immersion in hot water and cutting into the muscle adversely affect tenderization. Increasing the *scalding temperature* or the scalding time is observed to increase toughness, and decreasing the scalding temperature or scalding time is observed to increase tenderness. The *beating action* of mechanical feather-picking machines causes considerable toughening. *Cutting up* the bird into pieces before natural tenderization is completed reduces tenderness considerably. Therefore, it is recommended that any cutting be done after tenderization. *Rapid chilling* of poultry can also have a toughening effect. It is found that the tenderization process can be speeded up considerably by a patented *electrical stunning* process.

Poultry products are *highly perishable,* and thus they should be kept at the *lowest* possible temperature to maximize their shelf life. Studies have shown that the populations of certain bacteria double every 36 h at −2°C, every 14 h at 0°C, every 7 h at 5°C, and in less than 1 h at 25°C (Fig. 3-97). Studies have also shown that the total bacterial count on birds held at 2°C for 14 days is equivalent to that for birds held at 10°C for 5 days or 24°C for 1 day. It has also been found that birds held at −1°C had 8 days of additional shelf life over those held at 4°C.

The growth of microorganisms on the *surfaces* of the poultry causes the development of an *off odor* and *bacterial slime*. The higher the initial amount of bacterial contamination, the faster the sliming occurs. Therefore, good sanitation practices during processing such as cleaning the equipment frequently and washing the carcasses are as important as the storage temperature in extending shelf life.

Poultry must be frozen *rapidly* to assure a light, pleasing appearance. Poultry that is frozen slowly appears dark and develops large ice crystals that damage the tissue. The ice crystals formed during rapid freezing are small. Delaying freezing of poultry causes the ice crystals to become larger. Rapid freezing can be accomplished by forced air at temperatures of −23 to −40°C and velocities of 1.5 to 5 m/s in *air-blast tunnel freezers*. Most poultry is frozen this way. Also, the packaged birds freeze much faster on open shelves than they do in boxes. If poultry packages must be frozen in boxes, then it is very desirable to leave the boxes open or to cut holes in the boxes in the direction of airflow during freezing. For best results, the blast tunnel should be fully loaded across its cross section with even spacing between the products to assure uniform airflow around all sides of the packages. The freezing time of poultry as a function of refrigerated air temperature is given in Fig. 3-98. Thermal properties of poultry are given in Table 3-8.

Other freezing methods for poultry include sandwiching between *cold plates*, *immersion* into a refrigerated liquid such as glycol or calcium chloride

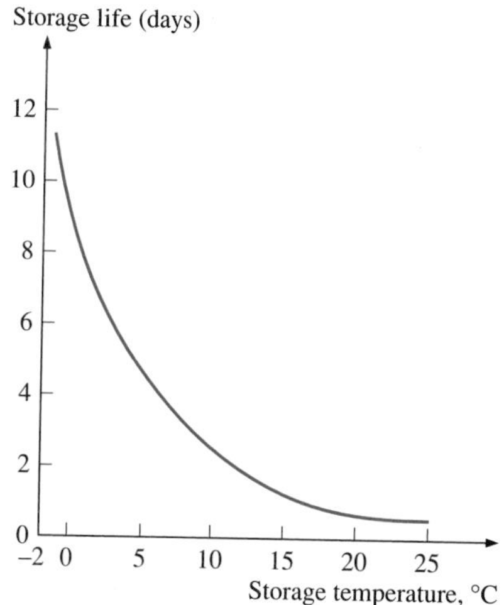

FIGURE 3-97
The storage life of fresh poultry decreases exponentially with increasing storage temperature.

FIGURE 3-98
The variation of freezing time of poultry with air temperature (from van der Berg and Lentz).

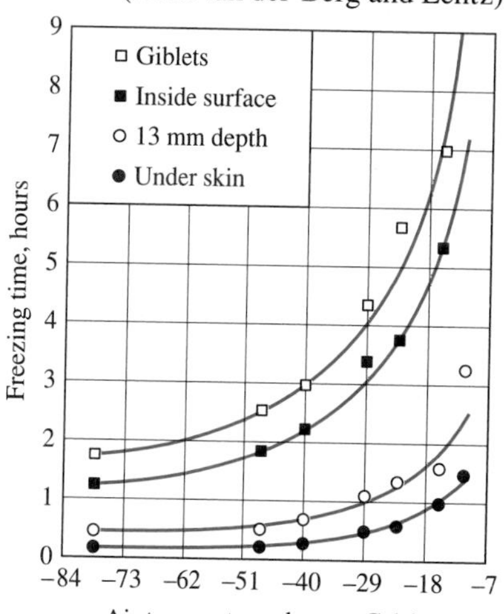

Note: Freezing time is the time required for temperature to fall from 0 to −4°C. The values are for 2.3 to 3.6 kg chickens with initial temperature of 0 to 2°C and with air velocity of 2.3 to 2.8 m/s.

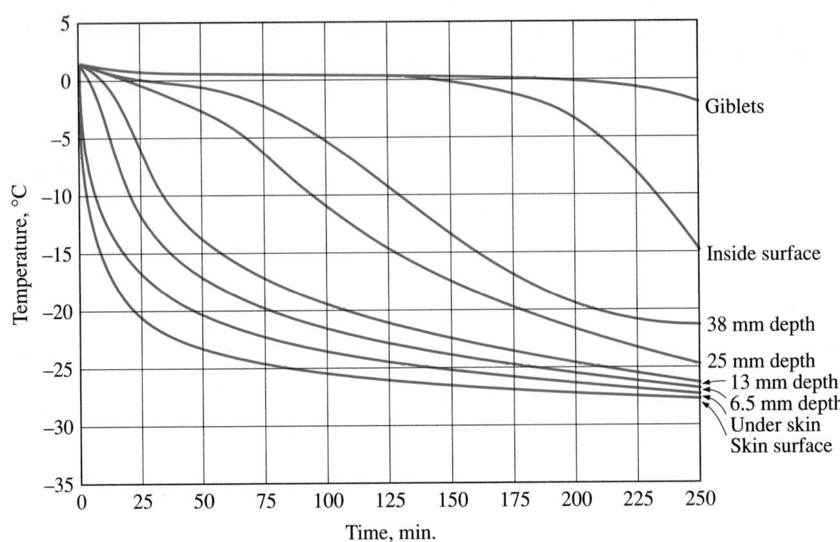

FIGURE 3-99

The variation of breast of 6.8 kg turkeys with depth during immersion cooling at −29°C (from van der Berg and Lentz).

TABLE 3-8

Thermal properties of poultry

Quantity	Typical value
Average density:	
Muscle	1070 kg/m^3
Skin	1030 kg/m^3
Specific heat:	
Above freezing	2.94 kJ/(kg · °C)
Below freezing	1.55 kJ/(kg · °C)
Freezing point	−2.8°C
Latent heat of fusion	247 kJ/kg
Thermal conductivity	[in W/(m · °C)]
Breast muscle	0.502 at 20°C
	1.384 at −20°C
	1.506 at −40°C
Dark muscle	1.557 at −40°C

brine, and *cryogenic cooling* with liquid nitrogen. Poultry can be frozen in several hours by cold plates. Very high freezing rates can be obtained by *immersing* the packaged birds into a low temperature brine. The freezing time of birds in −29°C brine can be as low as 20 min, depending on the size of the bird (Fig. 3-99). Also, immersion freezing produces a very appealing light appearance, and the high rate of heat transfer makes continuous line operation feasible. It also has lower initial and maintenance costs than forced air, but *leaks* into the packages through some small holes or cracks remain a concern. Sometimes liquid nitrogen is used to crust freeze the poultry products to −73°C. The freezing is then completed with air in a holding room at −23°C.

Properly packaged poultry products can be *stored* frozen for up to about a year at temperatures of −18°C or lower. The storage life drops considerably at higher (but still below-freezing) temperatures. Significant changes occur in flavor and juiciness when poultry is frozen for too long, and a stale rancid odor develops. Frozen poultry may become dehydrated and experience **freezer burn,** which may reduce the eye appeal of the product and cause toughening of the affected area. Dehydration and thus freezer burn can be controlled by *humidification, lowering* the storage temperature, and packaging the product in essentially *impermeable* film. The storage life can be extended by packing the poultry in an *oxygen-free* environment. The bacterial counts in precooked frozen products must be kept at safe levels since bacteria may not be destroyed completely during the reheating process at home.

Frozen poultry can be *thawed* in ambient air, water, refrigerator, or oven without any significant difference in taste. Big birds like turkey should be thawed safely by holding them in a refrigerator at 2 to 4°C for two to four days, depending on the size of the bird. They can also be thawed by immersing them into cool water in a large container for 4 to 6 h or by holding them in a paper bag. Care must be exercised to keep the bird's surface *cool* to minimize *microbiological growth* when thawing in air or water.

EXAMPLE 3-25 Freezing of Chicken in a Box

A supply of 50 kg of chicken at 6°C contained in a box is to be frozen to −18°C in a freezer. Determine the amount of heat that needs to be removed. The latent heat of the chicken is 247 kJ/kg, and its specific heat is 3.32 kJ/(kg · °C) above freezing and 1.77 kJ/(kg · °C) below freezing. The container box is 1.5 kg, and the specific heat of the box material is 1.4 kJ/(kg · °C). Also, the freezing temperature of chicken is −2.8°C.

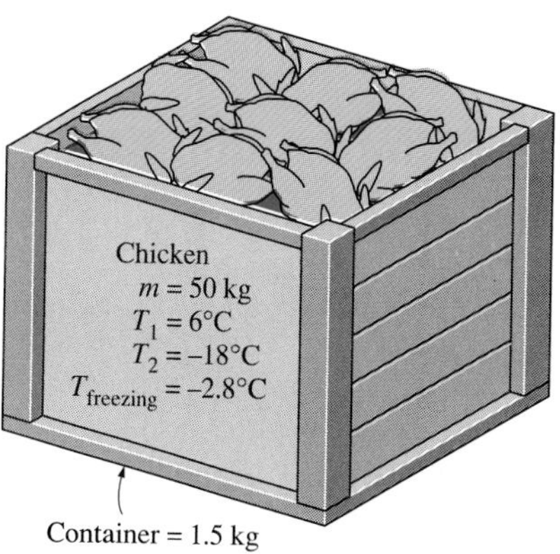

FIGURE 3-100

Schematic for Example 3-25.

Solution We take the chicken and the box they are in as the *system* (Fig. 3-100). This is a *closed system* since it involves a fixed mass (no mass transfer). We observe that the total amount of heat that needs to be removed (the cooling load of the freezer) is the sum of the latent heat and the sensible heats of the chicken before and after freezing, as well as the sensible heat of the box, and is determined below.

Assumptions **1** The energy change of the air in the box is negligible. **2** The thermal properties of fresh and frozen chicken are constant. **3** The entire water content of chicken freezes during the process. **4** The system is stationary and thus the kinetic and potential energy changes are zero, $\Delta KE = \Delta PE = 0$ and $\Delta E = \Delta U$. **5** There is no electrical, shaft, boundary, or any other kind of work involved.

Analysis Under the stated assumptions and observations, the energy balance on the system reduces to

$$\underbrace{E_{in} - E_{out}}_{\substack{\text{Net energy transfer} \\ \text{by heat, work, and mass}}} = \underbrace{\Delta E_{system}}_{\substack{\text{Change in internal, kinetic,} \\ \text{potential, etc., energies}}}$$

$$-Q_{out} = \Delta U_{chicken} + \Delta U_{box}$$

The total amount of heat that needs to be removed is the sum of the latent heat and the sensible heats of the chicken before and after freezing, as well as the sensible heat of the box:

Cooling fresh chicken from 6°C to −2.8°C:

$$\Delta U_{fresh\ chicken} = mC\Delta T = (50\text{ kg})[3.32\text{ kJ/(kg}\cdot{}^\circ\text{C)}](-2.8 - 6)\,{}^\circ\text{C} = -1461\text{ kJ}$$

Freezing chicken at −2.8°C:

$$\Delta U_{freezing} = mu_{latent} = (50\text{ kg})(-247\text{ kJ/kg}) = -12{,}350\text{ kJ}$$

Cooling frozen chicken from −2.8°C to −18°C:

$$\Delta U_{frozen\ chicken} = mC\Delta T = (50\text{ kg})[1.77\text{ kJ/(kg}\cdot{}^\circ\text{C)}][-18 - (-2.8)]{}^\circ\text{C} -1345\text{ kJ}$$

Cooling the box from 6°C to −18°C:

$$\Delta U_{box} = (mC\Delta T)_{box} = (1.5\text{kg})[1.4\text{ kJ/(kg}\cdot{}^\circ\text{C)}](-18 - 6){}^\circ\text{C} = -50\text{ kJ}$$

Therefore, the total amount of heat that needs to be removed is

$$\begin{aligned} Q_{out} &= -\Delta U_{chicken} - \Delta U_{box} \\ &= -(\Delta U_{fresh\ chicken} + \Delta U_{freezing} + \Delta U_{frozen\ chicken}) - \Delta U_{box} \\ &= 1461 + 12{,}350 + 1345 + 50 = \mathbf{15{,}206\ kJ} \end{aligned}$$

Discussion Note that most of the cooling load of the refrigeration system (81 percent of it) is due to the removal of the latent heat during the phase-change process. Also note that the cooling load due to the box is negligible (less than 1 percent) and can be ignored in calculations.

3-11 ■ SUMMARY

The first law of thermodynamics is essentially an expression of the conservation of energy principle. Energy can cross the boundaries of a closed system in the form of heat or work. If the energy transfer is due to a temperature difference between a system and its surroundings, it is *heat;* otherwise, it is *work*.

Heat is transferred in three different ways: conduction, convection, and radiation. *Conduction* is the transfer of energy from the more energetic particles of a substance to the adjacent less energetic ones as a result of interactions between the particles. *Convection* is the mode of energy transfer between a solid surface and the adjacent liquid or gas that is in motion, and it involves the combined effects of conduction and fluid motion. *Radiation* is the energy emitted by matter in the form of electromagnetic waves (or photons) as a result of the changes in the electronic configurations of the atoms or molecules. The three modes of heat transfer are expressed as

$$\dot{Q}_{\text{cond}} = -k_t A \frac{dT}{dx} \qquad \text{(W)}$$

$$\dot{Q}_{\text{conv}} = hA(T_s - T_f) \qquad \text{(W)}$$

$$\dot{Q}_{\text{rad}} = \varepsilon\sigma A(T_s^4 - T_{\text{surr}}^4) \qquad \text{(W)}$$

Various forms of work are expressed as follows:

Electrical work: $W_e = VI\,\Delta t$ (kJ)

Boundary work: $W_b = \int_1^2 P\,dV$ (kJ)

Gravitational work (= ΔPE): $W_g = mg(z_2 - z_1)$ (kJ)

Accelerational work (= ΔKE): $W_a = \frac{1}{2}m(\mathcal{V}_2^2 - \mathcal{V}_1^2)$ (kJ)

Shaft work: $W_{\text{sh}} = 2\pi n\tau$ (kJ)

Spring work: $W_{\text{spring}} = \frac{1}{2}k(x_2^2 - x_1^2)$ (kJ)

For the *polytropic process* (Pv^n = constant) of real gases, the boundary work can be expressed as

$$W_b = \frac{P_2V_2 - P_1V_1}{1 - n} \qquad (n \neq 1) \qquad \text{(kJ)}$$

The energy balances for *any system* undergoing *any process* can be expressed as

$$\underbrace{E_{\text{in}} - E_{\text{out}}}_{\substack{\text{Net energy transfer}\\ \text{by heat, work, and mass}}} = \underbrace{\Delta E_{\text{system}}}_{\substack{\text{Change in internal, kinetic,}\\ \text{potential, etc., energies}}} \qquad \text{(kJ)}$$

They can also be expressed in the *rate form* as

$$\underbrace{\dot{E}_{\text{in}} - \dot{E}_{\text{out}}}_{\substack{\text{Rate of net energy transfer}\\ \text{by heat, work, and mass}}} = \underbrace{\Delta \dot{E}_{\text{system}}}_{\substack{\text{Rate of change in internal, kinetic,}\\ \text{potential, etc., energies}}} \qquad \text{(kW)}$$

Taking heat transfer *to* the system and work done *by* the system to be positive quantities, the energy balance for a closed system can also be expressed as

$$Q - W = \Delta U + \Delta \text{KE} + \Delta \text{PE} \qquad \text{(kJ)}$$

where

$$W = W_{\text{other}} + W_b$$
$$\Delta U = m(u_2 - u_1)$$
$$\Delta \text{KE} = \tfrac{1}{2} m(\mathcal{V}_2^2 - \mathcal{V}_1^2)$$
$$\Delta \text{PE} = mg(z_2 - z_1)$$

For a constant-pressure process, $W_b + \Delta U = \Delta H$. Thus,

$$Q - W_{\text{other}} = \Delta H + \Delta \text{KE} + \Delta \text{PE} \qquad \text{(kJ)}$$

The amount of energy needed to raise theemperature of a unit mass of a substance by one degree is called the *specific heat at constant volume* C_v for a constant-volume process and the *specific heat at constant pressure* C_p for a constant-pressure process. They are defined as

$$C_v = \left(\frac{\partial u}{\partial T}\right)_v \quad \text{and} \quad C_p = \left(\frac{\partial h}{\partial T}\right)_p$$

h = Cp(T2−T1)
u = Cv(T2−T1)

For ideal gases u, h, C_v, and C_p are functions of temperature alone. The Δu and Δh of ideal gases can be expressed as

$$\Delta u = u_2 - u_1 = \int_1^2 C_v(T)\, dT \cong C_{v,\text{av}}(T_2 - T_1)$$

$$\Delta h = h_2 - h_1 = \int_1^2 C_p(T)\, dT \cong C_{p,\text{av}}(T_2 - T_1)$$

For ideal gases, C_v and C_p are related by

$$C_p = C_v + R \qquad [\text{kJ/(kg} \cdot \text{K)}]$$

where R is the gas constant. The *specific heat ratio* k is defined as

$$k = \frac{C_p}{C_v}$$

For *incompressible substances* (liquids and solids), both the constant-pressure and constant-volume specific heats are identical and denoted by C:

$$C_p = C_v = C \qquad [\text{kJ/(kg} \cdot \text{K)}]$$

The Δu and Δh of incompressible substances are given by

$$\Delta u = \int_1^2 C(T)\, dT \cong C_{av}(T_2 - T_1) \qquad (\text{kJ/kg})$$

$$\Delta h = \Delta u + v\, \Delta P \qquad (\text{kJ/kg})$$

The refrigeration and freezing of foods is a major application area of thermodynamics.

REFERENCES AND SUGGESTED READING

1. ASHRAE *Handbook of Fundamentals*. SI version. Atlanta, GA: American Society of Heating, Refrigerating, and Air-Conditioning Engineers, Inc., 1993.

2. ASHRAE *Handbook of Refrigeration*. SI version. Atlanta, GA: American Society of Heating, Refrigerating, and Air-Conditioning Engineers, Inc., 1994.

3. R. Balmer. *Thermodynamics*. St. Paul, MN: West Publishing, 1990.

4. A. Bejan. *Advanced Engineering Thermodynamics*. New York: John Wiley & Sons, 1988.

5. W. Z. Black and J. G. Hartley. *Thermodynamics*. New York: Harper and Row, 1985.

6. Y. A. Çengel. *Heat Transfer: A Practical Approach*. New York: McGraw-Hill, 1998.

7. H. Hillman. *Kitchen Science*. Mount Vernon, NY: Consumers Union, 1981.

8. J. B. Jones and G. A. Hawkins. *Engineering Thermodynamics*. 2nd ed. New York: John Wiley & Sons, 1986.

9. G. J. Van Wylen and R. E. Sonntag. *Fundamentals of Classical Thermodynamics*. 3rd ed. New York: John Wiley & Sons, 1985.

10. K. Wark. *Thermodynamics*. 5th ed. New York: McGraw-Hill, 1988.

PROBLEMS*

Heat Transfer and Work

3-1C In what forms can energy cross the boundaries of a system?

*Students are encouraged to answer *all* the concept "C" questions.

3-2C When is the energy crossing the boundaries of a system heat and when is it work?

3-3C What is an adiabatic process? What is an adiabatic system?

3-4C A gas in a piston-cylinder device is compressed, and as a result its temperature rises. Is this a heat or work interaction for the gas?

3-5C A room is heated by an iron that is left plugged in. Is this a heat or work interaction? Take the entire room, including the iron, as the system.

3-6C A room is heated as a result of solar radiation coming in through the windows. Is this a heat or work interaction for the room?

3-7C An insulated room is heated by burning candles. Is this a heat or work interaction? Take the entire room, including the candles, as the system.

3-8C What are point and path functions? Give some examples.

3-9C What are the mechanisms of heat transfer?

3-10C What is the caloric theory? When and why was it abandoned?

3-11C Does any of the energy of the sun reach the earth by conduction or convection?

3-12C Which is a better heat conductor, diamond or silver?

3-13C How does forced convection differ from natural convection?

3-14C Define emissivity and absorptivity. What is Kirchhoff's law of radiation?

3-15C What is a blackbody? How do real bodies differ from a blackbody?

3-16 The inner and outer surfaces of a 5-m × 6-m brick wall of thickness 30 cm and thermal conductivity 0.69 W/(m · °C) are maintained at temperatures of 20°C and 5°C, respectively. Determine the rate of heat transfer through the wall, in W.

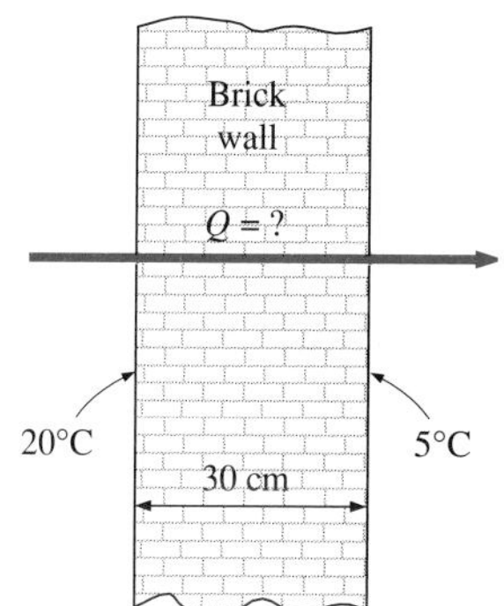

FIGURE P3-16

3-17 The inner and outer surfaces of a 0.5-cm-thick 2-m × 2-m window glass in winter are 10°C and 3°C, respectively. If the thermal conductivity of the glass is 0.78 W/(m · °C), determine the amount of heat loss, in kJ, through the glass over a period of 5 h. What would your answer be if the glass were 1-cm thick?

3-18 An aluminum pan whose thermal conductivity is 237 W/(m · °C) has a flat bottom whose diameter is 20 cm and thickness 0.4 cm. Heat is transferred steadily to boiling water in the pan through its bottom at a rate of 500 W. If the inner surface of the bottom of the pan is 105°C, determine the temperature of the outer surface of the bottom of the pan.

3-19 For heat transfer purposes, a standing man can be modeled as a 30-cm diameter, 170-cm long vertical cylinder with both the top and bottom surfaces insulated and with the side surface at an average temperature of 34°C. For a convection heat transfer coefficient of 15 W/(m^2 · °C), determine

the rate of heat loss from this man by convection in an environment at 20°C.
Answer: 336 W

3-20 A 5-cm-diameter spherical ball whose surface is maintained at a temperature of 70°C is suspended in the middle of a room at 20°C. If the convection heat transfer coefficient is 15 W/(m^2 · °C) and the emissivity of the surface is 0.8, determine the total rate of heat transfer from the ball.

3-21 Hot air at 80°C is blown over a 2-m × 4-m flat surface at 30°C. If the convection heat transfer coefficient is 55 W/(m^2 · °C), determine the rate of heat transfer from the air to the plate, in kW.

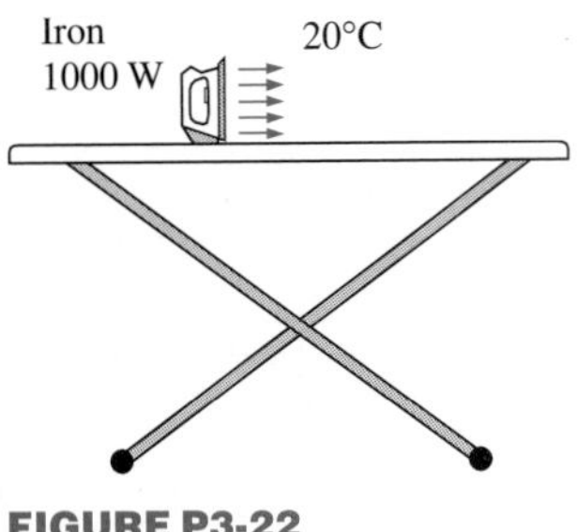

FIGURE P3-22

3-22 A 1000-W iron is left on the ironing board with its base exposed to the air at 20°C. The convection heat transfer coefficient between the base surface and the surrounding air is 35 W/(m^2 · °C). If the base has an emissivity of 0.6 and a surface area of 0.02 m^2, determine the temperature of the base of the iron.

3-23 A thin metal plate is insulated at the back and exposed to solar radiation at the front surface. The exposed surface of the plate has an absorptivity of 0.6 for solar radiation. If solar radiation is incident on the plate at a rate of 700 W/m^2 and the surrounding air temperature is 25°C, determine the surface temperature of the plate when the heat loss by convection equals the solar energy absorbed by the plate. Assume the convection heat transfer coefficient to be 50 W/(m^2 · °C), and disregard heat loss by radiation.

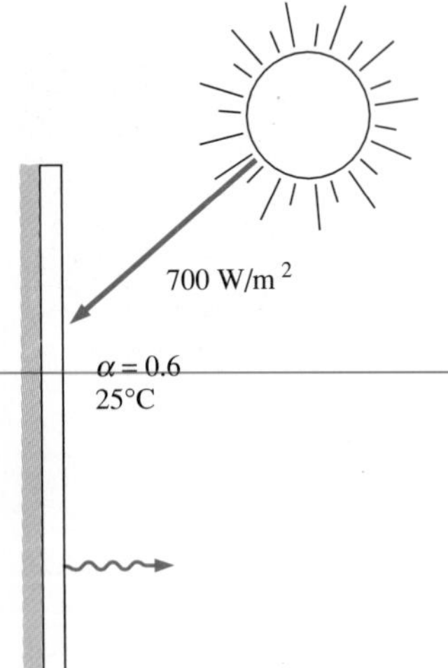

FIGURE P3-23

3-24 A 5-cm-external-diameter, 10-m-long hot water pipe at 80°C is losing heat to the surrounding air at 5°C by natural convection with a heat transfer coefficient of 25 W/(m^2 · °C). Determine the rate of heat loss from the pipe by natural convection, in kW.

3-25 The outer surface of a spacecraft in space has an emissivity of 0.8 and an absorptivity of 0.3 for solar radiation. If solar radiation is incident on the spacecraft at a rate of 1000 W/m^2, determine the surface temperature of the spacecraft when the radiation emitted equals the solar energy absorbed.

FIGURE P3-26

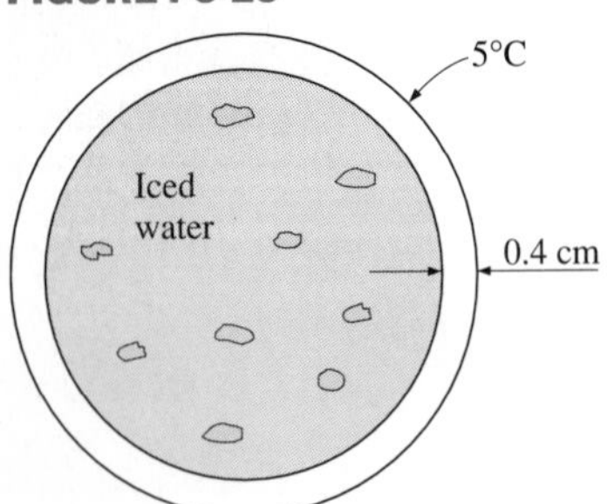

3-26 A hollow spherical iron container whose outer diameter is 20 cm and thickness is 0.4 cm is filled with iced water at 0°C. If the outer surface temperature is 5°C, determine the approximate rate of heat loss from the sphere, and the rate at which ice melts in the container.

3-27 The inner and outer glasses of a 2-m × 2-m double pane window are at 18°C and 6°C, respectively. If the 1-cm space between the two glasses is filled with still air, determine the rate of heat transfer through the window, in kW.

3-28 Two surfaces of a 2-cm-thick plate are maintained at 0°C and 100°C, respectively. If it is determined that heat is transferred through the plate at a rate of 500 W/m^2, determine its thermal conductivity.

Boundary Work

3-29C On a P-v diagram, what does the area under the process curve represent?

3-30C Is the boundary work associated with constant-volume systems always zero?

3-31C An ideal gas at a given state expands to a fixed final volume first at constant pressure and then at constant temperature. For which case is the work done greater?

3-32C Show that 1 kPa · m^3 = 1 kJ.

3-33 A mass of 5 kg of saturated water vapor at 200 kPa is heated at constant pressure until the temperature reaches 300°C. Calculate the work done by the steam during this process. *Answer:* 430.5 kJ

3-34 A frictionless piston-cylinder device initially contains 200 L of saturated liquid refrigerant-134a. The piston is free to move, and its mass is such that it maintains a pressure of 800 kPa on the refrigerant. The refrigerant is now heated until its temperature rises to 50°C. Calculate the work done during this process. *Answer:* 5227 kJ

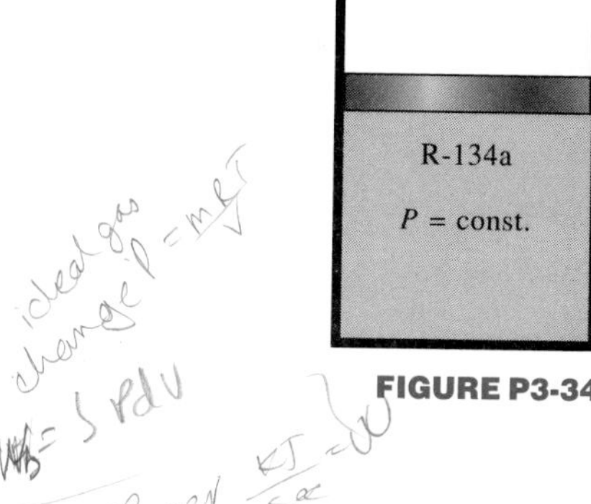

FIGURE P3-34

3-35E A frictionless piston-cylinder device contains 12 lbm of superheated water vapor at 60 psia and 500°F. Steam is now cooled at constant pressure until 70 percent of it, by mass, condenses. Determine the work done during this process.

3-36 A mass of 1.2 kg of air at 150 kPa and 12°C is contained in a gas-tight, frictionless piston-cylinder device. The air is now compressed to a final pressure of 600 kPa. During the process, heat is transferred from the air such that the temperature inside the cylinder remains constant. Calculate the work done during this process. *Answer:* −136.1 kJ

3-37 Nitrogen at an initial state of 300 K, 150 kPa, and 0.2 m^3 is compressed slowly in an isothermal process to a final pressure of 800 kPa. Determine the work done during this process.

3-38 A gas is compressed from an initial volume of 0.42 m^3 to a final volume of 0.12 m^3. During the quasi-equilibrium process, the pressure changes with volume according to the relation $P = aV + b$, where $a = -1200$ kPa/m^3 and $b = 600$ kPa. Calculate the work done during this process (*a*) by plotting the process on a P-V diagram and finding the area under the process curve and (*b*) by performing the necessary integrations.

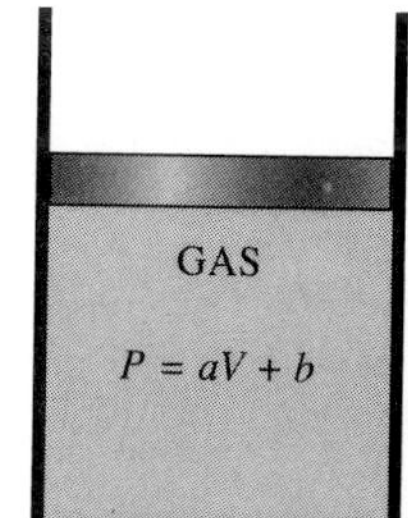

FIGURE P3-38

3-39E During an expansion process, the pressure of a gas changes from 15 to 100 psia according to the relation $P = aV + b$, where $a = 5$ psia/ft^3 and b is a constant. If the initial volume of the gas is 7 ft^3, calculate the work done during the process. *Answer:* 180.9 Btu

3-40 During some actual expansion and compression processes in piston-cylinder devices, the gases have been observed to satisfy the relationship $PV^n = C$, where n and C are constants. Calculate the work done when a gas expands from a state of 150 kPa and 0.03 m^3 to a final volume of 0.2 m^3 for the case of $n = 1.3$.

3-41 A frictionless piston-cylinder device contains 2 kg of nitrogen at 100 kPa and 300 K. Nitrogen is now compressed slowly according to the relation $PV^{1.4}$ = constant until it reaches a final temperature of 360 K. Calculate the work done during this process. *Answer:* -89.0 kJ

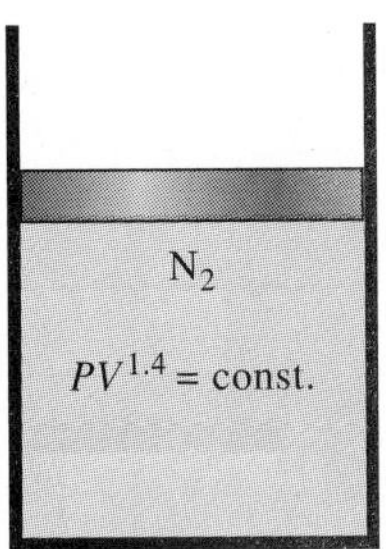

FIGURE P3-41

3-42 The equation of state of a gas is given as $\bar{v}(P + 10/\bar{v}^2) = R_uT$, where the units of $\bar{v}$ and P are m^3/kmol and kPa, respectively. Now 0.5 kmol of this gas is expanded in a quasi-equilibrium manner from 2 to 4 m^3 at a constant temperature of 300 K. Determine (*a*) the unit of the quantity 10 in the equation and (*b*) the work done during this isothermal expansion process.

3-43 Carbon dioxide contained in a piston-cylinder device is compressed from 0.3 to 0.1 m^3. During the process, the pressure and volume are related by $P = aV^{-2}$, where $a = 8$ kPa · m^6. Calculate the work done on the carbon dioxide during this process. *Answer:* -53.3 kJ

3-44E Hydrogen is contained in a piston-cylinder device at 14.7 psia and 15 ft^3. At this state, a linear spring ($F \propto x$) with a spring constant of 15,000 lbf/ft is touching the piston but exerts no force on it. The cross-sectional area of the piston is 3 ft^2. Heat is transferred to the hydrogen, causing it to expand until its volume doubles. Determine (*a*) the final pressure, (*b*) the total work done by the hydrogen, and (*c*) the fraction of this work done against the spring. Also, show the process on a *P-V* diagram.

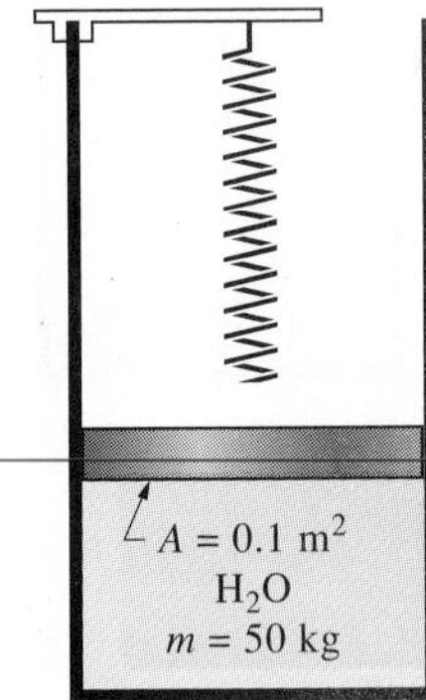

FIGURE P3-45

3-45 A piston-cylinder device contains 50 kg of water at 150 kPa and 25°C. The cross-sectional area of the piston is 0.1 m^2. Heat is now transferred to the water, causing part of it to evaporate and expand. When the volume reaches 0.2 m^3, the piston reaches a linear spring whose spring constant is 100 kN/m. More heat is transferred to the water until the piston rises 20 cm more. Determine (*a*) the final pressure and temperature and (*b*) the work done during this process. Also, show the process on a *P-V* diagram.
Answers: (*a*) 350 kPa, 138.88°C; (*b*) 27.5 kJ

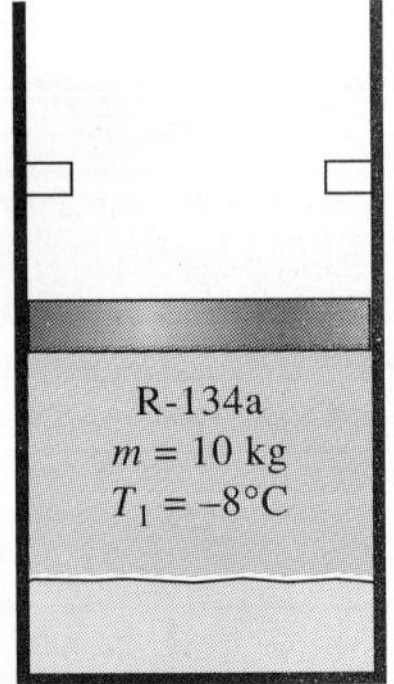

FIGURE P3-46

3-46 A piston-cylinder device with a set of stops contains 10 kg of refrigerant-134a. Initially, 8 kg of the refrigerant is in the liquid form, and the temperature is -8°C. Now heat is transferred slowly to the refrigerant until the piston hits the stops, at which point the volume is 400 L. Determine (*a*) the temperature when the piston first hits the stops and (*b*) the work done during this process. Also, show the process on a *P-V* diagram.
Answers: (*a*) -8°C, (*b*) 45.6 kJ

3-47 A frictionless piston-cylinder device contains 10 kg of saturated refrigerant-134a vapor at 50°C. The refrigerant is then allowed to expand isothermally by gradually decreasing the pressure in a quasi-equilibrium manner to a final value of 500 kPa. Determine the work done during this process (*a*) by using the experimental specific volume data from the tables and (*b*) by treating the refrigerant vapor as an ideal gas. Also, determine the error involved in the latter case.

3-48 Determine the boundary work done by a gas during an expansion process if the measured pressure and volume values at various states are measured to be 300 kPa, 1 L; 290 kPa, 1.1 L; 270 kPa, 1.2 L; 250 kPa, 1.4 L; 220 kPa, 1.7 L; and 200 kPa, 2 L.

Other Forms of Mechanical Work

3-49C A car is accelerated from rest to 85 km/h in 10 s. Would the work energy transferred to the car be different if it were accelerated to the same speed in 5 s?

3-50C Lifting a weight to a height of 20 m takes 20 s for one crane and 10 s for another. Is there any difference in the amount of work done on the weight by each crane?

3-51 Determine the work required to accelerate an 800-kg car from rest to 100 km/h on a level road. *Answer:* 308.6 kJ

3-52 Determine the work required to accelerate a 2000-kg car from 20 to 70 km/h on an uphill road with a vertical rise of 40 m.

3-53E Determine the torque applied to the shaft of a car that transmits 450 hp and rotates at a rate of 3000 rpm.

3-54 Determine the work required to deflect a linear spring with a spring constant of 70 kN/m by 20 cm from its rest position.

3-55 The engine of a 1500-kg automobile has a power rating of 75 kW. Determine the time required to accelerate this car from rest to a speed of 85 km/h at full power on a level road. Is your answer realistic?

3-56 A ski lift has a one-way length of 1 km and a vertical rise of 200 m. The chairs are spaced 20 m apart, and each chair can seat three people. The lift is operating at a steady speed of 10 km/h. Neglecting friction and air drag and assuming that the average mass of each loaded chair is 250 kg, determine the power required to operate this ski lift. Also estimate the power required to accelerate this ski lift in 5 s to its operating speed when it is first turned on.

3-57 Determine the power required for a 2000-kg car to climb a 100-m-long uphill road with a slope of 30° (from horizontal) in 10 s (*a*) at a constant velocity, (*b*) from rest to a final velocity of 30 m/s, and (*c*) from 35 m/s to a final velocity of 5 m/s. Disregard friction, air drag, and rolling resistance.
Answers: (*a*) 98.07 kW, (*b*) 188.07 kW, (*c*) −21.93 kW

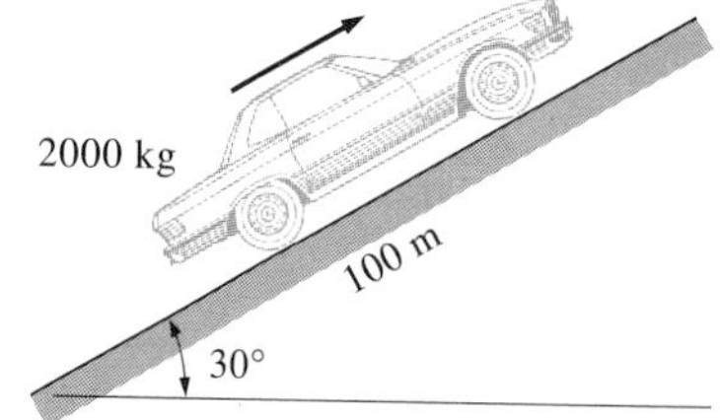

FIGURE P3-57

3-58 A damaged 1200-kg car is being towed by a truck. Neglecting the friction, air drag, and rolling resistance, determine the extra power required (*a*) for constant velocity on a level road, (*b*) for constant velocity of 50 km/h on a 30° (from horizontal) uphill road, and (*c*) to accelerate on a level road from stop to 90 km/h in 12 s. *Answers:* (*a*) 0, (*b*) 81.7 kW, (*c*) 31.25 kW

Closed-System Energy Analysis: General Systems

3-59C For a cycle, is the net work necessarily zero? For what kind of systems will this be the case?

3-60C Under what conditions is the relation $Q - W_{\text{other}} = H_2 - H_1$ valid for a closed system?

3-61C On a hot summer day, a student turns his fan on when he leaves his room in the morning. When he returns in the evening, will the room be warmer or cooler than the neighboring rooms? Why? Assume all the doors and windows are kept closed.

3-62C Consider two identical rooms, one with a refrigerator in it and the other without one. If all the doors and windows are closed, will the room that contains the refrigerator be cooler or warmer than the other room? Why?

3-63C Consider a can of soft drink that is dropped from the top of a tall building. Will the temperature of the soft drink increase as it falls, as a result of decreasing potential energy?

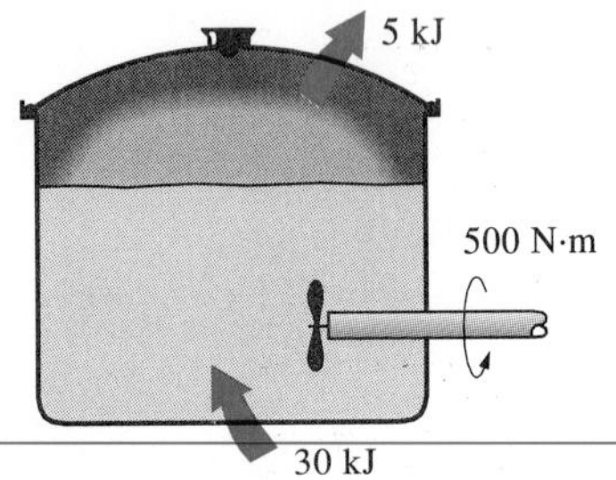

FIGURE P3-64

3-64 Water is being heated in a closed pan on top of a range while being stirred by a paddle wheel. During the process, 30 kJ of heat is transferred to the water, and 5 kJ of heat is lost to the surrounding air. The paddle-wheel work amounts to 500 N · m. Determine the final energy of the system if its initial energy is 10 kJ. *Answer:* 35.5 kJ

3-65E A vertical piston-cylinder device contains water and is being heated on top of a range. During the process, 50 Btu of heat is transferred to the water, and heat losses from the side walls amount to 8 Btu. The piston rises as a result of evaporation, and 5 Btu of boundary work is done. Determine the change in the energy of the water for this process. *Answer:* 37 Btu

3-66 Fill in the missing data for each of the following processes of a closed system between states 1 and 2. (Everything is in kJ.)

	Q_{in}	W_{out}	E_1	E_2	ΔE
(*a*)	18	6		35	
(*b*)	10			4	−15
(*c*)		12	3		32
(*d*)	25		14		10

3-67 Fill in the missing data for each of the following processes of a closed system between states 1 and 2. (Everything is in kJ.)

	Q_{in}	W_{out}	E_1	E_2	ΔE
(*a*)		18	6		20
(*b*)	5		20		35
(*c*)	25	10		40	
(*d*)	9			12	−15

3-68E A closed system undergoes a cycle consisting of two processes. During the first process, 40 Btu of heat is transferred to the system while the system does 60 Btu of work. During the second process, 45 Btu of work is done on the system.

(*a*) Determine the heat transfer during the second process.

(*b*) Calculate the net work and net heat transfer for the cycle.

3-69 A closed system undergoes a cycle consisting of three processes. During the first process, which is adiabatic, 50 kJ of work is done on the system. During the second process, 200 kJ of heat is transferred to the system while no work interaction takes place. And during the third process, the system does 90 kJ of work as it returns to its initial state.

(*a*) Determine the heat transfer during the last process.

(*b*) Determine the net work done during this cycle.

3-70 A classroom that normally contains 40 people is to be air-conditioned with window air-conditioning units of 5-kW rating. A person at rest may be assumed to dissipate heat at a rate of about 360 kJ/h. There are 10 light bulbs in the room, each with a rating of 100 W. The rate of heat transfer to the classroom through the walls and the windows is estimated to be 15,000 kJ/h. If the room air is to be maintained at a constant temperature of 21°C, determine the number of window air-conditioning units required.

Answer: 2 units

3-71 The lighting requirements of an industrial facility are being met by 700 40-W standard fluorescent lamps. The lamps are close to completing their service life and are to be replaced by their 34-W high-efficiency counterparts that operate on the existing standard ballasts. The standard and high-efficiency fluorescent lamps can be purchased in quantity at a cost of $1.77 and $2.26 each, respectively. The facility operates 2800 hours a year, and all of the lamps are kept on during operating hours. Taking the unit cost of electricity to be $0.08/kWh and the ballast factor to be 1.1 (i.e., ballasts consume 10 percent of the rated power of the lamps), determine how much energy and money will be saved per year as a result of switching to the high-efficiency fluorescent lamps. Also, determine the simple payback period.

3-72 The lighting needs of a storage room are being met by 6 fluorescent light fixtures, each fixture containing four lamps rated at 60 W each. All the lamps are on during operating hours of the facility, which are 6 A.M. to 6 P.M. 365 days a year. The storage room is actually used for an average of three hours a day. If the price of electricity is $0.08/kWh, determine the amount of energy and money that will be saved as a result of installing motion sensors. Also, determine the simple payback period if the purchase price of the sensor is $32 and it takes 1 hour to install it at a cost of $40.

3-73 A university campus has 200 classrooms and 400 faculty offices. The classrooms are equipped with 12 fluorescent light bulbs, each consuming 110 W, including the electricity used by the ballasts. The faculty offices, on average, have half as many light bulbs. The campus is open 240 days a year. The classrooms and faculty offices are not occupied an average of four hours a day, but the lights are kept on. If the unit cost of electricity is \$0.075/kWh, determine how much the campus will save a year if the lights in the classrooms and faculty offices are turned off during unoccupied periods.

Closed-System Energy Analysis: Saturation, Compressed Liquid, and Superheat Data

3-74 The radiator of a steam heating system has a volume of 20 L and is filled with superheated vapor at 300 kPa and 250°C. At this moment both the inlet and exit valves to the radiator are closed. Determine the amount of heat that will be transferred to the room when the steam pressure drops to 100 kPa. Also, show the process on a *P-v* diagram with respect to saturation lines. *Answer:* −33.4 kJ

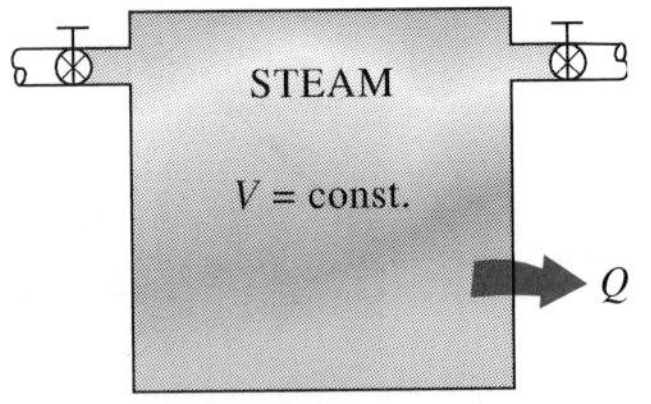

FIGURE P3-74

3-75 A 0.5-m³ rigid tank contains refrigerant-134a initially at 200 kPa and 40 percent quality. Heat is now transferred to the refrigerant until the pressure reaches 800 kPa. Determine (*a*) the mass of the refrigerant in the tank and (*b*) the amount of heat transferred. Also, show the process on a *P-v* diagram with respect to saturation lines.

3-76E A 20-ft³ rigid tank initially contains refrigerant-134a in the saturated vapor form at 120 psia. As a result of heat transfer from the refrigerant, the pressure drops to 30 psia. Show the process on a *P-v* diagram with respect to saturation lines, and determine (*a*) the final temperature, (*b*) the amount of refrigerant that has condensed, and (*c*) the heat transfer.

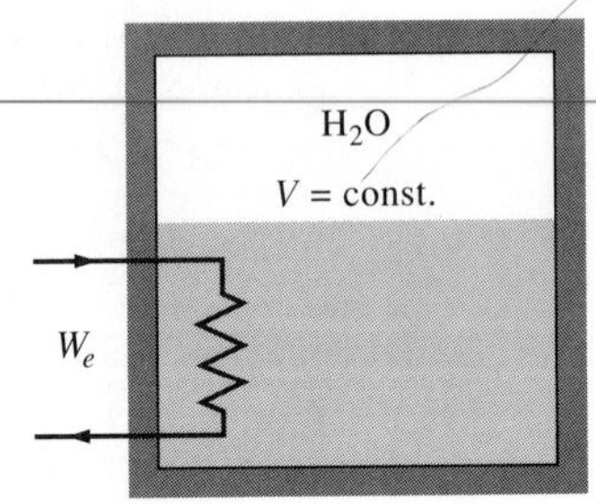

FIGURE P3-77

3-77 A well-insulated rigid tank contains 5 kg of a saturated liquid–vapor mixture of water at 100 kPa. Initially, three-quarters of the mass is in the liquid phase. An electric resistor placed in the tank is connected to a 110-V source, and a current of 8 A flows through the resistor when the switch is turned on. Determine how long it will take to vaporize all the liquid in the tank. Also, show the process on a *T-v* diagram with respect to saturation lines.

3-78 An insulated tank is divided into two parts by a partition. One part of the tank contains 2.5 kg of compressed liquid water at 60°C and 600 kPa while the other part is evacuated. The partition is now removed, and the water expands to fill the entire tank. Determine the final temperature of the water and the volume of the tank for a final pressure of 10 kPa.

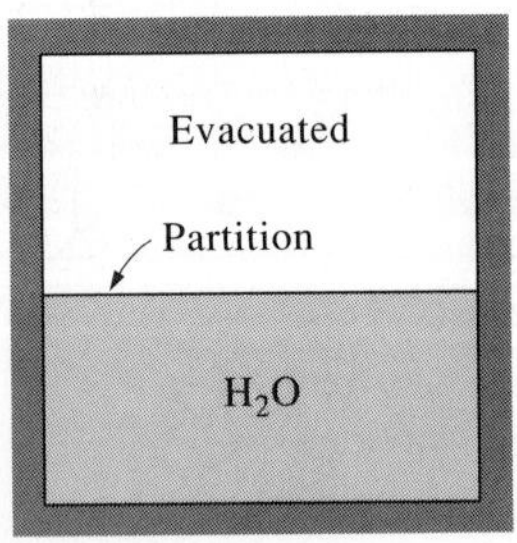

FIGURE P3-78

3-79 A piston-cylinder device contains 5 kg of refrigerant-134a at 800 kPa and 60°C. The refrigerant is now cooled at constant pressure until it exists as a liquid at 20°C. Determine the amount of heat loss and show the process on a *T-v* diagram with respect to saturation lines. *Answer:* 1089 kJ

3-80E A piston-cylinder device contains 0.5 lbm of water initially at 120 psia and 2 ft³. Now 200 Btu of heat is transferred to the water while its pressure is held constant. Determine the final temperature of the water. Also, show the process on a *T-v* diagram with respect to saturation lines.

3-81 An insulated piston-cylinder device contains 5 L of saturated liquid water at a constant pressure of 150 kPa. Water is stirred by a paddle wheel while a current of 8 A flows for 45 min through a resistor placed in the water. If one-half of the liquid is evaporated during this constant-pressure process and the paddle-wheel work amounts to 300 kJ, determine the voltage of the source. Also, show the process on a *P-v* diagram with respect to saturation lines. *Answer:* 230.9 V

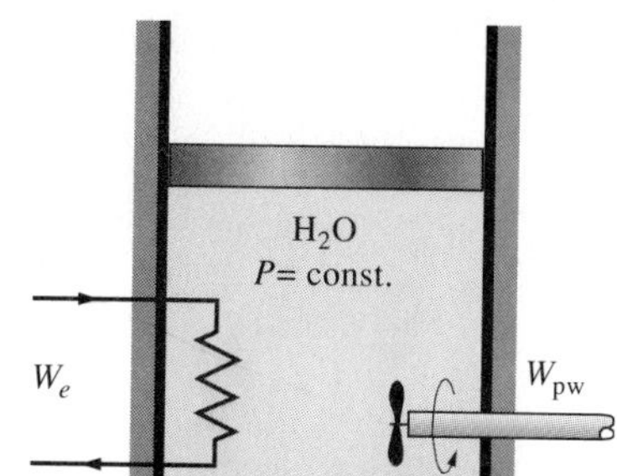

FIGURE P3-81

3-82 A piston-cylinder device contains steam initially at 1 MPa, 350°C, and 1.5 m^3. Steam is allowed to cool at constant pressure until it first starts condensing. Show the process on a *T-v* diagram with respect to saturation lines and determine (*a*) the mass of the steam, (*b*) the final temperature, and (*c*) the amount of heat transfer.

3-83 A piston-cylinder device initially contains steam at 200 kPa, 200°C, and 0.5 m^3. At this state, a linear spring ($F \propto x$) is touching the piston but exerts no force on it. Heat is now slowly transferred to the steam, causing the pressure and the volume to rise to 500 kPa and 0.6 m^3, respectively. Show the process on a *P-v* diagram with respect to saturation lines and determine (*a*) the final temperature, (*b*) the work done by the steam, and (*c*) the total heat transferred. *Answers:* (*a*) 1131°C, (*b*) 35 kJ, (*c*) 772 kJ

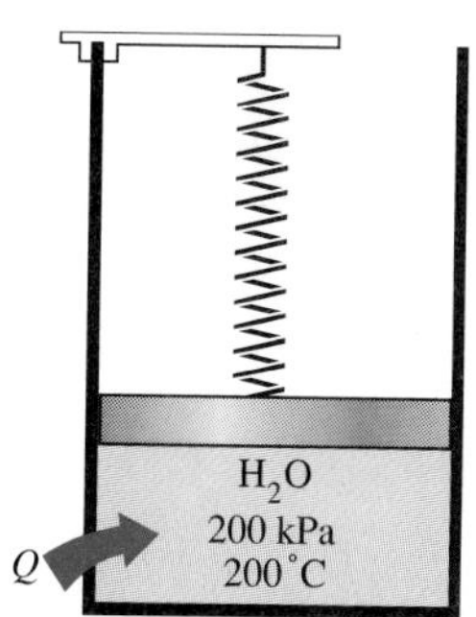

FIGURE P3-83

3-84 A piston-cylinder device initially contains 0.5 m^3 of saturated water vapor at 200 kPa. At this state, the piston is resting on a set of stops, and the mass of the piston is such that a pressure of 300 kPa is required to move it. Heat is now slowly transferred to the steam until the volume doubles. Show the process on a *P-v* diagram with respect to saturation lines and determine (*a*) the final temperature, (*b*) the work done during this process, and (*c*) the total heat transfer. *Answers:* (*a*) 878.90°C, (*b*) 150 kJ, (*c*) 875 kJ

3-85E A piston-cylinder device with a set of stops on the top contains 5 lbm of saturated liquid water at 20 psia. Heat is now transferred to the water, causing some of the liquid to evaporate and move the piston up. When the piston reaches the stops, the enclosed volume is 1.5 ft^3. More heat is transferred until the pressure is doubled. Show the process on a *P-v* diagram with respect to saturation lines. Determine (*a*) the amount of liquid at the final state, if any, (*b*) the final temperature, and (*c*) the total work and heat transfer.

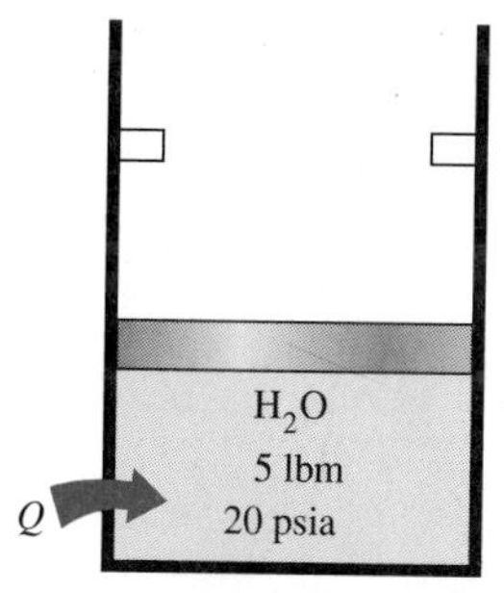

FIGURE P3-85E

3-86 Two rigid tanks are connected by a valve. Tank *A* contains 0.2 m^3 of water at 400 kPa and 80 percent quality. Tank *B* contains 0.5 m^3 of water at 200 kPa and 250°C. The valve is now opened, and the two tanks eventually come to the same state. Determine the pressure and the amount of heat transfer when the system reaches thermal equilibrium with the surroundings at 25°C. *Answers:* 3.169 kPa, −2170 kJ

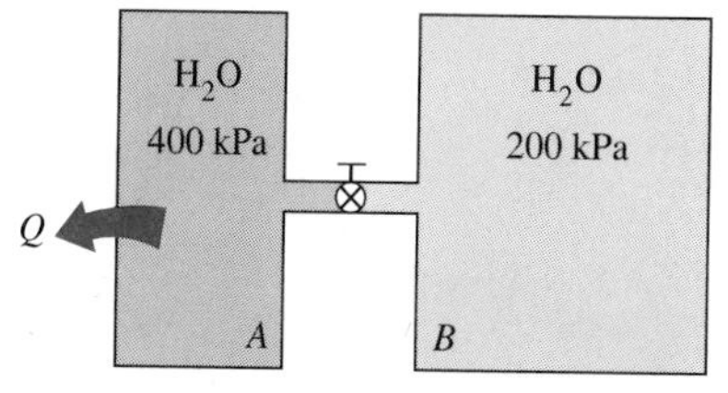

FIGURE P3-86

Specific Heats, Δu, and Δh of ideal Gases

3-87C Is the relation $\Delta U = mC_{v,\text{av}}\,\Delta T$ restricted to constant-volume processes only, or can it be used for any kind of process of an ideal gas? Answer the same question for $Q = mC_{v,\text{av}}\,\Delta T$.

3-88C Is the relation $\Delta H = mC_{p,\mathrm{av}}\,\Delta T$ restricted to constant-pressure processes only, or can it be used for any kind of process of an ideal gas?

3-89C Show that for an ideal gas $\overline{C}_p = \overline{C}_v + R_u$.

3-90C Is the energy required to heat air from 295 to 305 K the same as the energy required to heat it from 345 to 355 K? Assume the pressure remains constant in both cases.

3-91C In the relation $\Delta U = mC_v\Delta T$, what is the correct unit of C_v—kJ/(kg · °C) or kJ/(kg · K)?

3-92C A fixed mass of an ideal gas is heated from 50 to 80°C at a constant pressure of (*a*) 1 atm and (*b*) 3 atm. For which case do you think the energy required will be greater? Why?

3-93C A fixed mass of an ideal gas is heated from 50 to 80°C at a constant volume of (*a*) 1 m^3 and (*b*) 3 m^3. For which case do you think the energy required will be greater? Why?

3-94C A fixed mass of an ideal gas is heated from 50 to 80°C (*a*) at constant volume and (*b*) at constant pressure. For which case do you think the energy required will be greater? Why?

3-95 Determine the enthalpy change Δh of nitrogen, in kJ/kg, as it is heated from 600 to 1000 K, using (*a*) the empirical data for h from the nitrogen table (Table A-18), (*b*) the empirical specific heat equation as a function of temperature (Table A-2*c*), (*c*) the C_p value at the average temperature (Table A-2*b*), and (*d*) the C_p value at room temperature (Table A-2*a*). Also, determine the percentage error involved in each case.

Answers: (*a*) 448.6 kJ/kg; (*b*) 447.8 kJ/kg, 0.2 percent; (*c*) 448.4 kJ/kg, 0.04 percent; (*d*) 415.6 kJ/kg, 7.4 percent

3-96E Determine the enthalpy change Δh of oxygen, in Btu/lbm, as it is heated from 800 to 1500 R, using (*a*) the empirical data for h from the oxygen table (Table A-19E), (*b*) the empirical specific heat equation as a function of temperature (Table A-2E*c*), (*c*) the C_p value at the average temperature (Table A-2E*b*), and (*d*) the C_p value at room temperature (Table A-2E*a*).

Answers: (*a*) 169.2 Btu/lbm, (*b*) 170.1 Btu/lbm, (*c*) 178.5 Btu/lbm, (*d*) 153.3 Btu/lbm

3-97 Determine the internal energy change Δu of hydrogen, in kJ/kg, as it is heated from 400 to 1000 K, using (*a*) the empirical data for u from the hydrogen table (Table A-22), (*b*) the empirical specific heat equation as a function of temperature (Table A-2*c*), (*c*) the C_v value at average temperature (Table A-2*b*), and (*d*) the C_v value at room temperature (Table A-2*a*). Also, determine the percentage error involved in each case.

Closed-System Energy Analysis: Ideal Gases

3-98C Is it possible to compress an ideal gas isothermally in an adiabatic piston-cylinder device? Explain.

3-99E A rigid tank contains 20 lbm of air at 50 psia and 80°F. The air is now heated until its pressure doubles. Determine (*a*) the volume of the tank and (*b*) the amount of heat transfer. *Answers:* (*a*) 80 ft^3, (*b*) 2035 Btu

3-100 A 1-m^3 rigid tank contains hydrogen at 250 kPa and 500 K. The gas is now cooled until its temperature drops to 300 K. Determine (*a*) the final pressure in the tank and (*b*) the amount of heat transfer.

3-101 A 4-m × 5-m × 6-m room is to be heated by a baseboard resistance heater. It is desired that the resistance heater be able to raise the air temperature in the room from 7 to 23°C within 15 min. Assuming no heat losses from the room and an atmospheric pressure of 100 kPa, determine the required power of the resistance heater. Assume constant specific heats at room temperature. *Answer:* 1.91 kW

3-102 A 4-m × 5-m × 7-m room is heated by the radiator of a steam-heating system. The steam radiator transfers heat at a rate of 10,000 kJ/h, and a 100-W fan is used to distribute the warm air in the room. The rate of heat loss from the room is estimated to be about 5000 kJ/h. If the initial temperature of the room air is 10°C, determine how long it will take for the air temperature to rise to 20°C. Assume constant specific heats at room temperature.

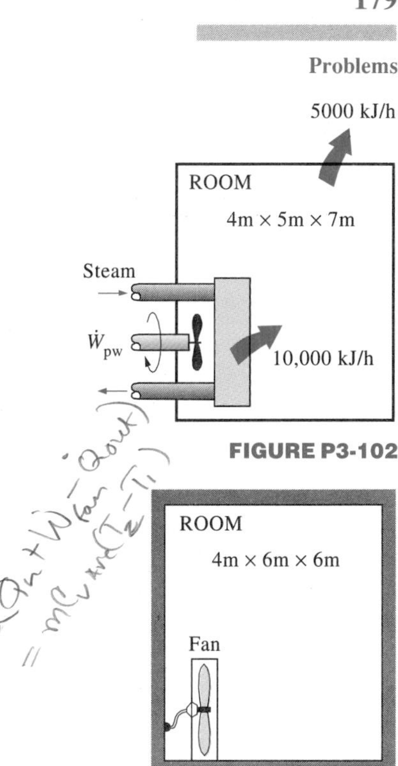

FIGURE P3-102

3-103 A student living in a 4-m × 6-m × 6-m dormitory room turns on her 150-W fan before she leaves the room on a summer day, hoping that the room will be cooler when she comes back in the evening. Assuming all the doors and windows are tightly closed and disregarding any heat transfer through the walls and the windows, determine the temperature in the room when she comes back 10 h later. Use specific heat values at room temperature, and assume the room to be at 100 kPa and 15°C in the morning when she leaves. *Answer:* 58.2°C

FIGURE P3-103

3-104E A 10-ft^3 tank contains oxygen initially at 14.7 psia and 80°F. A paddle wheel within the tank is rotated until the pressure inside rises to 20 psia. During the process 20 Btu of heat is lost to the surroundings. Determine the paddle-wheel work done. Neglect the energy stored in the paddle wheel.

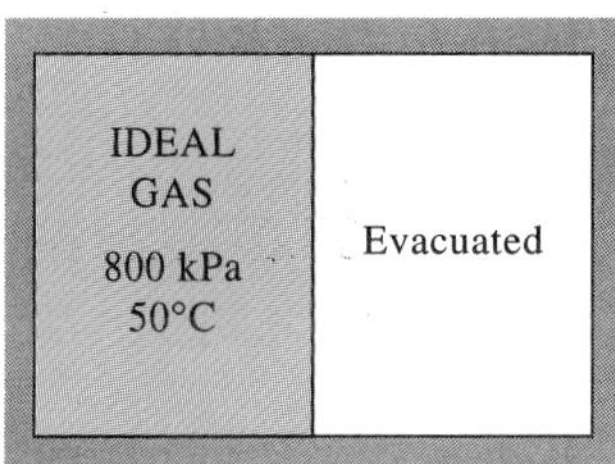

FIGURE P3-105

3-105 An insulated rigid tank is divided into two equal parts by a partition. Initially, one part contains 3 kg of an ideal gas at 800 kPa and 50°C, and the other part is evacuated. The partition is now removed, and the gas expands into the entire tank. Determine the final temperature and pressure in the tank.

3-106 A piston-cylinder device whose piston is resting on top of a set of stops initially contains 0.5 kg of helium gas at 100 kPa and 25°C. The mass of the piston is such that 500 kPa of pressure is required to raise it. How much heat must be transferred to the helium before the piston starts rising? *Answer:* 1857 kJ

3-107 An insulated piston-cylinder device contains 100 L of air at 400 kPa and 25°C. A paddle wheel within the cylinder is rotated until 15 kJ of work is done on the air while the pressure is held constant. Determine the final temperature of the air. Neglect the energy stored in the paddle wheel.

FIGURE P3-107

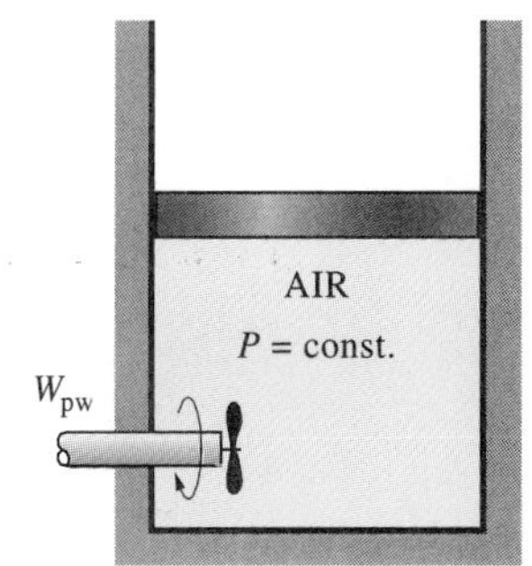

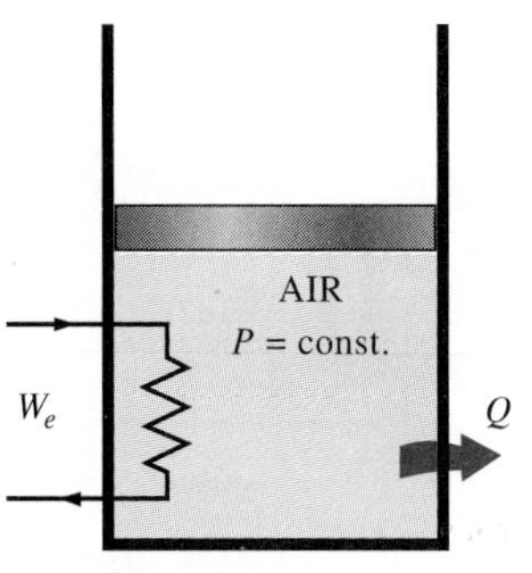

FIGURE P3-109

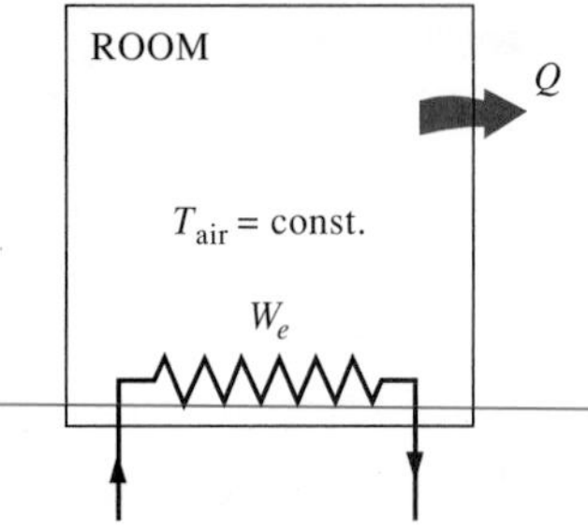

FIGURE P3-112

3-108E A piston-cylinder device contains 25 ft^3 of nitrogen at 50 psia and 700°F. Nitrogen is now allowed to cool at constant pressure until the temperature drops to 140°F. Using data from the nitrogen table, determine the heat transfer.

3-109 A mass of 15 kg of air in a piston-cylinder device is heated from 25 to 77°C by passing current through a resistance heater inside the cylinder. The pressure inside the cylinder is held constant at 300 kPa during the process, and a heat loss of 60 kJ occurs. Determine the electric energy supplied, in kWh. *Answer:* 0.235 kWh

3-110 An insulated piston-cylinder device initially contains 0.3 m^3 of carbon dioxide at 200 kPa and 27°C. An electric switch is turned on, and a 110-V source supplies current to a resistance heater inside the cylinder for a period of 10 min. The pressure is held constant during the process, while the volume is doubled. Determine the current that passes through the resistance heater.

3-111 A piston-cylinder device contains 0.8 kg of nitrogen initially at 100 kPa and 27°C. The nitrogen is now compressed slowly in a polytropic process during which $PV^{1.3}$ = constant until the volume is reduced by one-half. Determine the work done and the heat transfer for this process.

3-112 A room is heated by a baseboard resistance heater. When the heat losses from the room on a winter day amount to 8000 kJ/h, the air temperature in the room remains constant even though the heater operates continuously. Determine the power rating of the heater, in kW.

3-113E A piston-cylinder device contains 3 ft^3 of air at 60 psia and 150°F. Heat is transferred to the air in the amount of 40 Btu as the air expands isothermally. Determine the amount of boundary work done during this process.

3-114 A piston-cylinder device contains 5 kg of argon at 400 kPa and 30°C. During a quasi-equilibrium, isothermal expansion process, 15 kJ of boundary work is done by the system, and 3 kJ of paddle-wheel work is done on the system. Determine the heat transfer for this process. *Answer:* 12 kJ

3-115 A piston-cylinder device, whose piston is resting on a set of stops initially contains 3 kg of air at 200 kPa and 27°C. The mass of the piston is such that a pressure of 400 kPa is required to move it. Heat is now transferred to the air until its volume doubles. Determine the work done by the air and the total heat transferred to the air during this process. Also show the process on a P-v diagram. *Answers:* 516 kJ, 2674 kJ

3-116 A piston-cylinder device, with a set of stops on the top, initially contains 3 kg of air at 200 kPa and 27°C. Heat is now transferred to the air, and the piston rises until it hits the stops, at which point the volume is twice the initial volume. More heat is transferred until the pressure inside the cylinder also doubles. Determine the work done and the amount of heat transfer for this process. Also, show the process on a P-v diagram.

3-117 A rigid tank containing 0.4 m^3 of air at 400 kPa and 30°C is connected by a valve to a piston-cylinder device with zero clearance. The mass of the piston is such that a pressure of 200 kPa is required to raise the piston. The valve is now opened slightly, and air is allowed to flow into the cylinder until the pressure in the tank drops to 200 kPa. During this process, heat is exchanged with the surroundings such that the entire air remains at 30°C at all times. Determine the heat transfer for this process.

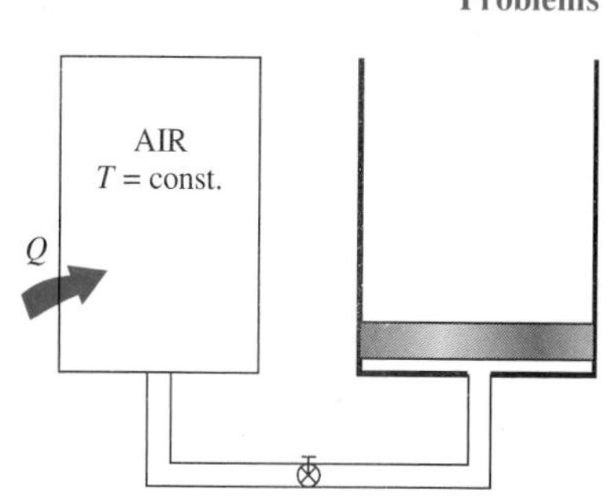

FIGURE P3-117

Closed-System Energy Analysis: Solids and Liquids

3-118 In a manufacturing facility, 5-cm-diameter brass balls [ρ = 8522 kg/m^3 and C_p = 0.385 kJ/(kg · °C)] initially at 120°C are quenched in a water bath at 50°C for a period of 2 minutes at a rate of 100 balls per minute. If the temperature of the balls after quenching is 74°C, determine the rate at which heat needs to be removed from the water in order to keep its temperature constant at 50°C.

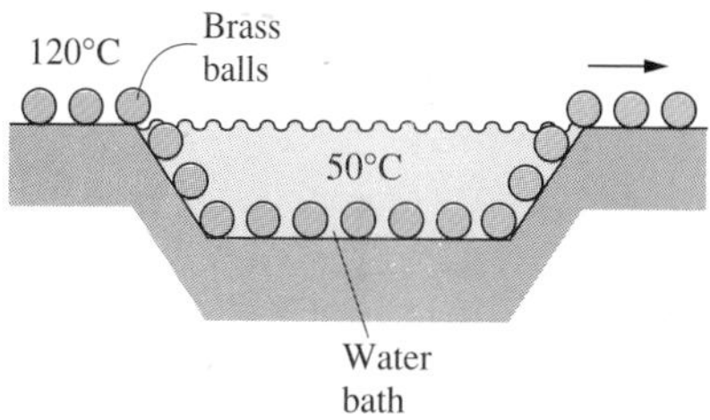

FIGURE P3-118

3-119 Repeat Prob. 3-118 for aluminum balls.

3-120E During a picnic on a hot summer day, all the cold drinks disappeared quickly, and the only available drinks were those at the ambient temperature of 75°F. In an effort to cool a 12-fluid-oz drink in a can, a person grabs the can and starts shaking it in the iced water of the chest at 32°F. Using the properties of water for the drink, determine the mass of ice that will melt by the time the canned drink cools to 45°F.

3-121 Consider a 1000-W iron whose base plate is made of 0.5-cm-thick aluminum alloy 2024-T6 [ρ = 2770 kg/m^3 and C_p = 875 J/(kg · °C)]. The base plate has a surface area of 0.03 m^2. Initially, the iron is in thermal equilibrium with the ambient air at 22°C. Assuming 85 percent of the heat generated in the resistance wires is transferred to the plate, determine the minimum time needed for the plate temperature to reach 140°C.

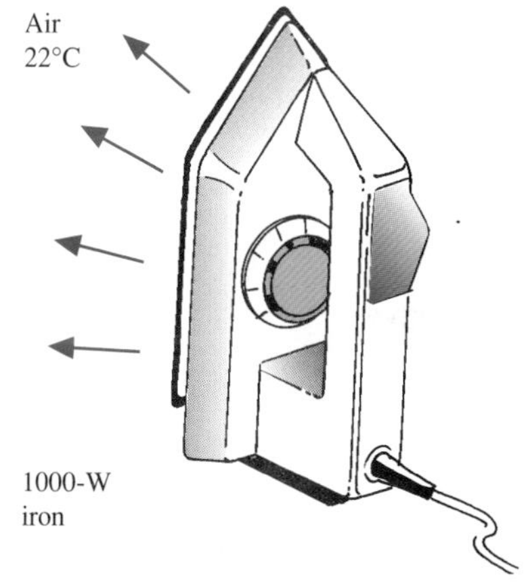

FIGURE P3-121

3-122 Stainless steel ball bearings [ρ = 8085 kg/m^3 and C_p = 0.480 kJ/(kg · °C)] having a diameter of 1.2 cm are to be quenched in water at a rate of 1400 per minute. The balls leave the oven at a uniform temperature of 900°C and are exposed to air at 30°C for a while before they are dropped into the water. If the temperature of the balls drops to 850°C prior to quenching, determine the rate of heat transfer from the balls to the air.

3-123 Carbon steel balls [ρ = 7833 kg/m^3 and C_p = 0.465 kJ/(kg · °C)] 8 mm in diameter are annealed by heating them first to 900°C in a furnace, and then allowing them to cool slowly to 100°C in ambient air at 35°C. If 2500 balls are to be annealed per hour, determine the total rate of heat transfer from the balls to the ambient air. *Answer:* 542 W

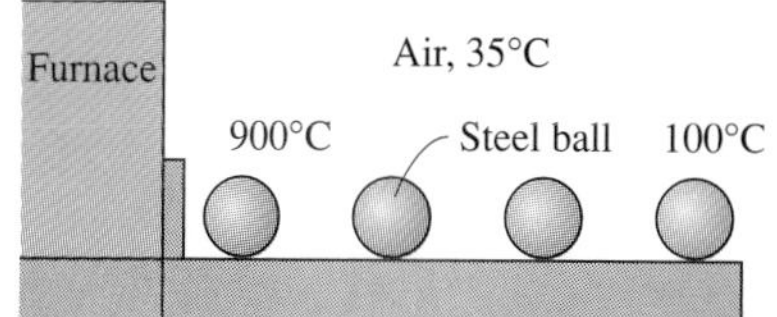

FIGURE P3-123

3-124 An electronic device dissipating 30 W has a mass of 20 g, a specific heat of 850 J/(kg · °C), and a surface area of 5 cm^2. The device is lightly used, and it is on for 5 minutes and then off for several hours, during which it cools to the ambient temperature of 25°C. Determine the highest possible temperature of the device at the end of the 5-min operating period. What would your answer be if the device were attached to a 0.2-kg aluminum heat sink? Assume the device and the heat sink to be nearly isothermal.

3-125 An ordinary egg can be approximated as a 5.5-cm-diameter sphere. The egg is initially at a uniform temperature of 8°C and is dropped into boiling water at 97°C. 'Taking the properties of the egg to be $\rho = 1020\ \text{kg/m}^3$ and $C_p = 3.32$ kJ/(kg · °C), determine how much heat is transferred to the egg by the time the average temperature of the egg rises to 70°C.

3-126E In a production facility, 1.2-in.-thick 2-ft × 2-ft square brass plates [$\rho = 532.5\ \text{lbm/ft}^3$ and $C_p = 0.091$ Btu/(lbm · °F)] that are initially at a uniform temperature of 75°F are heated by passing them through an oven at 1300°F at a rate of 300 per minute. If the plates remain in the oven until their average temperature rises to 1000°F, determine the rate of heat transfer to the plates in the furnace.

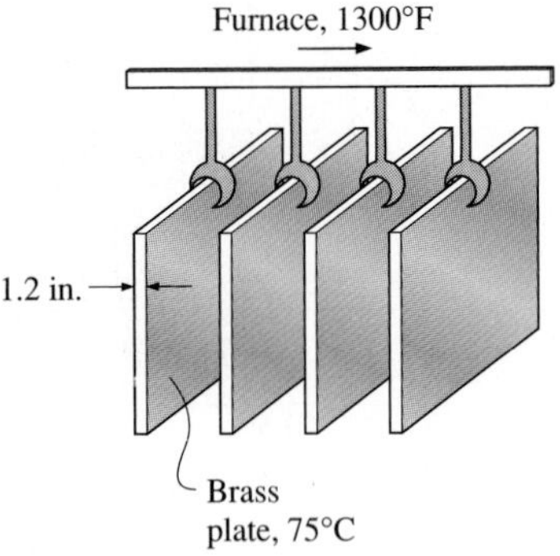

FIGURE P3-126E

3-127 Long cylindrical steel rods [$\rho = 7833\ \text{kg/m}^3$ and $C_p = 0.465$ kJ/(kg · °C)] of 10-cm diameter are heat-treated by drawing them at a velocity of 3 m/min through a 6-m-long oven maintained at 900°C. If the rods enter the oven at 30°C and leave at 700°C, determine the rate of heat transfer to the rods in the oven.

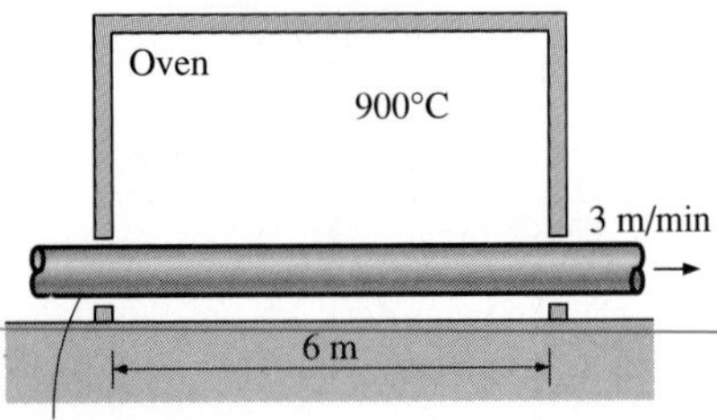

FIGURE P3-127

Refrigeration and Freezing of Foods

3-128C What are the common kinds of microorganisms? What undesirable changes do microorganisms cause in foods?

3-129C How does refrigeration prevent or delay the spoilage of foods? Why does freezing extend the storage life of foods for months?

3-130C What is the effect of cooking on the microorganisms in foods? Why is it important that the internal temperature of a roast in an oven be raised above 70°C?

3-131C What is the difference between the freezing injury and the chilling injury of fruits and vegetables?

3-132C How does the rate of freezing affect the size of the ice crystals that form during freezing and the quality of the frozen food products?

3-133C Which type of freezing is more likely to cause dehydration of foods: airblast freezing or cryogenic freezing?

3-134 Explain how the latent heat of fusion of food products whose water content is known is determined.

3-135C Whose specific heat is greater: apricots with a water content of 70 percent or apples with a water content of 82 percent?

3-136C Do carrots freeze at a fixed temperature or over a range of temperatures? Explain.

3-137C Consider 1 kg of cherries with a water content of 75 percent and 1 kg of roast beef also with a water content of 75 percent both at 5°C. Now both the cherries and the beef are completely frozen to −40°C. The heat removed from the cherries will be (*a*) much less than, (*b*) about equal to, or (*c*) much greater than the heat removed from the beef.

3-138C Why does a beef carcass lose up to 2 percent of its weight as it is cooled in the chilling room? How can this weight loss be minimized?

3-139C How does immersion chilling of poultry compare to forced-air chilling with respect to (*a*) cooling time, (*b*) moisture loss of poultry, and (*c*) microbial growth.

3-140C What is the heat of respiration of fruits and vegetables?

3-141C What is the operating principle of vacuum cooling of fruits and vegetables? How can the moisture loss of fruits and vegetables be minimized during vacuum cooling?

3-142C Why is it not recommended to store bananas below 13°C while it is highly recommended that apples be stored at −1°C?

3-143 A 35-kg box of beef at 6°C having a water content of 60 percent is to be frozen to a temperature of −20°C in 3 h. Determine (*a*) the total amount of heat that must be removed from the beef and (*b*) the average rate of heat removal from the beef.

3-144 A 50-kg box of sweet cherries at 8°C having a water content of 77 percent is to be frozen to a temperature of −20°C. Determine the total amount of heat that must be removed from the cherries.

FIGURE P3-144

3-145 Fresh strawberries with a water content of 88 percent (by mass) at 30°C are stored in 0.8-kg boxes made of nylon [C_p = 1.7 kJ/(kg · °C)]. Each box contains 23 kg of strawberries, and the strawberries are to be cooled to an average temperature of 4°C at a rate of 60 boxes per hour. Taking the average specific heat of the strawberries to be C_p = 3.89 kJ/(kg · °C) and the average rate of heat of respiration to be 210 mW/kg, determine the rate of heat removal from the strawberries and their boxes, in kJ/h. What would be the percent error involved if the strawberry boxes were ignored in the calculations?

3-146 Lettuce is to be vacuum cooled from the environment temperature of 24°C to a temperature of 2°C in 45 min in a 4-m outer-diameter insulated spherical vacuum chamber whose walls consist of 3-cm-thick urethane insulation [k = 0.020 W/(m · °C)] sandwiched between thin metal plates. The vacuum chamber contains 5000 kg of lettuce when loaded. Disregarding any heat transfer through the walls of the vacuum chamber, determine (*a*) the final pressure in the vacuum chamber and (*b*) the amount of moisture removed from the lettuce, in kg. It is claimed that the error involved in (*b*) due to neglecting heat transfer through the wall of the chamber is less than 2 percent. Is this a reasonable claim?
Answers: (*a*) 0.714 kPa, (*b*) 0.179 kg

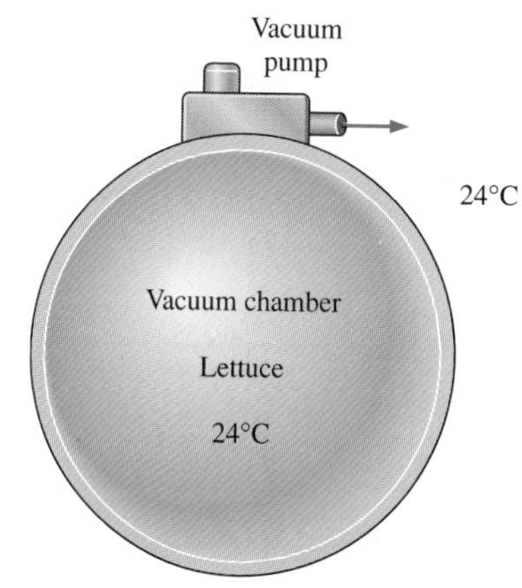

FIGURE P3-146

3-147 A supply of 40 kg of shrimp at 8°C contained in a box is to be frozen to −18°C in a freezer. Determine the amount of heat that needs to be

FIGURE P3-147

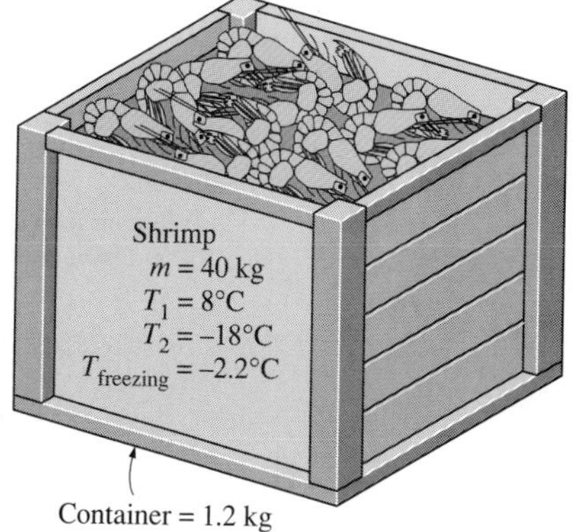

removed. The latent heat of the shrimp is 277 kJ/kg and its specific heat is 3.62 kJ/(kg · °C) above freezing and 1.89 kJ/(kg · °C) below freezing. The container box is 1.2 kg and is made up of polyethylene, whose specific heat is 2.3 kJ/(kg · °C). Also, the freezing temperature of shrimp is −2.2°C.

3-148E A cold storage room whose internal dimensions are 12 ft × 15 ft × 30 ft is maintained at 35°F. Under worst conditions, the amount of air infiltration is estimated to be 0.4 air change per hour at the ambient temperature of 90°F. Determine the total sensible infiltration load of this room under these conditions.

3-149E Orange juice [ρ = 62.4 lbm/ft^3, C_p = 0.900 Btu/(lbm · °F)] is to be transported from Florida to New York for a distance of 1250 miles in a 27-ft-long, 6.3-ft-external-diameter cylindrical tank whose walls consist of 0.15-ft-thick urethane insulation [k = 0.011 Btu/(h · ft · °F)] sandwiched between thin metal plates. The juice is precooled to 35°F before loading, and its temperature is not to exceed 46°F on arrival. The average ambient temperature can be as high as 92°F, and the effect of solar radiation and the radiation from hot pavement surfaces can be taken to be equivalent to a 12°F rise in ambient temperature. If the average velocity during transportation is 35 mph, determine if the orange juice can be transported without any refrigeration.

3-150 A large truck is to transport 30,000 kg of oranges precooled to 4°C under average ambient temperature of 27°C. The structure of the walls of the truck is such that the rate of heat transmission is UA = 80 W per °C temperature difference between the ambient and the oranges. From past experience, ambient air is estimated to enter the cargo space of the truck through the cracks at a rate of 4 L/s. Also, the average heat of respiration of the oranges at 4°C is 0.017 W/kg for this particular load. Disregarding any condensation and taking the density of air to be 1.15 kg/m^3, determine the refrigeration load of this truck and the amount of ice needed to meet the entire refrigeration need of the truck for a 15-h-long trip.

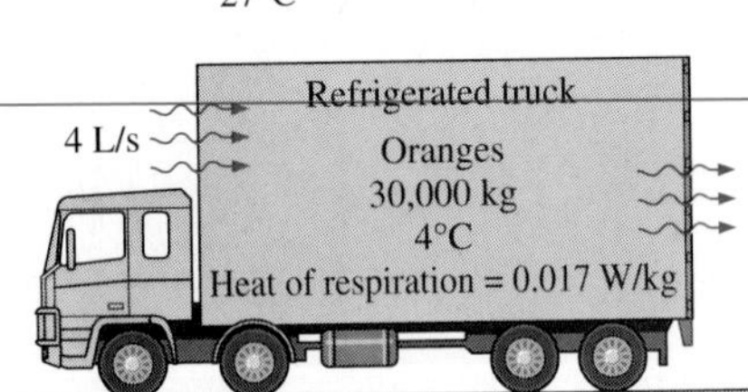

FIGURE P3-150

Review Problems

3-151 A well-insulated 4-m × 4-m × 5-m room initially at 10°C is heated by the radiator of a steam heating system. The radiator has a volume of 15 L and is filled with superheated vapor at 200 kPa and 200°C. At this moment both the inlet and the exit valves to the radiator are closed. A 120-W fan is used to distribute the air in the room. The pressure of the steam is observed to drop to 100 kPa after 30 min as a result of heat transfer to the room. Assuming constant specific heats for air at room temperature, determine the average temperature of air in 30 min. Assume the air pressure in the room remains constant at 100 kPa.

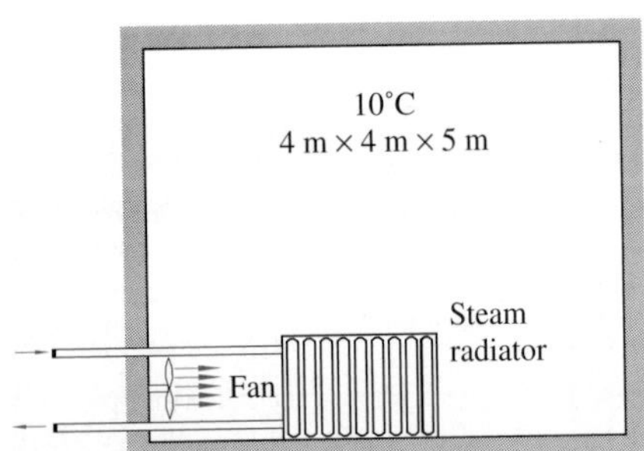

FIGURE P3-151

3-152 Consider a vertical elevator whose cabin has a total mass of 800 kg when fully loaded and 150 kg when empty. The weight of the elevator cabin is partially balanced by a 400-kg counterweight that is connected to the top of the cabin by cables that pass through a pulley located on top of the elevator well. Neglecting the weight of the cables and assuming the guide rails and the pulleys to be frictionless, determine (*a*) the power required while the fully loaded cabin is rising at a constant speed of 2 m/s and (*b*) the power required while the empty cabin is descending at a constant speed of 2 m/s.

What would your answer be to (*a*) if no counterweight were used? What would your answer be to (*b*) if a friction force of 1200 N has developed between the cabin and the guide rails?

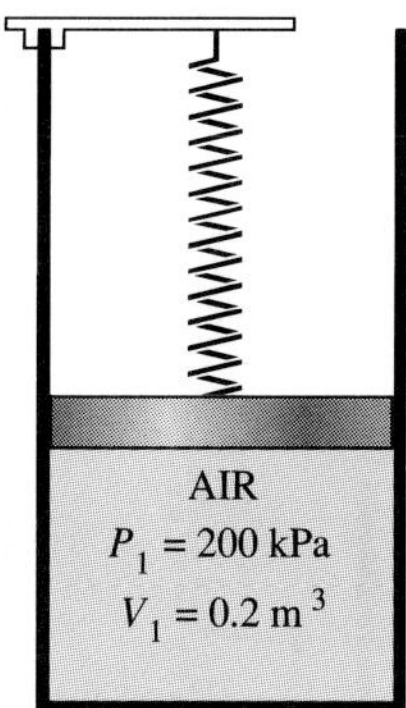

FIGURE P3-153

3-153 A frictionless piston-cylinder device initially contains air at 200 kPa and 0.2 m^3. At this state, a linear spring ($F \propto x$) is touching the piston but exerts no force on it. The air is now heated to a final state of 0.5 m^3 and 800 kPa. Determine (*a*) the total work done by the air and (*b*) the work done against the spring. Also, show the process on a *P-v* diagram.
Answers: (*a*) 150 kJ, (*b*) 90 kJ

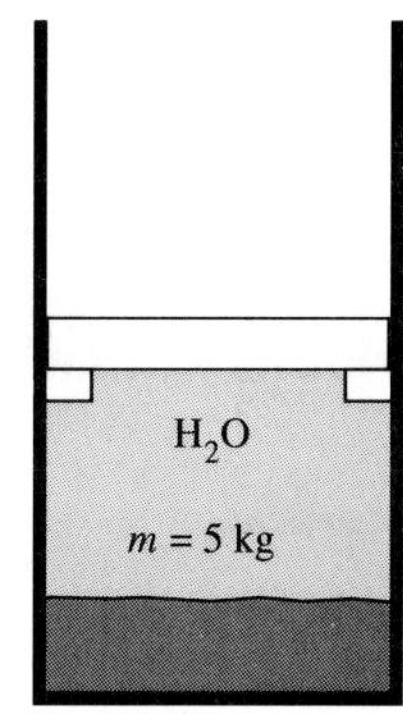

FIGURE P3-154

3-154 A mass of 5 kg of saturated liquid–vapor mixture of water is contained in a piston-cylinder device at 100 kPa. Initially, 2 kg of the water is in the liquid phase and the rest is in the vapor phase. Heat is now transferred to the water, and the piston, which is resting on a set of stops, starts moving when the pressure inside reaches 200 kPa. Heat transfer continues until the total volume increases by 20 percent. Determine (*a*) the initial and final temperatures, (*b*) the mass of liquid water when the piston first starts moving, and (*c*) the work done during this process. Also, show the process on a *P-v* diagram.

3-155E A spherical balloon contains 10 lbm of air at 30 psia and 800 R. The balloon material is such that the pressure inside is always proportional to the square of the diameter. Determine the work done when the volume of the balloon doubles as a result of heat transfer. *Answer:* 715.3 Btu

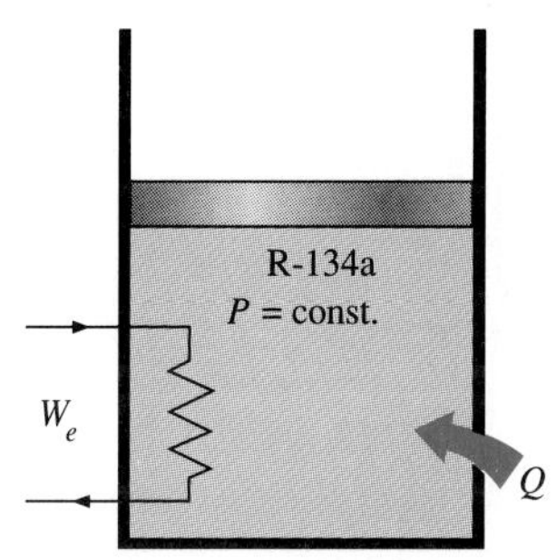

FIGURE P3-156

3-156 A mass of 12 kg of saturated refrigerant-134a vapor is contained in a piston-cylinder device at 200 kPa. Now 250 kJ of heat is transferred to the refrigerant at constant pressure while a 110-V source supplies current to a resistor within the cylinder for 6 min. Determine the current supplied if the final temperature is 70°C. Also, show the process on a *T-v* diagram with respect to the saturation lines. *Answer:* 15.7 A

3-157 A mass of 0.2 kg of saturated refrigerant-134a is contained in a piston-cylinder device at 200 kPa. Initially, 75 percent of the mass is in the liquid phase. Now heat is transferred to the refrigerant at constant pressure until the cylinder contains vapor only. Show the process on a *P-v* diagram with respect to saturation lines. Determine (*a*) the volume occupied by the refrigerant initially, (*b*) the work done, and (*c*) the total heat transfer.

FIGURE P3-158

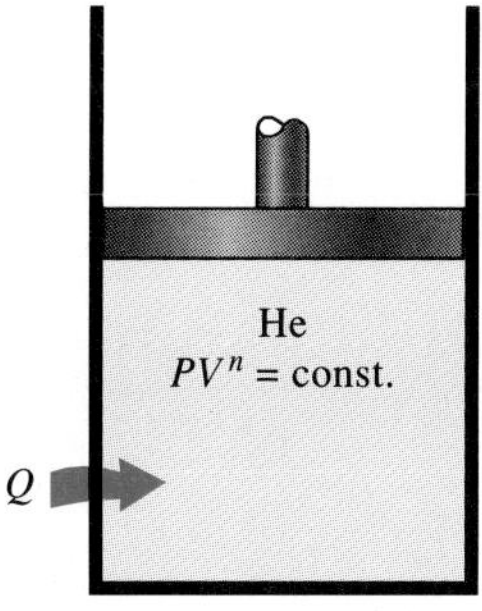

3-158 A piston-cylinder device contains helium gas initially at 150 kPa, 20°C, and 0.5 m^3. The helium is now compressed in a polytropic process (PV^n = constant) to 400 kPa and 140°C. Determine the heat loss or gain during this process. *Answer:* 11.2 kJ loss

3-159 A frictionless piston-cylinder device and a rigid tank initially contain 12 kg of an ideal gas each at the same temperature, pressure, and volume. It is desired to raise the temperatures of both systems by 15°C. Determine the amount of extra heat that must be supplied to the gas in the cylinder which is maintained at constant pressure to achieve this result. Assume the molar mass of the gas is 25.

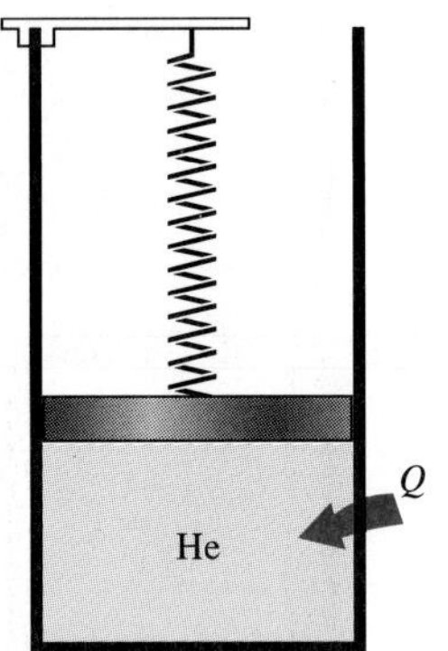

FIGURE P3-160

3-160 A piston-cylinder device contains 0.5 m^3 of helium gas initially at 100 kPa and 25°C. At this position, a linear spring is touching the piston but exerts no force on it. Heat is now transferred to helium until both the pressure and the volume triple. Determine (*a*) the work done and (*b*) the amount of heat transfer for this process. Also, show the process on a *P-v* diagram.

Answers: (*a*) W_{out} = 200 kJ, (*b*) Q_{in} = 801.4 kJ

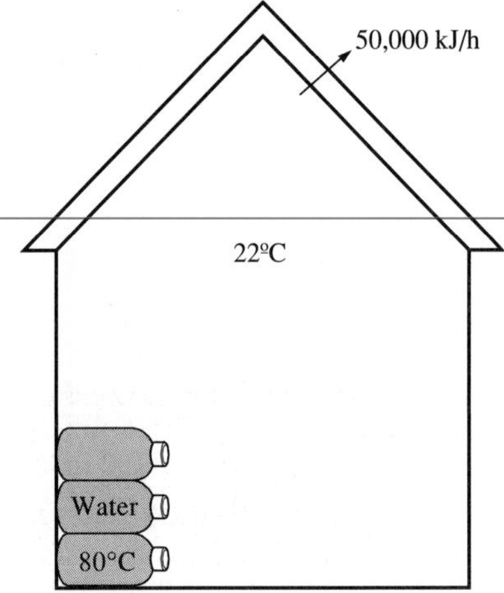

FIGURE P3-161

3-161 A passive solar house that is losing heat to the outdoors at an average rate of 50,000 kJ/h is maintained at 22°C at all times during a winter night for 10 h. The house is to be heated by 50 glass containers each containing 20 L of water that is heated to 80°C during the day by absorbing solar energy. A thermostat controlled 15-kW back-up electric resistance heater turns on whenever necessary to keep the house at 22°C. (*a*) How long did the electric heating system run that night? (*b*) How long would the electric heater run that night if the house incorporated no solar heating?

Answers: (*a*) 4.77 h, (*b*) 9.26 h

3-162 It is well known that wind makes cold air feel much colder as a result of the *windchill* effect, which is due to the increase in the convection heat transfer coefficient as a result of the increase in air velocity. The windchill effect is usually expressed in terms of the *windchill factor,* which is the difference between the actual air temperature and the equivalent calm-air temperature. For example, a windchill factor of 20°C for an actual air temperature of 5°C means that the windy air at 5°C feels as cold as the still air at −15°C. In other words, a person will lose as much heat to air at 5°C with a windchill factor of 20°C as he or she would in calm air at −15°C.

For heat transfer purposes, a standing man can be modeled as a 30-cm-diameter, 170-cm-long vertical cylinder with both the top and bottom surfaces insulated and with the side surface at an average temperature of 34°C. For a convection heat transfer coefficient of 15 W/($m^2 \cdot$ °C), determine the rate of heat loss from this man by convection in still air at 20°C. What would your answer be if the convection heat transfer coefficient was increased to 50 W/($m^2 \cdot$ °C) as a result of winds? What is the windchill factor in this case?

3-163 A 50-cm-long, 800-W electric resistance heating element whose diameter is 0.5 cm and surface temperature 120°C is immersed in 40 kg of water initially at 20°C. Determine how long it will take for this heater to raise the water temperature to 80°C. Also, determine the convection heat transfer coefficients at the beginning and at the end of the heating process.

3-164 One ton (1000 kg) of liquid water at 80°C is brought into a well-insulated and well-sealed 4-m × 5-m × 6-m room initially at 22°C and

100 kPa. Assuming constant specific heats for both air and water at room temperature, determine the final equilibrium temperature in the room.
Answer: 78.6°C

3-165 A 4-m × 5-m × 6-m room is to be heated by one ton (1000 kg) of liquid water contained in a tank that is placed in the room. The room is losing heat to the outside at an average rate of 10,000 kJ/h. The room is initially at 20°C and 100 kPa and is maintained at an average temperature of 20°C at all times. If the hot water is to meet the heating requirements of this room for a 24-h period, determine the minimum temperature of the water when it is first brought into the room. Assume constant specific heats for both air and water at room temperature.

3-166 Consider a well-insulated horizontal rigid cylinder that is divided into two compartments by a piston that is free to move but does not allow either gas to leak into the other side. Initially, one side of the piston contains 1 m^3 of N_2 gas at 500 kPa and 80°C while the other side contains 1 m^3 of He gas at 500 kPa and 25°C. Now thermal equilibrium is established in the cylinder as a result of heat transfer through the piston. Using constant specific heats at room temperature, determine the final equilibrium temperature in the cylinder. What would your answer be if the piston were not free to move?

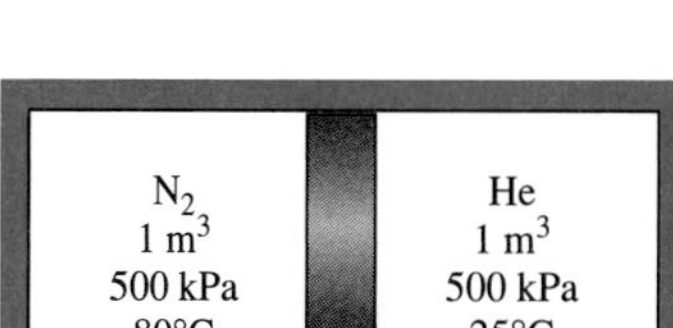

FIGURE P3-166

3-167 Repeat the problem above by assuming the piston is made of 5 kg of copper initially at the average temperature of the two gases on both sides.
Answer: 56°C

3-168 Catastrophic explosions of steam boilers in the 1800s and early 1900s resulted in hundreds of deaths, which prompted the development of the ASME Boiler and Pressure Vessel Code in 1915. Considering that the pressurized fluid in a vessel eventually reaches equilibrium with its surroundings shortly after the explosion, the work that a pressurized fluid would do if allowed to expand adiabatically to the state of the surroundings can be viewed as the *explosive energy* of the pressurized fluid. Because of the very short time period of the explosion and the apparent stability afterward, the explosion process can be considered to be adiabatic with no changes in kinetic and potential energies. The closed-system conservation of energy relation in this case reduces to $W_{in} = m(u_2 - u_1)$. Then the explosive energy E_{exp} becomes

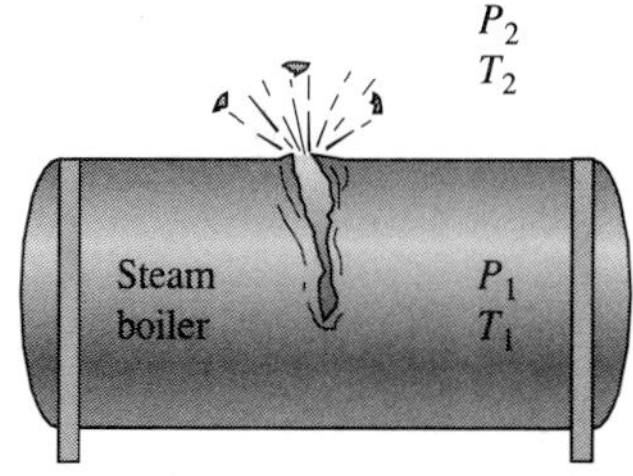

FIGURE P3-168

$$E_{exp} = m(u_1 - u_2)$$

where the subscripts 1 and 2 refer to the state of the fluid before and after the explosion, respectively. The specific explosion energy e_{exp} is usually expressed *per unit volume,* and it is obtained by dividing the quantity above by the total volume V of the vessel:

$$e_{exp} = \frac{u_1 - u_2}{v_1}$$

where v_1 is the specific volume of the fluid before the explosion.

Show that the specific explosion energy of an ideal gas with constant specific heats is

$$e_{exp} = \frac{P_1}{k - 1}\left(1 - \frac{T_2}{T_1}\right)$$

Also, determine the total explosion energy of 20 m^3 of air at 5 MPa and 100°C when the surroundings are at 20°C.

3-169 Using the relations in the problem above, determine the explosive energy of 20 m^3 of steam at 10 MPa and 500°C assuming the steam condenses and becomes a liquid at 25°C after the explosion. To how many kilograms of TNT is this explosive energy equivalent? The explosive energy of TNT is about 3250 kJ/kg.

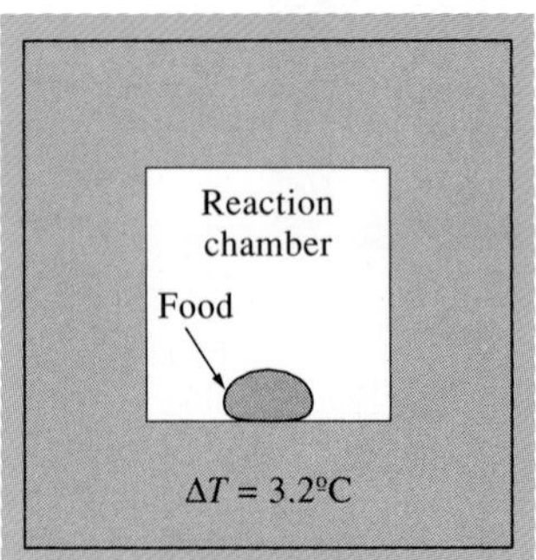

FIGURE P3-170

3-170 The energy content of a certain food is to be determined in a bomb calorimeter that contains 3 kg of water by burning a 2-g sample of it in the presence of 100 g of air in the reaction chamber. If the water temperature rises by 3.2°C when equilibrium is established, determine the energy content of the food, in kJ/kg, by neglecting the thermal energy stored in the reaction chamber and the energy supplied by the mixer. What is a rough estimate of the error involved in neglecting the thermal energy stored in the reaction chamber? *Answer:* 20,083 kJ/kg

3-171 A 68-kg man whose average body temperature is 38°C drinks 1 L of cold water at 3°C in an effort to cool down. Taking the average specific heat of the human body to be 3.6 kJ/(kg · °C), determine the drop in the average body temperature of this person under the influence of this cold water.

3-172 A 0.2-L glass of water at 20°C is to be cooled with ice to 5°C. Determine how much ice needs to be added to the water, in grams, if the ice is at (*a*) 0°C and (*b*) −8°C. Also determine how much water would be needed if the cooling is to be done with cold water at 0°C. The melting temperature and the heat of fusion of ice at atmospheric pressure are 0°C and 333.7 kJ/kg, respectively, and the density of water is 1 kg/L.

3-173 In order to cool 1 ton (1000 kg) of water at 20°C in a insulated tank, a person pours 80 kg of ice at −5°C into the water. Determine the final equilibrium temperature in the tank. The melting temperature and the heat of fusion of ice at atmospheric pressure are 0°C and 333.7 kJ/kg, respectively.
Answer: 12.4°C

FIGURE P3-175

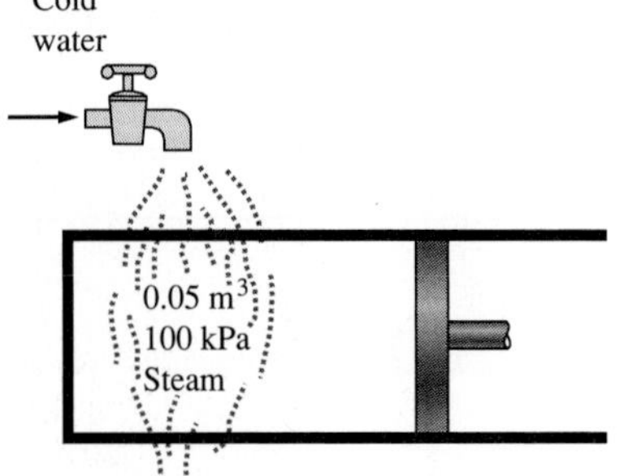

3-174 An insulated piston-cylinder device initially contains 0.01 m^3 of saturated liquid–vapor mixture with a quality of 0.2 at 100°C. Now some ice at 0°C is added to the cylinder. If the cylinder contains saturated liquid at 100°C when thermal equilibrium is established, determine the amount of ice added. The melting temperature and the heat of fusion of ice at atmospheric pressure are 0°C and 333.7 kJ/kg, respectively.

3-175 The early steam engines were driven by the atmospheric pressure acting on the piston fitted into a cylinder filled with saturated steam.

A vacuum was created in the cylinder by cooling the cylinder externally with cold water, and thus condensing the steam.

Consider a piston-cylinder device with a piston surface area of 0.1 m^2 initially filled with 0.05 m^3 of saturated water vapor at the atmospheric pressure of 100 kPa. Now cold water is poured outside the cylinder, and the steam inside starts condensing as a result of heat transfer to the cooling water outside. If the piston is stuck at its initial position, determine the friction force acting on the piston and the amount of heat transfer when the temperature inside the cylinder drops to 30°C.

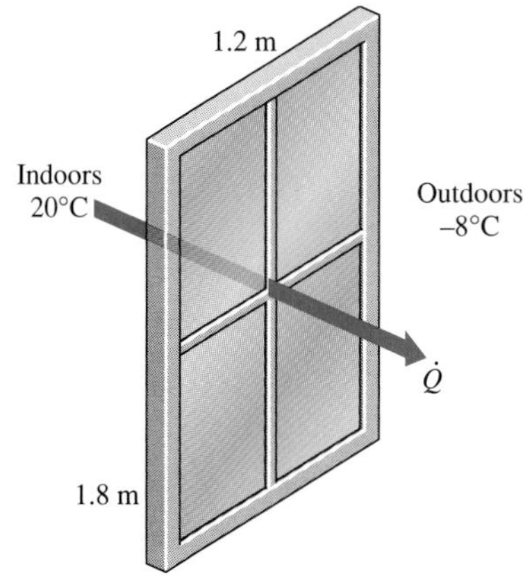

FIGURE P3-176

3-176 The rate of heat loss through a unit surface area of a window per unit temperature difference between the indoors and the outdoors is called the *U*-factor. The value of the *U*-factor ranges from about 1.25 W/(m^2 · °C) [or 0.22 But/(h · ft^2 · °F)] for low-*e* coated, argon-filled, quadruple-pane windows to 6.25 W/(m^2 · °C) [or 1.1 Btu/(h · ft^2 · °F)] for a single-pane window with aluminum frames. Determine the range for the rate of heat loss through a 1.2-m × 1.8-m window of a house that is maintained at 20°C when the outdoor air temperature is −8°C.

3-177 Consider a house in Atlanta, Georgia, that is maintained at 22°C and has a total of 20 m^2 of window area. The windows are double-door type with wood frames and metal spacers and have a *U*-factor of 2.5 W/(m^2 · °C) (see the previous problem for the definition of *U*-factor). The winter average temperature of Atlanta is 11.3°C. Determine the average rate of heat loss through the windows in winter.

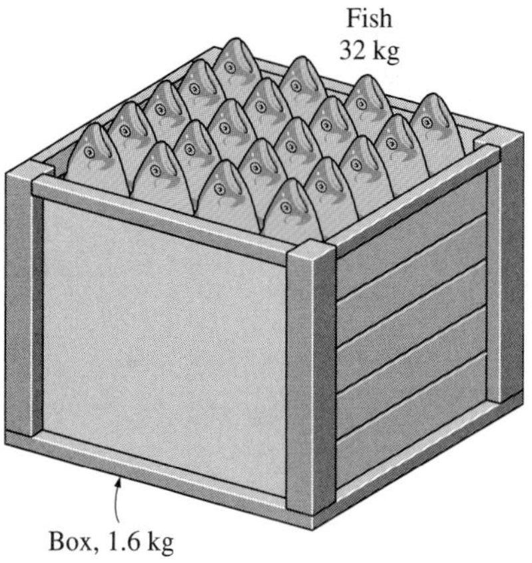

FIGURE P3-179

3-178E Wet broccoli is to be vacuum cooled from the environment temperature of 75°F to a temperature of 40°F in 1 h in a 15-ft outer-diameter insulated spherical vacuum chamber. The vacuum chamber contains 15,000 lbm of broccoli when loaded, but 2 percent of this mass is due to water that sticks to the surface of the broccoli during wetting. Disregarding any heat transfer through the walls of the vacuum chamber, determine the final mass of the broccoli after cooling.

3-179 A 1.6-kg box made of polypropylene [C_p = 1.9 kJ/(kg · °C)] contains 32 kg of haddock fish with a water content of 83 percent (by mass) at 16°C. The fish is to be frozen to an average temperature of −20°C in 4 h in its box. The specific heat of fish is 3.62 kJ/(kg · °C) above the freezing temperature of −2.2°C, and 1.89 kJ/(kg · °C) below the freezing temperature. Determine (*a*) the total amount of heat that must be removed from the fish and (*b*) the average rate of heat removal from the fish.

FIGURE P3-180

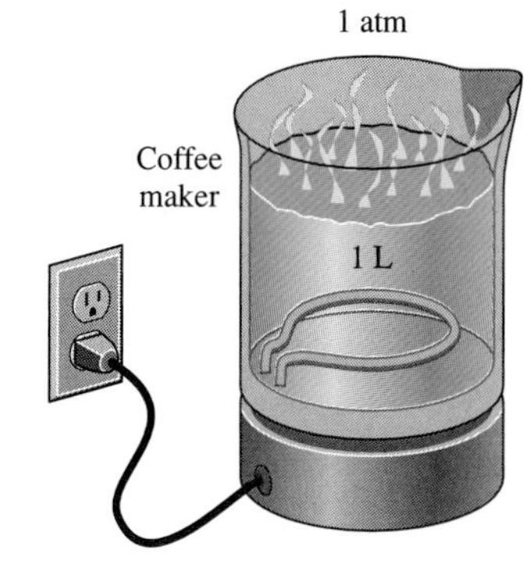

3-180 Water is boiled at sea level in a coffee maker equipped with an immersion-type electric heating element. The coffee maker contains 1 L of water when full. Once boiling starts, it is observed that half of the water in the coffee maker evaporates in 25 min. Determine the power rating of the electric heating element immersed in water. Also, determine how long it will take for this heater to raise the temperature of 1 L of cold water from 18°C to the boiling temperature.

3-181 In a gas-fired boiler, water is boiled at 150°C by hot gases flowing through 50-m-long, 5-cm-outer-diameter stainless steel pipes submerged in water. If the outer surface temperature of the pipes is 165°C and the average heat transfer coefficient is 6200 W/(m^2 · °C), determine (*a*) the rate of heat transfer from the hot gases to water and (*b*) the rate of evaporation of water.

3-182 Cold water enters a steam generator at 20°C and leaves as saturated vapor at 100°C. Determine the fraction of heat used in the steam generator to preheat the liquid water from 20°C to the saturation temperature of 100°C.

3-183 Cold water enters a steam generator at 20°C and leaves as saturated vapor at the boiler pressure. At what pressure will the amount of heat needed to preheat the water to saturation temperature be equal to the heat needed to vaporize the liquid at the boiler pressure?

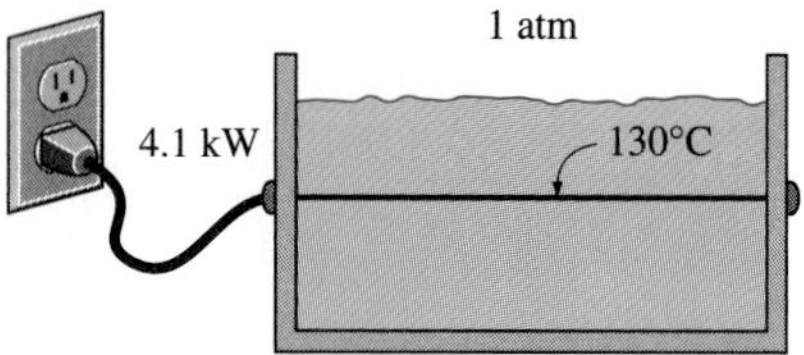

FIGURE P3-184

3-184 A 50-cm-long, 2-mm-diameter electric resistance wire submerged in water is used to determine the boiling heat transfer coefficient in water at 1 atm experimentally. The wire temperature is measured to be 130°C when a wattmeter indicates the electric power consumed to be 4.1 kW. Using Newton's law of cooling, determine the boiling heat transfer coefficient.

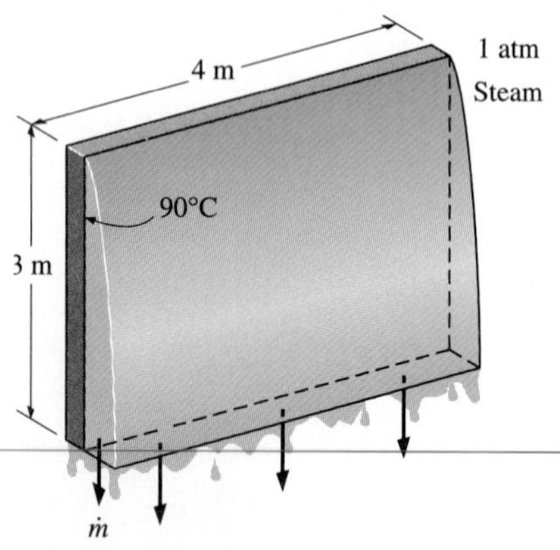

FIGURE P3-185

3-185 Saturated steam at 1 atm condenses on a 3-m-high, 4-m-wide vertical plate that is maintained at 90°C by circulating cooling water through the other side. If the average heat transfer coefficient is 5800 W/(m^2 · °C), determine (*a*) the rate of heat transfer by condensation to the plate and (*b*) the rate at which the condensate drips off the plate at the bottom.

3-186 Water is boiled at 100°C electrically by a 50-cm-long, 2-mm-diameter, 5-kW resistance wire. Determine (*a*) the rate of evaporation of water and (*b*) the heat flux at the surface of the wire.

FIGURE P3-186

3-187 In solar-heated buildings, energy is often stored as sensible heat in rocks, concrete, or water during the day for use at night. To minimize the storage space, it is desirable to use a material that can store a large amount of heat while experiencing a small temperature change. A large amount of heat can be stored essentially at constant temperature during a phase change process, and thus materials that change phase at about room temperature such as glaubers salt (sodium sulfate decahydrate), which has a melting point of 32°C and heat of fusion of 329 kJ/L, are very suitable for this purpose. Determine how much heat can be stored in a 5-m^3 storage space using (*a*) glaubers salt undergoing a phase change, (*b*) granite rocks with a heat capacity of 2.32 kJ/(kg · °C) and a temperature change of 20°C, and (*c*) water with a heat capacity of 4.00 kJ/(kg · °C) and a temperature change of 20°C.

3-188 Consider a well-insulated piston-cylinder device that contains 4 kg of liquid water and 1 kg of water vapor at 120°C and is maintained at constant pressure. Now a 5-kg copper block at 30°C is dropped into the cylinder. Determine the equilibrium temperature inside the cylinder once thermal equilibrium is established, and the mass of the water vapor at the final state.

3-189 The gage pressure of an automobile tire is measured to be 200 kPa before a trip and 220 kPa after the trip at a location where the atmospheric pressure is 90 kPa. Assuming the volume of the tire remains constant at 0.022 m^3, determine the temperature rise of air in the tire during the trip. Assume constant specific heats at room temperature.

Computer, Design, and Essay Problems

3-190 Write a computer program to express the variation of specific heat $\overline{C}_p$ of air with temperature as a third-degree polynomial, using the data in Table A-2*b*. Compare your result with that given in Table A-2*c*.

3-191 Find out how the specific heats of gases, liquids, and solids are determined in national laboratories. Describe the experimental apparatus and the procedures used.

3-192 Using information from the utility bills for the coldest month last year, estimate the average rate of heat loss from your house for that month. In your analysis, consider the contribution of the internal heat sources such as people, lights, and appliances. Identify the primary sources of heat losses from your house, and propose ways of improving the energy efficiency of your house.

3-193 Design an experiment complete with instrumentation to determine the specific heats of a gas using a resistance heater. Discuss how the experiment will be conducted, what measurements need to taken, and how the specific heats will be determined. What are the sources of error in your system? How can you minimize the experimental error?

3-194 Design an experiment complete with instrumentation to determine the specific heat of a liquid using a resistance heater. Discuss how the experiment will be conducted, what measurements need to taken, and how the specific heats will be determined. What are the sources of error in your system? How can you minimize the experimental error? How would you modify this system to determine the specific heat of a solid?

3-195 Design a reciprocating compressor capable of supplying compressed air at 800 kPa at a rate of 15 kg/min. Also specify the size of the electric motor capable of driving this compressor. The compressor is to operate at no more than 2000 rpm (revolutions per minute).

3-196 Conduct the following experiment to determine the heat transfer coefficient between an incandescent light bulb and the surrounding air using a 60-W light bulb. You will need an indoor–outdoor thermometer, which can be purchased for about $10 in a hardware store, and a metal glue. You will also need a piece of string and a ruler to calculate the surface area of the light bulb. First, measure the air temperature in the room, and then glue the tip of the thermocouple wire of the thermometer to the glass of the light bulb. Turn the light on and wait until the temperature reading stabilizes. The temperature reading will give the surface temperature of the light bulb. Assuming 10 percent of the rated power of the bulb is converted to light, calculate the heat transfer coefficient from Newton's law of cooling.

3-197 You are asked to design a heating system for a swimming pool that is 2-m-deep, 25-m-long, and 25-m wide. Your client desires that the heating system be large enough to raise the water temperature from 20°C to 30°C in 3 h. The rate of heat loss from the water to the air at the outdoor design conditions is determined to be 960 W/m^2, and the heater must also be able to maintain the pool at 30°C at those conditions. Heat losses to the ground are expected to be small and can be disregarded. The heater considered is a natural gas furnace whose efficiency is 80 percent. What heater size (in Btu/h input) would you recommend to your client?

3-198 A 1982 U.S. Department of Energy article (FS #204) states that a leak of one drip of hot water per second can cost $1.00 per month. Making reasonable assumptions about the drop size and the unit cost of energy, determine if this claim is reasonable.

The First Law of Thermodynamics: Control Volumes

CHAPTER 4

In Chap. 3, we discussed the energy interactions between a system and its surroundings, and the conservation of energy principle for closed (nonflow) systems. In this chapter, we extend the analysis to systems that involve mass flow across their boundaries, that is, *control volumes*. The conservation of energy equation for a general control volume can be rather involved and intimidating. Therefore, we treat the control volume energy analysis in two stages. First, we consider the *steady-flow process,* which is the model process for many engineering devices such as turbines, compressors, and heat exchangers. Second, we discuss the general unsteady-flow processes with particular emphasis on the *uniform-flow process,* which is the model process for commonly encountered charging and discharging processes.

4-1 ■ THERMODYNAMIC ANALYSIS OF CONTROL VOLUMES

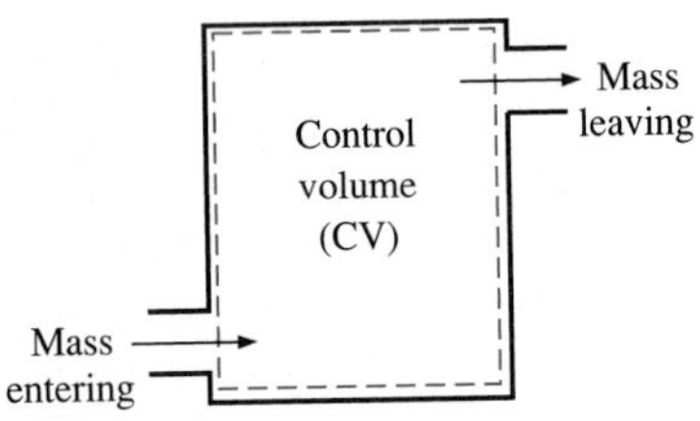

FIGURE 4-1

Mass may flow into and out of a control volume.

A large number of engineering problems involve mass flow in and out of a system and, therefore, are modeled as *control volumes* (Fig. 4-1). A water heater, a car radiator, a turbine, and a compressor all involve mass flow and should be analyzed as control volumes (open systems) instead of as control masses (closed systems). In general, *any arbitrary region in space* can be selected as a control volume. There are no concrete rules for the selection of control volumes, but the proper choice certainly makes the analysis much easier. If we were to analyze the flow of air through a nozzle, for example, a good choice for the control volume would be the region within the nozzle.

The boundaries of a control volume are called a *control surface,* and they can be real or imaginary. In the case of a nozzle, the inner surface of the nozzle forms the real part of the boundary, and the entrance and exit areas form the imaginary part, since there are no physical surfaces there (Fig. 4-2).

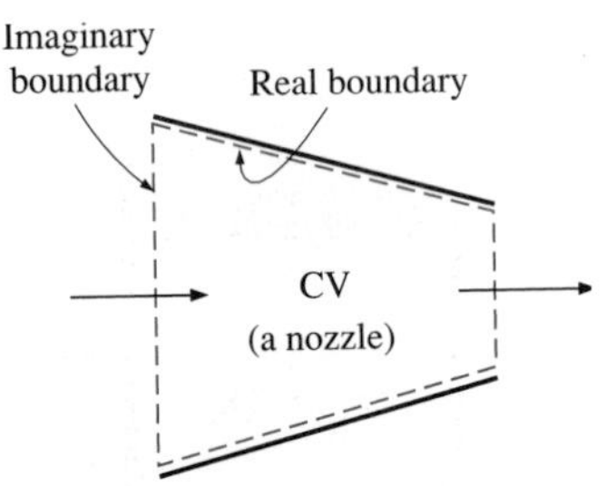

FIGURE 4-2

Real and imaginary boundaries of a control volume.

A control volume can be fixed in size and shape, as in the case of a nozzle, or it may involve a moving boundary, as shown in Fig. 4-3. Most control volumes, however, have fixed boundaries and thus do not involve any moving boundary work. A control volume may also involve heat and work interactions just as a closed system, in addition to mass interaction.

The terms *steady* and *uniform* are used extensively in this chapter, and thus it is important to have a clear understanding of their meanings. The term *steady* implies *no change with time.* The opposite of steady is *unsteady,* or *transient.* The term *uniform,* however, implies *no change with location* over a specified region. These meanings are consistent with their everyday use (steady girlfriend, uniform distribution, etc.).

An overview of the conservation of mass and the conservation of energy principles for control volumes is given below.

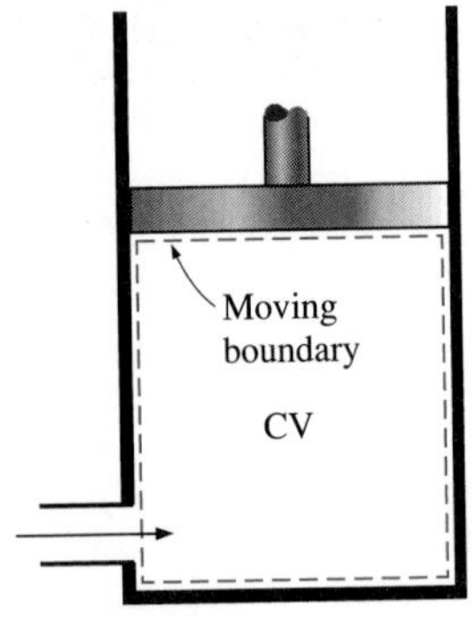

FIGURE 4-3

Some control volumes involve moving boundaries.

Conservation of Mass Principle

The conservation of mass is one of the most fundamental principles in nature. We are all familiar with this principle, and it is not difficult to understand. As the saying goes, you cannot have your cake and eat it, too! A person does not have to be an engineer to figure out how much vinegar-and-oil dressing he is going to have if he mixes 100 g of oil with 25 g of vinegar. Even chemical equations are balanced on the basis of the conservation of mass principle. When 16 kg of oxygen reacts with 2 kg of hydrogen, 18 kg of water is formed (Fig. 4-4). In an electrolysis process, the water will separate back to 2 kg of hydrogen and 16 kg of oxygen.

Mass, like energy, is a conserved property, and it cannot be created or destroyed. However, mass m and energy E can be converted to each other according to the famous formula proposed by Einstein:

$$E = mc^2$$

where c is the speed of light. This equation suggests that the mass of a system will change when its energy changes. However, for all energy interactions encountered in practice, with the exception of nuclear reactions, the change in mass is extremely small and cannot be detected by even the most sensitive devices. For example, when 1 kg of water is formed from oxygen and hydrogen, the amount of energy released is 15,879 kJ, which corresponds to a mass of 1.76×10^{-10} kg. A mass of this magnitude is beyond the accuracy required by practically all engineering calculations and thus can be disregarded.

For closed systems, the conservation of mass principle is implicitly used by requiring that the mass of the system remain constant during a process. For control volumes, however, mass can cross the boundaries, and so we must keep track of the amount of the mass entering and leaving the control volume (Fig. 4-5).

The **conservation of mass principle** can be expressed as: *net mass transfer to or from a system during a process is equal to the net change (increase or decrease) in the total mass of the system during that process.* That is,

$$\left(\begin{array}{c}\text{Total mass}\\ \text{entering the system}\end{array}\right) - \left(\begin{array}{c}\text{Total mass}\\ \text{leaving the system}\end{array}\right) = \left(\begin{array}{c}\text{Net change in mass}\\ \text{within the system}\end{array}\right)$$

or

$$m_{\text{in}} - m_{\text{out}} = \Delta m_{\text{system}} \qquad \text{(kg)} \qquad \text{(4-1)}$$

It can also be expressed in the *rate form* as

$$\dot{m}_{\text{in}} - \dot{m}_{\text{out}} = \Delta \dot{m}_{\text{system}} \qquad \text{(kg/s)} \qquad \text{(4-2)}$$

where $\dot{m}_{\text{in}}$ and $\dot{m}_{\text{out}}$ are the total rates of mass flow into and out of the system and $\Delta \dot{m}_{\text{system}}$ (or dm_{system}/dt) is the rate of change of mass within the system boundaries. The relations above are often referred to as the **mass balance** and are applicable to any kind of system undergoing any kind of process.

The conservation of mass principle is based on experimental observations and requires every bit of mass to be accounted for during a process.

A person who can balance a checkbook (by keeping track of deposits and withdrawals, or simply by observing the "conservation of money" principle) should have no difficulty in applying the conservation of mass principle to thermodynamic systems. The conservation of mass equation is often referred to as the *continuity equation* in fluid mechanics.

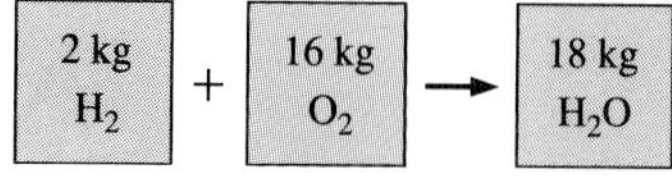

FIGURE 4-4
Mass is conserved even during chemical reactions.

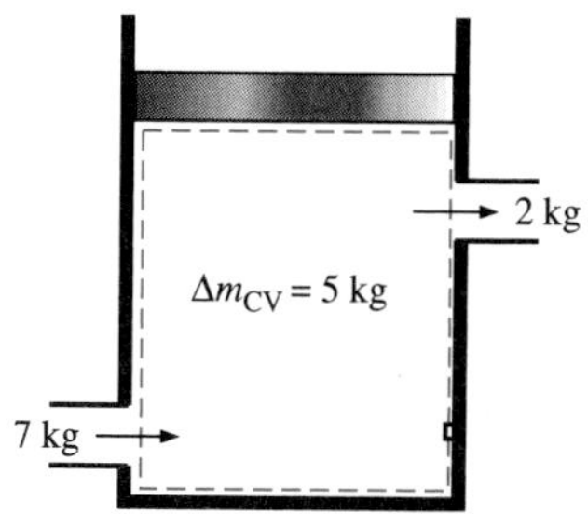

FIGURE 4-5
Conservation of mass principle for a control volume.

Mass and Volume Flow Rates

The amount of mass flowing through a cross section per unit time is called the **mass flow rate** and is denoted $\dot{m}$. As before, the dot over a symbol is used to indicate a quantity per unit time.

A liquid or a gas flows in and out of a control volume through pipes or ducts. The mass flow rate of a fluid flowing in a pipe or duct is proportional to the cross-sectional area A of the pipe or duct, the density ρ, and the velocity $\mathcal{V}$ of the fluid. The mass flow rate through a differential area dA can be expressed as

$$d\dot{m} = \rho \mathcal{V}_n \, dA \tag{4-3}$$

where $\mathcal{V}_n$ is the velocity component normal to dA. The mass flow rate through the entire cross-sectional area of the pipe or duct is obtained by integration:

$$\dot{m} = \int_A \rho \mathcal{V}_n \, dA \qquad \text{(kg/s)} \tag{4-4}$$

In most practical applications, the flow of a fluid through a pipe or duct can be approximated to be **one-dimensional flow.** That is, the properties can be assumed to vary in *one* direction only (the direction of flow). As a result, all properties are *uniform* at any cross-section normal to the flow direction, and the properties are assumed to have *bulk average values* over the cross-section. But the values of the properties at a cross section *may* change with time.

The one-dimensional-flow approximation has little impact on most properties of a fluid flowing in a pipe or duct such as temperature, pressure, and density since these properties usually remain constant over the cross section. But this is not the case for *velocity,* whose value varies from zero at the wall to a maximum at the center because of the viscous effects (friction between fluid layers). Under the one-dimensional-flow assumption, the velocity is assumed to be constant across the entire cross-section at some equivalent average value (Fig. 4-6). Then the integration in Eq. 4-4 can be performed for one-dimensional flow to yield

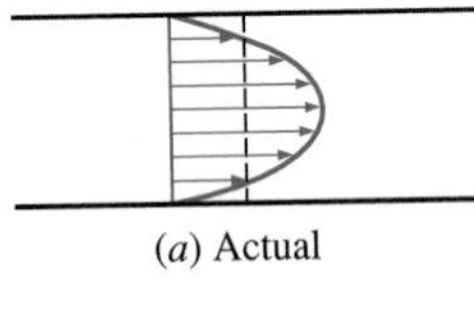

(*a*) Actual

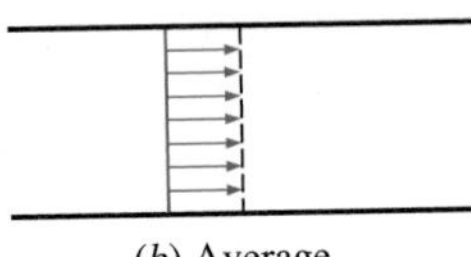

(*b*) Average

FIGURE 4-6
Actual and average velocity profiles for flow in a pipe (the mass flow rate is the same for both cases).

$$\dot{m} = \rho \mathcal{V}_{av} A \qquad \text{(kg/s)} \tag{4-5}$$

where

ρ = density, kg/m^3 (= $1/v$)
$\mathcal{V}_{av}$ = average fluid velocity normal to A, m/s
A = cross-sectional area normal to flow direction, m^2

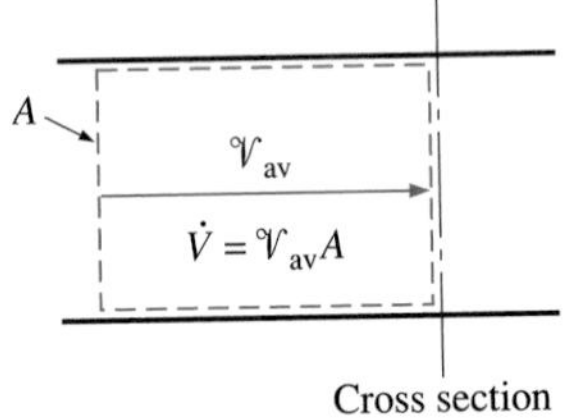

FIGURE 4-7
The volume flow rate is the volume of fluid flowing through a cross section per unit time.

The volume of the fluid flowing through a cross section per unit time is called the **volume flow rate** $\dot{V}$ (Fig. 4-7) and is given by

$$\dot{V} = \int_A \mathcal{V}_n \, dA = \mathcal{V}_{av} A \qquad \text{(m}^3\text{/s)} \tag{4-6}$$

The mass and volume flow rates are related by

$$\dot{m} = \rho \dot{V} = \frac{\dot{V}}{v} \tag{4-7}$$

This relation is analogous to $m = V/v$, which is the relation between the mass and the volume of a fluid in a container.

For simplicity, we drop the subscript on the average velocity. Unless otherwise stated, $\mathcal{V}$ denotes the average velocity in the flow direction. Also, A denotes the cross-sectional area normal to the flow direction.

Energy Balance for a Control Volume

In Chap. 3 we expressed the conservation of energy principle (or the energy balance) for *any system* undergoing *any process* as *the net change (increase or decrease) in the total energy of the system during a process is equal to the difference between the total energy entering and the total energy leaving the system during that process.* We also pointed out that energy can be transferred to or from a system in three forms—*heat, work,* and *mass flow*—and the total energy of a simple compressible system consists of internal, kinetic, and potential energies. Then the energy balance was written as

$$\underbrace{E_{in} - E_{out}}_{\substack{\text{Net energy transfer}\\ \text{by heat, work, and mass}}} = \underbrace{\Delta E_{system}}_{\substack{\text{Change in internal, kinetic,}\\ \text{potential, etc., energies}}} \quad \text{(kJ)} \qquad (4\text{-}8)$$

or, in the *rate form*, as

$$\underbrace{\dot{E}_{in} - \dot{E}_{out}}_{\substack{\text{Rate of net energy transfer}\\ \text{by heat, work, and mass}}} = \underbrace{\Delta \dot{E}_{system}}_{\substack{\text{Rate of change in internal, kinetic,}\\ \text{and potential, etc., energies}}} \quad \text{(kW)} \qquad (4\text{-}9)$$

In Chap. 3 we considered systems that involved only heat transfer and work as energy interactions (i.e., closed systems). In this chapter we extend the analysis to systems that involve mass flow across the system boundaries (i.e., to control volumes).

Mass flow into and out of a system serves as an additional mechanism to change the energy content of the system (Fig. 4-8). When mass enters a control volume, the energy of the control volume increases because the entering mass carries some energy with it. Likewise, when some mass leaves the control volume, the energy contained within the control volume decreases because the leaving mass takes *out* some energy with it. For example, when some hot water is taken out of a water heater and is replaced by the same amount of cold water, the energy content of the hot-water tank (the control volume) decreases as a result of this mass interaction.

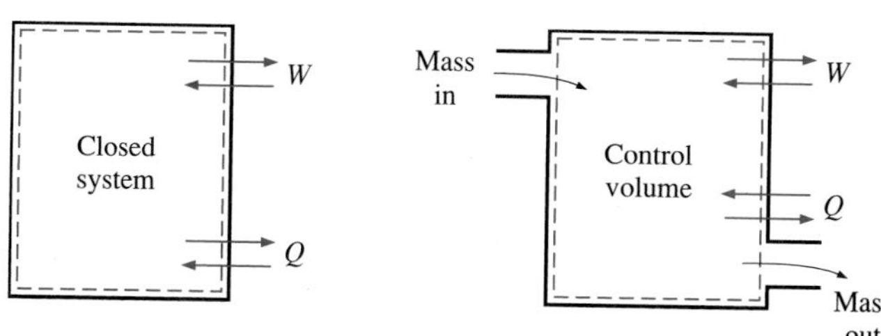

FIGURE 4-8
The energy content of a control volume can be changed by mass flow as well as heat and work interactions.

Heat transfer to or from a control volume should not be confused with the energy transported with mass into and out of a control volume. Remember that heat is the form of energy transferred as a result of a temperature difference between the control volume and the surroundings.

A control volume, like a closed system, may involve one or more forms of work at the same time (Fig. 4-9). If the boundary of the control volume is stationary, as is often the case, the moving boundary work is zero. Then the work term will involve, at most, shaft work and electrical work for simple compressible systems. As before, when the control volume is insulated, the heat transfer term becomes zero.

The energy required to push fluid into or out of a control volume is called the *flow work,* or *flow energy*. It is considered to be part of the energy transported with the fluid and is discussed below.

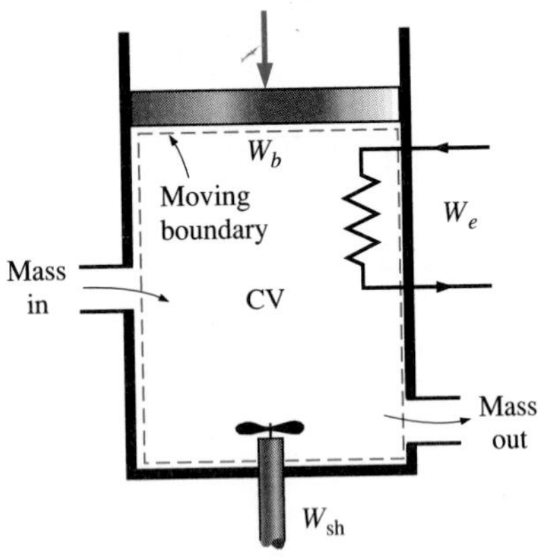

FIGURE 4-9
A control volume may involve boundary work in addition to electrical and shaft work.

Flow Work

Unlike closed systems, control volumes involve mass flow across their boundaries, and some work is required to push the mass into or out of the control volume. This work is known as the **flow work,** or **flow energy,** and is necessary for maintaining a continuous flow through a control volume.

To obtain a relation for flow work, consider a fluid element of volume V as shown in Fig. 4-10. The fluid immediately upstream will force this fluid element to enter the control volume; thus, it can be regarded as an imaginary piston. The fluid element can be chosen to be sufficiently small so that it has uniform properties throughout.

If the fluid pressure is P and the cross-sectional area of the fluid element is A (Fig. 4-11), the force applied on the fluid element by the imaginary piston is

$$F = PA$$

To push the entire fluid element into the control volume, this force must act through a distance L. Thus, the work done in pushing the fluid element across the boundary (i.e., the flow work) is

$$W_{\text{flow}} = FL = PAL = PV \qquad \text{(kJ)} \qquad \text{(4-10)}$$

The flow work per unit mass is obtained by dividing both sides of this equation by the mass of the fluid element:

$$w_{\text{flow}} = Pv \qquad \text{(kJ/kg)} \qquad \text{(4-11)}$$

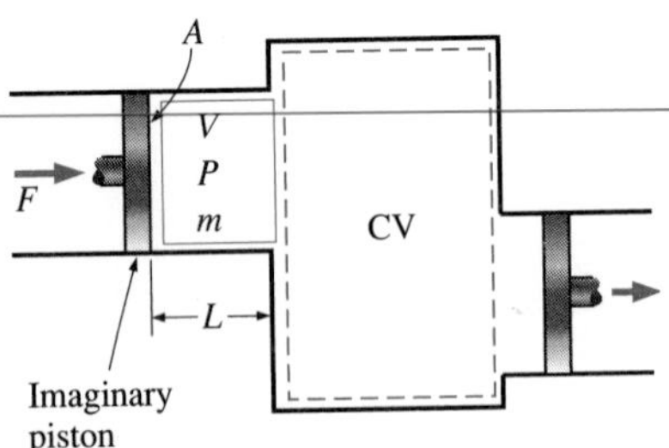

FIGURE 4-10
Schematic for flow work.

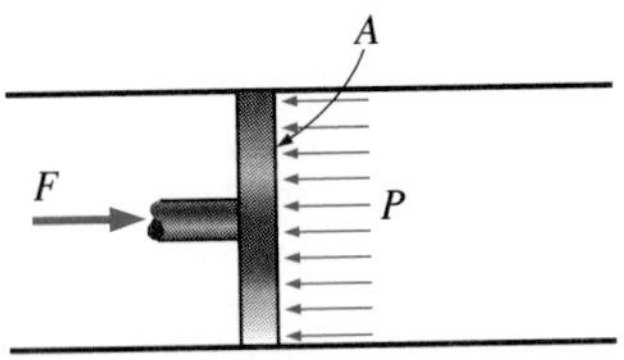

FIGURE 4-11
In the absence of acceleration, the force applied on a fluid by a piston is equal to the force applied on the piston by the fluid.

The flow work relation is the same whether the fluid is pushed into or out of the control volume (Fig. 4-12).

It is interesting that unlike other work quantities, flow work is expressed in terms of the properties. In fact, it is the product of two properties of the fluid. For that reason, some people view it as a *combination property* (like enthalpy) and refer to it as *flow energy, convected energy,* or *transport energy* instead of flow work. Others, however, argue rightfully that the product Pv represents energy for flowing fluids only and does not represent any form of energy for nonflow (closed) systems. Therefore, it should be treated as work.

This controversy is not likely to end, but it is comforting to know that both arguments yield the same result for the energy equation. In the discussions that follow, we consider the flow energy to be part of the energy of a flowing fluid, since this greatly simplifies the derivation of the energy equation for control volumes.

Total Energy of a Flowing Fluid

As we discussed in Chap. 1, the total energy of a simple compressible system consists of three parts: internal, kinetic, and potential energies (Fig. 4-13). On a unit-mass basis, it is expressed as

$$e = u + \text{ke} + \text{pe} = u + \frac{\mathcal{V}^2}{2} + gz \qquad \text{(kJ/kg)} \qquad (4\text{-}12)$$

where $\mathcal{V}$ is the velocity and z is the elevation of the system relative to some external reference point.

The fluid entering or leaving a control volume possesses an additional form of energy—the *flow energy Pv*, as discussed above. Then the total energy of a **flowing fluid** on a unit-mass basis (denoted θ) becomes

$$\theta = Pv + e = Pv + (u + \text{ke} + \text{pe})$$

But the combination $Pv + u$ has been previously defined as the enthalpy h. So the above relation reduces to

$$\theta = h + \text{ke} + \text{pe} = h + \frac{\mathcal{V}^2}{2} + gz \qquad \text{(kJ/kg)} \qquad (4\text{-}13)$$

Professor J. Kestin proposed in 1966 that the term θ be called **methalpy** (from *metaenthalpy,* which means *beyond enthalpy*).

By using the enthalpy instead of the internal energy to represent the energy of a flowing fluid, one does not need to be concerned about the flow work. The energy associated with pushing the fluid into or out of the control volume is automatically taken care of by enthalpy. In fact, this is the main reason for defining the property enthalpy. From now on, the energy of a fluid stream flowing into or out of a control volume is represented by Eq. 4-13, and no reference will be made to flow work or flow energy. Thus, the work term

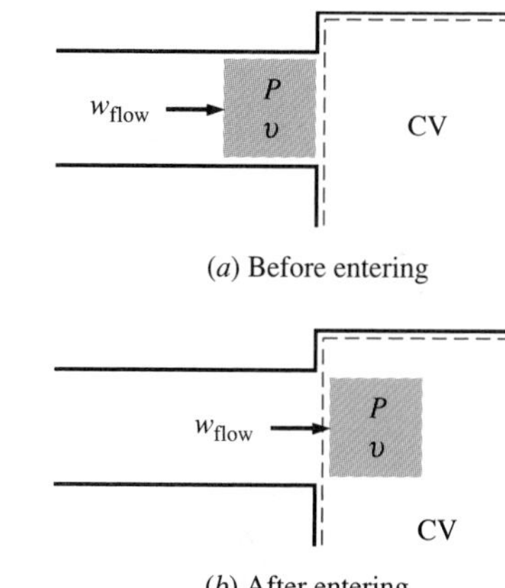

FIGURE 4-12

Flow work is the energy needed to push a fluid into or out of a control volume, and it is equal to Pv.

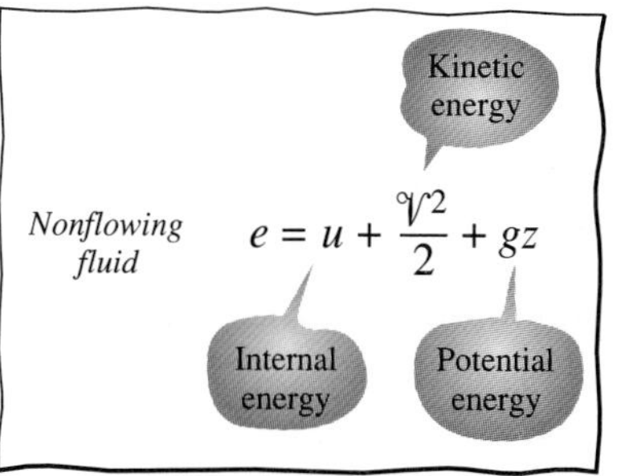

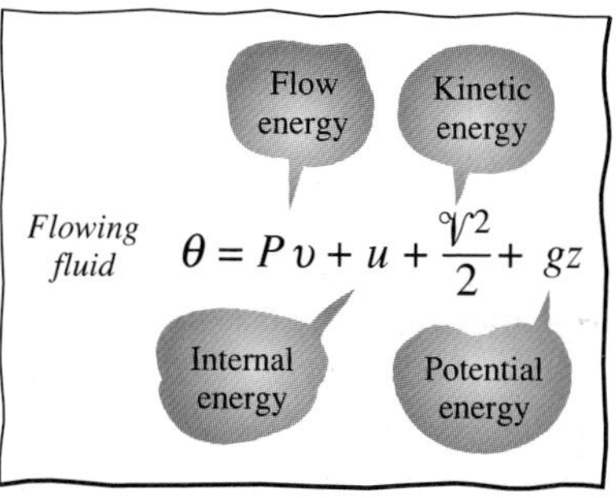

FIGURE 4-13

The total energy consists of three parts for a nonflowing fluid and four parts for a flowing fluid.

W in the control-volume energy equations will represent all forms of work (boundary, shaft, electrical, etc.) except flow work.

4-2 ■ THE STEADY-FLOW PROCESS

A large number of engineering devices such as turbines, compressors, and nozzles operate for long periods of time under the same conditions, and they are classified as *steady-flow devices.*

Processes involving steady-flow devices can be represented reasonably well by a somewhat idealized process, called the **steady-flow process.** A steady-flow process can be defined as *a process during which a fluid flows through a control volume steadily* (Fig. 4-14). That is, the fluid properties can change from point to point within the control volume, but at any fixed point they remain the same during the entire process. (Remember, *steady* means *no change with time.*) A steady-flow process is characterized by the following:

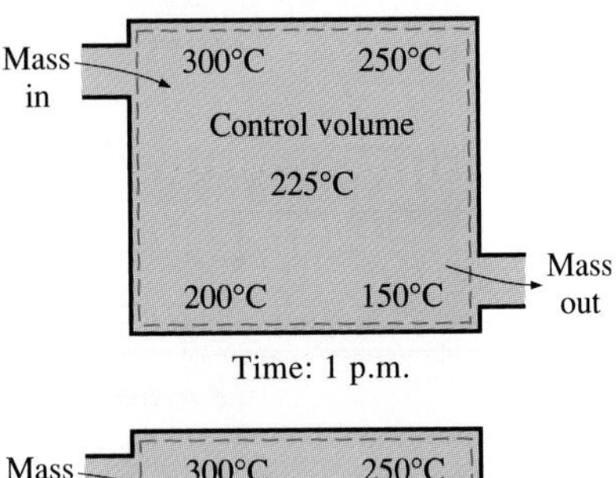

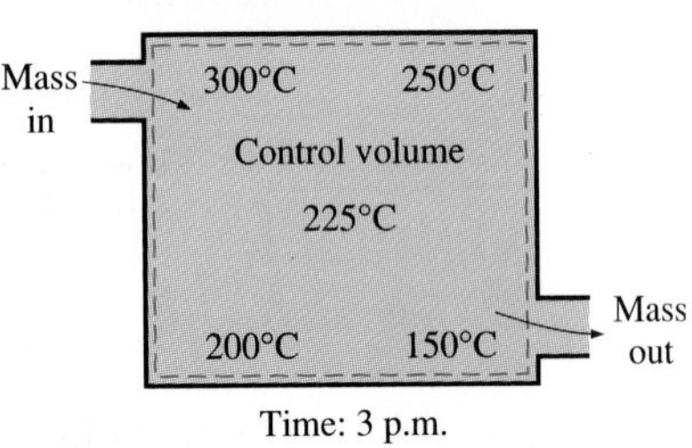

FIGURE 4-14

During a steady-flow process, fluid properties within the control volume may change with position, but not with time.

1. No properties (intensive or extensive) *within the control volume* change with time. Thus, the volume V, the mass m, and the total energy content E of the control volume remain constant during a steady flow process (Fig. 4-15). As a result, the boundary work is zero for steady flow systems (since V_{CV} = constant), and the total mass or energy entering the control volume must be equal to the total mass or energy leaving it (since m_{CV} = constant and E_{CV} = constant). These observations greatly simplify the analysis.

2. No properties change at the *boundaries* of the control volume with time. Thus, the fluid properties at an inlet or an exit will remain the same during the entire process. The properties may, however, be different at different inlets and exits. They may even vary over the cross section of an inlet or an exit. But all properties, including the velocity and elevation, must remain constant with time at a fixed position. It follows that the mass flow rate of the fluid at an opening must remain constant during a steady flow process (Fig. 4-16). As an added simplification, the fluid properties at an opening are usually considered to be uniform (at some average value) over the cross-section. Thus, the fluid properties at an inlet or exit may be specified by the average single values.

3. The *heat* and *work* interactions between a steady-flow system and its surroundings do not change with time. Thus, the power delivered by a system and the rate of heat transfer to or from a system remain constant during a steady flow process.

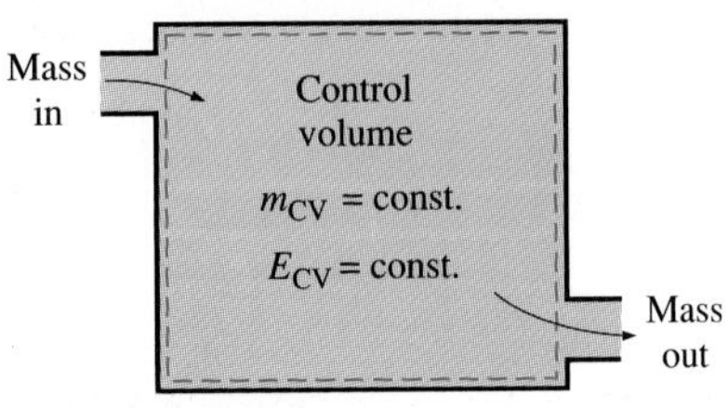

FIGURE 4-15

Under steady-flow conditions, the mass and energy contents of a control volume remain constant.

FIGURE 4-16

Under steady-flow conditions, the fluid properties at an inlet or exit remain constant (do not change with time).

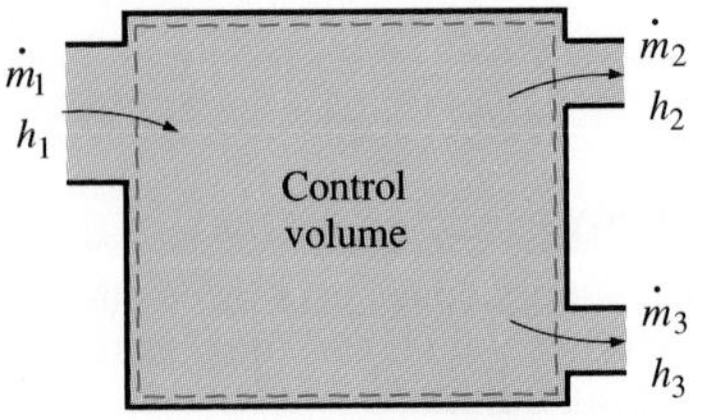

Some cyclic devices, such as reciprocating engines or compressors, do not satisfy any of the conditions stated above since the flow at the inlets and the exits will be pulsating and not steady. However, the fluid properties vary with time in a periodic manner, and the flow through these devices can still be analyzed as a steady flow process by using time-averaged values for the properties and the heat flow rates through the boundaries.

Steady-flow conditions can be closely approximated by devices that are intended for continuous operation such as turbines, pumps, boilers, condensers, and heat exchangers of steam power plants. The equations that are developed later in this section can be used for these and similar devices once the transient start-up period is completed and a steady operation is established.

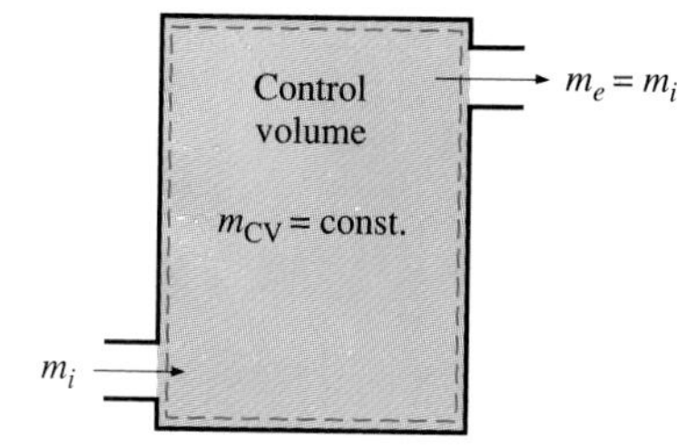

FIGURE 4-17

During a steady-flow process, the amount of mass entering the control volume equals the amount of mass leaving.

Conservation of Mass

During a steady-flow process, the total amount of mass contained within a control volume does not change with time (m_{CV} = constant). Then the conservation of mass principle requires that the total amount of mass entering a control volume equal the total amount of mass leaving it (Fig. 4-17). For a garden hose nozzle, for example, the amount of water entering the nozzle is equal to the amount of water leaving it in steady operation.

When dealing with steady-flow processes, we are not interested in the amount of mass that flows in and out of a device over time; instead, we are interested in the amount of mass flowing per unit time, that is, *the mass flow rate* $\dot{m}$. The **conservation of mass principle** for a general steady-flow system with multiple inlets and exits (Fig. 4-18) can be expressed in the rate form as

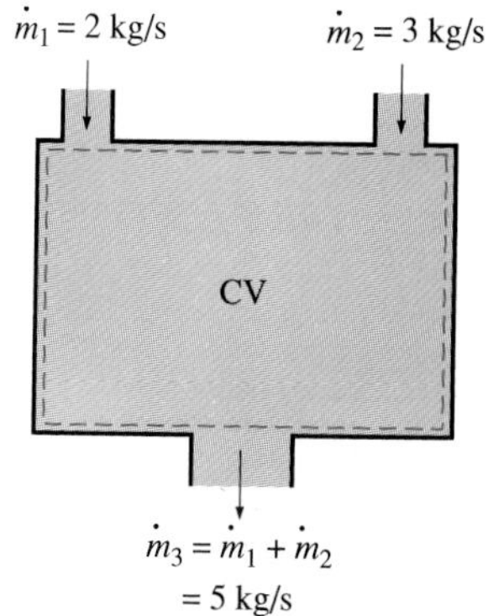

FIGURE 4-18

Conservation of mass principle for a two-inlet–one-exit steady-flow system.

$$\begin{pmatrix}\text{Total mass}\\ \text{entering CV}\\ \text{per unit time}\end{pmatrix} = \begin{pmatrix}\text{Total mass}\\ \text{leaving CV}\\ \text{per unit time}\end{pmatrix}$$

or

$$\sum \dot{m}_i = \sum \dot{m}_e \qquad \text{(kg/s)} \tag{4-14}$$

where the subscript *i* stands for *inlet* and *e* for *exit*. Most engineering devices such as nozzles, diffusers, turbines, compressors, and pumps involve a single stream (only one inlet and one exit). For these cases, we denote the inlet state by the subscript 1 and the exit state by the subscript 2. We also drop the summation signs. Then Eq. 4-14 reduces, for single-stream steady-flow systems, to

$$\dot{m}_1 = \dot{m}_2 \qquad \text{(kg/s)} \tag{4-15}$$

or

$$\rho_1 \mathcal{V}_1 A_1 = \rho_2 \mathcal{V}_2 A_2 \tag{4-16}$$

or

$$\frac{1}{v_1}\mathcal{V}_1 A_1 = \frac{1}{v_2}\mathcal{V}_2 A_2 \tag{4-17}$$

where

ρ = density, kg/m^3
v = specific volume, m^3/kg (= $1/\rho$)
$\mathcal{V}$ = average flow velocity in the flow direction, m/s
A = cross-sectional area normal to the flow direction, m^2

The reader is reminded that there is no such thing as a "conservation of volume" principle. Therefore, the volume flow rates ($\dot{V} = \mathcal{V}A$, m^3/s) into and out of a steady-flow device may be different. The volume flow rate at the exit of an air compressor will be much less than that at the inlet even though the mass flow rate of air through the compressor is constant (Fig. 4-19). This is due to the higher density of air at the compressor exit. For liquid flow, however, the volume flow rates, as well as the mass flow rates, remain constant since liquids are essentially incompressible (constant-density) substances. Water flow through the nozzle of a garden hose is a good example for the latter case.

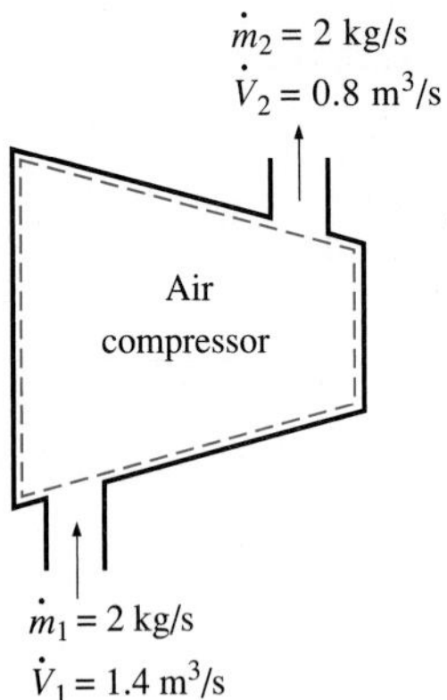

FIGURE 4-19
During a steady-flow process, volume flow rates are not necessarily conserved.

Energy Balance for Steady-Flow Systems

It was pointed out earlier that during a steady-flow process the total energy content of a control volume remains constant (E_{CV} = constant). That is, the change in the total energy of the control volume during such a process is zero ($\Delta E_{CV} = 0$). Thus, the amount of energy entering a control volume in all forms (heat, work, mass transfer) must be equal to the amount of energy leaving it for a steady-flow process. Then the rate form of the general energy balance reduces for a steady-flow process to

$$\underbrace{\dot{E}_{in} - \dot{E}_{out}}_{\substack{\text{Rate of net energy transfer} \\ \text{by heat, work, and mass}}} = \underbrace{\Delta \dot{E}_{system}^{\nearrow 0 \text{ (steady)}}}_{\substack{\text{Rate of change in internal, kinetic,} \\ \text{potential, etc., energies}}} = 0 \tag{4-18}$$

or

$$\underbrace{\dot{E}_{in}}_{\substack{\text{Rate of net energy transfer in} \\ \text{by heat, work, and mass}}} = \underbrace{\dot{E}_{out}}_{\substack{\text{Rate of net energy transfer out} \\ \text{by heat, work, and mass}}} \tag{4-19}$$

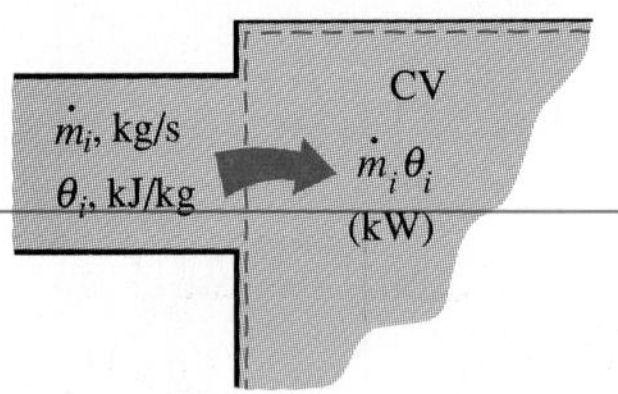

FIGURE 4-20
The product $\dot{m}_i\theta_i$ is the energy transported into the control volume by mass per unit time.

Noting that energy can be transferred by heat, work, and mass only, the energy balance above for a general steady-flow system can also be written as

$$\dot{Q}_{in} + \dot{W}_{in} + \sum \dot{m}_i\theta_i = \dot{Q}_{out} + \dot{W}_{out} + \sum \dot{m}_e\theta_e \tag{4-20}$$

or

$$\dot{Q}_{in} + \dot{W}_{in} + \underbrace{\sum \dot{m}_i\left(h_i + \frac{\mathcal{V}_i^2}{2} + gz_i\right)}_{\text{for each inlet}} = \dot{Q}_{out} + \dot{W}_{out} + \underbrace{\sum \dot{m}_e\left(h_e + \frac{\mathcal{V}_e^2}{2} + gz_e\right)}_{\text{for each exit}} \tag{4-21}$$

since the energy of a flowing fluid per unit mass is $\theta = h + \text{ke} + \text{pe} = h + \mathcal{V}^2/2 + gz$ (Fig. 4-20). The first law relation for steady-flow systems first appeared in 1859 in a German thermodynamics book written by Gustav Zeuner.

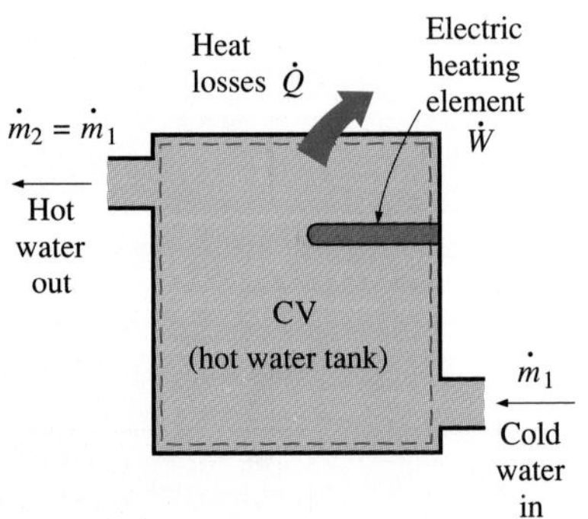

FIGURE 4-21
A water heater under steady operation.

Consider, for example, an ordinary electric hot-water heater under steady operation, as shown in Fig. 4-21. A cold-water stream with a mass flow rate $\dot{m}$ is continuously flowing into the water heater, and a hot-water stream of the same mass flow rate is continuously flowing out of it. The water heater (the control volume) is losing heat to the surrounding air at a rate of $\dot{Q}_{out}$, and the electric heating element is supplying electrical work (heating) to the water at a rate of $\dot{W}_{in}$. On the basis of the conservation of energy principle, we can say that the water stream will experience an increase in its total energy as it flows through the water heater that is equal to the electric energy supplied to the water minus the heat losses.

The energy balance (or the first law) relation given above is intuitive in nature and is easy to use when the magnitudes and directions of heat and work transfers are known. But when performing a general analytical study or solving a problem that involves an unknown heat or work interaction, we need to assume a direction for the heat or work interactions. In such cases, it is common practice to assume heat to be transferred *into the system* (heat input) at a rate of $\dot{Q}$, and power produced *by the system* (work output) at a rate of $\dot{W}$, and then solve the problem. The first law or energy balance relation in that case for a general steady-flow system becomes

$$\dot{Q} - \dot{W} = \underbrace{\sum \dot{m}_e \left(h_e + \frac{\mathcal{V}_e^2}{2} + gz_e \right)}_{\text{for each exit}} - \underbrace{\sum \dot{m}_i \left(h_i + \frac{\mathcal{V}_i^2}{2} + gz_i \right)}_{\text{for each inlet}} \qquad \text{(kW)} \qquad (4\text{-}22)$$

That is, the rate of heat transfer to a system minus power produced by the system is equal to the net change in the energy of the flow streams. Obtaining a negative quantity for $\dot{Q}$ or $\dot{W}$ simply means that the assumed direction for that quantity is wrong and should be reversed.

For single-stream (one-inlet–one-exit) systems, the summations over the inlets and the exits drop out, and the inlet and exit states in this case are denoted by subscripts 1 and 2, respectively, for simplicity. The mass flow rate through the entire control volume remains constant ($\dot{m}_1 = \dot{m}_2$) and is denoted $\dot{m}$. Then the energy balance for *single-stream steady-flow systems* becomes

$$\dot{Q} - \dot{W} = \dot{m}\left[h_2 - h_1 + \frac{\mathcal{V}_2^2 - \mathcal{V}_1^2}{2} + g(z_2 - z_1) \right] \qquad \text{(kW)} \qquad (4\text{-}23)$$

Dividing the equation above by $\dot{m}$ gives the energy balance on a unit-mass basis as

$$q - w = \Delta h + \Delta \text{ke} + \Delta \text{pe} = h_2 - h_1 + \frac{\mathcal{V}_2^2 - \mathcal{V}_1^2}{2} + g(z_2 - z_1) \qquad \text{(kJ/kg)} \qquad (4\text{-}24)$$

where $q = \dot{Q}/\dot{m}$ and $w = \dot{W}/\dot{m}$ are the heat transfer and work done per unit mass of the working fluid, respectively.

If the fluid experiences a negligible change in its kinetic and potential energies as it flows through the control volume (that is, $\Delta \text{ke} \cong 0$, $\Delta \text{pe} \cong 0$), then the energy equation for a single-stream steady-flow system reduces further to

$$q - w = \Delta h \qquad \text{(kJ/kg)} \qquad (4\text{-}25)$$

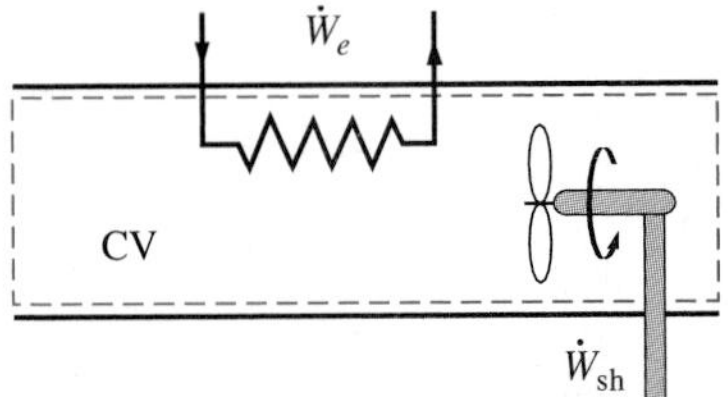

FIGURE 4-22
Under steady operation, shaft work and electrical work are the only forms of work a simple compressible system may involve.

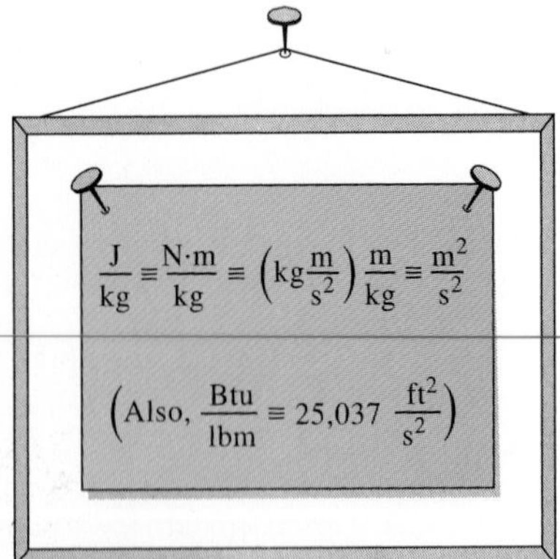

FIGURE 4-23
The units m^2/s^2 and J/kg are equivalent.

FIGURE 4-24
At very high velocities, even small changes in velocities may cause significant changes in the kinetic energy of the fluid.

$\mathcal{V}_1$ m/s	$\mathcal{V}_2$ m/s	Δke kJ/kg
0	40	1
50	67	1
100	110	1
200	205	1
500	502	1

The various terms appearing in the above equations are as follows:

$\dot{Q}$ = **rate of heat transfer between the control volume and its surroundings.** When the control volume is losing heat (as in the case of the water heater), $\dot{Q}$ is negative. If the control volume is well insulated (i.e., adiabatic), then $\dot{Q} = 0$.

$\dot{W}$ = **power.** For steady-flow devices, the volume of the control volume is constant; thus, there is no boundary work involved. The work required to push mass into and out of the control volume is also taken care of by using enthalpies for the energy of fluid streams instead of internal energies. Then $\dot{W}$ represents the remaining forms of work done per unit time (Fig. 4-22). Many steady-flow devices, such as turbines, compressors, and pumps, transmit power through a shaft, and $\dot{W}$ simply becomes the shaft power for those devices. If the control surface is crossed by electric wires (as in the case of an electric water heater), $\dot{W}$ will represent the electrical work done per unit time. If neither is present then $\dot{W} = 0$.

$\Delta h = h_{exit} - h_{inlet}$. The enthalpy change of a fluid can easily be determined by reading the enthalpy values at the exit and inlet states from the tables. For ideal gases, it may be approximated by $\Delta h = C_{p,av}(T_2 - T_1)$. Note that (kg/s)(kJ/kg) ≡ kW.

$\Delta ke = (\mathcal{V}_2^2 - \mathcal{V}_1^2)/2$. The unit of kinetic energy is m^2/s^2, which is equivalent to J/kg (Fig. 4-23). The enthalpy is usually given in kJ/kg. To add these two quantities, the kinetic energy should be expressed in kJ/kg. This is easily accomplished by dividing it by 1000.

A velocity of 45 m/s corresponds to a kinetic energy of only 1 kJ/kg, which is a very small value compared with the enthalpy values encountered in practice. Thus, the kinetic energy term at low velocities can be neglected. When a fluid stream enters and leaves a steady-flow device at about the same velocity ($\mathcal{V}_1 \cong \mathcal{V}_2$), the change in the kinetic energy is close to zero regardless of the velocity. Caution should be exercised at high velocities, however, since small changes in velocities may cause significant changes in kinetic energy (Fig. 4-24).

$\Delta pe = g(z_2 - z_1)$. A similar argument can be given for the potential energy term. A potential energy change of 1 kJ/kg corresponds to an elevation difference of 102 m. The elevation difference between the inlet and exit of most industrial devices such as turbines and compressors is well below this value, and the potential energy term is always neglected for these devices. The only time the potential energy term is significant is when a process involves pumping a fluid to high elevations. This is particularly true for systems involving negligible heat transfer.

4-3 ■ SOME STEADY-FLOW ENGINEERING DEVICES

Many engineering devices operate essentially under the same conditions for long periods of time. The components of a steam power plant (turbines,

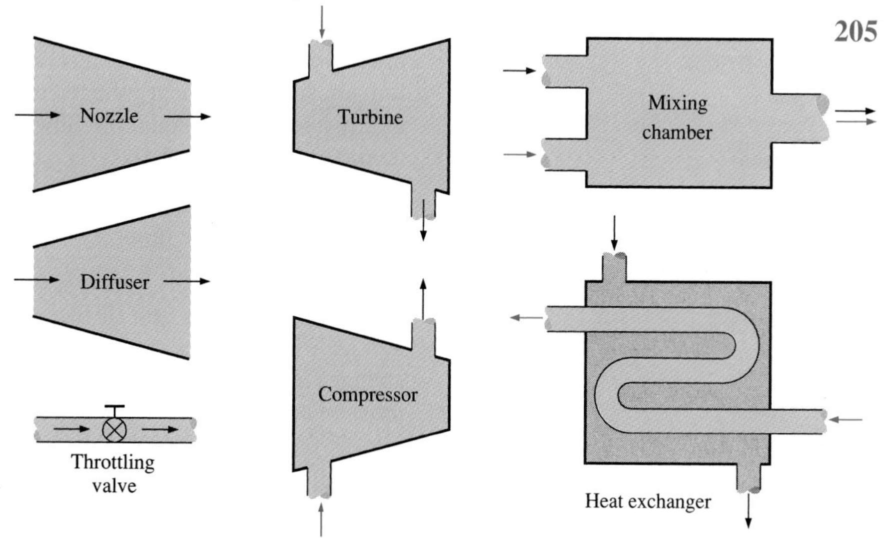

FIGURE 4-25
Steady-flow devices operate steadily for long periods.

compressors, heat exchangers, and pumps), for example, operate nonstop for months before the system is shut down for maintenance (Fig. 4-25). Therefore, these devices can be conveniently analyzed as steady-flow devices.

In this section, some common steady-flow devices are described, and the thermodynamic aspects of the flow through them are analyzed. The conservation of mass and the conservation of energy principles for these devices are illustrated with examples.

1 Nozzles and Diffusers

Nozzles and diffusers are commonly utilized in jet engines, rockets, spacecraft, and even garden hoses. A **nozzle** is a device that *increases the velocity of a fluid* at the expense of pressure. A **diffuser** is a device that *increases the pressure of a fluid* by slowing it down. That is, nozzles and diffusers perform opposite tasks. The cross-sectional area of a nozzle decreases in the flow direction for subsonic flows and increases for supersonic flows. The reverse is true for diffusers. The different behavior of the fluid for supersonic flows is explained in Chap. 16. Figure 4-26 shows a nozzle and a diffuser schematically.

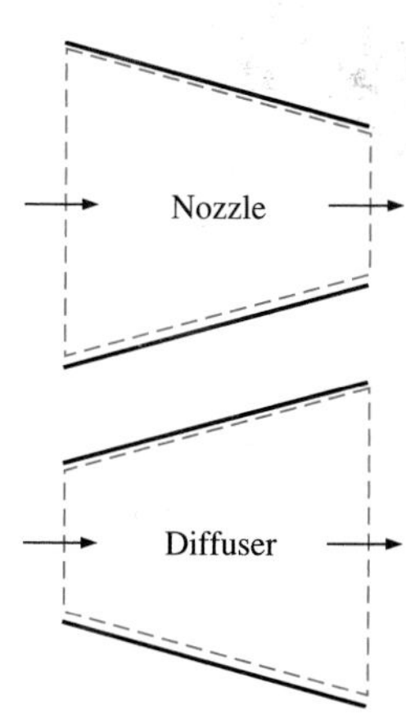

FIGURE 4-26
Schematic of a nozzle and diffuser for subsonic flows (velocities under the speed of sound).

The relative importance of the terms appearing in the energy equation for nozzles and diffusers is as follows:

$\dot{Q} \cong 0.$ The rate of heat transfer between the fluid flowing through a nozzle or a diffuser and the surroundings is usually very small, even when these devices are not insulated. This is mainly due to the fluid's having high velocities and thus not spending enough time in the device for any significant heat transfer to take place. Therefore, in the absence of any heat transfer data, the flow through nozzles and diffusers may be assumed to be adiabatic.

$\dot{W} = 0$. The work term for nozzles and diffusers is zero since these devices basically are properly shaped ducts and they involve no shaft or electric resistance wires.

$\Delta ke \neq 0$. Nozzles and diffusers usually involve very high velocities, and as a fluid passes through a nozzle or diffuser, it experiences large changes in its velocity (Fig. 4-27). Therefore, the kinetic energy changes must be accounted for in analyzing the flow through these devices.

$\Delta pe \cong 0$. The fluid usually experiences little or no change in its elevation as it flows through a nozzle or a diffuser, and therefore the potential energy term can be neglected.

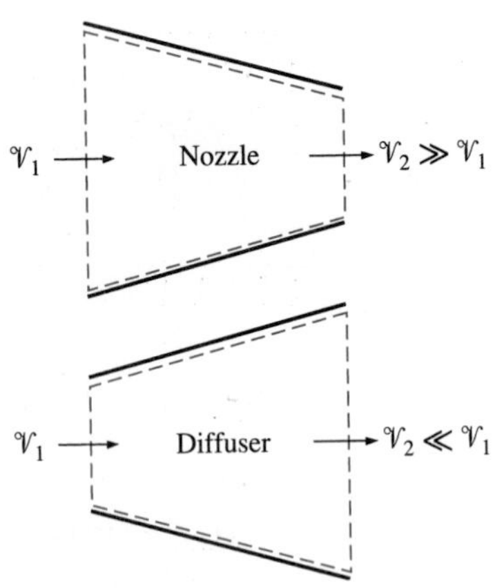

FIGURE 4-27
Nozzles and diffusers are shaped so that they cause large changes in fluid velocities and thus kinetic energies.

EXAMPLE 4-1 Deceleration of Air in a Diffuser

Air at 10°C and 80 kPa enters the diffuser of a jet engine steadily with a velocity of 200 m/s. The inlet area of the diffuser is 0.4 m². The air leaves the diffuser with a velocity that is very small compared with the inlet velocity. Determine (*a*) the mass flow rate of the air and (*b*) the temperature of the air leaving the diffuser.

Solution We take the *diffuser* as the system (Fig. 4-28), This is a *control volume* since mass crosses the system boundary during the process. We observe that there is only one inlet and one exit and thus $\dot{m}_1 = \dot{m}_2 = \dot{m}$.

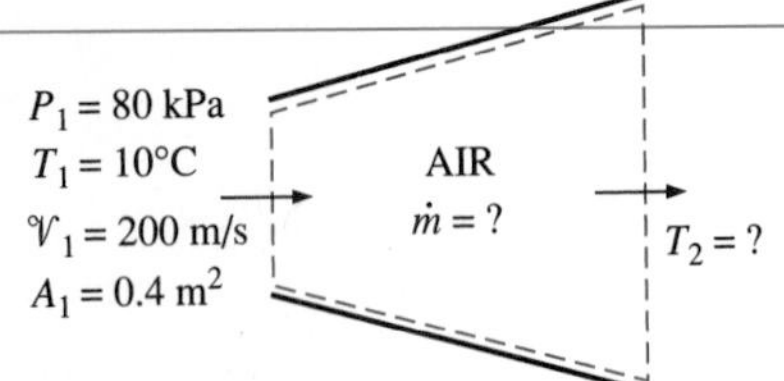

FIGURE 4-28
Schematic for Example 4-1.

Assumptions **1** This is a steady-flow process since there is no change with time at any point and thus $\Delta m_{CV} = 0$ and $\Delta E_{CV} = 0$. **2** Air is an ideal gas since it is at a high temperature and low pressure relative to its critical point values of −141°C and 3.77 MPa. **3** The potential energy change is zero, $\Delta pe = 0$. **4** Heat transfer is negligible. **5** Kinetic energy at the diffuser exit is negligible **6** There are no work interactions.

Analysis (*a*) To determine the mass flow rate, we need to find the specific volume of the air first. This is determined from the ideal-gas relation at the inlet conditions:

$$v_1 = \frac{RT_1}{P_1} = \frac{[0.287 \text{ kPa}\cdot\text{m}^3/(\text{kg}\cdot\text{K})](283 \text{ K})}{80 \text{ kPa}} = 1.015 \text{ m}^3/\text{kg}$$

Then from Eq. 4-5,

$$\dot{m} = \frac{1}{v_1}\mathcal{V}_1 A_1 = \frac{1}{1.015 \text{ m}^3/\text{kg}}(200 \text{ m/s})(0.4 \text{ m}^2) = \mathbf{78.8 \text{ kg/s}}$$

Since the flow is steady, the mass flow rate through the entire diffuser will remain constant at this value.

(*b*) Under stated assumptions and observations, the energy balance for this steady-flow system can be expressed in the rate form as

$$\underbrace{\dot{E}_{in} - \dot{E}_{out}}_{\substack{\text{Rate of net energy transfer} \\ \text{by heat, work, and mass}}} = \underbrace{\Delta \dot{E}_{system}}_{\substack{\text{Rate of change in internal, kinetic,} \\ \text{potential, etc., energies}}}\nearrow^{0\text{ (steady)}} = 0$$

$$\dot{E}_{in} = \dot{E}_{out}$$

$$\dot{m}\left(h_1 + \frac{\mathscr{V}_1^2}{2}\right) = \dot{m}\left(h_2 + \frac{\mathscr{V}_2^2}{2}\right) \qquad \text{(since } \dot{Q} \cong 0,\ \dot{W} = 0, \text{ and } \Delta pe \cong 0\text{)}$$

$$h_2 = h_1 - \frac{\mathscr{V}_2^2 - \mathscr{V}_1^2}{2}$$

The exit velocity of a diffuser is usually small compared with the inlet velocity ($\mathscr{V}_2 \ll \mathscr{V}_1$); thus, the kinetic energy at the exit can be neglected. The enthalpy of air at the diffuser inlet is determined from the air table (Table A-17) to be

$$h_1 = h_{@283\text{ K}} = 283.14 \text{ kJ/kg}$$

Substituting, we get

$$h_2 = 283.14 \text{ kJ/kg} - \frac{0 - (200 \text{ m/s})^2}{2}\left(\frac{1 \text{ kJ/kg}}{1000 \text{ m}^2/\text{s}^2}\right)$$
$$= 303.14 \text{ kJ/kg}$$

From Table A-17, the temperature corresponding to this enthalpy value is

$$T_2 = \mathbf{303.1\ K}$$

which shows that the temperature of the air increased by about 20°C as it was slowed down in the diffuser. The temperature rise of the air is mainly due to the conversion of kinetic energy to internal energy.

EXAMPLE 4-2 Acceleration of Steam in a Nozzle

Steam at 250 psia and 700°F steadily enters a nozzle whose inlet area is 0.2 ft². The mass flow rate of the steam through the nozzle is 10 lbm/s. Steam leaves the nozzle at 200 psia with a velocity of 900 ft/s. The heat losses from the nozzle per unit mass of the steam are estimated to be 1.2 Btu/lbm. Determine (*a*) the inlet velocity and (*b*) the exit temperature of the steam.

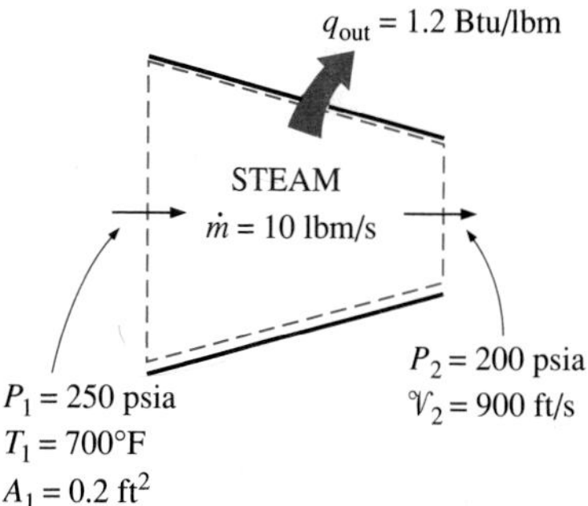

FIGURE 4-29
Schematic for Example 4-2.

Solution We take the *nozzle* as the system (Fig. 4-29). This is a *control volume* since mass crosses the system boundary during the process. We *observe* that there is only one inlet and one exit and thus $\dot{m}_1 = \dot{m}_2 = \dot{m}$.

Assumptions **1** This is a steady-flow process since there is no change with time at any point and thus $\Delta m_{CV} = 0$ and $\Delta E_{CV} = 0$. **2** There are no work interactions. **3** The potential energy change is zero, $\Delta pe = 0$.

Analysis (*a*) The inlet velocity is determined from Eq. 4-5. But first we need to determine the specific volume of the steam at the nozzle inlet:

$$\left.\begin{array}{l} P_1 = 250 \text{ psia} \\ T_1 = 700°\text{F} \end{array}\right\} \quad \begin{array}{l} v_1 = 2.688 \text{ ft}^3/\text{lbm} \\ h_1 = 1371.1 \text{ Btu/lbm} \end{array} \qquad \text{(Table A-6E)}$$

Then from Eq. 4-5.

$$\dot{m} = \frac{1}{v_1}\mathcal{V}_1 A_1$$

$$10 \text{ lbm/s} = \frac{1}{2.688 \text{ ft}^3/\text{lbm}}(\mathcal{V}_1)(0.2 \text{ ft}^2)$$

$$\mathcal{V}_1 = \mathbf{134.4 \text{ ft/s}}$$

(*b*) Under stated assumptions and observations, the energy balance for this steady-flow system can be expressed in the rate form as

$$\underbrace{\dot{E}_{in} - \dot{E}_{out}}_{\substack{\text{Rate of net energy transfer}\\ \text{by heat, work, and mass}}} = \underbrace{\Delta \dot{E}_{system}^{\nearrow 0 \text{ (steady)}}}_{\substack{\text{Rate of change in internal, kinetic,}\\ \text{potential, etc., energies}}} = 0$$

$$\dot{E}_{in} = \dot{E}_{out}$$

$$\dot{m}\left(h_1 + \frac{\mathcal{V}_1^2}{2}\right) = \dot{Q}_{out} + \dot{m}\left(h_2 + \frac{\mathcal{V}_2^2}{2}\right) \qquad (\text{since } \dot{W} \cong 0, \text{ and } \Delta pe \cong 0)$$

Dividing by the mass flow rate $\dot{m}$ and substituting, h_2 is determined to be

$$h_2 = h_1 - q_{out} - \frac{\mathcal{V}_2^2 - \mathcal{V}_1^2}{2}$$

$$= (1371.1 - 1.2) \text{ Btu/lbm} - \frac{(900 \text{ ft/s})^2 - (134.4 \text{ ft/s})^2}{2}\left(\frac{1 \text{ Btu/lbm}}{25{,}037 \text{ ft}^2/\text{s}^2}\right)$$

$$= 1354.1 \text{ Btu/lbm}$$

Then,

$$\left.\begin{aligned} P_2 &= 200 \text{ psia} \\ h_2 &= 1354.1 \text{ Btu/lbm} \end{aligned}\right\} \quad T_2 = \mathbf{661.9°F} \qquad \text{(Table A-6E)}$$

Therefore, the temperature of steam will drop by 38.1°F as it flows through the nozzle. This drop in temperature is mainly due to the conversion of internal energy to kinetic energy. (The heat loss is too small to cause any significant effect in this case.)

2 Turbines and Compressors

In steam, gas, or hydroelectric power plants, the device that drives the electric generator is the turbine. As the fluid passes through the turbine, work is done against the blades, which are attached to the shaft. As a result, the shaft rotates, and the turbine produces work. The work done in a turbine is positive since it is done by the fluid.

Compressors, as well as pumps and fans, are devices used to increase the pressure of a fluid. Work is supplied to these devices from an external source through a rotating shaft. Therefore, the work term for compressors is negative since work is done on the fluid. Even though these three devices function similarly, they do differ in the tasks they perform. A *fan* increases the pressure of a gas slightly and is mainly used to move a gas around. A *compressor* is capable of compressing the gas to very high pressures. *Pumps* work very much like compressors except that they handle liquids instead of gases.

For turbines and compressors, the relative magnitudes of the various terms appearing in the energy equation are as follows:

$\dot{Q} \cong 0$. The heat transfer for these devices is generally small relative to the shaft work unless there is intentional cooling (as in compressors). An estimated value based on the experimental studies can be used in the analysis, or the heat transfer may be neglected if there is no intentional cooling.

$\dot{W} \neq 0$. All of these devices involve rotating shafts crossing their boundaries; therefore, the work term is important. For turbines, $\dot{W}$ represents the power output; for pumps and compressors, it represents the power input.

$\Delta\text{pe} \cong 0$. The potential energy change that a fluid experiences as it flows through turbines, compressors, fans, and pumps is usually very small and is normally neglected.

$\Delta\text{ke} \cong 0$. The velocities involved with these devices, with the exception of turbines, are usually too low to cause any significant change in the kinetic energy. The fluid velocities encountered in most turbines are very high, and the fluid experiences a significant change in its kinetic energy. However, this change is usually very small relative to the change in enthalpy, and thus it is often disregarded.

EXAMPLE 4-3 Compressing Air by a Compressor

Air at 100 kPa and 280 K is compressed steadily to 600 kPa and 400 K. The mass flow rate of the air is 0.02 kg/s, and a heat loss of 16 kJ/kg occurs during the process. Assuming the changes in kinetic and potential energies are negligible, determine the necessary power input to the compressor.

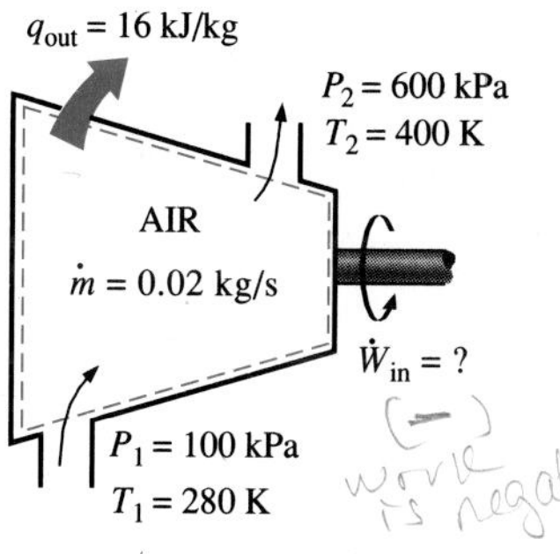

FIGURE 4-30

Schematic for Example 4-3.

Solution We take the *compressor* as the system (Fig. 4-30). This is a *control volume* since mass crosses the system boundary during the process. We observe that there is only one inlet and one exit and thus $\dot{m}_1 = \dot{m}_2 = \dot{m}$. Also, heat is lost from the system and work is supplied to the system.

Assumptions **1** This is a steady-flow process since there is no change with time at any point and thus $\Delta m_{CV} = 0$ and $\Delta E_{CV} = 0$. **2** Air is an ideal gas since it is at a high temperature and low pressure relative to its critical point values of $-141°C$ and 3.77 MPa. **3** The kinetic and potential energy changes are zero, $\Delta\text{pe} = \Delta\text{pe} = 0$.

Analysis Under stated assumptions and observations, the energy balance for this steady-flow system can be expressed in the rate form as

$$\underbrace{\dot{E}_{in} - \dot{E}_{out}}_{\text{Rate of net energy transfer by heat, work, and mass}} = \underbrace{\Delta \dot{E}_{system}}_{\text{Rate of change in internal, kinetic, potential, etc., energies}}^{\nearrow 0 \text{ (steady)}} = 0$$

$$\dot{E}_{in} = \dot{E}_{out}$$

$$\dot{W} + \dot{m}h_1 = \dot{Q}_{out} + \dot{m}h_2 \qquad (\text{since } \Delta ke = \Delta pe \cong 0)$$

$$\dot{W}_{in} = \dot{m}q_{out} + \dot{m}(h_2 - h_1)$$

The enthalpy of an ideal gas depends on temperature only, and the enthalpies of the air at the specified temperatures are determined from the air table (Table A-17) to be

$$h_1 = h_{@\,280\,K} = 280.13 \text{ kJ/kg}$$
$$h_2 = h_{@\,400\,K} = 400.98 \text{ kJ/kg}$$

Substituting, the power input to the compressor is determined to be

$$\dot{W}_{in} = (0.02 \text{ kg/s})(16 \text{ kJ/kg}) + (0.02 \text{ kg/s})(400.98 - 280.13 \text{ kJ/kg})$$
$$= \mathbf{2.74\ kW}$$

EXAMPLE 4-4 Power Generation by a Steam Turbine

The power output of an adiabatic steam turbine is 5 MW, and the inlet and the exit conditions of the steam are as indicated in Fig. 4-31.

(*a*) Compare the magnitudes of Δh, Δke, and Δpe.

(*b*) Determine the work done per unit mass of the steam flowing through the turbine.

(*c*) Calculate the mass flow rate of the steam.

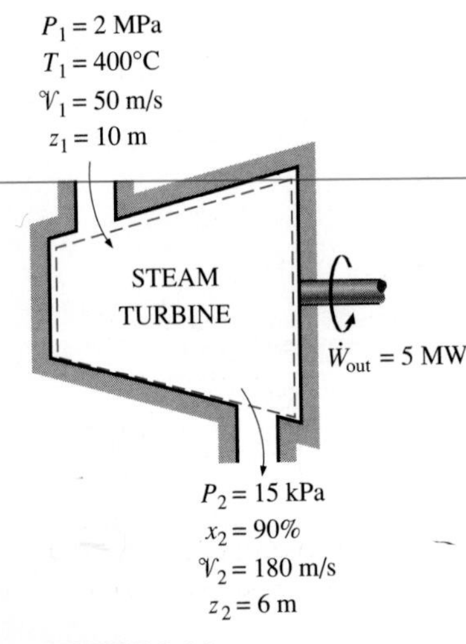

FIGURE 4-31
Schematic for Example 4-4.

Solution We take the *turbine* as the system (Fig. 4-31). This is a *control volume* since mass crosses the system boundary during the process. We observe that there is only one inlet and one exit and thus $\dot{m}_1 = \dot{m}_2 = \dot{m}$. Also, work is done by the system. The inlet and exit velocities and elevations are given, and thus the kinetic and potential energies are to be considered.

Assumptions **1** This is a steady-flow process since there is no change with time at any point and thus $\Delta m_{CV} = 0$ and $\Delta E_{CV} = 0$. **2** The system is adiabatic and thus there is no heat transfer.

Analysis (*a*) At the inlet, steam is in a superheated vapor state, and its enthalpy is

$$\left.\begin{matrix} P_1 = 2 \text{ MPa} \\ T_1 = 400°C \end{matrix}\right\} \quad h_1 = 3247.6 \text{ kJ/kg} \qquad \text{(Table A-6)}$$

At the turbine exit, we obviously have a saturated liquid–vapor mixture at 15-kPa pressure. The enthalpy at this state is

$$h_2 = h_f + x_2 h_{fg} = [225.94 + (0.9)(2373.1)] \text{ kJ/kg} = 2361.73 \text{ kJ/kg}$$

Then

$$\Delta h = h_2 - h_1 = (2361.73 - 3247.6)\text{ kJ/kg} = \mathbf{-885.87\text{ kJ/kg}}$$

$$\Delta ke = \frac{\mathcal{V}_2^2 - \mathcal{V}_1^2}{2} = \frac{(180\text{ m/s})^2 - (50\text{ m/s})^2}{2}\left(\frac{1\text{ kJ/kg}}{1000\text{ m}^2/\text{s}^2}\right) = \mathbf{14.95\text{ kJ/kg}}$$

$$\Delta pe = g(z_2 - z_1) = (9.807\text{ m/s}^2)[(6 - 10)\text{ m}]\left(\frac{1\text{ kJ/kg}}{1000\text{ m}^2/\text{s}^2}\right) = \mathbf{-0.04\text{ kJ/kg}}$$

Two observations can be made from the above results. First, the change in potential energy is insignificant in comparison to the changes in enthalpy and kinetic energy. This is typical for most engineering devices. Second, as a result of low pressure and thus high specific volume, the steam velocity at the turbine exit can be very high. Yet the change in kinetic energy is a small fraction of the change in enthalpy (less than 2 percent in our case) and is therefore often neglected.

(*b*) The energy balance for this steady-flow system can be expressed in the rate form as

$$\underbrace{\dot{E}_{in} - \dot{E}_{out}}_{\substack{\text{Rate of net energy transfer}\\ \text{by heat, work, and mass}}} = \underbrace{\Delta \dot{E}_{system}^{\nearrow 0\text{ (steady)}}}_{\substack{\text{Rate of change in internal, kinetic,}\\ \text{potential, etc., energies}}} = 0$$

$$\dot{E}_{in} = \dot{E}_{out}$$

$$\dot{m}(h_1 + \mathcal{V}_1^2/2 + gz_1) = \dot{W}_{out} + \dot{m}(h_2 + \mathcal{V}_2^2/2 + gz_2) \qquad (\text{since } \dot{Q} = 0)$$

Dividing by the mass flow rate $\dot{m}$ and substituting, the work done by the turbine per unit mass of the steam is determined to be

$$w_{out} = -\left[(h_2 - h_1) + \frac{\mathcal{V}_2^2 - \mathcal{V}_1^2}{2} + g(z_2 - z_1)\right] = -(\Delta h + \Delta ke + \Delta pe)$$

$$= -[-885.87 + 14.95 - 0.04]\text{ kJ/kg} = \mathbf{870.96\text{ kJ/kg}}$$

(*c*) The required mass flow rate for a 5-MW power output is

$$\dot{m} = \frac{\dot{W}}{w} = \frac{5000\text{ kJ/s}}{870.96\text{ kJ/kg}} = \mathbf{5.74\text{ kg/s}}$$

3 Throttling Valves

Throttling valves are *any kind of flow-restricting devices* that cause a significant pressure drop in the fluid. Some familiar examples are ordinary adjustable valves, capillary tubes, and porous plugs (Fig. 4-32). Unlike turbines, they produce a pressure drop without involving any work. The pressure drop in the fluid is often accompanied by *a large drop in temperature,* and for that reason throttling devices are commonly used in refrigeration and air-conditioning applications. The magnitude of the temperature drop (or, sometimes, the temperature rise) during a throttling process is governed by a property called the *Joule-Thomson coefficient,* which is discussed in Chap. 11.

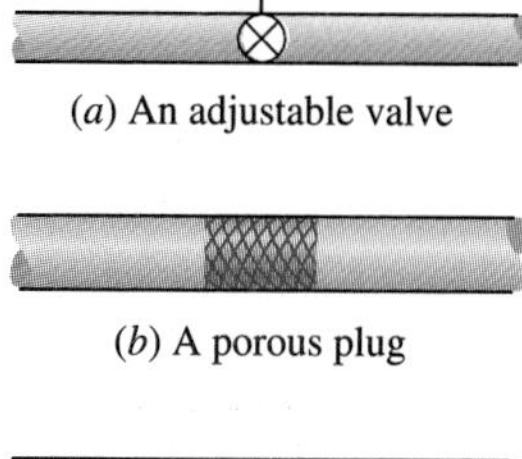

FIGURE 4-32
Throttling valves are devices that cause large pressure drops in the fluid.

Throttling valves are usually small devices, and the flow through them may be assumed to be adiabatic ($q \cong 0$) since there is neither sufficient time nor large enough area for any effective heat transfer to take place. Also, there is no work done ($w = 0$), and the change in potential energy, if any, is very small ($\Delta\text{pe} \cong 0$). Even though the exit velocity is often considerably higher than the inlet velocity, in many cases, the increase in kinetic energy is insignificant ($\Delta\text{ke} \cong 0$). Then the conservation of energy equation for this single-stream steady-flow device reduces to

$$h_2 \cong h_1 \qquad \text{(kJ/kg)} \qquad (4\text{-}26)$$

That is, enthalpy values at the inlet and exit of a throttling valve are the same. For this reason, a throttling valve is sometimes called an *isenthalpic device.*

To gain some insight into how throttling affects fluid properties, let us express Eq. 4-26 as follows:

$$u_1 + P_1 v_1 = u_2 + P_2 v_2$$

or

$$\text{Internal energy} + \text{Flow energy} = \text{Constant}$$

Thus the final outcome of a throttling process depends on which of the two quantities increases during the process. If the flow energy increases during the process ($P_2 v_2 > P_1 v_1$), it can do so at the expense of the internal energy. As a result, internal energy decreases, which is usually accompanied by a drop in temperature. If the product Pv decreases, the internal energy and the temperature of a fluid will increase during a throttling process. In the case of an ideal gas, $h = h(T)$, and thus the temperature has to remain constant during a throttling process (Fig. 4-33).

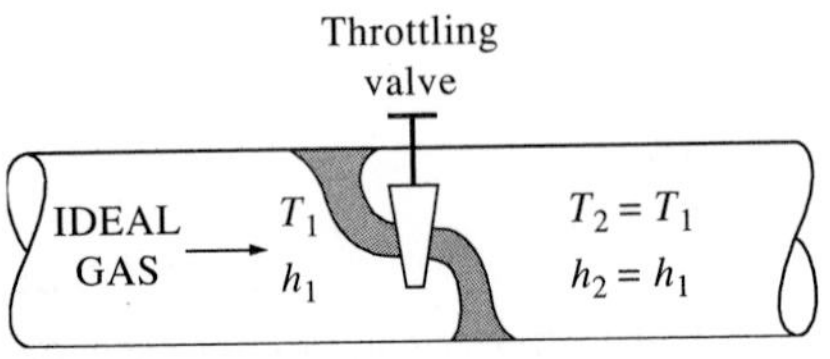

FIGURE 4-33

The temperature of an ideal gas does not change during a throttling (h = constant) process since $h = h(T)$.

EXAMPLE 4-5 Expansion of Refrigerant-134a in a Refrigerator

Refrigerant-134a enters the capillary tube of a refrigerator as saturated liquid at 0.8 MPa and is throttled to a pressure of 0.12 MPa. Determine the quality of the refrigerant at the final state and the temperature drop during this process.

Solution A capillary tube is a simple flow-restricting device that is commonly used in refrigeration applications to cause a large pressure drop in the refrigerant. Flow through a capillary tube is a throttling process; thus, the enthalpy of the refrigerant remains constant (Fig. 4-34).

At inlet: $\left.\begin{array}{l} P_1 = 0.8\text{ MPa} \\ \text{sat. liquid} \end{array}\right\}$ $\begin{array}{l} T_1 = T_{\text{sat @ 0.8 MPa}} = 31.33°\text{C} \\ h_1 = h_{f\text{ @ 0.8 MPa}} = 93.42\text{ kJ/kg} \end{array}$ (Table A-12)

At exit: $P_2 = 0.12\text{ MPa} \longrightarrow h_f = 21.32\text{ kJ/kg} \qquad T_{\text{sat}} = -22.36°\text{C}$

$(h_2 = h_1) \qquad h_g = 233.86\text{ kJ/kg}$

Obviously $h_f < h_2 < h_g$; thus, the refrigerant exists as a saturated mixture at the exit state. The quality at this state is determined from

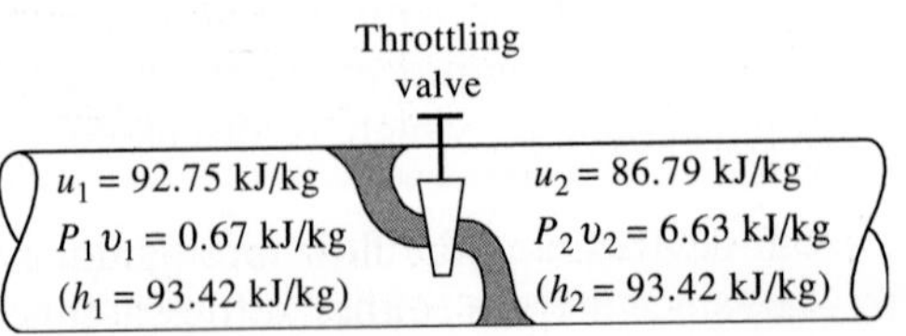

FIGURE 4-34

During a throttling process, the enthalpy (flow energy + internal energy) of a fluid remains constant. But internal and flow energies may be converted to each other.

$$x_2 = \frac{h_2 - h_f}{h_{fg}} = \frac{93.42 - 21.32}{233.86 - 21.32} = \mathbf{0.339}$$

Since the exit state is a saturated mixture at 0.12 MPa, the exit temperature must be the saturation temperature at this pressure, which is −22.36°C. Then the temperature change for this process becomes

$$\Delta T = T_2 - T_1 = (-22.36 - 31.33)°\text{C} = \mathbf{-53.69°C}$$

That is, the temperature of the refrigerant drops by 53.69°C during this throttling process. Notice that 33.9 percent of the refrigerant vaporizes during this throttling process, and the energy needed to vaporize this refrigerant is absorbed from the refrigerant itself.

4a Mixture Chambers

In engineering applications, mixing two streams of fluids is not a rare occurrence. The section where the mixing process takes place is commonly referred to as a **mixing chamber.** The mixing chamber does not have to be a distinct "chamber." An ordinary T-elbow or a Y-elbow in a shower, for example, serves as the mixing chamber for the cold- and hot-water streams (Fig. 4-35).

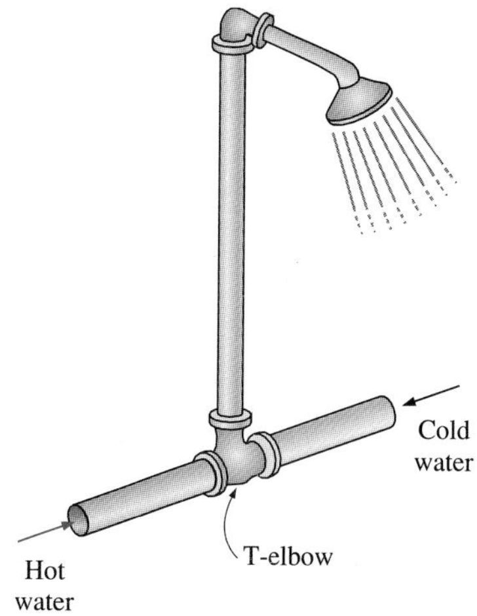

FIGURE 4-35

The T-elbow of an ordinary shower serves as the mixing chamber for the hot- and the cold-water streams.

The conservation of mass principle for a mixing chamber requires that the sum of the incoming mass flow rates equal the mass flow rate of the outgoing mixture.

Mixing chambers are usually well insulated ($q \cong 0$) and do not involve any kind of work ($w = 0$). Also, the kinetic and potential energies of the fluid streams are usually negligible (ke $\cong$ 0, pe $\cong$ 0). Then all there is left in the energy equation (Eq. 4-19) is the total energies of the incoming streams and the outgoing mixture. The conservation of energy principle requires that these two equal each other. Therefore, the conservation of energy equation becomes analogous to the conservation of mass equation for this case.

EXAMPLE 4-6 Mixing of Hot and Cold Waters in a Shower

Consider an ordinary shower where hot water at 140°F is mixed with cold water at 50°F. If it is desired that a steady stream of warm water at 110°F be supplied, determine the ratio of the mass flow rates of the hot to cold water. Assume the heat losses from the mixing chamber to be negligible and the mixing to take place at a pressure of 20 psia.

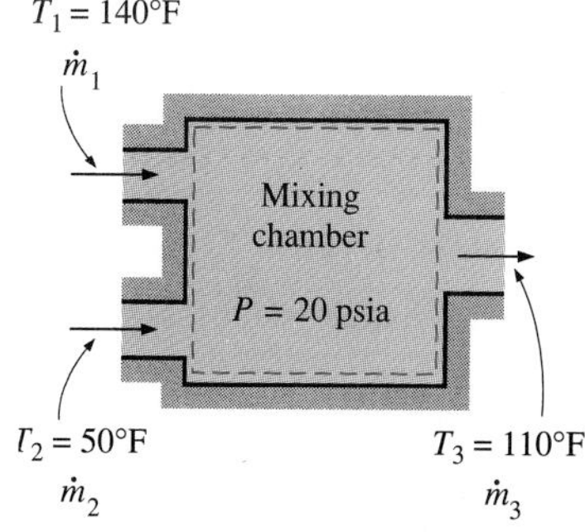

FIGURE 4-36

Schematic for Example 4-6.

Solution We take the *mixing chamber* as the system (Fig. 4-36). This is a *control volume* since mass crosses the system boundary during the process. We observe that there are two inlets and one exit.

Assumptions **1** This is a steady-flow process since there is no change with time at any point and thus $\Delta m_{CV} = 0$ and $\Delta E_{CV} = 0$. **2** The kinetic and potential energies are negligible, ke $\cong$ pe $\cong$ 0. **3** Heat losses from the system are negligible and thus $\dot{Q} \cong 0$. **4** There is no work interaction involved.

Analysis Under the stated assumptions and observations, the mass and energy balances for this steady-flow system can be expressed in the rate form as follows:

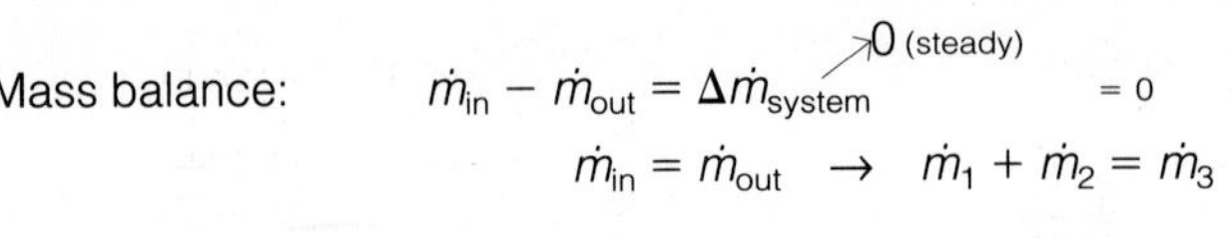

Mass balance: $\dot{m}_{in} - \dot{m}_{out} = \Delta \dot{m}_{system}^{\nearrow 0 \text{ (steady)}} = 0$

$$\dot{m}_{in} = \dot{m}_{out} \quad \rightarrow \quad \dot{m}_1 + \dot{m}_2 = \dot{m}_3$$

Energy balance:
$$\underbrace{\dot{E}_{in} - \dot{E}_{out}}_{\substack{\text{Rate of net energy transfer} \\ \text{by heat, work, and mass}}} = \underbrace{\Delta \dot{E}_{system}^{\nearrow 0 \text{ (steady)}}}_{\substack{\text{Rate of change in internal, kinetic,} \\ \text{potential, etc., energies}}} = 0$$

$$\dot{E}_{in} = \dot{E}_{out}$$

$$\dot{m}_1 h_1 + \dot{m}_2 h_2 = \dot{m}_3 h_3 \qquad (\text{since } \dot{Q} \cong 0,\ \dot{W} = 0,\ \text{ke} \cong \text{pe} \cong 0)$$

Combining the mass and energy balances,

$$\dot{m}_1 h_1 + \dot{m}_2 h_2 = (\dot{m}_1 + \dot{m}_2) h_3$$

Dividing this equation by $\dot{m}_2$ yields

$$y h_1 + h_2 = (y + 1) h_3$$

where $y = \dot{m}_1/\dot{m}_2$ is the desired mass flow rate ratio.

The saturation temperature of water at 20 psia is 227.96°F. Since the temperatures of all three streams are below this value ($T < T_{sat}$), the water in all three streams exists as a compressed liquid (Fig. 4-37). A compressed liquid can be approximated as a saturated liquid at the given temperature. Thus,

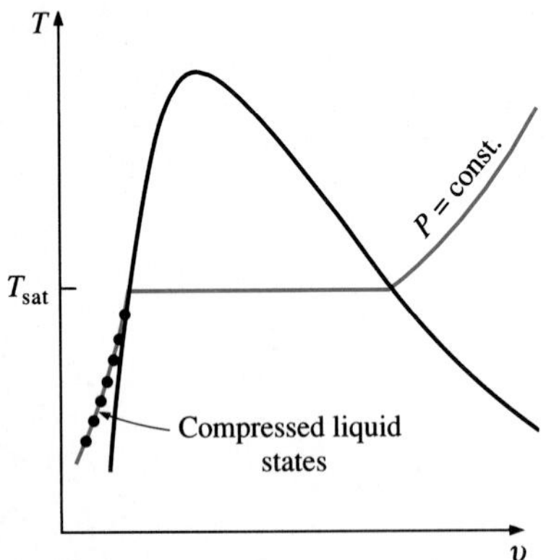

FIGURE 4-37
A substance exists as a compressed liquid at temperatures below the saturation temperatures at the given pressure.

$$h_1 \cong h_{f\,@\,140°F} = 107.96 \text{ Btu/lbm}$$
$$h_2 \cong h_{f\,@\,50°F} = 18.06 \text{ Btu/lbm}$$
$$h_3 \cong h_{f\,@\,110°F} = 78.02 \text{ Btu/lbm}$$

Solving for y and substituting yields

$$y = \frac{h_3 - h_2}{h_1 - h_3} = \frac{78.02 - 18.06}{107.96 - 78.02} = 2.0$$

Thus the mass flow rate of the hot water must be twice the mass flow rate of the cold water for the mixture to leave at 110°F.

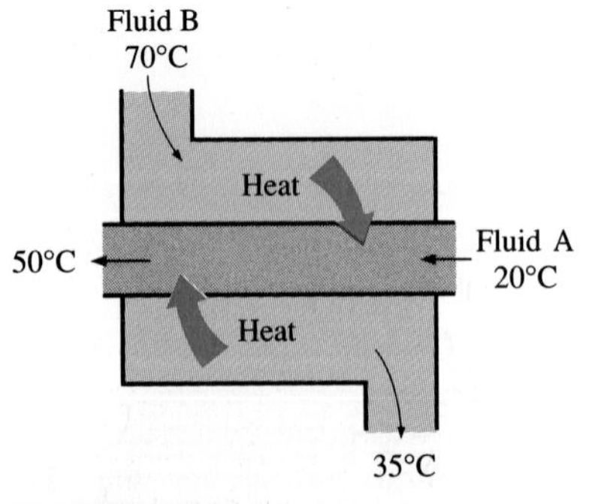

FIGURE 4-38
A heat exchanger can be as simple as two concentric pipes.

4b Heat Exchangers

As the name implies, **heat exchangers** are devices where two moving fluid streams exchange heat without mixing. Heat exchangers are widely used in various industries, and they come in various designs.

The simplest form of a heat exchanger is a *double-tube* (also called *tube-and-shell*) *heat exchanger,* shown in Fig. 4-38. It is composed of two concentric pipes of different diameters. One fluid flows in the inner pipe, and the other in the annular space between the two pipes. Heat is transferred from the hot fluid to the cold one through the wall separating them. Sometimes the inner tube makes a couple of turns inside the shell to increase the heat transfer area, and thus the rate of heat transfer. The mixing chambers discussed earlier are sometimes classified as *direct-contact* heat exchangers.

The conservation of mass principle for a heat exchanger in steady operation requires that the sum of the inbound mass flow rates equal the sum of

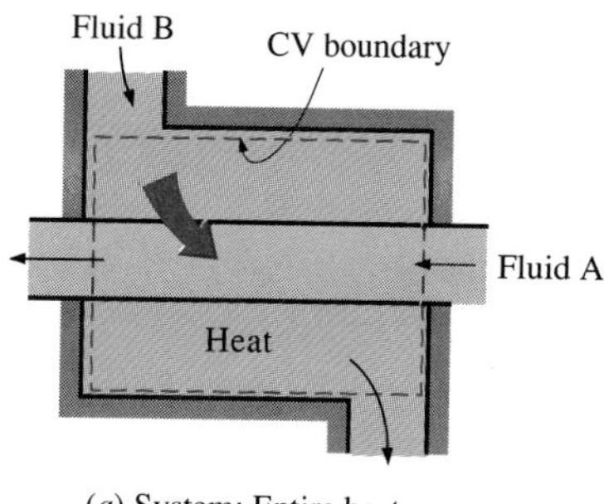

(a) System: Entire heat exchanger ($Q_{CV} = 0$)

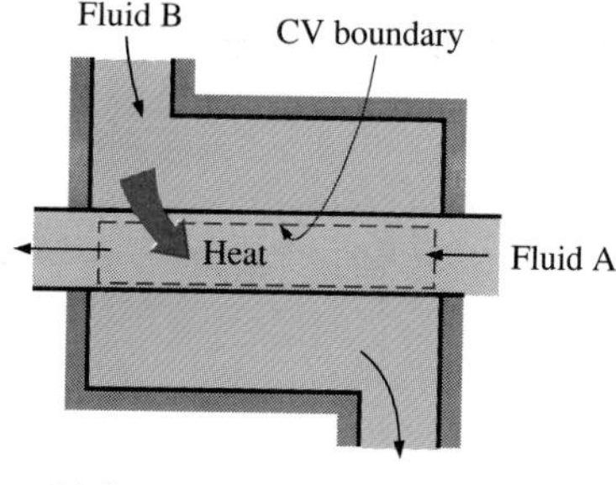

(a) System: Fluid A ($Q_{CV} \neq 0$)

FIGURE 4-39
The heat transfer associated with a heat exchanger may be zero or nonzero depending on how the system is selected.

the outbound mass flow rates. This principle can also be expressed as follows: *Under steady operation, the mass flow rate of each fluid stream flowing through a heat exchanger remains constant.*

Heat exchangers typically involve no work interactions ($w = 0$) and negligible kinetic and potential energy changes ($\Delta ke \cong 0$, $\Delta pe \cong 0$) for each fluid stream. The heat transfer rate associated with heat exchangers depends on how the control volume is selected. Heat exchangers are intended for heat transfer between two fluids *within* the device, and the outer shell is usually well insulated to prevent any heat loss to the surrounding medium.

When the entire heat exchanger is selected as the control volume, $\dot{Q}$ becomes zero, since the boundary for this case lies just beneath the insulation and little or no heat crosses the boundary (Fig. 4-39). If, however, only one of the fluids is selected as the control volume then heat will cross this boundary as it flows from one fluid to the other and $\dot{Q}$ will not be zero. In fact, $\dot{Q}$ in this case will be the rate of heat transfer between the two fluids.

EXAMPLE 4-7 Cooling of Refrigerant-134a by Water

Refrigerant-134a is to be cooled by water in a condenser. The refrigerant enters the condenser with a mass flow rate of 6 kg/min at 1 MPa and 70°C and leaves at 35°C. The cooling water enters at 300 kPa and 15°C and leaves at 25°C. Neglecting any pressure drops, determine (*a*) the mass flow rate of the cooling water required and (*b*) the heat transfer rate from the refrigerant to water.

Solution We take the *entire heat exchanger* as the system (Fig. 4-40). This is a *control volume* since mass crosses the system boundary during the process. In general, there are several possibilities for selecting the control volume for multiple-stream steady-flow devices, and the proper choice depends on the situation at hand. We observe that there are two fluid streams (and thus two inlets and two exits) but no mixing.

Assumptions **1** This is a steady-flow process since there is no change with time at any point and thus $\Delta m_{CV} = 0$ and $\Delta E_{CV} = 0$. **2** The kinetic and potential energies are negligible, ke $\cong$ pe $\cong$ 0. **3** Heat losses from the system are negligible and thus $\dot{Q} \cong 0$. **4** There is no work interaction.

Analysis (*a*) Under the stated assumptions and observations, the mass and energy balances for this steady-flow system can be expressed in the rate form as follows:

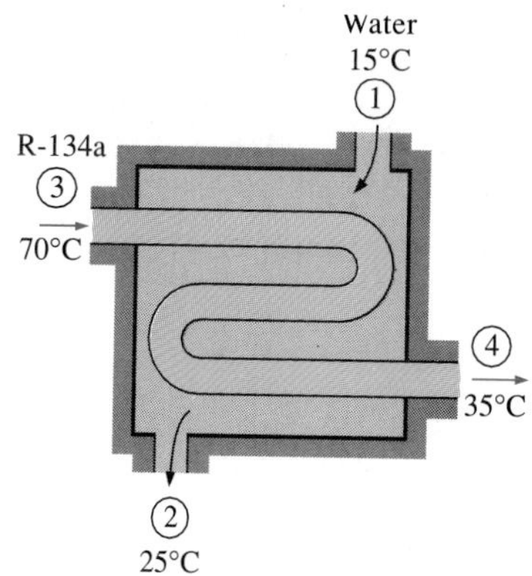

FIGURE 4-40
Schematic for Example 4-7.

Mass balance: $\dot{m}_{in} = \dot{m}_{out}$

for each fluid stream since there is no mixing. Thus,

$$\dot{m}_1 = \dot{m}_2 = \dot{m}_w$$
$$\dot{m}_3 = \dot{m}_4 = \dot{m}_R$$

Energy balance:

$$\underbrace{\dot{E}_{in} - \dot{E}_{out}}_{\substack{\text{Rate of net energy transfer}\\ \text{by heat, work, and mass}}} = \underbrace{\Delta \dot{E}_{system}}_{\substack{\text{Rate of change in internal, kinetic,}\\ \text{potential, etc., energies}}} \nearrow^{0\ \text{(steady)}} = 0$$

$$\dot{E}_{in} = \dot{E}_{out}$$

$$\dot{m}_1 h_1 + \dot{m}_3 h_3 = \dot{m}_2 h_2 + \dot{m}_4 h_4 \quad (\text{since } \dot{Q} \cong 0,\ \dot{W} = 0,\ \text{ke} \cong \text{pe} \cong 0)$$

Combining the mass and energy balances and rearranging give

$$\dot{m}_w(h_1 - h_2) = \dot{m}_R(h_4 - h_3)$$

Now we need to determine the enthalpies at all four states. Water exists as a compressed liquid at both the inlet and the exit since the temperatures at both locations are below the saturation temperature of water at 300 kPa (133.55°C). Approximating the compressed liquid as a saturated liquid at the given temperature, we have

$$h_1 \cong h_{f@15°C} = 62.99 \text{ kJ/kg}$$
$$h_2 \cong h_{f@25°C} = 104.89 \text{ kJ/kg} \qquad \text{(Table A-4)}$$

The refrigerant enters the condenser as a superheated vapor and leaves as a compressed liquid at 35°C. From refrigerant-134a tables,

$$\left.\begin{array}{l} P_3 = 1 \text{ MPa} \\ T_3 = 70°C \end{array}\right\} \quad h_3 = 302.34 \text{ kJ/kg} \qquad \text{(Table A-13)}$$

$$\left.\begin{array}{l} P_4 = 1 \text{ MPa} \\ T_4 = 35°C \end{array}\right\} \quad h_4 \cong h_{f@35°C} = 98.78 \text{ kJ/kg} \qquad \text{(Table A-11)}$$

Substituting, we find

$$\dot{m}_w(62.99 - 104.89) \text{ kJ/kg} = (6 \text{ kg/min}) [(- 302.34) \text{ kJ/kg}]$$
$$\dot{m}_w = \mathbf{29.15\ kg/min}$$

(*b*) To determine the heat transfer from the refrigerant to the water, we have to choose a control volume whose boundary lies on the path of the heat flow. We can choose the volume occupied by either fluid as our control volume. For no particular reason, we choose the volume occupied by the water. All the assumptions stated earlier apply, except that the heat flow is no longer zero. Then assuming heat to be transferred to water, the energy balance for this single-stream steady-flow system reduces to

$$\underbrace{\dot{E}_{in} - \dot{E}_{out}}_{\substack{\text{Rate of net energy transfer}\\ \text{by heat, work, and mass}}} = \underbrace{\Delta \dot{E}_{system}}_{\substack{\text{Rate of change in internal, kinetic,}\\ \text{potential, etc., energies}}} \nearrow^{0\ \text{(steady)}} = 0$$

$$\dot{E}_{in} = \dot{E}_{out}$$

$$\dot{Q}_{w,in} + \dot{m}_w h_1 = \dot{m}_w h_2$$

Rearranging and substituting,

$$\dot{Q}_{w,\text{in}} = \dot{m}_w(h_2 - h_1) = (29.15 \text{ kg/min})[(104.89 - 62.99) \text{ kJ/kg}]$$
$$= \mathbf{1221 \text{ kJ/min}}$$

Discussion Had we chosen the volume occupied by the refrigerant as the control volume (Fig. 4-41), we would have obtained the same result for $\dot{Q}_{R,\text{out}}$ since the heat gained by the water is equal to the heat lost by the refrigerant.

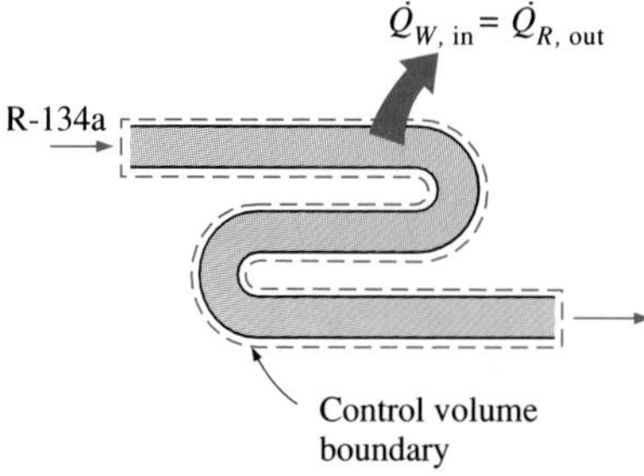

FIGURE 4-41

In a heat exchanger, the heat transfer depends on the choice of the control volume.

5 Pipe and Duct Flow

The transport of liquids or gases in pipes and ducts is of great importance in many engineering applications. Flow through a pipe or a duct usually satisfies the steady-flow conditions and thus can be analyzed as a steady-flow process. This, of course, excludes the transient start-up and shut-down periods. The control volume can be selected to coincide with the interior surfaces of the portion of the pipe or the duct that we are interested in analyzing.

When flow through pipes or ducts is analyzed, the following points should be considered:

$\dot{Q} \neq 0$. Under normal operating conditions, the amount of heat gained or lost by the fluid may be very significant, particularly if the pipe or duct is long (Fig. 4-42). Sometimes heat transfer is desirable and is the sole purpose of the flow. Water flow through the pipes in the furnace of a power plant, the flow of refrigerant in a freezer, and the flow in heat exchangers are some examples of this case. At other times, heat transfer is undesirable, and the pipes or ducts are insulated to prevent any heat loss or gain, particularly when the temperature difference between the flowing fluid and the surroundings is large. Heat transfer in this case is negligible.

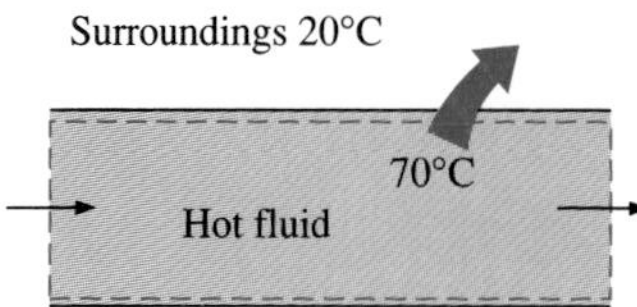

FIGURE 4-42

Heat losses from a hot fluid flowing through an uninsulated pipe or duct to the cooler environment may be very significant.

$\dot{W} \neq 0$. If the control volume involves a heating section (electric wires), a fan, or a pump (shaft), the work term interactions should be considered (Fig. 4-43). Of these, fan work is usually small and often neglected. If the control volume involves none of these work devices, the work term is zero.

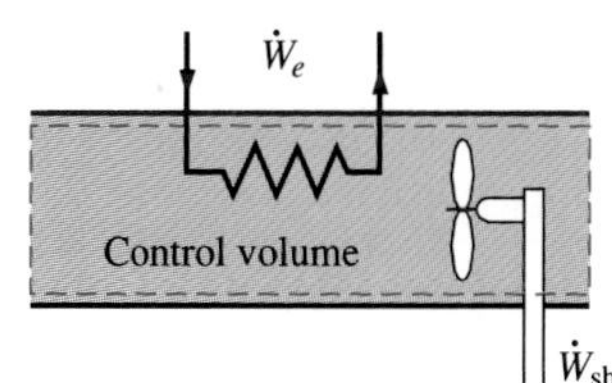

FIGURE 4-43

Pipe or duct flow may involve more than one form of work at the same time.

$\Delta\text{ke} \cong 0$. The velocities involved in pipe and duct flow are relatively low, and the kinetic energy changes are usually insignificant. This is particularly true when the pipe or duct diameter is constant and the heating effects are negligible. Kinetic energy changes may be significant, however, for gas flow in ducts with variable cross-sectional areas.

$\Delta\text{pe} \neq 0$. In pipes and ducts, the fluid may undergo a considerable elevation change. Thus, the potential energy term may be significant. This is particularly true for flow through insulated pipes and ducts where the heat transfer does not overshadow other effects.

EXAMPLE 4-8 Electric Heating of Air in a House

The electric heating systems used in many houses consist of a simple duct with resistance wires. Air is heated as it flows over resistance wires. Consider a 15-kW electric heating system. Air enters the heating section at 100 kPa and 17°C with a volume flow rate of 150 m³/min. If heat is lost from the air in the duct to the surroundings at a rate of 200 W, determine the exit temperature of air.

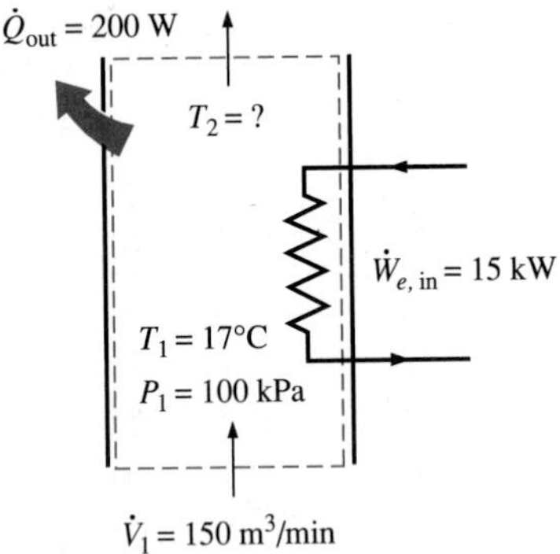

FIGURE 4-44
Schematic for Example 4-8.

Solution We take the *heating section portion of the duct* as the system (Fig. 4-44). This is a *control volume* since mass crosses the system boundary during the process. We observe that there is only one inlet and one exit and thus $\dot{m}_1 = \dot{m}_2 = \dot{m}$. Also, heat is lost from the system and electrical work is supplied to the system.

Assumptions **1** This is a steady-flow process since there is no change with time at any point and thus $\Delta m_{CV} = 0$ and $\Delta E_{CV} = 0$. **2** Air is an ideal gas since it is at a high temperature and low pressure relative to its critical point values of −141°C and 3.77 MPa. **3** The kinetic and potential energy changes are negligible, $\Delta ke \cong \Delta pe \cong 0$. **4** Constant specific heats at room temperature can be used for air.

Analysis At temperatures encountered in heating and air-conditioning applications, Δh can be replaced by $C_p\Delta T$ where C_p = 1.005 kJ/(kg · °C)—the value at room temperature—with negligible error (Fig. 4-45). Then the energy balance for this steady-flow system can be expressed in the rate form as

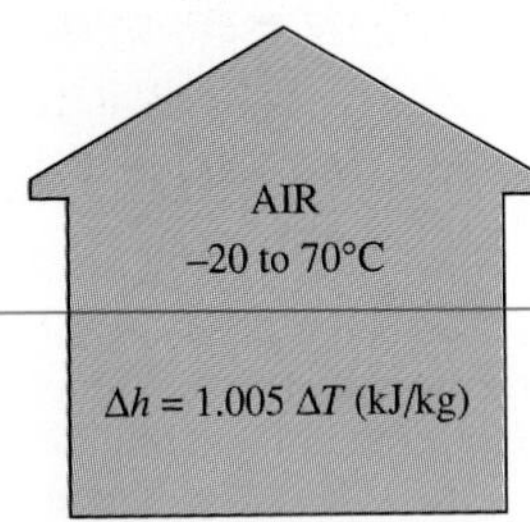

FIGURE 4-45
The error involved in $\Delta h = C_p\Delta T$, where C_p = 1.005 kJ/(kg · °C), is less than 0.5 percent for air in the temperature range −20 to 70°C.

$$\underbrace{\dot{E}_{in} - \dot{E}_{out}}_{\text{Rate of net energy transfer by heat, work, and mass}} = \underbrace{\Delta \dot{E}_{system}^{\nearrow 0 \text{ (steady)}}}_{\text{Rate of change in internal, kinetic, potential, etc., energies}} = 0$$

$$\dot{E}_{in} = \dot{E}_{out}$$

$$\dot{W}_{e,in} + \dot{m}h_1 = \dot{Q}_{out} + \dot{m}h_2 \qquad \text{(since } \Delta ke \cong \Delta pe \cong 0\text{)}$$

$$\dot{W}_{e,in} - \dot{Q}_{out} = \dot{m}C_p(T_2 - T_1)$$

From the ideal gas relation, the specific volume of air at the inlet of the duct is

$$v_1 = \frac{RT_1}{P_1} = \frac{[0.287 \text{ kPa} \cdot \text{m}^3/(\text{kg} \cdot \text{K})](290 \text{ K})}{100 \text{ kPa}} = 0.832 \text{ m}^3/\text{kg}$$

The mass flow rate of the air through the duct is determined from

$$\dot{m} = \frac{\dot{V}_1}{v_1} = \frac{150 \text{ m}^3/\text{min}}{0.832 \text{ m}^3/\text{kg}}\left(\frac{1 \text{ min}}{60 \text{ s}}\right) = 3.0 \text{ kg/s}$$

Substituting the known quantities, the exit temperature of the air is determined to be

$$(15 \text{ kJ/s}) - (0.2 \text{ kJ/s}) = (3 \text{ kg/s})[1.005 \text{ kJ/(kg} \cdot \text{°C)}](T_2 - 17)\text{°C}$$

$$T_2 = \mathbf{21.9°C}$$

EXAMPLE 4-9 Pumping Water from a Well

In rural areas, water is often extracted from underground by pumps. Consider an underground water source whose free surface is 60 m below ground level. The water is to be raised 5 m above the ground by a pump. The diameter of the pipe is 15 cm at the inlet and 20 cm at the exit. Neglecting any heat interaction with the surroundings and frictional heating effects, determine the power input to the pump required for a steady flow of water at a rate of 15 L/s (= 0.015 m^3/s).

Solution We take the *pipes and the pump* as the system (Fig. 4-46). This is a *control volume* since mass crosses the system boundary during the process. We observe that there is only one inlet and one exit and thus $\dot{m}_1 = \dot{m}_2 = \dot{m}$. Also, shaft work is supplied to the pump. The kinetic and potential energies can be significant in this case, and thus they will be considered.

Assumptions **1** This is a steady-flow process since there is no change with time at any point and thus $\Delta m_{CV} = 0$ and $\Delta E_{CV} = 0$. **2** Heat transfer is negligible. **3** Frictional heating effects are disregarded.

Analysis The density of liquid water at or about room temperature can be taken to be constant at 1000 kg/m^3 with negligible error. Then the mass flow rate and the flow velocities become

$$\dot{m} = \rho\dot{V} = (1000\ \text{kg/m}^3)(0.015\ \text{m}^3/\text{s}) = 15\ \text{kg/s}$$

$$\mathcal{V}_1 = \frac{\dot{m}}{\rho_1 A_1} = \frac{15\ \text{kg/s}}{(1000\ \text{kg/m}^3)[\pi(0.15\ \text{m})^2/4]} = 0.85\ \text{m/s}$$

$$\mathcal{V}_2 = \frac{\dot{m}}{\rho_2 A_2} = \frac{15\ \text{kg/s}}{(1000\ \text{kg/m}^3)[\pi(0.2\ \text{m})^2/4]} = 0.48\ \text{m/s}$$

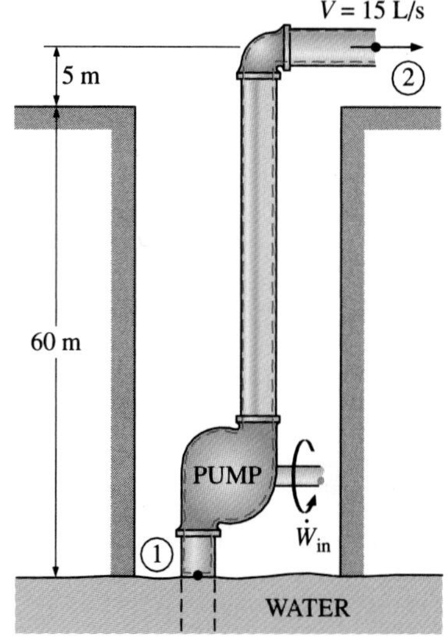

FIGURE 4-46
Schematic for Example 4-9.

As pointed out in Chap. 3, liquids can be treated as incompressible substances (v = constant). Thus, their enthalpy change can be expressed as

$$\begin{aligned} h_2 - h_1 &= (u_2 + P_2v_2) - (u_1 + P_1v_1) \\ &= (u_2 - u_1) + v(P_2 - P_1) \\ &= C(T_2 - T_1) + v(P_2 - P_1) \end{aligned}$$

since $\Delta u = C\Delta T$. In our case, $\Delta h = 0$ since there is no change in temperature ($T_2 = T_1$) and pressure ($P_2 = P_1 = P_{atm}$). (Note that we have atmospheric pressure at both the inlet and the exit.) Then the energy balance for this steady-flow system can be expressed in the rate form as

$$\underbrace{\dot{E}_{in} - \dot{E}_{out}}_{\substack{\text{Rate of net energy transfer} \\ \text{by heat, work, and mass}}} = \underbrace{\Delta \dot{E}_{system}}_{\substack{\text{Rate of change in internal, kinetic,} \\ \text{potential, etc., energies}}}^{\nearrow 0\ \text{(steady)}} = 0$$

$$\dot{E}_{in} = \dot{E}_{out}$$

$$\dot{W}_{sh,in} + \dot{m}\left(\frac{\mathcal{V}_1^2}{2} + gz_1\right) = \dot{m}\left(\frac{\mathcal{V}_2^2}{2} + gz_2\right) \qquad (\text{since } \dot{Q} = 0,\ \Delta h \cong 0)$$

$$\dot{W}_{e,in} = \dot{m}\left[\frac{\mathcal{V}_2^2 - \mathcal{V}_1^2}{2} + g(z_2 - z_1)\right]$$

Substituting gives

$$\begin{aligned} \dot{W}_{e,in} &= (15\ \text{kg/s})\left[\frac{(0.48\ \text{m/s})^2 - (0.85\ \text{m/s})^2}{2} + (9.8\ \text{m/s}^2)(65\ \text{m})\right] \\ &= (15\ \text{kg/s})(-0.246\ \text{m}^2/\text{s}^2 + 637.5\ \text{m}^2/\text{s}^2)\left(\frac{1\ \text{kJ/kg}}{1000\ \text{m}^2/\text{s}^2}\right) \\ &= \mathbf{9.55\ kW} \end{aligned}$$

Discussion This is the required pump work. It is interesting to note how small the kinetic energy term can be relative to the potential energy term when the process involves a liquid undergoing a considerable elevation change. This is typical for many actual processes. We should also note that the frictional losses for flow problems through pipes and ducts can be very significant. Therefore, in reality we would need a larger pump to overcome this extra resistance to flow. Frictional losses are treated in detail in fluid mechanics courses.

4-4 ■ UNSTEADY-FLOW PROCESSES

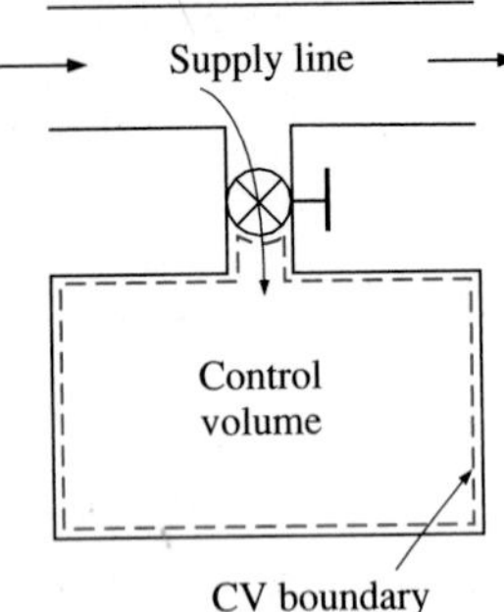

FIGURE 4-47
Charging of a rigid tank from a supply line is an unsteady-flow process since it involves changes within the control volume.

During a steady-flow process, no changes occur within the control volume; thus, one does not need to be concerned about what is going on within the boundaries. Not having to worry about any changes within the control volume with time greatly simplifies the analysis.

Many processes of interest, however, involve *changes* within the control volume with time. Such processes are called **unsteady-flow,** or **transient-flow processes.** The steady-flow relations developed in Sec. 4-2 are obviously not applicable to these processes. When an unsteady-flow process is analyzed, it is important to keep track of the mass and energy contents of the control volume as well as the energy interactions across the boundary.

Some familiar unsteady-flow processes are the charging of rigid vessels from supply lines (Fig. 4-47), discharging a fluid from a pressurized vessel, driving a gas turbine with pressurized air stored in a large container, inflating tires or balloons, and even cooking with an ordinary pressure cooker.

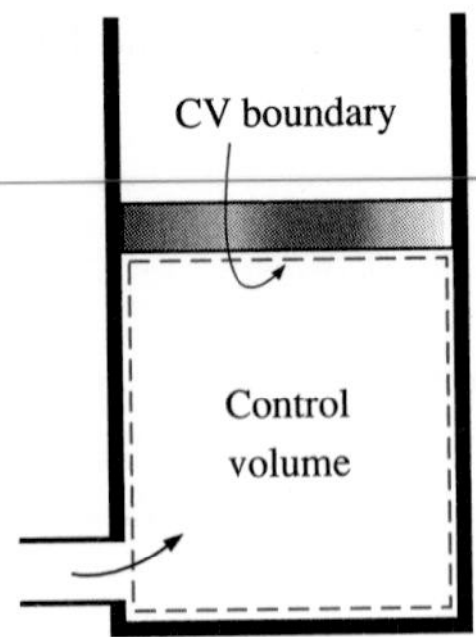

FIGURE 4-48
The shape and size of a control volume may change during an unsteady-flow process.

Unlike steady-flow processes, unsteady-flow processes start and end over some finite time period instead of continuing indefinitely. Therefore in this section, we deal with changes that occur over some time interval Δt instead of with the rate of changes (changes per unit time). An unsteady-flow system, in some respects, is similar to a closed system, except that the mass within the system boundaries does not remain constant during a process.

Another difference between steady- and unsteady-flow systems is that steady-flow systems are fixed in space, size, and shape. Unsteady-flow systems, however, are not (Fig. 4-48). They are usually stationary; that is, they are fixed in space, but they may involve moving boundaries and thus boundary work. Next we develop the mass and energy balance relations for general unsteady-flow processes.

Mass Balance

Unlike the case of steady-flow processes, the amount of mass within the control volume *does* change with time during an unsteady-flow process. The magnitude of change depends on the amounts of mass that enter and leave the control volume during the process. The mass balance for a system undergoing any process was expressed earlier in Eq. 4-1 as

$$m_{\text{in}} - m_{\text{out}} = \Delta m_{\text{system}}$$

where $\Delta m_{\text{system}} = m_{\text{final}} - m_{\text{initial}}$ is the change in the mass of the system during the process (Fig. 4-49). The mass balance for a control volume can also be expressed more explicitly as

$$\sum m_i - \sum m_e = (m_2 - m_1)_{\text{system}} \qquad \text{(kg)} \qquad (4\text{-}27)$$

where i = inlet; e = exit; 1 = initial state and 2 = final state of the control volume; and the summation signs are used to emphasize that all the inlets and exits are to be considered.

Often one or more terms in the equation above are zero. For example, $m_i = 0$ if no mass enters the control volume during the process, $m_e = 0$ if no mass leaves the control volume during the process, and $m_1 = 0$ if the control volume is initially evacuated.

The mass balance was also expressed in the rate form in Eq. 4-2 as

$$\dot{m}_{\text{in}} - \dot{m}_{\text{out}} = \Delta \dot{m}_{\text{system}} \qquad \text{(kg/s)}$$

where m_{in} and $\dot{m}_{\text{out}}$ are the total rates of mass flow into and out of the system and $\Delta \dot{m}_{\text{system}}$ is the rate of change of mass within the system boundaries. If the properties at the inlets and the exits as well as within the control volume are not uniform, then the rate form of the mass balance can be expressed as

$$\sum \int_{A_i} (\rho \mathcal{V}_n dA)_i - \sum \int_{A_e} (\rho \mathcal{V}_n dA)_e = \frac{d}{dt} \int_V (\rho dV)_{\text{CV}} \qquad (4\text{-}28)$$

to account for the variation of properties. The integration of $dm_{CV} = \rho dV$ on the right-hand side over the volume of the control volume gives the total mass contained within the control volume at time t.

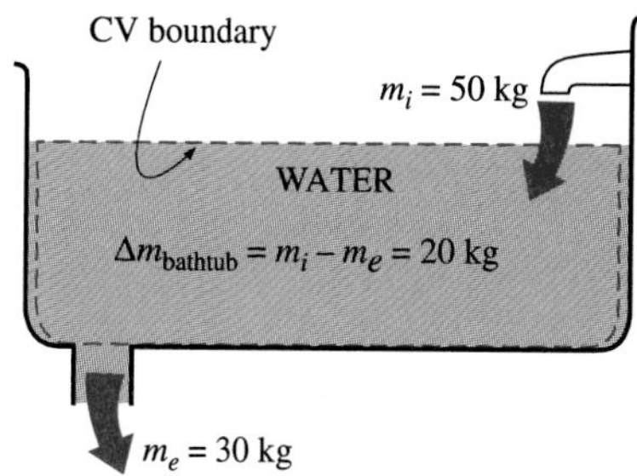

FIGURE 4-49

Conservation of mass principle for an ordinary bathtub.

Energy Balance

The energy content of a control volume changes with time during an unsteady-flow process. The magnitude of change depends on the amount of energy transfer across the system boundaries as heat and work as well as on the amount of energy transported into and out of the control volume by mass during the process. When analyzing an unsteady-flow process, we must keep track of the energy content of the control volume as well as the energies of the incoming and outgoing flow streams.

The general energy balance was given earlier in Eq. 4-8 as

$$\underbrace{E_{\text{in}} - E_{\text{out}}}_{\substack{\text{Net energy transfer} \\ \text{by heat, work, and mass}}} = \underbrace{\Delta E_{\text{system}}}_{\substack{\text{Change in internal, kinetic,} \\ \text{potential, etc., energies}}} \qquad \text{(kJ)}$$

which is applicable to any system undergoing any process. Noting that energy can be transferred by heat, work, and mass only, the energy balance above can also be written more explicitly as

$$Q_{\text{in}} + W_{\text{in}} + \sum m_i \theta_i = Q_{\text{out}} + W_{\text{out}} + \sum m_e \theta_e + \Delta E_{\text{system}} \qquad (4\text{-}29)$$

where $\theta = h + \text{ke} + \text{pe} = h + \mathcal{V}^2/2 + gz$ is the energy of a flowing fluid at any inlet or exit per unit mass and $\Delta E_{\text{system}} = (\Delta U + \Delta \text{KE} + \Delta \text{PE})_{\text{CV}}$.

When performing a general analytical study or solving a problem that involves unknown heat or work interactions, we need to assume a direction for the heat or work interactions. In such cases, it is common practice to assume the direction of net heat transfer Q to be *into the system* (heat input) and the direction of net work to be *out of the system* (work output), and then solve the problem. The first law or energy balance relation in that case becomes

$$Q - W = \sum m_e\theta_e - \sum m_i\theta_i + \Delta E_{\text{system}} \tag{4-30}$$

That is, the heat transferred to a system minus the work produced by the system is equal to the sum of the net change in the energy of the flow streams and the net change in the energy of the system itself. Obtaining a negative quantity for Q or W simply means that the assumed direction for that quantity is wrong and should be reversed.

The heat and work terms (Q and W) in the above equation can be determined by external measurements. The total energy of the control volume at the beginning and at the end of the process (E_1 and E_2) can also be determined by measuring the relevant properties of the substance at these two states. The total energy transported by mass into or out of the control volume ($m_i\theta_i$ and $m_e\theta_e$) however, is not as easy to determine since the properties of the mass at each inlet or exit may be changing with time as well as over the cross section. Thus, the only way to determine the energy transport through an opening as a result of mass flow is to consider sufficiently small differential masses δm that have uniform properties and to add their total energies.

The total energy of a flowing fluid of mass δm is $\theta\delta m$. Then the total energy transported by mass through an inlet or exit ($m_i\theta_i$ and $m_e\theta_e$) is obtained by integration. At an inlet, for example, it becomes

$$\Theta_i = m_i\theta_{i,\text{av}} = \int_{m_i}\left(h_i + \frac{\mathcal{V}_i^2}{2} + gz_i\right)\delta m_i \qquad \text{(kJ)} \tag{4-31}$$

Doing this at each inlet and exit and substituting into Eq. 4-30 give

$$Q - W = \sum\int_{m_e}\left(h_e + \frac{\mathcal{V}_e^2}{2} + gz_e\right)\delta m_e - \sum\int_{m_i}\left(h_i + \frac{\mathcal{V}_i^2}{2} + gz_i\right)\delta m_i + \Delta E_{\text{system}} \tag{4-32}$$

This equation can also be expressed in the rate form by dividing each term by Δt and taking the limit as $\Delta t \rightarrow 0$. It yields

$$\dot{Q} - \dot{W} = \sum \dot{m}_e\left(h_e + \frac{\mathcal{V}_e^2}{2} + gz_e\right) - \sum \dot{m}_i\left(h_i + \frac{\mathcal{V}_i^2}{2} + gz_i\right) + \frac{dE_{\text{system}}}{dt} \tag{4-33}$$

To perform the integrations in the above equations, one needs to know how the properties of the mass at the inlets and the exits change during a process.

Special Case: Uniform-Flow Processes

The unsteady-flow processes are, in general, difficult to analyze because the integrations in Eq. 4-32 are difficult to perform. Some unsteady-flow proc-

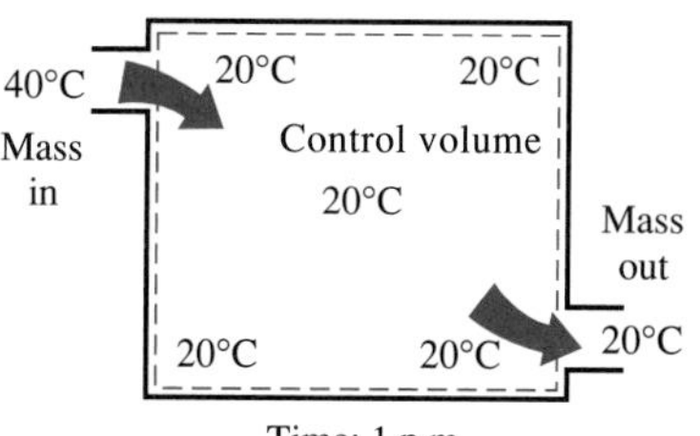

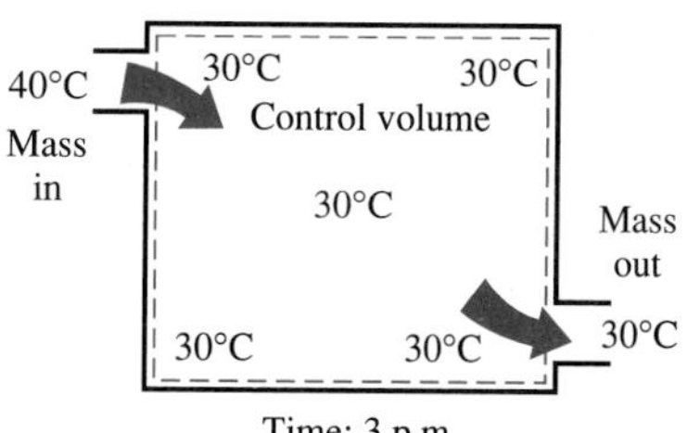

FIGURE 4-50

A control volume undergoing a uniform-flow process.

esses, however, can be represented reasonably well by another simplified model—the **uniform-flow process.** A uniform-flow process involves the following idealizations, which greatly simplify the analysis:

1. At any instant during the process, the state of the control volume is uniform (i.e., it is the same throughout). The state of the control volume may change with time, but it does so uniformly (Fig. 4-50). Consequently, the state of the mass leaving the control volume at any instant is the same as the state of the mass in the control volume at that instant. (This assumption is in contrast to the steady-flow assumption, which requires that the state of the control volume change with location but not with time.)

2. The fluid properties may differ from one inlet or exit to another, but the fluid flow at an inlet or exit is *uniform* and *steady*. That is, the properties do not change with time or position over the cross section of an inlet or exit. If they do, they are averaged and treated as constants for the entire process.

Under these idealizations, the integrations in Eq. 4-32 can easily be performed, and the conservation of energy equation for uniform-flow process becomes

$$Q - W = \sum m_e\left(h_e + \frac{\mathcal{V}_e^2}{2} + gz_e\right) - \sum m_i\left(h_i + \frac{\mathcal{V}_i^2}{2} + gz_i\right) + (m_2e_2 - m_1e_1)_{CV} \tag{4-34}$$

When the kinetic and potential energy changes associated with the control volume and fluid streams are negligible, Eq. 4-34 reduces to

$$Q - W = \sum m_e h_e - \sum m_i h_i + (m_2u_2 - m_1u_1)_{CV} \qquad \text{(kJ)} \tag{4-35}$$

Again, the various subscripts appearing in the above relations are i = inlet, e = exit, 1 = initial state, and 2 = final state of the control volume.

Notice that if no mass is entering or leaving the control volume ($m_i = m_e = 0$), the first two terms on the right-hand side of the above relation drop out and this equation reduces to the first-law relation for closed systems (Fig. 4-51).

Brief descriptions of the various terms appearing in the above equations are as follows:

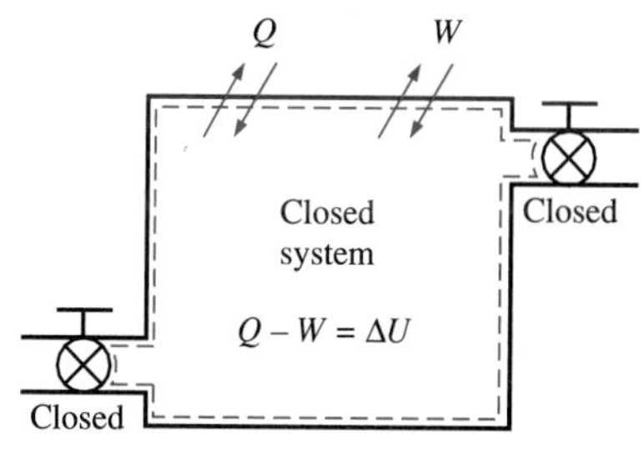

FIGURE 4-51

The energy equation of a uniform-flow system reduces to that of a closed system when all the inlets and exits are closed.

Q = **total heat transfer between the control volume and the surroundings during the process.** It is negative if heat is leaving the control volume and zero if the control volume is well insulated.

W = **total work associated with the control volume.** It may involve electrical work, shaft work, and even boundary work if the boundaries of the control volume move during the process (Fig. 4-52). It is zero for a control volume that involves no moving boundaries, shafts, or electric resistors.

m_e = **mass leaving the control volume.** It is zero if no mass leaves the control volume during the process.

m_i = **mass entering the control volume.** It is zero if no mass enters the control volume during a process.

$U_1 = m_1u_1$ = **total initial internal energy of the control volume.** It is zero for a control volume that is initially evacuated.

$U_2 = m_2u_2$ = **total final energy of the control volume.**

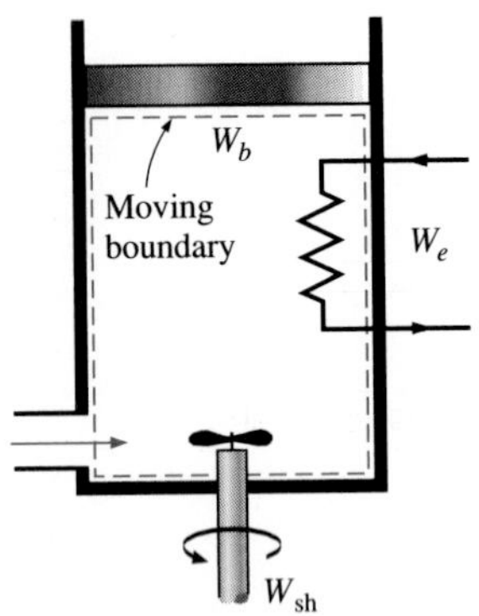

FIGURE 4-52
A uniform-flow system may involve electrical, shaft, and boundary work all at once.

Even though both the steady-flow and uniform-flow processes are somewhat idealized, many actual processes can be approximated reasonably well by one of these with satisfactory results. The degree of satisfaction depends upon the desired accuracy and the degree of validity of the assumptions made.

Engineers often find themselves in a position of having to choose between a quick, simple analysis with simplifying assumptions, at the expense of some accuracy, and an accurate, in-depth analysis with minimal assumptions, at the expense of time and extra effort. The right choice depends on the situation at hand.

EXAMPLE 4-10 Charging of a Rigid Tank by Steam

A rigid, insulated tank that is initially evacuated is connected through a valve to a supply line that carries steam at 1 MPa and 300°C. Now the valve is opened, and steam is allowed to flow slowly into the tank until the pressure reaches 1 MPa, at which point the valve is closed. Determine the final temperature of the steam in the tank.

Solution We take the *tank* as the system (Fig. 4-53). This is a *control volume* since mass crosses the system boundary during the process. We observe that this is an unsteady-flow process since changes occur within the control volume. The control volume is initially evacuated and thus $m_1 = 0$ and $m_1u_1 = 0$. Also, there is one inlet and no exits for mass flow.

Assumptions **1** This process can be analyzed as a *uniform flow process* since the properties of the steam entering the control volume remain constant during the entire process. **2** The kinetic and potential energies of the streams are negligible, ke ≅ pe ≅ 0. **3** The tank is stationary and thus its kinetic and potential energy changes are zero; that is, $\Delta KE = \Delta PE = 0$ and $\Delta E_{system} = \Delta U_{system}$. **4** There are no boundary, electrical, or shaft work interactions involved. **5** The tank is well insulated and thus there is no heat transfer.

Analysis Noting that microscopic energies of flowing and nonflowing fluids are represented by enthalpy *h* and internal energy *u*, respectively, the mass and energy balances for this uniform-flow system can be expressed as

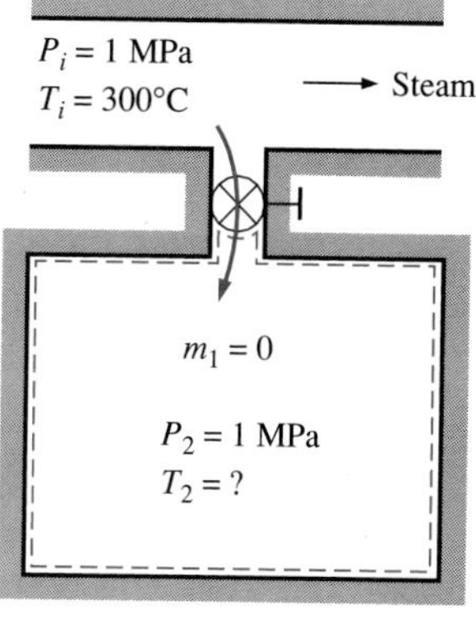

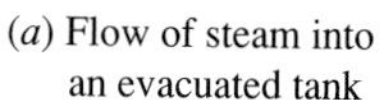
(*a*) Flow of steam into an evacuated tank

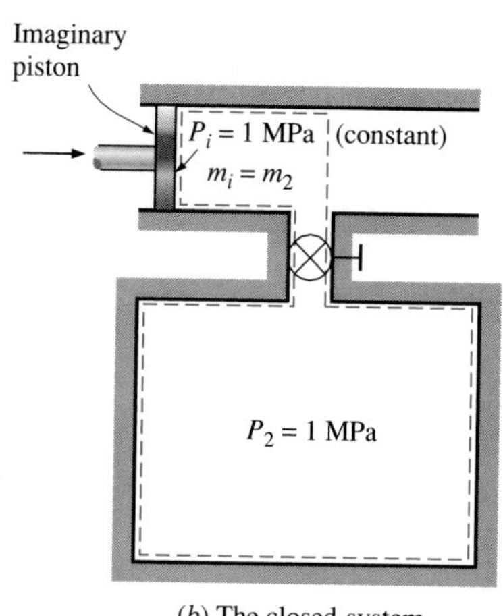

(*b*) The closed-system equivalence

FIGURE 4-53
Schematic for Example 4-10.

Mass balance: $m_{in} - m_{out} = \Delta m_{system} \rightarrow m_i = m_2 - m_1^{\nearrow 0} = m_2$

Energy balance:

$$\underbrace{E_{in} - E_{out}}_{\substack{\text{Net energy transfer} \\ \text{by heat, work, and mass}}} = \underbrace{\Delta E_{system}}_{\substack{\text{Change in internal, kinetic,} \\ \text{potential, etc., energies}}}$$

$$m_i h_i = m_2 u_2 \qquad \text{(since } W = Q = 0, \text{ ke} \cong \text{pe} \cong 0, m_1 = 0)$$

Combining the mass and energy balances gives

$$u_2 = h_i$$

That is, the final internal energy of the steam in the tank is equal to the enthalpy of the steam entering the tank. The enthalpy of the steam at the inlet state is

$$\left.\begin{array}{l} P_i = 1 \text{ MPa} \\ T_i = 300°\text{C} \end{array}\right\} \quad h_i = 3051.2 \text{ kJ/kg} \qquad \text{(Table A-6)}$$

which is equal to u_2. Since we now know two properties at the final state, it is fixed and the temperature at this state is determined from the same table to be

$$\left.\begin{array}{l} P_2 = 1 \text{ MPa} \\ u_2 = 3051.2 \text{ kJ/kg} \end{array}\right\} \quad T_2 = \mathbf{456.2°C}$$

Discussion Note that the temperature of the steam in the tank has increased by 156.2°C. This result may be surprising at first, and you may be wondering where the energy to raise the temperature of the steam came from. The answer lies in the enthalpy term $h = u + Pv$. Part of the energy represented by enthalpy is the flow energy Pv, and this flow energy is converted to sensible internal energy once the flow ceases to exist in the control volume, and it shows up as an increase in temperature (Fig. 4-54).

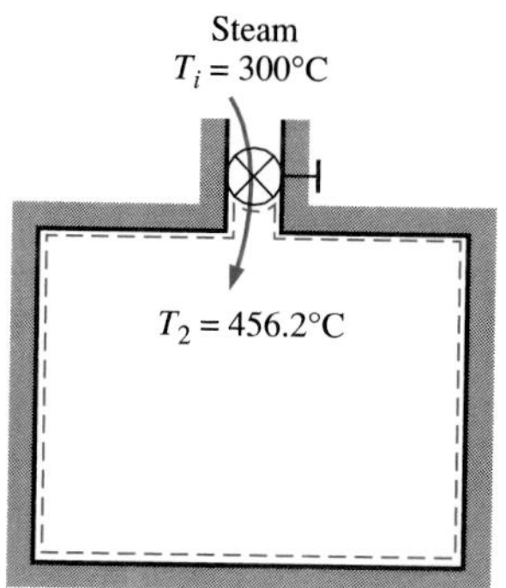

FIGURE 4-54
The temperature of steam rises from 300 to 456.2°C as it enters a tank as a result of flow energy being converted to internal energy.

Alternative solution This problem can also be solved by considering the region within the tank and the mass that is destined to enter the tank as a closed system, as shown in Fig. 4-53*b*. Since no mass crosses the boundaries, viewing this as a closed system is appropriate.

During the process, the steam upstream (the imaginary piston) will push the enclosed steam in the supply line into the tank at a constant pressure of 1 MPa. Then the boundary work done during this process is

$$W_{b,\,in} = -\int_1^2 P_i dV = -P_i(V_2 - V_1) = -P_i[V_{tank} - (V_{tank} + V_i)] = P_i V_i$$

where V_i is the volume occupied by the steam before it enters the tank and P_i is the pressure at the moving boundary (the imaginary piston face). The energy balance for the closed system gives

$$\underbrace{E_{in} - E_{out}}_{\substack{\text{Net energy transfer}\\ \text{by heat, work, and mass}}} = \underbrace{\Delta E_{system}}_{\substack{\text{Change in internal, kinetic,}\\ \text{potential, etc., energies}}}$$

$$W_{b,\,in} = \Delta U$$

$$m_i P_i v_i = m_2 u_2 - m_i u_i$$

$$u_2 = u_i + P_i v_i = h_i$$

since the initial state of the system is simply the line conditions of the steam. This result is identical to the one obtained with the uniform-flow analysis. Once again, the temperature rise is caused by the so-called flow energy or flow work, which is the energy required to push the substance into the tank.

EXAMPLE 4-11 Cooking with a Pressure Cooker

A pressure cooker is a pan that cooks food much faster than ordinary pans by maintaining a higher pressure and temperature during cooking. The pressure inside the pan is controlled by a pressure regulator (the petcock) that keeps the pressure at a constant level by periodically allowing some steam to escape, thus preventing any excess pressure buildup.

A certain pressure cooker has a volume of 6 L and an operating pressure of 75 kPa gage. Initially, it contains 1 kg of water. Heat is supplied to the pressure cooker at a rate of 500 W for 30 min after the operating pressure is reached. Assuming an atmospheric pressure of 100 kPa, determine (*a*) the temperature at which cooking takes place and (*b*) the amount of water left in the pressure cooker at the end of the process.

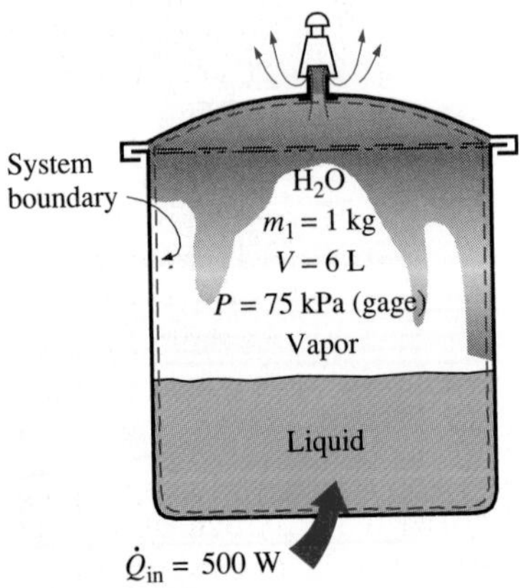

System boundary

H_2O

$m_1 = 1$ kg

$V = 6$ L

$P = 75$ kPa (gage)

Vapor

Liquid

$\dot{Q}_{in} = 500$ W

FIGURE 4-55

Schematic for Example 4-11.

Solution We take the *pressure cooker* as the system (Fig. 4-55). This is a *control volume* since mass crosses the system boundary during the process. We observe that this is an unsteady-flow process since changes occur within the control volume. Also, there is one exit and no inlets for mass flow.

Assumptions **1** This process can be analyzed as a *uniform-flow process* since the properties of the steam leaving the control volume remain constant during the entire cooking process. **2** The kinetic and potential energies of the streams are negligible, ke ≅ pe ≅ 0. **3** The pressure cooker is stationary and thus its kinetic and potential energy changes are zero; that is, $\Delta KE = \Delta PE = 0$ and $\Delta E_{system} = \Delta U_{system}$. **4** The pressure (and thus temperature) in the pressure cooker remains constant. **5** Steam leaves as a saturated vapor at the cooker pressure. **6** There are no boundary, electrical, or shaft work interactions involved. **7** Heat is transferred to the cooker at a constant rate.

Analysis (*a*) The absolute pressure within the cooker is

$$P_{abs} = P_{gage} + P_{atm} = 75 \text{ kPa} + 100 \text{ kPa} = 175 \text{ kPa}$$

Since saturation conditions exist in the cooker at all times (Fig. 4-56), the cooking temperature must be the saturation temperature corresponding to this pressure. From Table A-5, it is

$$T = T_{sat @ 175 \text{ kPa}} = \mathbf{116.06°C}$$

which is about 16°C higher than the ordinary cooking temperature.

(*b*) Noting that the microscopic energies of flowing and nonflowing fluids are represented by enthalpy *h* and internal energy *u*, respectively, the mass and energy balances for this uniform-flow system can be expressed as

Mass balance:

$$m_{in} - m_{out} = \Delta m_{system} \quad \rightarrow \quad -m_e = (m_2 - m_1)_{CV} \quad \text{or} \quad m_e = (m_1 - m_2)_{CV}$$

Energy balance:

$$\underbrace{E_{in} - E_{out}}_{\substack{\text{Net energy transfer}\\ \text{by heat, work, and mass}}} = \underbrace{\Delta E_{system}}_{\substack{\text{Change in internal, kinetic,}\\ \text{potential, etc., energies}}}$$

$$Q_{in} - m_e h_e = (m_2 u_2 - m_1 u_1)_{CV} \qquad (\text{since } W = 0, \text{ ke} \cong \text{pe} \cong 0)$$

Combining the mass and energy balances gives

$$Q_{in} = (m_1 - m_2)h_e + (m_2 u_2 - m_1 u_1)_{CV}$$

The amount of heat transfer during this process is found from

$$Q_{in} = \dot{Q}_{in}\,\Delta t = (0.5 \text{ kJ/s})(30 \times 60 \text{ s}) = 900 \text{ kJ}$$

Steam leaves the pressure cooker as saturated vapor at 175 kPa at all times (Fig. 4-57). Thus,

$$h_e = h_{g @ 175 \text{ kPa}} = 2700.6 \text{ kJ/kg}$$

The initial internal energy is found after the quality is determined:

$$v_1 = \frac{V}{m_1} = \frac{0.006 \text{ m}^3}{1 \text{ kg}} = 0.006 \text{ m}^3\text{/kg}$$

$$x_1 = \frac{v_1 - v_f}{v_{fg}} = \frac{0.006 - 0.001}{1.004 - 0.001} = 0.005$$

Thus,

$$u_1 = u_f + x_1 u_{fg} = 486.8 + (0.005)(2038.1) \text{ kJ/kg} = 497.0 \text{ kJ/kg}$$

and

$$U_1 = m_1 u_1 = (1 \text{ kg})(497 \text{ kJ/kg}) = 497 \text{ kJ}$$

The mass of the system at the final state is $m_2 = V/v_2$. Substituting this into the energy equation yields

$$Q_{in} = \left(m_1 - \frac{V}{v_2}\right)h_e + \left(\frac{V}{v_2}u_2 - m_1 u_1\right)$$

There are two unknowns in this equation, u_2 and v_2. Thus we need to relate them to a single unknown before we can determine these unknowns. Assuming there is still some liquid water left in the cooker at the final state (i.e., saturation conditions exist), v_2 and u_2 can be expressed as

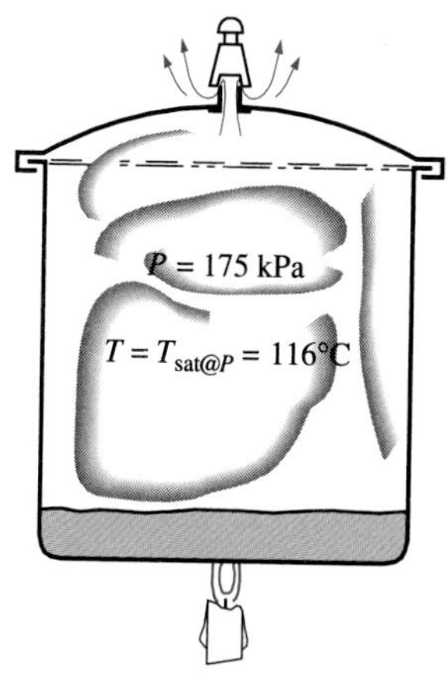

FIGURE 4-56
As long as there is liquid in a pressure cooker, the saturation conditions exist and the temperature remains constant at the saturation temperature.

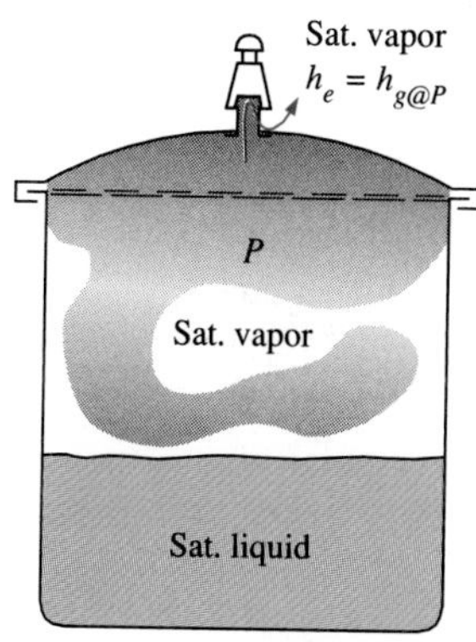

FIGURE 4-57
In a pressure cooker, the enthalpy of the exiting steam is $h_{g@P}$ (enthalpy of the saturated vapor at the given pressure).

$$v_2 = v_f + x_2 v_{fg} = 0.001 + x_2(1.004 - 0.001)\ \mathrm{m^3/kg}$$
$$u_2 = u_f + x_2 u_{fg} = 486.8 + x_2(2038.1)\ \mathrm{kJ/kg}$$

Notice that during a boiling process at constant pressure, the properties of each phase remain constant (only the amounts change). When these expressions are substituted into the above energy equation, x_2 becomes the only unknown, and it is determined to be

$$x_2 = 0.009$$

Thus,

$$v_2 = 0.001 + (0.009)(1.004 - 0.0001)\ \mathrm{m^3/kg} = 0.010\ \mathrm{m^3/kg}$$

and

$$m_2 = \frac{V}{v_2} = \frac{0.006\ \mathrm{m^3}}{0.01\ \mathrm{m^3/kg}} = \mathbf{0.6\ kg}$$

Therefore, after 30 min there is 0.6 kg water (liquid + vapor) left in the pressure cooker.

4-5 ■ SUMMARY

A control volume differs from a closed system in that it involves mass transfer. Mass carries energy with it, and thus the mass and energy content of a system change when mass enters or leaves. The mass and energy balances for *any system* undergoing *any process* can be expressed as

$$m_{in} - m_{out} = \Delta m_{system} \qquad \text{(kg)}$$

$$\underbrace{E_{in} - E_{out}}_{\substack{\text{Net energy transfer}\\ \text{by heat, work, and mass}}} = \underbrace{\Delta E_{system}}_{\substack{\text{Change in internal, kinetic,}\\ \text{potential, etc., energies}}} \qquad \text{(kJ)}$$

They can also be expressed in the *rate form* as

$$\dot{m}_{in} - \dot{m}_{out} = \Delta \dot{m}_{system} \qquad \text{(kg/s)}$$

$$\underbrace{\dot{E}_{in} - \dot{E}_{out}}_{\substack{\text{Rate of net energy transfer}\\ \text{by heat, work, and mass}}} = \underbrace{\Delta \dot{E}_{system}}_{\substack{\text{Rate of change in internal, kinetic,}\\ \text{potential, etc., energies}}} \qquad \text{(kW)}$$

Mass flow through a cross section per unit time is called the *mass flow rate* and is denoted $\dot{m}$. It is expressed as

$$\dot{m} = \rho \mathcal{V}_{av} A \qquad \text{(kg/s)}$$

where

ρ = density, kg/m³ (= $1/v$)
$\mathcal{V}_{av}$ = average fluid velocity normal to A, m/s
A = cross-sectional area, m²

The fluid volume flowing through a cross section per unit time is called the *volume flow rate* $\dot{V}$. It is given by

$$\dot{V} = \int_A \mathcal{V}_n \, dA = \mathcal{V}_{av} A \qquad (\mathrm{m}^3/\mathrm{s})$$

The mass and volume flow rates are related by

$$\dot{m} = \rho \dot{V} = \frac{\dot{V}}{v}$$

Thermodynamic processes involving control volumes can be considered in two groups: steady-flow processes and unsteady-flow processes. During a *steady-flow process,* the fluid flows through the control volume steadily, experiencing no change with time at a fixed position. The mass and energy content of the control volume remain constant during a steady-flow process. Taking heat transfer *to* the system and work done *by* the system to be positive quantities, the conservation of mass and energy equations for steady-flow processes are expressed as

$$\sum \dot{m}_i = \sum \dot{m}_e \qquad (\mathrm{kg/s})$$

$$\dot{Q} - \dot{W} = \underbrace{\sum \dot{m}_e \left(h_e + \frac{\mathcal{V}_e^2}{2} + gz_e \right)}_{\text{for each exit}} - \underbrace{\sum \dot{m}_i \left(h_i + \frac{\mathcal{V}_i^2}{2} + gz_i \right)}_{\text{for each inlet}} \qquad (\mathrm{kW})$$

where the subscript i stands for inlet and e for exit. These are the most general forms of the equations for steady-flow processes. For single-stream (one-inlet–one-exit) systems such as nozzles, diffusers, turbines, compressors, and pumps, they simplify to

$$\dot{m}_1 = \dot{m}_2 \qquad (\mathrm{kg/s})$$

or

$$\frac{1}{v_1} \mathcal{V}_1 A_1 = \frac{1}{v_2} \mathcal{V}_2 A_2$$

and

$$\dot{Q} - \dot{W} = \dot{m} \left[h_2 - h_1 + \frac{\mathcal{V}_2^2 - \mathcal{V}_1^2}{2} + g(z_2 - z_1) \right] \qquad (\mathrm{kW})$$

In the above relations, subscripts 1 and 2 denote the inlet and exit states, respectively.

During a uniform-flow process, the state of the control volume may change with time, but it may do so uniformly. Also, the fluid properties at the inlets and the exits are assumed to remain constant during the entire process. The conservation of energy equation for a uniform-flow process reduces to

$$Q - W = \sum m_e \left(h_e + \frac{\mathcal{V}_e^2}{2} + gz_e \right) - \sum m_i \left(h_i + \frac{\mathcal{V}_i^2}{2} + gz_i \right) + (m_2 e_2 - m_1 e_1)_{\mathrm{CV}}$$

When the kinetic and potential energy changes associated with the control volume and the fluid streams are negligible, it simplifies to

$$Q - W = \sum m_e h_e - \sum m_i h_i + (m_2 u_2 - m_1 u_1)_{\mathrm{CV}} \qquad (\mathrm{kJ})$$

REFERENCES AND SUGGESTED READING

1. A. Bejan. *Advanced Engineering Thermodynamics.* New York: John Wiley & Sons, 1988.

2. W. Z. Black and J. G. Hartley. *Thermodynamics.* New York: Harper & Row, 1985.

3. J. R. Howell and R. O. Buckius. *Fundamentals of Engineering Thermodynamics.* New York: McGraw-Hill, 1987.

4. J. B. Jones and G. A. Hawkins. *Engineering Thermodynamics.* 2nd ed. New York: John Wiley & Sons, 1986.

5. W. C. Reynolds and H. C. Perkins. *Engineering Thermodynamics.* 2nd ed. New York: McGraw-Hill, 1977.

6. G. J. Van Wylen and R. E. Sonntag. *Fundamentals of Classical Thermodynamics.* 3rd ed. New York: John Wiley & Sons, 1985.

7. K. Wark. *Thermodynamics.* 5th ed. New York: McGraw-Hill, 1988.

PROBLEMS*

General Control Volume Analysis

4-1C Define mass and volume flow rates. How do they differ?

4-2C What are the different mechanisms for transferring energy to or from a control volume?

4-3C What is flow energy? Do fluids at rest possess any flow energy?

4-4C How do the energies of flowing fluids and a fluid at rest compare? Name the specific forms of energy associated with each case.

Steady-Flow Processes

4-5C When is the flow through a control volume steady?

4-6C How is a steady-flow system characterized?

4-7C Can a steady-flow system involve boundary work?

Nozzles and Diffusers

4-8C A diffuser is an adiabatic device that decreases the kinetic energy of the fluid by slowing it down. What happens to this *lost* kinetic energy?

4-9C The kinetic energy of a fluid increases as it is accelerated in an adiabatic nozzle. Where does this energy come from?

4-10C Is heat transfer to or from the fluid desirable as it flows through a nozzle? How will heat transfer affect the fluid velocity at the nozzle exit?

---

*Students are encouraged to answer *all* the concept "C" questions.

4-11 Air enters an adiabatic nozzle steadily at 300 kPa, 200°C, and 30 m/s and leaves at 100 kPa and 180 m/s. The inlet area of the nozzle is 80 cm^2. Determine (*a*) the mass flow rate through the nozzle, (*b*) the exit temperature of the air, and (*c*) the exit area of the nozzle.
Answers: (*a*) 0.5304 kg/s, (*b*) 184.60°C, (*c*) 38.7 cm^2

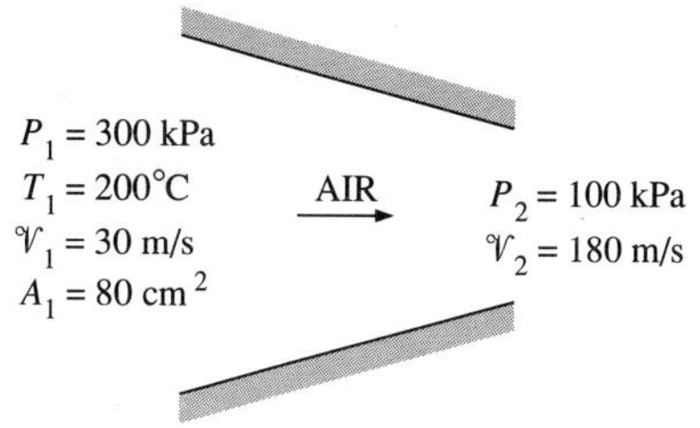

P_1 = 300 kPa
T_1 = 200°C
$\mathcal{V}_1$ = 30 m/s
A_1 = 80 cm^2
AIR
P_2 = 100 kPa
$\mathcal{V}_2$ = 180 m/s

FIGURE P4-11

4-12 Steam at 5 MPa and 500°C enters a nozzle steadily with a velocity of 80 m/s, and it leaves at 2 MPa and 400°C. The inlet area of the nozzle is 50 cm^2, and heat is being lost at a rate of 90 kJ/s. Determine (*a*) the mass flow rate of the steam, (*b*) the exit velocity of the steam, and (*c*) the exit area of the nozzle.

4-13 Carbon dioxide enters an adiabatic nozzle steadily at 1 MPa and 500°C with a mass flow rate of 6000 kg/h and leaves at 100 kPa and 450 m/s. The inlet area of the nozzle is 40 cm^2. Determine (*a*) the inlet velocity and (*b*) the exit temperature. *Answers:* (*a*) 60.8 m/s, (*b*) 685.8 K

4-14E Air enters a nozzle steadily at 50 psia, 140°F, and 150 ft/s and leaves at 14.7 psia and 900 ft/s. The heat loss from the nozzle is estimated to be 6.5 Btu/lbm of air flowing. The inlet area of the nozzle is 0.1 ft^2. Determine (*a*) the exit temperature of air and (*b*) the exit area of the nozzle.
Answers: (*a*) 441.7 R, (*b*) 0.0417 ft^2

4-15 Refrigerant-134a at 700 kPa and 100°C enters an adiabatic nozzle steadily with a velocity of 20 m/s and leaves at 300 kPa and 30°C. Determine (*a*) the exit velocity and (*b*) the ratio of the inlet to exit area A_1/A_2.

4-16 Steam at 3 MPa and 400°C enters an adiabatic nozzle steadily with a velocity of 40 m/s and leaves at 2.5 MPa and 300 m/s. Determine (*a*) the exit temperature and (*b*) the ratio of the inlet to exit area A_1/A_2.

P_1 = 3 MPa
T_1 = 400°C
$\mathcal{V}_1$ = 40 m/s
STEAM
P_2 = 2.5 MPa
$\mathcal{V}_2$ = 300 m/s

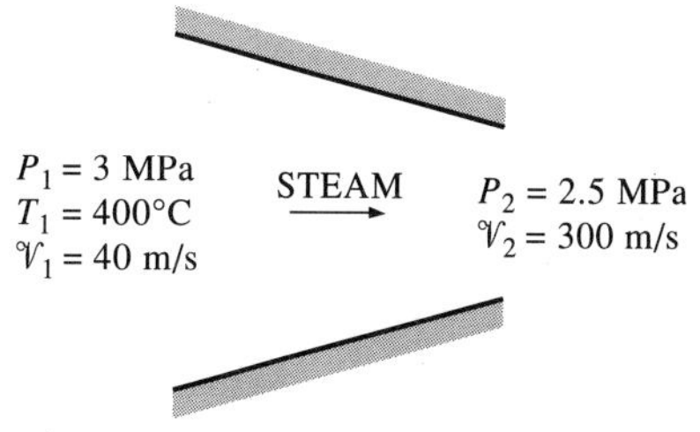

FIGURE P4-16

4-17 Air at 600 kPa and 500 K enters an adiabatic nozzle that has an inlet-to-exit area ratio of 2:1 with a velocity of 120 m/s and leaves with a velocity of 380 m/s. Determine (*a*) the exit temperature and (*b*) the exit pressure of the air. Answers: (*a*) 436.5 K, (*b*) 330.8 kPa

4-18 Air at 80 kPa and 127°C enters an adiabatic diffuser steadily at a rate of 6000 kg/h and leaves at 100 kPa. The velocity of the airstream is decreased from 230 to 30 m/s as it passes through the diffuser. Find (*a*) the exit temperature of the air and (*b*) the exit area of the diffuser.

4-19E Air at 13 psia and 20°F enters an adiabatic diffuser steadily with a velocity of 600 ft/s and leaves with a low velocity at a pressure of 14.5 psia. The exit area of the diffuser is 5 times the inlet area. Determine (*a*) the exit temperature and (*b*) the exit velocity of the air.

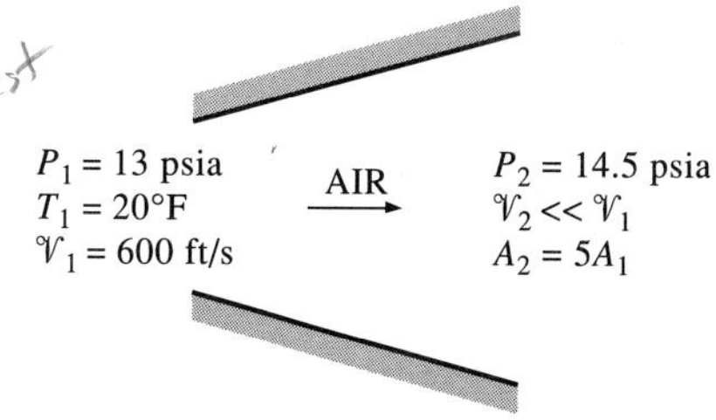

P_1 = 13 psia
T_1 = 20°F
$\mathcal{V}_1$ = 600 ft/s
AIR
P_2 = 14.5 psia
$\mathcal{V}_2 << \mathcal{V}_1$
$A_2 = 5A_1$

FIGURE P4-19E

4-20 Air at 80 kPa, 27°C, and 220 m/s enters a diffuser at a rate of 2.5 kg/s and leaves at 42°C. The exit area of the diffuser is 400 cm^2. The air is estimated to lose heat at a rate of 18 kJ/s during this process. Determine (*a*) the exit velocity and (*b*) the exit pressure of the air.
Answers: (*a*) 62.0 m/s, (*b*) 91.1 kPa

4-21 Nitrogen gas at 60 kPa and 7°C enters an adiabatic diffuser steadily with a velocity of 200 m/s and leaves at 85 kPa and 22°C. Determine (*a*) the exit velocity of the nitrogen and (*b*) the ratio of the inlet to exit area A_1/A_2.

4-22 Refrigerant-134a enters a diffuser steadily as saturated vapor at 700 kPa with a velocity of 140 m/s, and it leaves at 800 kPa and 40°C. The refrigerant is gaining heat at a rate of 3 kJ/s as it passes through the diffuser. If the exit area is 80 percent greater than the inlet area, determine (*a*) the exit velocity and (*b*) the mass flow rate of the refrigerant.

Answers: (*a*) 71.7 m/s, (*b*) 0.655 kg/s

Turbines and Compressors

4-23C Consider an adiabatic turbine operating steadily. Does the work output of the turbine have to be equal to the decrease in the energy of the steam flowing through it?

4-24C Consider a steam turbine operating steadily. Would you expect the temperatures at the turbine inlet and exit to be the same?

4-25C Consider an air compressor operating steadily. Would you expect the air density to be the same at the compressor inlet and exit?

4-26C Consider an air compressor operating steadily. How would you compare the volume flow rates of the air at the compressor inlet and exit?

4-27C Will the temperature of air rise as it is compressed by an adiabatic compressor? Why?

4-28C Somebody proposes the following system to cool a house in the summer: Compress the regular outdoor air, let it cool back to the outdoor temperature, pass it through a turbine, and discharge the cold air leaving the turbine into the house. From a thermodynamic point of view, is the proposed system sound?

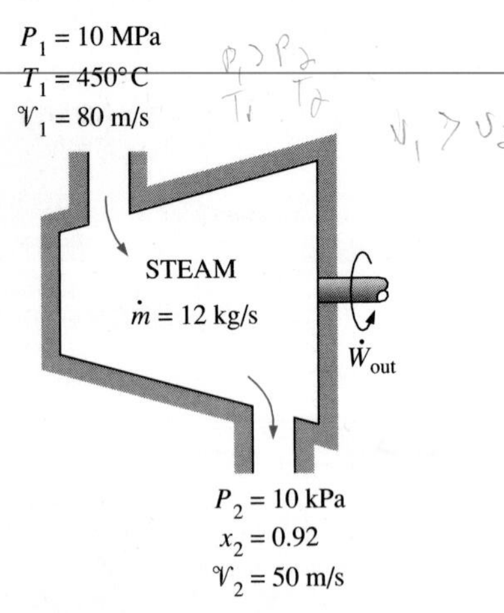

FIGURE P4-29

4-29 Steam flows steadily through an adiabatic turbine. The inlet conditions of the steam are 10 MPa, 450°C, and 80 m/s, and the exit conditions are 10 kPa, 92 percent quality, and 50 m/s. The mass flow rate of the steam is 12 kg/s. Determine (*a*) the change in kinetic energy, (*b*) the power output, and (*c*) the turbine inlet area.

Answers: (*a*) −1.95 kJ/kg, (*b*) 10.2 MW, (*c*) 0.00446 m^2

4-30 Steam enters an adiabatic turbine at 10 MPa and 400°C and leaves at 20 kPa with a quality of 90 percent. Neglecting the changes in kinetic and potential energies, determine the mass flow rate required for a power output of 5 MW. *Answer:* 6.919 kg/s

4-31E Steam flows steadily through a turbine at a rate of 45,000 lbm/h, entering at 1000 psia and 900°F and leaving at 5 psia as saturated vapor. If the power generated by the turbine is 4 MW, determine the rate of heat loss from the steam.

4-32 Steam enters an adiabatic turbine at 10 MPa and 500°C at a rate of 3 kg/s and leaves at 20 kPa. If the power output of the turbine is 2 MW, determine the temperature of the steam at the turbine exit. Neglect kinetic energy changes. *Answer:* 110.8°C

4-33 Argon gas enters steadily an adiabatic turbine at 900 kPa and 450°C with a velocity of 80 m/s and leaves at 150 kPa with a velocity of 150 m/s. The inlet area of the turbine is 60 cm². If the power output of the turbine is 250 kW, determine the exit temperature of the argon. *Answer:* 267°C

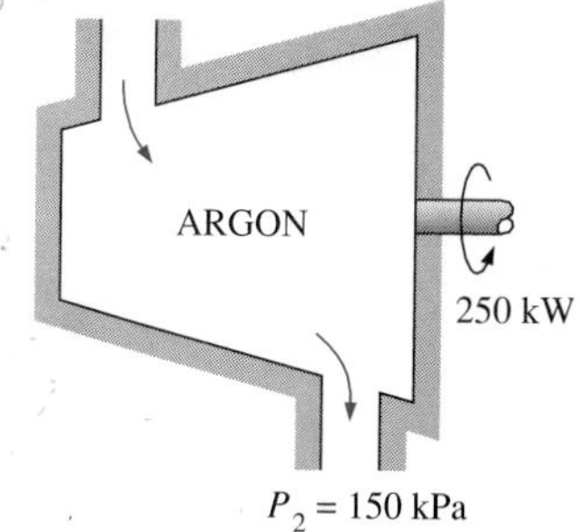

FIGURE P4-33

4-34E Air flows steadily through an adiabatic turbine, entering at 150 psia, 900°F, and 350 ft/s and leaving at 20 psia, 300°F, and 700 ft/s. The inlet area of the turbine is 0.1 ft². Determine (*a*) the mass flow rate of the air and (*b*) the power output of the turbine.

4-35 Refrigerant-134a enters an adiabatic compressor as saturated vapor at −20°C and leaves at 0.7 MPa and 70°C. The mass flow rate of the refrigerant is 1.2 kg/s. Determine (*a*) the power input to the compressor and (*b*) the volume flow rate of the refrigerant at the compressor inlet.

4-36 Air enters the compressor of a gas-turbine plant at ambient conditions of 100 kPa and 25°C with a low velocity and exits at 1 MPa and 347°C with a velocity of 90 m/s. The compressor is cooled at a rate of 1500 kJ/min, and the power input to the compressor is 250 kW. Determine the mass flow rate of air through the compressor. *Answer:* 0.680 kg/s

4-37E Air is compressed from 14.7 psia and 60°F to a pressure of 150 psia while being cooled at a rate of 10 Btu/lbm by circulating water through the compressor casing. The volume flow rate of the air at the inlet conditions is 5000 ft³/min, and the power input to the compressor is 700 hp. Determine (*a*) the mass flow rate of the air and (*b*) the temperature at the compressor exit. *Answers:* (*a*) 6.36 lbm/s, (*b*) 781 R

4-38 Helium is to be compressed from 120 kPa and 310 K to 700 kPa and 430 K. A heat loss of 20 kJ/kg occurs during the compression process. Neglecting kinetic energy changes, determine the power input required for a mass flow rate of 90 kg/min.

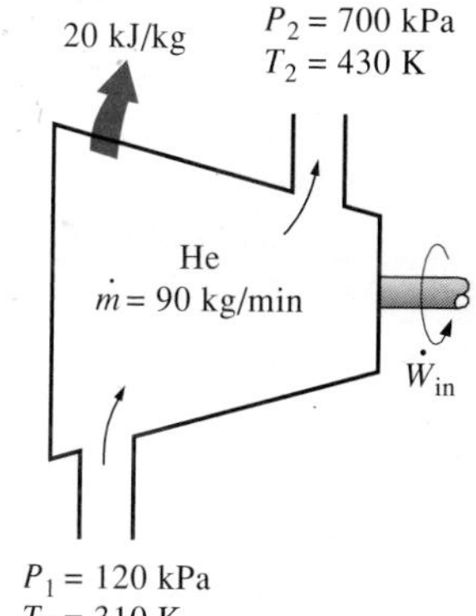

FIGURE P4-38

4-39 Carbon dioxide enters an adiabatic compressor at 100 kPa and 300 K at a rate of 0.5 kg/s and leaves at 600 kPa and 450 K. Neglecting kinetic energy changes, determine (*a*) the volume flow rate of the carbon dioxide at the compressor inlet and (*b*) the power input to the compressor.
Answers: (*a*) 0.28 m³/s, (*b*) 68.8 kW

Throttling Valves

4-40C Why are throttling devices commonly used in refrigeration and air-conditioning applications?

4-41C During a throttling process, the temperature of a fluid drops from 30 to −20°C. Can this process occur adiabatically?

4-42C Would you expect the temperature of air to drop as it undergoes a steady-flow throttling process?

4-43C Would you expect the temperature of a liquid to change as it is throttled? How?

4-44 Refrigerant-134a is throttled from the saturated liquid state at 800 kPa to a pressure of 140 kPa. Determine the temperature drop during this process and the final specific volume of the refrigerant.
Answers: 50.1°C, 0.0454 m^3/kg

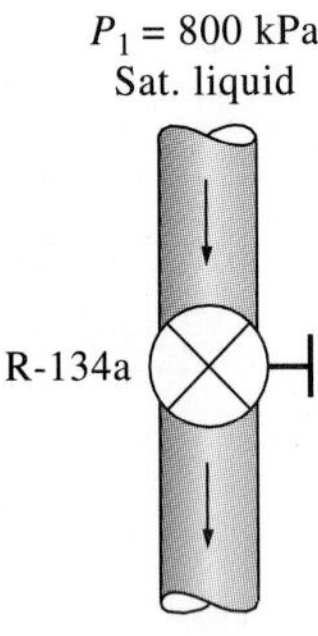

FIGURE P4-44

4-45 Refrigerant-134a at 800 kPa and 25°C is throttled to a temperature of −20°C. Determine the pressure and the internal energy of the refrigerant at the final state. *Answers:* 133 kPa, 78.8 kJ/kg

4-46 A well-insulated valve is used to throttle steam from 8 MPa and 500°C to 6 MPa. Determine the final temperature of the steam.
Answer: 490.1°C

4-47E Air at 200 psia and 90°F is throttled to the atmospheric pressure of 14.7 psia. Determine the final temperature of the air.

Mixing Chambers and Heat Exchangers

4-48C When two fluid streams are mixed in a mixing chamber, can the mixture temperature be lower than the temperature of both streams? Explain.

4-49C Consider a steady-flow mixing process. Under what conditions will the energy transported into the control volume by the incoming streams be equal to the energy transported out of it by the outgoing stream?

4-50C Consider a steady-flow heat exchanger involving two different fluid streams. Under what conditions will the amount of heat lost by one fluid be equal to the amount of heat gained by the other?

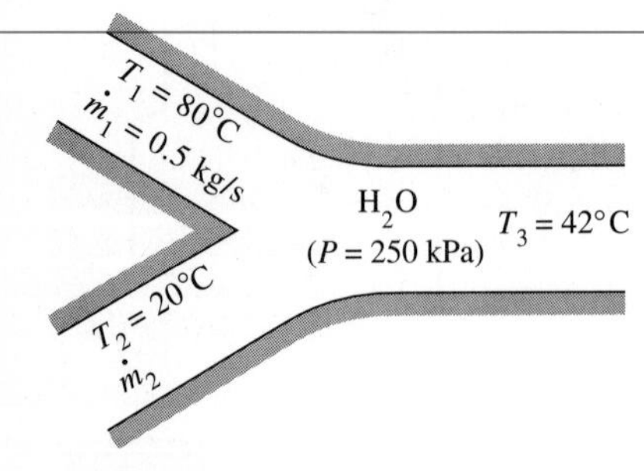

FIGURE P4-51

4-51 A hot-water stream at 80°C enters a mixing chamber with a mass flow rate of 0.5 kg/s where it is mixed with a stream of cold water at 20°C. If it is desired that the mixture leave the chamber at 42°C, determine the mass flow rate of the cold-water stream. Assume all the streams are at a pressure of 250 kPa. *Answer:* 0.864 kg/s

4-52 Liquid water at 300 kPa and 20°C is heated in a chamber by mixing it with superheated steam at 300 kPa and 300°C. Cold water enters the chamber at a rate of 1.8 kg/s. If the mixture leaves the mixing chamber at 60°C, determine the mass flow rate of the superheated steam required.
Answer: 0.107 kg/s

FIGURE P4-53

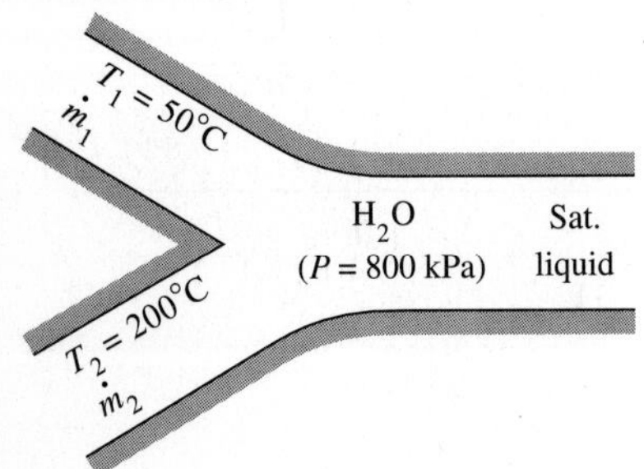

4-53 In steam power plants, open feedwater heaters are frequently utilized to heat the feedwater by mixing it with steam bled off the turbine at some intermediate stage. Consider an open feedwater heater that operates at a pressure of 800 kPa. Feedwater at 50°C and 800 kPa is to be heated with superheated steam at 200°C and 800 kPa. In an ideal feedwater heater, the mixture leaves the heater as saturated liquid at the feedwater pressure. Determine the ratio of the mass flow rates of the feedwater and the superheated vapor for this case. *Answer:* 4.14

4-54E Water at 50°F and 50 psia is heated in a chamber by mixing it with saturated water vapor at 50 psia. If both streams enter the mixing chamber at the same mass flow rate, determine the temperature and the quality of the exiting stream. *Answers:* 281°F, 0.374

4-55 A stream of refrigerant-134a at 1 MPa and 12°C is mixed with another stream at 1 MPa and 60°C. If the mass flow rate of the cold stream is twice that of the hot one, determine the temperature and the quality of the exit stream.

4-56 Refrigerant-134a at 1 MPa and 80°C is to be cooled to 1 MPa and 30°C in a condenser by air. The air enters at 100 kPa and 27°C with a volume flow rate of 800 m^3/min and leaves at 95 kPa and 60°C. Determine the mass flow rate of the refrigerant. *Answer:* 139 kg/min

4-57E Air enters the evaporator section of a window air conditioner at 14.7 psia and 90°F with a volume flow rate of 200 ft^3/min. Refrigerant-134a at 20 psia with a quality of 30 percent enters the evaporator at a rate of 4 lbm/min and leaves as saturated vapor at the same pressure. Determine (*a*) the exit temperature of the air and (*b*) the rate of heat transfer from the air.

4-58 Refrigerant-134a at 800 kPa, 70°C, and 8 kg/min is cooled by water in a condenser until it exists as a saturated liquid at the same pressure. The cooling water enters the condenser at 300 kPa and 15°C and leaves at 30°C at the same pressure. Determine the mass flow rate of the cooling water required to cool the refrigerant. *Answer:* 27.0 kg/min

4-59E In a steam heating system, air is heated by being passed over some tubes through which steam flows steadily. Steam enters the heat exchanger at 30 psia and 400°F at a rate of 15 lbm/min and leaves at 25 psia and 212°F. Air enters at 14.7 psia and 80°F and leaves at 130°F. Determine the volume flow rate of air at the inlet.

4-60 Steam enters the condenser of a steam power plant at 20 kPa and a quality of 95 percent with a mass flow rate of 20,000 kg/h. It is to be cooled by water from a nearby river by circulating the water through the tubes within the condenser. To prevent thermal pollution, the river water is not allowed to experience a temperature rise above 10°C. If the steam is to leave the condenser as saturated liquid at 20 kPa, determine the mass flow rate of the cooling water required. *Answer:* 17,866 kg/min

4-61 Steam is to be condensed in the condenser of a steam power plant at a temperature of 50°C (h_{fg} = 2305 kJ/kg) with cooling water [C_p = 4.20 kJ/(kg · °C)] from a nearby lake, which enters the tubes of the condenser at 18°C at a rate of 101 kg/s and leaves at 27°C. Determine the rate of condensation of the steam in the condenser. *Answer:* 1.65 kg/s

4-62 A heat exchanger is to heat water [C_p = 4.18 kJ(kg · °C)] from 25°C to 60°C at a rate of 0.2 kg/s. The heating is to be accomplished by geothermal

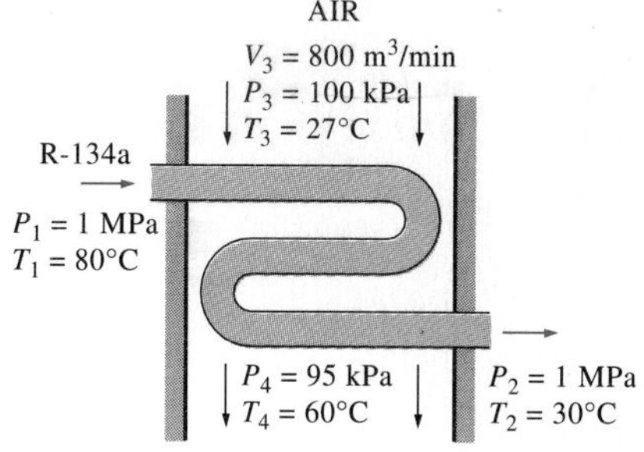

FIGURE P4-56

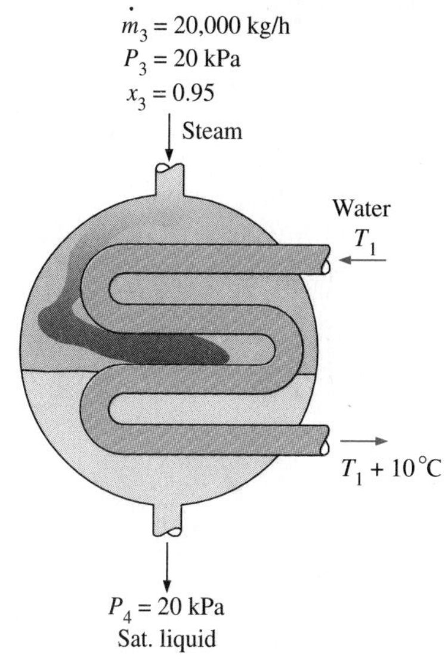

FIGURE P4-60

FIGURE P4-61

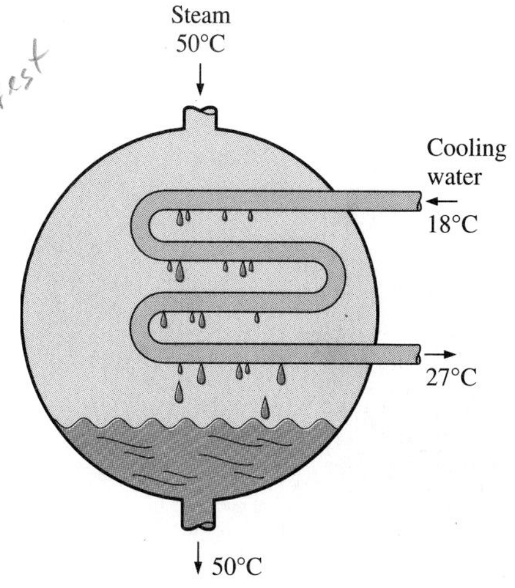

water [C_p = 4.31 kJ/(kg · °C)] available at 140°C at a mass flow rate of 0.3 kg/s. The inner tube is thin-walled and has a diameter of 0.8 cm. Determine the rate of heat transfer in the heat exchanger and the exit temperature of geothermal water.

4-63 A heat exchanger is to cool ethylene glycol [C_p = 2.56 kJ/(kg · °C)] flowing at a rate of 2 kg/s from 80°C to 40°C by water [(C_p = 4.18 kJ/(kg · °C)] that enters at 20°C and leaves at 55°C. Determine (*a*) the rate of heat transfer and (*b*) the mass flow rate of water.

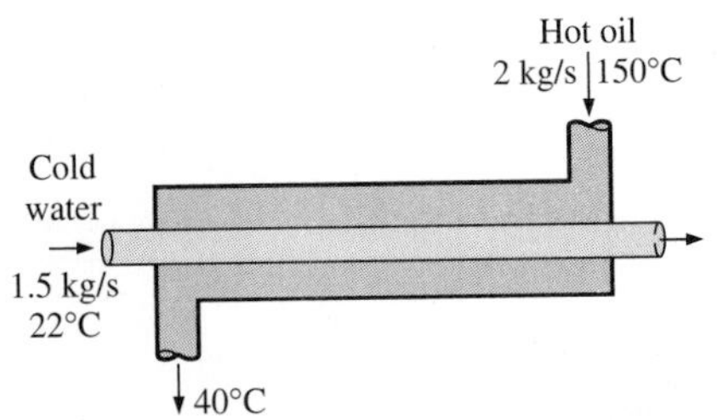

FIGURE P4-64

4-64 A thin-walled double-pipe counter-flow heat exchanger is to be used to cool oil [C_p = 2.20 kJ/(kg · °C)] from 150°C to 40°C at a rate of 2 kg/s by water [C_p = 4.18 kJ/(kg · °C)] that enters at 22°C at a rate of 1.5 kg/s. Determine the rate of heat transfer in the heat exchanger and the exit temperature of water.

4-65 Cold water [C_p = 4.18 kJ/(kg · °C)] leading to a shower enters a thin-walled double-pipe counter-flow heat exchanger at 15°C at a rate of 0.25 kg/s and is heated to 45°C by hot water [C_p = 4.19 kJ/(kg · °C)] that enters at 100°C at a rate of 3 kg/s. Determine the rate of heat transfer in the heat exchanger and the exit temperature of the hot water.

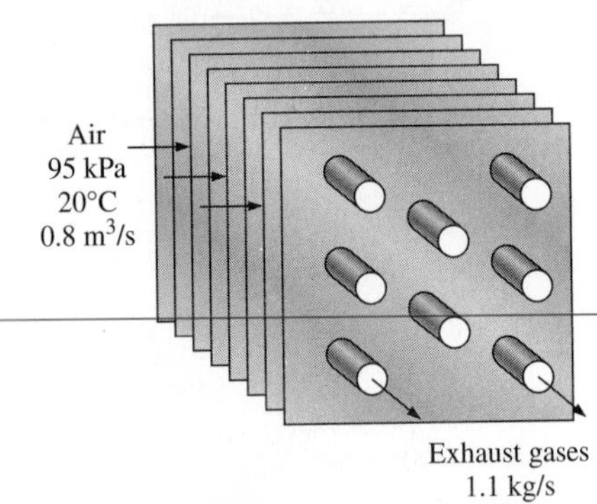

FIGURE P4-66

4-66 Air [C_p = 1.005 kJ/(kg · °C)] is to be preheated by hot exhaust gases in a cross-flow heat exchanger before it enters the furnace. Air enters the heat exchanger at 95 kPa and 20°C at a rate of 0.8 m^3/s. The combustion gases [C_p = 1.10 kJ/(kg · °C)] enter at 180°C at a rate of 1.1 kg/s and leave at 95°C. Determine the rate of heat transfer to the air and its outlet temperature.

4-67 A well-insulated shell-and-tube heat exchanger is used to heat water [C_p = 4.18 kJ/(kg · °C)] in the tubes from 20°C to 70°C at a rate of 4.5 kg/s. Heat is supplied by hot oil [C_p = 2.30 kJ/(kg · °C)] that enters the shell side at 170°C at a rate of 10 kg/s. Determine the rate of heat transfer in the heat exchanger and the exit temperature of oil.

4-68E Steam is to be condensed on the shell side of a heat exchanger at 90°F (h_{fg} = 1043 Btu/lbm). Cooling water [C_p = 1.0 Btu/(lbm · °F)] enters the tubes at 60°F at a rate of 115.3 lbm/s and leaves at 73°F. Assuming the heat exchanger to be well-insulated, determine the rate of heat transfer in the heat exchanger and the rate of condensation of the steam.

FIGURE P4-69

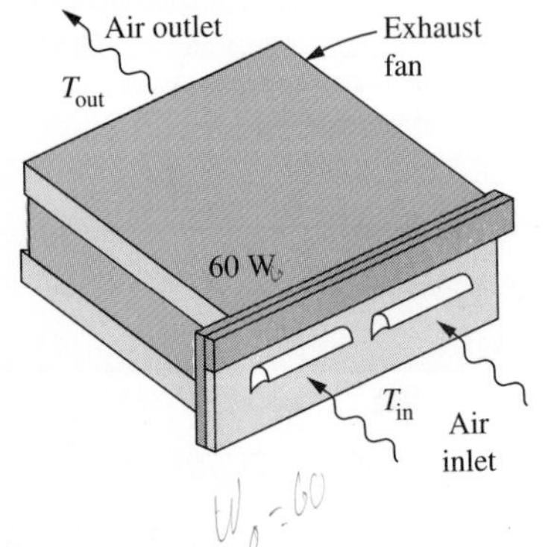

Pipe and Duct Flow

4-69 A desktop computer is to be cooled by a fan. The electronic components of the computer consume 60 W of power under full load conditions. The computer is to operate in environments at temperatures up to 45°C and at elevations up to 3400 m where the average atmospheric pressure is 66.63 kPa. The exit temperature of air is not to exceed 60°C to meet the reliability requirements. Also, the average velocity of air is not to exceed 110 m/min at the exit of the computer case where the fan is installed to keep the noise level down. Determine the flow rate of the fan that needs to be installed and the diameter of the casing of the fan.

4-70 Repeat Prob. 4-69 for a computer that consumes 100 W of power.

4-71E Water enters the tubes of a cold plate at 95°F with an average velocity of 60 ft/min and leaves at 105°F. The diameter of the tubes is 0.25 in. Assuming 15 percent of the heat generated is dissipated from the components to the surroundings by convection and radiation, and the remaining 85 percent is removed by the cooling water, determine the amount of heat generated by the electronic devices mounted on the cold plate. *Answer:* 263 W

4-72 A sealed electronic box is to be cooled by tap water flowing through the channels on two of its sides. It is specified that the temperature rise of the water not exceed 4°C. The power dissipation of the box is 2 kW, which is removed entirely by water. If the box operates 24 hours a day, 365 days a year, determine the mass flow rate of water flowing through the box and the amount of cooling water used per year.

4-73 Repeat Prob. 4-72 for a power dissipation of 3 kW.

4-74 A long roll of 2-m-wide and 0.5-cm-thick 1-Mn manganese steel plate [ρ = 785 kg/m^3 and C_p = 0.454 kJ/(kg · °C)] coming off a furnace at 820°C is to be quenched in an oil bath at 45°C to a temperature of 51.1°C. If the metal sheet is moving at a steady velocity of 10 m/min, determine the required rate of heat removal from the oil to keep its temperature constant at 45°C.
Answer: 437 kW

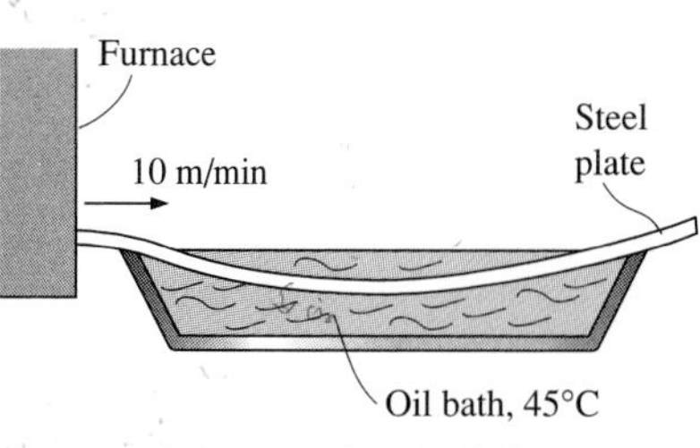

FIGURE P4-74

4-75 The components of an electronic system dissipating 180 W are located in a 1.4-m-long horizontal duct whose cross section is 20 cm × 20 cm. The components in the duct are cooled by forced air that enters the duct at 30°C and 1 atm at a rate of 0.6 m^3/min and leaves at 40°C. Determine the rate of heat transfer from the outer surfaces of the duct to the ambient.
Answer: 64.3 W

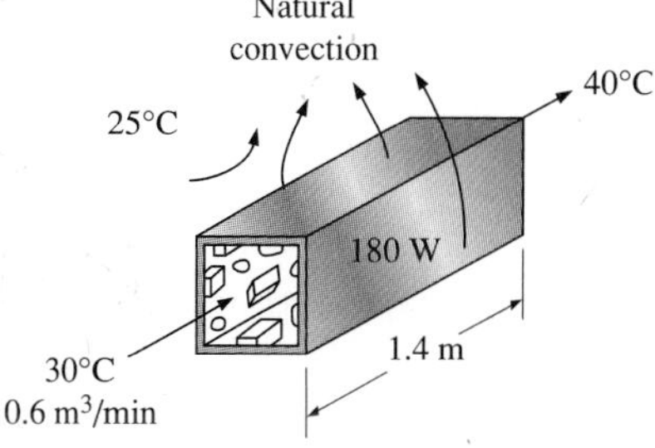

FIGURE P4-75

4-76 Repeat Prob. 4-75 for a circular horizontal duct of diameter 10 cm.

4-77E The hot water needs of a household are to be met by heating water at 55°F to 200°F by a parabolic solar collector at a rate of 4 lbm/s. Water flows through a 1.25-in.-diameter thin aluminum tube whose outer surface is black-anodized in order to maximize its solar absorption ability. The centerline of the tube coincides with the focal line of the collector, and a glass sleeve is placed outside the tube to minimize the heat losses. If solar energy is transferred to water at a net rate of 350 Btu/h per ft length of the tube, determine the required length of the parabolic collector to meet the hot water requirements of this house.

4-78 Consider a hollow-core printed circuit board 12 cm high and 18 cm long, dissipating a total of 20 W. The width of the air gap in the middle of the PCB is 0.25 cm. If the cooling air enters the 12-cm-wide core at 32°C at a rate of 0.8 L/s, determine the average temperature at which the air leaves the hollow core. *Answer:* 53.7°C

4-79 A computer cooled by a fan contains eight PCBs, each dissipating 10 W power. The height of the PCBs is 12 cm and the length is 18 cm. The cooling air is supplied by a 25-W fan mounted at the inlet. If the temperature rise of air as it flows through the case of the computer is not to exceed 10°C, determine (*a*) the flow rate of the air that the fan needs to deliver and (*b*) the fraction of the temperature rise of air that is due to the heat generated by the fan and its motor. *Answers:* (*a*) 0.0104 kg/s, (*b*) 31 percent

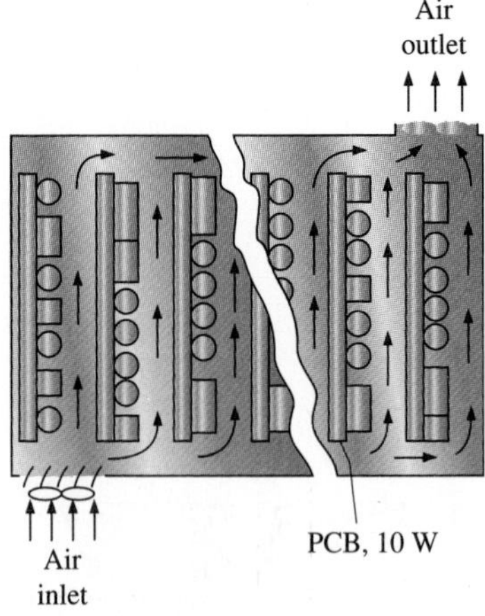

FIGURE P4-79

4-80 Hot water at 90°C enters a 15-m section of a cast iron pipe whose inner diameter is 4 cm at an average velocity of 0.8 m/s. The outer surface of the pipe is exposed to the cold air at 10°C in a basement. If water leaves the basement at 88°C, determine the rate of heat loss from the water.

4-81 A 5-m × 6-m × 8-m room is to be heated by an electric resistance heater placed in a short duct in the room. Initially, the room is at 15°C, and the local atmospheric pressure is 98 kPa. The room is losing heat steadily to the outside at a rate of 200 kJ/min. A 200-W fan circulates the air steadily through the duct and the electric heater at an average mass flow rate of 50 kg/min. The duct can be assumed to be adiabatic, and there is no air leaking in or out of the room. If it takes 15 min for the room air to reach an average temperature of 25°C, find (*a*) the power rating of the electric heater and (*b*) the temperature rise that the air experiences each time it passes through the heater.

4-82 A house has an electric heating system that consists of a 300-W fan and an electric resistance heating element placed in a duct. Air flows steadily through the duct at a rate of 0.6 kg/s and experiences a temperature rise of 5°C. The rate of heat loss from the air in the duct is estimated to be 400 W. Determine the power rating of the electric resistance heating element.
Answer: 3.12 kW

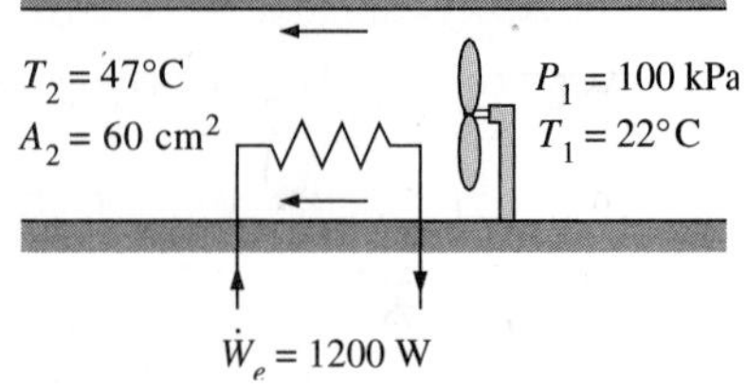

FIGURE 4-83

4-83 A hair dryer is basically a duct in which a few layers of electric resistors are placed. A small fan pulls the air in and forces it through the resistors where it is heated. Air enters a 1200-W hair dryer at 100 kPa and 22°C and leaves at 47°C. The cross-sectional area of the hair dryer at the exit is 60 cm^2. Neglecting the power consumed by the fan and the heat losses through the walls of the hair dryer, determine (*a*) the volume flow rate of air at the inlet and (*b*) the velocity of the air at the exit.
Answers: (*a*) 0.0404 m^3/kg, (*b*) 7.31 m/s

4-84 The ducts of an air heating system pass through an unheated area. As a result of heat losses, the temperature of the air in the duct drops by 4°C. If the mass flow rate of air is 120 kg/min, determine the rate of heat loss from the air to the cold environment.

4-85E Air enters the duct of an air-conditioning system at 15 psia and 50°F at a volume flow rate of 450 ft^3/min. The diameter of the duct is 10 in., and heat is transferred to the air in the duct from the surroundings at a rate of 2 Btu/s. Determine (*a*) the velocity of the air at the duct inlet and (*b*) the temperature of the air at the exit.

4-86 Water is heated in an insulated, constant-diameter tube by a 7-kW electric resistance heater. If the water enters the heater steadily at 15°C and leaves at 70°C, determine the mass flow rate of water.

4-87 Water is to be pumped from a well to the top of a 200-m-tall building. There is a 15-kW pump available in the basement, and the water surface level in the well is 40 m below ground level. Neglecting any heat transfer and frictional effects, determine the maximum flow rate of water that can be maintained by this pump.

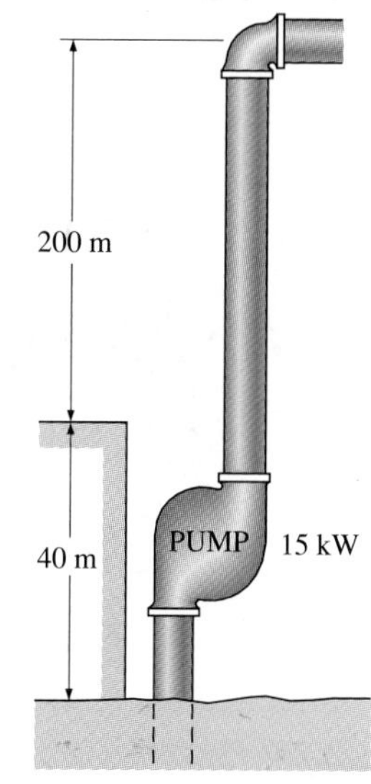

FIGURE P4-87

4-88 Steam enters a long, horizontal pipe with an inlet diameter of $D_1 =$ 12 cm at 1 MPa and 250°C with a velocity of 2 m/s. Further downstream, the conditions are 800 kPa and 200°C, and the diameter is $D_2 = 10$ cm. Determine (*a*) the mass flow rate of the steam and (*b*) the rate of heat transfer.

Answers: (*a*) 0.0972 kg/s, (*b*) −10.04 kJ/s

4-89E A 5-hp pump is used to raise the elevation of a lake's water by 75 ft from the free surface of the lake. The pipe inlet is 6 ft below the free surface. The temperature of water increases by 0.1°F during this process as a result of the frictional effects. Neglecting any heat transfer and kinetic energy changes, determine the mass flow rate of the water.

4-90 The free surface of the water in a well is 20 m below the ground level. This water is to be pumped steadily to an elevation of 30 m above the ground level. Neglecting any heat transfer, kinetic energy changes, and frictional effects, determine the power input to the pump required for a steady flow of water at a rate of 1.5 m^3/min. *Answer:* 12.3 kW

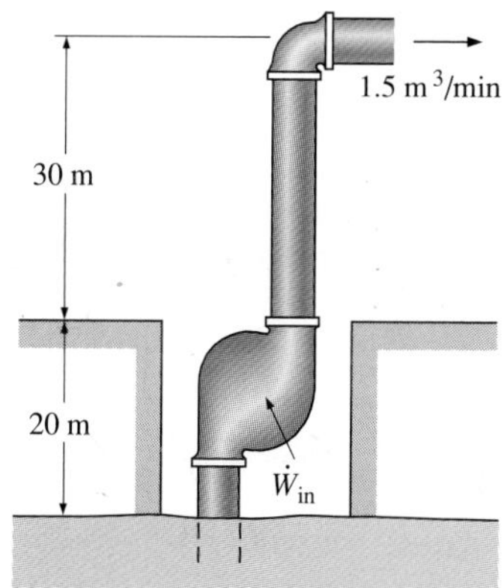

FIGURE P4-90

Unsteady-Flow Processes

4-91C Does the amount of mass entering a control volume have to be equal to the amount of mass leaving during an unsteady-flow process? How about the amounts of energy entering and leaving?

4-92C Under what conditions can an unsteady-flow process be approximated as a uniform-flow process?

4-93C Can a uniform-flow system involve boundary work?

4-94C The valve of an initially evacuated, adiabatic rigid tank is opened, and air at 30°C flows in. When the pressure inside the tank reaches atmospheric pressure, the air temperature in the tank increases to 150°C. Explain what caused this temperature rise.

4-95C When a can that contains a refrigerant at 500 kPa and 25°C is slightly opened and refrigerant is allowed to escape, a layer of ice forms outside the can. Explain how that happens.

4-96C The valve of an insulated rigid vessel containing air at a high pressure is slightly opened, allowing some air to escape. Will the temperature of air in the tank change during this process? How?

Charging Processes

4-97 Consider a 5-L evacuated rigid bottle that is surrounded by the atmosphere at 100 kPa and 17°C. A valve at the neck of the bottle is now opened and the atmospheric air is allowed to flow into the bottle. The air trapped in the bottle eventually reaches thermal equilibrium with the atmosphere as a result of heat transfer through the wall of the bottle. The valve remains open during the process so that the trapped air also reaches mechanical equilibrium with the atmosphere. Determine the net heat transfer through the wall of the bottle during this filling process. *Answer:* $Q_{out} = 0.5$ kJ

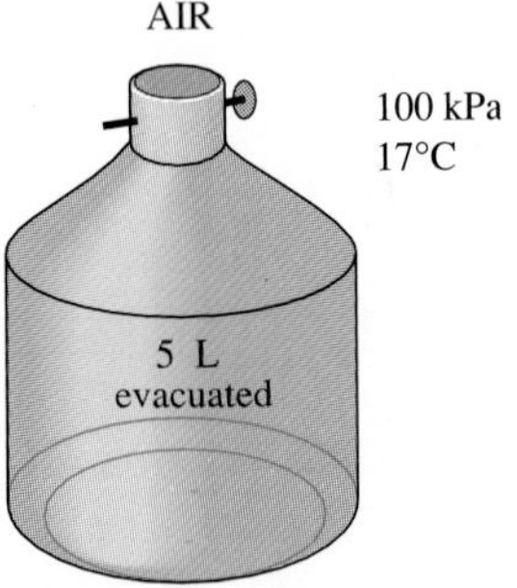

FIGURE P4-97

4-98 An insulated rigid tank is initially evacuated. A valve is opened, and atmospheric air at 95 kPa and 17°C enters the tank until the pressure in the tank reaches 95 kPa, at which point the valve is closed. Determine the final temperature of the air in the tank. Assume constant specific heats.
Answer: 406 K

4-99 A 2-m^3 rigid tank initially contains air at 100 kPa and 22°C. The tank is connected to a supply line through a valve. Air is flowing in the supply line at 600 kPa and 22°C. The valve is opened, and air is allowed to enter the tank until the pressure in the tank reaches the line pressure, at which point the valve is closed. A thermometer placed in the tank indicates that the air temperature at the final state is 77°C. Determine (*a*) the mass of air that has entered the tank and (*b*) the amount of heat transfer.
Answers: (*a*) 9.58 kg, (*b*) $Q_{out} = 339.4$ kJ

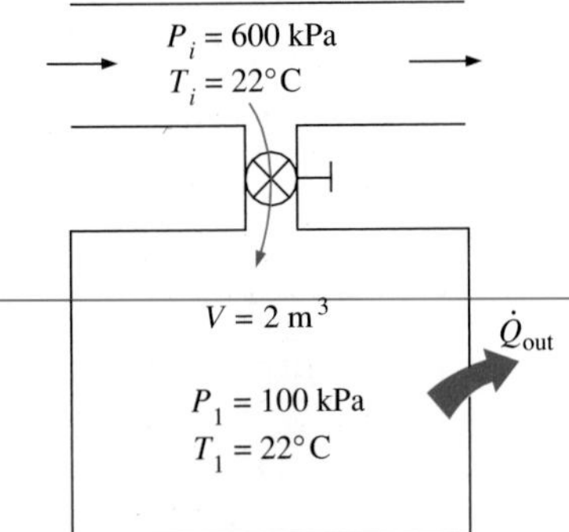

FIGURE P4-99

4-100 A 0.2-m^3 rigid tank initially contains refrigerant-134a at 8°C. At this state, 60 percent of the mass is in the vapor phase, and the rest is in the liquid phase. The tank is connected by a valve to a supply line where refrigerant at 1 MPa and 120°C flows steadily. Now the valve is opened slightly, and the refrigerant is allowed to enter the tank. When the pressure in the tank reaches 800 kPa, the entire refrigerant in the tank exists in the vapor phase only. At this point the valve is closed. Determine (*a*) the final temperature in the tank, (*b*) the mass of refrigerant that has entered the tank, and (*c*) the heat transfer between the system and the surroundings.

4-101E A 4-ft^3 rigid tank initially contains saturated water vapor at 250°F. The tank is connected by a valve to a supply line that carries steam at 160 psia and 400°F. Now the valve is opened, and steam is allowed to enter the tank. Heat transfer takes place with the surroundings such that the temperature in the tank remains constant at 250°F at all times. The valve is closed when it is observed that one-half of the volume of the tank is occupied by liquid water. Find (*a*) the final pressure in the tank, (*b*) the amount of steam that has entered the tank, and (*c*) the amount of heat transfer.
Answers: (*a*) 29.82 psia, (*b*) 117.5 lbm, (*c*) −117,539 Btu

FIGURE P4-102

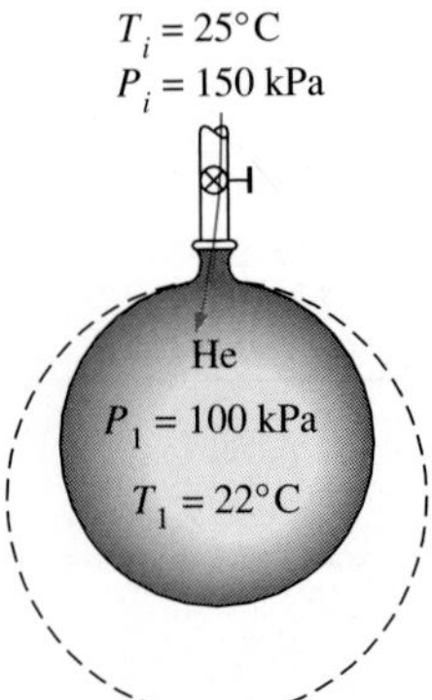

4-102 A balloon initially contains 65 m^3 of helium gas at atmospheric conditions of 100 kPa and 22°C. The balloon is connected by a valve to a large reservoir that supplies helium gas at 150 kPa and 25°C. Now the valve is opened, and helium is allowed to enter the balloon until the pressure equilibrium with the helium at the supply line is reached. The material of the balloon is such that its volume increases linearly with pressure. If no heat transfer takes place during this process, determine the final temperature in the balloon. *Answer:* 320.2 K

4-103 A vertical piston-cylinder device initially contains 0.01 m^3 of steam at 200°C. The mass of the frictionless piston is such that it maintains a constant pressure of 500 kPa inside. Now steam at 1 MPa and 350°C is allowed to enter the cylinder from a supply line until the volume inside doubles. Neglecting any heat transfer that may have taken place during the process, determine (*a*) the final temperature of the steam in the cylinder and (*b*) the amount of mass that has entered. *Answers:* (*a*) 262.6°C, (*b*) 0.0176 kg

4-104 An insulated, vertical piston-cylinder device initially contains 10 kg of water, 8 kg of which is in the vapor phase. The mass of the piston is such that it maintains a constant pressure of 300 kPa inside the cylinder. Now steam at 0.5 MPa and 350°C is allowed to enter the cylinder from a supply line until all the liquid in the cylinder has vaporized. Determine (*a*) the final temperature in the cylinder and (*b*) the mass of the steam that has entered.
Answers: (*a*) 133.6°C, (*b*) 9.78 kg

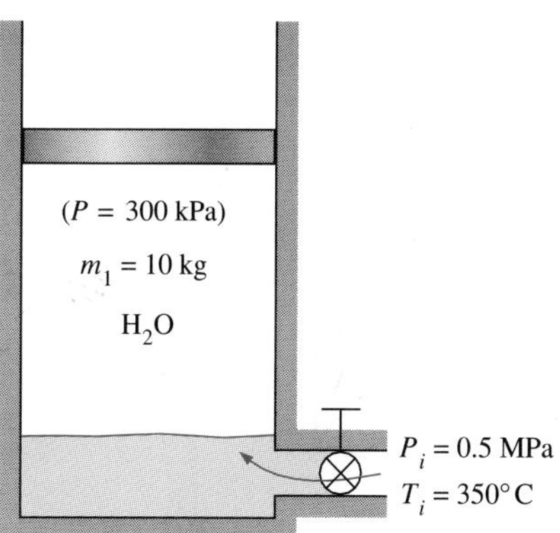

FIGURE P4-104

4-105 A 0.1-m^3 rigid tank initially contains refrigerant-134a at 1 MPa and 100 percent quality. The tank is connected by a valve to a supply line that carries refrigerant at 1.2 MPa and 30°C. Now the valve is opened, and the refrigerant is allowed to enter the tank. The valve is closed when it is observed that the tank contains saturated liquid at 1.2 MPa. Determine (*a*) the mass of the refrigerant that has entered the tank and (*b*) the amount of heat transfer. *Answers:* (*a*) 107.1 kg, (*b*) 1825 kJ

Discharging Processes

4-106 A 0.3-m^3 rigid tank is filled with saturated liquid water at 200°C. A valve at the bottom of the tank is opened, and liquid is withdrawn from the tank. Heat is transferred to the water such that the temperature in the tank remains constant. Determine the amount of heat that must be transferred by the time one-half of the total mass has been withdrawn.

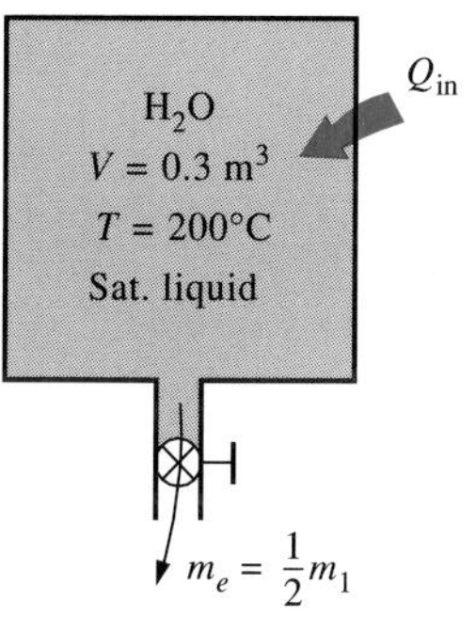

FIGURE P4-106

4-107 A 0.1-m^3 rigid tank contains saturated refrigerant-134a at 800 kPa. Initially, 40 percent of the volume is occupied by liquid and the rest by vapor. A valve at the bottom of the tank is now opened, and liquid is withdrawn from the tank. Heat is transferred to the refrigerant such that the pressure inside the tank remains constant. The valve is closed when no liquid is left in the tank and vapor starts to come out. Determine the total heat transfer for this process. *Answer:* 267.6 kJ

4-108E A 4-ft^3 rigid tank contains saturated refrigerant-134a at 100 psia. Initially, 20 percent of the volume is occupied by liquid and the rest by vapor. A valve at the top of the tank is now opened, and vapor is allowed to escape slowly from the tank. Heat is transferred to the refrigerant such that the pressure inside the tank remains constant. The valve is closed when the last drop of liquid in the tank is vaporized. Determine the total heat transfer for this process.

FIGURE P4-108E

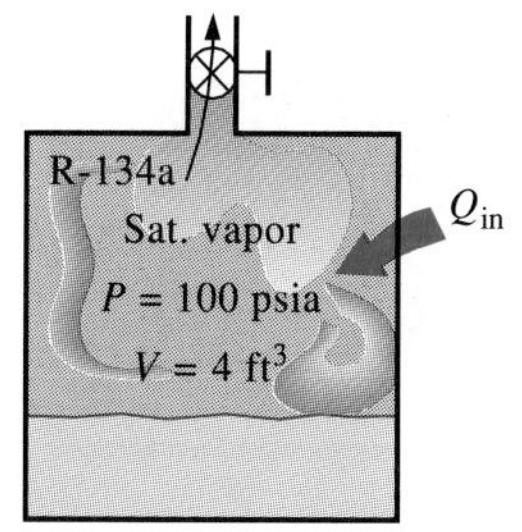

4-109 A 0.2-m^3 rigid tank equipped with a pressure regulator contains steam at 2 MPa and 300°C. The steam in the tank is now heated. The regulator keeps the steam pressure constant by letting out some steam, but the temperature inside rises. Determine the amount of heat transferred when the steam temperature reaches 500°C.

4-110 A 4-L pressure cooker has an operating pressure of 175 kPa. Initially, one-half of the volume is filled with liquid and the other half with vapor. If it is desired that the pressure cooker not run out of liquid water for 1 h, determine the highest rate of heat transfer allowed.

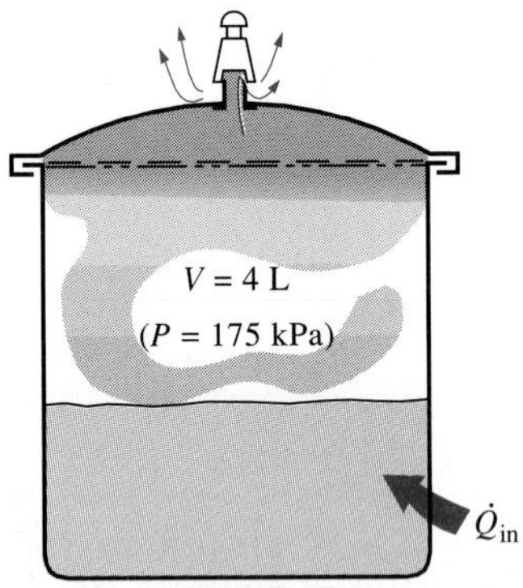

FIGURE P4-110

4-111 An insulated 0.08-m^3 tank contains helium at 2 MPa and 80°C. A valve is now opened, allowing some helium to escape. The valve is closed when one-half of the initial mass has escaped. Determine the final temperature and pressure in the tank. *Answers:* 225 K, 637 kPa

4-112E An insulated 60-ft^3 rigid tank contains air at 75 psia and 120°F. A valve connected to the tank is now opened, and air is allowed to escape until the pressure inside drops to 30 psia. The air temperature during this process is maintained constant by an electric resistance heater placed in the tank. Determine the electrical work done during this process.

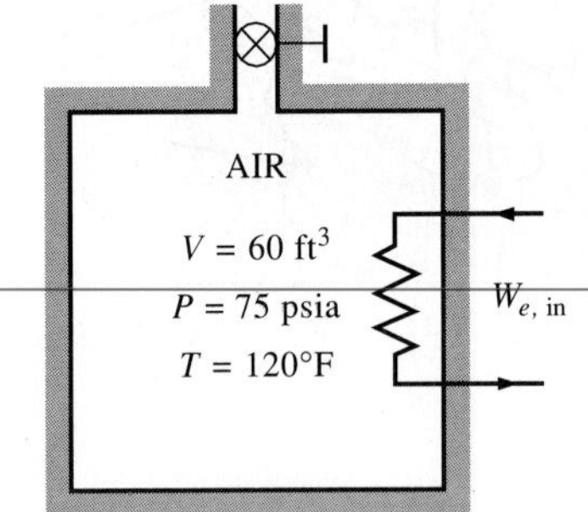

FIGURE P4-112E

4-113 A vertical piston-cylinder device initially contains 0.2 m^3 of air at 20°C. The mass of the piston is such that it maintains a constant pressure of 300 kPa inside. Now a valve connected to the cylinder is opened, and air is allowed to escape until the volume inside the cylinder is decreased by one-half. Heat transfer takes place during the process so that the temperature of the air in the cylinder remains constant. Determine (*a*) the amount of air that has left the cylinder and (*b*) the amount of heat transfer.
Answers: (*a*) 0.357 kg, (*b*) 0

4-114 A balloon initially contains 10 m^3 of helium gas at 150 kPa and 27°C. Now a valve is opened, and helium is allowed to escape slowly until the pressure inside drops to 100 kPa, at which point the valve is closed. During this process the volume of the balloon decreases by 15 percent. The balloon material is such that the volume of the balloon changes linearly with pressure in this range. If the heat transfer during this process is negligible, find (*a*) the final temperature of the helium in the balloon and (*b*) the amount of helium that has escaped.

FIGURE P4-115

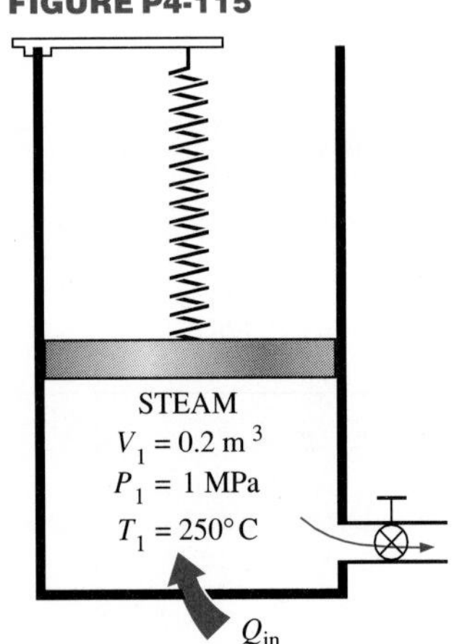

4-115 A vertical piston-cylinder device initially contains 0.2 m^3 of steam at 1 MPa and 250°C. A linear spring at this point applies full force to the piston. A valve connected to the cylinder is now opened, and steam is allowed to escape. As the piston moves down, the spring unwinds, and at the final state the pressure drops to 800 kPa and the volume to 0.1 m^3. If at the final state the cylinder contains saturated vapor only, determine (*a*) the initial and final masses in the cylinder and (*b*) the amount and direction of any heat transfer.

4-116 A vertical piston-cylinder device initially contains 0.3 m^3 of steam at 250°C. The mass of the piston is such that it maintains a constant pressure of 300 kPa. Now a valve is opened and steam is allowed to escape. Heat transfer takes place during the process so that the temperature inside remains constant. If the final volume is 0.1 m^3, determine (*a*) the amount of steam that has escaped and (*b*) the amount of heat transfer. *Answers:* (*a*) 0.251 kg, (*b*) 0

Review Problems

4-117 Consider two identical buildings: one in Los Angeles, California, where the atmospheric pressure is 101 kPa and the other in Denver, Colorado, where the atmospheric pressure is 83 kPa. Both buildings are maintained at 21°C, and the infiltration rate for both buildings is 1.2 air changes per hour (ACH). That is, the entire air in the building is replaced completely by the outdoor air 1.2 times per hour on a day when the outdoor temperature at both locations is 10°C. Disregarding latent heat, determine the ratio of the heat losses by infiltration at the two cities.

4-118 The ventilating fan of the bathroom of a building has a volume flow rate of 30 L/s and runs continuously. The building is located in San Francisco, California, where the average winter temperature is 12.2°C, and is maintained at 22°C at all times. The building is heated by electricity whose unit cost is \$0.09/kWh. Determine the amount and cost of the heat "vented out" per month in winter.

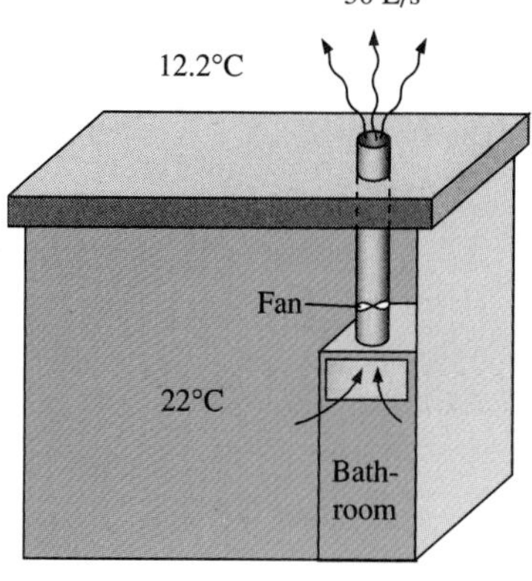

FIGURE P4-118

4-119 Consider a large classroom on a hot summer day with 150 students, each dissipating 60 W of sensible heat. All the lights, with 4.0 kW of rated power, are kept on. The room has no external walls, and thus heat gain through the walls and the roof is negligible. Chilled air is available at 15°C, and the temperature of the return air is not to exceed 25°C. Determine the required flow rate of air, in kg/s, that needs to be supplied to the room to keep the average temperature of the room constant. *Answer:* 1.45 kg/s

4-120 A typical full-carlot-capacity banana room contains 36 pallets of bananas. Each pallet consists of 24 boxes, and thus the room stores 864 boxes of bananas. A box holds an average of 19 kg of bananas and is made of 2.3 kg of fiberboard. The specific heats of banana and the fiberboard are 3.55 kJ/(kg · °C) and 1.7 kJ/(kg · °C), respectively. The peak heat of respiration of bananas is 0.3 W/kg. The bananas are cooled at a rate of 0.4°C/h. The rate of heat gain through the walls and other surfaces of the room is estimated to be 1800 kJ/h. If the temperature rise of refrigerated air is not to exceed 2.0°C as it flows thorough the room, determine the minimum flow rate of air needed. Take the density and specific heat of air to be 1.2 kg/m^3 and 1.0 kJ/(kg · °C), respectively.

4-121 The chilling room of a meat plant is 15 m × 18 m × 5.5 m in size and has a capacity of 350 beef carcasses. The power consumed by the fans and the lights in the chilling room are 22 and 2 kW, respectively, and the room gains heat through its envelope at a rate of 11 kW. The carcasses have an average mass of 280 kg, and enter the chilling room at 35°C after they are washed to facilitate evaporative cooling and are cooled to 16°C in 12 h. The air enters the chilling room at −2.2°C and leaves at 0.5°C. Determine (*a*) the refrigeration load of the chilling room and (*b*) the volume flow rate of air. The average specific heats of beef carcasses and air are 3.14 and 1.0 kJ/(kg · °C), respectively. Also, the density of air can be taken to be 1.28 kg/m³.

4-122 Chickens with an average mass of 2.2 kg and average specific heat of 3.54 kJ/(kg · °C) are to be cooled by chilled water that enters a continuous-flow-type immersion chiller at 0.5°C. Chickens are dropped into the chiller at a uniform temperature of 15°C at a rate of 500 chickens per hour and are cooled to an average temperature of 3°C before they are taken out. The chiller gains heat from the surroundings at a rate of 200 kJ/h. Determine (*a*) the rate of heat removal from the chickens, in kW, and (*b*) the mass flow rate of water, in kg/s, if the temperature rise of water is not to exceed 2°C.

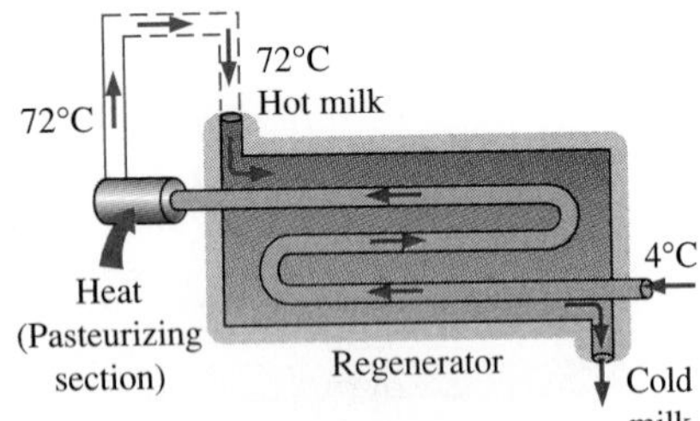

FIGURE P4-123

4-123 In a dairy plant, milk at 4°C is pasteurized continuously at 72°C at a rate of 12 L/s for 24 hours a day and 365 days a year. The milk is heated to the pasteurizing temperature by hot water heated in a natural-gas-fired boiler that has an efficiency of 82 percent. The pasteurized milk is then cooled by cold water at 18°C before it is finally refrigerated back to 4°C. To save energy and money, the plant installs a regenerator that has an effectiveness of 82 percent. If the cost of natural gas is $0.52/therm (1 therm = 105,500 kJ), determine how much energy and money the regenerator will save this company per year.

4-124E A refrigeration system is being designed to cool eggs [ρ = 67.4 lbm/ft³ and C_p = 0.80 Btu/(lbm · °F) with an average mass of 0.14 lbm from an initial temperature of 90°F to a final average temperature of 50°F by air at 34°F at a rate of 10,000 eggs per hour. Determine (*a*) the rate of heat removal from the eggs, in Btu/h and (*b*) the required volume flow rate of air, in ft³/h, if the temperature rise of air is not to exceed 10°F.

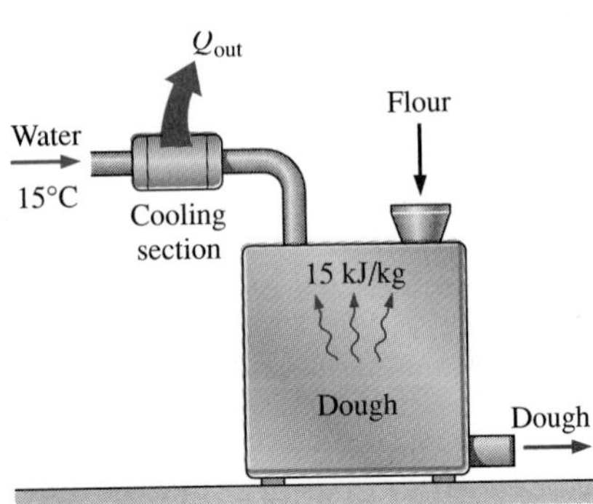

FIGURE P4-125

4-125 The heat of hydration of dough, which is 15 kJ/kg, will raise its temperature to undesirable levels unless some cooling mechanism is utilized. A practical way of absorbing the heat of hydration is to use refrigerated water when kneading the dough. If a recipe calls for mixing 2 kg of flour with 1 kg of water, and the temperature of the city water is 15°C, determine the temperature to which the city water must be cooled before mixing in order for the water to absorb the entire heat of hydration when the water temperature rises to 15°C. Take the specific heats of the flour and the water to be 1.76 and 4.18 kJ/(kg · °C), respectively. *Answer:* 4.2°C

4-126 A glass bottle washing facility uses a well-agitated hot water bath at 55°C that is placed on the ground. The bottles enter at a rate of 800 per minute at an ambient temperature of 20°C and leave at the water temperature. Each bottle has a mass of 150 g and removes 0.2 g of water as it leaves the bath wet. Make-up water is supplied at 15°C. Disregarding any heat losses from the outer surfaces of the bath, determine the rate at which (*a*) water and (*b*) heat must be supplied to maintain steady operation.

4-127 Repeat Prob. 4-126 for a water bath temperature of 50°C.

4-128 Long aluminum wires of diameter 3 mm [ρ = 2702 kg/m^3 and C_p = 0.896 kJ/(kg · °C)] are extruded at a temperature of 350°C and are cooled to 50°C in atmospheric air at 30°C. If the wire is extruded at a velocity of 10 m/min, determine the rate of heat transfer from the wire to the extrusion room.

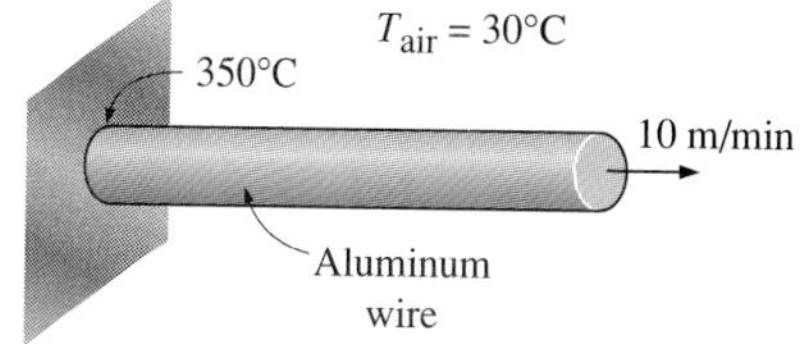

FIGURE P4-128

4-129 Repeat Prob. 4-128 for a copper wire [ρ = 8950 kg/m^3 and C_p = 0.383 kJ/(kg · °C)].

4-130 Turkeys with an average mass of 7.5 kg and average specific heat of 3.28 kJ/(kg · °C) are to be cooled by chilled water that enters a continuous-flow-type immersion chiller at 0.6°C. Turkeys are dropped into the chiller at a uniform temperature of 14°C at a rate of 200 turkeys per hour and are cooled to an average temperature of 4°C before they are taken out. The chiller gains heat from the surroundings at a rate of 350 kJ/h. Determine (*a*) the rate of heat removal from the turkeys, in kW, and (*b*) the mass flow rate of water, in kg/s, if the temperature rise of water is not to exceed 2.5°C.

4-131 Steam at 40°C condenses on the outside of a 4-mm-long, 3-cm-diameter thin horizontal copper tube by cooling water that enters the tube at 25°C at an average velocity of 2 m/s and leaves at 35°C. Determine the rate of condensation of steam and the average overall heat transfer coefficient between the steam and the cooling water based on the outside surface area of the tube.

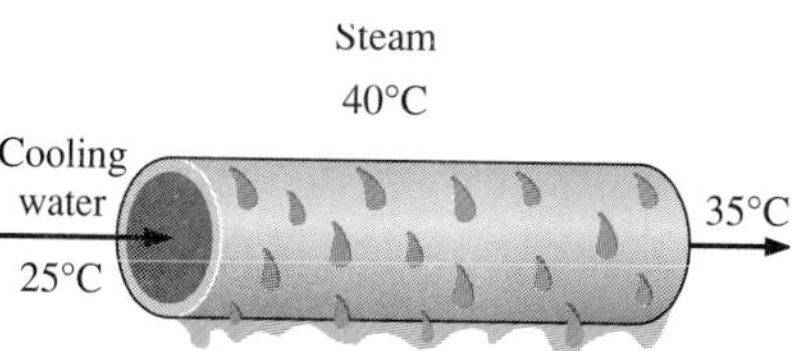

FIGURE P4-131

4-132E The condenser of a steam power plant operates at a pressure of 0.95 psia. The condenser consists of 144 horizontal tubes arranged in a 12 × 12 square array. Steam condenses on the outer surfaces of the tubes whose inner and outer diameters are 1 in. and 1.2 in., respectively. If steam is to be condensed at a rate of 6800 lbm/h and the temperature rise of the cooling water is limited to 8°F, determine (*a*) the rate of heat transfer from the steam to the cooling water and (*b*) the average velocity of the cooling water through the tubes.

4-133 Saturated refrigerant-134a vapor at 30°C is to be condensed as it flows in a 1-cm-diameter horizontal tube at a rate of 0.1 kg/min. Determine the rate of heat transfer from the refrigerant. What would your answer be if the condensed refrigerant is cooled to 20°C?

4-134E The average atmospheric pressure in Spokane, Washington (elevation = 2350 ft), is 13.5 psia, and the average winter temperature is 36.5°F. The pressurization test of a 9-ft-high, 3,000-ft^2 older home revealed that the seasonal average infiltration rate of the house is 2.2 air changes per hour (ACH). That is, the entire air in the house is replaced completely 2.2 times per hour by the outdoor air. It is suggested that the infiltration rate of the house can be reduced by half to 1.1 ACH by winterizing the doors and the windows. If the house is heated by natural gas whose unit cost is \$0.62/therm and the heating season can be taken to be six months, determine how much the home owner will save from the heating costs per year by this winterization project. Assume the house is maintained at 72°F at all times and the efficiency of the furnace is 0.65. Also assume the latent heat load during the heating season to be negligible.

4-135 Determine the rate of sensible heat loss from a building due to infiltration if the outdoor air at −10°C and 90 kPa enters the building at a rate of 35 L/s when the indoors is maintained at 22°C.

4-136 The maximum flow rate of standard shower heads is about 3.5 gpm (13.3 L/min) and can be reduced to 2.75 gpm (10.5 L/min) by switching to low-flow shower heads that are equipped with flow controllers. Consider a family of four, with each person taking a five-minute shower every morning. City water at 15°C is heated to 55°C in an electric water heater and tempered to 42°C by cold water at the T-elbow of the shower before being routed to the shower heads. Assuming a constant specific heat of 4.18 kJ/(kg · °C) for water, determine (*a*) the ratio of the flow rates of the hot and cold water as they enter the T-elbow and (*b*) the amount of electricity that will be saved per year, in kWh, by replacing the standard shower heads by the low-flow ones.

4-137 A fan is powered by a 0.5-hp motor and delivers air at a rate of 130 m^3/min. Determine the highest value for the average velocity of air mobilized by the fan. Take the density of air to be 1.18 kg/m^3.

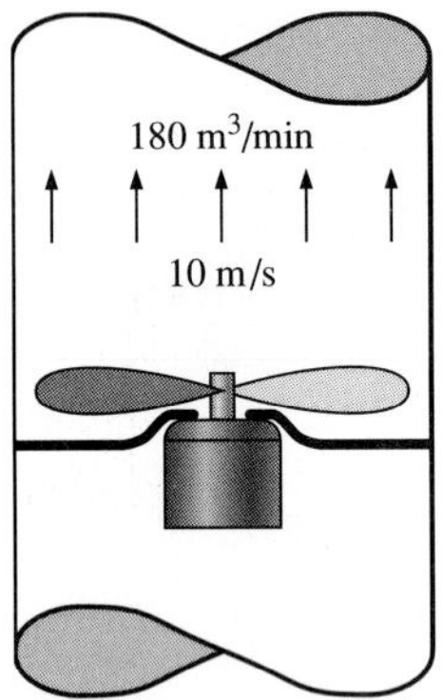

FIGURE P4-138

4-138 An air-conditioning system requires airflow at the main supply duct at a rate of 180 m^3/min. The average velocity of air in the circular duct is not to exceed 10 m/s to avoid excessive vibration and pressure drops. Assuming the fan converts 70 percent of the electrical energy it consumes into kinetic energy of air, determine the size of the electric motor needed to drive the fan and the diameter of the main duct. Take the density of air to be 1.20 kg/m^3.

4-139 Consider an evacuated rigid bottle of volume V that is surrounded by the atmosphere at pressure P_0 and temperature T_0. A valve at the neck of the bottle is now opened and the atmospheric air is allowed to flow into the bottle. The air trapped in the bottle eventually reaches thermal equilibrium with the atmosphere as a result of heat transfer through the wall of the bottle. The valve remains open during the process so that the trapped air also reaches mechanical equilibrium with the atmosphere. Determine the net heat transfer through the wall of the bottle during this filling process in terms of the properties of the system and the surrounding atmosphere.

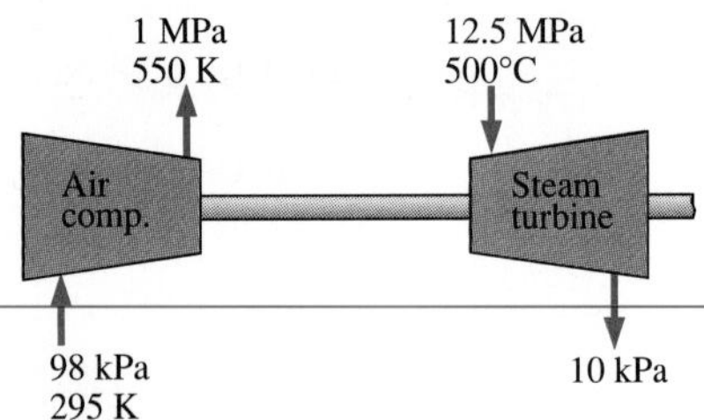

FIGURE P4-140

4-140 An adiabatic air compressor is to be powered by a direct-coupled adiabatic steam turbine that is also driving a generator. Steam enters the turbine at 12.5 MPa and 500°C at a rate of 25 kg/s and exits at 10 kPa and a quality of 0.92. Air enters the compressor at 98 kPa and 295 K at a rate of 10 kg/s and exits at 1 MPa and 550 K. Determine the net power delivered to the generator by the turbine.

FIGURE P4-141

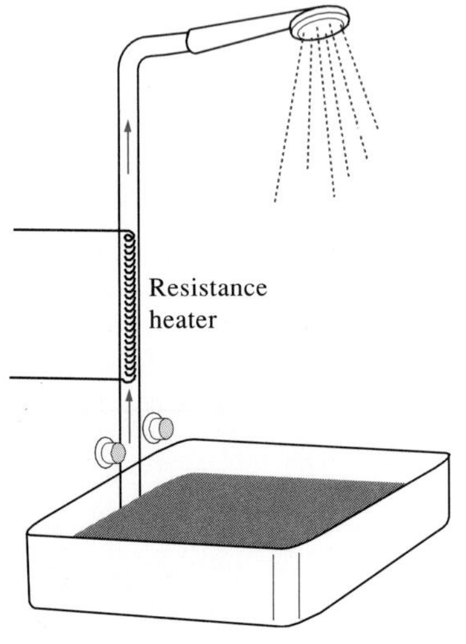

4-141 Water flows through a shower head steadily at a rate of 10 L/min. An electric resistance heater placed in the water pipe heats the water from 16°C to 43°C. Taking the density of water to be 1 kg/L, determine the electric power input to the heater, in kW.

In an effort to conserve energy, it is proposed to pass the drained warm water at a temperature of 39°C through a heat exchanger to preheat the incoming cold water. If the heat exchanger has an effectiveness of 0.50 (that is, it recovers only half of the energy that can possibly be transferred from the drained water to incoming cold water), determine the electric power input required in this case. If the price of the electric energy is 8.5 ¢/kWh, determine how much money is saved during a 10 min shower as a result of installing this heat exchanger.

4-142 Steam enters a turbine steadily at 10 MPa and 550°C with a velocity of 60 m/s and leaves at 25 kPa with a quality of 95 percent. A heat loss of 30 kJ/kg occurs during the process. The inlet area of the turbine is 150 cm^2, and the exit area is 1400 cm^2. Determine (*a*) the mass flow rate of the steam, (*b*) the exit velocity, and (*c*) the power output.

4-143E Refrigerant-134a enters an adiabatic compressor at 15 psia and 20°F with a volume flow rate of 10 ft^3/s and leaves at a pressure of 120 psia. The power input to the compressor is 60 hp. Find (*a*) the mass flow rate of the refrigerant and (*b*) the exit temperature.

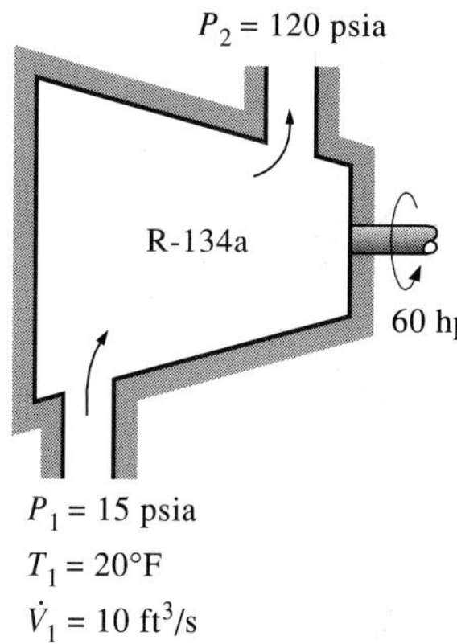

FIGURE P4-143E

4-144 In large gas-turbine power plants, air is preheated by the exhaust gases in a heat exchanger called the *regenerator* before it enters the combustion chamber. Air enters the regenerator at 1 MPa and 550 K at a mass flow rate of 800 kg/min. Heat is transferred to the air at a rate of 3200 kJ/s. Exhaust gases enter the regenerator at 140 kPa and 800 K and leave at 130 kPa and 600 K. Treating the exhaust gases as air, determine (*a*) the exit temperature of the air and (*b*) the mass flow rate of exhaust gases.
Answers: (*a*) 775 K, (*b*) 14.9 kg/s

4-145 In large steam power plants, the feedwater is frequently heated in a closed feedwater heater by using steam extracted from the turbine at some stage. Steam enters the feedwater heater at 1 MPa and 200°C and leaves as saturated liquid at the same pressure. Feedwater enters the heater at 2.5 MPa and 50°C and leaves at 10°C below the exit temperature of the steam. Determine the ratio of the mass flow rates of the extracted steam and the feedwater.

4-146 A building with an internal volume of 400 m^3 is to be heated by a 30-kW electric resistance heater placed in the duct inside the building. Initially, the air in the building is at 14°C, and the local atmospheric pressure is 95 kPa. The building is losing heat to the surroundings at a steady rate of 450 kJ/min. Air is forced to flow through the duct and the heater steadily by a 250-W fan, and it experiences a temperature rise of 5°C each time it passes through the duct, which may be assumed to be adiabatic.

(*a*) How long will it take for the air inside the building to reach an average temperature of 24°C?

(*b*) Determine the average mass flow rate of air through the duct.
Answers: (*a*) 146 s, (*b*) 6.02 kg/s

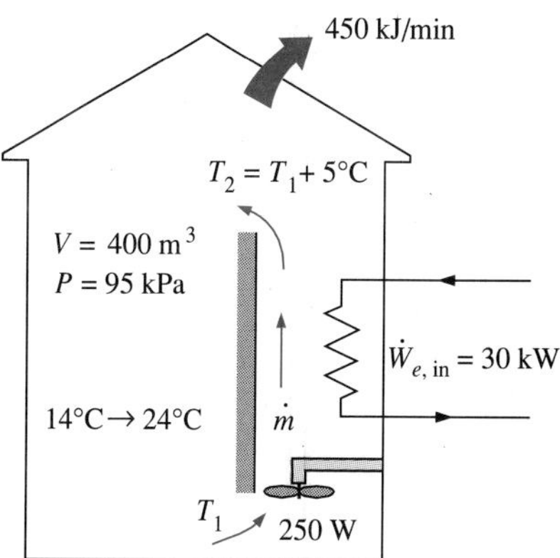

FIGURE P4-146

4-147 It is proposed to have a water heater that consists of an insulated pipe of 5-cm diameter and an electric resistor inside. Cold water at 15°C enters the heating section steadily at a rate of 30 L/min. If water is to be heated to

50°C, determine (*a*) the power rating of the resistance heater and (*b*) the average velocity of the water in the pipe.

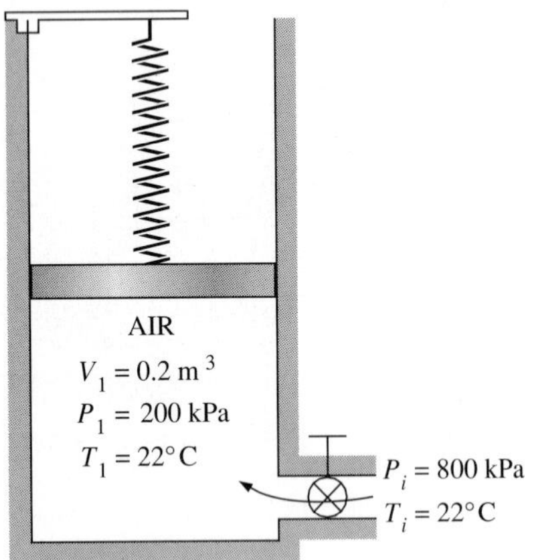

FIGURE P4-148

4-148 An insulated vertical piston-cylinder device initially contains 0.2 m^3 of air at 200 kPa and 22°C. At this state, a linear spring touches the piston but exerts no force on it. The cylinder is connected by a valve to a line that supplies air at 800 kPa and 22°C. The valve is opened, and air from the high-pressure line is allowed to enter the cylinder. The valve is turned off when the pressure inside the cylinder reaches 600 kPa. If the enclosed volume inside the cylinder doubles during this process, determine (*a*) the mass of air that entered the cylinder and (*b*) the final temperature of the air inside the cylinder.

4-149 Pressurized air stored in a 10,000-m^3 cave at 500 kPa and 400 K is to be used to drive a turbine at times of high demand for electric power. If the turbine exit conditions are 100 kPa and 300 K, determine the amount of work delivered by the turbine when the air pressure in the cave drops to 300 kPa. Assume both the cave and the turbine to be adiabatic.
Answer: 980.8 kJ

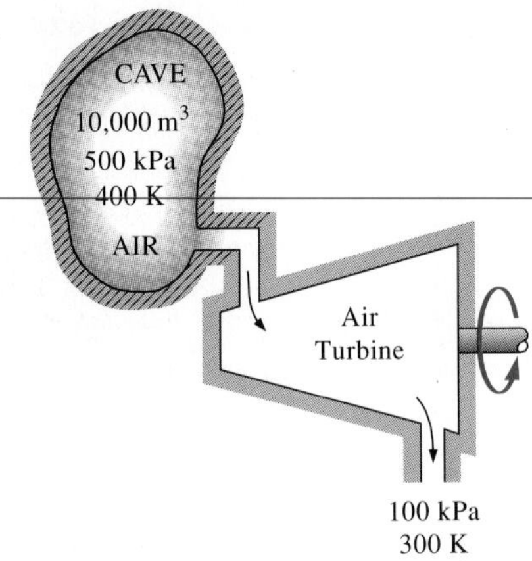

FIGURE P4-149

4-150 The velocity of a liquid flowing in a circular pipe of radius R varies from zero at the wall to a maximum at the pipe center. The velocity distribution in the pipe can be represented as $V(r)$, where r is the radial distance from the pipe center. Based on the definition of mass flow rate $\dot{m}$, obtain a relation for the average velocity in terms of $V(r)$, R, and r.

4-151E Steam at 14.7 psia and 320°F enters a diffuser with a velocity of 500 ft/s and leaves as saturated vapor at 240°F with a velocity of 100 ft/s. The exit area of the diffuser is 120 in^2. Determine (*a*) the mass flow rate of the steam, (*b*) the rate of heat transfer, and (*c*) the inlet area of the diffuser.
Answers: (*a*) 5.1 lbm/s, (*b*) 235.8 Btu/s loss, (*c*) 46.1 in^2

FIGURE P4-151E

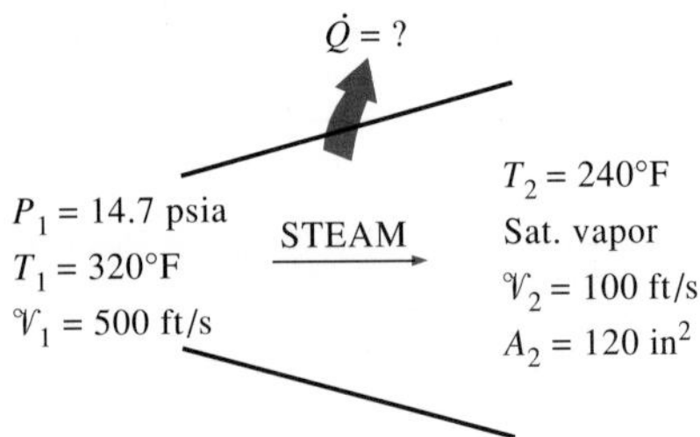

4-152 A 5-L pressure cooker has an operating pressure of 200 kPa. Initially, 20 percent of the volume is occupied by liquid and the rest by vapor. The cooker is placed on a heating unit that supplies heat to the water inside at a rate of 400 W. Determine how long it will take for the liquid in the pressure cooker to be depleted (i.e., the cooker contains only saturated vapor at the final state). *Answer:* 1.44 h

4-153 A spherical balloon initially contains 25 m^3 of helium gas at 20°C and 150 kPa. A valve is now opened, and the helium is allowed to escape slowly. The valve is closed when the pressure inside the balloon drops to the atmospheric pressure of 100 kPa. The elasticity of the balloon material is such that the pressure inside the balloon during the process varies with volume according to the relation $P = a + bV$, where $a = -100$ kPa and b is a constant. Disregarding any heat transfer, determine (*a*) the final temperature in the balloon and (*b*) the mass of helium that has escaped.
Answers: (*a*) 249.7 K, (*b*) 2.306 kg

4-154 Write a computer program to solve Prob. 4-153, using a stepwise approach. Use (*a*) 5, (*b*) 20, and (*c*) 50 increments for pressure between the initial value of 150 kPa and the final value of 100 kPa. Take the starting point of the first step to be the initial state of the helium (150 kPa, 20°C, and 25 m^3). The starting point of the second step is the state of the helium at the end of the first step, and so on. Compare your results with those obtained by using the uniform-flow approximation (i.e., a one-step solution).

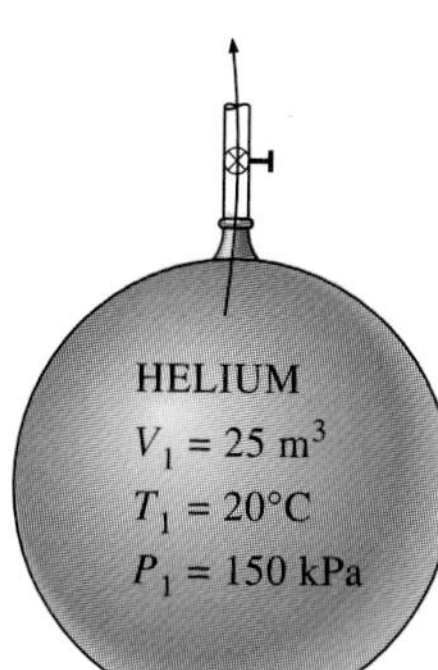

FIGURE P4-153

4-155 Using a thermometer and a tape measure only, explain how you can determine the average velocity of air at the exit of your hair dryer at its highest power setting.

4-156 Design a 1200 W electric hair dryer such that the air temperature and velocity in the dryer will not exceed 50°C and 3 m/s, respectively.

4-157 Design an electric hot water heater for a family of four in your area. The maximum water temperature in the tank and the power consumption are not to exceed 60°C and 4 kW, respectively. There are two showers in the house, and the flow rate of water through each of the shower heads is about 10 L/min. Each family member takes a 5-min shower every morning. Explain why a hot water tank is necessary, and determine the proper size of the tank for this family.

4-158 Write an essay on the classification of heat exchangers. Describe the operation of each type of heat exchanger, and discuss how different types differ from each other.

4-159 A manufacturing facility requires saturated steam at 120°C at a rate of 1.2 kg/min. Design an electric steam boiler for this purpose under the following constraints:

- The boiler will be in cylindrical shape with a height-to-diameter ratio of 1.5. The boiler can be horizontal or vertical.
- A commercially available plug-in type electrical heating element made of mechanically polished stainless steel will be used. The diameter of the heater can be between 0.5 cm and 3 cm. Also, the heat flux at the surface of the heater cannot exceed 150 kW/m^2.
- Half of the volume of the boiler should be occupied by steam, and the boiler should be large enough to hold enough water for a 2-h supply of steam. Also, the boiler will be well-insulated.

You are to specify the following: (1) The height and inner diameter of the tank, (2) the length, diameter, power rating, and surface temperature of the electric heating element, and (3) the maximum rate of steam production during short periods (less than 30 min) of overload conditions, and how it can be accomplished.

4-160 Repeat Prob. 4-159 for a boiler that produces steam at 150°C at a rate of 2.5 kg/min.

4-161 Design a scalding unit for slaughtered chicken to loosen their feathers before they are routed to feather-picking machines with a capacity of 1200 chickens per hour under the following conditions:

The unit will be of an immersion type filled with hot water at an average temperature of 53°C at all times. Chicken with an average mass of 2.2 kg and an average temperature of 36°C will be dipped into the tank, held in the water for 1.5 min, and taken out by a slow-moving conveyor. The chicken is expected to leave the tank 15 percent heavier as a result of the water that sticks to its surface. The center-to-center distance between chickens in any direction will be at least 30 cm. The tank can be as wide as 3 m and as high as 60 cm. The water is to be circulated through and heated by a natural gas furnace, but the temperature rise of water will not exceed 5°C as it passes through the furnace. The water loss is to be made up by the city water at an average temperature of 16°C. The walls and the floor of the tank are well-insulated. The unit operates 24 h a day and 6 days a week. Assuming reasonable values for the average properties, recommend reasonable values for (*a*) the mass flow rate of the make-up water that must be supplied to the tank, (*b*) the rate of heat transfer from the water to the chicken, in kW, (*c*) the size of the heating system in kJ/h, and (*d*) the operating cost of the scalding unit per month for a unit cost of \$0.56/therm of natural gas (1 therm = 105,500 kJ).

The Second Law of Thermodynamics

CHAPTER 5

To this point, we have focused our attention on the first law of thermodynamics, which requires that energy be conserved during a process. In this chapter, we introduce the second law of thermodynamics, which asserts that processes occur in a certain direction and that energy has quality as well as quantity. A process cannot take place unless it satisfies both the first and second laws of thermodynamics. In this chapter, the thermal energy reservoirs, reversible and irreversible processes, heat engines, refrigerators, and heat pumps are introduced first. Various statements of the second law are followed by a discussion of perpetual-motion machines and the absolute thermodynamic temperature scale. The Carnot cycle is introduced next, and the Carnot principles, idealized Carnot heat engines, refrigerators, and heat pumps are examined. Finally, energy conservation associated with the use of household refrigerators is discussed.

5-1 ■ INTRODUCTION TO THE SECOND LAW OF THERMODYNAMICS

FIGURE 5-1

A cup of hot coffee does not get hotter in a cooler room.

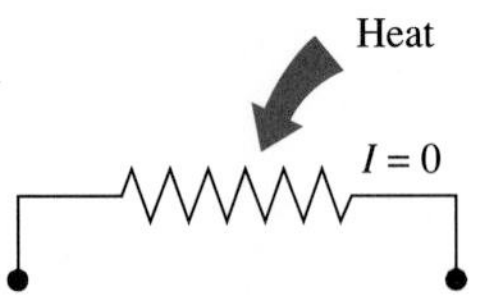

FIGURE 5-2

Transferring heat to a wire will not generate electricity.

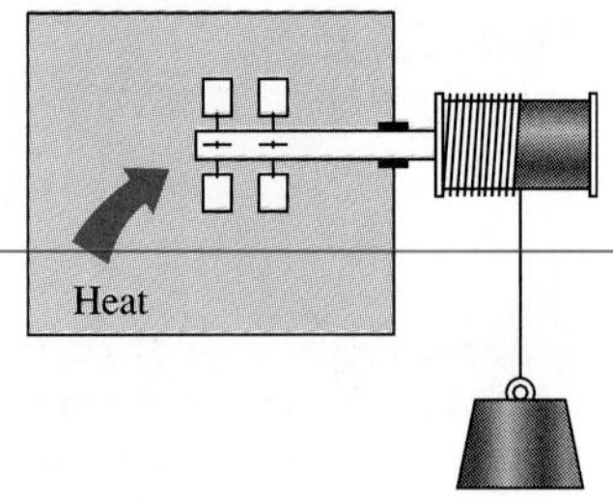

FIGURE 5-3

Transferring heat to a paddle wheel will not cause it to rotate.

FIGURE 5-4

Processes occur in a certain direction, and not in the reverse direction.

In the preceding two chapters, we applied the *first law of thermodynamics,* or the *conservation of energy principle,* to processes involving closed and open systems. As pointed out repeatedly in those chapters, energy is a conserved property, and no process is known to have taken place in violation of the first law of thermodynamics. Therefore, it is reasonable to conclude that a process must satisfy the first law to occur. However, as explained below, satisfying the first law alone does not ensure that the process will actually take place.

It is common experience that a cup of hot coffee left in a cooler room eventually cools off (Fig. 5-1). This process satisfies the first law of thermodynamics since the amount of energy lost by the coffee is equal to the amount gained by the surrounding air. Now let us consider the reverse process—the hot coffee getting even hotter in a cooler room as a result of heat transfer from the room air. We all know that this process never takes place. Yet, doing so would not violate the first law as long as the amount of energy lost by the air is equal to the amount gained by the coffee.

As another familiar example, consider the heating of a room by the passage of current through an electric resistor (Fig. 5-2). Again, the first law dictates that the amount of electric energy supplied to the resistance wires be equal to the amount of energy transferred to the room air as heat. Now let us attempt to reverse this process. It will come as no surprise that transferring some heat to the wires will not cause an equivalent amount of electric energy to be generated in the wires, even though doing so would not violate the first law.

Finally, consider a paddle-wheel mechanism that is operated by the fall of a mass (Fig. 5-3). The paddle wheel rotates as the mass falls and stirs a fluid within an insulated container. As a result, the potential energy of the mass decreases, and the internal energy of the fluid increases in accordance with the conservation of energy principle. However, the reverse process, raising the mass by transferring heat from the fluid to the paddle wheel, does not occur in nature, although doing so would not violate the first law of thermodynamics.

It is clear from the above arguments that processes proceed in a *certain direction* and not in the reverse direction (Fig. 5-4). The first law places no restriction on the direction of a process, but satisfying the first law does not ensure that that process will actually occur. This inadequacy of the first law to identify whether a process can take place is remedied by introducing another general principle, the *second law of thermodynamics.* We show later in this chapter that the reverse processes discussed above violate the second law of thermodynamics. This violation is easily detected with the help of a property, called *entropy,* defined in the next chapter. *A process will not occur unless it satisfies both the first and the second laws of thermodynamics* (Fig. 5-5).

There are numerous valid statements of the second law of thermodynamics. Two such statements are presented and discussed later in this chapter in relation to some engineering devices that operate on cycles.

The use of the second law of thermodynamics is not limited to identifying the direction of processes, however. The second law also asserts that energy has *quality* as well as quantity. The first law is concerned with the quantity of energy and the transformations of energy from one form to another with no regard to its quality. Preserving the quality of energy is a major concern to engineers, and the second law provides the necessary means to determine the quality as well as the degree of degradation of energy during a process. As discussed later in this chapter, more of high-temperature energy can be converted to work, and thus it has a higher quality than the same amount of energy at a lower temperature.

The second law of thermodynamics is also used in determining the *theoretical limits* for the performance of commonly used engineering systems, such as heat engines and refrigerators, as well as predicting the *degree of completion* of chemical reactions.

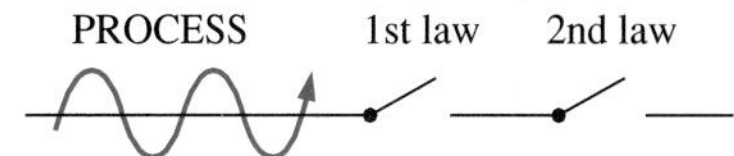

FIGURE 5-5

A process must satisfy both the first and second laws of thermodynamics to proceed.

5-2 ■ THERMAL ENERGY RESERVOIRS

In the development of the second law of thermodynamics, it is very convenient to have a hypothetical body with a relatively large *thermal energy capacity* (mass × specific heat) that can supply or absorb finite amounts of heat without undergoing any change in temperature. Such a body is called a **thermal energy reservoir,** or just a **reservoir.** In practice, large bodies of water such as oceans, lakes, and rivers as well as the atmospheric air can be modeled accurately as thermal energy reservoirs because of their large thermal energy storage capabilities or thermal masses (Fig. 5-6). The *atmosphere,* for example, does not warm up as a result of heat losses from residential buildings in winter. Likewise, megajoules of waste energy dumped in large rivers by power plants do not cause any significant change in water temperature.

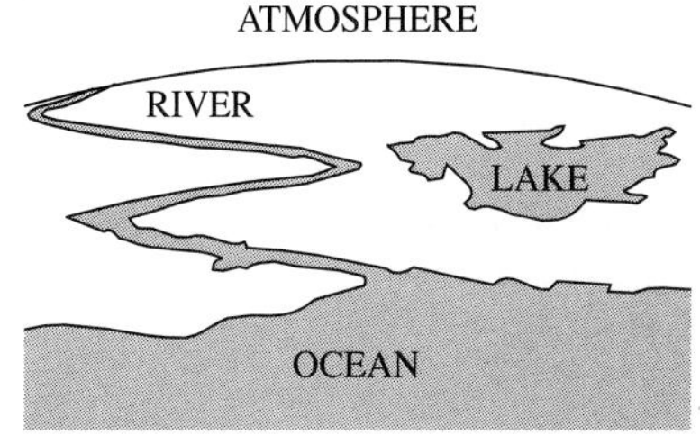

FIGURE 5-6

Bodies with relatively large thermal masses can be modeled as thermal energy reservoirs.

A *two-phase system* can be modeled as a reservoir also since it can absorb and release large quantities of heat while remaining at constant temperature. Another familiar example of a thermal energy reservoir is the *industrial furnace*. The temperatures of most furnaces are carefully controlled, and they are capable of supplying large quantities of thermal energy as heat in an essentially isothermal manner. Therefore, they can be modeled as reservoirs.

A body does not actually have to be very large to be considered a reservoir. Any physical body whose thermal energy capacity is large relative to the amount of energy it supplies or absorbs can be modeled as one. The air in a room, for example, can be treated as a reservoir in the analysis of the heat dissipation from a TV set in the room, since the amount of heat transfer from the TV set to the room air is not large enough to have a noticeable effect on the room air temperature.

A reservoir that supplies energy in the form of heat is called a **source,** and one that absorbs energy in the form of heat is called a **sink** (Fig. 5-7). Thermal energy reservoirs are often referred to as **heat reservoirs** since they supply or absorb energy in the form of heat.

FIGURE 5-7

A source supplies energy in the form of heat, and a sink absorbs it.

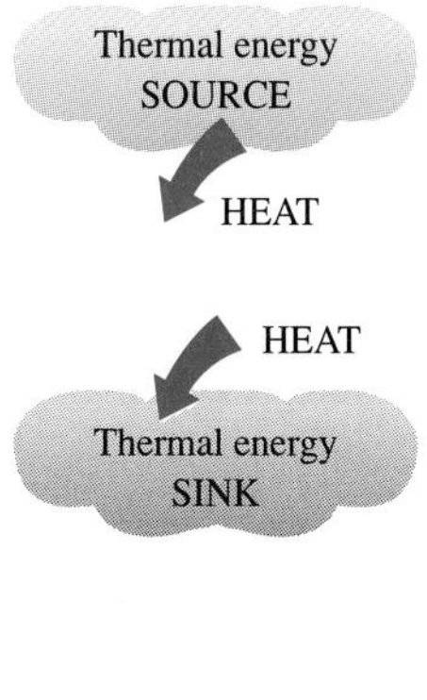

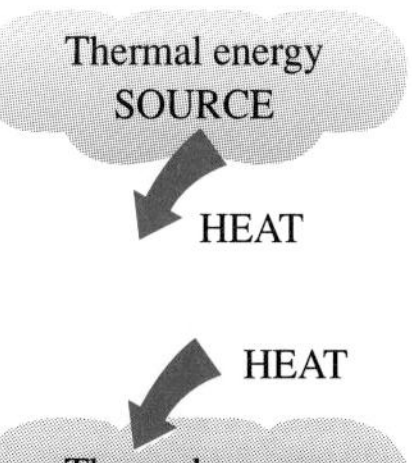

Heat transfer from industrial sources to the environment is of major concern to environmentalists as well as to engineers. Irresponsible management of waste energy can significantly increase the temperature of portions of the environment, causing what is called *thermal pollution*. If it is not carefully controlled, thermal pollution can seriously disrupt marine life in lakes and rivers. However, by careful design and management, the waste energy dumped into large bodies of water can be used to improve the quality of marine life by keeping the local temperature increases within safe and desirable levels.

5-3 ■ HEAT ENGINES

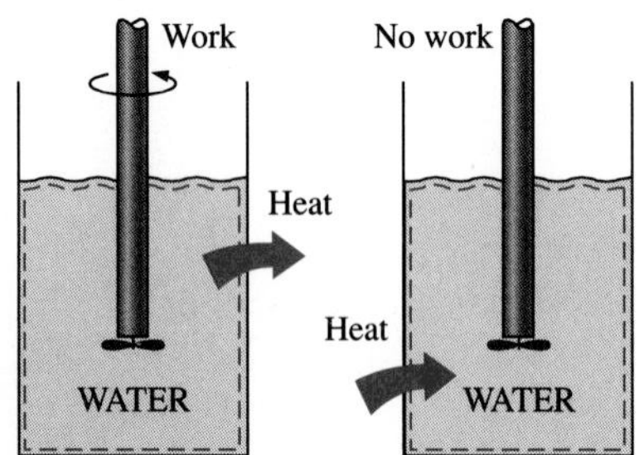

FIGURE 5-8

Work can always be converted to heat directly and completely, but the reverse is not true.

As pointed out in Sec. 5-1, work can easily be converted to other forms of energy, but converting other forms of energy to work is not that easy. The mechanical work done by the shaft shown in Fig. 5-8, for example, is first converted to the internal energy of the water. This energy may then leave the water as heat. We know from experience that any attempt to reverse this process will fail. That is, transferring heat to the water will not cause the shaft to rotate. From this and other observations, we conclude that work can be converted to heat directly and completely, but converting heat to work requires the use of some special devices. These devices are called **heat engines.**

Heat engines differ considerably from one another, but all can be characterized by the following (Fig. 5-9):

1. They receive heat from a high-temperature source (solar energy, oil furnace, nuclear reactor, etc.).
2. They convert part of this heat to work (usually in the form of a rotating shaft).
3. They reject the remaining waste heat to a low-temperature sink (the atmosphere, rivers, etc.).
4. They operate on a cycle.

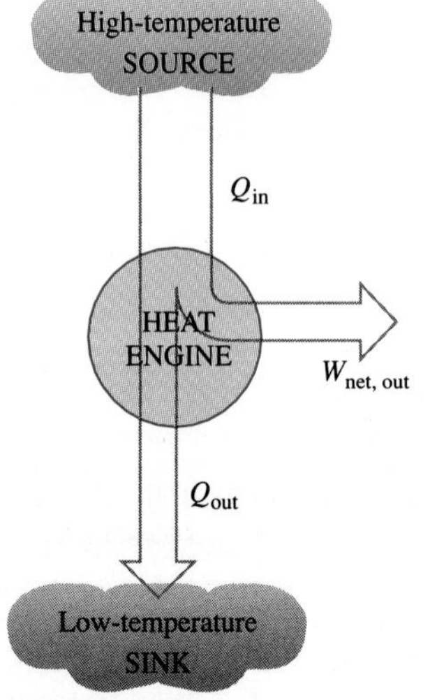

FIGURE 5-9

Part of the heat received by a heat engine is converted to work, while the rest is rejected to a sink.

Heat engines and other cyclic devices usually involve a fluid to and from which heat is transferred while undergoing a cycle. This fluid is called the **working fluid.**

The term *heat engine* is often used in a broader sense to include work-producing devices that do not operate in a thermodynamic cycle. Engines that involve internal combustion such as gas turbines and car engines fall into this category. These devices operate in a mechanical cycle but not in a thermodynamic cycle since the working fluid (the combustion gases) does not undergo a complete cycle. Instead of being cooled to the initial temperature, the exhaust gases are purged and replaced by fresh air-and-fuel mixture at the end of the cycle.

The work-producing device that best fits into the definition of a heat engine is the *steam power plant,* which is an external-combustion engine. That is, combustion takes place outside the engine, and the thermal energy

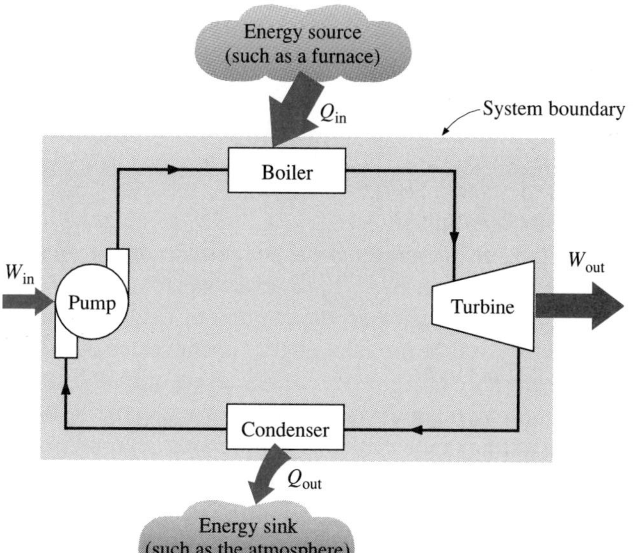

FIGURE 5-10
Schematic of a steam power plant.

released during this process is transferred to the steam as heat. The schematic of a basic steam power plant is shown in Fig. 5-10. This is a rather simplified diagram, and the discussion of actual steam power plants with all their complexities is left to Chap. 9. The various quantities shown on this figure are as follows:

Q_{in} = amount of heat supplied to steam in boiler from a high-temperature source (furnace)
Q_{out} = amount of heat rejected from steam in condenser to a low-temperature sink (the atmosphere, a river, etc.)
W_{out} = amount of work delivered by steam as it expands in turbine
W_{in} = amount of work required to compress water to boiler pressure

Notice that the directions of the heat and work interactions are indicated by the subscripts *in* and *out*. Therefore, all four quantities described above are always *positive*.

The net work output of this power plant is simply the difference between the total work output of the plant and the total work input (Fig. 5-11):

$$W_{net,out} = W_{out} - W_{in} \qquad \text{(kJ)} \qquad (5\text{-}1)$$

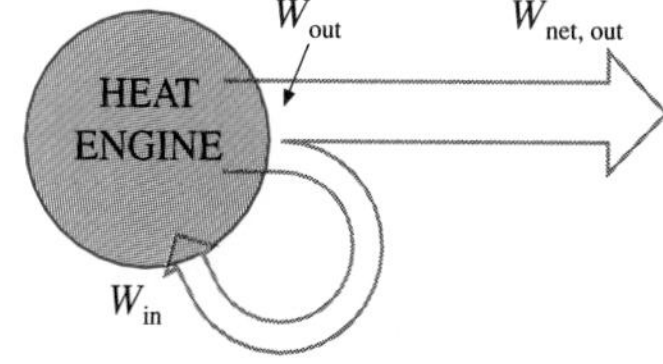

FIGURE 5-11
A portion of the work output of a heat engine is consumed internally to maintain continuous operation.

The net work can also be determined from the heat transfer data alone. The four components of the steam power plant involve mass flow in and out, and therefore they should be treated as open systems. These components, together with the connecting pipes, however, always contain the same fluid (not counting the steam that may leak out, of course). No mass enters or leaves this combination system, which is indicated by the shaded area on Fig. 5-11; thus, it can be analyzed as a closed system. Recall that for a closed system undergoing a cycle, the change in internal energy ΔU is zero, and therefore the net work output of the system is also equal to the net heat transfer to the system:

$$W_{\text{net, out}} = Q_{\text{in}} - Q_{\text{out}} \quad \text{(kJ)} \tag{5-2}$$

Thermal Efficiency

In Eq. 5-2, Q_{out} represents the magnitude of the energy wasted in order to complete the cycle. But Q_{out} is never zero; thus, the net work output of a heat engine is always less than the amount of heat input. That is, only part of the heat transferred to the heat engine is converted to work. *The fraction of the heat input that is converted to net work output is a measure of the performance of a heat engine and is called the* **thermal efficiency** η_{th} (Fig. 5-12).

Performance or efficiency, in general, can be expressed in terms of the desired output and the required input as (Fig. 5-13)

$$\text{Performance} = \frac{\text{Desired output}}{\text{Required input}} \tag{5-3}$$

For heat engines, the desired output is the net work output, and the required input is the amount of heat supplied to the working fluid. Then the thermal efficiency of a heat engine can be expressed as

$$\text{Thermal efficiency} = \frac{\text{Net work output}}{\text{Total heat input}} \tag{5-4}$$

or

$$\eta_{\text{th}} = \frac{W_{\text{net, out}}}{Q_{\text{in}}}$$

It can also be expressed as

$$\eta_{\text{th}} = 1 - \frac{Q_{\text{out}}}{Q_{\text{in}}} \tag{5-5}$$

since $W_{\text{net, out}} = Q_{\text{in}} - Q_{\text{out}}$.

Cyclic devices of practical interest such as heat engines, refrigerators, and heat pumps operate between a high-temperature medium (or reservoir) at temperature T_H and a low-temperature medium (or reservoir) at temperature T_L. To bring uniformity to the treatment of heat engines, refrigerators, and heat pumps, we define the following two quantities:

Q_H = magnitude of heat transfer between the cyclic device and the high-temperature medium at temperature T_H

Q_L = magnitude of heat transfer between the cyclic device and the low-temperature medium at temperature T_L

Notice that both Q_L and Q_H are defined as *magnitudes* and therefore are positive quantities. The direction of Q_H and Q_L is easily determined by in-

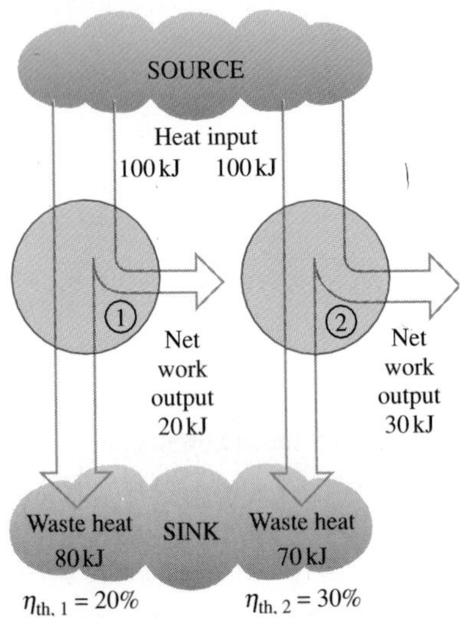

FIGURE 5-12
Some heat engines perform better than others (convert more of the heat they receive to work).

FIGURE 5-13
The definition of performance is not limited to thermodynamics only.

spection. Then the net work output and thermal efficiency relations for any heat engine (shown in Fig. 5-14) can also be expressed as

$$W_{\text{net, out}} = Q_H - Q_L$$

and

$$\eta_{\text{th}} = \frac{W_{\text{net, out}}}{Q_H}$$

or

$$\eta_{\text{th}} = 1 - \frac{Q_L}{Q_H} \qquad (5\text{-}6)$$

The thermal efficiency of a heat engine is always less than unity since both Q_L and Q_H are defined as positive quantities.

Thermal efficiency is a measure of how efficiently a heat engine converts the heat that it receives to work. Heat engines are built for the purpose of converting heat to work, and engineers are constantly trying to improve the efficiencies of these devices since increased efficiency means less fuel consumption and thus lower fuel bills and less pollution.

The thermal efficiencies of work-producing devices are relatively low. Ordinary spark-ignition automobile engines have a thermal efficiency of about 25 percent. That is, an automobile engine converts, at an average, about 25 percent of the chemical energy of the gasoline to mechanical work. This number is about 35 percent for diesel engines and large gas-turbine plants and 50 percent for large combined gas-steam power plants. Thus, even with the most efficient heat engines available today, about one-half of the energy supplied ends up in the rivers, lakes, or the atmosphere as waste or useless energy (Fig. 5-15).

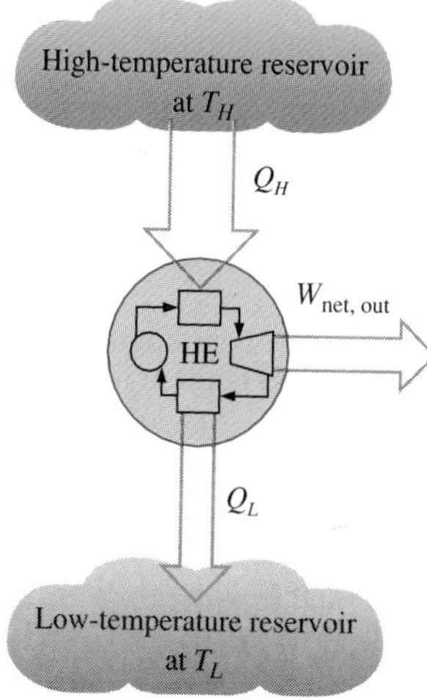

FIGURE 5-14
Schematic of a heat engine.

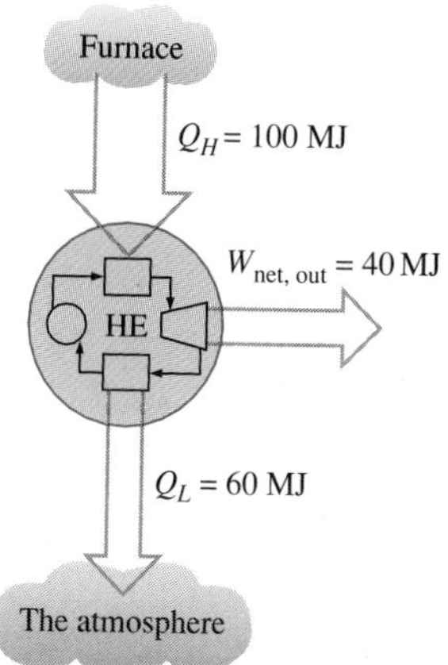

FIGURE 5-15
Even the most efficient heat engines reject most of the heat they receive as waste heat.

Can We Save Q_{out}?

In a steam power plant, the condenser is the device where large quantities of waste heat is rejected to rivers, lakes, or the atmosphere. Then one may ask, can we not just take the condenser out of the plant and save all that waste energy? The answer to this question is, unfortunately, a firm *no* for the simple reason that without a heat rejection process in a condenser, the cycle cannot be completed. (Cyclic devices such as steam power plants cannot run continuously unless the cycle is completed.) This is demonstrated below with the help of a simple heat engine.

Consider the simple heat engine shown in Fig. 5-16 that is used to lift weights. It consists of a piston-cylinder device with two sets of stops. The working fluid is the gas contained within the cylinder. Initially, the gas temperature is 30°C. The piston, which is loaded with the weights, is resting on top of the lower stops. Now 100 kJ of heat is transferred to the gas in the cylinder from a source at 100°C, causing it to expand and to raise the loaded piston until the piston reaches the upper stops, as shown in the figure. At this point, the load is removed, and the gas temperature is observed to be 90°C.

The work done on the load during this expansion process is equal to the increase in its potential energy, say 15 kJ. Even under ideal conditions

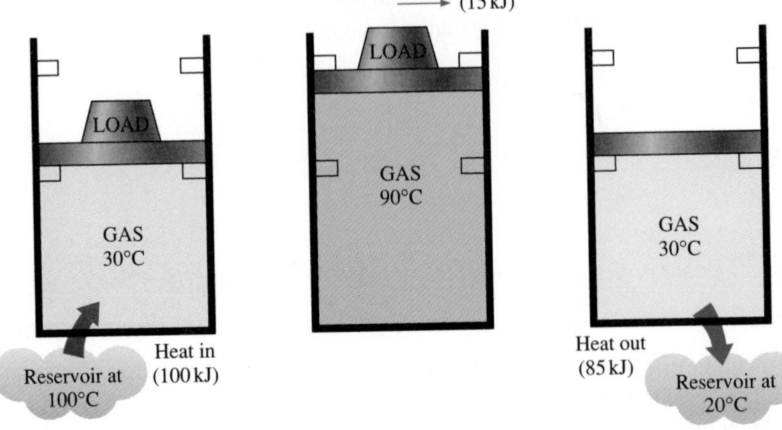

FIGURE 5-16

A heat-engine cycle cannot be completed without rejecting some heat to a low-temperature sink.

(weightless piston, no friction, no heat losses, and quasi-equilibrium expansion), the amount of heat supplied to the gas is greater than the work done since part of the heat supplied is used to raise the temperature of the gas.

Now let us try to answer the following question: *Is it possible to transfer the 85 kJ of excess heat at 90°C back to the reservoir at 100°C for later use?* If it is, then we will have a heat engine that can have a thermal efficiency of 100 percent under ideal conditions. The answer to this question is again *no,* for the very simple reason that heat always flows from a high-temperature medium to a low-temperature one, and never the other way around. Therefore, we cannot cool this gas from 90 to 30°C by transferring heat to a reservoir at 100°C. Instead, we have to bring the system into contact with a low-temperature reservoir, say at 20°C, so that the gas can return to its initial state by rejecting its 85 kJ of excess energy as heat to this reservoir. This energy cannot be recycled, and it is properly called *waste energy.*

We conclude from the above discussion that every heat engine must *waste* some energy by transferring it to a low-temperature reservoir in order to complete the cycle, even under idealized conditions. The requirement that a heat engine exchange heat with at least two reservoirs for continuous operation forms the basis for the Kelvin-Planck expression of the second law of thermodynamics discussed later in this section.

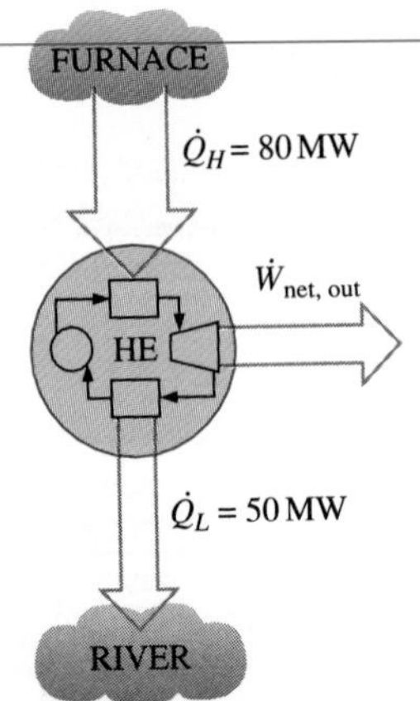

FIGURE 5-17
Schematic for Example 5-1.

EXAMPLE 5-1 Net Power Production of a Heat Engine

Heat is transferred to a heat engine from a furnace at a rate of 80 MW. If the rate of waste heat rejection to a nearby river is 50 MW, determine the net power output and the thermal efficiency for this heat engine.

Solution A schematic of the heat engine is given in Fig. 5-17. The furnace serves as the high-temperature reservoir for this heat engine and the river as the low-temperature reservoir.

Assumptions Heat losses through the pipes and other components are negligible.

Analysis The given quantities can be expressed in rate form as

$$\dot{Q}_H = 80 \text{ MW and } \dot{Q}_L = 50 \text{ MW}$$

The net power output of this heat engine is

$$\dot{W}_{\text{net, out}} = \dot{Q}_H - \dot{Q}_L = (80 - 50) \text{ MW} = 30 \text{ MW}$$

Then the thermal efficiency is easily determined to be

$$\eta_{\text{th}} = \frac{\dot{W}_{\text{net, out}}}{\dot{Q}_H} = \frac{30 \text{ MW}}{80 \text{ MW}} = \mathbf{0.375} \text{ (or 37.5\%)}$$

That is, the heat engine converts 37.5 percent of the heat it receives to work.

EXAMPLE 5-2 Fuel Consumption Rate of a Car

A car engine with a power output of 65 hp has a thermal efficiency of 24 percent. Determine the fuel consumption rate of this car if the fuel has a heating value of 19,000 Btu/lbm (that is, 19,000 Btu of energy is released for each lbm of fuel burned).

Solution A schematic of the car engine is given in Fig. 5-18. The car engine is powered by converting 24 percent of the chemical energy released during the combustion process to work.

Assumptions The power output of the car is constant.

Analysis The amount of energy input required to produce a power output of 65 hp is determined from the definition of thermal efficiency to be

$$\dot{Q}_H = \frac{\dot{W}_{\text{net, out}}}{\eta_{\text{th}}} = \frac{65 \text{ hp}}{0.24}\left(\frac{2545 \text{ Btu/h}}{1 \text{ hp}}\right) = 689{,}262 \text{ Btu/h}$$

To supply energy at this rate, the engine must burn fuel at a rate of

$$\dot{m} = \frac{689{,}262 \text{ Btu/h}}{19{,}000 \text{ Btu/lbm}} = \mathbf{36.3 \text{ lbm/h}}$$

since 19,000 Btu of thermal energy is released for each lbm of fuel burned.

$\dot{m}_{\text{fuel}}$

Combustion chamber

$\dot{Q}_H$

CAR ENGINE (idealized)

$\dot{W}_{\text{net, out}} = 65 \text{ hp}$

$\dot{Q}_L$

Atmosphere

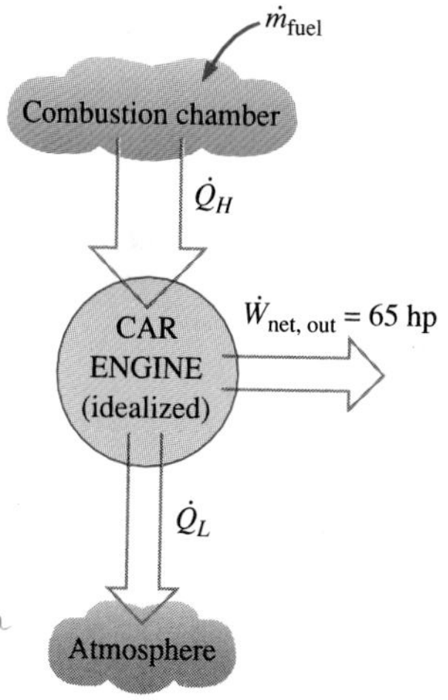

FIGURE 5-18
Schematic for Example 5-2.

The Second Law of Thermodynamics: Kelvin-Planck Statement

We have demonstrated earlier with reference to the heat engine shown in Fig. 5-16 that, even under ideal conditions, a heat engine must reject some heat to a low-temperature reservoir in order to complete the cycle. That is, no heat engine can convert all the heat it receives to useful work. This limitation on the thermal efficiency of heat engines forms the basis for the Kelvin-Planck statement of the second law of thermodynamics, which is expressed as follows:

> *It is impossible for any device that operates on a cycle to receive heat from a single reservoir and produce a net amount of work.*

That is, a heat engine must exchange heat with a low-temperature sink as well as a high-temperature source to keep operating. The Kelvin-Planck statement can also be expressed as *no heat engine can have a thermal efficiency of 100 percent* (Fig. 5-19), or as *for a power plant to operate, the working fluid must exchange heat with the environment as well as the furnace.*

Note that the impossibility of having a 100 percent efficient heat engine is not due to friction or other dissipative effects. It is a limitation that applies to both the idealized and the actual heat engines. Later in this chapter, we develop a relation for the maximum thermal efficiency of a heat engine. We also demonstrate that this maximum value depends on the reservoir temperatures only.

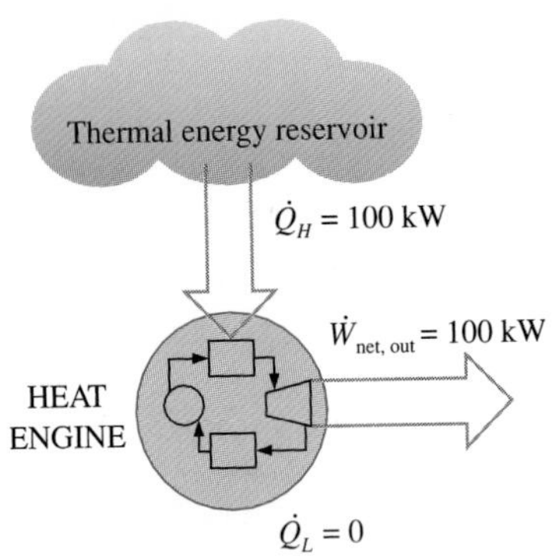

FIGURE 5-19
A heat engine that violates the Kelvin-Planck statement of the second law.

5-4 ■ ENERGY CONVERSION EFFICIENCIES

Efficiency is one of the most frequently used terms in thermodynamics, and it indicates how well an energy conversion or transfer process is accomplished. The *thermal efficiency* of a heat engine, for example, is the fraction of the thermal energy a heat engine converts to work. Efficiency is also one of the most frequently misused terms in thermodynamics and a source of misunderstandings. This is because efficiency is often used without being properly defined first. Below we will clarify this further, and define some efficiencies commonly used in practice.

If you are shopping for a water heater, a knowledgeable salesperson will tell you that the efficiency of a conventional electric water heater is about 90 percent (Fig. 5-20). You may find this confusing, since the heating elements of electric water heaters are resistance heaters, and the efficiency of all resistance heaters is 100 percent as they convert all the electrical energy they consume into heat. A knowledgeable salesperson will clarify this by explaining that the heat losses from the hot water tank to the surrounding air amount to 10 percent of the electrical energy consumed, and the **efficiency of a water heater** is defined as the ratio of the *energy delivered to the house by hot water* to the *energy supplied to the water heater.* A clever salesperson may even talk you into buying a more expensive water heater with thicker insulation that has an efficiency of 94 percent. If you are a knowledgeable consumer and have access to natural gas, you will probably purchase a gas water heater whose efficiency is only 55 percent since a gas unit costs about the same as an electric unit to purchase and install, but the annual energy cost of a gas unit will be less than half of that of an electric unit at national average electricity and gas prices.

FIGURE 5-20
Typical efficiencies of conventional and high-efficiency electric and natural gas water heaters.

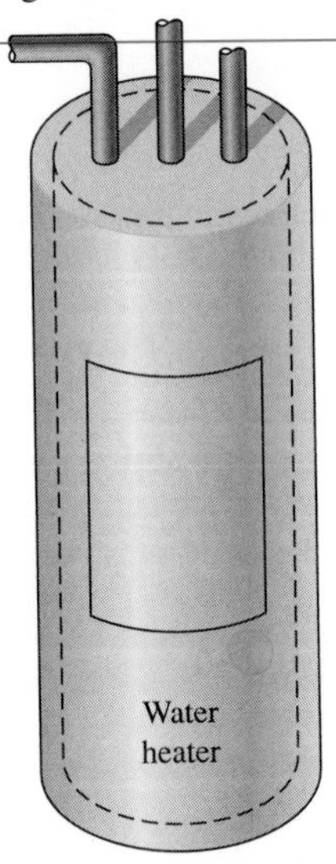

Type	Efficiency
Gas, conventional	55%
Gas, high efficiency	62%
Elect., conventional	90%
Elect., high-efficiency	94%

Perhaps you are wondering how the efficiency for a gas water heater is defined, and why it is much lower than the efficiency of an electric heater. As a general rule, the efficiency of equipment that involves the combustion of a fuel is based on the **heating value of the fuel,** which is *the amount of heat released when a specified amount of fuel (usually a unit mass) at room temperature is completely burned and the combustion products are cooled to the room temperaturer* (Figure 5-21). Then the performance of combustion equipment can be characterized by **combustion efficiency,** defined as

$$\eta_{\text{combustion}} = \frac{Q}{\text{HV}} = \frac{\text{Amount of heat released during combustion}}{\text{Heating value of the fuel burned}} \tag{5-7}$$

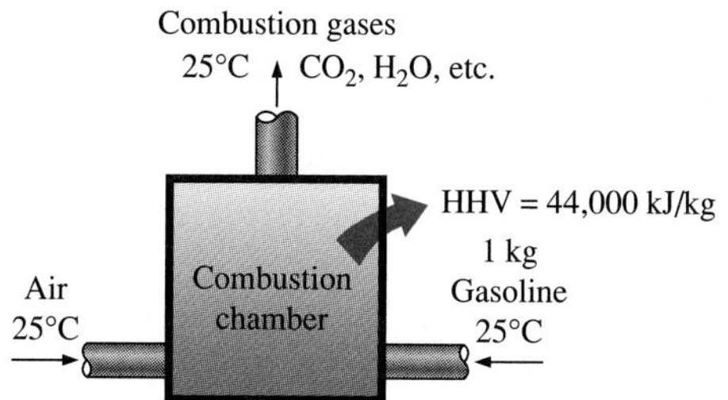

FIGURE 5-21
The definition of the heating value of gasoline.

A combustion efficiency of 100 percent indicates that the fuel is burned completely and the stack gases leave the combustion chamber at room temperature, and thus the amount of heat released during a combustion process is equal to the heating value of the fuel.

Most fuels contain hydrogen, which forms water when burned, and the heating value of a fuel will be different, depending on whether the water in combustion products is in the liquid or vapor form. The heating value is called the *lower heating value,* or LHV, when the water leaves as a vapor, and the *higher heating value,* or HHV, when the water in the combustion gases is completely condensed and thus the heat of vaporization is also recovered. The difference between these two heating values is equal to the product of the amount of water and the enthalpy of vaporization of water at room temperature. For example, the lower and higher heating values of gasoline are 44,000 kJ/kg and 47,300 kJ/kg, respectively. An efficiency definition should make it clear whether it is based on the higher or lower heating value of the fuel. Efficiencies of cars and jet engines are normally based on *lower heating values* since water normally leaves as a vapor in the exhaust gases, and it is not practical to try to recuperate the heat of vaporization. Efficiencies of furnaces, on the other hand, are based on *higher heating values.*

The efficiency of space heating systems of residential and commercial buildings is usually expressed in terms of the **annual fuel utilization efficiency,** or **AFUE,** which accounts for the combustion efficiency as well as other losses such as heat losses to unheated areas and start-up and cool-down losses. The AFUE of most new heating systems is close to 85 percent, although the AFUE of some old heating systems is under 60 percent. The AFUE of some new high-efficiency furnaces exceeds 96 percent, but the high cost of such furnaces cannot be justified for locations with mild to moderate winters. Such high efficiencies are achieved by reclaiming most of the heat in the flue gases, condensing the water vapor, and discharging the flue gases at temperatures as low as 38°C (or 100°F) instead of about 200°C (or 400°F) for the conventional models.

For *car engines,* the work output is understood to be the power delivered by the crankshaft. But for power plants, the work output can be the mechanical power at the turbine exit, or the electrical power output of the generator.

A generator is a device that converts mechanical energy to electrical energy, and the effectiveness of a generator is characterized by the **generator efficiency,** which is the ratio of the *electrical power output* to the *mechanical power input.* The *thermal efficiency* of a power plant, which is of primary interest in thermodynamics, is usually defined as the ratio of the shaft work output of the turbine to the heat input to the working fluid. The effects of other factors are incorporated by defining an **overall efficiency** for the power plant as the ratio of the *net electric power output* to the *rate of fuel energy input*. That is,

$$\eta_{\text{overall}} = \eta_{\text{combustion}}\,\eta_{\text{thermal}}\,\eta_{\text{generator}} = \frac{\dot{W}_{\text{net, electric}}}{\text{HHV} \times \dot{m}_{\text{fuel}}} \tag{5-8}$$

The overall efficiencies are about 25–28 percent for gasoline automotive engines, 34–38 percent for diesel engines, and 40–60 percent for large power plants.

Electrical energy is commonly converted to *rotating mechanical energy* by electric motors to drive fans, compressors, robot arms, car starters, and so forth. The effectiveness of this conversion process is characterized by the **motor efficiency** η_{motor}, which is the ratio of the *mechanical energy output* of the motor to the *electrical energy input*. The full-load motor efficiencies range from about 35 percent for small motors to over 96 percent for large high-efficiency motors. The difference between the electrical energy consumed and the mechanical energy delivered is dissipated as waste heat.

We are all familiar with the conversion of electrical energy to *light* by incandescent light bulbs, fluorescent tubes, and high-intensity discharge lamps. The efficiency for the conversion of electricity to light can be defined as the ratio of the energy converted to light to the electrical energy consumed. For example, common incandescent light bulbs convert about 10 percent of the electrical energy they consume to light; the rest of the energy consumed is dissipated as heat, which adds to the cooling load of the air conditioner in summer. However, it is more common to express the effectiveness of this conversion process by **lighting efficacy,** which is defined as the *amount of light output in lumens per W of electricity consumed*.

TABLE 5-1

The efficacy of different lighting systems

Type of lighting	Efficacy, lumens/W
Combustion	
Candle	0.2
Incandescent	
Ordinary	5–20
Halogen	15–25
Fluorescent	
Ordinary	40–60
High output	70–90
Compact	50–80
High intensity discharge	
Mercury vapor	50–60
Metal halide	55–125
High-pressure sodium	100–150
Low-pressure sodium	up to 200

The efficacy of different lighting systems is given in Table 5-1. Note that a compact fluorescent light bulb produces about four times as much light as an incandescent light bulb per W, and thus a 15-W fluorescent bulb can replace a 60-W incandescent light bulb (Fig. 5-22). Also, a compact fluorescent bulb last about 10,000 h, which is 10 times as long as an incandescent bulb, and it plugs directly into the socket of an incandescent lamp. Therefore, despite their higher initial cost, compact fluorescents reduce the lighting costs considerably through reduced electricity consumption. Sodium-filled high intensity discharge lamps provide the most efficient lighting, but their use is limited to outdoor use because of their yellowish light.

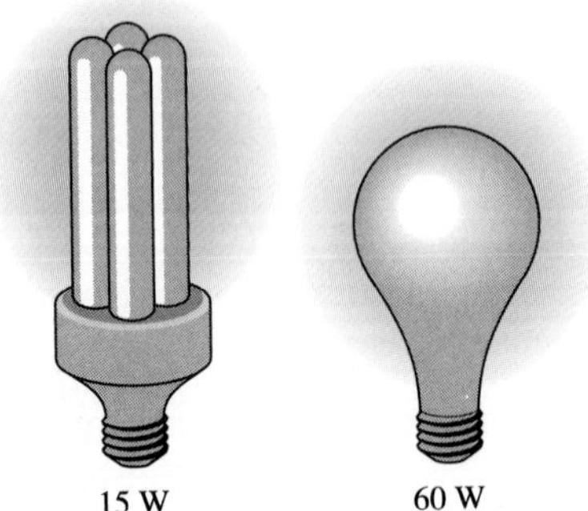

FIGURE 5-22
A 15-W compact fluorescent lamp provides as much light as a 60-W incandescent lamp.

We can also define efficiency for cooking appliances since they convert electrical or chemical energy to heat for cooking. The **efficiency of a cooking appliance** can be defined as the ratio of the *useful energy transferred to the food* to the *energy consumed by the appliance* (Fig. 5-23). Electric ranges are more efficient than gas ranges, but it is much cheaper to cook with natural gas than with electricity because of the lower unit cost of natural gas (Table 5-2).

$$\text{Efficiency} = \frac{\text{Energy utilized}}{\text{Energy supplied to appliance}}$$

$$= \frac{3\text{ kWh}}{5\text{ kWh}} = 0.60$$

FIGURE 5-23

The efficiency of a cooking appliance represents the fraction of the energy supplied to the appliance that is transferred to the food.

TABLE 5-2

Energy costs of cooking a casserole with different appliances*
[from A. Wilson and J. Morril, *Consumer Guide to Home Energy Savings,* Washington, D.C.: American Council for an Energy-Efficient Economy, 1996, p. 192.]

Cooking appliance	Cooking temperature	Cooking time	Energy used	Cost of energy
Electric oven	350°F (177°C)	1 h	2.0 kWh	$0.16
Convection oven (elect.)	325°F (163°C)	45 min	1.39 kWh	$0.11
Gas oven	350°F (177°C)	1 h	0.112 therm	$0.07
Frying pan	420°F (216°C)	1 h	0.9 kWh	$0.07
Toaster oven	425°F (218°C)	50 min	0.95 kWh	$0.08
Crockpot	200°F (93°C)	7 h	0.7 kWh	$0.06
Microwave oven	"High"	15 min	0.36 kWh	$0.03

*Assumes a unit cost of $0.08/kWh for electricity and $0.60/therm for gas.

The cooking efficiency depends on user habits as well as the individual appliances. Convection and microwave ovens are inherently more efficient than conventional ovens. On average, convection ovens save about *one-third* and microwave ovens save about *two-thirds* of the energy used by conventional ovens. The cooking efficiency can be increased by using the smallest oven for baking, using a pressure cooker, using a crockpot for stews and soups, using the smallest pan that will do the job, using the smaller heating element for small pans on electric ranges, using flat-bottomed pans on electric burners to assure good contact, keeping burner drip pans clean and shiny, defrosting frozen foods in the refrigerator before cooking, avoiding preheating unless it is necessary, keeping the pans covered during cooking, using timers and thermometers to avoid overcooking, using the self-cleaning feature of ovens right after cooking, and keeping inside surfaces of microwave ovens clean.

Using energy-efficient appliances and practicing energy conservation measures help our pocketbooks by reducing our utility bills. It will also help the **environment** by reducing the amount of pollutants emitted to the atmosphere during the combustion of fuel at home or at the power plants where electricity is generated. The combustion of *each therm of natural gas* produces 6.4 kg of carbon dioxide, which causes global climate change; 4.7 g of nitrogen oxides and 0.54 g of hydrocarbons, which cause smog; 2.0 g of carbon monoxide, which is toxic; and 0.030 g of sulfur dioxide, which causes acid rain. Each therm of natural gas saved eliminates the emission of these pollutants while saving \$0.60 for the average consumer in the United States. Each kWh of electricity saved will save 0.4 kg of coal and 1.0 kg of CO_2 and 15 g of SO_2 from a coal power plant.

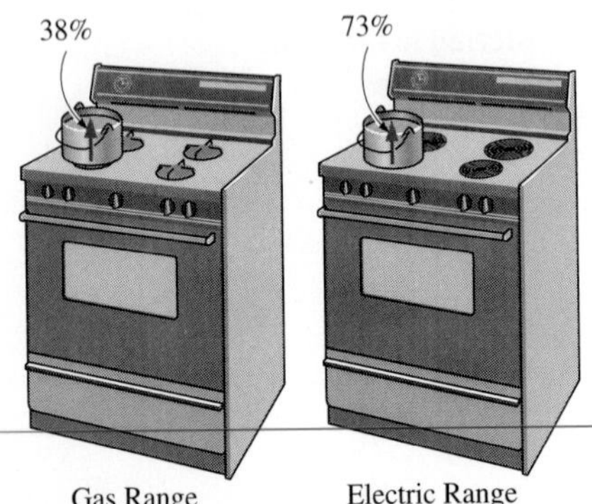

FIGURE 5-24

Schematic of the 73 percent efficient electric heating unit and 38 percent efficient gas burner discussed in Example 5-3.

EXAMPLE 5-3 Cost of Cooking with Electric and Gas Ranges

The efficiency of cooking appliances affects the internal heat gain from them since an inefficient appliance consumes a greater amount of energy for the same task, and the excess energy consumed shows up as heat in the living space. The efficiency of open burners is determined to be 73 percent for electric units and 38 percent for gas units (Fig. 5-24). Consider a 2-kW electric burner in an area where the unit costs of electricity and natural gas are \$0.09/kWh and \$0.55/therm, respectively. Determine the rate of energy consumption by the burner and the unit cost of utilized energy for both electric and gas burners.

Solution This example is to demonstrate the economics of electric and gas ranges.

Analysis The efficiency of the electric heater is given to be 73 percent. Therefore, a burner that consumes 2 kW of electrical energy will supply

$$\dot{Q}_{\text{utilized}} = (\text{Energy input}) \times (\text{Efficiency}) = (2\text{ kW})(0.73) = \mathbf{1.46\ kW}$$

of useful energy. The unit cost of utilized energy is inversely proportional to the efficiency, and is determined from

$$\text{Cost of utilized energy} = \frac{\text{Cost of energy input}}{\text{Efficiency}} = \frac{\$0.09/\text{kWh}}{0.73} = \mathbf{\$0.123/kWh}$$

Noting that the efficiency of a gas burner is 38 percent, the energy input to a gas burner that supplies utilized energy at the same rate (1.46 kW) is

$$\dot{Q}_{\text{input, gas}} = \frac{\dot{Q}_{\text{utilized}}}{\text{Efficiency}} = \frac{1.46\text{ kW}}{0.38} = \mathbf{3.84\ kW} \qquad (= 13{,}100\text{ Btu/h})$$

since 1 kW = 3412 Btu/h. Therefore, a gas burner should have a rating of at least 13,100 Btu/h to perform as well as the electric unit.

Noting that 1 therm = 29.3 kWh, the unit cost of utilized energy in the case of a gas burner is determined to be

$$\text{Cost of utilized energy} = \frac{\text{Cost of energy input}}{\text{Efficiency}} = \frac{\$0.55/29.3\text{ kWh}}{0.38} = \mathbf{\$0.049/kWh}$$

Discussion The cost of utilized gas is less than half of the unit cost of utilized electricity. Therefore, despite its higher efficiency, cooking with an electric burner will cost more than twice as much compared to a gas burner in this case. This explains why cost-conscious consumers always ask for gas appliances, and it is not wise to use electricity for heating purposes.

5-5 ■ REFRIGERATORS AND HEAT PUMPS

We all know from experience that heat flows in the direction of decreasing temperature, that is, from high-temperature mediums to low-temperature ones. This heat transfer process occurs in nature without requiring any devices. The reverse process, however, cannot occur by itself. The transfer of heat from a low-temperature medium to a high-temperature one requires special devices called **refrigerators.**

Refrigerators, like heat engines, are cyclic devices. The working fluid used in the refrigeration cycle is called a **refrigerant.** The most frequently used refrigeration cycle is the *vapor-compression refrigeration cycle,* which involves four main components: a compressor, a condenser, an expansion valve, and an evaporator, as shown in Fig. 5-25.

The refrigerant enters the compressor as a vapor and is compressed to the condenser pressure. It leaves the compressor at a relatively high temperature and cools down and condenses as it flows through the coils of the condenser by rejecting heat to the surrounding medium. It then enters a capillary tube where its pressure and temperature drop drastically due to the throttling effect. The low-temperature refrigerant then enters the evaporator, where it evaporates by absorbing heat from the refrigerated space. The cycle is completed as the refrigerant leaves the evaporator and reenters the compressor.

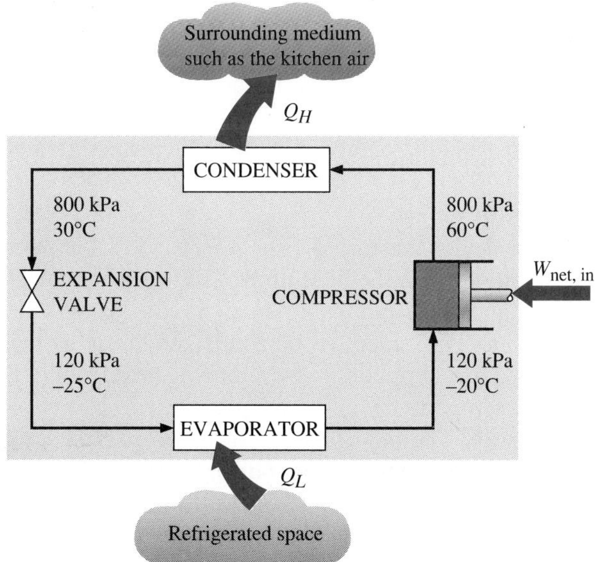

FIGURE 5-25

Basic components of a refrigeration system and typical operating conditions.

In a household refrigerator, the freezer compartment where heat is picked up by the refrigerant serves as the evaporator, and the coils behind the refrigerator where heat is dissipated to the kitchen air serve as the condenser.

A refrigerator is shown schematically in Fig. 5-26. Here Q_L is the magnitude of the heat removed from the refrigerated space at temperature T_L, Q_H is the magnitude of the heat rejected to the warm environment at temperature T_H, and $W_{net, in}$ is the net work input to the refrigerator. As discussed before, Q_L and Q_H represent magnitudes and thus are positive quantities.

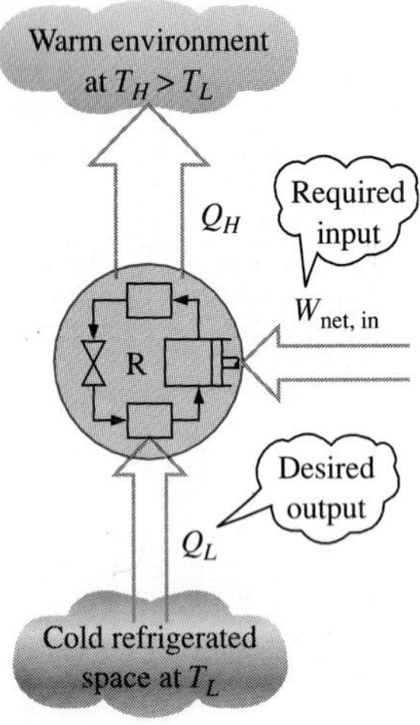

FIGURE 5-26
The objective of a refrigerator is to remove Q_L from the cooled space.

Coefficient of Performance

The *efficiency* of a refrigerator is expressed in terms of the **coefficient of performance** (COP), denoted by COP_R. The objective of a refrigerator is to remove heat (Q_L) from the refrigerated space. To accomplish this objective, it requires a work input of $W_{net, in}$. Then the COP of a refrigerator can be expressed as

$$COP_R = \frac{\text{Desired output}}{\text{Required input}} = \frac{Q_L}{W_{net, in}} \tag{5-9}$$

This relation can also be expressed in rate form by replacing Q_L by $\dot{Q}_L$ and $W_{net, in}$ by $\dot{W}_{net, in}$.

The conservation of energy principle for a cyclic device requires that

$$W_{net, in} = Q_H - Q_L \qquad \text{(kJ)} \tag{5-10}$$

Then the COP relation can also be expressed as

$$COP_R = \frac{Q_L}{Q_H - Q_L} = \frac{1}{Q_H/Q_L - 1} \tag{5-11}$$

Notice that the value of COP_R can be *greater than unity*. That is, the amount of heat removed from the refrigerated space can be greater than the amount of work input. This is in contrast to the thermal efficiency, which can never be greater than 1. In fact, one reason for expressing the efficiency of a refrigerator by another term—the coefficient of performance—is the desire to avoid the oddity of having efficiencies greater than unity.

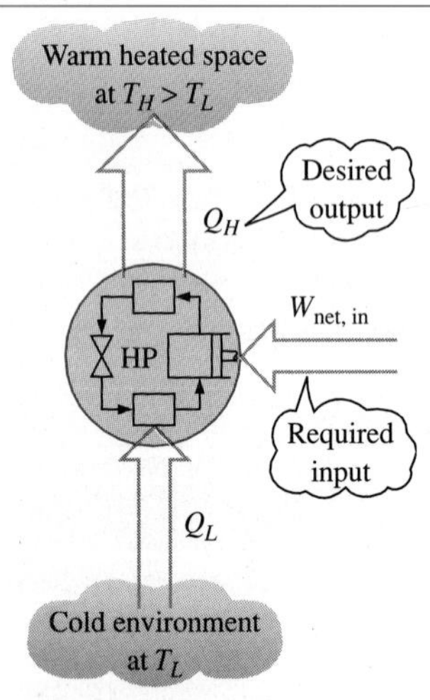

FIGURE 5-27
The objective of a heat pump is to supply heat Q_H into the warmer space.

Heat Pumps

Another device that transfers heat from a low-temperature medium to a high-temperature one is the **heat pump,** shown schematically in Fig. 5-27. Refrigerators and heat pumps operate on the same cycle but differ in their objectives. The objective of a refrigerator is to maintain the refrigerated space at a low temperature by removing heat from it. Discharging this heat to a higher-temperature medium is merely a necessary part of the operation, not the purpose. The objective of a heat pump, however, is to maintain a heated space at a high temperature. This is accomplished by absorbing heat from a low-temperature source, such as well water or cold outside air in winter, and supplying this heat to the high-temperature medium such as a house (Fig. 5-28).

An ordinary refrigerator that is placed in the window of a house with its door open to the cold outside air in winter will function as a heat pump since it will try to cool the outside by absorbing heat from it and rejecting this heat into the house through the coils behind it (Fig. 5-29).

The measure of performance of a heat pump is also expressed in terms of the **coefficient of performance** COP_{HP}, defined as

$$COP_{HP} = \frac{\text{Desired output}}{\text{Required input}} = \frac{Q_H}{W_{net,\,in}} \quad (5\text{-}12)$$

which can also be expressed as

$$COP_{HP} = \frac{Q_H}{Q_H - Q_L} = \frac{1}{1 - Q_L/Q_H} \quad (5\text{-}13)$$

A comparison of Eqs. 5-9 and 5-12 reveals that

$$COP_{HP} = COP_R + 1 \quad (5\text{-}14)$$

for fixed values of Q_L and Q_H. This relation implies that the coefficient of performance of a heat pump is always greater than unity since COP_R is a positive quantity. That is, a heat pump will function, at worst, as a resistance heater, supplying as much energy to the house as it consumes. In reality, however, part of Q_H is lost to the outside air through piping and other devices, and COP_{HP} may drop below unity when the outside air temperature is too low. When this happens, the system usually switches to a resistance heating mode. Most heat pumps in operation today have a seasonally averaged COP of 2 to 3.

Most existing heat pumps use the cold outside air as the heat source in winter, and they are referred to as *air-source heat pumps*. The COP of such heat pumps currently approaches 3.0 at design conditions. Air-source heat pumps are not appropriate for cold climates since their efficiency drops considerably when temperatures are below the freezing point. In such cases, geothermal (also called ground-source) heat pumps that use the ground as the heat source can be used. Geothermal heat pumps require the burial of pipes in the ground 1 to 2 m deep. Such heat pumps are more expensive to install, but they are also more efficient (up to 45 percent more efficient than air-source heat pumps). The COP of ground-source heat pumps currently approaches 4.0.

Air conditioners are basically refrigerators whose refrigerated space is a room or a building instead of the food compartment. A window air conditioning unit cools a room by absorbing heat from the room air and discharging it to the outside. The same air conditioning unit can be used as a heat pump in winter by installing it backwards. In this mode, the unit will pick up heat from the cold outside and deliver it to the room. Air conditioning systems that are equipped with proper controls and a reversing valve operate as air conditioners in summer and as heat pumps in winter.

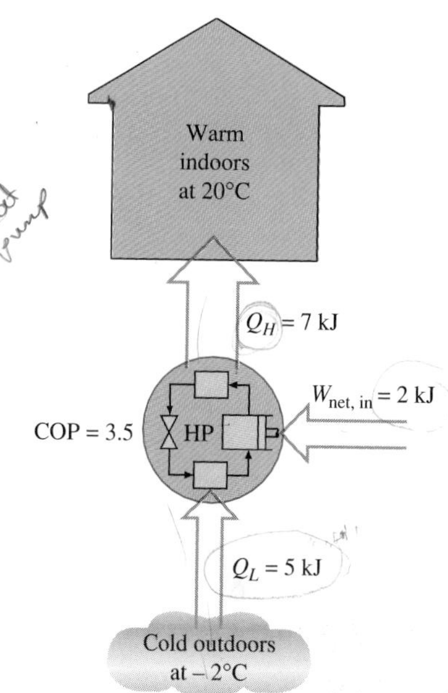

FIGURE 5-28
The work supplied to a heat pump is used to extract energy from the cold outdoors and carry it into the warm indoors.

FIGURE 5-29
When installed backwards, an air conditioner will function as a heat pump.

The performance of refrigerators and air conditioners in the United States is often expressed in terms of the **Energy Efficiency Rating** (EER), which is the amount of heat removed from the cooled space in Btu's for 1 Wh (watt-hour) of electricity consumed. Considering that 1 kWh = 3412 Btu and thus 1 Wh = 3.412 Btu, a unit that removes 1 kWh of heat from the cooled space for each kWh of electricity it consumes (COP = 1) will have an EER of 3.412. Therefore, the relation between EER and COP is

$$\text{EER} = 3.412\ \text{COP}_\text{R}$$

Most air conditioners have an EER between 8 and 12 (a COP of 2.3 to 3.5). A high-efficiency heat pump recently manufactured by the Trane Company using a reciprocating variable-speed compressor is reported to have a COP of 3.3 in the heating mode and an EER of 16.9 (COP of 5.0) in the air-conditioning mode. Variable-speed compressors and fans allow the unit to operate at maximum efficiency for varying heating/cooling needs and weather conditions as determined by a microprocessor. In the air-conditioning mode, for example, they operate at higher speeds on hot days and at lower speeds on cooler days, enhancing both efficiency and comfort.

The EER or COP of a refrigerator decreases with decreasing refrigeration temperature. Therefore, it is not economical to refrigerate to a lower temperature than needed. The COPs of refrigerators are in the range of 2.5–3.0 for cutting and preparation rooms; 2.3–2.6 for meat, deli, dairy, and produce; 1.2–1.5 for frozen foods; and 1.0–1.2 for ice cream units. Note that the COP of freezers is about half of the COP of meat refrigerators, and thus it will cost twice as much to cool the meat products with refrigerated air that is cold enough to cool frozen foods. It is good energy conservation practice to use separate refrigeration systems to meet different refrigeration needs.

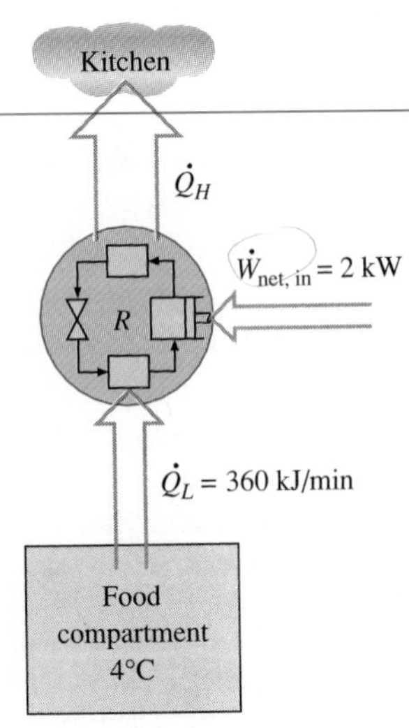

FIGURE 5-30
Schematic for Example 5-4.

EXAMPLE 5-4 Heat Rejection by a Refrigerator

The food compartment of a refrigerator, shown in Fig. 5-30, is maintained at 4°C by removing heat from it at a rate of 360 kJ/min. If the required power input to the refrigerator is 2 kW, determine (*a*) the coefficient of performance of the refrigerator and (*b*) the rate of heat rejection to the room that houses the refrigerator.

Solution The power consumption of a refrigerator is given. The COP and the rate of heat rejection are to be determined.

Assumptions Steady operating conditions exist.

Analysis (*a*) The coefficient of performance of a refrigerator is defined by Eq. 5-9, which can be expressed in rate form as

$$\text{COP}_\text{R} = \frac{\dot{Q}_L}{\dot{W}_{\text{net, in}}} = \frac{360\ \text{kJ/min}}{2\ \text{kW}}\left(\frac{1\ \text{kW}}{60\ \text{kJ/min}}\right) = \mathbf{3}$$

That is, 3 kJ of heat is removed from the refrigerated space for each kJ of work supplied.

(*b*) The rate at which heat is discharged to the room that houses the refrigerator is determined from the conservation of energy relation for cyclic devices (Eq. 5-10), expressed in rate form as

$$\dot{Q}_H = \dot{Q}_L + \dot{W}_{net,\,in} = 360\text{ kJ/min} + (2\text{ kW})\left(\frac{60\text{ kJ/min}}{1\text{ kW}}\right) = \mathbf{480\ kJ/min}$$

Discussion Notice that both the energy removed from the refrigerated space as heat and the energy supplied to the refrigerator as electrical work eventually show up in the room air and become part of the internal energy of the air. This demonstrates that energy can change from one form to another, can move from one place to another, but is never destroyed during a process.

EXAMPLE 5-5 Heating a House by a Heat Pump

A heat pump is used to meet the heating requirements of a house and maintain it at 20°C. On a day when the outdoor air temperature drops to −2°C, the house is estimated to lose heat at a rate of 80,000 kJ/h. If the heat pump under these conditions has a COP of 2.5, determine (*a*) the power consumed by the heat pump and (*b*) the rate at which heat is absorbed from the cold outdoor air.

Solution The COP of a heat pump is given. The power consumption and the rate of heat absorption are to be determined.

Assumptions Steady operating conditions exist.

Analysis (*a*) The power consumed by this heat pump, shown in Fig. 5-31, can be determined from the definition of the coefficient of performance of a heat pump (Eq. 5-12), expressed in the rate form as

$$\dot{W}_{net,\,in} = \frac{\dot{Q}_H}{\text{COP}_{HP}} = \frac{80{,}000\text{ kJ/h}}{2.5} = \mathbf{32{,}000\ kJ/h\ (or\ 8.9\ kW)}$$

(*b*) The house is losing heat at a rate of 80,000 kJ/h. If the house is to be maintained at a constant temperature of 20°C, the heat pump must deliver heat to the house at the same rate, that is, at a rate of 80,000 kJ/h. Then the rate of heat transfer from the outdoor air is determined from the conservation of energy principle for a cyclic device (Eq. 5-10):

$$\dot{Q}_L = \dot{Q}_H - \dot{W}_{net,\,in} = (80{,}000 - 32{,}000)\text{ kJ/h} = \mathbf{48{,}000\ kJ/h}$$

Discussion Note that 48,000 of the 80,000 kJ/h heat delivered to the house is actually extracted from the cold outdoor air. Therefore, we are paying only for the 32,000-kJ/h energy that is supplied as electrical work to the heat pump. If we were to use an electric resistance heater instead, we would have to supply the entire 80,000 kJ/h to the resistance heater as electric energy. This would mean a heating bill that is 2.5 times higher. This explains the popularity of heat pumps as heating systems and why they are preferred to simple electric resistance heaters despite their considerably higher initial cost.

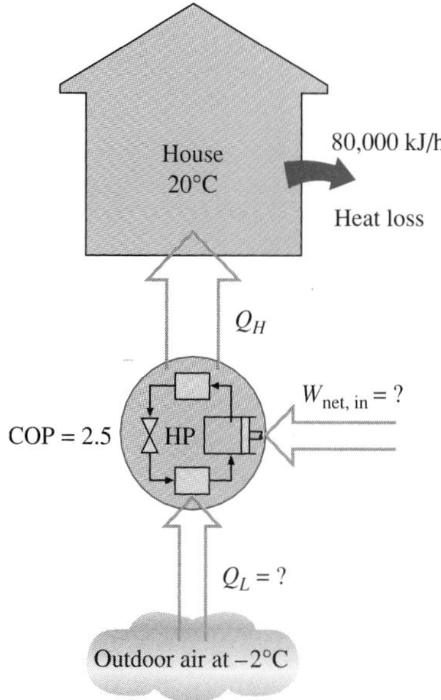

FIGURE 5-31
Schematic for Example 5-5.

The Second Law of Thermodynamics: Clausius Statement

There are two classical statements of the second law—the Kelvin-Planck statement, which is related to heat engines and discussed in the preceding section, and the Clausius statement, which is related to refrigerators or heat pumps. The Clausius statement is expressed as follows:

Clausius statement

It is impossible to construct a device that operates in a cycle and produces no effect other than the transfer of heat from a lower- temperature body to a higher-temperature body.

It is common knowledge that heat does not, of its own volition, flow from a cold medium to a warmer one. The Clausius statement does not imply that a cyclic device that transfers heat from a cold medium to a warmer one is impossible to construct. In fact, this is precisely what a common household refrigerator does. It simply states that a refrigerator will not operate unless its compressor is driven by an external power source, such as an electric motor (Fig. 5-32). This way, the net effect on the surroundings involves the consumption of some energy in the form of work, in addition to the transfer of heat from a colder body to a warmer one. That is, it leaves a trace in the surroundings. Therefore, a household refrigerator is in complete compliance with the Clausius statement of the second law.

Both the Kelvin-Planck and the Clausius statements of the second law are negative statements, and a negative statement cannot be proved. Like any other physical law, the second law of thermodynamics is based on experimental observations. To date, no experiment has been conducted that contradicts the second law, and this should be taken as sufficient evidence of its validity.

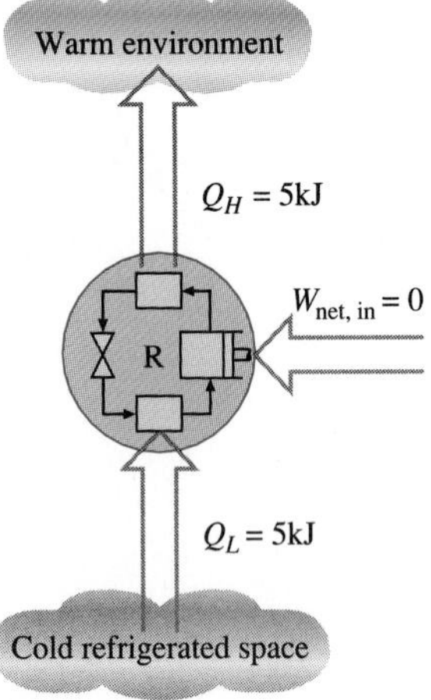

FIGURE 5-32
A refrigerator that violates the Clausius statement of the second law.

Equivalence of the Two Statements

The Kelvin-Planck and the Clausius statements are equivalent in their consequences, and either statement can be used as the expression of the second law of thermodynamics. Any device that violates the Kelvin-Planck statement also violates the Clausius statement, and vice versa. This can be demonstrated as follows:

Consider the heat-engine-refrigerator combination shown in Fig. 5-33*a*, operating between the same two reservoirs. The heat engine is assumed to

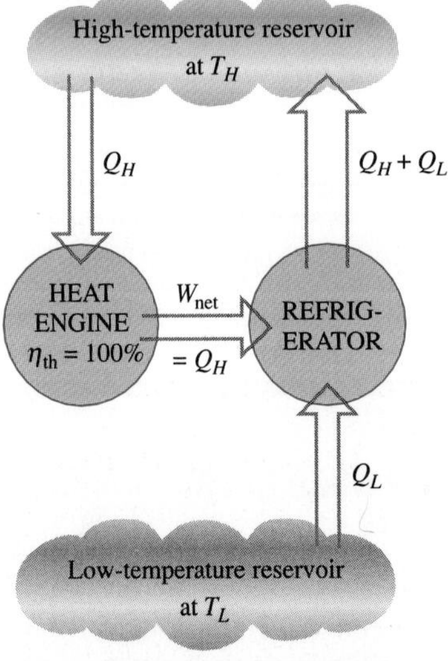

(*a*) A refrigerator which is powered by a 100% efficient heat engine

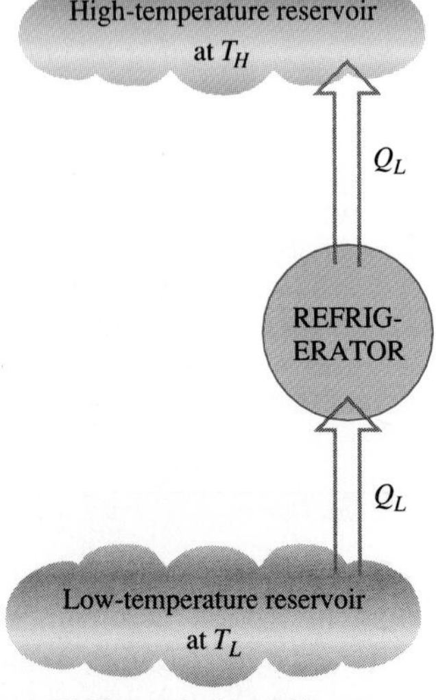

(*b*) The equivalent refrigerator

FIGURE 5-33
Proof that the violation of the Kelvin-Planck statement leads to the violation of the Clausius statement.

have, in violation of the Kelvin-Planck statement, a thermal efficiency of 100 percent, and therefore it converts all the heat Q_H it receives to work W. This work is now supplied to a refrigerator that removes heat in the amount of Q_L from the low-temperature reservoir and rejects heat in the amount of $Q_L + Q_H$ to the high-temperature reservoir. During this process, the high-temperature reservoir recieves a net amount of heat Q_L (the difference between $Q_L + Q_H$ and Q_H). Thus, the combination of these two devices can be viewed as a refrigerator, as shown in Fig. 5-33*b*, that transfers heat in an amount of Q_L from a cooler body to a warmer one without requiring any input from outside. This is clearly a violation of the Clausius statement. Therefore, a violation of the Kelvin-Planck statement results in the violation of the Clausius statement.

It can also be shown in a similar manner that a violation of the Clausius statement leads to the violation of the Kelvin-Planck statement. Therefore, the Clausius and the Kelvin-Planck statements are two equivalent expressions of the second law of thermodynamics.

5-6 ■ PERPETUAL-MOTION MACHINES

We have repeatedly stated that a process cannot take place unless it satisfies both the first and second laws of thermodynamics. Any device that violates either law is called a **perpetual-motion machine,** and despite numerous attempts, no perpetual-motion machine is known to have worked. But this has not stopped inventors from trying to create new ones.

A device that violates the first law of thermodynamics (by *creating* energy) is called a **perpetual-motion machine of the first kind** (PMM1), and a device that violates the second law of thermodynamics is called a **perpetual-motion machine of the second kind** (PMM2).

Consider the steam power plant shown in Fig. 5-34. It is proposed to heat the steam by resistance heaters placed inside the boiler, instead of by the energy supplied from fossil or nuclear fuels. Part of the electricity generated by the plant is to be used to power the resistors as well as the pump. The rest of the electric energy is to be supplied to the electric network as the net work

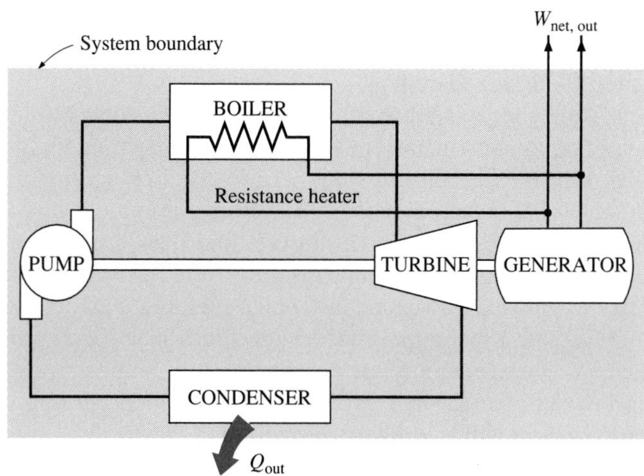

FIGURE 5-34
A perpetual-motion machine that violates the first law of thermodynamics (PMM1).

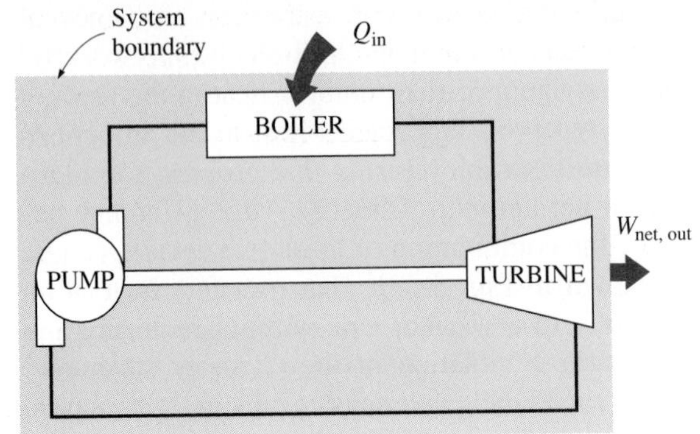

FIGURE 5-35

A perpetual-motion machine that violates the second law of thermodynamics (PMM2).

output. The inventor claims that once the system is started, this power plant will produce electricity indefinitely without requiring any energy input from the outside.

Well, here is an invention that could solve the world's energy problem—if it works, of course. A careful examination of this invention reveals that the system enclosed by the shaded area is continuously supplying energy to the outside at a rate of $\dot{Q}_{out} + \dot{W}_{net,out}$ without receiving any energy. That is, this system is creating energy at a rate of $\dot{Q}_{out} + \dot{W}_{net,out}$, which is clearly a violation of the first law. Therefore, this wonderful device is nothing more than a PMM1 and does not warrant any further consideration.

Now let us consider another novel idea by the same inventor. Convinced that energy cannot be created, the inventor suggests the following modification that will greatly improve the thermal efficiency of that power plant without violating the first law. Aware that more than one-half of the heat transferred to the steam in the furnace is discarded in the condenser to the environment, the inventor suggests getting rid of this wasteful component and sending the steam to the pump as soon as it leaves the turbine, as shown in Fig. 5-35. This way, all the heat transferred to the steam in the boiler will be converted to work, and thus the power plant will have a theoretical efficiency of 100 percent. The inventor realizes that some heat losses and friction between the moving components are unavoidable and that these effects will hurt the efficiency somewhat, but still expects the efficiency to be no less than 80 percent (as opposed to 40 percent in actual power plants) for a carefully designed system.

Well, the possibility of doubling the efficiency would certainly be very tempting to plant managers and, if not properly trained, they would probably give this idea a chance, since intuitively they see nothing wrong with it. A student of thermodynamics, however, will immediately label this device as a PMM2, since it works on a cycle and does a net amount of work while exchanging heat with a single reservoir (the furnace) only. It satisfies the first law but violates the second law, and therefore it will not work.

Countless perpetual-motion machines have been proposed throughout history, and many more are being proposed. Some proposers have even gone so far as to patent their inventions, only to find out that what they actually have in their hands is a worthless piece of paper.

Some perpetual-motion machine inventors were very successful in fund raising. For example, a Philadelphia carpenter named J. W. Kelly collected millions of dollars between 1874 and 1898 from investors in his *hydro-pneumatic-pulsating-vacu-engine,* which supposedly could push a railroad train 3000 miles on one liter of water. Of course, it never did. After his death in 1898, the investigators discovered that the demonstration machine was powered by a hidden motor. Recently a group of investors was set to invest $2.5 million into a mysterious *energy augmentor,* which multiplied whatever power it took in, but their lawyer wanted an expert opinion first. Confronted by the scientists, the "inventor" fled the scene without even attempting to run his demo machine.

Tired of applications for perpetual-motion machines, the U.S. Patent Office decreed in 1918 that it would no longer even consider any perpetual-motion applications. But several such patent applications were still filed, and some made it through the patent office undetected. Some applicants whose patent applications were denied sought legal action. For example, in 1982 the U.S. Patent Office dismissed as just another perpetual-motion machine a huge device that involves several hundred kilograms of rotating magnets and kilometers of copper wire that is supposed to be generating more electricity than it is consuming from a battery pack. But the inventor challenged the decision, and in 1985 the National Bureau of Standards finally tested the machine just to certify that it is battery-operated. But it did not convince the inventor that his machine will not work.

The proposers of perpetual-motion machines generally have innovative minds, but they usually lack formal engineering training, which is very unfortunate. No one is immune from being deceived by an innovative perpetual-motion machine. But, as the saying goes, if something sounds too good to be true, it probably is.

5-7 ■ REVERSIBLE AND IRREVERSIBLE PROCESSES

The second law of thermodynamics states that no heat engine can have an efficiency of 100 percent. Then one may ask, What is the highest efficiency that a heat engine *can* possibly have? Before we can answer this question, we need to define an idealized process first, which is called the *reversible process.*

The processes that were discussed in Sec. 5-1 occurred in a certain direction. Once having taken place, these processes cannot reverse themselves spontaneously and restore the system to its initial state. For this reason, they are classified as *irreversible processes.* Once a cup of hot coffee cools, it will not heat up retrieving the heat it lost from the surroundings. If it could, the surroundings, as well as the system (coffee), would be restored to their original condition, and this would be a reversible process.

A **reversible process** is defined as a *process that can be reversed without leaving any trace on the surroundings* (Fig. 5-36). That is, both the system *and* the surroundings are returned to their initial states at the end of the reverse process. This is possible only if the net heat *and* net work exchange between the system and the surroundings is zero for the combined (original

FIGURE 5-36
Two familiar reversible processes.

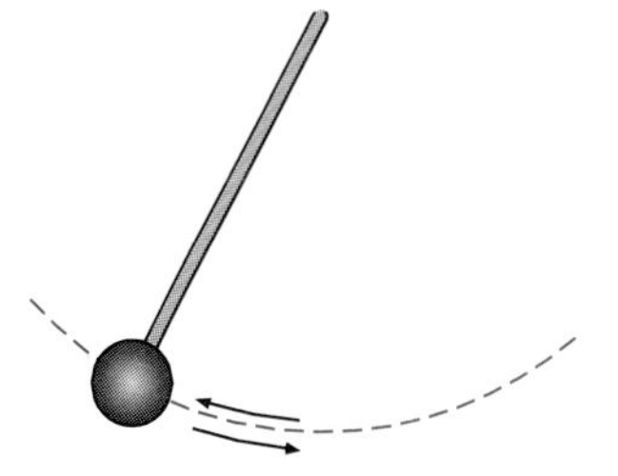

(*a*) Frictionless pendulum

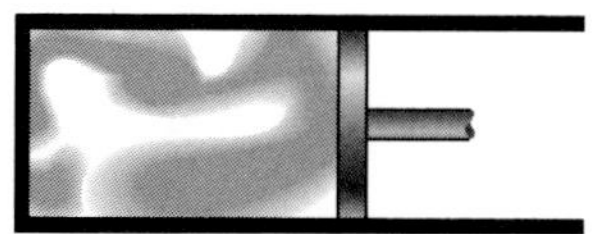

(*b*) Quasi-equilibrium expansion and compression of a gas

and reverse) process. Processes that are not reversible are called **irreversible processes.**

It should be pointed out that a system can be restored to its initial state following a process, regardless of whether the process is reversible or irreversible. But for reversible processes, this restoration is made without leaving any net change on the surroundings, whereas for irreversible processes, the surroundings usually do some work on the system and therefore will not return to their original state.

Reversible processes actually do not occur in nature. They are merely *idealizations* of actual processes. Reversible processes can be approximated by actual devices, but they can never be achieved. That is, all the processes occurring in nature are irreversible. You may be wondering, then, *why* we are bothering with such fictitious processes. There are two reasons. First, they are easy to analyze, since a system passes through a series of equilibrium states during a reversible process; second, they serve as idealized models to which actual processes can be compared.

In daily life, the concepts of Mr. Right and Ms. Right are also idealizations, just like the concept of a reversible (perfect) process. People who insist on finding Mr. or Ms. Right to settle down are bound to remain Mr. or Ms. Single for the rest of their lives. The possibility of finding the perfect prospective mate is no higher than the possibility of finding a perfect (reversible) process. Likewise, a person who insists on perfection in friends is bound to have no friends.

Engineers are interested in reversible processes because work-producing devices such as car engines and gas or steam turbines *deliver the most work,* and work-consuming devices such as compressors, fans, and pumps *require the least work* when reversible processes are used instead of irreversible ones (Fig. 5-37).

Reversible processes can be viewed as *theoretical limits* for the corresponding irreversible ones. Some processes are more irreversible than others. We may never be able to have a reversible process, but we may certainly approach it. The more closely we approximate a reversible process, the more work delivered by a work-producing device or the less work required by a work-consuming device.

The concept of reversible processes leads to the definition of the **second-law efficiency** for actual processes, which is the degree of approximation to the corresponding reversible processes. This enables us to compare the performance of different devices that are designed to do the same task on the

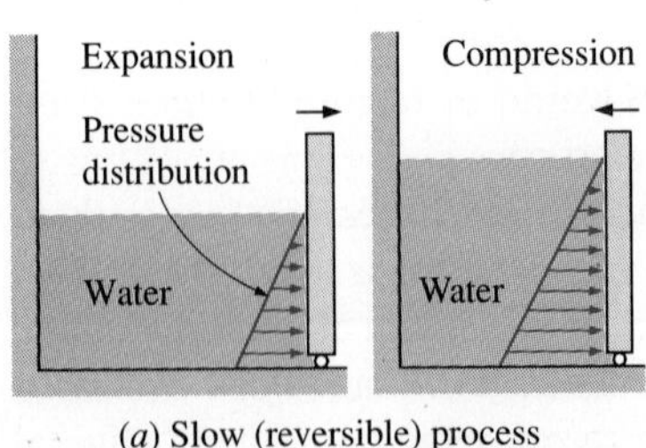

(*a*) Slow (reversible) process

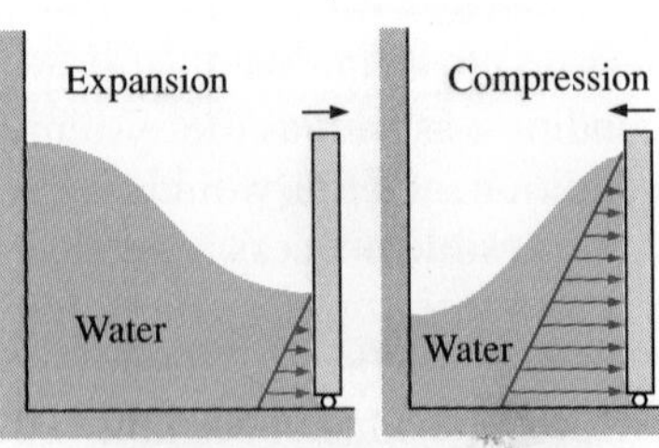

(*b*) Fast (irreversible) process

FIGURE 5-37
Reversible processes deliver the most and consume the least work.

basis of their efficiencies. The better the design, the lower the irreversibilities and the higher the second-law efficiency.

Irreversibilities

The factors that cause a process to be irreversible are called **irreversibilities.** They include friction, unrestrained expansion, mixing of two gases, heat transfer across a finite temperature difference, electric resistance, inelastic deformation of solids, and chemical reactions. The presence of any of these effects renders a process irreversible. A reversible process involves none of these. Some of the frequently encountered irreversibilities are discussed briefly below.

Friction

Friction is a familiar form of irreversibility associated with bodies in motion. When two bodies in contact are forced to move relative to each other (a piston in a cylinder, for example, as shown in Fig. 5-38), a friction force that opposes the motion develops at the interface of these two bodies, and some work is needed to overcome this friction force. The energy supplied as work is eventually converted to heat during the process and is transferred to the bodies in contact, as evidenced by a temperature rise at the interface. When the direction of the motion is reversed, the bodies will be restored to their original position, but the interface will not cool, and heat will not be converted back to work. Instead, more of the work will be converted to heat while overcoming the friction forces that also oppose the reverse motion. Since the system (the moving bodies) and the surroundings cannot be returned to their original states, this process is irreversible. Therefore, any process that involves friction is irreversible. The larger the friction forces involved, the more irreversible the process is.

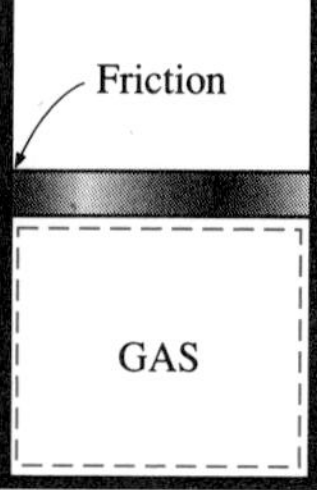

FIGURE 5-38
Friction renders a process irreversible.

Friction does not always involve two solid bodies in contact. It is also encountered between a fluid and solid and even between the layers of a fluid moving at different velocities. A considerable fraction of the power produced by a car engine is used to overcome the friction (the drag force) between the air and the external surfaces of the car, and it eventually becomes part of the internal energy of the air. It is not possible to reverse this process and recover that lost power, even though doing so would not violate the conservation of energy principle.

Non-Quasi-Equilibrium Expansion and Compression

In Chap. 1, we defined a quasi-equilibrium process as one during which the system remains infinitesimally close to a state of equilibrium at all times. Consider a frictionless adiabatic piston-cylinder device that contains a gas. Now the piston is pushed into the cylinder, compressing the gas. If the piston velocity is not very high, the pressure and the temperature will increase uniformly throughout the gas. Since the system is always maintained at a state close to equilibrium, this is a quasi-equilibrium process.

Now the external force on the piston is slightly decreased, allowing the gas to expand. The expansion process will also be *quasi-equilibrium* if the gas is allowed to expand slowly. When the piston returns to its original position, all the boundary ($P\ dV$) work done on the gas during compression is returned to the surroundings during expansion. That is, the net work for the combined process is zero. Also, there has been no heat transfer involved during this process, and thus both the system and the surroundings will return to their initial states at the end of the reverse process. Therefore, the slow frictionless adiabatic expansion or compression of a gas is a reversible process.

Now let us repeat this adiabatic process in a *non-quasi-equilibrium* manner, as shown in Fig. 5-39. If the piston is pushed in very rapidly, the gas molecules near the piston face will not have sufficient time to escape, and they will pile up in front of the piston. This will raise the pressure near the piston face, and as a result, the pressure there will be higher than the pressure in other parts of the cylinder. The nonuniformity of pressure will render this process non-quasi-equilibrium. The actual boundary work is a function of pressure at the piston face. Because of this higher pressure value at the piston face, a non-quasi-equilibrium compression process will require a larger work input than the corresponding quasi-equilibrium one. When the process is reversed by letting the gas expand rapidly, the gas molecules in the cylinder will not be able to follow the piston as fast, thus creating a low-pressure region before the piston face. Because of this low-pressure value at the piston face, a non-quasi-equilibrium process will deliver less work than a corresponding reversible one. Consequently, the work done by the gas during expansion is less than the work done by the surroundings on the gas during compression, and thus the surroundings have a net work deficit. When the piston returns to its initial position, the gas will have excess internal energy, equal in magnitude to the work deficit of the surroundings.

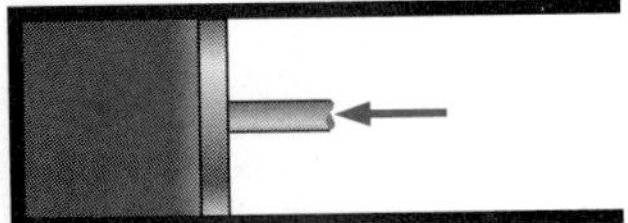
(*a*) Fast compression

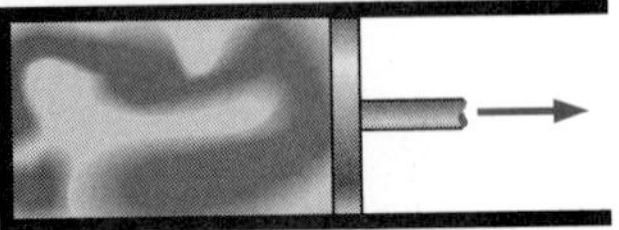
(*b*) Fast expansion

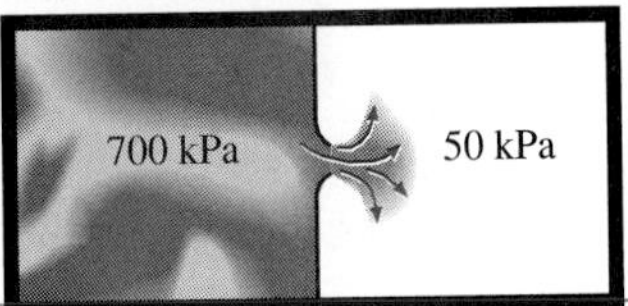

(*c*) Unrestrained expansion

FIGURE 5-39

Irreversible compression and expansion processes.

The system can easily be returned to its initial state by transferring this excess internal energy to the surroundings as heat. But the only way the surroundings can be returned to their initial condition is by completely converting this heat to work, which can only be done by a heat engine that has an efficiency of 100 percent. This, however, is impossible to do, even theoretically, since it would violate the second law of thermodynamics. Since only the system, not both the system and the surroundings, can be returned to its initial state, we conclude that the adiabatic non-quasi-equilibrium expansion or compression processes are irreversible.

Another example of non-quasi-equilibrium expansion processes is the unrestrained expansion of a gas separated from a vacuum by a membrane, as shown in Fig. 5-39c. When the membrane is ruptured, the gas fills the entire tank. The only way to restore the system to its original state is to compress it to its initial volume, while transferring heat from the gas until it reaches its initial temperature. From the conservation of energy considerations, it can easily be shown that the amount of heat transferred from the gas equals the amount of work done on the gas by the surroundings. The restoration of the surroundings involves conversion of this heat completely to work, which would violate the second law. Therefore, unrestrained expansion of a gas is an irreversible process.

Heat Transfer

Another form of irreversibility familiar to us all is heat transfer through a finite temperature difference. Consider a can of cold soda left in a warm room (Fig. 5-40). Heat will flow from the warmer room air to the cooler soda. The only way this process can be reversed and the soda restored to its original temperature is to provide refrigeration, which requires some work input. At the end of the reverse process, the soda will be restored to its initial state, but the surroundings will not be. The internal energy of the surroundings will increase by an amount equal in magnitude to the work supplied to the refrigerator. The restoration of the surroundings to the initial state can be done only by converting this excess internal energy completely to work, which is impossible to do without violating the second law. Since only the system, not both the system and the surroundings, can be restored to its initial condition, heat transfer through a finite temperature difference is an irreversible process.

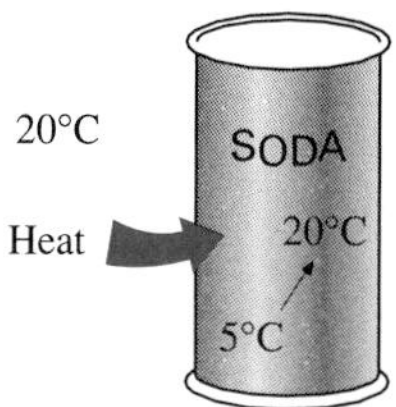

(*a*) An irreversible heat transfer process

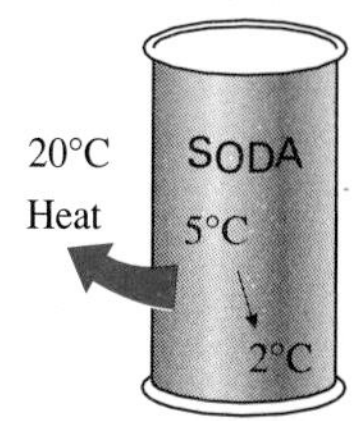

(*b*) An impossible heat transfer process

FIGURE 5-40

(*a*) Heat transfer through a temperature difference is irreversible, and (*b*) the reverse process is impossible.

Heat transfer can occur only when there is a temperature difference between a system and its surroundings. Therefore, it is physically impossible to have a reversible heat transfer process. But a heat transfer process becomes less and less irreversible as the temperature difference between the two bodies approaches zero. Then heat transfer through a differential temperature difference dT can be considered to be reversible. As dT approaches zero, the process can be reversed in direction (at least theoretically) without requiring any refrigeration. Notice that reversible heat transfer is a conceptual process and cannot be duplicated in the real world.

The smaller the temperature difference between two bodies, the smaller the heat transfer rate will be. Any significant heat transfer through a small temperature difference will require a very large surface area and a very long time. Therefore, even though approaching reversible heat transfer is desirable from a thermodynamic point of view, it is impractical and not economically feasible.

Internally and Externally Reversible Processes

A process is an interaction between a system and its surroundings, and a reversible process involves no irreversibilities associated with either of them.

A process is called **internally reversible** if no irreversibilities occur within the boundaries of the system during the process. During an internally reversible process, a system proceeds through a series of equilibrium states, and when the process is reversed, the system passes through exactly the same equilibrium states while returning to its initial state. That is, the paths of the forward and reverse processes coincide for an internally reversible process. The quasi-equilibrium process discussed earlier is an example of an internally reversible process.

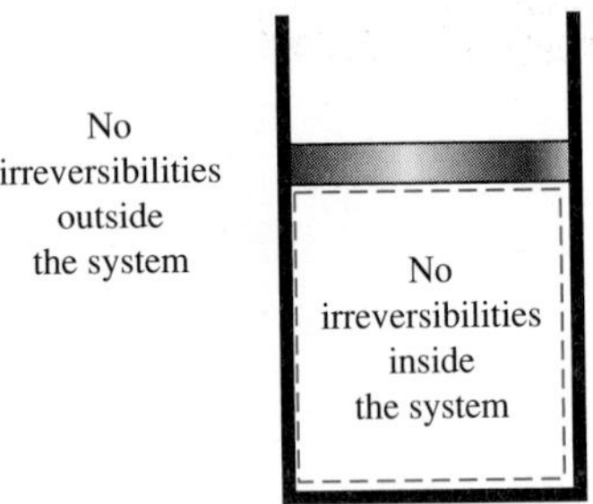

FIGURE 5-41
A reversible process involves no internal and external irreversibilities.

A process is called **externally reversible** if no irreversibilities occur outside the system boundaries during the process. Heat transfer between a reservoir and a system is an externally reversible process if the surface of contact between the system and the reservoir is at the temperature of the reservoir.

A process is called **totally reversible,** or simply **reversible,** if it involves no irreversibilities within the system or its surroundings (Fig. 5-41). A totally reversible process involves no heat transfer through a finite temperature difference, no non-quasi-equilibrium changes, and no friction or other dissipative effects.

As an example, consider the transfer of heat to two identical systems that are undergoing a constant-pressure (thus constant-temperature) phase-change process, as shown in Fig. 5-42. Both processes are internally reversible, since both take place isothermally and both pass through exactly the same equilibrium states. The first process shown is externally reversible also, since heat transfer for this process takes place through an infinitesimal temperature difference dT. The second process, however, is externally irreversible, since it involves heat transfer through a finite temperature difference ΔT.

5-8 ■ THE CARNOT CYCLE

We mentioned earlier that heat engines are cyclic devices and that the working fluid of a heat engine returns to its initial state at the end of each cycle. Work is done by the working fluid during one part of the cycle and on the working fluid during another part. The difference between these two is the net work delivered by the heat engine. The efficiency of a heat-engine cycle greatly depends on how the individual processes that make up the cycle are executed. The net work, thus the cycle efficiency, can be maximized by using processes that require the least amount of work and deliver the most, that is, by using *reversible processes.* Therefore, it is no surprise that the most efficient cycles are reversible cycles, that is, cycles that consist entirely of reversible processes.

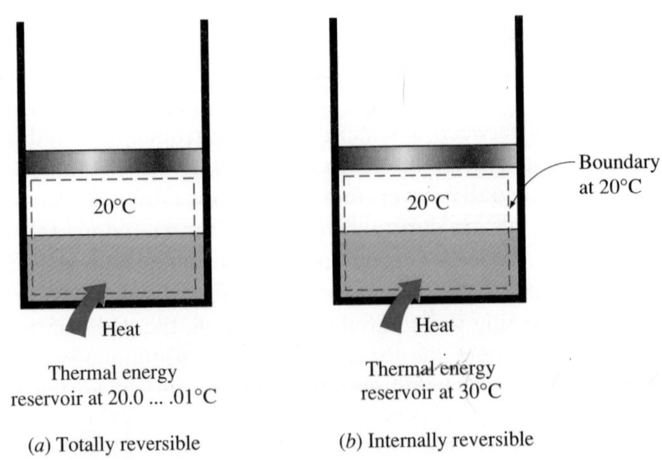

FIGURE 5-42
Totally and internally reversible heat transfer processes.

Reversible cycles cannot be achieved in practice because the irreversibilities associated with each process cannot be eliminated. However, reversible cycles provide upper limits on the performance of real cycles. Heat engines and refrigerators that work on reversible cycles serve as models to which actual heat engines and refrigerators can be compared. Reversible cycles also serve as starting points in the development of actual cycles and are modified as needed to meet certain requirements.

Probably the best known reversible cycle is the **Carnot cycle,** first proposed in 1824 by French engineer Sadi Carnot. The theoretical heat engine that operates on the Carnot cycle is called the **Carnot heat engine.** The Carnot cycle is composed of four reversible processes—two isothermal and two adiabatic—and it can be executed either in a closed or a steady-flow system.

Consider a closed system that consists of a gas contained in an adiabatic piston-cylinder device, as shown in Fig. 5-43. The insulation of the cylinder head is such that it may be removed to bring the cylinder into contact with reservoirs to provide heat transfer. The four reversible processes that make up the Carnot cycle are as follows:

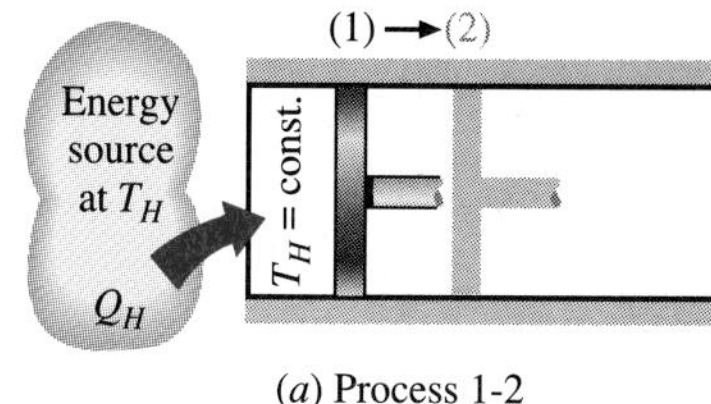

(a) Process 1-2

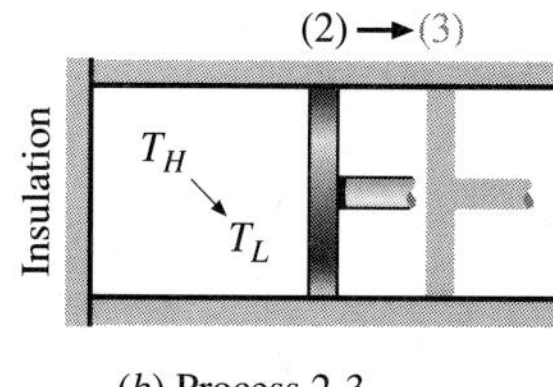

(b) Process 2-3

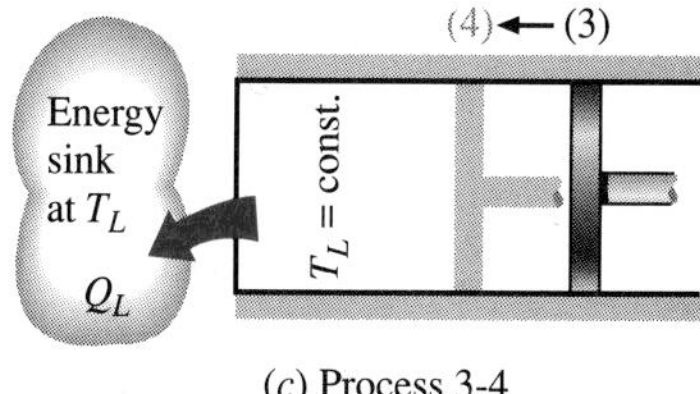

(c) Process 3-4

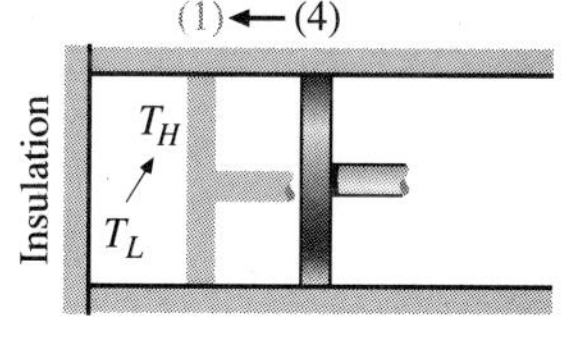

(d) Process 4-1

FIGURE 5-43
Execution of the Carnot cycle in a closed system.

Reversible Isothermal Expansion (process 1-2, T_H = constant). Initially (state 1), the temperature of the gas is T_H and the cylinder head is in close contact with a source at temperature T_H. The gas is allowed to expand slowly, doing work on the surroundings. As the gas expands, the temperature of the gas tends to decrease. But as soon as the temperature drops by an infinitesimal amount dT, some heat flows from the reservoir into the gas, raising the gas temperature to T_H. Thus, the gas temperature is kept constant at T_H. Since the temperature difference between the gas and the reservoir never exceeds a differential amount dT, this is a reversible heat transfer process. It continues until the piston reaches position 2. The amount of total heat transferred to the gas during this process is Q_H.

Reversible Adiabatic Expansion (process 2-3, temperature drops from T_H to T_L). At state 2, the reservoir that was in contact with the cylinder head is removed and replaced by insulation so that the system becomes adiabatic. The gas continues to expand slowly, doing work on the surroundings until its temperature drops from T_H to T_L (state 3). The piston is assumed to be frictionless and the process to be quasi-equilibrium, so the process is reversible as well as adiabatic.

Reversible Isothermal Compression (process 3-4, T_L = constant). At state 3, the insulation at the cylinder head is removed, and the cylinder is brought into contact with a sink at temperature T_L. Now the piston is pushed inward by an external force, doing work on the gas. As the gas is compressed, its temperature tends to rise. But as soon as it rises by an infinitesimal amount dT, heat flows from the gas to the sink, causing the gas temperature to drop to T_L. Thus, the gas temperature is maintained constant at T_L. Since the temperature difference between the gas and the sink never exceeds a differential amount dT, this is a reversible heat transfer process. It continues until the piston reaches state 4. The amount of heat rejected from the gas during this process is Q_L.

Reversible Adiabatic Compression (process 4-1, temperature rises from T_L to T_H). State 4 is such that when the low-temperature reservoir is removed, the insulation is put back on the cylinder head, and the gas is compressed in a reversible manner, the gas returns to its initial state (state 1). The temperature rises from T_L to T_H during this reversible adiabatic compression process, which completes the cycle.

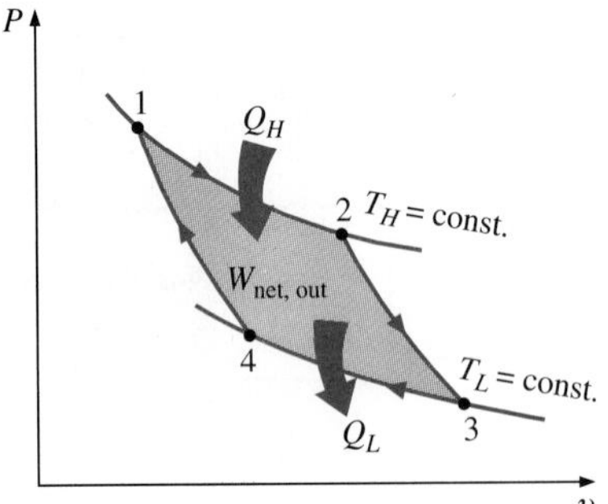

FIGURE 5-44

P-v diagram of the Carnot cycle.

The *P-v* diagram of this cycle is shown in Fig. 5-44. Remembering that on a *P-v* diagram the area under the process curve represents the boundary work for quasi-equilibrium (internally reversible) processes, we see that the area under curve 1-2-3 is the work done by the gas during the expansion part of the cycle, and the area under curve 3-4-1 is the work done on the gas during the compression part of the cycle. The area enclosed by the path of the cycle (area 1-2-3-4-1) is the difference between these two and represents the net work done during the cycle.

Notice that if we acted stingily and compressed the gas at state 3 adiabatically instead of isothermally in an effort *to save* Q_L, we would end up back at state 2, retracing the process path 3-2. By doing so we would save Q_L, but we would not be able to obtain any net work output from this engine. This illustrates once more the necessity of a heat engine exchanging heat with at least two reservoirs at different temperatures to operate in a cycle and produce a net amount of work.

The Carnot cycle can also be executed in a steady-flow system. It is discussed in Chap. 8 in conjunction with other power cycles.

Being a reversible cycle, the Carnot cycle is the most efficient cycle operating between two specified temperature limits. Even though the Carnot cycle cannot be achieved in reality, the efficiency of actual cycles can be improved by attempting to approximate the Carnot cycle more closely.

FIGURE 5-45

P-v diagram of the reversed Carnot cycle.

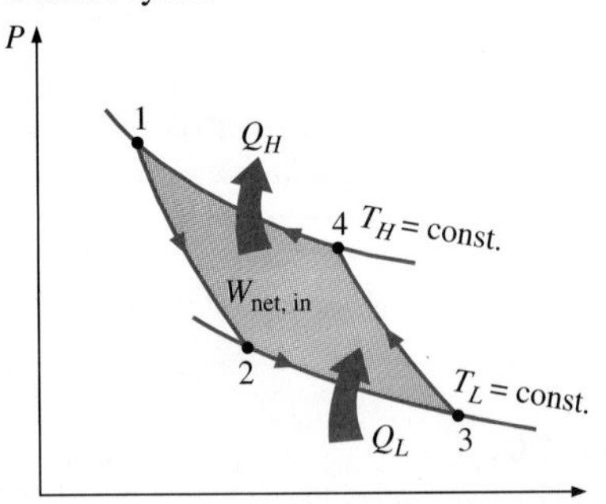

The Reversed Carnot Cycle

The Carnot heat-engine cycle described above is a totally reversible cycle. Therefore, all the processes that comprise it can be *reversed,* in which case it becomes the **Carnot refrigeration cycle.** This time, the cycle remains exactly the same, except that the directions of any heat and work interactions are reversed: Heat in the amount of Q_L is absorbed from the low-temperature reservoir, heat in the amount of Q_H is rejected to a high-temperature reservoir, and a work input of $W_{net, in}$ is required to accomplish all this.

The *P-v* diagram of the reversed Carnot cycle is the same as the one given for the Carnot cycle, except that the directions of the processes are reversed, as shown in Fig. 5-45.

The second law of thermodynamics puts limits on the operation of cyclic devices as expressed by the Kelvin-Planck and Clausius statements. A heat engine cannot operate by exchanging heat with a single reservoir, and a refrigerator cannot operate without a net work input from an external source.

We can draw valuable conclusions from these statements. Two conclusions pertain to the thermal efficiency of reversible and irreversible (i.e., actual) heat engines, and they are known as the **Carnot principles** (Fig. 5-46), expressed as follows:

1. *The efficiency of an irreversible heat engine is always less than the efficiency of a reversible one operating between the same two reservoirs.*
2. *The efficiencies of all reversible heat engines operating between the same two reservoirs are the same.*

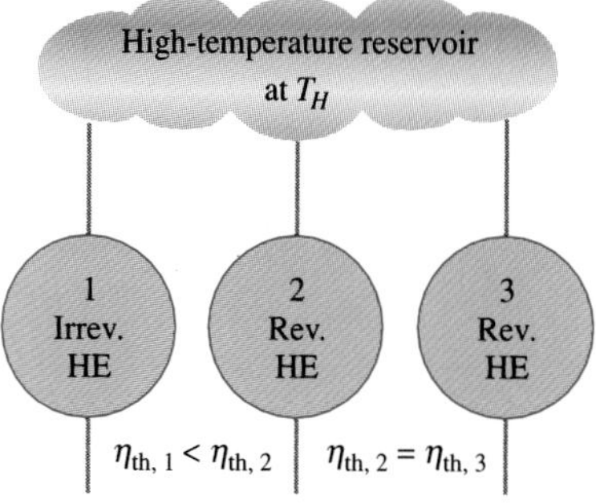

FIGURE 5-46
The Carnot principles.

These two statements can be proved by demonstrating that the violation of either statement results in the violation of the second law of thermodynamics.

To prove the first statement, consider two heat engines operating between the same reservoirs, as shown in Fig. 5-47. One engine is reversible and the other is irreversible. Now each engine is supplied with the same amount of heat Q_H. The amount of work produced by the reversible heat engine is W_{rev}, and the amount produced by the irreversible one is W_{irrev}.

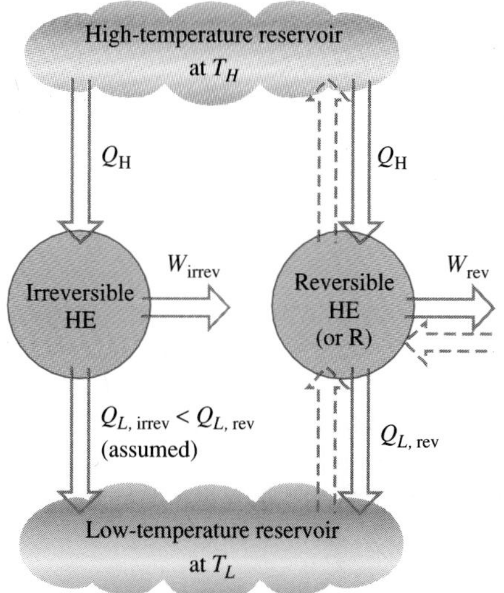

(*a*) A reversible and an irreversible heat engine operating between the same two reservoirs (the reversible heat engine is then reversed to run as a refrigerator)

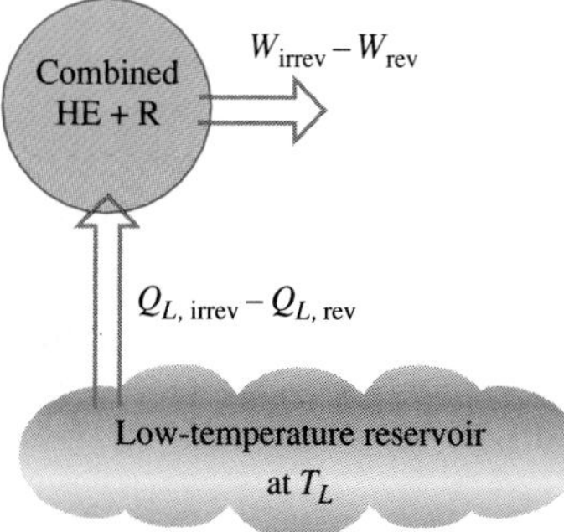

(*b*) The equivalent combined system

FIGURE 5-47
Proof of the first Carnot principle.

In violation of the first Carnot principle, we assume that the irreversible heat engine is more efficient than the reversible one (that is, $\eta_{\text{th, irrev}} > \eta_{\text{th, rev}}$) and thus delivers more work than the reversible one. Now let the reversible heat engine be reversed and operate as a refrigerator. This refrigerator will receive a work input of W_{rev} and reject heat to the high-temperature reservoir. Since the refrigerator is rejecting heat in the amount of Q_H to the high-temperature reservoir and the irreversible heat engine is receiving the same amount of heat from this reservoir, the net heat exchange for this reservoir is zero. Thus, it could be eliminated by having the refrigerator discharge Q_H directly into the irreversible heat engine.

Now considering the refrigerator and the irreversible engine together, we have an engine that produces a net work in the amount of $W_{\text{irrev}} - W_{\text{rev}}$ while exchanging heat with a single reservoir—a violation of the Kelvin-Planck statement of the second law. Therefore, our initial assumption that $\eta_{\text{th, irrev}} > \eta_{\text{th, rev}}$ is incorrect. Then we conclude that no heat engine can be more efficient than a reversible heat engine operating between the same reservoirs.

The second Carnot principle can also be proved in a similar manner. This time, let us replace the irreversible engine by another reversible engine that is more efficient and thus delivers more work than the first reversible engine. By following through the same reasoning as above, we will end up having an engine that produces a new amount of work while exchanging heat with a single reservoir, which is a violation of the second law. Therefore we conclude that no reversible heat engine can be more efficient than a reversible one operating between the same two reservoirs, regardless of how the cycle is completed or the kind of working fluid used.

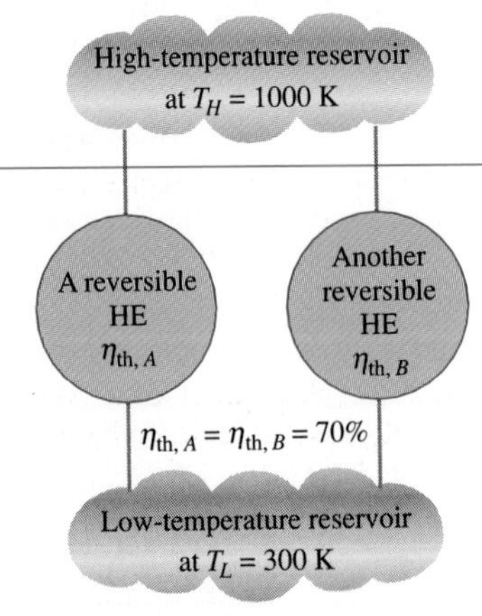

FIGURE 5-48

All reversible heat engines operating between the same two reservoirs have the same efficiency (the second Carnot principle).

5-10 ■ THE THERMODYNAMIC TEMPERATURE SCALE

A temperature scale that is independent of the properties of the substances that are used to measure temperature is called a **thermodynamic temperature scale.** Such a temperature scale offers great conveniences in thermodynamic calculations, and its derivation is given below using some reversible heat engines.

The second Carnot principle discussed in Sec. 5-9 states that all reversible heat engines have the same thermal efficiency when operating between the same two reservoirs (Fig. 5-48). That is, the efficiency of a reversible engine is independent of the working fluid employed and its properties, the way the cycle is executed, or the type of reversible engine used. Since energy reservoirs are characterized by their temperatures, the thermal efficiency of reversible heat engines is a function of the reservoir temperatures only. That is,

$$\eta_{\text{th, rev}} = g(T_H, T_L)$$

or

$$\frac{Q_H}{Q_L} = f(T_H, T_L) \tag{5-15}$$

since $\eta_{\text{th}} = 1 - Q_L/Q_H$. In these relations T_H and T_L are the temperatures of the high- and low-temperature reservoirs, respectively.

The functional form of $f(T_H, T_L)$ can be developed with the help of the three reversible heat engines shown in Fig. 5-49. Engines A and C are supplied with the same amount of heat Q_1 from the high-temperature reservoir at T_1. Engine C rejects Q_3 to the low-temperature reservoir at T_3. Engine B receives the heat Q_2 rejected by engine A at temperature T_2 and rejects heat in the amount of Q_3 to a reservoir at T_3.

The amounts of heat rejected by engines B and C must be the same since engines A and B can be combined into one reversible engine operating between the same reservoirs as engine C and thus the combined engine will have the same efficiency as engine C. Since the heat input to engine C is the same as the heat input to the combined engines A and B, both systems must reject the same amount of heat.

Applying Eq. 5-15 to all three engines separately, we obtain

$$\frac{Q_1}{Q_2} = f(T_1, T_2), \qquad \frac{Q_2}{Q_3} = f(T_2, T_3) \qquad \text{and} \qquad \frac{Q_1}{Q_3} = f(T_1, T_3)$$

Now consider the identity

$$\frac{Q_1}{Q_3} = \frac{Q_1}{Q_2}\frac{Q_2}{Q_3}$$

which corresponds to

$$f(T_1, T_3) = f(T_1, T_2) \cdot f(T_2, T_3)$$

A careful examination of this equation reveals that the left-hand side is a function of T_1 and T_3, and therefore the right-hand side must also be a function of T_1 and T_3 only, and not T_2. That is, the value of the product on the right-hand side of this equation is independent of the value of T_2. This condition will be satisfied only if the function f has the following form:

$$f(T_1, T_2) = \frac{\phi(T_1)}{\phi(T_2)} \qquad \text{and} \qquad f(T_2, T_3) = \frac{\phi(T_2)}{\phi(T_3)}$$

so that $\phi(T_2)$ will cancel from the product of $f(T_1, T_2)$ and $f(T_2, T_3)$, yielding

$$\frac{Q_1}{Q_3} = f(T_1, T_3) = \frac{\phi(T_1)}{\phi(T_3)} \tag{5-16}$$

This relation is much more specific than Eq. 5-15 for the functional form of Q_1/Q_3 in terms of T_1 and T_3.

For a reversible heat engine operating between two reservoirs at temperatures T_H and T_L, Eq. 5-16 can be written as

$$\frac{Q_H}{Q_L} = \frac{\phi(T_H)}{\phi(T_L)} \tag{5-17}$$

This is the only requirement that the second law places on the ratio of heat flows to and from the reversible heat engines. Several functions $\phi(T)$ will satisfy this equation, and the choice is completely arbitrary. Lord Kelvin first proposed taking $\phi(T) = T$ to define a thermodynamic temperature scale as (Fig. 5-50)

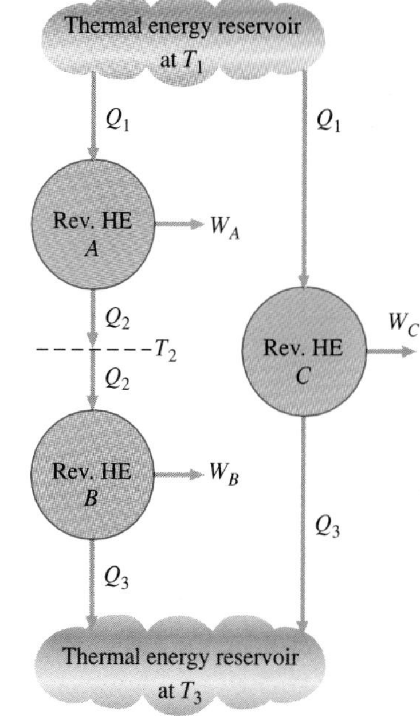

FIGURE 5-49
The arrangement of heat engines used to develop the thermodynamic temperature scale.

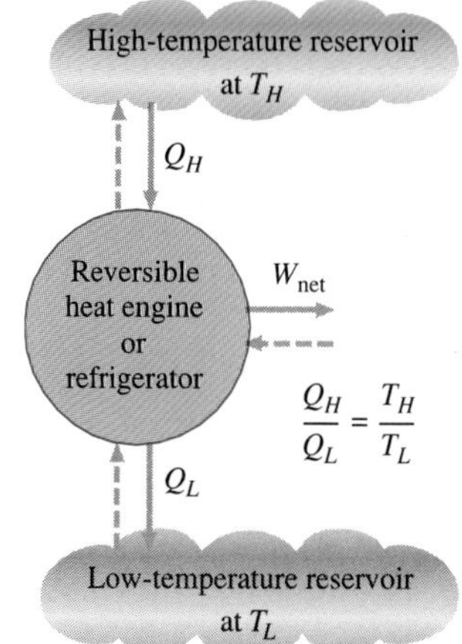

FIGURE 5-50
For reversible cycles, the heat transfer ratio Q_H/Q_L can be replaced by the absolute temperature ratio T_H/T_L.

$$\left(\frac{Q_H}{Q_L}\right)_{\text{rev}} = \frac{T_H}{T_L} \tag{5-18}$$

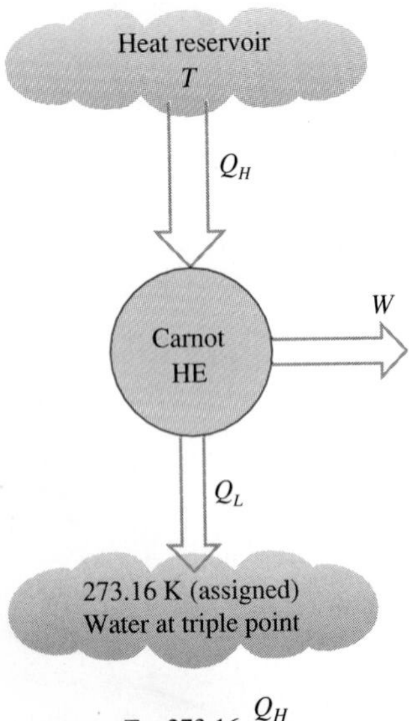

$$T = 273.16\,\frac{Q_H}{Q_L}$$

FIGURE 5-51
A conceptual experimental setup to determine thermodynamic temperatures on the Kelvin scale by measuring heat transfers Q_H and Q_L.

This temperature scale is called the **Kelvin scale,** and the temperatures on this scale are called **absolute temperatures.** On the Kelvin scale, the temperature ratios depend on the ratios of heat transfer between a reversible heat engine and the reservoirs and are independent of the physical properties of any substance. On this scale, temperatures vary between zero and infinity.

The thermodynamic temperature scale is not completely defined by Eq. 5-18 since it gives us only a ratio of absolute temperatures. We also need to know the magnitude of a kelvin. At the International Conference on Weights and Measures held in 1954, the triple point of water (the state at which all three phases of water exist in equilibrium) was assigned the value 273.16 K (Fig. 5-51). The *magnitude of a kelvin* is defined as 1/273.16 of the temperature interval between absolute zero and the triple-point temperature of water. The magnitudes of temperature units on the Kelvin and Celsius scales are identical (1 K ≡ 1°C). The temperatures on these two scales differ by a constant 273.15:

$$T(°\text{C}) = T(\text{K}) - 273.15 \tag{5-19}$$

Even though the thermodynamic temperature scale is defined with the help of the reversible heat engines, it is not possible, nor is it practical, to actually operate such an engine to determine numerical values on the absolute temperature scale. Absolute temperatures can be measured accurately by other means, such as the constant-volume ideal-gas thermometer discussed in Chap. 1 together with extrapolation techniques. The validity of Eq. 5-18 can be demonstrated from physical considerations for a reversible cycle using an ideal gas as the working fluid.

5-11 ■ THE CARNOT HEAT ENGINE

The hypothetical heat engine that operates on the reversible Carnot cycle is called the **Carnot heat engine.** The thermal efficiency of any heat engine, reversible or irreversible, is given by Eq. 5-6 as

$$\eta_{\text{th}} = 1 - \frac{Q_L}{Q_H}$$

efficiency

where Q_H is heat transferred to the heat engine from a high-temperature reservoir at T_H, and Q_L is heat rejected to a low-temperature reservoir at T_L. For reversible heat engines, the heat transfer ratio in the above relation can be replaced by the ratio of the absolute temperatures of the two reservoirs, as given by Eq. 5-18. Then the efficiency of a Carnot engine, or any reversible heat engine, becomes

Carnot efficiency For reversible Heat Engine

$$\eta_{\text{th, rev}} = 1 - \frac{T_L}{T_H} \tag{5-20}$$

This relation is often referred to as the **Carnot efficiency,** since the Carnot heat engine is the best known reversible engine. *This is the highest efficiency a heat engine operating between the two thermal energy reservoirs at temperatures T_L and T_H can have* (Fig. 5-52). All irreversible (i.e., actual) heat engines operating between these temperature limits (T_L and T_H) will have lower efficiencies. An actual heat engine cannot reach this maximum theoretical efficiency value because it is impossible to completely eliminate all the irreversibilities associated with the actual cycle.

Note that T_L and T_H in Eq. 5-20 are *absolute temperatures.* Using °C or °F for temperatures in this relation will give results grossly in error.

The thermal efficiencies of actual and reversible heat engines operating between the same temperature limits compare as follows (Fig. 5-53):

$$\eta_{th} \begin{cases} < \eta_{th,rev} & \text{irreversible heat engine} \\ = \eta_{th,rev} & \text{reversible heat engine} \\ > \eta_{th,rev} & \text{impossible heat engine} \end{cases} \tag{5-21}$$

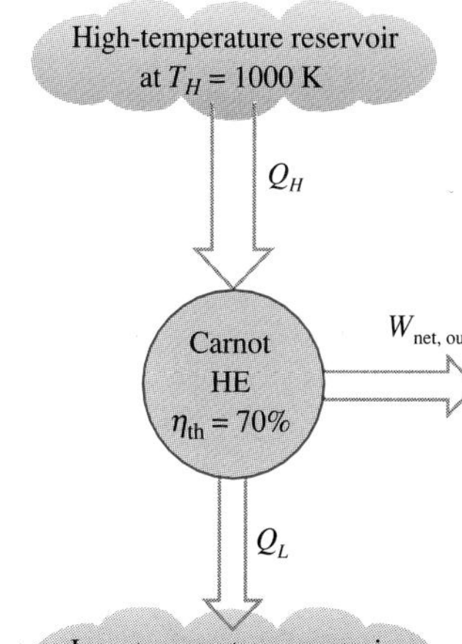

FIGURE 5-52

The Carnot heat engine is the most efficient of all heat engines operating between the same high- and low-temperature reservoirs.

Most work-producing devices (heat engines) in operation today have efficiencies under 40 percent, which appear low relative to 100 percent. However, when the performance of actual heat engines is assessed, the efficiencies should not be compared to 100 percent; instead, they should be compared to the efficiency of a reversible heat engine operating between the same temperature limits—because this is the true theoretical upper limit for the efficiency, not 100 percent.

The maximum efficiency of a steam power plant operating between T_H = 750 K and T_L = 300 K is 60 percent, as determined from Eq. 5-20. Compared with this value, an actual efficiency of 40 percent does not seem so bad, even though there is still plenty of room for improvement.

It is obvious from Eq. 5-20 that the efficiency of a Carnot heat engine increases as T_H is increased, or as T_L is decreased. This is to be expected since as T_L decreases, so does the amount of heat rejected, and as T_L approaches zero, the Carnot efficiency approaches unity. This is also true for actual heat engines. *The thermal efficiency of actual heat engines can be maximized by*

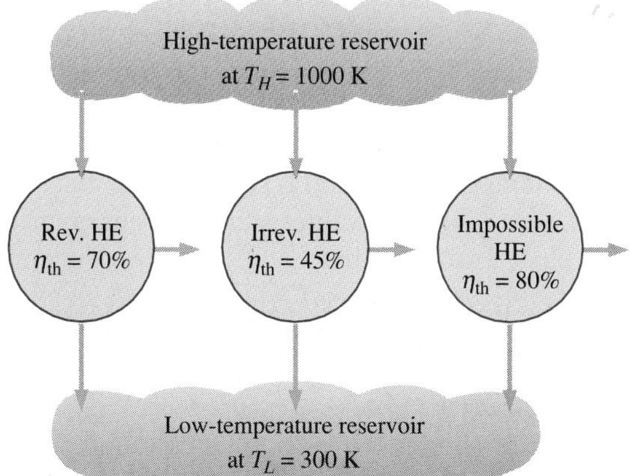

FIGURE 5-53

No heat engine can have a higher efficiency than a reversible heat engine operating between the same high- and low-temperature reservoirs.

supplying heat to the engine at the highest possible temperature (limited by material strength) *and rejecting heat from the engine at the lowest possible temperature* (limited by the temperature of the cooling medium such as rivers, lakes, or the atmosphere).

EXAMPLE 5-6 Analysis of a Carnot Heat Engine

A Carnot heat engine, shown in Fig. 5-54, receives 500 kJ of heat per cycle from a high-temperature source at 652°C and rejects heat to a low-temperature sink at 30°C. Determine (*a*) the thermal efficiency of this Carnot engine and (*b*) the amount of heat rejected to the sink per cycle.

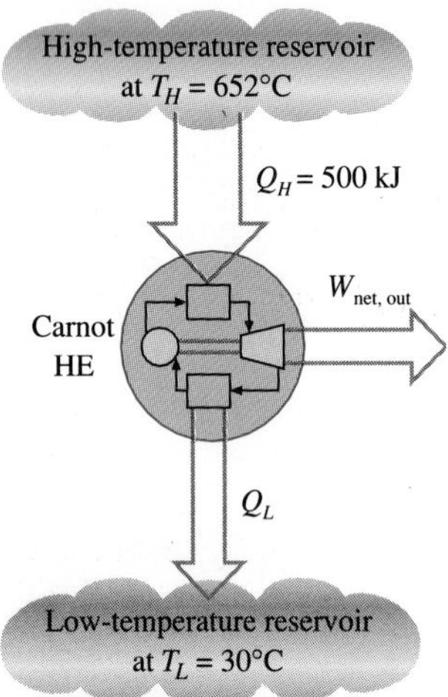

FIGURE 5-54
Schematic for Example 5-6.

Solution The heat supplied to a Carnot heat engine is given. The thermal efficiency and the heat rejected are to be determined.

Analysis (*a*) The Carnot heat engine is a reversible heat engine, and so its efficiency can be determined from Eq. 5-20 to be

$$\eta_{\text{th},C} = \eta_{\text{th,rev}} = 1 - \frac{T_L}{T_H} = 1 - \frac{(30 + 273)\text{ K}}{(652 + 273)\text{ K}} = \mathbf{0.672}$$

That is, this Carnot heat engine converts 67.2 percent of the heat it receives to work.

(*b*) The amount of heat rejected Q_L by this reversible heat engine is easily determined from Eq. 5-18 to be

$$Q_{L,\text{rev}} = \frac{T_L}{T_H} Q_{H,\text{rev}} = \frac{(30 + 273)\text{ K}}{(652 + 273)\text{ K}}(500\text{ kJ}) = \mathbf{163.8\text{ kJ}}$$

Discussion Note that this Carnot heat engine rejects to a low-temperature sink 163.8 kJ of the 500 kJ of heat it receives during each cycle.

The Quality of Energy

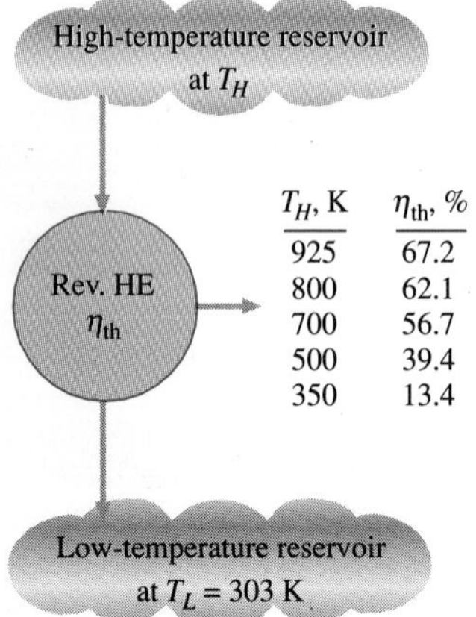

T_H, K	η_{th}, %
925	67.2
800	62.1
700	56.7
500	39.4
350	13.4

FIGURE 5-55
The fraction of heat that can be converted to work as a function of source temperature (for T_L = 303 K).

The Carnot heat engine in Example 5-6 receives heat from a source at 925 K and converts 67.2 percent of it to work while rejecting the rest (32.8 percent) to a sink at 303 K. Now let us examine how the thermal efficiency varies with the source temperature when the sink temperature is held constant.

The thermal efficiency of a Carnot heat engine that rejects heat to a sink at 303 K is evaluated at various source temperatures using Eq. 5-20 and is listed in Fig. 5-55. Clearly the thermal efficiency decreases as the source temperature is lowered. When heat is supplied to the heat engine at 500 instead of 925 K, for example, the thermal efficiency drops from 67.2 to 39.4 percent. That is, the fraction of heat that can be converted to work drops to 39.4 percent when the temperature of the source drops to 500 K. When the source temperature is 350 K, this fraction becomes a mere 13.4 percent.

These efficiency values show that energy has **quality** as well as quantity. It is clear from the thermal efficiency values in Fig. 5-55 that *more of the high-temperature thermal energy can be converted to work. Therefore, the higher the temperature, the higher the quality of the energy* (Fig. 5-56).

Large quantities of solar energy, for example, can be stored in large bodies of water called *solar ponds* at about 350 K. This stored energy can then be supplied to a heat engine to produce work (electricity). However, the efficiency of solar pond power plants is very low (under 5 percent) because of the low quality of the energy stored in the source, and the construction and maintenance costs are relatively high. Therefore, they are not competitive even though the energy supply of such plants is free. The temperature (and thus the quality) of the solar energy stored could be raised by utilizing concentrating collectors, but the equipment cost in that case becomes very high.

Work is a more valuable form of energy than heat since 100 percent of work can be converted to heat, but only a fraction of heat can be converted to work. When heat is transferred from a high-temperature body to a lower-temperature one, it is degraded since less of it now can be converted to work. For example, if 100 kJ of heat is transferred from a body at 1000 K to a body at 300 K, at the end we will have 100 kJ of thermal energy stored at 300 K, which has no practical value. But if this conversion is made through a heat engine, up to $1 - 300/1000 = 70$ percent of it could be converted to work, which is a more valuable form of energy. Thus 70 kJ of work potential is wasted as a result of this heat transfer, and energy is degraded. The degradation of energy during a process is discussed more fully in Chap. 7.

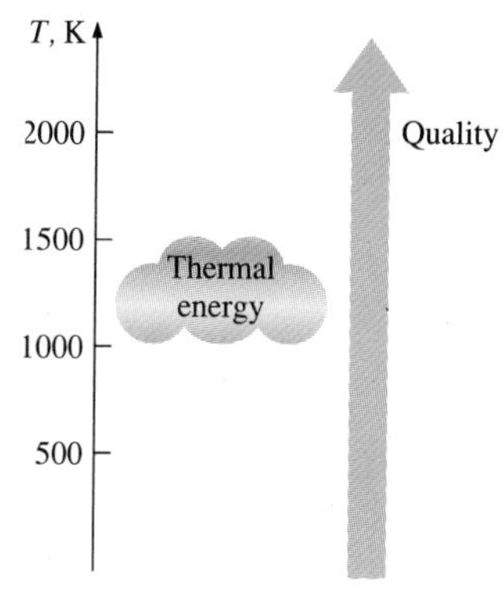

FIGURE 5-56
The higher the temperature of the thermal energy, the higher its quality.

Quantity versus Quality in Daily Life

At times of energy crisis, we are bombarded with speeches and articles on how to "conserve" energy. Yet we all know that the *quantity* of energy is already conserved. What is not conserved is the *quality* of energy, or the work potential of energy. Wasting energy is synonymous to converting it to a less useful form. One unit of high-quality energy can be more valuable than three units of lower-quality energy. For example, a finite amount of heat energy at high temperature is more attractive to power plant engineers than a vast amount of heat energy at low temperature, such as the energy stored in the upper layers of the oceans at tropical climates.

As part of our culture, we seem to be fascinated by quantity, and little attention is given to quality. But quantity alone cannot give the whole picture, and we need to consider quality as well. That is, we need to look at something from both the first- and second-law points of view when evaluating something, even in nontechnical areas. Below we present some ordinary events and show their relevance to the second law of thermodynamics.

Those who shop for a diamond engagement ring quickly realize that not all diamonds are created equal. For example, a half carat fine-quality diamond ring is more valuable (and thus more expensive) than a full carat standard-quality diamond ring. Thus, bigger is not necessarily better.

As another example, consider two students Andy and Wendy. Andy has 10 friends who never miss his parties and are always around during fun times. But they seem to be busy when Andy needs their help. Wendy, on the other hand, has five friends. But they are never too busy for her, and she can count on them at times of need. Let us now try to answer the question, *Who has more friends?* From the first law point of view, which considers quantity only, it is obvious that Andy has more friends. But from the second-law point of view, which considers quality as well, there is no doubt that Wendy is the one with more friends.

Another example with which most people will identify is the multi-billion-dollar diet industry, which is primarily based on the first law of thermodynamics. But considering that 90 percent of the people who lose weight gain it back quickly, with interest, suggests that the first law alone does not give the whole picture. This is also confirmed by recent work that shows that calories that come from fat are more likely to be stored as fat than the calories that come from carbohydrates and protein. A Stanford study found that body weight was related to fat calories consumed and not calories per se. A Harvard study found no correlation between calories eaten and degree of obesity. A major Cornell University survey involving 6500 people in nearly all provinces of China found that the Chinese eat more—gram for gram, calorie for calorie—than Americans do, but they weigh less, with less body fat. Studies indicate that the metabolism rates and hormone levels change noticeably in the mid 30s. Some researchers concluded that prolonged dieting teaches a body to survive on fewer calories, making it more *fuel efficient*. This probably explains why the dieters gain more weight than they lost once they go back to their normal eating levels.

People who seem to be eating whatever they want, whenever they want, are living proof that the calorie-counting technique (the first law) leaves many questions on dieting unanswered. Obviously, more research focused on the second-law effects of dieting is needed before we can fully understand the weight-gain and weight-loss process.

Those holding academic positions know well that "quantity" is still used as the basis in the annual merit reviews. That is, faculty members are judged on the basis of the *number* of their papers and the *dollar amount* of their research grants. Skillful faculty members play this game well by diluting their work over two or more papers, instead of publishing their findings in one sterling paper. The result is a huge increase in the number of publications, but little increase in new knowledge. The emphasis on numbers (the *quantity*) encourages faculty to publish many papers of questionable quality instead of a few high-quality ones, and discourages them from undertaking long-term high-risk studies. The issue of declining quality of publications is frequently raised, but no serious effort is made to find a workable solution. Maybe it is time that publications and other scholarly works are examined from the second-law point of view to assess their value realistically.

It is tempting to judge things on the basis of their *quantity* instead of their *quality* since assessing quality is much more difficult than assessing quantity. However, assessments made on the basis of quantity only (the first law) may be grossly inadequate and misleading. The above discussions on the quality of energy and the examples that follow indicate that there is a need for a further in-depth study of the second law of thermodynamics. This is what we will do in the following two chapters.

5-12 ■ THE CARNOT REFRIGERATOR AND HEAT PUMP

A refrigerator or a heat pump that operates on the reversed Carnot cycle is called a **Carnot refrigerator,** or a **Carnot heat pump.** The coefficient of performance of any refrigerator or heat pump, reversible or irreversible, is given by Eqs. 5-11 and 5-13 as

$$\text{COP}_\text{R} = \frac{1}{Q_H/Q_L - 1} \quad \text{and} \quad \text{COP}_\text{HP} = \frac{1}{1 - Q_L/Q_H}$$

where Q_L is the amount of heat absorbed from the low-temperature medium and Q_H is the amount of heat rejected to the high-temperature medium. The COPs of all reversible (such as Carnot) refrigerators or heat pumps can be determined by replacing the heat transfer ratios in the above relations by the ratios of the absolute temperatures of the high- and low-temperature media, as expressed by Eq. 5-18. Then the COP relations for reversible refrigerators and heat pumps become

$$\text{COP}_\text{R, rev} = \frac{1}{T_H/T_L - 1} \tag{5-22}$$

and

$$\text{COP}_\text{HP, rev} = \frac{1}{1 - T_L/T_H} \tag{5-23}$$

These are the highest coefficients of performance that a refrigerator or a heat pump operating between the temperature limits of T_L and T_H can have. All actual refrigerators or heat pumps operating between these temperature limits (T_L and T_H) will have lower coefficients of performance (Fig. 5-57).

The coefficients of performance of actual and reversible (such as Carnot) refrigerators operating between the same temperature limits can be compared as follows:

$$\text{COP}_\text{R} \begin{cases} < \text{COP}_\text{R, rev} & \text{irreversible refrigerator} \\ = \text{COP}_\text{R, rev} & \text{reversible refrigerator} \\ > \text{COP}_\text{R, rev} & \text{impossible refrigerator} \end{cases} \tag{5-24}$$

A similar relation can be obtained for heat pumps by replacing all values of COP_R in Eq. 5-24 by COP_HP.

The COP of a reversible refrigerator or heat pump is the maximum theoretical value for the specified temperature limits. Actual refrigerators or

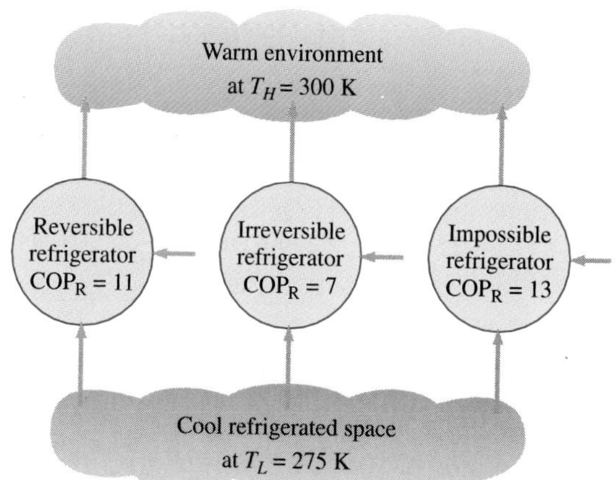

FIGURE 5-57

No refrigerator can have a higher COP than a reversible refrigerator operating between the same temperature limits.

heat pumps may approach these values as their designs are improved, but they can never reach them.

As a final note, the COPs of both the refrigerators and the heat pumps decrease as T_L decreases. That is, it requires more work to absorb heat from lower-temperature media. As the temperature of the refrigerated space approaches zero, the amount of work required to produce a finite amount of refrigeration approaches infinity and COP_R approaches zero.

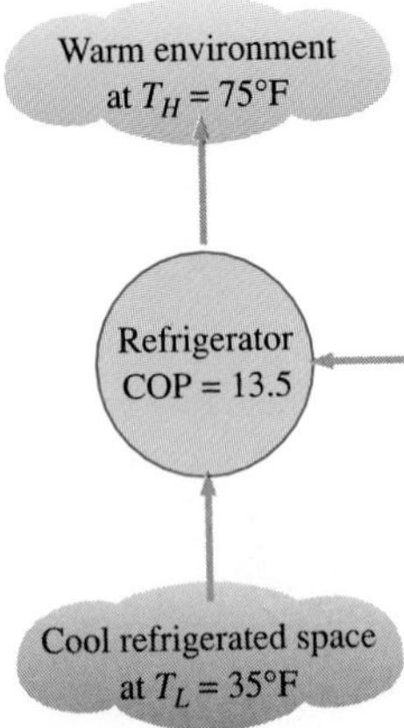

FIGURE 5-58
Schematic for Example 5-7.

EXAMPLE 5-7 A Questionable Claim for a Refrigerator

An inventor claims to have developed a refrigerator that maintains the refrigerated space at 35°F while operating in a room where the temperature is 75°F and that has a COP of 13.5. Is this claim reasonable?

Solution An extraordinary claim made for the performance of a refrigerator is to be evaluated.

Assumptions Steady operating conditions exist.

Analysis The performance of this refrigerator (shown in Fig. 5-58) can be evaluated by comparing it with a Carnot or any other reversible refrigerator operating between the same temperature limits:

$$COP_{R,\,max} = COP_{R,\,rev} = \frac{1}{T_H/T_L - 1}$$

$$= \frac{1}{(75 + 460\text{ R})/(35 + 460\text{ R}) - 1} = 12.4$$

This is the highest COP a refrigerator can have when removing heat from a cool medium at 35°F to a warmer medium at 75°F. Since the COP claimed by the inventor is above this maximum value, the claim is *false*.

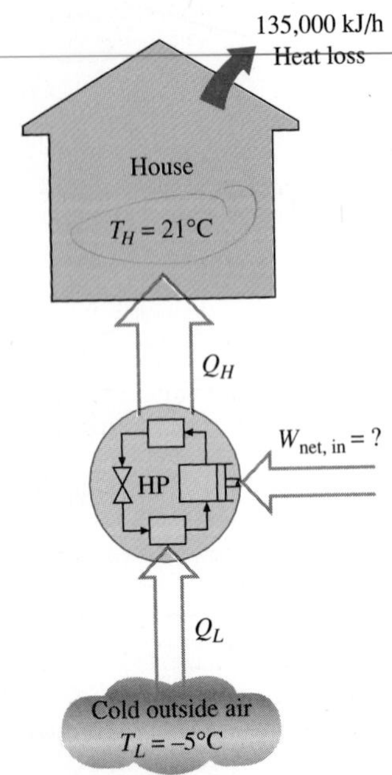

FIGURE 5-59
Schematic for Example 5-8.

EXAMPLE 5-8 Heating a House by a Carnot Heat Pump

A heat pump is to be used to heat a house during the winter, as shown in Fig. 5-59. The house is to be maintained at 21°C at all times. The house is estimated to be losing heat at a rate of 135,000 kJ/h when the outside temperature drops to −5°C. Determine the minimum power required to drive this heat pump.

Solution To maintain the house at a fixed temperature, the heat pump must supply the house as much heat as it is losing. That is, the heat pump must reject heat to the house (the high-temperature medium) at a rate of $\dot{Q}_H$ = 135,000 kJ/h = 37.5 kW.

Assumptions Steady operating conditions exist.

Analysis The power requirements will be minimum if a reversible heat pump is used to do the job. The COP of a reversible heat pump operating between the house (T_H = 21 + 273 = 294 K) and the outside air (T_L = −5 + 273 = 268 K) is

$$COP_{HP,\,rev} = \frac{1}{1 - T_L/T_H} = \frac{1}{1 - (268\text{ K}/294\text{ K})} = 11.3$$

Then the required power input to this reversible heat pump is determined from the definition of the COP, Eq. 5-12:

$$\dot{W}_{net,\,in} = \frac{\dot{Q}_H}{COP_{HP}} = \frac{37.5\text{ kw}}{11.3} = \mathbf{3.32\ kW}$$

Discussion This heat pump can meet the heating requirements of this house by consuming electric power at a rate of 3.32 kW only. If this house were to be heated by electric resistance heaters instead, the power consumption rate would jump up to 11.3 times to 37.5 kW. This is because in resistance heaters the electric energy is converted to heat at a one-to-one ratio. With a heat pump, however, energy is absorbed from the outside and carried to the inside using a refrigeration cycle that consumes only 3.32 kW. Notice that the heat pump does not create energy. It merely transports it from one medium (the cold outdoors) to another (the warm indoors).

5-13 ■ HOUSEHOLD REFRIGERATORS

Refrigerators to preserve perishable foods have long been one of the essential appliances in a household. They have proven to be highly durable and reliable, providing satisfactory service for over 15 years. A typical household refrigerator is actually a combination refrigerator-freezer since it has a freezer compartment to make ice and to store frozen food.

Today's refrigerators use much less energy as a result of using *smaller* and *higher-efficiency* motors and compressors, *better insulation materials, larger coil surface areas,* and *better door seals* (Fig. 5-60). At an average electricity rate of 8.3 cents per kWh, an average refrigerator costs about $72 a year to run, which is half the annual operating cost of a refrigerator 20 years ago. Replacing a 20-year-old, 18-ft^3 refrigerator with a new energy-efficient model will save over 1000 kWh of electricity per year. For the environment, this means a reduction of over 1 ton of CO_2, which is the leading cause of global climate change, and over 10 kg of SO_2, which is the leading cause of acid rain.

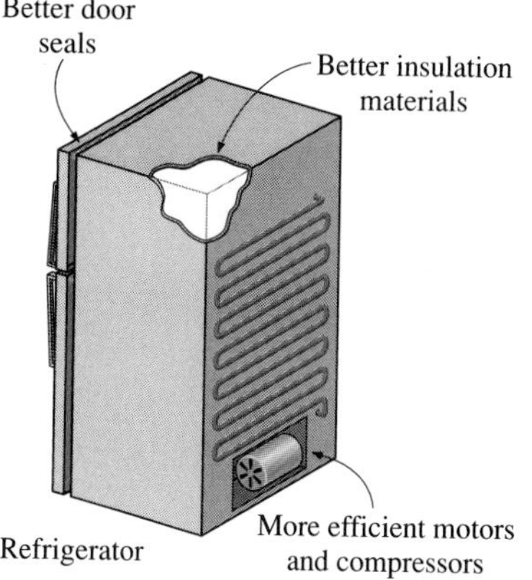

FIGURE 5-60

Today's refrigerators are much more efficient because of the improvements in technology and manufacturing.

Despite the improvements made in several areas during the past 100 years in household refrigerators, the basic *vapor-compression refrigeration cycle* has remained unchanged. The alternative *absorption refrigeration* and *thermoelectric refrigeration* systems are currently more expensive and less efficient, and they have found limited use in some specialized applications (Table 5-3).

TABLE 5-3

Typical operating efficiencies of some refrigeration systems for a freezer temperature of −18°C and ambient temperature of 32°C.

Type of refrigeration system	Coefficient of performance
Vapor-compression	1.3
Absorption refrigeration	0.4
Thermoelectric refrigeration	0.1

A household refrigerator is designed to maintain the freezer section at −18°C (0°F) and the refrigerator section at 3°C (37°F). Lower freezer temperatures increase energy consumption without improving the storage life of frozen foods significantly. Different temperatures for the storage of specific foods can be maintained in the refrigerator section by using *special-purpose* compartments.

Practically all full-size refrigerators have a large *air-tight* drawer for leafy vegetables and fresh fruits to seal in moisture and to protect them from the drying effect of cool air circulating in the refrigerator. A covered *egg compartment* in the lid extends the life of eggs by slowing down the moisture loss from the eggs. It is common for refrigerators to have a special warmer compartment for *butter* in the door to maintain butter at spreading temperature. The compartment also isolates butter and prevents it from absorbing *odors* and *tastes* from other food items. Some upscale models have a temperature-controlled *meat compartment* maintained at −0.5°C (31°F), which keeps meat at the lowest possible temperature without freezing it, and thus extending its storage life. The more expensive models come with an automatic *icemaker* located in the freezer section that is connected to the water line, as well as automatic ice and chilled-water dispensers. A typical icemaker can produce 2 to 3 kg of ice per day and store 3 to 5 kg of ice in a removable ice storage container.

Household refrigerators consume from about 90 W to 600 W of electrical energy when running and are designed to perform satisfactorily in environments at up to 43°C (110°F). Refrigerators run intermittently, as you may have noticed, running about 30 percent of the time under normal use in a house at 25°C (77°F).

For specified external dimensions, a refrigerator is desired to have *maximum* food storage volume, *minimum* energy consumption, and the *lowest* possible cost to the consumer. The total food storage volume has been increased over the years without an increase in the external dimensions by using thinner but more effective insulation and minimizing the space occupied by the compressor and the condenser. Switching from the fiberglass insulation [k = 0.032–0.040 W/(m · °C)] to expanded-in-place urethane foam insulation [k = 0.019 W/(m · °C)] made it possible to reduce the wall thickness of the refrigerator by almost half, from about 90 mm to 48 mm for the freezer section and from about 70 mm to 40 mm for the refrigerator section. The rigidity and bonding action of the foam also provide additional structural support. However, the entire shell of the refrigerator must be carefully sealed to prevent any water leakage or moisture migration into the insulation since moisture degrades the effectiveness of insulation.

The size of the compressor and the other components of a refrigeration system are determined on the basis of the anticipated heat load (or refrigeration load), which is the rate of heat flow into the refrigerator. The heat load consists of the *predictable part,* such as heat transfer through the walls and door gaskets of the refrigerator, fan motors, and defrost heaters (Fig. 5-61), and the *unpredictable part,* which depends on the user habits such as opening the door, making ice, and loading the refrigerator. The amount of *energy* consumed by the refrigerator can be minimized by practicing good *conservation measures* as discussed below.

1. *Open the refrigerator door the fewest times possible* for the shortest duration possible. Each time the refrigerator door is opened, the cool air inside is replaced by the warmer air outside, which needs to be cooled. Keeping the refrigerator or freezer full will save energy by reducing the amount of cold air that can escape each time the door is opened.

2. *Cool the hot foods* to room temperature first before putting them into the refrigerator. Moving a hot pan from the oven directly into the refrigerator not only wastes energy by making the refrigerator work longer, but it also causes the nearby perishable foods to spoil by creating a warm environment in its immediate surroundings (Fig. 5-62).

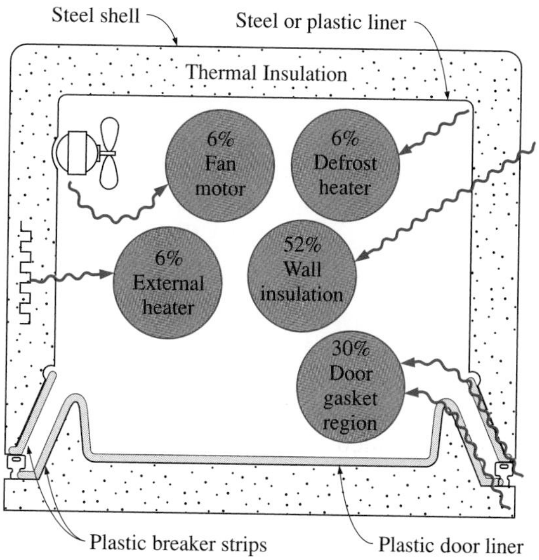

Steel shell
Steel or plastic liner
Thermal Insulation
6% Fan motor
6% Defrost heater
6% External heater
52% Wall insulation
30% Door gasket region
Plastic breaker strips
Plastic door liner

FIGURE 5-61

The cross section of a refrigerator showing the relative magnitudes of various effects that constitute the predictable heat load (from ASHRAE *Handbook of Refrigeration*, Chap. 48, Fig. 2).

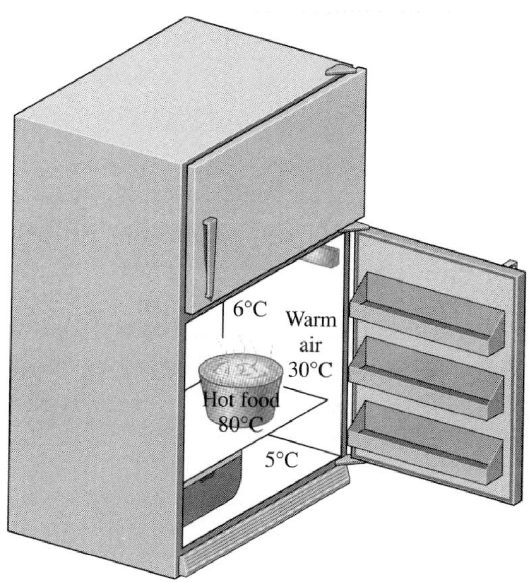

FIGURE 5-62

Putting hot foods into the refrigerator without cooling them first not only wastes energy but also could spoil the foods nearby.

3. *Clean the condenser coils* behind the refrigerator. The dust and grime that collect on the coils act as insulation that slows down heat dissipation through them. Cleaning the coils a couple of times a year with a damp cloth or a vacuum cleaner will improve cooling ability of the refrigerator while cutting down the power consumption by a few percent. Sometimes a fan is used to force-cool the condensers of large or built-in refrigerators, and the strong air motion keeps the coils clean.

4. *Check the door gasket* for air leaks. This can be done by placing a flashlight into the refrigerator, turning off the kitchen lights, and looking for light leaks. Heat transfer through the door gasket region accounts for almost

one-third of the regular heat load of the refrigerators, and thus any defective door gaskets must be repaired immediately.

5. *Avoid unnecessarily low temperature settings.* The recommended temperatures for freezers and refrigerators are −18°C (0°F) and 3°C (37°F), respectively. Setting the freezer temperature below −18°C adds significantly to the energy consumption but does not add much to the storage life of frozen foods. Keeping temperatures 6°C (or 10°F) below recommended levels can increase the energy use by as much as 25 percent.

6. *Avoid excessive ice build-up* on the interior surfaces of the evaporator. The ice layer on the surface acts as insulation and slows down heat transfer from the freezer section to the refrigerant. The refrigerator should be defrosted by manually turning off the temperature control switch when the ice thickness exceeds a few millimeters.

Defrosting is done automatically in no-frost refrigerators by supplying heat to the evaporator by a 300-W to 1000-W resistance heater or by hot refrigerant gas, periodically for short periods. The water is then drained to a pan outside where it is evaporated using the heat dissipated by the condenser. The no-frost evaporators are basically finned tubes subjected to air flow circulated by a fan. Practically all the frost collects on fins, which are the coldest surfaces, leaving the exposed surfaces of the freezer section and the frozen food frost-free.

7. *Use the power-saver switch* that controls the heating coils and prevents condensation on the outside surfaces in humid environments. The low-wattage heaters are used to raise the temperature of the outer surfaces of the refrigerator at critical locations above the dew point in order to avoid water droplets forming on the surfaces and sliding down. Condensation is most likely to occur in summer in hot and humid climates in homes without air-conditioning. The moisture formation on the surfaces is undesirable since it may cause the painted finish of the outer surface to deteriorate and it may wet the kitchen floor. About 10 percent of the total energy consumed by the refrigerator can be saved by turning this heater off and keeping it off unless there is visible condensation on the outer surfaces.

8. *Do not block the air flow passages* to and from the condenser coils of the refrigerator. The heat dissipated by the condenser to the air is carried away by air that enters through the bottom and sides of the refrigerator and leaves through the top. Any blockage of this natural convection air circulation path by large objects such as several cereal boxes on top of the refrigerator will degrade the performance of the condenser and thus the refrigerator (Fig. 5-63).

FIGURE 5-63

The condenser coils of a refrigerator must be cleaned periodically, and the airflow passages must not be blocked to maintain high performance.

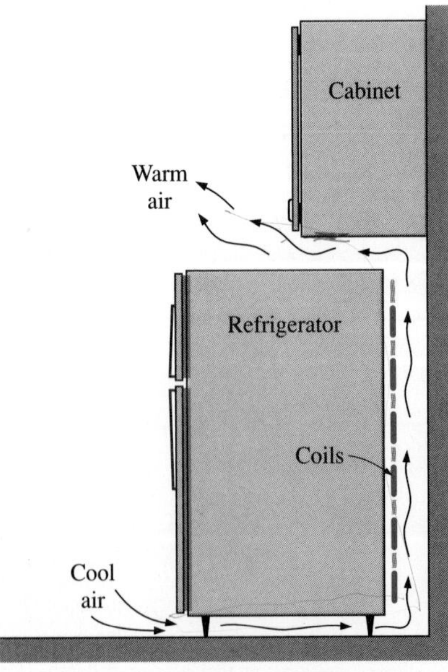

These and other commonsense conservation measures will result in a reduction in the energy and maintenance costs of a refrigerator as well as an extended trouble-free life of the device.

EXAMPLE 5-9 Malfunction of a Refrigerator Light Switch

The interior lighting of refrigerators is provided by incandescent lamps whose switches are actuated by the opening of the refrigerator door. Consider a refrigerator whose 40-W light bulb remains on continuously as a result of a malfunction of the switch (Fig. 5-64). If the refrigerator has a coefficient of performance of 1.3 and the cost of electricity is 8 cents per kWh, determine the increase in the energy consumption of the refrigerator and its cost per year if the switch is not fixed.

Solution The light bulb consumes 40 W of power when it is on, and thus adds 40 W to the heat load of the refrigerator.

Assumptions The life of the light bulb is more than 1 year.

Analysis Noting that the coefficient of performance of the refrigerator is 1.3, the power consumed by the refrigerator to remove the heat generated by the light bulb is determined from

$$\dot{W}_{\text{refrig}} = \frac{\dot{Q}_{\text{refrig}}}{\text{COP}_{\text{R}}} = \frac{40 \text{ W}}{1.3} = 30.8 \text{ W}$$

Therefore, the total additional power consumed by the refrigerator is

$$\dot{W}_{\text{total, additional}} = \dot{W}_{\text{light}} + \dot{W}_{\text{refrig}} = 40 + 30.8 = 70.8 \text{ W}$$

The total number of hours in a year is

$$\text{Annual hours} = (365 \text{ days/yr})(24 \text{ h/day}) = 8760 \text{ h/yr}$$

Assuming the refrigerator is opened 20 times a day for an average of 30 s, the light would normally be on for

$$\begin{aligned}\text{Normal operating hours} &= (20 \text{ times/day})(30 \text{ s/time})(1 \text{ h/3600 s})(365 \text{ days/yr})\\ &= 61 \text{ h/yr}\end{aligned}$$

Then the additional hours the light remains on as a result of the malfunction becomes

$$\begin{aligned}\text{Additional operating hours} &= \text{Annual hours} - \text{Normal operating hours}\\ &= 8760 - 61 = 8699 \text{ h/yr}\end{aligned}$$

Therefore, the additional electric power consumption and its cost per year are

$$\begin{aligned}\text{Additional power consumption} &= \dot{W}_{\text{total, additional}} \times (\text{Additional operating hours})\\ &= (0.0708 \text{ kW})(8699 \text{ h/yr}) = \mathbf{616 \text{ kWh/yr}}\end{aligned}$$

and

$$\begin{aligned}\text{Additional power cost} &= (\text{Additional power consumption})(\text{Unit cost})\\ &= (616 \text{ kWh/yr})(\$0.08/\text{kWh}) = \mathbf{\$49.3/yr}\end{aligned}$$

Discussion Note that not repairing the switch will cost the homeowner about $50 a year. This is alarming when we consider that at $0.08/kWh, a typical refrigerator consumes about $70 worth of electricity a year.

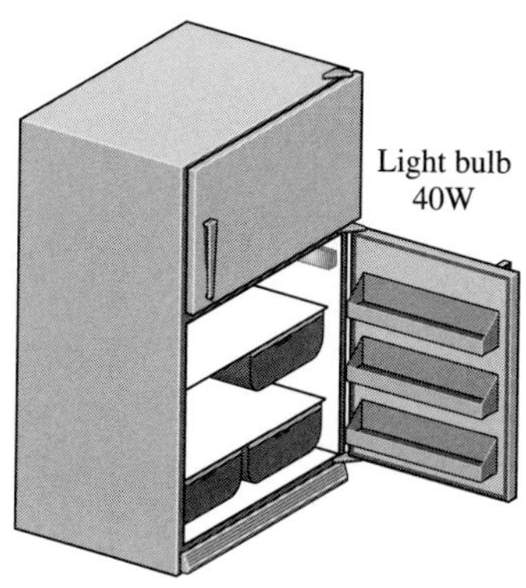

FIGURE 5-64
Schematic for Example 5-9.

5-14 ■ SUMMARY

The *second law of thermodynamics* states that processes occur in a certain direction, not in any direction. A process will not occur unless it satisfies both the first and the second laws of thermodynamics. Bodies that can absorb or reject finite amounts of heat isothermally are called *thermal energy reservoirs* or *heat reservoirs*.

Work can be converted to heat directly, but heat can be converted to work only by some devices called heat engines. The *thermal efficiency* of a heat engine is defined as

$$\eta_{\text{th}} = \frac{W_{\text{net, out}}}{Q_H} = 1 - \frac{Q_L}{Q_H}$$

where $W_{\text{net, out}}$ is the net work output of the heat engine, Q_H is the amount of heat supplied to the engine, and Q_L is the amount of heat rejected by the engine.

Refrigerators and heat pumps are devices that absorb heat from low-temperature media and reject it to higher-temperature ones. The performance of a refrigerator or a heat pump is expressed in terms of the *coefficient of performance,* which is defined as

$$\text{COP}_{\text{R}} = \frac{Q_L}{W_{\text{net, in}}} = \frac{1}{Q_H/Q_L - 1}$$

$$\text{COP}_{\text{HP}} = \frac{Q_H}{W_{\text{net, in}}} = \frac{1}{1 - Q_L/Q_H}$$

The *Kelvin-Planck statement* of the second law of thermodynamics states that no heat engine can produce a net amount of work while exchanging heat with a single reservoir only. The *Clausius statement* of the second law states that no device can transfer heat from a cooler body to a warmer one without leaving an effect on the surroundings.

Any device that violates the first or the second law of thermodynamics is called a *perpetual-motion machine.*

A process is said to be *reversible* if both the system and the surroundings can be restored to their original conditions. Any other process is *irreversible.* The effects such as friction, non-quasi-equilibrium expansion or compression, and heat transfer through a finite temperature difference render a process irreversible and are called *irreversibilities.*

The *Carnot cycle* is a reversible cycle that is composed of four reversible processes, two isothermal and two adiabatic. The *Carnot principles* state that the thermal efficiencies of all reversible heat engines operating between the same two reservoirs are the same, and that no heat engine is more efficient than a reversible one operating between the same two reservoirs. These statements form the basis for establishing a *thermodynamic temperature scale* related to the heat transfers between a reversible device and the high- and low-temperature reservoirs by

$$\left(\frac{Q_H}{Q_L}\right)_{\text{rev}} = \frac{T_H}{T_L}$$

Therefore, the Q_H/Q_L ratio can be replaced by T_H/T_L for reversible devices, where T_H and T_L are the absolute temperatures of the high- and low-temperature reservoirs, respectively.

A heat engine that operates on the reversible Carnot cycle is called a *Carnot heat engine.* The thermal efficiency of a Carnot heat engine, as well as all other reversible heat engines, is given by

$$\eta_{\text{th, rev}} = 1 - \frac{T_L}{T_H}$$

This is the maximum efficiency a heat engine operating between two reservoirs at temperatures T_H and T_L can have.

The COPs of reversible refrigerators and heat pumps are given in a similar manner as

$$\text{COP}_{\text{R, rev}} = \frac{1}{T_H/T_L - 1}$$

and

$$\text{COP}_{\text{HP, rev}} = \frac{1}{1 - T_L/T_H}$$

Again, these are the highest COPs a refrigerator or a heat pump operating between the temperature limits of T_H and T_L can have.

REFERENCES AND SUGGESTED READING

1. W. Z. Black and J. G. Hartley. *Thermodynamics.* New York: Harper & Row, 1985.

2. J. R. Howell and R. O. Buckius. *Fundamentals of Engineering Thermodynamics.* New York: McGraw-Hill, 1987.

3. D. Stewart. "Wheels Go Round and Round, but Always Run Down." November 1986, Smithsonian, pp. 193–208.

4. G. J. Van Wylen and R. E. Sonntag. *Fundamentals of Classical Thermodynamics.* 3rd ed. New York: John Wiley & Sons, 1985.

5. K. Wark. *Thermodynamics.* 5th ed. New York: McGraw-Hill, 1988.

PROBLEMS*

Introduction to the Second Law of Thermodynamics

5-1C A mechanic claims to have developed a car engine that runs on water instead of gasoline. What is your response to this claim?

5-2C Describe an imaginary process that satisfies the first law but violates the second law of thermodynamics.

5-3C Describe an imaginary process that satisfies the second law but violates the first law of thermodynamics.

5-4C Describe an imaginary process that violates both the first and the second laws of thermodynamics.

*Students are encouraged to answer all the concept "C" questions.

5-5C An experimentalist claims to have raised the temperature of a small amount of water to 150°C by transferring heat from high-pressure steam at 120°C. Is this a reasonable claim? Why? Assume no refrigerator or heat pump is used in the process.

Thermal Energy Reservoirs

5-6C Consider the energy dissipated by a computer in a room. What is a suitable choice for a thermal energy reservoir?

5-7C What is a thermal energy reservoir? Give some examples.

5-8C Consider the process of baking potatoes in a conventional oven. Can the hot air in the oven be treated as a thermal energy reservoir? Explain.

5-9C Consider the energy generated by a TV set. What is a suitable choice for a thermal energy reservoir?

Heat Engines and Thermal Efficiency

5-10C Is it possible for a heat engine to operate without rejecting any waste heat to a low-temperature reservoir? Explain.

5-11C What are the characteristics of all heat engines?

5-12C Describe two ways to determine the net work output of a heat engine.

5-13C Consider the process of baking potatoes in a conventional oven. How would you define the efficiency of the oven for this baking process?

5-14C Consider a pan of water being heated (*a*) by placing it on an electric range and (*b*) by placing a heating element in the water. Which method is a more efficient way of heating water? Explain.

5-15C Which is a more efficient way of converting electricity to light—using a light bulb or using a fluorescent tube?

5-16C Baseboard heaters are basically electric resistance heaters and are frequently used in space heating. A homeowner claims that her 5-year-old baseboard heaters have a conversion efficiency of 100 percent. Is this claim in violation of any thermodynamic laws? Explain.

5-17C What is the Kelvin-Planck expression of the second law of thermodynamics?

5-18C Does a heat engine that has a thermal efficiency of 100 percent necessarily violate (*a*) the first law and (*b*) the second law of thermodynamics? Explain.

5-19C In the absence of any friction and other irreversibilities, can a heat engine have an efficiency of 100 percent? Explain.

5-20C Are the efficiencies of all the work-producing devices, including the hydroelectric power plants, limited by the Kelvin-Planck statement of the second law? Explain.

5-21 An 800-MW steam power plant, which is cooled by a nearby river, has a thermal efficiency of 40 percent. Determine the rate of heat transfer to the river water. Will the actual heat transfer rate be higher or lower than this value? Why?

5-22 A steam power plant receives heat from a furnace at a rate of 280 GJ/h. Heat losses to the surrounding air from the steam as it passes through the pipes and other components are estimated to be about 8 GJ/h. If the waste heat is transferred to the cooling water at a rate of 145 GJ/h, determine (*a*) net power output and (*b*) the thermal efficiency of this power plant.
Answers: (*a*) 35.3 MW, (*b*) 45.4 percent

5-23E A car engine with a power output of 95 hp has a thermal efficiency of 28 percent. Determine the rate of fuel consumption if the heating value of the fuel is 19,000 Btu/lbm.

5-24 A steam power plant with a power output of 150 MW consumes coal at a rate of 60 tons/h. If the heating value of the coal is 30,000 kJ/kg, determine the thermal efficiency of this plant (1 ton ≡ 1000 kg).
Answer: 30.0 percent

5-25 An automobile engine consumes fuel at a rate of 20 L/h and delivers 60 kW of power to the wheels. If the fuel has a heating value of 44,000 kJ/kg and a density of 0.8 g/cm^3, determine the efficiency of this engine.
Answer: 30.7 percent

5-26E Solar energy stored in large bodies of water, called solar ponds, is being used to generate electricity. If such a solar power plant has an efficiency of 4 percent and a net power output of 300 kW, determine the average value of the required solar energy collection rate, in Btu/h.

5-27 The United States produces about 55 percent of its electricity from coal, and the efficiency of conventional coal-fired power plants is about 34 percent. The combined power rating of the power plants in the United States is about 500,000 MW. Assuming steady power generation at 400 MW, determine the amount of heat rejected by the coal-fired power plants in the United States per year, in kJ.

5-28 The Department of Energy projects that between the years 1995 and 2010, the United States will need to build new power plants to generate an additional 150,000 MW of electricity to meet the increasing demand for electric power. One possibility is to build coal-fired power plants, which cost \$1300 per kW to construct and have an efficiency of 34 percent. Another possibility is to use the clean-burning Integrated Gasification Combined Cycle (IGCC) plants where the coal is subjected to heat and pressure to gasify it while removing sulfur and particulate matter from it. The gaseous coal is then burned in a gas turbine, and part of the waste heat from the exhaust gases is recovered to generate steam for the steam turbine. Currently the construction of IGCC plants costs about \$1500 per kW, but their efficiency is about 45 percent. The average heating value of the coal is about 28,000,000 kJ per ton (that is, 28,000,000 kJ of heat is released when 1 ton of coal is

FIGURE P5-30

burned). If the IGCC plant is to recover its cost difference from fuel savings in five years, determine what the price of coal should be in $ per ton.

5-29 Repeat Prob. 5-28 for a simple payback period of three years instead of five years.

5-30 Wind energy has been used since 4000 BC to power sailboats, grind grain, pump water for farms, and, more recently, generate electricity. In the United States alone, more than 6 million small windmills, most of them under 5 hp, have been used since the 1850s to pump water. Small windmills have been used to generate electricity since 1900, but the development of modern wind turbines occurred only recently in response to the energy crises in the early 1970s. The cost of wind power has dropped an order of magnitude from about $0.50/kWh in the early 1980s to about $0.05/kWh in the mid 1990s, which is about the price of electricity generated at coal-fired power plants. Areas with an average wind speed of 6 m/s (or 14 mph) are potential sites for economical wind power generation. Commercial wind turbines generate from 100 kW to 3.2 MW of electric power each at peak design conditions. The blade span (or rotor) diameter of the 3.2 MW wind turbine built by Boeing Engineering is 320 ft (97.5 m). The rotation speed of rotors of wind turbines is usually under 40 rpm (under 20 rpm for large turbines). Altamont Pass in California is the world's largest windfarm with 15,000 modern wind turbines. This farm and two others in California produced 2.8 billion kWh of electricity in 1991, which is enough power to meet the electricity needs of San Francisco. Many wind turbines currently in operation have just two blades. This is because at tip speeds of 100 to 200 mph, the efficiency of the two bladed turbine approaches the theoretical maximum, and the increase in the efficiency by adding a third or fourth blade is so little that they do not justify the added cost and weight. Take the density of air to be 1.20 kg/m^3.

Consider a wind turbine with an 80-m-diameter rotor that is rotating at 20 rpm under steady winds at an average velocity of 30 km/h. Assuming the turbine has an efficiency of 35 percent (i.e., it converts 35 percent of the kinetic energy of the wind to electricity), determine (*a*) the power produced, in kW; (*b*) the tip speed of the blade, in km/h; and (*c*) the revenue generated by the wind turbine per year if the electric power produced is sold to the utility at $0.06/kWh.

5-31 Repeat Prob. 5-30 for an average wind velocity of 25 km/h.

5-32E An Ocean Thermal Energy Conversion (OTEC) power plant built in Hawaii in 1987 was designed to operate between the temperature limits of 86°F at the ocean surface and 41°F at a depth of 2100 ft. About 13,300 gpm of cold sea water was pumped from deep ocean through a 40-in-diameter pipe as the cooling medium or heat sink. If the cooling water rises by 6°C, determine the highest thermal efficiency that can be obtained by this power plant, and the maximum amount of power that can be generated.

Energy Conversion Efficiencies

5-33 Consider a 3-kW hooded electric open burner in an area where the unit costs of electricity and natural gas are $0.07/kWh and $0.60/therm, respectively. The efficiency of open burners can be taken to be 73 percent for electric burners and 38 percent for gas burners. Determine the rate of energy consumption by the burner and the unit cost of utilized energy for both electric and gas burners.

5-34 A 75-hp motor that has an efficiency of 91.0 percent is worn out and is replaced by a high-efficiency motor that has an efficiency of 95.4 percent. Determine the reduction in the heat gain of the room due to higher efficiency under full-load conditions.

5-35 A 60-hp electric car is powered by an electric motor mounted in the engine compartment. If the motor has an average efficiency of 88 percent, determine the rate of heat supply by the motor to the engine compartment at full load.

5-36 A 75-hp motor that has an efficiency of 91.0 percent is worn out and is to be replaced by a high-efficiency motor that has an efficiency of 95.4 percent. The motor operates 4368 hours a year at a load factor of 0.75. Taking the cost of electricity to be \$0.08/kWh, determine the amount of energy and money saved as a result of installing the high-efficiency motor instead of the standard motor. Also, determine the simple payback period if the purchase prices of the standard and high-efficiency motors are \$5449 and \$5520, respectively.

5-37E The steam requirements of a manufacturing facility are being met by a boiler whose rated heat input is 3.6×10^6 Btu/h. The combustion efficiency of the boiler is measured to be 0.7 by a hand-held flue gas analyzer. After tuning up the boiler, the combustion efficiency rises to 0.8. The boiler operates 1500 hours a year intermittently. Taking the unit cost of energy to be \$4.35/10^6 Btu, determine the annual energy and cost savings as a result of tuning up the boiler.

5-38 The space heating of a facility is accomplished by natural gas heaters that are 80 percent efficient. The compressed air needs of the facility are met by a large liquid-cooled compressor. The coolant of the compressor is cooled by air in a liquid-to-air heat exchanger whose airflow section is 1.0 m high and 1.0 m wide. During typical operation, the air is heated from 20°C to 52°C as it flows through the heat exchanger. The average velocity of air on the inlet side is measured to be 3 m/s. The compressor operates 20 hours a day and 5 days a week throughout the year. Taking the heating season to be 6 months (26 weeks) and the cost of the natural gas to be \$0.50/therm (1 therm = 105,500 kJ), determine how much money will be saved by diverting the compressor waste heat into the facility during the heating season.

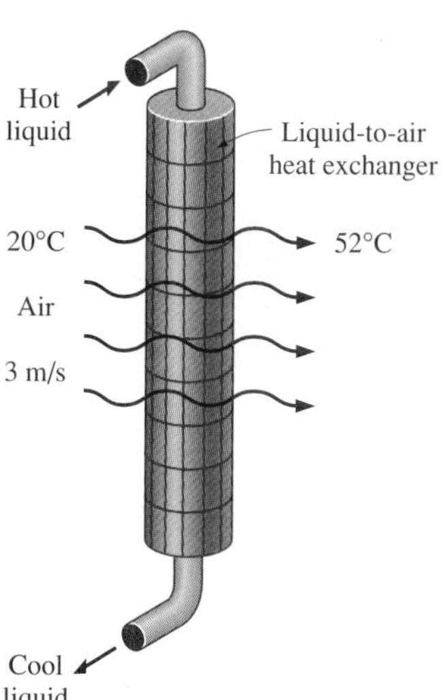

FIGURE P5-38

5-39 An exercise room has 8 weight-lifting machines that have no motors and 4 treadmills each equipped with a 2.5-hp motor. The motors operate at an average load factor of 0.7, at which their efficiency is 0.77. During peak evening hours, all 12 pieces of exercising equipment are used continuously, and there are also two people doing light exercises while waiting in line for one piece of the equipment. Determine the rate of heat gain of the exercise room from people and the equipment at peak load conditions.

5-40 Consider a classroom for 40 students and one instructor, each generating heat at a rate of 100 W. Lighting is provided by 18 fluorescent light bulbs, 40 W each, and the ballasts consume an additional 10 percent. Determine the rate of internal heat generation in this classroom when it is fully occupied.

5-41 A room is cooled by circulating chilled water through a heat exchanger located in a room. The air is circulated through the heat exchanger by a 0.25-hp fan. Typical efficiency of small electric motors driving 0.25-hp equipment is 54 percent. Determine the rate of heat supply by the fan–motor assembly to the room.

Refrigerators and Heat Pumps

5-42C What is the difference between a refrigerator and a heat pump?

5-43C What is the difference between a refrigerator and an air conditioner?

5-44C In a refrigerator, heat is transferred from a lower-temperature medium (the refrigerated space) to a higher-temperature one (the kitchen air). Is this a violation of the second law of thermodynamics? Explain.

5-45C A heat pump is a device that absorbs energy from the cold outdoor air and transfers it to the warmer indoors. Is this a violation of the second law of thermodynamics? Explain.

5-46C Define the coefficient of performance of a refrigerator in words. Can it be greater than unity?

5-47C Define the coefficient of performance of a heat pump in words. Can it be greater than unity?

5-48C A heat pump that is used to heat a house has a COP of 2.5. That is, the heat pump delivers 2.5 kWh of energy to the house for each 1 kWh of electricity it consumes. Is this a violation of the first law of thermodynamics? Explain.

5-49C A refrigerator has a COP of 1.5. That is, the refrigerator removes 1.5 kWh of energy from the refrigerated space for each 1 kWh of electricity it consumes. Is this a violation of the first law of thermodynamics? Explain.

5-50C What is the Clausius expression of the second law of thermodynamics?

5-51C Show that the Kelvin-Planck and the Clausius expressions of the second law are equivalent.

5-52 A household refrigerator with a COP of 1.8 removes heat from the refrigerated space at a rate of 90 kJ/min. Determine (*a*) the electric power consumed by the refrigerator and (*b*) the rate of heat transfer to the kitchen air. *Answers:* (*a*) 0.83 kW, (*b*) 140 kJ/min

5-53 An air conditioner removes heat steadily from a house at a rate of 750 kJ/min while drawing electric power at a rate of 6 kW. Determine (*a*) the COP of this air conditioner and (*b*) the rate of heat transfer to the outside air. *Answers:* (*a*) 2.08, (*b*) 1110 kJ/min

5-54 A household refrigerator runs one-fourth of the time and removes heat from the food compartment at an average rate of 1200 kJ/h. If the COP of the refrigerator is 2.5, determine the power the refrigerator draws when running.

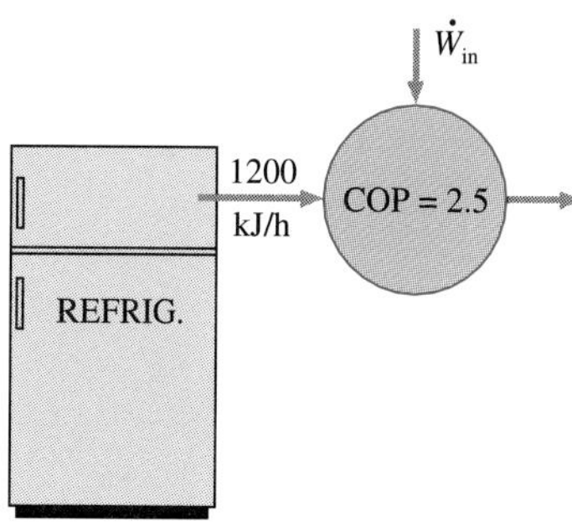

FIGURE P5-54

5-55E Water enters an ice machine at 55°F and leaves as ice at 25°F. If the COP of the ice machine is 2.4 during this operation, determine the required power input for an ice production rate of 20 lbm/h. (169 Btu of energy needs to be removed from each lbm of water at 55°F to turn it into ice at 25°F.)

5-56 A household refrigerator that has a power input of 450 W and a COP of 2.5 is to cool five large watermelons, 10 kg each, to 8°C. If the watermelons are initially at 20°C, determine how long it will take for the refrigerator to cool them. The watermelons can be treated as water whose specific heat is 4.2 kJ/(kg · °C). Is your answer realistic or optimistic? Explain.
Answer: 2240 s

5-57 When a man returns to his well-sealed house on a summer day, he finds that the house is at 32°C. He turns on the air conditioner, which cools the entire house to 20°C in 15 min. If the COP of the air-conditioning system is 2.5, determine the power drawn by the air conditioner. Assume the entire mass within the house is equivalent to 800 kg of air for which C_v = 0.72 kJ/(kg · °C) and C_p = 1.0 kJ/(kg · °C).

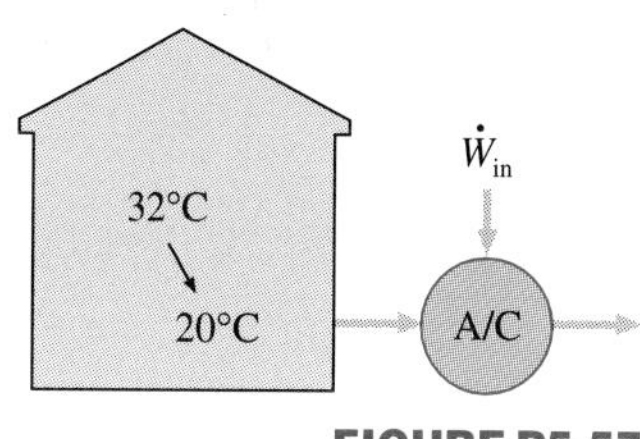

FIGURE P5-57

5-58E Determine the COP of a refrigerator that removes heat from the food compartment at a rate of 8000 kJ/h for each kW of power it consumes. Also, determine the rate of heat rejection to the outside air.

5-59 Determine the COP of a heat pump that supplies energy to a house at a rate of 8000 kJ/h for each kW of electric power it draws. Also, determine the rate of energy absorption from the outdoor air.
Answers: 2.22, 4400 kJ/h

5-60 A house that was heated by electric resistance heaters consumed 1200 kWh of electric energy in a winter month. If this house were heated instead by a heat pump that has an average COP of 2.4, determine how much money the homeowner would have saved that month. Assume a price of 8.5¢/kWh for electricity.

5-61E A heat pump with a COP of 2.5 supplies energy to a house at a rate of 60,000 Btu/h. Determine (*a*) the electric power drawn by the heat pump and (*b*) the rate of heat removal from the outside air.
Answers: (*a*) 9.43 hp, (*b*) 36,000 Btu/h

5-62 A heat pump used to heat a house runs about one-third of the time. The house is losing heat at an average rate of 15,000 kJ/h. If the COP of the heat pump is 3.5, determine the power the heat pump draws when running.

5-63 A heat pump is used to maintain a house at a constant temperature of 23°C. The house is losing heat to the outside air through the walls and the

FIGURE P5-63

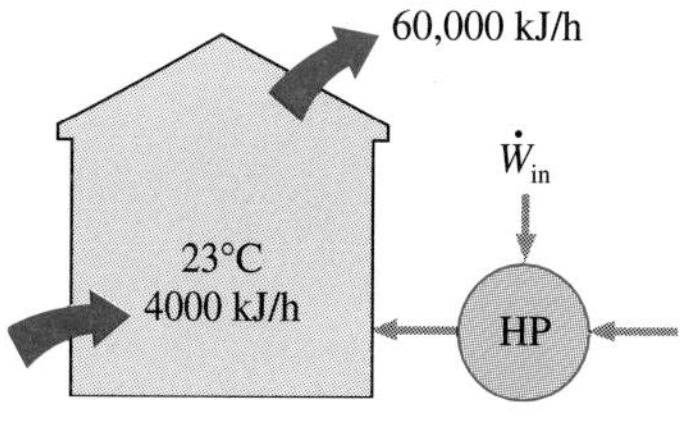

60,000 kJ/h
$\dot{W}_{in}$
23°C
4000 kJ/h
HP

windows at a rate of 60,000 kJ/h while the energy generated within the house from people, lights, and appliances amounts to 4000 kJ/h. For a COP of 2.5, determine the required power input to the heat pump. *Answer:* 6.22 kW

5-64 Consider an office room that is being cooled adequately by a 12,000 Btu/h window air conditioner. Now it is decided to convert this room into a computer room by installing several computers, terminals, and printers with a total rated power of 3.5 kW. The facility has several 4000 Btu/h air conditioners in storage that can be installed to meet the additional cooling requirements. Assuming a usage factor of 0.4 (i.e., only 40 percent of the rated power will be consumed at any given time) and additional occupancy of four people, each generating heat at a rate of 100 W, determine how many of these air conditioners need to be installed to the room.

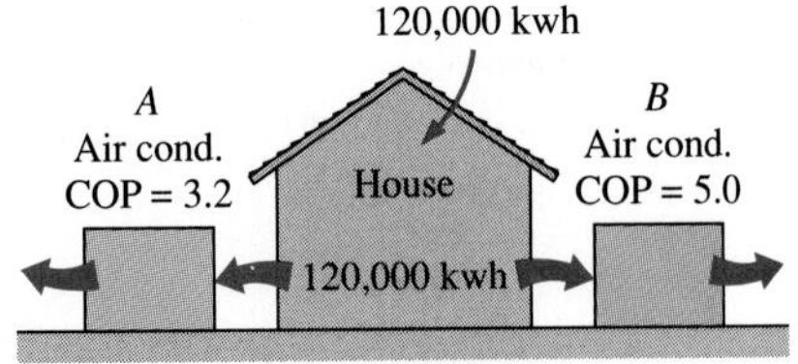

FIGURE P5-65

5-65 Consider a building whose annual air-conditioning load is estimated to be 120,000 kWh in an area where the unit cost of electricity is \$0.10/kWh. Two air conditioners are considered for the building. Air conditioner A has a seasonal average Cop of 3.2 and costs \$5500 to purchase and install. Air conditioner B has a seasonal average COP of 5.0 and costs \$7000 to purchase and install. All else being equal, determine which air conditioner is a better buy.

Perpetual-Motion Machines

5-66C An inventor claims to have developed a resistance heater that supplies 1.2 kWh of energy to a room for each kWh of electricity it consumes. Is this a reasonable claim, or has the inventor developed a perpetual-motion machine? Explain.

5-67C It is common knowledge that the temperature of air rises as it is compressed. An inventor thought about using this high-temperature air to heat buildings. He used a compressor driven by an electric motor. The inventor claims that the compressed hot-air system is 25 percent more efficient than a resistance heating system that provides an equivalent amount of heating. Is this claim valid, or is this just another perpetual-motion machine? Explain.

Reversible and Irreversible Processes

5-68C A cold canned drink is left in a warmer room where its temperature rises as a result of heat transfer. Is this a reversible process? Explain.

5-69C A hot baked potato is left on a table where it cools to the room temperature. Is this a reversible or an irreversible process? Explain.

5-70C Why are engineers interested in reversible processes even though they can never be achieved?

5-71C Air is compressed from 20°C and 100 kPa to 300°C and 800 kPa first in a reversible manner and then in an irreversible manner. Which case do you think will require more work input?

5-72C Why does a non-quasi-equilibrium compression process require a larger work input than the corresponding quasi-equilibrium one?

5-73C Why does a non-quasi-equilibrium expansion process deliver less work than the corresponding quasi-equilibrium one?

5-74C How do you distinguish between internal and external irreversibilities?

5-75C Is a reversible expansion or compression process necessarily quasi-equilibrium? Is a quasi-equilibrium expansion or compression process necessarily reversible? Explain.

The Carnot Cycle and Carnot Principles

5-76C What are the four processes that make up the Carnot cycle?

5-77C What are the two statements known as the Carnot principles?

5-78C Somebody claims to have developed a new reversible heat-engine cycle that has a higher theoretical efficiency than the Carnot cycle operating between the same temperature limits. How do you evaluate this claim?

5-79C Somebody claims to have developed a new reversible heat-engine cycle that has the same theoretical efficiency as the Carnot cycle operating between the same temperature limits. Is this a reasonable claim?

5-80C Is it possible to develop (*a*) an actual and (*b*) a reversible heat-engine cycle that is more efficient than a Carnot cycle operating between the same temperature limits? Explain.

Carnot Heat Engines

5-81C Is there any way to increase the efficiency of a Carnot heat engine other than by increasing T_H or decreasing T_L?

5-82C Consider two actual power plants operating with solar energy. Energy is supplied to one plant from a solar pond at 80°C and to the other from concentrating collectors that raise the water temperature to 600°C. Which of these power plants will have a higher efficiency? Explain.

5-83 A Carnot heat engine operates between a source at 1000 K and a sink at 300 K. If the heat engine is supplied with heat at a rate of 800 kJ/min, determine (*a*) the thermal efficiency and (*b*) the power output of this heat engine. *Answers:* (*a*) 70 percent, (*b*) 9.33 kW

5-84 A Carnot heat engine receives 500 kJ of heat from a source of unknown temperature and rejects 200 kJ of it to a sink at 17°C. Determine (*a*) the temperature of the source and (*b*) the thermal efficiency of the heat engine.

5-85 A heat engine operates between a source at 550°C and a sink at 25°C. If heat is supplied to the heat engine at a steady rate of 1200 kJ/min, determine the maximum power output of this heat engine.

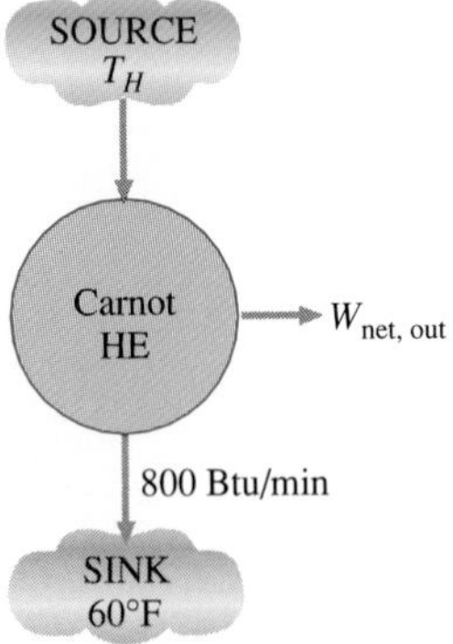

FIGURE P5-86E

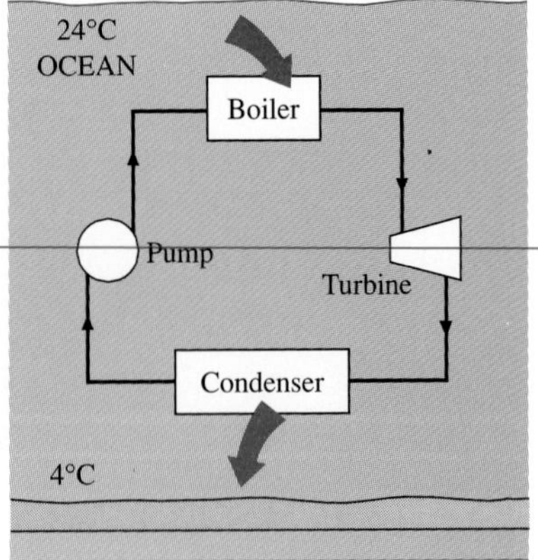

FIGURE P5-87

5-86E A heat engine is operating on a Carnot cycle and has a thermal efficiency of 55 percent. The waste heat from this engine is rejected to a nearby lake at 60°F at a rate of 800 Btu/min. Determine (*a*) the power output of the engine and (*b*) the temperature of the source.

Answers: (*a*) 23.1 hp, (*b*) 1155.6 R

5-87 In tropical climates, the water near the surface of the ocean remains warm throughout the year as a result of solar energy absorption. In the deeper parts of the ocean, however, the water remains at a relatively low temperature since the sun's rays cannot penetrate very far. It is proposed to take advantage of this temperature difference and construct a power plant that will absorb heat from the warm water near the surface and reject the waste heat to the cold water a few hundred meters below. Determine the maximum thermal efficiency of such a plant if the water temperatures at the two respective locations are 24 and 4°C.

5-88 An innovative way of power generation involves the utilization of geothermal energy—the energy of hot water that exists naturally underground—as the heat source. If a supply of hot water at 140°C is discovered at a location where the environmental temperature is 20°C, determine the maximum thermal efficiency a geothermal power plant built at that location can have.

Answer: 29.1 percent

5-89 An inventor claims to have developed a heat engine that receives 800 kJ of heat from a source at 400 K and produces 250 kJ of net work while rejecting the waste heat to a sink at 300 K. Is this a reasonable claim? Why?

5-90E An experimentalist claims that, based on his measurements, a heat engine receives 300 Btu of heat from a source of 900 R, converts 160 Btu of it to work, and rejects the rest as waste heat to a sink at 540 R. Are these measurements reasonable? Why?

Carnot Refrigerators and Heat Pumps

5-91C How can we increase the COP of a Carnot refrigerator?

5-92C What is the highest COP that a refrigerator operating between temperature levels T_L and T_H can have?

5-93C In an effort to conserve energy in a heat-engine cycle, somebody suggests incorporating a refrigerator that will absorb some of the waste energy Q_L and transfer it to the energy source of the heat engine. Is this a smart idea? Explain.

5-94C It is well established that the thermal efficiency of a heat engine increases as the temperature at which heat is rejected from the heat engine T_L decreases. In an effort to increase the efficiency of a power plant, somebody suggests refrigerating the cooling water before it enters the condenser, where heat rejection takes place. Would you be in favor of this idea? Why?

5-95C It is well known that the thermal efficiency of heat engines increases as the temperature of the energy source increases. In an attempt to improve the efficiency of a power plant, somebody suggests transferring heat from the available energy source to a higher-temperature medium by a heat pump before energy is supplied to the power plant. What do you think of this suggestion? Explain.

5-96 A Carnot refrigerator operates in a room in which the temperature is 25°C and consumes 2 kW of power when operating. If the food compartment of the refrigerator is to be maintained at 3°C, determine the rate of heat removal from the food compartment.

5-97 A refrigerator is to remove heat from the cooled space at a rate of 300 kJ/min to maintain its temperature at −8°C. If the air surrounding the refrigerator is at 25°C, determine the minimum power input required for this refrigerator. *Answer:* 0.623 kW

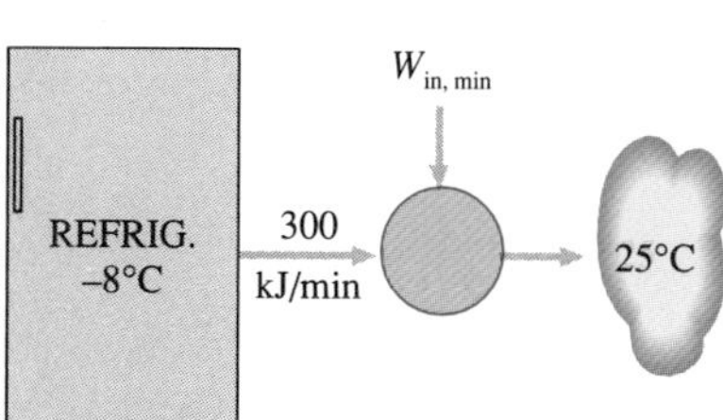

FIGURE P5-97

5-98 An air-conditioning system operating on the reversed Carnot cycle is required to transfer heat from a house at a rate of 750 kJ/min, to maintain its temperature at 20°C. If the outdoor air temperature is 35°C, determine the power required to operate this air-conditioning system. *Answer:* 0.64 kW

5-99E An air-conditioning system is used to maintain a house at 70°F when the temperature outside is 90°F. If this air-conditioning system draws 5 hp of power when operating, determine the maximum rate of heat removal from the house that it can provide.

5-100 A Carnot refrigerator operates in a room in which the temperature is 25°C. The refrigerator consumes 500 W of power when operating and has a COP of 4.5. Determine (*a*) the rate of heat removal from the refrigerated space and (*b*) the temperature of the refrigerated space.
Answers: (*a*) 133 kJ/min, (*b*) −29.2°C

5-101 An inventor claims to have developed a refrigeration system that removes heat from the closed region at −5°C and transfers it to the surrounding air at 22°C while maintaining a COP of 8.2. Is this claim reasonable? Why?

5-102 During an experiment conducted in a room at 25°C, a laboratory assistant measures that a refrigerator that draws 2 kW of power has removed 30,000 kJ of heat from the refrigerated space, which is maintained at −30°C. The running time of the refrigerator during the experiment was 20 min. Determine if these measurements are reasonable.

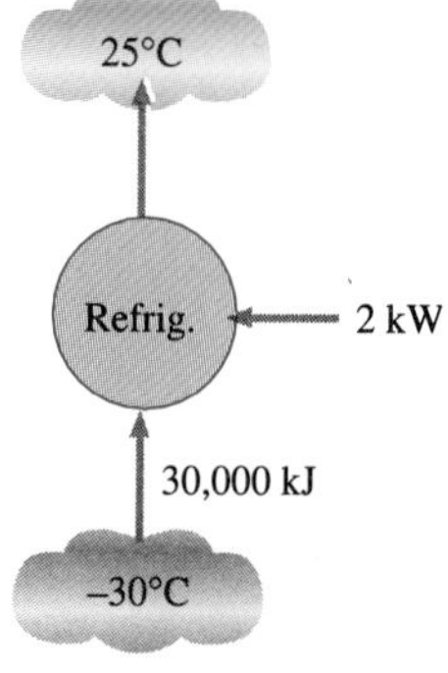

FIGURE P5-102

5-103E An air-conditioning system is used to maintain a house at 75°F when the temperature outside is 95°F. The house is gaining heat through the walls and the windows at a rate of 750 Btu/min, and the heat generation rate within the house from people, lights, and appliances amounts to 150 Btu/min. Determine the minimum power input required for this air-conditioning system. *Answer:* 0.79 hp

5-104 A heat pump is used to heat a house and maintain it at 20°C. On a winter day when the outdoor air temperature is −5°C, the house is estimated to lose heat at a rate of 75,000 kJ/h. Determine the minimum power required to operate this heat pump.

5-105 A heat pump is used to maintain a house at 22°C by extracting heat from the outside air on a day when the outside air temperature is 2°C. The house is estimated to lose heat at a rate of 110,000 kJ/h, and the heat pump consumes 8 kW of electric power when running. Is this heat pump powerful enough to do the job?

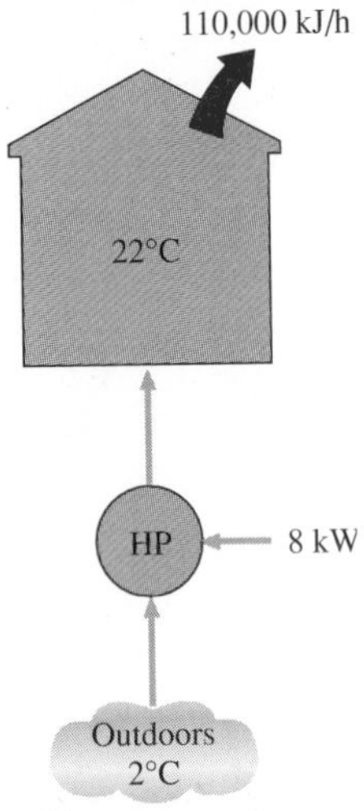

FIGURE P5-105

5-106 The structure of a house is such that it loses heat at a rate of 5400 kJ/h per °C difference between the indoors and outdoors. A heat pump that requires a power input of 6 kW is used to maintain this house at 21°C. Determine the lowest outdoors temperature for which the heat pump can meet the heating requirements of this house. *Answer:* −13.3°C

5-107 The performance of a heat pump degrades (i.e., its COP decreases) as the temperature of the heat source decreases. This makes using heat pumps at locations with severe weather conditions unattractive. Consider a house that is heated and maintained at 20°C by a heat pump during the winter. What is the maximum COP for this heat pump if heat is extracted from the outdoor air at (*a*) 10°C, (*b*) −5°C, and (*c*) −30°C?

5-108E A heat pump is to be used for heating a house in winter. The house is to be maintained at 78°F at all times. When the temperature outdoors drops to 25°F, the heat losses from the house are estimated to be 80,000 Btu/h. Determine the minimum power required to run this heat pump if heat is extracted from (*a*) the outdoor air at 25°F and (*b*) the well water at 50°F.

5-109 A Carnot heat pump is to be used to heat a house and maintain it at 20°C in winter. On a day when the average outdoor temperature remains at about 2°C, the house is estimated to lose heat at a rate of 82,000 kJ/h. If the heat pump consumes 8 kW of power while operating, determine (*a*) how long the heat pump ran on that day; (*b*) the total heating costs, assuming an average price of 8.5¢/kWh for electricity; and (*c*) the heating cost for the same day if resistance heating is used instead of a heat pump. *Answers:* (*a*) 4.19 h, (*b*) $2.85, (*c*) $46.47

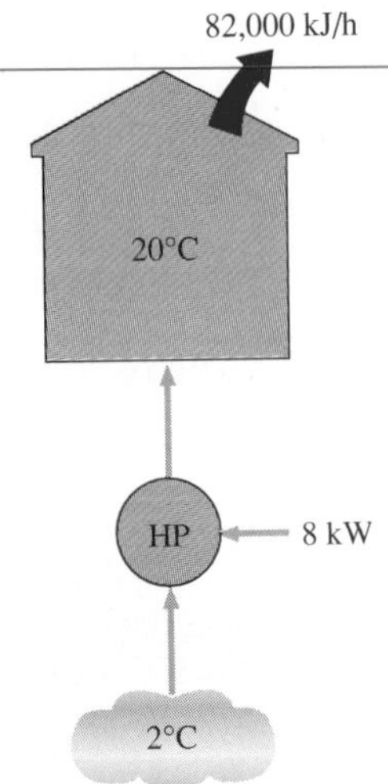

FIGURE P5-109

5-110 A Carnot heat engine receives heat from a reservoir at 900°C at a rate of 800 kJ/min and rejects the waste heat to the ambient air at 27°C. The entire work output of the heat engine is used to drive a refrigerator that removes heat from the refrigerated space at −5°C and transfers it to the same ambient air at 27°C. Determine (*a*) the maximum rate of heat removal from the refrigerated space and (*b*) the total rate of heat rejection to the ambient air. *Answers:* (*a*) 4982 kJ/min, (*b*) 5782 kJ

5-111E A Carnot heat engine receives heat from a reservoir at 1700°F at a rate of 700 Btu/min and rejects the waste heat to the ambient air at 80°F. The entire work output of the heat engine is used to drive a refrigerator that removes heat from the refrigerated space at 20°F and transfers it to the same ambient air at 80°F. Determine (*a*) the maximum rate of heat removal from the refrigerated space and (*b*) the total rate of heat rejection to the ambient air. *Answers:* (*a*) 4262 Btu/min, (*b*) 4962 Btu/min

5-112C Someone proposes that the refrigeration system of a supermarket be overdesigned so that the entire air-conditioning needs of the store can be met by refrigerated air without installing any air-conditioning system. What do you think of this proposal?

5-113C Someone proposes that the entire refrigerator/freezer requirements of a store be met using a large freezer that supplies sufficient cold air at $-20°C$ instead of installing separate refrigerators and freezers. What do you think of this proposal?

5-114C Explain how you can reduce the energy consumption of your household refrigerator.

5-115C Why is it important to clean the condenser coils of a household refrigerator a few times a year? Also, why is it important not to block airflow through the condenser coils?

5-116C Why are today's refrigerators much more efficient than those built in the past?

5-117 The "Energy Guide" label of a refrigerator states that the refrigerator will consume \$83 worth of electricity per year under normal use if the cost of electricity is \$0.08/kWh. If the electricity consumed by the light bulb is negligible and the refrigerator consumes 300 W when running, determine the fraction of the time the refrigerator will run.

5-118 The interior lighting of refrigerators is usually provided by incandescent lamps whose switches are actuated by the opening of the refrigerator door. Consider a refrigerator whose 40-W light bulb remains on about 60 h per year. It is proposed to replace the light bulb by an energy-efficient bulb that consumes only 18 W but costs \$25 to purchase and install. If the refrigerator has a coefficient of performance of 1.3 and the cost of electricity is 8 cents per kWh, determine if the energy savings of the proposed light bulb justify its cost.

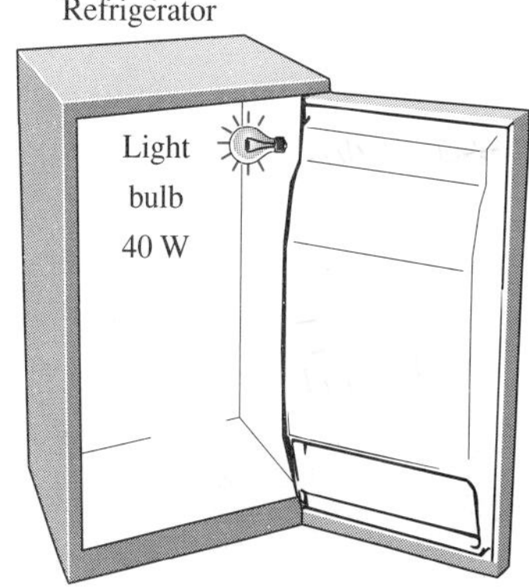

FIGURE P5-118

5-119 It is commonly recommended that hot foods be cooled first to room temperature by simply waiting a while before they are put into the refrigerator to save energy. Despite this commonsense recommendation, a person keeps cooking a large pan of stew twice a week and putting the pan into the refrigerator while it is still hot, thinking that the money saved is probably too little. But he says he can be convinced if you can show that the money saved is significant. The average mass of the pan and its contents is 5 kg. The average temperature of the kitchen is 20°C, and the average temperature of the food is 95°C when it is taken off the stove. The refrigerated space is maintained at 3°C, and the average specific heat of the food and the pan can be taken to be 3.9 kJ/(kg · °C). If the refrigerator has a coefficient of performance of 1.2 and the cost of electricity is 10 cents per kWh, determine how much this person will save a year by waiting for the food to cool to room temperature before putting it into the refrigerator.

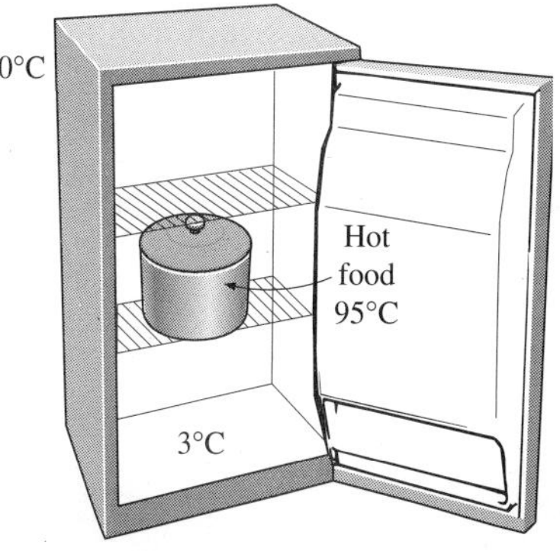

FIGURE P5-119

5-120 It is often stated that the refrigerator door should be opened as few times as possible for the shortest duration of time to save energy. Consider a household refrigerator whose interior volume is 0.9 m^3 and average internal temperature is 4°C. At any given time, one-third of the refrigerated space is occupied by food items, and the remaining 0.6 m^3 is filled with air. The average temperature and pressure in the kitchen are 20°C and 95 kPa, respectively. Also, the moisture contents of the air in the kitchen and the refrigerator are 0.010 and 0.004 kg per kg of air, respectively, and thus 0.006 kg of water vapor is condensed and removed for each kg of air that enters. The refrigerator door is opened an average of 8 times a day, and each time half of the air volume in the refrigerator is replaced by the warmer kitchen air. If the refrigerator has a coefficient of performance of 1.4 and the cost of electricity is 7.5 cents per kWh, determine the cost of the energy wasted per year as a result of opening the refrigerator door. What would your answer be if the kitchen air were very dry and thus a negligible amount of water vapor condensed in the refrigerator?

Review Problems

5-121 Consider a Carnot heat-engine cycle executed in a steady-flow system using steam as the working fluid. The cycle has a thermal efficiency of 30 percent, and steam changes from saturated liquid to saturated vapor at 300°C during the heat addition process. If the mass flow rate of the steam is 5 kg/s, determine the net power output of this engine, in kW.

5-122 A heat pump with a COP of 2.4 is used to heat a house. When running, the heat pump consumes 8 kW of electric power. If the house is losing heat to the outside at an average rate of 40,000 kJ/h and the temperature of the house is 3°C when the heat pump is turned on, determine how long it will take for the temperature in the house to rise to 22°C. Assume the house is well sealed (i.e., no air leaks) and take the entire mass within the house (air, furniture, etc.) to be equivalent to 2000 kg of air.

5-123 An old gas turbine has an efficiency of 17 percent and develops a power output of 6000 kW. Determine the fuel consumption rate of this gas turbine, in L/min, if the fuel has a heating value of 46,000 kJ/kg and a density of 0.8 g/cm^3.

5-124 Show that $COP_{HP} = COP_R + 1$ when both the heat pump and the refrigerator have the same Q_L and Q_H values.

5-125 An air-conditioning system is used to maintain a house at a constant temperature of 20°C. The house is gaining heat from outdoors at a rate of 20,000 kJ/h, and the heat generated in the house from the people, lights, and appliances amounts to 8000 kJ/h. For a COP of 2.5, determine the required power input to this air-conditioning system. *Answer:* 3.11 kW

5-126 Consider a Carnot heat-engine cycle executed in a closed system using 0.01 kg of refrigerant-134a as the working fluid. The cycle has a thermal efficiency of 15 percent, and the refrigerant-134a changes from saturated liquid to saturated vapor at 70°C during the heat addition process. Determine the net work output of this engine, in kJ.

5-127 A heat pump with a COP of 3.2 is used to heat a house. When running, the heat pump consumes power at a rate of 5 kW. If the temperature in the house is 7°C when the heat pump is turned on, how long will it take for the heat pump to raise the temperature of the house to 22°C? Is this answer realistic or optimistic? Explain. Assume the entire mass within the house (air, furniture, etc.) is equivalent to 1500 kg of air for which $C_v = 0.72$ kJ/(kg · °C) and $C_p = 1.0$ kJ/(kg · °C). *Answer:* 1012 s

5-128 A promising method of power generation involves collecting and storing solar energy in large artificial lakes a few meters deep, called solar ponds. Solar energy is absorbed by all parts of the pond, and the water temperature rises everywhere. The top part of the pond, however, loses to the atmosphere much of the heat it absorbs, and as a result, its temperature drops. This cool water serves as insulation for the bottom part of the pond and helps trap the energy there. Usually, salt is planted at the bottom of the pond to prevent the rise of this hot water to the top. A power plant that uses an organic fluid, such as alcohol, as the working fluid can be operated between the top and the bottom portions of the pond. If the water temperature is 35°C near the surface and 80°C near the bottom of the pond, determine the maximum thermal efficiency that this power plant can have. Is it realistic to use 35 and 80°C for temperatures in the calculations? Explain. *Answer:* 12.7 percent

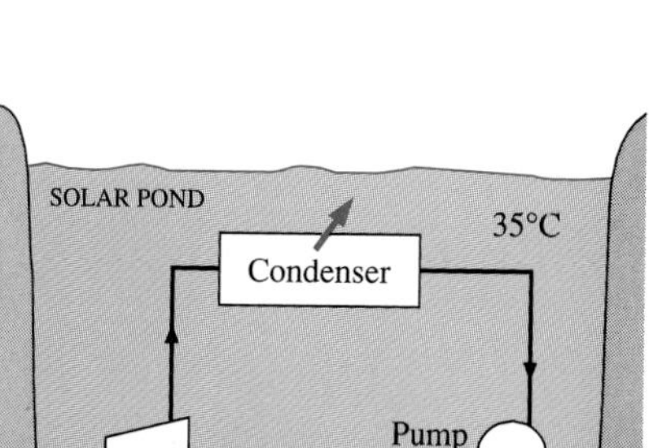

FIGURE P5-128

5-129 Consider a Carnot heat-engine cycle executed in a closed system using 0.0103 kg of steam as the working fluid. It is known that the maximum absolute temperature in the cycle is twice the minimum absolute temperature, and the net work output of the cycle is 25 kJ. If the steam changes from saturated vapor to saturated liquid during the heat rejection process, determine the temperature of the steam during the heat rejection process, in °C.

5-130 Consider a Carnot refrigeration cycle executed in a closed system in the saturated liquid–vapor mixture region using 0.96 kg of refrigerant-134a as the working fluid. It is known that the maximum absolute temperature in the cycle is 1.2 times the minimum absolute temperature, and the net work input to the cycle is 22 kJ. If the refrigerant changes from saturated liquid to saturated vapor during the heat rejection process, determine the minimum pressure in the cycle, in kPa.

5-131 Consider two Carnot heat engines operating in series. The first engine receives heat from the reservoir at 1200 K and rejects the waste heat to another reservoir at temperature *T*. The second engine receives this energy rejected by the first one, converts some of it to work, and rejects the rest to a reservoir at 300 K. If the thermal efficiencies of both engines are the same, determine the temperature *T*. *Answer:* 600 K

5-132 The COP of a refrigerator decreases as the temperature of the refrigerated space is decreased. That is, removing heat from a medium at a very low temperature will require a large work input. Determine the minimum work input required to remove 1 kJ of heat from liquid helium at 3 K when the outside temperature is 300 K. *Answer:* 99 kJ

5-133E A Carnot heat pump is used to heat and maintain a residential building at 75°F. An energy analysis of the house reveals that it loses heat at a rate of 2500 Btu/h per °F temperature difference between the indoors and the outdoors. For an outdoor temperature of 35°F, determine (*a*) the coefficient of performance and (*b*) the required power input to the heat pump.
Answers: (*a*) 13.4, (*b*) 2.93 hp

5-134 A Carnot heat engine receives heat at 750 K and rejects the waste heat to the environment at 300 K. The entire work output of the heat engine is used to drive a Carnot refrigerator that removes heat from the cooled space at −15°C at a rate of 400 kJ/min and rejects it to the same environment at 300 K. Determine (*a*) the rate of heat supplied to the heat engine and (*b*) the total rate of heat rejection to the environment.

5-135 A heat engine operates between two reservoirs at 800 and 20°C. One-half of the work output of the heat engine is used to drive a Carnot heat pump that removes heat from the cold surroundings at 2°C and transfers it to a house maintained at 22°C. If the house is losing heat at a rate of 95,000 kJ/h, determine the minimum rate of heat supply to the heat engine required to keep the house at 22°C.

5-136 Consider a Carnot refrigeration cycle executed in a closed system in the saturated liquid–vapor mixture region using 0.8 kg of refrigerant-134a as the working fluid. The maximum and the minimum temperatures in the cycle are 20°C and −10°C, respectively. It is known that the refrigerant is saturated liquid at the end of the heat rejection process, and the net work input to the cycle is 12 kJ. Determine the fraction of the mass of the refrigerant that vaporizes during the heat addition process, and the pressure at the end of the heat rejection process.

5-137 Consider a Carnot heat-pump cycle executed in a steady-flow system in the saturated liquid–vapor mixture region using refrigerant-134a flowing at a rate of 0.264 kg/s as the working fluid. It is known that the maximum absolute temperature in the cycle is 1.15 times the minimum absolute temperature, and the net power input to the cycle is 5 kW. If the refrigerant changes from saturated vapor to saturated liquid during the heat rejection process, determine the ratio of the maximum to minimum pressures in the cycle.

5-138 A Carnot heat engine is operating between a source at T_H and a sink at T_L. If it is desired to double the thermal efficiency of this engine, what should the new source temperature be? Assume the sink temperature is held constant.

5-139 When discussing Carnot engines, it is assumed that the engine is in thermal equilibrium with the source and the sink during the heat addition and heat rejection processes, respectively. That is, it is assumed that $T_H^* = T_H$ and $T_L^* = T_L$ so that there is no external irreversibility. In that case, the thermal efficiency of the Carnot engine is $\eta_C = 1 - T_L/T_H$.

FIGURE P5-139

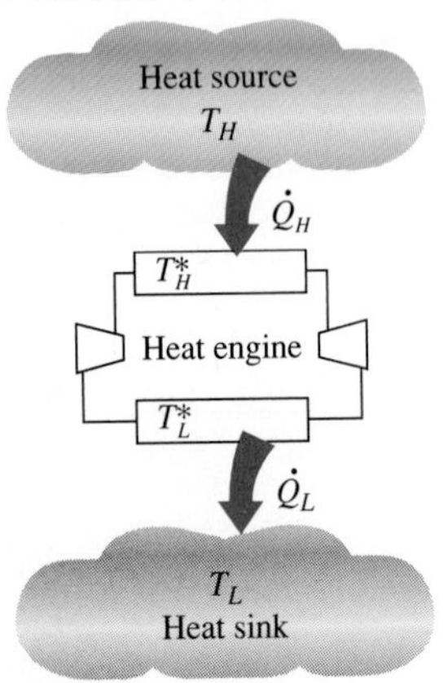

In reality, however, we must maintain a reasonable temperature difference between the two heat transfer media in order to have an acceptable heat transfer rate through a finite heat exchanger surface area. The heat transfer rates in that case can be expressed as

$$\dot{Q}_H = (hA)_H(T_H - T_H^*)$$
$$\dot{Q}_L = (hA)_L(T_L^* - T_L)$$

where h and A are the heat transfer coefficient and heat transfer surface area, respectively. When the values of h, A, T_H, and T_L are fixed, show that the power output will be a maximum when

$$\frac{T_L^*}{T_H^*} = \left(\frac{T_L}{T_H}\right)^{1/2}$$

Also, show that the maximum net power output in this case is

$$\dot{W}_{C,\max} = \frac{(hA)_H T_H}{1 + (hA)_H/(hA)_L}\left[1 - \left(\frac{T_L}{T_H}\right)^{1/2}\right]^2$$

5-140 Consider a homeowner who is replacing his 20-year-old natural gas furnace that has an efficiency of 60 percent. The homeowner is considering a conventional furnace that has an efficiency of 82 percent and costs \$1600 and a high-efficiency furnace that has an efficiency of 95 percent and costs \$2700. The home owner would like to buy the high-efficiency furnace if the savings from the natural gas pay for the additional cost in less than 8 years. If the home owner presently pays \$1100 a year for heating, determine if he should buy the conventional or high-efficiency model.

5-141 Replacing incandescent lights with energy-efficient fluorescent lights can reduce the lighting energy consumption to one-fourth of what it was before. The energy consumed by the lamps is eventually converted to heat, and thus switching to energy-efficient lighting also reduces the cooling load in summer but increases the heating load in winter. Consider a building that is heated by a natural gas furnace with an efficiency of 80 percent and cooled by an air conditioner with a COP of 3.5. If electricity costs \$0.08/kWh and natural gas costs \$0.70/therm, and the annual heating load of the building is roughly equal to the annual cooling load, determine if efficient lighting will increase or decrease the total heating and cooling costs of the building.

5-142 The cargo space of a refrigerated truck whose inner dimensions are 12 m × 2.3 m × 3.5 m is to be precooled from 25°C to an average temperature of 5°C. The construction of the truck is such that a transmission heat gain occurs at a rate of 80 W/°C. If the ambient temperature is 25°C, determine how long it will take for a system with a refrigeration capacity of 8 kW to precool this truck.

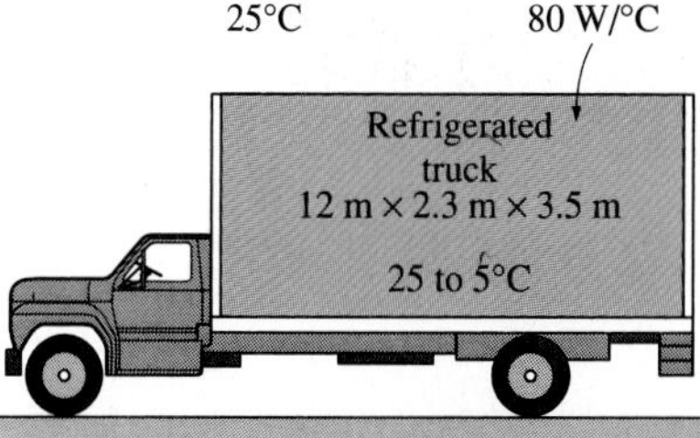

FIGURE P5-142

5-143 A refrigeration system is to cool bread loaves with an average mass of 450 g from 22°C to −10°C at a rate of 500 loaves per hour by refrigerated air at −30°C. Taking the average specific and latent heats of bread to be 2.93 kJ/(kg · °C) and 109.3 kJ/kg, respectively, determine (*a*) the rate of heat removal from the breads, in kJ/h; (*b*) the required volume flow rate of air, in m^3/h, if the temperature rise of air is not to exceed 8°C; and (*c*) the size of the compressor of the refrigeration system, in kW, for a COP of 1.2 for the refrigeration system.

5-144 The drinking water needs of a production facility with 20 employees is to be met by a bobbler type water fountain. The refrigerated water fountain is to cool water from 22°C to 8°C and supply cold water at a rate of 0.4 L per hour per person. The outer diameter of the water reservoir where water is cooled and stored is 20 cm, and its height is 25 cm. The outer surface temperature of the reservoir is 17°C, and heat is transferred to the reservoir from the surroundings at 25°C with a heat transfer coefficient of 10 W/($m^2 \cdot$ °C). If the COP of the refrigeration system is 2.9, determine the size of the compressor, in W, that will be suitable for the refrigeration system of this water cooler.

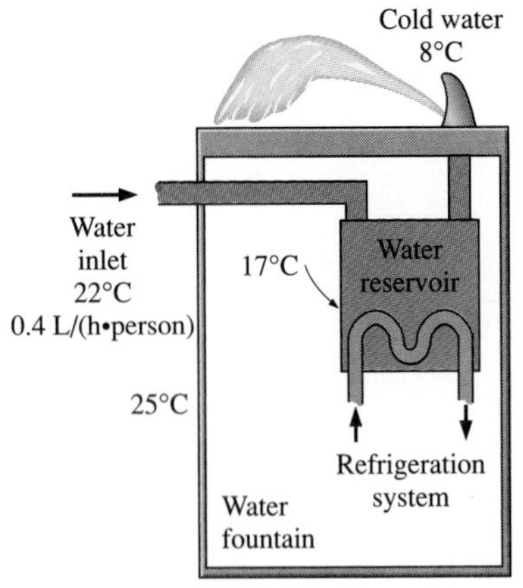

FIGURE P5-144

5-145 The "Energy Guide" label on a washing machine indicates that the washer will use $85 worth of hot water if the water is heated by an electric water heater at an electricity rate of $0.082/kWh. If the water is heated from 15°C to 55°C, determine how many liters of hot water an average family uses per week. Disregard the electricity consumed by the washer, and take the efficiency of the electric water heater to be 90 percent.

5-146 The "Energy Guide" label on a washing machine indicates that the washer will use $33 worth of hot water if the water is heated by a gas water heater at a natural gas rate of $0.605/therm. If the water is heated from 60°F to 130°F, determine how many liters of hot water an average family uses per week. Disregard the electricity consumed by the washer, and take the efficiency of the gas water heater to be 55 percent.

5-147 A typical electric water heater has an efficiency of 90 percent and costs $390 a year to operate at a unit cost of electricity of $0.08/kWh. A typical heat pump–powered water heater has a COP of 2.2 but costs about $800 more to install. Determine how many years it will take for the heat pump water heater to pay for its cost differential from the energy it saves.

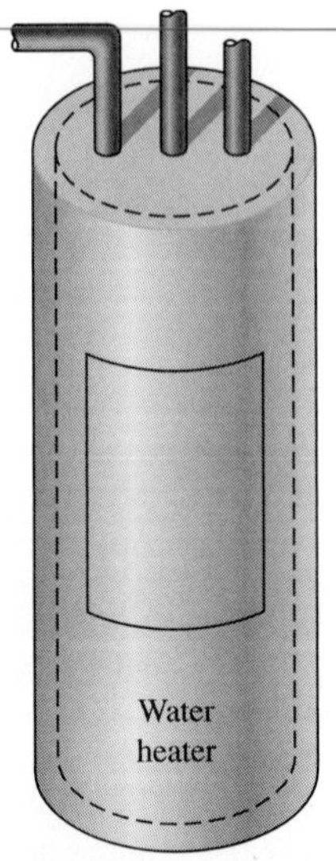

FIGURE P5-147

5-148E The energy contents, unit costs, and typical conversion efficiencies of various energy sources for use in water heaters are given as follows: 1025 Btu/ft^3, $0.0060/$ft^3$, and 55 percent for natural gas; 138,700 Btu/gal, $1.15 gal, and 55 percent for heating oil; and 1 kWh/kWh $0.084/kWh, and 90 percent for resistance electric heaters, respectively. Determine the lowest-cost energy source for water heaters.

5-149E A homeowner is considering the following heating systems for heating his house. Electric resistance heating with $0.09/kWh and 1 kWh = 3412 Btu, gas heating with $0.062/therm and 1 therm = 105,500 kJ, and oil heating with $1.25/gal and 1 gal of oil = 138,500 kJ. Assuming efficiencies of 100 percent for the electric furnace and 87 percent for the gas and oil furnaces, determine the heating system with the lowest energy cost.

5-150 A homeowner is trying to decide between a high-efficiency natural gas furnace with an efficiency of 97 percent and a ground-source heat pump with a COP of 3.7. The unit costs of electricity and natural gas are $0.092/kWh and $0.71/therm (1 therm = 105,500 kJ). Determine which system will have a lower energy cost.

5-151 The maximum flow rate of a standard shower head is about 3.5 gpm (13.3 L/min) and can be reduced to 2.75 gpm (10.5 L/min) by switching to a low-flow shower head that is equipped with flow controllers. Consider a family of four, with each person taking a 5-minute shower every morning. City water at 15°C is heated to 55°C in an oil water heater whose efficiency is 65 percent and then tempered to 42°C by cold water at the T-elbow of the shower before being routed to the shower head. The price of heating oil is $1.20/gal and its heating value is 146,300 kJ/gal. Assuming a constant specific heat of 4.18 kJ/(kg · °C) for water, determine the amount of oil and money saved per year by replacing the standard shower heads by the low-flow ones.

5-152 A typical household pays about $1200 a year on energy bills, and the U.S. Department of Energy estimates that 46 percent of this energy is used for heating and cooling, 15 percent for heating water, 15 percent for refrigerating and freezing, and the remaining 24 percent for lighting, cooking, and running other appliances. The heating and cooling costs of a poorly insulated house can be reduced by up to 30 percent by adding adequate insulation. If the cost of insulation is $200, determine how long it will take for the insulation to pay for itself from the energy it saves.

5-153 The kitchen, bath, and other ventilation fans in a house should be used sparingly since these fans can discharge a houseful of warmed or cooled air in just one hour. Consider a 200 m^2 house whose ceiling height is 2.4 m. The house is heated by a 96 percent efficient gas heater and is maintained at 22°C and 92 kPa. If the unit cost of natural gas is $0.60/therm (1 therm = 105,500 kJ), determine the cost of energy "vented out" by the fans in 1 h. Assume the average outdoor temperature during the heating season to be 5°C.

5-154 Repeat Prob. 5-153 for the air-conditioning cost in a dry climate for an outdoor temperature of 32°C. Assume the COP of the air-conditioning system to be 3.2, and the unit cost of electricity to be $0.10/kWh.

5-155 The U.S. Department of Energy estimates that up to 10 percent of the energy use of a house can be saved by caulking and weatherstripping doors and windows to reduce air leaks at a cost of about $50 for materials for an average home with 12 windows and 2 doors. Caulking and weatherstripping every gas-heated home properly would save enough energy to heat about 4 million homes. The savings can be increased by installing storm windows. Determine how long it will take for the caulking and weatherstripping to pay for itself from the energy they save for a house whose annual energy use is $1100.

5-156 The U.S. Department of Energy estimates that 570,000 barrels of oil would be saved per day if every household in the United States lowered the thermostat setting in winter by 6°F (3.3°C). Assuming the average heating season to be 180 days and the cost of oil to be $20/barrel, determine how much money would be saved per year.

Computer, Design, and Essay Problems

5-157 Write a computer program to determine the maximum work that can be extracted from a pond containing 10^5 kg of water at 350 K when the temperature of the surroundings is 300 K. Notice that the temperature of water in the pond will be gradually decreasing as energy is extracted from it; therefore, the efficiency of the engine will be decreasing. Use temperature intervals of (*a*) 5 K, (*b*) 2 K, and (*c*) 1 K until the pond temperature drops to 300 K. Also solve this problem exactly by integration, and compare the results.

5-158 Find out the prices of heating oil, natural gas, and electricity in your area, and determine the cost of each per kWh of energy supplied to the house as heat. Go through your utility bills and determine how much money you spent for heating last January. Also determine how much your January heating bill would be for each of the heating systems if you had the latest and most efficient system installed.

5-159 Prepare a report on the heating systems available in your area for residential buildings. Discuss the advantages and disadvantages of each system and compare their initial and operating costs. What are the important factors in the selection of a heating system? Give some guidelines. Identify the conditions under which each heating system would be the best choice in your area.

5-160 The performance of a cyclic device is defined as the ratio of the desired output to the required input, and this definition can be extended to nontechnical fields. For example, your performance in this course can be viewed as the grade you earn relative to the effort you put in. If you have been investing a lot of time in this course and your grades do not reflect it, you are performing poorly. In that case, perhaps you should try to find out the underlying cause and how to correct the problem. Give three other definitions of performance from nontechnical fields and discuss them.

5-161 Devise a Carnot heat engine using steady-flow components, and describe how the Carnot cycle is executed in that engine. What happens when the directions of heat and work interactions are reversed?

5-162 When was the concept of the heat pump conceived and by whom? When was the first heat pump built, and when were the heat pumps first mass-produced?

5-163 Your neighbor lives in a 2500-square-foot (about 250 m^2) older house heated by natural gas. The current gas heater was installed in the early 1970s and has an efficiency (called the Annual Fuel Utilization Efficiency rating, or AFUE) of 65 percent. It is time to replace the furnace, and the neighbor is trying to decide between a conventional furnace that has an efficiency of 80 percent and costs $1500 and a high-efficiency furnace that has an efficiency of 95 percent and costs $2500. Your neighbor offered to pay you $100 if you help him make the right decision. Considering the weather data, typical heating loads, and the price of natural gas in your area, make a recommendation to your neighbor based on a convincing economic analysis.

5-164 Using a thermometer, measure the temperature of the main food compartment of your refrigerator, and check if it is between 1 and 4°C. Also, measure the temperature of the freezer compartment, and check if it is at the recommended value of −18°C.

5-165 Using a timer (or watch) and a thermometer, conduct the following experiment to determine the rate of heat gain of your refrigerator. First make sure that the door of the refrigerator is not opened for at least a few hours so that steady operating conditions are established. Start the timer when the refrigerator stops running and measure the time Δt_1 it stays off before it kicks in. Then measure the time Δt_2 it stays on. Noting that the heat removed during Δt_2 is equal to the heat gain of the refrigerator during $\Delta t_1 + \Delta t_2$ and using the power consumed by the refrigerator when it is running, determine the average rate of heat gain for your refrigerator, in W. Take the COP (coefficient of performance) of your refrigerator to be 1.3 if it is not available.

5-166 Design a hydrocooling unit that can cool fruits and vegetables from 30°C to 5°C at a rate of 20,000 kg/h under the following conditions:

The unit will be of flood type, which will cool the products as they are conveyed into the channel filled with water. The products will be dropped into the channel filled with water at one end and be picked up at the other end. The channel can be as wide as 3 m and as high as 90 cm. The water is to be circulated and cooled by the evaporator section of a refrigeration system. The refrigerant temperature inside the coils is to be −2°C, and the water temperature is not to drop below 1°C and not to exceed 6°C.

Assuming reasonable values for the average product density, specific heat, and porosity (the fraction of air volume in a box), recommend reasonable values for (*a*) the water velocity through the channel and (*b*) the refrigeration capacity of the refrigeration system.

Entropy: A Measure of Disorder

CHAPTER 6

In Chap. 5, we introduced the second law of thermodynamics and applied it to cycles and cyclic devices. In this chapter, we apply the second law to processes. The first law of thermodynamics deals with the property *energy* and the conservation of it. The second law leads to the definition of a new property called *entropy*. Entropy is a somewhat abstract property, and it is difficult to give a physical description of it. Entropy is best understood and appreciated by studying its uses in commonly encountered engineering processes, and this is what we intend to do.

This chapter starts with a discussion of the Clausius inequality, which forms the basis for the definition of entropy, and continues with the increase of entropy principle. Unlike energy, entropy is a nonconserved property, and there is no such thing as a *conservation of entropy principle*. Next, the entropy changes that take place during processes for pure substances, incompressible substances, and ideal gases are discussed, and a special class of idealized processes, called *isentropic processes*, are examined. Then, the reversible steady-flow work and the isentropic efficiencies of various engineering devices such as turbines and compressors are considered. Finally, entropy balance is introduced and applied to various systems.

6-1 ■ ENTROPY

The second law of thermodynamics often leads to expressions that involve inequalities. An irreversible (i.e., actual) heat engine, for example, is less efficient than a reversible one operating between the same two thermal energy reservoirs. Likewise, an irreversible refrigerator or a heat pump has a lower coefficient of performance (COP) than a reversible one operating between the same temperature limits. Another important inequality that has major consequences in thermodynamics is the **Clausius inequality.** It was first stated by the German physicist R. J. E. Clausius (1822-1888), one of the founders of thermodynamics, and is expressed as

$$\oint \frac{\delta Q}{T} \leq 0$$

That is, *the cyclic integral of* $\delta Q/T$ *is always less than or equal to zero.* This inequality is valid for all cycles, reversible or irreversible. The symbol $\oint$ (integral symbol with a circle in the middle) is used to indicate that the integration is to be performed over the entire cycle. Any heat transfer to or from a system can be considered to consist of differential amounts of heat transfer. Then the cyclic integral of $\delta Q/T$ can be viewed as the sum of all these differential amounts of heat transfer divided by the absolute temperature at the boundary.

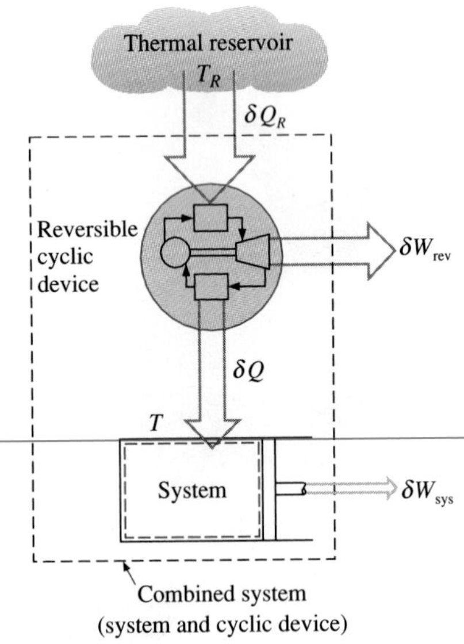

FIGURE 6-1

The system considered in the development of Clausius inequality.

To demonstrate the validity of Clausius inequality, consider a system connected to a thermal energy reservoir at a constant absolute temperature of T_R through a *reversible* cyclic device (Fig. 6-1). The cyclic device receives heat δQ_R from the reservoir and supplies heat δQ to the system whose absolute temperature at that part of the boundary is T (a variable) while producing work δW_{rev}. The system produces work δW_{sys} as a result of this heat transfer. Applying the energy balance to the combined system identified by dashed lines yields

$$\delta W_C = \delta Q_R - dE_C$$

where δW_C is the total work of the combined system ($\delta W_{rev} + \delta W_{sys}$) and dE_C is the change in the total energy of the combined system. Considering that the cyclic device is a *reversible* one, we have (Eq. 5-18)

$$\frac{\delta Q_R}{T_R} = \frac{\delta Q}{T}$$

where the sign of δQ is determined with respect to the system (positive if *to* the system and negative if *from* the system) and the sign of δQ_R is determined with respect to the reversible cyclic device. Eliminating δQ_R from the two relations above yields

$$\delta W_C = T_R \frac{\delta Q}{T} - dE_C$$

We now let the system undergo a cycle while the cyclic device undergoes an integral number of cycles. Then the relation above becomes

$$W_C = T_R \oint \frac{\delta Q}{T}$$

since the cyclic integral of energy (the net change in the energy, which is a property, during a cycle) is zero. Here W_C is the cyclic integral of δW_C, and it represents the net work for the combined cycle.

It appears that the combined system is exchanging heat with a single thermal energy reservoir while involving (producing or consuming) work W_C during a cycle. On the basis of the Kelvin-Planck statement of the second law, which states that *no system can produce a net amount of work while operating in a cycle and exchanging heat with a single thermal energy reservoir,* we reason that W_C cannot be a work output, and thus it cannot be a positive quantity. Considering that T_R is an absolute temperature and thus a positive quantity, we must have

$$\oint \frac{\delta Q}{T} \leq 0 \qquad (6\text{-}1)$$

which is the *Clausius inequality.* This inequality is valid for all thermodynamic cycles, reversible or irreversible, including the refrigeration cycles.

If no irreversibilities occur within the system as well as the reversible cyclic device, then the cycle undergone by the combined system will be internally reversible. As such, it can be reversed. In the reversed cycle case, all the quantities will have the same magnitude but the opposite sign. Therefore, the work W_C, which could not be a positive quantity in the regular case, cannot be a negative quantity in the reversed case. Then it follows that $W_{C,\text{int rev}} = 0$ since it cannot be a positive or negative quantity, and therefore

$$\oint \left(\frac{\delta Q}{T}\right)_{\text{int rev}} = 0 \qquad (6\text{-}2)$$

for internally reversible cycles. Thus we conclude that *the equality in the Clausius inequality holds for totally or just internally reversible cycles and the inequality for the irreversible ones.*

To develop a relation for the definition of entropy, let us examine Eq. 6-2 more closely. Here we have a quantity whose cyclic integral is zero. Let us think for a moment what kind of quantities can have this characteristic. We know that the cyclic integral of *work* is not zero. (It is a good thing that it is not. Otherwise, heat engines that work on a cycle such as steam power plants would produce zero net work.) Neither is the cyclic integral of heat.

Now consider the volume occupied by a gas in a piston-cylinder device undergoing a cycle, as shown in Fig. 6-2. When the piston returns to its initial position at the end of a cycle, the volume of the gas also returns to its initial value. Thus the net change in volume during a cycle is zero. This is also expressed as

FIGURE 6-2
The net change in volume (a property during a cycle is always zero).

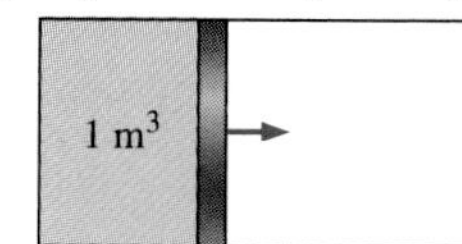

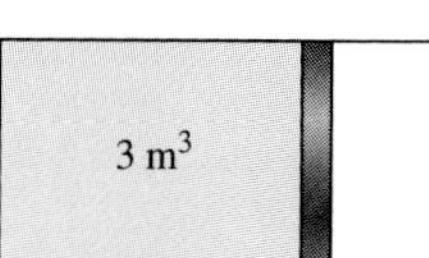

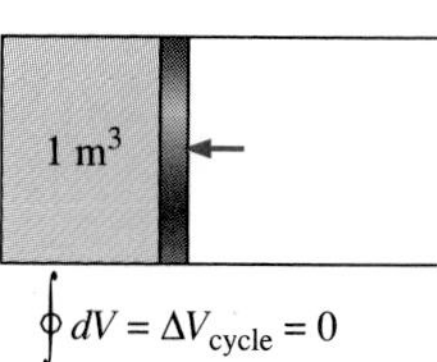

$\oint dV = \Delta V_{\text{cycle}} = 0$

$$\oint dV = 0 \tag{6-3}$$

That is, the cyclic integral of volume (or any other property) is zero. Conversely, a quantity whose cyclic integral is zero depends on the *state* only and not the process path, and thus it is a property. Therefore, the quantity $(\delta Q/T)_{\text{int rev}}$ must represent a property in the differential form.

Clausius realized in 1865 that he had discovered a new thermodynamic property, and he chose to name this property **entropy.** It is designated S and is defined as

$$dS = \left(\frac{\delta Q}{T}\right)_{\text{int rev}} \quad \text{(kJ/K)} \tag{6-4}$$

Entropy is an extensive property of a system and sometimes is referred to as *total entropy*. Entropy per unit mass, designated s, is an intensive property and has the unit kJ/(kg · K). The term *entropy* is generally used to refer to both total entropy and entropy per unit mass since the context usually clarifies which one is meant.

The entropy change of a system during a process can be determined by integrating Eq. 6-4 between the initial and the final states:

$$\Delta S = S_2 - S_1 = \int_1^2 \left(\frac{\delta Q}{T}\right)_{\text{int rev}} \quad \text{(kJ/K)} \tag{6-5}$$

Notice that we have actually defined the *change* in entropy instead of entropy itself, just as we defined the change in energy instead of the energy itself when we developed the first-law relation for closed systems in Chap. 3. Absolute values of entropy are determined on the basis of the third law of thermodynamics, which is discussed later in this chapter. Engineers are usually concerned with the *changes* in entropy. Therefore, the entropy of a substance can be assigned a zero value at some arbitrarily selected reference state, and the entropy values at other states can be determined from Eq. 6-5 by choosing state 1 to be the reference state ($S = 0$) and state 2 to be the state at which entropy is to be determined.

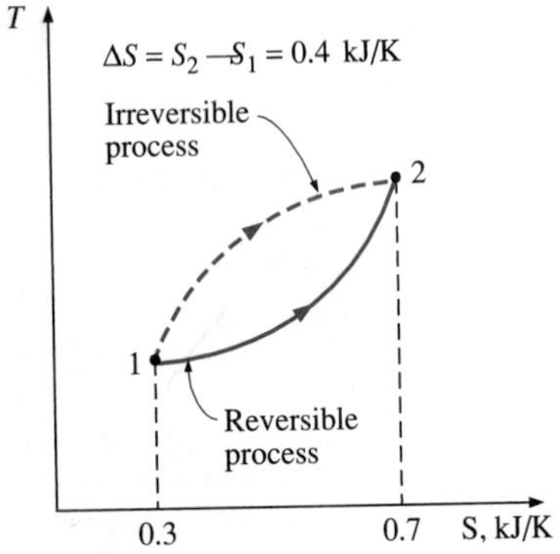

To perform the integration in Eq. 6-5, one needs to know the relation between Q and T during a process. This relation is often not available, and the integral in Eq. 6-5 can be performed for a few cases only. For the majority of cases we have to rely on tabulated data for entropy.

Note that entropy is a property, and like all other properties, it has fixed values at fixed states. Therefore, the entropy change ΔS between two specified states is the same no matter what path, reversible or irreversible, is followed during a process (Fig. 6-3).

FIGURE 6-3
The entropy change between two specified states is the same whether the process is reversible or irreversible.

Also note that the integral of $\delta Q/T$ will give us the value of entropy change *only if* the integration is carried out along an *internally reversible* path between the two states. The integral of $\delta Q/T$ along an irreversible path is not a property, and in general, different values will be obtained when the integration is carried out along different irreversible paths. Therefore, even for irreversible processes, the entropy change should be determined by carrying out this integration along some convenient *imaginary* internally reversible path between the specified states.

A Special Case: Internally Reversible Isothermal Heat Transfer Processes

We pointed out in Chap. 5 that isothermal heat transfer processes are internally reversible. Therefore, the entropy change of a system during an internally reversible isothermal heat transfer process can be determined by performing the integration in Eq. 6-5:

$$\Delta S = \int_1^2 \left(\frac{\delta Q}{T}\right)_{\text{int rev}} = \int_1^2 \left(\frac{\delta Q}{T_0}\right)_{\text{int rev}} = \frac{1}{T_0}\int_1^2 (\delta Q)_{\text{int rev}}$$

which reduces to

$$\Delta S = \frac{Q}{T_0} \qquad \text{(kJ/K)} \qquad (6\text{-}6)$$

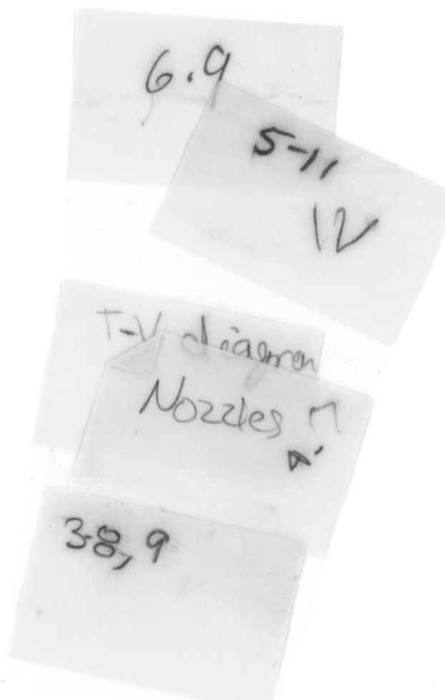

where T_0 is the constant absolute temperature of the system and Q is the heat transfer for the internally reversible process. Equation 6-6 is particularly useful for determining the entropy changes of thermal energy reservoirs that can absorb or supply heat indefinitely at a constant temperature.

Notice that the entropy change of a system during an internally reversible isothermal process can be positive or negative, depending on the direction of heat transfer. Heat transfer to a system will increase the entropy of a system, whereas heat transfer from a system will decrease it. In fact, losing heat is the only way the entropy of a system can be decreased.

EXAMPLE 6-1 Entropy Change during an Isothermal Process

A piston-cylinder device contains a liquid–vapor mixture of water at 300 K. During a constant pressure process, 750 kJ of heat is transferred to the water. As a result, part of the liquid in the cylinder vaporizes. Determine the entropy change of the water during this process.

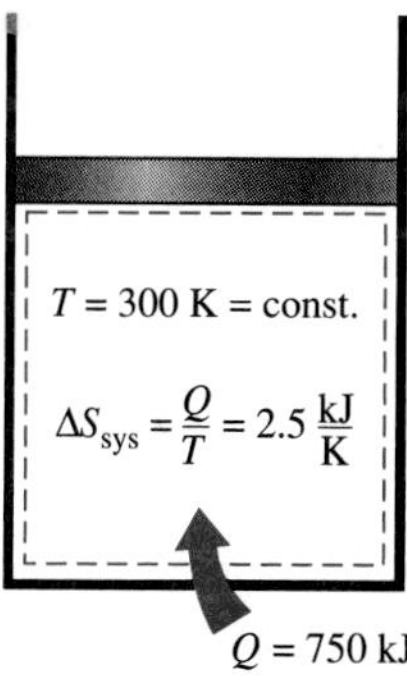

FIGURE 6-4
Schematic for Example 6-1.

Solution We take the *entire water* (liquid + vapor) in the cylinder as the system (Fig. 6-4). This is a *closed system* since no mass crosses the system boundary during the process. We note that the temperature of the system remains constant at 300 K during this process since the temperature of a pure substance remains constant at the saturation value during a phase change process at constant pressure.

Assumptions No irreversibilities occur within the system boundaries during the process.

Analysis The system undergoes an internally reversible, isothermal process, and thus its entropy change can be determined directly from Eq. 6-6 to be

$$\Delta S_{\text{sys, isothermal}} = \frac{Q}{T_{\text{sys}}} = \frac{750 \text{ kJ}}{300 \text{ K}} = \mathbf{2.5\ kJ/K}$$

Discussion Note that the entropy change of the system is positive, as expected, since heat transfer is *to* the system.

6-2 ■ THE INCREASE OF ENTROPY PRINCIPLE

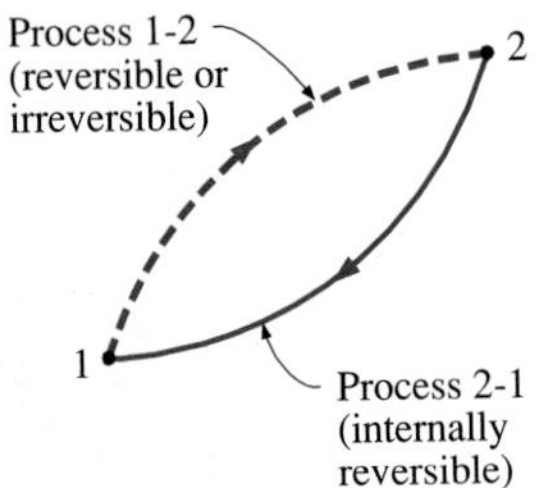

FIGURE 6-5
A cycle composed of a reversible and an irreversible process.

Consider a cycle that is made up of two processes: process 1-2, which is arbitrary (reversible or irreversible), and process 2-1, which is internally reversible, as shown in Fig. 6-5. From the Clausius inequality,

$$\oint \frac{\delta Q}{T} \leq 0$$

or

$$\int_1^2 \frac{\delta Q}{T} + \int_2^1 \left(\frac{\delta Q}{T}\right)_{\text{int rev}} \leq 0$$

The second integral in the above relation is readily recognized as the entropy change $S_1 - S_2$. Therefore,

$$\int_1^2 \frac{\delta Q}{T} + S_1 - S_2 \leq 0$$

which can be rearranged as

$$S_2 - S_1 \geq \int_1^2 \frac{\delta Q}{T} \tag{6-7}$$

Equation 6-7 can be viewed as a mathematical statement of the second law of thermodynamics for a fixed mass. It can also be expressed in differential form as

$$dS \geq \frac{\delta Q}{T} \tag{6-8}$$

where the equality holds for an internally reversible process and the inequality for an irreversible process. We may conclude from these equations that *the entropy change of a closed system during an irreversible process is greater than the integral of $\delta Q/T$ evaluated for that process. In the limiting case of a reversible process, these two quantities become equal*. We again emphasize that T in the above relations is the *absolute temperature* at the *boundary* where the differential heat δQ is transferred between the system and the surroundings.

The quantity $\Delta S = S_2 - S_1$ represents the *entropy change* of the system. For a reversible process, it becomes equal to $\int_1^2 \delta Q/T$, which represents the *entropy transfer* with heat.

The inequality sign in the relations above is a constant reminder that the entropy change of a closed system during an irreversible process is always greater than the entropy transfer. That is, some entropy is *generated* or *created* during an irreversible process, and this generation is due entirely to the presence of irreversibilities. The entropy generated during a process is called **entropy generation** and is denoted by S_{gen}. Noting that the difference between the entropy change of a closed system and the entropy transfer is equal to entropy generation, Eq. 6-8 can be rewritten as an equality as

$$\Delta S_{sys} = S_2 - S_1 = \int_1^2 \frac{\delta Q}{T} + S_{gen} \tag{6-9}$$

Note that the entropy generation S_{gen} is always a *positive* quantity or zero. Its value depends on the process, and thus it is *not* a property of the system. Also, in the absence of any entropy transfer, the entropy change of a system is equal to the entropy generation.

Equation 6-7 has far-reaching implications in thermodynamics. For an isolated system (or simply an adiabatic closed system), the heat transfer is zero, and Eq. 6-7 reduces to

$$\Delta S_{isolated} \geq 0 \tag{6-10}$$

This equation can be expressed as *the entropy of an isolated system during a process always increases or, in the limiting case of a reversible process, remains constant.* In other words, it *never* decreases. This is known as the **increase of entropy principle.** Note that in the absence of any heat transfer, entropy change is due to irreversibilities only, and their effect is always to increase entropy.

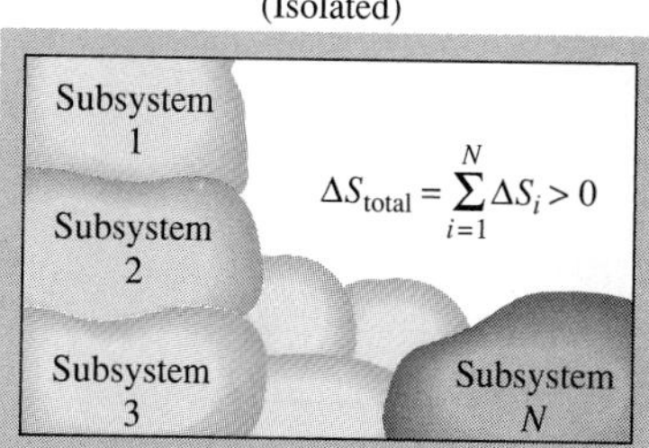

FIGURE 6-6
The entropy change of an isolated system is the sum of the entropy changes of its components, and is never less than zero.

Entropy is an extensive property, and thus the total entropy of a system is equal to the sum of the entropies of the parts of the system. An isolated system may consist of any number of subsystems (Fig. 6-6). A system and its surroundings, for example, constitute an isolated system since both can be enclosed by a sufficiently large arbitrary boundary across which there is no heat, work, or mass transfer (Fig. 6-7). Therefore, a system and its surroundings can be viewed as the two subsystems of an isolated system, and the entropy change of this isolated system during a process is the sum of the entropy changes of the system and its surroundings, which is equal to the entropy generation since an isolated system involves no entropy transfer. That is,

$$S_{gen} = \Delta S_{total} = \Delta S_{sys} + \Delta S_{surr} \geq 0 \tag{6-11}$$

where the equality holds for reversible processes and the inequality for irreversible ones. Note that ΔS_{surr} refers to the change in the entropy of the surroundings as a result of the occurrence of the process under consideration.

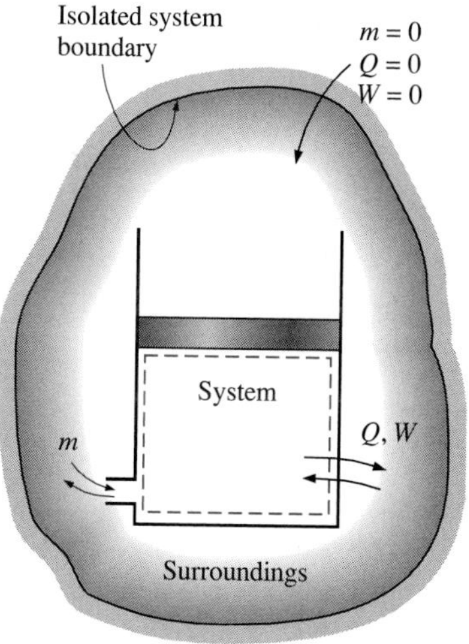

FIGURE 6-7
A system and its surroundings form an isolated system.

Since no actual process is truly reversible, we can conclude that some entropy is generated during a process, and therefore the entropy of the universe, which can be considered to be an isolated system, is continuously increasing. The more irreversible a process, the larger the entropy generated during that process. No entropy is generated during reversible processes ($S_{gen} = 0$).

Entropy increase of the universe is a major concern not only to engineers but also to philosophers and theologians since entropy is viewed as a measure of the disorder (or "mixed-up-ness") in the universe.

The increase of entropy principle does not imply that the entropy of a system cannot decrease. The entropy change of a system *can* be negative during a process (Fig. 6-8), but entropy generation cannot. The increase of entropy principle can be summarized as follows:

$$S_{gen} \begin{cases} > 0 & \text{Irreversible process} \\ = 0 & \text{Reversible process} \\ < 0 & \text{Impossible process} \end{cases}$$

This relation serves as a criterion in determining whether a process is reversible, irreversible, or impossible.

Things in nature have a tendency to change until they attain a state of equilibrium. The increase of entropy principle dictates that the entropy of an isolated system will increase until the entropy of the system reaches a *maximum* value. At that point, the system is said to have reached an equilibrium state since the increase of entropy principle prohibits the system from undergoing any change of state that will result in a decrease in entropy.

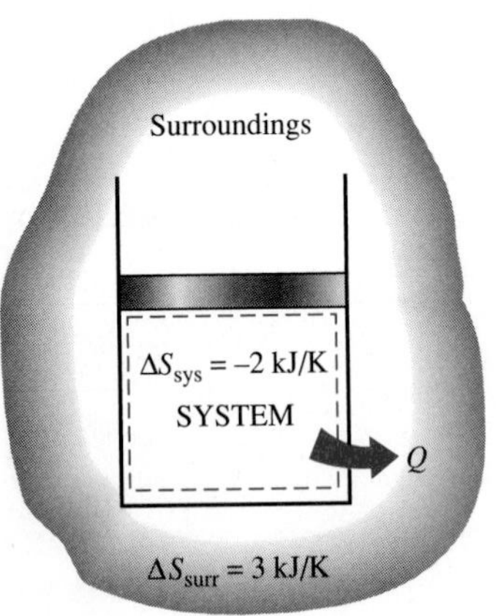

$S_{gen} = \Delta S_{total} = \Delta S_{sys} + \Delta S_{surr} = 1 \text{kJ/K}$

FIGURE 6-8

The entropy change of a system can be negative, but the entropy generation cannot.

Some Remarks about Entropy

In light of the preceding discussions, we can draw the following conclusions:

1. Processes can occur in a *certain* direction only, not in *any* direction. A process must proceed in the direction that complies with the increase of entropy principle, that is, $S_{gen} \geq 0$. A process that violates this principle is impossible. This principle often forces chemical reactions to come to a halt before reaching completion.

2. Entropy is a *nonconserved property,* and there is *no* such thing as the *conservation of entropy principle*. Entropy is conserved during the idealized reversible processes only and increases during *all* actual processes. Therefore, the entropy of the universe is continuously increasing.

3. The performance of engineering systems is degraded by the presence of irreversibilities, and *entropy generation* is a measure of the magnitudes of the irreversibilities present during that process. The greater the extent of irreversibilities, the greater the entropy generation. Therefore, entropy generation can be used as a quantitative measure of irreversibilities associated with a process. It is also used to establish criteria for the performance of engineering devices. This point is illustrated further in the following example.

FIGURE 6-9

Schematic for Example 6-2.

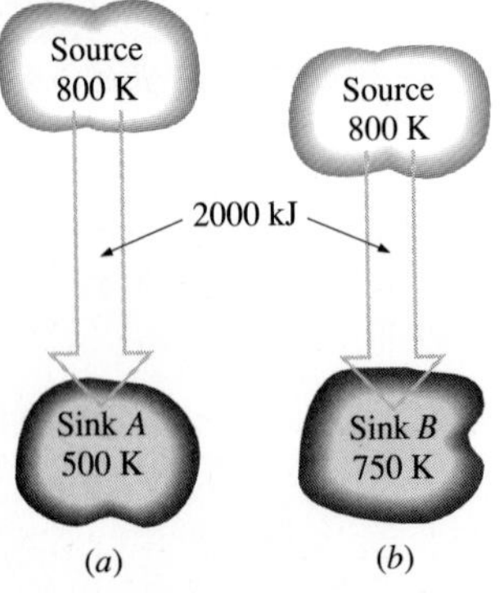

EXAMPLE 6-2 Entropy Generation during Heat Transfer Processes

A heat source at 800 K loses 2000 kJ of heat to a sink at (*a*) 500 K and (*b*) 750 K. Determine which heat transfer process is more irreversible.

Solution A sketch of the reservoirs is shown in Fig. 6-9. Both cases involve heat transfer through a finite temperature difference, and therefore both are irreversible. The magnitude of the irreversibility associated with each process can be determined by calculating the total entropy change for each case. The total entropy change for a heat transfer process involving two reservoirs (a source and a sink) is the sum of the entropy changes of each reservoir since the two reservoirs form an adiabatic system.

Or do they? The problem statement gives the impression that the two reservoirs are in direct contact during the heat transfer process. But this cannot be the case since the temperature at a point can have only one value, and thus it cannot be 800 K on one side of the point of contact and 500 K on the other side. In other words, the temperature function cannot have a jump discontinuity. Therefore, it is reasonable to assume that the two reservoirs are separated by a partition through which the temperature drops from 800 K on one side to 500 K (or 750 K) on the other. Therefore, the entropy change of the partition should also be considered when evaluating the total entropy change for this process. However, considering that entropy is a property and the values of properties depend on the state of a system, we can argue that the entropy change of the partition is zero since the partition appears to have undergone a *steady* process and thus experienced no change in its properties at any point. We base this argument on the fact that the temperature on both sides of the partition and thus throughout remained constant during this process. Therefore, we are justified to assume that $\Delta S_{\text{partition}} \cong 0$ since the entropy (as well as the energy) content of the partition essentially remained constant during this process.

The entropy change for each reservoir can be determined from Eq. 6-6 since each reservoir undergoes an internally reversible, isothermal process.

(*a*) For the heat transfer process to a sink at 500 K.

$$\Delta S_{\text{source}} = \frac{Q_{\text{source}}}{T_{\text{source}}} = \frac{-2000 \text{ kJ}}{800 \text{ K}} = -2.5 \text{ kJ/K}$$

$$\Delta S_{\text{sink}} = \frac{Q_{\text{sink}}}{T_{\text{sink}}} = \frac{+2000 \text{ kJ}}{500 \text{ K}} = +4.0 \text{ kJ/K}$$

and $$S_{\text{gen}} = \Delta S_{\text{total}} = \Delta S_{\text{source}} + \Delta S_{\text{sink}} = (-2.5 + 4.0) \text{ kJ/K} = \mathbf{+1.5 \text{ kJ/K}}$$

Therefore, 1.5 kJ/K of entropy is generated during this process. Noting that both reservoirs have undergone internally reversible processes, the entire entropy generation took place in the partition.

(*b*) Repeating the calculations in part (*a*) for a sink temperature of 750 K, we obtain

$$\Delta S_{\text{source}} = -2.5 \text{ kJ/K}$$

$$\Delta S_{\text{sink}} = +2.7 \text{ kJ/K}$$

and $$S_{\text{gen}} = \Delta S_{\text{total}} = (-2.5 + 2.7) \text{ kJ/K} = \mathbf{+0.2 \text{ kJ/K}}$$

The total entropy change for the process in part (*b*) is smaller, and therefore it is less irreversible. This is expected since the process in (*b*) involves a smaller temperature difference and thus a smaller irreversibility.

Discussion The irreversibilities associated with both processes could be eliminated by operating a Carnot heat engine between the source and the sink. For this case it can be easily shown that $\Delta S_{\text{total}} = 0$.

6-3 ■ ENTROPY CHANGE OF PURE SUBSTANCES

Entropy is a property, and thus the value of entropy of a system is fixed once the state of the system is specified. Specifying two intensive independent properties fixes the state of a simple compressible system, and thus the value of entropy, as well as the values of other properties at that state. Starting with its defining relation (Eq. 6-4), the entropy change of a substance can be expressed in terms of other properties (see Sec. 6-7). But in general, these relations are too complicated and are not practical to use for hand calculations. Therefore, using a suitable reference state, the entropies of substances are evaluated from measurable property data following rather involved computations, and the results are tabulated in the same manner as the other properties such as *v*, *u*, and *h* (Fig. 6-10).

The entropy values in the property tables are given relative to an arbitrary reference state. In steam tables the entropy of saturated liquid s_f at 0.01°C is assigned the value of zero. For refrigerant-134a, the zero value is assigned to saturated liquid at −40°C. The entropy values become negative at temperatures below the reference value.

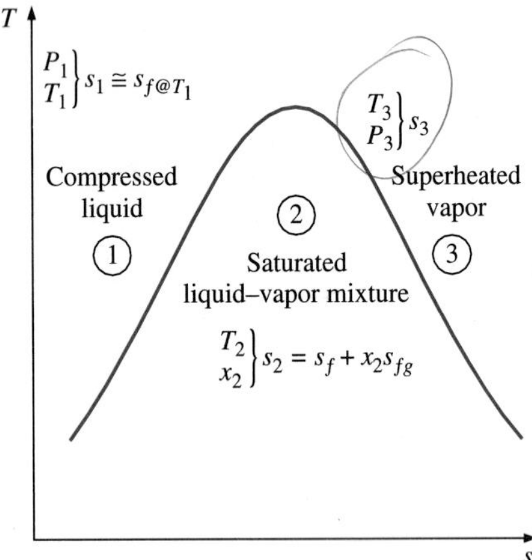

FIGURE 6-10
The entropy of a pure substance is determined from the tables, just as for any other property.

The value of entropy at a specified state is determined just like any other property. In the compressed liquid and superheated vapor regions, it can be obtained directly from the tables at the specified state. In the saturated mixture region, it is determined from

$$s = s_f + xs_{fg} \qquad [\text{kJ/(kg} \cdot \text{K)}]$$

where x is the quality and s_f and s_{fg} values are listed in the saturation tables. In the absence of compressed liquid data, the entropy of the compressed liquid can be approximated by the entropy of the saturated liquid at the given temperature:

$$s_{@T,P} \cong s_{f@T} \qquad [\text{kJ/(kg} \cdot \text{K)}]$$

The entropy change of a specified mass m (such as a closed system) during a process is simply

$$\Delta S = m\Delta s = m(s_2 - s_1) \qquad (\text{kJ/K}) \qquad (6\text{-}12)$$

which is the difference between the entropy values at the final and initial states.

When studying the second law aspects of processes, entropy is commonly used as a coordinate on diagrams such as the *T-s* and *h-s* diagrams. The general characteristics of the *T-s* diagram of pure substances are shown in Fig. 6-11 using data for water. Notice from this diagram that the constant volume lines are steeper than the constant pressure lines and the constant pressure lines are parallel to the constant temperature lines in the saturated liquid–vapor mixture region. Also, the constant pressure lines almost coincide with the saturated liquid line in the compressed liquid region. The *T-s* diagram will be used commonly in this and the forthcoming chapters.

EXAMPLE 6-3 Entropy Change of a Substance in a Tank

A rigid tank contains 5 kg of refrigerant-134a initially at 20°C and 140 kPa. The refrigerant is now cooled while being stirred until its pressure drops to 100 kPa. Determine the entropy change of the refrigerant during this process.

Solution We take the refrigerant in the tank as the *system* (Fig. 6-12). This is a *closed system* since no mass crosses the system boundary during the process.

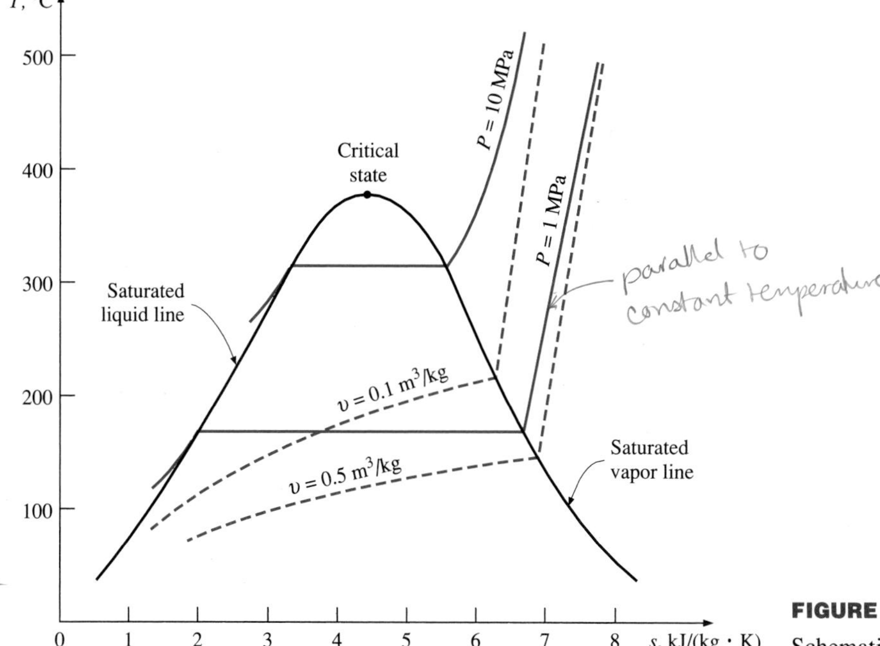

FIGURE 6-11

Schematic of the *T-s* diagram for water.

We note that the change in entropy of a substance during a process is simply the difference between the entropy values at the final and initial states. The initial state of the refrigerant is completely specified.

Assumptions The volume of the tank is constant and thus $v_2 = v_1$.

Analysis Recognizing that the specific volume remains constant during this process, the properties of the refrigerant at both states are determined to be

State 1:
$$\left.\begin{aligned} P_1 &= 140 \text{ kPa} \\ T_1 &= 20°\text{C} \end{aligned}\right\} \quad \begin{aligned} s_1 &= 1.0532 \text{ kJ/(kg} \cdot \text{K)} \\ v_1 &= 0.1652 \text{ m}^3\text{/kg} \end{aligned}$$

State 2:
$$\left.\begin{aligned} P_2 &= 100 \text{ kPa} \\ (v_2 &= v_1) \end{aligned}\right\} \quad \begin{aligned} v_f &= 0.0007258 \text{ m}^3\text{/kg} \\ v_g &= 0.1917 \text{ m}^3\text{/kg} \end{aligned}$$

The refrigerant is a saturated liquid–vapor mixture at the final state since $v_f < v_2 < v_g$ at 100 kPa pressure. Therefore, we need to determine the quality first:

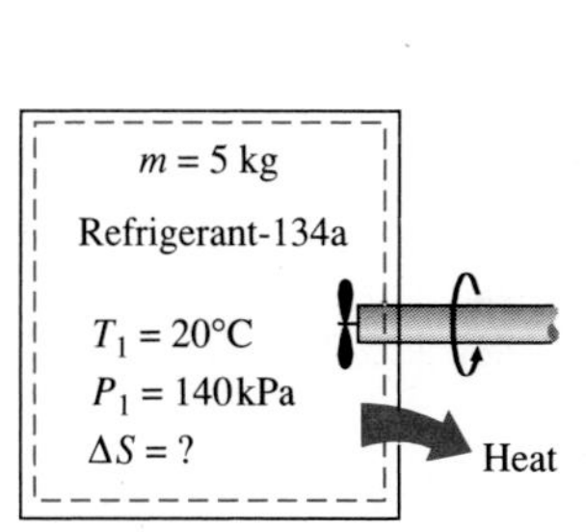

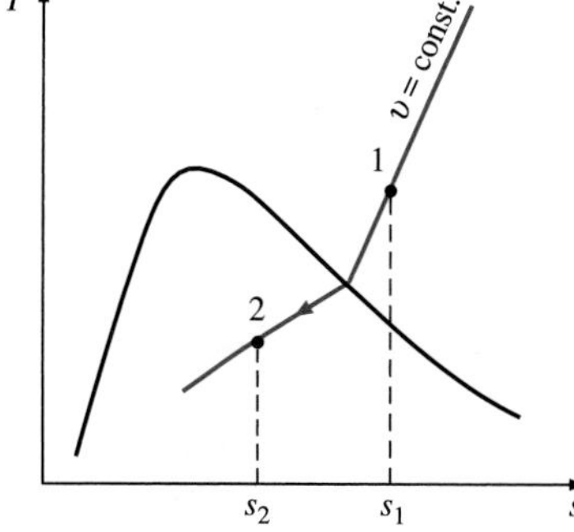

FIGURE 6-12

Schematic and *T-s* diagram for Example 6-3.

$$x_2 = \frac{v_2 - v_f}{v_{fg}} = \frac{0.1652 - 0.0007258}{0.1917 - 0.0007258} = 0.861$$

Thus,

$$s_2 = s_f + x_2 s_{fg} = 0.0678 + (0.861)(0.9395 - 0.0678) = 0.8183 \text{ kJ/(kg} \cdot \text{K)}$$

Then the entropy change of the refrigerant during this process is determined from

$$\Delta S = m(s_2 - s_1) = (5 \text{ kg})[(0.8183 - 1.0532 \text{ kJ/(kg} \cdot \text{K)}]$$
$$= \mathbf{-1.174 \text{ kJ/K}}$$

Discussion The negative sign indicates that the entropy of the system is decreasing during this process. This is not a violation of the second law, however, since it is the *entropy generation* S_{gen} that cannot be negative.

EXAMPLE 6-4 Entropy Change during a Constant Pressure Process

A piston-cylinder device initially contains 3 lbm of liquid water at 20 psia and 70°F. The water is now heated at constant pressure by the addition of 3450 Btu of heat. Determine the entropy change of the water during this process.

Solution We take the water in the cylinder as the *system* (Fig. 6-13). This is a *closed system* since no mass crosses the system boundary during the process. We note that a piston-cylinder device typically involves a moving boundary and thus boundary work W_b. Also, heat is transferred to the system and electrical work W_e is supplied to the system.

Assumptions **1** The tank is stationary and thus the kinetic and potential energy changes are zero, $\Delta KE = \Delta PE = 0$. **2** The process is quasi-equilibrium. **3** The pressure remains constant during the process and thus $P_2 = P_1$.

Analysis Water exists as a compressed liquid at the initial state since its pressure is greater than the saturation pressure of 0.3632 psia at 70°F. By approximating the compressed liquid as a saturated liquid at the given temperature, the properties at the initial state are

State 1: $\left.\begin{aligned} P_1 &= 20 \text{ psia} \\ T_1 &= 70°\text{F} \end{aligned}\right\}$ $\begin{aligned} s_1 &\cong s_{f@70°F} = 0.07463 \text{ Btu/(lbm} \cdot \text{R)} \\ h_1 &\cong h_{f@70°F} = 38.09 \text{ Btu/lbm} \end{aligned}$

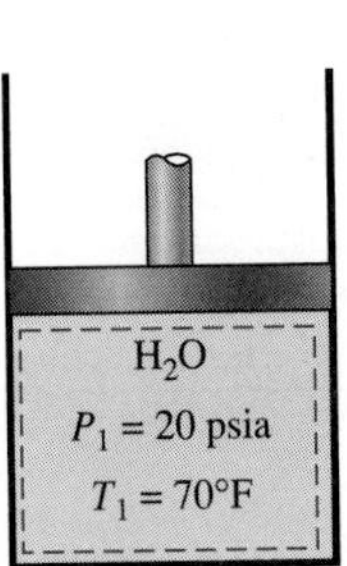

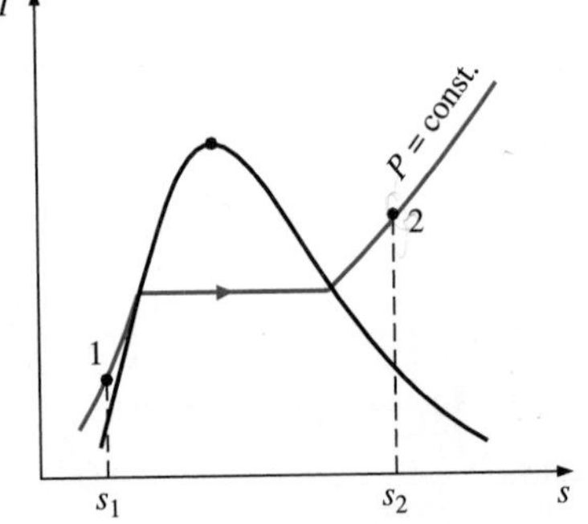

FIGURE 6-13
Schematic and *T-s* diagram for Example 6-4.

At the final state, the pressure is still 20 psia, but we need one more property to fix the state. This property is determined from the energy balance applied to the system,

$$\underbrace{E_{in} - E_{out}}_{\substack{\text{Net energy transfer} \\ \text{by heat, work, and mass}}} = \underbrace{\Delta E_{system}}_{\substack{\text{Change in internal, kinetic,} \\ \text{potential, etc., energies}}}$$

$$Q_{in} - W_b = \Delta U$$

$$Q_{in} = \Delta H = m(h_2 - h_1)$$

$$3450 \text{ Btu} = (3 \text{ lbm})(h_2 - 38.09 \text{ Btu/lbm})$$

$$h_2 = 1188.1 \text{ Btu/lbm}$$

since $\Delta U + W_b = \Delta H$ for a constant pressure quasi-equilibrium process. Then,

State 2: $\left.\begin{array}{l} P_2 = 20 \text{ psia} \\ h_2 = 1188.1 \text{ Btu/lbm} \end{array}\right\}$ $\begin{array}{l} s_2 = 1.7759 \text{ Btu/(lbm} \cdot \text{R)} \\ \text{(Table A-6E, interpolation)} \end{array}$

Therefore, the entropy change of water during this process is

$$\Delta S = m(s_2 - s_1) = (3 \text{ lbm})(1.7759 - 0.07463) \text{ Btu/(lbm} \cdot \text{R)}$$
$$= \mathbf{5.1038 \text{ Btu/R}}$$

6-4 ■ ISENTROPIC PROCESSES

We mentioned earlier that the entropy of a fixed mass can be changed by (1) heat transfer and (2) irreversibilities. Then it follows that the entropy of a fixed mass will not change during a process that is *internally reversible* and *adiabatic*. A process during which the entropy remains constant is called an **isentropic process.** An isentropic process is characterized by

Isentropic process: $\Delta s = 0$ or $s_2 = s_1$ [kJ/(kg · K)] (6-13)

That is, a substance will have the same entropy value at the end of the process as it does at the beginning if the process is carried out in an isentropic manner. We will develop some very useful isentropic relations for ideal gases later in this chapter.

Many engineering systems or devices such as pumps, turbines, nozzles, and diffusers are essentially adiabatic in their operation, and they perform best when the irreversibilities, such as the friction associated with the process, are minimized. Therefore, an isentropic process can serve as an appropriate model for actual processes. Also, isentropic processes enable us to define efficiencies for processes to compare the actual performance of these devices to the performance under idealized conditions. This should be sufficient motivation for studying the isentropic processes.

It should be recognized that a *reversible adiabatic* process is necessarily isentropic ($s_2 = s_1$), but an *isentropic* process is not necessarily a reversible

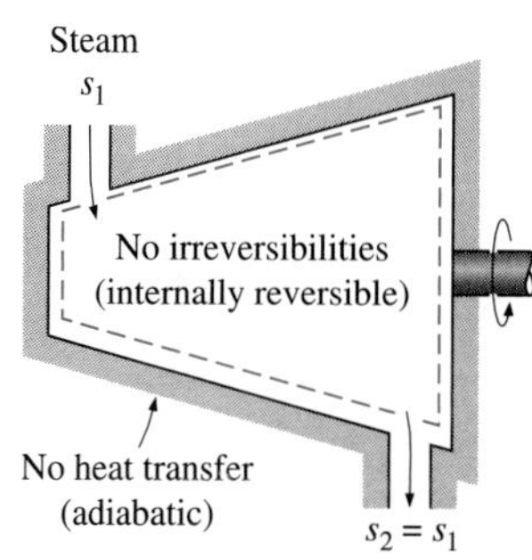

FIGURE 6-14

During an internally reversible, adiabatic (isentropic) process, the entropy of a system remains constant.

adiabatic process. (The entropy increase of a substance during a process as a result of irreversibilities may be offset by a decrease in entropy as a result of heat losses, for example.) However, the term *isentropic process* is customarily used in thermodynamics to imply an *internally reversible, adiabatic process.*

Not necessarily

EXAMPLE 6-5 Isentropic Expansion of Steam in a Turbine

Steam enters an adiabatic turbine at 5 MPa and 450°C and leaves at a pressure of 1.4 MPa. Determine the work output of the turbine per unit mass of steam flowing through the turbine if the process is reversible and the changes in kinetic and potential energies are negligible.

Solution We take the *turbine* as the system (Fig. 6-15). This is a *control volume* since mass crosses the system boundary during the process. We note that there is only one inlet and one exit, and thus $\dot{m}_1 = \dot{m}_2 = \dot{m}$.

Assumptions **1** This is a steady-flow process since there is no change with time at any point and thus $\Delta m_{CV} = 0$, $\Delta E_{CV} = 0$, and $\Delta S_{CV} = 0$. **2** The process is reversible. **3** Kinetic and potential energies are negligible. **4** The turbine is adiabatic and thus there is no heat flow.

Analysis The power output of the turbine is determined from the rate form of the energy balance,

$$\underbrace{\dot{E}_{in} - \dot{E}_{out}}_{\substack{\text{Rate of net energy transfer}\\ \text{by heat, work, and mass}}} = \underbrace{\Delta \dot{E}_{system}^{\nearrow 0 \text{ (steady)}}}_{\substack{\text{Rate of change in internal, kinetic,}\\ \text{potential, etc., energies}}} = 0$$

$$\dot{E}_{in} = \dot{E}_{out}$$

$$\dot{m}h_1 = \dot{W}_{out} + \dot{m}h_2 \qquad (\text{since } \dot{Q} = 0, \text{ ke} \cong \text{pe} \cong 0)$$

$$\dot{W}_{out} = \dot{m}(h_1 - h_2)$$

The inlet state is completely specified since two properties are given. But only one property (pressure) is given at the final state, and we need one more property to fix it. The second property comes from the observation that the process is reversible and adiabatic, and thus isentropic. Therefore, $s_2 = s_1$, and

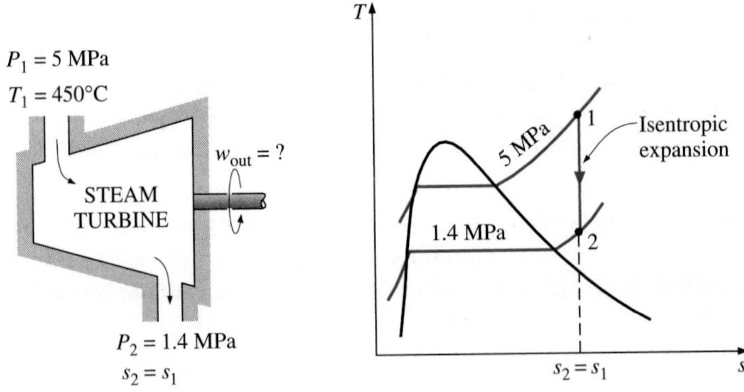

FIGURE 6-15
Schematic and *T-s* diagram for Example 6-5.

State 1: $\left.\begin{array}{l} P_1 = 5 \text{ MPa} \\ T_1 = 450°\text{C} \end{array}\right\}$ $\begin{array}{l} h_1 = 3316.2 \text{ kJ/kg} \\ s_1 = 6.8186 \text{ kJ/(kg} \cdot \text{K)} \end{array}$

State 2: $\left.\begin{array}{l} P_2 = 1.4 \text{ MPa} \\ s_2 = s_1 \end{array}\right\}$ $h_2 = 2966.6 \text{ kJ/kg}$

Then the work output of the turbine per unit mass of the steam flowing through it becomes

$$w_{out} = h_1 - h_2 = 3316.2 - 2966.6 = \mathbf{349.6\ kJ/kg}$$

6-5 ■ WHAT IS ENTROPY?

It is clear from the previous discussion that entropy is a useful property and serves as a valuable tool in the second-law analysis of engineering devices. But this does not mean that we know and understand entropy well. Because we do not. In fact, we cannot even give an adequate answer to the question, What is entropy? Not being able to describe entropy fully, however, does not take anything away from its usefulness. In Chap. 1, we could not define *energy* either, but it did not interfere with our understanding of energy transformations and the conservation of energy principle. Granted, entropy is not a household word like energy. But with continued use, our understanding of entropy will deepen, and our appreciation of it will grow. The discussion below will shed some light on the physical meaning of entropy by considering the microscopic nature of matter.

Entropy can be viewed as a measure of *molecular disorder,* or *molecular randomness.* As a system becomes more disordered, the positions of the molecules become less predictable and the entropy increases. Thus, it is not surprising that the entropy of a substance is lowest in the solid phase and highest in the gas phase (Fig. 6-16). In the solid phase, the molecules of a substance continually oscillate about their equilibrium positions, but they cannot move relative to each other, and their position at any instant can be predicted with good certainty. In the gas phase, however, the molecules move about at random, collide with each other, and change direction, making it extremely difficult to predict accurately the microscopic state of a system at any instant. Associated with this molecular chaos is a high value of entropy.

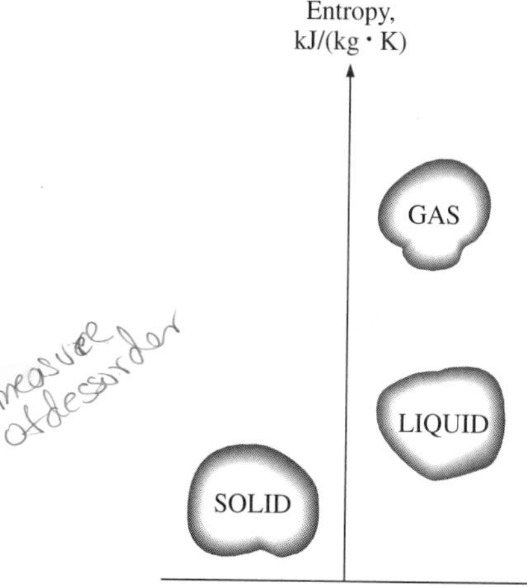

FIGURE 6-16

The level of molecular disorder (entropy) of a substance increases as it melts or evaporates.

When viewed microscopically (from a statistical thermodynamics point of view), an isolated system that appears to be at a state of equilibrium may exhibit a high level of activity because of the continual motion of the molecules. To each state of macroscopic equilibrium there corresponds a large number of possible microscopic states or molecular configurations. The entropy of a system is related to the total number of possible microscopic states of that system, called *thermodynamic probability p,* by the **Boltzmann relation,** expressed as

$$S = k \ln p \tag{6-14}$$

where $k = 1.3806 \times 10^{-23}$ kJ/(kmol · K) is the **Boltzmann constant.** Therefore, from a microscopic point of view, the entropy of a system increases whenever the molecular randomness or uncertainty (i.e., molecular probability) of a system increases. Thus, entropy is a measure of molecular disorder, and the molecular disorder of an isolated system increases anytime it undergoes a process.

FIGURE 6-17

Disorganized energy does not create much useful effect, no matter how large it is.

Molecules in the gas phase possess a considerable amount of kinetic energy. But we know that no matter how large their kinetic energies are, the gas molecules will not rotate a paddle wheel inserted into the container and produce work. This is because the gas molecules, and the energy they possess, are disorganized. Probably the number of molecules trying to rotate the wheel in one direction at any instant is equal to the number of molecules that are trying to rotate it in the opposite direction, causing the wheel to remain motionless. Therefore, we cannot extract any useful work directly from disorganized energy (Fig. 6-17).

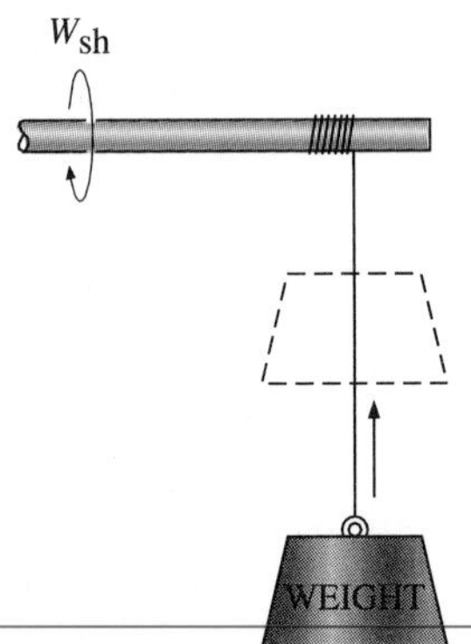

FIGURE 6-18

In the absence of friction, raising a weight by a rotating shaft does not create any disorder (entropy), and thus energy is not degraded during this process.

Now consider a rotating shaft shown in Fig. 6-18. This time the energy of the molecules is completely organized since the molecules of the shaft are rotating in the same direction together. This organized energy can readily be used to perform useful tasks such as raising a weight or generating electricity. Being an organized form of energy, work is free of disorder or randomness and thus free of entropy. *There is no entropy transfer associated with energy transfer as work.* Therefore, in the absence of any friction, the process of raising a weight by a rotating shaft (or a flywheel) will not produce any entropy. Any process that does not produce a net entropy is reversible, and thus the process described above can be reversed by lowering the weight. Therefore, energy is not degraded during this process, and no potential to do work is lost.

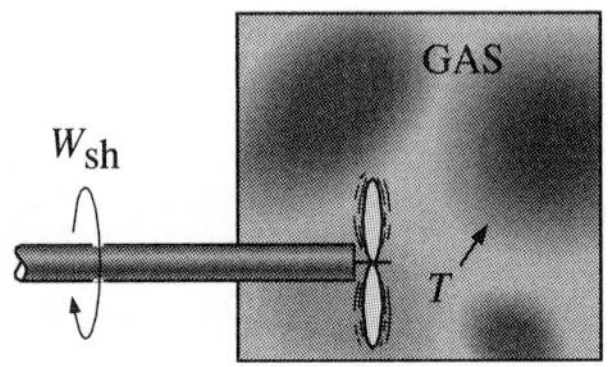

FIGURE 6-19

The paddle-wheel work done on a gas increases the level of disorder (entropy) of the gas, and thus energy is degraded during this process.

Instead of raising a weight, let us operate the paddle wheel in a container filled with a gas, as shown in Fig. 6-19. The paddle-wheel work in this case will be converted to the internal energy of the gas, as evidenced by a rise in gas temperature, creating a higher level of molecular chaos and disorder in the container. This process is quite different from raising a weight since the organized paddle-wheel energy is now converted to a highly disorganized form of energy, which cannot be converted back to the paddle wheel as the rotational kinetic energy. Only a portion of this energy can be converted to work by partially reorganizing it through the use of a heat engine. Therefore, energy is degraded during this process, the ability to do work is reduced, molecular disorder is produced, and associated with all this is an increase in entropy.

The *quantity* of energy is always preserved during an actual process (the first law), but the *quality* is bound to decrease (the second law). This decrease in quality is always accompanied by an increase in entropy. As an example, consider the transfer of 10 kJ of energy as heat from a hot medium to a cold one. At the end of the process, we will still have the 10 kJ of energy, but at a lower temperature and thus at a lower quality.

Heat is, in essence, a form of *disorganized energy,* and some disorganization (entropy) will flow with heat (Fig. 6-20). As a result, the entropy and the level of molecular disorder or randomness of the hot body will decrease with the entropy and the level of molecular disorder of the cold body will increase. The second law requires that the increase in entropy of the cold body be greater than the decrease in entropy of the hot body, and thus the net entropy of the combined system (the cold body and the hot body) increases. That is, the combined system is at a state of greater disorder at the final state. Thus we can conclude that processes can occur only in the direction of increased overall entropy or molecular disorder. That is, the entire universe is getting more and more chaotic every day.

From a statistical point of view, entropy is a measure of molecular randomness, that is, the uncertainty about the positions of molecules at any instant. Even in the solid phase, the molecules of a substance continually oscillate, creating an uncertainty about their position. These oscillations, however, fade as the temperature is decreased, and the molecules become completely motionless at absolute zero. This represents a state of ultimate molecular order (and minimum energy). Therefore, *the entropy of a pure crystalline substance at absolute zero temperature is zero* since there is no uncertainty about the state of the molecules at that instant (Fig. 6-21). This statement is known as the **third law of thermodynamics.** The third law of thermodynamics provides an absolute reference point for the determination of entropy. The entropy determined relative to this point is called **absolute entropy,** and it is extremely useful in the thermodynamic analysis of chemical reactions. Notice that the entropy of a substance that is not pure crystalline (such as a solid solution) is not zero at absolute zero temperature. This is because more than one molecular configuration exist for such substances, which introduces some uncertainty about the microscopic state of the substance.

The concept of entropy as a measure of disorganized energy can also be applied to other areas. Iron molecules, for example, create a magnetic field around themselves. In ordinary iron, molecules are randomly aligned, and they cancel each other's magnetic effect. When iron is treated and the molecules are realigned, however, that piece of iron turns into a piece of magnet, creating a powerful magnetic field around it.

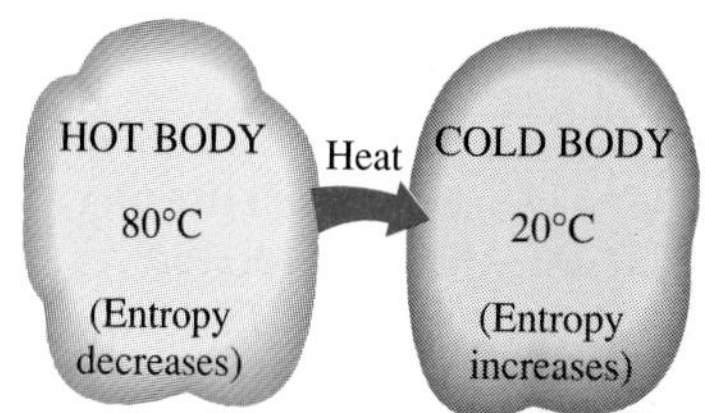

FIGURE 6-20
During a heat transfer process, the net disorder (entropy) increases. (The increase in the disorder of the cold body more than offsets the decrease in the disorder of the hot body.)

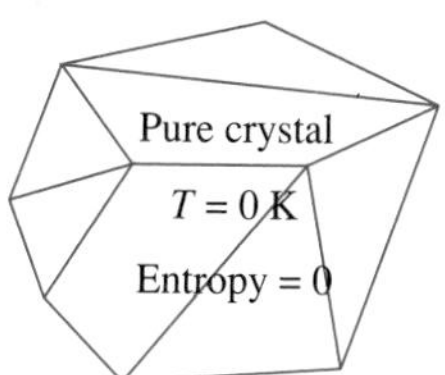

FIGURE 6-21
A pure substance at absolute zero temperature is in perfect order, and its entropy is zero (the third law of thermodynamics).

Entropy and Entropy Generation in Daily Life

Entropy can be viewed as a measure of disorder or disorganization in a system. Likewise, entropy generation can be viewed as a measure of disorder or disorganization generated during a process. The concept of entropy is not used in daily life nearly as extensively as the concept of energy, even though entropy is readily applicable to various aspects of daily life. The extension of the entropy concept to nontechnical fields is not a novel idea. It has been the topic of several articles, and even some books. Below we present several ordinary events and show their relevance to the concept of entropy and entropy generation.

Efficient people lead low-entropy (highly organized) lives. They have a place for everything (minimum uncertainty), and it takes minimum energy for them to locate something. Inefficient people, on the other hand, are disorganized and lead high-entropy lives. It takes them minutes (if not hours) to find something they need, and they are likely to create a bigger disorder as they are searching since they will probably conduct the search in a disorganized manner (Fig. 6-22). People leading high-entropy lifestyles are always on the run, and never seem to catch up.

FIGURE 6-22
The use of entropy (disorganization, uncertainty) is not limited to thermodynamics.

You probably noticed (with frustration) that some people seem to *learn* fast and remember well what they learn. We can call this type of learning organized or low-entropy learning. These people make a conscientious effort to file the new information properly by relating it to their existing knowledge base and creating a solid information network in their minds. On the other hand, people who throw the information into their minds as they study, with no effort to secure it, may *think* they are learning. They are bound to discover otherwise when they need to locate the information, for example during a test. It is not easy to retrieve information from a database that is, in a sense, in the gas phase. Students who have blackouts during tests should reexamine their study habits.

A *library* with a good shelving and indexing system can be viewed as a low-entropy library because of the high level of organization. Likewise, a library with a poor shelving and indexing system can be viewed as a high-entropy library because of the high level of disorganization. A library with no indexing system is like no library, since a book is of no value if it cannot be found.

Consider two identical buildings, each containing one million books. In the first building, the books are *piled* on top of each other, whereas in the second building they are *highly organized, shelved, and indexed* for easy reference. There is no doubt about which building a student will prefer to go to for checking out a certain book. Yet, some may argue from the first law point of view that these two buildings are equivalent since the mass and energy content of the two buildings are identical, despite the high level of disorganization (entropy) in the first building. This example illustrates that any realistic comparisons should involve the second law point of view.

Two *textbooks* that seem to be identical because both cover basically the same topics and present the same information may actually be *very* different depending on *how* they cover the topics. After all, two seemingly identical cars are not so identical if one goes only half as many miles as the other one on the same amount of fuel. Likewise, two seemingly identical books are not so identical if it takes twice as long to learn a topic from one of them as it does from the other. Thus, comparisons made on the basis of the first law only may be highly misleading.

Having a disorganized (high-entropy) *army* is like having no army at all. It is no coincidence that the command centers of any armed forces are among the primary targets during a war. One army that consists of ten divisions is ten times more powerful than ten armies each consisting of a single division. Likewise, one country that consists of ten states is more powerful than ten countries, each consisting of a single state. The *United States* would not be such a powerful country if there were fifty independent countries in its place instead of a single country with fifty states. The new European common market has the potential to be a new economic superpower. The old cliché "divide and conquer" can be rephrased as "increase the entropy and conquer."

We know that mechanical friction is always accompanied by entropy generation, and thus reduced performance. We can generalize this to daily life: *friction in the work place* with fellow workers is bound to generate entropy, and thus adversely affect performance. It will result in reduced productivity. Hopefully, someday we will be able to come up with some procedures to quantify entropy generated during nontechnical activities, and maybe even pinpoint its primary sources and magnitude.

We also know that *unrestrained expansion* (or explosion) and uncontrolled electron exchange (chemical reactions) generate entropy and are highly irreversible. Likewise, unrestrained opening of the mouth to scatter angry words is highly irreversible since this generates entropy, and it can cause considerable damage. A person who gets up in anger is bound to sit down at a loss.

6-6 ■ PROPERTY DIAGRAMS INVOLVING ENTROPY

Property diagrams serve as great visual aids in the thermodynamic analysis of processes. We have used *P-v* and *T-v* diagrams extensively in previous chapters in conjunction with the first law of thermodynamics. In the second-law analysis, it is very helpful to plot the processes on diagrams for which one of the coordinates is entropy. The two diagrams commonly used in the second-law analysis are the *temperature-entropy* and the *enthalpy-entropy* diagrams.

1 The *T-s* Diagram

Consider the defining equation of entropy (Eq. 6-4). It can be rearranged as

$$\delta Q_{\text{int rev}} = T\,dS \qquad \text{(kJ)} \qquad (6\text{-}15)$$

As shown in Fig. 6-23, $\delta Q_{\text{rev int}}$ corresponds to a differential area on a *T-S* diagram. The total heat transfer during an internally reversible process is determined by integration to be

$$Q_{\text{int rev}} = \int_1^2 T\,dS \qquad \text{(kJ)} \qquad (6\text{-}16)$$

which corresponds to the area under the process curve on a *T-S* diagram. Therefore, we conclude that *the area under the process curve on a T-S diagram represents heat transfer during an internally reversible process.* This is somewhat analogous to reversible boundary work being represented by the area under the process curve on a *P-V* diagram. Note that the area under the process curve represents heat transfer for processes that are internally (or totally) reversible. The area has no meaning for irreversible processes.

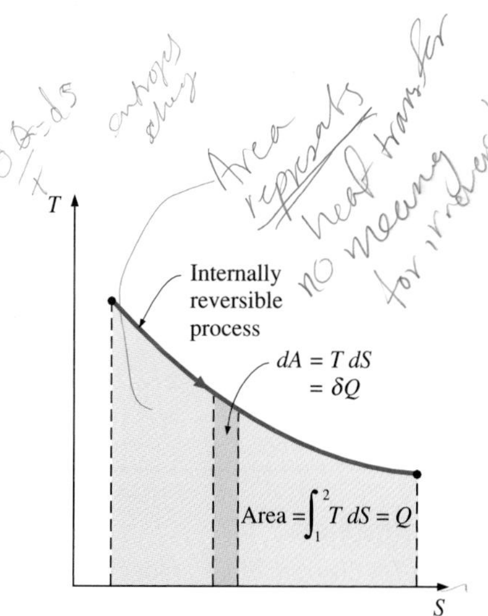

FIGURE 6-23

On a *T-S* diagram, the area under the process curve represents the heat transfer for internally reversible processes.

Equations 6-15 and 6-16 can also be expressed on a unit-mass basis as

$$\delta q_{\text{int rev}} = T\,ds \qquad \text{(kJ/kg)} \qquad (6\text{-}17)$$

and

$$q_{\text{int rev}} = \int_1^2 T\,ds \qquad \text{(kJ/kg)} \qquad (6\text{-}18)$$

To perform the integrations in Eqs. 6-16 and 6-18, one needs to know the relationship between T and s during a process. One special case for which these integrations can be performed easily is the *internally reversible isothermal process*. It yields

$$Q_{\text{int rev}} = T_0\,\Delta S \qquad \text{(kJ)} \qquad (6\text{-}19)$$

or

$$q_{\text{int rev}} = T_0\,\Delta s \qquad \text{(kJ/kg)} \qquad (6\text{-}20)$$

where T_0 is the constant temperature and ΔS is the entropy change of the system during the process.

In the relations above, T is the absolute temperature, which is always positive. Therefore, heat transfer during internally reversible processes is positive when entropy increases and negative when entropy decreases. An isentropic process on a *T-s* diagram is easily recognized as a *vertical-line segment*. This is expected since an isentropic process involves no heat transfer, and therefore the area under the process path must be zero (Fig. 6-24). The *T-s* diagrams serve as valuable tools for visualizing the second-law aspects of processes and cycles, and thus they are frequently used in thermodynamics. The *T-s* diagram of water is given in the appendix in Fig. A-9.

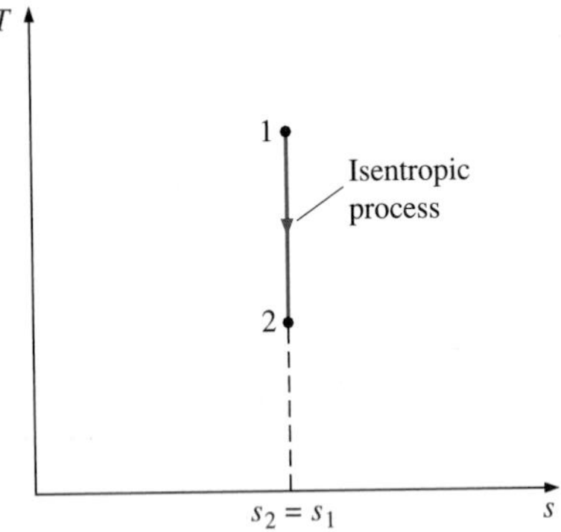

FIGURE 6-24
During an internally reversible, adiabatic (isentropic) process, the entropy of a system remains constant.

EXAMPLE 6-6 The *T-S* Diagram of the Carnot Cycle

Show the Carnot cycle on a *T-S* diagram and indicate the areas that represent the heat supplied Q_H, heat rejected Q_L, and the net work output $W_{\text{net, out}}$ on this diagram.

Solution You will recall from Chap. 5 that the Carnot cycle is made up of two reversible isothermal (T = constant) processes and two isentropic (s = constant) processes. These four processes form a rectangle on a *T-S* diagram, as shown in Fig. 6-25.

On a *T-S* diagram, the area under the process curve represents the heat transfer for that process. Thus the area *A*12*B* represents Q_H, the area *A*43*B* represents Q_L, and the difference between these two (the area in color) represents the net work since

$$W_{\text{net, out}} = Q_H - Q_L$$

Therefore, the area enclosed by the path of a cycle (area 1234) on a *T-S* diagram represents the net work. Recall from Chap. 3 that the area enclosed by the path of a cycle also represents the net work on a *P-V* diagram.

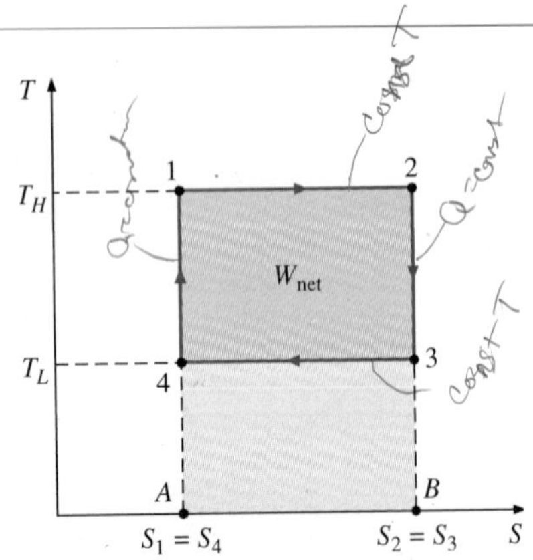

FIGURE 6-25
The *T-S* diagram of a Carnot cycle (Example 6-6).

The *h-s* Diagram

Another diagram commonly used in engineering is the enthalpy-entropy diagram, which is quite valuable in the analysis of steady-flow devices such as

turbines, compressors, and nozzles. The coordinates of an *h-s* diagram represent two properties of major interest: enthalpy, which is a primary property in the first-law analysis of the steady-flow devices, and entropy, which is the property that accounts for irreversibilities during adiabatic processes. In analyzing the steady flow of steam through an adiabatic turbine, for example, the vertical distance between the inlet and the exit states (Δh) is a measure of the work output of the turbine, and the horizontal distance (Δs) is a measure of the irreversibilities associated with the process (Fig. 6-26).

The *h-s* diagram is also called a **Mollier diagram** after the German scientist R. Mollier (1863–1935). An *h-s* diagram is given in the appendix for steam in Fig. A-10. The general features of an *h-s* diagram are illustrated in Fig. 6-27. On an *h-s* diagram, the constant-temperature lines are straight in the saturated liquid–vapor mixture region. They become almost horizontal in the superheated vapor region, particularly at low pressures. This is not surprising since steam approaches ideal-gas behavior as it moves away from the saturation region, and for ideal gases the enthalpy is a function of temperature only.

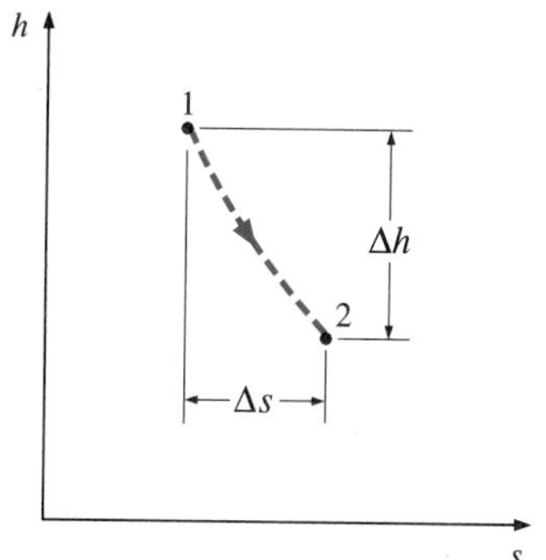

FIGURE 6-26

For adiabatic steady-flow devices, the vertical distance Δh on an *h-s* diagram is a measure of work, and the horizontal distance Δs is a measure of irreversibilities.

6-7 ■ THE $T\,ds$ RELATIONS

Earlier in this chapter, it was shown that the quantity $(\delta Q/T)_{\text{int rev}}$ corresponds to a differential change in a property, called *entropy*. The entropy change for a process, then, was evaluated by integrating $\delta Q/T$ along some imaginary

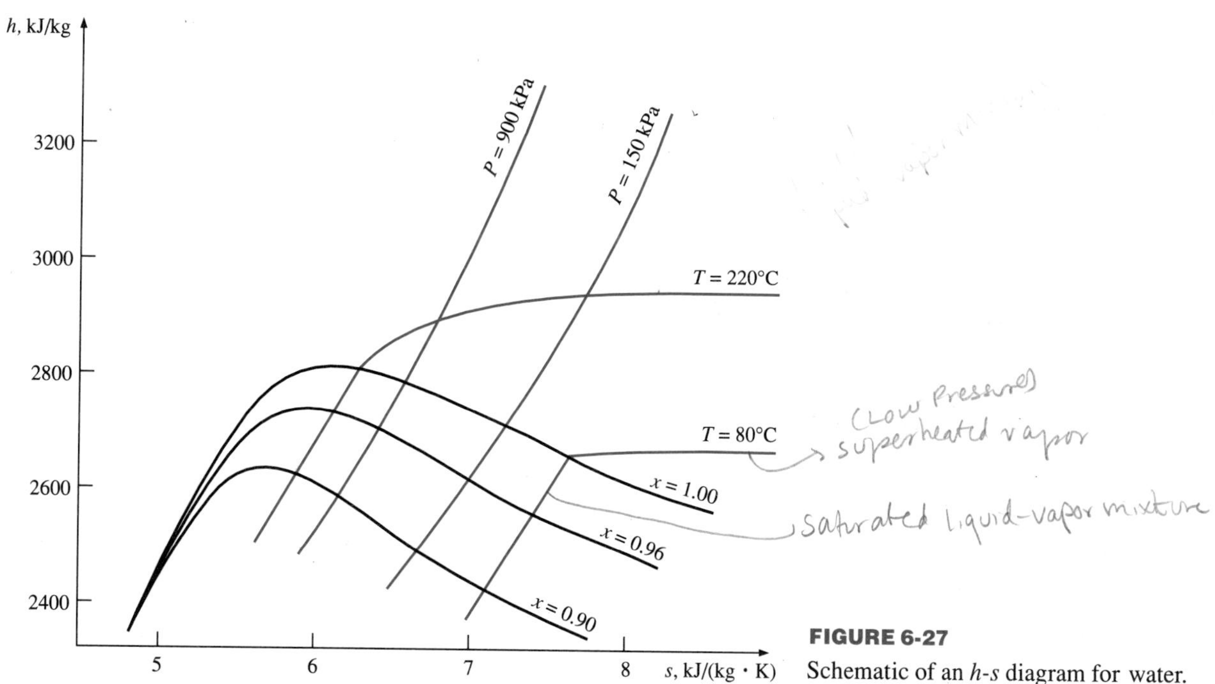

FIGURE 6-27

Schematic of an *h-s* diagram for water.

internally reversible path between the actual end states (Eq. 6-5). For isothermal internally reversible processes, this integration is straightforward. But when the temperature varies during the process, we have to have a relation between δQ and T to perform this integration. Finding such relations is what we intend to do in this section.

The differential form of the conservation of energy equation for a closed stationary system (a fixed mass) containing a simple compressible substance (Eq. 3-40) can be expressed for an internally reversible process as

$$\delta Q_{\text{int rev}} - \delta W_{\text{int rev, out}} = dU \tag{6-21}$$

But

$$\delta Q_{\text{int rev}} = T\,dS \qquad \text{(Eq. 6-15)}$$

$$\delta W_{\text{int rev, out}} = P\,dV$$

Thus,

$$T\,dS = dU + P\,dV \tag{6-22}$$

or

$$T\,ds = du + P\,dv \tag{6-23}$$

per unit mass. This equation is known as the first *T ds,* or *Gibbs, equation.* Notice that the only type of work interaction a simple compressible system may involve as it undergoes an internally reversible process is the quasi-equilibrium boundary work.

The second *T ds* equation is obtained by eliminating *du* from Eq. 6-23 by using the definition of enthalpy ($h = u + Pv$):

$$\left.\begin{array}{lll} h = u + Pv & \longrightarrow & dh = du + P\,dv + v\,dP \\ \text{(Eq. 6-23)} & \longrightarrow & T\,ds = du + P\,dv \end{array}\right\} \quad T\,ds = dh - v\,dP \tag{6-24}$$

Equations 6-23 and 6-24 are extremely valuable since they relate entropy changes of a system to the changes in other properties. Unlike Eq. 6-4, they are property relations and therefore are independent of the type of the processes.

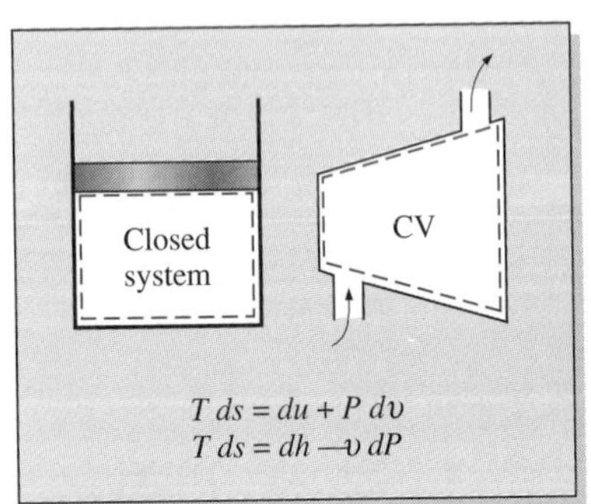

FIGURE 6-28

The *T ds* relations are valid for both reversible and irreversible processes and for both closed and open systems.

The *T ds* relations above are developed with an internally reversible process in mind since the entropy change between two states must be evaluated along a reversible path. But the results obtained are valid for both reversible and irreversible processes since entropy is a property and the change in a property between two states is independent of the type of process the system undergoes. Equations 6-23 and 6-24 are relations between the properties of a unit mass of a simple compressible system as it undergoes a change of state, and they are applicable whether the change occurs in a closed or an open system (Fig. 6-28).

Explicit relations for differential changes in entropy are obtained by solving for *ds* in Eqs. 6-23 and 6-24:

$$ds = \frac{du}{T} + \frac{P\,dv}{T} \tag{6-25}$$

and

$$ds = \frac{dh}{T} - \frac{v\,dP}{T} \tag{6-26}$$

The entropy change during a process can be determined by integrating either of these equations between the initial and the final states. To perform these integrations, however, we must know the relationship between du or dh and the temperature (such as $du = C_v\, dT$ and $dh = C_p\, dT$ for ideal gases) as well as the equation of state for the substance (such as the ideal-gas equation of state $Pv = RT$). For substances for which such relations exist, the integration of Eq. 6-25 or 6-26 is straightforward. This is done later in this chapter. For other substances, we have to rely on tabulated data.

The $T\, ds$ relations for nonsimple systems, that is, systems that involve more than one mode of quasi-equilibrium work, can be obtained in a similar manner by including all the relevant quasi-equilibrium work modes.

6-8 ■ ENTROPY CHANGE OF LIQUIDS AND SOLIDS

We mentioned in Chap. 3 that liquids and solids can be approximated as *incompressible substances* since their specific volumes remain nearly constant during a process. Thus, $dv \cong 0$ for liquids and solids, and Eq. 6-26 for this case reduces to

$$ds = \frac{du}{T} = \frac{C\, dT}{T} \tag{6-27}$$

since $C_p = C_v = C$ and $du = C\, dT$ for incompressible substances. Then the entropy change during a process is determined by integration to be

Liquids, solids: $$s_2 - s_1 = \int_1^2 C(T)\frac{dT}{T} \cong C_{av} \ln \frac{T_2}{T_1} \quad \text{[kJ/(kg·K)]} \tag{6-28}$$

where C_{av} is the *average* specific heat of the substance over the given temperature range. Note that the entropy change of a truly incompressible substance depends on temperature only and is independent of pressure.

Equation 6-28 can be used to determine the entropy changes of solids and liquids with reasonable accuracy. However, for liquids that expand considerably with temperature, it may be necessary to consider the effects of volume change in calculations. This is especially the case when the temperature change is large.

A relation for isentropic processes of liquids and solids is obtained by setting the entropy change relation above equal to zero. It gives

Isentropic: $$s_2 - s_1 = C_{av} \ln \frac{T_2}{T_1} = 0 \quad \rightarrow \quad T_2 = T_1 \tag{6-29}$$

That is, the temperature of a truly incompressible substance remains constant during an isentropic process. Therefore, the isentropic process of an incompressible substance is also isothermal. This behavior is closely approximated by liquids and solids.

TABLE 6-1

Properties of liquid methane

Temp., T K	Pressure, P MPa	Density, ρ kg/m^3	Enthalpy, h kJ/kg	Entropy, s kJ/(kg · K)	Specific heat, C_p kJ/(kg · K)
110	0.5	425.3	208.3	4.878	3.476
	1.0	425.8	209.0	4.875	3.471
	2.0	426.6	210.5	4.867	3.460
	5.0	429.1	215.0	4.844	3.432
120	0.5	410.4	243.4	5.185	3.551
	1.0	411.0	244.1	5.180	3.543
	2.0	412.0	245.4	5.171	3.528
	5.0	415.2	249.6	5.145	3.486

EXAMPLE 6-7 Effect of Density of a Liquid on Entropy

Liquid methane is commonly used in various cryogenic applications. The critical temperature of methane is 191 K (or −82°C), and thus methane must be maintained below 191 K to keep it in liquid phase. The properties of liquid methane at various temperatures and pressures are given in Table 6-1. Determine the entropy change of liquid methane as it undergoes a process from 110 K and 1 MPa to 120 K and 5 MPa (*a*) using actual data for methane and (*b*) approximating liquid methane as an incompressible substance. What is the error involved in the latter case?

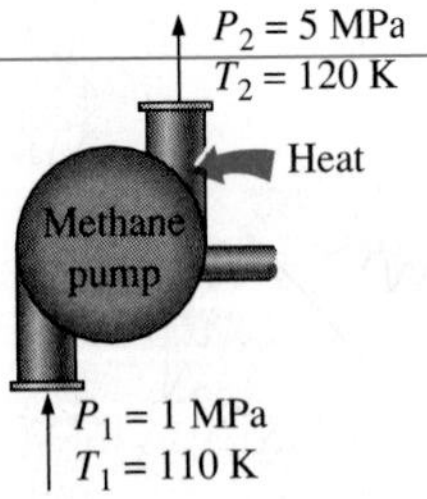

$P_2 = 5$ MPa
$T_2 = 120$ K
Heat
Methane pump
$P_1 = 1$ MPa
$T_1 = 110$ K

FIGURE 6-29
Schematic for Example 6-7.

Solution (*a*) We consider a unit mass of liquid methane (Fig. 6-29). The entropies of the methane at the initial and final states are determined directly from Table 6-1 to be

State 1: $\left.\begin{array}{l} P_1 = 1 \text{ MPa} \\ T_1 = 110 \text{ K} \end{array}\right\} \quad \begin{array}{l} s_1 = 4.875 \text{ kJ/(kg} \cdot \text{K)} \\ C_{p1} = 3.471 \text{ kJ/(kg} \cdot \text{K)} \end{array}$

State 2: $\left.\begin{array}{l} P_2 = 5 \text{ MPa} \\ T_2 = 120 \text{ K} \end{array}\right\} \quad \begin{array}{l} s_2 = 5.145 \text{ kJ/(kg} \cdot \text{K)} \\ C_{p2} = 3.486 \text{ kJ/(kg} \cdot \text{K)} \end{array}$

Therefore,

$$\Delta s = s_2 - s_1 = 5.145 - 4.875 = \mathbf{0.270\ kJ/(kg \cdot K)}$$

(*b*) Approximating liquid methane as an incompressible substance, its entropy change is determined to be

$$\Delta s = C_{av} \ln \frac{T_2}{T_1} = [3.4785 \text{ kJ/(kg} \cdot \text{K)}] \ln \frac{120 \text{ K}}{110 \text{ K}} = \mathbf{0.303\ kJ/(kg \cdot K)}$$

since

$$C_{p,\text{av}} = \frac{C_{p1} + C_{p2}}{2} = \frac{3.471 + 3.486}{2} = 3.4785 \text{ kJ/(kg} \cdot \text{K)}$$

Therefore, the error involved in approximating liquid methane as an incompressible substance is

$$\text{Error} = \frac{|\Delta s_{\text{actual}} - \Delta s_{\text{ideal}}|}{\Delta s_{\text{actual}}} = \frac{|0.270 - 0.303|}{0.270} = \mathbf{0.122\ (or\ 12.2\%)}$$

Discussion This result is not surprising since the density of liquid methane changes during this process from 425.8 to 415.2 kg/m^3 (about 3 percent), which makes us question the validity of the incompressible substance assumption. Still, this assumption enables us to obtain reasonably accurate results with less effort, which proves to be very convenient in the absence of compressed liquid data.

EXAMPLE 6-8 Economics of Replacing a Valve by a Turbine

A cryogenic manufacturing facility handles liquid methane at 115 K and 5 MPa at a rate of 0.280 m^3/s . A process requires dropping the pressure of liquid methane to 1 MPa, which is done by throttling the liquid methane by passing it through a flow resistance such as a valve. A recently hired engineer proposes to replace the throttling valve by a turbine in order to produce power while dropping the pressure to 1 MPa. Using data from Table 6-1, determine the maximum amount of power that can be produced by such a turbine. Also, determine how much this turbine will save the facility from electricity usage costs per year if the turbine operates continuously (8760 hr/yr) and the facility pays $0.075/kWh for electricity.

Solution We take the *turbine* as the system (Fig. 6-30). This is a *control volume* since mass crosses the system boundary during the process. We note that there is only one inlet and one exit and thus $\dot{m}_1 = \dot{m}_2 = \dot{m}$.

FIGURE 6-30
A 1.0 MW liquified natural gas (LNG) turbine with 95-cm turbine runner diameter being installed in a cryogenic test facility (courtesy of Ebara International Corporation, Cryodynamics Division, Sparks, Nevada).

Assumptions **1** This is a steady-flow process since there is no change with time at any point and thus $\Delta m_{CV} = 0$, $\Delta E_{CV} = 0$, and $\Delta S_{CV} = 0$. **2** The turbine is adiabatic and thus there is no heat transfer. **3** The process is reversible. **4** Kinetic and potential energies are negligible.

Analysis The assumptions above are reasonable since a turbine is normally well-insulated and it must involve no irreversibilities for best performance and thus *maximum* power production. Therefore, the process through the turbine must be *reversible adiabatic* or *isentropic.* Then, $s_2 = s_1$ and

State 1: $\left.\begin{array}{l} P_1 = 5\text{ MPa} \\ T_1 = 115\text{ K} \end{array}\right\}$ $\begin{array}{l} h_1 = 232.3\text{ kJ/kg} \\ s_1 = 4.9945\text{ kJ/(kg}\cdot\text{K)} \\ \rho_1 = 422.5\text{ kg/s} \end{array}$

State 2: $\left.\begin{array}{l} P_2 = 1\text{ MPa} \\ s_2 = s_1 \end{array}\right\}$ $h_2 = 222.6\text{ kJ/kg}$

Also, the mass flow rate of liquid methane is

$$\dot{m} = \rho_1 \dot{V}_1 = (422.5\text{ kg/m}^3)(0.280\text{ m}^3\text{/s}) = 118.3\text{ kg/s}$$

Then the power output of the turbine is determined from the rate form of the energy balance to be

$$\underbrace{\dot{E}_{in} - \dot{E}_{out}}_{\substack{\text{Rate of net energy transfer} \\ \text{by heat, work, and mass}}} = \underbrace{\Delta \dot{E}_{system}^{\nearrow 0 \text{ (steady)}}}_{\substack{\text{Rate of change in internal, kinetic,} \\ \text{potential, etc., energies}}} = 0$$

$$\dot{E}_{in} = \dot{E}_{out}$$

$$\dot{m}h_1 = \dot{W}_{out} + \dot{m}h_2 \qquad (\text{since } \dot{Q} = 0, \text{ ke} \cong \text{pe} \cong 0)$$

$$\begin{aligned} \dot{W}_{out} &= \dot{m}(h_1 - h_2) \\ &= (118.3 \text{ kg/s})(232.3 - 222.6) \text{ kJ/kg} \\ &= \mathbf{1148 \text{ kW}} \end{aligned}$$

For continuous operation (365 × 24 = 8760 h), the amount of power produced per year will be

$$\begin{aligned} \text{Annual power production} &= \dot{W}_{out} \times \Delta t = (1148 \text{ kW})(8760 \text{ h/yr}) \\ &= 1.0056 \times 10^7 \text{ kWh/yr} \end{aligned}$$

At \$0.075/kWh, the amount of money this turbine will save the facility becomes

$$\begin{aligned} \text{Annual power savings} &= (\text{Annual power production})(\text{Unit cost of power}) \\ &= (1.0056 \times 10^7 \text{ kWh/yr})(\$0.075/\text{kWh}) \\ &= \mathbf{\$754{,}200/yr} \end{aligned}$$

That is, this turbine can save the facility \$754,200 a year by simply taking advantage of the potential that is currently being wasted by a throttling valve, and the engineer who made this observation should be rewarded.

Discussion This example shows the importance of the property entropy since it enabled us to quantify the work potential that is being wasted. In practice, the turbine will not be isentropic, and thus the power produced will be less. The analysis above gave us the upper limit. An actual turbine-generator assembly can utilize about 80 percent of the potential and produce more than 900 kW of power while saving the facility more than \$600,000 a year.

It can also be shown that the temperature of methane will drop to 113.9 K (a drop of 1.1 K) during the isentropic expansion process in the turbine instead of remaining constant at 115 K as would be the case if methane were assumed to be an incompressible substance. The temperature of methane would rise to 116.6 K (a rise of 1.6 K) during the throttling process.

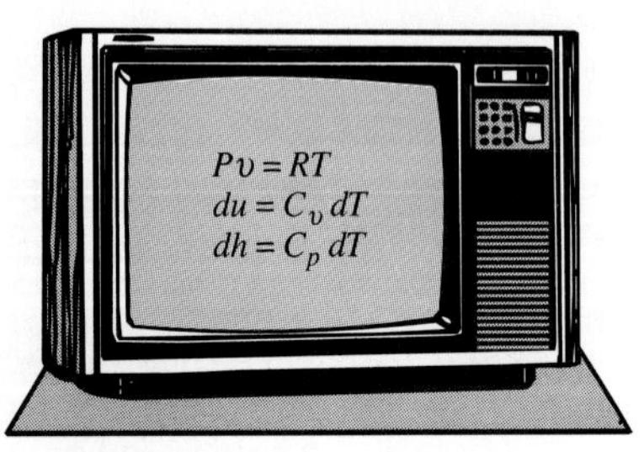

FIGURE 6-31
A broadcast from channel IG.

6-9 ■ THE ENTROPY CHANGE OF IDEAL GASES

An expression for the entropy change of an ideal gas can be obtained from Eq. 6-25 or 6-26 by employing the property relations for ideal gases (Fig. 6-31). By substituting $du = C_v\, dT$ and $P = RT/v$ into Eq. 6-25, the differential entropy change of an ideal gas becomes

$$ds = C_v \frac{dT}{T} + R \frac{dv}{v} \tag{6-30}$$

The entropy change for a process is obtained by integrating this relation between the end states:

$$s_2 - s_1 = \int_1^2 C_v(T)\frac{dT}{T} + R\ln\frac{v_2}{v_1} \tag{6-31}$$

A second relation for the entropy change of an ideal gas is obtained in a similar manner by substituting $dh = C_p\,dT$ and $v = RT/P$ into Eq. 6-26 and integrating. The result is

$$s_2 - s_1 = \int_1^2 C_p(T)\frac{dT}{T} - R\ln\frac{P_2}{P_1} \tag{6-32}$$

The specific heats of ideal gases, with the exception of monatomic gases, depend on temperature, and the integrals in Eqs. 6-31 and 6-32 cannot be performed unless the dependence of C_v and C_p on temperature is known. Even when the $C_v(T)$ and $C_p(T)$ functions are available, performing long integrations every time entropy change is calculated is not practical. Then two reasonable choices are left: either perform these integrations by simply assuming constant specific heats or evaluate those integrals once and tabulate the results. Both approaches are presented below.

1 Constant Specific Heats: Approximate Analysis

Assuming constant specific heats for ideal gases is a common approximation, and we used this assumption before on several occasions. It usually simplifies the analysis greatly, and the price we pay for this convenience is some loss in accuracy. The magnitude of the error introduced by this assumption depends on the situation at hand. For example, for monatomic ideal gases such as helium, the specific heats are independent of temperature, and therefore the constant-specific-heat assumption introduces no error. For ideal gases whose specific heats vary almost linearly in the temperature range of interest, the possible error is minimized by using specific-heat values evaluated at the average temperature (Fig. 6-32). The results obtained in this way usually are sufficiently accurate for most ideal gases if the temperature range is not greater than a few hundred degrees.

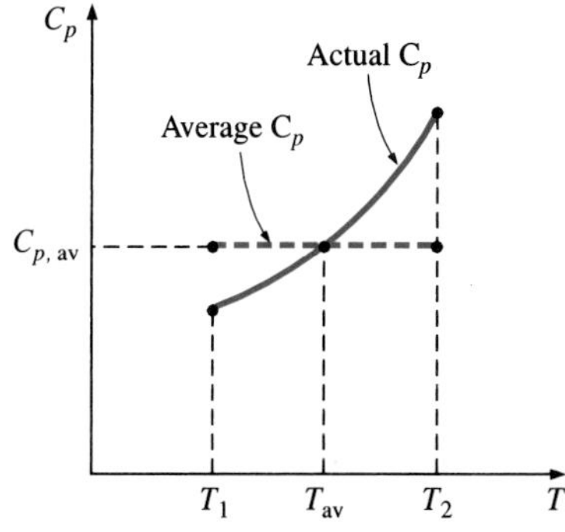

FIGURE 6-32
Under the constant-specific-heat assumption, the specific heat is assumed to be constant at some average value.

The entropy-change relations for ideal gases under the constant-specific-heat assumption are easily obtained by replacing $C_v(T)$ and $C_p(T)$ in Eqs. 6-31 and 6-32 by $C_{v,\text{av}}$ and $C_{p,\text{av}}$, respectively, and performing the integrations. We obtain

$$s_2 - s_1 = C_{v,\text{av}}\ln\frac{T_2}{T_1} + R\ln\frac{v_2}{v_1} \quad [\text{kJ/(kg}\cdot\text{K)}] \tag{6-33}$$

and

$$s_2 - s_1 = C_{p,\text{av}}\ln\frac{T_2}{T_1} - R\ln\frac{P_2}{P_1} \quad [\text{kJ/(kg}\cdot\text{K)}] \tag{6-34}$$

Entropy changes can also be expressed on a unit-mole basis by multiplying these relations by molar mass:

$$\bar{s}_2 - \bar{s}_1 = \bar{C}_{v,\text{av}} \ln \frac{T_2}{T_1} + R_u \ln \frac{v_2}{v_1} \qquad \text{[kJ/(kmol} \cdot \text{K)]} \qquad (6\text{-}35)$$

and

$$\bar{s}_2 - \bar{s}_1 = \bar{C}_{p,\text{av}} \ln \frac{T_2}{T_1} - R_u \ln \frac{P_2}{P_1} \qquad \text{[kJ/(kmol} \cdot \text{K)]} \qquad (6\text{-}36)$$

2 Variable Specific Heats: Exact Analysis

When the temperature change during a process is large and the specific heats of the ideal gas vary nonlinearly within the temperature range, the assumption of constant specific heats may lead to considerable errors in entropy-change calculations. For those cases, the variation of specific heats with temperature should be properly accounted for by utilizing accurate relations for the specific heats as a function of temperature. The entropy change during a process is then determined by substituting these $C_v(T)$ or $C_p(T)$ relations into Eq. 6-31 or 6-32 and performing the integrations.

Instead of performing these laborious integrals each time we have a new process, it is convenient to perform these integrals once and tabulate the results. For this purpose, we choose absolute zero as the reference temperature and define a function $s°$ as

$$s° = \int_0^T C_p(T) \frac{dT}{T} \qquad (6\text{-}37)$$

According to this definition, $s°$ is a function of temperature alone, and its value is zero at absolute zero temperature. The values of $s°$ are calculated at various temperatures from Eq. 6-37, and the results are tabulated in the appendix as a function of temperature for several ideal gases. Given this definition, the integral Eq. 6-32 becomes

$$\int_1^2 C_p(T) \frac{dT}{T} = s_2° - s_1° \qquad (6\text{-}38)$$

where $s_2°$ is the value of $s°$ at T_2 and $s_1°$ is the value at T_1. Thus,

$$s_2 - s_1 = s_2° - s_1° - R \ln \frac{P_2}{P_1} \qquad \text{[kJ/(kg} \cdot \text{K)]} \qquad (6\text{-}39)$$

It can also be expressed on a unit-mole basis as

$$\bar{s}_2 - \bar{s}_1 = \bar{s}_2° - \bar{s}_1° - R_u \ln \frac{P_2}{P_1} \qquad \text{[kJ/(kmol} \cdot \text{K)]} \qquad (6\text{-}40)$$

T, K	$s°(T)$, kJ/(kg · K)
.	.
.	.
.	.
300	1.70203
310	1.73498
320	1.76690
.	.
.	.
.	.

(Table A-17)

FIGURE 6-33

The entropy of an ideal gas depends on both T and P. The function $s°$ represents only the temperature-dependent part of entropy.

Note that unlike internal energy and enthalpy, the entropy of an ideal gas varies with specific volume or pressure as well as the temperature. Therefore, entropy cannot be tabulated as a function of temperature alone. The $s°$ values in the tables account for the temperature dependence of entropy (Fig. 6-33). The variation of entropy with pressure is accounted for by the last term in

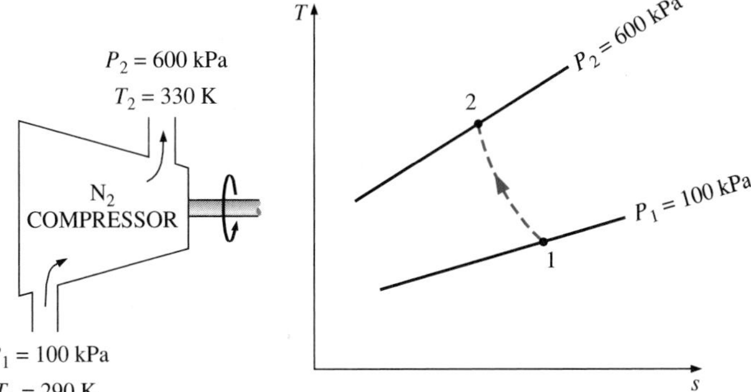

FIGURE 6-34

Schematic and *T-s* diagram for Example 6-9.

Eq. 6-39. Another relation for entropy change can be developed based on Eq. 6-31, but this would require the definition of another function and tabulation of its values, which is not practical.

EXAMPLE 6-9 Entropy Change of an Ideal Gas

Nitrogen gas is compressed from an initial state of 100 kPa and 17°C to a final state of 600 kPa and 57°C. Determine the entropy change of the nitrogen during this compression process by using (*a*) property values from the nitrogen table and (*b*) average specific heats.

Solution A sketch of the system and the *T-s* diagram for the process are given in Fig. 6-34. We note that both the initial and the final states of the nitrogen are completely specified.

Assumptions Nitrogen is an ideal gas since it is at a high temperature and low pressure relative to its critical point values of −147°C and 3.99 MPa. Therefore, entropy change relations developed under the ideal gas assumption are applicable.

Analysis (*a*) The properties of the nitrogen gas in the nitrogen table (Table A-18) are given on a unit-mole basis. Therefore, it is more convenient to determine the entropy change of the nitrogen by using Eq. 6-40 and then convert the results to the desired units (Fig. 6-35). Reading $\bar{s}^\circ$ values at given temperatures and substituting, we find

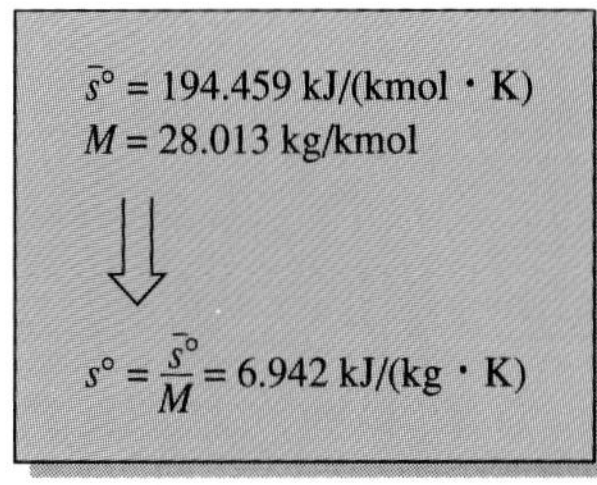

FIGURE 6-35

Properties per unit mole and properties per unit mass are related to each other through the molar mass of the substance.

$$\bar{s}_2 - \bar{s}_1 = \bar{s}_2^\circ - \bar{s}_1^\circ - R_u \ln \frac{P_2}{P_1}$$

$$= [(194.459 - 190.695) \text{ kJ/(kmol} \cdot \text{K)}] - [8.314 \text{ kJ/(kmol} \cdot \text{K)}] \ln \frac{600 \text{ kPa}}{100 \text{ kPa}}$$

$$= -11.133 \text{ kJ/(kmol} \cdot \text{K)}$$

Then $$s_2 - s_1 = \frac{\bar{s}_2 - \bar{s}_1}{M} = \frac{-11.133 \text{ kJ/(kmol} \cdot \text{K)}}{28.013 \text{ kg/kmol}} = \mathbf{-0.3974 \text{ kJ/(kg} \cdot \text{K)}}$$

(*b*) The entropy change of the nitrogen during this process can also be determined approximately from Eq. 6-34 by using a C_p value at the average temperature of 37°C (Table A-2*b*) and treating it as a constant:

$$s_2 - s_1 = C_{p,\text{av}} \ln \frac{T_2}{T_1} - R \ln \frac{P_2}{P_1}$$

$$= [1.0394 \text{ kJ/(kg} \cdot \text{K)}] \ln \frac{330 \text{ K}}{290 \text{ K}} - [0.297 \text{ kJ/(kg} \cdot \text{K)}] \ln \frac{600 \text{ kPa}}{100 \text{ kPa}}$$

$$= \mathbf{-0.3978 \text{ kJ/(kg} \cdot \text{K)}}$$

Discussion The two results above are almost identical since the change in temperature during this process is relatively small. When the temperature change is large, however, they may differ significantly. For those cases, Eq. 6-40 should be used instead of Eq. 6-34 since it accounts for the variation of specific heats with temperature.

Isentropic Processes of Ideal Gases

Several relations for the isentropic processes of ideal gases can be obtained by setting the entropy-change relations developed above equal to zero. Again, this is done first for the case of constant specific heats and then for the case of variable specific heats.

Constant Specific Heats: Approximate Treatment

When the constant-specific-heat assumption is valid, the isentropic relations for ideal gases are obtained by setting Eqs. 6-33 and 6-34 equal to zero. From Eq. 6-33,

$$\ln \frac{T_2}{T_1} = -\frac{R}{C_v} \ln \frac{v_2}{v_1}$$

which can be rearranged as

$$\ln \frac{T_2}{T_1} = \ln \left(\frac{v_1}{v_2}\right)^{R/C_v} \qquad (6\text{-}41)$$

or

$$\left(\frac{T_2}{T_1}\right)_{s=\text{const.}} = \left(\frac{v_1}{v_2}\right)^{k-1} \qquad \text{(ideal gas)} \qquad (6\text{-}42)$$

since $R = C_p - C_v$, $k = C_p/C_v$, and thus $R/C_v = k - 1$.

Equation 6-42 is the *first isentropic relation* for ideal gases under the constant-specific-heat assumption. The *second isentropic relation* is obtained in a similar manner from Eq. 6-34 with the following result:

$$\left(\frac{T_2}{T_1}\right)_{s=\text{const.}} = \left(\frac{P_2}{P_1}\right)^{(k-1)/k} \qquad \text{(ideal gas)} \qquad (6\text{-}43)$$

The *third isentropic relation* is obtained by substituting Eq. 6-43 into Eq. 6-42 and simplifying:

$$\left(\frac{P_2}{P_1}\right)_{s=\text{const.}} = \left(\frac{v_1}{v_2}\right)^k \qquad \text{(ideal gas)} \qquad (6\text{-}44)$$

Equations 6-42 through 6-44 can also be expressed in a compact form as

$$Tv^{k-1} = \text{constant} \qquad (6\text{-}45)$$

$$TP^{(1-k)/k} = \text{constant} \qquad \text{(ideal gas)} \qquad (6\text{-}46)$$

$$Pv^k = \text{constant} \qquad (6\text{-}47)$$

The specific heat ratio k, in general, varies with temperature, and thus an average k value for the given temperature range should be used.

Note that the ideal gas isentropic relations above, as the name implies, are strictly valid for isentropic processes only when the constant-specific-heat assumption is appropriate (Fig. 6-36).

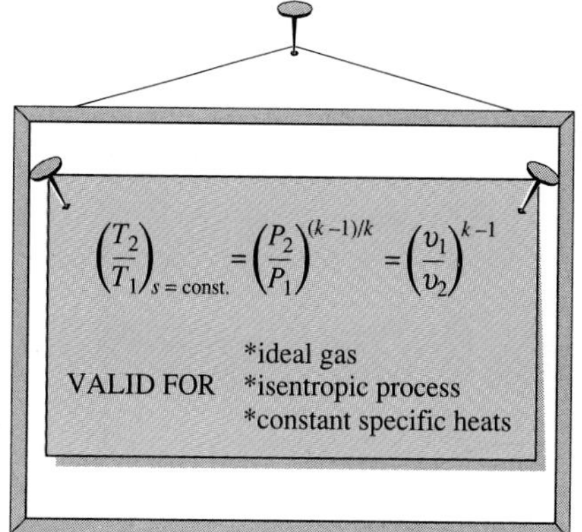

FIGURE 6-36

The isentropic relations of ideal gases are valid for the isentropic processes of ideal gases only.

Variable Specific Heats: Exact Treatment

When the constant-specific-heat assumption is not appropriate, the isentropic relations developed above will yield results that are not quite accurate. For such cases, we should use an isentropic relation obtained from Eq. 6-39 that accounts for the variation of specific heats with temperature. Setting this equation equal to zero gives

$$0 = s_2^\circ - s_1^\circ - R \ln \frac{P_2}{P_1}$$

or

$$s_2^\circ = s_1^\circ + R \ln \frac{P_2}{P_1} \qquad (6\text{-}48)$$

where s_2° is the s° value at the end of the isentropic process.

Relative Pressure and Relative Specific Volume

Equation 6-48 provides an accurate way of evaluating property changes of ideal gases during isentropic processes since it accounts for the variation of specific heats with temperature. However, it involves tedious iterations when the volume ratio is given instead of the pressure ratio. This is quite an inconvenience in optimization studies, which usually require numerous repetitive calculations. To remedy this deficiency, we define two new dimensionless quantities associated with isentropic processes.

The definition of the first is based on Eq. 6-48, which can be rearranged as

$$\frac{P_2}{P_1} = \exp\frac{s_2^\circ - s_1^\circ}{R}$$

or

$$\frac{P_2}{P_1} = \frac{\exp(s_2^\circ/R)}{\exp(s_1^\circ/R)}$$

The quantity $\exp(s^\circ/R)$ is defined as the **relative pressure** P_r. With this definition, the above relation becomes

$$\left(\frac{P_2}{P_1}\right)_{s=\text{const.}} = \frac{P_{r2}}{P_{r1}} \tag{6-49}$$

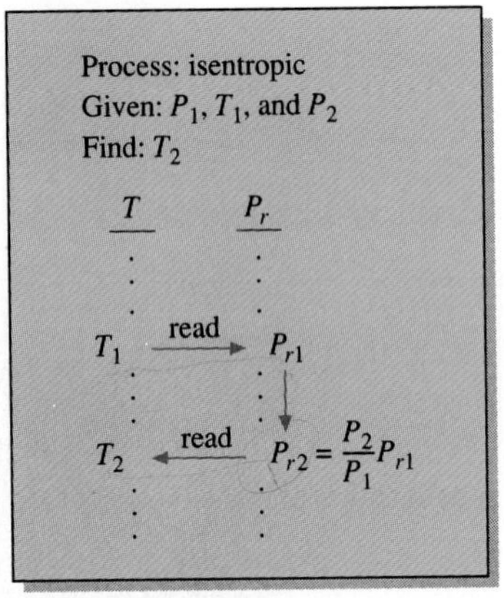

Note that the relative pressure P_r is a *dimensionless* quantity that is a function of temperature only since s° depends on temperature alone. Therefore, values of P_r can be tabulated against temperature. This is done for air in Table A-17. The use of P_r data is illustrated in Fig. 6-37.

Sometimes specific volume ratios are given instead of pressure ratios. This is particularly the case when automotive engines are analyzed. In such cases, one needs to work with volume ratios. Therefore, we define another quantity related to specific volume ratios for isentropic processes. This is done by utilizing the ideal-gas relation and Eq. 6-49:

$$\frac{P_1 v_1}{T_1} = \frac{P_2 v_2}{T_2} \longrightarrow \frac{v_2}{v_1} = \frac{T_2}{T_1}\frac{P_1}{P_2} = \frac{T_2}{T_1}\frac{P_{r1}}{P_{r2}} = \frac{T_2/P_{r2}}{T_1/P_{r1}}$$

Process: isentropic
Given: P_1, T_1, and P_2
Find: T_2

T		P_r
T_1	read →	P_{r1}
T_2	← read	$P_{r2} = \frac{P_2}{P_1} P_{r1}$

FIGURE 6-37
The use of P_r data for calculating the final temperature during an isentropic process.

The quantity T/P_r is a function of temperature only and is defined as **relative specific volume** v_r. Thus,

$$\left(\frac{v_2}{v_1}\right)_{s=\text{const.}} = \frac{v_{r2}}{v_{r1}} \tag{6-50}$$

Equations 6-49 and 6-50 are strictly valid for isentropic processes of ideal gases only. They account for the variation of specific heats with temperature and therefore give more accurate results than Eqs. 6-42 through 6-47. The values of P_r and v_r are listed for air in Table A-17.

EXAMPLE 6-10 Isentropic Compression of Air in a Car Engine

Air is compressed in a car engine from 22°C and 95 kPa in a reversible and adiabatic manner. If the compression ratio V_1/V_2 of this piston-cylinder device is 8, determine the final temperature of the air.

Solution A sketch of the system and the *T-s* diagram for the process are given in Fig. 6-38. We note that the process is reversible and adiabatic.

Assumptions At specified conditions, air can be treated as an ideal gas since it is at a high temperature and low pressure relative to its critical point values of −141°C and 3.77 MPa. Therefore, the isentropic relations developed earlier for ideal gases are applicable.

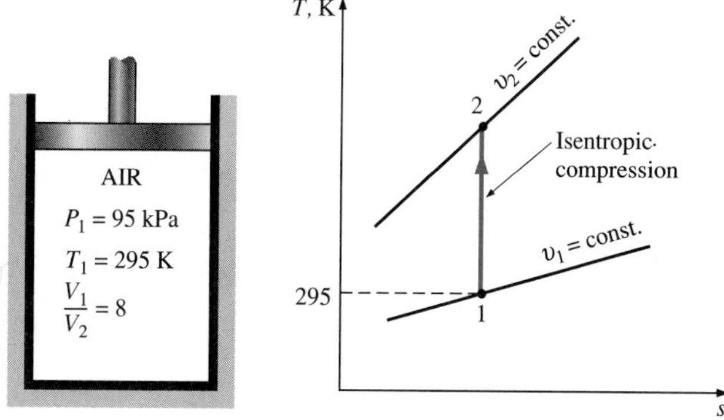

FIGURE 6-38

Schematic and *T-s* diagram for Example 6-10.

Analysis This process is easily recognized as being isentropic since it is both reversible and adiabatic. The final temperature for this isentropic process can be determined from Eq. 6-50 with the help of relative specific volume data (Table A-17), as illustrated in Fig. 6-39.

For closed systems: $$\frac{V_2}{V_1} = \frac{v_2}{v_1}$$

At $T_1 = 295$ K: $$v_{r1} = 647.9$$

From Eq. 6-50: $$v_{r2} = v_{r1}\left(\frac{v_2}{v_1}\right) = (647.9)\left(\frac{1}{8}\right) = 80.99 \longrightarrow T_2 = \mathbf{662.7\ K}$$

Therefore, the temperature of air will increase by 367.7°C during this process.

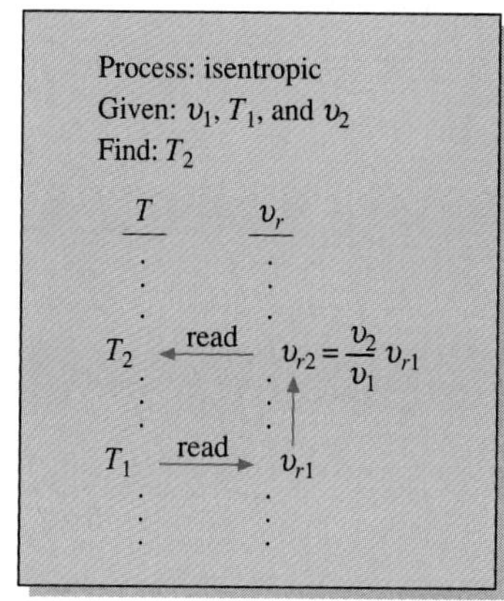

FIGURE 6-39

The use of v_r data for calculating the final temperature during an isentropic process (Example 6-10).

Alternative solution The final temperature could also be determined from Eq. 6-42 by assuming constant specific heats for air:

$$\left(\frac{T_2}{T_1}\right)_{s=\text{const.}} = \left(\frac{v_1}{v_2}\right)^{k-1}$$

The specific heat ratio *k* also varies with temperature, and we need to use the value of *k* corresponding to the average temperature. However, the final temperature is not given, and so we cannot determine the average temperature in advance. For such cases, calculations can be started with a *k* value at the initial or the anticipated average temperature. This value could be refined later, if necessary, and the calculations can be repeated. We know that the temperature of the air will rise considerably during this adiabatic compression process, so we *guess* that the average temperature will be about 450 K. The *k* value at this anticipated average temperature is determined from Table A-2*b* to be 1.391. Then the final temperature of air becomes

$$T_2 = (295\text{ K})(8)^{1.391-1} = 665.2\text{ K}$$

This will give an average temperature value of 480.1 K, which is sufficiently close to the assumed value of 450 K. Therefore, it is not necessary to repeat the calculations by using the *k* value at this average temperature.

The result obtained by assuming constant specific heats for this case is in error by about 0.4 percent, which is rather small. This is not surprising since the temperature change of air is relatively small (only a few hundred degrees) and

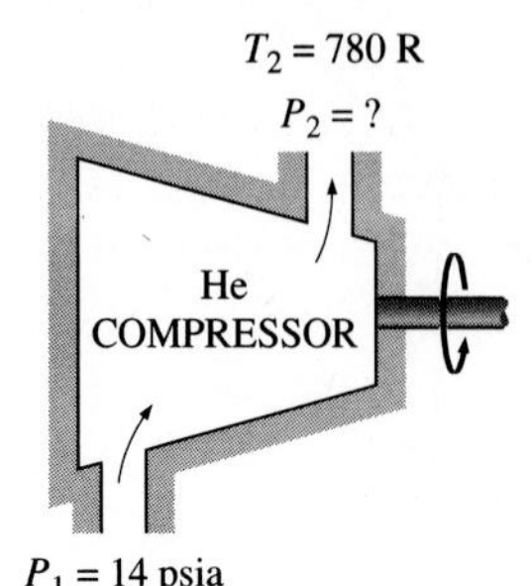

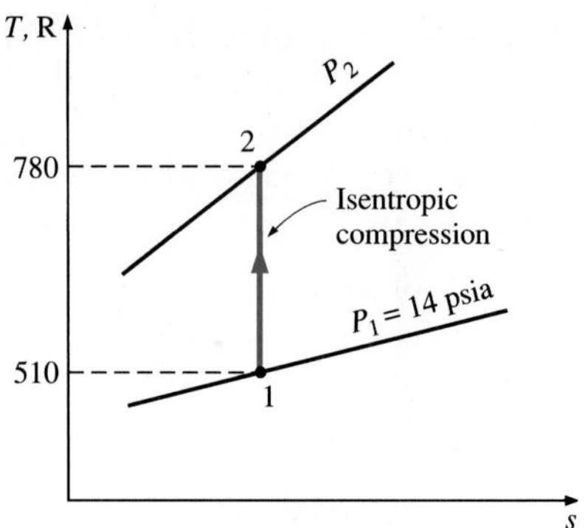

FIGURE 6-40
Schematic and *T-s* diagram for Example 6-11.

the specific heats of air vary almost linearly with temperature in this temperature range.

EXAMPLE 6-11 Isentropic Compression of an Ideal Gas

Helium gas is compressed in an adiabatic compressor from an initial state of 14 psia and 50°F to a final temperature of 320°F in a reversible manner. Determine the exit pressure of helium.

Solution A sketch of the system and the *T-s* diagram for the process are given in Fig. 6-40. We note that the process is reversible and adiabatic.

Assumptions At specified conditions, helium can be treated as an ideal gas since it is at a high temperature relative to its critical point value of −450°F. Therefore, the isentropic relations developed earlier for ideal gases are applicable.

Analysis The specific heat ratio *k* of helium is 1.667 and is independent of temperature in the region where it behaves as an ideal gas. Thus the final pressure of helium can be determined from Eq. 6-43:

$$P_2 = P_1\left(\frac{T_2}{T_1}\right)^{k/(k-1)} = (14 \text{ psia})\left(\frac{780 \text{ R}}{510 \text{ R}}\right)^{1.667/0.667} = \mathbf{40.5 \text{ psia}}$$

6-10 ■ REVERSIBLE STEADY-FLOW WORK

The work done during a process depends on the path followed as well as on the properties at the end states. In Chap. 3, we discussed reversible (quasi-equilibrium) moving boundary work associated with closed systems and expressed it in terms of the fluid properties as

$$W_b = \int_1^2 P\,dV$$

We mentioned that the quasi-equilibrium work interactions lead to the maximum work output for work-producing devices and the minimum work input for work-consuming devices.

It would also be very insightful to express the work associated with steady-flow devices in terms of fluid properties.

Taking the positive direction of work to be from the system (work output), the energy balance for a steady-flow device undergoing an internally reversible process can be expressed in differential form as

$$\delta q_{rev} - \delta w_{rev} = dh + d\text{ke} + d\text{pe}$$

But

$$\left.\begin{aligned} \delta q_{rev} &= T\,ds \qquad &\text{(Eq. 6-17)} \\ T\,ds &= dh - v\,dP \qquad &\text{(Eq. 6-24)} \end{aligned}\right\} \quad \delta q_{rev} = dh - v\,dP$$

Substituting this into the relation above and canceling dh yield

$$-\delta w_{rev} = v\,dP + d\text{ke} + d\text{pe}$$

Integrating, we find

$$w_{rev} = -\int_1^2 v\,dP - \Delta\text{ke} - \Delta\text{pe} \qquad \text{(kJ/kg)} \qquad (6\text{-}51)$$

When the changes in kinetic and potential energies are negligible, this equation reduces to

$$w_{rev} = -\int_1^2 v\,dP \qquad \text{(kJ/kg)} \qquad (6\text{-}52)$$

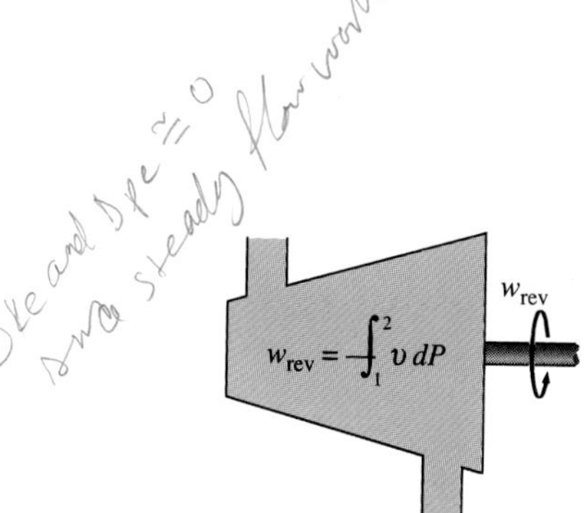

(*a*) Steady-flow system

Equations 6-51 and 6-52 are relations for the *reversible work output* associated with an internally reversible process in a steady-flow device. They will give a negative result when work is done on the system. To avoid the negative sign, Eq. 6-51 can be written for work input to steady-flow devices such as compressors and pumps as

$$w_{rev,in} = \int_1^2 v\,dP + \Delta\text{ke} + \Delta\text{pe} \qquad (6\text{-}53)$$

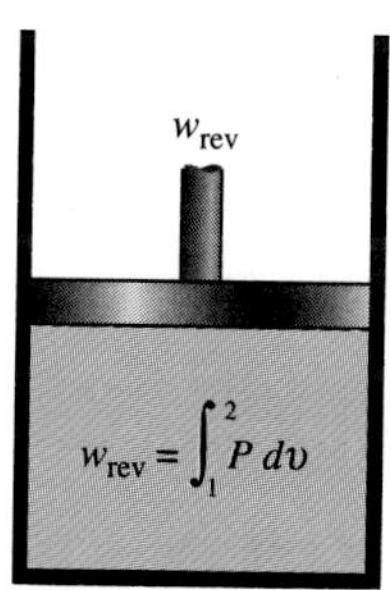

(*b*) Closed system

FIGURE 6-41

Reversible work relations for steady-flow and closed systems.

The resemblance between the $v\,dP$ in these relations and $P\,dv$ is striking. They should not be confused with each other, however, since $P\,dv$ is associated with reversible boundary work in closed systems (Fig. 6-41).

Obviously, one needs to know v as a function of P for the given process to perform the integration in Eq. 6-51. When the working fluid is an *incompressible fluid,* the specific volume v remains constant during the process and can be taken out of the integration. Then Eq. 6-51 simplifies to

$$w_{rev} = -v(P_2 - P_1) - \Delta\text{ke} - \Delta\text{pe} \qquad \text{(kJ/kg)} \qquad (6\text{-}54)$$

For the steady flow of a liquid through a device that involves no work interactions (such as a nozzle or a pipe section), the work term is zero, and the equation above can be expressed as

$$v(P_2 - P_1) + \frac{\mathcal{V}_2^2 - \mathcal{V}_1^2}{2} + g(z_2 - z_1) = 0 \qquad \text{(6-55)}$$

which is known as the **Bernoulli equation** in fluid mechanics. It is developed for an internally reversible process and thus is applicable to incompressible fluids that involve no irreversibilities such as friction or shock waves. This equation can be modified, however, to incorporate these effects.

Equation 6-52 has far-reaching implications in engineering regarding devices that produce or consume work steadily such as turbines, compressors, and pumps. It is obvious from this equation that the reversible steady-flow work is closely associated with the specific volume of the fluid flowing through the device. *The larger the specific volume, the larger the reversible work produced or consumed by the steady-flow device* (Fig. 6-42). This conclusion is equally valid for actual steady-flow devices. Therefore, every effort should be made to keep the specific volume of a fluid as small as possible during a compression process to minimize the work input and as large as possible during an expansion process to maximize the work output.

$$W = -\int_1^2 v\,dP$$

$$W = -\int_1^2 v\,dP$$

$$W = -\int_1^2 v\,dP$$

FIGURE 6-42
The larger the specific volume, the greater the work produced (or consumed) by a steady-flow device.

In steam or gas power plants, the pressure rise in the pump or compressor is equal to the pressure drop in the turbine if we disregard the pressure losses in various other components. In steam power plants, the pump handles liquid, which has a very small specific volume, and the turbine handles vapor, whose specific volume is many times larger. Therefore, the work output of the turbine is much larger than the work input to the pump. This is one of the reasons for the overwhelming popularity of steam power plants in electric power generation.

If we were to compress the steam exiting the turbine back to the turbine inlet pressure before cooling it first in the condenser in order to "save" the heat rejected, we would have to supply all the work produced by the turbine back to the compressor. In reality, the required work input would be even greater than the work output of the turbine because of the irreversibilities present in both processes.

In gas power plants, the working fluid (typically air) is compressed in the gas phase, and a considerable portion of the work output of the turbine is consumed by the compressor. As a result, a gas power plant delivers less net work per unit mass of the working fluid.

EXAMPLE 6-12 Compressing a Substance in the Liquid vs. Gas Phases

Determine the compressor work input required to compress steam isentropically from 100 kPa to 1 MPa, assuming that the steam exists as (*a*) saturated liquid and (*b*) saturated vapor at the initial state.

Solution We take the turbine and then the pump as the system. Both are control volumes since mass crosses the boundary. Sketches of the pump and the turbine together with the *T-s* diagram are given in Fig. 6-43.

Assumptions **1** Steady operating conditions exist. **2** Kinetic and potential energy changes are negligible. **3** The process is given to be isentropic.

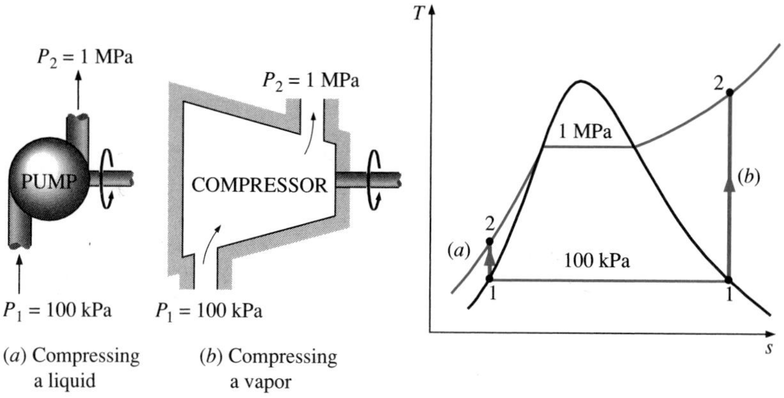

FIGURE 6-43
Schematic and *T-s* diagram for Example 6-12.

Analysis (*a*) In this case, steam is a saturated liquid initially, and its specific volume is

$$v_1 = v_{f\,@\,100\,\text{kPa}} = 0.001043\ \text{m}^3/\text{kg} \qquad \text{(Table A-5)}$$

which remains essentially constant during the process. Thus,

$$\begin{aligned} w_{\text{rev, in}} &= \int_1^2 v\,dP \cong v_1(P_2 - P_1) \\ &= (0.001043\ \text{m}^3/\text{kg})[(1000 - 100)\ \text{kPa}]\left(\frac{1\ \text{kJ}}{1\ \text{kPa}\cdot\text{m}^3}\right) \\ &= \mathbf{0.94\ kJ/kg} \end{aligned}$$

(*b*) This time, steam is a saturated vapor initially and remains a vapor during the entire compression process. Since the specific volume of a gas changes considerably during a compression process, we need to know how *v* varies with *P* to perform the integration in Eq. 6-53. This relation, in general, is not readily available. But for an isentropic process, it is easily obtained from the second *T ds* relation by setting $ds = 0$:

$$\left.\begin{aligned} T\,ds &= dh - v\,dP \quad \text{(Eq. 6-24)} \\ ds &= 0 \quad \text{(isentropic process)} \end{aligned}\right\} \quad v\,dP = dh$$

Thus,

$$w_{\text{rev, in}} = \int_1^2 v\,dP = \int_1^2 dh = h_2 - h_1$$

This result could also be obtained from the energy balance relation for an isentropic steady-flow process. Next we determine the enthalpies:

State 1:
$$\left.\begin{aligned} P_1 &= 100\ \text{kPa} \\ &\text{(sat. vapor)} \end{aligned}\right\} \quad \begin{aligned} h_1 &= 2675.5\ \text{kJ/kg} \\ s_1 &= 7.3594\ \text{kJ/(kg}\cdot\text{K)} \end{aligned} \qquad \text{(Table A-5)}$$

State 2:
$$\left.\begin{aligned} P_2 &= 1\text{MPa} \\ s_2 &= s_1 \end{aligned}\right\} \quad h_2 = 3195.5\ \text{kJ/kg} \qquad \text{(Table A-6)}$$

Thus,

$$w_{\text{rev, in}} = (3195.5 - 2675.5)\ \text{kJ/kg} = \mathbf{520\ kJ/kg}$$

Discussion Note that compressing steam in the vapor form would require over 500 times more work than compressing it in the liquid form between the same pressure limits.

Proof that Steady-Flow Devices Deliver the Most and Consume the Least Work when the Process Is Reversible

We have shown in Chap. 5 that cyclic devices (heat engines, refrigerators, and heat pumps) deliver the most work and consume the least when reversible processes are used. Now we will demonstrate that this is also the case for individual devices such as turbines and compressors in steady operation.

Consider two steady-flow devices, one reversible and the other irreversible, operating between the same inlet and exit states. Again taking heat transfer to the system and work done by the system to be positive quantities, the energy balance for each of these devices can be expressed in the differential form as

Actual: $$\delta q_{act} - \delta w_{act} = dh + dke + dpe$$

Reversible: $$\delta q_{rev} - \delta w_{rev} = dh + dke + dpe$$

The right-hand sides of these two equations are identical since both devices are operating between the same end states. Thus,

$$\delta q_{act} - \delta w_{act} = \delta q_{rev} - \delta w_{rev}$$

or

$$\delta w_{rev} - \delta w_{act} = \delta q_{rev} - \delta q_{act}$$

But

$$\delta q_{rev} = T\,ds$$

Substituting this relation into the equation above and dividing each term by T, we obtain

$$\frac{\delta w_{rev} - \delta w_{act}}{T} = ds - \frac{\delta q_{act}}{T} \geq 0$$

since

$$ds \geq \frac{\delta q_{act}}{T} \qquad \text{(see Eq. 6-8)}$$

Also, T is the absolute temperature, which is always positive. Thus,

$$\delta w_{rev} \geq \delta w_{act}$$

or

$$w_{rev} \geq w_{act}$$

Thus work-producing devices such as turbines (w is positive) deliver more work, and work-consuming devices such as pumps and compressors (w is negative) require less work when they operate reversibly (Fig. 6-44).

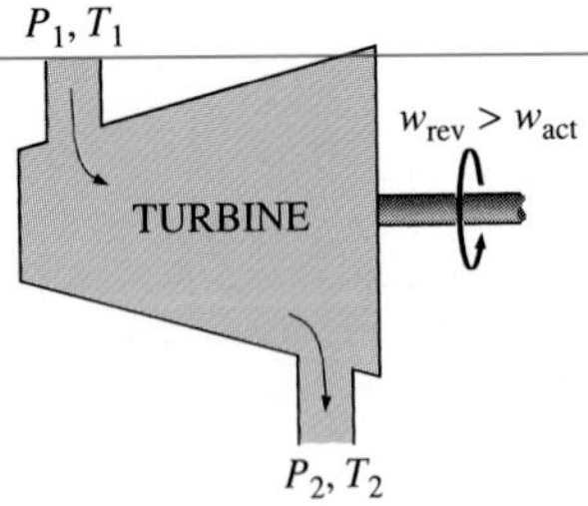

P_1, T_1; $w_{rev} > w_{act}$; TURBINE; P_2, T_2

FIGURE 6-44

A reversible turbine delivers more work than an irreversible one if both operate between the same end states.

6-11 ■ MINIMIZING THE COMPRESSOR WORK

We have shown in Sec. 6-10 that the work input to a compressor is minimized when the compression process is executed in an internally reversible manner. When the changes in kinetic and potential energies are negligible, the compressor work is given by (Eq. 6-53)

$$w_{rev,in} = \int_1^2 v\,dP \qquad (6\text{-}56)$$

Obviously one way of minimizing the compressor work is to approach an internally reversible process as much as possible by minimizing the irreversibilities such as friction, turbulence, and non-quasi-equilibrium compression. The extent to which this can be accomplished is limited by economic considerations. A second (and more practical) way of reducing the compressor work is to keep the specific volume of the gas as small as possible during the compression process. This is done by maintaining the temperature of the gas as low as possible during compression since the specific volume of a gas is proportional to temperature. Therefore, reducing the work input to a compressor requires that the gas be cooled as it is compressed.

To have a better understanding of the effect of cooling during the compression process, we compare the work input requirements for three kinds of processes: *an isentropic process* (involves no cooling), *a polytropic process* (involves some cooling), and *an isothermal process* (involves maximum cooling). Assuming all three processes are executed between the same pressure levels (P_1 and P_2) in an internally reversible manner and the gas behaves as an ideal gas ($Pv = RT$), we see that the compression work is determined by performing the integration in Eq. 6-56 for each case, with the following results:

Isentropic (Pv^k = constant):

$$w_{\text{comp, in}} = \frac{kR(T_2 - T_1)}{k - 1} = \frac{kRT_1}{k - 1}\left[\left(\frac{P_2}{P_1}\right)^{(k-1)/k} - 1\right] \qquad \text{(6-57}a\text{)}$$

Polytropic (Pv^n = constant):

$$w_{\text{comp, in}} = \frac{nR(T_2 - T_1)}{n - 1} = \frac{nRT_1}{n - 1}\left[\left(\frac{P_2}{P_1}\right)^{(n-1)/n} - 1\right] \qquad \text{(6-57}b\text{)}$$

Isothermal (Pv = constant):

$$w_{\text{comp, in}} = RT \ln \frac{P_2}{P_1} \qquad \text{(6-57}c\text{)}$$

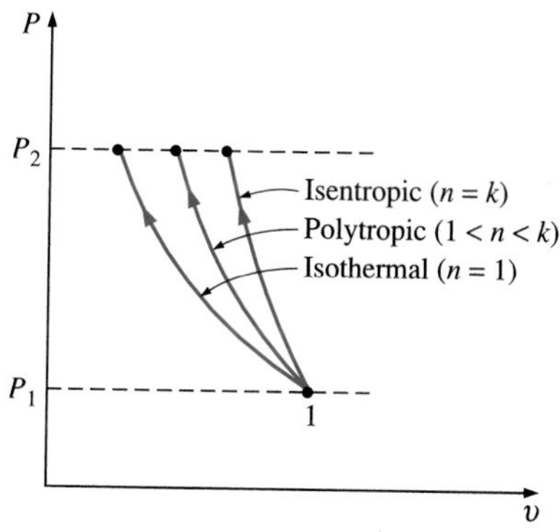

FIGURE 6-45

P-v diagrams of isentropic, polytropic, and isothermal compression processes between the same pressure limits.

The three processes are plotted on a *P-v* diagram in Fig. 6-45 for the same inlet state and exit pressure. On a *P-v* diagram, the area to the left of the process curve is the integral of $v\,dP$. Thus it is a measure of the steady-flow compression work. It is interesting to observe from this diagram that of the three internally reversible cases considered, the adiabatic compression (Pv^k = constant) requires the maximum work and the isothermal compression (T = constant or Pv = constant) requires the minimum. The work input requirement for the polytropic case (Pv^n = constant) is between these two and decreases as the polytropic exponent n is decreased, by increasing the heat rejection during the compression process. If sufficient heat is removed, the value of n approaches unity and the process becomes isothermal. One

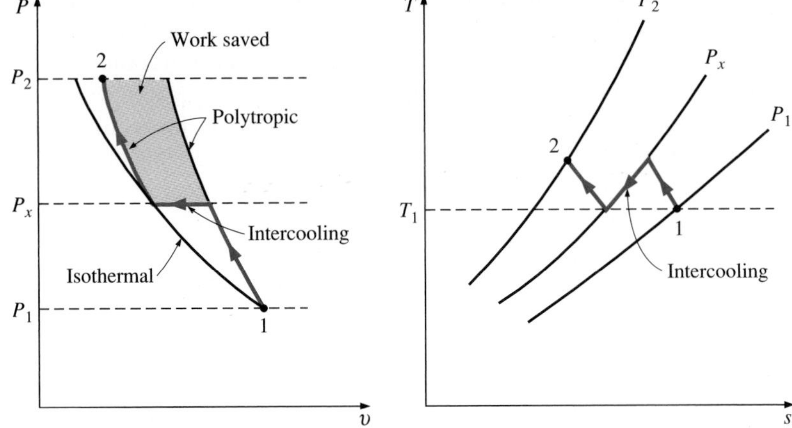

FIGURE 6-46
P-v and *T-s* diagrams for a two-stage steady-flow compression process.

common way of cooling the gas during compression is to use cooling jackets around the casing of the compressors.

Multistage Compression with Intercooling

It is clear from the above arguments that cooling a gas as it is compressed is desirable since this reduces the required work input to the compressor. However, often it is not possible to have effective cooling through the casing of the compressor, and it becomes necessary to use other techniques to achieve effective cooling. One such technique is **multistage compression with intercooling,** where the gas is compressed in stages and cooled between each stage by passing it through a heat exchanger called an *intercooler.* Ideally, the cooling process takes place at constant pressure, and the gas is cooled to the initial temperature T_1 at each intercooler. Multistage compression with intercooling is especially attractive when a gas is to be compressed to very high pressures.

The effect of intercooling on compressor work is graphically illustrated on *P-v* and *T-s* diagrams in Fig. 6-46 for a two-stage compressor. The gas is compressed in the first stage from P_1 to an intermediate pressure P_x, cooled at constant pressure to the initial temperature T_1, and compressed in the second stage to the final pressure P_2. The compression processes, in general, can be modeled as polytropic (Pv^n = constant) where the value of n varies between k and 1. The colored area on the *P-v* diagram represents the work saved as a result of two-stage compression with intercooling. The process paths for single-stage isothermal and polytropic processes are also shown for comparison.

The size of the colored area (the saved work input) varies with the value of the intermediate pressure P_x, and it is of practical interest to determine the conditions under which this area is maximized. The total work input for a two-stage compressor is the sum of the work inputs for each stage of compression, as determined from Eq. 6-57*b*:

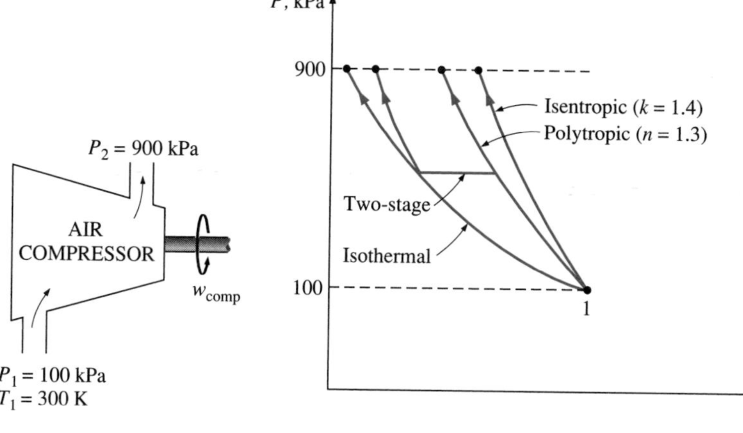

FIGURE 6-47
Schematic and *P-v* diagram for Example 6-13.

$$w_{\text{comp, in}} = w_{\text{comp I, in}} + w_{\text{comp II, in}}$$
$$= \frac{nRT_1}{n-1}\left[\left(\frac{P_x}{P_1}\right)^{(n-1)/n} - 1\right] + \frac{nRT_1}{n-1}\left[\left(\frac{P_2}{P_x}\right)^{(n-1)/n} - 1\right] \qquad (6\text{-}58)$$

The only variable in this equation is P_x. The P_x value that will minimize the total work is determined by differentiating this expression with respect to P_x and setting the resulting expression equal to zero. It yields

$$P_x = (P_1P_2)^{1/2} \quad \text{or} \quad \frac{P_x}{P_1} = \frac{P_2}{P_x} \qquad (6\text{-}59)$$

That is, *to minimize compression work during two-stage compression, the pressure ratio across each stage of the compressor must be the same*. When this condition is satisfied, the compression work at each stage becomes identical, that is, $w_{\text{comp I, in}} = w_{\text{comp II, in}}$.

EXAMPLE 6-13 Work Input for Various Compression Processes

Air is compressed steadily by a reversible compressor from an inlet state of 100 kPa and 300 K to an exit pressure of 900 kPa. Determine the compressor work per unit mass for (*a*) isentropic compression with $k = 1.4$, (*b*) polytropic compression with $n = 1.3$, (*c*) isothermal compression, and (*d*) ideal two-stage compression with intercooling with a polytropic exponent of 1.3.

Solution We take the compressor to be the system. This is a control volume since mass crosses the boundary. A sketch of the system and the *T-s* diagram for the process are given in Fig. 6-47.

Assumptions **1** Steady operating conditions exist. **2** At specified conditions, air can be treated as an ideal gas since it is at a high temperature and low pressure relative to its critical point values of 141°C and 3.77 MPa. **3** Kinetic and potential energy changes are negligible.

Analysis The steady-flow compression work for all these four cases is determined by using the relations developed earlier in this section:

(*a*) Isentropic compression with $k = 1.4$ (Eq. 6-57*a*):

$$w_{\text{comp, in}} = \frac{kRT_1}{k-1}\left[\left(\frac{P_2}{P_1}\right)^{(k-1)/k} - 1\right]$$

$$= \frac{(1.4)[0.287\ \text{kJ/(kg}\cdot\text{K)}](300\ \text{K})}{1.4-1}\left[\left(\frac{900\ \text{kPa}}{100\ \text{kPa}}\right)^{(1.4-1)/1.4} - 1\right]$$

$$= \mathbf{263.2\ kJ/kg}$$

(*b*) Polytropic compression with $n = 1.3$ (Eq. 6-57*b*):

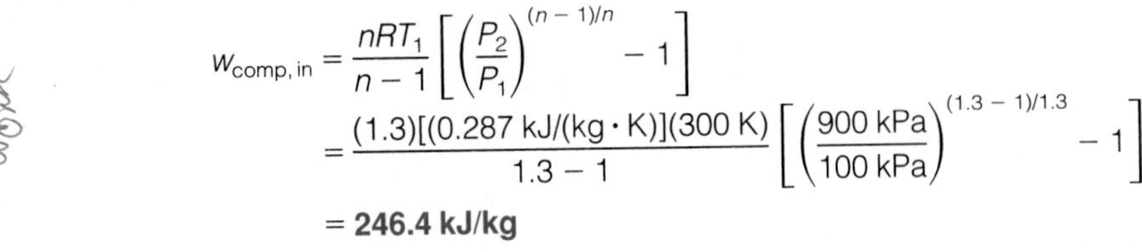

$$w_{\text{comp, in}} = \frac{nRT_1}{n-1}\left[\left(\frac{P_2}{P_1}\right)^{(n-1)/n} - 1\right]$$

$$= \frac{(1.3)[(0.287\ \text{kJ/(kg}\cdot\text{K)}](300\ \text{K})}{1.3-1}\left[\left(\frac{900\ \text{kPa}}{100\ \text{kPa}}\right)^{(1.3-1)/1.3} - 1\right]$$

$$= \mathbf{246.4\ kJ/kg}$$

(*c*) Isothermal compression (Eq. 6-57*c*):

$$w_{\text{comp, in}} = RT \ln\frac{P_2}{P_1} = [0.287\ \text{kJ/(kg}\cdot\text{K)}](300\ \text{K}) \ln\frac{100\ \text{kPa}}{900\ \text{kPa}}$$

$$= \mathbf{189.2\ kJ/kg}$$

(*d*) Ideal two-stage compression with intercooling ($n = 1.3$): In this case, the pressure ratio across each stage is the same, and its value is determined from Eq. 6-59:

$$P_x = (P_1P_2)^{1/2} = [(100\ \text{kPa})(900\ \text{kPa})]^{1/2} = 300\ \text{kPa}$$

The compressor work across each stage is also the same. Thus the total compressor work is twice the compression work for a single stage:

$$w_{\text{comp, in}} = 2w_{\text{comp I, in}} = 2\frac{nRT_1}{n-1}\left[\left(\frac{P_x}{P_1}\right)^{(n-1)/n} - 1\right]$$

$$= \frac{2(1.3)[0.287\ \text{kJ/(kg}\cdot\text{K)}](300\ \text{K})}{1.3-1}\left[\left(\frac{300\ \text{kPa}}{100\ \text{kPa}}\right)^{(1.3-1)/1.3} - 1\right]$$

$$= \mathbf{215.3\ kJ/kg}$$

Discussion Of all four cases considered, the isothermal compression requires the minimum work and the isentropic compression the maximum. The compressor work is decreased when two stages of polytropic compression are utilized instead of just one. As the number of compressor stages is increased, the compressor work approaches the value obtained for the isothermal case.

6-12 ■ REDUCING THE COST OF COMPRESSED AIR

Compressed air at gage pressures of 550 kPa to 1000 kPa (80 to 150 psig) is commonly used in industrial facilities to perform a wide variety of tasks such as *cleaning, operating pneumatic equipment,* and even *refrigeration*. It is often referred to as the *fourth utility* after electricity, water, and natural gas or oil. In production facilities, there is a widespread waste of energy associated with compressed-air systems and a general lack of awareness about the opportunities to conserve energy. A considerable portion of the energy waste associated with compressed-air systems can be avoided by following some commonsense measures. In this section we discuss the energy losses associated with compressed-air systems and their costs to manufacturers. We also show how to reduce the cost of compressed air in existing facilities by making some modifications with attractive payback periods. With the exception of a few compressors that are driven by natural gas engines, all compressors are driven by electric motors (Fig. 6-48).

FIGURE 6-48
A 1250-hp compressor assembly (courtesy of Dresser Rand Company, Painted Post, NY).

Some primitive methods of producing an *air blast* to keep the fire in furnaces alive, such as air-threading bags and the Chinese wind box, date back at least to 2000 BC. The *water trompe* that compresses air by the fall of water in a tube to blow forges (metal heat shops) is believed to have been in use by 150 BC. In 1650, Otto van Guericke made great improvements in both the compressor and vacuum pump. In 1683, Papin proposed using compressed air to *transmit power* over long distances. In 1829, William Mann received a patent for multistage compression of air. In 1830, Thilorier was recognized for compressing gases to high pressures in *stages*. In 1890, Edward Rix transmitted power with air several miles to operate lifting machines in the North Star mine near Grass Valley, California, by using a compressor driven by Pelton wheels. In 1872, *cooling* was adapted to increase efficiency by spraying water directly into the cylinder through the air inlet valves. This "wet compression" was abandoned later because of the problems it caused. The cooling then was accomplished externally by *water jacketing* the cylinders. The first large-scale compressor used in the United States was a four-cylinder unit built in 1866 for use in the Hoosac tunnel. The cooling was first accomplished by water injection into the cylinder, and later by running a stream of water over the cylinder. Major advances in recent compressor technology are due to Burleigh, Ingersoll, Sergeant, Rand, and Clayton, among others.

The compressors used range from a few hp to more than 10,000 hp in size, and they are among the major energy-consuming equipment in most manufacturing facilities. Manufacturers are quick to identify energy (and thus money) losses from *hot surfaces* and to insulate those surfaces. But somehow they are not so sensitive when it comes to saving *compressed air* since they view air as being free, and the only time the air leaks and dirty air filters get some attention is when the air and pressure losses interfere with the normal operation of the plant. However, paying attention to the compressed-air system and practicing some simple conservation measures can result in considerable energy and cost savings for the plants.

The hissing of *air leaks* can sometimes be heard even in high-noise manufacturing facilities. *Pressure drops* at end-use points in the order of 40 percent of the compressor-discharged pressure are not uncommon. Yet a common response to such a problem is the installation of a larger compressor instead of checking the system and finding out what the problem is. The latter corrective action is usually taken only after the larger compressor also fails to eliminate the problem. The energy wasted in compressed-air systems because of poor installation and maintenance can account for up to 50 percent of the energy consumed by the compressor, and about half of this amount can be saved by simple measures.

The cost of electricity to operate a compressor for one year can exceed the purchase price of the compressor. This is especially the case for larger compressors operating two or three shifts. For example, operating a 125-hp compressor powered by a 90-percent efficient electric motor at full load for 6000 hours a year at \$0.085/kWh will cost \$52,820 a year in electricity cost, which greatly exceeds the purchase and installation cost of a typical unit (Fig. 6-49).

Compressor: 125 hp = 93.21 kW
Operating hours: 6000 h/yr
Unit cost of electricity: \$0.085/kWh
Motor efficiency: 0.90

Annual energy usage: 621,417 kWh
Annual electricity cost: \$52,820/yr

FIGURE 6-49
The cost of electricity to operate a compressor for one year can exceed the purchase price of the compressor.

Below we describe some procedures to reduce the cost of compressed air in industrial facilities and quantify the energy and cost savings associated with them. Once the compressor power wasted is determined, the *annual energy* (usually electricity) and *cost savings* can be determined from

$$\text{Energy savings} = (\text{Power saved})(\text{Operating hours})/\eta_{\text{motor}} \quad (6\text{-}60)$$

and

$$\text{Cost savings} = (\text{Energy savings})(\text{Unit cost of energy}) \quad (6\text{-}61)$$

where η_{motor} is the efficiency of the motor driving the compressor and the unit cost of energy is usually expressed in dollars per kWh (1 kWh = 3600 kJ).

1 Repairing Air Leaks on Compressed-Air Lines

Air leaks are the greatest single cause of energy loss in manufacturing facilities associated with compressed-air systems. It takes energy to compress the air, and thus the loss of compressed air is a loss of energy for the facility. A compressor must work harder and longer to make up for the lost air and must use more energy in the process. Several studies at plants have revealed that up to 40 percent of the compressed air is lost through leaks. Eliminating the air leaks totally is impractical, and a leakage rate of 10 percent is considered acceptable.

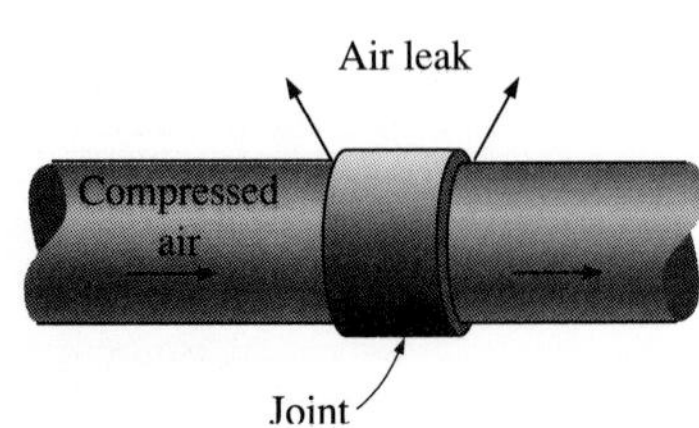

FIGURE 6-50
Air leaks commonly occur at joints and connections.

Air leaks, in general, occur at the *joints, flange connections, elbows, reducing bushes, sudden expansions, valve systems, filters, hoses, check valves, relief valves, extensions,* and the *equipment* connected to the compressed-air lines (Fig. 6-50). Expansion and contraction as a result of thermal cycling and vibration are common causes of loosening at the joints, and thus air leaks. Therefore, it is a good practice to *check* the joints for tightness and to *tighten* them periodically. Air leaks also commonly occur at the points of end use or where the compressed-air lines are connected to the equipment that operates on compressed air. Because of the frequent opening and closing of the compressed-air lines at these points, the gaskets wear out quickly, and they need to be replaced periodically.

There are many ways of detecting air leaks in a compressed-air system. Perhaps the simplest way of detecting a large air leak is to listen for it. The high velocity of the air escaping the line produces a *hissing sound* that is difficult not to notice except in environments with a high noise level. Another way of detecting air leaks, especially small ones, is to test the suspected area with *soap water* and to watch for soap bubbles. This method is obviously not practical for a large system with many connections. A modern way of checking for air leaks is to use an acoustic leak detector, which consists of a directional microphone, amplifiers, audio filters, and digital indicators.

A practical way of quantifying the air leaks in a production facility in its entirety is to conduct a *pressure drop test*. The test is conducted by stopping all the operations that use compressed air and by shutting down the compressors and closing the pressure relief valve, which relieves pressure automatically if the compressor is equipped with one. This way, any pressure drop in the compressed-air lines is due to the cumulative effects of air leaks. The drop in pressure in the system with time is observed, and the test is conducted until the pressure drops by an amount that can be measured accurately, usually 0.5 atm. The time it takes for the pressure to drop by this amount is measured, and the decay of pressure as a function of time is recorded. The total volume of the compressed-air system, including the compressed-air tanks, the headers, accumulators, and the primary compressed-air lines, is calculated. Ignoring the small lines will make the job easier and will cause the result to be more conservative. The rate of air leak can be determined using the ideal gas equation of state.

The amount of *mechanical energy wasted* as a unit mass of air escapes through the leaks is equivalent to the actual amount of energy it takes to compress it, and is determined from Eq. 6-57, modified as (Fig. 6-51)

$$w_{\text{comp, in}} = \frac{w_{\text{reversible comp, in}}}{\eta_{\text{comp}}} = \frac{nRT_1}{\eta_{\text{comp}}(n-1)}\left[\left(\frac{P_2}{P_1}\right)^{(n-1)/n} - 1\right] \quad (6\text{-}62)$$

For energy used in compressor

where n is the polytropic compression exponent ($n = 1.4$ when the compression is isentropic and $1 < n < 1.4$ when there is intercooling) and η_{comp} is the

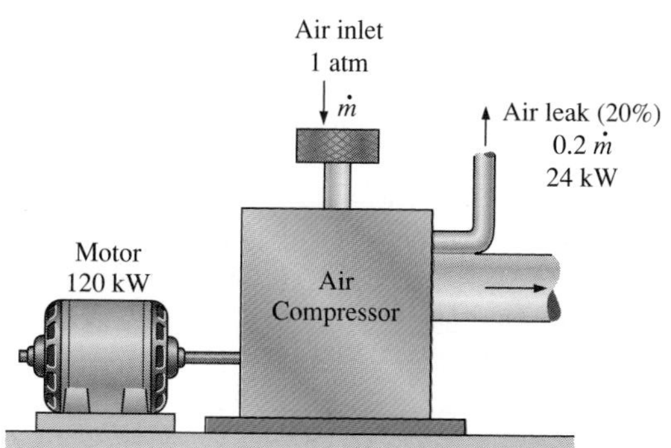

FIGURE 6-51

The energy wasted as compressed air escapes through the leaks is equivalent to the energy it takes to compress it.

compressor efficiency (discussed in the next section), whose value usually ranges between 0.7 and 0.9.

Using compressible-flow theory (see Chap. 16), it can be shown that whenever the line pressure is above 2 atm, which is usually the case, the velocity of air at the leak site must be equal to the local *speed of sound*. Then the **mass flow rate of air** through a leak of minimum cross-sectional area A becomes

$$\dot{m}_{\text{air}} = C_{\text{discharge}}\left(\frac{2}{k+1}\right)^{1/(k-1)} \frac{P_{\text{line}}}{RT_{\text{line}}} A \sqrt{kR\left(\frac{2}{k+1}\right)T_{\text{line}}} \qquad \text{(6-63)}$$

where k is the specific heat ratio ($k = 1.4$ for air) and $C_{\text{discharge}}$ is a discharge (or loss) coefficient that accounts for imperfections in flow at the leak site. Its value ranges from about 0.60 for an orifice with sharp edges to 0.97 for a well-rounded circular hole. The air-leak sites are imperfect in shape, and thus the discharge coefficient can be taken to be 0.65 in the absence of actual data. Also, T_{line} and P_{line} are the temperature and pressure in the compressed-air line, respectively.

Once $\dot{m}_{\text{air}}$ and $w_{\text{comp, in}}$ are available, the **power wasted** by the leaking compressed air (or the power saved by repairing the leak) is determined from

$$\text{Power saved} = \text{Power wasted} = \dot{m}_{\text{air}} w_{\text{comp, in}} \qquad \text{(6-64)}$$

EXAMPLE 6-14 Energy and Cost Savings by Fixing Air Leaks

The compressors of a production facility maintain the compressed-air lines at a (gage) pressure of 700 kPa at sea level where the atmospheric pressure is 101 (Fig. 6-52). The average temperature of air is 20°C at the compressor inlet and 24°C in the compressed-air lines. The facility operates 4200 hours a year, and the average price of electricity is \$0.078/kWh. Taking the compressor efficiency to be 0.8, the motor efficiency to be 0.92, and the discharge coefficient to be 0.65, determine the energy and money saved per year by sealing a leak equivalent to a 3-mm-diameter hole on the compressed-air line.

Solution We note that the absolute pressure is the sum of the gage and atmospheric pressures.

Assumptions **1** Steady operating conditions exist. **2** Air is an ideal gas. **3** Pressure losses in the compressed air lines are negligible.

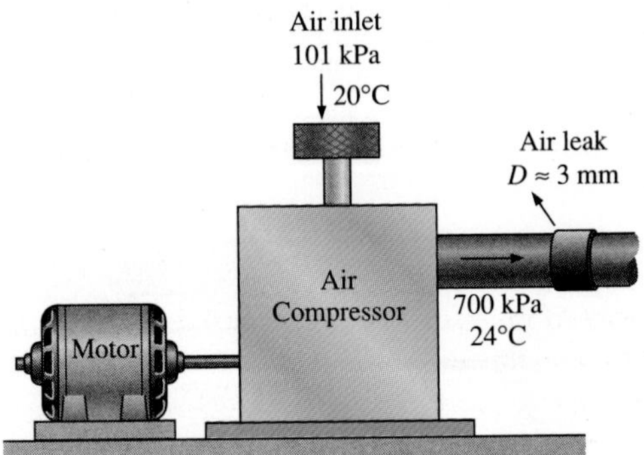

FIGURE 6-52
Schematic for Example 6-14.

Analysis The work needed to compress a unit mass of air at 20°C from the atmospheric pressure of 101 kPa to 700 + 101 = 801 kPa is

$$w_{\text{comp, in}} = \frac{nRT_1}{\eta_{\text{comp}}(n-1)}\left[\left(\frac{P_2}{P_1}\right)^{(n-1)/n} - 1\right]$$

$$= \frac{(1.4)[0.287\text{ kJ/(kg}\cdot\text{K)}](293\text{ K})}{(0.8)(1.4-1)}\left[\left(\frac{801\text{ kPa}}{101\text{ kPa}}\right)^{0.4/1.4} - 1\right] = 296.9\text{ kJ/kg}$$

The cross-sectional area of the 3-mm-diameter hole is

$$A = \pi D^2/4 = \pi(3\times10^{-3}\text{ m})^2/4 = 7.069\times10^{-6}\text{ m}^2$$

Noting that the line conditions are 297 K and 801 kPa, the mass flow rate of the air leaking through the hole is determined to be

$$\dot{m}_{\text{air}} = C_{\text{discharge}}\left(\frac{2}{k+1}\right)^{1/(k-1)}\frac{P_{\text{line}}}{RT_{\text{line}}}A\sqrt{kR\left(\frac{2}{k+1}\right)T_{\text{line}}}$$

$$= (0.65)\left(\frac{2}{1.4+1}\right)^{1/(1.4-1)}\frac{801\text{ kPa}}{[0.287\text{ kPa}\cdot\text{m}^3/(\text{kg}\cdot\text{K})]\,(297\text{ K})}(7.069\times10^{-6}\text{ m}^2)$$

$$\times\sqrt{(1.4)[0.287\text{ kJ/(kg}\cdot\text{K)}]\left(\frac{1000\text{ m}^2/\text{s}^2}{1\text{ kJ/kg}}\right)\left(\frac{2}{1.4+1}\right)(297\text{ K})}$$

$$= 0.008632\text{ kg/s}$$

Then the power wasted by the leaking compressed air becomes

$$\text{Power wasted} = \dot{m}_{\text{air}}\,w_{\text{comp, in}}$$
$$= (0.008632\text{ kg/s})(296.9\text{ kJ/kg})$$
$$= 2.563\text{ kW}$$

The compressor operates 4200 hours a year, and the motor efficiency is 0.92. Then the annual energy and cost savings resulting from repairing this leak are determined to be

Energy savings = (Power saved)(Operating hours)/η_{motor}
= (2.563 kW)(4200 h/yr)/0.92
= **11,700 kWh/yr**

Cost savings = (Energy savings)(Unit cost of energy)
= (11,700 kWh/yr)($0.078/kWh)
= **$913/yr**

Discussion Note that the facility will save 11,700 kWh of electricity worth $913 a year when this air leak is fixed. This is a substantial amount for a single leak whose equivalent diameter is 3 mm.

Installing High Efficiency Motors

Practically all compressors are powered by electric motors, and the *electrical energy* a motor draws for a specified power output is *inversely proportional* to its efficiency. Electric motors cannot convert the electrical energy they consume into mechanical energy completely, and the ratio of the mechanical

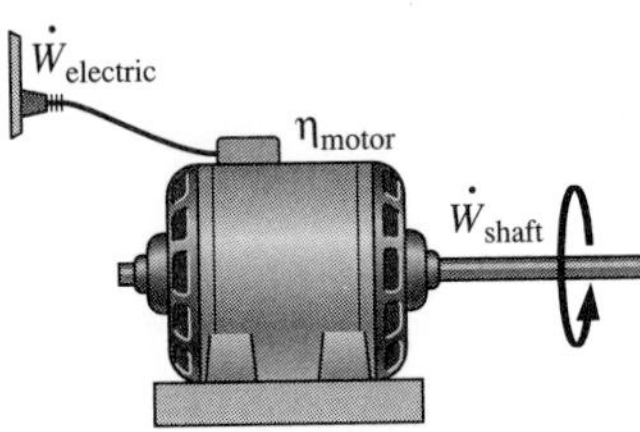

Motor efficiency η_{motor}	Electrical power consumed per kW of mechanical (shaft) power output, $\dot{W}_{electric} = \dot{W}_{shaft}/\eta_{motor}$
100%	1.00 kW
90	1.11
80	1.25
70	1.43
60	1.67
50	2.00
40	2.50
30	3.33
20	5.00
10	10.00

FIGURE 6-53
The electrical energy consumed by a motor is inversely proportional to its efficiency.

power supplied to the electrical power consumed during operation is called the **motor efficiency,** η_{motor}. Therefore, the electric power consumed by the motor and the mechanical (shaft) power supplied to the compressor are related to each other by (Fig. 6-53)

$$\dot{W}_{electric} = \dot{W}_{comp}/\eta_{motor} \tag{6-65}$$

For example, assuming no transmission losses, a motor that is 80 percent efficient will draw 1/0.8 = 1.25 kW of electric power for each kW of shaft power it delivers to the compressor, whereas a motor that is 95 percent efficient will draw only 1/0.95 = 1.05 kW to deliver 1 kW. Therefore, high-efficiency motors cost less to operate than their standard counterparts, but they also usually cost more to purchase. However, the energy savings usually make up for the price differential during the first few years. This is especially true for large compressors that operate more than one regular shift. The *electric power saved* by replacing the existing standard motor of efficiency $\eta_{standard}$ by a high-efficiency one of efficiency $\eta_{efficient}$ is determined from

$$\begin{aligned}\dot{W}_{electric,\,saved} &= \dot{W}_{electric,\,standard} - \dot{W}_{electric,\,efficient}\\ &= \dot{W}_{comp}\,(1/\eta_{standard} - 1/\eta_{efficient})\\ &= \text{(Rated power)(Load factor)}(1/\eta_{standard} - 1/\eta_{efficient})\end{aligned} \tag{6-66}$$

where *rated power* is the nominal power of the motor listed on its label (the power the motor delivers at full load) and the *load factor* is the fraction of the rated power at which the motor normally operates. Then the annual energy savings as a result of replacing a motor by a high-efficiency motor instead of a comparable standard one is

$$\text{Energy savings} = \dot{W}_{electric,\,saved} \times \text{Annual operating hours} \tag{6-67}$$

The efficiencies of motors used to power compressors usually range from about 70 percent to over 96 percent. The portion of electric energy not converted to mechanical energy is converted to heat. The amount of heat generated by the motors may reach high levels, especially at part load, and it may cause overheating if not dissipated effectively. It may also cause the air temperature in the compressor room to rise to undesirable levels. For example, a 90-percent-efficient 100-kW motor generates as much heat as a 10-kW resistance heater in the confined space of the compressor room, and it contributes greatly to the heating of the air in the room. If this heated air is not vented properly, and the air into the compressor is drawn from inside the compressor room, the performance of the compressor will also decline, as explained later.

Important considerations in the selection of a motor for a compressor are the operating profile of the compressor (i.e., the variation of the load with time), and the efficiency of the motor at part-load conditions. The part-load efficiency of a motor is as important as the full-load efficiency if the compressor is expected to operate at part load during a significant portion of the

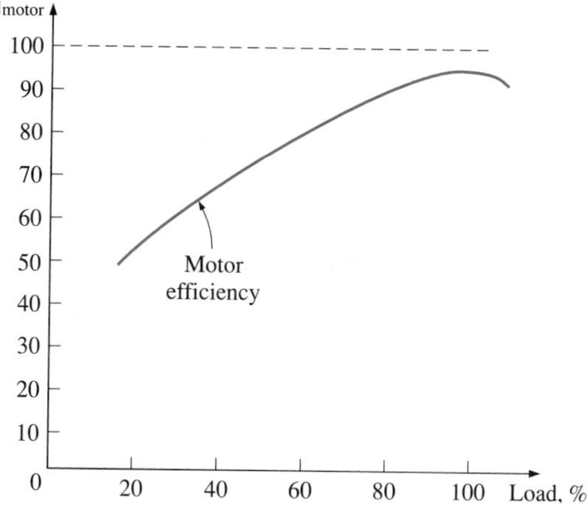

FIGURE 6-54

The efficiency of an electric motor decreases at part load.

total operating time. Also, it is well-known that the efficiency of a motor decreases with decreasing load. For example, the efficiency of a motor may drop from 90 percent at full load to 87 percent at half load and 80 percent at quarter load (Fig. 6-54). The efficiency of another motor of similar specifications, on the other hand, may drop from 91 percent at full load to 75 percent at quarter load. The first motor is obviously better suited for a situation in which a compressor is expected to operate at quarter load during a significant portion of the time. The efficiency at part-load conditions can be improved greatly by installing variable voltage controllers if it is economical to do so. Also, *oversizing* a motor just to be on the safe side and to have some excess power just in case is a bad practice since this will cause the motor to operate almost always at *part load* and thus at a *lower efficiency*. Besides, motors can handle occasional overloading well without any problems.

Using a Smaller Motor at Full Capacity

We tend to purchase *larger equipment* than needed for reasons like having a safety margin or anticipated future expansion, and compressors are no exception. The uncertainties in plant operation are partially responsible for opting for a larger compressor, since it is preferred to have an oversized compressor than an undersized one. Sometimes compressors that have several times the required capacity are purchased with the perception that the extra capacity may be needed some day. The result is a compressor that runs intermittently at full load, or one that runs continuously at part load.

A compressor that operates at part load also causes the motor to operate less efficiently since the efficiency of an electric motor decreases as the point of operation shifts down from its rated power, as explained above. The result is a motor that consumes more electricity per unit power delivered, and thus a more expensive operation. The operating costs can be reduced by switching to a smaller motor that runs at rated power and thus at a higher efficiency.

3 Using Outside Air for Compressor Intake

We have pointed out earlier that the power consumed by a compressor is proportional to the *specific volume,* which is proportional to the *absolute temperature* of the gas at a given pressure. It is also clear from Eq. 6-62 that the compressor work is directly proportional to the *inlet temperature* of air. Therefore, the lower the inlet temperature of the air, the smaller the compressor work. Then the *power reduction factor,* which is the fraction of compressor power reduced as a result of taking intake air from the outside, becomes

$$f_{\text{reduction}} = \frac{W_{\text{comp, inside}} - W_{\text{comp, outside}}}{W_{\text{comp, inside}}} = \frac{T_{\text{inside}} - T_{\text{outside}}}{T_{\text{inside}}} = 1 - \frac{T_{\text{outside}}}{T_{\text{inside}}} \tag{6-68}$$

where T_{inside} and T_{outside} are the absolute temperatures (in K or R) of the ambient air inside and outside the facility, respectively. Thus, reducing the absolute inlet temperature by 5 percent, for example, will reduce the compressor power input by 5 percent. As a rule of thumb, for a specified amount of compressed air, the power consumption of the compressor decreases (or, for a fixed power input, the amount of compressed air increases) by 1 percent for each 3°C drop in the temperature of the inlet air to the compressor.

Compressors are usually located inside the production facilities or in adjacent shelters specifically built outside these facilities. The intake air is normally drawn from inside the building or the shelter. But in many locations the air temperature in the building is higher than the outside air temperature, because of space heaters in the winter and the heat given up by a large number of mechanical and electrical equipment as well as the furnaces year round. The temperature rise in the shelter is also due to the heat dissipation from the compressor and its motor. The outside air is generally *cooler* and thus *denser* than the air in the compressor room even on hot summer days. Therefore, it is advisable to install an *intake duct* to the compressor inlet so that the air is supplied directly from the outside of the building instead of the inside, as shown in Fig. 6-55. This will reduce the energy consumption of the compressor since it takes less energy to compress a specified amount of cool air than the same amount of warm air. Compressing the warm air in a building in winter also wastes the energy used to heat the air.

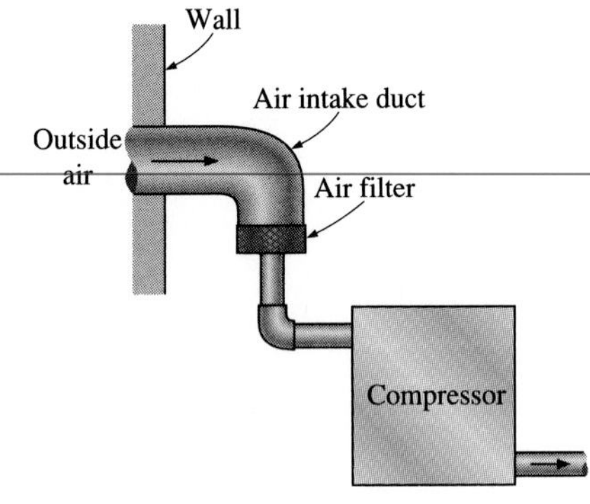

FIGURE 6-55

The power consumption of a compressor can be reduced by taking in air from the outside.

4 Reducing the Air Pressure Setting

Another source of energy waste in compressed-air systems is compressing the air to a higher pressure than required by the air-driven equipment since it takes more energy to compress air to a higher pressure. In such cases considerable energy savings can be realized by determining the minimum required pressure and then reducing the air pressure control setting on the compressor accordingly. This can be done on both screw-type and reciprocating compressors by simply adjusting the pressure setting to match the needs.

The amount of energy it takes to compress a unit mass of air is determined from Eq. 6-62. We note from that relation that the higher the pressure

P_2 at the compressor exit, the larger the work required for compression. Reducing the exit pressure of the compressor to $P_{2,\text{reduced}}$ will reduce the power input requirements of the compressor by a factor of

$$f_{\text{reduction}} = \frac{w_{\text{comp, current}} - w_{\text{comp, reduced}}}{w_{\text{comp, current}}} = 1 - \frac{(P_{2,\text{reduced}}/P_1)^{(n-1)/n} - 1}{(P_2/P_1)^{(n-1)/n} - 1} \quad (6\text{-}69)$$

A power reduction (or savings) factor of $f_{\text{reduction}} = 0.08$, for example, indicates that the power consumption of the compressor is reduced by 8 percent as a result of reducing the pressure setting.

Some applications require slightly compressed air. In such cases, the need can be met by a blower instead of a compressor. Considerable energy can be saved in this manner since a blower requires a small fraction of the power needed by a compressor for a specified mass flow rate.

EXAMPLE 6-15 Reducing the Pressure Setting to Reduce Cost

The compressed-air requirements of a plant located at 1400-m elevation is being met by a 75-hp compressor that takes in air at the local atmospheric pressure of 85.6 kPa and the average temperature of 15°C and compresses it to 900 kPa gage (Fig. 6-56). The plant is currently paying $12,000 a year in electricity costs to run the compressor. An investigation of the compressed-air system and the equipment using the compressed air reveals that compressing the air to 800 kPa is sufficient for this plant. Determine how much money will be saved as a result of reducing the pressure of the compressed air.

Solution The cost savings associated with pressure reduction of a compressor are to be determined.

Assumptions Air is an ideal gas.

Analysis The fraction of energy saved as a result of reducing the pressure setting of the compressor is

$$\begin{aligned} f_{\text{reduction}} &= 1 - \frac{(P_{2,\text{reduced}}/P_1)^{(n-1)/n} - 1}{(P_2/P_1)^{(n-1)/n} - 1} \\ &= 1 - \frac{(885.6/85.6)^{(1.4-1)/1.4} - 1}{(985.6/85.6)^{(1.4-1)/1.4} - 1} = 0.060 \end{aligned}$$

That is, reducing the pressure setting will reduce the energy consumed by the compressor by about 6 percent. Then,

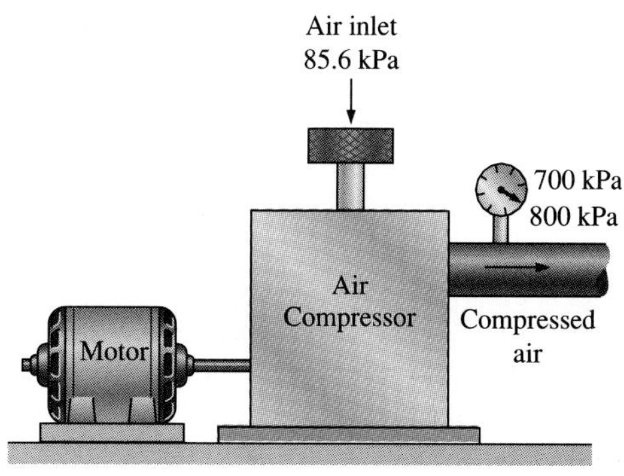

FIGURE 6-56
Schematic for Example 6-15.

$$\text{Cost savings} = (\text{Current cost}) f_{\text{reduction}} = (\$12{,}000/\text{yr})(0.06) = \mathbf{\$720/yr}$$

Therefore, reducing the pressure setting by 100 kPa will result in annual savings of $720 in this case.

There are also other ways to reduce the cost of compressed air in industrial facilities. An obvious way is *turning the compressor off* during nonproduction periods such as lunch hours, nights, and even weekends. A considerable amount of power can be wasted during this stand-by mode. This is especially the case for screw-type compressors since they consume up to 85 percent of their rated power in this mode. The reciprocating compressors are not immune from this deficiency, however, since they also must cycle on and off because of the air leaks present in the compressed-air lines. The system can be shut down manually during nonproduction periods to save energy, but installing a timer (with manual override) is preferred to do this automatically since it is human nature to put things off when the benefits are not obvious or immediate.

The compressed air is sometimes cooled considerably below its dew point in *refrigerated dryers* in order to condense and remove a large fraction of the water vapor in the air as well as other noncondensable gases such as oil vapors. The temperature of air rises considerably as it is compressed, sometimes exceeding 250°C at compressor exit when compressed adiabatically to just 700 kPa. Therefore, it is desirable to cool air after compression in order to minimize the amount of power consumed by the refrigeration system, just as it is desirable to let the hot food in a pan cool to the ambient temperature before putting it into the refrigerator. The cooling can be done by either ambient air or water, and the heat picked up by the cooling medium can be used for space heating, feedwater heating, or process-related heating.

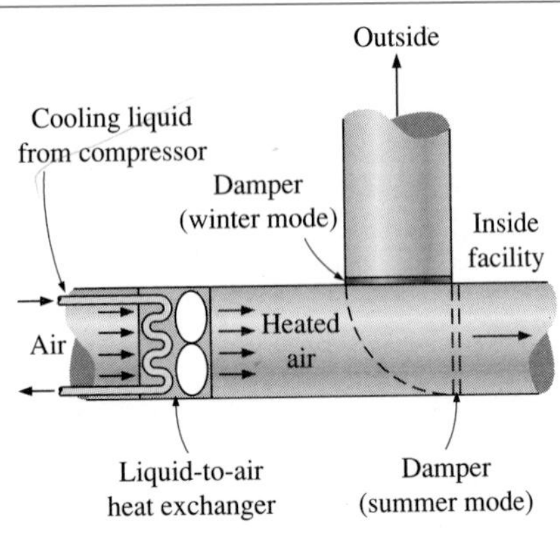

FIGURE 6-57

Waste heat from a compressor can be used to heat a building in winter.

Compressors are commonly cooled directly by air or by circulating a liquid such as oil or water through them in order to minimize the power consumption. The heat picked up by the oil or water is usually rejected to the ambient in a liquid-to-air heat exchanger. This *heat rejected* usually amounts to 60 to 90 percent of the power input, and thus it represents a huge amount of energy that can be used for a useful purpose such as *space heating* in winter, *preheating* the air or water in a furnace, or other process-related purposes (Fig. 6-57). For example, assuming 80 percent of the power input is converted to heat, a 150-hp compressor can reject as much heat as a 90-kW electric resistance heater or a 400,000-Btu/h natural gas heater when operating at full load. Thus, the proper utilization of the waste heat from a compressor can result in significant energy and cost savings.

6-13 ■ ISENTROPIC EFFICIENCIES OF STEADY-FLOW DEVICES

We mentioned on several occasions that irreversibilities inherently accompany all actual processes and that their effect is always to downgrade the performance of devices. In engineering analysis, it would be very desirable to have some parameters that would enable us to quantify the degree of degradation of energy in these devices. In Chap. 5 we did this for cyclic devices, such as heat engines and refrigerators, by comparing the actual cycles to the idealized ones, such as the Carnot cycle. A cycle that was composed entirely of reversible processes served as the *model cycle* to which the actual cycles could be compared. This idealized model cycle enabled us to determine the theoretical limits of performance for cyclic devices under specified conditions and to examine how the performance of actual devices suffered as a result of irreversibilities.

Now we extend the analysis to discrete engineering devices working under steady-flow conditions, such as turbines, compressors, and nozzles, and we examine the degree of degradation of energy in these devices as a result of irreversibilities. But first we need to define an ideal process that will serve as a model for the actual processes.

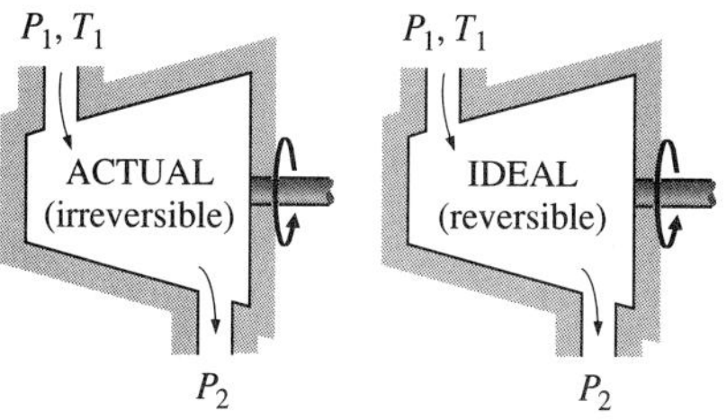

FIGURE 6-58

The isentropic process involves no irreversibilities and serves as the ideal process for adiabatic devices.

Although some heat transfer between these devices and the surrounding medium is unavoidable, most steady-flow devices are intended to operate under adiabatic conditions. Therefore, the model process for these devices should be an adiabatic one. Furthermore, an ideal process should involve no irreversibilities since the effect of irreversibilities is always to downgrade the performance of engineering devices. Thus, the ideal process that can serve as a suitable model for most steady-flow devices is the *isentropic* process (Fig. 6-58).

The more closely the actual process approximates the idealized isentropic process, the better the device will perform. Thus, it would be desirable to have a parameter that expresses quantitatively how efficiently an actual device approximates an idealized one. This parameter is the **isentropic or adiabatic efficiency,** which is a measure of the deviation of actual processes from the corresponding idealized ones.

Adiabatic efficiencies are defined differently for different devices since each device is set up to perform different tasks. Below we define the adiabatic efficiencies of turbines, compressors, and nozzles by comparing the actual performance of these devices to their performance under isentropic conditions for the same inlet state and exit pressure.

Isentropic Efficiency of Turbines

For a turbine under steady operation, the inlet state of the working fluid and the exhaust pressure are fixed. Therefore, the ideal process for an adiabatic turbine is an isentropic process between the inlet state and the exhaust pressure. The desired output of a turbine is the work produced, and the **isentropic efficiency of a turbine** is defined as *the ratio of the actual work output of the turbine to the work output that would be achieved if the process between the inlet state and the exit pressure were isentropic:*

$$\eta_T = \frac{\text{Actual turbine work}}{\text{Isentropic turbine work}} = \frac{w_a}{w_s} \tag{6-70}$$

Usually the changes in kinetic and potential energies associated with a fluid stream flowing through a turbine are small relative to the change in enthalpy and can be neglected. Then the work output of an adiabatic turbine simply becomes the change in enthalpy, and the above relation for this case can be expressed as

$$\eta_T \cong \frac{h_1 - h_{2a}}{h_1 - h_{2s}} \tag{6-71}$$

where h_{2a} and h_{2s} are the enthalpy values at the exit state for actual and isentropic processes, respectively. The actual and isentropic processes in a turbine are illustrated in Fig. 6-59.

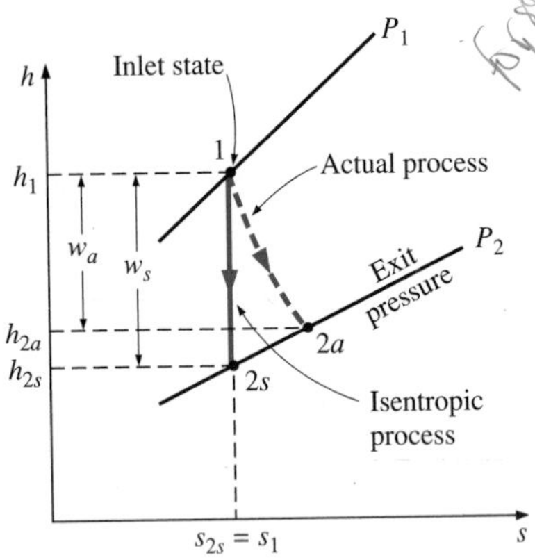

FIGURE 6-59
The *h-s* diagram for the actual and isentropic processes of an adiabatic turbine.

The value of η_T greatly depends on the design of the individual components that make up the turbine. Well-designed, large turbines have isentropic efficiencies above 90 percent. For small turbines, however, it may drop even below 70 percent. The value of the isentropic efficiency of a turbine is determined by measuring the actual work output of the turbine and by calculating the isentropic work output for the measured inlet conditions and the exit pressure. This value may then be used conveniently in the design of power plants.

EXAMPLE 6-16 Isentropic Efficiency of a Steam Turbine

Steam enters an adiabatic turbine steadily at 3 MPa and 400°C and leaves at 50 kPa and 100°C. If the power output of the turbine is 2 MW, determine (*a*) the isentropic efficiency of the turbine and (*b*) the mass flow rate of the steam flowing through the turbine.

Solution A sketch of the system and the *T-s* diagram of the process are given in Fig. 6-60.

Assumptions **1** Steady operating conditions exist. **2** The changes in kinetic and potential energies are negligible. **3** The turbine is adiabatic.

Analysis (*a*) The enthalpies at various states are

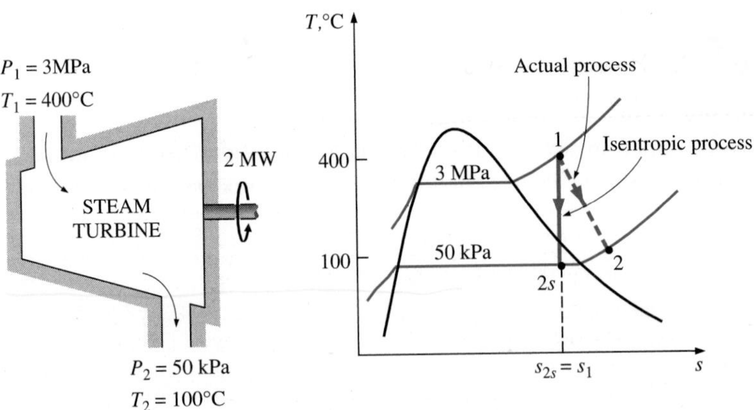

FIGURE 6-60
Schematic and *T-s* diagram for Example 6-16.

State 1: $\left.\begin{array}{l} P_1 = 3 \text{ MPa} \\ T_1 = 400°\text{C} \end{array}\right\}$ $\begin{array}{l} h_1 = 3230.9 \text{ kJ/kg} \\ s_1 = 6.9212 \text{ kJ/(kg·K)} \end{array}$ (Table A-6)

State 2a: $\left.\begin{array}{l} P_{2a} = 50 \text{ kPa} \\ T_{2a} = 100°\text{C} \end{array}\right\}$ $h_{2a} = 2682.5 \text{ kJ/kg}$ (Table A-6)

The exit enthalpy of the steam for the isentropic process h_{2s} is determined from the requirement that the entropy of the steam remain constant ($s_{2s} = s_1$):

State 2s: $\begin{array}{l} P_{2s} = 50 \text{ kPa} \\ (s_{2s} = s_1) \end{array} \longrightarrow \begin{array}{l} s_f = 1.0910 \text{ kJ/(kg·K)} \\ s_g = 7.5939 \text{ kJ/(kg·K)} \end{array}$ (Table A-5)

Obviously, at the end of the isentropic process steam will exist as a saturated liquid–vapor mixture since $s_f < s_{2s} < s_g$. Thus we need to find the quality at state 2*s* first:

$$x_{2s} = \frac{s_{2s} - s_f}{s_{fg}} = \frac{6.9212 - 1.0910}{6.5029} = 0.897$$

and $$h_{2s} = h_f + x_{2s}h_{fg} = 340.49 + 0.897\,(2305.4) = 2407.4 \text{ kJ/kg}$$

By substituting these enthalpy values into Eq. 6-71, the isentropic efficiency of this turbine is determined to be

$$\eta_T \cong \frac{h_1 - h_{2a}}{h_1 - h_{2s}} = \frac{3230.9 - 2682.5}{3230.9 - 2407.4} = \mathbf{0.667, \text{ or } 66.7\%}$$

(*b*) The mass flow rate of steam through this turbine is determined from the energy balance for steady-flow systems:

$$\dot{E}_{in} = \dot{E}_{out}$$

$$\dot{m}h_1 = \dot{W}_{a,\text{out}} + \dot{m}h_{2a}$$

$$\dot{W}_{a,\text{out}} = \dot{m}(h_1 - h_{2a})$$

$$2 \text{ MW}\left(\frac{1000 \text{ kJ/s}}{1 \text{ MW}}\right) = \dot{m}(3230.9 - 2682.5) \text{ kJ/kg}$$

$$\dot{m} = \mathbf{3.65 \text{ kg/s}}$$

Isentropic Efficiencies of Compressors and Pumps

The **isentropic efficiency of a compressor** is defined as *the ratio of the work input required to raise the pressure of a gas to a specified value in an isentropic manner to the actual work input*:

$$\eta_C = \frac{\text{Isentropic compressor work}}{\text{Actual compressor work}} = \frac{w_s}{w_a} \qquad (6\text{-}72)$$

Notice that the isentropic compressor efficiency is defined with the *isentropic work input in the numerator* instead of in the denominator. This is because w_s is a smaller quantity than w_a, and this definition prevents η_C from becoming greater than 100 percent, which would falsely imply that the actual compressors performed better than the isentropic ones. Also notice that the inlet conditions and the exit pressure of the gas are the same for both the actual and the isentropic compressor.

When the changes in kinetic and potential energies of the gas being compressed are negligible, the work input to an adiabatic compressor becomes equal to the change in enthalpy, and Eq. 6-72 for this case becomes

$$\eta_C \cong \frac{h_{2s} - h_1}{h_{2a} - h_1} \tag{6-73}$$

where h_{2a} and h_{2s} are the enthalpy values at the exit state for actual and isentropic compression processes, respectively, as illustrated in Fig. 6-61. Again, the value of η_C greatly depends on the design of the compressor. Well-designed compressors have isentropic efficiencies that range from 75 to 85 percent.

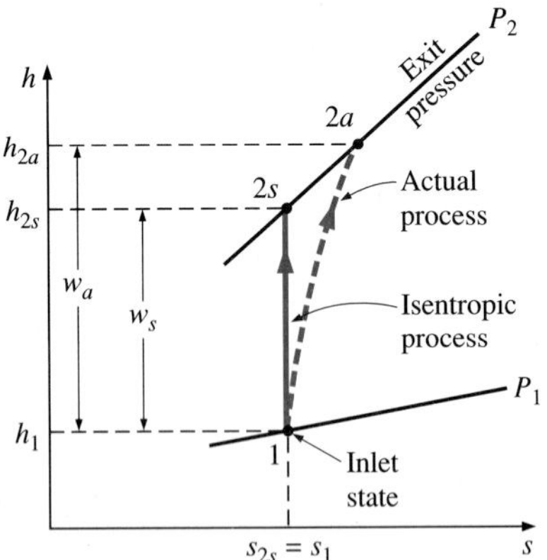

FIGURE 6-61

The *h-s* diagram of the actual and isentropic processes of an adiabatic compressor.

When the changes in potential and kinetic energies of a liquid are negligible, the isentropic efficiency of a pump is defined similarly as

$$\eta_P = \frac{w_s}{w_a} = \frac{v(P_2 - P_1)}{h_{2a} - h_1} \tag{6-74}$$

When no attempt is made to cool the gas as it is compressed, the actual compression process is nearly adiabatic and the reversible adiabatic (i.e., isentropic) process serves well as the ideal process. But sometimes *compressors are cooled intentionally* by utilizing fins or a water jacket placed around the casing to reduce the work input requirements (Fig. 6-62). In this case, the isentropic process is not suitable as the model process since the device is no longer adiabatic and the isentropic compressor efficiency defined above is meaningless. A realistic model process for compressors that are intentionally cooled during the compression process is the *reversible isothermal process*. Then we can conveniently define an **isothermal efficiency** for such cases by comparing the actual process to a reversible isothermal one:

$$\eta_C = \frac{w_t}{w_a} \tag{6-75}$$

where w_t and w_a are the required work inputs to the compressor for the reversible isothermal and actual cases, respectively.

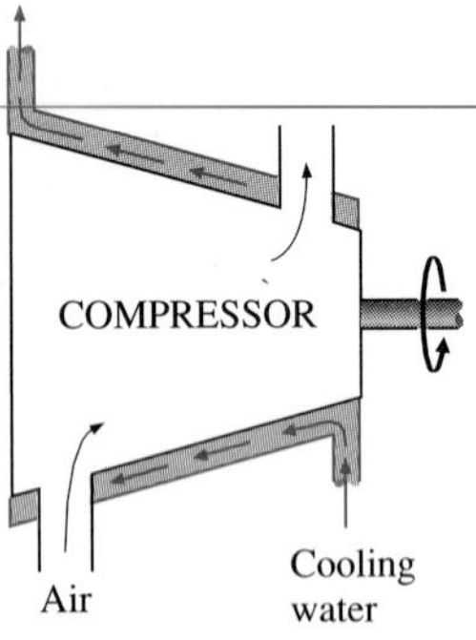

FIGURE 6-62

Compressors are sometimes intentionally cooled to minimize the work input.

EXAMPLE 6-17 Effect of Efficiency on Compressor Power Input

Air is compressed by an adiabatic compressor from 100 kPa and 12°C to a pressure of 800 kPa at a steady rate of 0.2 kg/s. If the isentropic efficiency of the compressor is 80 percent, determine (*a*) the exit temperature of air and (*b*) the required power input to the compressor.

Solution A sketch of the system and the *T-s* diagram of the process are given in Fig. 6-63.

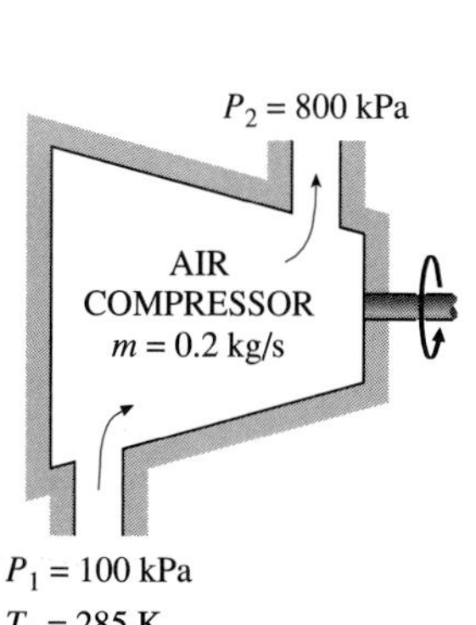

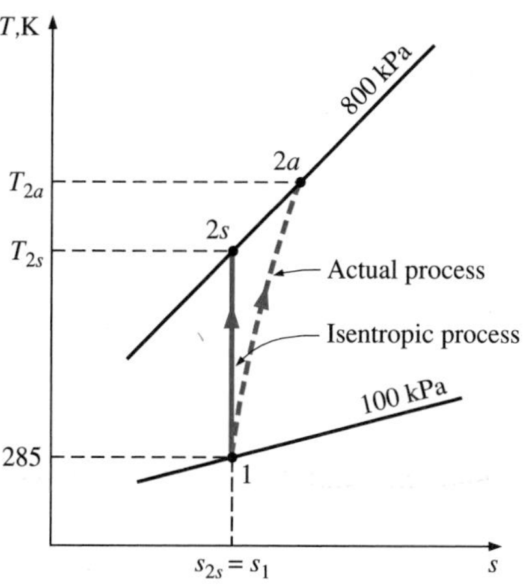

FIGURE 6-63

Schematic and *T-s* diagram for Example 6-17.

Assumptions **1** Steady operating conditions exist. **2** Air is an ideal gas. **3** The changes in kinetic and potential energies are negligible. **4** The compressor is adiabatic.

Analysis (*a*) We know only one property (pressure) at the exit state, and we need to know one more to fix the state and thus determine the exit temperature. The property that can be determined with minimal effort in this case is h_{2a} since the adiabatic efficiency of the compressor is given.

The enthalpy of an ideal gas is a function of temperature only, and h_1 is easily determined from the air table at the inlet temperature:

$$T_1 = 285 \text{ K} \longrightarrow \begin{array}{l} h_1 = 285.14 \text{ kJ/kg} \\ (P_{r1} = 1.1584) \end{array} \qquad \text{(Table A-17)}$$

Now we need to determine h_{2s}, the enthalpy of the air at the end of the isentropic compression process. This is done by using one of the isentropic relations of ideal gases, such as Eq. 6-49:

$$P_{r2} = P_{r1}\left(\frac{P_2}{P_1}\right) = 1.1584\left(\frac{800 \text{ kPa}}{100 \text{ kPa}}\right) = 9.2672$$

and $$P_{r2} = 9.2672 \longrightarrow h_{2s} = 517.05 \text{ kJ/kg} \qquad \text{(Table A-17)}$$

Substituting the known quantities into Eq. 6-73, we have

$$\eta_C \cong \frac{h_{2s} - h_1}{h_{2a} - h_1} \longrightarrow 0.80 = \frac{(517.05 - 285.14) \text{ kJ/kg}}{h_{2a} - 285.14) \text{ kJ/kg}}$$

Thus, $$h_{2a} = 575.03 \text{ kJ/kg} \longrightarrow T_{2a} = \mathbf{569.5\ K} \qquad \text{(Table A-17)}$$

(*b*) The required power input to the compressor is determined from the energy balance for steady-flow devices,

$$\dot{E}_{in} = \dot{E}_{out}$$

$$\dot{m}h_1 + \dot{W}_{a,in} = \dot{m}h_{2a}$$

$$\dot{W}_{a,in} = \dot{m}(h_{2a} - h_1)$$

$$= (0.2\ \text{kg/s})[(575.03 - 285.14)\ \text{kJ/kg}]$$

$$= \mathbf{58.0\ kW}$$

Discussion Notice that in determining the power input to the compressor, we used h_{2a} instead of h_{2s} since h_{2a} is the actual enthalpy of the air as it exits the compressor. The quantity h_{2s} is a hypothetical enthalpy value that the air would have if the process were isentropic.

Isentropic Efficiency of Nozzles

Nozzles are essentially adiabatic devices and are used to accelerate a fluid. Therefore, the isentropic process serves as a suitable model for nozzles. The **isentropic efficiency of a nozzle** is defined as *the ratio of the actual kinetic energy of the fluid at the nozzle exit to the kinetic energy value at the exit of an isentropic nozzle for the same inlet state and exit pressure.* That is,

$$\eta_N = \frac{\text{Actual KE at nozzle exit}}{\text{Isentropic KE at nozzle exit}} = \frac{V_{2a}^2}{V_{2s}^2} \quad \text{(6-76)}$$

Note that the exit pressure is the same for both the actual and isentropic processes, but the exit state is different.

Nozzles involve no work interactions, and the fluid experiences little or no change in its potential energy as it flows through the device. If, in addition, the inlet velocity of the fluid is small relative to the exit velocity, the energy balance for this steady-flow device reduces to

$$h_1 = h_{2a} + \frac{V_{2a}^2}{2}$$

Then the adiabatic efficiency of the nozzle can be expressed in terms of enthalpies as

$$\eta_N \cong \frac{h_1 - h_{2a}}{h_1 - h_{2s}} \quad \text{(6-77)}$$

where h_{2a} and h_{2s} are the enthalpy values at the nozzle exit for the actual and isentropic processes, respectively (Fig. 6-64). Isentropic efficiencies of nozzles are typically above 90 percent, and nozzle efficiencies above 95 percent are not uncommon.

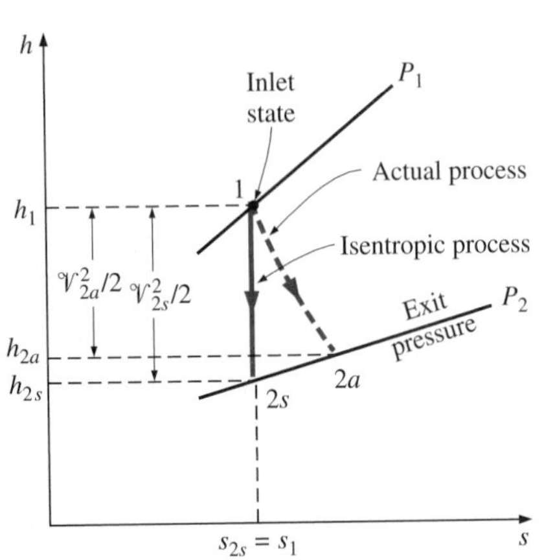

FIGURE 6-64

The *h-s* diagram of the actual and isentropic processes of an adiabatic nozzle.

EXAMPLE 6-18 Effect of Efficiency on Nozzle Exit Velocity

Air at 200 kPa and 950 K enters an adiabatic nozzle at low velocity and is discharged at a pressure of 80 kPa. If the isentropic efficiency of the nozzle is 92 percent, determine (*a*) the maximum possible exit velocity, (*b*) the exit temperature, and (*c*) the actual velocity of the air. Assume constant specific heats for air.

Solution A sketch of the system and the *T-s* diagram of the process are given in Fig. 6-65.

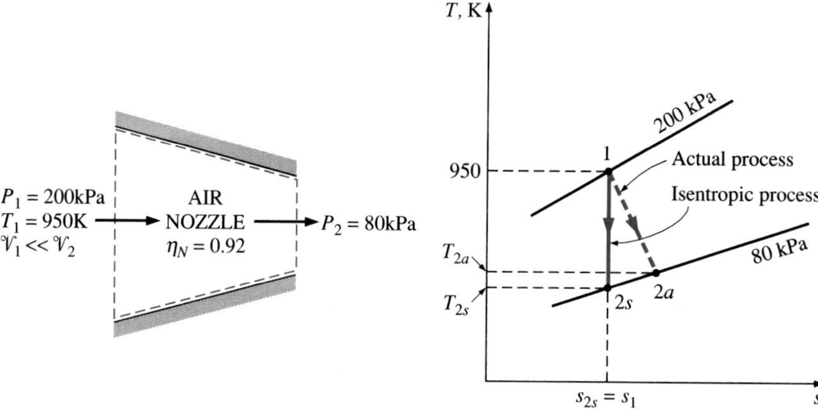

FIGURE 6-65

Schematic and *T-s* diagram for Example 6-18.

Assumptions **1** Steady operating conditions exist. **2** Air is an ideal gas. **3** The inlet kinetic energy is negligible. **4** The nozzle is adiabatic.

Analysis The temperature of air will drop during this acceleration process because some of its internal energy is converted to kinetic energy. This problem can be solved accurately by using property data from the air table. But we will assume constant specific heats (thus sacrifice some accuracy) to demonstrate their use. Let us guess that the average temperature of the air will be about 800 K. Then the average values of C_p and k at this anticipated average temperature are determined from Table A-2*b* to be $C_p = 1.099$ kJ/(kg · K) and $k = 1.354$.

(*a*) The exit velocity of the air will be a maximum when the process in the nozzle involves no irreversibilities. The exit velocity in this case is determined from the steady-flow energy equation. But first we need to determine the exit temperature. For the isentropic process of an ideal gas with constant specific heats, the temperatures and pressures are related by Eq. 6-43:

$$\frac{T_{2s}}{T_1} = \left(\frac{P_{2s}}{P_1}\right)^{(k-1)/k}$$

or

$$T_{2s} = T_1\left(\frac{P_{2s}}{P_1}\right)^{(k-1)/k} = (950 \text{ K})\left(\frac{80 \text{ kPa}}{200 \text{ kPa}}\right)^{0.354/1.354} = 748 \text{ K}$$

This will give an average temperature of 849 K, which is somewhat higher than the assumed average temperature (800 K). This result could be refined by reevaluating the k value at 749 K and repeating the calculations, but it is not warranted since the two average temperatures are sufficiently close (doing so would change the temperature by only 1.5 K, which is not significant).

Now we can determine the isentropic exit velocity of the air from the energy balance for this isentropic steady-flow process:

$$e_{in} = e_{out}$$

$$h_1 + \frac{\mathcal{V}_1^2}{2}^{\nearrow 0} = h_{2s} + \frac{\mathcal{V}_{2s}^2}{2}$$

or

$$\mathcal{V}_{2s} = \sqrt{2(h_{2s} - h_1)} = \sqrt{2C_{p,\text{av}}(T_1 - T_{2s})}$$

$$= \sqrt{2[1.099 \text{ kJ/(kg}\cdot\text{K)}][(950 - 748) \text{ K}]\left(\frac{1000 \text{ m}^2/\text{s}^2}{1 \text{ kJ/kg}}\right)}$$

$$= \mathbf{666 \text{ m/s}}$$

(*b*) The actual exit temperature of the air will be higher than the isentropic exit temperature evaluated above, and it is determined from Eq. 6-77. For constant specific heats,

$$\eta_N \cong \frac{h_1 - h_{2a}}{h_1 - h_{2s}} = \frac{C_{p,\text{av}}(T_1 - T_{2a})}{C_{p,\text{av}}(T_1 - T_{2s})}$$

or

$$0.92 = \frac{950 - T_{2a}}{950 - 748} \longrightarrow T_{2a} = \mathbf{764\ K}$$

That is, the temperature will be 16 K higher at the exit of the actual nozzle as a result of irreversibilities such as friction. It represents a loss since this rise in temperature comes at the expense of kinetic energy (Fig. 6-66).

(*c*) The actual exit velocity of air can be determined from the definition of isentropic efficiency (Eq. 6-76):

$$\eta_N = \frac{\mathcal{V}_{2a}^2}{\mathcal{V}_{2s}^2} \longrightarrow \mathcal{V}_{2a} = \sqrt{\eta_N \mathcal{V}_{2s}^2} = \mathbf{639\ m/s}$$

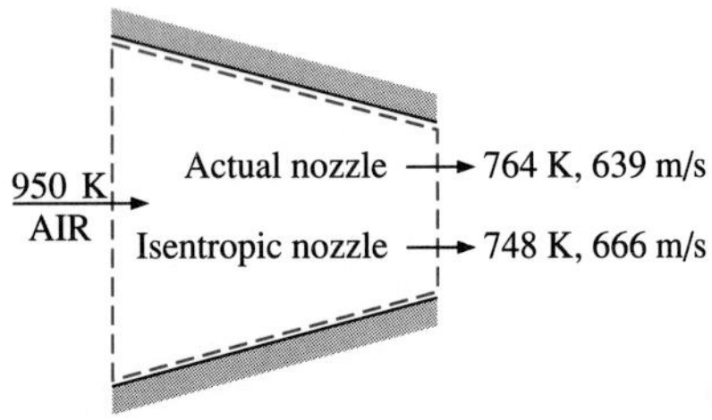

FIGURE 6-66

A substance leaves actual nozzles at a higher temperature (thus a lower velocity) as a result of friction.

6-14 ■ ENTROPY BALANCE

The property *entropy* is a measure of molecular disorder or randomness of a system, and the second law of thermodynamics states that entropy can be created but it cannot be destroyed. Therefore, the entropy change of a system during a process is greater than the entropy transfer by an amount equal to the entropy generated during the process within the system, and the *increase of entropy principle* for any system is expressed as (Fig. 6-67)

$$\begin{pmatrix}\text{Total}\\ \text{entropy}\\ \text{entering}\end{pmatrix} - \begin{pmatrix}\text{Total}\\ \text{entropy}\\ \text{leaving}\end{pmatrix} + \begin{pmatrix}\text{Total}\\ \text{entropy}\\ \text{generated}\end{pmatrix} = \begin{pmatrix}\text{Change in the}\\ \text{total entropy}\\ \text{of the system}\end{pmatrix}$$

or

$$S_{\text{in}} - S_{\text{out}} + S_{\text{gen}} = \Delta S_{\text{system}} \tag{6-78}$$

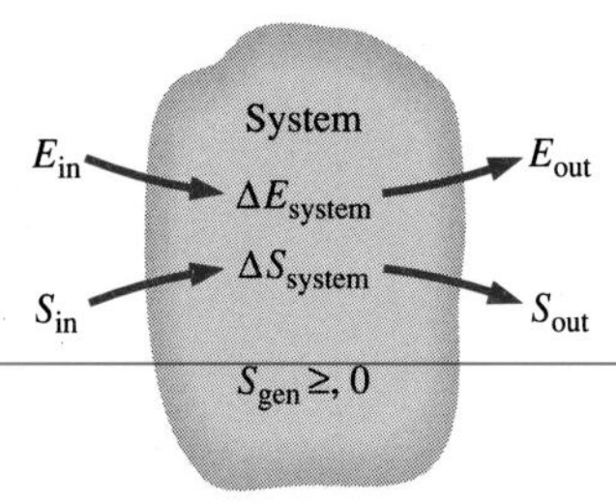

$\Delta E_{\text{system}} = E_{\text{in}} - E_{\text{out}}$
$\Delta S_{\text{system}} = S_{\text{in}} - S_{\text{out}} + S_{\text{gen}}$

FIGURE 6-67

Energy and entropy balances for a system.

which is a verbal statement of Eq. 6-9. This relation is often referred to as the **entropy balance** and is applicable to any kind of system undergoing any kind of process. The entropy balance relation above can be stated as: *the entropy change of a system during a process is equal to the net entropy transfer through the system boundary and the entropy generated within the system as a result of irreversibilities.* Next we discuss the various terms in that relation.

Entropy Change of a System, ΔS_{system}

Despite the reputation of entropy as being vague and abstract and the intimidation associated with it, entropy balance is actually easier to deal with than energy balance since, unlike energy, entropy does not exist in various forms. Therefore, the determination of entropy change of a system during a process involves evaluating entropy of the system at the beginning and at the end of the process and taking their difference. That is,

Entropy change = Entropy at final state − Entropy at initial state

or

$$\Delta S_{\text{system}} = S_{\text{final}} - S_{\text{initial}} = S_2 - S_1 \qquad (6\text{-}79)$$

Note that entropy is a property, and the value of a property does not change unless the state of the system changes. Therefore, the entropy change of a system is zero if the state of the system does not change during the process. For example, the entropy change of steady-flow devices such as nozzles, compressors, turbines, pumps, and heat exchangers is zero during steady operation.

When the properties of the system are not uniform, the entropy of the system can be determined by integration from

$$S_{\text{system}} = \int s\,\delta m = \int_V s\rho\, dV \qquad (6\text{-}80)$$

where V is the volume of the system and ρ is density.

Mechanisms of Entropy Transfer, S_{in} and S_{out}

Entropy can be transferred to or from a system by two mechanisms: *heat transfer* and *mass flow* (in contrast, energy is transferred by work also). Entropy transfer is recognized at the system boundary as it crosses the boundary, and it represents the entropy gained or lost by a system during a process. The only form of entropy interaction associated with a fixed mass or closed system is *heat transfer,* and thus the entropy transfer for an adiabatic closed system is zero.

1 Heat Transfer

Heat is, in essence, a form of disorganized energy, and some disorganization (entropy) will flow with heat. Heat transfer to a system increases the entropy of that system and thus the level of molecular disorder or randomness, and heat transfer from a system decreases it. In fact, heat rejection is the only way the entropy of a fixed mass can be decreased. The ratio of the heat transfer Q at a location to the absolute temperature T at that location is called the *entropy flow* or *entropy transfer* and is expressed as (Fig. 6-68)

Entropy transfer by heat transfer: $$S_{\text{heat}} = \frac{Q}{T} \qquad (T = \text{constant}) \qquad (6\text{-}81)$$

The quantity Q/T represents the entropy transfer accompanied by heat transfer, and the direction of entropy transfer is the same as the direction of heat transfer since absolute temperature T is always a positive quantity.

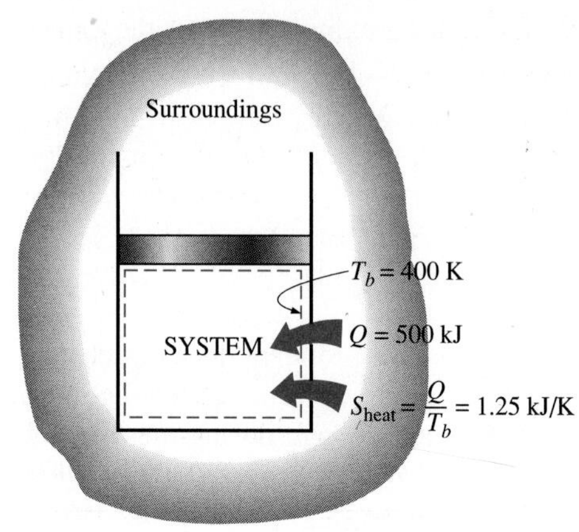

FIGURE 6-68
Heat transfer is always accompanied by entropy transfer in the amount of Q/T where T is the boundary temperature.

When the temperature T is not constant, the entropy transfer during a process 1-2 can be determined by integration (or by summation if appropriate) as

$$S_{heat} = \int_1^2 \frac{\delta Q}{T} \cong \sum \frac{Q_k}{T_k} \tag{6-82}$$

where Q_k is the heat transfer through the boundary at temperature T_k at location k.

When two systems are in contact, the entropy transfer from the warmer system is equal to the entropy transfer into the cooler one at the point of contact. That is, no entropy can be created or destroyed at the boundary since the boundary has no thickness and occupies no volume.

Note that **work** is entropy-free, and no entropy is transferred by work. Energy is transferred by both heat and work, whereas entropy is transferred only by heat. That is,

Entropy transfer by work: $\qquad S_{work} = 0 \qquad$ (zero) $\qquad$ (6-83)

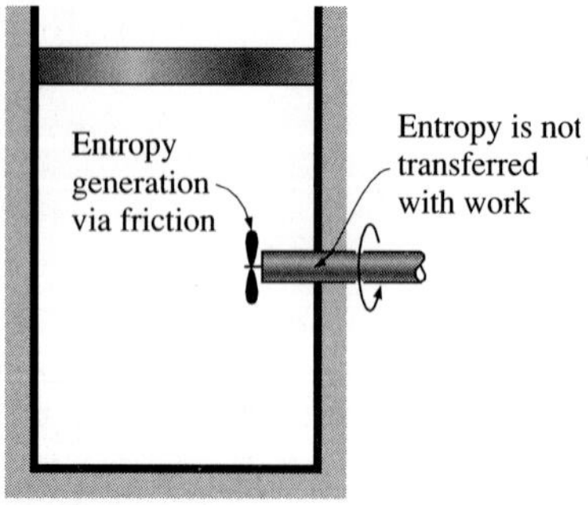

FIGURE 6-69
No entropy accompanies work as it crosses the system boundary. But entropy may be generated within the system as work energy is dissipated into a less useful form of energy.

The first law of thermodynamics makes no distinction between heat transfer and work; it considers them as *equals*. The distinction between heat transfer and work is brought out by the second law: *an energy interaction that is accompanied by entropy transfer is heat transfer, and an energy interaction that is not accompanied by entropy transfer is work*. That is, no entropy is exchanged during a work interaction between a system and its surroundings. Thus, only *energy* is exchanged during work interaction whereas both *energy* and *entropy* are exchanged during heat transfer (Fig. 6-69).

2 Mass flow

Mass contains entropy as well as energy, and the entropy and energy contents of a system are proportional to the mass. (When the mass of a system is doubled, so are the entropy and energy contents of the system.) Both entropy and energy are carried into or out of a system by streams of matter, and the rates of entropy and energy transport into or out of a system are proportional to the mass flow rate. Closed systems do not involve any mass flow and thus any entropy transfer by mass. When a mass in the amount of m enters or leaves a system, entropy in the amount of ms, where s is the specific entropy (entropy per unit mass entering or leaving), accompanies it (Fig. 6-70). That is,

Entropy transfer by mass flow: $$S_{\text{mass}} = ms \qquad (6\text{-}84)$$

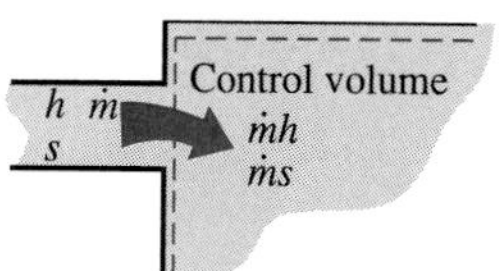

FIGURE 6-70
Mass contains entropy as well as energy, and thus mass flow into or out of system is always accompanied by energy and entropy transfer.

Therefore, the entropy of a system increases by ms when mass in the amount of m enters and decreases by the same amount when the same amount of mass at the same state leaves the system. When the properties of the mass change during the process, the entropy transfer by mass flow can be determined by integration from

$$\dot{S}_{\text{mass}} = \int_{A_c} s\rho \mathcal{V}_n \, dA_c \quad \text{and} \quad S_{\text{mass}} = \int s\,\delta m = \int_{\Delta t} \dot{S}_{\text{mass}}\, dt \qquad (6\text{-}85)$$

where A_c is the cross-sectional area of the flow and $\mathcal{V}_n$ is the local velocity normal to dA_c.

3 Entropy Generation, S_{gen}

Irreversibilities such as friction, mixing, chemical reactions, heat transfer through a finite temperature difference, unrestrained expansion, non-quasi-equilibrium compression, or expansion always cause the entropy of a system to increase, and entropy generation is a measure of the entropy created by such effects during a process.

For a *reversible process* (a process that involves no irreversibilities), the entropy generation is zero and thus the *entropy change* of a system is equal to the *entropy transfer*. Therefore, the entropy balance relation in the reversible case becomes analogous to the energy balance relation, which states that *energy change* of a system during a process is equal to the *energy transfer* during that process. However, note that the energy change of a system equals the energy transfer for *any* process, but the entropy change of a system equals the entropy transfer only for a *reversible* process.

The entropy transfer by heat Q/T is zero for adiabatic systems, and the entropy transfer by mass ms is zero for systems that involve no mass flow across their boundary (i.e., closed systems).

Entropy balance for *any system* undergoing *any process* can be expressed more explicitly as

$$\underbrace{S_{\text{in}} - S_{\text{out}}}_{\substack{\text{Net entropy transfer}\\ \text{by heat and mass}}} + \underbrace{S_{\text{gen}}}_{\substack{\text{Entropy}\\ \text{generation}}} = \underbrace{\Delta S_{\text{system}}}_{\substack{\text{Change}\\ \text{in entropy}}} \qquad (\text{kJ/K}) \qquad (6\text{-}86)$$

or, in the **rate form**, as

$$\underbrace{\dot{S}_{in} - \dot{S}_{out}}_{\substack{\text{Rate of net entropy transfer}\\ \text{by heat and mass}}} + \underbrace{\dot{S}_{gen}}_{\substack{\text{Rate of entropy}\\ \text{generation}}} = \underbrace{\Delta \dot{S}_{system}}_{\substack{\text{Rate of change}\\ \text{of entropy}}} \qquad (\text{kW/K}) \qquad (6\text{-}87)$$

where the rates of entropy transfer by heat transferred at a rate of $\dot{Q}$ and mass flowing at a rate of $\dot{m}$ are $\dot{S}_{heat} = \dot{Q}/T$ and $\dot{S}_{mass} = \dot{m}s$. The entropy balance can also be expressed on a **unit-mass basis** as

$$(s_{in} - s_{out}) + s_{gen} = \Delta s_{system} \qquad [\text{kJ/(kg} \cdot \text{K)}] \qquad (6\text{-}88)$$

where all the quantities are expressed per unit mass of the system. Note that for a *reversible process,* the entropy generation term S_{gen} drops out from all of the relations above.

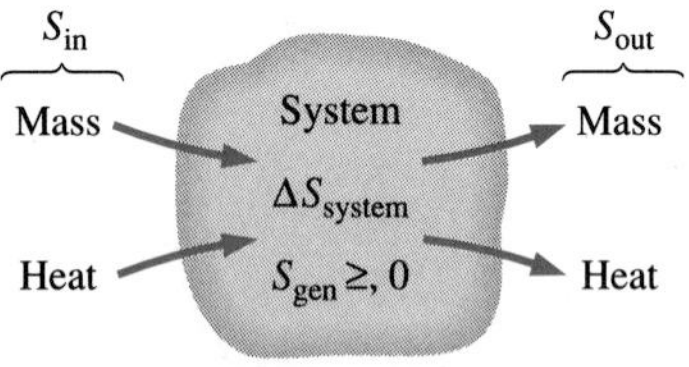

FIGURE 6-71
Mechanisms of entropy transfer for a general system.

The term S_{gen} represents the entropy generation *within the system boundary* only (Fig. 6-71), and not the entropy generation that may occur outside the system boundary during the process as a result of external irreversibilities. Therefore, a process for which $S_{gen} = 0$ is *internally reversible,* but not necessarily *totally* reversible. The *total* entropy generated during a process can be determined by applying the entropy balance to an *extended system* that includes the system itself and its immediate surroundings where external irreversibilities might be occurring (Fig. 6-72). Also, the entropy change in this case is equal to the sum of the entropy change of the system and the entropy change of the immediate surroundings. Note that under steady conditions, the state and thus the entropy of the immediate surroundings (let us call it the "buffer zone") at any point will not change during the process, and the entropy change of the buffer zone will be zero. The entropy change of the buffer zone, if any, is usually small relative to the entropy change of the system, and thus it is usually disregarded.

When evaluating the entropy transfer between an extended system and the surroundings, the boundary temperature of the extended system is simply taken to be the *environment temperature.*

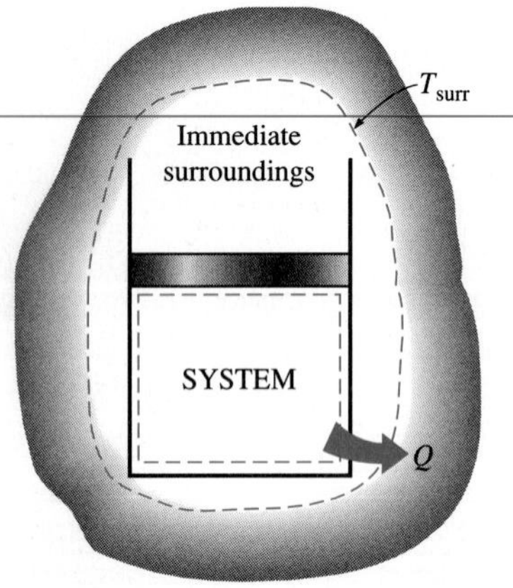

FIGURE 6-72
Entropy generation outside system boundaries can be accounted for by writing an entropy balance on an extended system that includes the system and its immediate surroundings.

Closed Systems

A closed system involves *no mass flow* across its boundaries, and its entropy change is simply the difference between the initial and final entropies of the system. The *entropy change* of a closed system is due to the *entropy transfer* accompanying heat transfer and the *entropy generation* within the system boundaries. Taking the positive direction of heat transfer to be *to* the system, the general entropy balance relation (Eq. 6-86) can be expressed for a closed system as

$$\textit{Closed system:} \quad \sum \frac{Q_k}{T_k} + S_{gen} = \Delta S_{system} = S_2 - S_1 \qquad (\text{kJ/K}) \qquad (6\text{-}89)$$

The entropy balance relation above can be stated as: *the entropy change of a closed system during a process is equal to the sum of the net entropy trans-*

ferred through the system boundary by heat transfer and the entropy generated within the system boundaries.

For an *adiabatic process* ($Q = 0$), the entropy transfer term in the above relation drops out and the entropy change of the closed system becomes equal to the entropy generation within the system boundaries. That is,

Adiabatic closed system: $$S_{\text{gen}} = \Delta S_{\text{adiabatic system}} \qquad (6\text{-}90)$$

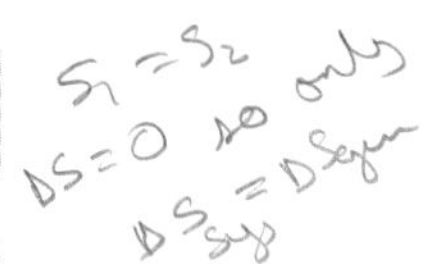

Noting that any closed system and its surroundings can be treated as an adiabatic system and the total entropy change of a system is equal to the sum of the entropy changes of its parts, the entropy balance for a closed system and its surroundings can be written as

System + Surroundings: $$S_{\text{gen}} = \sum \Delta S = \Delta S_{\text{system}} + \Delta S_{\text{surroundings}} \qquad (6\text{-}91)$$

where $\Delta S_{\text{system}} = m(s_2 - s_1)$ and the entropy change of the surroundings can be determined from $\Delta S_{\text{surr}} = Q_{\text{surr}}/T_{\text{surr}}$ if its temperature is constant. At initial stages of studying entropy and entropy transfer, it is more instructive to start with the general form of the entropy balance (Eq. 6-86) and to simplify it for the problem under consideration. The specific relations above are convenient to use after a certain degree of intuitive understanding of the material is achieved.

Control Volumes

The entropy balance relations for control volumes differ from those for closed systems in that they involve one more mechanism of entropy exchange: *mass flow across the boundaries.* As mentioned earlier, mass possesses entropy as well as energy, and the amounts of these two extensive properties are proportional to the amount of mass (Fig. 6-73).

Taking the positive direction of heat transfer to be *to* the system, the general entropy balance relations (Eqs. 6-86 and 6-87) can be expressed for control volumes as

$$\sum \frac{Q_k}{T_k} + \sum m_i s_i - \sum m_e s_e + S_{\text{gen}} = (S_2 - S_1)_{\text{CV}} \qquad (\text{kJ/K}) \qquad (6\text{-}92)$$

or, in the rate form, as

$$\sum \frac{\dot{Q}_k}{T_k} + \sum \dot{m}_i s_i - \sum \dot{m}_e s_e + \dot{S}_{\text{gen}} = \Delta \dot{S}_{\text{CV}} \qquad (\text{kW/K}) \qquad (6\text{-}93)$$

The entropy balance relation above can be stated as: *the rate of entropy change within the control volume during a process is equal to the sum of the rate of entropy transfer through the control volume boundary by heat transfer, the net rate of entropy transfer into the control volume by mass flow, and the rate of entropy generation within the boundaries of the control volume as a result of irreversibilities.*

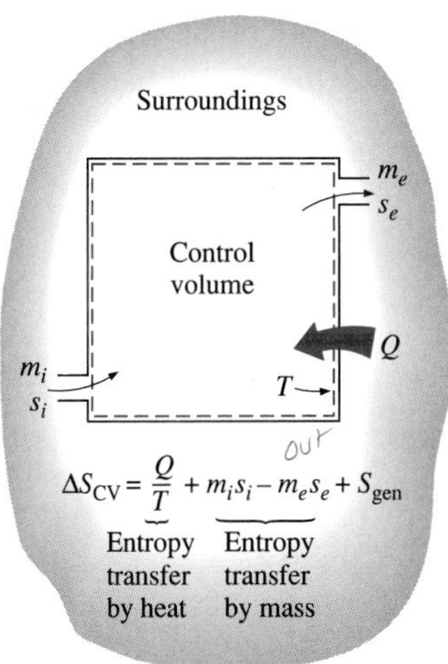

FIGURE 6-73

The entropy of a control volume changes as a result of mass flow as well as heat flow.

Most control volumes encountered in practice such as turbines, compressors, nozzles, diffusers, heat exchangers, pipes, and ducts operate steadily, and thus they experience no change in their entropy. Therefore, the entropy balance relation for a general **steady-flow process** can be obtained from Eq. 6-93 by setting $\Delta \dot{S}_{CV} = 0$ and rearranging to give

$$\textit{Steady-flow:} \qquad \dot{S}_{gen} = \sum \dot{m}_e s_e - \sum \dot{m}_i s_i - \sum \frac{\dot{Q}_k}{T_k} \qquad (6\text{-}94)$$

For *single-stream* (one inlet and one exit) steady-flow devices, the entropy balance relation simplifies to

$$\textit{Steady-flow, single-stream:} \qquad \dot{S}_{gen} = \dot{m}(s_e - s_i) - \sum \frac{\dot{Q}_k}{T_k} \qquad (6\text{-}95)$$

For the case of an *adiabatic* single-stream device, the entropy balance relation further simplifies to

$$\textit{Steady-flow, single-stream, adiabatic:} \qquad \dot{S}_{gen} = \dot{m}(s_e - s_i) \qquad (6\text{-}96)$$

which indicates that the specific entropy of the fluid must increase as it flows through an adiabatic device since $\dot{S}_{gen} \geq 0$ (Fig. 6-74). If the flow through the device is *reversible* and *adiabatic,* then the entropy will remain constant, $s_e = s_i$, regardless of the changes in other properties.

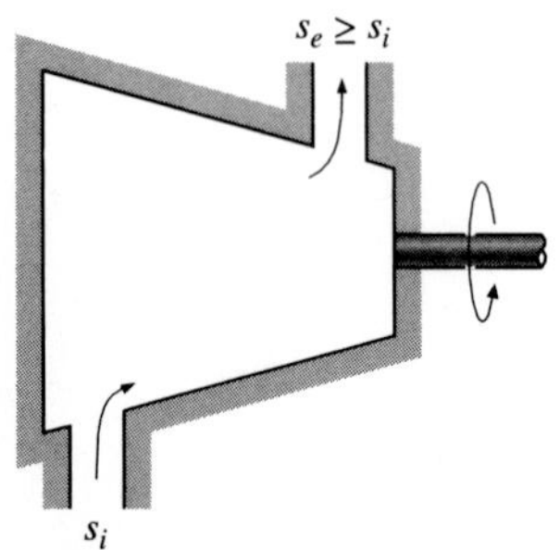

FIGURE 6-74
The entropy of a substance always increases (or remains constant in the case of a reversible process) as it flows through a single-stream, adiabatic, steady-flow device.

EXAMPLE 6-19 Entropy Generation in a Wall

Consider steady heat flow through a 5-m × 6-m brick wall of a house of thickness 30 cm and thermal conductivity 0.69 W/(m · °C). On a day when the temperature of the outdoors is 0°C, the house is maintained at 27°C. The temperatures of the inner and outer surfaces of the brick wall are measured to be 20°C and 5°C, respectively. Determine the rate of heat transfer through the wall, the rate of entropy generation in the wall, and the rate of total entropy generation associated with this heat transfer process.

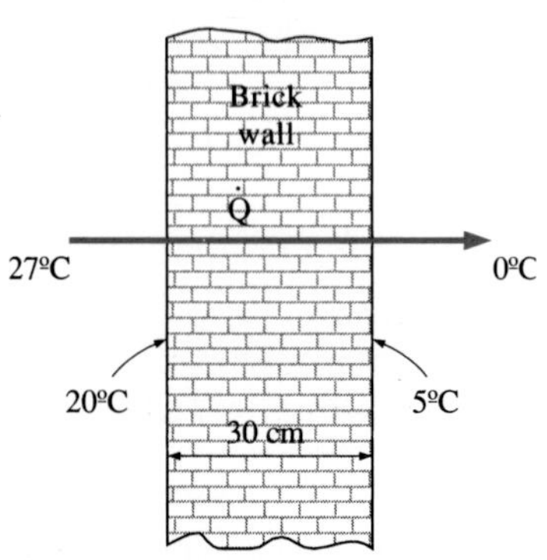

FIGURE 6-75
Schematic for Example 6-19.

Solution We first take the *wall* as the system (Fig. 6-75). This is a *closed system* since no mass crosses the system boundary during the process. We note that the entropy change of the wall is zero during this process since the state and thus the entropy of the wall do not change anywhere in the wall. Heat and entropy are entering from one side of the wall, and leaving from the other side.

Assumptions **1** The process is steady, and thus the rate of heat transfer through the wall is constant. **2** Heat conduction through the wall is one-dimensional. **3** The thermal conductivity is constant.

Analysis Knowing the wall surface temperatures, the rate of heat transfer through the wall is determined from Fourier's law of heat conduction to be

$$\dot{Q} = kA\left(\frac{\Delta T}{L}\right)_{\text{wall}} = [0.69\ \text{W/(m}\cdot{}^\circ\text{C)}][(5 \times 6)\ \text{m}^2]\frac{(20-5)^\circ\text{C}}{0.3\ \text{m}} = \mathbf{1035\ W}$$

The rate form of the entropy balance for the wall simplifies to

$$\underbrace{\dot{S}_{in} - \dot{S}_{out}}_{\substack{\text{Rate of net entropy transfer}\\ \text{by heat and mass}}} + \underbrace{\dot{S}_{gen}}_{\substack{\text{Rate of entropy}\\ \text{generation}}} = \underbrace{\Delta \dot{S}_{system}^{\nearrow 0}}_{\substack{\text{Rate of change}\\ \text{of entropy}}}$$

$$\left(\frac{\dot{Q}}{T}\right)_{in} - \left(\frac{\dot{Q}}{T}\right)_{out} + \dot{S}_{gen} = 0$$

$$\frac{1035 \text{ W}}{293 \text{ K}} - \frac{1035 \text{ W}}{278 \text{ K}} + \dot{S}_{gen} = 0$$

Therefore, the rate of entropy generation in the wall is

$$\dot{S}_{gen,\,wall} = \mathbf{0.191\ W/K}$$

Note that entropy transfer by heat at any location is Q/T at that location, and the direction of entropy transfer is the same as the direction of heat transfer.

To determine the rate of total entropy generation during this heat transfer process, we extend the system to include the regions on both sides of the wall that experience a temperature change. Then one side of the system boundary becomes room temperature while the other side becomes the temperature of the outdoors. The entropy balance for this *extended system* (System + Immediate surroundings) will be the same as that given above, except the two boundary temperatures will be 300 and 273 K instead of 293 and 278 K, respectively. Then the rate of total entropy generation becomes

$$\frac{1035 \text{ W}}{300 \text{ K}} - \frac{1035 \text{ W}}{273 \text{ K}} + \dot{S}_{gen,\,total} = 0 \quad \rightarrow \quad \dot{S}_{gen,\,total} = \mathbf{0.341\ W/K}$$

Discussion Note that the entropy change of this extended system is also zero since the state of air does not change at any point during the process. The differences between the two entropy generations is 0.150 W/K, and it represents the entropy generated in the air layers on both sides of the wall. The entropy generation in this case is entirely due to irreversible heat transfer through a finite temperature difference.

EXAMPLE 6-20 Entropy Generation during a Throttling Process

Steam at 7 MPa and 450°C is throttled in a valve to a pressure of 3 MPa during a steady-flow process. Determine the entropy generated during this process and check if the increase of entropy principle is satisfied.

Solution We take the throttling valve as the *system* (Fig. 6-76). This is a *control volume* since mass crosses the system boundary during the process. We note that there is only one inlet and one exit and thus $\dot{m}_1 = \dot{m}_2 = \dot{m}$. Also, the enthalpy of a fluid remains nearly constant during a throttling process and thus $h_2 \cong h_1$.

Assumptions **1** This is a steady-flow process since there is no change with time at any point and thus $\Delta m_{CV} = 0$, $\Delta E_{CV} = 0$, and $\Delta S_{CV} = 0$. **2** Heat transfer to or from the valve is negligible. **3** The kinetic and potential energy changes are negligible, $\Delta \text{ke} = \Delta \text{pe} = 0$.

Analysis Noting that $h_2 = h_1$, the entropy of the steam at the inlet and the exit states is determined from the steam tables to be

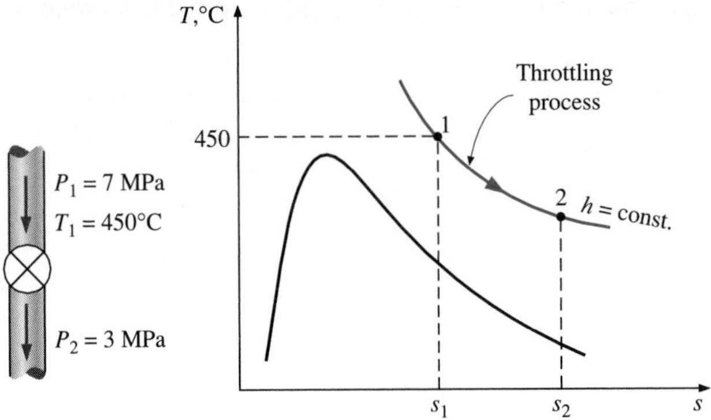

FIGURE 6-76
Schematic for Example 6-20.

State 1: $\left.\begin{array}{l} P_1 = 7 \text{ MPa} \\ T_1 = 450°\text{C} \end{array}\right\}$ $\begin{array}{l} h_1 = 3287.1 \text{ kJ/kg} \\ s_1 = 6.6327 \text{ kJ/(kg} \cdot \text{K)} \end{array}$

State 2: $\left.\begin{array}{l} P_2 = 3 \text{ MPa} \\ h_2 = h_1 \end{array}\right\}$ $s_2 = 6.9919 \text{ kJ/(kg} \cdot \text{K)}$

Then the entropy generation per unit mass of the steam is determined from the entropy balance applied to the throttling valve,

$$\underbrace{\dot{S}_{\text{in}} - \dot{S}_{\text{out}}}_{\substack{\text{Rate of net entropy transfer} \\ \text{by heat and mass}}} + \underbrace{\dot{S}_{\text{gen}}}_{\substack{\text{Rate of entropy} \\ \text{generation}}} = \underbrace{\Delta \dot{S}_{\text{system}}}_{\substack{\text{Rate of change} \\ \text{of entropy}}}^{\nearrow 0 \text{ (steady)}}$$

$$\dot{m}s_1 - \dot{m}s_2 + \dot{S}_{\text{gen}} = 0$$

$$\dot{S}_{\text{gen}} = \dot{m}(s_2 - s_1)$$

Dividing by mass flow rate and substituting gives

$$s_{\text{gen}} = s_2 - s_1 = 6.9919 - 6.6327 = \mathbf{0.3592\ kJ/(kg \cdot K)}$$

This is the amount of entropy generated per unit mass of steam as it is throttled from the inlet state to the final pressure, and it is caused by unrestrained expansion. The increase of entropy principle is obviously satisfied during this process since the entropy generation is positive.

EXAMPLE 6-21 Entropy Generated when a Hot Block Is Dropped in a Lake

A 50-kg block of iron casting at 500 K is thrown into a large lake that is at a temperature of 285 K. The iron block eventually reaches thermal equilibrium with the lake water. Assuming an average specific heat of 0.45 kJ/(kg · K) for the iron, determine (*a*) the entropy change of the iron block, (*b*) the entropy change of the lake water, and (*c*) the entropy generated during this process.

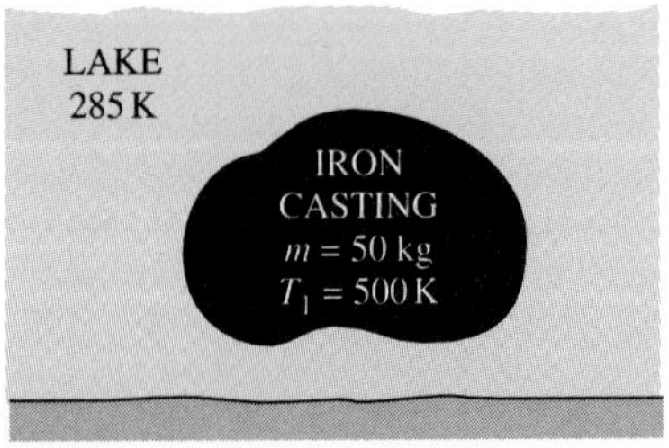

LAKE 285 K

IRON CASTING $m = 50$ kg $T_1 = 500$ K

FIGURE 6-77
Schematic for Example 6-21.

Solution We take the *iron casting* as the system (Fig. 6-77). This is a *closed system* since no mass crosses the system boundary during the process.

To determine the entropy change for the iron block and for the lake, first we need to know the final equilibrium temperature. Given that the thermal energy capacity of the lake is very large relative to that of the iron block, the lake will absorb all the heat rejected by the iron block without experiencing any change in its temperature. Therefore, the iron block will cool to 285 K during this process while the lake temperature remains constant at 285 K.

Assumptions **1** Both the water and the iron block are incompressible substances. **2** Constant specific heats can be used for the water and the iron. **3** The kinetic and potential energy changes of the iron are negligible, $\Delta KE = \Delta PE = 0$ and thus $\Delta E = \Delta U$. **4** There are no work interactions.

Analysis (*a*) Approximating the iron block as an incompressible substance, its entropy change can be determined from

$$\begin{aligned}\Delta S_{\text{iron}} &= m(s_2 - s_1) = mC_{\text{av}} \ln \frac{T_2}{T_1} \\ &= (50 \text{ kg})[0.45 \text{ kJ/(kg} \cdot \text{K)}] \ln \frac{285 \text{ K}}{500 \text{ K}} \\ &= \mathbf{-12.65 \text{ kJ/K}}\end{aligned}$$

(*b*) The temperature of the lake water remains constant during this process at 285 K. Also, the amount of heat transfer from the iron block to the lake is determined from an energy balance on the iron block to be

$$\underbrace{E_{\text{in}} - E_{\text{out}}}_{\substack{\text{Net energy transfer} \\ \text{by heat, work, and mass}}} = \underbrace{\Delta E_{\text{system}}}_{\substack{\text{Change in internal, kinetic,} \\ \text{potential, etc., energies}}}$$

$$-Q_{\text{out}} = \Delta U = mC_{\text{av}}(T_2 - T_1)$$

or $\quad Q_{\text{out}} = mC_{\text{av}}(T_1 - T_2) = (50 \text{ kg})[0.45 \text{ kJ/(kg} \cdot \text{K)}](500 - 285) \text{ K} = 4838 \text{ kJ}$

Then the entropy change of the lake becomes

$$\Delta S_{\text{lake}} = \frac{Q_{\text{lake}}}{T_{\text{lake}}} = \frac{+4838 \text{ kJ}}{285 \text{ K}} = \mathbf{16.97 \text{ kJ/K}}$$

(*c*) The entropy generated during this process can be determined by applying an entropy balance on an *extended system* that includes the iron block and its immediate surroundings so that the boundary temperature of the extended system is at 285 K at all times:

$$\underbrace{S_{\text{in}} - S_{\text{out}}}_{\substack{\text{Net entropy transfer} \\ \text{by heat and mass}}} + \underbrace{S_{\text{gen}}}_{\substack{\text{Entropy} \\ \text{generation}}} = \underbrace{\Delta S_{\text{system}}}_{\substack{\text{Change} \\ \text{in entropy}}}$$

$$-\frac{Q_{\text{out}}}{T_b} + S_{\text{gen}} = \Delta S_{\text{system}}$$

or $$S_{\text{gen}} = \frac{Q_{\text{out}}}{T_b} + \Delta S_{\text{system}} = \frac{4838 \text{ kJ}}{285 \text{ K}} - (12.65 \text{ kJ/K}) = \mathbf{4.32 \text{ kJ/K}}$$

Discussion The entropy generated can also be determined by taking the iron block and the entire lake as the system, which is an isolated system, and applying an entropy balance. An isolated system involves no heat or entropy transfer, and thus the entropy generation in this case becomes equal to the total entropy change,

$$S_{\text{gen}} = \Delta S_{\text{total}} = \Delta S_{\text{system}} + \Delta S_{\text{lake}} = -12.65 + 16.97 = \mathbf{4.32 \text{ kJ/K}}$$

which is the same result obtained above.

EXAMPLE 6-22 Entropy Generation in a Mixing Chamber

Water at 20 psia and 50°F enters a mixing chamber at a rate of 300 lbm/min where it is mixed steadily with steam entering at 20 psia and 240°F. The mixture leaves the chamber at 20 psia and 130°F, and heat is lost to the surrounding air at 70°F at a rate of 180 Btu/min. Neglecting the changes in kinetic and potential energies, determine the rate of entropy generation during this process.

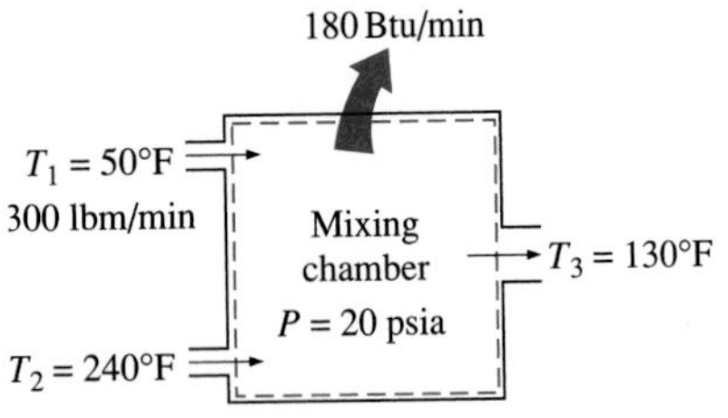

FIGURE 6-78
Schematic for Example 6-22.

Solution We take the *mixing chamber* as the system (Fig. 6-78). This is a *control volume* since mass crosses the system boundary during the process. We note that there are two inlets and one exit.

Assumptions **1** This is a steady-flow process since there is no change with time at any point and thus $\Delta m_{CV} = 0$, ΔE_{CV}, and $\Delta S_{CV} = 0$. **2** There are no work interactions involved. **3** The kinetic and potential energies are negligible, ke ≅ pe ≅ 0.

Analysis Under the stated assumptions and observations, the mass and energy balances for this steady-flow system can be expressed in the rate form as follows:

Mass balance: $$\dot{m}_{in} - \dot{m}_{out} = \Delta \dot{m}_{system}^{\nearrow 0 \text{ (steady)}} = 0 \rightarrow \dot{m}_1 + \dot{m}_2 = \dot{m}_3$$

Energy balance: $$\underbrace{\dot{E}_{in} - \dot{E}_{out}}_{\substack{\text{Rate of net energy transfer} \\ \text{by heat, work, and mass}}} = \underbrace{\Delta \dot{E}_{system}^{\nearrow 0 \text{ (steady)}}}_{\substack{\text{Rate of change in internal, kinetic,} \\ \text{potential, etc., energies}}} = 0$$

$$\dot{E}_{in} = \dot{E}_{out}$$

$$\dot{m}_1 h_1 + \dot{m}_2 h_2 = \dot{m}_3 h_3 + \dot{Q}_{out} \qquad (\text{since } \dot{W} = 0, \text{ ke} \cong \text{pe} \cong 0)$$

Combining the mass and energy balances gives

$$\dot{Q}_{out} = \dot{m}_1 h_1 + \dot{m}_2 h_2 - (\dot{m}_1 + \dot{m}_2) h_3$$

The desired properties at the specified states are determined from the steam tables to be

State 1: $$\left.\begin{array}{l} P_1 = 20 \text{ psia} \\ T_1 = 50°F \end{array}\right\} \begin{array}{l} h_1 \cong h_{f@50°F} = 18.06 \text{ Btu/lbm} \\ s_1 \cong s_{f@50°F} = 0.03607 \text{ Btu/(lbm}\cdot\text{R)} \end{array}$$

State 2: $$\left.\begin{array}{l} P_2 = 20 \text{ psia} \\ T_2 = 240°F \end{array}\right\} \begin{array}{l} h_2 = 1162.3 \text{ Btu/lbm} \\ s_2 = 1.7405 \text{ Btu/(lbm}\cdot\text{R)} \end{array}$$

State 3: $$\left.\begin{array}{l} P_3 = 20 \text{ psia} \\ T_3 = 130°F \end{array}\right\} \begin{array}{l} h_3 \cong h_{f@130°F} = 97.98 \text{ Btu/lbm} \\ s_3 \cong s_{f@130°F} = 0.18172 \text{ Btu/(lbm}\cdot\text{R)} \end{array}$$

Substituting,

$$180 \text{ Btu/min} = [300 \times 18.06 + \dot{m}_2 \times 1162.3 - (300 + \dot{m}_2) \times 97.98] \text{ Btu/min}$$

which gives

$$\dot{m}_2 = 22.7 \text{ kg/min}$$

The rate of entropy generation during this process can be determined by applying the rate form of the entropy balance on an *extended system* that includes the mixing chamber and its immediate surroundings so that the boundary temperature of the extended system is 70°F = 530 R:

$$\underbrace{\dot{S}_{in} - \dot{S}_{out}}_{\substack{\text{Rate of net entropy transfer} \\ \text{by heat and mass}}} + \underbrace{\dot{S}_{gen}}_{\substack{\text{Rate of entropy} \\ \text{generation}}} = \underbrace{\Delta \dot{S}_{system}^{\nearrow 0 \text{ (steady)}}}_{\substack{\text{Rate of change} \\ \text{of entropy}}}$$

$$\dot{m}_1 s_1 + \dot{m}_2 s_2 - \dot{m}_3 s_3 - \frac{\dot{Q}_{out}}{T_b} + \dot{S}_{gen} = 0$$

Substituting, the rate of entropy generation is determined to be

$$\dot{S}_{gen} = \dot{m}_3 s_3 - \dot{m}_1 s_1 - \dot{m}_2 s_2 + \frac{\dot{Q}_{out}}{T_b}$$

$$= (322.7 \times 0.18172 - 300 \times 0.03607 - 22.7 \times 1.7405) \text{ Btu/(min} \cdot \text{R)} + \frac{180 \text{ Btu/min}}{530 \text{ R}}$$

$$= \mathbf{8.65\ Btu/(min \cdot R)}$$

Discussion Note that entropy is generated during this process at a rate of 8.65 Btu/(min · R). This entropy generation is caused by the mixing of two fluid streams (an irreversible process) and the heat transfer between the mixing chamber and the surroundings through a finite temperature difference (another irreversible process).

EXAMPLE 6-23 Entropy Generation Associated with Heat Transfer

A frictionless piston-cylinder device contains a saturated mixture of water at 100°C. During a constant-pressure process, 600 kJ of heat is transferred to the surrounding air at 25°C. As a result, part of the water vapor contained in the cylinder condenses. Determine (*a*) the entropy change of the water and (*b*) the total entropy generation during this heat transfer process.

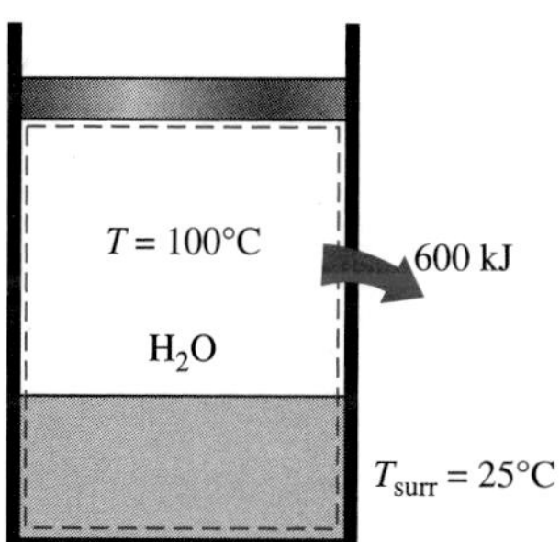

FIGURE 6-79
Schematic for Example 6-23.

Solution We first take the *water in the cylinder* as the system (Fig. 6-79). This is a *closed system* since no mass crosses the system boundary during the process. We note that the pressure and thus the temperature of water in the cylinder remain constant during this process. Also, the entropy of the system decreases during the process because of heat loss.

Assumptions **1** There are no irreversibilities involved within the system boundaries, and thus the process is internally reversible. **2** The water temperature remains constant at 100°C everywhere, including the boundaries.

Analysis (*a*) Noting that water undergoes an internally reversible isothermal process, its entropy change can be determined from

$$\Delta S_{system} = \frac{Q}{T_{system}} = \frac{-600 \text{ kJ}}{(100 + 273 \text{ K})} = \mathbf{-1.61\ kJ/K}$$

(*b*) To determine the total entropy generation during this process, we consider the *extended system,* which includes the water, the piston-cylinder device, and the region immediately outside the system that experiences a temperature change so that the entire boundary of the extended system is at the surrounding temperature of 25°C. The entropy balance for this *extended system* (system + immediate surroundings) yields

$$\underbrace{S_{in} - S_{out}}_{\substack{\text{Net entropy transfer} \\ \text{by heat and mass}}} + \underbrace{S_{gen}}_{\substack{\text{Entropy} \\ \text{generation}}} = \underbrace{\Delta S_{system}}_{\substack{\text{Change} \\ \text{in entropy}}}$$

$$-\frac{Q_{out}}{T_b} + S_{gen} = \Delta S_{system}$$

or $$S_{gen} = \frac{Q_{out}}{T_b} + \Delta S_{system} = \frac{600 \text{ kJ}}{(25 + 273) \text{ K}} + (-1.61 \text{ kJ/K}) = \mathbf{0.40 \text{ kJ/K}}$$

The entropy generation in this case is entirely due to irreversible heat transfer through a finite temperature difference.

Note that the entropy change of this extended system is equivalent to the entropy change of water since the piston-cylinder device and the immediate surroundings do not experience any change of state at any point, and thus any change in any property, including entropy.

Discussion For the sake of argument, consider the reverse process (i.e., the transfer of 600 kJ of heat from the surrounding air at 25°C to saturated water at 100°C) and see if the increase of entropy principle can detect the impossibility of this process. This time, heat transfer will be to the water (heat gain instead of heat loss), and thus the entropy change of water will be +1.61 kJ/K. Also, the entropy transfer at the boundary of the extended system will have the same magnitude but opposite direction. This will result in an entropy generation of −0.4 kJ/K. The negative sign for the entropy generation indicates that the reverse process is *impossible.*

To complete the discussion, let us consider the case where the surrounding air temperature is a differential amount below 100°C (say 99.999 . . . 9°C) instead of being 25°C. This time, heat transfer from the saturated water to the surrounding air will take place through a differential temperature difference rendering this process *reversible.* It can be shown that $S_{gen} = 0$ for this process.

Remember that reversible processes are idealized processes, and they can be approached but never reached in reality.

Entropy Generation Associated with a Heat Transfer Process

In the example above it is determined that 0.4 kJ/K of entropy is generated during the heat transfer process, but it is not clear where exactly the entropy generation takes place, and how. To pinpoint the location of entropy generation, we need to be more precise about the description of the system, its surroundings, and the system boundary.

In that example, we assumed both the system and the surrounding air to be isothermal at 100°C and 25°C, respectively. This assumption is reasonable if both fluids are well mixed. The inner surface of the wall must also be at 100°C while the outer surface is at 25°C since two bodies in physical contact must have the same temperature at the point of contact. Considering that entropy transfer with heat transfer Q through a surface at constant temperature T is Q/T, the entropy transfer from the water into the wall is $Q/T_{sys} =$ 1.61 kJ/K. Likewise, entropy transfer from the outer surface of the wall into

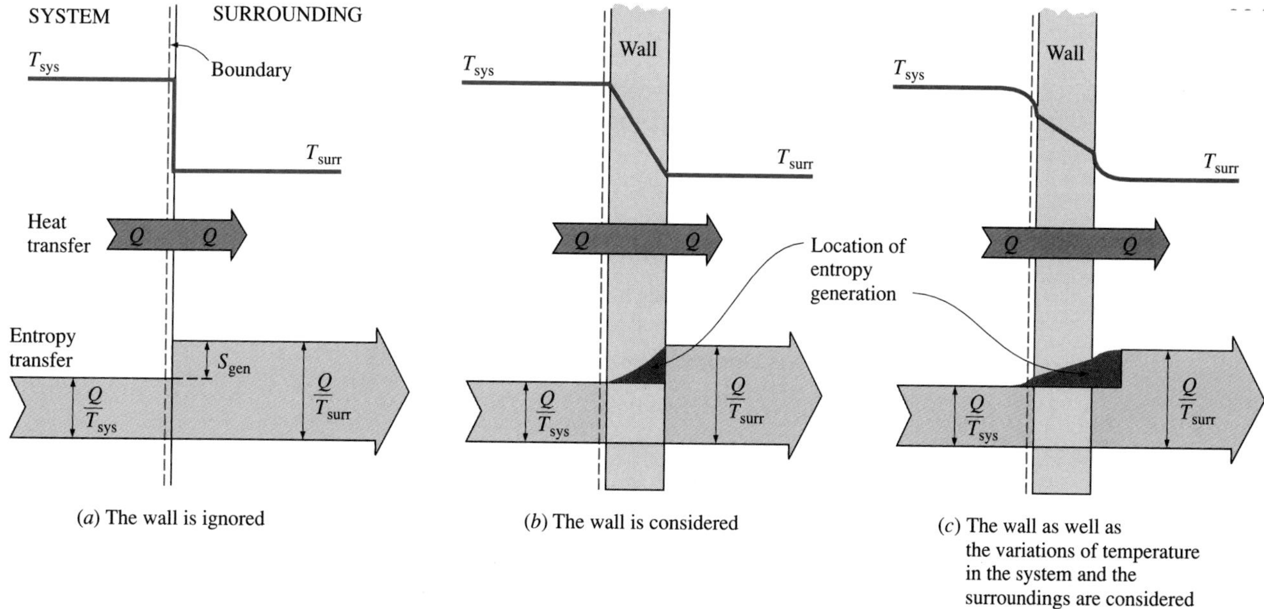

(*a*) The wall is ignored

(*b*) The wall is considered

(*c*) The wall as well as the variations of temperature in the system and the surroundings are considered

FIGURE 6-80

Graphical representation of entropy generation during a heat transfer process through a finite temperature difference.

the surrounding air is $Q/T_{surr} = 2.01$ kJ/K. Obviously, entropy in the amount of $2.01 - 1.61 = 0.4$ kJ/K is generated in the wall, as illustrated in Fig. 6-80*b*.

Identifying the location of entropy generation enables us to determine whether a process is internally reversible. A process is internally reversible if no entropy is generated within the system boundaries. Therefore, the heat transfer process discussed in the example above is internally reversible if the inner surface of the wall is taken as the system boundary, and thus the system excludes the container wall. If the system boundary is taken to be the outer surface of the container wall, then the process is no longer internally reversible since the wall, which is the site of entropy generation, is now part of the system.

For thin walls, it is very tempting to ignore the mass of the wall and to regard the wall as the boundary between the system and the surroundings. This seemingly harmless choice hides the site of the entropy generation from view and is a source of confusion. The temperature in this case drops suddenly from T_{sys} to T_{surr} at the boundary surface, and confusion arises as to which temperature to use in the relation Q/T for entropy transfer at the boundary.

Note that if the system and the surrounding air are not isothermal as a result of insufficient mixing, then part of the entropy generation will occur in both the system and the surrounding air in the vicinity of the wall, as shown in Fig. 6-80*c*.

6-15 ■ SUMMARY

The second law of thermodynamics leads to the definition of a new property called *entropy,* which is a quantitative measure of microscopic disorder for a system. The definition of entropy is based on the *Clausius inequality,* given by

$$\oint \frac{\delta Q}{T} \leq 0 \qquad \text{(kJ/K)}$$

where the equality holds for internally or totally reversible processes and the inequality for irreversible processes. Any quantity whose cyclic integral is zero is a property, and entropy is defined as

$$dS = \left(\frac{\delta Q}{T}\right)_{\text{int rev}} \qquad \text{(kJ/K)}$$

For the special case of an internally reversible, isothermal process, it gives

$$\Delta S = \frac{Q}{T_0} \qquad \text{(kJ/K)}$$

The inequality part of the Clausius inequality combined with the definition of entropy yields an inequality known as the *increase of entropy principle,* expressed as

$$S_{\text{gen}} \geq 0 \qquad \text{(kJ/K)}$$

where S_{gen} is the *entropy generated* during the process. Entropy change is caused by heat transfer, mass flow, and irreversibilities. Heat transfer to a system increases the entropy, and heat transfer from a system decreases it. The effect of irreversibilities is always to increase the entropy.

Entropy is a property, and it can be expressed in terms of more familiar properties through the $T\,ds$ relations, expressed as

$$T\,ds = du + P\,dv$$

and

$$T\,ds = dh - v\,dP$$

These two relations have many uses in thermodynamics and serve as the starting point in developing entropy-change relations for processes. The successful use of $T\,ds$ relations depends on the availability of property relations. Such relations do not exist for a general pure substance but are available for incompressible substances (solids, liquids) and ideal gases.

The *entropy-change* and *isentropic relations* for a process can be summarized as follows:

1. *Pure substances:*

Any process: $\Delta s = s_2 - s_1$ [kJ/(kg · K)]

Isentropic process: $s_2 = s_1$

2. *Incompressible substances:*

Any process: $s_2 - s_1 = C_{av} \ln \frac{T_2}{T_1}$ [kJ/(kg · K)]

Isentropic process: $T_2 = T_1$

3. *Ideal gases:*

a. Constant specific heats (approximate treatment):

Any process:

$$s_2 - s_1 = C_{v,av} \ln \frac{T_2}{T_1} + R \ln \frac{v_2}{v_1} \qquad \text{[kJ/(kg · K)]}$$

and $$s_2 - s_1 = C_{p,av} \ln \frac{T_2}{T_1} - R \ln \frac{P_2}{P_1} \qquad \text{[kJ/(kg · K)]}$$

Or, on a unit-mole basis,

$$\bar{s}_2 - \bar{s}_1 = \overline{C}_{v,av} \ln \frac{T_2}{T_1} + R_u \ln \frac{v_2}{v_1} \qquad \text{[kJ/(kmol · K)]}$$

and $$\bar{s}_2 - \bar{s}_1 = \overline{C}_{p,av} \ln \frac{T_2}{T_1} - R_u \ln \frac{P_2}{P_1} \qquad \text{[kJ/(kmol · K)]}$$

Isentropic process:

$$\left(\frac{T_2}{T_1}\right)_{s=\text{const.}} = \left(\frac{v_1}{v_2}\right)^{k-1}$$

$$\left(\frac{T_2}{T_1}\right)_{s=\text{const.}} = \left(\frac{P_2}{P_1}\right)^{(k-1)/k}$$

$$\left(\frac{P_2}{P_1}\right)_{s=\text{const.}} = \left(\frac{v_1}{v_2}\right)^{k}$$

b. Variable specific heats (exact treatment):

Any process:

$$s_2 - s_1 = s_2^\circ - s_1^\circ - R \ln \frac{P_2}{P_1} \qquad \text{[kJ/(kg · K)]}$$

or $$\bar{s}_2 - \bar{s}_1 = \bar{s}_2^\circ - \bar{s}_1^\circ - R_u \ln \frac{P_2}{P_1} \qquad \text{[kJ/(kmol · K)]}$$

Isentropic process:

$$s_2^\circ = s_1^\circ + R \ln \frac{P_2}{P_1} \qquad \text{[kJ/(kg · K)]}$$

$$\left(\frac{P_2}{P_1}\right)_{s=\text{const.}} = \frac{P_{r2}}{P_{r1}}$$

$$\left(\frac{v_2}{v_1}\right)_{s=\text{const.}} = \frac{v_{r2}}{v_{r1}}$$

where P_r is the *relative pressure* and v_r is the *relative specific volume*. The function $s°$ depends on temperature only.

The *steady-flow work* for a reversible process can be expressed in terms of the fluid properties as

$$w_{\text{rev}} = -\int_1^2 v\,dP - \Delta\text{ke} - \Delta\text{pe} \qquad \text{(kJ/kg)}$$

For incompressible substances (v = constant) it simplifies to

$$w_{\text{rev}} = -v(P_2 - P_1) - \Delta\text{ke} - \Delta\text{pe} \qquad \text{(kJ/kg)}$$

The work done during a steady-flow process is proportional to the specific volume. Therefore, v should be kept as small as possible during a compression process to minimize the work input and as large as possible during an expansion process to maximize the work output.

The reversible work inputs to a compressor compressing an ideal gas from T_1, P_1 to P_2 in an isentropic (Pv^k = constant), polytropic (Pv^n = constant), or isothermal (Pv = constant) manner, are determined by integration for each case with the following results:

Isentropic: $$w_{\text{comp, in}} = \frac{kR(T_2 - T_1)}{k-1} = \frac{kRT_1}{k-1}\left[\left(\frac{P_2}{P_1}\right)^{(k-1)/k} - 1\right] \qquad \text{(kJ/kg)}$$

Polytropic: $$w_{\text{comp, in}} = \frac{nR(T_2 - T_1)}{n-1} = \frac{nRT_1}{n-1}\left[\left(\frac{P_2}{P_1}\right)^{(n-1)/n} - 1\right] \qquad \text{(kJ/kg)}$$

Isothermal: $$w_{\text{comp, in}} = RT\ln\frac{P_2}{P_1} \qquad \text{(kJ/kg)}$$

The work input to a compressor can be reduced by using multistage compression with intercooling. For maximum savings from the work input, the pressure ratio across each stage of the compressor must be the same.

Most steady-flow devices operate under adiabatic conditions, and the ideal process for these devices is the isentropic process. The parameter that describes how efficiently a device approximates a corresponding isentropic device is called *isentropic* or *adiabatic efficiency*. It is expressed for turbines, compressors, and nozzles as follows:

$$\eta_T = \frac{\text{Actual turbine work}}{\text{Isentropic turbine work}} = \frac{w_a}{w_s} \cong \frac{h_1 - h_{2a}}{h_1 - h_{2s}}$$

$$\eta_C = \frac{\text{Isentropic compressor work}}{\text{Actual compressor work}} = \frac{w_s}{w_a} \cong \frac{h_{2s} - h_1}{h_{2a} - h_1}$$

$$\eta_N = \frac{\text{Actual KE at nozzle exit}}{\text{Isentropic KE at nozzle exit}} = \frac{\mathcal{V}_{2a}^2}{\mathcal{V}_{2s}^2} \cong \frac{h_1 - h_{2a}}{h_1 - h_{2s}}$$

In the relations above, h_{2a} and h_{2s} are the enthalpy values at the exit state for actual and isentropic processes, respectively.

The entropy balance for any system undergoing any process can be expressed in the general form as

$$\underbrace{S_{\text{in}} - S_{\text{out}}}_{\substack{\text{Net entropy transfer}\\ \text{by heat and mass}}} + \underbrace{S_{\text{gen}}}_{\substack{\text{Entropy}\\ \text{generation}}} = \underbrace{\Delta S_{\text{system}}}_{\substack{\text{Change in}\\ \text{entropy}}} \qquad \text{(kJ/K)}$$

or, in the *rate form*, as

$$\underbrace{\dot{S}_{\text{in}} - \dot{S}_{\text{out}}}_{\substack{\text{Rate of net entropy transfer}\\ \text{by heat and mass}}} + \underbrace{\dot{S}_{\text{gen}}}_{\substack{\text{Rate of entropy}\\ \text{generation}}} = \underbrace{\Delta \dot{S}_{\text{system}}}_{\substack{\text{Rate of change}\\ \text{of entropy}}} \qquad \text{(kJ/K)}$$

For a general *steady-flow process* it simplifies to

$$\dot{S}_{\text{gen}} = \sum \dot{m}_e s_e - \sum \dot{m}_i s_i - \sum \frac{\dot{Q}_k}{T_k}$$

REFERENCES AND SUGGESTED READING

1. A. Bejan. *Advanced Engineering Thermodynamics.* New York: John Wiley & Sons, 1988.

2. A. Bejan. *Entropy Generation through Heat and Fluid Flow.* New York: John Wiley & Sons–Interscience, 1982.

3. W. Z. Black and J. G. Hartley. *Thermodynamics.* New York: Harper & Row, 1985.

4. Y. Cerci, Y. A. Çengel, and R. H. Turner. "Reducing the Cost of Compressed Air in Industrial Facilities." *International Mechanical Engineering Congress and Exposition,* San Francisco, California, November 12–17, 1995.

5. W. F. E. Feller. *Air Compressors: Their Installation, Operation, and Maintenance,* New York: McGraw-Hill, 1944.

6. M. S. Moran and H. N. Shapiro. *Fundamentals of Engineering Thermodynamics,* New York: John Wiley & Sons, 1988.

7. D. W. Nutter, A. J. Britton, and W. M. Heffington. "Conserve Energy to Cut Operating Costs." *Chemical Engineering,* September 1993, pp. 126–37.

8. J. Rifkin. *Entropy.* New York: The Viking Press, 1980.

9. E. M. Talbott. *Compressed Air Systems: A Guidebook on Energy and Cost Savings.* 2nd ed. Liburn, GA: Fairmont Press, 1993.

10. G. J. Van Wylen and R. E. Sonntag. *Fundamentals of Classical Thermodynamics.* 3rd ed. New York: John Wiley & Sons, 1985.

PROBLEMS*

Entropy and the Increase of Entropy Principle

6-1C Does the temperature in the Clausius inequality relation have to be absolute temperature? Why?

6-2C Does a cycle for which $\oint \delta Q > 0$ violate the Clausius inequality? Why?

6-3C Is a quantity whose cyclic integral is zero necessarily a property?

6-4C Does the cyclic integral of heat have to be zero (i.e., does a system have to reject as much heat as it receives to complete a cycle)? Explain.

6-5C Does the cyclic integral of work have to be zero (i.e., does a system have to produce as much work as it consumes to complete a cycle)? Explain.

6-6C A system undergoes a process between two fixed states first in a reversible manner and then in an irreversible manner. For which case is the entropy change greater? Why?

6-7C Is the value of the integral $\int_1^2 \delta Q/T$ the same for all processes between states 1 and 2? Explain.

6-8C Is the value of the integral $\int_1^2 \delta Q/T$ the same for all reversible processes between states 1 and 2? Why?

6-9C To determine the entropy change for an irreversible process between states 1 and 2, should the integral $\int_1^2 \delta Q/T$ be performed along the actual process path or an imaginary reversible path? Explain.

6-10C Is an isothermal process necessarily internally reversible? Explain your answer with an example.

6-11C How do the values of the integral $\int_1^2 \delta Q/T$ compare for a reversible and irreversible process between the same end states?

6-12C The entropy of a hot baked potato decreases as it cools. Is this a violation of the increase of entropy principle? Explain.

6-13C Is it possible to create entropy? Is it possible to destroy it?

6-14C A piston-cylinder device contains helium gas. During a reversible, isothermal process, the entropy of the helium will (*never, sometimes, always*) increase.

---

*Students are encouraged to answer *all* the concept "C" questions.

6-15C A piston-cylinder device contains nitrogen gas. During a reversible, adiabatic process, the entropy of the nitrogen will (*never, sometimes, always*) increase.

6-16C A piston-cylinder device contains superheated steam. During an actual adiabatic process, the entropy of the steam will (*never, sometimes, always*) increase.

6-17C The entropy of steam will (*increase, decrease, remain the same*) as it flows through an actual adiabatic turbine.

6-18C The entropy of the working fluid of the ideal Carnot cycle (*increases, decreases, remains the same*) during the isothermal heat addition process.

6-19C The entropy of the working fluid of the ideal Carnot cycle (*increases, decreases, remains the same*) during the isothermal heat rejection process.

6-20C During a heat transfer process, the entropy of a system (*always, sometimes, never*) increases.

6-21C Is it possible for the entropy change of a closed system to be zero during an irreversible process? Explain.

6-22C What three different mechanisms can cause the entropy of a control volume to change?

6-23C Steam is accelerated as it flows through an actual adiabatic nozzle. The entropy of the steam at the nozzle exit will be (*greater than, equal to, less than*) the entropy at the nozzle inlet.

6-24C Consider a person who organizes his room, and thus decreases the entropy of the room. Does this process violate the second law of thermodynamics?

6-25C Consider a fruit tree that makes highly organized fruits out of the water and highly disorganized soil, and thus decreases the entropy of its locality. Does this process violate the second law of thermodynamics?

6-26C Consider an army unit whose soldiers are walking around at random in a field. Suddenly an order is issued and the soldiers align in a highly organized manner, decreasing the entropy. Does this process violate the second law of thermodynamics? Explain.

6-27 A rigid tank contains an ideal gas at 40°C that is being stirred by a paddle wheel. The paddle wheel does 200 kJ of work on the ideal gas. It is observed that the temperature of the ideal gas remains constant during this process as a result of heat transfer between the system and the surroundings at 25°C. Determine the entropy change of the ideal gas.

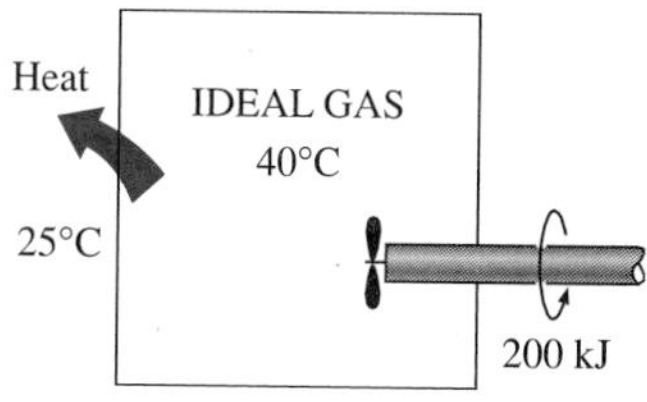

FIGURE P6-27

6-28 Air is compressed by a 8-kW compressor from P_1 to P_2. The air temperature is maintained constant at 25°C during this process as a result of heat transfer to the surrounding medium at 10°C. Determine the rate of entropy change of the air. State the assumptions made in solving this problem.
Answers: −0.0268 kW/K

6-29 During the isothermal heat addition process of a Carnot cycle, 900 kJ of heat is added to the working fluid from a source at 400°C. Determine (*a*) the entropy change of the working fluid, (*b*) the entropy change of the source, and (*c*) the total entropy generation for the process.

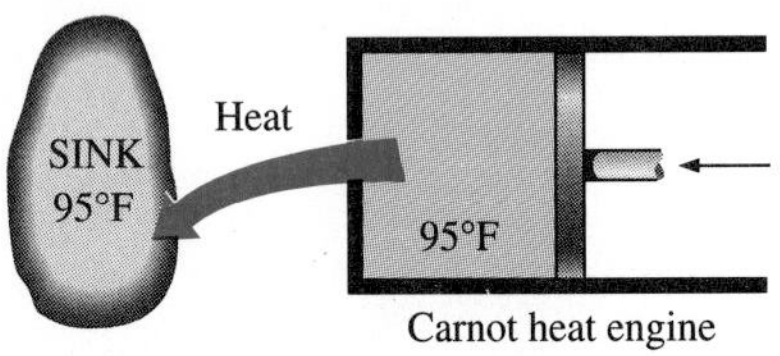

FIGURE P6-30E

6-30E During the isothermal heat rejection process of a Carnot cycle, the working fluid experiences an entropy change of −0.7 Btu/R. If the temperature of the energy sink is 95°F, determine (*a*) the amount of heat transfer, (*b*) the entropy change of the sink, and (*c*) the total entropy change for this process. *Answers:* (*a*) 388.5 Btu, (*b*) 0.7 Btu/R, (*c*) 0

6-31 Refrigerant-134a enters the coils of the evaporator of a refrigeration system as a saturated liquid–vapor mixture at a pressure of 200 kPa. The refrigerant absorbs 120 kJ of heat from the cooled space, which is maintained at −5°C, and leaves as saturated vapor at the same pressure. Determine (*a*) the entropy change of the refrigerant, (*b*) the entropy change of the cooled space, and (*c*) the total entropy generation for this process.

Entropy Changes of Pure Substances

6-32C Is a process that is internally reversible and adiabatic necessarily isentropic? Explain.

6-33 The radiator of a steam heating system has a volume of 20 L and is filled with superheated water vapor at 200 kPa and 200°C. At this moment both the inlet and the exit valves to the radiator are closed. After a while the temperature of the steam drops to 80°C as a result of heat transfer to the room air. Determine the entropy change of the steam during this process, in kJ/K. *Answer:* −0.0806 kJ/K

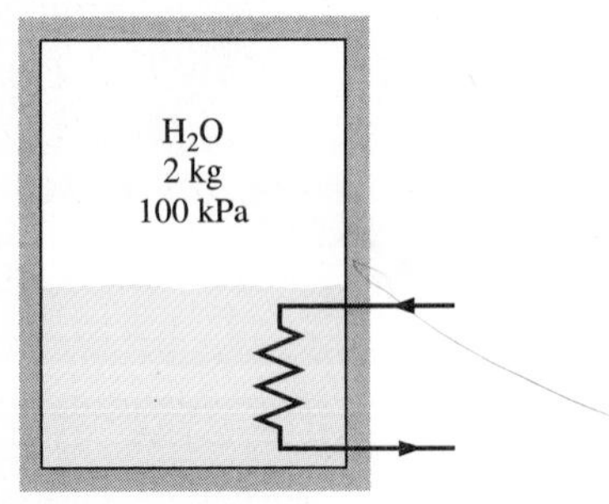

FIGURE P6-35

6-34 A 0.5-m^3 rigid tank contains refrigerant-134a initially at 200 kPa and 40 percent quality. Heat is transferred now to the refrigerant from a source at 35°C until the pressure rises to 400 kPa. Determine (*a*) the entropy change of the refrigerant, (*b*) the entropy change of the heat source, and (*c*) the total entropy change for this process. *Answers:* (*a*) 3.883 kJ/K, (*b*) −3.441 kJ/K, (*c*) 0.422 kJ/K

6-35 A well-insulated rigid tank contains 2 kg of a saturated liquid–vapor mixture of water at 100 kPa. Initially, three-quarters of the mass is in the liquid phase. An electric resistance heater placed in the tank is now turned on and kept on until all the liquid in the tank is vaporized. Determine the entropy change of the steam during this process. *Answer:* 8.0962 kJ/K

FIGURE P6-36

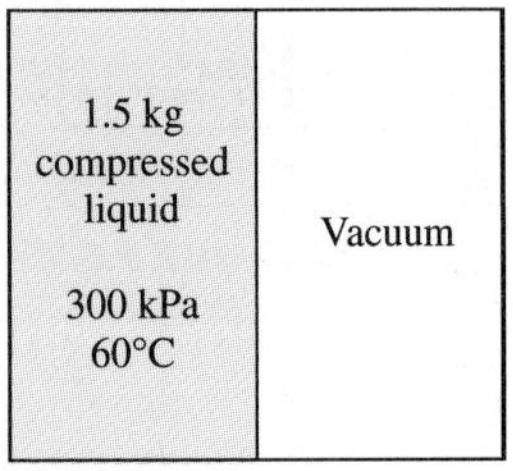

1.5 kg compressed liquid
300 kPa
60°C
Vacuum

6-36 A rigid tank is divided into two equal parts by a partition. One part of the tank contains 1.5 kg of compressed liquid water at 300 kPa and 60°C while the other part is evacuated. The partition is now removed, and the water expands to fill the entire tank. Determine the entropy change of water during this process, if the final pressure in the tank is 15 kPa. *Answer:* −0.1134 kJ/K

6-37E A piston-cylinder device contains 3 lbm of refrigerant-134a at 120 psia and 120°F. The refrigerant is now cooled at constant pressure until it exists as a liquid at 90°F. Determine the entropy change of the refrigerant during this process.

6-38 An insulated piston-cylinder device contains 5 L of saturated liquid water at a constant pressure of 150 kPa. An electric resistance heater inside the cylinder is now turned on, and electrical work is done on the steam in the amount of 2200 kJ. Determine the entropy change of the water during this process, in kJ/K. *Answer:* 5.72 kJ/K

6-39 An insulated piston-cylinder device contains 0.01 m^3 of saturated refrigerant-134a vapor at 0.8-MPa pressure. The refrigerant is now allowed to expand in a reversible manner until the pressure drops to 0.4 MPa. Determine (*a*) the final temperature in the cylinder and (*b*) the work done by the refrigerant.

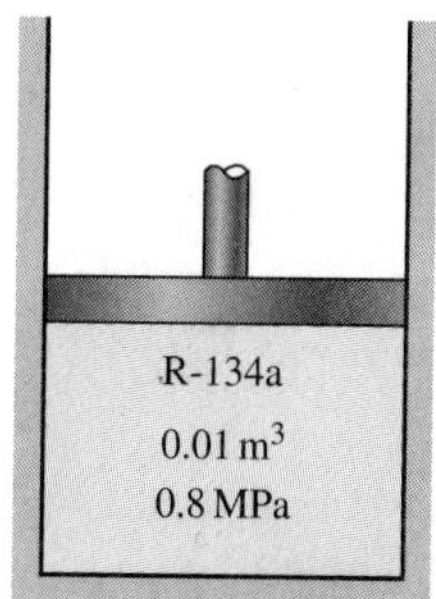

FIGURE P6-40

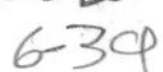

6-40 Refrigerant-134a enters an adiabatic compressor as saturated vapor at 140 kPa a rate of 2 m^3/min and is compressed to a pressure of 700 kPa. Determine the minimum power that must be supplied to the compressor.

6-41E Steam enters an adiabatic turbine at 800 psia and 900°F and leaves at a pressure of 40 psia. Determine the maximum amount of work that can be delivered by this turbine.

6-42 A heavily insulated piston-cylinder device contains 0.05 m^3 of steam at 300 kPa and 150°C. Steam is now compressed in a reversible manner to a pressure of 1 MPa. Determine the work done on the steam during this process.

6-43 A piston-cylinder device contains 0.5 kg of saturated water vapor at 200°C. Heat is now transferred to steam, and steam expands reversibly and isothermally to a final pressure of 800 kPa. Determine the heat transferred and the work done during this process.

Entropy Change of Incompressible Substances

6-44C Consider two solid blocks, one hot and the other cold, brought into contact in an adiabatic container. After a while, thermal equilibrium is established in the container as a result of heat transfer. The first law requires that the amount of energy lost by the hot solid be equal to the energy gained by the cold one. Does the second law require that the decrease in entropy of the hot solid be equal to the increase in entropy of the cold one?

6-45 A 50-kg copper block initially at 80°C is dropped into an insulated tank that contains 120 L of water at 25°C. Determine the final equilibrium temperature and the total entropy change for this process.

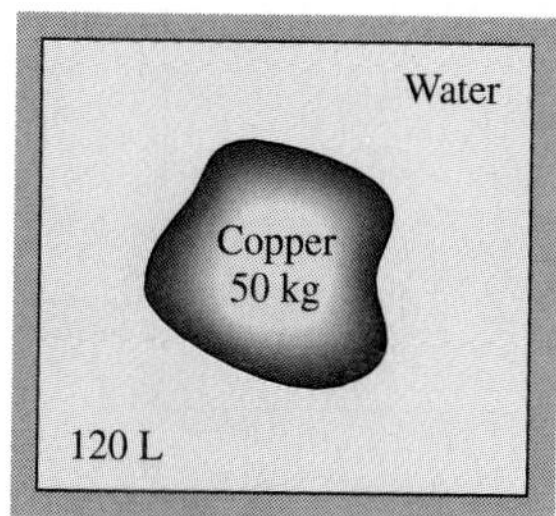

FIGURE P6-45

6-46 A 5-kg iron block initially at 350°C is quenched in an insulated tank that contains 100 kg of water at 22°C. Assuming the water that vaporizes during the process condenses back in the tank, determine the amount of entropy generated during this process.

6-47 A 20-kg aluminum block initially at 200°C is brought into contact with a 20-kg block of iron at 100°C in an insulated enclosure. Determine the final equilibrium temperature and the total entropy change for this process. *Answers:* 168.4°C, 0.169 kJ/K

6-48 A 50-kg iron block and a 20-kg copper block, both initially at 80°C, are dropped into a large lake at 15°C. Thermal equilibrium is established after a while as a result of heat transfer between the blocks and the lake water. Determine the total entropy generation for this process.

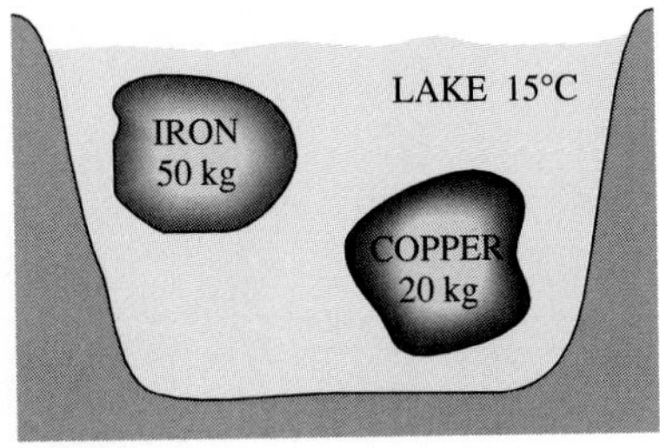

FIGURE P6-48

Entropy Change of Ideal Gases

6-49C Prove that the two relations for entropy change of ideal gases under the constant-specific-heat assumption (Eqs. 6-33 and 6-34) are equivalent.

6-50C Starting with the second $T\,ds$ relation (Eq. 6-26), obtain Eq. 6-34 for the entropy change of ideal gases under the constant-specific-heat assumption.

6-51C What does the function s° in the ideal-gas tables represent?

6-52C Some properties of ideal gases such as internal energy and enthalpy vary with temperature only [that is, $u = u(T)$ and $h = h(T)$]. Is this also the case for entropy?

6-53C Starting with Eq. 6-34, obtain Eq. 6-43.

6-54C What are P_r and v_r called? Is their use limited to isentropic processes? Explain.

6-55C Can the entropy of an ideal gas change during an isentropic process?

6-56C An ideal gas undergoes a process between two specified temperatures, first at constant pressure and then at constant volume. For which case will the ideal gas experience a larger entropy change? Explain.

6-57 Oxygen gas is compressed in a piston-cylinder device from an initial state of 0.8 m^3/kg and 25°C to a final state of 0.1 m^3/kg and 287°C. Determine the entropy change of the oxygen during this process, assuming (*a*) constant specific heats and (*b*) variable specific heats.

6-58 A 0.5-m^3 insulated rigid tank contains 0.9 kg of carbon dioxide at 100 kPa. Now paddle-wheel work is done on the system until the pressure in the tank rises to 120 kPa. Determine the entropy change of carbon dioxide during this process in kJ/K. Assume constant specific heats. *Answer:* 0.108 kJ/K

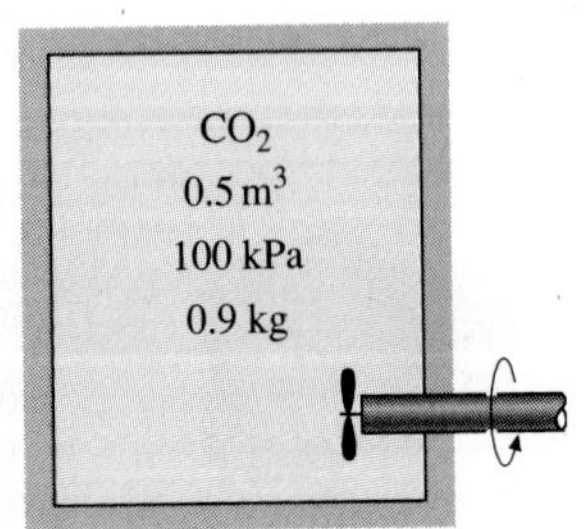

FIGURE P6-58

6-59 An insulated piston-cylinder device initially contains 300 L of air at 120 kPa and 17°C. Air is now heated for 15 min by a 200-W resistance heater placed inside the cylinder. The pressure of air is maintained constant during this process. Determine the entropy change of air, assuming (*a*) constant specific heats and (*b*) variable specific heats.

6-60 A piston-cylinder device contains 1.2 kg of nitrogen gas at 120 kPa and 27°C. The gas is now compressed slowly in a polytropic process during which $PV^{1.3}$ = constant. The process ends when the volume is reduced by one-half. Determine the entropy change of nitrogen during this process.
Answer: −0.0615 kJ/K

6-61E A mass of 8 lbm of helium undergoes a process from an initial state of 50 ft³/lbm and 80°F to a final state of 10 ft³/lbm and 200°F. Determine the entropy change of helium during this process, assuming (*a*) the process is reversible and (*b*) the process is irreversible.

6-62 Air is compressed in a piston-cylinder device from 90 kPa and 20°C to 400 kPa in a reversible isothermal process. Determine (*a*) the entropy change of air and (*b*) the work done.

6-63 Air is compressed steadily by a 5-kW compressor from 100 kPa and 17°C to 600 kPa and 167°C at a rate of 1.6 kg/min. During this process, some heat transfer takes place between the compressor and the surrounding medium at 17°C. Determine the rate of entropy change of air during this process. *Answers:* −0.0025 kW/K

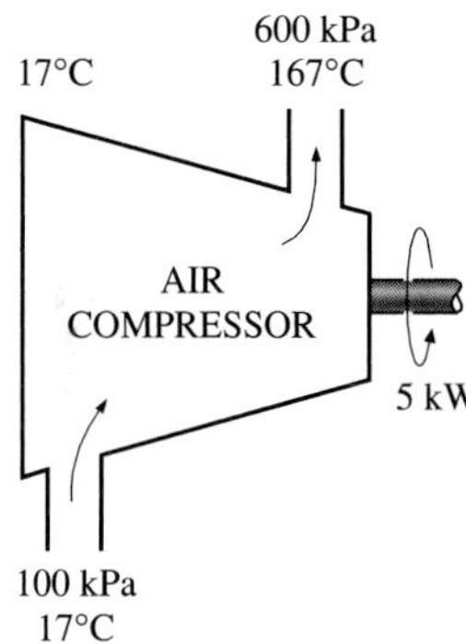

FIGURE P6-63

6-64 An insulated rigid tank is divided into two equal parts by a partition. Initially, one part contains 5 kmol of an ideal gas at 400 kPa and 50°C, and the other side is evacuated. The partition is now removed, and the gas fills the entire tank. Determine the total entropy change during this process.
Answer: 28.81 kJ/K

6-65 Air is compressed in a piston-cylinder device from 100 kPa and 17°C to 800 kPa in a reversible, adiabatic process. Determine the final temperature and the work done during this process, assuming (*a*) constant specific heats and (*b*) variable specific heats for air.
Answers: (*a*) 525.3 K, 171.1 kJ/kg; (*b*) 522.4 K, 169.3 kJ/kg

6-66 Helium gas is compressed from 100 kPa and 30°C to 500 kPa in a reversible, adiabatic process. Determine the final temperature and the work done, assuming the process takes place (*a*) in a piston-cylinder device and (*b*) in a steady-flow compressor.

6-67 An insulated, rigid tank contains 4 kg of argon gas at 450 kPa and 30°C. A valve is now opened, and argon is allowed to escape until the pressure inside drops to 150 kPa. Assuming the argon remaining inside the tank has undergone a reversible, adiabatic process, determine the final mass in the tank. *Answer:* 2.07 kg

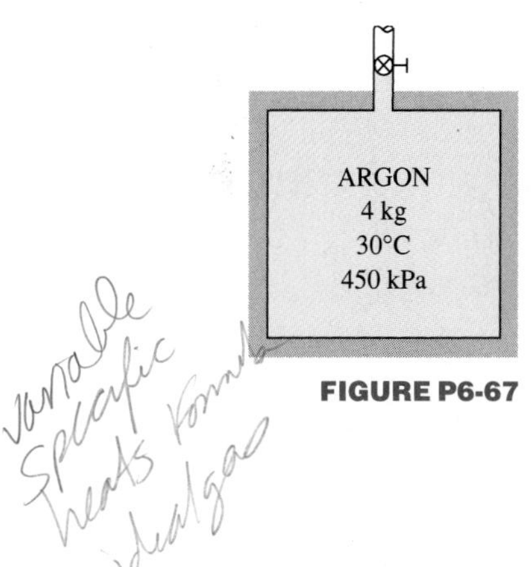

FIGURE P6-67

6-68E Air enters an adiabatic nozzle at 60 psia, 540°F, and 200 ft/s and exits at 12 psia. Assuming air to be an ideal gas with variable specific heats and disregarding any irreversibilities, determine the exit velocity of the air.

6-69 Air enters a nozzle steadily at 280 kPa and 77°C with a velocity of 50 m/s and exits at 85 kPa and 320 m/s. The heat losses from the nozzle to the surrounding medium at 20°C are estimated to be 3.2 kJ/kg. Determine (*a*) the exit temperature and (*b*) the total entropy change for this process.

Reversible Steady-Flow Work

6-70C In large compressors, the gas is frequently cooled while being compressed to reduce the power consumed by the compressor. Explain how cooling the gas during a compression process reduces the power consumption.

6-71C The turbines in steam power plants operate essentially under adiabatic conditions. A plant engineer suggests to end this practice. She proposes to run cooling water through the outer surface of the casing to cool the steam as it flows through the turbine. This way, she reasons, the entropy of the steam will decrease, the performance of the turbine will improve, and as a result the work output of the turbine will increase. How would you evaluate this proposal?

6-72C It is well known that the power consumed by a compressor can be reduced by cooling the gas during compression. Inspired by this, somebody proposes to cool the liquid as it flows through a pump, in order to reduce the power consumption of the pump. Would you support this proposal? Explain.

6-73 Water enters the pump of a steam power plant as saturated liquid at 20 kPa at a rate of 20 kg/s and exits at 6 MPa. Neglecting the changes in kinetic and potential energies and assuming the process to be reversible, determine the power input to the pump.

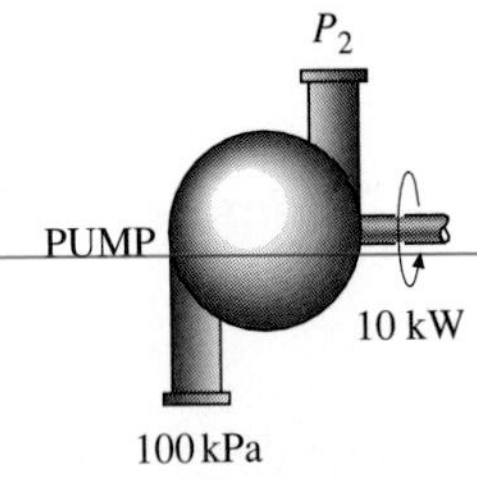

FIGURE P6-74

6-74 Liquid water enters a 10-kW pump at 100-kPa pressure at a rate of 5 kg/s. Determine the highest pressure the liquid water can have at the exit of the pump. Neglect the kinetic and potential energy changes of water, and take the specific volume of water to be 0.001 m^3/kg. *Answer:* 2100 kPa

6-75E Saturated refrigerant-134a vapor at 20 psia is compressed reversibly in an adiabatic compressor to 120 psia. Determine the work input to the compressor. What would your answer be if the refrigerant were first condensed at constant pressure before it was compressed?

6-76 Consider a steam power plant that operates between the pressure limits of 10 MPa and 20 kPa. Steam enters the pump as saturated liquid and leaves the turbine as saturated vapor. Determine the ratio of the work delivered by the turbine to the work consumed by the pump. Assume the entire cycle to be reversible and the heat losses from the pump and the turbine to be negligible.

6-77 Liquid water at 120 kPa enters a 15-kW pump where its pressure is raised to 3 MPa. If the elevation difference between the exit and the inlet levels is 10 m, determine the highest mass flow rate of liquid water this pump can handle. Neglect the kinetic energy change of water, and take the specific volume of water to be 0.001 m^3/kg.

6-78E Helium gas is compressed from 14 psia and 70°F to 120 psia at a rate of 5 ft^3/s. Determine the power input to the compressor, assuming the compression process to be (*a*) isentropic, (*b*) polytropic with $n = 1.2$, and (*c*) isothermal, and (*d*) ideal two-stage polytropic with $n = 1.2$.

6-79 Nitrogen gas is compressed from 80 kPa and 27°C to 480 kPa by a 10-kW compressor. Determine the mass flow rate of nitrogen through the compressor, assuming the compression process to be (*a*) isentropic, (*b*) polytropic with $n = 1.3$, (*c*) isothermal, and (*d*) ideal two-stage polytropic with $n = 1.3$.

Answers: (*a*) 0.048 kg/s, (*b*) 0.05 kg/s, (*c*) 0.063 kg/s, (*d*) 0.058 kg/s

6-80 The compression stages in the axial compressor of the industrial gas turbine are close coupled, making intercooling very impractical. To cool the air in such compressors and to reduce the compression power, it is proposed to spray water mist with drop size on the order of 5 microns into the air stream as it is compressed and to cool the air continuously as the water evaporates. Although the collision of water droplets with turbine blades is a concern, experience with steam turbines indicates that they can cope with water droplet concentrations of up to 14 percent. Assuming air is compressed isentropically at a rate of 2 kg/s from 300 K and 100 kPa to 1200 kPa and the water is injected at a temperature of 20°C at a rate of 0.2 kg/s, determine the reduction in the exit temperature of the compressed air and the compressor power saved. Assume the water vaporizes completely before leaving the compressor, and assume an average mass flow rate of 2.1 kg/s throughout the compressor.

6-81 Reconsider Prob. 6-80. The water-injected compressor is used in a gas turbine power plant. It is claimed that the power output of a gas turbine will increase because of the increase in the mass flow rate of the gas (air + water vapor) through the turbine. Do you agree?

Reducing the Cost of Compressed Air

6-82 Compressed air is one of the key utilities in manufacturing facilities, and the total installed power of compressed-air systems in the United States is estimated to be about 20 million horsepower. Assuming the compressors to operate at full load during one-third of the time on average and the average motor efficiency to be 88 percent, determine how much energy and money will be saved per year if the energy consumed by compressors is reduced by 5 percent as a result of implementing some conservation measures. Take the unit cost of electricity to be \$0.07/kWh.

6-83 The energy used to compress air in the United States is estimated to exceed one-half quadrillion (0.5×10^{15}) kJ per year. It is also estimated that 10 to 40 percent of the compressed air is lost through leaks. Assuming, on average, 20 percent of the compressed air is lost through air leaks and the unit cost of electricity is \$0.07/kWh, determine the amount and cost of electricity wasted per year due to air leaks.

6-84 The compressed-air requirements of a plant at sea level are being met by a 125-hp compressor that takes in air at the local atmospheric pressure of 101.3 kPa and the average temperature of 15°C and compresses it to 900 kPa. An investigation of the compressed-air system and the equipment using the compressed air reveals that compressing the air to 750 kPa is sufficient for this plant. The compressor operates 3500 hours a year at 75 percent of the rated load and is driven by an electric motor that has an efficiency of 88 percent. Taking the price of electricity to be \$0.085/kWh, determine the amount of energy and money saved as a result of reducing the pressure of the compressed air.

6-85 A 150-hp compressor in an industrial facility is housed inside the production area where the average temperature during operating hours is 25°C. The average temperature outdoors during the same hours is 10°C. The compressor operates 6000 hours a year at 85 percent of rated load and is driven by an electric motor that has an efficiency of 90 percent. Taking the price of electricity to be \$0.08/kWh, determine the amount of energy and money saved as a result of drawing outside air to the compressor instead of using the inside air.

6-86 The compressed-air requirements of a plant are being met by a 100-hp screw compressor that runs at full load during 40 percent of the time and idles the rest of the time during operating hours. The compressor consumes 35 percent of the rated power when idling and 90 percent of the power when compressing air. The annual operating hours of the facility are 3800 hours, and the unit cost of electricity is \$0.075/kWh.

It is determined that the compressed-air requirements of the facility during 60 percent of the time can be met by a 25-hp reciprocating compressor that consumes 95 percent of the rated power when compressing air and no power when not compressing air. It is estimated that the 25-hp compressor runs 85 percent of the time. The efficiencies of the motors of the large and the small compressors at or near full load are 0.90 and 0.88, respectively. The efficiency of the large motor at 35 percent load is 0.82. Determine the amount of energy and money saved as a result of switching to the 25-hp compressor during 60 percent of the time.

6-87 The compressed-air requirements of a plant are being met by a 125-hp screw compressor. The facility stops production for one hour every day, including weekends, for lunch break, but the compressor is kept operating. The compressor consumes 35 percent of the rated power when idling, and the unit cost of electricity is \$0.09/kWh. Determine the amount of energy and money saved per year as a result of turning the compressor off during lunch break. Take the efficiency of the motor at part load to be 84 percent.

6-88 The compressed-air requirements of a plant are met by a 150-hp compressor equipped with an intercooler, an aftercooler, and a refrigerated dryer. The plant operates 4800 hours a year, but the compressor is estimated to be compressing air during only one-third of the operating hours, that is, 1600 hours a year. The compressor is either idling or is shut off the rest of the time. Temperature measurements and calculations indicate that 40 percent of the energy input to the compressor is removed from the compressed air as heat in the aftercooler. The COP of the refrigeration unit is 3.5, and the cost of electricity is \$0.08/kWh. Determine the amount of the energy and money saved per year as a result of cooling the compressed air before it enters the refrigerated dryer.

6-89 The 1800-rpm, 150-hp motor of a compressor is burned out and is to be replaced by either a standard motor that has a full-load efficiency of 93.0 percent and costs $9031 or a high-efficiency motor that has an efficiency of 96.2 percent and costs $10,942. The compressor operates 4368 hours a year at full load, and its operation at part load is negligible. If the cost of electricity is $0.085/kWh, determine the amount of energy and money this facility will save by purchasing the high-efficiency motor instead of the standard motor. Also, determine if the savings from the high-efficiency motor justify the price differential if the expected life of the motor is 12 years. Ignore any possible rebates from the local power company.

6-90 The space heating of a facility is accomplished by natural gas heaters that are 80 percent efficient. The compressed air needs of the facility are met by a large liquid-cooled compressor. The coolant of the compressor is cooled by air in a liquid-to-air heat exchanger whose airflow section is 1.0-m high and 1.0-m wide. During typical operation, the air is heated from 20°C to 52°C as it flows through the heat exchanger. The average velocity of air on the inlet side is measured to be 3 m/s. The compressor operates 20 hours a day and 5 days a week throughout the year. Taking the heating season to be 6 months (26 weeks) and the cost of the natural gas to be $0.50/therm (1 therm = 100,000 Btu = 105,500 kJ), determine how much money will be saved by diverting the compressor waste heat into the facility during the heating season.

6-91 The compressors of a production facility maintain the compressed-air lines at a (gage) pressure of 850 kPa at 1400-m elevation, where the atmospheric pressure is 85.6 kPa. The average temperature of air is 15°C at the compressor inlet and 25°C in the compressed-air lines. The facility operates 5200 hours a year, and the average price of electricity is $0.072/kWh. Taking the compressor efficiency to be 0.8, the motor efficiency to be 0.93, and the discharge coefficient to be 0.65, determine the energy and money saved per year by sealing a leak equivalent to a 5-mm-diameter hole on the compressed-air line.

Isentropic Efficiencies of Steady-Flow Devices

6-92C Describe the ideal process for an (*a*) adiabatic turbine, (*b*) adiabatic compressor, and (*c*) adiabatic nozzle, and define the isentropic efficiency for each device.

6-93C Is the isentropic process a suitable model for compressors that are cooled intentionally? Explain.

6-94C On a *T-s* diagram, does the actual exit state (state 2) of an adiabatic turbine have to be on the right-hand side of the isentropic exit state (state 2*s*)? Why?

6-95 Steam enters an adiabatic turbine at 8 MPa and 500°C with a mass flow rate of 3 kg/s and leaves at 30 kPa. The isentropic efficiency of the turbine is 0.90. Neglecting the kinetic energy change of the steam, determine

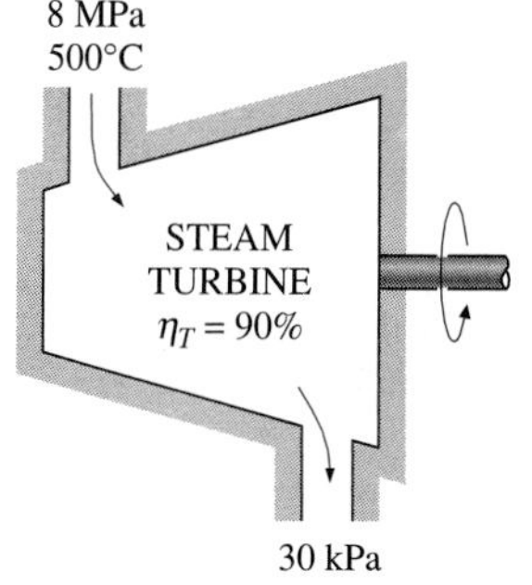

FIGURE P6-95

(*a*) the temperature at the turbine exit and (*b*) the power output of the turbine.
Answers: (*a*) 69.1°C, (*b*) 3052 kW

6-96 Steam enters an adiabatic turbine at 6 MPa, 600°C, and 80m/s and leaves at 50 kPa, 100°C, and 140 m/s. If the power output of the turbine is 5 MW, determine (*a*) the mass flow rate of the steam flowing through the turbine and (*b*) the isentropic efficiency of the turbine.
Answers: (*a*) 5.16 kg/s, (*b*) 83.7 percent

6-97 Argon gas enters an adiabatic turbine at 800°C and 1.5 MPa at a rate of 80 kg/min and exhausts at 200 kPa. If the power output of the turbine is 370 kW, determine the isentropic efficiency of the turbine.

6-98E Combustion gases enter an adiabatic gas turbine at 1540°F and 120 psia and leave at 60 psia with a low velocity. Treating the combustion gases as air and assuming an isentropic efficiency of 86 percent, determine the work output of the turbine. *Answer:* 75.2 Btu/lbm

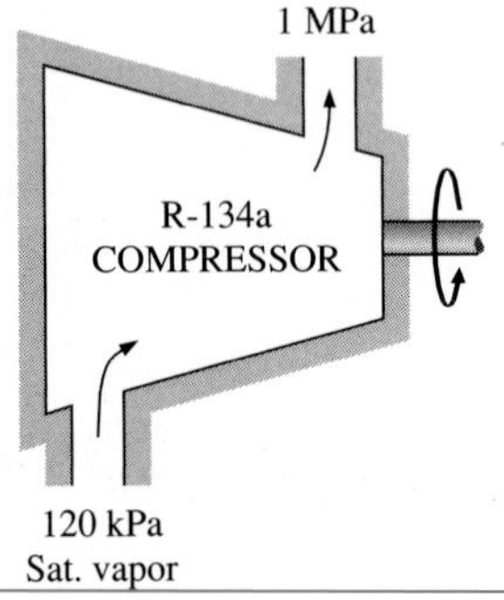

FIGURE P6-99

6-99 Refrigerant-134a enters an adiabatic compressor as saturated vapor at 120 kPa at a rate of 0.3 m^3/min and exits at 1-MPa pressure. If the isentropic efficiency of the compressor is 80 percent, determine (*a*) the temperature of the refrigerant at the exit of the compressor and (*b*) the power input, in kW. Also, show the process on a *T-s* diagram with respect to saturation lines.
Answers: (*a*) 57.7°C, (*b*) 1.70 kW

6-100 Air enters an adiabatic compressor at 100 kPa and 17°C at a rate of 1.2 m^3/s, and it exits at 257°C. The compressor has an isentropic efficiency of 84 percent. Neglecting the changes in kinetic and potential energies, determine (*a*) the exit pressure of air and (*b*) the power required to drive the compressor.

6-101 Air is compressed by an adiabatic compressor from 95 kPa and 27°C to 600 kPa and 277°C. Assuming variable specific heats and neglecting the changes in kinetic and potential energies, determine (*a*) the isentropic efficiency of the compressor and (*b*) the exit temperature of air if the process were reversible. *Answers:* (*a*) 81.9 percent, (*b*) 505.5 K

6-102E Argon gas enters an adiabatic compressor at 20 psia and 90°F with a velocity of 60 ft/s, and it exits at 200 psia and 240 ft/s. If the isentropic efficiency of the compressor is 80 percent, determine (*a*) the exit temperature of the argon and (*b*) the work input to the compressor.

6-103 Carbon dioxide enters an adiabatic compressor at 100 kPa and 300 K at a rate of 0.5 kg/s and exits at 600 kPa and 450 K. Neglecting the kinetic energy changes, determine (*a*) the isentropic efficiency of the compressor and (*b*) the rate of entropy generation during this process.

FIGURE P6-105

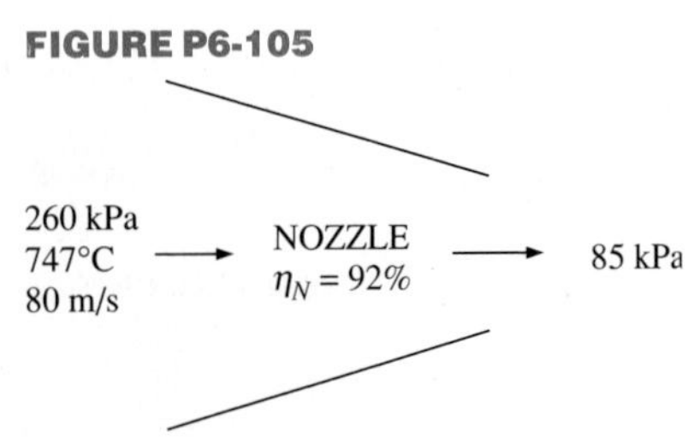

6-104E Air enters an adiabatic nozzle at 60 psia and 1020°F with low velocity and exits at 800 ft/s. If the isentropic efficiency of the nozzle is 90 percent, determine the exit temperature and pressure of the air.

6-105 Hot combustion gases enter the nozzle of a turbojet engine at 260 kPa, 747°C, and 80 m/s, and they exit at a pressure of 85 kPa. Assuming an isentropic efficiency of 92 percent and treating the combustion gases as air, determine (*a*) the exit velocity and (*b*) the exit temperature.

Answers: (*a*) 828.2 m/s, (*b*) 786.3 K

Entropy Balance

6-106 Consider a family of four, with each person taking a five-minute shower every morning. The average flow rate through the shower head is 12 L/min. City water at 15°C is heated to 55°C in an electric water heater and tempered to 42°C by cold water at the T-elbow of the shower before being routed to the shower head. Determine the amount of entropy generated by this family per year as a result of taking daily showers.

6-107 Steam is to be condensed in the condenser of a steam power plant at a temperature of 50°C (h_{fg} = 2305 kJ/kg) with cooling water [C_p = 4.20 kJ/(kg · °C)] from a nearby lake, which enters the tubes of the condenser at 18°C at a rate of 101 kg/s and leaves at 27°C. Assuming the condenser to be perfectly insulated, determine (*a*) the rate of condensation of the steam and (*b*) the rate of entropy generation in the condenser.

Answers: (*a*) 1.65 kg/s, (*b*) 1.15 kW/K

6-108 A well-insulated heat exchanger is to heat water [C_p = 4.18 kJ/(kg · °C)] from 25°C to 60°C at a rate of 0.2 kg/s. The heating is to be accomplished by geothermal water [C_p = 4.31 kJ/(kg · °C)] available at 140°C at a mass flow rate of 0.3 kg/s. The inner tube is thin-walled and has a diameter of 0.8 cm. Determine (*a*) the rate of heat transfer and (*b*) the rate of entropy generation in the heat exchanger.

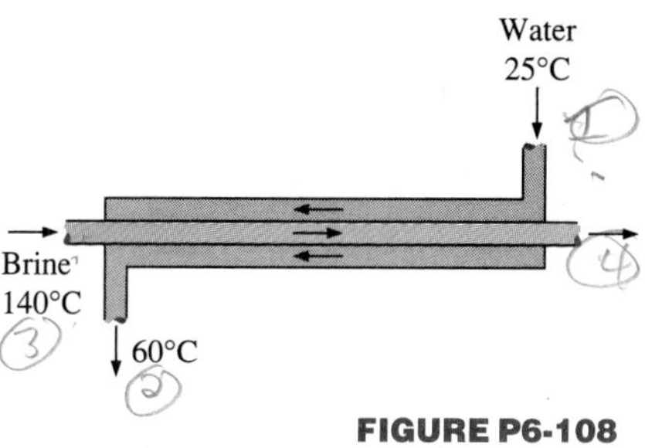

FIGURE P6-108

6-109 An adiabatic heat exchanger is to cool ethylene glycol [C_p = 2.56 kJ/(kg · °C)] flowing at a rate of 2 kg/s from 80°C to 140°C by water [C_p = 4.18 kJ/(kg · °C)] that enters at 20°C and leaves at 55°C. Determine (*a*) the rate of heat transfer and (*b*) the rate of entropy generation in the heat exchanger.

6-110 A well-insulated, thin-walled, double-pipe, counter-flow heat exchanger is to be used to cool oil [C_p = 2.20 kJ/(kg · °C)] from 150°C to 40°C at a rate of 2 kg/s by water [C_p = 4.18 kJ/(kg · °C)] that enters at 22°C at a rate of 1.5 kg/s. The diameter of the tube is 2.5 cm, and its length is 6 m. Determine (*a*) the rate of heat transfer and (*b*) the rate of entropy generation in the heat exchanger.

6-111 Cold water [C_p = 4.18 kJ/(kg · °C)] leading to a shower enters a well-insulated, thin-walled, double-pipe, counter-flow heat exchanger at 15°C at a rate of 0.25 kg/s and is heated to 45°C by hot water [C_p = 4.19 kJ/(kg · °C)] that enters at 100°C at a rate of 3 kg/s. Determine (*a*) the rate of heat transfer and (*b*) the rate of entropy generation in the heat exchanger.

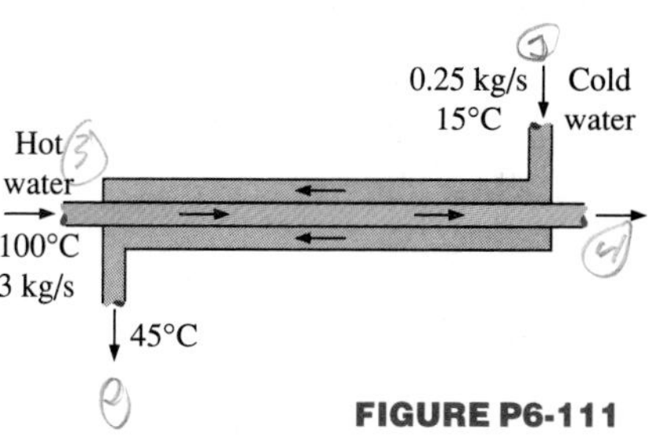

FIGURE P6-111

6-112 Air [C_p = 1.005 kJ/(kg · °C)] is to be preheated by hot exhaust gases in a cross-flow heat exchanger before it enters the furnace. Air enters the heat exchanger at 95 kPa and 20°C at a rate of 0.8 m^3/S. The combustion gases [C_p = 1.10 kJ/(kg · °C)] enter at 180°C at a rate of 1.1 kg/s and leave at

95°C. Determine the rate of heat transfer to the air and the outlet temperature of the air.

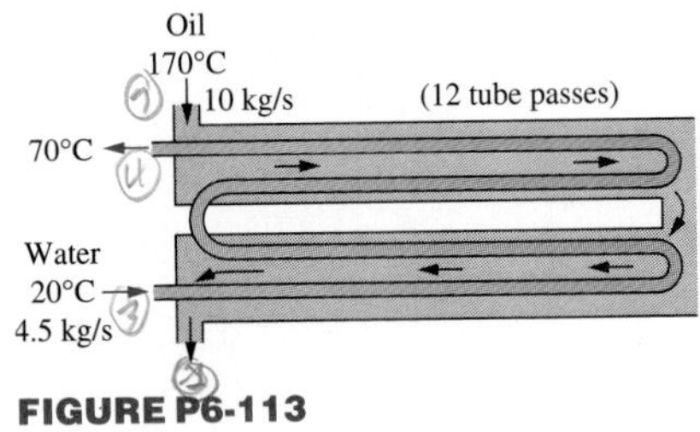

FIGURE P6-113

6-113 A well-insulated, shell-and-tube heat exchanger is used to heat water [C_p = 4.18 kJ/(kg · °C)] in the tubes from 20°C to 70°C at a rate of 4.5 kg/s. Heat is supplied by hot oil [C_p = 2.30 kJ/(kg · °C)] that enters the shell side at 170°C at a rate of 10 kg/s. Disregarding any heat loss from the heat exchanger, determine (*a*) the exit temperature of the oil and (*b*) the rate of entropy generation in the heat exchanger.

6-114E Steam is to be condensed on the shell side of a heat exchanger at 90°F (h_{fg} = 1043 Btu/lbm). Cooling water [C_p = 1.0 Btu/(lbm · °F)] enters the tubes at 60°F at a rate of 115.3 lbm/s and leaves at 73°F. Assuming the heat exchanger to be well-insulated, determine (*a*) the rate of heat transfer in the heat exchanger and (*b*) the rate of entropy generation in the heat exchanger.

6-115 Chickens with an average mass of 2.2 kg and average specific heat of 3.54 kJ/(kg · °C) are to be cooled by chilled water that enters a continuous-flow-type immersion chiller at 0.5°C and leaves at 2.5°C. Chickens are dropped into the chiller at a uniform temperature of 15°C at a rate of 500 chickens per hour and are cooled to an average temperature of 3°C before they are taken out. The chiller gains heat from the surroundings at 25°C at a rate of 200 kJ/h. Determine (*a*) the rate of heat removal from the chickens, in kW, and (*b*) the rate of entropy generation during this chilling process.

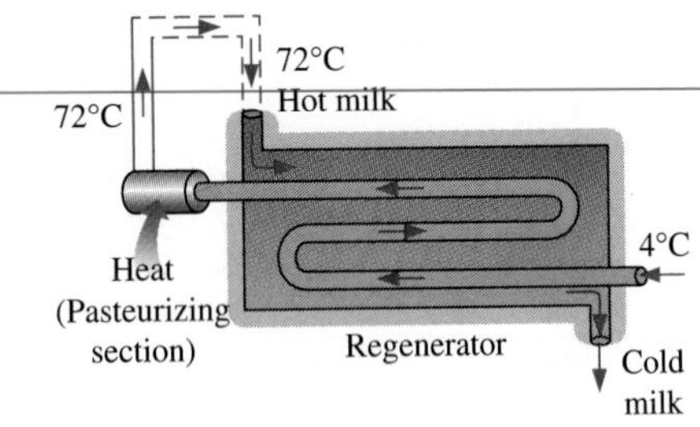

FIGURE P6-116

6-116 In a dairy plant, milk at 4°C is pasteurized continuously at 72°C at a rate of 12 L/s for 24 hours a day and 365 days a year. The milk is heated to the pasteurizing temperature by hot water heated in a natural-gas-fired boiler that has an efficiency of 82 percent. The pasteurized milk is then cooled by cold water at 18°C before it is finally refrigerated back to 4°C. To save energy and money, the plant installs a regenerator that has an effectiveness of 82 percent. If the cost of natural gas is $0.52/therm (1 therm = 105,500 kJ), determine how much energy and money the regenerator will save this company per year and the annual reduction in entropy generation.

6-117 Stainless-steel ball bearings [ρ = 8085 kg/m^3 and C_p = 0.480 kJ/(kg · °C)] having a diameter of 1.2 cm are to be quenched in water at a rate of 1400 per minute. The balls leave the oven at a uniform temperature of 900°C and are exposed to air at 30°C for a while before they are dropped into the water. If the temperature of the balls drops to 850°C prior to quenching, determine (*a*) the rate of heat transfer from the balls to the air and (*b*) the rate of entropy generation due to heat loss from the balls to the air.

FIGURE P6-118

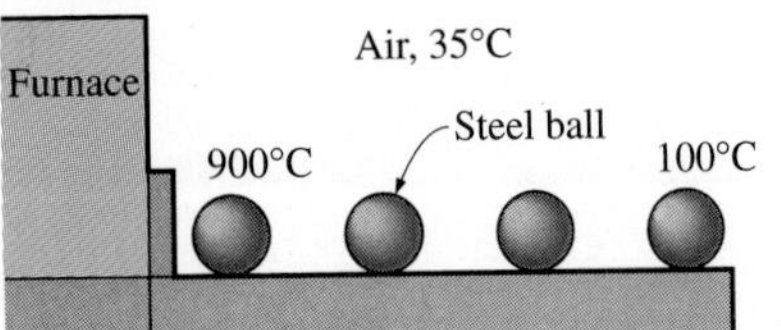

6-118 Carbon-steel balls [ρ = 7833 kg/m^3 and C_p = 0.465 kJ/(kg · °C)] 8 mm in diameter are annealed by heating them first to 900°C in a furnace and then allowing them to cool slowly to 100°C in ambient air at 35°C. If 2500 balls are to be annealed per hour, determine (*a*) the rate of heat transfer from the balls to the air and (*b*) the rate of entropy generation due to heat loss from the balls to the air. *Answers:* (*a*) 542 W, (*b*) 1.082 W/K

6-119 An ordinary egg can be approximated as a 5.5-cm-diameter sphere. The egg is initially at a uniform temperature of 8°C and is dropped into boiling water at 97°C. Taking the properties of the egg to be ρ = 1020 kg/m^3 and C_p = 3.32 kJ/(kg · °C), determine how much heat is transferred to the egg by the time the average temperature of the egg rises to 70°C and the amount of entropy generation associated with this heat transfer process.

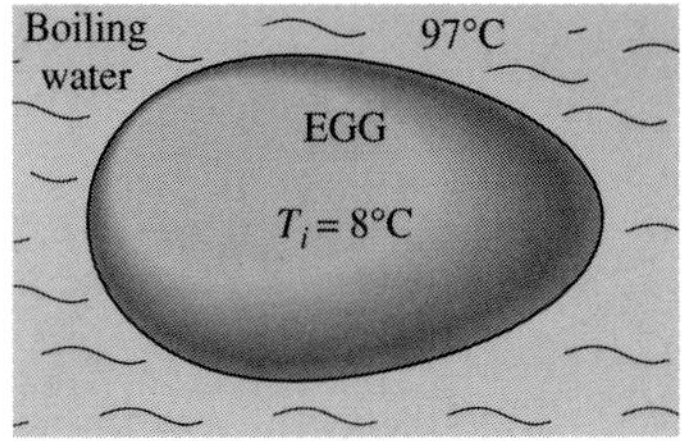

FIGURE P6-119

6-120E In a production facility, 1.2-in.-thick, 2-ft × 2-ft square brass plates [ρ = 532.5 lbm/ft^3 and C_p = 0.091 Btu/(lbm · °F)] that are initially at a uniform temperature of 75°F are heated by passing them through an oven at 1300°F at a rate of 300 per minute. If the plates remain in the oven until their average temperature rises to 1000°F, determine the rate of heat transfer to the plates in the furnace and the rate of entropy generation associated with this heat transfer process.

6-121 Long cylindrical steel rods [ρ = 7833 kg/m^3 and C_p = 0.465 kJ/(kg · °C)] of 10-cm diameter are heat treated by drawing them at a velocity of 3 m/min through a 6-m-long oven maintained at 900°C. If the rods enter the oven at 30°C and leave at 700°C, determine (*a*) the rate of heat transfer to the rods in the oven and (*b*) the rate of entropy generation associated with this heat transfer process.

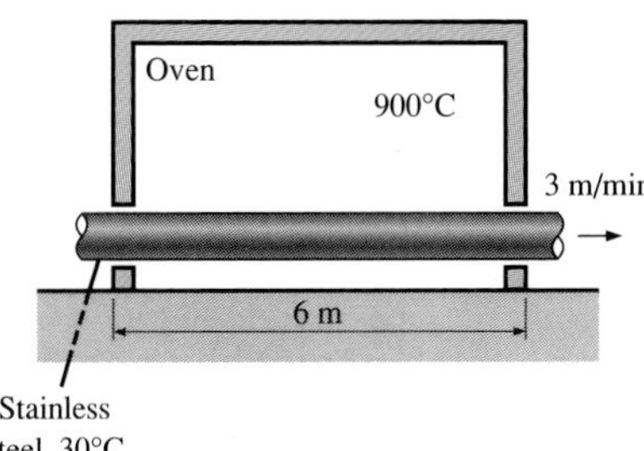

FIGURE P6-121

6-122 The inner and outer surfaces of a 5-m × 6-m brick wall of thickness 30 cm and thermal conductivity 0.69 W/(m · °C) are maintained at temperatures of 20°C and 5°C, respectively. Determine (*a*) the rate of heat transfer through the wall, in W, and (*b*) the rate of entropy generation within the wall.

6-123 For heat transfer purposes, a standing man can be modeled as a 30-cm diameter, 170-cm long vertical cylinder with both the top and bottom surfaces insulated and with the side surface at an average temperature of 34°C. For a convection heat transfer coefficient of 15 W/(m^2 · °C), determine the rate of heat loss from this man by convection in an environment at 20°C. Also, determine the rate of entropy transfer from the body of this person accompanying heat transfer, in W/K.

6-124 A 1000 W iron is left on the ironing board with its base exposed to the air at 20°C. The convection heat transfer coefficient between the base surface and the surrounding air is 80 W/(m^2 · °C). If the base has a surface area of 0.02 m^2, determine (*a*) the temperature of the base of the iron and (*b*) the rate of entropy generation during this process in steady operation. How much of this entropy generation occurs within the iron? Disregard heat transfer by radiation.

6-125E A frictionless piston-cylinder device contains saturated liquid water at 20-psia pressure. Now 600 Btu of heat is transferred to water from a source at 900°F, and part of the liquid vaporizes at constant pressure. Determine the total entropy generated during this process, in Btu/R.

6-126E Steam enters a diffuser at 20 psia and 240°F with a velocity of 900 ft/s and exits as saturated vapor at 240°F and 100 ft/s. The exit area of the diffuser is 1 ft^2. Determine (*a*) the mass flow rate of the steam and (*b*) the rate of entropy generation during this process. Assume an ambient temperature of 77°F.

6-127 Steam expands in a turbine steadily at a rate of 25,000 kg/h, entering at 8 MPa and 450°C and leaving at 50 kPa as saturated vapor. If the power generated by the turbine is 4 MW, determine the rate of entropy generation for this process. Assume the surrounding medium is at 25°C.
Answer: 8.38 kW/K

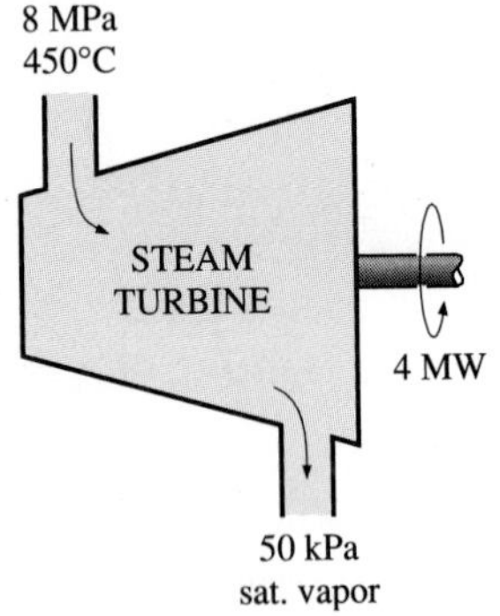

FIGURE P6-127

6-128 A hot-water stream at 70°C enters an adiabatic mixing chamber with a mass flow rate of 1.8 kg/s, where it is mixed with a stream of cold water at 20°C. If the mixture leaves the chamber at 42°C, determine (*a*) the mass flow rate of the cold water and (*b*) the rate of entropy generation during this adiabatic mixing process. Assume all the streams are at a pressure of 200 kPa.

6-129 Liquid water at 200 kPa and 20°C is heated in a chamber by mixing it with superheated steam at 200 kPa and 300°C. Liquid water enters the mixing chamber at a rate of 2.5 kg/s, and the chamber is estimated to lose heat to the surrounding air at 25°C at a rate of 600 kJ/min. If the mixture leaves the mixing chamber at 200 kPa and 60°C, determine (*a*) the mass flow rate of the superheated steam and (*b*) the rate of entropy generation during this mixing process. *Answers:* (*a*) 0.152 kg/s, (*b*) 0.297 kW/K

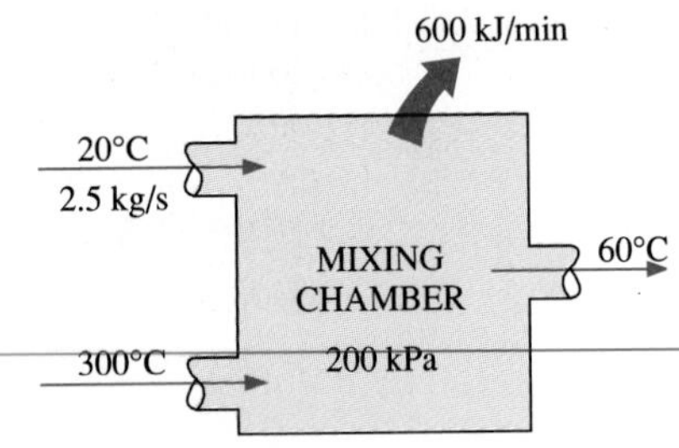

FIGURE P6-129

6-130 A 0.4-m^3 rigid tank is filled with saturated liquid water at 200°C. A valve at the bottom of the tank is now opened, and one-half of the total mass is withdrawn from the tank in the liquid form. Heat is transferred to water from a source at 250°C so that the temperature in the tank remains constant. Determine (*a*) the amount of heat transfer and (*b*) the total entropy change for this process.

6-131E An iron block of unknown mass at 185°F is dropped into an insulated tank that contains 0.8 ft^3 of water at 70°F. At the same time, a paddle wheel driven by a 200-W motor is activated to stir the water. Thermal equilibrium is established after 10 min with a final temperature of 75°F. Determine the mass of the iron block and the entropy generated during this process.

6-132E Air enters a compressor at ambient conditions of 15 psia and 60°F with a low velocity and exits at 150 psia, 620°F, and 350 ft/s. The compressor is cooled by the ambient air at 60°F at a rate of 1500 Btu/min. The power input to the compressor is 400 hp. Determine (*a*) the mass flow rate of air and (*b*) the rate of entropy generation.

6-133 Steam enters an adiabatic nozzle at 3 MPa and 400°C with a velocity of 70 m/s and exits at 2 MPa and 320 m/s. If the nozzle has an inlet area of 7 cm^2, determine (*a*) the exit temperature and (*b*) the rate of entropy generation for this process. *Answers:* (*a*) 370.4°C, (*b*) 0.0517 kW/K

Review Problems

6-134 Demonstrate the validity of the Clausius inequality using a reversible and an irreversible heat engine operating between the same two thermal energy reservoirs at constant temperatures of T_L and T_H.

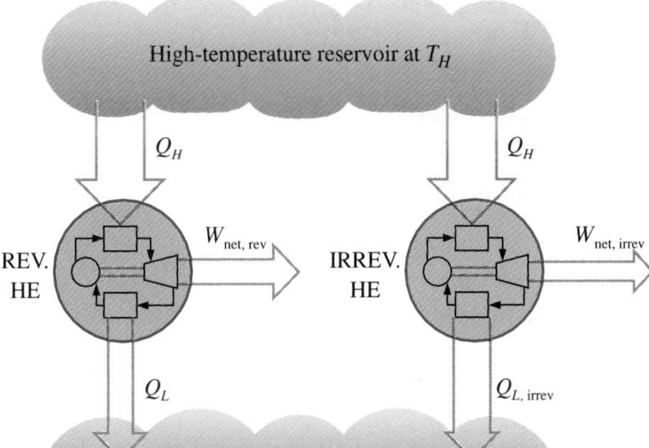

FIGURE P6-134

6-135 Show that the difference between the reversible steady-flow work and reversible moving boundary work is equal to the flow energy.

6-136 The inner and outer surfaces of a 0.5-cm-thick, 2-m × 2-m window glass in winter are 10°C and 3°C, respectively. If the thermal conductivity of the glass is 0.78 W/(m · °C), determine the amount of heat loss, in kJ, through the glass over a period of 5 h. Also, determine the amount of entropy generated during this process within the glass.

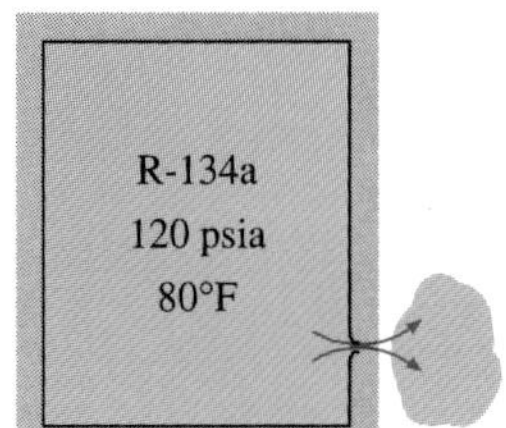

FIGURE P6-137E

6-137E A 0.8-ft^3 well-insulated rigid can initially contains refrigerant-134a at 120 psia and 80°F. Now a crack develops in the can, and the refrigerant starts to leak out slowly. Assuming the refrigerant remaining in the can has undergone a reversible, adiabatic process, determine the final mass in the can when the pressure drops to 30 psia.

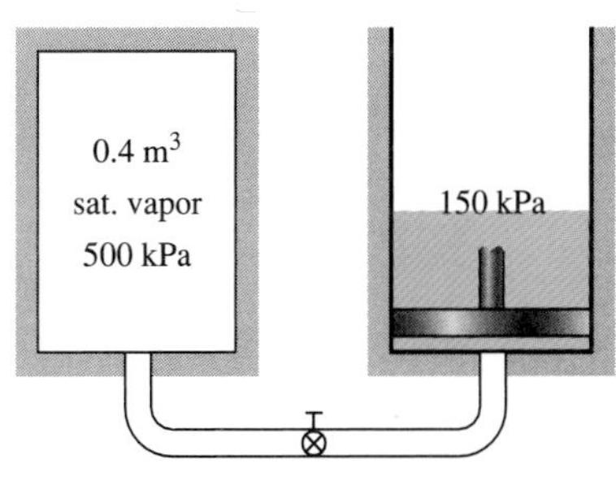

FIGURE P6-138

6-138 An insulated tank containing 0.4 m^3 of saturated water vapor at 500 kPa is connected to an initially evacuated, insulated piston-cylinder device. The mass of the piston is such that a pressure of 150 kPa is required to raise it. Now the valve is opened slightly, and part of the steam flows to the cylinder, raising the piston. This process continues until the pressure in the tank drops to 150 kPa. Assuming the steam that remains in the tank to have undergone a reversible adiabatic process, determine the final temperature (*a*) in the rigid tank and (*b*) in the cylinder.

6-139 Two rigid tanks are connected by a valve. Tank *A* is insulated and contains 0.2 m^3 of steam at 400 kPa and 80 percent quality. Tank *B* is uninsulated and contains 3 kg of steam at 200 kPa and 250°C. The valve is now opened, and steam flows from tank *A* to tank *B* until the pressure in tank *A* drops to 300 kPa. During this process 600 kJ of heat is transferred from tank *B* to the surroundings at 0°C. Assuming the steam remaining inside tank *A* to have undergone a reversible adiabatic process, determine (*a*) the final temperature in each tank and (*b*) the entropy generated during this process.

Answers: (*a*) 133.55°C, 113.0°C; (*b*) 0.912 kJ/K

FIGURE P6-139

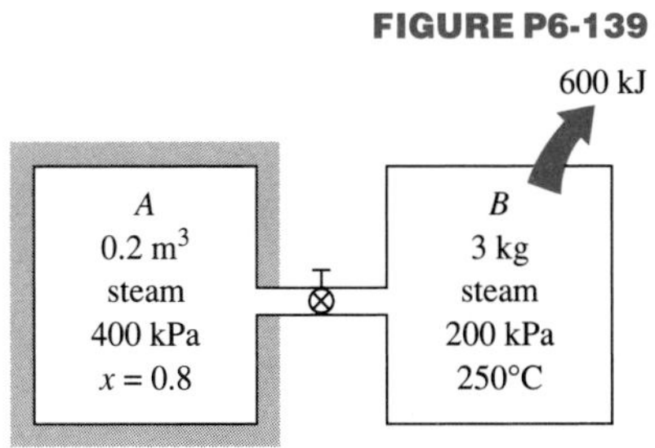

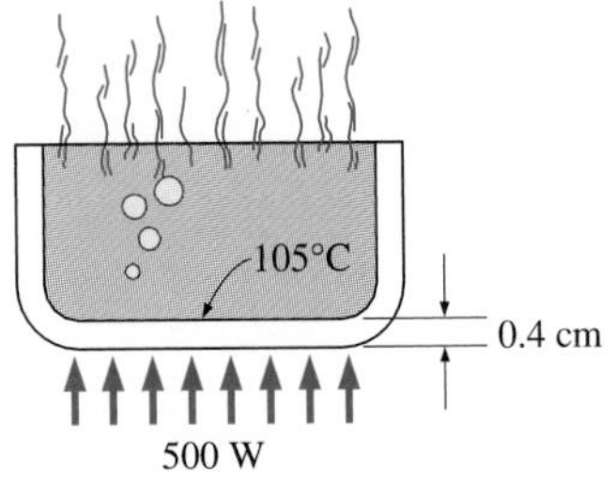

FIGURE P6-140

6-140 An aluminum pan [k_t = 237 W/(m · K)] has a flat bottom whose diameter is 20 cm and thickness is 0.4 cm. Heat is transferred steadily to boiling water in the pan through its bottom at a rate of 500 W. If the inner surface of the bottom of the pan is 105°C, determine (*a*) the temperature of the outer surface of the bottom of the pan and (*b*) the rate of entropy generation within bottom of the pan, in W/K.

6-141 A 50-cm-long, 800-W electric resistance heating element whose diameter is 0.5 cm is immersed in 40 kg of water initially at 20°C. Assuming the water container is well-insulated, determine how long it will take for this heater to raise the water temperature to 80°C. Also, determine the entropy generated during this process, in kJ/K.

6-142 A 5-cm-external-diameter, 10-m-long hot water pipe at 80°C is losing heat to the surrounding air at 5°C by natural convection with a heat transfer coefficient of 25 W/(m^2 · °C). Determine the rate of heat loss from the pipe by natural convection, in W, and the rate of entropy generation in the surrounding air, in W/K.

6-143 In large steam power plants, the feedwater is frequently heated in closed feedwater heaters, which are basically heat exchangers, by steam extracted from the turbine at some stage. Steam enters the feedwater heater at 1 MPa and 200°C and leaves as saturated liquid at the same pressure. Feedwater enters the heater at 2.5 MPa and 50°C and leaves 10°C below the exit temperature of the steam. Neglecting any heat losses from the outer surfaces of the heater, determine (*a*) the ratio of the mass flow rates of the extracted steam and the feedwater heater and (*b*) the total entropy change for this process per unit mass of the feedwater.

6-144E A 3-ft^3 rigid tank initially contains refrigerant-134a at 120 psia and 100 percent quality. The tank is connected by a valve to a supply line that carries refrigerant-134a at 160 psia and 80°F. The valve is now opened, allowing the refrigerant to enter the tank, and is closed when it is observed that the tank contains only saturated liquid at 140 psia. Determine (*a*) the mass of the refrigerant that entered the tank, (*b*) the amount of heat transfer with the surroundings at 120°F, and (*c*) the entropy generated during this process.

FIGURE P6-146

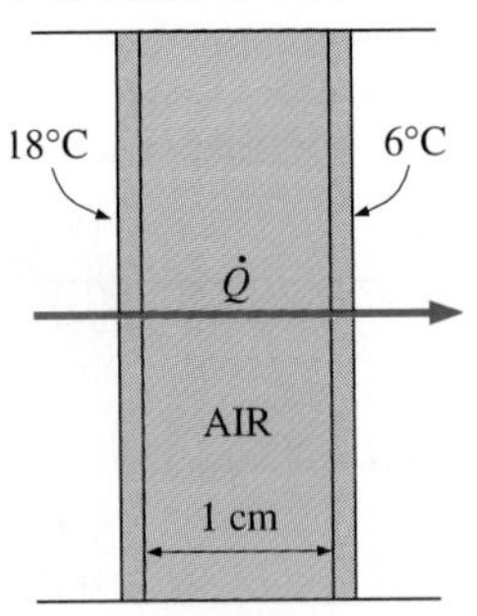

6-145 During a heat transfer process, the entropy change of incompressible substances, such as liquid water, can be determined from $\Delta S = mC_{av} \ln(T_2/T_1)$. Show that for thermal energy reservoirs, such as large lakes, this relation reduces to $\Delta S = Q/T$.

6-146 The inner and outer glasses of a 2-m × 2-m double-pane window are at 18°C and 6°C, respectively. If the 1-cm space between the two glasses is filled with still air [k_t = 0.026 W/(m · K)] and the glasses are very nearly isothermal, determine the rate of heat transfer through the window, in W. Also, determine the rates of entropy transfer through both sides of the window and the rate of entropy generation within the window, in W/K.

6-147 A well-insulated 4-m × 4-m × 5-m room initially at 10°C is heated by the radiator of a steam heating system. The radiator has a volume of 15 L and is filled with superheated vapor at 200 kPa and 200°C. At this moment both the inlet and the exit valves to the radiator are closed. A 120-W fan is used to distribute the air in the room. The pressure of the steam is observed to drop to 100 kPa after 30 min as a result of heat transfer to the room. Assuming constant specific heats for air at room temperature, determine (*a*) the average temperature of air in 30 min, (*b*) the entropy change of the steam, (*c*) the entropy change of the air in the room, and (*d*) the entropy generated during this process, in kJ/K. Assume the air pressure in the room remains constant at 100 kPa at all times.

6-148 A passive solar house that is losing heat to the outdoors at 3°C at an average rate of 50,000 kJ/h is maintained at 22°C at all times during a winter night for 10 h. The house is to be heated by 50 glass containers, each containing 20 L of water that is heated to 80°C during the day by absorbing solar energy. A thermostat controlled 15 kW back-up electric resistance heater turns on whenever necessary to keep the house at 22°C. Determine how long the electric heating system was on that night and the amount of entropy generated during the night.

6-149E A 15-ft^3 steel container that has a mass of 40 lbm when empty is filled with liquid water. Initially, both the steel tank and the water are at 120°F. Now heat is transferred, and the entire system cools to the surrounding air temperature of 70°F. Determine the total entropy generated during this process.

6-150 One ton (1000 kg) of liquid water at 80°C is brought into a well-insulated and well-sealed 4-m × 5-m × 6-m room initially at 22°C and 100 kPa. Assuming constant specific heats for both air and water at room temperature, determine (*a*) the final equilibrium temperature in the room and (*b*) the entropy generated during this process, in kJ/K.

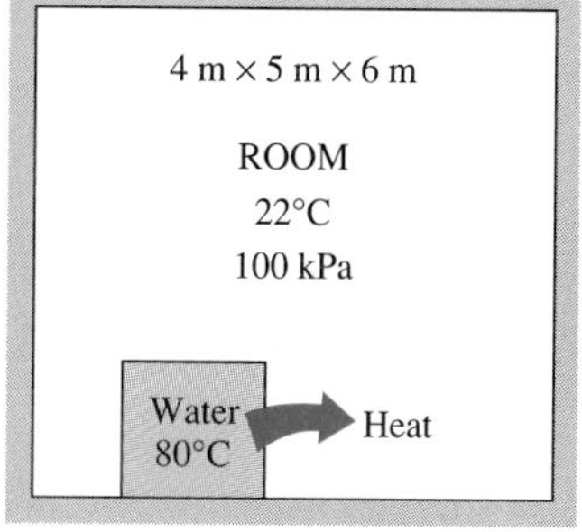

FIGURE P6-150

6-151E A piston-cylinder device initially contains 15 ft^3 of helium gas at 25 psia and 70°F. Helium is now compressed in a polytropic process (PV^n = constant) to 70 psia and 300°F. Determine (*a*) the entropy change of helium, (*b*) the entropy change of the surroundings, and (*c*) whether this process is reversible, irreversible, or impossible. Assume the surroundings are at 70°F. *Answers:* (*a*) −0.016 Btu/R, (*b*) 0.019 Btu/R, (*c*) irreversible

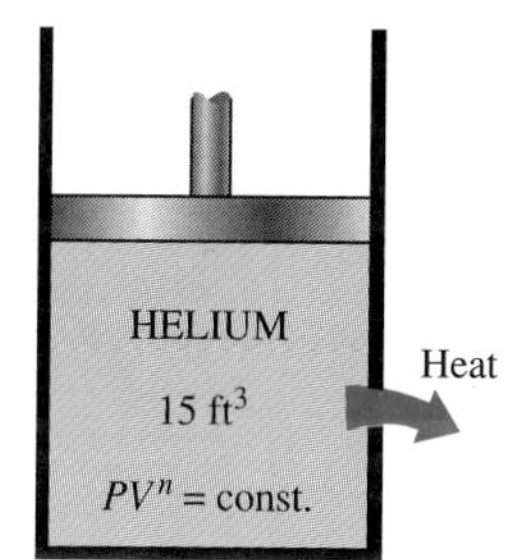

FIGURE P6-151E

6-152 Air is compressed steadily by a compressor from 100 kPa and 17°C to 700 kPa at a rate of 2 kg/min. Determine the minimum power input required if the process is (*a*) adiabatic and (*b*) isothermal. Assume air to be an ideal gas with constant specific heats, and neglect the changes in kinetic and potential energies. *Answers:* (*a*) 7.21 kW, (*b*) 5.4 kW.

6-153 Air enters the evaporator section of a window air conditioner at 100 kPa and 27°C with a volume flow rate of 6 m^3/min. The refrigerant-134a at 120 kPa with a quality of 0.3 enters the evaporator at a rate of 2 kg/min and leaves as saturated vapor at the same pressure. Determine the exit temperature of the air and the rate of entropy generation for this process,

FIGURE P6-153

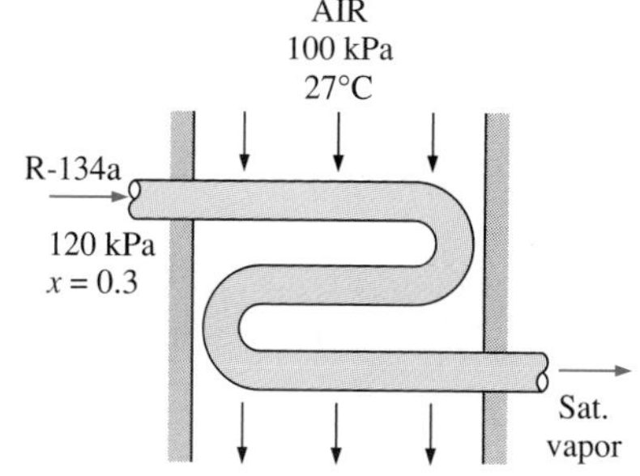

assuming (*a*) the outer surfaces of the air conditioner are insulated and (*b*) heat is transferred to the evaporator of the air conditioner from the surrounding medium at 32°C at a rate of 30 kJ/min.
Answers: (*a*) −5.7°C, 0.00194 kW/K, (*b*) −1.4°C, 0.00216 kW/K

6-154 A 4-m × 5-m × 6-m well-sealed room is to be heated by one ton of liquid water contained in a tank that is placed in the room. The room is losing heat to the outside air at 5°C at an average rate of 10,000 kJ/h. The room is initially at 20°C and 100 kPa and is maintained at a temperature of 20°C at all times. If the hot water is to meet the heating requirements of this room for a 24-h period, determine (*a*) the minimum temperature of the water when it is first brought into the room and (*b*) the entropy generated during a 24-h period. Assume constant specific heats for both air and water at room temperature.

6-155 Consider a well-insulated horizontal rigid cylinder that is divided into two compartments by a piston that is free to move but does not allow either gas to leak into the other side. Initially, one side of the piston contains 1 m^3 of N_2 gas at 500 kPa and 80°C while the other side contains 1 m^3 of He gas at 500 kPa and 25°C. Now thermal equilibrium is established in the cylinder as a result of heat transfer through the piston. Using constant specific heats at room temperature, determine (*a*) the final equilibrium temperature in the cylinder and (*b*) the entropy generation during this process. What would your answer be if the piston were not free to move?

6-156 Repeat the problem above by assuming the piston is made of 5 kg of copper initially at the average temperature of the two gases on both sides.

6-157 An insulated 5-m^3 rigid tank contains air at 500 kPa and 57°C. A valve connected to the tank is now opened, and air is allowed to escape until the pressure inside drops to 200 kPa. The air temperature during this process is maintained constant by an electric resistance heater placed in the tank. Determine (*a*) the electrical work done during this process and (*b*) the total entropy change for this process.
Answers: (*a*) −1500 kJ, (*b*) 4.47 kJ/K

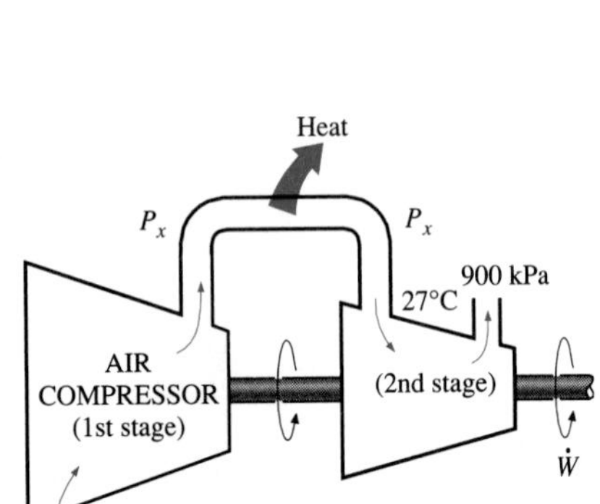

FIGURE P6-158

6-158 Air enters a two-stage compressor at 100 kPa and 27°C and is compressed to 900 kPa. The pressure ratio across each stage is the same, and the air is cooled to the initial temperature between the two stages. Assuming the compression process to be isentropic, determine the power input to the compressor for a mass flow rate of 0.02 kg/s. What would your answer be if only one stage of compression were used? *Answers:* 4.44 kW, 5.26 kW

6-159 Consider a three-stage isentropic compressor with two intercoolers that cool the gas to the initial temperature between the stages. Determine the two intermediate pressures (P_x and P_y) in terms of inlet and exit pressures (P_1 and P_2) that will minimize the work input to the compressor.
Answers: $P_x = (P_1^2 P_2)^{1/3}$, $P_y = (P_1 P_2^2)^{1/3}$

6-160 In order to cool 1-ton (1000 kg) of water at 20°C in an insulated tank, a person pours 80 kg of ice at −5°C into the water. Determine (*a*) the final equilibrium temperature in the tank and (*b*) the entropy generation during this process. The melting temperature and the heat of fusion of ice at atmospheric pressure are 0°C and 333.7 kJ/kg.

6-161 An insulated piston-cylinder device initially contains 0.01 m^3 of saturated liquid–vapor mixture of water with a quality of 0.2 at 100°C. Now some ice at −5°C is dropped into the cylinder. If the cylinder contains saturated liquid at 100°C when thermal equilibrium is established, determine (*a*) the amount of ice added and (*b*) the entropy generation during this process. The melting temperature and the heat of fusion of ice at atmospheric pressure are 0°C and 333.7 kJ/kg.

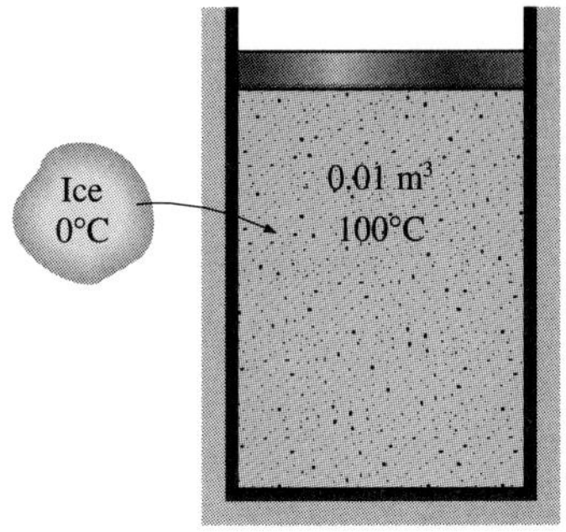

FIGURE P6-161

6-162 Steam at 7 MPa and 500°C enters a two-stage adiabatic turbine at a rate of 15 kg/s. Ten percent of the steam is extracted at the end of the first stage at a pressure of 1 MPa for other use. The remainder of the steam is further expanded in the second stage and leaves the turbine at 50 kPa. Determine the power output of the turbine, assuming (*a*) the process is reversible and (*b*) the turbine has an isentropic efficiency of 88 percent.
Answers: (*a*) 14,928 kW, (*b*) 13,136 kW

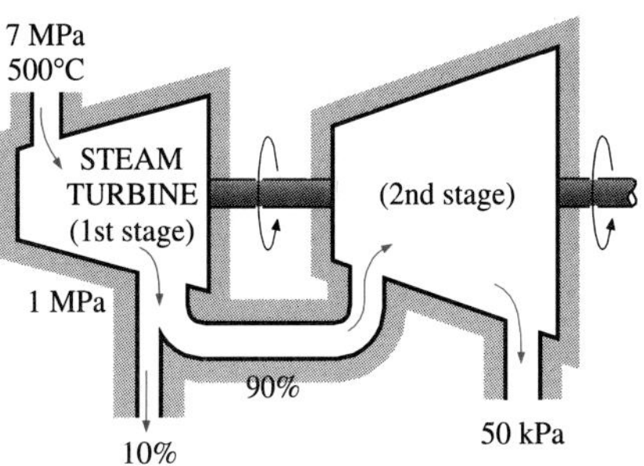

FIGURE P6-162

6-163 Steam enters a two-stage adiabatic turbine at 8 MPa and 500°C. It expands in the first stage to a pressure of 2 MPa. Then steam is reheated at constant pressure to 500°C before it is expanded in a second stage to a pressure of 100 kPa. The work output of the turbine is 40 MW. Assuming an isentropic efficiency of 84 percent for each stage of the turbine, determine the required mass flow rate of steam. Also, show the process on a *T-s* diagram with respect to saturation lines. *Answer:* 41.0 kg/s

6-164 Refrigerant-134a at 140 kPa and −10°C is compressed by an adiabatic 0.5-kW compressor to an exit state of 700 kPa and 60°C. Neglecting the changes in kinetic and potential energies, determine (*a*) the isentropic efficiency of the compressor, (*b*) the volume flow rate of the refrigerant at the compressor inlet, in L/min, and (*c*) the maximum volume flow rate at the inlet conditions that this adiabatic 0.5-kW compressor can handle without violating the second law.

6-165E Helium gas enters a nozzle whose isentropic efficiency is 94 percent with a low velocity, and it exits at 14 psia, 180°F, and 1000 ft/s. Determine the pressure and temperature at the nozzle inlet.

6-166 Consider a 5-L evacuated rigid bottle that is surrounded by the atmosphere at 100 kPa and 17°C. A valve at the neck of the bottle is now opened and the atmospheric air is allowed to flow into the bottle. The air trapped in the bottle eventually reaches thermal equilibrium with the atmosphere as a result of heat transfer through the wall of the bottle. The valve remains open during the process so that the trapped air also reaches mechanical equilibrium with the atmosphere. Determine the net heat transfer through the wall of the bottle and the entropy generation during this filling process.
Answers: 0.5 kJ, 0.0017 kJ/K

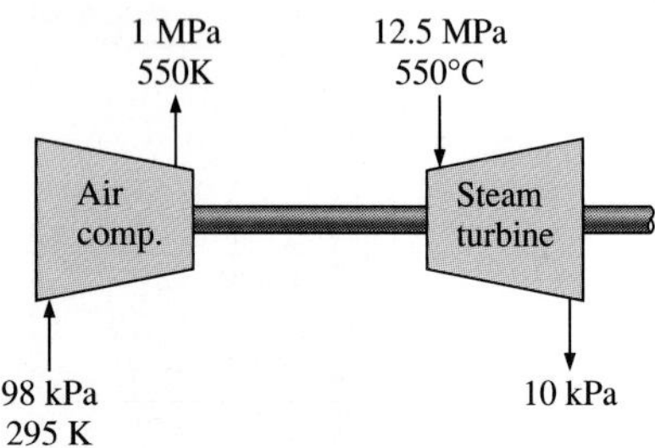

FIGURE P6-167

6-167 An adiabatic air compressor is to be powered by a direct-coupled adiabatic steam turbine that is also driving a generator. Steam enters the turbine at 12.5 MPa and 500°C at a rate of 25 kg/s and exits at 10 kPa and a quality of 0.92. Air enters the compressor at 98 kPa and 295 K at a rate of 10 kg/s and exits at 1 MPa and 550 K. Determine the net power delivered to the generator by the turbine and the rate of entropy generation within the turbine and the compressor during this process.

6-168 (*a*) Water flows through a shower head steadily at a rate of 10 L/min. An electric resistance heater placed in the water pipe heats the water from 16°C to 43°C. Taking the density of water to be 1 kg/L, determine the electric power input to the heater, in kW, and the rate of entropy generation during this process, in kW/K.

(*b*) In an effort to conserve energy, it is proposed to pass the drained warm water at a temperature of 39°C through a heat exchanger to preheat the incoming cold water. If the heat exchanger has an effectiveness of 0.50 (that is, it recovers only half of the energy that can possibly be transferred from the drained water to incoming cold water), determine the electric power input required in this case and the reduction in the rate of entropy generation in the resistance heating section.

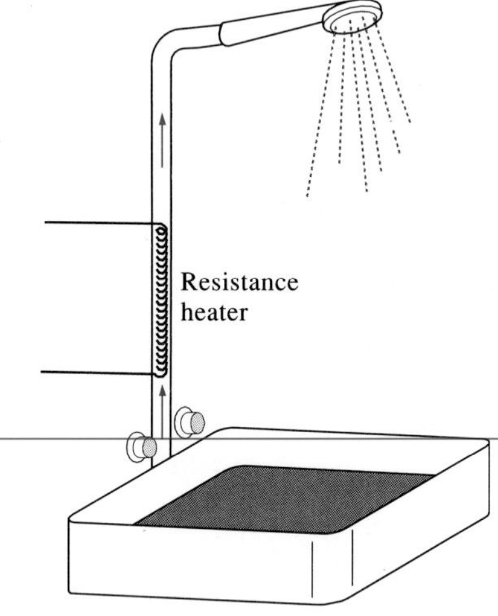

FIGURE P6-168

6-169 Consider two bodies of identical mass m and specific heat C used as thermal reservoirs (source and sink) for a heat engine. The first body is initially at an absolute temperature T_1 while the second one is at a lower absolute temperature T_2. Heat is transferred from the first body to the heat engine, which rejects the waste heat to the second body. The process continues until the final temperatures of the two bodies T_f become equal. Show that $T_f = \sqrt{T_1 T_2}$ when the heat engine produces the maximum possible work.

FIGURE P6-169

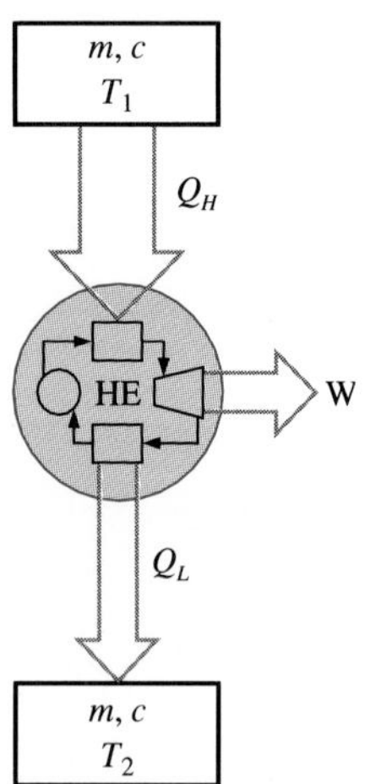

6-170 The explosion of a hot water tank in a school in Spencer, Oklahoma, in 1982 killed 7 people while injuring 33 others. Although the number of such explosions has decreased dramatically since the development of the ASME Pressure Vessel Code, which requires the tanks to be designed to withstand four times the normal operating pressures, they still occur as a result of the failure of the pressure relief valves and thermostats. When a tank filled with a high-pressure and high-temperature liquid ruptures, the sudden drop of the pressure of the liquid to the atmospheric level causes part of the liquid to flash into vapor, and thus to experience a huge rise in its volume. The resulting pressure wave that propagates rapidly can cause considerable damage.

Considering that the pressurized liquid in the tank eventually reaches equilibrium with its surroundings shortly after the explosion, the work that a pressurized liquid would do if allowed to expand reversibly and adiabatically to the pressure of the surroundings can be viewed as the *explosive energy* of the pressurized liquid. Because of the very short time period of the explosion and the apparent calm afterward, the explosion process can be considered to be adiabatic with no changes in kinetic and potential energies and no mixing with the air.

Consider a 100-L hot-water tank that has a working pressure of 0.5 MPa. As a result of some malfunction, the pressure in the tank rises to 2 MPa, at which point the tank explodes. Taking the atmospheric pressure to be 100 kPa and assuming the liquid in the tank to be saturated at the time of explosion, determine the total explosion energy of the tank in terms of the TNT equivalence. (The explosion energy of TNT is about 3250 kJ/kg, and 5 kg of TNT can cause total destruction of unreinforced structures within about a 7-m radius.) *Answer:* 2.467 kg TNT

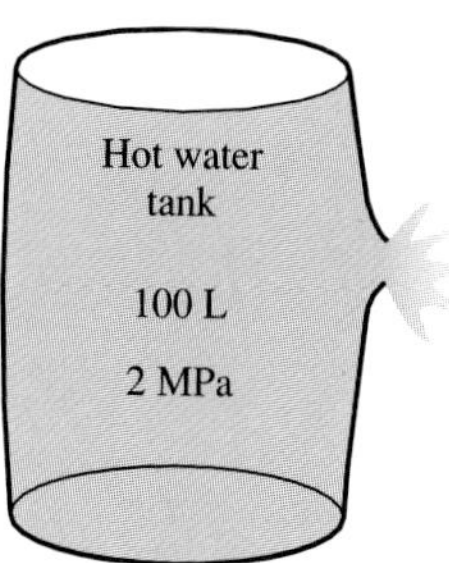

FIGURE P6-170

6-171 Using the arguments in the problem above, determine the total explosion energy of a 0.2-L canned drink that explodes at a pressure of 1 MPa. To how many kg of TNT is this explosion energy equivalent?

Computer, Design, and Essay Problems

6-172 Write a computer program to determine the work input to a multistage compressor for a given set of inlet and exit pressures for any number of stages. Assume that the pressure ratio across each stage is identical and the compression process is polytropic. List and plot the compressor work against the number of stages for $P_1 = 100$ kPa, $T_1 = 17°C$, $P_2 = 800$ kPa, and $n = 1.35$ for air. Based on this chart, can you justify using compressors with 3 or more stages?

6-173 It is well known that the temperature of a gas rises while it is compressed as a result of the energy input in the form of compression work. At high compression ratios, the air temperature may rise above the autoignition temperature of some hydrocarbons, including some lubricating oil. Therefore, the presence of some lubricating oil vapor in high-pressure air raises the possibility of an explosion, creating a fire hazard. The concentration of the oil within the compressor is usually too low to create a real danger. However, the oil that collects on the inner walls of exhaust piping of the compressor may cause an explosion. Such explosions have largely been eliminated by using the proper lubricating oils, carefully designing the equipment, intercooling between compressor stages, and keeping the system clean.

A compressor is to be designed for an industrial application in Los Angeles. If the compressor exit temperature is not to exceed 250°C for safety consideration, determine the maximum allowable compression ratio that is safe for all possible weather conditions for that area.

6-174 Identify the major sources of entropy generation in your house and propose ways of reducing them.

6-175 Obtain the following information about a power plant that is closest to your town: the net power output; the type and amount of fuel; the power consumed by the pumps, fans, and other auxiliary equipment; stack gas losses; temperatures at several locations; and the rate of heat rejection at the condenser. Using these and other relevant data, determine the rate of entropy generation in that power plant.

6-176 Think about all of your activities all day yesterday. List three of the activities during which you contributed considerably to the entropy increase of the universe. Explain why those activities are irreversible and how they generate entropy.

Exergy: A Measure of Work Potential

CHAPTER 7

The increased awareness that the world's energy resources are limited has caused some governments to reexamine their energy policies and take drastic measures in eliminating waste. It has also sparked interest in the scientific community to take a closer look at the energy conversion devices and to develop new techniques to better utilize the existing limited resources. The first law of thermodynamics deals with the *quantity* of energy and asserts that energy cannot be created or destroyed. This law merely serves as a necessary tool for the bookkeeping of energy during a process and offers no challenges to the engineer. The second law, however, deals with the *quality* of energy. More specifically, it is concerned with the degradation of energy during a process, the entropy generation, and the lost opportunities to do work; and it offers plenty of space for improvement.

The second law of thermodynamics has proved to be a very powerful tool in the optimization of complex thermodynamic systems. In this chapter, we examine the performance of engineering devices in light of the second law of thermodynamics. We start our discussions with the introduction of *exergy* (also called *availability*), which is the maximum useful work that could be obtained from the system at a given state in a specified environment, and we continue with the *reversible work,* which is the maximum useful work that can be obtained as a system undergoes a process between two specified states. Next we discuss the *irreversibility* (also called the *exergy destruction* or *lost work*), which is the wasted work potential during a process as a result of irreversibilities, and we define a *second-law efficiency.* We then develop the *exergy balance* relation and apply it to closed systems and control volumes.

7-1 ■ EXERGY: WORK POTENTIAL OF ENERGY

When a new energy source, such as a geothermal well, is discovered, the first thing the explorers do is estimate the amount of energy contained in the source. This information alone, however, is of little value in deciding whether to build a power plant on that site. What we really need to know is the *work potential* of the source—that is, the amount of energy we can extract as useful work. The rest of the energy will eventually be discarded as waste energy and is not worthy of our consideration. Thus, it would be very desirable to have a property to enable us to determine the useful work potential of a given amount of energy at some specified state. This property is *exergy,* which is also called the *availability* or *available energy.*

The work potential of the energy contained in a system at a specified state is simply the maximum useful work that can be obtained from the system. You will recall that the work done during a process depends on the initial state, the final state, and the process path. That is,

$$\text{Work} = f(\text{initial state, process path, final state})$$

In an exergy analysis, the *initial state* is specified, and thus it is not a variable. The work output is maximized when the process between two specified states is executed in a *reversible manner,* as shown in Chap. 6. Therefore, all the irreversibilities are disregarded in determining the work potential. Finally, the system must be in the *dead state* at the end of the process to maximize the work output.

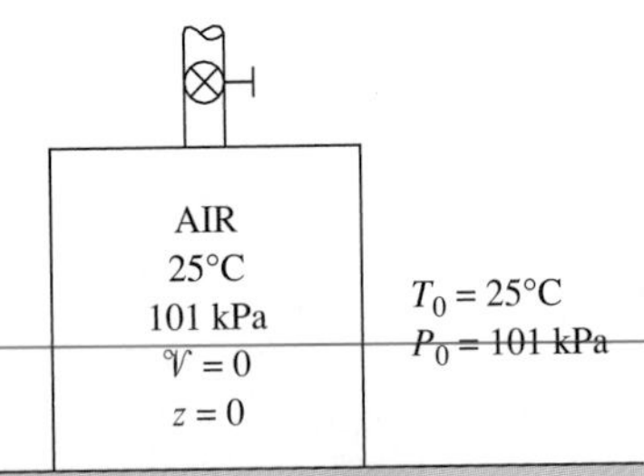

FIGURE 7-1

A system that is in equilibrium with the environment it is in is said to be at the dead state.

A system is said to be in the **dead state** when it is in thermodynamic equilibrium with the environment it is in (Fig. 7-1). At the dead state, a system is at the temperature and pressure of its environment (in thermal and mechanical equilibrium); it has no kinetic or potential energy relative to the environment (zero velocity and zero elevation above a reference level); and it does not react with the environment (chemically inert). Also, there are no unbalanced magnetic, electrical and surface tension effects between the system and its surroundings, if these are relevant to the situation at hand. The properties of a system at the dead state are denoted by subscript zero, for example, P_0, T_0, h_0, u_0, and s_0. Unless specified otherwise, the dead-state temperature and pressure are assumed to be T_0 = 25°C (77°F) and P_0 = 1 atm (101.325 kPa or 14.7 psia). A system has zero availability at the dead state (Fig. 7-2).

FIGURE 7-2

At the dead state, the useful work potential (exergy) of a system is zero.

Distinction should be made between the *surroundings, immediate surroundings,* and the *environment.* By definition, **surroundings** are everything outside the system boundaries. The **immediate surroundings** refer to the portion of the surroundings that is affected by the process, and **environment** refers to the region beyond the immediate surroundings whose properties are not affected by the process at any point. Therefore, any irreversibilities during a process occur within the system and its immediate surroundings, and the environment is free of any irreversibilities. When analyzing the cooling of a hot baked potato in a room at 25°C, for example, the warm air that surrounds the potato is the immediate surroundings, and the remaining part of the room air at 25°C is the environment. Note that the temperature of the immediate surroundings changes from the temperature of the potato at the boundary to the environment temperature of 25°C (Fig. 7-3).

The notion that a system must go to the dead state at the end of the process to maximize the work output can be explained as follows: If the system temperature at the final state is greater than (or less than) the temperature of the environment it is in, we can always produce additional work by running a heat engine between these two temperature levels. If the final pressure is greater than (or less than) the pressure of the environment, we can still obtain work by letting the system expand to the pressure of the environment. If the final velocity of the system is not zero, we can catch that extra kinetic energy by a turbine and convert it to rotating shaft work, and so on. No work can be produced from a system that is initially at the dead state. The atmosphere around us contains a tremendous amount of energy. However, the atmosphere is in the dead state, and the energy it contains has no work potential (Fig. 7-4).

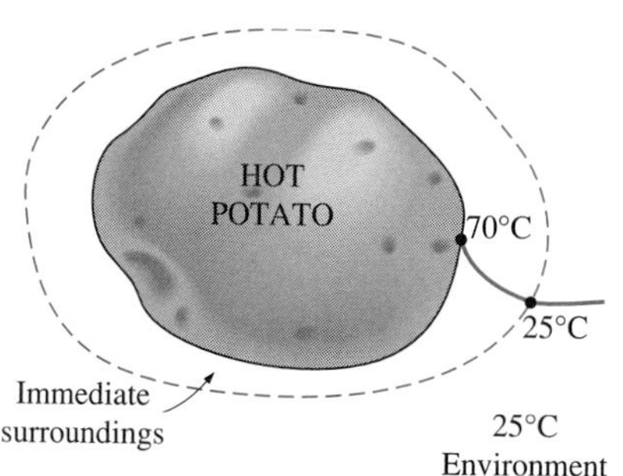

FIGURE 7-3
The immediate surroundings of a hot potato are simply the temperature gradient zone of the air next to the potato.

Therefore, we conclude that a *system will deliver the maximum possible work as it undergoes a reversible process from the specified initial state to the state of its environment, that is, the dead state* (Fig. 7-4). This represents the *useful work potential* of the system at the specified state and is called **exergy.** It is important to realize that exergy does not represent the amount of work that a work-producing device will actually deliver upon installation. Rather, it represents the *upper limit on the amount of work a device can deliver without violating any thermodynamic laws.* There will always be a difference, large or small, between exergy and the actual work delivered by a device. This difference represents the room engineers have for improvement.

FIGURE 7-4
The atmosphere contains a tremendous amount of energy, but no exergy.

Note that the exergy of a system at a specified state depends on the conditions of the environment (the dead state) as well as the properties of the system. Therefore, availability is a property of the *system–environment combination* and not of the system alone. Altering the environment is another way of increasing exergy, but it is definitely not an easy alternative.

The term *availability* was made popular in the United States by the M.I.T. School of Engineering in the 1940s. Today, an equivalent term, *exergy,* introduced in Europe in the 1950s, has found global acceptance partly because it is shorter, it rhymes with energy and entropy, and it can be adapted without requiring translation. In this text the preferred term is *exergy.* The reader should be aware that some authors define *exergy* and *availability* slightly differently.

FIGURE 7-5
Schematic for Example 7-1.

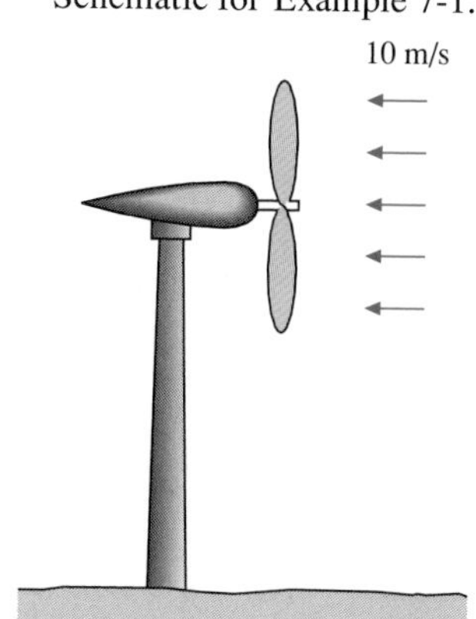

EXAMPLE 7-1 Maximum Power Generation by a Windmill

A windmill with a 12-m-diameter rotor, as shown in Fig. 7-5, is to be installed at a location where the wind is blowing steadily at an average velocity of 10 m/s. Determine the maximum power that can be generated by the windmill.

Solution The air flowing with the wind has the same properties as the stagnant atmospheric air except that it possesses a velocity and thus some kinetic energy.

Assumptions Air is at standard conditions of 1 atm and 25°C, and thus its density is 1.18 kg/m^3.

Analysis This air will reach the dead state when it is brought to a complete stop. Therefore, the exergy of the blowing air is simply the kinetic energy it possesses:

$$\text{Exergy} = ke_1 = \frac{\mathcal{V}_1^2}{2} = \frac{(100\text{ m/s})^2}{2}\left(\frac{1\text{ kJ/kg}}{1000\text{ m}^2/\text{s}^2}\right) = 0.05\text{ kJ/kg}$$

That is, every unit mass of air flowing at a velocity of 10 m/s has a work potential of 0.05 kJ/kg. In other words, a perfect windmill will bring the air to a complete stop and capture that 0.05 kJ/kg of work potential. To determine the maximum power, we need to know the amount of air passing through the rotor of the windmill per unit time, that is, the mass flow rate, which is determined to be

$$\dot{m} = \rho A\mathcal{V}_1 = \rho\frac{\pi D^2}{2}\mathcal{V}_1 = (1.18\text{ kg/m}^3)\frac{\pi(12\text{ m})^2}{4}(10\text{ m/s}) = 1335\text{ kg/s}$$

Thus,

$$\text{Maximum power} = \dot{m}(ke_1) = (1335\text{ kg/s})(0.05\text{ kJ/kg}) = \mathbf{66.7\text{ kW}}$$

This is the maximum power available to the windmill. Assuming a conversion efficiency of 25 percent, an actual windmill will convert 16.7 kW to electricity. Notice that the work potential for this case is equal to the entire kinetic energy of the air. This is because kinetic energy itself is a form of "mechanical" energy.

Discussion It should be noted that although the entire kinetic energy of the wind is available for power production, Betz's law states that the power output of a wind machine will be at maximum when the wind is slowed to one-third of its initial velocity. Therefore, for maximum power (and thus minimum cost per installed power), the highest efficiency of a wind turbine is about 59 percent. In practice, the actual efficiency ranges between 20 and 40 percent, and is about 35 percent for most wind turbines.

Wind power is suitable for harvesting when there are steady winds with an average velocity of at least 6 m/s (or 14 mph). Recent improvements in wind turbine design have brought the cost of generating wind power to about 5 cents per kWh at some locations, which is well below the average price of 8 cents per kWh the consumers are paying for electricity in the United States.

EXAMPLE 7-2 Exergy Transfer from a Furnace

Consider a large furnace that can supply heat at a temperature of 2000 R at a steady rate of 8000 Btu/s. Determine the exergy (work potential) of this energy. Assume an environment temperature of 77°F.

Solution The furnace in this example can be modeled as a heat reservoir that supplies heat indefinitely at a constant temperature. The exergy of this heat energy is its useful work potential, that is, the maximum possible amount of work that can be extracted from it. This corresponds to the amount of work that a reversible heat engine operating between the furnace and the environment can produce.

Analysis The thermal efficiency of this reversible heat engine is

$$\eta_{th,max} = \eta_{th,rev} = 1 - \frac{T_L}{T_H} = 1 - \frac{T_0}{T_H} = 1 - \frac{537\text{ R}}{2000\text{ R}} = 0.732\text{ (or 73.2\%)}$$

That is, a heat engine can convert, at best, 73.2 percent of the heat received from this furnace to work. Thus, the exergy of this furnace is equivalent to the power produced by the reversible heat engine:

$$\dot{W}_{max} = \dot{W}_{rev} = \eta_{th,rev}\,\dot{Q}_{in} = (0.732)(3000\text{ Btu/s}) = \mathbf{2196\ Btu/s}$$

Discussion Notice that 26.8 percent of the heat transferred from the furnace is not available for doing work. The portion of energy that cannot be converted to work is called **unavailable energy** (Fig. 7-6). Unavailable energy is simply the difference between the total energy of a system at a specified state and the exergy of that energy.

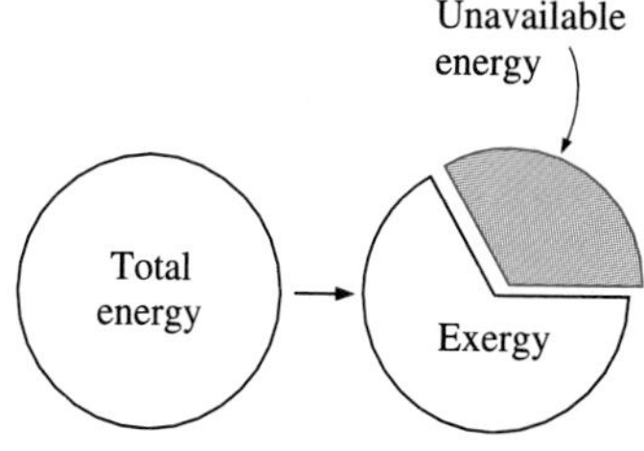

FIGURE 7-6
Unavailable energy is the portion of energy that cannot be converted to work by even a reversible heat engine.

7-2 ■ REVERSIBLE WORK AND IRREVERSIBILITY

The property exergy serves as a valuable tool in determining the quality of energy and comparing the work potentials of different energy sources or systems. The evaluation of exergy alone, however, is not sufficient for studying engineering devices operating between two fixed states. This is because when evaluating exergy, the final state is always assumed to be the *dead state,* which is hardly ever the case for actual engineering systems. The adiabatic efficiencies discussed in Chap. 6 are also of limited use because the exit state of the model (isentropic) process is not the same as the actual exit state.

In this section, we describe two quantities that are related to the actual initial and final states of processes and serve as valuable tools in the thermodynamic analysis of components or systems. These two quantities are the *reversible work* and *irreversibility* (or *exergy destruction*). But first we examine the **surroundings work,** which is the work done by or against the surroundings during a process.

The work done by work-producing devices is not always entirely in a usable form. For example, when a gas in a piston-cylinder device expands, part of the work done by the gas is used to push the atmospheric air out of the way of the piston (Fig. 7-7). This work, which cannot be recovered and utilized for any useful purpose, is equal to the atmospheric pressure P_0 times the volume change of the system,

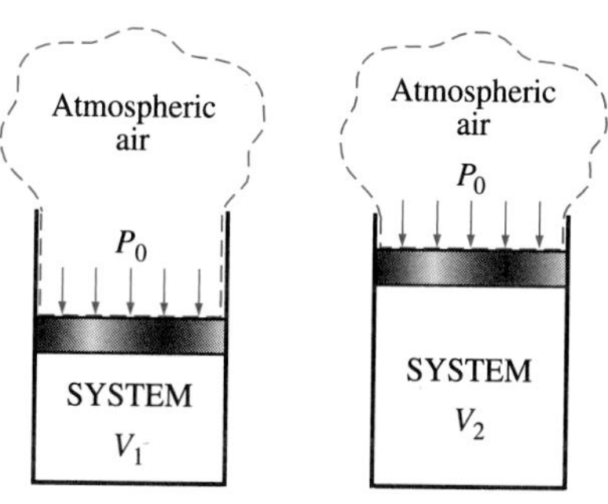

FIGURE 7-7
As a closed system expands, some work needs to be done to push the atmospheric air out of the way (W_{surr}).

$$W_{surr} = P_0(V_2 - V_1) \tag{7-1}$$

The difference between the actual work W and the surroundings work W_{surr} is called the **useful work** W_u:

$$W_u = W - W_{surr} = W - P_0(V_2 - V_1) \tag{7-2}$$

When a system is expanding and doing work, part of the work done is used to overcome the atmospheric pressure, and thus W_{surr} represents a loss. When a system is compressed, however, the atmospheric pressure helps the compression process, and thus W_{surr} represents a gain.

Note that the work done by or against the atmospheric pressure has significance only for systems whose volume changes during the process (i.e., systems that involve moving boundary work). It has no significance for cyclic devices and systems whose boundaries remain fixed during a process such as rigid tanks and steady-flow devices (turbines, compressors, nozzles, heat exchangers, etc.), as shown in Fig. 7-8.

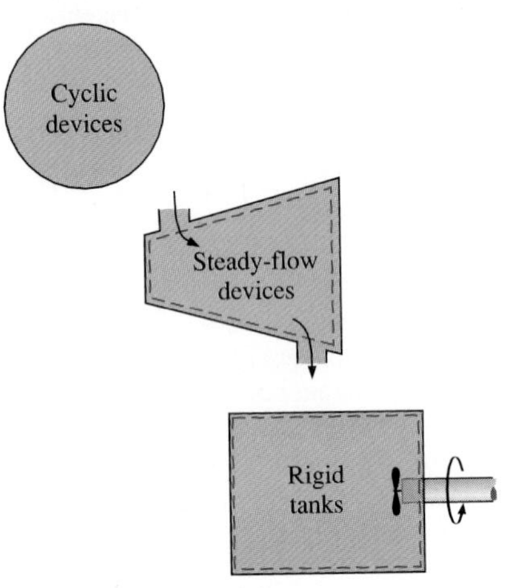

FIGURE 7-8
For constant-volume systems, the total actual and useful works are identical ($W_u = W$).

Reversible work W_{rev} is defined as *the maximum amount of useful work that can be produced (or the minimum work that needs to be supplied) as a system undergoes a process between the specified initial and final states.* This is the useful work output (or input) obtained (or expended) when the process between the initial and final states is executed in a totally reversible manner. That is, any heat transfer between the system and the surroundings must take place reversibly, and no irreversibilities should be present within the system during the process. When the final state is the dead state, the reversible work equals exergy. For processes that require work, reversible work represents the minimum amount of work necessary to carry out that process. For convenience in presentation, the term *work* is used to denote both work and power throughout this chapter.

Any difference between the reversible work W_{rev} and the useful work W_u is due to the irreversibilities present during the process, and this difference is called **irreversibility** I. It is expressed as (Fig. 7-9)

$$I = W_{rev,out} - W_{u,out} \quad \text{or} \quad I = W_{u,in} - W_{rev,in} \tag{7-3}$$

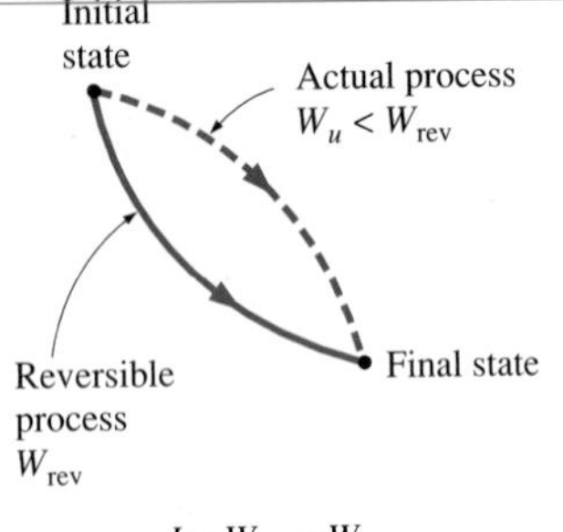

FIGURE 7-9
The difference between reversible work and actual useful work is the irreversibility.

The irreversibility is equivalent to the *exergy destroyed,* discussed in Sec. 7-5. For a totally reversible process, the actual and reversible work terms are identical, and thus the irreversibility is zero. This is expected since totally reversible processes generate no entropy. Irreversibility is a *positive quantity* for all actual (irreversible) processes since $W_{rev} \geq W_u$ for work-producing devices and $W_{rev} \leq W_u$ for work-consuming devices.

Irreversibility can be viewed as the *wasted work potential* or the *lost opportunity* to do work. It represents the energy that could have been converted to work but was not. The smaller the irreversibility associated with a process, the greater the work that will be produced (or the smaller the work that will be consumed). The performance of a system can be improved by minimizing the irreversibility associated with it.

EXAMPLE 7-3 The Rate of Irreversibility of a Heat Engine

A heat engine receives heat from a source at 1200 K at a rate of 500 kJ/s and rejects the waste heat to a medium at 300 K (Fig. 7-10). The power output of the heat engine is 180 kW. Determine the reversible power and the irreversibility rate for this process.

Solution The reversible power for this process is the amount of power that a reversible heat engine, such as a Carnot heat engine, would produce when operating between the same temperature limits.

Analysis Reversible power is determined by using the definition of thermal efficiency for a reversible heat-engine cycle:

$$\dot{W}_{rev} = \eta_{th,\,rev}\dot{Q}_{in} = \left(1 - \frac{T_{sink}}{T_{source}}\right)\dot{Q}_{in} = \left(1 - \frac{300\text{K}}{1200\text{ K}}\right)(500\text{ kW}) = \mathbf{375\text{ kW}}$$

This is the maximum power that can be produced by a heat engine operating between the specified temperature limits and receiving heat at the specified rate. This would also represent the *available power* if 300 K were the lowest temperature available for heat rejection.

The irreversibility rate is the difference between the reversible power (maximum power that could have been produced) and the useful power output:

$$\dot{I} = \dot{W}_{rev,\,out} - \dot{W}_{u,\,out} = 375 - 180 = \mathbf{195\text{ kW}}$$

Discussion Note that 195 kW of power potential is wasted during this process as a result of irreversibilities. Notice that the 500 − 375 = 125 kW of heat rejected to the sink is not available for converting to work and thus is not part of the irreversibility.

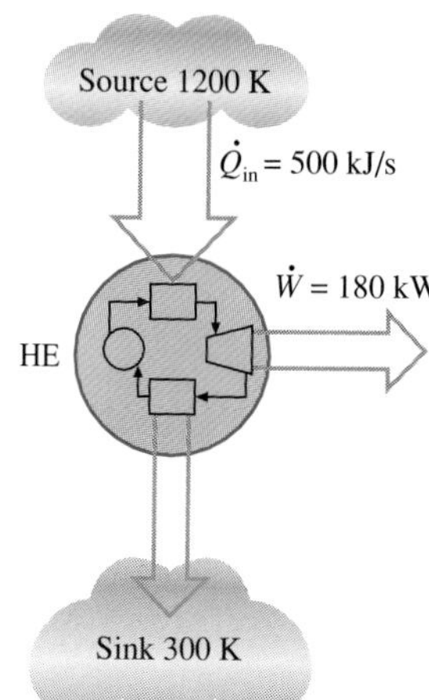

FIGURE 7-10
Schematic for Example 7-3.

EXAMPLE 7-4 Irreversibility during the Cooling of an Iron Block

A 500-kg iron block shown in Fig. 7-11 is initially at 200°C and is allowed to cool to 27°C by transferring heat to the surrounding air at 27°C. Determine the reversible work and the irreversibility for this process.

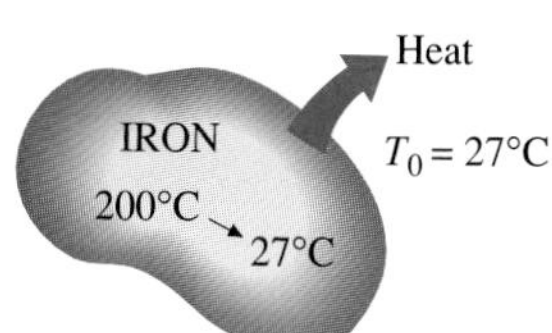

FIGURE 7-11
Schematic for Example 7-4.

Solution We take the *iron block* as the system. This is a *closed system* since no mass crosses the system boundary. We note that heat is lost from the system.

Assumptions **1** The kinetic and potential energies are negligible. **2** The process involves no work interactions.

Analysis It probably came as a surprise to you that we are asking to find the "reversible work" for a process that does not involve any work interactions. Well, even if no attempt is made to produce work during this process, the potential to do work still exists, and the reversible work is a quantitative measure of this potential.

The reversible work in this case is determined by considering a series of imaginary reversible heat engines operating between the source (at a variable temperature T) and the sink (at a constant temperature T_0), as shown in Fig. 7-12. Summing their work output:

$$\delta W_{rev} = \eta_{th,\,rev}Q_{in} = \left(1 - \frac{T_{sink}}{T_{source}}\right)\delta Q_{in} = \left(1 - \frac{T_0}{T}\right)\delta Q_{in}$$

and

$$W_{rev} = \int\left(1 - \frac{T_0}{T}\right)\delta Q_{in}$$

The source temperature T changes from T_1 = 200°C = 473 K to T_0 = 27°C = 300 K during this process. A relation for the differential heat transfer from the iron block can be obtained from the differential form of the energy balance applied on the iron block,

$$\underbrace{\delta E_{in} - \delta E_{out}}_{\substack{\text{Net energy transfer}\\ \text{by heat, work, and mass}}} = \underbrace{dE_{system}}_{\substack{\text{Change in internal, kinetic,}\\ \text{potential, etc., energies}}}$$

$$-\delta Q_{out} = dU = mC_{av}\,dT$$

Then,

$$\delta Q_{in,\,heat\,engine} = \delta Q_{out,\,system} = -mC_{av}\,dT$$

FIGURE 7-12
An irreversible heat transfer process can be made reversible by the use of a reversible heat engine.

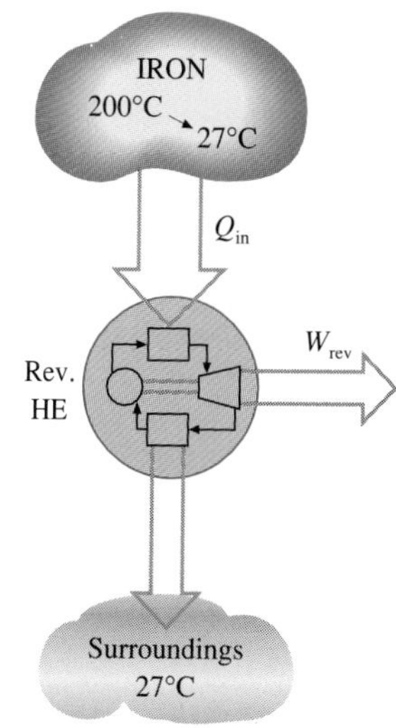

since heat transfers from the iron and to the heat engine are equal in magnitude and opposite in direction. Substituting and performing the integration, the reversible work is determined to be

$$W_{rev} = \int_{T_1}^{T_0}\left(1 - \frac{T_0}{T}\right)(-\,mC_{av}\,dT) = mC_{av}\,(T_1 - T_0) - mC_{av}\,T_0 \ln\frac{T_1}{T_0}$$

$$= (500\text{ kg})[0.45\text{ kJ/(kg}\cdot\text{K)}]\left[(473 - 300)\text{K} - (300\text{K})\ln\frac{473\text{ K}}{300\text{ K}}\right]$$

$$= \mathbf{8191\ kJ}$$

where the specific heat value is obtained from Table A-3. The first term in the above equation $[Q = mC_{av}(T_1 - T_0) = 38{,}925\text{ kJ}]$ is the total heat transfer from the iron block to the heat engine. The reversible work for this problem is found to be 8191 kJ, which means that 8191 (21 percent) of the 38,925 kJ of heat transferred from the iron block to the ambient air *could* have been converted to work. If the specified ambient temperature of 27°C is the lowest available environment temperature, the reversible work determined above also represents the exergy, which is the maximum work potential of the sensible energy contained in the iron block.

The irreversibility for this process is determined from its definition,

$$I = W_{rev} - W_u = 8191 - 0 = \mathbf{8191\ kJ}$$

Notice that the reversible work (the work potential) and irreversibility (the wasted work potential) are the same for this case since the entire work potential is wasted. The source of irreversibility in this process is the heat transfer through a finite temperature difference.

EXAMPLE 7-5 Heating Potential of a Hot Iron Block

The iron block discussed in Example 7-4 is to be used to maintain a house at 27°C when the outdoor temperature is 5°C. Determine the maximum amount of heat that can be supplied to the house as the iron cools to 27°C.

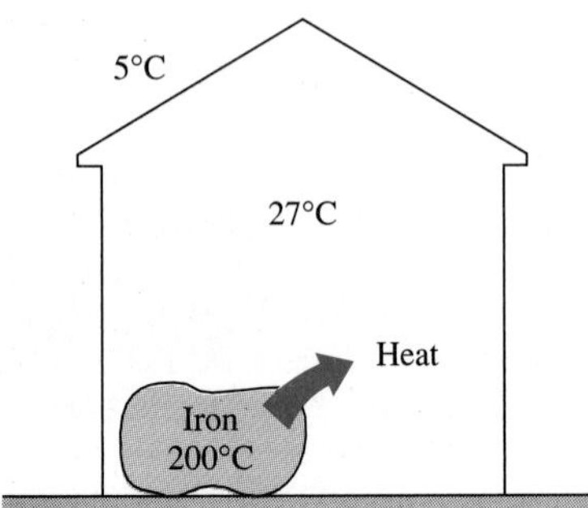

FIGURE 7-13
Schematic for Example 7-5.

Solution Probably the first thought that comes to mind to make the most use of the energy stored in the iron block is to take it inside and let it cool in the house, as shown in Fig. 7-13, transferring its sensible energy as heat to the indoors air (provided that it meets the approval of the household, of course). The iron block can keep "losing" heat until its temperature drops to the indoor temperature of 27°C, transferring a total of 38,925 kJ of heat. Since we utilized the entire energy of the iron block available for heating without wasting a single kilojoule, it seems like we have a 100-percent-efficient operation, and nothing can beat this, right? Well, not quite.

In Example 7-4 we determined that this process has an irreversibility of 8191 kJ, which implies that things are not as "perfect" as they seem. A "perfect" process is one that involves "zero" irreversibility. The irreversibility in this process is associated with the heat transfer through a finite temperature difference that can be eliminated by running a reversible heat engine between the iron block and the indoor air. This heat engine will produce (as determined in Example 7-4) 8191 kJ of work and reject the remaining 38,925 − 8191 = 30,734 kJ of heat to the house. Now we managed to eliminate the irreversibility and ended up with 8191 kJ of work. What can we do with this work? Well, at worst we can convert it to heat by running a paddle wheel, for example, creating an equal amount of irreversibility. Or we can supply this work to a heat pump that will transport heat from the outdoors at 5°C to the indoors at 27°C. Such a heat pump, if reversible, will have a coefficient of performance of

$$\text{COP}_{\text{HP}} = \frac{1}{1 - T_L/T_H} = \frac{1}{1 - (278\text{ K})/(300\text{ K})} = 13.6$$

That is, this heat pump can supply the house with 13.6 times the energy it consumes as work. In our case, it will consume the 8191 kJ of work and deliver 8191 × 13.6 = 111,398 kJ of heat to the house. Therefore, the hot iron block has the potential of supplying

$$(30{,}734 + 111{,}398)\text{ kJ} = \mathbf{142{,}132\ kJ}$$

of heat to the house. The irreversibility for this process is zero, and this is *the best* we can do under the specified conditions. A similar argument can be given for the electric heating of residential or commercial buildings.

Now try to answer the following question: What would happen if the heat engine were operated between the iron block and the outside air instead of the house until the temperature of the iron block fell to 27°C? Would the amount of heat supplied to the house still be 142,132 kJ? Here is a hint: The initial and final states in both cases are the same, and the irreversibility for both cases is zero.

7-3 ■ SECOND-LAW EFFICIENCY η_{II}

In Chap. 5 we defined the *thermal efficiency* and the *coefficient of performance* for devices as a measure of their performance. They were defined on the basis of the first law only, and they are sometimes referred to as the *first-law efficiencies*. The first law efficiency (also known as the *conversion efficiency*), however, makes no reference to the best possible performance, and thus it may be misleading.

Consider two heat engines, both having a thermal efficiency of 30 percent, as shown in Fig. 7-14. One of the engines (engine *A*) is supplied with heat from a source at 600 K, and the other one (engine *B*) from a source at 1000 K. Both engines reject heat to a medium at 300 K. At first glance, both engines seem to convert to work the same fraction of heat that they receive; thus they are performing equally well. When we take a second look at these engines in light of the second law of thermodynamics, however, we see a totally different picture. These engines, at best, can perform as reversible (Carnot) engines, in which case their efficiencies would be

$$\eta_{\text{rev},A} = \left(1 - \frac{T_L}{T_H}\right)_A = 1 - \frac{300\text{ K}}{600\text{ K}} = 50\%$$

$$\eta_{\text{rev},B} = \left(1 - \frac{T_L}{T_H}\right)_B = 1 - \frac{300\text{ K}}{1000\text{ K}} = 70\%$$

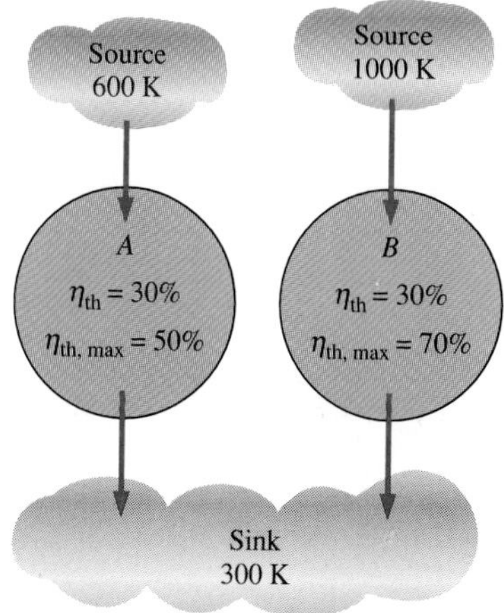

FIGURE 7-14

Now it is becoming apparent that engine *B* has a greater work potential available to it (70 percent of the heat supplied as compared to 50 percent for engine *A*), and thus should do a lot better than engine *A*. Therefore, we can say that engine B is performing poorly relative to engine A even though both have the same thermal efficiency.

It is obvious from this example that the first-law efficiency alone is not a realistic measure of performance of engineering devices. To overcome this deficiency, we define a **second-law efficiency** η_{II} as the ratio of the actual thermal efficiency to the maximum possible (reversible) thermal efficiency under the same conditions (Fig. 7-15):

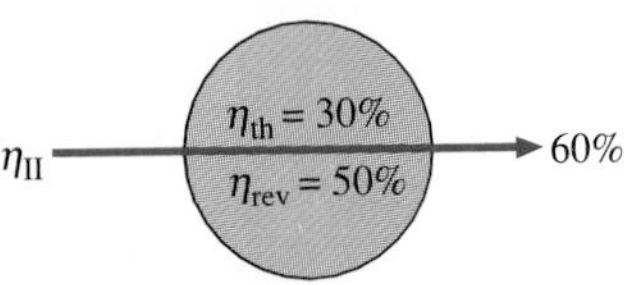

FIGURE 7-15
Second-law efficiency is a measure of the performance of a device relative to its performance under reversible conditions.

$$\eta_{II} = \frac{\eta_{th}}{\eta_{th,rev}} \quad \text{(heat engines)} \tag{7-4}$$

Based on this definition, the second-law efficiencies of the two heat engines discussed above are

$$\eta_{II,A} = \frac{0.30}{0.50} = 0.60 \quad \text{and} \quad \eta_{II,B} = \frac{0.30}{0.70} = 0.43$$

That is, engine A is converting 60 percent of the available work potential to useful work. This ratio is only 43 percent for engine B.

The second-law efficiency can also be expressed as the ratio of the useful work output and the maximum possible (reversible) work output:

$$\eta_{II} = \frac{W_u}{W_{rev}} \quad \text{(work-producing devices)} \tag{7-5}$$

This definition is more general since it can be applied to processes (in turbines, piston-cylinder devices, etc.) as well as to cycles. Note that the second-law efficiency cannot exceed 100 percent (Fig. 7-16).

FIGURE 7-16
The second-law efficiency of all reversible devices is 100 percent.

We can also define a second-law efficiency for work-consuming noncyclic (such as compressors) and cyclic (such as refrigerators) devices as the ratio of the minimum (reversible) work input to the useful work input:

$$\eta_{II} = \frac{W_{rev}}{W_u} \quad \text{(work-consuming devices)} \tag{7-6}$$

For cyclic devices such as refrigerators and heat pumps, it can also be expressed in terms of the coefficients of performance as

$$\eta_{II} = \frac{\text{COP}}{\text{COP}_{rev}} \quad \text{(refrigerators and heat pumps)} \tag{7-7}$$

Again, because of the way we defined the second-law efficiency, its value cannot exceed 100 percent. In the above relations, the reversible work W_{rev} should be determined by using the same initial and final states as in the actual process.

The definitions above for the second-law efficiency do not apply to devices that are not intended to produce or consume work. Therefore, we need a more general definition. But there is no agreement on a general definition of the second-law efficiency, and thus a person may encounter different definitions for the same device. The second-law efficiency is intended to serve as a measure of approximation to reversible operation, and thus its value should change from zero in the worst case (complete destruction of exergy) to one in the best case (no destruction of exergy). With this in mind, we define the second-law efficiency of a system during a process as

$$\eta_{II} = \frac{\text{Exergy recovered}}{\text{Exergy supplied}} = 1 - \frac{\text{Exergy destroyed}}{\text{Exergy supplied}} \tag{7-8}$$

Therefore, when determining the second-law efficiency, the first thing we need to do is determine how much exergy or work potential is consumed during a process. In a reversible operation, we should be able to recover entirely the exergy supplied during the process, and the irreversibility in this case should be zero. The second-law efficiency will be zero when we recover none of the exergy supplied to the system. Note that the exergy can be supplied or recovered at various amounts in various forms such as heat, work, kinetic energy, potential energy, internal energy, and enthalpy. Sometimes there are differing (though valid) opinions on what constitutes supplied exergy, and this causes differing definitions for second-law efficiency. But at all times, the exergy recovered and the exergy destroyed (the irreversibility) must add up to the exergy supplied. Also, we need to define the system precisely in order to identify correctly any interactions between the system and its surroundings.

For a *heat engine,* the exergy supplied is the decrease in the exergy of the heat transferred to the engine, which is the difference between the exergy of the heat supplied and the exergy of the heat rejected. (The exergy of the heat rejected at the temperature of the surroundings is zero.) The net work output is the recovered exergy.

For a *refrigerator* or *heat pump,* the exergy supplied is the work input W since the work supplied to a cyclic device is entirely available. The recovered exergy is the exergy of the heat transferred to the high-temperature medium (which is the reversible work) for a heat pump, and the exergy of the heat transferred from the low-temperature medium for a refrigerator.

For a heat exchanger with two unmixed fluid streams, normally the exergy is the decrease in the exergy of the higher-temperature fluid stream, and the exergy recovered is the increase in the exergy of the lower-temperature fluid stream. This is discussed further in Sec. 7-9.

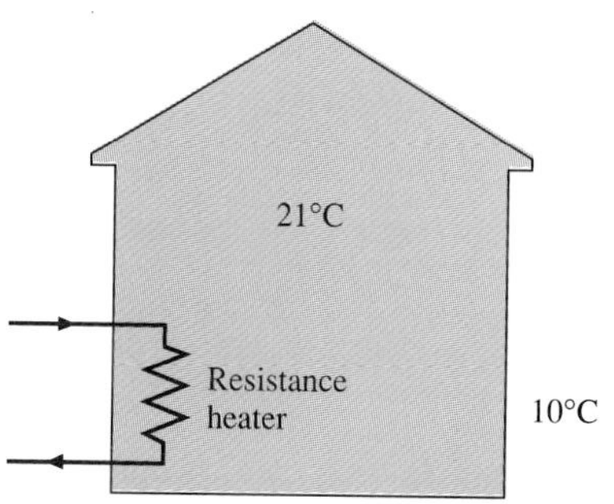

FIGURE 7-17
Schematic for Example 7-6.

EXAMPLE 7-6 Second-Law Efficiency of Resistance Heaters

A dealer advertises that he has just received a shipment of electric resistance heaters for residential buildings that have an efficiency of 100 percent, as shown in Fig. 7-17. Assuming an indoor temperature of 21°C and outdoor temperature of 10°C, determine the second-law efficiency of these heaters.

Solution Obviously the efficiency that the dealer is referring to is the first-law efficiency, meaning that for each unit of electric energy (work) consumed, the heater will supply the house with 1 unit of energy (heat). That is, the advertised heater has a COP of 1.

At the specified conditions, a reversible heat pump would have a coefficient of the performance of

$$\mathrm{COP_{HP,\,rev}} = \frac{1}{1 - T_L/T_H} = \frac{1}{1 - (283\ \mathrm{K})/(294\ \mathrm{K})} = 26.7$$

That is, it would supply the house with 26.7 units of heat (extracted from the cold outside air) for each unit of electric energy it consumes.

The second-law efficiency of this resistance heater is determined from Eq. 7-7 to be

$$\eta_{\mathrm{II}} = \frac{\mathrm{COP}}{\mathrm{COP_{rev}}} = \frac{1.0}{26.7} = \mathbf{0.037\ or\ 3.7\%}$$

which does not look so impressive. The dealer will not be happy to see this value. Considering the high price of electricity, a consumer will probably be better off with a "less" efficient gas heater.

7-4 ■ EXERGY ASSOCIATED WITH ke, pe, u, Pv, and h

We mentioned earlier that *energy* can exist in numerous forms such as kinetic, potential, internal, flow, and enthalpy (the sum of internal and flow energies), and their sum constitutes the total energy of a system. *Exergy* is the useful work potential of energy, and thus the exergy of a system is the sum of the exergies of different forms of energy it contains. Therefore, the determination of the exergy of a system requires the identification of the different forms of energy contained in the system together with the exergy relations for each form of energy. Once they are available, the exergy of a system at a specified state is determined by simply adding the exergies of each form of energy the system possesses. Below we will develop relations for the exergy content of different energy forms associated with simple compressible systems such as the kinetic energy, potential energy, internal energy, flow energy, and enthalpy in an environment at temperature T_0 and pressure P_0.

Exergy (Work Potential) Associated with Kinetic Energy, *ke*

Kinetic energy is a form of *mechanical energy,* and thus is can be converted to work entirely. Therefore, the *work potential* or *exergy* of the kinetic energy of a system is equal to the kinetic energy itself regardless of the temperature and pressure of the environment. That is,

Exergy of kinetic energy: $$x_{ke} = ke = \frac{\mathcal{V}^2}{2} \qquad \text{(kJ/kg)} \qquad (7\text{-}9)$$

where $\mathcal{V}$ is the velocity of the system relative to the environment.

Exergy Associated with Potential Energy, *pe*

Potential energy is also a form of *mechanical energy,* and thus it can also be converted to work entirely. Therefore, the *work potential* or *exergy* of the potential energy of a system is equal to the potential energy itself regardless of the temperature and pressure of the environment (Fig. 7-18). That is,

$$\textit{Exergy of potential energy:} \qquad x_{pe} = pe = gz \qquad \text{(kJ/kg)} \qquad (7\text{-}10)$$

where g is the gravitational acceleration and z is the elevation of the system relative to a reference level in the environment.

Therefore, the exergies of kinetic and potential energies are equal to themselves, and they are entirely available for work. But the internal energy u and enthalpy h of a system are not entirely available for work, as shown below.

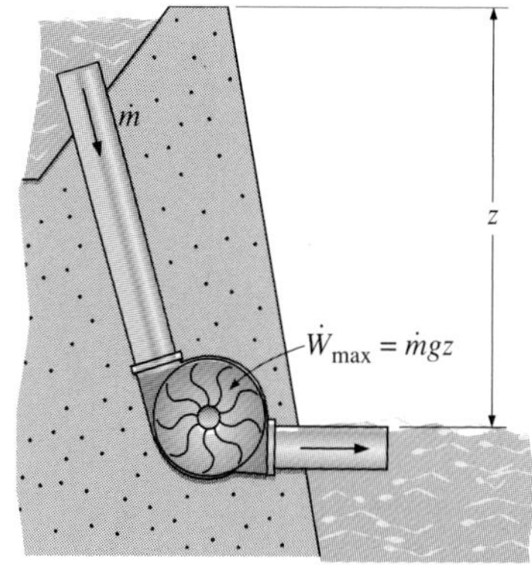

FIGURE 7-18
The *work potential* or *exergy* of potential energy is equal to the potential energy itself.

Exergy (Work Potential) Associated with Internal Energy, *u*

In general, internal energy consists of *sensible, latent, chemical,* and *nuclear* energies. But in the absence of any chemical or nuclear reactions, the chemical and nuclear energies can be disregarded and the internal energy can be considered to consist of only sensible and latent energies that can be transferred to or from a system as *heat* whenever there is a temperature difference across the system boundary. The second law of thermodynamics states that heat cannot be converted to work entirely, and thus the work potential of internal energy must be less than the internal energy itself. But how much less?

To answer that question, we need to consider a stationary closed system at a specified state that undergoes a *reversible* process to the state of the environment (that is, the final temperature and pressure of the system should be T_0 and P_0, respectively). The useful work delivered during this process is the exergy of the system at its initial state, which is equivalent to the exergy of the internal energy of the system since the only form of energy a stationary simple compressible closed system possesses is internal energy (Fig. 7-19).

Consider a piston-cylinder device that contains a fluid of mass m at temperature T and pressure P. The system (the mass inside the cylinder) has a volume V, internal energy U, and entropy S. The system is now allowed to undergo a differential change of state during which the volume changes by a differential amount dV and heat is transferred in the differential amount of δQ. Taking the direction of heat and work transfers to be *from* the system (heat and work outputs), the energy balance for the system during this differential process can be expressed as

$$\underbrace{\delta E_{in} - \delta E_{out}}_{\substack{\text{Net energy transfer} \\ \text{by heat, work and mass}}} = \underbrace{dE_{system}}_{\substack{\text{Change in internal, kinetic,} \\ \text{potential, etc., energies}}}$$

$$-\delta Q - \delta W = dU \qquad (7\text{-}11)$$

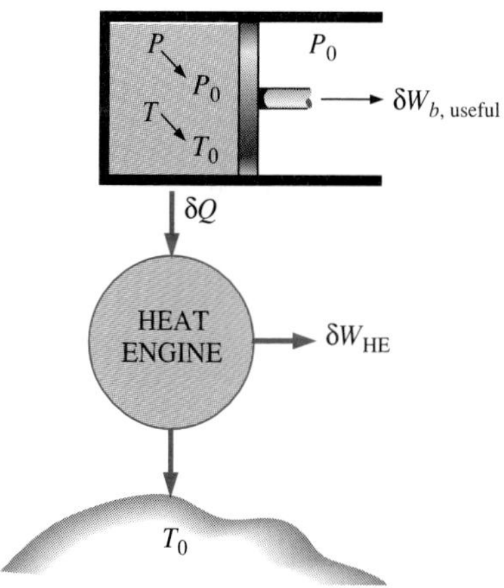

FIGURE 7-19
The *exergy* of a specified mass at a specified state is the useful work that can be produced as it undergoes a reversible process to the state of the environment.

since the only form of energy the system contains is *internal energy,* and the only forms of energy transfer a fixed mass can involve are heat and work. Also, the only form of work a simple compressible system can involve during a reversible process is the boundary work, which is given to be $\delta W = P\,dV$ when the direction of work is taken to be from the system (otherwise it would be $-P\,dV$). The pressure P in the $P\,dV$ expression is the absolute pressure, which is measured from absolute zero. Any useful work delivered by a piston-cylinder device is due to the pressure above the atmospheric level. Therefore,

$$\delta W = P\,dV = (P - P_0)dV + P_0\,dV = \delta W_{b,\text{useful}} + P_0\,dV \qquad \text{(7-12)}$$

A reversible process cannot involve any heat transfer through a finite temperature difference, and thus any heat transfer between the system at temperature T and its surroundings at T_0 must occur through a reversible heat engine. Noting that $dS = \delta Q/T$ for a reversible process, and the thermal efficiency of a reversible heat engine operating between the temperatures of T and T_0 is $\eta_{\text{th}} = 1 - T_0/T$, the differential work produced by the engine as a result of this heat transfer is

$$\delta W_{\text{HE}} = \left(1 - \frac{T_0}{T}\right)\delta Q = \delta Q - \frac{T_0}{T}\delta Q = \delta Q + T_0\,dS \;\rightarrow\; \delta Q = \delta W_{\text{HE}} - T_0\,dS \qquad \text{(7-13)}$$

Substituting the δW and δQ expressions in Eqs. 7-12 and 7-13 into the energy balance relation (Eq. 7-11) gives, after rearranging,

$$\delta W_{\text{total useful}} = \delta W_{\text{HE}} + \delta W_{b,\text{useful}} = -dU - P_0\,dV + T_0\,dS$$

Integrating from the given state (no subscript) to the dead state (0 subscript) we obtain

$$W_{\text{total useful}} = (U - U_0) + P_0(V - V_0) - T_0(S - S_0) \qquad \text{(7-14)}$$

where $W_{\text{total useful}}$ is the total useful work delivered as the system undergoes a reversible process from the given state to the dead state, which is *exergy* by definition. Then the *useful work potential* or *exergy* of internal energy can be expressed on a unit-mass basis as

Exergy of internal energy: $\quad x_u = \phi = (u - u_0) + P_0(v - v_0) - T_0(s - s_0) \qquad \text{(7-15)}$

where u_0, v_0, and s_0 are the properties of the *system* evaluated at the dead state. Note that the exergy of the internal energy of a system that is at the dead state is zero since $u = u_0$, $v = v_0$, and $s = s_0$ at that state.

The exergy of internal energy (or a closed system) is either *positive* or *zero.* It is never negative. Even a medium at *low temperature* ($T < T_0$) and/or *low pressure* ($P < P_0$) contains exergy since a cold medium can serve as the heat sink to a heat engine that absorbs heat from the environment at T_0, and an evacuated space makes it possible for the atmospheric pressure to move a piston and do work (Fig. 7-20).

FIGURE 7-20

The *exergy* of a cold medium is also a *positive* quantity since work can be produced by transferring heat to it.

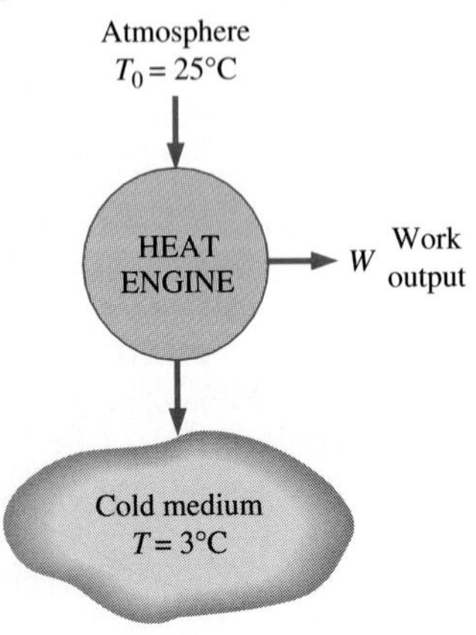

Exergy (Work Potential) Associated with Flow Work, *Pv*

The *flow work* or *flow energy* was defined in Chap. 4 as the energy needed to push a unit mass of a fluid into or out of a control volume, or to maintain flow in a pipe or duct, and was determined to be

$$w_{\text{flow}} = Pv \qquad \text{(kJ/kg)}$$

where v is the specific volume of the fluid, which is equivalent to the *volume change* of a unit mass of the fluid as it is displaced during flow. The flow work is essentially the boundary work done by the fluid on the fluid downstream, and thus the exergy of flow work is equivalent to the exergy of the boundary work, which is the boundary work in excess of the work done against the atmospheric air at P_0 to displace it by a volume v (Fig. 7-21). Noting that the flow work is Pv and the work done against the atmosphere is P_0v, the *useful work potential* or *exergy* of flow work can be expressed as

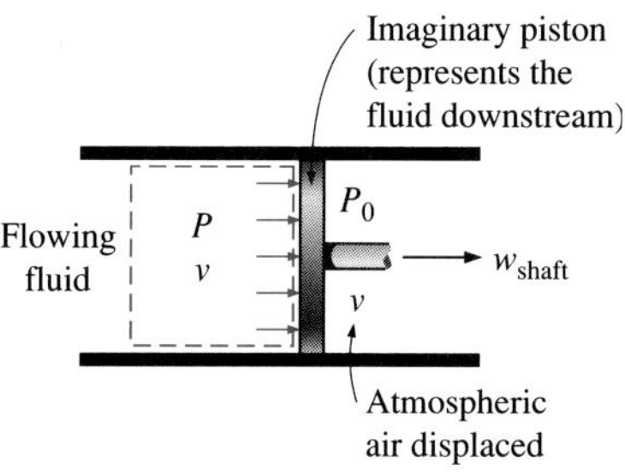

FIGURE 7-21
The exergy of flow work is the useful work that would be delivered by an imaginary piston in the flow section.

Exergy of flow work: $$x_{Pv} = Pv - P_0v = (P - P_0)v \qquad \text{(kJ/kg)} \qquad (7\text{-}16)$$

Therefore, the exergy of flow work is obtained by replacing the pressure P in the flow work relation by the pressure in excess of the atmospheric pressure, $P - P_0$. Note that the exergy of the flow work of a flowing fluid at the atmospheric pressure is *zero*. Also note that the exergy of flow work is *negative* when the pressure of the flowing fluid is less than the atmospheric pressure, which means that work must be done on the flowing fluid to bring it to the state of the environment (the dead state, which is the state of zero exergy).

Exergy (Work Potential) Associated with Enthalpy, *h*

We could determine the exergy content of enthalpy by following the approach used to determine the exergy content of internal energy by considering a unit mass in a flow stream at a specified state with negligible kinetic and potential energies that undergoes a process to the dead state in a reversible manner. The useful work delivered during this process would be the exergy of the stream at its initial state, which is equivalent to the exergy associated with the enthalpy of the fluid stream. Instead, we leave this as an exercise to the student, and determine the exergy of enthalpy directly by using its definition.

Enthalpy was defined as the sum of the internal and flow energies, $h = u + Pv$. Therefore, the exergy associated with enthalpy is simply the sum of the exergies of its components (Fig. 7-22). That is,

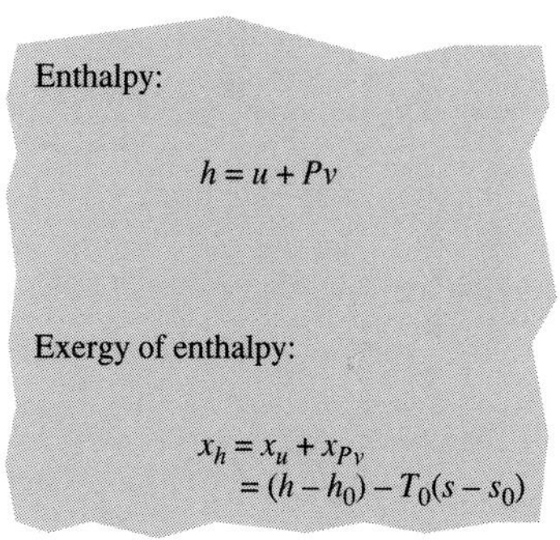

FIGURE 7-22
The *exergy* of enthalpy is the sum of the exergies of the internal energy and flow energy.

$$\begin{aligned} x_h &= x_u + x_{Pv} \\ &= [(u - u_0) + P_0(v - v_0) - T_0(s - s_0)] + (P - P_0)v \\ &= (u + Pv) - (u_0 + P_0v_0) - T_0(s - s_0) \end{aligned}$$

or

Exergy of enthalpy: $x_h = (h - h_0) - T_0(s - s_0)$ (kJ/kg) (7-17)

Here h_0 and s_0 are the enthalpy and entropy of the *fluid* at the dead state. Note that the exergy of flow energy is negative at a pressure below the atmospheric pressure, and thus the exergy of enthalpy can be negative at subatmospheric pressures.

7-5 ■ EXERGY CHANGE OF A SYSTEM

The property *exergy* is the work potential of a system in a specified environment and represents the maximum amount of useful work that can be obtained as the system is brought to equilibrium with the environment. Unlike energy, the value of exergy depends on the state of the environment as well as the state of the system. Therefore, exergy is a combination property. The exergy of a system that is in equilibrium with the environment it is in is zero. The state of the environment is referred to as the "dead state" since the system is practically "dead" (cannot do any work) from a thermodynamic point of view when it reaches that state.

In this section we will limit the discussion to **thermo-mechanical exergy,** and thus disregard any mixing and chemical reactions. Therefore, a system at this "restricted dead state" will be at the temperature and pressure of the environment and it will have no kinetic or potential energies relative to the environment. However, it may have a different chemical composition than the environment. Exergy associated with different chemical compositions and chemical reactions is discussed in later chapters.

Below we develop relations for the exergies and exergy changes for a fixed mass and a flow stream.

Exergy of a Fixed Mass: Nonflow (or Closed System) Exergy

For simple compressible systems, the total energy of a specified *nonflowing* mass is the sum of its internal, kinetic, and potential energies. Therefore, the exergy of such a system is simply the sum of the exergies of its internal, kinetic, and potential energies. That is, $x_{\text{nonflow}} = x_u + x_{\text{ke}} + x_{\text{pe}}$. This is called the **nonflow** (or **closed system**) **exergy** ϕ and is expressed on a unit-mass basis as

Nonflow exergy:

$$\begin{aligned}\phi &= (u - u_0) + P_0(v - v_0) - T_0(s - s_0) + \frac{\mathcal{V}^2}{2} + gz \\ &= (e - e_0) + P_0(v - v_0) - T_0(s - s_0)\end{aligned} \quad (7\text{-}18)$$

Then the *exergy change* of a unit mass as it undergoes a process from state 1 to state 2 becomes

$$\Delta\phi = \phi_2 - \phi_1 = (u_2 - u_1) + P_0(v_2 - v_1) - T_0(s_2 - s_1) + \frac{V_2^2 - V_1^2}{2} + g(z_2 - z_1)$$
$$= (e_2 - e_1) + P_0(v_2 - v_1) - T_0(s_2 - s_1) \quad (7\text{-}19)$$

The determination of exergy change of a system during a process involves evaluating the exergy of the system at the beginning and at the end of the process and taking their difference. That is,

Exergy change = Exergy at final state − Exergy at initial state

or

$$\Delta X = X_2 - X_1 = m(\phi_2 - \phi_1) = (E_2 - E_1) + P_0(V_2 - V_1) - T_0(S_2 - S_1)$$
$$= (U_2 - U_1) + P_0(V_2 - V_1) - T_0(S_2 - S_1) + m\frac{V_2^2 - V_1^2}{2} + mg(z_2 - z_1) \quad (7\text{-}20)$$

For *stationary* closed systems, the kinetic and potential energy terms drop out.

When the properties of a system are not uniform, the exergy of the system can be determined by integration from

$$X_{\text{system}} = \int \phi \delta m = \int_V \phi\rho\, dV \quad (7\text{-}21)$$

where V is the volume of the system and ρ is density.

Note that exergy is a property, and the value of a property does not change unless the *state* changes. Therefore, the *exergy change* of a system is *zero* if the state of the system or the environment does not change during the process. For example, the exergy change of steady-flow devices such as nozzles, compressors, turbines, pumps, and heat exchangers in a given environment is zero during steady operation.

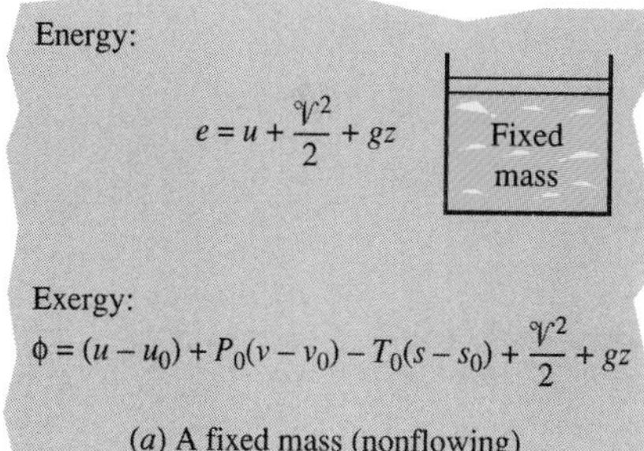

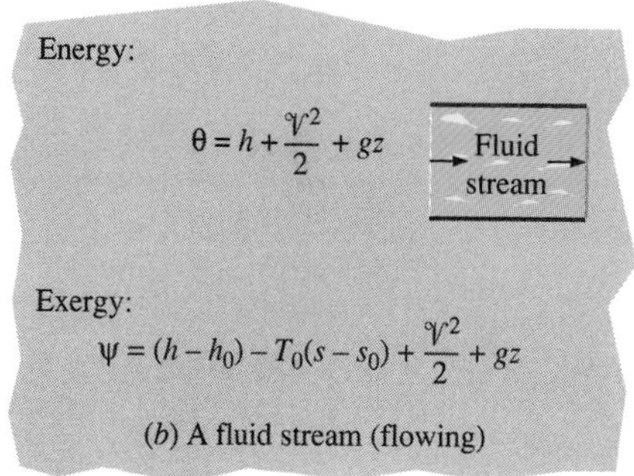

FIGURE 7-23
The *energy* and *exergy* contents of (*a*) a fixed mass and (*b*) a fluid stream.

Exergy of a Flow Stream: Flow (or Stream) Exergy

The total energy of a *flowing* fluid is the sum of the enthalpy, kinetic energy, and potential energy. Therefore, the exergy of such a fluid is simply the sum of the exergies of its enthalpy, kinetic energy, and potential energy. That is, $x_{\text{flow}} = x_h + x_{\text{ke}} + x_{\text{pe}}$. This is called the **flow** (or **stream**) **exergy** ψ, and is expressed on a unit-mass basis as (Fig. 7-23)

Flow exergy:
$$\psi = (h - h_0) - T_0(s - s_0) + \frac{V^2}{2} + gz \quad (7\text{-}22)$$

Then the *exergy change* of a fluid stream as it undergoes a process from state 1 to state 2 becomes

$$\Delta\psi = \psi_2 - \psi_1 = (h_2 - h_1) - T_0(s_2 - s_1) + \frac{V_2^2 - V_1^2}{2} + g(z_2 - z_1) \quad (7\text{-}23)$$

For fluid streams with negligible kinetic and potential energies, the kinetic and potential energy terms drop out.

Note that the *exergy change* of a closed system or a fluid stream represents the *maximum* amount of useful work that can be done (or the *minimum* amount of useful work that needs to be supplied if it is negative) as the system changes from state 1 to state 2 in a specified environment, and represents the *reversible work* W_{rev}. It is independent of the type of process executed, the kind of system used, and the nature of energy interactions with the surroundings. Also note that the exergy of a closed system cannot be negative, but the exergy of a flow stream can at pressures below the environment pressure P_0.

EXAMPLE 7-7 Work Potential of Compressed Air in a Tank

A 200-m^3 rigid tank contains compressed air at 1 MPa and 300 K. Determine how much work can be obtained from this air if the environment conditions are 100 kPa and 300 K.

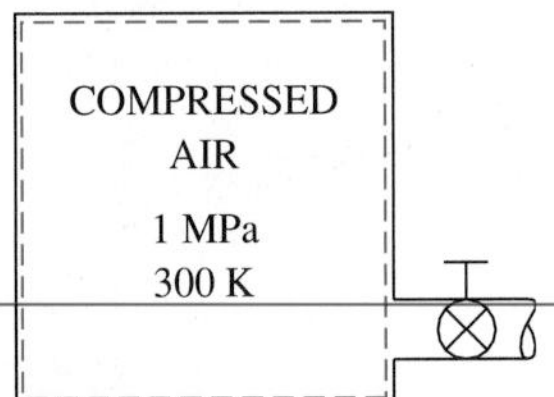

FIGURE 7-24
Schematic for Example 7-7.

Solution We take the *rigid tank* as the system (Fig. 7-24). This is a *closed system* since no mass crosses the system boundary during the process. Here the question is the work potential of a fixed mass, which is the nonflow exergy by definition.

Assumptions **1** Air is an ideal gas since it is at a high temperature and low pressure relative to its critical point values of −141°C and 3.77 MPa. **2** The kinetic and potential energies are negligible.

Analysis Taking the state of the air in the tank to be state 1 and noting that $T_1 = T_0 = 300$ K, the mass of air in the tank is determined to be

$$m_1 = \frac{P_1 V}{RT_1} = \frac{(1000 \text{ kPa})(200 \text{ m}^3)}{[0.287 \text{ kPa}\cdot\text{m}^3/(\text{kg}\cdot\text{K})](300\text{K})} = 2323 \text{ kg}$$

The exergy content of the compressed air can be determined from

$$X_1 = m\phi_1 = m\left[(u_1 - u_0)^{\nearrow 0} + P_0(v_1 - v_0) - T_0(s_1 - s_0) + \frac{V_1^2}{2}^{\nearrow 0} + gz_1^{\nearrow 0}\right]$$
$$= m[P_0(v_1 - v_0) - T_0(s_1 - s_0)]$$

We note that

$$P_0(v_1 - v_0) = P_0\left(\frac{RT_1}{P_1} - \frac{RT_0}{P_0}\right) = RT_0\left(\frac{P_0}{P_1} - 1\right) \qquad (\text{since } T_1 = T_0)$$

$$T_0(s_1 - s_0) = T_0\left(C_p \ln\frac{T_1}{T_0}^{\nearrow 0} - R\ln\frac{P_2}{P_0}\right) = -RT_0 \ln\frac{P_1}{P_0} \qquad (\text{since } T_1 = T_0)$$

Therefore,

$$\phi_1 = RT_0\left(\frac{P_0}{P_2} - 1\right) + RT_0 \ln\frac{P_1}{P_0} = RT_0\left(\ln\frac{P_1}{P_0} + \frac{P_0}{P_1} - 1\right)$$
$$= [0.287\ \text{kJ/(kg}\cdot\text{K)}](300\text{K})\left(\ln\frac{1000\ \text{kPa}}{100\ \text{kPa}} + \frac{100\ \text{kPa}}{1000\ \text{kPa}} - 1\right)$$
$$= 120.76\ \text{kJ/kg}$$

and

$$X_1 = m_1\phi_1 = (2323\ \text{kg})(120.76\ \text{kJ/kg}) = \mathbf{280{,}525\ kJ}$$

Discussion The work potential of the system is 280,525 kJ, and thus a maximum of 280,525 kJ of useful work can be obtained from the compressed air stored in the tank in the specified environment.

EXAMPLE 7-8 Exergy Change during a Compression Process

Refrigerant-134a is to be compressed from 0.14 MPa and −10°C to 0.8 MPa and 50°C steadily by a compressor. Taking the environment conditions to be 20°C and 95 kPa, determine the exergy change of the refrigerant during this process and the minimum work input that needs to be supplied to the compressor per unit mass of the refrigerant.

Solution We take the *compressor* as the system (Fig. 7-25). This is a *control volume* since mass crosses the system boundary during the process. Here the question is the exergy change of a fluid stream, which is the change in the flow exergy ψ.

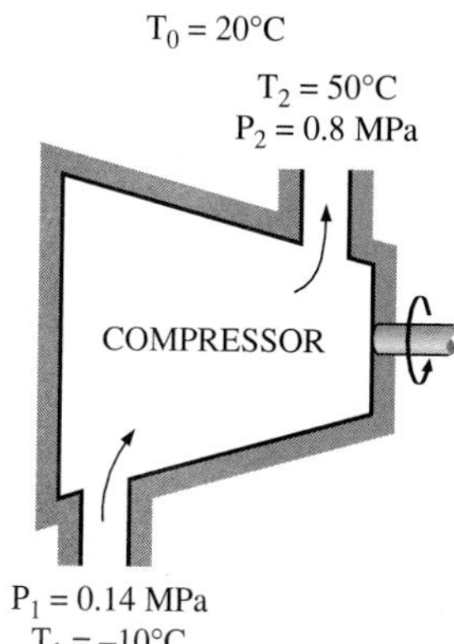

FIGURE 7-25
Schematic for Example 7-8.

Assumptions **1** Steady operating conditions exist. **2** The kinetic and potential energies are negligible.

Analysis The properties of the refrigerant at the inlet and the exit states are

Inlet state: $\left.\begin{array}{l} P_1 = 0.14\ \text{MPa} \\ T_1 = -10°\text{C} \end{array}\right\}$ $\begin{array}{l} h_1 = 243.40\ \text{kJ/kg} \\ s_1 = 0.9606\ \text{kJ/(kg}\cdot\text{K)} \end{array}$

Exit state: $\left.\begin{array}{l} P_2 = 0.8\ \text{MPa} \\ T_2 = 50°\text{C} \end{array}\right\}$ $\begin{array}{l} h_2 = 284.39\ \text{kJ/kg} \\ s_2 = 0.9711\ \text{kJ/(kg}\cdot\text{K)} \end{array}$

The exergy change of the refrigerant during this compression process is determined directly from Eq. 7-23 to be

$$\Delta\psi = \psi_2 - \psi_1 = (h_2 - h_1) - T_0(s_2 - s_1) + \frac{V_2^2 - V_1^2}{2}\nearrow^0 + g(z_2 - z_1)\nearrow^0$$
$$= (h_2 - h_1) - T_0(s_2 - s_1)$$
$$= (284.39 - 243.40)\ \text{kJ/kg} - (293.15\ \text{K})[(0.9711 - 0.9606)\text{kJ/(kg}\cdot\text{K)}]$$
$$= \mathbf{37.9\ kJ/kg}$$

Therefore, the exergy of the refrigerant will increase during compression by 37.9 kJ/kg.

The exergy change of a system in a specified environment represents the reversible work in that environment, which is the minimum work input required for work-consuming devices such as compressors. Therefore, the increase in exergy of the refrigerant is equal to the minimum work that needs to be supplied to the compressor,

$$W_{\text{in, min}} = \psi_2 - \psi_1 = \mathbf{37.9\ kJ/kg}$$

Note that if the compressed refrigerant at 0.8 MPa and 50°C were to be expanded to 0.14 MPa and −10°C in a turbine in the same environment in a reversible manner, 37.9 kJ/kg of work would be produced.

7-6 ■ EXERGY TRANSFER BY HEAT, WORK, AND MASS

Exergy, like energy, can be transferred to or from a system in three forms: *heat, work,* and *mass flow*. Exergy transfer is recognized at the system boundary as exergy crosses it, and it represents the exergy gained or lost by a system during a process. The only two forms of exergy interactions associated with a fixed mass or closed system are *heat transfer* and *work*.

Exergy Transfer by Heat Transfer, Q

HEAT SOURCE

Temperature: T

Energy content: E

Exergy $= \left(1 - \frac{T_0}{T}\right)E$

T_0

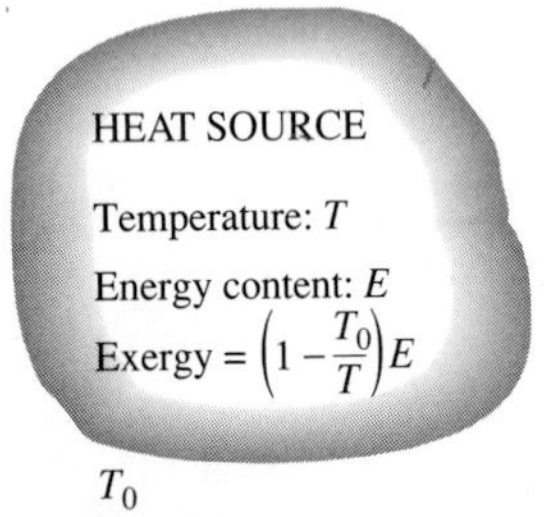

FIGURE 7-26

The Carnot efficiency $\eta_c = 1 - T_0/T$ represents the fraction of the energy of a heat source at temperature T that can be converted to work in an environment at temperature T_0.

You will recall from Chap 5. that the work potential of the energy of a heat source at temperature T is the maximum work that can be obtained from that source in an environment at temperature T_0 and is equivalent to the work produced by a Carnot heat engine operating between the source and the environment. Therefore, the Carnot efficiency $\eta_C = 1 - T_0/T$ represents the fraction of energy of a heat source at temperature T that can be converted to work (Fig. 7-26). For example, only 70 percent of the energy in a heat source at $T = 1000$ K can be converted to work in an environment at $T_0 = 300$ K. In other words, the exergy content of each kJ of the energy in that source is 0.7 kJ.

Heat is a form of disorganized energy, and thus only a portion of it can be converted to work, which is a form of organized energy (the second law). We can always produce work from heat at a temperature above the environment temperature by transferring it to a heat engine that rejects the waste heat to the environment. Therefore, heat transfer is always accompanied by exergy transfer. Heat transfer Q at a location at absolute temperature T is always accompanied by *exergy transfer* X_{heat} in the amount of

Exergy transfer by heat:
$$X_{\text{heat}} = \left(1 - \frac{T_0}{T}\right)Q \quad \text{(kJ)} \qquad (7\text{-}24)$$

This relation gives the exergy transfer accompanying heat transfer Q whether T is greater than or less than T_0. When $T > T_0$, heat transfer to a system increases the exergy of that system and heat transfer from a system decreases it. But the opposite is true when $T < T_0$. In this case, the heat transfer Q is the heat rejected to the cold medium (the waste heat), and it should not be confused with the heat supplied by the environment at T_0. The exergy transferred with heat is zero when $T = T_0$ at the point of transfer.

Perhaps you are wondering what happens when $T < T_0$. That is, what if we have a medium that is at a lower temperature than the environment? In this case it is conceivable that we can run a heat engine between the environment and the "cold" medium, and thus a cold medium offers us an opportunity to produce work. But this time the environment serves as the heat source and the cold medium as the heat sink. In this case, the relation above gives the negative of the exergy transfer associated with the heat Q transferred to the cold medium. For example, for $T = 100$ K and a heat transfer of $Q = 1$ kJ to the medium, Eq. 7-24 gives $X_{heat} = (1 - 300/100)(1 \text{ kJ}) = -2$ kJ, which means that the exergy of the cold medium decreases by 2 kJ. It also means that this exergy can be recovered, and the cold medium–environment combination has the potential to produce 2 units of work for each unit of heat rejected to the cold medium at 100 K. That is, a Carnot heat engine operating between $T_0 = 300$ K and $T = 100$ K will produce 2 units of work while rejecting 1 unit of heat for each 3 units of heat it receives from the environment.

When $T > T_0$, the exergy and heat transfer are in the same direction. That is, both the exergy and energy content of the medium to which heat is transferred increase. But when $T < T_0$ (cold medium), the exergy and heat transfer are in opposite directions. That is, the energy of the cold medium increases as a result of heat transfer, but its exergy decreases. The exergy of the cold medium eventually becomes zero when its temperature reaches T_0. Equation 7-24 can also be viewed as the *exergy of thermal energy Q* at temperature T.

When the temperature T at the location where heat transfer is taking place is not constant, the exergy transfer accompanying heat transfer is determined by integration to be

$$X_{heat} = \int \left(1 - \frac{T_0}{T}\right) \delta Q \tag{7-25}$$

Note that heat transfer through a finite temperature difference is irreversible, and some entropy is generated as a result. The entropy generation is always accompanied by exergy destruction, as illustrated in Fig. 7-27. Also note that *heat transfer Q* at a location at temperature T is always accompanied by *entropy transfer* in the amount of Q/T and *exergy transfer* in the amount of $(1 - T_0/T)Q$.

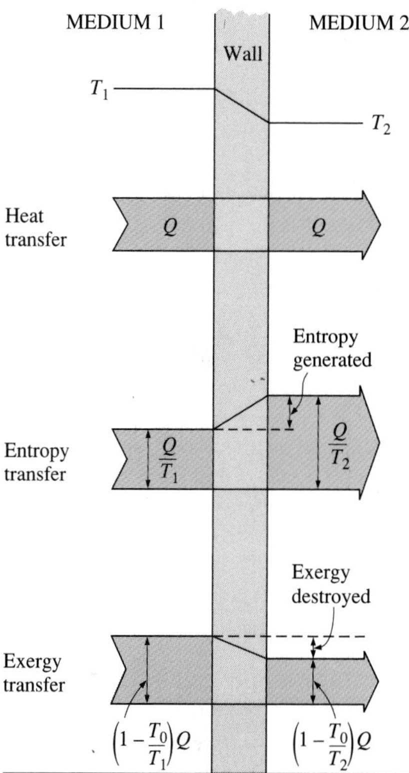

FIGURE 7-27
The transfer and destruction of exergy during a heat transfer process through a finite temperature difference.

Exergy Transfer by Work, W

Exergy is the useful work potential, and the exergy transfer by work can simply be expressed as

$$\textit{Exergy transfer by work: } X_{work} = \begin{cases} W - W_{surr} & \text{(for boundary work)} \\ W & \text{(for other forms of work)} \end{cases} \tag{7-26}$$

where $W_{surr} = P_0(V_2 - V_1)$, P_0 is atmospheric pressure, and V_1 and V_2 are the initial and final volumes of the system. Therefore, the exergy transfer with work such as shaft work and electrical work is equal to the work W itself. In the case of a system that involves boundary work, such as a piston-cylinder device, the work done to push the atmospheric air out of the way during expansion cannot be transferred, and thus it must be subtracted. Also, during a compression process, part of the work is done by the atmospheric air, and thus we need to supply less useful work from an external source.

To clarify this point further, consider a vertical cylinder fitted with a weightless and frictionless piston (Fig. 7-28). The cylinder is filled with a gas that is maintained at the atmospheric pressure P_0 at all times. Heat is now transferred to the system and the gas in the cylinder expands. As a result, the piston rises and boundary work is done. However, this work cannot be used for any useful purpose since it is just enough to push the atmospheric air aside. (If we connect the piston to an external load to extract some useful work, the pressure in the cylinder will have to rise above P_0 to beat the resistance offered by the load.) When the gas is cooled, the piston moves down, compressing the gas. Again, no work is needed from an external source to accomplish this compression process. Thus we conclude that the work done by or against the atmosphere is not available for any useful purpose, and thus should be excluded from available work.

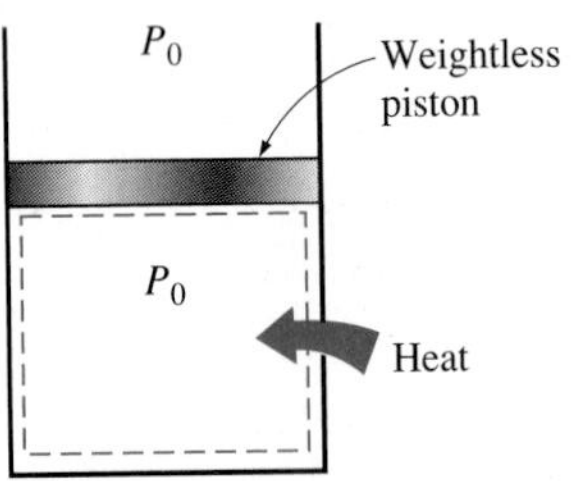

FIGURE 7-28

There is no useful work transfer associated with boundary work when the pressure of the system is maintained constant at atmospheric pressure.

Exergy Transfer by Mass, *m*

Mass contains *exergy* as well as energy and entropy, and the exergy, energy, and entropy contents of a system are proportional to mass. Also, the rates of exergy entropy, and energy transport into or out of a system are proportional to the mass flow rate. Mass flow is a mechanism to transport exergy, entropy, and energy into or out of a system. When mass in the amount of m enters or leaves a system, exergy in the amount of $m\psi$, where $\psi = (h - h_0) - T_0(s - s_0) + \mathcal{V}^2/2 + gz$, accompanies it. That is,

Exergy transfer by mass:
$$X_{\text{mass}} = m\psi \qquad (7\text{-}27)$$

Therefore, the exergy of a system increases by $m\psi$ when mass in the amount of m enters, and decreases by the same amount when the same amount of mass at the same state leaves the system (Fig. 7-29).

When the properties of the fluid change during the process, the exergy transfer by mass flow can be determined by integration from

$$\dot{X}_{\text{mass}} = \int_{A_c} \psi\rho\mathcal{V}_n \, dA_c \qquad \text{and} \qquad X_{\text{mass}} = \int \psi\delta m = \int_{\Delta t} \dot{X}_{\text{mass}} \, dt \qquad (7\text{-}28)$$

where A_c is the cross-sectional area of the flow and $\mathcal{V}_n$ is the local velocity normal to dA_c.

Note that exergy transfer by heat X_{heat} is zero for adiabatic systems, and the exergy transfer by mass X_{mass} is zero for systems that involve no mass flow across their boundary (i.e., closed systems). The total exergy transfer is zero for isolated systems since they involve no heat, work, or mass transfer.

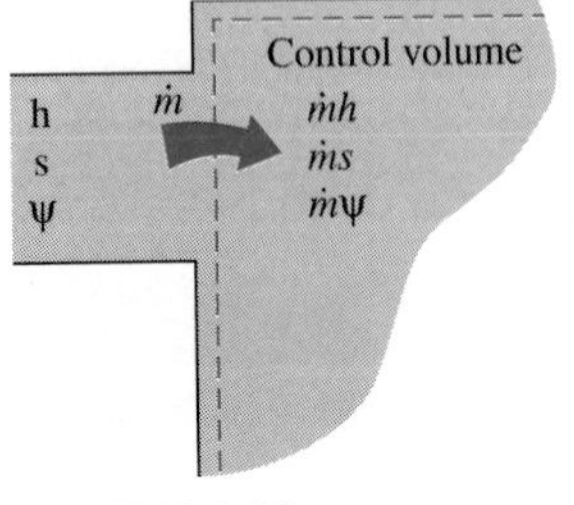

FIGURE 7-29

Mass contains energy, entropy, and exergy, and thus mass flow into or out of a system is accompanied by energy, entropy, and exergy transfer.

7-7 ■ THE DECREASE OF EXERGY PRINCIPLE AND EXERGY DESTRUCTION

In Chap. 3 we presented the *conservation of energy principle* and indicated that energy cannot be created or destroyed. In Chap. 6 we established the *increase of entropy principle,* which can be regarded as one of the statements of the second law, and indicated that entropy can be created but cannot be destroyed. That is, entropy generation S_{gen} must be positive (actual processes) or zero (reversible processes), but it cannot be negative. Now we are about to establish an alternative statement of the second law of thermodynamics, called the *decrease of exergy principle,* which is the counterpart of the increase of entropy principle.

Consider an *isolated system* shown in Fig. 7-30. By definition, no heat, work, or mass can cross the boundaries of an isolated system, and thus there is no energy and entropy transfer. Then the *energy* and *entropy* balances for an isolated system can be expressed as

Energy balance: $E_{in}^{\nearrow 0} - E_{out}^{\nearrow 0} = \Delta E_{system} \rightarrow 0 = E_2 - E_1$

Entropy balance: $S_{in}^{\nearrow 0} - S_{out}^{\nearrow 0} + S_{gen} = \Delta S_{system} \rightarrow S_{gen} = S_2 - S_1$

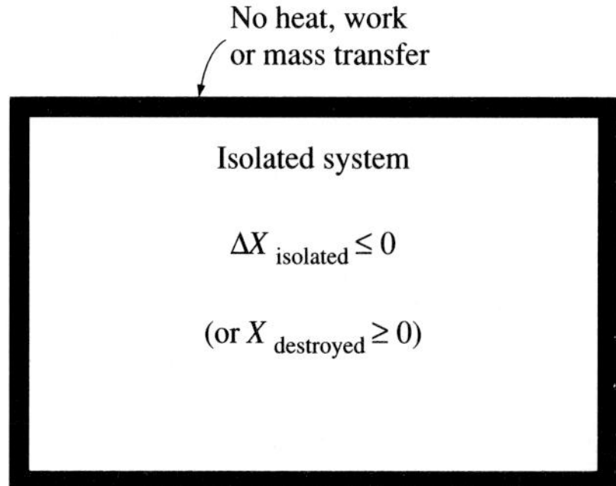

FIGURE 7-30
The isolated system considered in the development of the decrease of exergy principle.

Multiplying the second relation by T_0 and subtracting it from the first one gives

$$-T_0 S_{gen} = E_2 - E_1 - T_0(S_2 - S_1) \qquad (7\text{-}29)$$

From Eq. 7-20 we have

$$\begin{aligned} X_2 - X_1 &= (E_2 - E_1) + P_0(V_2 - V_1)^{\nearrow 0} - T_0(S_2 - S_1) \\ &= (E_2 - E_1) - T_0(S_2 - S_1) \end{aligned} \qquad (7\text{-}30)$$

since $V_2 = V_1$ for an isolated system (it cannot involve any moving boundary and thus any boundary work). Combining Eqs. 7-29 and 7-30 gives

$$-T_0 S_{gen} = X_2 - X_1 \leq 0 \qquad (7\text{-}31)$$

since T_0 is the absolute temperature of the environment and thus a positive quantity, $S_{gen} \geq 0$, and thus $T_0 S_{gen} \geq 0$. Then we conclude that

$$\Delta X_{isolated} = (X_2 - X_1)_{isolated} \leq 0 \qquad (7\text{-}32)$$

This equation can be expressed as *the exergy of an isolated system during a process always decreases or, in the limiting case of a reversible process, remains constant.* In other words, it *never* increases and *exergy is destroyed* during an actual process. This is known as the **decrease of exergy principle.** For an isolated system, the decrease in exergy equals exergy destroyed.

Exergy Destruction

Irreversibilities such as friction, mixing, chemical reactions, heat transfer through a finite temperature difference, unrestrained expansion, non-quasi-equilibrium compression, or expansion always *generate entropy,* and anything that generates entropy always *destroys exergy.* The **exergy destroyed** is proportional to the entropy generated, as can be seen from Eq. 7-31, and is expressed as

$$X_{\text{destroyed}} = T_0 S_{\text{gen}} \geq 0 \tag{7-33}$$

Note that exergy destroyed is a *positive quantity* for any actual process and becomes *zero* for a reversible process. Exergy destroyed represents the lost work potential and is also called the *irreversibility* or *lost work.*

Equations 7-32 and 7-33 for the decrease of exergy and the exergy destruction are applicable to *any kind of system* (closed or open) undergoing *any kind of process* since any system and its surroundings can be enclosed by a sufficiently large arbitrary boundary across which there is no heat, work, and mass transfer, and thus any system and its surroundings constitute an *isolated system.*

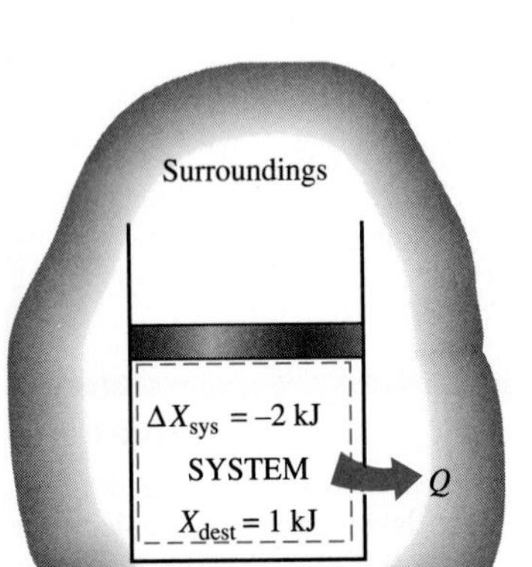

FIGURE 7-31

The exergy change of a system can be negative, but the exergy destruction cannot.

No actual process is truly reversible, and thus some exergy is destroyed during a process. Therefore, the exergy of the universe, which can be considered to be an isolated system, is continuously decreasing. The more irreversible a process is, the larger the exergy destruction during that process. No exergy is destroyed during a reversible process ($X_{\text{destroyed,rev}} = 0$).

The decrease of exergy principle does not imply that the exergy of a system cannot increase. The exergy change of a system *can* be positive or negative during a process (Fig. 7-31), but exergy destroyed cannot be negative. The decrease of exergy principle can be summarized as follows:

$$X_{\text{destroyed}} \begin{cases} > 0 & \text{Irreversible process} \\ = 0 & \text{Reversible process} \\ < 0 & \text{Impossible process} \end{cases} \tag{7-34}$$

This relation serves as an alternative criterion to determine whether a process is reversible, irreversible, or impossible.

7-8 ■ EXERGY BALANCE: Closed Systems

The nature of exergy is opposite to that of entropy in that exergy can be *destroyed,* but it cannot be created. Therefore, the *exergy change* of a system during a process is less than the *exergy transfer* by an amount equal to the *exergy destroyed* during the process within the system boundaries. Then the *decrease of exergy principle* can be expressed as (Fig. 7-32)

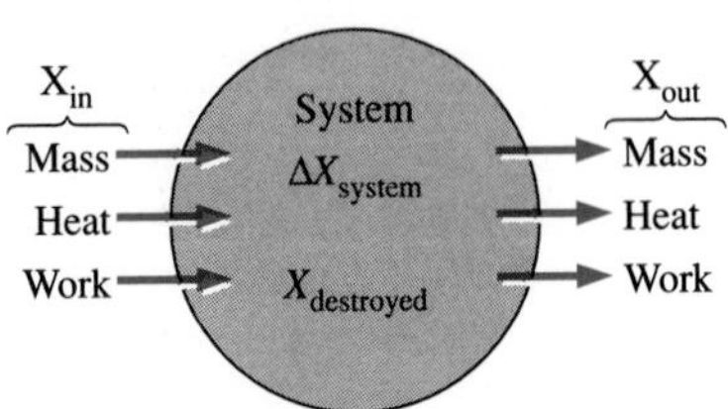

FIGURE 7-32

Mechanisms of exergy transfer for a general system.

$$\begin{pmatrix}\text{Total}\\ \text{exergy}\\ \text{entering}\end{pmatrix} - \begin{pmatrix}\text{Total}\\ \text{exergy}\\ \text{leaving}\end{pmatrix} - \begin{pmatrix}\text{Total}\\ \text{exergy}\\ \text{destroyed}\end{pmatrix} = \begin{pmatrix}\text{Change in the}\\ \text{total exergy}\\ \text{of the system}\end{pmatrix}$$

or

$$X_{\text{in}} - X_{\text{out}} - X_{\text{destroyed}} = \Delta X_{\text{system}} \tag{7-35}$$

This relation is referred to as the **exergy balance** and can be stated as *the exergy change of a system during a process is equal to the difference between* *the net exergy transfer through the system boundary and the exergy destroyed within the system boundaries as a result of irreversibilities (or entropy generation).*

We mentioned earlier that exergy can be transferred to or from a system by heat, work, and mass transfer. Then the exergy balance for *any system* undergoing *any process* can be expressed more explicitly as

$$\textit{General:}\quad \underbrace{X_{in} - X_{out}}_{\substack{\text{Net exergy transfer}\\ \text{by heat, work and mass}}} \underbrace{- X_{destroyed}}_{\substack{\text{Exergy}\\ \text{destruction}}} = \underbrace{\Delta X_{system}}_{\substack{\text{Change}\\ \text{in exergy}}} \quad \text{(kJ)} \qquad (7\text{-}36)$$

or, in the **rate form,** as

$$\textit{General, rate form:}\quad \underbrace{\dot{X}_{in} - \dot{X}_{out}}_{\substack{\text{Rate of net exergy transfer}\\ \text{by heat, work, and mass}}} \underbrace{- \dot{X}_{destroyed}}_{\substack{\text{Rate of exergy}\\ \text{destruction}}} = \underbrace{\Delta \dot{X}_{system}}_{\substack{\text{Rate of change}\\ \text{of exergy}}} \quad \text{(kW)} \qquad (7\text{-}37)$$

where the rates of exergy transfer by heat, work, and mass are expressed as $\dot{X}_{heat} = (1 - T_0/T)\dot{Q}$, $\dot{X}_{work} = \dot{W}_{useful}$, and $\dot{X}_{mass} = \dot{m}\psi$, respectively, and $\Delta\dot{X}_{system} = dX_{system}/dt$. The exergy balance can also be expressed per unit mass as

$$\textit{General, unit-mass basis:}\quad (x_{in} - x_{out}) - x_{destroyed} = \Delta x_{system} \quad \text{(kJ/kg)} \qquad (7\text{-}38)$$

where all the quantities are expressed per unit mass of the system. Note that for a *reversible process,* the exergy destruction term $X_{destroyed}$ drops out from all of the relations above. Also, it is usually more convenient to find the entropy generation S_{gen} first, and then to evaluate the exergy destroyed directly from Eq. 7-33. That is,

$$X_{destroyed} = T_0 S_{gen} \quad \text{or} \quad \dot{X}_{destroyed} = T_0 \dot{S}_{gen} \qquad (7\text{-}39)$$

When the environment conditions P_0 and T_0 and the end states of the system are specified, the exergy change of the system $\Delta X_{system} = X_2 - X_1$ can be determined directly from Eq. 7-20 regardless of how the process is executed. However, the determination of the exergy transfers by heat, work, and mass requires a knowledge of these interactions.

A *closed system* does not involve any mass flow and thus any exergy transfer by it. Taking the positive direction of heat transfer to be to the system and the positive direction of work transfer to be from the system, the exergy balance for a closed system can be expressed more explicitly as (Fig. 7-33)

$$\textit{Closed system:}\quad X_{heat} - X_{work} - X_{destroyed} = \Delta X_{system} \qquad (7\text{-}40)$$

or

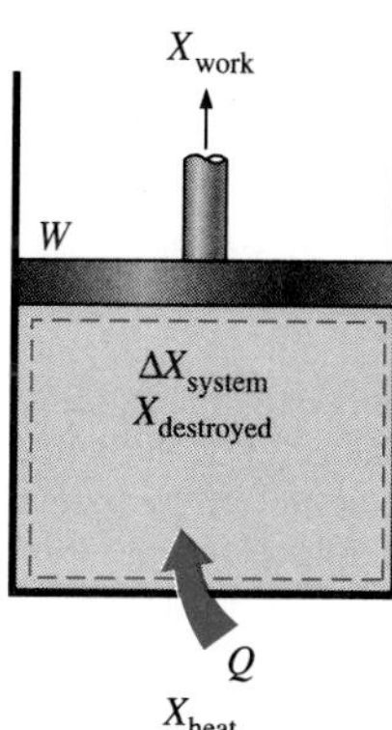

$X_{heat} - X_{work} - X_{destroyed} = \Delta X_{system}$

FIGURE 7-33

Exergy balance for a closed system when the direction of heat transfer is taken to be to the system and the direction of work from the system.

$$\textit{Closed system:} \quad \sum\left(1-\frac{T_0}{T_k}\right)Q_k - [W - P_0(V_2 - V_1)] - T_0 S_{gen} = X_2 - X_1 \tag{7-41}$$

where Q_k is the heat transfer through the boundary at temperature T_k at location k. Dividing the equation above by the time interval Δt and taking the limit as $\Delta t \to 0$ gives the *rate form* of the exergy balance for a closed system,

$$\textit{Rate form:} \quad \sum\left(1-\frac{T_0}{T_k}\right)\dot{Q}_k - \left(\dot{W} - P_0\frac{dV_{system}}{dt}\right) - T_0\dot{S}_{gen} = \frac{dX_{system}}{dt} \tag{7-42}$$

Note that the relations above for a closed system are developed by taking the heat transfer to a system and work done by the system to be positive quantities. Therefore, heat transfer from the system and work done on the system will be taken to be negative quantities when using those relations.

The exergy balance relations presented above can be used to determine the *reversible work* W_{rev} by setting the exergy destruction term equal to zero. The work W in that case becomes the reversible work. That is, $W = W_{rev}$ when $X_{destroyed} = T_0 S_{gen} = 0$.

Note that $X_{destroyed}$ represents the exergy destroyed *within the system boundary* only, and not the exergy destruction that may occur outside the system boundary during the process as a result of external irreversibilities. Therefore, a process for which $X_{destroyed} = 0$ is *internally reversible* but not necessarily *totally* reversible. The *total* exergy destroyed during a process can be determined by applying the exergy balance to an *extended system* that includes the system itself and its immediate surroundings where external irreversibilities might be occurring (Fig. 7-34). Also, the exergy change in this case is equal to the sum of the exergy changes of the system and the *exergy change* of the immediate surroundings. Note that under steady conditions, the state and thus the exergy of the immediate surroundings (the "buffer zone") at any point will not change during the process, and thus the exergy change of the buffer zone will be zero. When evaluating the exergy transfer between an extended system and the environment, the boundary temperature of the extended system is simply taken to be the environment temperature T_0.

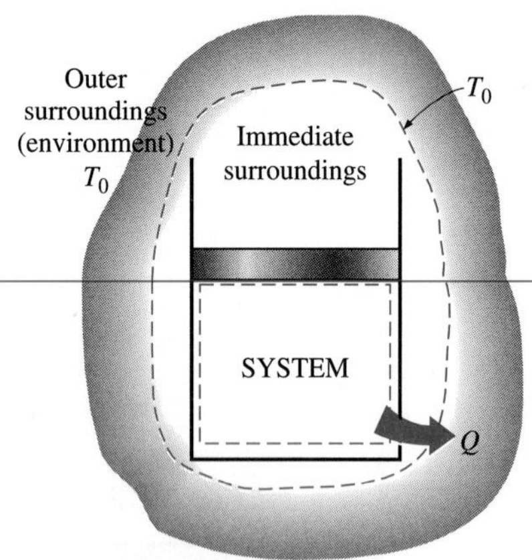

FIGURE 7-34

Exergy destroyed outside system boundaries can be accounted for by writing an exergy balance on the extended system that includes the system and its immediate surroundings.

For a *reversible process,* the *entropy generation* and thus the *exergy destruction* are *zero,* and the exergy balance relation in this case becomes analogous to the energy balance relation. That is, the exergy change of the system becomes equal to the exergy transfer.

Note that the *energy change* of a system equals the *energy transfer* for *any* process, but the *exergy change* of a system equals the *exergy transfer* only for a *reversible* process. The *quantity* of energy is always preserved during an actual process (the first law), but the *quality* is bound to decrease (the second law). This decrease in quality is always accompanied by an increase in entropy and a decrease in exergy. When 10 kJ of heat is transferred from a hot medium to a cold one, for example, we will still have 10 kJ of energy at the end of the process, but at a lower temperature, and thus at a lower quality and at a lower potential to do work.

EXAMPLE 7-9 General Exergy Balance for Closed Systems

Starting with energy and entropy balances, derive the general exergy balance relation for a closed system (Eq. 7-41).

Solution We consider a general closed system (a fixed mass) that is free to exchange heat and work with its surroundings (Fig. 7-35). The system undergoes a process from state 1 to state 2. Taking the direction of heat transfer to be *to* the system and the direction of work transfer to be *from* the system, the energy and entropy balances for this closed system can be expressed as

Energy balance: $E_{in} - E_{out} = \Delta E_{system} \quad \rightarrow \quad Q - W = E_2 - E_1$

Entropy balance: $S_{in} - S_{out} + S_{gen} = \Delta S_{system} \quad \rightarrow \quad \int_1^2 \left(\frac{\delta Q}{T}\right)_{boundary} + S_{gen} = S_2 - S_1$

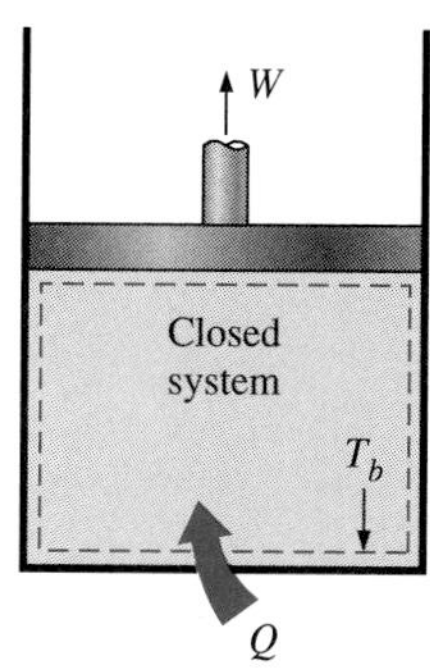

FIGURE 7-35
A general closed system considered in Example 7-9.

Multiplying the second relation by T_0 and subtracting it from the first one gives

$$Q - T_0\int_1^2 \left(\frac{\delta Q}{T}\right)_{boundary} - W - T_0 S_{gen} = E_2 - E_1 - T_0(S_2 - S_1)$$

But the heat transfer for the process 1-2 can be expressed as $Q = \int_1^2 \delta Q$ and the right side of the above equation is, from Eq. 7-20, $(X_2 - X_1) - P_0(V_2 - V_1)$. Thus,

$$\int_1^2 \delta Q - T_0\int_1^2 \left(\frac{\delta Q}{T}\right)_{boundary} - W - T_0 S_{gen} = X_2 - X_1 - P_0(V_2 - V_1)$$

Letting T_b denote the boundary temperature and rearranging give

$$\int_1^2 \left(1 - \frac{T_0}{T_b}\right)\delta Q - [W - P_0(V_2 - V_1)] - T_0 S_{gen} = X_2 - X_1 \qquad (7\text{-}43)$$

which is equivalent to Eq. 7-41 for the exergy balance except that the integration is replaced by summation in that equation for convenience. This completes the proof.

Discussion Note that the exergy balance relation above is obtained by adding the energy and entropy balance relations, and thus it is not an independent equation. However, it can be used in place of the entropy balance relation as an alternative second law expression in exergy analysis.

EXAMPLE 7-10 Exergy Destruction during Heat Conduction

Consider steady heat flow through a 5-m × 6-m brick wall of a house of thickness 30 cm and thermal conductivity 0.69 W/(m · °C). On a day when the temperature of the outdoors is 0°C, the house is maintained at 27°C. The temperatures of the inner and outer surfaces of the brick wall are measured to be 20°C and 5°C, respectively. Determine the rate of heat transfer through the wall, the rate of exergy destruction in the wall, and the rate of total exergy destruction associated with this heat transfer process.

FIGURE 7-36
Schematic for Example 7-10.

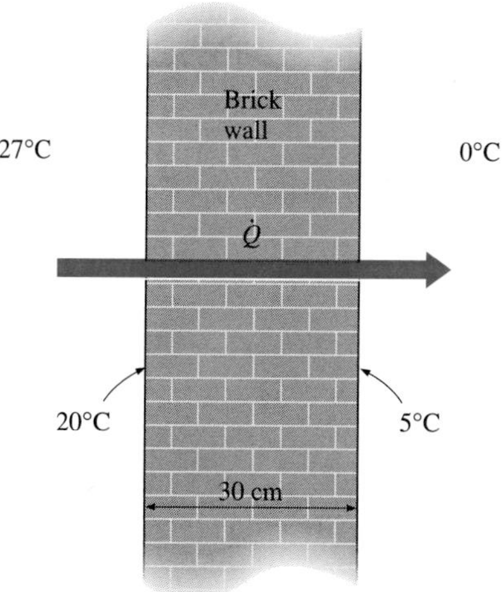

Solution We first take the *wall* as the system (Fig. 7-36). This is a *closed system* since no mass crosses the system boundary during the process. We note that heat and exergy are entering from one side of the wall and leaving from the other side.

Assumptions **1** The process is steady, and thus the rate of heat transfer through the wall is constant. **2** The exergy change of the wall is zero during this process since the state and thus the exergy of the wall do not change anywhere in the wall. **3** Heat conduction through the wall is one-dimensional.

Analysis Knowing the wall surface temperatures, the rate of heat transfer through the wall is determined from Fourier's law of heat conduction to be

$$\dot{Q} = kA\left(\frac{\Delta T}{L}\right)_{\text{wall}} = [0.69\ \text{W/(m}\cdot{}^\circ\text{C)}][(5\times 6)\text{m}^2]\frac{(20-5)^\circ\text{C}}{0.3\ \text{m}} = \mathbf{1035\ W}$$

Applying the rate form of the exergy balance to the wall gives

$$\underbrace{\dot{X}_{\text{in}} - \dot{X}_{\text{out}}}_{\substack{\text{Rate of net exergy transfer}\\ \text{by heat, work, and mass}}} - \underbrace{\dot{X}_{\text{destroyed}}}_{\substack{\text{Rate of exergy}\\ \text{destruction}}} = \underbrace{\Delta\dot{X}_{\text{system}}}_{\substack{\text{Rate of change}\\ \text{of exergy}}}^{\nearrow 0\ \text{(steady)}} = 0$$

$$\dot{Q}\left(1-\frac{T_0}{T}\right)_{\text{in}} - \dot{Q}\left(1-\frac{T_0}{T}\right)_{\text{out}} - \dot{X}_{\text{destroyed}} = 0$$

$$(1035\ \text{W})\left(1-\frac{273\ \text{K}}{293\ \text{K}}\right) - (1035\ \text{W})\left(1-\frac{273\ \text{K}}{278\ \text{K}}\right) - \dot{X}_{\text{destroyed}} = 0$$

Solving, the rate of exergy destruction in the wall is determined to be

$$\dot{X}_{\text{destroyed}} = (1035\ \text{W})\left(1-\frac{273\ \text{K}}{293\ \text{K}}\right) - (1035\ \text{W})\left(1-\frac{273\ \text{K}}{278\ \text{K}}\right) = \mathbf{52.0\ W}$$

Note that exergy transfer with heat at any location is $(1 - T_0/T)Q$ at that location, and the direction of exergy transfer is the same as the direction of heat transfer.

To determine the rate of total exergy destruction during this heat transfer process, we extend the system to include the regions on both sides of the wall that experience a temperature change. Then one side of the system boundary becomes room temperature while the other side, the temperature of the outdoors. The exergy balance for this *extended system* (system + immediate surroundings) will be the same as that given above, except the two boundary temperatures will be 300 and 273 K instead of 293 and 278 K, respectively. Then the rate of total exergy destruction becomes

$$\dot{X}_{\text{destroyed, total}} = (1035\ \text{W})\left(1-\frac{273\ \text{K}}{300\ \text{K}}\right) - (1035\ \text{W})\left(1-\frac{273\ \text{K}}{273\ \text{K}}\right) = \mathbf{93.2\ W}$$

Note that the exergy change of this extended system is also zero since the state of air at any point does not change during the process. The difference between the two exergy destructions is 41.2 W and represents the exergy destroyed in the air layers on both sides of the wall. The exergy destruction in this case is entirely due to irreversible heat transfer through a finite temperature difference.

Discussion This problem was solved in Chap. 6 for entropy generation. We could have determined the exergy destroyed by simply multiplying the entropy generations by the environment temperature of $T_0 = 273$ K.

EXAMPLE 7-11 Exergy Destruction during Expansion of Steam

A piston-cylinder device contains 0.05 kg of steam at 1 MPa and 300°C. The steam now expands to a final state of 200 kPa and 150°C, doing work. Heat losses from the system to the surroundings are estimated to be 2 kJ during this process. Assuming the surroundings to be at T_0 = 25°C and P_0 = 100 kPa, determine (*a*) the exergy of the steam at the initial and the final states, (*b*) the exergy change of the steam, (*c*) the exergy destroyed, and (*d*) the second-law efficiency for the process.

Solution We take the *steam* contained within the piston-cylinder device as the system (Fig. 7-37). This is a *closed system* since no mass crosses the system boundary during the process. We note that boundary work is done by the system and heat is lost from the system during the process.

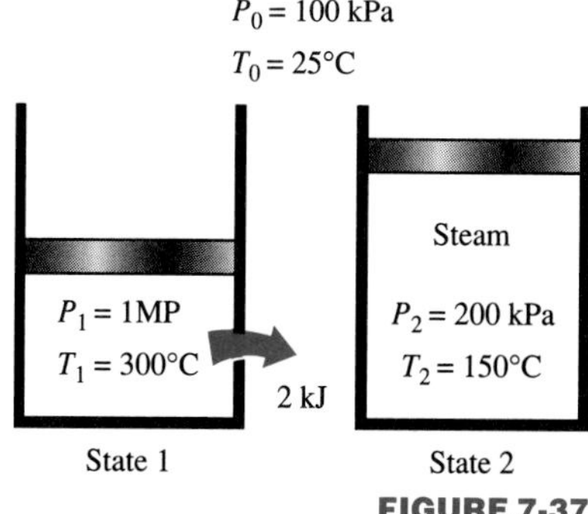

FIGURE 7-37

Schematic for Example 7-11.

Assumptions The kinetic and potential energies are negligible.

Analysis (*a*) First let us determine the properties of the steam at the initial and final states as well as the state of the surroundings:

State 1: $\left.\begin{array}{l} P_1 = 1\text{ MPa} \\ T_1 = 300°\text{C} \end{array}\right\}$ $\begin{array}{l} u_1 = 2793.2\text{ kJ/kg} \\ v_1 = 0.2579\text{ m}^3\text{/kg} \\ s_1 = 7.1229\text{ kJ(kg}\cdot\text{K)} \end{array}$ (Table A-6)

State 2: $\left.\begin{array}{l} P_2 = 200\text{ kPa} \\ T_2 = 150°\text{C} \end{array}\right\}$ $\begin{array}{l} u_2 = 2576.9\text{ kJ/kg} \\ v_2 = 0.9596\text{ m}^3\text{/kg} \\ s_2 = 7.2795\text{ kJ/(kg}\cdot\text{K)} \end{array}$ (Table A-6)

Dead state: $\left.\begin{array}{l} P_0 = 100\text{ kPa} \\ T_0 = 25°\text{C} \end{array}\right\}$ $\begin{array}{l} u_0 \cong u_{f @ 25°\text{C}} = 104.88\text{ kJ/kg} \\ v_0 \cong v_{f @ 25°\text{C}} = 0.0010\text{ m}^3\text{/kg} \\ s_0 \cong s_{f @ 25°\text{C}} = 0.3674\text{ kJ/(kg}\cdot\text{K)} \end{array}$ (Table A-4)

The exergies of the system at the initial state X_1 and the final state X_2 are determined from Eq. 7-15 to be

$$\begin{aligned} X_1 &= m[(u_1 - u_0) - T_0(s_1 - s_0) + P_0(v_1 - v_0)] \\ &= (0.05\text{ kg})\{(2793.2 - 104.88)\text{ kJ/kg} \\ &\quad - (298\text{ K})[(7.1229 - 0.3674)\text{ kJ/(kg}\cdot\text{K)}] \\ &\quad + (100\text{ kPa})[(0.2579 - 0.0010)\text{ m}^3\text{/kg})\}[\text{kJ/(kPa}\cdot\text{m}^3)] \\ &= \mathbf{35.0\text{ kJ}} \end{aligned}$$

and

$$\begin{aligned} X_2 &= m[(u_2 - u_0) - T_0(s_2 - s_0) + P_0(v_2 - v_0)] \\ &= (0.05\text{ kg})\{(2576.9 - 104.88)\text{ kJ/kg} \\ &\quad - (298\text{ K})[(7.2795 - 0.3674)\text{ kJ/(kg}\cdot\text{K)}] \\ &\quad + (100\text{ kPa})[(0.9596 - 0.0010)\text{ m}^3\text{/kg}]\}[\text{kJ/(kPa}\cdot\text{m}^3)] \\ &= \mathbf{25.4\text{ kJ}} \end{aligned}$$

That is, the system (steam) initially has an exergy (useful work potential) of 35 kJ, which drops to 25.4 kJ at the end of the process. In other words, if the system were allowed to undergo a reversible process from the initial state to the state of the environment (the dead state), it would produce 35 kJ of useful work in that environment.

(*b*) The exergy change for a process is simply the difference between the exergy at the initial and final states of the process, Eq. 7-20:

$$\Delta X = X_2 - X_1 = 25.4 - 35.0 = \mathbf{-9.6\ kJ}$$

That is, if the process between states 1 and 2 were executed in a reversible manner, the system would deliver 9.6 kJ of useful work.

(*c*) The exergy destroyed during this process can be determined from the exergy balance applied on the *extended system* (system + immediate surroundings) whose boundary is at the environment temperature of T_0 (so that there is no exergy transfer accompanying heat transfer to or from the environment),

$$\underbrace{X_{in} - X_{out}}_{\substack{\text{Net exergy transfer}\\ \text{by heat, work, and mass}}} - \underbrace{X_{destroyed}}_{\substack{\text{Exergy}\\ \text{destruction}}} = \underbrace{\Delta X_{system}}_{\substack{\text{Change}\\ \text{in exergy}}}$$

$$-X_{work,\,out} - X_{heat,\,out}\nearrow^{0} - X_{destroyed} = X_2 - X_1$$

$$X_{destroyed} = X_1 - X_2 - W_{u,\,out}$$

where $W_{u,\,out}$ is the useful boundary work delivered as the system expands. By writing an energy balance on the system, the total boundary work done during the process is determined to be

$$\underbrace{E_{in} - E_{out}}_{\substack{\text{Net energy transfer}\\ \text{by heat, work, and mass}}} = \underbrace{\Delta E_{system}}_{\substack{\text{Change in internal, kinetic,}\\ \text{potential, etc., energies}}}$$

$$-Q_{out} - W_{b,\,out} = \Delta U$$

$$W_{b,\,out} = -Q_{out} - \Delta U = -Q_{out} - m(u_2 - u_1)$$

$$= -(2\text{ kJ}) - (0.05\text{ kg})(2576.9 - 2793.2)\text{ kJ/kg}$$

$$= 8.8\text{ kJ}$$

This is the total boundary work done by the system, including the work done against the atmosphere to push the atmospheric air out of the way during the expansion process. The useful work is the difference between the two:

$$W_u = W - W_{surr} = W_{b,\,out} - P_0(V_2 - V_1) = W_{b,\,out} - P_0 m(v_2 - v_1)$$

$$= 8.8\text{ kJ} - (100\text{ kPa})(0.05\text{ kg})[(0.9596 - 0.2579)\text{ m}^3\text{/kg}]\left(\frac{1\text{ kJ}}{1\text{ kPa}\cdot\text{m}^3}\right)$$

$$= 5.3\text{ kJ}$$

Substituting, the exergy destroyed is determined to be

$$X_{destroyed} = X_1 - X_2 - W_{u,\,out} = 35.0 - 25.4 - 5.3 = \mathbf{4.3\ kJ}$$

That is, 4.3 kJ of work potential is wasted during this process. In other words, an additional 4.3 kJ of energy *could have been* converted to work during this process, but was not.

The irreversibility could also be determined from

$$I = T_0 S_{gen} = T_0\left[m(s_2 - s_1) + \frac{Q_{surr}}{T_0}\right]$$

$$= (298\text{ K})\left\{(0.05\text{ kg})[(7.2795 - 7.1229)\text{ kJ/(kg}\cdot\text{K)}] + \frac{2\text{ kJ}}{298\text{ K}}\right\}$$

$$= 4.3\text{ kJ}$$

which is the same result obtained before.

(*d*) Noting that the decrease in the exergy of the steam is the exergy supplied and the useful work output is the exergy recovered, the second-law efficiency for this process can be determined from

$$\eta_{II} = \frac{\text{Exergy recovered}}{\text{Exergy supplied}} = \frac{W_u}{X_1 - X_2} = \frac{5.3}{35.0 - 25.4} = \mathbf{0.552 \text{ or } 55.2\%}$$

That is, 44.8 percent of the work potential of the steam is wasted during this process.

EXAMPLE 7-12 Exergy Destroyed during Mixing of a Gas

An insulated rigid tank contains 2 lbm of air at 20 psia and 70°F. A paddle wheel inside the tank is now rotated by an external power source until the temperature in the tank rises to 130°F. If the surrounding air is at $T_0 = 70°F$, determine (*a*) the exergy destroyed and (*b*) the reversible work for this process.

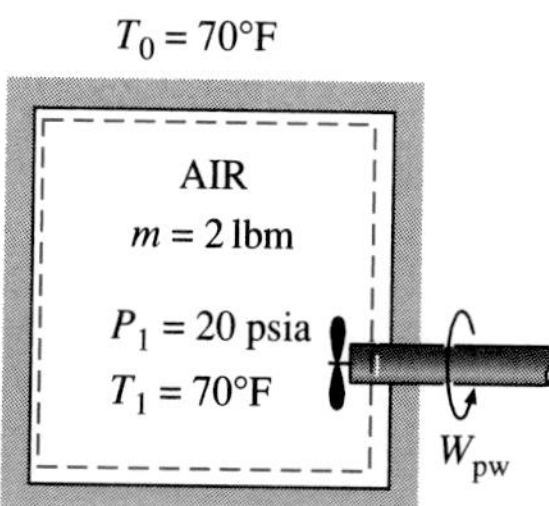

FIGURE 7-38
Schematic for Example 7-12.

Solution We take the *air* contained within the tank as the system (Fig. 7-38). This is a *closed system* since no mass crosses the system boundary during the process. We note that shaft work is done on the system.

Assumptions **1** Air at about atmospheric conditions can be treated as an ideal gas with constant specific heats at room temperature. **2** The kinetic and potential energies are negligible. **3** The volume of a rigid tank is constant, and thus there is no boundary work and $v_2 = v_1$. **4** The tank is well-insulated and thus there is no heat transfer.

Analysis (*a*) The exergy destroyed during a process can be determined from an exergy balance, or directly from $X_{destroyed} = T_0 S_{gen}$. We will use the second approach since it is usually easier. But first we determine the entropy generated from an entropy balance,

$$\underbrace{S_{in} - S_{out}}_{\substack{\text{Net entropy transfer}\\\text{by heat and mass}}} + \underbrace{S_{gen}}_{\substack{\text{Entropy}\\\text{generation}}} = \underbrace{\Delta S_{system}}_{\substack{\text{Change}\\\text{in entropy}}}$$

$$0 + S_{gen} = \Delta S_{system} = m\left(C_v \ln\frac{T_2}{T_1} + R\ln\frac{V_2}{V_1}^{\nearrow 0}\right)$$

$$S_{gen} = mC_v \ln\frac{T_2}{T_1}$$

Taking $C_v = 0.172$ Btu/(lbm · °F) and substituting, the exergy destroyed becomes

$$X_{\text{destroyed}} = T_0 S_{\text{gen}} = T_0 m C_v \ln \frac{T_2}{T_1}$$

$$= (530\text{ R})(2\text{ lbm})[0.172\text{ Btu/(lbm}\cdot°\text{F)}] \ln \frac{590\text{ R}}{530\text{ R}}$$

$$= \mathbf{19.6\ Btu}$$

(*b*) The reversible work, which represents the minimum work input $W_{\text{rev, in}}$ in this case, can be determined from the exergy balance by setting the exergy destruction equal to zero,

$$\underbrace{X_{\text{in}} - X_{\text{out}}}_{\substack{\text{Net exergy transfer}\\ \text{by heat, work, and mass}}} - \underbrace{X_{\text{destroyed}}^{\nearrow 0 \text{ (reversible)}}}_{\substack{\text{Exergy}\\ \text{destruction}}} = \underbrace{\Delta X_{\text{system}}}_{\substack{\text{Change}\\ \text{in exergy}}}$$

$$W_{\text{rev, in}} = X_2 - X_1$$

$$= (E_2 - E_1) + P_0(V_2 - V_1)^{\nearrow 0} - T_0(S_2 - S_1)$$

$$= (U_2 - U_1) - T_0(S_2 - S_1)$$

since $\Delta KE = \Delta PE = 0$ and $V_2 = V_1$. Noting that $T_0(S_2 - S_1) = T_0 \Delta S_{\text{system}} =$ 19.6 Btu, the reversible work becomes

$$W_{\text{rev, in}} = mC_v(T_2 - T_1) - T_0(S_2 - S_1)$$

$$= (2\text{ lbm})[0.172\text{ Btu/(lbm}\cdot°\text{F)}](130 - 70)°\text{F} - 19.6\text{ Btu}$$

$$= 20.6 - 19.6\text{ Btu}$$

$$= \mathbf{1.0\ Btu}$$

Therefore, a work input of just 1.0 Btu would be sufficient to accomplish this process (raise the temperature of air in the tank from 70 to 130°F) if all the irreversibilities were eliminated.

Discussion The solution is complete at this point. But to gain some physical insight, we will set the stage for a discussion. First, let us determine the actual work (the paddle-wheel work W_{pw}) done during this process. Applying the energy balance on the system,

$$\underbrace{E_{\text{in}} - E_{\text{out}}}_{\substack{\text{Net energy transfer}\\ \text{by heat, work, and mass}}} = \underbrace{\Delta E_{\text{system}}}_{\substack{\text{Change in internal, kinetic,}\\ \text{potential, etc., energies}}}$$

$$W_{\text{pw, in}} = \Delta U = 20.6\text{ Btu} \qquad [\text{from part } (b)]$$

since the system is adiabatic ($Q = 0$) and involves no moving boundaries ($W_b = 0$). Also, the value of ΔU is determined in part (*b*) above.

To put the information into perspective, 20.6 Btu of work is consumed during the process, 19.6 Btu of irreversibility is generated, and the reversible work input for the process is 1.0 Btu. What does all this mean? It simply means that we could have created the same effect on the closed system (raising its temperature to 130°F at constant volume) by consuming 1.0 Btu of work only instead of 20.6 Btu, and thus saving 19.6 Btu of work from going to waste. This would have been accomplished by a reversible heat pump.

To prove what we have just said, consider a Carnot heat pump that absorbs heat from the surroundings at $T_0 = 530$ R and transfers it to the air in the rigid tank until the air temperature T rises from 530 to 590 R, as shown in Fig. 7-39. The system (the air in the rigid tank) involves no direct work interactions in this case, and the heat supplied to the system can be expressed in differential form as

FIGURE 7-39
The same effect on the system can be accomplished by a reversible heat pump that consumes only 1 Btu of work.

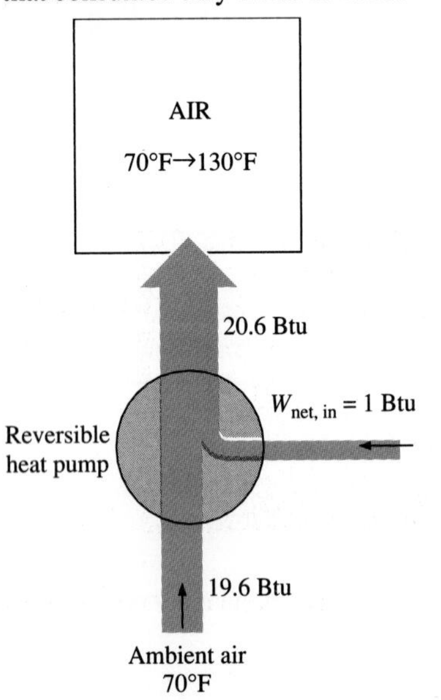

$$\delta Q_H = dU = mC_v\,dT$$

The coefficient of performance of a reversible heat pump is given by

$$\text{COP}_{\text{HP}} = \frac{\delta Q_H}{\delta W_{\text{net, in}}} = \frac{1}{1 - T_0/T}$$

Thus

$$\delta W_{\text{net, in}} = \frac{\delta Q_H}{\text{COP}_{\text{HP}}} = \left(1 - \frac{T_0}{T}\right) mC_v\,dT$$

Integrating, we get

$$\begin{aligned} W_{\text{net, in}} &= \int_1^2 \left(1 - \frac{T_0}{T}\right) mC_v\,dT \\ &= mC_{v,\text{av}}(T_2 - T_1) - T_0 mC_{v,\text{av}} \ln\frac{T_2}{T_1} \\ &= (20.6 - 19.6)\text{ Btu} = 1.0\text{ Btu} \end{aligned}$$

The first term on the right-hand side of the final expression above is recognized as ΔU and the second term as the exergy destroyed, whose values were determined earlier. By substituting those values, the total work input to the heat pump is determined to be 1.0 Btu, proving our claim. Notice that the system is still supplied with 20.6 Btu of energy; all we did in the latter case is replace the 19.6 Btu of valuable work by an equal amount of "useless" energy captured from the surroundings.

Discussion It is also worth mentioning that the exergy of the system as a result of 20.6 Btu of paddle-wheel work done on it has increased by 1.0 Btu only, that is, by the amount of the reversible work. In other words, if the system were returned to its initial state, it would produce, at most, 1.0 Btu of work.

EXAMPLE 7-13 Dropping a Hot Iron Block into Water

A 5-kg block initially at 350°C is quenched in an insulated tank that contains 100 kg of water at 30°C. Assuming the water that vaporizes during the process condenses back in the tank and the surroundings are at 20°C and 100 kPa, determine (*a*) the final equilibrium temperature, (*b*) the exergy of the combined system at the initial and the final states, and (*c*) the wasted work potential during this process.

Solution We take the entire contents of the tank, *water + iron block,* as the *system* (Fig. 7-40). This is a *closed system* since no mass crosses the system boundary during the process. We note that the volume of a rigid tank is constant, and thus there is no boundary work.

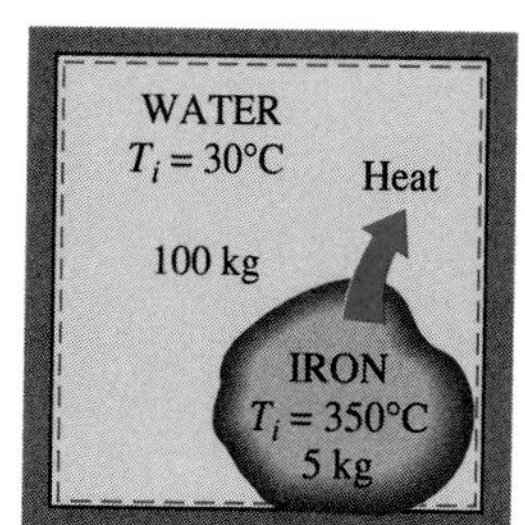

FIGURE 7-40
Schematic for Example 7-13.

Assumptions **1** Both water and the iron block are incompressible substances. **2** Constant specific heats at room temperature can be used for both the water and the iron. **3** The system is stationary and thus the kinetic and potential energy changes are zero, $\Delta KE = \Delta PE = 0$ and $\Delta E = \Delta U$. **4** There are no electrical, shaft, or other forms of work involved. **5** The system is well-insulated and thus there is no heat transfer.

Analysis (*a*) Noting that no energy enters or leaves the system during the process, the application of the energy balance gives.

$$\underbrace{E_{in} - E_{out}}_{\substack{\text{Net energy transfer} \\ \text{by heat, work, and mass}}} = \underbrace{\Delta E_{system}}_{\substack{\text{Change in internal, kinetic,} \\ \text{potential, etc., energies}}}$$

$$0 = \Delta U$$

$$0 = (\Delta U)_{iron} + (\Delta U)_{water}$$

$$0 = [mC(T_f - T_i)]_{iron} + [mC(T_f - T_i)]_{water}$$

By using the specific-heat values for water and iron at room temperature (from Table A-3*a*), the final equilibrium temperature T_f becomes

$$0 = (5\text{ kg})[0.45\text{ kJ/(kg}\cdot{}^\circ\text{C)}](T_f - 350^\circ\text{C}) + (100\text{ kg})[4.184\text{ kJ/(kg}\cdot{}^\circ\text{C)}](T_f - 30^\circ\text{C})$$

which yields

$$T_f = \mathbf{31.7^\circ C}$$

(*b*) Exergy X is an extensive property, and the exergy of a composite system at a specified state is the sum of the exergies of the components of that system at that state. It is determined from Eq. 7-15, which for an incompressible substance reduces to

$$X = (U - U_0) - T_0(S - S_0) + P_0(V - V_0 \overset{\nearrow 0}{)}$$

$$= mC(T - T_0) - T_0 mC \ln\frac{T}{T_0} + 0$$

$$= mC\left(T - T_0 - T_0 \ln\frac{T}{T_0}\right)$$

where T is the absolute temperature at the specified state and T_0 is the absolute temperature of the surroundings. At the initial state,

$$X_{1,\text{ iron}} = (5\text{ kg})[0.45\text{ kJ/(kg}\cdot\text{K)}]\left[(623 - 293)\text{ K} - (293\text{ K})\ln\frac{623}{293}\right]$$

$$= 245.2\text{ kJ}$$

$$X_{1,\text{ water}} = (100\text{ kg})[4.184\text{ kJ/(kg}\cdot\text{K)}]\left[(303 - 293)\text{ K} - (293\text{ K})\ln\frac{303}{293}\right]$$

$$= 69.8\text{ kJ}$$

$$X_{1,\text{ total}} = X_{1,\text{ iron}} + X_{1,\text{ water}} = (245.2 + 69.8)\text{kJ} = \mathbf{315\text{ kJ}}$$

Similarly, the exergy at the final state is

$$X_{2,\text{ iron}} = 0.5\text{ kJ}$$

$$X_{2,\text{ water}} = 95.2\text{ kJ}$$

$$X_{2,\text{ total}} = X_{2,\text{ iron}} + X_{2,\text{ water}} = (0.5 + 95.2)\text{ kJ} = \mathbf{95.7\text{ kJ}}$$

That is, the exergy of the combined system (water + iron) decreased from 315 to 95.7 kJ as a result of this irreversible heat transfer process.

(*c*) The wasted work potential is equivalent to the exergy destroyed, which can be determined from $X_{destroyed} = T_0 S_{gen}$ or by performing an exergy balance on the system. The second approach is more convenient in this case since the initial and final exergies of the system are already evaluated.

$$\underbrace{X_{in} - X_{out}}_{\substack{\text{Net exergy transfer}\\ \text{by heat, work, and mass}}} - \underbrace{X_{destroyed}}_{\substack{\text{Exergy}\\ \text{destruction}}} = \underbrace{\Delta X_{system}}_{\substack{\text{Change}\\ \text{in exergy}}}$$

$$0 - X_{destroyed} = X_2 - X_1$$

$$X_{destroyed} = X_1 - X_2 = 315 - 95.7 = \mathbf{219.3\ kJ}$$

Discussion Note that 219.3 kJ of work could have been produced as the iron was cooled from 350 to 31.7°C and water was heated from 30 to 31.7°C, but was not.

EXAMPLE 7-14 Exergy Destruction during Heat Transfer to a Gas

A frictionless piston-cylinder device, shown in Fig. 7-41, initially contains 0.01 m^3 of argon gas at 400 K and 350 kPa. Heat is now transferred to the argon from a furnace at 1200 K, and the argon expands isothermally until its volume is doubled. No heat transfer takes place between the argon and the surrounding atmospheric air, which is at $T_0 = 300$ K and $P_0 = 100$ kPa. Determine (*a*) the useful work output, (*b*) the exergy destroyed and (*c*) the reversible work for this process.

Solution We take the *argon gas* contained within the piston-cylinder device as the system (Fig. 7-41). This is a *closed system* since no mass crosses the system boundary during the process. We note that heat is transferred to the system from a source at 1200 K, but there is no heat exchange with the environment at 300 K. Also, the temperature of the system remains constant during the expansion process, and its volume doubles, that is, $T_2 = T_1$ and $V_2 = 2V_1$.

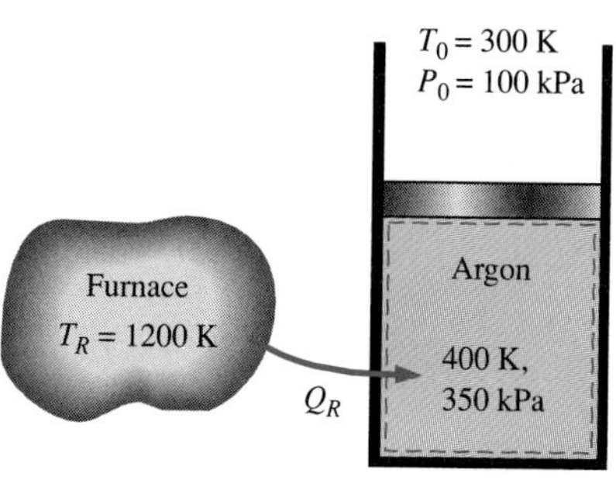

FIGURE 7-41
Schematic for Example 7-14.

Assumptions **1** Argon at specified conditions can be treated as an ideal gas since it is well above its critical temperature of 151 K. **2** The kinetic and potential energies are negligible.

Analysis (*a*) The only work interaction involved during this isothermal process is the quasi-equilibrium boundary work, which is determined from

$$W = W_b = \int_1^2 P\,dV = P_1V_1 \ln\frac{V_2}{V_1} = (350\text{ kPa})(0.01\text{ m}^3)\ln\frac{0.02\text{ m}^3}{0.01\text{ m}^3}$$
$$= 2.43\text{ kPa}\cdot\text{m}^3 = 2.43\text{ kJ}$$

This is the total boundary work done by the argon gas. Part of this work is done against the atmospheric pressure P_0 to push the air out of the way, and it cannot be used for any useful purpose. It is determined from Eq. 7-1:

$$W_{surr} = P_0(V_2 - V_1) = (100\text{ kPa})[(0.02 - 0.01)\text{m}^3]\left(\frac{1\text{ kJ}}{1\text{ kPa}\cdot\text{m}^3}\right) = 1\text{ kJ}$$

The useful work is the difference between these two:

$$W_u = W - W_{surr} = 2.43 - 1 = \mathbf{1.43\ kJ}$$

That is, 1.43 kJ of the work done is available for creating a useful effect such as rotating a shaft.

Also, the heat transfer from the furnace to the system is determined from an energy balance on the system to be

$$\underbrace{E_{in} - E_{out}}_{\substack{\text{Net energy transfer}\\ \text{by heat, work, and mass}}} = \underbrace{\Delta E_{system}}_{\substack{\text{Change in internal, kinetic,}\\ \text{potential, etc., energies}}}$$

$$Q_{in} - W_{b,\text{out}} = \Delta U = mC_v\Delta T^{\nearrow 0} = 0$$

$$Q_{in} = W_{b,\text{out}} = 2.43 \text{ kJ}$$

(*b*) The exergy destroyed during a process can be determined from an exergy balance, or directly from $X_{destroyed} = T_0 S_{gen}$. We will use the second approach since it is usually easier. But first we determine the entropy generation by applying an entropy balance on an *extended system* (system + immediate surroundings), which includes the temperature gradient zone between the cylinder and the furnace so that the temperature at the boundary where heat transfer occurs is $T_R = 1200$ K. This way, the entropy generation associated with the heat transfer is included. Also, the entropy change of the argon gas can be determined from Q/T_{sys} since its temperature remains constant.

$$\underbrace{S_{in} - S_{out}}_{\substack{\text{Net entropy transfer}\\ \text{by heat and mass}}} + \underbrace{S_{gen}}_{\substack{\text{Entropy}\\ \text{generation}}} = \underbrace{\Delta E_{system}}_{\substack{\text{Change}\\ \text{in entropy}}}$$

$$\frac{Q}{T_R} + S_{gen} = \Delta S_{system} = \frac{Q}{T_{sys}}$$

Therefore,

$$S_{gen} = \frac{Q}{T_{sys}} - \frac{Q}{T_R} = \frac{2.43 \text{ kJ}}{400 \text{ K}} - \frac{2.43 \text{ kJ}}{1200 \text{ K}} = 0.00405 \text{ kJ}$$

and $$X_{destroyed} = T_0 S_{gen} = (300 \text{ K})(0.00405 \text{ kJ/K}) = \mathbf{1.22 \text{ kJ}}$$

(*c*) The reversible work, which represents the maximum useful work that could be produced $W_{rev,\text{out}}$, can be determined from the exergy balance by setting the exergy destruction equal to zero,

$$\underbrace{X_{in} - X_{out}}_{\substack{\text{Net exergy transfer}\\ \text{by heat, work, and mass}}} - \underbrace{X_{destroyed}^{\nearrow 0 \text{ (reversible)}}}_{\substack{\text{Exergy}\\ \text{destruction}}} = \underbrace{\Delta X_{system}}_{\substack{\text{Change}\\ \text{in exergy}}}$$

$$\left(1 - \frac{T_0}{T_b}\right)Q - W_{rev,\text{out}} = X_2 - X_1$$

$$= (E_2 - E_1) + P_0(V_2 - V_1) - T_0(S_2 - S_1)$$

$$= 0 + W_{surr} - T_0\frac{Q}{T_{sys}}$$

since $\Delta KE = \Delta PE = 0$ and $\Delta U = 0$ (the change in internal energy of an ideal gas is zero during an isothermal process), and $\Delta S_{sys} = Q/T_{sys}$ for isothermal processes in the absence of any irreversibilities. Then,

$$W_{\text{rev, out}} = T_0 \frac{Q}{T_{\text{sys}}} - W_{\text{surr}} + \left(1 - \frac{T_0}{T_R}\right)Q$$
$$= (300 \text{ K}) \frac{2.43 \text{ kJ}}{400 \text{ K}} - (1 \text{ kJ}) + \left(1 - \frac{300 \text{ K}}{1200 \text{ K}}\right)(2.43 \text{ kJ})$$
$$= \mathbf{2.65 \text{ kJ}}$$

Therefore, the useful work output would be 2.65 kJ instead of 1.43 kJ if the process were executed in a totally reversible manner.

Alternative approach The reversible work could also be determined by applying the basics only, without resorting to exergy balance. This is done by replacing the irreversible portions of the process by reversible ones that will create the same effect on the system. The useful work output of this idealized process (between the actual end states) is the reversible work.

The only irreversibility the actual process involves is the heat transfer between the system and the furnace through a finite temperature difference. This irreversibility can be eliminated by operating a reversible heat engine between the furnace at 1200 K and the surroundings at 300 K. When 2.43 kJ of heat is supplied to this heat engine, it will have a work output of

$$W_{\text{HE}} = \eta_{\text{rev}} Q_H = \left(1 - \frac{T_L}{T_H}\right) Q_H = \left(1 - \frac{300 \text{ K}}{1200 \text{ K}}\right)(2.43 \text{ kJ}) = 1.82 \text{ kJ}$$

The 2.43 kJ of heat that was transferred to the system from the source is now extracted from the surrounding air at 300 K by a reversible heat pump that will require a work input of

$$W_{\text{HP, in}} = \frac{Q_H}{\text{COP}_{\text{HP}}} = \left[\frac{Q_H}{T_H/T_H - T_L)}\right]_{\text{HP}} = \frac{2.43 \text{ kJ}}{(400 \text{ K})/[(400 - 300) \text{ K}]} = 0.61 \text{ kJ}$$

Then the net work output of this reversible process (i.e., the reversible work) becomes

$$W_{\text{rev}} = W_u + W_{\text{HE}} - W_{\text{HP, in}} = 1.43 + 1.82 - 0.61 = 2.64 \text{ kJ}$$

which is practically identical to the result obtained earlier. Also, the exergy destroyed is the difference between the reversible work and the useful work, and is determined to be

$$X_{\text{destroyed}} = W_{\text{rev, out}} - W_{u,\text{ out}} = 2.65 - 1.43 = 1.22 \text{ kJ}$$

which is identical to the result obtained earlier.

7-9 ■ EXERGY BALANCE: Control Volumes

The exergy balance relations for control volumes differ from those for closed systems in that they involve one more mechanism of exergy transfer: *mass flow across the boundaries*. As mentioned earlier, mass possesses exergy as well as energy and entropy, and the amounts of these three extensive properties are proportional to the amount of mass (Fig. 7-42). Again taking the positive direction of heat transfer to be to the system and the positive direction of work transfer to be from the system, the general exergy balance

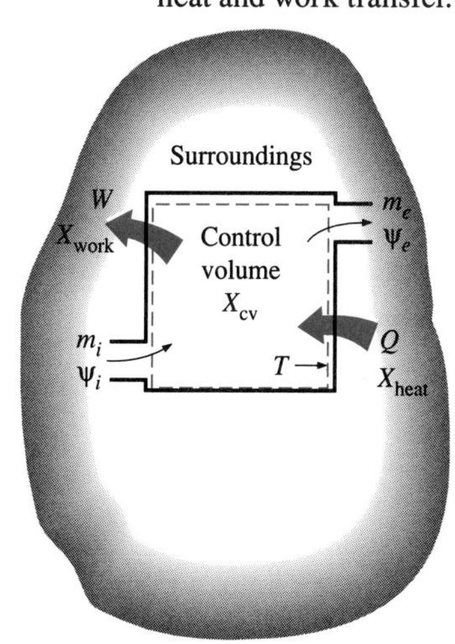

FIGURE 7-42
Exergy is transferred into or out of a control volume by mass as well as heat and work transfer.

relations (Eqs. 7-36 and 7-37) can be expressed for a control volume more explicitly as

$$X_{\text{heat}} - X_{\text{work}} + X_{\text{mass, in}} - X_{\text{mass, out}} - X_{\text{destroyed}} = (X_2 - X_1)_{\text{CV}} \quad (7\text{-}44)$$

or

$$\sum\left(1 - \frac{T_0}{T_k}\right)Q_k - [W - P_0(V_2 - V_1)] + \sum m_i\psi_i - \sum m_e\psi_e - X_{\text{destroyed}} = (X_2 - X_1)_{\text{CV}} \quad (7\text{-}45)$$

where the subscripts are i = inlet, e = exit, 1 = initial state, and 2 = final state of the control volume. It can also be expressed in the **rate form** as

$$\sum\left(1 - \frac{T_0}{T_k}\right)\dot{Q}_k - \left(\dot{W} - P_0\frac{dV_{\text{CV}}}{dt}\right) + \sum \dot{m}_i\psi_i - \sum \dot{m}_e\psi_e - \dot{X}_{\text{destroyed}} = \frac{dX_{\text{CV}}}{dt} \quad (7\text{-}46)$$

The exergy balance relation above can be stated as *the rate of exergy change within the control volume during a process is equal to the rate of net exergy transfer through the control volume boundary by heat, work, and mass flow minus the rate of exergy destruction within the boundaries of the control volume as a result of irreversibilities.*

When the initial and final states of the control volume are specified, the exergy change of the control volume can be determined from $X_2 - X_1 = m_2\phi_2 - m_1\phi_1$.

Exergy Balance for Steady-Flow Systems

Most control volumes encountered in practice such as turbines, compressors, nozzles, diffusers, heat exchangers, pipes, and ducts operate steadily, and thus they experience no changes in their mass, energy, entropy, and exergy contents as well as their volumes. Therefore, $dV_{\text{CV}}/dt = 0$ and $dX_{\text{CV}}/dt = 0$ for such systems, and the amount of exergy entering a steady-flow system in all forms (heat, work, mass transfer) must be equal to the amount of exergy leaving plus the exergy destroyed. Then the rate form of the general exergy balance (Eq. 7-46) reduces for a **steady-flow process** to (Fig. 7-43)

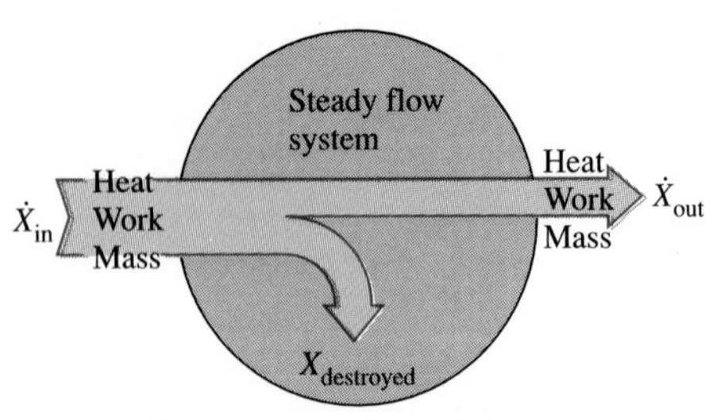

FIGURE 7-43
The exergy transfer to a steady-flow system is equal to the exergy transfer from it plus the exergy destruction within the system.

$$\textit{Steady-flow:}\quad \sum\left(1 - \frac{T_0}{T_k}\right)\dot{Q}_k - \dot{W} + \sum \dot{m}_i\psi_i - \sum \dot{m}_e\psi_e - \dot{X}_{\text{destroyed}} = 0 \quad (7\text{-}47)$$

For a *single-stream* (one-inlet, one-exit) steady-flow device, the relation above further reduces to

$$\textit{Single-stream:}\quad \sum\left(1 - \frac{T_0}{T_k}\right)\dot{Q}_k - \dot{W} + \dot{m}(\psi_1 - \psi_2) - \dot{X}_{\text{destroyed}} = 0 \quad (\text{kW}) \quad (7\text{-}48)$$

where $\dot{m}$ is the mass flow rate and the change in the flow exergy is given by Eq. 7-23 as

$$\psi_1 - \psi_2 = (h_1 - h_2) - T_0(s_1 - s_2) + \frac{V_1^2 - V_2^2}{2} + g(z_1 - z_2)$$

Dividing Eq. 7-48 by $\dot{m}$ gives the exergy balance on a *unit-mass basis* as

$$\textit{Per-unit mass:} \quad \sum\left(1 - \frac{T_0}{T_k}\right)q_k - w + (\psi_1 - \psi_2) - x_{\text{destroyed}} = 0 \qquad \text{(kJ/kg)} \quad (7\text{-}49)$$

where $q = \dot{Q}/\dot{m}$ and $w = \dot{W}/\dot{m}$ are the heat transfer and work done per unit mass of the working fluid, respectively.

For the case of an *adiabatic* single-stream device with no work interactions, the exergy balance relation further simplifies to $\dot{X}_{\text{destroyed}} = \dot{m}(\psi_1 - \psi_2)$, which indicates that the specific exergy of the fluid must decrease as it flows through a work-free adiabatic device or remain the same ($\psi_2 = \psi_1$) in the limiting case of a reversible process regardless of the changes in other properties of the fluid.

Reversible Work, W_{rev}

The exergy balance relations presented above can be used to determine the reversible work W_{rev} by setting the exergy destroyed equal to zero. The work W in that case becomes the reversible work. That is,

$$\textit{General:} \qquad W = W_{\text{rev}} \qquad \text{when} \qquad X_{\text{destroyed}} = 0 \qquad (7\text{-}50)$$

For example, the reversible power for a single-stream steady-flow device is, from Eq. 7-48,

$$\textit{Single stream:} \quad \dot{W}_{\text{rev}} = \dot{m}(\psi_1 - \psi_2) + \sum\left(1 - \frac{T_0}{T_k}\right)\dot{Q}_k \quad \text{(kW)} \qquad (7\text{-}51)$$

which reduces for an adiabatic device to

$$\textit{Adiabatic, single stream:} \quad \dot{W}_{\text{rev}} = \dot{m}(\psi_1 - \psi_2) \qquad (7\text{-}52)$$

Note that the exergy destroyed is zero only for a reversible process, and reversible work represents the maximum work output for work-producing devices such as turbines and the minimum work input for work-consuming devices such as compressors.

Second-Law Efficiency of Steady-Flow Devices, η_{II}

The *second-law efficiency* of various steady-flow devices can be determined from its general definition, η_{II} = (Exergy recovered)/(Exergy supplied). When the changes in kinetic and potential energies are negligible, the second-law efficiency of an *adiabatic turbine* can be determined from

$$\eta_{\text{II,turb}} = \frac{w}{w_{\text{rev}}} = \frac{h_1 - h_2}{\psi_1 - \psi_2} \qquad \text{or} \qquad \eta_{\text{II,turb}} = 1 - \frac{T_0 s_{\text{gen}}}{\psi_1 - \psi_2} \qquad (7\text{-}53)$$

where $s_{\text{gen}} = s_2 - s_1$. For an *adiabatic compressor* with negligible kinetic and potential energies, the second-law efficiency becomes

$$\eta_{\text{II,comp}} = \frac{w_{\text{rev,in}}}{w_{\text{in}}} = \frac{\psi_2 - \psi_1}{h_2 - h_1} \quad \text{or} \quad \eta_{\text{II,comp}} = 1 - \frac{T_0 s_{\text{gen}}}{h_2 - h_1} \tag{7-54}$$

where again $s_{\text{gen}} = s_2 - s_1$.

For an adiabatic *heat exchanger* with two unmixed fluid streams (Fig. 7-44), the exergy supplied is the decrease in the exergy of the hot stream, and the exergy recovered is the increase in the exergy of the cold stream, provided that the cold stream is not at a lower temperature than the surroundings. Then the second-law efficiency of the heat exchanger becomes

$$\eta_{\text{II,HX}} = \frac{\dot{m}_{\text{cold}}(\psi_4 - \psi_3)}{\dot{m}_{\text{hot}}(\psi_1 - \psi_2)} \quad \text{or} \quad \eta_{\text{II,HX}} = 1 - \frac{T_0 \dot{S}_{\text{gen}}}{\dot{m}_{\text{hot}}(\psi_1 - \psi_2)} \tag{7-55}$$

where $\dot{S}_{\text{gen}} = \dot{m}_{\text{hot}}(s_2 - s_1) + \dot{m}_{\text{cold}}(s_4 - s_3)$. Perhaps you are wondering what happens if the heat exchanger is not adiabatic; that is, it is losing some heat to its surroundings at T_0. If the temperature of the boundary (the outer surface of the heat exchanger) T_b is equal T_0, the definition above still holds (except the entropy generation term needs to be modified if the second definition is used). But if $T_b > T_0$, then the exergy of the lost heat at the boundary should be included in the recovered exergy. Although no attempt is made in practice to utilize this exergy and it is allowed to be destroyed, the heat exchanger should not be held responsible for this destruction, which occurs outside its boundaries. If we are interested in the exergy destroyed during the process, not just within the boundaries of the device, then it makes sense to consider an *extended system* that includes the immediate surroundings of the device such that the boundaries of the new enlarged system are at T_0. The second-law efficiency of the extended system will reflect the effects of the irreversibilities that occur within and just outside the device.

An interesting situation arises when the temperature of the cold stream remains below the temperature of the surroundings at all times. In that case the exergy of the cold stream actually decreases instead of increasing. The second-law efficiency becomes zero in this case since none of the exergy supplied, which is the sum of the exergies of both streams, is recovered.

For an adiabatic *mixing chamber* where a hot stream 1 is mixed with a cold stream 2 ($T_2 > T_0$), forming a mixture 3, the exergy supplied can be viewed as the decrease in the exergy of the hot stream, and the exergy recovered as the increase in the exergy of the cold stream. Then the second-law efficiency of the mixing chamber becomes

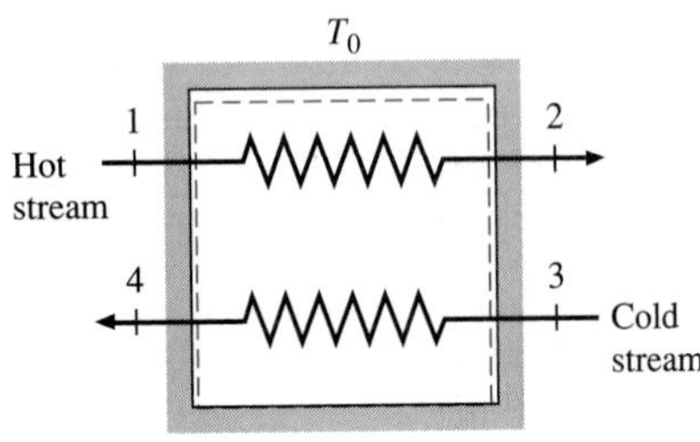

FIGURE 7-44
A heat exchanger with two unmixed fluid streams.

$$\eta_{\text{II,mix}} = \frac{\dot{m}_2(\psi_3 - \psi_2)}{\dot{m}_1(\psi_1 - \psi_3)} \quad \text{or} \quad \eta_{\text{II,mix}} = 1 - \frac{T_0 \dot{S}_{\text{gen}}}{\dot{m}_1(\psi_1 - \psi_3)} \tag{7-56}$$

where $\dot{S}_{\text{gen}} = \dot{m}_3 s_3 - \dot{m}_2 s_2 - \dot{m}_1 s_1$.

EXAMPLE 7-15 Second-Law Analysis of a Steam Turbine

Steam enters a turbine steadily at 3 MPa and 450°C at a rate of 8 kg/s and exits at 0.2 MPa and 150°C. The steam is losing heat to the surrounding air at 100 kPa and 25°C at a rate of 2300 kW, and the kinetic and potential energy changes are negligible. Determine (*a*) the actual power output, (*b*) the maximum possible power output (the reversible power), (*c*) the second-law efficiency, (*d*) the exergy destroyed, and (*e*) the exergy of the steam at the inlet conditions.

Solution We take the *turbine* as the system (Fig. 7-45). This is a *control volume* since mass crosses the system boundary during the process. We note that there is only one inlet and one exit and thus $\dot{m}_1 = \dot{m}_2 = \dot{m}$. Also, heat is lost to the surrounding air and work is done by the system.

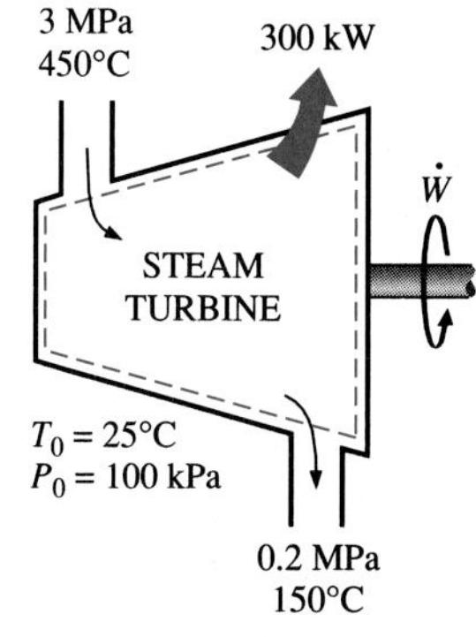

FIGURE 7-45
Schematic for Example 7-15.

Assumptions **1** This is a steady-flow process since there is no change with time at any point and thus $\Delta m_{CV} = 0$, $\Delta E_{CV} = 0$, and $\Delta X_{CV} = 0$. **2** The kinetic and potential energies are negligible.

Analysis The properties of the steam at the inlet and exit states and the state of the environment are

Inlet state: $\left.\begin{array}{l} P_1 = 3 \text{ MPa} \\ T_1 = 450°\text{C} \end{array}\right\}$ $\begin{array}{l} h_1 = 3344.0 \text{ kJ/kg} \\ s_1 = 7.0834 \text{ kJ/(kg} \cdot \text{K)} \end{array}$ (Table A-6)

Exit state: $\left.\begin{array}{l} P_2 = 0.2 \text{ MPa} \\ T_2 = 150°\text{C} \end{array}\right\}$ $\begin{array}{l} h_2 = 2768.8 \text{ kJ/kg} \\ s_2 = 7.2795 \text{ kJ/(kg} \cdot \text{K)} \end{array}$ (Table A-6)

Dead state: $\left.\begin{array}{l} P_0 = 100 \text{ kPa} \\ T_0 = 25°\text{C} \end{array}\right\}$ $\begin{array}{l} h_0 \cong h_{f\,@\,25°\text{C}} = 104.89 \text{ kJ/kg} \\ s_0 \cong s_{f\,@\,25°\text{C}} = 0.3674 \text{ kJ/(kg} \cdot \text{K)} \end{array}$ (Table A-4)

(*a*) The actual power output of the turbine is determined from the rate form of the energy balance,

$$\underbrace{\dot{E}_{in} - \dot{E}_{out}}_{\text{Rate of net energy transfer by heat, work, and mass}} = \underbrace{\Delta \dot{E}_{system}^{\nearrow 0 \text{ (steady)}}}_{\text{Rate of change in internal, kinetic, potential, etc., energies}} = 0$$

$$\dot{E}_{in} = \dot{E}_{out}$$

$$\dot{m}h_1 = \dot{W}_{out} + \dot{Q}_{out} + \dot{m}h_2 \qquad (\text{since ke} \cong \text{pe} \cong 0)$$

$$\dot{W}_{out} = \dot{m}(h_1 - h_2) - \dot{Q}_{out}$$
$$= (8 \text{ kg/s})[(3344.0 - 2768.8)\text{kJ/kg}] - 300 \text{ kW}$$
$$= \mathbf{4302 \text{ kW}}$$

(*b*) The maximum power output (reversible power) is determined from the rate form of the exergy balance applied on the *extended system* (system + immediate surroundings), whose boundary is at the environment temperature of T_0, and by setting the exergy destruction term equal to zero,

$$\underbrace{\dot{X}_{in} - \dot{X}_{out}}_{\text{Rate of net exergy transfer by heat, work, and mass}} - \underbrace{\dot{X}_{destroyed}^{\nearrow 0 \text{ (reversible)}}}_{\text{Rate of exergy destruction}} = \underbrace{\Delta \dot{X}_{system}^{\nearrow 0 \text{ (steady)}}}_{\text{Rate of change of exergy}} = 0$$

$$\dot{X}_{in} = \dot{X}_{out}$$

$$\dot{m}\psi_1 = \dot{W}_{rev,out} + \dot{X}_{heat}^{\nearrow 0} + \dot{m}\psi_2$$

$$\dot{W}_{rev,out} = \dot{m}(\psi_1 - \psi_2)$$

$$= \dot{m}[(h_1 - h_2) - T_0(s_1 - s_2) - \Delta \text{ke}^{\nearrow 0} - \Delta \text{pe}^{\nearrow 0}]$$

Note that exergy transfer with heat is zero when the temperature at the point of transfer is the environment temperature T_0. Substituting,

$$\dot{W}_{rev,\,out} = (8\text{ kg/s})[(3344.0 - 2768.8)\text{ kJ/kg} - (298\text{ K})(7.0834 - 7.2795)\text{ kJ/(kg}\cdot\text{K)}]$$
$$= \mathbf{4660\ kW}$$

(*c*) The second-law efficiency of a turbine is the ratio of the actual work delivered to the reversible work,

$$\eta_{II} = \frac{\dot{W}_{out}}{\dot{W}_{rev,\,out}} = \frac{4302\text{ kW}}{4660\text{ kW}} = \mathbf{0.923\ (or\ 92.3\%)}$$

That is, 7.7 percent of the work potential is wasted during this process.

(*d*) The difference between the reversible work and the actual useful work is the exergy destroyed, which is determined to be

$$\dot{X}_{destroyed} = \dot{W}_{rev,\,out} - \dot{W}_{out} = 4660 - 4302 = \mathbf{358\ kW}$$

That is, the potential to produce useful work is wasted at a rate of 358 kW during this process. The exergy destroyed could also be determined by first calculating the net entropy generated $\dot{S}_{gen}$ during the process.

(*e*) the exergy (maximum work potential) of the steam at the inlet conditions is simply the stream exergy, and is determined from

$$\psi_1 = (h_1 - h_0) - T_0(s_1 - s_0) + \frac{\mathcal{V}_1^2}{2}^{\nearrow 0} + gz_1^{\nearrow 0}$$
$$= (h_1 - h_0) - T_0(s_1 - s_0)$$
$$= (3344.0 - 104.89)\text{kJ/kg} - (298\text{ K})[(7.0834 - 0.3674)\text{kJ/(kg}\cdot\text{K)}]$$
$$= \mathbf{1238\ kJ/kg}$$

That is, not counting the kinetic and potential energies, every kilogram of the steam entering the turbine has a work potential of 1238 kJ. This corresponds to a power potential of (8 kg/s)(1238 kJ/kg) = 9904 kW. Obviously, the turbine is converting 4302/9904 = 43.4 percent of the available work potential of the steam to work.

EXAMPLE 7-16 Exergy Destroyed during Mixing of Fluid Streams

Water at 20 psia and 50°F enters a mixing chamber at a rate of 300 lbm/min, where it is mixed steadily with steam entering at 20 psia and 240°F. The mixture leaves the chamber at 20 psia and 130°F, and heat is being lost to the surrounding air at T_0 = 70°F at a rate of 180 Btu/min. Neglecting the changes in kinetic and potential energies, determine the reversible work and exergy destroyed for this process.

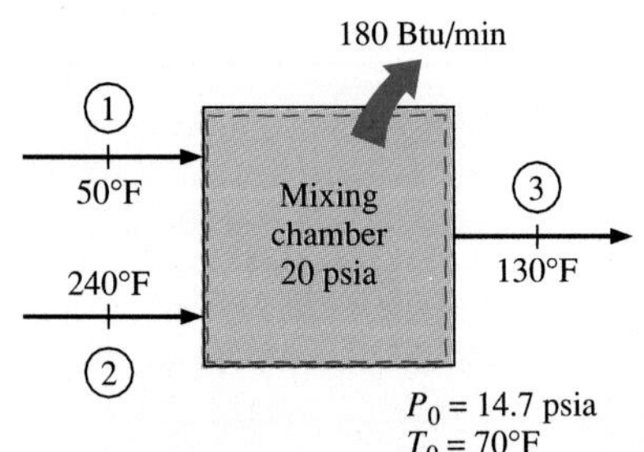

FIGURE 7-46
Schematic for Example 7-16.

Solution This is a steady-flow process, which was discussed in Example 6-22 with regard to entropy generation. A sketch of the mixing chamber and the system boundaries is given in Fig. 7-46. By applying the conservation of mass and the conservation of energy equations, the mass flow rate of the steam was determined in Example 6-22 to be $\dot{m}_2$ = 22.7 lbm/min.

The maximum power output (reversible power) is determined from the rate form of the exergy balance applied on the *extended system* (system + immediate surroundings), whose boundary is at the environment temperature of T_0, and by setting the exergy destruction term equal to zero,

$$\underbrace{\dot{X}_{in} - \dot{X}_{out}}_{\substack{\text{Rate of net exergy transfer}\\ \text{by heat, work, and mass}}} \underbrace{- \dot{X}_{destroyed}^{\nearrow 0 \text{ (reversible)}}}_{\substack{\text{Rate of exergy}\\ \text{destruction}}} = \underbrace{\Delta \dot{X}_{system}^{\nearrow 0 \text{ (steady)}}}_{\substack{\text{Rate of change}\\ \text{of exergy}}} = 0$$

$$\dot{X}_{in} = \dot{X}_{out}$$

$$\dot{m}_1\psi_1 + \dot{m}_2\psi_2 = \dot{W}_{rev,out} + \dot{X}_{heat}^{\nearrow 0} + \dot{m}_3\psi_3$$

$$\dot{W}_{rev,out} = \dot{m}_1\psi_1 + \dot{m}_2\psi_2 - \dot{m}_3\psi_3$$

Note that exergy transfer by heat is zero when the temperature at the point of transfer is the environment temperature T_0, and the kinetic and potential energies are negligible. Therefore,

$$\begin{aligned}\dot{W}_{rev,out} &= \dot{m}_1(h_1 - T_0s_1) + \dot{m}_2(h_2 - T_0s_2) - \dot{m}_3(h_3 - T_0s_3)\\ &= (300 \text{ lbm/min})\{18.06 \text{ Btu/lbm} - (530 \text{ R})[0.03607 \text{ Btu/(lbm}\cdot\text{R)}]\}\\ &\quad + (22.7 \text{ lbm/min})\{1162.3 \text{ Btu/lbm} - (530 \text{ R})[1.7405 \text{ Btu/(lbm}\cdot\text{R)}]\}\\ &\quad - (322.7 \text{ lbm/min})\{97.98 \text{ Btu/lbm} - (530\text{R})[0.18172 \text{ Btu/(lbm}\cdot\text{R)}]\}\\ &= \mathbf{4588.7 \text{ Btu/min}}\end{aligned}$$

That is, we could have produced work at a rate of 4588.7 Btu/min if we ran a heat engine between the hot and the cold fluid streams instead of allowing them to mix directly.

The exergy destroyed is determined from

$$\dot{X}_{destroyed} = \dot{W}_{rev,out} - \dot{W}_u^{\nearrow 0} = T_0\dot{S}_{gen}$$

Thus,
$$\dot{X}_{destroyed} = \dot{W}_{rev,out} = \mathbf{4588.7 \text{ Btu/min}}$$

since there is no actual work produced during the process (Fig. 7-47).

The entropy generation rate for this process was determined in Example 6-22 to be $\dot{S}_{gen} = 8.65$ Btu/(min · R). Thus the exergy destroyed could also be determined from the second part of the above equation:

$$\dot{X}_{destroyed} = T_0\dot{S}_{gen} = (530 \text{ R})[8.65 \text{ Btu/(min}\cdot\text{R)}] = 4584.5 \text{ Btu/min}$$

The slight difference between the two results is due to roundoff error.

FIGURE 7-47
For systems that involve no actual work, the reversible work and irreversibility are identical.

EXAMPLE 7-17 Charging a Compressed Air Storage System

A 200-m^3 rigid tank initially contains atmospheric air at 100 kPa and 300 K and is to be used as a storage vessel for compressed air at 1 MPa and 300K. Compressed air is to be supplied by a compressor that takes in atmospheric air at P_0 = 100 kPa and T_0 = 300 K. Determine the minimum work requirement for this process.

FIGURE 7-48
Schematic for Example 7-17.

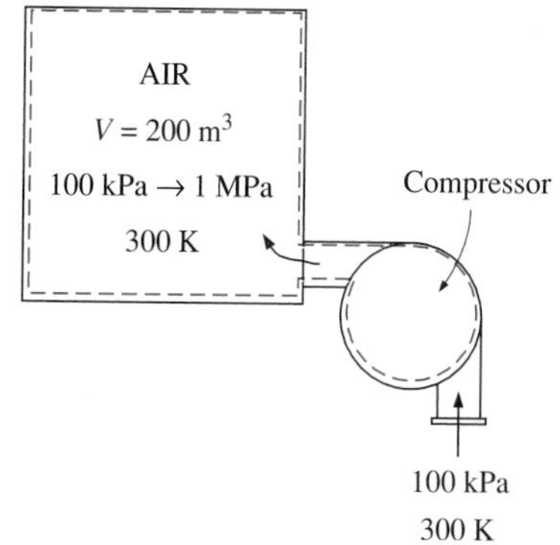

Solution We take the *rigid tank combined with the compressor* as the system (Fig. 7-48). This is a *control volume* since mass crosses the system boundary during the process. We note that this is an unsteady-flow process since the mass content of the system will change as the tank is charged. Also, there is only one inlet and no exit.

Assumptions **1** Air is an ideal gas since it is at a high temperature and low pressure relative to its critical point values of −141°C and 3.77 MPa. **2** The kinetic and potential energies are negligible. **3** The properties of air at the inlet remain constant during the entire charging process.

Analysis The minimum work required for a process is the *reversible work,* which can be determined from the exergy balance applied on the *extended system* (system + immediate surroundings) whose boundary is at the environment temperature of T_0 (so that there is no exergy transfer accompanying heat transfer to or from the environment) and by setting the exergy destruction term equal to zero,

$$\underbrace{X_{in} - X_{out}}_{\substack{\text{Net exergy transfer}\\ \text{by heat, work, and mass}}} \quad \underbrace{- X_{destroyed}^{\nearrow 0 \text{ (reversible)}}}_{\substack{\text{Exergy}\\ \text{destruction}}} = \underbrace{\Delta X_{system}}_{\substack{\text{Change}\\ \text{in exergy}}}$$

$$\dot{X}_{in} - \dot{X}_{out} = X_2 - X_1$$

$$W_{rev,in} + m_1\psi_1^{\nearrow 0} = m_2\phi_2 - m_1\phi_1^{\nearrow 0}$$

$$W_{rev,in} = \dot{m}_2\phi_2$$

Note that $\phi_1 = \psi_1 = 0$ since the initial air in the tank and the air entering are at the state of the environment, and the exergy of a substance at the state of the environment is zero. The final mass of air and the exergy of the pressurized air in the tank at the end of the process are determined to be

$$m_2 = \frac{P_2V}{RT_2} = \frac{(1000 \text{ kPa})(200 \text{ m}^3)}{[0.287(\text{kPa}\cdot\text{m}^3)/(\text{kg}\cdot\text{K})](300 \text{ K})} = 2323 \text{ kg}$$

$$\phi_2 = (u_2 - u_0)^{\nearrow 0 (\text{since } T_2 = T_0)} + P_0(v_2 - v_0) - T_0(s_2 - s_0) + \frac{\mathcal{V}_2^2}{2}^{\nearrow 0} + gz_2^{\nearrow 0}$$

$$= P_0(v_2 - v_0) - T_0(s_2 - s_0)$$

We note that

$$P_0(v_2 - v_0) = P_0\left(\frac{RT_2}{P_2} - \frac{RT_0}{P_0}\right) = RT_0\left(\frac{P_0}{P_2} - 1\right) \qquad (\text{since } T_2 = T_0)$$

$$T_0(s_2 - s_0) = T_0\left(C_p \ln\frac{T_2}{T_0}^{\nearrow 0} - R\ln\frac{P_2}{P_0}\right) = -RT_0 \ln\frac{P_2}{P_0} \qquad (\text{since } T_2 = T_0)$$

Therefore,

$$\phi_2 = RT_0\left(\frac{P_0}{P_2} - 1\right) + RT_0\ln\frac{P_2}{P_0} = RT_0\left(\ln\frac{P_2}{P_0} + \frac{P_0}{P_2} - 1\right)$$

$$= [0.287 \text{ kJ/(kg}\cdot\text{K)}](300 \text{ K})\left(\ln\frac{1000 \text{ kPa}}{100 \text{ kPa}} + \frac{100 \text{ kPa}}{1000 \text{ kPa}} - 1\right)$$

$$= 120.76 \text{ kJ/kg}$$

and

$$W_{rev,in} = m_2\phi_2 = (2323 \text{ kg})(120.76 \text{ kJ/kg}) = \mathbf{280{,}525 \text{ kJ}}$$

That is, a minimum of 280,525 kJ of work input is required to fill the tank with compressed air at 300 K and 1 MPa. In reality, the required work input will be greater by an amount equal to the exergy destruction during the process. Compare this to the result of Example 7-7. What can you conclude?

7-10 ■ SECOND-LAW ASPECTS OF DAILY LIFE

Thermodynamics is a fundamental natural science that deals with various aspects of energy, and even nontechnical people have a basic understanding of energy and the first law of thermodynamics since there is hardly any aspect of life that does not involve the transfer or transformation of energy in different forms. All the *dieters,* for example, base their lifestyle on the conservation of energy principle. Although the first-law aspects of thermodynamics are readily understood and easily accepted by most people there is not a public awareness about the second law of thermodynamics, and the second-law aspects are not fully appreciated even by people with technical backgrounds. This causes some students to view the second law as something that is of theoretical interest rather than an important and practical engineering tool. As a result, students show little interest in a detailed study of the second law of thermodynamics. This is unfortunate because the students end up with a one-sided view of thermodynamics and miss the balanced, complete picture.

Many *ordinary events* that go unnoticed can serve as excellent vehicles to convey important concepts of thermodynamics. Below we will attempt to demonstrate the relevance of the second-law concepts such as exergy, reversible work, irreversibility, and the second-law efficiency to various aspects of daily life using examples with which even nontechnical people can identify. Hopefully, this will enhance our understanding and appreciation of the second law of thermodynamics and encourage us to use it more often in technical and even nontechnical areas. The critical reader is reminded that the concepts presented below are *soft* and *difficult to quantize,* and that they are offered here to stimulate interest in the study of the second law of thermodynamics and to enhance our understanding and appreciation of it.

The second-law concepts are implicitly used in various aspects of daily life. Many successful people seem to make extensive use of them without even realizing it. There is growing awareness that quality plays as important a role as quantity in even ordinary daily activities. The following appeared in an article in the *Reno Gazette-Journal* on March 3, 1991:

> Dr. Held considers himself a survivor of the tick-tock conspiracy. About four years ago, right around his 40th birthday, he was putting in 21-hour days—working late, working out, taking care of his three children and getting involved in sports. He got about four or five hours of sleep a night. . . . "Now I'm in bed by 9:30 and I'm up by 6," he says. "I get twice as much done as I used to. I don't have to do things twice or read things three times before I understand them."

The statement above has a strong relevance to the second-law discussions. It indicates that the problem is not how much time we have (the first law), but, rather, how effectively we use it (the second law). For a person to get *more done in less time* is no different than for a car to go *more miles on less fuel*.

In thermodynamics, *reversible work* for a process is defined as the maximum useful work output (or minimum work input) for that process. It is the useful work that a system would deliver (or consume) during a process between two specified states if that process is executed in a reversible (perfect) manner. The difference between the reversible work and the actual useful work is due to imperfections and is called *irreversibility* (the wasted work potential). For the special case of the final state being the dead state or the state of the surroundings, the reversible work becomes a maximum and is called the *exergy* of the system at the initial state. The irreversibility for a reversible or perfect process is zero.

The *exergy* of a person in daily life can be viewed as the best job that person can do under the most favorable conditions. The *reversible work* in daily life, on the other hand, can be viewed as the best job a person can do under some specified conditions. Then the difference between the reversible work and the actual work done under those conditions can be viewed as the *irreversibility* or the *exergy destroyed*. In engineering systems, we try to identify the major sources of irreversibilities and minimize them in order to maximize performance. In daily life, a person should do just that to maximize his or her performance.

The exergy of a person at a given time and place can be viewed as the maximum amount of work he or she can do at that time and place. Exergy is certainly difficult to quantify because of the interdependence of physical and intellectual capabilities of a person. The ability to perform physical and intellectual tasks simultaneously complicates things even further. *Schooling* and *training* obviously increase the exergy of a person. *Aging* decreases the physical exergy. Unlike most mechanical things, the exergy of human beings is a function of time, and the physical and/or intellectual exergy of a person goes to waste if it is not utilized at the time. A barrel of oil loses nothing from its exergy if left unattended for 40 years. But a person will lose much of his or her entire exergy during that time period if he or she just sits back.

A hard-working farmer, for example, may make full use of his *physical exergy* but very little use of his *intellectual exergy*. That farmer, for example, could learn a foreign language or a science by listening to some educational tapes at the same time he is doing his physical work. This is also true for people who spend considerable time in the car commuting to work. It is hoped that some day we will be able to do exergy analysis for people and irreversibility analysis for their activities. Such an analysis will point out the way for people to minimize their irreversibility, and get more done in less time. Mainframe computers can perform several tasks at once. Why shouldn't human beings be able to do the same?

Children are born with different levels of *exergies* (talents) in different areas. Giving aptitude tests to children at an early age is simply an attempt to uncover the extent of their "hidden' exergies, or talents. The children are

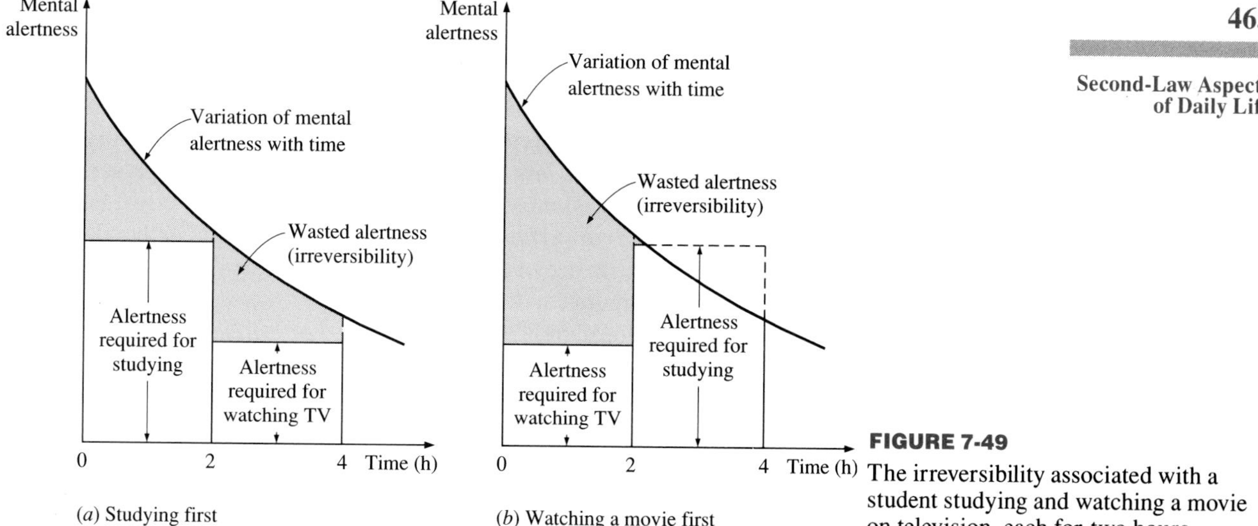

FIGURE 7-49
The irreversibility associated with a student studying and watching a movie on television, each for two hours.

then directed to areas in which they have the greatest exergy. As adults, they are more likely to perform at high levels without stretching the limits if they are naturally fit to be in that area.

We can view the level of *alertness* of a person as his or her *exergy* for intellectual affairs. When a person is well-rested, the degree of alertness, and thus intellectual exergy, is at a maximum and this exergy decreases with time as the person gets tired, as illustrated in Fig. 7-49. Different tasks in daily life require different levels of intellectual exergy, and the difference between available and required alertness can be viewed as the *wasted alertness* or *irreversibility*. To minimize irreversibility, there should be a close match between available alertness and required alertness.

Consider a well-rested student who is planning to spend her next 4 h studying and watching a 2-h-long movie. From the *first-law* point of view, it makes no difference in what order these tasks are performed. But from the *second-law* point of view, it makes a lot of difference. Of these two tasks, studying requires more intellectual alertness than watching a movie does, and thus it makes thermodynamic sense to study first when the alertness is high and to watch the movie later when the alertness is lower, as shown in the figure. A student who does it backwards will waste a lot of alertness while watching the movie, as illustrated in Fig. 7-49, and she will have to keep going back and forth while studying because of insufficient alertness, thus getting less done in the same time period.

In thermodynamics, *the first-law efficiency* (or thermal efficiency) of a heat engine is defined as the ratio of net work output to total heat input. That is, it is the fraction of the heat supplied that is converted to net work. In general, the first-law efficiency can be viewed as the ratio of the desired output to the required input. The first-law efficiency makes no reference to the *best possible performance*, and thus the first-law efficiency alone is not a realistic measure of performance. To overcome this deficiency, we defined the second-law efficiency, which is a measure of actual performance relative to the best possible performance under the same conditions. For heat engines, the second-law efficiency is defined as the ratio of the actual thermal efficiency to the maximum possible (reversible) thermal efficiency under the same conditions.

In daily life, the *first-law efficiency* or *performance* of a person can be viewed as the accomplishment of that person relative to the effort he or she puts in. The *second-law efficiency* of a person, on the other hand, can be viewed as the performance of that person relative to the best possible performance under the circumstances.

Happiness is closely related to the *second-law efficiency.* Small children are probably the happiest human beings because there is so little they can do, but they do it so well, considering their limited capabilities. That is, children have very high second-law efficiencies in their daily lives. The term "full life" also refers to second-law efficiency. A person is considered to have a full life, and thus a very high second-law efficiency, if he or she has utilized all of his or her abilities to the limit during a lifetime.

Even a person with some disabilities will have to put in considerably more effort to accomplish what a physically fit person accomplishes. Yet, despite accomplishing less with more effort, the person with disabilities who gives an impressive performance will probably get more praise. Thus we can say that this person with disabilities had a low first-law efficiency (accomplishing little with a lot of effort) but a very high second-law efficiency (accomplishing as much as possible under the circumstances).

I have only just a minute,
Only 60 seconds in it,
Forced upon me - can't refuse it
Didn't seek it, didn't choose it.
But it is up to me to use it.
I must suffer if I lose it.
Give account if I abuse it,
Just a tiny little minute -
But eternity is in it.

Benjamin E. Mays, President
Morehouse College, Atlanta, GA

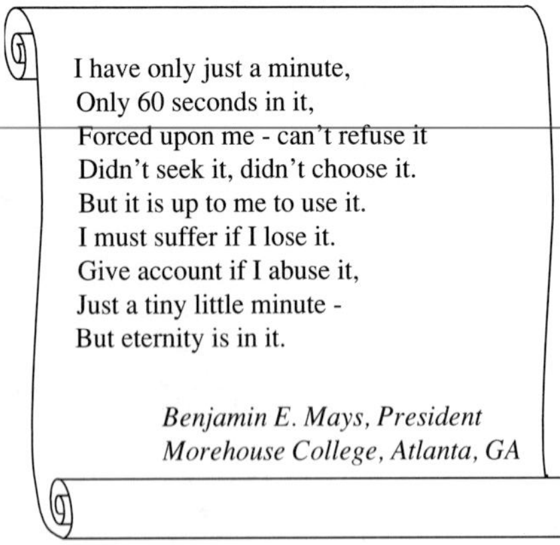

FIGURE 7-50

A poetic expression of exergy and exergy destruction.

In daily life, exergy can also be viewed as the *opportunities that we have* and the exergy destruction as the *opportunities wasted.* Time is the biggest asset that we have, and the time wasted is the wasted opportunity to do something useful (Fig. 7-50).

The examples above show that several *parallels* can be drawn between the supposedly abstract concepts of thermodynamics related to the second law and daily life, and that the second-law concepts can be used in daily life as frequently and authoritatively as the first-law (or conservation) concepts. Relating the *abstract concepts* of thermodynamics to *ordinary events* of life benefits both engineers and social scientists: it helps *engineers* to have a clearer picture of those concepts and to understand them better, and it enables *social scientists* to use these concepts to describe and formulate some social or psychological phenomena better and with more precision. This is like mathematics and sciences being used in support of each other: abstract mathematical concepts are best understood using examples from sciences, and scientific phenomena are best described and formulated with the help of mathematics.

The arguments presented in this section are exploratory in nature, and they are hoped to initiate some interesting discussions and research that may lead into better understanding of performance in various aspects of daily life. The second law may eventually be used to determine quantitatively the most effective way to improve the quality of life and performance in daily life, as it is presently used to improve the performance of engineering systems.

7-11 ■ SUMMARY

The energy content of the universe is constant, just as its mass content is. Yet at times of crisis we are bombarded with speeches and articles on how to "conserve" energy. As engineers, we know that energy is already conserved. What is not concerned is *exergy,* which is the useful work potential of the energy. Once the exergy is wasted, it can never be recovered. When we use energy (to heat our homes for example), we are not destroying any energy; we are merely converting it to a less useful form, a form of less exergy.

The useful work potential of a system at the specified state is called *exergy*. Exergy is a property and is associated with the state of the system and the environment. A system that is in equilibrium with its surroundings has zero exergy and is said to be at the *dead state*. The exergy of the thermal energy of thermal reservoirs is equivalent to the work output of a Carnot heat engine operating between the reservoir and the environment.

Reversible work W_{rev} is defined as the maximum amount of useful work that can be produced (or the minimum work that needs to be supplied) as a system undergoes a process between the specified initial and final states. This is the useful work output (or input) obtained when the process between the initial and final states is executed in a totally reversible manner. The difference between the reversible work W_{rev} and the useful work W_u is due to the irreversibilities present during the process and is called the *ireversibility I*. It is equivalent to the *exergy destroyed* and is expressed as

$$I = X_{destroyed} = T_0 S_{gen} = W_{rev,out} - W_{u,out} = W_{u,in} - W_{rev,in}$$

where S_{gen} is the entropy generated during the process. For a totally reversible process, the useful and reversible work terms are identical and thus irreversibility is zero. Exergy destroyed represents the lost work potential and is also called the *wasted work* or *lost work*.

The *second-law efficiency* is a measure of the performance of a device relative to the performance under reversible conditions for the same end states and is given by

$$\eta_{II} = \frac{\eta_{th}}{\eta_{th,rev}} = \frac{W_u}{W_{rev}}$$

for heat engines and other work-producing devices and

$$\eta_{II} = \frac{COP}{COP_{rev}} = \frac{W_{rev}}{W_u}$$

for refrigerators, heat pumps, and other work-consuming devices. In general, the second-law efficiency is expressed as

$$\eta_{II} = \frac{\text{Exergy recovered}}{\text{Exergy supplied}} = 1 - \frac{\text{Exergy destroyed}}{\text{Exergy supplied}}$$

The exergy of various forms of energy are

Exergy of kinetic energy: $x_{ke} = \text{ke} = \dfrac{\mathcal{V}^2}{2}$

Exergy of potential energy: $x_{pe} = \text{pe} = gz$

Exergy of internal energy: $x_u = (u - u_0) + P_0(v - v_0) - T_0(s - s_0)$

Exergy of flow energy: $x_{Pv} = Pv - P_0v = (P - P_0)v$

Exergy of enthalpy: $x_h = (h - h_0) - T_0(s - s_0)$

The exergies of a fixed mass (nonflow exergy) and of a flow stream are expressed as

$$\textit{Nonflow exergy:}\quad \phi = (u - u_0) + P_0(v - v_0) - T_0(s - s_0) + \frac{\mathcal{V}^2}{2} + gz$$
$$= (e - e_0) + P_0(v - v_0) - T_0(s - s_0)$$

$$\textit{Flow exergy:}\quad \psi = (h - h_0) - T_0(s - s_0) + \frac{\mathcal{V}^2}{2} + gz$$

Then the *exergy change* of a fixed mass or fluid stream as it undergoes a process from state 1 to state 2 is given by

$$\Delta X = X_2 - X_1 = m(\phi_2 - \phi_1) = (E_2 - E_1) + P_0(V_2 - V_1) - T_0(S_2 - S_1)$$
$$= (U_2 - U_1) + P_0(V_2 - V_1) - T_0(S_2 - S_1) + m\frac{\mathcal{V}_2^2 - \mathcal{V}_1^2}{2} + mg(z_2 - z_1)$$

$$\Delta\psi = \psi_2 - \psi_1 = (h_2 - h_1) - T_0(s_2 - s_1) + \frac{\mathcal{V}_2^2 - \mathcal{V}_1^2}{2} + g(z_2 - z_1)$$

Exergy can be transferred by heat, work, and mass flow, and exergy transfer accompanied by heat, work, and mass transfer are given by

Exergy transfer by heat: $X_{heat} = \left(1 - \dfrac{T_0}{T}\right)Q$

Exergy transfer by work: $X_{work} = \begin{cases} W - W_{surr} & \text{(for boundary work)} \\ W & \text{(for other forms of work)} \end{cases}$

Exergy transfer by mass: $X_{mass} = m\psi$

The exergy of an isolated system during a process always decreases or, in the limiting case of a reversible process, remains constant. This is known as the *decrease of exergy principle* and is expressed as

$$\Delta X_{isolated} = (X_2 - X_1)_{isolated} \leq 0$$

Exergy balance for *any system* undergoing *any process* can be expressed as

General:

$$\underbrace{X_{in} - X_{out}}_{\substack{\text{Net exergy transfer}\\ \text{by heat, work, and mass}}} - \underbrace{X_{destroyed}}_{\substack{\text{Exergy}\\ \text{destruction}}} = \underbrace{\Delta X_{system}}_{\substack{\text{Change}\\ \text{in exergy}}}$$

General, rate form:

$$\underbrace{\dot{X}_{in} - \dot{X}_{out}}_{\substack{\text{Rate of net exergy transfer}\\ \text{by heat, work, and mass}}} - \underbrace{\dot{X}_{destroyed}}_{\substack{\text{Rate of exergy}\\ \text{destruction}}} = \underbrace{\Delta \dot{X}_{system}}_{\substack{\text{Rate of change}\\ \text{of exergy}}}$$

General, unit-mass basis: $(x_{in} - x_{out}) - x_{destroyed} = \Delta x_{system}$

where

$$\dot{X}_{heat} = (1 - T_0/T)\dot{Q}$$
$$\dot{X}_{work} = \dot{W}_{useful}$$
$$\dot{X}_{mass} = \dot{m}\psi$$
$$\Delta\dot{X}_{system} = dX_{system}/dt$$

For a *reversible process,* the exergy destruction term $X_{destroyed}$ drops out. Taking the positive direction of heat transfer to be to the system and the positive direction of work transfer to be from the system, the general exergy balance relations can be expressed more explicitly as

$$\sum\left(1 - \frac{T_0}{T_k}\right)Q_k - [W - P_0(V_2 - V_1)] + \sum m_i\psi_i - \sum m_e\psi_e - X_{destroyed} = X_2 - X_1$$

$$\sum\left(1 - \frac{T_0}{T_k}\right)\dot{Q}_k - \left(\dot{W} - P_0\frac{dV_{CV}}{dt}\right) + \sum \dot{m}_i\psi_i - \sum \dot{m}_e\psi_e - \dot{X}_{destroyed} = \frac{dX_{CV}}{dt}$$

where the subscripts are i = inlet, e = exit, 1 = initial state, and 2 = final state of the system.

REFERENCES AND SUGGESTED READING

1. J. E. Ahern. *The Exergy Method of Energy Systems Analysis.* New York: John Wiley & Sons, 1980.

2. A. Bejan. *Advanced Engineering Thermodynamics.* New York: John Wiley & Sons, 1988.

3. A Bejan. *Entropy Generation through Heat and Fluid Flow.* New York: John Wiley & Sons, 1982.

4. Y. A. Çengel. "A Unified and Intuitive Approach to Teaching Thermodynamics." ASME International Congress and Exposition, Atlanta, Georgia, November 17–22, 1996.

5. M. S. Moran and H. N. Shapiro. *Fundamentals of Engineering Thermodynamics.* 3rd ed. New York: John Wiley & Sons, 1996.

6. G. J. Van Wylen and R. E. Sonntag. *Fundamentals of Classical Thermodynamics.* 3rd ed. New York: John Wiley & Sons, 1985.

7. K. Wark. *Thermodynamics.* 5th ed. New York: McGraw-Hill, 1988.

PROBLEMS*

Exergy, Irreversibility, Reversible Work, and Second-Law Efficiency

7-1C What is the difference between reversible work and exergy?

7-2C How does reversible work differ from useful work?

7-3C What is wind? How is it caused? Why is it a form of energy? How can you express wind energy per unit mass of air? Do you agree with the claim that wind is a form of solar energy?

7-4C Under what conditions does the reversible work equal irreversibility for a process?

7-5S What final state will maximize the work output of a device?

7-6C Is the exergy of a system different in different environments?

7-7C How does useful work differ from actual work? For what kind of systems are these two identical?

7-8C Are unavailable energy and irreversibility basically the same thing? If not, how do they differ?

7-9C Consider a process that involves no irreversibilities. Will the actual useful work for that process be equal to the reversible work?

7-10C Consider two geothermal wells whose energy contents are estimated to be the same. Will the exergies of these wells necessarily be the same? Explain.

7-11C Consider two systems that are at the same pressure as the environment. The first system is at the same temperature as the environment, whereas the second system is at a lower temperature than the environment. How would you compare the exergies of these two systems?

7-12C Consider an environment of zero absolute pressure (such as outer space). How will the actual work and the useful work compare in that environment?

7-13C What is the second-law efficiency? How does it differ from the first-law efficiency?

7-14C Does a power plant that has a higher thermal efficiency necessarily have a higher second-law efficiency than one with a lower thermal efficiency? Explain.

7-15C Does a refrigerator that has a higher COP necessarily have a higher second-law efficiency than one with a lower COP? Explain.

7-16C Can a process for which the reversible work is zero be reversible? Can it be irreversible? Explain.

---

*Students are encouraged to answer *all* the concept "C" questions.

7-17C Consider a steam turbine operating between two specified states. As the design of the turbine improves, will the actual work increase? How about the reversible work? Explain.

7-18C Consider a process during which no entropy is generated ($S_{gen} = 0$). Does the irreversibility for this process have to be zero?

7-19 The electric power needs of a community are to be met by windmills with 10-m-diameter rotors. The windmills are to be located where the wind is blowing steadily at an average velocity of 12 m/s. Determine the minimum number of windmills that need to be installed if the required power output is 400 kW.

7-20 One method of meeting the extra electric power demand at peak periods is to pump some water from a large body of water (such as a lake) to a water reservoir at a higher elevation at times of low demand and to generate electricity at times of high demand by letting this water run down and rotate a turbine (i.e., convert the electric energy to potential energy and then back to electric energy). For an energy storage capacity of 5×10^6 kWh, determine the minimum amount of water that needs to be stored at an average elevation (relative to the ground level) of 75 m. *Answer:* 2.45×10^{10} kg

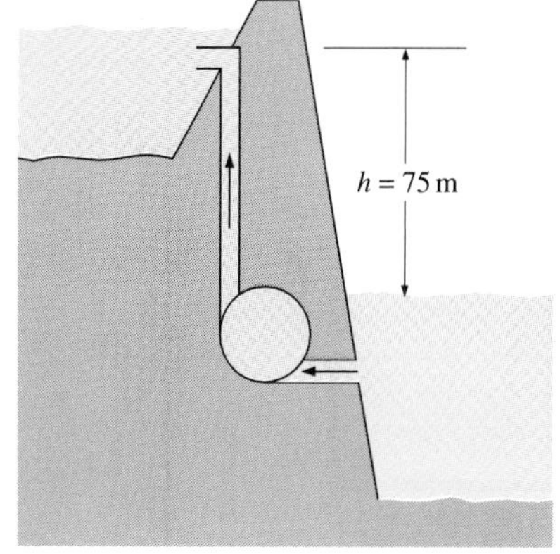

FIGURE P7-20

7-21 Consider a thermal energy reservoir at 1500 K that can supply heat at a rate of 150,000 kJ/h. Determine the exergy of this supplied energy, assuming an environmental temperature of 25°C.

7-22 A heat engine receives heat from a source at 1500 K at a rate of 700 kJ/s, and it rejects the waste heat to a medium at 320 K. The measured power output of the heat engine is 320 kW, and the lowest naturally occurring environment temperature is 25°C. Determine (*a*) the reversible power, (*b*) the rate of irreversibility, and (*c*) the second-law efficiency of this heat engine.
Answers: (*a*) 550.7 kW, (*b*) 230.7 kW, (*c*) 58.1 percent

7-23E A heat engine that rejects waste heat to a sink at 530 R has a thermal efficiency of 36 percent and a second-law efficiency of 60 percent. Determine the temperature of the source that supplies heat to this engine.
Answer: 1325 R

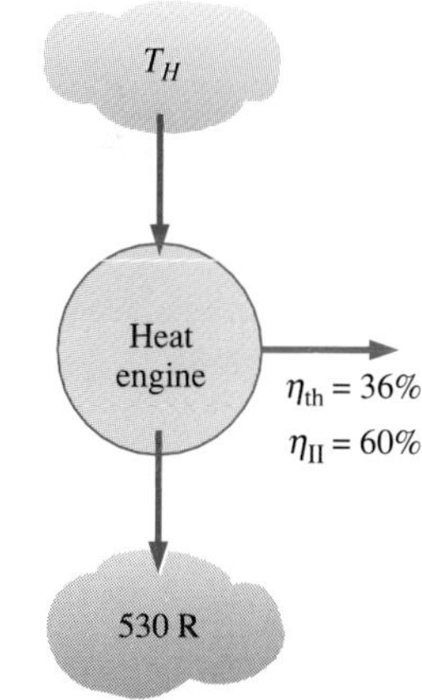

FIGURE P7-23E

7-24 How much of the 100 kJ of thermal energy at 600 K can be converted to useful work? Assume the environment to be at 27°C.

7-25 A heat engine that receives heat from a furnace at 1100°C and rejects waste heat to a river at 20°C has a thermal efficiency of 35 percent. Determine the second-law efficiency of this power plant.

7-26 A house that is losing heat at a rate of 80,000 kJ/h when the outside temperature drops to 15°C is to be heated by electric resistance heaters. If the house is to be maintained at 22°C at all times, determine the reversible work input for this process and the irreversibility. *Answers:* 0.53 kW, 21.69 kW

7-27E A freezer is maintained at 20°F by removing heat from it at a rate of 75 Btu/min. The power input to the freezer is 0.70 hp, and the surrounding

air is at 75°F. Determine (*a*) the reversible power, (*b*) the irreversibility, and (*c*) the second-law efficiency of this freezer.
Answers: (*a*) 0.20 hp, (*b*) 0.50 hp, (*c*) 28.9 percent

7-28 Show that the power produced by a wind turbine is proportional to the cube of the wind velocity and to the square of the blade span diameter.

Second-Law Analysis of Closed Systems

7-29C Is a process during which no entropy is generated ($S_{gen} = 0$) necessarily reversible?

7-30C Can a system have a higher second-law efficiency than the first-law efficiency during a process? Give examples.

7-31 A piston-cylinder device initially contains 2 L of air at 100 kPa and 25°C. Air is now compressed to a final state of 600 kPa and 150°C. The useful work input is 1.2 kJ. Assuming the surroundings are at 100 kPa and 25°C, determine (*a*) the exergy of the air at the initial and the final states, (*b*) the minimum work that must be supplied to accomplish this compression process, and (*c*) the second-law efficiency of this process.
Answers: (*a*) 0, 0.171 kJ; (*b*) 0.171 kJ; (*c*) 14.3 percent

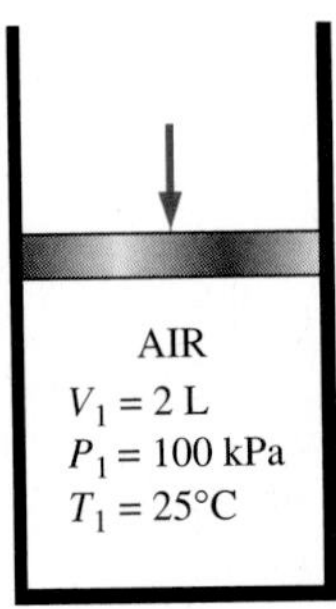

FIGURE P7-31

7-32 A piston-cylinder device contains 3 kg of refrigerant-134a at 0.8 MPa and 50°C. The refrigerant is now cooled at constant pressure until it exists as a liquid at 30°C. If the surroundings are at 100 kPa and 30°C, determine (*a*) the exergy of the refrigerant at the initial and the final states and (*b*) the exergy destroyed during this process.

7-33 The radiator of a steam heating system has a volume of 20 L and is filled with superheated water vapor at 200 kPa and 200°C. At this moment both the inlet and the exit valves to the radiator are closed. After a while it is observed that the temperature of the steam drops to 80°C as a result of heat transfer to the room air, which is at 21°C. Assuming the surroundings to be at 0°C, determine (*a*) the amount of heat transfer to the room and (*b*) the maximum amount of heat that can be supplied to the room if this heat from the radiator is supplied to a heat engine that is driving a heat pump. Assume the heat engine operates between the radiator and the surroundings.
Answers: (*a*) 30.3 kJ, (*b*) 114.8 kJ

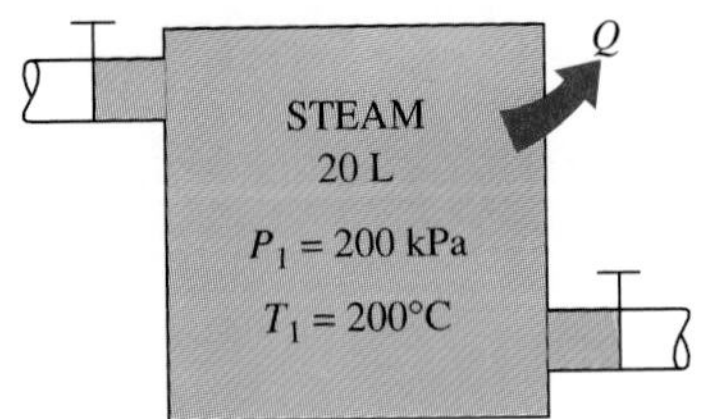

FIGURE P7-33

7-34E A well-insulated rigid tank contains 4 lbm of saturated liquid–vapor mixture of water at 15 psia. Initially, three-quarters of the mass is in the liquid phase. An electric resistance heater placed in the tank is turned on and kept on until all the liquid in the tank is vaporized. Assuming the surroundings to be at 75°F and 14.7 psia, determine (*a*) the exergy destruction and (*b*) the second-law efficiency for this process.

7-35 A rigid tank is divided into two equal parts by a partition. One part of the tank contains 1.5 kg of compressed liquid water at 300 kPa and 60°C and the other side is evacuated. Now the partition is removed, and the water expands to fill the entire tank. If the final pressure in the tank is 15 kPa, determine the exergy destroyed during this process. Assume the surroundings to be at 25°C and 100 kPa. *Answer:* 3.66 kJ

7-36 An insulated piston-cylinder device contains 2 L of saturated liquid water at a constant pressure of 150 kPa. An electric resistance heater inside the cylinder is turned on, and electrical work is done on the water in the amount of 2200 kJ. Assuming the surroundings to be at 25°C and 100 kPa, determine (*a*) the minimum work with which this process could be accomplished and (*b*) the exergy destroyed during this process.
Answers: (*a*) 437.8 kJ, (*b*) 1704.5 kJ

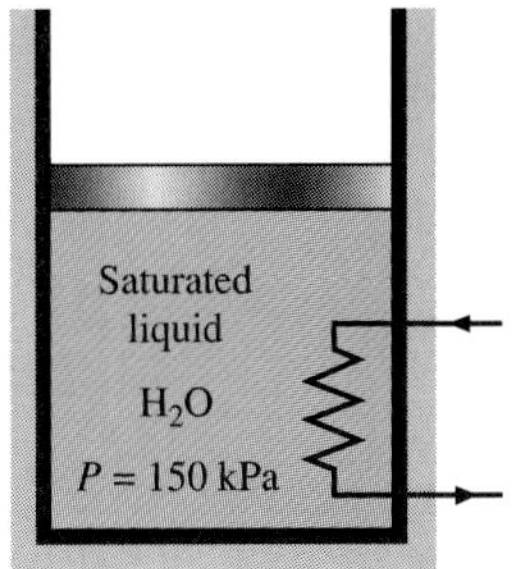

FIGURE P7-36

7-37 An insulated piston-cylinder device contains 0.05 m^3 of saturated refrigerant-134a vapor at 0.8 MPa pressure. The refrigerant is now allowed to expand in a reversible manner until the pressure drops to 0.4 MPa. Determine the change in the exergy of the refrigerant during this process and the reversible work. Assume the surroundings to be at 25°C and 100 kPa.

7-38E Oxygen gas is compressed in a piston-cylinder device from an initial state of 12 ft^3/lbm and 75°F to a final state of 1.5 ft^3/lbm and 525°F. Determine the reversible work input and the increase in the exergy of the oxygen during this process. Assume the surroundings to be at 14.7 psia and 75°F.
Answers: (*a*) 60.7 Btu/lbm, 60.7 Btu/lbm

7-39 A 1.2-m^3 insulated rigid tank contains 2.13 kg of carbon dioxide at 100 kPa. Now paddle-wheel work is done on the system until the pressure in the tank rises to 120 kPa. Determine (*a*) the actual paddle-wheel work done during this process and (*b*) the minimum paddle-wheel work with which this process (between the same end states) could be accomplished.
Answers: (*a*) 87.0 kJ, (*b*) 7.66 kJ

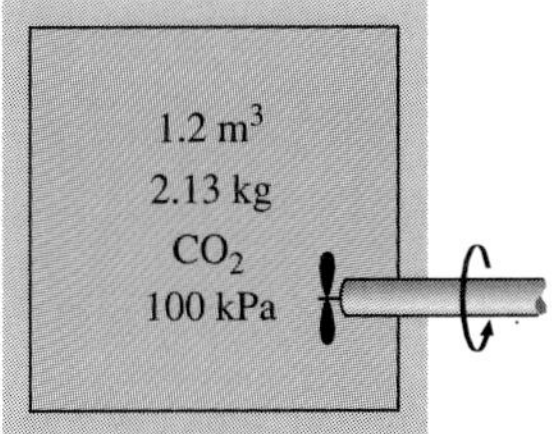

FIGURE P7-39

7-40 An insulated piston-cylinder device initially contains 30 L of air at 120 kPa and 27°C. Air is now heated for 5 min by a 50-W resistance heater placed inside the cylinder. The pressure of air is maintained constant during this process, and the surroundings are at 27°C and 100 kPa. Determine the exergy destroyed during this process. *Answer:* 9.9 kJ

7-41 A mass of 5 kg of helium undergoes a process from an initial state of 3 m^3/kg and 15°C to a final state of 0.5 m^3/kg and 80°C. Assuming the surroundings to be at 25°C, determine the increase in the useful work potential of the helium during this process.

7-42 An insulated rigid tank is divided into two equal parts by a partition. Initially, one part contains 5 kg of argon gas at 300 kPa and 70° C, and the other side is evacuated. The partition is now removed, and the gas fills the entire tank. Assuming the surroundings to be at 25°C, determine the exergy destroyed during this process. *Answer:* 215 kJ

7-43E A 70-lbm copper block initially at 200°F is dropped into an insulated tank that contains 1 ft^3 of water at 75°F. Determine (*a*) the final equilibrium temperature and (*b*) the work potential wasted during this process. Assume the surroundings to be at 75°F.

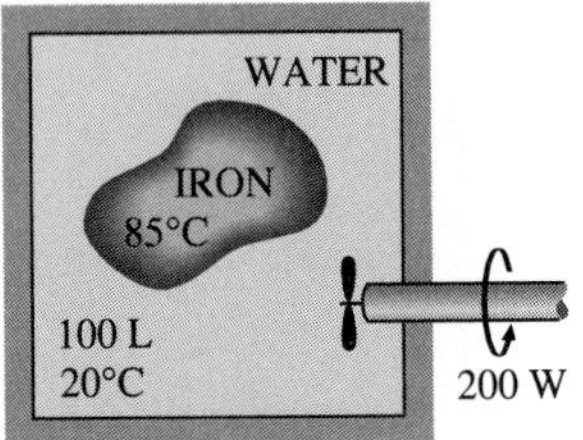

FIGURE P7-45

7-44 A 20-kg aluminum block initially at 200°C is brought into contact with a 20-kg block of iron at 100°C in an insulated enclosure. Assuming the surroundings to be at 25°C, determine (*a*) the final equilibrium temperature and (*b*) the exergy destroyed during this process.

7-45 An iron block of unknown mass at 85°C is dropped into an insulated tank that contains 100 L of water at 20°C. At the same time, a paddle wheel driven by a 200-W motor is activated to stir the water. It is observed that thermal equilibrium is established after 20 min with a final temperature of 24°C. Assuming the surroundings to be at 20°C, determine (*a*) the mass of the iron block and (*b*) the exergy destroyed during this process.

Answers: (*a*) 52.2 kJ, (*b*) 376 kJ

7-46 A 50-kg iron block and a 20-kg copper block, both initially at 80°C, are dropped into a large lake at 15°C. Thermal equilibrium is established after a while as a result of heat transfer between the blocks and the lake water. Assuming the surroundings to be at 20°C, determine the amount of work that could have been produced if the entire process was executed in a reversible manner.

7-47E A 8-ft^3 rigid tank contains refrigerant-134a at 30 psia and 40 percent quality. Heat is transferred now to the refrigerant from a source at 120°F until the pressure rises to 60 psia. Assuming the surroundings to be at 75°F, determine (*a*) the amount of heat transfer between the source and the refrigerant and (*b*) the exergy destroyed during this process.

7-48 Chickens with an average mass of 2.2 kg and average specific heat of 3.54 kJ/(kg · °C) are to be cooled by chilled water that enters a continuous-flow-type immersion chiller at 0.5°C and leaves at 2.5°C. Chickens are dropped into the chiller at a uniform temperature of 15°C at a rate of 500 chickens per hour and are cooled to an average temperature of 3°C before they are taken out. The chiller gains heat from the surroundings at a rate of 200 kJ/h. Determine (*a*) the rate of heat removal from the chicken, in kW, and (*b*) the rate of exergy destruction during this chilling process.

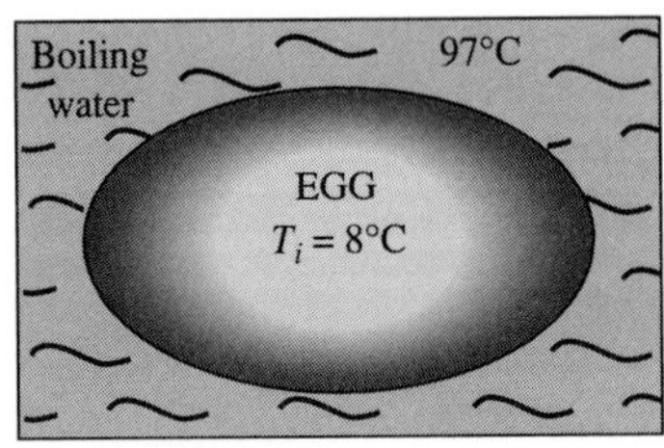

FIGURE P7-49

7-49 An ordinary egg can be approximated as a 5.5-cm-diameter sphere. The egg is initially at a uniform temperature of 8°C and is dropped into boiling water at 97°C. Taking the properties of egg to be ρ = 1020 kg/m^3 and C_p = 3.32 kJ/(kg · °C), determine how much heat is transferred to the egg by the time the average temperature of the egg rises to 70°C and the amount of exergy destruction associated with this heat transfer process.

FIGURE P7-51

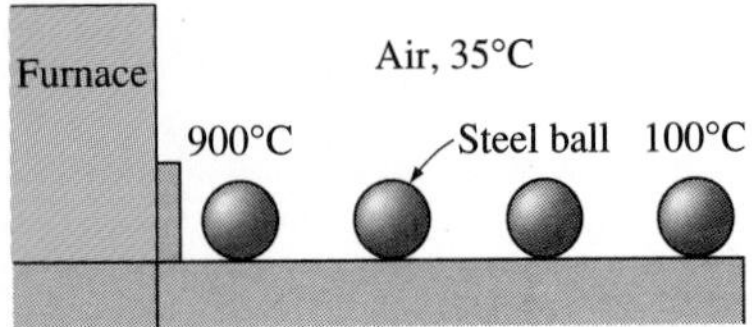

7-50 Stainless steel ball bearings [ρ = 8085 kg/m^3 and C_p = 0.480 kJ/(kg · °C)] having a diameter of 1.2 cm are to be quenched in water at a rate of 1400 per minute. The balls leave the oven at a uniform temperature of 900°C and are exposed to air at 30°C for a while before they are dropped into the water. If the temperature of the balls drops to 850°C prior to quenching, determine (*a*) the rate of heat transfer from the balls to the air and (*b*) the rate of exergy destruction due to heat loss from the balls to the air.

7-51 Carbon steel balls [ρ = 7833 kg/m^3 and C_p = 0.465 kJ/(kg · °C)] 8 mm in diameter are annealed by heating them first to 900°C in a furnace and then allowing them to cool slowly to 100°C in ambient air at 35°C. If 2500 balls are to be annealed per hour, determine (*a*) the rate of heat transfer from the balls to the air and (*b*) the rate of exergy destruction due to heat loss from the balls to the air. *Answers:* (*a*) 542 W, (*b*) 333 W

Second-Law Analysis of Control Volumes

7-52 Steam is throttled from 10 MPa and 450°C to 8 MPa. Determine the wasted work potential during this throttling process. Assume the surroundings to be at 25°C. *Answer:* 27.3 kJ/kg

7-53 Air is compressed steadily by an 8-kW compressor from 100 kPa and 17°C to 600 kPa and 167°C at a rate of 2.1 kg/min. Neglecting the changes in kinetic and potential energies, determine (*a*) the increase in the exergy of the air and (*b*) the rate of exergy destroyed during this process. Assume the surroundings to be at 17°C.

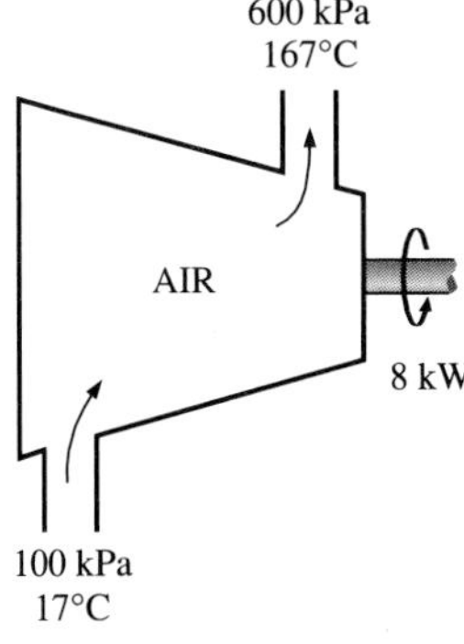

FIGURE P7-53

7-54 Refrigerant-134a at 1 MPa and 100°C is throttled to a pressure of 0.8 MPa. Determine the reversible work and exergy destroyed during this throttling process. Assume the surroundings to be at 25°C.

7-55 Air enters a nozzle steadily at 300 kPa and 87°C with a velocity of 50 m/s and exits at 95 kPa and 300 m/s. The heat loss from the nozzle to the surrounding medium at 17°C is estimated to be 4 kJ/kg. Determine (*a*) the exit temperature and (*b*) the exergy destroyed during this process.
Answers: (*a*) 39.5°C, (*b*) 58.4 kJ/kg

7-56 Steam enters a diffuser at 10 kPa and 50°C with a velocity of 300 m/s and exits as saturated vapor at 50°C and 50 m/s. The exit area of the diffuser is 2 m^2. Determine (*a*) the mass flow rate of the steam and (*b*) the wasted work potential during this process. Assume the surroundings to be at 25°C.

7-57E Air is compressed steadily by a compressor from 14.7 psia and 60°F to 100 psia and 480°F at a rate of 15 lbm/min. Assuming the surroundings to be at 60°F, determine the minimum power input to the compressor. Assume air to be an ideal gas with variable specific heats, and neglect the changes in kinetic and potential energies.

7-58 Steam enters an adiabatic turbine at 6 MPa, 600°C, and 80 m/s and leaves at 50 kPa, 100°C, and 140 m/s. If the power output of the turbine is 5 MW, determine (*a*) the reversible power output and (*b*) the second-law efficiency of the turbine. Assume the surroundings to be at 25°C.
Answers: (*a*) 5.81 MW, (*b*) 86.1 percent

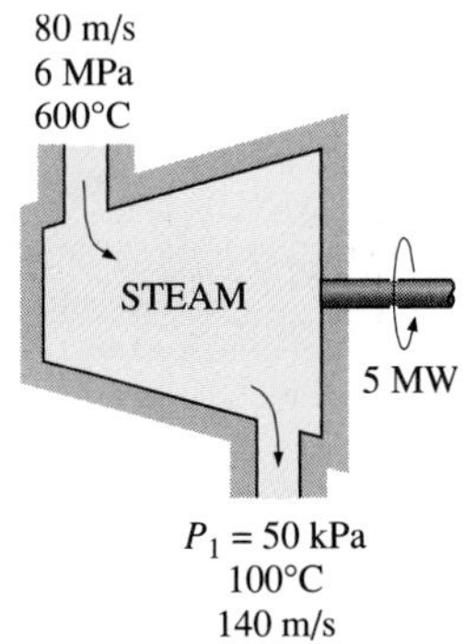

FIGURE P7-58

7-59 Steam is throttled from 9 MPa and 500°C to a pressure of 7 MPa. Determine the decrease in exergy of the steam during this process. Assume the surroundings to be at 25°C. *Answer:* 32.0 kJ/kg

7-60 Combustion gases enter a gas turbine at 900°C, 800 kPa, and 100 m/s and leave at 650°C, 400 kPa, and 220 m/s. Taking $C_p = 1.15$ kJ/(kg · °C) and $k = 1.3$ for the combustion gases, determine (*a*) the exergy of the combustion gases at the turbine inlet and (*b*) the work output of the turbine under reversible conditions. Assume the surroundings to be at 25°C and 100 kPa.

7-61E Refrigerant-134a enters an adiabatic compressor as saturated vapor at 20 psia at a rate of 20 ft^3/min and exits at 100 psia pressure. If the adiabatic efficiency of the compressor is 80 percent, determine (*a*) the actual power input and (*b*) the second-law efficiency of the compressor. Assume the surroundings to be at 75°F. *Answers:* (*a*) 3.77 hp, (*b*) 81.2 percent

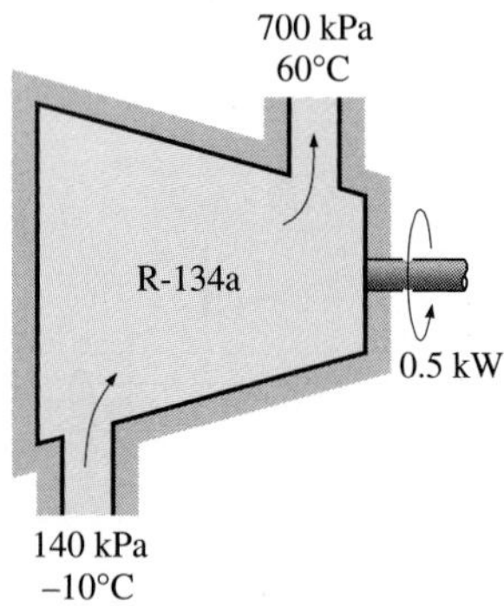

FIGURE P7-62

7-62 Refrigerant-134a at 140 kPa and −10°C is compressed by an adiabatic 0.5-kW compressor to an exit state of 700 kPa and 60°C. Neglecting the changes in kinetic and potential energies and assuming the surroundings to be at 25°C, determine (*a*) the adiabatic efficiency and (*b*) the second-law efficiency of the compressor.

7-63 Air is compressed by a compressor from 95 kPa and 27°C to 600 kPa and 277°C at a rate of 0.06 kg/s. Neglecting the changes in kinetic and potential energies and assuming the surroundings to be at 25°C, determine the reversible power input for this process. *Answer:* 13.7 kW

7-64 Argon gas enters an adiabatic compressor at 120 kPa and 30°C with a velocity of 20 m/s and exits at 1.2 MPa, 530°C, and 80 m/s. The inlet area of the compressor is 130 cm^2. Assuming the surroundings to be at 25°C, determine the reversible power input and irreversibility.
Answers: 116 kW, 3.80 kW

7-65 Steam expands in a turbine steadily at a rate of 15,000 kg/h, entering at 8 MPa and 450°C and leaving at 50 kPa as saturated vapor. Assuming the surroundings to be at 100 kPa and 25°C, determine (*a*) the power potential of the steam at the inlet conditions and (*b*) the power output of the turbine if there were no irreversibilities present.
Answers: (*a*) 5513 kW, (*b*) 3899 kW

7-66E Air enters a compressor at ambient conditions of 15 psia and 60°F with a low velocity and exits at 150 psia, 620°F, and 350 ft/s. The compressor is cooled by the ambient air at 60°F at a rate of 1500 Btu/min. The power input to the compressor is 400 hp. Determine (*a*) the mass flow rate of air and (*b*) the portion of the power input that is used just to overcome the irreversibilities.

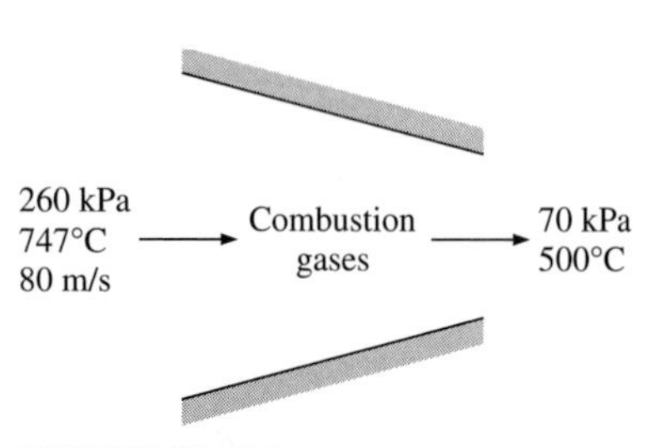

FIGURE P7-67

7-67 Hot combustion gases enter the nozzle of a turbojet engine at 260 kPa, 747°C, and 80 m/s and exit at 70 kPa and 500°C. Assuming the nozzle to be adiabatic and the surroundings to be at 17°C, determine (*a*) the exit velocity and (*b*) the decrease in the exergy of the gases. Take $k = 1.3$ and $C_p = 1.15$ kJ/(kg · °C) for the combustion gases.

7-68 Steam is usually accelerated in the nozzle of a turbine before it strikes the turbine blades. Steam enters an adiabatic nozzle at 7 MPa and 500°C with a velocity of 70 m/s and exits at 5 MPa and 450°C. Assuming the surroundings to be at 25°C, determine (*a*) the exit velocity of the steam, (*b*) the adiabatic efficiency, and (*c*) the exergy destroyed within the nozzle.

7-69 Carbon dioxide enters a compressor at 100 kPa and 300 K at a rate of 0.2 kg/s and exits at 600 kPa and 450 K. Determine the power input to the compressor if the process involved no irreversibilities. Assume the surroundings to be at 25°C. *Answer:* −25.5 kW

7-70E A hot-water stream at 160°F enters an adiabatic mixing chamber with a mass flow rate of 4 lbm/s, where it is mixed with a stream of cold water at 70°F. If the mixture leaves the chamber at 110°F, determine (*a*) the mass flow rate of the cold water and (*b*) the exergy destroyed during this adiabatic mixing process. Assume all the streams are at a pressure of 50 psia and the surroundings are at 75°F. *Answers:* (*a*) 5.0 lbm, (*b*) 14.6 Btu/s

7-71 Liquid water at 200 kPa and 20°C is heated in a chamber by mixing it with superheated steam at 200 kPa and 300°C. Liquid water enters the mixing chamber at a rate of 2.5 kg/s, and the chamber is estimated to lose heat to the surrounding air at 25°C at a rate of 600 kJ/min. If the mixture leaves the mixing chamber at 200 kPa and 60°C, determine (*a*) the mass flow rate of the superheated steam and (*b*) the wasted work potential during this mixing process.

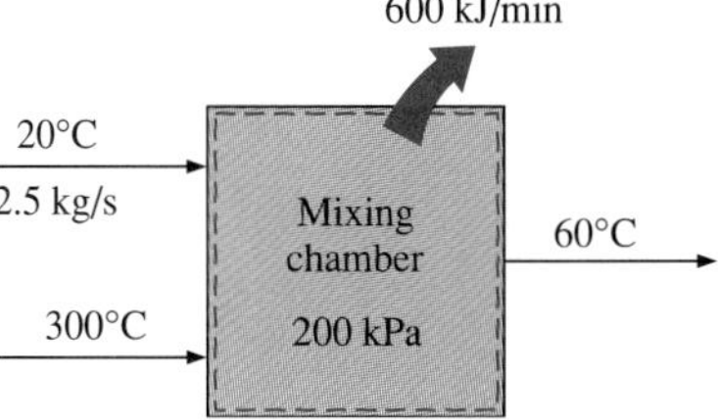

FIGURE P7-71

7-72 Air enters the evaporator section of a window air conditioner at 100 kPa and 27°C with a volume flow rate of 6 m^3/min. Refrigerant-134a at 120 kPa with a quality of 0.3 enters the evaporator at a rate of 2 kg/min and leaves as saturated vapor at the same pressure. Determine the exit temperature of the air and the exergy destruction for this process, assuming (*a*) heat is transferred to the evaporator of the air conditioner from the surrounding medium at 32°C at a rate of 30 kJ/min and (*b*) the outer surfaces of the air conditioner are insulated.

7-73 A 0.1-m^3 rigid tank initially contains refrigerant-134a at 1 MPa and 100 percent quality. The tank is connected by a valve to a supply line that carries refrigerant-134a at 1.4 MPa and 30°C. The valve is now opened, allowing the refrigerant to enter the tank, and it is closed when the tank contains only saturated liquid at 1.2 MPa. The refrigerant exchanges heat with its surroundings at 50°C and 100 kPa during this process. Determine (*a*) the mass of the refrigerant that entered the tank and (*b*) the exergy destroyed during this process.

7-74 A 0.4-m^3 rigid tank is filled with saturated liquid water at 200°C. A valve at the bottom of the tank is now opened, and one-half of the total mass is withdrawn from the tank in liquid form. Heat is transferred to water from a source of 250°C so that the temperature in the tank remains constant. Determine (*a*) the amount of heat transfer and (*b*) the reversible work and exergy destruction for this process. Assume the surroundings to be at 25°C and 100 kPa. *Answers:* (*a*) 3077 kJ; (*b*) 183.6 kJ, 183.6 kJ

7-75E An insulated 150-ft^3 rigid tank contains air at 75 psia and 140°F. A valve connected to the tank is opened, and air is allowed to escape until the pressure inside drops to 30 psia. The air temperature during this process is maintained constant by a electric resistance heater placed in the tank. Determine (*a*) the electrical work done during this process and (*b*) the exergy destruction. Assume the surroundings to be at 70°F.
Answers: (*a*) 1249 Btu, (*b*) 1068 Btu

7-76 A 0.1-m^3 rigid tank contains saturated refrigerant-134a at 800 kPa. Initially, 20 percent of the volume is occupied by liquid and the rest by vapor. A valve at the bottom of the tank is opened, and liquid is withdrawn from the tank. Heat is transferred to the refrigerant from a source at 50°C so that the pressure inside the tank remains constant. The valve is closed when no liquid is left in the tank and vapor starts to come out. Assuming the surroundings to be at 25°C, determine (*a*) the final mass in the tank and (*b*) the reversible work associated with this process. *Answers:* (*a*) 4.57 kg, (*b*) 6.24 kJ

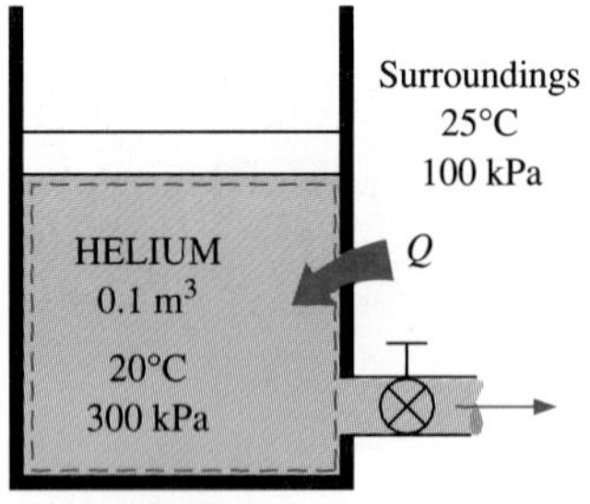

FIGURE P7-77

7-77 A vertical piston-cylinder device initially contains 0.1 m^3 of helium at 20°C. The mass of the piston is such that it maintains a constant pressure of 300 kPa inside. A valve is now opened, and helium is allowed to escape until the volume inside the cylinder is decreased by one-half. Heat transfer takes place between the helium and its surroundings at 25°C and 100 kPa so that the temperature of helium in the cylinder remains constant. Determine (*a*) the maximum work potential of the helium at the initial state and (*b*) the exergy destroyed during this process.

7-78 A 0.2-m^3 rigid tank initially contains saturated refrigerant-134a vapor at 1 MPa. The tank is connected by a valve to a supply line that carries refrigerant-134a at 1.4 MPa and 60°C. The valve is now opened, and the refrigerant is allowed to enter the tank. The valve is closed when one-half of the volume of the tank is filled with liquid and the rest with vapor at 1.2 MPa. The refrigerant exchanges heat during this process with the surroundings at 25°C. Determine (*a*) the amount of heat transfer and (*b*) the exergy destruction associated with this process.

7-79 An insulated vertical piston-cylinder device initially contains 15 kg of water, 12 kg of which is in the vapor phase. The mass of the piston is such that it maintains a constant pressure of 300 kPa inside the cylinder. Now steam at 1 MPa and 400°C is allowed to enter the cylinder from a supply line until all the liquid in the cylinder is vaporized. Assuming the surroundings to be at 25°C and 100 kPa, determine (*a*) the amount of steam that has entered and (*b*) the exergy destroyed during this process.
Answers: (*a*) 12.05 kg, (*b*) 3057 kJ

7-80 Consider a family of four, with each person taking a 5-minute shower every morning. The average flow rate through the shower head is 12 L/min. City water at 15°C is heated to 55°C in an electric water heater and tempered to 42°C by cold water at the T-elbow of the shower before being routed to the shower head. Determine the amount of exergy destroyed by this family per year as a result of taking daily showers.

FIGURE P7-82

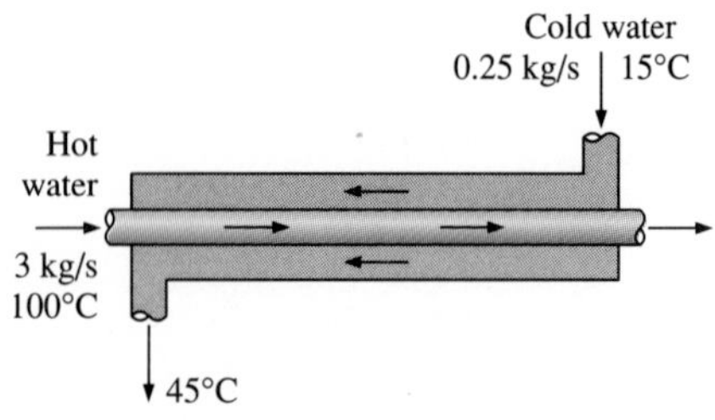

7-81 Ambient air at 100 kPa and 300 K is compressed isentropically in a steady-flow device to 1 MPa. Determine (*a*) the work input to the compressor, (*b*) the exergy of the air at the compressor exit, and (*c*) the exergy of compressed air after it is cooled to 300 K at 1 MPa pressure.

7-82 Cold water [C_p = 4.18 kJ/(kg · °C)] leading to a shower enters a well-insulated, thin-walled, double-pipe, counter-flow heat exchanger at 15°C at a rate of 0.25 kg/s and is heated to 45°C by hot water [C_p = 4.19 kJ/(kg · °C)]

that enters at 100°C at a rate of 3 kg/s. Determine (*a*) the rate of heat transfer and (*b*) the rate of exergy destruction in the heat exchanger.

7-83 Air [C_p = 1.005 kJ/(kg · °C)] is to be preheated by hot exhaust gases in a cross-flow heat exchanger before it enters the furnace. Air enters the heat exchanger at 95 kPa and 20°C at a rate of 0.8 m³/s. The combustion gases [C_p = 110 kJ/(kg · °C)] enter at 180°C at a rate of 1.1 kg/s and leave at 95°C. Determine the rate of heat transfer to the air and the outlet temperature of the air.

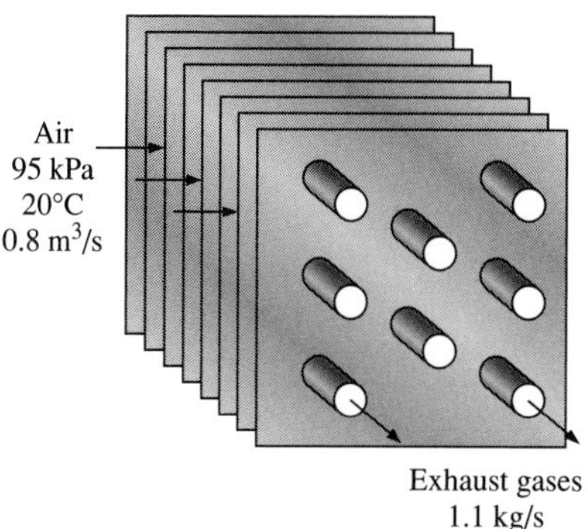

FIGURE P7-83

7-84 A well-insulated shell-and-tube heat exchanger is used to heat water [C_p = 4.18 kJ/(kg · °C)] in the tubes from 20°C to 70°C at a rate of 4.5 kg/s. Heat is supplied by hot oil [C_p = 2.30 kJ/(kg · °C)] that enters the shell side at 170°C at a rate of 10 kg/s. Disregarding any heat loss from the heat exchanger, determine (*a*) the exit temperature of oil and (*b*) the rate of exergy destruction in the heat exchanger.

7-85E Steam is to be condensed on the shell side of a heat exchanger at 90°F (h_{fg} = 1043 Btu/lbm) at a rate of 20 lbm/s. Cooling water [C_p = 1.0 Btu/(lbm · °F)] enters the tubes at 60°F at a rate of 115.3 lbm/s and leaves at 73°F. Assuming the heat exchanger to be well-insulated, determine (*a*) the rate of heat transfer in the heat exchanger and (*b*) the rate of exergy destruction in the heat exchanger.

Review Problems

7-86 The inner and outer surfaces of a 5-m × 6-m brick wall of thickness 30 cm and thermal conductivity 0.69 W/(m · °C) are maintained at temperatures of 20°C and 5°C, respectively. Determine (*a*) the rate of heat transfer through the wall and (*b*) the rate of exergy destruction associated with this process, both in W. Take T_0 = 0°C.

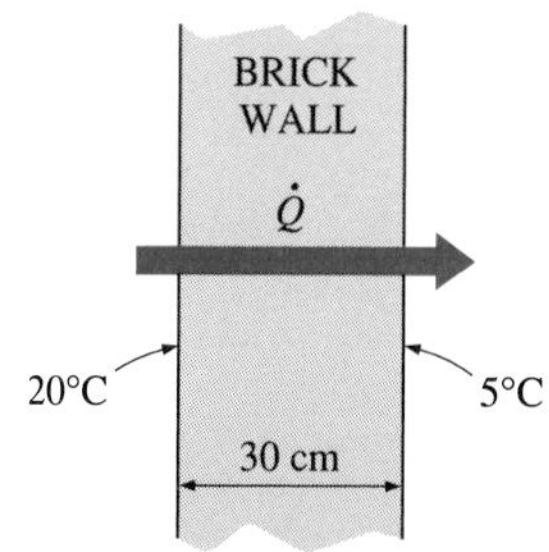

FIGURE P7-86

7-87 A 1000-W iron is left on the ironing board with its base exposed to the air at 20°C. The convection heat transfer coefficient between the base surface and the surrounding air is 80W/(m² · °C). If the base has a surface area of 0.02 m², determine (*a*) the temperature of the base of the iron and (*b*) the rate of exergy destruction for this process as a result of heat transfer, in W, in steady operation. Take T_0 = 20°C.

7-88 One method of passive solar heating is to stack gallons of liquid water inside the buildings and expose them to the sun. The solar energy stored in the water during the day is released at night to the room air, providing some heating. Consider a house that is maintained at 22°C and whose heating is assisted by a 500-L water storage system. If the water is heated to 45°C during the day, determine the amount of heating this water will provide to the house at night. Assuming an outside temperature of 5°C, determine the exergy destruction associated with this process.
Answers: (*a*) 48,070 kJ, (*b*) 1768 kJ

7-89 The inner and outer surfaces of a 0.5-cm-thick, 2-m × 2-m window glass in winter are 10°C and 3°C, respectively. If the thermal conductivity of the glass is 0.78W/(m · °C), determine the amount of heat loss, in kJ, through the glass over a period of 5 h. Also, determine the exergy destruction associated with this process. Take T_0 = 5°C.

7-90 An aluminum pan [k_t = 237 W/(m · K)] has a flat bottom whose diameter is 20 cm and thickness is 0.4 cm. Heat is transferred steadily to boiling water in the pan through its bottom at a rate of 500 W. If the inner surface of the bottom of the pan is 105°C, determine (*a*) the temperature of the outer surface of the bottom of the pan and (*b*) the rate of exergy destruction within the bottom of the pan during this process, in W. Take T_0 = 25°C.

7-91 A crater lake has a base area of 20,000 m^2, and the water it contains is 12 m deep. The ground surrounding the crater is nearly flat and is 140 m below the base of the lake. Determine the maximum amount of electrical work, in kWh, that can be generated by feeding this water to a hydroelectric power plant. *Answer:* (*a*) 95,600 kWh

7-92E A refrigerator has a second-law efficiency of 45 percent, and heat is removed from the refrigerated space at a rate of 200 Btu/min. If the space is maintained at 35°F while the surrounding air temperature is 75°F, determine the power input to the refrigerator.

7-93 Writing the first- and second-law relations and simplifying, obtain the reversible work relation for a closed system that exchanges heat with the surrounding medium at T_0 in the amount of Q_0 as well as a heat reservoir at T_R in the amount of Q_R. (*Hint:* Eliminate Q_0 between the two equations.)

7-94 Writing the first- and second-law relations and simplifying, obtain the reversible work relation for a steady-flow system that exchanges heat with the surrounding medium at T_0 in the amount of $\dot{Q}_0$ as well as a thermal reservoir at T_R at a rate of $\dot{Q}_R$. (*Hint:* Eliminate $\dot{Q}_0$ between the two equations.)

7-95 Writing the first- and second-law relations and simplifying, obtain the reversible work relation for a uniform-flow system that exchanges heat with the surrounding medium at T_0 in the amount of Q_0 as well as a heat reservoir at T_R in the amount of Q_R. (*Hint:* Eliminate Q_0 between the two equations.)

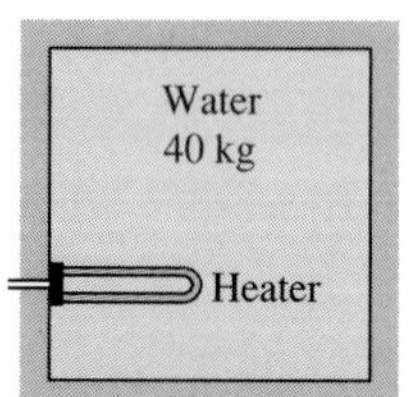

FIGURE P7-96

7-96 A 50-cm-long, 800-W electric resistance heating element whose diameter is 0.5 cm is immersed in 40 kg of water initially at 20°C. Assuming the water container is well-insulated, determine how long it will take for this heater to raise the water temperature to 80°C. Also, determine the minimum work input required and exergy destruction for this process, in kJ. Take T_0 = 20°C.

7-97 A 5-cm-external-diameter, 10-m-long hot water pipe at 80°C is losing heat to the surrounding air at 5°C by natural convection with a heat transfer coefficient of 25 W/(m^2 · °C). Determine the rate of heat loss from the pipe by natural convection and the rate at which the work potential is wasted during this process as a result of this heat loss.

FIGURE P7-98

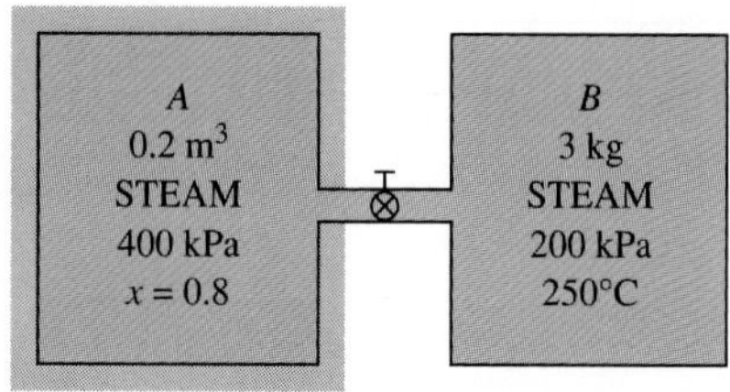

7-98 Two rigid tanks are connected by a valve. Tank *A* is insulated and contains 0.2 m^3 of steam at 400 kPa and 80 percent quality. Tank *B* is uninsulated and contains 3 kg of steam at 200 kPa and 250°C. The valve is now

opened, and steam flows from tank A to tank B until the pressure in tank A drops to 300 kPa. During this process 600 kJ of heat is transferred from tank B to the surroundings at 0°C. Assuming the steam remaining inside tank A to have undergone a reversible adiabatic process, determine (*a*) the final temperature in each tank and (*b*) the work potential wasted during this process.

7-99E A piston-cylinder device initially contains 15 ft^3 of helium gas at 25 psia and 70°F. Helium is now compressed in a polytropic process (PV^n = constant) to 70 psia and 300°F. Assuming the surroundings to be at 14.7 psia and 70°F, determine (*a*) the actual useful work consumed and (*b*) the minimum useful work input needed for this process.

Answers: (*a*) 36 Btu, (*b*) 34.3 Btu

7-100 A well-insulated 4-m × 4-m × 5-m room initially at 10°C is heated by the radiator of a steam heating system. The radiator has a volume of 15 L and is filled with superheated vapor at 200 kPa and 200°C. At this moment both the inlet and the exit valves to the radiator are closed. A 120-W fan is used to distribute the air in the room. The pressure of the steam is observed to drop to 100 kPa after 30 min as a result of heat transfer to the room. Assuming constant specific heats for air at room temperature, determine (*a*) the average temperature of room air in 30 min, (*b*) the entropy change of the steam, (*c*) the entropy change of the air in the room, and (*d*) the exergy destruction for this process, in kJ. Assume the air pressure in the room remains constant at 100 kPa at all times, and take T_0 = 10°C.

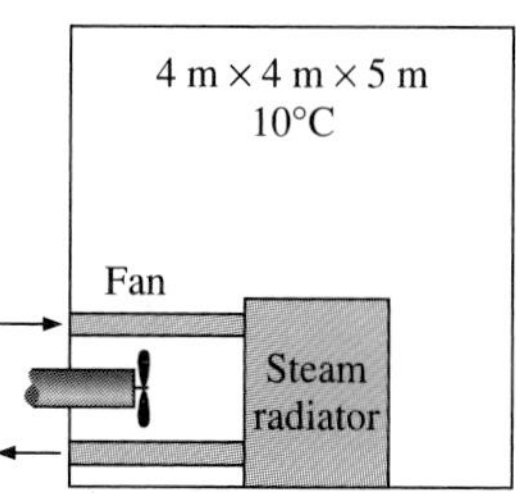

FIGURE P7-100

7-101 A passive solar house that is losing heat to the outdoors at 3°C at an average rate of 50,000 kJ/h is maintained at 22°C at all times during a winter night for 10 h. The house is to be heated by 50 glass containers, each containing 20 L of water that is heated to 80°C during the day by absorbing solar energy. A thermostat-controlled 15-kW back-up electric resistance heater turns on whenever necessary to keep the house at 22°C. Determine (*a*) how long the electric heating system was on that night, (*b*) the exergy destruction, and (*c*) the minimum work input required for that night, in kJ.

7-102 Steam at 7 MPa and 500°C enters a two-stage adiabatic turbine at a rate of 15 kg/s. Ten percent of the steam is extracted at the end of the first stage at a pressure of 1 MPa for other use. The remainder of the steam is further expanded in the second stage and leaves the turbine at 50 kPa. If the turbine has an adiabatic efficiency of 88 percent, determine the wasted power potential during this process as a result of irreversibilities. Assume the surroundings to be at 25°C.

7-103 Steam enters a two-stage adiabatic turbine at 8 MPa and 500°C. It expands in the first stage to a state of 2 MPa and 350°C. Steam is then reheated at constant pressure to a temperature of 500°C before it is routed to the second stage, where it exits at 30 kPa and a quality of 97 percent. The work output of the turbine is 5 MW. Assuming the surroundings to be at 25°C, determine the reversible power output and the rate of exergy destruction within this turbine. *Answers:* (*a*) 5463 kW, 463 kW

FIGURE P7-103

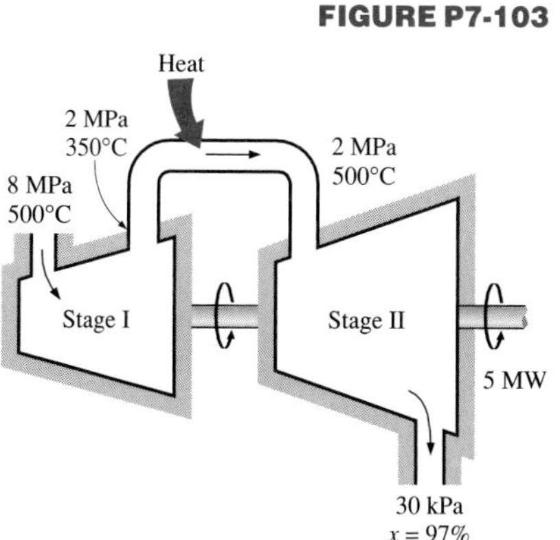

7-104 One ton (1000 kg) of liquid water at 80°C is brought into a well-insulated and well-sealed 4-m × 5-m × 6-m room initially at 22°C and 100 kPa. Assuming constant specific heats for both the air and water at room temperature, determine (*a*) the final equilibrium temperature in the room, (*b*) the exergy destruction, (*c*) the maximum amount of work that can be produced during this process, in kJ. Take T_0 = 15°C.

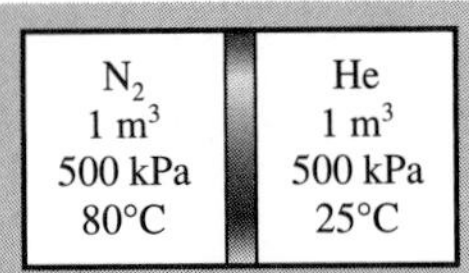

FIGURE P-105

7-105 Consider a well-insulated horizontal rigid cylinder that is divided into two compartments by a piston that is free to move but does not allow either gas to leak into the other side. Initially, one side of the piston contains 1 m³ of N_2 gas at 500 kPa and 80°C while the other side contains 1 m³ of He gas at 500 kPa and 25°C. Now thermal equilibrium is established in the cylinder as a result of heat transfer through the piston. Using constant specific heats at room temperature, determine (*a*) the final equilibrium temperature in the cylinder and (*b*) the wasted work potential during this process. What would your answer be if the piston were not free to move? Take T_0 = 25°C.

7-106 Repeat the problem above by assuming the piston is made of 5 kg of copper initially at the average temperature of the two gases on both sides.

7-107E Argon gas enters an adiabatic turbine at 1500°F and 200 psia at a rate of 40 lbm/min and exhausts at 30 psia. If the power output of the turbine is 95 hp, determine (*a*) the adiabatic efficiency and (*b*) the second-law efficiency of the turbine. Assume the surroundings to be at 77°F.

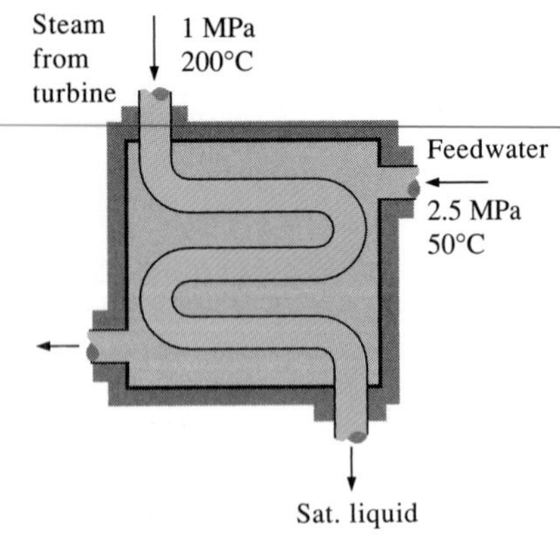

FIGURE P7-108

7-108 In large steam power plants, the feedwater is frequently heated in closed feedwater heaters, which are basically heat exchangers, by steam extracted from the turbine at some stage. Steam enters the feedwater heater at 1 MPa and 200°C and leaves as saturated liquid at the same pressure. Feedwater enters the heater at 2.5 MPa and 50°C and leaves 10°C below the exit temperature of the steam. Neglecting any heat losses from the outer surfaces of the heater, determine (*a*) the ratio of the mass flow rates of the extracted steam and the feedwater heater and (*b*) the reversible work for this process per unit mass of the feedwater. Assume the surroundings to be at 25°C.
Answers: (*a*) 0.247, (*b*) 63.5 kJ/kg

7-109 In order to cool 1 ton (1000 kg) of water at 20°C in an insulated tank, a person pours 80 kg of ice at −5°C into the water. Determine (*a*) the final equilibrium temperature in the tank and (*b*) the exergy destroyed during this process. The melting temperature and the heat of fusion of ice at atmospheric pressure are 0°C and 333.7 kJ/kg, respectively. Take T_0 = 25°C.

FIGURE P7-110

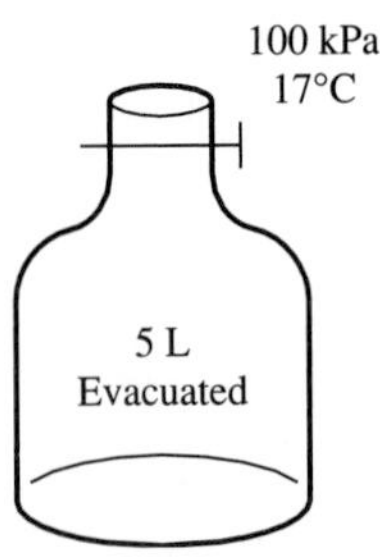

7-110 Consider a 5-L evacuated rigid bottle that is surrounded by the atmosphere at 100 kPa and 17°C. A valve at the neck of the bottle is now opened and the atmospheric air is allowed to flow into the bottle. The air trapped in the bottle eventually reaches thermal equilibrium with the atmosphere as a result of heat transfer through the wall of the bottle. The valve remains open during the process so that the trapped air also reaches mechanical equilibrium with the atmosphere. Determine the net heat transfer through the wall of the bottle and the exergy destroyed during this filling process.

7-111 Two constant-volume tanks, each filled with 20 kg of air, have temperatures of 1000 K and 300 K. A heat engine placed between the two tanks extracts heat from the high-temperature tank, produces work, and rejects heat to the low-temperature tank. Determine the maximum work that can be produced by the heat engine and the final temperatures of the tanks. Assume constant specific heats at room temperature.

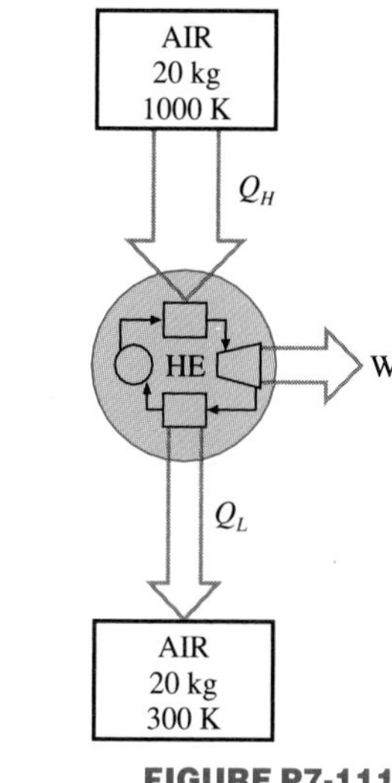

FIGURE P7-111

7-112 Two constant-pressure devices, each filled with 20 kg of air, have temperatures of 1000 K and 300 K. A heat engine placed between the two devices extracts heat from the high-temperature device, produces work, and rejects heat to the low-temperature device. Determine the maximum work that can be produced by the heat engine and the final temperatures of the devices. Assume constant specific heats at room temperature.

7-113 A 4-L pressure cooker has an operating pressure of 175 kPa. Initially, one-half of the volume is filled with liquid water and the other half by water vapor. The cooker is now placed on top of a 500-W electrical heating unit that is kept on for 30 min. Assuming the surroundings to be at 25°C and 100 kPa, determine (*a*) the amount of water that remained in the cooker and (*b*) the exergy destruction associated with the entire process, including the conversion of electric energy to heat energy.

Answers: (*a*) 1.487 kg, (*b*) 690 kg

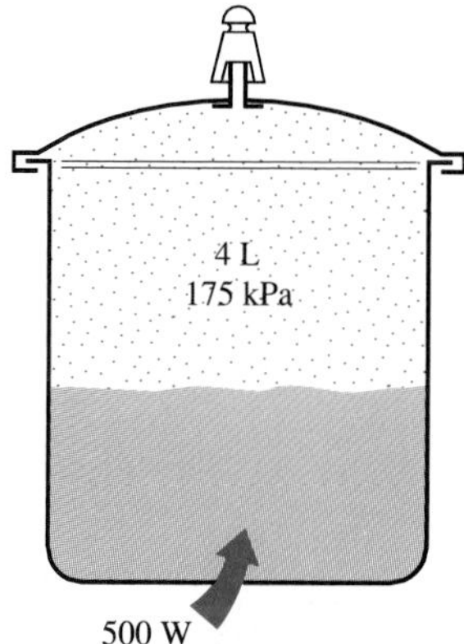

FIGURE P7-113

7-114 What would your answer to Prob. 7-113 be if heat were supplied to the pressure cooker from a heat source at 180°C instead of the electrical heating unit?

7-115 A constant-volume tank contains 20 kg of nitrogen at 1000 K, and a constant-pressure device contains 10 kg of argon at 300 K. A heat engine placed between the tank and device extracts heat from the high-temperature tank, produces work, and rejects heat to the low-temperature device. Determine the maximum work that can be produced by the heat engine and the final temperatures of the nitrogen and argon. Assume constant specific heats at room temperature.

FIGURE P7-115

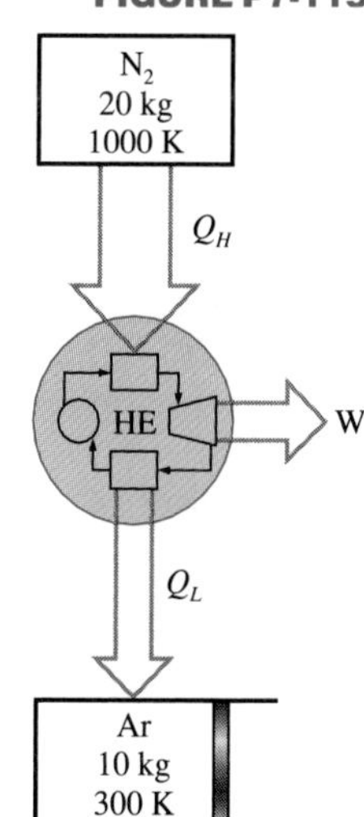

7-116 A constant-volume tank has a temperature of 800 K and a constant-pressure device has a temperature of 290 K. Both the tank and device are filled with 20 kg of air. A heat engine placed between the tank and device receives heat from the high-temperature tank, produces work, and rejects heat to the low-temperature device. Determine the maximum work that can be produced by the heat engine and the final temperatures of the tank and device. Assume constant specific heats at room temperature.

7-117 Can closed-system exergy be negative? How about flow exergy: Explain using an incompressible substance as an example.

7-118 Obtain a relation for the second-law efficiency of a heat engine that receives heat Q_H from a source at temperature T_H and rejects heat Q_L to a sink at T_L, which is higher than T_0 (the temperature of the surroundings), while producing work in the amount of W.

7-119E In a production facility, 1.2-in-thick, 2-ft × 2-ft square brass plates [ρ = 532.5 lbm/ft³ and C_p = 0.091 Btu/(lbm · °F)] that are initially at a uniform temperature of 75°F are heated by passing them through an oven at 1300°F at a rate of 300 per minute. If the plates remain in the oven until their average temperature rises to 1000°F, determine the rate of heat transfer to the plates in the furnace and the rate of exergy destruction associated with this heat transfer process.

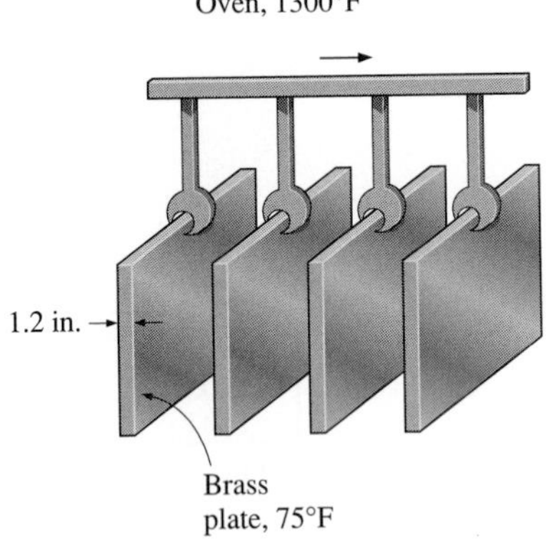

FIGURE P7-119E

7-120 Long cylindrical steel rods [ρ = 7833 kg/m³ and C_p = 0.465 kJ/(kg · °C)] of 10-cm diameter are heat-treated by drawing them at a velocity of 3 m/min through a 6-m-long oven maintained at 900°C. If the rods enter the oven at 30°C and leave at 700°C, determine (*a*) the rate of heat transfer to the rods in the oven and (*b*) the rate of exergy destruction associated with this heat transfer process.

7-121 Steam is to be condensed in the condenser of a steam power plant at a temperature of 50°C (h_{fg} = 2305 kJ/kg) with cooling water [C_p = 4.20 kJ/(kg · °C)] from a nearby lake that enters the tubes of the condenser at 18°C at a rate of 101 kg/s and leaves at 27°C. Assuming the condenser to be perfectly insulated, determine (*a*) the rate of condensation of the steam and (*b*) the rate of exergy destruction in the condenser.

Answers: (*a*) 1.65 kg, (*b*) 335 kW

7-122 A well-insulated heat exchanger is to heat water [C_p = 4.18 kJ/(kg · °C)] from 25°C to 60°C at a rate of 0.2 kg/s. The heating is to be accomplished by geothermal water [C_p = 4.31 kJ/(kg · °C)] available at 140°C at a mass flow rate of 0.3 kg/s. The inner tube is thin-walled and has a diameter of 0.8 cm. Determine (*a*) the rate of heat transfer and (*b*) the rate of exergy destruction in the heat exchanger.

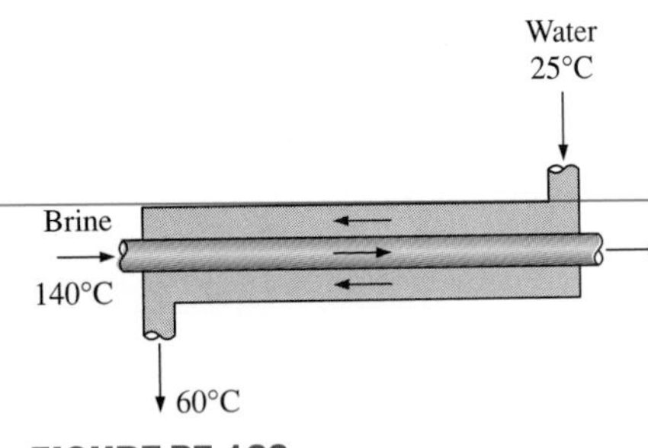

FIGURE P7-122

7-123 An adiabatic heat exchanger is to cool ethylene glycol [C_p = 2.56 kJ/(kg · °C)] flowing at a rate of 2 kg/s from 80°C to 140°C by water [C_p = 4.18 kJ/(kg · °C)] that enters at 20°C and leaves at 55°C. Determine (*a*) the rate of heat transfer and (*b*) the rate of exergy destruction in the heat exchanger.

7-124 A well-insulated, thin-walled, counter-flow heat exchanger is to be used to cool oil [C_p = 2.20 kJ/(kg · °C)] from 150°C to 40°C at a rate of 2 kg/s by water [C_p = 4.18 kJ/(kg · °C)] that enters at 22°C at a rate of 1.5 kg/s. The diameter of the tube is 2.5 cm, and its length is 6 m. Determine (*a*) the rate of heat transfer and (*b*) the rate of exergy destruction in the heat exchanger.

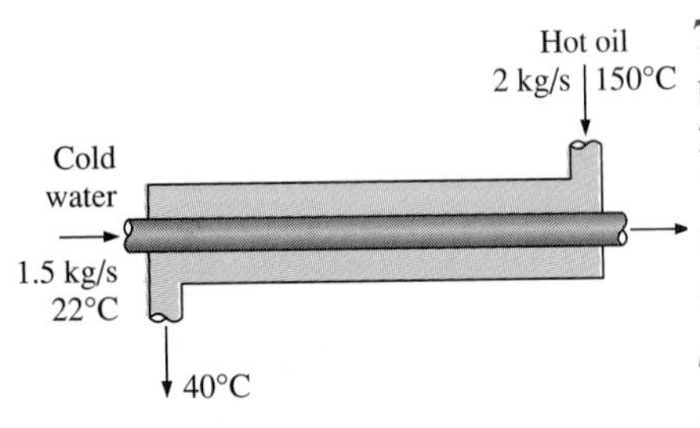

FIGURE P7-124

7-125 In a dairy plant, milk at 4°C is pasteurized continuously at 72°C at a rate of 12 L/s for 24 hours a day and 365 days a year. The milk is heated to the pasteurizing temperature by hot water heated in a natural gas-fired boiler having an efficiency of 82 percent. The pasteurized milk is then cooled by cold water at 18°C before it is finally refrigerated back to 4°C. To save energy and money, the plant installs a regenerator that has an effectiveness of 82 percent. If the cost of natural gas is $0.52/therm (1 therm = 105,500 kJ), determine how much energy and money the regenerator will save this company per year and the annual reduction in exergy destruction.

Computer, Design, and Essay Problems

7-126 The concept of second-law efficiency is based on comparing the performance of the system to the best possible performance under identical conditions, and this concept can be extended to nontechnical fields. For example, the second-law efficiency of a person in expressing an idea verbally or in writing can be viewed as the ratio of the minimum number of words necessary to express the idea to the actual number of words used. Give five examples of second-law efficiency from nontechnical fields not mentioned in this chapter.

7-127 Obtain the following information about a power plant that is closest to your town: the net power output; the type and amount of fuel used; the power consumed by the pumps, fans, and other auxiliary equipment; stack gas losses; temperatures at several locations; and the rate of heat rejection at the condenser. Using these and other relevant data, determine the rate of irreversibility in that power plant.

7-128 Make a list of what you have done during the last 24 h and compare it to what you could have done during that period. The quantification of this comparison can be viewed as your second-law efficiency for that day. If you have done everything you could have done, and did it right, you can claim that you had a "full" day, and your second-law efficiency for that day is 100 percent. What do you think your second-law efficiency for the last 24 h has been?

7-129 Human beings are probably the most capable creatures, and they have a high level of physical intellectual, emotional, and spiritual potentials or exergies. Unfortunately people make little use of their exergies, letting most of their exergies go to waste. Draw four exergy versus time charts, and plot your physical, intellectual, emotional, and spiritual exergies on each of these charts for a 24-h period using your best judgment based on your experience. On these four charts, plot your respective exergies that you have utilized during the last 24 h. Compare the two plots on each chart and determine if you are living a "full" life or if you are wasting your life away. Can you think of any ways to reduce the mismatch between your exergies and your utilization of them?

7-130 Consider a student who takes classes for 4 h a day and works for another 4 h in the library. Keeping in mind the things that a typical engineering student does in a normal day, prepare a time schedule of activities for that student for a 24-h period such that the student's daily life involves minimum irreversibilities.

7-131 Consider natural gas, electric resistance, and heat pump heating systems. For a specified heating load, which one of these systems will do the job with the least irreversibility? Explain.

7-132 The domestic hot water systems involve a high level of irreversibility and thus they have low second-law efficiencies. The water in these systems is heated from about 15°C to about 60°C, and most of the hot water is mixed with cold water to reduce its temperature to 45°C or even lower before it is used for any useful purpose such as taking a shower or washing clothes at a warm setting. The water is discarded at about the same temperature at which it was used and replaced by fresh cold water at 15°C. Redesign a typical residential hot water system such that the irreversibility is greatly reduced. Draw a sketch of your proposed design.

Gas Power Cycles

CHAPTER

Two important areas of application for thermodynamics are power generation and refrigeration. Both are usually accomplished by systems that operate on a thermodynamic cycle. Thermodynamic cycles can be divided into two general categories: *power cycles,* which are discussed in this and the next chapter, and *refrigeration cycles,* which are discussed in Chap. 10.

The devices or systems used to produce a net power output are often called *engines,* and the thermodynamic cycles they operate on are called *power cycles.* The devices or systems used to produce a refrigeration effect are called *refrigerators, air conditioners,* or *heat pumps,* and the cycles they operate on are called *refrigeration cycles.*

Thermodynamic cycles can also be categorized as *gas cycles* and *vapor cycles,* depending on the *phase* of the working fluid. In gas cycles, the working fluid remains in the gaseous phase throughout the entire cycle, whereas in vapor cycles the working fluid exists in the vapor phase during one part of the cycle and in the liquid phase during another part.

Thermodynamic cycles can be categorized yet another way: *closed* and *open cycles.* In closed cycles, the working fluid is returned to the initial state at the end of the cycle and is recirculated. In open cycles, the working fluid is renewed at the end of each cycle instead of being recirculated. In automobile engines, the combustion gases are exhausted and replaced by fresh air–fuel mixture at the end of each cycle. The engine operates on a mechanical cycle, but the working fluid does not go through a complete thermodynamic cycle.

Heat engines are categorized as *internal combustion* and *external combustion engines,* depending on how the heat is supplied to the working fluid. In external combustion engines (such as steam power plants), energy is supplied to the working fluid from an external source such as a furnace, a geothermal well, a nuclear reactor, or even the sun. In internal combustion engines (such as automobile engines), this is done by burning the fuel within the system boundaries. In this chapter, various gas power cycles are analyzed under some simplifying assumptions.

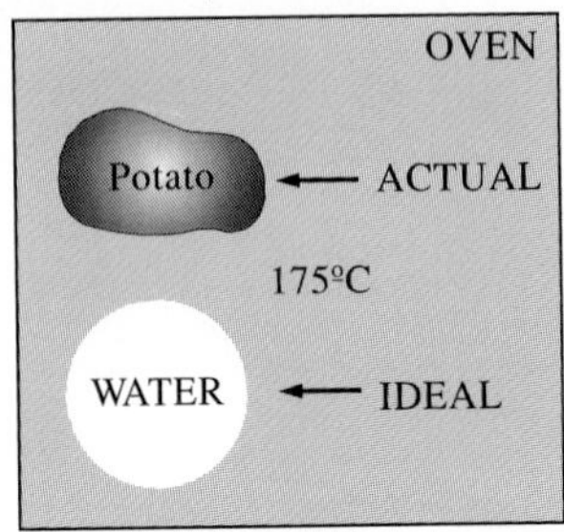

FIGURE 8-1

Modeling is a powerful engineering tool that provides great insight and simplicity at the expense of some loss in accuracy.

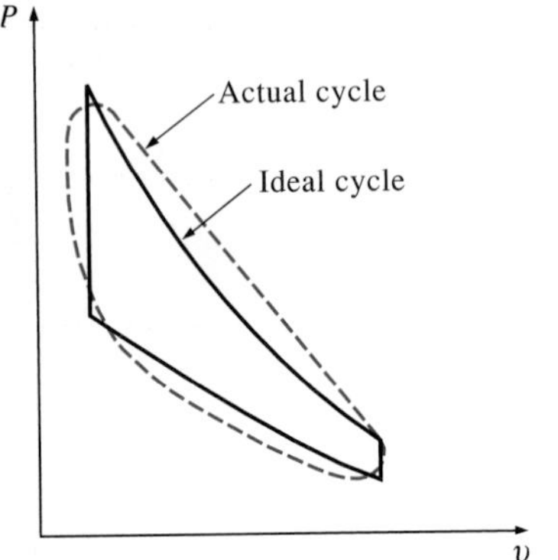

FIGURE 8-2

The analysis of many complex processes can be reduced to a manageable level by utilizing some idealizations.

8-1 ■ BASIC CONSIDERATIONS IN THE ANALYSIS OF POWER CYCLES

Most power-producing devices operate on cycles, and the study of power cycles is an exciting and important part of thermodynamics. The cycles encountered in actual devices are difficult to analyze because of the presence of complicating effects, such as friction, and the absence of sufficient time for establishment of the equilibrium conditions during the cycle. To make an analytical study of a cycle feasible, we have to keep the complexities at a manageable level and utilize some idealizations (Fig. 8-1). When the actual cycle is stripped off all the internal irreversibilities and complexities, we end up with a cycle that resembles the actual cycle closely but is made up totally of internally reversible processes. Such a cycle is called an **ideal cycle** (Fig. 8-2).

A simple idealized model enables engineers to study the effects of the major parameters that dominate the cycle without getting bogged down in the details. The cycles discussed in this chapter are somewhat idealized, but they still retain the general characteristics of the actual cycles they represent. The conclusions reached from the analysis of ideal cycles are also applicable to actual cycles. The thermal efficiency of the Otto cycle, the ideal cycle for spark-ignition automobile engines, for example, increases with the compression ratio. This is also the case for actual automobile engines. The numerical values obtained from the analysis of an ideal cycle, however, are not necessarily representative of the actual cycles, and care should be exercised in their interpretation (Fig. 8-3). The simplified analysis presented in this chapter for various power cycles of practical interest may also serve as the starting point for a more in-depth study.

Heat engines are designed for the purpose of converting other forms of energy (usually in the form of heat) to work, and their performance is expressed in terms of the **thermal efficiency** η_{th}, which is the ratio of the net work produced by the engine to the total heat input:

$$\eta_{th} = \frac{W_{net}}{Q_{in}} \quad \text{or} \quad \eta_{th} = \frac{w_{net}}{q_{in}} \qquad (8\text{-}1a, b)$$

It was pointed out in Chap. 6 that heat engines that operate on a totally reversible cycle, such as the Carnot cycle, have the highest thermal efficiency

FIGURE 8-3

Care should be exercised in the interpretation of the results from ideal cycles.

of all heat engines operating between the same temperature levels. That is, nobody can develop a cycle more efficient than the *Carnot cycle.* Then the following question arises naturally: If the Carnot cycle is the best possible cycle, why do we not use it as the model cycle for all the heat engines instead of bothering with several so-called *ideal* cycles? The answer to this question is hardware-related. Most cycles encountered in practice differ significantly from the Carnot cycle, which makes it unsuitable as a realistic model. Each ideal cycle discussed in this chapter is related to a specific work-producing device and is an *idealized* version of the actual cycle.

The ideal cycles are *internally reversible,* but, unlike the Carnot cycle, they are not necessarily externally reversible. That is, they may involve irreversibilities external to the system such as heat transfer through a finite temperature difference. Therefore, the thermal efficiency of an ideal cycle, in general, is less than that of a totally reversible cycle operating between the same temperature limits. However, it is still considerably higher than the thermal efficiency of an actual cycle because of the idealizations utilized.

The idealizations and simplifications commonly employed in the analysis of power cycles can be summarized as follows:

1. The cycle does not involve any *friction.* Therefore, the working fluid does not experience any pressure drop as it flows in pipes or devices such as heat exchangers.

2. All expansion and compression processes take place in a *quasi-equilibrium* manner (Fig. 8-4).

3. The pipes connecting the various components of a system are well insulated, and *heat transfer* through them is negligible.

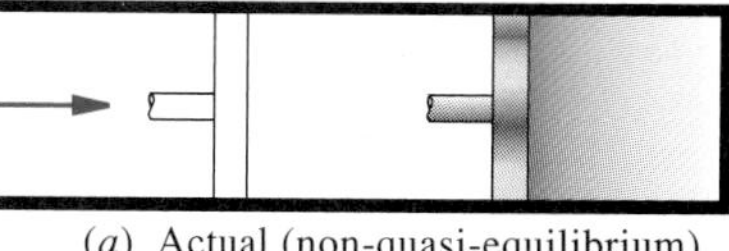

(*a*) Actual (non-quasi-equilibrium) compression

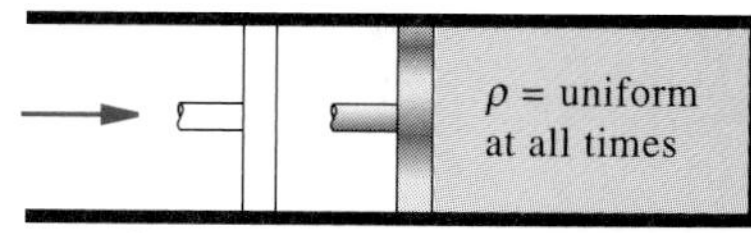

(*b*) Ideal (quasi-equilibrium) compression

FIGURE 8-4

All compression and expansion processes in ideal cycles are assumed to be quasi-equilibrium (internally reversible).

Neglecting the changes in *kinetic* and *potential energies* of the working fluid is another commonly utilized simplification in the analysis of power cycles. This is a reasonable assumption since in devices that involve shaft work, such as turbines, compressors, and pumps, the kinetic and potential energy terms are usually very small relative to the other terms in the energy equation. Fluid velocities encountered in devices such as condensers, boilers, and mixing chambers are typically low, and the fluid streams experience little change in their velocities, again making kinetic energy changes negligible. The only devices where the changes in kinetic energy are significant are the nozzles and diffusers, which are specifically designed to create large changes in velocity.

In the preceding chapters, *property diagrams* such as the *P-v* and *T-s* diagrams have served as valuable aids in the analysis of thermodynamic processes. On both the *P-v* and *T-s* diagrams, the area enclosed by the process curves of a cycle represents the net work produced during the cycle (Fig. 8-5), which is also equivalent to the net heat transfer for that cycle. The *T-s* diagram is particularly useful as a visual aid in the analysis of ideal power cycles. An ideal power cycle does not involve any internal irreversibilities, and so the only effect that can change the entropy of the working fluid during a process is heat transfer.

On a *T-s* diagram, a *heat-addition* process proceeds in the direction of increasing entropy, a *heat-rejection* process proceeds in the direction of

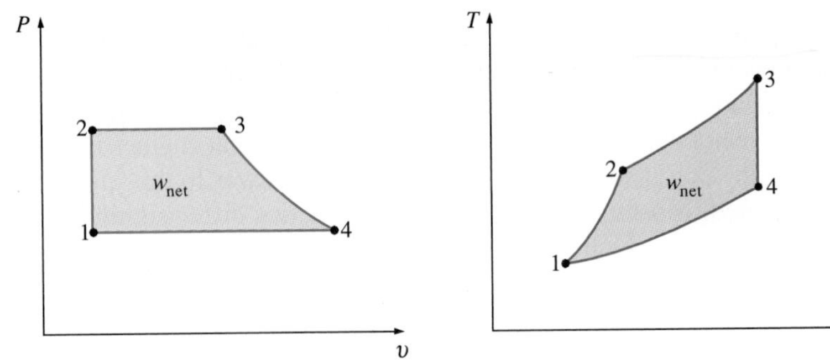

FIGURE 8-5

On both *P-v* and *T-s* diagrams, the area enclosed by the process curve represents the net work of the cycle.

decreasing entropy, and an *isentropic* (internally reversible, adiabatic) process proceeds at constant entropy. The area under the process curve on a *T-s* diagram represents the heat transfer for that process. The area under the heat addition process on a *T-s* diagram is a geometric measure of the total heat supplied during the cycle q_{in}, and the area under the heat rejection process is a measure of the total heat rejected q_{out}. The difference between these two (the area enclosed by the cyclic curve) is the net heat transfer, which is also the net work produced during the cycle. Therefore, on a *T-s* diagram, the ratio of the area enclosed by the cyclic curve to the area under the heat-addition process curve represents the thermal efficiency of the cycle. *Any modification that will increase the ratio of these two areas will also improve the thermal efficiency of the cycle.*

Although the working fluid in an ideal power cycle operates on a closed loop, the type of individual processes that comprises the cycle depends on the individual devices used to execute the cycle. In the Rankine cycle, which is the ideal cycle for steam power plants, the working fluid flows through a series of steady-flow devices such as the turbine and condenser, whereas in the Otto cycle, which is the ideal cycle for the spark-ignition automobile engine, the working fluid is alternately expanded and compressed in a piston-cylinder device. Therefore, equations pertaining to steady-flow systems should be used in the analysis of the Rankine cycle, and equations pertaining to closed systems should be used in the analysis of the Otto cycle.

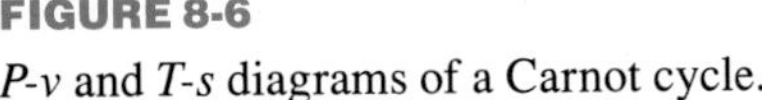

FIGURE 8-6

P-v and *T-s* diagrams of a Carnot cycle.

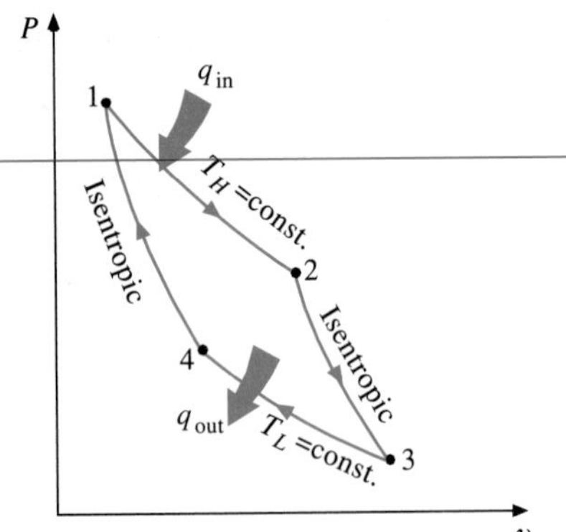

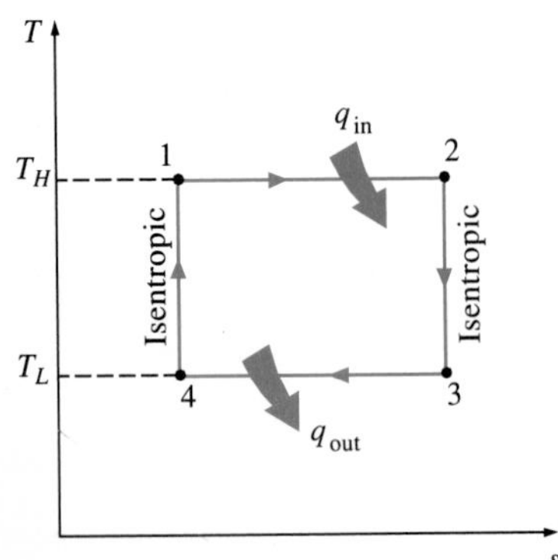

8-2 ■ THE CARNOT CYCLE AND ITS VALUE IN ENGINEERING

The Carnot cycle, which was introduced and discussed in Chap. 5, is composed of four totally reversible processes: isothermal heat addition, isentropic expansion, isothermal heat rejection, and isentropic compression. The *P-v* and *T-s* diagrams of a Carnot cycle are replotted in Fig. 8-6. The Carnot cycle can be executed in a closed system (a piston-cylinder device) or a steady-flow system (utilizing two turbines and two compressors, as shown in Fig. 8-7), and either a gas or a vapor can be utilized as the working fluid. The Carnot cycle is the most efficient cycle that can be executed between a heat source at temperature T_H and a sink at temperature T_L, and its thermal efficiency is expressed as

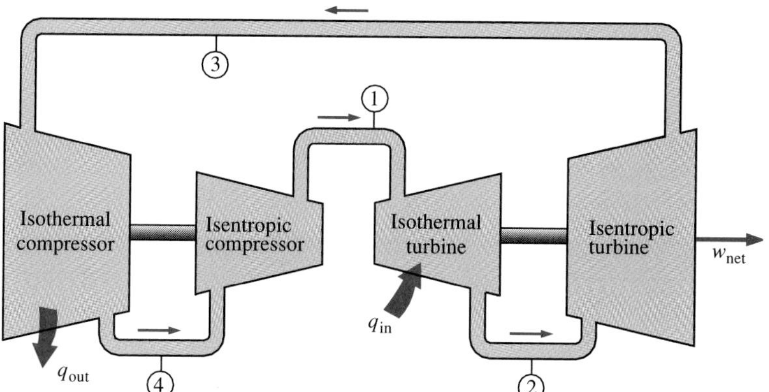

FIGURE 8-7
A steady-flow Carnot engine.

$$\eta_{th,\,Carnot} = 1 - \frac{T_L}{T_H} \tag{8-2}$$

Reversible isothermal heat transfer is very difficult to achieve in reality because it would require very large heat exchangers and it would take a very long time (a power cycle in a typical engine is completed in a fraction of a second). Therefore, it is not practical to build an engine that would operate on a cycle that closely approximates the Carnot cycle.

The real value of the Carnot cycle comes from its being a standard against which the actual or the ideal cycles can be compared. The thermal efficiency of the Carnot cycle is a function of the sink and source temperatures only, and the thermal efficiency relation for the Carnot cycle (Eq. 8-2) conveys an important message that is equally applicable to both ideal and actual cycles: *Thermal efficiency increases with an increase in the average temperature at which heat is supplied to the system or with a decrease in the average temperature at which heat is rejected from the system.*

The source and sink temperatures that can be used in practice are not without limits, however. The highest temperature in the cycle is limited by the maximum temperature that the components of the heat engine, such as the piston or the turbine blades, can withstand. The lowest temperature is limited by the temperature of the cooling medium utilized in the cycle such as a lake, a river, or the atmospheric air.

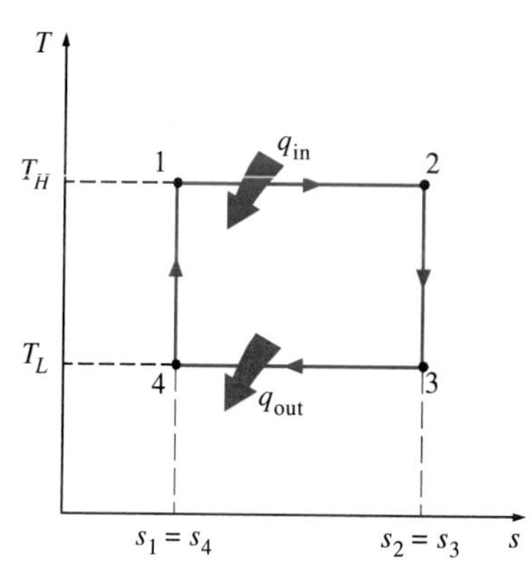

FIGURE 8-8
T-s diagram for Example 8-1.

EXAMPLE 8-1 Derivation of the Efficiency of the Carnot Cycle

Show that the thermal efficiency of a Carnot cycle operating between the temperature limits of T_H and T_L is solely a function of these two temperatures and is given by Eq. 8-2.

Solution The *T-s* diagram of a Carnot cycle is redrawn in Fig. 8-8. All four processes that comprise the Carnot cycle are reversible, and thus the area under each process curve represents the heat transfer for that process. Heat is transferred to the system during process 1-2 and rejected during process 3-4. Therefore, the amount of heat input and heat output for the cycle can be expressed as

$$q_{in} = T_H(s_2 - s_1) \qquad \text{and} \qquad q_{out} = T_L(s_3 - s_4) = T_L(s_2 - s_1)$$

const T for carnot

since processes 2-3 and 4-1 are isentropic, and thus $s_2 = s_3$ and $s_4 = s_1$. Substituting these into Eq. 8-1*b*, we see that the thermal efficiency of a Carnot cycle is

$$\eta_{th} = \frac{w_{net}}{q_{in}} = 1 - \frac{q_{out}}{q_{in}} = 1 - \frac{T_L(s_2 - s_1)}{T_H(s_2 - s_1)} = 1 - \frac{T_L}{T_H}$$

which is the desired result. Notice that the thermal efficiency of a Carnot cycle is independent of the type of the working fluid used (an ideal gas, steam, etc.) or whether the cycle is executed in a closed or steady-flow system.

8-3 ■ AIR-STANDARD ASSUMPTIONS

In gas power cycles, the working fluid remains a gas throughout the entire cycle. Spark-ignition automobile engines, diesel engines, and conventional gas turbines are familiar examples of devices that operate on gas cycles. In all these engines, energy is provided by burning a fuel within the system boundaries. That is, they are *internal combustion engines*. Because of this combustion process, the composition of the working fluid changes from air and fuel to combustion products during the course of the cycle. However, considering that air is predominantly nitrogen that undergoes hardly any chemical reactions in the combustion chamber, the working fluid closely resembles air at all times.

Even though internal combustion engines operate on a mechanical cycle (the piston returns to its starting position at the end of each revolution), the working fluid does not undergo a complete thermodynamic cycle. It is thrown out of the engine at some point in the cycle (as exhaust gases) instead of being returned to the initial state. Working on an open cycle is the characteristic of all internal combustion engines.

The actual gas power cycles are rather complex. To reduce the analysis to a manageable level, we utilize the following approximations, commonly known as the **air-standard assumptions:**

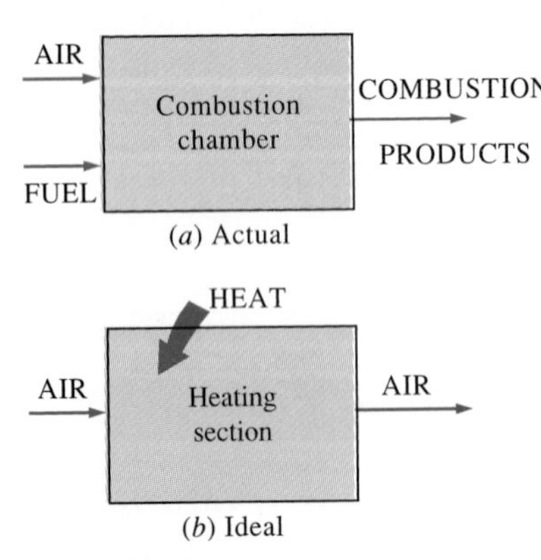

FIGURE 8-9
The combustion process is replaced by a heat-addition process in ideal cycles.

1. The working fluid is air, which continuously circulates in a closed loop and always behaves as an ideal gas.
2. All the processes that make up the cycle are internally reversible.
3. The combustion process is replaced by a heat-addition process from an external source (Fig. 8-9).
4. The exhaust process is replaced by a heat rejection process that restores the working fluid to its initial state.

Another assumption that is often utilized to simplify the analysis even more is that the air has constant specific heats whose values are determined at *room temperature* (25°C, or 77°F). When this assumption is utilized, the air-standard assumptions are called the **cold-air-standard assumptions.** A cycle for which the air-standard assumptions are applicable is frequently referred to as an **air-standard cycle.**

The air-standard assumptions stated above provide considerable simplification in the analysis without significantly deviating from the actual cycles. This simplified model enables us to study qualitatively the influence of major parameters on the performance of the actual engines.

Brief Overview of Reciprocating Engines

Despite its simplicity, the reciprocating engine (basically a piston-cylinder device) is one of the rare inventions that has proved to be very versatile and to have a wide range of applications. It is the powerhouse of the vast majority of automobiles, trucks, light aircraft, ships, and electric power generators, as well as many other devices.

The basic components of a reciprocating engine are shown in Fig. 8-10. The piston reciprocates in the cylinder between two fixed positions called the **top dead center** (TDC)—the position of the piston when it forms the smallest volume in the cylinder—and the **bottom dead center** (BDC)—the position of the piston when it forms the largest volume in the cylinder. The distance between the TDC and the BDC is the largest distance that the piston can travel in one direction, and it is called the **stroke** of the engine. The diameter of the piston is called the **bore.** The air or air–fuel mixture is drawn into the cylinder through the **intake valve,** and the combustion products are expelled from the cylinder through the **exhaust valve.**

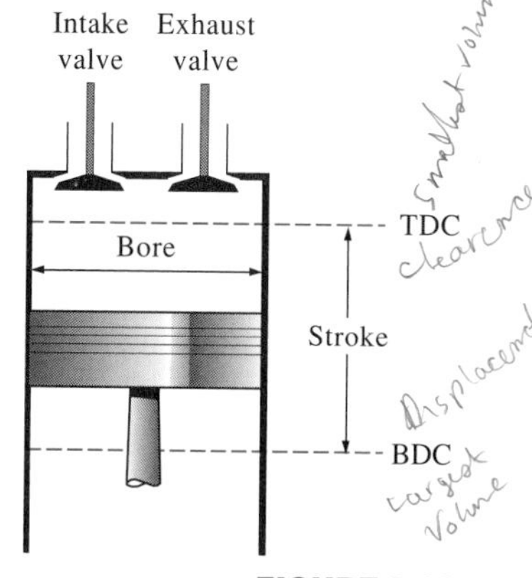

FIGURE 8-10

Nomenclature for reciprocating engines.

The minimum volume formed in the cylinder when the piston is at TDC is called the **clearance volume** (Fig. 8-11). The volume displaced by the piston as it moves between TDC and BDC is called the **displacement volume.** The ratio of the maximum volume formed in the cylinder to the minimum (clearance) volume is called the **compression ratio** r of the engine:

$$r = \frac{V_{max}}{V_{min}} = \frac{V_{BDC}}{V_{TDC}} \tag{8-3}$$

Notice that the compression ratio is a *volume ratio* and should not be confused with the pressure ratio.

Another term frequently used in conjunction with reciprocating engines is the **mean effective pressure** (MEP). It is a fictitious pressure that, if it acted on the piston during the entire power stroke, would produce the same

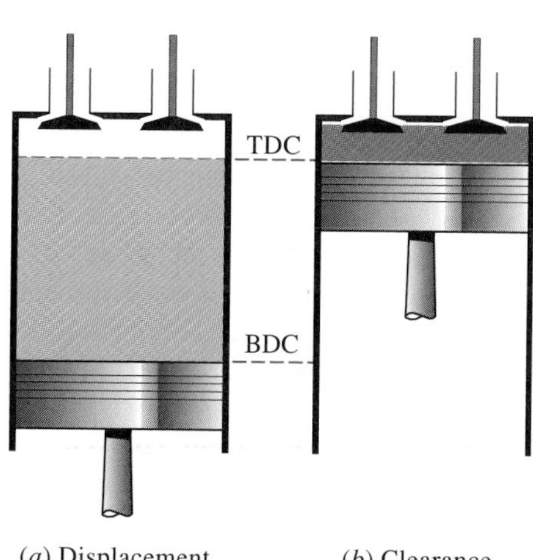

(*a*) Displacement volume

(*b*) Clearance volume

FIGURE 8-11

Displacement and clearance volumes of a reciprocating engine.

amount of net work as that produced during the actual cycle (Fig. 8-12). That is,

$$W_{net} = \text{MEP} \times \text{Piston area} \times \text{Stroke} = \text{MEP} \times \text{Displacement volume}$$

or

$$\text{MEP} = \frac{W_{net}}{V_{max} - V_{min}} = \frac{w_{net}}{v_{max} - v_{min}} \qquad \text{(kPa)} \qquad (8\text{-}4)$$

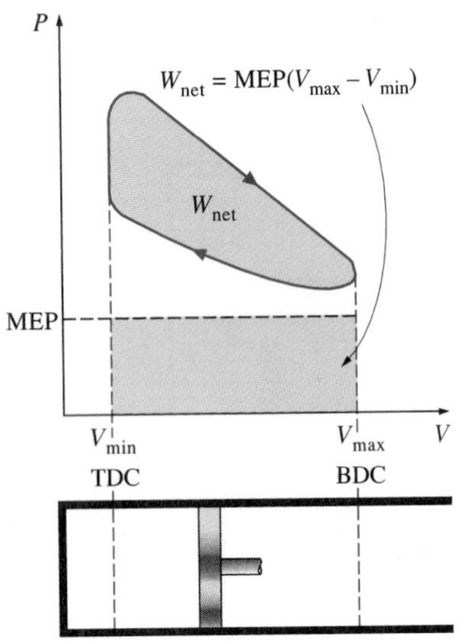

FIGURE 8-12
The net work output of a cycle is equivalent to the product of the mean effective pressure and the displacement volume.

The mean effective pressure can be used as a parameter to compare the performances of reciprocating engines of equal size. The engine with a larger value of MEP will deliver more net work per cycle and thus will perform better.

Reciprocating engines are classified as **spark-ignition** (SI) **engines** or **compression-ignition** (CI) **engines,** depending on how the combustion process in the cylinder is initiated. In SI engines, the combustion of the air–fuel mixture is initiated by a spark plug. In CI engines, the air–fuel mixture is self-ignited as a result of compressing the mixture above its self-ignition temperature. In the next two sections, we discuss the *Otto* and *Diesel cycles,* which are the ideal cycles for the SI and CI reciprocating engines, respectively.

8-5 ■ OTTO CYCLE: The Ideal Cycle for Spark-Ignition Engines

The Otto cycle is the ideal cycle for spark-ignition reciprocating engines. It is named after Nikolaus A. Otto, who built a successful four-stroke engine in 1876 in Germany using the cycle proposed by Frenchman Beau de Rochas in 1862. In most spark-ignition engines, the piston executes four complete strokes (two mechanical cycles) within the cylinder, and the crankshaft completes two revolutions for each thermodynamic cycle. These engines are called **four-stroke** internal combustion engines. A schematic of each stroke as well as a *P-v* diagram for an actual four-stroke spark-ignition engine is given in Fig. 8-13(*a*).

Initially, both the intake and the exhaust valves are closed, and the piston is at its lowest position (BDC). During the *compression stroke,* the piston moves upward, compressing the air–fuel mixture. Shortly before the piston reaches its highest position (TDC), the spark plug fires and the mixture ignites, increasing the pressure and temperature of the system. The high-pressure gases force the piston down, which in turn forces the crankshaft to rotate, producing a useful work output during the *expansion* or *power stroke.* At the end of this stroke, the piston is at its lowest position (the completion of the first mechanical cycle), and the cylinder is filled with combustion products. Now the piston moves upward one more time, purging the exhaust gases through the exhaust valve (the *exhaust stroke*), and down a second time, drawing in fresh air–fuel mixture through the intake valve (the *intake stroke*). Notice that the pressure in the cylinder is slightly above the atmospheric value during the exhaust stroke and slightly below during the intake stroke.

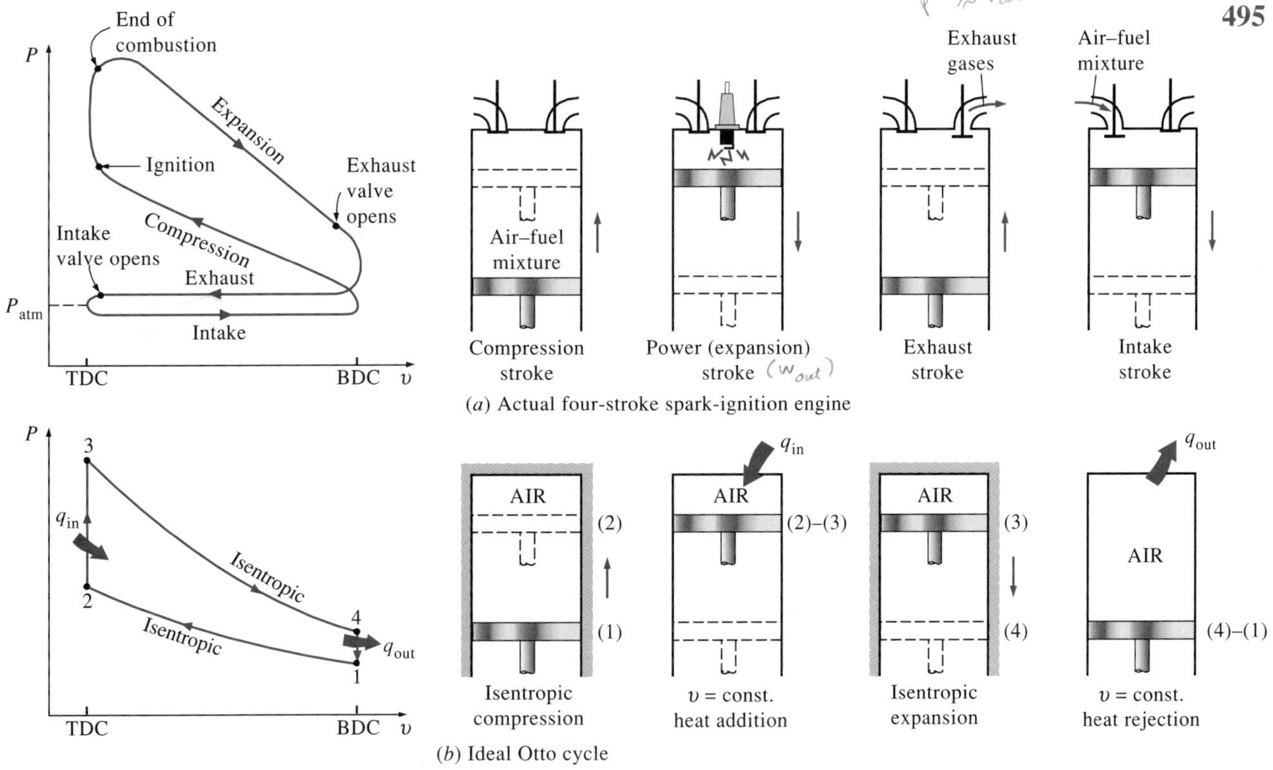

(*a*) Actual four-stroke spark-ignition engine

(*b*) Ideal Otto cycle

FIGURE 8-13

Actual and ideal cycles in spark-ignition engines and their *P-v* diagrams.

In **two-stroke engines,** all four functions described above are executed in just two strokes: the power stroke and the compression stroke. In these engines, the crankcase is sealed, and the outward motion of the piston is used to slightly pressurize the air–fuel mixture in the crankcase, as shown in Fig. 8-14. Also, the intake and exhaust valves are replaced by openings in the lower portion of the cylinder wall. During the latter part of the power stroke, the piston uncovers first the exhaust port, allowing the exhaust gases to be partially expelled, and then the intake port, allowing the fresh air–fuel mixture to rush in and drive most of the remaining exhaust gases out of the cylinder. This mixture is then compressed as the piston moves upward during the compression stroke and is subsequently ignited by a spark plug.

The two-stroke engines are generally less efficient than their four-stroke counterparts because of the incomplete expulsion of the exhaust gases and the partial expulsion of the fresh air–fuel mixture with the exhaust gases. However, they are relatively simple and inexpensive, and they have high power-to-weight and power-to-volume ratios, which make them suitable for applications requiring small size and weight such as for motorcycles, chain saws, and lawn mowers.

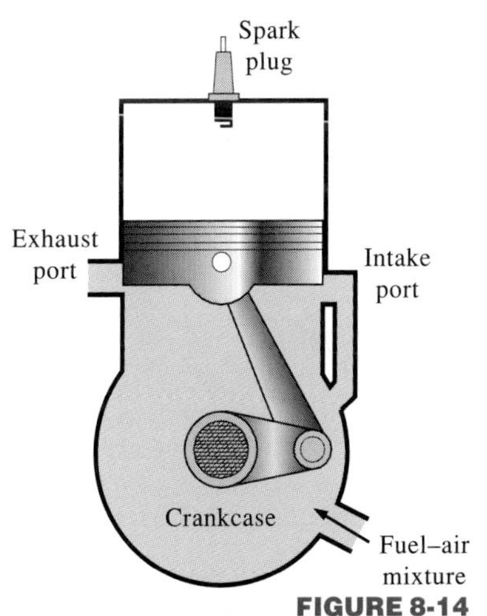

FIGURE 8-14

Schematic of a two-stroke reciprocating engine.

Advances in several technologies—such as direct fuel injection, stratified charge combustion, and electronic controls—brought about a renewed interest in two-stroke engines that can offer high performance and fuel economy while satisfying the future stringent emission requirements. For a given weight and displacement, a well-designed two-stroke engine can provide significantly more power than its four-stroke counterpart because two-stroke engines produce power on every engine revolution instead of every other one. In the new two-stroke engines under development, the highly atomized fuel spray that is injected into the combustion chamber toward the end of the compression stroke burns much more completely. The fuel is sprayed after the exhaust valve is closed, which prevents unburned fuel from being ejected into the atmosphere. With stratified combustion, the flame that is initiated by igniting a small amount of the rich fuel–air mixture near the spark plug propagates through the combustion chamber filled with a much leaner mixture, and this results in much cleaner combustion. Also, the advances in electronics have made it possible to ensure the optimum operation under varying engine load and speed conditions. Major car companies have research programs underway on two-stroke engines which are expected to make a comeback in the future.

The thermodynamic analysis of the actual four-stroke or two-stroke cycles described above is not a simple task. However, the analysis can be simplified significantly if the air-standard assumptions are utilized. The resulting cycle, which closely resembles the actual operating conditions, is the ideal **Otto cycle.** It consists of four internally reversible processes:

1-2	Isentropic compression
2-3	Constant volume heat addition
3-4	Isentropic expansion
4-1	Constant volume heat rejection

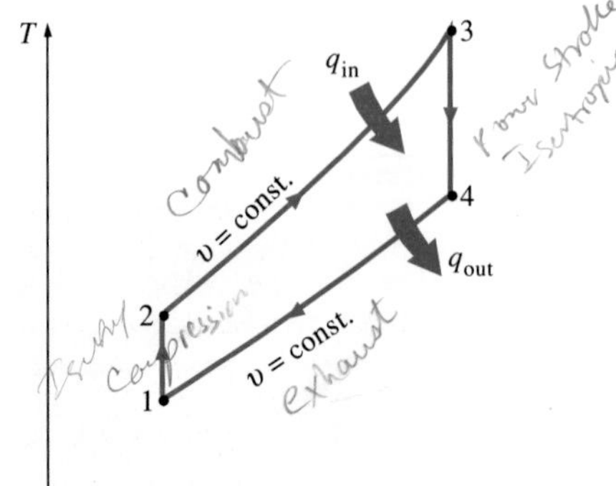

FIGURE 8-15
T-s diagram for the ideal Otto cycle.

The execution of the Otto cycle in a piston-cylinder device together with a *P-v* diagram is illustrated in Fig. 8-13*b*. The *T-s* diagram of the Otto cycle is given in Fig. 8-15.

The Otto cycle is executed in a closed system, and disregarding the changes in kinetic and potential energies, the first-law relation for any of the processes is expressed, on a unit-mass basis, as

$$(q_{in} - q_{out}) + (w_{in} - w_{out}) = \Delta u \qquad \text{(kJ/kg)} \tag{8-5}$$

No work is involved during the two heat transfer processes since both take place at constant volume. Therefore, heat transfer to and from the working fluid can be expressed as

$$q_{in} = u_3 - u_2 = C_v(T_3 - T_2) \tag{8-6a}$$

and

$$q_{out} = u_4 - u_1 = C_v(T_4 - T_1) \tag{8-6b}$$

Then the thermal efficiency of the ideal Otto cycle under the cold air standard assumptions becomes

$$\eta_{\text{th, Otto}} = \frac{w_{\text{net}}}{q_{\text{in}}} = 1 - \frac{q_{\text{out}}}{q_{\text{in}}} = 1 - \frac{T_4 - T_1}{T_3 - T_2} = 1 - \frac{T_1(T_4/T_1 - 1)}{T_2(T_3/T_2 - 1)}$$

Processes 1-2 and 3-4 are isentropic, and $v_2 = v_3$ and $v_4 = v_1$. Thus,

$$\frac{T_1}{T_2} = \left(\frac{v_2}{v_1}\right)^{k-1} = \left(\frac{v_3}{v_4}\right)^{k-1} = \frac{T_4}{T_3} \qquad (8\text{-}7)$$

Substituting these equations into the thermal efficiency relation and simplifying give

$$\eta_{\text{th, Otto}} = 1 - \frac{1}{r^{k-1}} \qquad (8\text{-}8)$$

where
$$r = \frac{V_{\text{max}}}{V_{\text{min}}} = \frac{V_1}{V_2} = \frac{v_1}{v_2} \qquad (8\text{-}9)$$

is the compression ratio and k is the specific heat ratio C_p/C_v.

Equation 8-8 shows that under the cold-air-standard assumptions, the thermal efficiency of an ideal Otto cycle depends on the compression ratio of the engine and the specific heat ratio of the working fluid (if different from air). The thermal efficiency of the ideal Otto cycle increases with both the compression ratio and the specific heat ratio. This is also true for actual spark-ignition internal combustion engines. A plot of thermal efficiency versus the compression ratio is given in Fig. 8-16 for $k = 1.4$, which is the specific heat ratio value of air at room temperature. For a given compression ratio, the thermal efficiency of an actual spark-ignition engine will be less than that of an ideal Otto cycle because of the irreversibilities, such as friction, and other factors such as incomplete combustion.

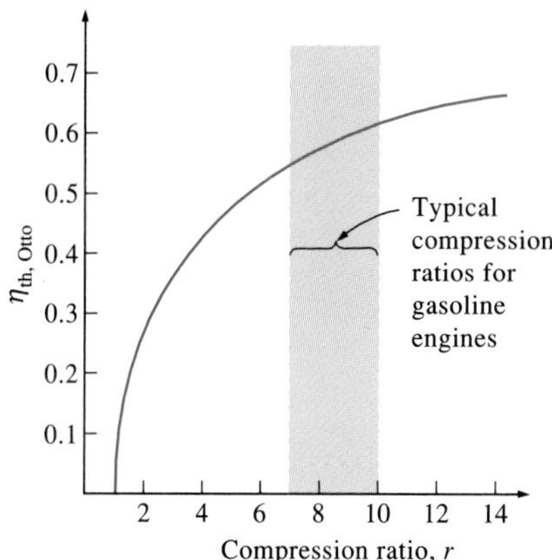

FIGURE 8-16
Thermal efficiency of the ideal Otto cycle as a function of compression ratio ($k = 1.4$).

We can observe from Fig. 8-16 that the thermal efficiency curve is rather steep at low compression ratios but flattens out starting with a compression ratio value of about 8. Therefore, the increase in thermal efficiency with the compression ratio is not that pronounced at high compression ratios. Also, when high compression ratios are used, the temperature of the air–fuel mixture rises above the autoignition temperature of the fuel (the temperature at which the fuel ignites without the help of a spark) during the combustion process, causing an early and rapid burn of the fuel at some point or points ahead of the flame front, followed by almost instantaneous inflammation of the end gas (Fig. 8-17). This premature ignition of the fuel, called **autoignition,** produces an audible noise, which is called **engine knock.** Autoignition in spark-ignition engines cannot be tolerated because it hurts performance and can cause engine damage. The requirement that autoignition not be allowed places an upper limit on the compression ratios that can be used in spark-ignition internal combustion engines.

Improvement of the thermal efficiency of gasoline engines by utilizing higher compression ratios (up to about 12) without facing the autoignition problem has been made possible by using gasoline blends that have good antiknock characteristics, such as gasoline mixed with tetraethyl lead.

FIGURE 8-17
At high compression ratios, the air–fuel mixture temperature rises above the self-ignition temperature of the fuel during the compression process.

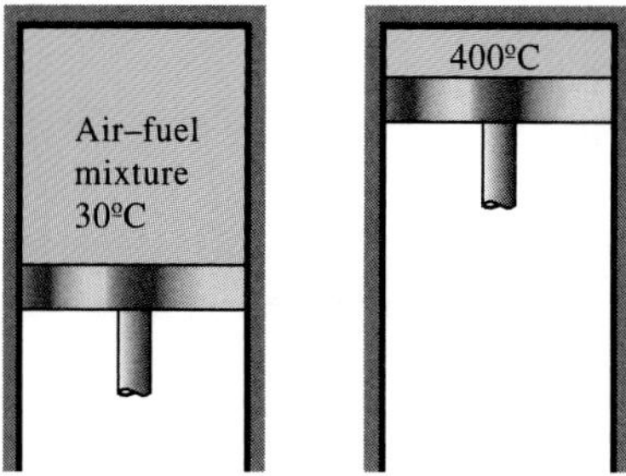

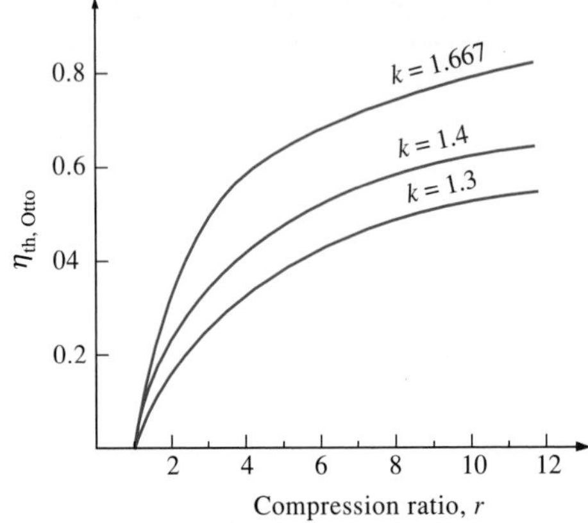

FIGURE 8-18
The thermal efficiency of the Otto cycle increases with the specific heat ratio k of the working fluid.

Tetraethyl lead has been added to gasoline since the 1920s because it is the cheapest method of raising the *octane rating*, which is a measure of the engine knock resistance of a fuel. Leaded gasoline, however, has a very undesirable side effect: it forms compounds during the combustion process that are hazardous to health and pollute the environment. In an effort to combat air pollution, the government adopted a policy in the mid-1970s that resulted in the eventual phase-out of leaded gasoline. Unable to use lead, the refiners developed other, more elaborate techniques to improve the antiknock characteristics of gasoline. Most cars made since 1975 have been designed to use unleaded gasoline, and the compression ratios had to be lowered to avoid engine knock. The thermal efficiency of car engines has decreased somewhat as a result of decreased compression ratios. But, owing to the improvements in other areas (reduction in overall automobile weight, improved aerodynamic design, etc.), today's cars have better fuel economy and consequently get more miles per gallon of fuel. This is an example of how engineering decisions involve compromises, and efficiency is only one of the considerations in final design.

The second parameter affecting the thermal efficiency of an ideal Otto cycle is the specific heat ratio k. For a given compression ratio, an ideal Otto cycle using a monatomic gas (such as argon or helium, $k = 1.667$) as the working fluid will have the highest thermal efficiency. The specific heat ratio k, and thus the thermal efficiency of the ideal Otto cycle, decreases as the molecules of the working fluid get larger (Fig. 8-18). At room temperature it is 1.4 for air, 1.3 for carbon dioxide, and 1.2 for ethane. The working fluid in actual engines contains larger molecules such as carbon dioxide, and the specific heat ratio decreases with temperature, which is one of the reasons that the actual cycles have lower thermal efficiencies than the ideal Otto cycle. The thermal efficiencies of actual spark-ignition engines range from about 25 to 30 percent.

EXAMPLE 8-2 The Ideal Otto Cycle

An ideal Otto cycle has a compression ratio of 8. At the beginning of the compression process, the air is at 100 kPa and 17°C, and 800 kJ/kg of heat is transferred to air during the constant-volume heat-addition process. Accounting

ideal
air standard

for the variation of specific heats of air with temperature, determine (*a*) the maximum temperature and pressure that occur during the cycle, (*b*) the net work output, (*c*) the thermal efficiency, and (*d*) the mean effective pressure for the cycle.

Solution The *P-v* diagram of the ideal Otto cycle described is shown in Fig. 8-19. We note that the air contained in the cylinder forms a closed system.

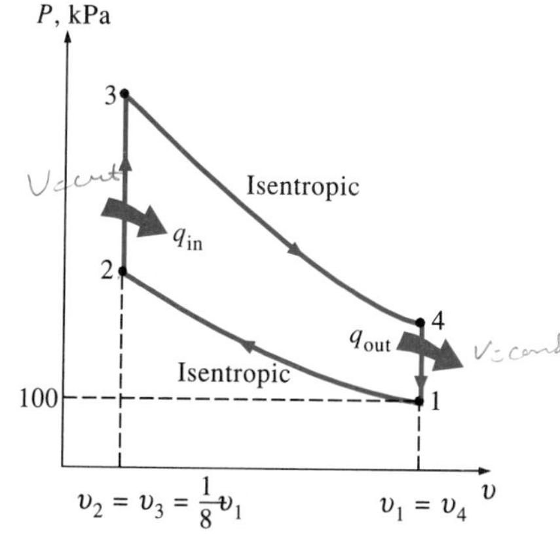

FIGURE 8-19
P-v diagram for the Otto cycle discussed in Example 8-2.

Assumptions **1** The air-standard assumptions are applicable. **2** Kinetic and potential energy changes are negligible. **3** The variation of specific heats with temperature is to be accounted for.

Analysis (*a*) The maximum temperature and pressure in an Otto cycle occur at the end of the constant-volume heat-addition process (state 3). But first we need to determine the temperature and pressure of air at the end of the isentropic compression process (state 2), using data from Table A-17:

$$T_1 = 290 \text{ K} \longrightarrow u_1 = 206.91 \text{ kJ/kg}$$
$$v_{r1} = 676.1$$

Process 1-2 (isentropic compression of an ideal gas):

$$\frac{v_{r2}}{v_{r1}} = \frac{v_2}{v_1} = \frac{1}{r} \longrightarrow v_{r2} = \frac{v_{r1}}{r} = \frac{676.1}{8} = 84.51 \longrightarrow T_2 = 652.4 \text{ K}$$
$$u_2 = 475.11 \text{ kJ/kg}$$

$$\frac{P_2 v_2}{T_2} = \frac{P_1 v_1}{T_1} \longrightarrow P_2 = P_1\left(\frac{T_2}{T_1}\right)\left(\frac{v_1}{v_2}\right)$$
$$= (100 \text{ kPa})\left(\frac{652.4 \text{ K}}{290 \text{ K}}\right)(8) = 1799.7 \text{ kPa}$$

Process 2-3 (v = constant heat addition):

$$q_{in} = u_3 - u_2$$
$$800 \text{ kJ/kg} = u_3 - 475.11 \text{ kJ/kg}$$
$$u_3 = 1275.11 \text{ kJ/kg} \longrightarrow T_3 = \mathbf{1575.1 \text{ K}}$$
$$v_{r3} = 6.108$$

$$\frac{P_3 v_3}{T_3} = \frac{P_2 v_2}{T_2} \longrightarrow P_3 = P_2\left(\frac{T_3}{T_2}\right)\left(\frac{v_2}{v_3}\right)$$
$$= (1.7997 \text{ MPa})\left(\frac{1575.1 \text{ K}}{652.1 \text{ K}}\right)(1) = \mathbf{4.347 \text{ MPa}}$$

(*b*) The net work output for the cycle is determined either by finding the boundary ($P\,dV$) work involved in each process by integration and adding them or by finding the net heat transfer that is equivalent to the net work done during the cycle. We take the latter approach. But first we need to find the internal energy of the air at state 4:

Process 3-4 (isentropic expansion of an ideal gas):

$$\frac{v_{r4}}{v_{r3}} = \frac{v_4}{v_3} = r \longrightarrow v_{r4} = rv_{r3} = (8)(6.108) = 48.864 \longrightarrow T_4 = 795.6 \text{ K}$$
$$u_4 = 588.74 \text{ kJ/kg}$$

Process 4-1 (v = constant heat rejection):

$$-q_{out} = u_1 - u_4 \longrightarrow q_{out} = u_4 - u_1$$
$$q_{out} = 588.74 - 206.91 = 381.83 \text{ kJ/kg}$$

Thus, $w_{net} = q_{net} = q_{in} - q_{out} = 800 - 381.83 = \mathbf{418.17\ kJ/kg}$

(*c*) The thermal efficiency of the cycle is determined from its definition, Eq. 8-1:

$$\eta_{th} = \frac{w_{net}}{q_{in}} = \frac{418.17 \text{ kJ/kg}}{800 \text{ kJ/kg}} = \mathbf{0.523 \text{ or } 52.3\%}$$

Under the cold-air-standard assumptions (constant specific heat values at room temperature), the thermal efficiency would be (Eq. 8-8)

$$\eta_{th,\text{Otto}} = 1 - \frac{1}{r^{k-1}} = 1 - r^{1-k} = 1 - (8)^{1-1.4} = 0.565 \text{ or } 56.5\%$$

which is considerably different from the value obtained above. Therefore, care should be exercised in utilizing the cold-air standard assumptions.

(*d*) The mean effective pressure is determined from its definition, Eq. 8-4:

$$\text{MEP} = \frac{w_{net}}{v_1 - v_2} = \frac{w_{net}}{v_1 - v_1/r} = \frac{w_{net}}{v_1(1 - 1/r)}$$

where $v_1 = \frac{RT_1}{P_1} = \frac{[0.287 \text{ kPa}\cdot\text{m}^3/(\text{kg}\cdot\text{K})](290 \text{ K})}{100 \text{ kPa}} = 0.832 \text{ m}^3/\text{kg}$

Thus, $\text{MEP} = \frac{451.8 \text{ kJ/kg}}{(0.832 \text{ m}^3/\text{kg})(1 - \frac{1}{8})}\left(\frac{1 \text{ kPa}\cdot\text{m}^3}{1 \text{ kJ}}\right) = \mathbf{574.4\ kPa}$

Therefore, a constant pressure of 574.4 kPa during the power stroke would produce the same net work output as the entire cycle.

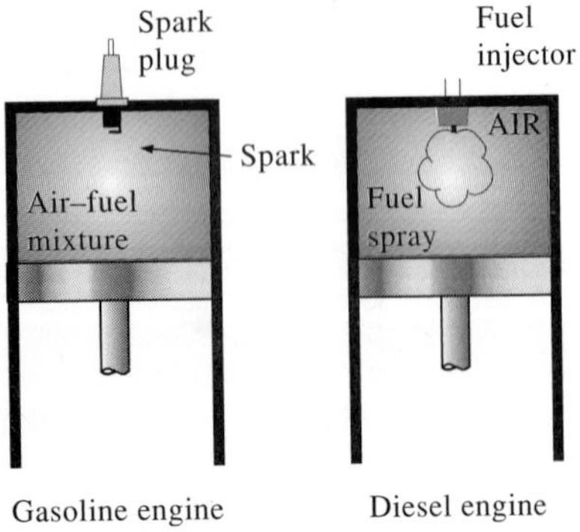

Spark plug | Fuel injector | Spark | AIR | Air–fuel mixture | Fuel spray

Gasoline engine   Diesel engine

FIGURE 8-20
In diesel engines, the spark plug is replaced by a fuel injector, and only air is compressed during the compression process.

8-6 ■ DIESEL CYCLE: The Ideal Cycle for Compression-Ignition Engines

The Diesel cycle is the ideal cycle for CI reciprocating engines. The CI engine, first proposed by Rudolph Diesel in the 1890s, is very similar to the SI engine discussed in the last section, differing mainly in the method of initiating combustion. In spark-ignition engines (also known as *gasoline engines*), the air–fuel mixture is compressed to a temperature that is below the autoignition temperature of the fuel, and the combustion process is initiated by firing a spark plug. In CI engines (also known as *diesel engines*), the air is compressed to a temperature that is above the autoignition temperature of the fuel, and combustion starts on contact as the fuel is injected into this hot air. Therefore, the spark plug and carburetor are replaced by a fuel injector in diesel engines (Fig. 8-20).

In gasoline engines, a mixture of air and fuel is compressed during the compression stroke, and the compression ratios are limited by the onset of autoignition or engine knock. In diesel engines, only air is compressed during the compression stroke, eliminating the possibility of autoignition. Therefore, diesel engines can be designed to operate at much higher compression ratios, typically between 12 and 24. Not having to deal with the problem of autoignition has another benefit: many of the stringent requirements placed on the gasoline can now be removed, and fuels that are less refined (thus less expensive) can be used in diesel engines.

The fuel injection process in diesel engines starts when the piston approaches TDC and continues during the first part of the power stroke. Therefore, the combustion process in these engines takes place over a longer interval. Because of this longer duration, the combustion process in the ideal Diesel cycle is approximated as a constant-pressure heat-addition process. In fact, this is the only process where the Otto and the Diesel cycles differ. The remaining three processes are the same for both ideal cycles. That is, process 1-2 is isentropic compression, 3-4 is isentropic expansion, and 4-1 is constant-volume heat rejection. The similarity between the two cycles is also apparent from the *P-v* and *T-s* diagrams of the Diesel cycle, shown in Fig. 8-21.

Noting that the Diesel cycle is executed in a piston-cylinder device, which forms a closed system, the amount of heat transferred to the working fluid at constant pressure and rejected from it at constant volume can be expressed as

$$q_{in} - w_{b,out} = u_3 - u_2 \longrightarrow q_{in} = P_2(v_3 - v_2) + (u_3 - u_2)$$
$$= h_3 - h_2 = C_p(T_3 - T_2) \quad (8\text{-}10a)$$

and

$$-q_{out} = u_1 - u_4 \longrightarrow q_{out} = u_4 - u_1 = C_v(T_4 - T_1) \quad (8\text{-}10b)$$

Then the thermal efficiency of the ideal Diesel cycle under the cold-air-standard assumptions becomes

$$\eta_{th,Diesel} = \frac{w_{net}}{q_{in}} = 1 - \frac{q_{out}}{q_{in}} = 1 - \frac{T_4 - T_1}{k(T_3 - T_2)} = 1 - \frac{T_1(T_4/T_1 - 1)}{kT_2(T_3/T_2 - 1)}$$

We now define a new quantity, the **cutoff ratio r_c,** as the ratio of the cylinder volumes after and before the combustion process:

$$r_c = \frac{V_3}{V_2} = \frac{v_3}{v_2} \quad (8\text{-}11)$$

Utilizing this definition and the isentropic ideal-gas relations for processes 1-2 and 3-4, we see that the thermal efficiency relation reduces to

$$\eta_{th,Diesel} = 1 - \frac{1}{r^{k-1}}\left[\frac{r_c^k - 1}{k(r_c - 1)}\right] \quad (8\text{-}12)$$

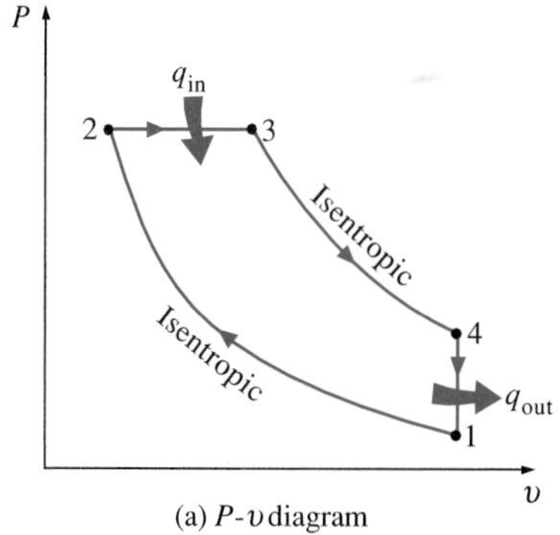

(a) *P-v* diagram

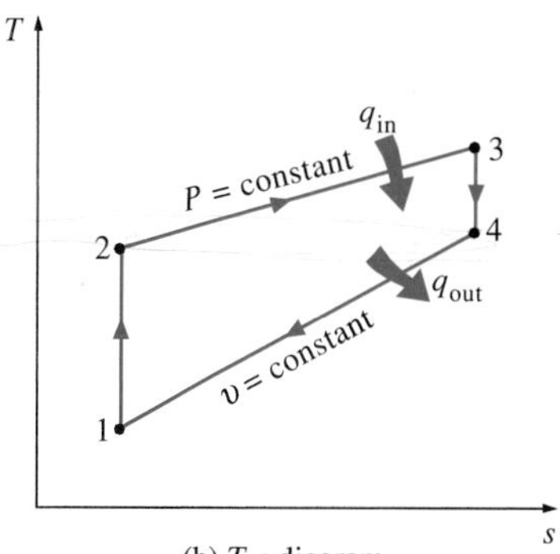

(b) *T-s* diagram

FIGURE 8-21

T-s and *P-v* diagrams for the ideal Diesel cycle.

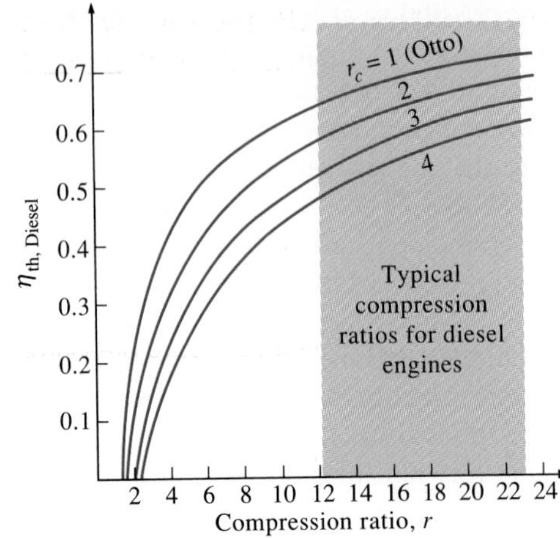

FIGURE 8-22

Thermal efficiency of the ideal Diesel cycle as a function of compression and cutoff ratios (k = 1.4).

where r is the compression ratio defined by Eq. 8-9. Looking at Eq. 8-12 carefully, one would notice that under the cold-air-standard assumptions, the efficiency of a Diesel cycle differs from the efficiency of an Otto cycle by the quantity in the brackets. This quantity is always greater than 1. Therefore,

$$\eta_{th,\,Otto} > \eta_{th,\,Diesel} \tag{8-13}$$

when both cycles operate on the same compression ratio. Also, as the cutoff ratio decreases, the efficiency of the Diesel cycle increases (Fig. 8-22). For the limiting case of r_c = 1, the quantity in the brackets becomes unity (can you prove it?), and the efficiencies of the Otto and Diesel cycles become identical. Remember, though, that diesel engines operate at much higher compression ratios and thus are usually more efficient than the spark-ignition (gasoline) engines. The diesel engines also burn the fuel more completely since they usually operate at lower revolutions per minute than spark-ignition engines. Thermal efficiencies of large diesel engines range from about 35 to 40 percent.

The higher efficiency and lower fuel costs of diesel engines make them the clear choice in applications requiring relatively large amounts of power, such as in locomotive engines, emergency power generation units, large ships, and heavy trucks. As an example of how large a diesel engine can be, a 12-cylinder diesel engine built in 1964 by the Fiat Corporation of Italy had a normal power output of 25,200 hp (18.8 MW) at 122 rpm, a cylinder bore of 90 cm, and a stroke of 91 cm.

FIGURE 8-23

P-v diagram of an ideal dual cycle.

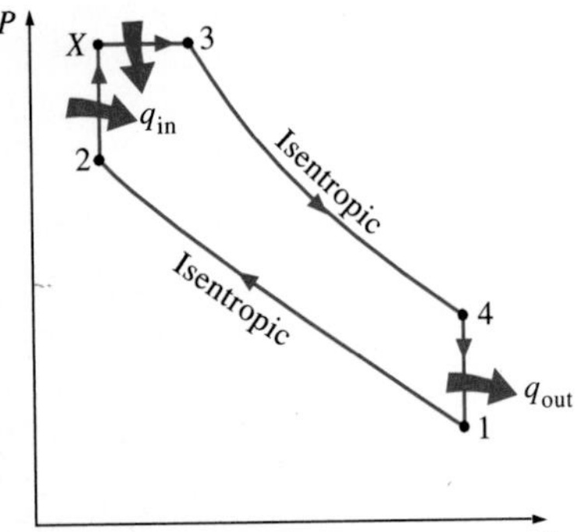

Approximating the combustion process in internal combustion engines as a constant-volume or a constant-pressure heat-addition process is overly simplistic and not quite realistic. Probably a better (but slightly more complex) approach would be to model the combustion process in both gasoline and diesel engines as a combination of two heat-transfer processes, one at constant volume and the other at constant pressure. The ideal cycle based on this concept is called the **dual cycle,** and a P-v diagram for it is given in Fig. 8-23. The relative amounts of heat transferred during each process can be adjusted to approximate the actual cycle more closely. Note that both the Otto and the Diesel cycles can be obtained as special cases of the dual cycle.

EXAMPLE 8-3 The Ideal Diesel Cycle

An ideal Diesel cycle with air as the working fluid has a compression ratio of 18 and a cutoff ratio of 2. At the beginning of the compression process, the working fluid is at 14.7 psia, 80°F, and 117 in³. Utilizing the cold-air-standard assumptions, determine (*a*) the temperature and pressure of the air at the end of each process, (*b*) the net output and the thermal efficiency, and (*c*) the mean effective pressure.

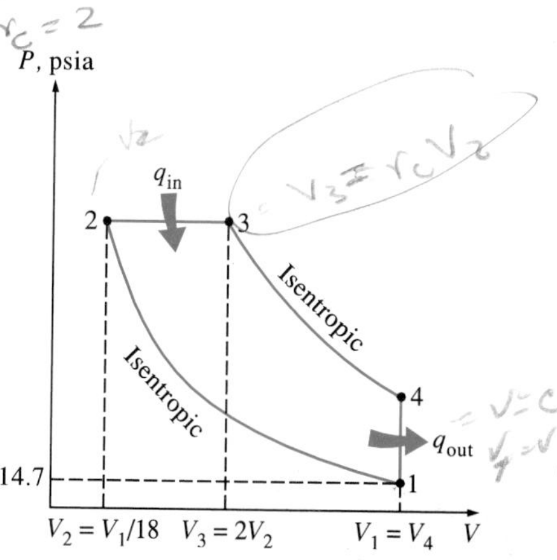

FIGURE 8-24

P-v diagram for the ideal Diesel cycle discussed in Example 8-3.

Solution The *P-v* diagram of the ideal Diesel cycle described is shown in Fig. 8-24. We note that the air contained in the cylinder forms a closed system.

Assumptions **1** The cold-air-standard assumptions are applicable and thus air can be assumed to have constant specific heats at room temperature. **2** Kinetic and potential energy changes are negligible.

Analysis The gas constant of air is $R = 0.06855$ Btu/(lbm · R), and its specific heats at room temperature are $C_p = 0.240$ Btu/(lbm · R) and $C_v = 0.171$ Btu/(lbm · R) (Table A-2E*a*).

(*a*) The temperature and pressure values at the end of each process can be determined by utilizing the ideal-gas isentropic relations for processes 1-2 and 3-4. But first we determine the volumes at the end of each process from the definitions of the compression ratio and the cutoff ratio:

$$V_2 = \frac{V_1}{r} = \frac{117 \text{ in}^3}{18} = 6.5 \text{ in}^3$$

$$V_3 = r_c V_2 = (2)(6.5 \text{ in}^3) = 13 \text{ in}^3$$

$$V_4 = V_1 = 117 \text{ in}^3$$

Process 1-2 (isentropic compression of an ideal gas, constant specific heats):

$$T_2 = T_1\left(\frac{V_1}{V_2}\right)^{k-1} = (540 \text{ R})(18)^{1.4-1} = \mathbf{1716\ R}$$

$$P_2 = P_1\left(\frac{V_1}{V_2}\right)^{k} = (14.7 \text{ psia})(18)^{1.4} = \mathbf{841\ psia}$$

Process 2-3 (P = constant heat addition to an ideal gas):

$$P_3 = P_2 = \mathbf{841\ psia}$$

$$\frac{P_2V_2}{T_2} = \frac{P_3V_3}{T_3} \longrightarrow T_3 = T_2\left(\frac{V_3}{V_2}\right) = (1716 \text{ R})(2) = \mathbf{3432\ R}$$

Process 3-4 (isentropic expansion of an ideal gas, constant specific heats):

$$T_4 = T_3\left(\frac{V_3}{V_4}\right)^{k-1} = (3432 \text{ R})\left(\frac{13 \text{ in}^3}{117 \text{ in}^3}\right)^{1.4-1} = \mathbf{1425\ R}$$

$$P_4 = P_3\left(\frac{V_3}{V_4}\right)^{k} = (841 \text{ psia})\left(\frac{13 \text{ in}^3}{117 \text{ in}^3}\right)^{1.4} = \mathbf{38.8\ psia}$$

(*b*) The net work for a cycle is equivalent to the net heat transfer, that is, the difference between the total heat supplied and the total heat rejected. But first we find the mass of air:

$$m = \frac{P_1V_1}{RT_1} = \frac{(14.7 \text{ psia})(117 \text{ in}^3)}{[0.3704 \text{ psia}\cdot\text{ft}^3/(\text{lbm}\cdot\text{R})](540 \text{ R})}\left(\frac{1 \text{ ft}^3}{1728 \text{ in}^3}\right)$$

$$= 0.00498 \text{ lbm}$$

Process 2-3 is a constant-pressure heat-addition process, for which the boundary work and Δu terms can be combined into Δh. Thus,

$$Q_{in} = m(h_3 - h_2) = mC_p(T_3 - T_2)$$

$$= (0.00498 \text{ lbm})[0.240 \text{ Btu}/(\text{lbm}\cdot\text{R})][(3432 - 1716) \text{ R}]$$

$$= 2.051 \text{ Btu}$$

Process 4-1 is a constant-volume heat-rejection process (it involves no work interactions), and the amount of heat rejected is

$$Q_{out} = m(u_4 - u_1) = mC_v(T_4 - T_1)$$

$$= (0.00498 \text{ lbm})[0.171 \text{ Btu}/(\text{lbm}\cdot\text{R})][(1425 - 540) \text{ R}]$$

$$= 0.758 \text{ Btu}$$

Thus, $\quad W_{net} = Q_{in} - Q_{out} = 2.051 - 0.758 = \mathbf{1.293 \text{ Btu}}$

Then the thermal efficiency becomes

$$\eta_{th} = \frac{W_{net}}{Q_{in}} = \frac{1.293 \text{ Btu}}{2.051 \text{ Btu}} = \mathbf{0.630 \text{ or } 63.0\%}$$

The thermal efficiency of this Diesel cycle under the cold-air-standard assumptions could also be determined from Eq. 8-12.

(*c*) The mean effective pressure is determined from its definition, Eq. 8-4:

$$\text{MEP} = \frac{W_{net}}{V_{max} - V_{min}} = \frac{W_{net}}{V_1 - V_2} = \frac{1.293 \text{ Btu}}{(117 - 6.5) \text{ in}^3}\left(\frac{778.17 \text{ lbf}\cdot\text{ft}}{1 \text{ Btu}}\right)\left(\frac{12 \text{ in.}}{1 \text{ ft}}\right)$$

$$= \mathbf{109.3 \text{ psia}}$$

Therefore, a constant pressure of 109.3 psia during the power stroke would produce the same net work output as the entire Diesel cycle.

8-7 ■ STIRLING AND ERICSSON CYCLES

The ideal Otto and Diesel cycles discussed in the preceding sections are composed entirely of internally reversible processes and thus are internally reversible cycles. These cycles are not totally reversible, however, since they involve heat transfer through a finite temperature difference during the non-isothermal heat-addition and rejection processes, which are irreversible. Therefore, the thermal efficiency of an Otto or Diesel engine will be less than that of a Carnot engine operating between the same temperature limits.

Consider a heat engine operating between a high-temperature reservoir at T_H and a low-temperature reservoir at T_L. For the heat-engine cycle to be totally reversible, the temperature difference between the working fluid and the thermal energy source (or sink) should never exceed a differential amount dT during any heat-transfer process. That is, both the heat-addition and heat-rejection processes during the cycle must take place isothermally, one at a temperature of T_H and the other at a temperature of T_L. This is precisely what happens in a Carnot cycle.

There are two other cycles that involve an isothermal heat-addition process at T_H and an isothermal heat-rejection process at T_L: the *Stirling cycle* and the *Ericsson cycle*. They differ from the Carnot cycle in that the two isentropic processes are replaced by two constant-volume regeneration processes in the Stirling cycle and by two constant-pressure regeneration processes in the Ericsson cycle. Both cycles utilize **regeneration,** a process during which heat is transferred to a thermal energy storage device (called a *regenerator*) during one part of the cycle and is transferred back to the working fluid during another part of the cycle (Fig. 8-25).

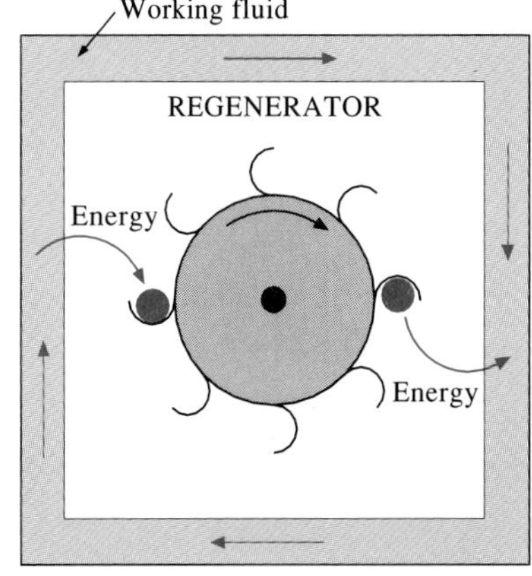

FIGURE 8-25

A regenerator is a device that borrows energy from the working fluid during one part of the cycle and pays it back (without interest) during another part.

Figure 8-26(*b*) shows the *T-s* and *P-v* diagrams of the **Stirling cycle,** which is made up of four totally reversible processes:

1-2	T = *constant* expansion (heat addition from the external source)
2-3	v = *constant* regeneration (internal heat transfer from the working fluid to the regenerator)
3-4	T = *constant* compression (heat rejection to the external sink)
4-1	v = *constant* regeneration (internal heat transfer from the regenerator back to the working fluid)

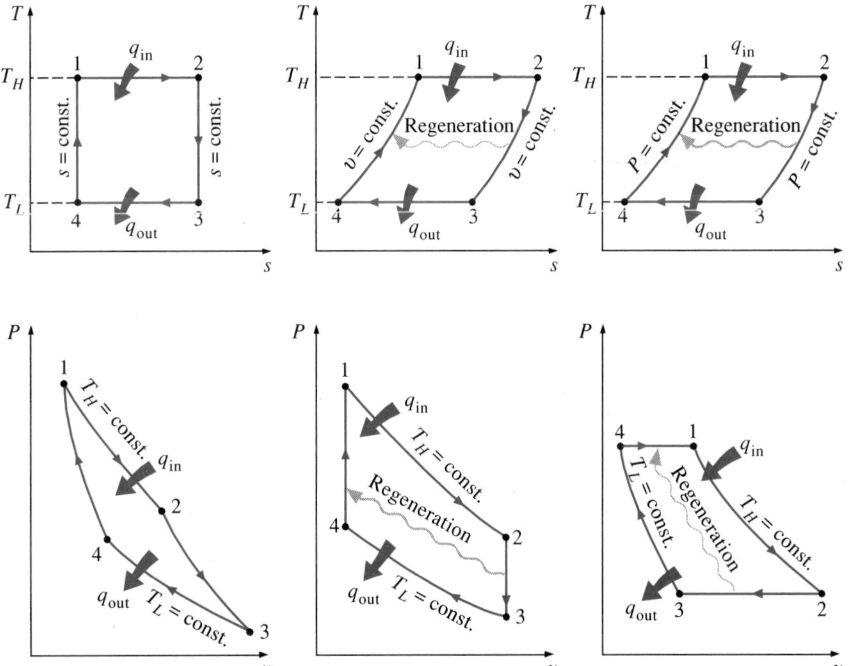

FIGURE 8-26

T-s and *P-v* diagrams of Carnot, Stirling, and Ericsson cycles.

The execution of the Stirling cycle requires rather innovative hardware. The actual Stirling engines, including the original one patented by Robert Stirling, are very heavy and complicated. To spare the reader the complexities, the execution of the Stirling cycle in a closed system is explained with the help of the hypothetical engine shown in Fig. 8-27.

This system consists of a cylinder with two pistons on each side and a regenerator in the middle. The regenerator can be a wire or a ceramic mesh or any kind of porous plug with a high thermal mass (mass times specific heat). It is used for the temporary storage of thermal energy. The mass of the working fluid contained within the regenerator at any instant is considered negligible.

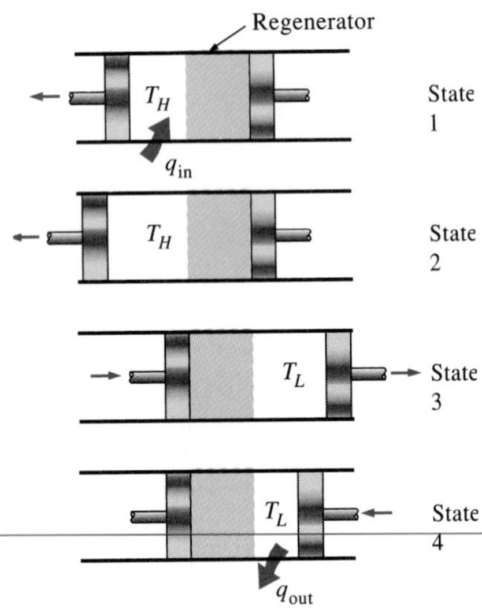

FIGURE 8-27
The execution of the Stirling cycle.

Initially, the left chamber houses the entire working fluid (a gas), which is at a high temperature and pressure. During *process 1-2,* heat is transferred to the gas at T_H from a source atT_H. As the gas expands isothermally, the left piston moves outward, doing work, and the gas pressure drops. During *process 2-3,* both pistons are moved to the right at the same rate (to keep the volume constant) until the entire gas is forced into the right chamber. As the gas passes through the regenerator, heat is transferred to the regenerator and the gas temperature drops from T_H to T_L. For this heat transfer process to be reversible, the temperature difference between the gas and the regenerator should not exceed a differential amount dT at any point. Thus, the temperature of the regenerator will be T_H at the left end and T_L at the right end of the regenerator when state 3 is reached. During *process 3-4,* the right piston is moved inward, compressing the gas. Heat is transferred from the gas to a sink at temperature T_L so that the gas temperature remains constant at T_L while the pressure rises. Finally, during *process 4-1,* both pistons are moved to the left at the same rate (to keep the volume constant), forcing the entire gas into the left chamber. The gas temperature rises from T_L to T_H as it passes through the regenerator and picks up the thermal energy stored there during process 2-3. This completes the cycle.

Notice that the second v = constant process takes place at a smaller volume than the first one, and the net heat transfer to the regenerator during a cycle is zero. That is, the amount of energy stored in the regenerator during process 2-3 is equal to the amount picked up by the gas during process 4-1.

The *T-s* and *P-v* diagrams of the **Ericsson cycle** are shown in Fig. 8-26*c*. The Ericsson cycle is very much like the Stirling cycle, except that the two constant-volume processes are replaced by two constant-pressure processes.

A steady-flow system operating on an Ericsson cycle is shown in Fig. 8-28. Here the isothermal expansion and compression processes are executed in a compressor and a turbine, respectively, and a counter-flow heat exchanger serves as a regenerator. Hot and cold fluid streams enter the heat exchanger from opposite ends, and heat transfer takes place between the two streams. In the ideal case, the temperature difference between the two fluid streams does not exceed a differential amount at any point, and the cold fluid stream leaves the heat exchanger at the inlet temperature of the hot stream.

Both the Stirling and Ericsson cycles are totally reversible, as is the Carnot cycle, and so according to the Carnot principle, all three cycles will have the same thermal efficiency when operating between the same temperature limits:

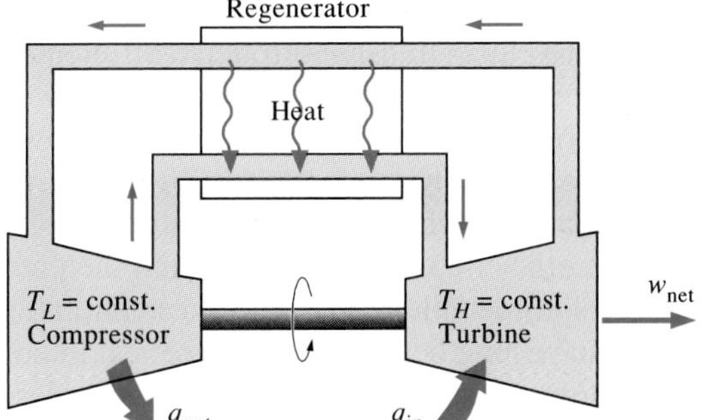

FIGURE 8-28
A steady-flow Ericsson engine.

$$\eta_{th,\,Stirling} = \eta_{th,\,Ericsson} = \eta_{th,\,Carnot} = 1 - \frac{T_L}{T_H} \tag{8-14}$$

This is proved for the Carnot cycle in Example 8-1 and can be proved in a similar manner for both the Stirling and Ericsson cycles.

EXAMPLE 8-4 Thermal Efficiency of the Ericsson Cycle

Using an ideal gas as the working fluid, show that the thermal efficiency of an Ericsson cycle is identical to the efficiency of a Carnot cycle operating between the same temperature limits.

Solution Heat is transferred to the working fluid isothermally from an external source at temperature T_H during process 1-2, and it is rejected again isothermally to an external sink at temperature T_L during process 3-4. For a reversible isothermal process, heat transfer is related to the entropy change by

$$q = T\,\Delta s$$

The entropy change of an ideal gas during an isothermal process is given by

$$\Delta s = C_p \ln \overset{0}{\cancel{\frac{T_e}{T_i}}} - R \ln \frac{P_e}{P_i} = -R \ln \frac{P_e}{P_i}$$

Then the amount of heat input and heat output can be expressed, on a unit-mass basis, as

$$q_{in} = T_H(s_2 - s_1) = T_H\left(-R \ln \frac{P_2}{P_1}\right) = RT_H \ln \frac{P_1}{P_2}$$

and

$$q_{out} = -T_L(s_4 - s_3) = -T_L\left(-R \ln \frac{P_4}{P_3}\right) = RT_L \ln \frac{P_4}{P_3}$$

Then the thermal efficiency of the Ericsson cycle becomes

$$\eta_{th,\,Ericsson} = 1 - \frac{q_{out}}{q_{in}} = 1 - \frac{RT_L \ln (P_4/P_3)}{RT_H \ln (P_1/P_2)} = 1 - \frac{T_L}{T_H}$$

since $P_1 = P_4$ and $P_3 = P_2$. Notice that this result is independent of whether the cycle is executed in a closed or steady-flow system.

Stirling and Ericsson cycles are difficult to achieve in practice because they involve heat transfer through a differential temperature difference in all components including the regenerator. This would require providing infinitely large surface areas for heat transfer or allowing an infinitely long time for the process. Neither is practical. In reality, all heat transfer processes will take place through a finite temperature difference, the regenerator will not have an efficiency of 100 percent, and the pressure losses in the regenerator will be considerable. Because of these limitations, both Stirling and Ericsson cycles have long been of only theoretical interest. However, there is renewed interest in engines that operate on these cycles because of their potential for higher efficiency and better emission control. The Ford Motor Company, General Motors Corporation, and the Phillips Research Laboratories of the Netherlands have successfully developed Stirling engines suitable for trucks, buses, and even automobiles. More research and development are needed before these engines can compete with the gasoline or diesel engines.

Both the Stirling and the Ericsson engines are *external combustion* engines. That is, the fuel in these engines is burned outside the system, as opposed to gasoline or diesel engines, where the fuel is burned inside the cylinder.

External combustion offers several advantages. First, a variety of fuels can be used as a source of thermal energy. Second, there is more time for combustion, and thus the combustion process is more complete, which means less air pollution and more energy extraction from the fuel. Third, these engines operate on closed cycles, and thus a working fluid that has the most desirable characteristics (stable, chemically inert, high thermal conductivity) can be utilized as the working fluid. Hydrogen and helium are two gases commonly employed in these engines.

Despite the physical limitations and impracticalities associated with them, both the Stirling and Ericsson cycles give a strong message to design engineers: *Regeneration can increase efficiency.* It is no coincidence that modern gas-turbine and steam power plants make extensive use of regeneration. In fact, the Brayton cycle with intercooling, reheating, and regeneration, which is utilized in large gas-turbine power plants and discussed later in this chapter, closely resembles the Ericsson cycle.

8-8 ■ BRAYTON CYCLE: The Ideal Cycle for Gas-Turbine Engines

The Brayton cycle was first proposed by George Brayton for use in the reciprocating oil-burning engine that he developed around 1870. Today, it is used for gas turbines only where both the compression and expansion processes take place in rotating machinery. Gas turbines usually operate on an *open cycle,* as shown in Fig. 8-29. Fresh air at ambient conditions is drawn into the compressor, where its temperature and pressure are raised. The high-pressure air proceeds into the combustion chamber, where the fuel is burned at constant pressure. The resulting high-temperature gases then enter the turbine, where they expand to the atmospheric pressure, thus producing

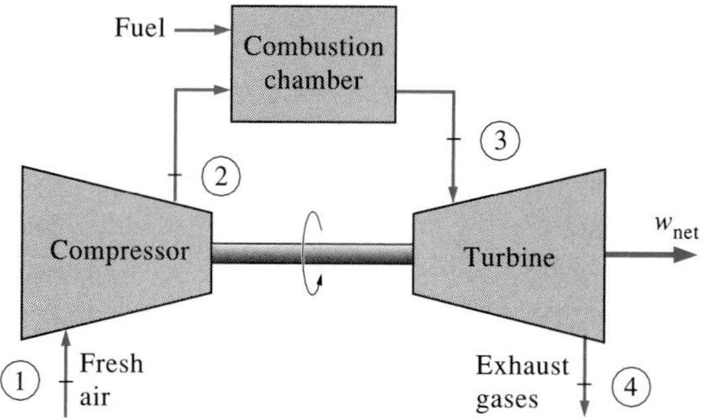

FIGURE 8-29
An open-cycle gas-turbine engine.

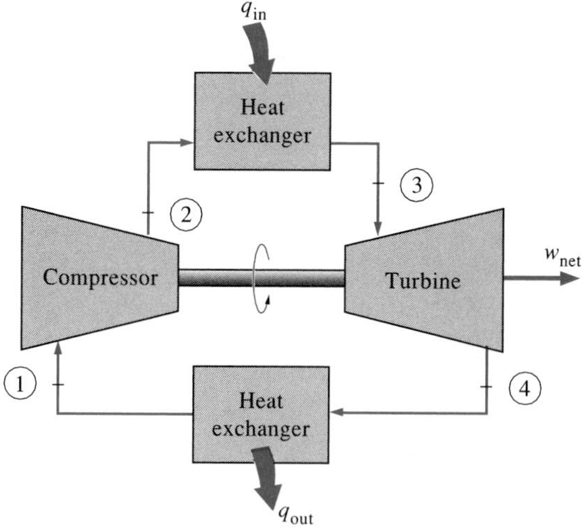

FIGURE 8-30
A closed-cycle gas-turbine engine.

power. The exhaust gases leaving the turbine are thrown out (not recirculated), causing the cycle to be classified as an open cycle.

The open gas-turbine cycle described above can be modeled as a *closed cycle,* as shown in Fig. 8-30, by utilizing the air-standard assumptions. Here the compression and expansion processes remain the same, but the combustion process is replaced by a constant-pressure heat-addition process from an external source, and the exhaust process is replaced by a constant-pressure heat-rejection process to the ambient air. The ideal cycle that the working fluid undergoes in this closed loop is the **Brayton cycle,** which is made up of four internally reversible processes:

1-2 Isentropic compression (in a compressor)

2-3 Constant pressure heat addition

3-4 Isentropic expansion (in a turbine)

4-1 Constant pressure heat rejection

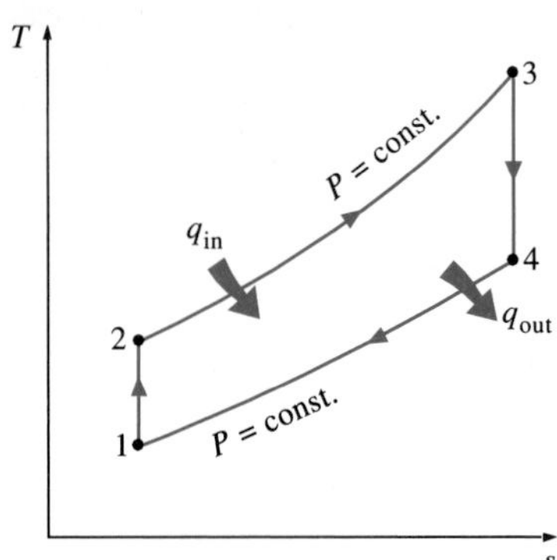

(a) *T-s* diagram

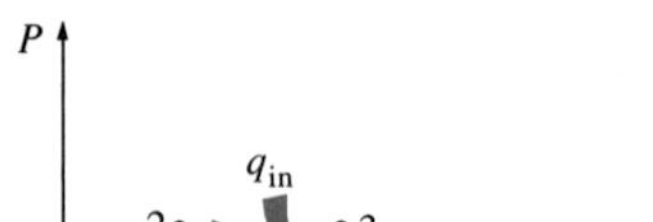

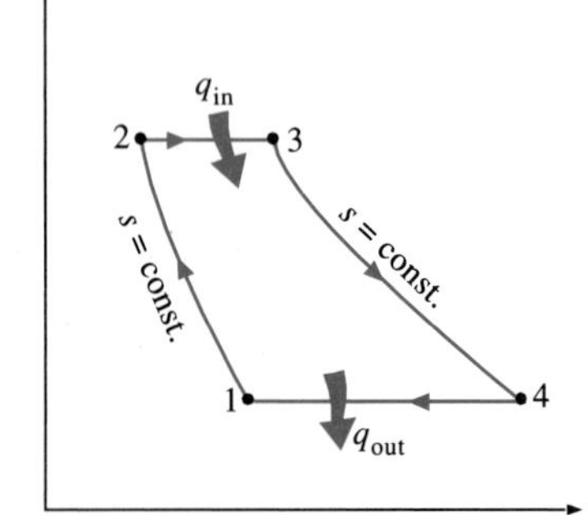

(b) *P-v* diagram

FIGURE 8-31

T-s and *P-v* diagrams for the ideal Brayton cycle.

The *T-s* and *P-v* diagrams of an ideal Brayton cycle are shown in Fig. 8-31. Notice that all four processes of the Brayton cycle are executed in steady-flow devices; thus, they should be analyzed as steady-flow processes. When the changes in kinetic and potential energies are neglected, the energy balance for a steady-flow process can be expressed, on a unit-mass basis, as

$$(q_{in} - q_{out}) + (w_{in} - w_{out}) = h_{exit} - h_{inlet} \tag{8-15}$$

Therefore, heat transfers to and from the working fluid are

$$q_{in} = h_3 - h_2 = C_p(T_3 - T_2) \tag{8-16a}$$

and

$$q_{out} = h_4 - h_1 = C_p(T_4 - T_1) \tag{8-16b}$$

Then the thermal efficiency of the ideal Brayton cycle under the cold air standard assumptions becomes

$$\eta_{th,\,Brayton} = \frac{w_{net}}{q_{in}} = 1 - \frac{q_{out}}{q_{in}} = 1 - \frac{C_p(T_4 - T_1)}{C_p(T_3 - T_2)} = 1 - \frac{T_1(T_4/T_1 - 1)}{T_2(T_3/T_2 - 1)}$$

Processes 1-2 and 3-4 are isentropic, and $P_2 = P_3$ and $P_4 = P_1$. Thus,

$$\frac{T_2}{T_1} = \left(\frac{P_2}{P_1}\right)^{(k-1)/k} = \left(\frac{P_3}{P_4}\right)^{(k-1)/k} = \frac{T_3}{T_4}$$

Substituting these equations into the thermal efficiency relation and simplifying give

$$\eta_{th,\,Brayton} = 1 - \frac{1}{r_p^{(k-1)/k}} \tag{8-17}$$

where

$$r_p = \frac{P_2}{P_1} \tag{8-18}$$

FIGURE 8-32

Thermal efficiency of the ideal Brayton cycle as a function of the pressure ratio.

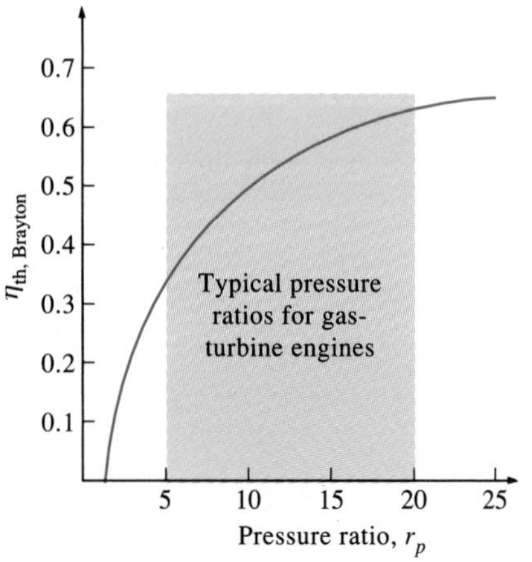

is the **pressure ratio** and k is the specific heat ratio. Equation 8-17 shows that under the cold-air-standard assumptions, the thermal efficiency of an ideal Brayton cycle depends on the pressure ratio of the gas turbine and the specific heat ratio of the working fluid (if different from air). The thermal efficiency increases with both of these parameters, which is also the case for actual gas turbines. A plot of thermal efficiency versus the pressure ratio is given in Fig. 8-32 for $k = 1.4$, which is the specific-heat-ratio value of air at room temperature.

The highest temperature in the cycle occurs at the end of the combustion process (state 3), and it is limited by the maximum temperature that the turbine blades can withstand. This also limits the pressure ratios that can be used in the cycle. For a fixed turbine inlet temperature T_3, the net work output per cycle increases with the pressure ratio, reaches a maximum, and then starts to decrease, as shown in Fig. 8-33. Therefore, there should be a compromise between the pressure ratio (thus the thermal efficiency) and the net work output. With less work output per cycle, a larger mass flow rate (thus a larger system) is needed to maintain the same power output which may not be economical. In most common designs, the pressure ratio of gas turbines ranges from about 11 to 16.

The air in gas turbines performs two important functions: It supplies the necessary oxidant for the combustion of the fuel, and it serves as a coolant to keep the temperature of various components within safe limits. The second function is accomplished by drawing in more air than is needed for the complete combustion of the fuel. In gas turbines, an air–fuel mass ratio of 50 or above is not uncommon. Therefore, in a cycle analysis, treating the combustion gases as air will not cause any appreciable error. Also, the mass flow rate through the turbine will be greater than that through the compressor, the difference being equal to the mass flow rate of the fuel. Thus, assuming a constant mass flow rate throughout the cycle will yield conservative results for open-loop gas-turbine engines.

The two major application areas of gas-turbine engines are *aircraft propulsion* and *electric power generation*. When it is used for aircraft propulsion, the gas turbine produces just enough power to drive the compressor and a small generator to power the auxiliary equipment. The high-velocity exhaust gases are responsible for producing the necessary thrust to propel the aircraft. Gas turbines are also used as stationary power plants to generate electricity as stand-alone units or in conjunction with steam power plants on the high-temperature side. In these plants, the exhaust gases of the gas turbine serve as the heat source for the steam. The gas-turbine cycle can also be executed as a closed cycle for use in nuclear power plants. This time the working fluid is not limited to air, and a gas with more desirable characteristics (such as helium) can be used.

The majority of the Western world's naval fleets already use gas-turbine engines for propulsion and electric power generation. The General Electric LM2500 gas turbines used to power ships have a simple-cycle thermal efficiency of 37 percent. The new General Electric WR-21 gas turbines equipped with intercooling and regeneration have a thermal efficiency of 43 percent and produce 21.6 MW (29,040 hp). The regeneration also reduces the exhaust temperature from 600°C (1100°F) to 350°C (650°F). Air is compressed to 3 atm before it enters the intercooler. Compared to steam-turbine and diesel-propulsion systems, the gas turbine offers greater power for a given size and weight, high reliability, long life, and more convenient operation. The engine start-up time has been reduced from 4 h required for a typical steam-propulsion system to less than 2 min for a gas turbine. Many modern marine propulsion systems use gas turbines together with diesel engines because of the high fuel consumption of simple-cycle gas-turbine engines. In combined diesel and gas-turbine systems, diesel is used to provide for efficient low-power and cruise operation, and gas turbine is used when high speeds are needed.

In gas-turbine power plants, the ratio of the compressor work to the turbine work, called the **back work ratio,** is very high (Fig. 8-34). Usually more than one-half of the turbine work output is used to drive the compressor. The situation is even worse when the adiabatic efficiencies of the compressor and the turbine are low. This is quite in contrast to steam power plants, where the back work ratio is only a few percent. This is not surprising, however,

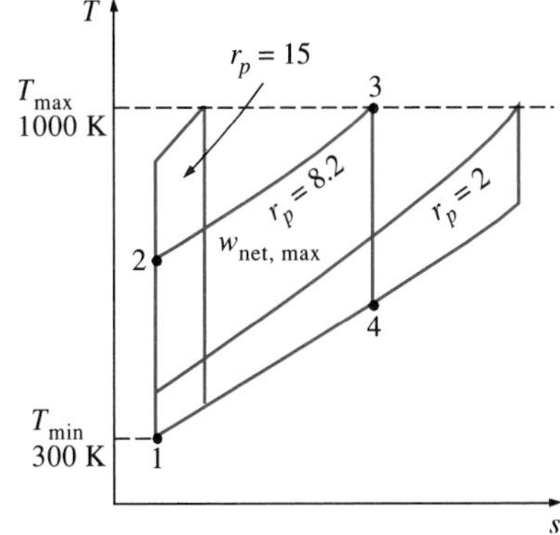

FIGURE 8-33

For fixed values of T_{min} and T_{max}, the net work of the Brayton cycle first increases with the pressure ratio, then reaches a maximum at $r_p = (T_{max}/T_{min})^{k/[2(k-1)]}$, and finally decreases.

FIGURE 8-34

The fraction of the turbine work used to drive the compressor is called the back work ratio.

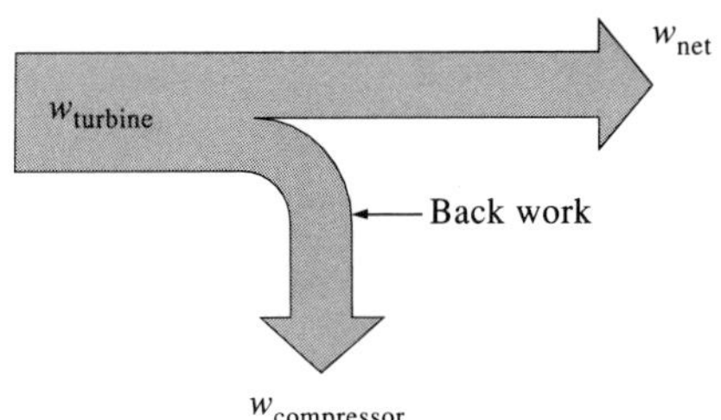

since a liquid is compressed in steam power plants instead of a gas, and the reversible steady-flow work is proportional to the specific volume of the working fluid.

A power plant with a high back work ratio requires a larger turbine to provide the additional power requirements of the compressor. Therefore, the turbines used in gas-turbine power plants are larger than those used in steam power plants of the same net power output.

Development of Gas Turbines

The gas turbine has experienced phenomenal progress and growth since its first successful development in the 1930s. The early gas turbines built in the 1940s and even 1950s had simple-cycle efficiencies of about 17 percent because of the low compressor and turbine efficiencies and low turbine inlet temperatures due to metallurgical limitations of those times. Therefore, gas turbines found only limited use despite their versatility and their ability to burn a variety of fuels. The efforts to improve the cycle efficiency concentrated in three areas:

1. **Increasing the turbine inlet (or firing) temperatures** This has been the primary approach taken to improve gas-turbine efficiency. The turbine inlet temperatures have increased steadily from about 540°C (1000°F) in the 1940s to 1425°C (2600°F) today. These increases were made possible by the development of new materials and the innovative cooling techniques for the critical components such as coating the turbine blades with ceramic layers and cooling the blades with the discharge air from the compressor. Maintaining high turbine inlet temperatures with air-cooling technique requires the combustion temperature to be higher to compensate for the cooling effect of the cooling air. But higher combustion temperatures increase the amount of nitrogen oxides (NO_x), which are responsible for the formation of ozone at ground level and smog. Using steam as the coolant allowed an increase in the turbine inlet temperatures by 200°F without an increase in the combustion temperature. Steam is also a much more effective heat transfer medium than air.

2. **Increasing the efficiencies of turbo-machinery components** The performance of early turbines suffered greatly from the inefficiencies of turbines and compressors. But the advent of computers and advanced techniques for computer-aided design made it possible to design these components aerodynamically with minimal losses. The increased efficiencies of the turbines and compressors resulted in a significant increase in the cycle efficiency.

3. **Adding modifications to the basic cycle** The simple-cycle efficiencies of early gas turbines were practically doubled by incorporating intercooling, regeneration (or recuperation), and reheating, discussed in the next two sections. These improvements, of course, come at the expense of increased initial and operation costs, and they cannot be justified unless the decrease in fuel costs offsets the increase in other costs. The relatively low fuel prices, the

general desire in the industry to minimize installation costs, and the tremendous increase in the simple-cycle efficiency to about 40 percent left little desire for opting for these modifications.

The first gas turbine for an electric utility was installed in 1949 in Oklahoma as part of a combined-cycle power plant. It was built by General Electric and produced 3.5 MW of power. Gas turbines installed until the mid-1970s suffered from low efficiency and poor reliability. In the past, the base-load electric power generation was dominated by large coal and nuclear power plants. However, there has been a historic shift toward natural gas–fired gas turbines because of their higher efficiencies, lower capital costs, shorter installation times, and better emission characteristics, and the abundance of natural gas supplies, and more and more electric utilities are using gas turbines for base-load power production as well as for peaking. The construction costs for gas-turbine power plants are roughly half that of comparable conventional fossil-fuel steam power plants, which were the primary base-load power plants until the early 1980s. More than half of all power plants to be installed in the foreseeable future are forecast to be gas-turbine or combined gas–steam turbine types.

A gas turbine manufactured by General Electric in the early 1990s had a pressure ratio of 13.5 and generated 135.7 MW of net power at a thermal efficiency of 33 percent in simple-cycle operation. A more recent gas turbine manufactured by General Electric uses a turbine inlet temperature of 1425°C (2600°F) and produces up to 282 MW while achieving a thermal efficiency of 39.5 percent in the simple-cycle mode. A 1.3-ton small-scale gas turbine labeled OP-16, built by the Dutch firm Opra Optimal Radial Turbine, can run on gas or liquid fuel and can replace a 16-ton diesel engine. It has a pressure ratio of 6.5 and produces up to 2 MW of power. Its efficiency is 26 percent in the simple-cycle operation, which rises to 37 percent when equipped with a regenerator.

EXAMPLE 8-5 The Simple Ideal Brayton Cycle

A stationary power plant operating on an ideal Brayton cycle has a pressure ratio of 8. The gas temperature is 300 K at the compressor inlet and 1300 K at the turbine inlet. Utilizing the air-standard assumptions, determine (*a*) the gas temperature at the exits of the compressor and the turbine, (*b*) the back work ratio, and (*c*) the thermal efficiency.

Solution The *T-s* diagram of the ideal Brayton cycle described is shown in Fig. 8-35. We note that the components involved in the Brayton cycle are steady-flow devices.

Assumptions **1** Steady operating conditions exist. **2** The air-standard assumptions are applicable. **3** Kinetic and potential energy changes are negligible. **4** The variation of specific heats with temperature is to be accounted for.

Analysis (*a*) The air temperatures at the compressor and turbine exits are determined by applying the energy equation to processes 1-2 and 3-4:

FIGURE 8-35

T-s diagram for the Brayton cycle discussed in Example 8-5.

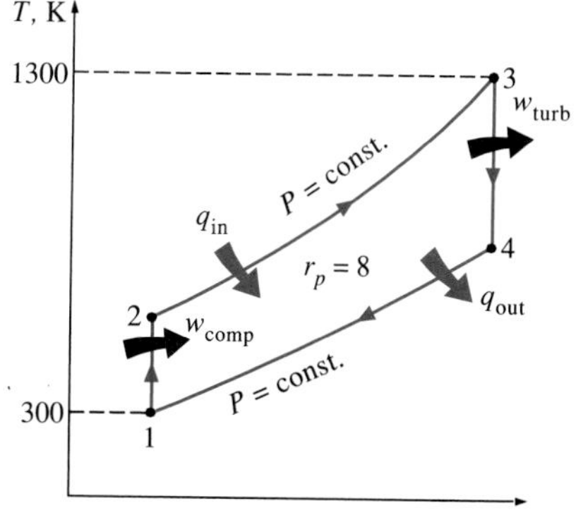

Process 1-2 (isentropic compression of an ideal gas);

$$T_1 = 300 \text{ K} \longrightarrow h_1 = 300.19 \text{ kJ/kg}$$
$$P_{r1} = 1.386$$

$$P_{r2} = \frac{P_2}{P_1} P_{r1} = (8)(1.386) = 11.09 \longrightarrow T_2 = \mathbf{540\ K} \quad \text{(at compressor exit)}$$
$$h_2 = 544.35 \text{ kJ/kg}$$

Process 3-4 (isentropic expansion of an ideal gas):

$$T_3 = 1300 \text{ K} \longrightarrow h_3 = 1395.97 \text{ kJ/kg}$$
$$P_{r3} = 330.9$$

$$P_{r4} = \frac{P_4}{P_3} P_{r3} = \left(\frac{1}{8}\right)(330.9) = 41.36 \longrightarrow T_4 = \mathbf{770\ K} \quad \text{(at turbine exit)}$$
$$h_4 = 789.11 \text{ kJ/kg}$$

(*b*) To find the back work ratio, we need to find the work input to the compressor and the work output of the turbine:

$$w_{\text{comp, in}} = h_2 - h_1 = 544.35 - 300.19 = 244.16 \text{ kJ/kg}$$
$$w_{\text{turb, out}} = h_3 - h_4 = 1395.97 - 789.11 = 606.86 \text{ kJ/kg}$$

Thus, $$\text{Back work ratio } r_{bw} = \frac{w_{\text{comp, in}}}{w_{\text{turb, out}}} = \frac{244.16 \text{ kJ/kg}}{606.86 \text{ kJ/kg}} = \mathbf{0.402}$$

That is, 40.2 percent of the turbine work output is used just to drive the compressor.

(*c*) The thermal efficiency of the cycle is the ratio of the net power output to the total heat input:

$$q_{\text{in}} = h_3 - h_2 = 1395.97 - 544.35 = 851.62 \text{ kJ/kg}$$
$$w_{\text{net}} = w_{\text{out}} - w_{\text{in}} = 606.86 - 244.16 = 362.7 \text{ kJ/kg}$$

Thus, $$\eta_{\text{th}} = \frac{w_{\text{net}}}{q_{\text{in}}} = \frac{362.7 \text{ kJ/kg}}{851.62 \text{ kJ/kg}} = \mathbf{0.426 \text{ or } 42.6\%}$$

The thermal efficiency could also be determined from

$$\eta_{\text{th}} = 1 - \frac{q_{\text{out}}}{q_{\text{in}}}$$

where $$q_{\text{out}} = h_4 - h_1 = 789.11 - 300.19 = 488.92 \text{ kJ/kg}$$

Discussion Under the cold-air-standard assumptions (constant specific heats, values at room temperature), the thermal efficiency would be, from Eq. 8-17,

$$\eta_{\text{th, Brayton}} = 1 - \frac{1}{r_p^{(k-1)/k}} = 1 - \frac{1}{8^{(1.4-1)/1.4}} = \mathbf{0.448}$$

which is sufficiently close to the value obtained by accounting for the variation of specific heats with temperature.

Deviation of Actual Gas-Turbine Cycles from Idealized Ones

The actual gas-turbine cycle differs from the ideal Brayton cycle on several accounts. For one thing, some pressure drop during the heat-addition and rejection processes is inevitable. More importantly, the actual work input to the compressor will be more, and the actual work output from the turbine will be less because of irreversibilities. The deviation of actual compressor and turbine behavior from the idealized isentropic behavior can be accurately accounted for by utilizing the adiabatic efficiencies of the turbine and compressor, defined as

$$\eta_C = \frac{w_s}{w_a} \cong \frac{h_1 - h_{2s}}{h_1 - h_{2a}} \qquad (8\text{-}19)$$

and

$$\eta_T = \frac{w_a}{w_s} \cong \frac{h_3 - h_{4a}}{h_3 - h_{4s}} \qquad (8\text{-}20)$$

where states $2a$ and $4a$ are the actual exit states of the compressor and the turbine, respectively, and $2s$ and $4s$ are the corresponding states for the isentropic case, as illustrated in Fig. 8-36. The effect of the turbine and compressor efficiencies on the thermal efficiency of the gas-turbine engines is illustrated below with an example.

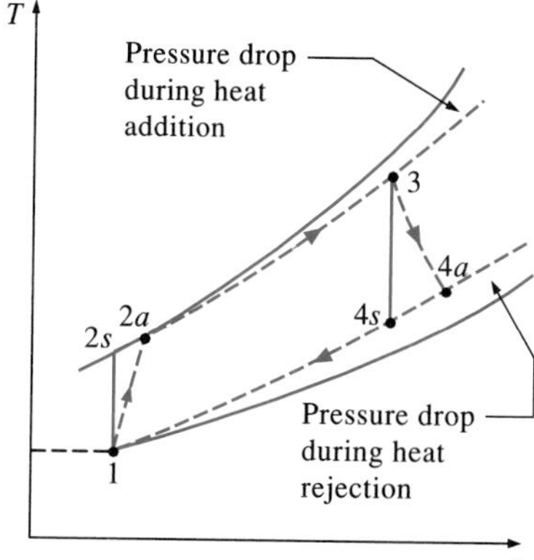

FIGURE 8-36

The deviation of an actual gas-turbine cycle from the ideal Brayton cycle as a result of irreversibilities.

EXAMPLE 8-6 An Actual Gas-Turbine Cycle

Assuming a compressor efficiency of 80 percent and a turbine efficiency of 85 percent, determine (*a*) the back work ratio, (*b*) the thermal efficiency, and (*c*) the turbine exit temperature of the gas-turbine power plant discussed in Example 8-5.

Solution (*a*) The *T-s* diagram of the cycle is shown in Fig. 8-37. The actual compressor work and turbine work are determined by using the definitions of compressor and turbine efficiencies, Eqs. 8-19 and 8-20:

Compressor: $$w_{\text{comp, in}} = \frac{w_s}{\eta_C} = \frac{244.16 \text{ kJ/kg}}{0.80} = -305.20 \text{ kJ/kg}$$

Turbine: $$w_{\text{turb, out}} = \eta_T w_s = (0.85)(606.86 \text{ kJ/kg}) = 515.83 \text{ kJ/kg}$$

Thus, $$r_{bw} = \frac{w_{\text{comp, in}}}{w_{\text{turb, out}}} = \frac{305.20 \text{ kJ/kg}}{515.83 \text{ kJ/kg}} = \mathbf{0.592}$$

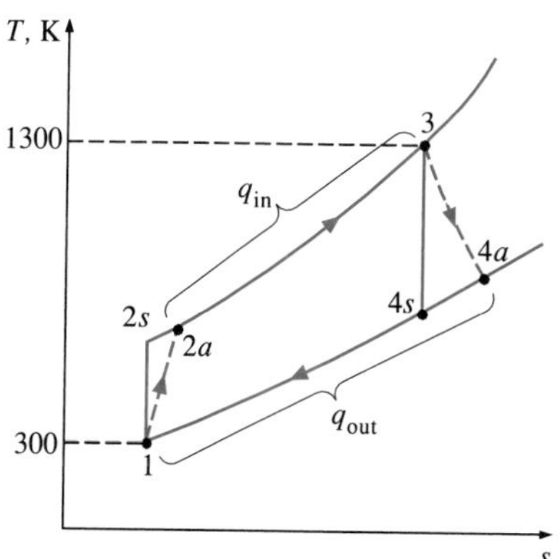

FIGURE 8-37

T-s diagram of the gas-turbine cycle discussed in Example 8-6.

That is, the compressor is now consuming 59.2 percent of the work produced by the turbine (up from 40.2 percent). This increase is due to the irreversibilities that occur within the compressor and the turbine.

(*b*) In this case, air will leave the compressor at a higher temperature and enthalpy, which are determined to be

$$w_{\text{comp, in}} = h_{2a} - h_1 \longrightarrow h_{2a} = h_1 + w_{\text{comp, in}}$$
$$= 300.19 + 305.20$$
$$= 605.39 \text{ kJ/kg} \qquad (\text{and } T_{2a} = 598 \text{ K})$$

Thus, $$q_{\text{in}} = h_3 - h_{2a} = 1395.97 - 605.39 = 790.58 \text{ kJ/kg}$$
$$w_{\text{net}} = w_{\text{out}} - w_{\text{in}} = 515.83 - 305.20 = 210.63 \text{ kJ/kg}$$

and $$\eta_{\text{th}} = \frac{w_{\text{net}}}{q_{\text{in}}} = \frac{210.63 \text{ kJ/kg}}{790.58 \text{ kJ/kg}} = \mathbf{0.266 \text{ or } 26.6\%}$$

That is, the irreversibilities occurring within the turbine and compressor caused the thermal efficiency of the gas turbine cycle to drop from 42.6 to 26.6 percent. This example shows how sensitive the performance of a gas-turbine power plant is to the efficiencies of the compressor and the turbine. In fact, gas-turbine thermal efficiencies did not reach competitive values until significant improvements were made in the design of gas turbines and compressors.

(*c*) The air temperature at the turbine exit is determined from an energy balance on the turbine:

$$w_{\text{turb}, a} = h_3 - h_{4a} \longrightarrow h_{4a} = h_3 - w_{\text{turb}, a}$$
$$= 1395.97 - 515.83$$
$$= 880.14 \text{ kJ/kg}$$

Then, from Table A-17,

$$T_{4a} = \mathbf{853 \text{ K}}$$

This value is considerably higher than the air temperature at the compressor exit (T_{2a} = 598 K), which suggests the use of regeneration to reduce the heat input requirements.

8-9 ■ THE BRAYTON CYCLE WITH REGENERATION

In gas-turbine engines, the temperature of the exhaust gas leaving the turbine is often considerably higher than the temperature of the air leaving the compressor. Therefore, the high-pressure air leaving the compressor can be heated by transferring heat to it from the hot exhaust gases in a counter-flow heat exchanger, which is also known as a *regenerator* or a *recuperator.* A sketch of the gas-turbine engine utilizing a regenerator and the *T-s* diagram of the new cycle are shown in Figs. 8-38 and 8-39, respectively.

The thermal efficiency of the Brayton cycle increases as a result of regeneration since the portion of energy of the exhaust gases that is normally rejected to the surroundings is now used to preheat the air entering the combustion chamber. This, in turn, decreases the heat input (thus fuel) requirements for the same net work output. Note, however, that the use of a regenerator is recommended only when the turbine exhaust temperature is higher than the compressor exit temperature. Otherwise, heat will flow in the reverse direction (*to* the exhaust gases), decreasing the efficiency. This situation is encountered in gas-turbine engines operating at very high pressure ratios.

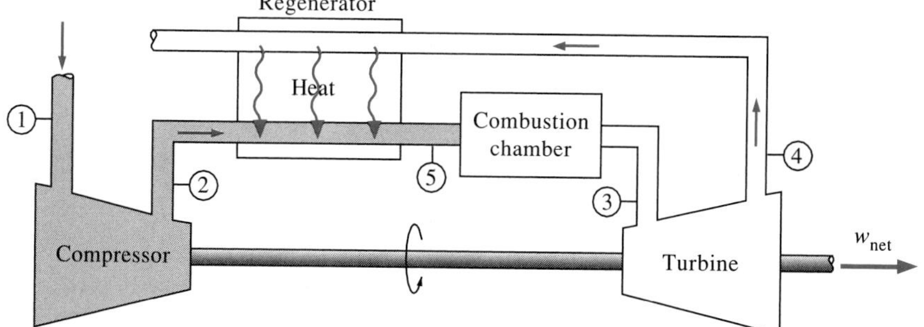

FIGURE 8-38
A gas-turbine engine with regenerator.

The highest temperature occurring within the regenerator is T_4, the temperature of the exhaust gases leaving the turbine and entering the regenerator. Under no conditions can the air be preheated in the regenerator to a temperature above this value. Air normally leaves the regenerator at a lower temperature, T_5. In the limiting (ideal) case, the air will exit the regenerator at the inlet temperature of the exhaust gases T_4. Assuming the regenerator to be well insulated and any changes in kinetic and potential energies to be negligible, the actual and maximum heat transfers from the exhaust gases to the air can be expressed as

$$q_{\text{regen, act}} = h_5 - h_2 \quad (8\text{-}21)$$

and

$$q_{\text{regen, max}} = h_{5'} - h_2 = h_4 - h_2 \quad (8\text{-}22)$$

The extent to which a regenerator approaches an ideal regenerator is called the **effectiveness** ε and is defined as

$$\varepsilon = \frac{q_{\text{regen, act}}}{q_{\text{regen, max}}} = \frac{h_5 - h_2}{h_4 - h_2} \quad (8\text{-}23)$$

When the cold-air-standard assumptions are utilized, it reduces to

$$\varepsilon \cong \frac{T_5 - T_2}{T_4 - T_2} \quad (8\text{-}24)$$

A regenerator with a higher effectiveness will obviously save a greater amount of fuel since it will preheat the air to a higher temperature prior to combustion. However, achieving a higher effectiveness requires the use of a larger regenerator, which carries a higher price tag and causes a larger pressure drop. Therefore, the use of a regenerator with a very high effectiveness cannot be justified economically unless the savings from the fuel costs exceed the additional expenses involved. The effectiveness of most regenerators used in practice is below 0.85.

Under the cold-air-standard assumptions, the thermal efficiency of an ideal Brayton cycle with regeneration is

$$\eta_{\text{th, regen}} = 1 - \left(\frac{T_1}{T_3}\right)(r_p)^{(k-1)/k} \quad (8\text{-}25)$$

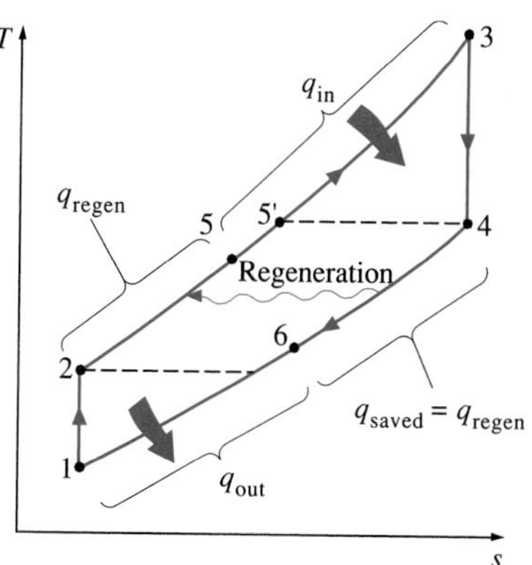

FIGURE 8-39
T-s diagram of a Brayton cycle with regeneration.

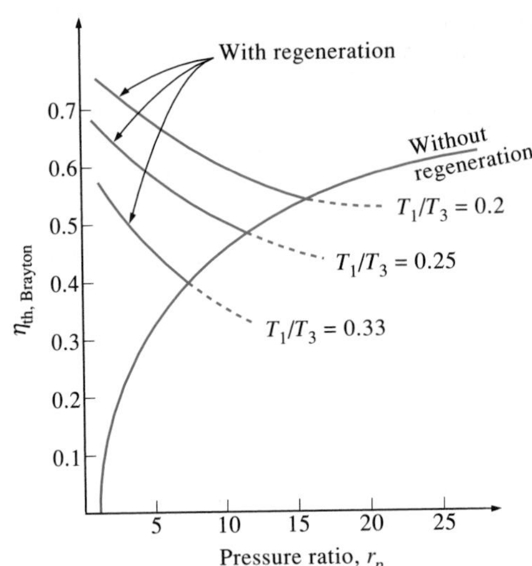

FIGURE 8-40

Thermal efficiency of the ideal Brayton cycle with and without regeneration.

Therefore, the thermal efficiency of an ideal Brayton cycle with regeneration depends on the ratio of the minimum to maximum temperatures as well as the pressure ratio. The thermal efficiency is plotted in Fig. 8-40 for various pressure ratios and minimum-to-maximum temperature ratios. This figure shows that regeneration is most effective at lower pressure ratios and low minimum-to-maximum temperature ratios.

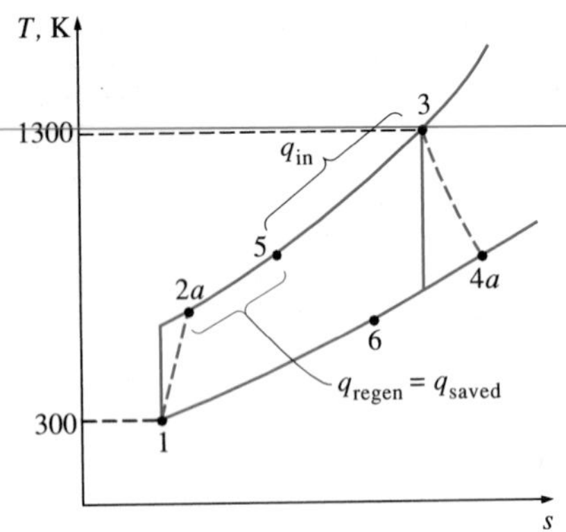

FIGURE 8-41

T-s diagram of the regenerative Brayton cycle described in Example 8-7.

EXAMPLE 8-7 Actual Gas-Turbine Cycle with Regeneration

Determine the thermal efficiency of the gas-turbine power plant described in Example 8-6 if a regenerator having an effectiveness of 80 percent is installed.

Solution The *T-s* diagram of the cycle is shown in Fig. 8-41. We first determine the enthalpy of the air at the exit of the regenerator, using the definition of effectiveness:

$$\varepsilon = \frac{h_5 - h_{2a}}{h_{4a} - h_{2a}}$$

$$0.80 = \frac{(h_5 - 605.39)\ \text{kJ/kg}}{(880.14 - 605.39)\ \text{kJ/kg}} \longrightarrow h_5 = 825.19\ \text{kJ/kg}$$

Thus, $\quad q_{in} = h_3 - h_5 = (1395.97 - 825.19)\ \text{kJ/kg} = 570.78\ \text{kJ/kg}$

This represents a savings of 219.8 kJ/kg from the heat input requirements. The addition of a regenerator (assumed to be frictionless) does not affect the net work output of the plant. Thus,

$$\eta_{th} = \frac{w_{net}}{q_{in}} = \frac{210.63\ \text{kJ/kg}}{570.78\ \text{kJ/kg}} = \mathbf{0.369\ or\ 36.9\%}$$

Discussion Note that the thermal efficiency of the power plant has gone up from 26.6 to 36.9 percent as a result of installing a regenerator that helps to recuperate some of the excess energy of the exhaust gases.

8-10 ■ THE BRAYTON CYCLE WITH INTERCOOLING, REHEATING, AND REGENERATION

The net work of a gas-turbine cycle is the difference between the turbine work output and the compressor work input, and it can be increased by either decreasing the compressor work or increasing the turbine work, or both. It was shown in Chap. 6 that the work required to compress a gas between two specified pressures can be decreased by carrying out the compression process in stages and cooling the gas in between (Fig. 8-42)—that is, using *multistage compression with intercooling.* As the number of stages is increased, the compression process becomes nearly isothermal at the compressor inlet temperature, and the compression work decreases.

Likewise, the work output of a turbine operating between two pressure levels can be increased by expanding the gas in stages and reheating it in between—that is, utilizing *multistage expansion with reheating.* This is accomplished without raising the maximum temperature in the cycle. As the number of stages is increased, the expansion process becomes nearly isothermal. The foregoing argument is based on a simple principle: The steady-flow compression or expansion work is proportional to the specific volume of the fluid. Therefore, the specific volume of the working fluid should be as low as possible during a compression process and as high as possible during an expansion process. This is precisely what intercooling and reheating accomplish.

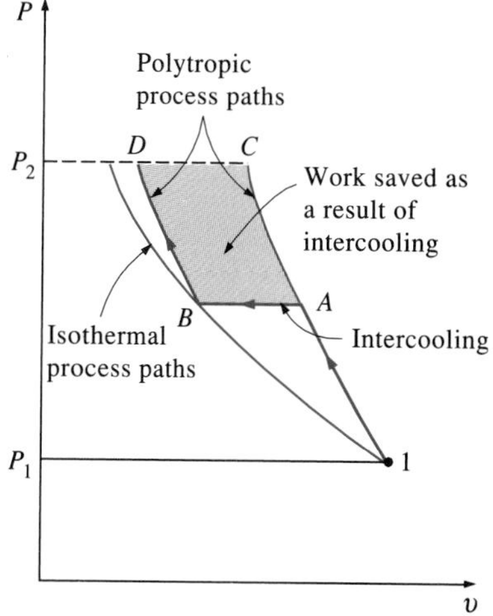

FIGURE 8-42
Comparison of work inputs to a single-stage compressor (1*AC*) and a two-stage compressor with intercooling (1*ABD*).

Combustion in gas turbines typically occurs at four times the amount of air needed for complete combustion to avoid excessive temperatures. Therefore, the exhaust gases are rich in oxygen, and reheating can be accomplished by simply spraying additional fuel into the exhaust gases between two expansion states.

The working fluid leaves the compressor at a lower temperature, and the turbine at a higher temperature, when intercooling and reheating are utilized. This makes regeneration more attractive since a greater potential for regeneration exists. Also, the gases leaving the compressor can be heated to a higher temperature before they enter the combustion chamber because of the higher temperature of the turbine exhaust.

A schematic of the physical arrangement and the *T-s* diagram of an ideal two-stage gas-turbine cycle with intercooling, reheating, and regeneration are shown in Figs. 8-43 and 8-44. The gas enters the first stage of the compressor at state 1, is compressed isentropically to an intermediate pressure P_2, is cooled at constant pressure to state 3 ($T_3 = T_1$), and is compressed in the second stage isentropically to the final pressure P_4. At state 4 the gas enters the regenerator, where it is heated to T_5 at constant pressure. In an ideal regenerator, the gas will leave the regenerator at the temperature of the turbine exhaust, that is, $T_5 = T_9$. The primary heat addition (or combustion) process takes place between states 5 and 6. The gas enters the first stage of the turbine at state 6 and expands isentropically to state 7, where it enters the reheater. It is reheated at constant pressure to state 8 ($T_8 = T_6$), where it enters the second stage of the turbine. The gas exits the turbine at state 9 and enters the regenerator, where it is cooled to state 10 at constant pressure. The cycle

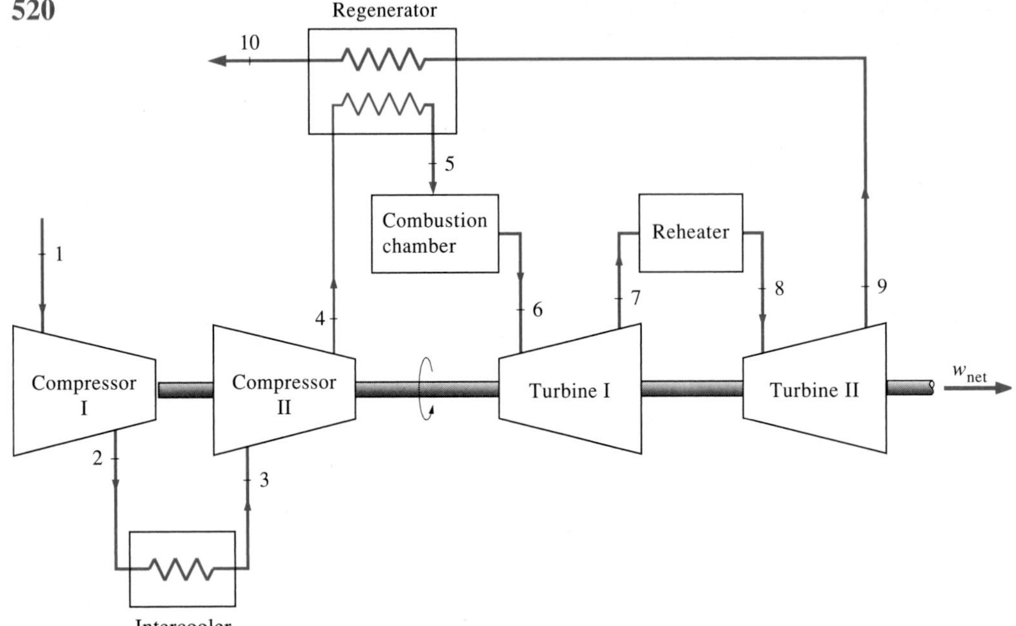

FIGURE 8-43

A gas-turbine engine with two-stage compression with intercooling, two-stage expansion with reheating, and regeneration.

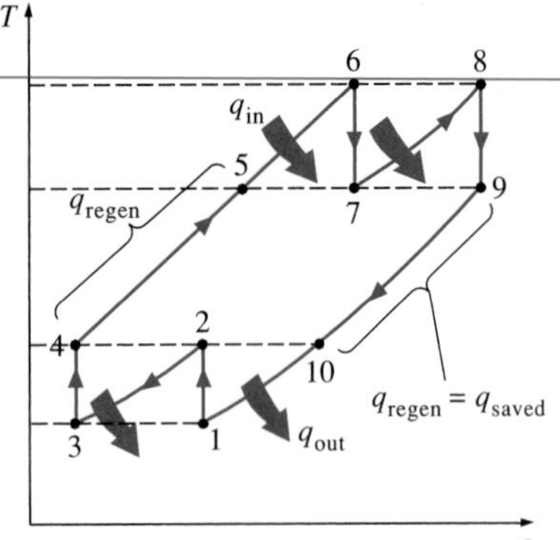

FIGURE 8-44

T-s diagram of an ideal gas-turbine cycle with intercooling, reheating, and regeneration.

is completed by cooling the gas to the initial state (or purging the exhaust gases).

It was shown in Chap. 6 that the work input to a two-stage compressor is minimized when equal pressure ratios are maintained across each stage. It can be shown that this procedure also maximizes the turbine work output. Thus, for optimum operation we have

$$\frac{P_2}{P_1} = \frac{P_4}{P_3} \quad \text{and} \quad \frac{P_6}{P_7} = \frac{P_8}{P_9} \tag{8-26}$$

In the analysis of the actual gas-turbine cycles, the irreversibilities that are present within the compressor, the turbine, and the regenerator as well as the pressure drops in the heat exchangers should be taken into consideration.

The back work ratio of a gas-turbine cycle improves as a result of intercooling and reheating. However, this does not mean that the thermal efficiency will also improve. The fact is, intercooling and reheating will always decrease the thermal efficiency unless they are accompanied by regeneration. This is because intercooling decreases the average temperature at which heat is added, and reheating increases the average temperature at which heat is rejected. This is also apparent from Fig. 8-44. Therefore, in gas-turbine power plants, intercooling and reheating are always used in conjunction with regeneration.

If the number of compression and expansion stages is increased, the ideal gas-turbine cycle with intercooling, reheating, and regeneration will approach the Ericsson cycle, as illustrated in Fig. 8-45, and the thermal efficiency will approach the theoretical limit (the Carnot efficiency). However, the contribution of each additional stage to the thermal efficiency is less and less, and the use of more than two or three stages cannot be justified economically.

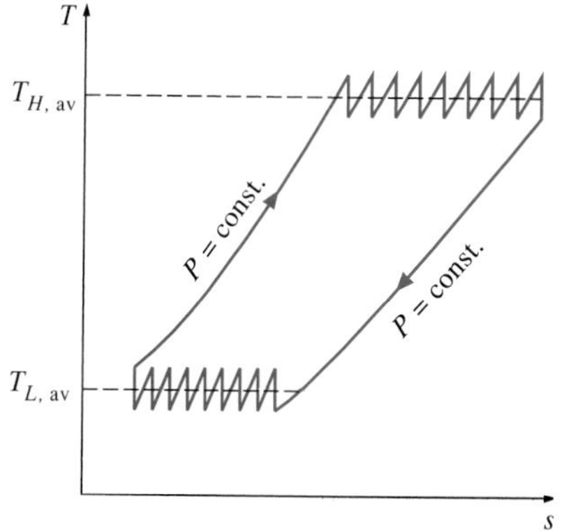

FIGURE 8-45

As the number of compression and expansion stages increases, the gas-turbine cycle with intercooling, reheating, and regeneration approaches the Ericsson cycle.

EXAMPLE 8-8 A Gas Turbine with Reheating and Intercooling

An ideal gas-turbine cycle with two stages of compression and two stages of expansion has an overall pressure ratio of 8. Air enters each stage of the compressor at 300 K and each stage of the turbine at 1300 K. Determine the back work ratio and the thermal efficiency of this gas-turbine cycle, assuming (*a*) no regenerators and (*b*) an ideal regenerator with 100 percent effectiveness. Compare the results with those obtained in Example 8-5.

Solution The *T-s* diagram of the ideal gas turbine cycle described is shown in Fig. 8-46. We note that the cycle involves two stages of expansion, two stages of compression, and regeneration.

Assumptions **1** Steady operating conditions exist. **2** The air-standard assumptions are applicable. **3** Kinetic and potential energy changes are negligible.

Analysis For two-stage compression and expansion, the work input is minimized and the work output is maximized when both stages of the compressor and the turbine have the same pressure ratio, as shown in Chap. 6. Thus,

$$\frac{P_2}{P_1} = \frac{P_4}{P_3} = \sqrt{8} = 2.83 \qquad \text{and} \qquad \frac{P_6}{P_7} = \frac{P_8}{P_9} = \sqrt{8} = 2.83$$

Air enters each stage of the compressor at the same temperature, and each stage has the same adiabatic efficiency (100 percent in this case). Therefore, the temperature (and enthalpy) of the air at the exit of each compression stage will be the same. A similar argument can be given for the turbine. Thus,

At inlets: $T_1 = T_3,\quad h_1 = h_3 \quad \text{and} \quad T_6 = T_8,\quad h_6 = h_8$

At exits: $T_2 = T_4,\quad h_2 = h_4 \quad \text{and} \quad T_7 = T_9,\quad h_7 = h_9$

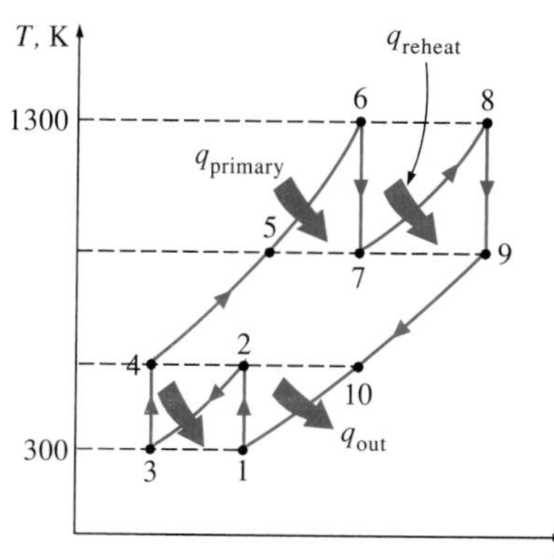

FIGURE 8-46

T-s diagram of the gas-turbine cycle discussed in Example 8-8.

Under these conditions, the work input to each stage of the compressor will be the same, and so will the work output from each stage of the turbine.

(*a*) In the absence of any regeneration, the back work ratio and the thermal efficiency are determined by using data from Table A-17 as follows:

$$T_1 = 300 \text{ K} \longrightarrow h_1 = 300.19 \text{ kJ/kg}$$
$$P_{r1} = 1.386$$

$$P_{r2} = \frac{P_2}{P_1} P_{r1} = \sqrt{8}\,(1.386) = 3.92 \longrightarrow T_2 = 403.3 \text{ K}$$
$$h_2 = 404.33 \text{ kJ/kg}$$

$$T_6 = 1300 \text{ K} \longrightarrow h_6 = 1395.97 \text{ kJ/kg}$$
$$P_{r6} = 330.9$$

$$P_{r7} = \frac{P_7}{P_6} P_{r6} = \frac{1}{\sqrt{8}}(330.9) = 117.0 \longrightarrow T_7 = 1006.4 \text{ K}$$
$$h_7 = 1053.35 \text{ kJ/kg}$$

Then

$$w_{\text{comp, in}} = 2(w_{\text{comp, in, I}}) = 2(h_2 - h_1) = 2(404.33 - 300.19) = 208.28 \text{ kJ/kg}$$
$$w_{\text{turb, out}} = 2(w_{\text{turb, out, I}}) = 2(h_6 - h_7) = 2(1395.97 - 1053.35) = 685.24 \text{ kJ/kg}$$
$$w_{\text{net}} = w_{\text{turb, out}} - w_{\text{comp, in}} = 685.24 - 208.28 = 476.96 \text{ kJ/kg}$$
$$q_{\text{in}} = q_{\text{primary}} + q_{\text{reheat}} = (h_6 - h_4) + h_8 - h_7)$$
$$= (1395.97 - 404.33) + (1395.97 - 1053.35) = 1334.19 \text{ kJ/kg}$$

Thus, $$r_{\text{bw}} = \frac{w_{\text{comp, in}}}{w_{\text{turb, out}}} = \frac{208.28 \text{ kJ/kg}}{685.24 \text{ kJ/kg}} = \mathbf{0.304 \text{ or } 30.4\%}$$

and $$\eta_{\text{th}} = \frac{w_{\text{net}}}{q_{\text{in}}} = \frac{476.96 \text{ kJ/kg}}{1334.19 \text{ kJ/kg}} = \mathbf{0.357 \text{ or } 35.7\%}$$

A comparison of these results with those obtained in Example 8-5 (single-stage compression and expansion) reveals that multistage compression with intercooling and multistage expansion with reheating improve the back work ratio (it drops from 40.2 to 30.4 percent) but hurt the thermal efficiency (it drops from 42.6 to 35.7 percent). Therefore, intercooling and reheating are not recommended in gas-turbine power plants unless they are accompanied by regeneration.

(*b*) The addition of an ideal regenerator (no pressure drops, 100 percent effectiveness) does not affect the compressor work and the turbine work. Therefore, the net work output and the back work ratio of an ideal gas-turbine cycle will be identical whether there is a regenerator or not. A regenerator, however, reduces the heat input requirements by preheating the air leaving the compressor, using the hot exhaust gases leaving the turbine. In an ideal regenerator, the compressed air is heated to the turbine exit temperature T_9 before it enters the combustion chamber. Thus, under the air-standard assumptions, $h_5 = h_7 = h_9$.

The heat input and the thermal efficiency in this case are.

$$q_{\text{in}} = q_{\text{primary}} + q_{\text{reheat}} = (h_6 - h_5) + (h_8 - h_7)$$
$$= (1395.97 - 1053.35) + (1395.97 - 1053.35) = 685.24 \text{ kJ/kg}$$

and $$\eta_{\text{th}} = \frac{w_{\text{net}}}{q_{\text{in}}} = \frac{476.96 \text{ kJ/kg}}{685.24 \text{ kJ/kg}} = \mathbf{0.696 \text{ or } 69.6\%}$$

That is, the thermal efficiency almost doubles as a result of regeneration compared to the no-regeneration case. The overall effect of two-stage compression and expansion with intercooling, reheating, and regeneration on the thermal efficiency is an increase of 63 percent. As the number of compression and expansion stages is increased, the cycle will approach the Ericsson cycle, and the thermal efficiency will approach

$$\eta_{\text{th, Ericsson}} = \eta_{\text{th, Carnot}} = 1 - \frac{T_L}{T_H} = 1 - \frac{300 \text{ K}}{1300 \text{ K}} = 0.769$$

Adding a second stage increases the thermal efficiency from 42.6 to 69.6 percent, an increase of 27 percentage points. This is a significant increase in efficiency, and usually it is well worth the extra cost associated with the second stage. Adding more stages, however (no matter how many), can increase the efficiency an additional 7.3 percentage points at most, and usually cannot be justified economically.

8-11 ■ IDEAL JET-PROPULSION CYCLES

Gas-turbine engines are widely used to power aircraft because they are light and compact and have a high power-to-weight ratio. Aircraft gas turbines operate on an open cycle called a **jet-propulsion cycle.** The ideal jet-propulsion cycle differs from the simple ideal Brayton cycle in that the gases are not expanded to the ambient pressure in the turbine. Instead, they are expanded to a pressure such that the power produced by the turbine is just sufficient to drive the compressor and the auxiliary equipment, such as a small generator and hydraulic pumps. That is, the net work output of a jet-propulsion cycle is zero. The gases that exit the turbine at a relatively high pressure are subsequently accelerated in a nozzle to provide the thrust to propel the aircraft (Fig. 8-47). Also, aircraft gas turbines operate at higher pressure ratios (typically between 10 and 25), and the fluid passes through a diffuser first, where it is decelerated and its pressure is increased before it enters the compressor.

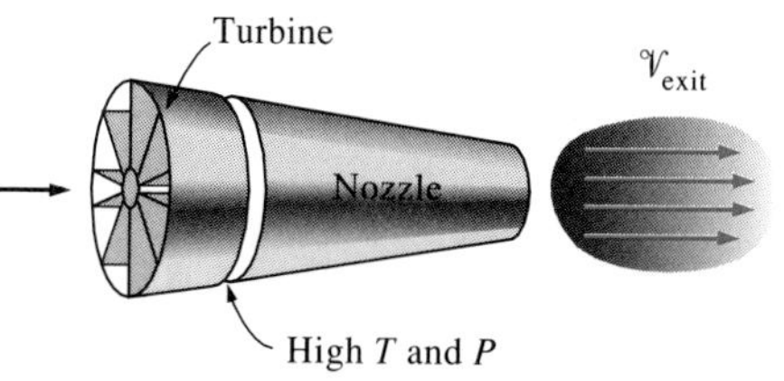

FIGURE 8-47
In jet engines, the high-temperature and high-pressure gases leaving the turbine are accelerated in a nozzle to provide thrust.

Aircraft are propelled by accelerating a fluid in the opposite direction to motion. This is accomplished by either slightly accelerating a large mass of fluid (*propeller-driven engine*) or greatly accelerating a small mass of fluid (*jet* or *turbojet engine*) or both (*turboprop engine*).

A schematic of a turbojet engine and the *T-s* diagram of the ideal turbojet cycle are shown in Fig. 8-48. The pressure of air rises slightly as it is decelerated in the diffuser. Air is compressed in the compressor. It is mixed with fuel in the combustion chamber, where the mixture is burned at constant pressure. The high-pressure and high-temperature combustion gases partially

FIGURE 8-48
Basic components of a turbojet engine and the *T-s* diagram for the ideal turbojet cycle. [*Source: The Aircraft Gas Turbine Engine and Its Operation*. © United Aircraft Corporation (now United Technologies Corp.), 1951, 1974.]

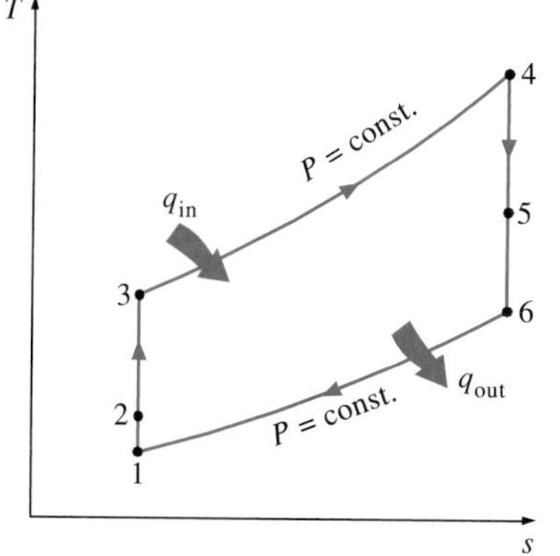

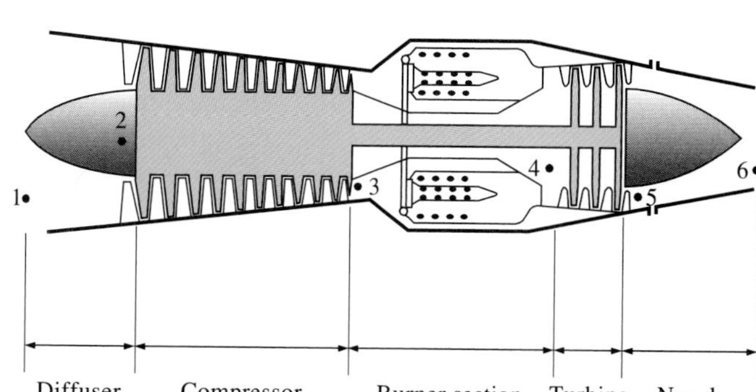

expand in the turbine, producing enough power to drive the compressor and other equipment. Finally, the gases expand in a nozzle to the ambient pressure and leave the aircraft at a high velocity.

In the ideal case, the turbine work is assumed to equal the compressor work. Also, the processes in the diffuser, the compressor, the turbine, and the nozzle are assumed to be isentropic. In the analysis of actual cycles, however, the irreversibilities associated with these devices should be considered. The effect of the irreversibilities is to reduce the thrust that can be obtained from a turbojet engine.

The **thrust** developed in a turbojet engine is the unbalanced force that is caused by the difference in the momentum of the low-velocity air entering the engine and the high-velocity exhaust gases leaving the engine, and it is determined from Newton's second law. The pressures at the inlet and the exit of a turbojet engine are identical (the ambient pressure); thus, the net thrust developed by the engine is

$$F = (\dot{m}\mathcal{V})_{\text{exit}} - (\dot{m}\mathcal{V})_{\text{inlet}} = \dot{m}(\mathcal{V}_{\text{exit}} - \mathcal{V}_{\text{inlet}}) \quad \text{(N)} \tag{8-27}$$

where $\mathcal{V}_{\text{exit}}$ is the exit velocity of the exhaust gases and $\mathcal{V}_{\text{inlet}}$ is the inlet velocity of the air, both relative to the aircraft. Thus, for an aircraft cruising in still air, $\mathcal{V}_{\text{inlet}}$ is the aircraft velocity. In reality, the mass flow rates of the gases at the engine exit and the inlet are different, the difference being equal to the combustion rate of the fuel. But the air–fuel mass ratio used in jet-propulsion engines is usually very high, making this difference very small. Thus, $\dot{m}$ in Eq. 8-27 is taken as the mass flow rate of air through the engine. For an aircraft cruising at a constant speed, the thrust is used to overcome the air drag, and the net force acting on the body of the aircraft is zero. Commercial airplanes save fuel by flying at higher altitudes during long trips since the air at higher altitudes is thinner and exerts a smaller drag force on aircraft.

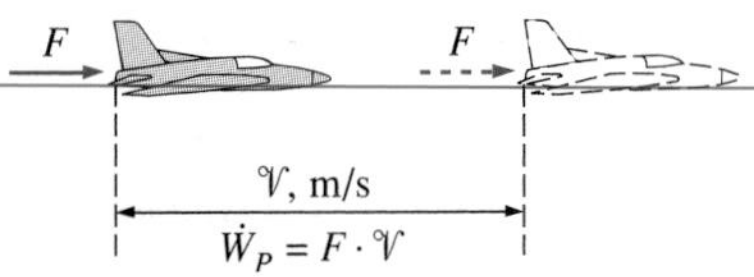

FIGURE 8-49
Propulsive power is the thrust acting on the aircraft through a distance per unit time.

The power developed from the thrust of the engine is called the **propulsive power** $\dot{W}_P$, which is the *propulsive force* (*thrust*) times the *distance* this force acts on the aircraft per unit time, that is, the thrust times the aircraft velocity (Fig. 8-49):

$$\dot{W}_P = (F)\mathcal{V}_{\text{aircraft}} = \dot{m}(\mathcal{V}_{\text{exit}} - \mathcal{V}_{\text{inlet}})\mathcal{V}_{\text{aircraft}} \quad \text{(kW)} \tag{8-28}$$

The net work developed by a turbojet engine is zero. Thus, we cannot define the efficiency of a turbojet engine in the same way as stationary gas-turbine engines. Instead, we should use the general definition of efficiency, which is the ratio of the desired output to the required input. The desired output in a turbojet engine is the *power produced* to propel the aircraft $\dot{W}_P$, and the required input is the *thermal energy of the fuel* released during the combustion process $\dot{Q}_{\text{in}}$. The ratio of these two quantities is called the **propulsive efficiency** and is given by

$$\eta_P = \frac{\text{Propulsive power}}{\text{Energy input rate}} = \frac{\dot{W}_P}{\dot{Q}_{\text{in}}} \tag{8-29}$$

Propulsive efficiency is a measure of how efficiently the energy released during the combustion process is converted to propulsive energy. The remaining part of the energy released will show up as the kinetic energy of the exhaust gases relative to a fixed point on the ground and as an increase in the enthalpy of the air leaving the engine.

EXAMPLE 8-9 The Ideal Jet Propulsion Cycle

A turbojet aircraft flies with a velocity of 850 ft/s at an altitude where the air is at 5 psia and −40°F. The compressor has a pressure ratio of 10, and the temperature of the gases at the turbine inlet is 2000°F. Air enters the compressor at a rate of 100 lbm/s. Utilizing the cold-air-standard assumptions, determine (*a*) the temperature and pressure of the gases at the turbine exit, (*b*) the velocity of the gases at the nozzle exit, and (*c*) the propulsive efficiency of the cycle.

Solution The *T-s* diagram of the ideal jet propulsion cycle described is shown in Fig. 8-50. We note that the components involved in the jet-propulsion cycle are steady-flow devices.

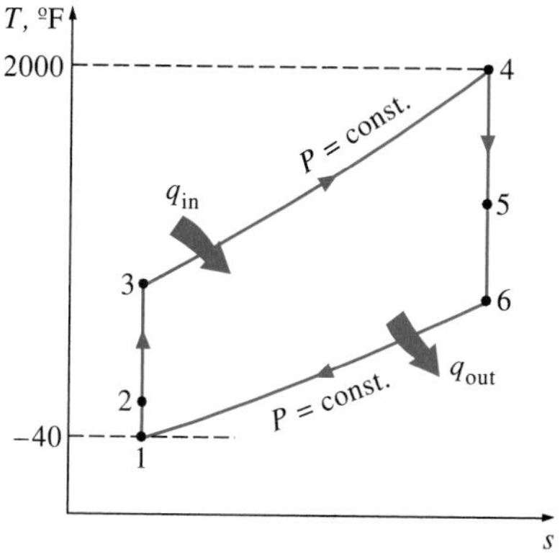

FIGURE 8-50
T-s diagram for the turbojet cycle described in Example 8-9.

Assumptions **1** Steady operating conditions exist. **2** The cold-air-standard assumptions are applicable and thus air can be assumed to have constant specific heats at room temperature [C_p = 0.240 Btu/(lbm · °F) and k = 1.4]. **3** Kinetic and potential energies are negligible, except at the diffuser inlet and the nozzle exit. **4** The turbine work output is equal to the compressor work input.

Analysis (*a*) Before we can determine the temperature and pressure at the turbine exit, we need to find the temperatures and pressures at other states:

Process 1-2 (isentropic compression of an ideal gas in a diffuser): For convenience, we can assume that the aircraft is stationary and the air is moving toward the aircraft at a velocity of $\mathcal{V}_1$ = 850 ft/s. Ideally, the air will leave the diffuser with a negligible velocity ($\mathcal{V}_2 \cong 0$):

$$h_2 + \frac{\mathcal{V}_2^2}{2}^{\nearrow 0} = h_1 + \frac{\mathcal{V}_1^2}{2}$$

$$0 = C_p(T_2 - T_1) - \frac{\mathcal{V}_1^2}{2}$$

$$T_2 = T_1 + \frac{\mathcal{V}_1^2}{2C_p}$$

$$= 420\text{ R} + \frac{(850\text{ ft/s})^2}{2[0.240\text{ Btu/(lbm}\cdot\text{R)}]}\left(\frac{1\text{ Btu/lbm}}{25{,}037\text{ ft}^2/\text{s}^2}\right)$$

$$= 480.1\text{ R}$$

$$P_2 = P_1\left(\frac{T_2}{T_1}\right)^{k/(k-1)} = (5\text{ psia})\left(\frac{480.1\text{ R}}{420\text{ R}}\right)^{1.4/(1.4-1)} = 8.0\text{ psia}$$

Process 2-3 (isentropic compression of an ideal gas in a compressor):

$$P_3 = (r_p)(P_2) = (10)(8.0\text{ psia}) = 80\text{ psia } (= P_4)$$

$$T_3 = T_2\left(\frac{P_3}{P_2}\right)^{(k-1)/k} = (480.1\text{ R})(10)^{(1.4-1)/1.4} = 926.9\text{ R}$$

Process 4-5 (isentropic expansion of an ideal gas in a turbine): Neglecting the kinetic energy changes across the compressor and the turbine and assuming the turbine work to be equal to the compressor work, we find the temperature and pressure at the turbine exit to be

$$w_{\text{comp, in}} = w_{\text{turb, out}}$$
$$h_3 - h_2 = h_4 - h_5$$
$$C_p(T_3 - T_2) = C_p(T_4 - T_5)$$
$$T_5 = T_4 - T_3 + T_2 = 2460 - 926.9 + 480.1 = \mathbf{2013.2\ R}$$
$$P_5 = P_4\left(\frac{T_5}{T_4}\right)^{k/(k-1)} = (80\text{ psia})\left(\frac{2013.2\text{ R}}{2460\text{ R}}\right)^{1.4/(1.4-1)} = \mathbf{39.7\ psia}$$

(*b*) To find the air velocity at the nozzle exit, we need to first determine the nozzle exit temperature and then apply the steady-flow energy equation.

Process 5-6 (isentropic expansion of an ideal gas in a nozzle):

$$T_6 = T_5\left(\frac{P_6}{P_5}\right)^{(k-1)/k} = (2013.2\text{ R})\left(\frac{5\text{ psia}}{39.7\text{ psia}}\right)^{(1.4-1)/1.4} = 1113.8\text{ R}$$

$$h_6 + \frac{\mathcal{V}_6^2}{2} = h_5 + \frac{\mathcal{V}_5^2}{2}\ (\to 0)$$

$$0 = C_p(T_6 - T_5) + \frac{\mathcal{V}_6^2}{2}$$

$$\mathcal{V}_6 = \sqrt{2C_p(T_5 - T_6)}$$
$$= \sqrt{2[0.240\text{ Btu/(lbm}\cdot\text{R)}][(2013.2 - 1113.8)\text{ R}]\left(\frac{25{,}037\text{ ft}^2/\text{s}^2}{1\text{ Btu/lbm}}\right)}$$
$$= \mathbf{3287.7\ ft/s}$$

(*c*) The propulsive, efficiency of a turbojet engine is the ratio of the propulsive power developed $\dot{W}_p$ to the total heat transfer rate to the working fluid:

$$\dot{W}_P = \dot{m}(\mathcal{V}_{\text{exit}} - \mathcal{V}_{\text{inlet}})\mathcal{V}_{\text{aircraft}}$$
$$= (100\text{ lbm/s})[(3287.7 - 850)\text{ ft/s}](850\text{ ft/s})\left(\frac{1\text{ Btu/lbm}}{25{,}037\text{ ft}^2/\text{s}^2}\right)$$
$$= 8276\text{ Btu/s} \qquad (11{,}707\text{ hp})$$

$$\dot{Q}_{\text{in}} = \dot{m}(h_4 - h_3) = \dot{m}C_p(T_4 - T_3)$$
$$= (100\text{ lbm/s})[0.240\text{ Btu/(lbm}\cdot\text{R)}][(2460 - 926.9)\text{ R}]$$
$$= 36{,}794\text{ Btu/s}$$

$$\eta_p = \frac{\dot{W}_P}{\dot{Q}_{\text{in}}} = \frac{8276\text{ Btu/s}}{36{,}794\text{ Btu/s}} = \mathbf{22.5\%}$$

That is, 22.5 percent of the energy input is used to propel the aircraft and to overcome the drag force exerted by the air.

Discussion For those who are wondering what happened to the rest of the energy, here is a brief account:

$$\dot{KE}_{out} = \dot{m}\frac{\mathcal{V}_g^2}{2} = (100 \text{ lbm/s})\left\{\frac{[(3287.7 - 850) \text{ ft/s}]^2}{2}\right\}\left(\frac{1 \text{ Btu/lbm}}{25{,}037 \text{ ft}^2/\text{s}^2}\right)$$
$$= 11{,}867 \text{ Btu/s} \qquad (32.2\%)$$
$$\dot{Q}_{out} = \dot{m}(h_6 - h_1) = \dot{m}C_p(T_6 - T_1)$$
$$= (100 \text{ lbm/s})[0.24 \text{ Btu/(lbm}\cdot\text{R)}][(1113.8 - 420) \text{ R}]$$
$$= 16{,}651 \text{ Btu/s} \qquad (45.3\%)$$

Thus, 32.2 percent of the energy shows up as excess kinetic energy (kinetic energy of the gases relative to a fixed point on the ground). Notice that for the highest propulsion efficiency, the velocity of the exhaust gases relative to the ground $\mathcal{V}_g$ should be zero. That is, the exhaust gases should leave the nozzle at the velocity of the aircraft. The remaining 45.3 percent of the energy shows up as an increase in enthalpy of the gases leaving the engine. These last two forms of energy eventually become part of the internal energy of the atmospheric air (Fig. 8-51).

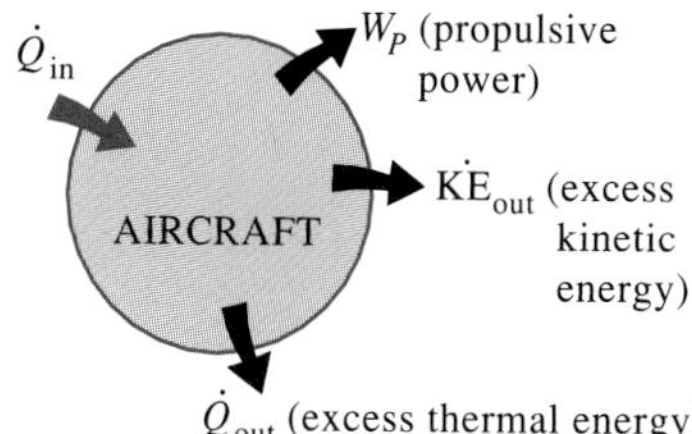

FIGURE 8-51
Energy supplied to an aircraft (from the burning of a fuel) manifests itself in various forms.

Modifications to Turbojet Engines

The first airplanes built were all propeller-driven, with propellers powered by engines essentially identical to automobile engines. The major breakthrough in commercial aviation occurred with the introduction of the turbojet engine in 1952. Both propeller-driven engines and jet-propulsion-driven engines have their own strengths and limitations, and several attempts have been made to combine the desirable characteristics of both in one engine. Two such modifications are the *propjet engine* and the *turbofan engine.*

The most widely used engine in aircraft propulsion is the **turbofan** (or *fanjet*) engine wherein a large fan driven by the turbine forces a considerable amount of air through a duct (cowl) surrounding the engine, as shown in Figs. 8-52 and 8-53. The fan exhaust leaves the duct at a higher velocity, enhancing the total thrust of the engine significantly. A turbofan engine is

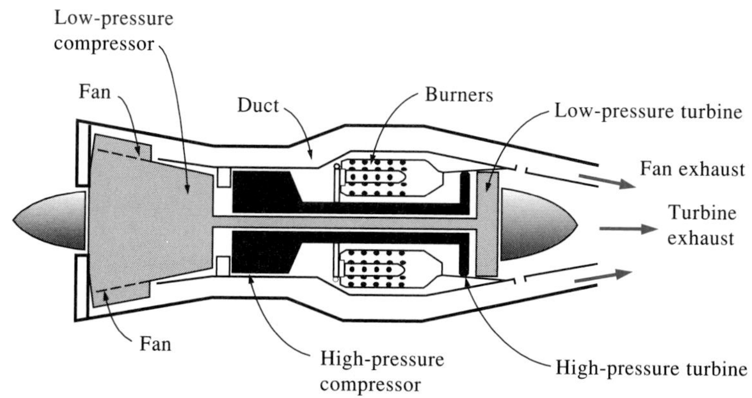

FIGURE 8-52
A turbofan engine. [*Source: The Aircraft Gas Turbine and Its Operation.* © United Aircraft Corporation (now United Technologies Corporation), 1951, 1974.]

FIGURE 8-53

A turbofan engine. (Courtesy of Allied-Signal Aerospace Company, Garrett Engine Division.)

based on the principle that for the same power, a large volume of slower-moving air will produce more thrust than a small volume of fast-moving air. The first commercial turbofan engine was successfully tested in 1955.

The turbofan engine on an airplane can be distinguished from the less-efficient turbojet engine by its fat cowling covering the large fan. All the thrust of a turbojet engine is due to the exhaust gases leaving the engine at about twice the speed of sound. In a turbofan engine, the high-speed exhaust gases are mixed with the lower-speed air, which results in a considerable reduction in noise.

New cooling techniques have resulted in considerable increases in efficiencies by allowing gas temperatures at the burner exit to reach over 1500°C, which is more than 100°C above the melting point of the turbine blade materials. Turbofan engines deserve most of the credit for the success of jumbo jets that weigh almost 400,000 kg and are capable of carrying over 400 passengers for up to 10,000 km at speeds over 950 km/h with less fuel per passenger mile.

The ratio of the mass flow rate of air bypassing the combustion chamber to that of air flowing through it is called the *bypass ratio*. The first commercial high-bypass-ratio engines had a bypass ratio of five. Increasing the bypass ratio of a turbofan engine increases thrust. Thus, it makes sense to remove the cowl from the fan. The result is a **propjet** engine, as shown in Fig. 8-54. Turbofan and propjet engines differ primarily in their bypass ratios: 5 or 6 for turbofans and as high as 100 for propjets. As a general rule, propellers are more efficient than jet engines, but they are limited to low-speed and low-altitude operation since their efficiency decreases at high speeds and altitudes. The old propjet engines (*turboprops*) were limited to speeds of about Mach 0.62 and to altitudes of around 9100 m. The new propjet engines (*propfans*) under development are expected to achieve speeds of about Mach 0.82

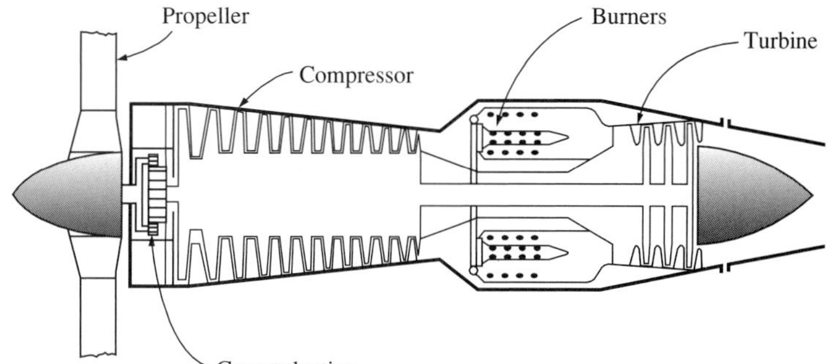

FIGURE 8-54

A turboprop engine. [*Source: The Aircraft Gas Turbine Engine and Its Operation*. © United Aircraft Corporation (now United Technologies Corporation), 1951, 1974.]

and altitudes of about 12,200 m. Commercial airplanes of medium size and range propelled by propfans are expected to fly as high and as fast as the planes propelled by turbofans, and to do so on less fuel.

Another modification that is popular in military aircraft is the addition of an **afterburner** section between the turbine and the nozzle. Whenever a need for extra thrust arises, such as for short takeoffs or combat conditions, additional fuel is injected into the oxygen-rich combustion gases leaving the turbine. As a result of this added energy, the exhaust gases leave at a higher velocity, providing a greater thrust.

A **ramjet** engine is a properly shaped duct with no compressor or turbine, as shown in Fig. 8-55, and is sometimes used for high-speed propulsion of missiles and aircraft. The pressure rise in the engine is provided by the ram effect of the incoming high-speed air being rammed against a barrier. Therefore, a ramjet engine needs to be brought to a sufficiently high speed by an external source before it can be fired.

The ramjet performs best in aircraft flying above Mach 2 or 3 (two or three times the speed of sound). In a ramjet, the air is slowed down to about Mach 0.2, fuel is added to the air and burned at this low velocity, and the combustion gases are expended and accelerated in a nozzle.

A **scramjet** engine is essentially a ramjet in which air flows through at supersonic speeds (above the speed of sound). Ramjets that convert to scramjet configurations at speeds above Mach 6 are successfully tested at speeds of about Mach 8.

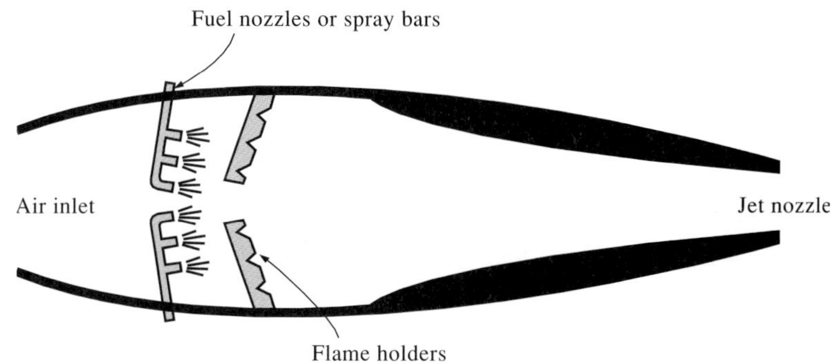

FIGURE 8-55

A ramjet engine. [*Source: The Aircraft Gas Turbine Engine and Its Operation*. © United Aircraft Corporation (now United Technologies Corporation), 1951, 1974.]

Finally, a **rocket** is a device where a solid or liquid fuel and an oxidizer react in the combustion chamber. The high-pressure combustion gases are then expanded in a nozzle. The gases leave the rocket at very high velocities, producing the thrust to propel the rocket.

8-12 ■ SECOND-LAW ANALYSIS OF GAS POWER CYCLES

The ideal Carnot, Ericsson, and Stirling cycles are *totally reversible*; thus they do not involve any irreversibilities. The ideal Otto, Diesel, and Brayton cycles, however, are only *internally reversible,* and they may involve irreversibilities external to the system. A second-law analysis of these cycles will reveal where the largest irreversibilities occur and where to start improvements.

Relations for *exergy* and *irreversibility* (or *exergy destruction*) for both closed and steady-flow systems are developed in Chap. 7. The irreversibility for a closed system can be expressed as

$$I = T_0 S_{\text{gen}} = T_0(\Delta S_{\text{sys}} - S_{\text{in}} + S_{\text{out}}) = T_0\left[(S_2 - S_1)_{\text{sys}} - \frac{Q_{\text{in}}}{T_{b,\text{in}}} + \frac{Q_{\text{out}}}{T_{b,\text{out}}}\right] \quad \text{(kJ)} \tag{8-30}$$

where $T_{b,\text{in}}$ and $T_{b,\text{out}}$ are the temperatures of the system boundary where heat is transferred into and out of the system, respectively. A similar relation for steady-flow systems can be expressed, in rate form, as

$$\dot{I} = T_0 \dot{S}_{\text{gen}} = T_0(\dot{S}_{\text{out}} - \dot{S}_{\text{in}}) = T_0\left(\sum \dot{m}_e s_e - \sum \dot{m}_i s_i - \frac{\dot{Q}_{\text{in}}}{T_{b,\text{in}}} + \frac{\dot{Q}_{\text{out}}}{T_{b,\text{out}}}\right) \quad \text{(kW)} \tag{8-31}$$

or, on a unit-mass basis for a one-inlet, one-exit steady-flow device, as

$$i = T_0 s_{\text{gen}} = T_0\left(s_e - s_i - \frac{q_{\text{in}}}{T_{b,\text{in}}} + \frac{q_{\text{out}}}{T_{b,\text{out}}}\right) \quad \text{(kJ/kg)} \tag{8-32}$$

where subscripts i and e denote the inlet and exit states for the process, respectively.

The irreversibility of a *cycle* is the sum of the irreversibilities of the processes that compose that cycle. The irreversibility of a cycle can also be determined without tracing the individual processes by considering the entire cycle as a single process and using one of the relations above. Entropy is a property, and its value depends on the state only. For a cycle, reversible or actual, the initial and the final states are identical; thus $s_e = s_i$. Therefore, the irreversibility of a cycle depends on the magnitude of the heat transfer with the high- and low-temperature reservoirs involved and on their temperatures. It can be expressed on a unit-mass basis as

$$i = T_0\left(\sum \frac{q_{out}}{T_{b,out}} - \sum \frac{q_{in}}{T_{b,in}}\right) \qquad \text{(kJ/kg)} \qquad (8\text{-}33)$$

For a cycle that involves heat transfer only with a source at T_H and a sink at T_L, the irreversibility becomes

$$i = T_0\left(\frac{q_{out}}{T_L} - \frac{q_{in}}{T_H}\right) \qquad \text{(kJ/kg)} \qquad (8\text{-}34)$$

The exergies of a closed system ϕ and a fluid stream ψ at any state can be determined from

$$\phi = (u - u_0) - T_0(s - s_0) + P_0(v - v_0) + \frac{\mathcal{V}^2}{2} + gz \qquad \text{(kJ/kg)} \qquad (8\text{-}35)$$

and $$\psi = (h - h_0) - T_0(s - s_0) + \frac{\mathcal{V}^2}{2} + gz \qquad \text{(kJ/kg)} \qquad (8\text{-}36)$$

where subscript "0" denotes the state of the surroundings. The reversible work for any process can be determined by finding the exergy change of the working fluid during that process.

EXAMPLE 8-10 Second-Law Analysis of an Otto Cycle

Determine the irreversibility of the Otto cycle (all four processes as well as the cycle) discussed in Example 8-2, assuming that heat is transferred to the working fluid from a source at 1700 K and heat is rejected to the surroundings at 290 K. Also, determine the exergy of the exhaust gases when they are purged.

Solution In Example 8-2, various quantities of interest were given or determined to be

$$\begin{aligned} r &= 8 & P_2 &= 1.7997 \text{ MPa} \\ T_0 &= 290 \text{ K} & P_3 &= 4.347 \text{ MPa} \\ T_1 &= 290 \text{ K} & q_{in} &= 800 \text{ kJ/kg} \\ T_2 &= 652.4 \text{ K} & q_{out} &= 381.83 \text{ kJ/kg} \\ T_3 &= 1575.1 \text{ K} & w_{net} &= 418.17 \text{ kJ/kg} \end{aligned}$$

Processes 1-2 and 3-4 are isentropic ($s_1 = s_2$, $s_3 = s_4$) and therefore do not involve any internal or external irreversibilities; that is, $i_{12} = 0$ and $i_{34} = 0$.

Processes 2-3 and 4-1 are constant-volume heat-addition and -rejection processes, respectively, and are internally reversible. But the heat transfer between the working fluid and the source or the sink takes place through a finite temperature difference, rendering both processes irreversible. The irreversibility associated with each process is determined from Eq. 8-32. But first we need to determine the entropy change of air during these processes:

$$s_3 - s_2 = s_3^\circ - s_2^\circ - R \ln \frac{P_3}{P_2}$$

$$= (3.5045 - 2.4975)\ \text{kJ/(kg}\cdot\text{K)} - [0.287\ \text{kJ/(kg}\cdot\text{K)}] \ln \frac{4.347\ \text{MPa}}{1.7997\ \text{MPa}}$$

$$= 0.7539\ \text{kJ/(kg}\cdot\text{K)}$$

Also, $\qquad q_{in} = 800\ \text{kJ/kg} \qquad \text{and} \qquad T_{source} = 1700\ \text{K}$

Thus,

$$i_{23} = T_0\left[(s_3 - s_2)_{sys} - \frac{q_{in}}{T_{source}}\right]$$

$$= (290\ \text{K})\left[0.7539\ \text{kJ/(kg}\cdot\text{K)} - \frac{800\ \text{kJ/kg}}{1700\ \text{K}}\right]$$

$$= 82.2\ \text{kJ/kg}$$

For process 4-1, $s_1 - s_4 = s_2 - s_3 = -0.7539$ kJ/(kg · K), $q_{R,41} = -q_{out} = 381.83$ kJ/kg, and $T_{sink} = 290$ K. Thus,

$$i_{41} = T_0\left[(s_1 - s_4)_{sys} + \frac{q_{out}}{T_{sink}}\right)$$

$$= (290\ \text{K})\left[-0.7539\ \text{kJ/(kg}\cdot\text{K)} + \frac{381.83\ \text{kJ/kg}}{290\ \text{K}}\right]$$

$$= 163.2\ \text{kJ/kg}$$

Therefore, the irreversibility of the cycle is

$$i_{cycle} = i_{12} + i_{23} + i_{34} + i_{41}$$

$$= 0 + 82.2\ \text{kJ/kg} + 0 + 163.2\ \text{kJ/kg}$$

$$= \mathbf{245.4\ kJ/kg}$$

The irreversibility of the cycle could also be determined from Eq. 8-34. Notice that the largest irreversibility in the cycle occurs during the heat-rejection process. Therefore, any attempt to reduce the irreversibilities should start with this process.

Disregarding any kinetic and potential energies, the exergy (work potential) of the working fluid before it is purged (state 4) is determined from Eq. 8-35:

$$\phi_4 = (u_4 - u_0) - T_0(s_4 - s_0) + P_0(v_4 - v_0)$$

where

$$s_4 - s_0 = s_4 - s_1 = 0.7539\ \text{kJ/(kg}\cdot\text{K)}$$
$$u_4 - u_0 = u_4 - u_1 = q_{out} = 381.83\ \text{kJ/kg}$$
$$v_4 - v_0 = v_4 - v_1 = 0$$

Thus, $\quad \phi_4 = 381.83\ \text{kJ/kg} - (290\ \text{K})[0.7539\ \text{kJ/(kg}\cdot\text{K)}] + 0 = 163.2\ \text{kJ/kg}$

which is equivalent to the irreversibility for process 4-1. (Why?)

Discussion Note that 163.2 kJ/kg of work could be obtained from the exhaust gases if they were brought to the state of the surroundings in a reversible manner.

8-13 ■ SUMMARY

A cycle during which a net amount of work is produced is called a *power cycle,* and a power cycle during which the working fluid remains a gas throughout is called a *gas power cycle.* The most efficient cycle operating between a heat source at temperature T_H and a sink at temperature T_L is the Carnot cycle, and its thermal efficiency is given by

$$\eta_{\text{th, Carnot}} = 1 - \frac{T_L}{T_H}$$

The actual gas cycles are rather complex. The approximations used to simplify the analysis are known as the *air-standard assumptions.* Under these assumptions, all the processes are assumed to be internally reversible; the working fluid is assumed to be air, which behaves as an ideal gas; and the combustion and exhaust processes are replaced by heat-addition and heat-rejection processes, respectively. The air-standard assumptions are called *cold-air-standard assumptions* if, in addition, air is assumed to have constant specific heats at room temperature.

In reciprocating engines, the *compression ratio r* and the *mean effective pressure* MEP are defined as

$$r = \frac{V_{\text{max}}}{V_{\text{min}}} = \frac{V_{\text{BDC}}}{V_{\text{TDC}}}$$

$$\text{MEP} = \frac{w_{\text{net}}}{v_{\text{max}} - v_{\text{min}}} \qquad \text{(kPa)}$$

The *Otto cycle* is the ideal cycle for the spark-ignition reciprocating engines, and it consists of four internally reversible processes: isentropic compression, constant volume heat addition, isentropic expansion, and constant volume heat rejection. Under cold-air-standard assumptions, the thermal efficiency of the ideal Otto cycle is

$$\eta_{\text{th, Otto}} = 1 - \frac{1}{r^{k-1}}$$

where r is the compression ratio and k is the specific heat ratio C_p/C_v.

The *Diesel cycle* is the ideal cycle for the compression-ignition reciprocating engines. It is very similar to the Otto cycle, except that the constant volume heat-addition process is replaced by a constant pressure heat-addition process. Its thermal efficiency under cold-air-standard assumptions is

$$\eta_{\text{th, Diesel}} = 1 - \frac{1}{r^{k-1}}\left[\frac{r_c^k - 1}{k(r_c - 1)}\right]$$

where r_c is the *cutoff ratio,* defined as the ratio of the cylinder volumes after and before the combustion process.

Stirling and *Ericsson cycles* are two totally reversible cycles that involve an isothermal heat-addition process at T_H and an isothermal heat-rejection process at T_L. They differ from the Carnot cycle in that the two isentropic processes are replaced by two constant volume regeneration processes in the

Stirling cycle and by two constant pressure regeneration processes in the Ericsson cycle. Both cycles utilize *regeneration*, a process during which heat is transferred to a thermal energy storage device (called a *regenerator*) during one part of the cycle that is then transferred back to the working fluid during another part of the cycle.

The ideal cycle for modern gas-turbine engines is the *Brayton cycle*, which is made up of four internally reversible processes: isentropic compression, constant pressure heat addition, isentropic expansion, and constant pressure heat rejection. Under cold-air-standard assumptions, its thermal efficiency is

$$\eta_{\text{th, Brayton}} = 1 - \frac{1}{r_p^{(k-1)/k}}$$

where $r_p = P_{max}/P_{min}$ is the pressure ratio and k is the specific heat ratio. The thermal efficiency of the simple Brayton cycle increases with the pressure ratio.

The deviation of the actual compressor and the turbine from the idealized isentropic ones can be accurately accounted for by utilizing their adiabatic efficiencies, defined as

$$\eta_C = \frac{w_s}{w_a} \cong \frac{h_1 - h_{2s}}{h_1 - h_{2a}}$$

and

$$\eta_T = \frac{w_a}{w_s} \cong \frac{h_3 - h_{4a}}{h_3 - h_{4s}}$$

where states 1 and 3 are the inlet states, $2a$ and $4a$ are the actual exit states, and $2s$ and $4s$ are the isentropic exit states.

In gas-turbine engines, the temperature of the exhaust gas leaving the turbine is often considerably higher than the temperature of the air leaving the compressor. Therefore, the high-pressure air leaving the compressor can be heated by transferring heat to it from the hot exhaust gases in a counter-flow heat exchanger, which is also known as a *regenerator*. The extent to which a regenerator approaches an ideal regenerator is called the *effectiveness* ε and is defined as

$$\varepsilon = \frac{q_{\text{regen, act}}}{q_{\text{regen, max}}}$$

Under cold-air-standard assumptions, the thermal efficiency of an ideal Brayton cycle with regeneration becomes

$$\eta_{\text{th, regen}} = 1 - \left(\frac{T_1}{T_3}\right)(r_p)^{(k-1)/k}$$

where T_1 and T_3 are the minimum and maximum temperatures, respectively, in the cycle.

The thermal efficiency of the Brayton cycle can also be increased by utilizing *multistage compression with intercooling, regeneration, and multistage expansion with reheating*. The work input to the compressor is minimized when equal pressure ratios are maintained across each stage. This procedure also maximizes the turbine work output.

Gas-turbine engines are widely used to power aircraft because they are light and compact and have a high power-to-weight ratio. The ideal *jet-propulsion cycle* differs from the simple ideal Brayton cycle in that the gases are partially expanded in the turbine. The gases that exit the turbine at a relatively high pressure are subsequently accelerated in a nozzle to provide the thrust needed to propel the aircraft.

The *net thrust* developed by the engine is

$$F = \dot{m}(\mathcal{V}_{\text{exit}} - \mathcal{V}_{\text{inlet}}) \qquad \text{(N)}$$

where $\dot{m}$ is the mass flow rate of gases, $\mathcal{V}_{\text{exit}}$ is the exit velocity of the exhaust gases, and $\mathcal{V}_{\text{inlet}}$ is the inlet velocity of the air, both relative to the aircraft.

The power developed from the thrust of the engine is called the *propulsive power* $\dot{W}_P$, and it is given by

$$\dot{W}_P = \dot{m}(\mathcal{V}_{\text{exit}} - \mathcal{V}_{\text{inlet}})\mathcal{V}_{\text{aircraft}} \qquad \text{(kW)}$$

Propulsive efficiency is a measure of how efficiently the energy released during the combustion process is converted to propulsive energy, and it is defined as

$$\eta_P = \frac{\text{Propulsive power}}{\text{Energy input rate}} = \frac{\dot{W}_P}{\dot{Q}_{\text{in}}}$$

For an ideal cycle that involves heat transfer only with a source at T_H and a sink at T_L, the irreversibility or exergy destruction is determined to be

$$i = T_0\left(\frac{q_{\text{out}}}{T_L} - \frac{q_{\text{in}}}{T_H}\right) \qquad \text{(kJ/kg)}$$

REFERENCES AND SUGGESTED READING

1. W. Z. Black and J. G. Hartley. *Thermodynamics.* New York: Harper & Row, 1985.

2. V. D. Chase. "Propfans: A New Twist for the Propeller." *Mechanical Engineering,* November 1986, pp. 47–50.

3. R. A. Harmon. "The Keys to Cogeneration and Combined Cycles." *Mechanical Engineering,* February 1988, pp. 64–73.

4. B. V. Karlekar. *Thermodynamics for Engineers.* Englewood Cliffs, NJ: Prentice Hall, 1983.

5. L. C. Lichty. *Combustion Engine Processes.* New York: McGraw-Hill, 1967.

6. D. C. Look, Jr., and H. J. Sauer, Jr. *Engineering Thermodynamics.* Boston: PWS Engineering, 1986.

7. H. McIntosh. "Jumbo Jet." *10 Outstanding Achievements 1964–1989.* Washington, D.C.: National Academy of Engineering, 1989, pp. 30–33.

8. W. Siuru. "Two-stroke Engines: Cleaner and Meaner." *Mechanical Engineering.* June 1990, pp. 66–69.

9. C. F. Taylor. *The Internal Combustion Engine in Theory and Practice.* Cambridge, MA: M.I.T. Press, 1968.

10. G. J. Van Wylen and R. E. Sonntag. *Fundamentals of Classical Thermodynamics.* 3rd ed. New York: John Wiley & Sons, 1985.

11. K. Wark. *Thermodynamics.* 5th ed. New York: McGraw-Hill, 1988.

PROBLEMS*

Actual and Ideal Cycles, Carnot Cycle, Air-Standard Assumptions, Reciprocating Engines

8-1C How do gas power cycles differ from vapor power cycles?

8-2C Why is the Carnot cycle not suitable as an ideal cycle for all power-producing cyclic devices?

8-3C How does the thermal efficiency of an ideal cycle, in general, compare to that of a Carnot cycle operating between the same temperature limits?

8-4C What does the area enclosed by the cycle represent on a *P-v* diagram? How about on a *T-s* diagram?

8-5C What is the difference between air-standard assumptions and the cold-air-standard assumptions?

8-6C Do internal combustion engines operate on a closed or an open cycle? Why?

8-7C How are the combustion and exhaust processes modeled under the air-standard assumptions?

8-8C What are the air-standard assumptions?

8-9C What is the difference between the clearance volume and the displacement volume of reciprocating engines?

8-10C Define the compression ratio for reciprocating engines.

8-11C How is the mean effective pressure for reciprocating engines defined?

8-12C Can the mean effective pressure of an automobile engine in operation be less than the atmospheric pressure?

*Students are encouraged to answer *all* the concept "C" questions.

8-13C As a car gets older, will its compression ratio change? How about the mean effective pressure?

8-14C What is the difference between spark-ignition and compression-ignition engines?

8-15C Define the following terms related to reciprocating engines: stroke, bore, top dead center, and clearance volume.

8-16 An air-standard cycle with variable specific heats is executed in a closed system and is composed of the following four processes:

1-2 Isentropic compression from 100 kPa and 27°C to 800 kPa
2-3 v = *constant* heat addition to 1800 K
3-4 Isentropic expansion to 100 kPa
4-1 P = *constant* heat rejection to initial state

(*a*) Show the cycle on *P-v* and *T-s* diagrams.
(*b*) Calculate the net work output per unit mass.
(*c*) Determine the thermal efficiency.

8-17 An air-standard cycle is executed in a closed system and is composed of the following four processes:

1-2 Isentropic compression from 100 kPa and 27°C to 1 MPa
2-3 P = *constant* heat addition in amount of 2840 kJ/kg
3-4 v = *constant* heat rejection to 100 kPa
4-1 P = *constant* heat rejection to initial state

(*a*) Show the cycle on *P-v* and *T-s* diagrams.
(*b*) Calculate the maximum temperature in the cycle.
(*c*) Determine the thermal efficiency.

Assume constant specific heats at room temperature.
Answers: (*b*) 3405.1 K, (*c*) 21.1 percent

8-18E An air-standard cycle with variable specific heats is executed in a closed system and is composed of the following four processes:

1-2 v = *constant* heat addition from 14.7 psia and 80°F in the amount of 300 Btu/lbm
2-3 P = *constant* heat addition to 3200 R
3-4 Isentropic expansion to 14.7 psia
4-1 P = *constant* heat rejection to initial state

(*a*) Show the cycle on *P-v* and *T-s* diagrams.
(*b*) Calculate the total heat input per unit mass.
(*c*) Determine the thermal efficiency.
Answers: (*b*) 612.4 Btu/lbm, (*c*) 24.2 percent

8-19E Repeat Prob. 8-18E using constant specific heats at room temperature.

8-20 An air-standard cycle is executed in a closed system with 0.001 kg of air and consists of the following three processes:

1-2 Isentropic compression from 100 kPa and 27°C to 1 MPa
2-3 P = *constant* heat addition in the amount of 1.84 kJ

3-1 $P = c_1 v + c_2$ heat rejection to initial state (c_1 and c_2 are constants)

(*a*) Show the cycle on *P-v* and *T-s* diagrams.
(*b*) Calculate the heat rejected.
(*c*) Determine the thermal efficiency.

Assume constant specific heats at room temperature.
Answers: (*b*) 1.422 kJ, (*c*) 22.7 percent

8-21 An air-standard cycle with variable specific heats is executed in a closed system with 0.003 kg of air and consists of the following three processes:

1-2 v = *constant* heat addition from 95 kPa and 17°C to 380 kPa
2-3 Isentropic expansion to 95 kPa
3-1 P = *constant* heat rejection to initial state

(*a*) Show the cycle on *P-v* and *T-s* diagrams.
(*b*) Calculate the net work per cycle, in kJ.
(*c*) Determine the thermal efficiency.

8-22 Repeat Prob. 8-21 using constant specific heats at room temperature.

8-23 Consider a Carnot cycle executed in a closed system with 0.004 kg of air. The temperature limits of the cycle are 300 and 1000 K, and the minimum and maximum pressures that occur during the cycle are 20 and 1800 kPa. Assuming constant specific heats, determine the net work output per cycle.

8-24 An air-standard Carnot cycle is executed in a closed system between the temperature limits of 350 and 1200 K. The pressures before and after the isothermal compression are 150 and 300 kPa, respectively. If the net work output per cycle is 0.5 kJ, determine (*a*) the maximum pressure in the cycle, (*b*) the heat transfer to air, and (*c*) the mass of air. Assume variable specific heats for air. *Answers:* (*a*) 30,013 kPa, (*b*) 0.706 kJ, (*c*) 0.00296 kg

8-25 Repeat Prob. 8-24 using helium as the working fluid.

Otto Cycle

8-26C What four processes make up the ideal Otto cycle?

8-27C How do the efficiencies of the ideal Otto cycle and the Carnot cycle compare for the same temperature limits? Explain.

8-28C How is the rpm (revolutions per minute) of an actual four-stroke gasoline engine related to the number of thermodynamic cycles? What would your answer be for a two-stroke engine?

8-29C Are the processes that make up the Otto cycle analyzed as closed-system or steady-flow processes? Why?

8-30C How does the thermal efficiency of an ideal Otto cycle change with the compression ratio of the engine and the specific heat ratio of the working fluid?

8-31C Why are high compression ratios not used in spark-ignition engines?

8-32C An ideal Otto cycle with a specified compression ratio is executed using (*a*) air, (*b*) argon, and (*c*) ethane as the working fluid. For which case will the thermal efficiency be the highest? Why?

8-33C What is the difference between fuel-injected gasoline engines and diesel engines?

8-34C An ideal Otto cycle has a compression ratio of 8. At the beginning of the compression process, air is at 95 kPa and 27°C, and 750 kJ/kg of heat is transferred to air during the constant-volume heat-addition process. Taking into account the variation of specific heats with temperature, determine (*a*) the pressure and temperature at the end of the heat-addition process, (*b*) the net work output, (*c*) the thermal efficiency, and (*d*) the mean effective pressure for the cycle.

Answers: (*a*) 3898 kPa, 1539 K; (*b*) 392.4 kJ/kg; (*c*) 52.3 percent; (*d*) 495 kPa

8-35 Repeat Prob. 8-34 using constant specific heats at room temperature.

8-36 The compression ratio of an air-standard Otto cycle is 9.5. Prior to the isentropic compression process, the air is at 100 kPa, 17°C, and 600 cm^3. The temperature at the end of the isentropic expansion process is 800 K. Using specific heat values at room temperature, determine (*a*) the highest temperature and pressure in the cycle; (*b*) the amount of heat transferred, in kJ; (*c*) the thermal efficiency; and (*d*) the mean effective pressure.

Answers: (*a*) 1969 K, 6449 kPa; (*b*) 0.65 kJ; (*c*) 59.4 percent; (*d*) 719 kPa

8-37 Repeat Prob. 8-36, but replace the isentropic expansion process by a polytropic expansion process with the polytropic exponent $n = 1.35$.

8-38E An ideal Otto cycle with air as the working fluid has a compression ratio of 8. The minimum and maximum temperatures in the cycle are 540 and 2200 R. Accounting for the variation of specific heats with temperature, determine (*a*) the amount of heat transferred to the air during the heat-addition process, (*b*) the thermal efficiency, and (*c*) the thermal efficiency of a Carnot cycle operating between the same temperature limits.

Answers: (*a*) 198.15 Btu/lbm, (*b*) 53.5 percent, (*c*) 75.5 percent

8-39E Repeat Prob. 8-38E using argon as the working fluid.

Diesel Cycle

8-40C What is the dual cycle? How does it differ from the Otto and Diesel cycles?

8-41C How does a diesel engine differ from a gasoline engine?.

8-42C How does the ideal Diesel cycle differ from the ideal Otto cycle?

8-43C For a specified compression ratio, is a diesel or gasoline engine more efficient?

8-44C Do diesel or gasoline engines operate at higher compression ratios? Why?

8-45C What is the cutoff ratio? How does it affect the thermal efficiency of a Diesel cycle?

8-46 An air-standard Diesel cycle has a compression ratio of 16 and a cutoff ratio of 2. At the beginning of the compression process, air is at 95 kPa and 27°C. Accounting for the variation of specific heats with temperature, determine (*a*) the temperature after the heat-addition process, (*b*) the thermal efficiency, and (*c*) the mean effective pressure.
Answers: (*a*) 1724.8 K, (*b*) 56.3 percent, (*c*) 675.9 kPa

8-47 Repeat Prob. 8-46 using constant specific heats at room temperature.

8-48E An air-standard Diesel cycle has a compression ratio of 18.2. Air is at 80°F and 14.7 psia at the beginning of the compression process and at 3400 R at the end of the heat-addition process. Accounting for the variation of specific heats with temperature, determine (*a*) the cutoff ratio, (*b*) the heat rejection per unit mass, and (*c*) the thermal efficiency.
Answers: (*a*) 2.09, (*b*) 216.3 Btu/lbm, (*c*) 57.3 percent

8-49E Repeat Prob. 8-48E using constant specific heats at room temperature.

8-50 An ideal diesel engine has a compression ratio of 20 and uses air as the working fluid. The state of air at the beginning of the compression process is 95 kPa and 20°C. If the maximum temperature in the cycle is not to exceed 2200 K, determine (*a*) the thermal efficiency and (*b*) the mean effective pressure. Assume constant specific heats for air at room temperature.
Answers: (*a*) 63.5 percent, (*b*) 933 kPa

8-51 Repeat Prob. 8-50, but replace the isentropic expansion process by polytropic expansion process with the polytropic exponent $n = 1.35$.

8-52 A four-cylinder 4.5-L diesel engine that operates on an ideal Diesel cycle has a compression ratio of 17 and a cutoff ratio of 2.2. Air is at 27°C and 97 kPa at the beginning of the compression process. Using the cold-air-standard assumptions, determine how much power the engine will deliver at 1500 rpm.

8-53 Repeat Prob. 8-52 using nitrogen as the working fluid.

8-54 The compression ratio of an ideal dual cycle is 14. Air is at 100 kPa and 300 K at the beginning of the compression process and at 2200 K at the end of the heat-addition process. Heat transfer to air takes place partly at constant volume and partly at constant pressure, and it amounts to 1520.4 kJ/kg. Assuming variable specific heats for air, determine (*a*) the fraction of heat transferred at constant volume and (*b*) the thermal efficiency of the cycle.

8-55 Repeat Prob. 8-54 using constant specific heats at room temperature. Is the constant specific heat assumption reasonable in this case?

Stirling and Ericsson Cycles

8-56C Consider the ideal Otto, Stirling, and Carnot cycles operating between the same temperature limits. How would you compare the thermal efficiencies of these three cycles?

8-57C Consider the ideal Diesel, Ericsson, and Carnot cycles operating between the same temperature limits. How would you compare the thermal efficiencies of these three cycles?

8-58C What cycle is composed of two isothermal and two constant-volume processes?

8-59C How does the ideal Ericsson cycle differ from the Carnot cycle?

8-60C How is regeneration accomplished in the ideal Ericsson cycle?

8-61C Name three advantages that external combustion engines have over internal combustion engines.

8-62E An ideal Ericsson engine using helium as the working fluid operates between temperature limits of 550 and 3000 R and pressure limits of 25 and 200 psia. Assuming a mass flow rate of 8 lbm/s, determine (*a*) the thermal efficiency of the cycle, (*b*) the heat transfer rate in the regenerator, and (*c*) the power delivered.

8-63 Consider an ideal Ericsson cycle with air as the working fluid executed in a steady-flow system. Air is at 27°C and 120 kPa at the beginning of the isothermal compression process, during which 150 kJ/kg of heat is rejected. Heat transfer to air occurs at 1200 K. Determine (*a*) the maximum pressure in the cycle, (*b*) the net work output per unit mass of air, and (*c*) the thermal efficiency of the cycle.

Answers: (*a*) 658.2 kPa, (*b*) 450 kJ/kg, (*c*) 75 percent

8-64 An ideal Stirling engine using helium as the working fluid operates between temperature limits of 300 and 2000 K and pressure limits of 100 kPa and 2 MPa. Assuming the mass of the helium used in the cycle is 1.5 kg, determine (*a*) the thermal efficiency of the cycle, (*b*) the amount of heat transfer in the regenerator, and (*c*) the work output per cycle.

Ideal and Actual Gas-Turbine (Brayton) Cycles

8-65C Why are the back work ratios relatively high in gas-turbine engines?

8-66C What four processes make up the simple ideal Brayton cycle?

8-67C For fixed maximum and minimum temperatures, what is the effect of the pressure ratio on (*a*) the thermal efficiency and (*b*) the net work output of a simple ideal Brayton cycle?

8-68C Why are gas turbines operated at very high air–fuel mass ratios?

8-69C Should the processes that make up the Brayton cycle be analyzed as closed-system or steady-flow processes? Why?

8-70C What is the back work ratio? What are typical back work ratio values for gas-turbine engines?

8-71C How can the irreversibilities in the turbine and compressor of gas-turbine engines be properly accounted for?

8-72C How do the inefficiencies of the turbine and the compressor affect (*a*) the back work ratio and (*b*) the thermal efficiency of a gas-turbine engine?

8-73E A simple ideal Brayton cycle with air as the working fluid has a pressure ratio of 10. The air enters the compressor at 520 R and the turbine at 2000 R. Accounting for the variation of specific heats with temperature, determine (*a*) the air temperature at the compressor exit, (*b*) the back work ratio, and (*c*) the thermal efficiency.

8-74 A simple Brayton cycle using air as the working fluid has a pressure ratio of 8. The minimum and maximum temperatures in the cycle are 310 and 1160 K. Assuming an adiabatic efficiency of 75 percent for the compressor and 82 percent for the turbine, determine (*a*) the air temperature at the turbine exit, (*b*) the net work output, and (*c*) the thermal efficiency.

8-75 Repeat Prob. 8-74 using constant specific heats at room temperature.

8-76 Air is used as the working fluid in a simple ideal Brayton cycle that has a pressure ratio of 12, a compressor inlet temperature of 300 K, and a turbine inlet temperature of 1000 K. Determine the required mass flow rate of air for a net power output of 30 MW, assuming both the compressor and the turbine have an isentropic efficiency of (*a*) 100 percent and (*b*) 80 percent. Assume constant specific heats at room temperature.
Answers: (*a*) 150.7 kg/s, (*b*) 1581 kg/s

8-77 A stationary gas-turbine power plant operates on a simple ideal Brayton cycle with air as the working fluid. The air enters the compressor at 95 kPa and 290 K and the turbine at 760 kPa and 1100 K. Heat is transferred to air at a rate of 50,000 kJ/s. Determine the power delivered by this plant (*a*) assuming constant specific heats at room temperature and (*b*) accounting for the variation of specific heats with temperature.

8-78 Air enters the compressor of a gas-turbine engine at 300 K and 100 kPa, where it is compressed to 700 kPa and 580 K. Heat is transferred to air in the amount of 950 kJ/kg before it enters the turbine. For a turbine efficiency of 86 percent, determine (*a*) the fraction of the turbine work output used to drive the compressor and (*b*) the thermal efficiency. Assume variable specific heats for air.

8-79 Repeat Prob. 8-78 using constant specific heats at room temperature.

8-80E A gas-turbine power plant operates on a simple Brayton cycle with air as the working fluid. The air enters the turbine at 120 psia and 2000 R and leaves at 15 psia and 1200 R. Heat is rejected to the surroundings at a rate of 6400 Btu/s, and air flows through the cycle at a rate of 40 lbm/s. Assuming a compressor efficiency of 80 percent, determine the net power output of the plant. Account for the variation of specific heats with temperature. *Answer:* 3373 kW

8-81E For what compressor efficiency will the gas-turbine power plant in Prob. 8-80E produce zero net work?

8-82 A gas-turbine power plant operates on the simple Brayton cycle with air as the working fluid and delivers 15 MW of power. The minimum and maximum temperatures in the cycle are 310 and 900 K, and the pressure of air at the compressor exit is 8 times the value at the compressor inlet. Assuming an adiabatic efficiency of 80 percent for the compressor and 86 percent for the turbine, determine the mass flow rate of air through the cycle. Account for the variation of specific heats with temperature.

8-83 Repeat Prob. 8-82 using constant specific heats at room temperature.

Brayton Cycle with Regeneration

8-84C How does regeneration affect the efficiency of a Brayton cycle, and how does it accomplish it?

8-85C Somebody claims that at very high pressure ratios, the use of regeneration actually decreases the thermal efficiency of a gas-turbine engine. Is there any truth in this claim? Explain.

8-86C Define the effectiveness of a regenerator used in gas-turbine cycles.

8-87C In an ideal regenerator, is the air leaving the compressor heated to the temperature at (*a*) turbine inlet, (*b*) turbine exit, (*c*) slightly above turbine exit?

8-88C In 1903, Aegidius Elling of Norway designed and built an 11-hp gas turbine that used steam injection between the combustion chamber and the turbine to cool the combustion gases to a safe temperature for the materials available at the time. Currently there are several gas-turbine power plants that use steam injection to augment power and improve thermal efficiency. For example, the thermal efficiency of the General Electric LM5000 gas turbine is reported to increase from 35.8 percent in simple-cycle operation to 43 percent when steam injection is used. Explain why steam injection increases the power output and the efficiency of gas turbines. Also, explain how you would obtain the steam.

8-89E The idea of using gas turbines to power automobiles was conceived in the 1930s, and considerable research was done in the 1940s and 1950s to develop automotive gas turbines by major automobile manufacturers such as the Chrysler and Ford corporations in the United States and Rover in the United Kingdom. The world's first gas-turbine-powered automobile, the 200-hp Rover Jet 1, was built in 1950 in the United Kingdom. This was followed by the production of the Plymouth Sport Coupe by Chrysler in 1954 under the leadership of G. J. Huebner. Several hundred gas-turbine-powered Plymouth cars were built in the early 1960s for demonstration purposes and were loaned to a select group of people to gather field experience. The users had no complaints other than slow acceleration. But the cars were never mass-produced because of the high production (especially material) costs and the failure to satisfy the provisions of the 1966 Clean Air Act.

A gas-turbine-powered Plymouth car built in 1960 had a turbine inlet temperature of 1700°F, a pressure ratio of 4, and a regenerator effectiveness of 0.9. Using isentropic efficiencies of 80 percent for both the compressor and the turbine, determine the thermal efficiency of this car. Also, determine the mass flow rate of air for a net power output of 135 hp. Assume the ambient air to be at 540 R and 14.5 psia.

8-90 The 7FA gas turbine manufactured by General Electric is reported to have an efficiency of 35.9 percent in the simple-cycle mode and to produce 159 MW of net power. The pressure ratio is 14.7, the turbine inlet temperature is 1288°C, and the exhaust temperature is 589°C. The mass flow rate through the turbine is 1,536,000 kg/h. Taking the ambient conditions to be 20°C and 100 kPa, determine the isentropic efficiency of the turbine and the compressor. Also, determine the thermal efficiency of this gas turbine if a regenerator with an effectiveness of 80 percent is added.

8-91 An ideal Brayton cycle with regeneration has a pressure ratio of 10. Air enters the compressor at 300 K and the turbine at 1200 K. If the effectiveness of the regenerator is 100 percent, determine the net work output and the thermal efficiency of the cycle. Account for the variation of specific heats with temperature.

8-92 Repeat Prob. 8-91 using constant specific heats at room temperature.

8-93 A Brayton cycle with regeneration using air as the working fluid has a pressure ratio of 8. The minimum and maximum temperatures in the cycle are 310 and 1150 K. Assuming an adiabatic efficiency of 75 percent for the compressor and 82 percent for the turbine and an effectiveness of 65 percent for the regenerator, determine (*a*) the air temperature at the turbine exit, (*b*) the net work output, and (*c*) the thermal efficiency.

Answers: (*a*) 763 K, (*b*) 101.64 kJ/kg, (*c*) 21.0 percent

8-94 A stationary gas-turbine power plant operates on an ideal regenerative Brayton cycle (ε = 100 percent) with air as the working fluid. Air enters the compressor at 95 kPa and 290 K and the turbine at 760 kPa and 1100 K. Heat is transferred to air from an external source at a rate of 60,000 kJ/s. Determine the power delivered by this plant (*a*) assuming constant specific heats for air at room temperature and (*b*) accounting for the variation of specific heats with temperature.

8-95 Air enters the compressor of a regenerative gas-turbine engine at 300 K and 100 kPa, where it is compressed to 800 kPa and 580 K. The regenerator has an effectiveness of 72 percent, and the air enters the turbine at 1200 K. For a turbine efficiency of 86 percent, determine (*a*) the amount of heat transfer in the regenerator and (*b*) the thermal efficiency. Assume variable specific heats for air.

Answers: (*a*) 152.5 kJ/kg, (*b*) 36.0 percent

8-96 Repeat Prob. 8-95 using constant specific heats at room temperature.

8-97 Repeat Prob. 8-95 for a regenerator effectiveness of 70 percent.

Brayton Cycle with Intercooling, Reheating, and Regeneration

8-98C Under what modifications will the ideal simple gas-turbine cycle approach the Ericsson cycle?

8-99C The single-stage compression process of an ideal Brayton cycle without regeneration is replaced by a multistage compression process with intercooling between the same pressure limits. As a result of this modification,

(*a*) Does the compressor work increase, decrease, or remain the same?
(*b*) Does the back work ratio increase, decrease, or remain the same?
(*c*) Does the thermal efficiency increase, decrease, or remain the same?

8-100C The single-stage expansion process of an ideal Brayton cycle without regeneration is replaced by a multistage expansion process with reheating between the same pressure limits. As a result of this modification,

(*a*) Does the turbine work increase, decrease, or remain the same?
(*b*) Does the back work ratio increase, decrease, or remain the same?
(*c*) Does the thermal efficiency increase, decrease, or remain the same?

8-101C A simple ideal Brayton cycle without regeneration is modified to incorporate multistage compression with intercooling and multistage expansion with reheating, without changing the pressure or temperature limits of the cycle. As a result of these two modifications,

(*a*) Does the net work output increase, decrease, or remain the same?
(*b*) Does the back work ratio increase, decrease, or remain the same?
(*c*) Does the thermal efficiency increase, decrease, or remain the same?
(*d*) Does the heat rejected increase, decrease, or remain the same?

8-102C A simple ideal Brayton cycle is modified to incorporate multistage compression with intercooling, multistage expansion with reheating, and regeneration without changing the pressure limits of the cycle. As a result of these modifications,

(*a*) Does the net work output increase, decrease, or remain the same?
(*b*) Does the back work ratio increase, decrease, or remain the same?
(*c*) Does the thermal efficiency increase, decrease, or remain the same?
(*d*) Does the heat rejected increase, decrease, or remain the same?

8-103C For a specified pressure ratio, why does multistage compression with intercooling decrease the compressor work, and multistage expansion with reheating increase the turbine work?

8-104C In an ideal gas-turbine cycle with intercooling, reheating, and regeneration, as the number of compression and expansion stages is increased, the cycle thermal efficiency approaches (*a*) 100 percent, (*b*) the Otto cycle efficiency, or (*c*) the Carnot cycle efficiency.

8-105 Consider an ideal gas-turbine cycle with two stages of compression and two stages of expansion. The pressure ratio across each stage of the compressor and turbine is 3. The air enters each stage of the compressor at 300 K and each stage of the turbine at 1200 K. Determine the back work ratio and the thermal efficiency of the cycle, assuming (*a*) no regenerator is used and (*b*) a regenerator with 75 percent effectiveness is used. Use constant specific heats at room temperature.

8-106 Repeat Prob. 8-105, assuming an efficiency of 80 percent for each compressor stage and an efficiency of 85 percent for each turbine stage.

8-107 Consider a regenerative gas-turbine power plant with two stages of compression and two stages of expansion. The overall pressure ratio of the cycle is 9. The air enters each stage of the compressor at 300 K and each stage of the turbine at 1200 K. Accounting for the variation of specific heats with temperature, determine the minimum mass flow rate of air needed to develop a net power output of 30 MW. *Answer:* 68.1 kg/s

8-108 Repeat Prob. 8-107 using argon as the working fluid.

Jet-Propulsion Cycles

8-109C How does the ideal jet-propulsion cycle differ from the ideal Brayton cycle?

8-110C What is the function of the nozzle in turbojet engines?

8-111C What is propulsive power? How is it related to thrust?

8-112C What is propulsive efficiency? How is it determined?

8-113C Is the effect of turbine and compressor irreversibilities of a turbojet engine to reduce (*a*) the net work, (*b*) the thrust, or (*c*) the fuel consumption rate?

8-114E A turbojet is flying with a velocity of 900 ft/s at an altitude of 20,000 ft, where the ambient conditions are 7 psia and 10°F. The pressure ratio across the compressor is 13, and the temperature at the turbine inlet is 2400 R. Assuming ideal operation for all components and constant specific heats for air at room temperature, determine (*a*) the pressure at the turbine exit, (*b*) the velocity of the exhaust gases, and (*c*) the propulsive efficiency.

8-115E Repeat Prob. 8-114E accounting for the variation of specific heats with temperature.

8-116 A turbojet aircraft is flying with a velocity of 320 m/s at an altitude of 9150 m, where the ambient conditions are 32 kPa and −32°C. The pressure ratio across the compressor is 12, and the temperature at the turbine inlet is 1400 K. Air enters the compressor at a rate of 40 kg/s, and the jet fuel has a heating value of 42,700 kJ/kg. Assuming ideal operation for all components and constant specific heats for air at room temperature, determine (*a*) the velocity of the exhaust gases, (*b*) the propulsive power developed, and (*c*) the rate of fuel consumption.

8-117 Repeat Prob. 8-116 using a compressor efficiency of 80 percent and a turbine efficiency of 85 percent.

8-118 Consider an aircraft powered by a turbojet engine that has a pressure ratio of 12. The aircraft is stationary on the ground, held in position by its brakes. The ambient air is at 27°C and 95 kPa and enters the engine at a rate of 10 kg/s. The jet fuel has a heating value of 42,700 kJ/kg, and it is burned completely at a rate of 0.2 kg/s. Neglecting the effect of the diffuser and disregarding the slight increase in mass at the engine exit as well as the inefficiencies of engine components, determine the force that must be applied on the brakes to hold the plane stationary. *Answer:* 9088 N

8-119 Air at 7°C enters a turbojet engine at a rate of 20 kg/s and at a velocity of 300 m/s (relative to the engine). Air is heated in the combustion chamber at a rate 20,000 kJ/s and it leaves the engine at 427°C. Determine the thrust produced by this turbojet engine. (*Hint:* Choose the entire engine as your control volume.)

Second-Law Analysis of Gas Power Cycles

8-120 Determine the total irreversibility associated with the Otto cycle described in Prob. 8-34, assuming a source temperature of 2000 K and a sink temperature of 300 K. Also, determine the exergy at the end of the power stroke. *Answers:* 245.12 kJ/kg, 145.2 kJ/kg

8-121 Determine the total irreversibility associated with the Diesel cycle described in Prob. 8-46, assuming a source temperature of 2000 K and a sink temperature of 300 K. Also, determine the exergy at the end of the isentropic compression process. *Answers:* 292.7 kJ/kg, 348.6 kJ/kg

8-122E Determine the irreversibility associated with the heat rejection process of the Diesel cycle described in Prob. 8-48E, assuming a source temperature of 3500 R and a sink temperature of 540 R. Also, determine the exergy at the end of the isentropic expansion process.

8-123 Calculate the irreversibility associated with each of the processes of the Brayton cycle described in Prob. 8-74, assuming a source temperature of 1400 K and a sink temperature of 310 K.

8-124 Determine the total irreversibility associated with the Brayton cycle described in Prob. 8-92, assuming a source temperature of 1400 K and a sink temperature of 300 K. Also, determine the exergy of the exhaust gases at the exit of the regenerator.

8-125 Determine the irreversibility associated with each of the processes of the Brayton cycle described in Prob. 8-93, assuming a source temperature of 1260 K and a sink temperature of 310 K. Also, determine the exergy of the exhaust gases at the exit of the regenerator. Take $P_{exhaust} = P_0 = 100$ kPa.

Review Problems

8-126 A four-stroke turbocharged V-16 diesel engine built by GE Transportation Systems to power fast trains produces 4000 hp at 1050 rpm. Determine the amount of power produced per cylinder per (*a*) mechanical cycle and (*b*) thermodynamic cycle.

8-127 Consider a simple ideal Brayton cycle operating between the temperature limits of 290 K and 1500 K. Using constant specific heats at room temperature, determine the pressure ratio for which the compressor and the turbine exit temperatures of air are equal.

8-128 An air-standard cycle with variable coefficients is executed in a closed system and is composed of the following four processes:

1-2 $v =$ *constant* heat addition from 100 kPa and 27°C to 300 kPa
2-3 $P =$ *constant* heat addition to 1027°C
3-4 Isentropic expansion to 100 kPa
4-1 $P =$ *constant* heat rejection to initial state

(*a*) Show the cycle on *P-v* and *T-s* diagrams.
(*b*) Calculate the net work output per unit mass.
(*c*) Determine the thermal efficiency.

8-129 Repeat Prob. 8-128 using constant specific heats at room temperature.

8-130 An air-standard cycle with variable specific heats is executed in a closed system with 0.002 kg of air, and it consists of the following three processes:

1-2 Isentropic compression from 100 kPa and 27°C to 700 kPa
2-3 $P =$ *constant* heat addition to initial specific volume
3-1 $v =$ *constant* heat rejection to initial state

(*a*) Show the cycle on *P-v* and *T-s* diagrams.
(*b*) Calculate the maximum temperature in the cycle.
(*c*) Determine the thermal efficiency.

Answers: (*b*) 2100 K, (*c*) 15.8 percent

8-131 Repeat Prob. 8-130 using constant specific heats at room temperature.

8-132 A Carnot cycle is executed in a closed system and uses 0.002 kg of air as the working fluid. The cycle efficiency is 70 percent, and the lowest temperature in the cycle is 300 K. The pressure at the beginning of the isentropic expansion is 700 kPa, and at the end of the isentropic compression it is 1 MPa. Determine the net work output per cycle.

8-133 A four-cylinder spark-ignition engine has a compression ratio of 8, and each cylinder has a maximum volume of 0.6 L. At the beginning of the compression process, the air is at 98 kPa and 17°C, and the maximum temperature in the cycle is 1800 K. Assuming the engine to operate on the ideal Otto cycle, determine (*a*) the amount of heat supplied per cylinder, (*b*) the thermal efficiency, and (*c*) the number of revolutions per minute required for a net power output of 60 kW. Assume variable specific heats for air.

8-134 An ideal Otto cycle has a compression ratio of 9.2 and uses air as the working fluid. At the beginning of the compression process, air is at 98 kPa and 27°C. The pressure is doubled during the constant-volume heat-addition process. Accounting for the variation of specific heats with temperature, determine (*a*) the amount of heat transferred to the air, (*b*) the net work output, (*c*) the thermal efficiency, and (*d*) the mean effective pressure for the cycle.

8-135 Repeat Prob. 8-134 using constant specific heats at room temperature.

8-136 Consider an engine operating on the ideal Diesel cycle with air as the working fluid. The volume of the cylinder is 1200 cm^3 at the beginning of the compression process, 75 cm^3 at the end, and 150 cm^3 after the heat-addition process. Air is at 17°C and 100 kPa at the beginning of the compression process. Determine (*a*) the pressure at the beginning of the heat-rejection process; (*b*) the net work per cycle, in kJ; and (*c*) the mean effective pressure.

8-137 Repeat Prob. 8-136 using argon as the working fluid.

8-138E An ideal dual cycle has a compression ratio of 12 and uses air as the working fluid. At the beginning of the compression process, air is at 14.7 psia and 90°F, and occupies a volume of 75 in^3. During the heat-addition process, 0.3 Btu of heat is transferred to air at constant volume and 1.1 Btu at constant pressure. Using constant specific heats evaluated at room temperature, determine the thermal efficiency of the cycle.

8-139 Consider an ideal Stirling cycle using air as the working fluid. Air is at 350 K and 200 kPa at the beginning of the isothermal compression process, and heat is supplied to air from a source at 1600 K in the amount of 800 kJ/kg. Determine (*a*) the maximum pressure in the cycle and (*b*) the net work output per unit mass of air. *Answers:* (*a*) 5218 kPa, (*b*) 625 kJ/kg

8-140 Consider a simple ideal Brayton cycle with air as the working fluid. The pressure ratio of the cycle is 6, and the minimum and maximum temperatures are 300 and 1300 K, respectively. Now the pressure ratio is doubled without changing the minimum and maximum temperatures in the cycle. Determine the change in (*a*) the net work output per unit mass and (*b*) the thermal efficiency of the cycle as a result of this modification. Assume variable specific heats for air. *Answers:* (*a*) 41.5 kJ/kg, (*b*) 10.6 percent

8-141 Repeat Prob. 8-140 using constant specific heats at room temperature.

8-142 Helium is used as the working fluid in a Brayton cycle with regeneration. The pressure ratio of the cycle is 8, the compressor inlet temperature is 300 K, and the turbine inlet temperature is 1800 K. The effectiveness of the regenerator is 75 percent. Determine the thermal efficiency and the required mass flow rate of helium for a net power output of 30 MW, assuming both the compressor and the turbine have an isentropic efficiency of (*a*) 100 percent and (*b*) 80 percent.

8-143 A gas-turbine engine with regeneration operates with two stages of compression and two stages of expansion. The pressure ratio across each stage of the compressor and turbine is 3.5. The air enters each stage of the compressor at 300 K and each stage of the turbine at 1200 K. The compressor and turbine efficiencies are 78 and 86 percent, respectively, and the effectiveness of the regenerator is 72 percent. Determine the back work ratio and the thermal efficiency of the cycle, assuming constant specific heats for air at room temperature. *Answers:* 53.2 percent, 39.2 percent

8-144 Repeat Prob. 8-143 using helium as the working fluid.

8-145 Consider the ideal regenerative Brayton cycle. Determine the pressure ratio that maximizes the thermal efficiency of the cycle and compare this value with the pressure ratio that maximizes the cycle net work. For the same maximum-to-minimum temperature ratios, explain why the pressure ratio for maximum efficiency is less than the pressure ratio for maximum work.

8-146 Consider an ideal gas-turbine cycle with one stage of compression and two stages of expansion and regeneration. The pressure ratio across each turbine stage is the same. The high-pressure turbine exhaust gas enters the regenerator and then enters the low-pressure turbine for expansion to the compressor inlet pressure. Determine the thermal efficiency of this cycle as a function of the compressor pressure ratio and the high-pressure turbine to compressor inlet temperature ratio. Compare your result with the efficiency of the standard regenerative cycle.

Computer, Design, and Essay Problems

8-147 Write a computer program to study the effect of variable specific heats on the thermal efficiency of the ideal Otto cycle using air as the working fluid. At the beginning of the compression process, air is at 100 kPa and 300 K. Use the equation in Table A-2*c* to account for the variation of specific heats with temperature. Determine the percentage of error involved in using constant specific heat values at room temperature for the following combinations of compression ratios and maximum cycle temperatures: r = 7, 8, 9, 10, 11, 12 and T_{max} = 1200, 1400, 1600, 1800, 2000, 2500 K.

8-148 Write a computer program to determine the effects of pressure ratio, maximum cycle temperature, and compressor and turbine inefficiencies on the net work output per unit mass and the thermal efficiency of a simple Brayton cycle. Assume the working fluid is air that is at 100 kPa and 300 K at the compressor inlet. Also, assume constant specific heats for air at room temperature. Determine the net work output and the thermal efficiency for all combinations of the following parameters:

Pressure ratio: 5, 8, 14

Maximum cycle temperature: 800, 1200, 1600 K

Compressor adiabatic efficiency: 80, 100 percent

Turbine adiabatic efficiency: 80, 100 percent

Draw conclusions from the results.

8-149 Repeat Prob. 8-148 using helium as the working fluid.

8-150 Repeat Prob. 8-148 by considering the variation of specific heats of air with temperature. Use the specific-heat expressions for air given in Table A-2(*c*).

8-151 Write a computer program to determine the effects of pressure ratio, maximum cycle temperature, regenerator effectiveness, and compressor and turbine efficiencies on the net work output per unit mass and on the thermal efficiency of a regenerative Brayton cycle. Assume the working fluid is air that is at 100 kPa and 300 K at the compressor inlet. Also, assume constant specific heats for air at room temperature. Determine the net work output and the thermal efficiency for all combinations of the following parameters:

Pressure ratio: 5, 8, 14

Maximum cycle temperature: 1000, 1400, 1600 K

Compressor adiabatic efficiency: 80, 100 percent

Turbine adiabatic efficiency: 80, 100 percent

Regenerator effectiveness: 70, 80 percent

8-152 Repeat Prob. 8-151 using helium as the working fluid.

8-153 Repeat Prob. 8-151 by considering the variation of specific heats of air with temperature. Use the specific-heat expressions for air given in Table A-2*c*.

8-154 Write a computer program to determine the effect of the number of compression and expansion stages on the thermal efficiency of an ideal regenerative Brayton cycle with multistage compression and expansion. Assume that the overall pressure ratio of the cycle is 12, and the air enters each stage of the compressor at 300 K and each stage of the turbine at 1200 K. Using constant specific heats for air at room temperature, determine the thermal efficiency of the cycle by varying the number of stages from 1 to 20. Plot the thermal efficiency versus the number of stages. Compare your results to the efficiency of an Ericsson cycle operating between the same temperature limits.

8-155 Repeat Prob. 8-154 using helium as the working fluid.

8-156 Design a closed-system air-standard gas power cycle composed of three processes and having a minimum thermal efficiency of 20 percent. The processes may be isothermal, isobaric, isochoric, isentropic, polytropic, or pressure as a linear function of volume. Prepare an engineering report describing your design, showing the system, *P-v* and *T-s* diagrams, and sample calculations.

8-157 Design a closed-system air-standard gas power cycle composed of three processes and having a minimum thermal efficiency of 20 percent. The processes may be isothermal, isobaric, isochoric, isentropic, polytropic, or pressure as a linear function of volume; however, the Otto, Diesel, Ericsson, and Stirling cycles may not be used. Prepare an engineering report describing your design, showing the system, *P-v* and *T-s* diagrams, and sample calculations.

8-158 Write an essay on the most recent developments on the two-stroke engines, and find out when we might be seeing cars powered by two-stroke engines in the market. Why do the major car manufacturers have a renewed interest in two-stroke engines?

8-159 In response to concerns about the environment, some major car manufacturers are currently marketing electric cars. Write an essay on the advantages and disadvantages of electric cars, and discuss when it is advisable to purchase an electric car instead of a traditional internal combustion car.

8-160 Intense research is underway to develop adiabatic engines that require no cooling of the engine block. Such engines are based on ceramic materials because of the ability of such materials to withstand high temperatures. Write an essay on the current status of adiabatic engine development. Also determine the highest possible efficiencies with these engines, and compare them to the highest possible efficiencies of current engines.

8-161 Since its introduction in 1903 by Aegidius Elling of Norway, steam injection between the combustion chamber and the turbine is used even in some modern gas turbines currently in operation to cool the combustion gases to a metallurgical-safe temperature while increasing the mass flow rate through the turbine. Currently there are several gas-turbine power plants that use steam injection to augment power and improve thermal efficiency.

Consider a gas-turbine power plant whose pressure ratio is 8. The isentropic efficiencies of the compressor and the turbine are 80 percent, and there is a regenerator with an effectiveness of 70 percent. When the mass flow rate of air through the compressor is 40 kg/s, the turbine inlet temperature becomes 1700 K. But the turbine inlet temperature is limited to 1500 K, and thus steam injection into the combustion gases is being considered. However, to avoid the complexities associated with steam injection, it is proposed to use excess air (that is, to take in much more air than needed for complete combustion) to lower the combustion and thus turbine inlet temperature while increasing the mass flow rate and thus power output of the turbine. Evaluate this proposal, and compare the thermodynamic performance of "high air flow" to that of a "steam-injection" gas-turbine power plant under the following design conditions: the ambient air is at 100 kPa and 25°C, adequate water supply is available at 20°C, and the amount of fuel supplied to the combustion chamber remains constant.

8-162 Using the enclosed software, determine the effects of compression ratio on the net work output and the thermal efficiency of the Otto cycle for a maximum cycle temperature of 2000 K. Take the working fluid to be air that is at 100 kPa and 300 K at the beginning of the compression process, and assume variable specific heats. Vary the compression ratio from 6 to 15 with an increment of 1. Tabulate and plot your results against the compression ratio.

8-163 Using the enclosed software, determine the effects of pressure ratio on the net work output and the thermal efficiency of the Brayton cycle for a maximum cycle temperature of 1800 K. Take the working fluid to be air that is at 100 kPa and 300 K at the beginning of the compression process, and assume variable specific heats. Vary the pressure ratio from 5 to 24 with an increment of 1. Tabulate and plot your results against the pressure ratio. At what pressure ratio does the net work output become a maximum? At what pressure ratio does the thermal efficiency become a maximum?

8-164 Repeat Prob. 8-163 assuming adiabatic efficiencies of 85 percent for both the turbine and the compressor.

8-165 Repeat Prob. 8-147 using the enclosed software.

8-166 Repeat Prob. 8-148 using the enclosed software.

8-167 Repeat Prob. 8-149 using the enclosed software.

Vapor and Combined Power Cycles

CHAPTER

9

In Chap. 8 we discussed gas power cycles for which the working fluid remains a gas throughout the entire cycle. In this chapter, we consider *vapor power cycles* in which the working fluid is alternatively vaporized and condensed. We also consider power generation coupled with process heating called *cogeneration*.

The continued quest for higher thermal efficiencies has resulted in some innovative modifications to the basic vapor power cycle. Among these, we discuss the *reheat* and *regenerative cycles* as well as power cycles that consist of two separate cycles known as *binary cycles* and *combined cycles* where the heat rejected by one fluid is used as the heat input to another fluid operating at a lower temperature.

Steam is the most common working fluid used in vapor power cycles because of its many desirable characteristics, such as low cost, availability, and high enthalpy of vaporization. Other working fluids used include sodium, potassium, and mercury for high-temperature applications and some organic fluids such as benzene and the freons for low-temperature applications. The majority of this chapter is devoted to the discussion of steam power plants, which produce most of the electric power in the world today.

Steam power plants are commonly referred to as *coal plants, nuclear plants,* or *natural gas plants,* depending on the type of fuel used to supply heat to the steam. But the steam goes through the same basic cycle in all of them. Therefore, all can be analyzed in the same manner.

9-1 ■ THE CARNOT VAPOR CYCLE

We have mentioned many times that the Carnot cycle is the most efficient cycle operating between two specified temperature levels. Thus it is natural to look at the Carnot cycle first as a prospective ideal cycle for vapor power plants. If we could, we would certainly adopt it as the ideal cycle. But as explained below, the Carnot cycle is not a suitable model for power cycles. Throughout the discussions, we assume *steam* to be the working fluid since it is the working fluid predominantly used in vapor power cycles.

Consider a steady-flow *Carnot cycle* executed within the saturation dome of a pure substance, as shown in Fig. 9-1*a*. The fluid is heated reversibly and isothermally in a boiler (process 1-2), expanded isentropically in a turbine (process 2-3), condensed reversibly and isothermally in a condenser (process 3-4), and compressed isentropically by a compressor to the initial state (process 4-1).

Several impracticalities are associated with this cycle:

1. Isothermal heat transfer to or from a two-phase system is not difficult to achieve in practice since maintaining a constant pressure in the device will automatically fix the temperature at the saturation value. Therefore, processes 1-2 and 3-4 can be approached closely in actual boilers and condensers. Limiting the heat transfer processes to two-phase systems, however, severely limits the maximum temperature that can be used in the cycle (it has to remain under the critical-point value, which is 374°C for water). Limiting the maximum temperature in the cycle also limits the thermal efficiency. Any attempt to raise the maximum temperature in the cycle will involve heat transfer to the working fluid in a single phase, which is not easy to accomplish isothermally.

2. The isentropic expansion process (process 2-3) can be approximated closely by a well-designed turbine. However, the quality of the steam decreases during this process, as shown on the *T-s* diagram in Fig. 9-1*a*. Thus the turbine will have to handle steam with low quality, that is, steam with a high moisture content. The impingement of liquid droplets on the turbine blades causes erosion and is a major source of wear. Thus steam with qualities

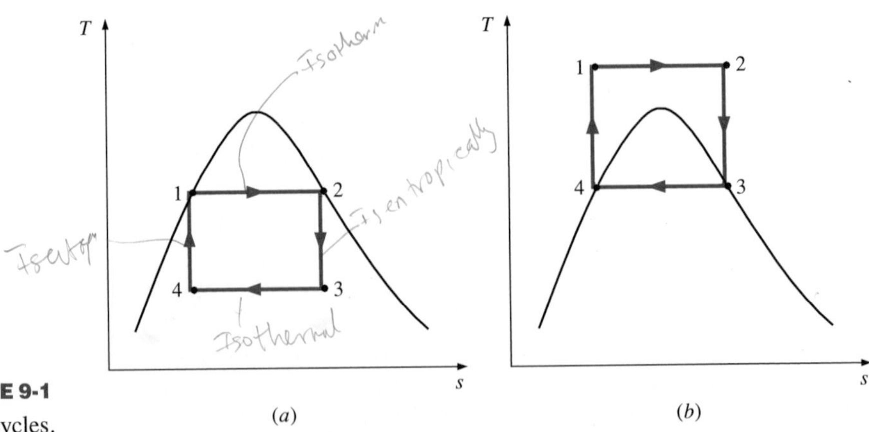

FIGURE 9-1
T-s diagram of two Carnot vapor cycles.

less than about 90 percent cannot be tolerated in the operation of power plants. This problem could be eliminated by using a working fluid with a very steep saturated vapor line.

3. The isentropic compression process (process 4-1) involves the compression of a liquid–vapor mixture to a saturated liquid. There are two difficulties associated with this process. First, it is not easy to control the condensation process so precisely as to end up with the desired quality at state 4. Second, it is not practical to design a compressor that will handle two phases.

Some of these problems could be eliminated by executing the Carnot cycle in a different way, as shown in Fig. 9-1*b*. This cycle, however, presents other problems such as isentropic compression to extremely high pressures and isothermal heat transfer at variable pressures. Thus we conclude that the Carnot cycle cannot be approximated in actual devices and is not a realistic model for vapor power cycles.

9-2 ■ RANKINE CYCLE: The Ideal Cycle for Vapor Power Cycles

Many of the impracticalities associated with the Carnot cycle can be eliminated by superheating the steam in the boiler and condensing it completely in the condenser, as shown schematically on a *T-s* diagram in Fig. 9-2. The cycle that results is the **Rankine cycle,** which is the ideal cycle for vapor power plants. The ideal Rankine cycle does not involve any internal irreversibilities and consists of the following four processes:

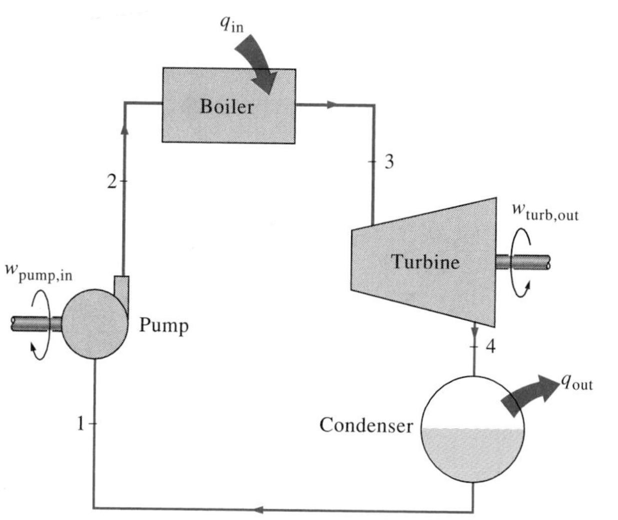

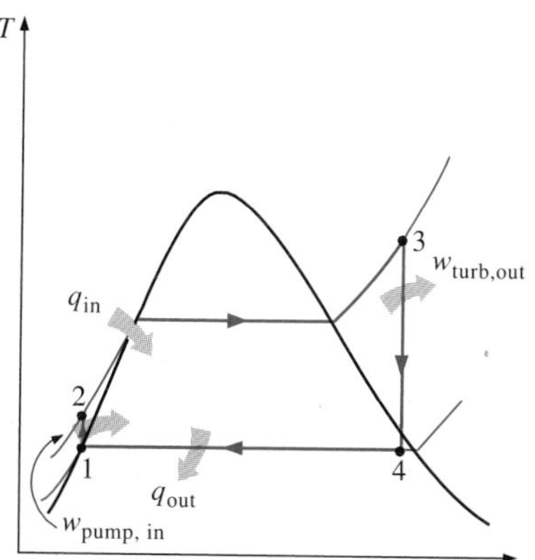

FIGURE 9-2
The simple ideal Rankine cycle.

1-2 Isentropic compression in a pump

2-3 Constant pressure heat addition in a boiler

3-4 Isentropic expansion in a turbine

4-1 Constant pressure heat rejection in a condenser

Water enters the *pump* at state 1 as saturated liquid and is compressed isentropically to the operating pressure of the boiler. The water temperature increases somewhat during this isentropic compression process due to a slight decrease in the specific volume of the water. The vertical distance between states 1 and 2 on the *T-s* diagram is greatly exaggerated for clarity. (If water were truly incompressible, would there be a temperature change at all during this process?)

Water enters the *boiler* as a compressed liquid at state 2 and leaves as a superheated vapor at state 3. The boiler is basically a large heat exchanger where the heat originating from combustion gases, nuclear reactors, or other sources is transferred to the water essentially at constant pressure. The boiler, together with the section where the steam is superheated (the superheater), is often called the *steam generator*.

The superheated vapor at state 3 enters the *turbine*, where it expands isentropically and produces work by rotating the shaft connected to an electric generator. The pressure and the temperature of the steam drop during this process to the values at state 4, where steam enters the *condenser*. At this state, steam is usually a saturated liquid–vapor mixture with a high quality. Steam is condensed at constant pressure in the condenser, which is basically a large heat exchanger, by rejecting heat to a cooling medium such as a lake, a river, or the atmosphere. Steam leaves the condenser as saturated liquid and enters the pump, completing the cycle. In areas where water is precious, the power plants are cooled by air instead of water. This method of cooling, which is also used in car engines, is called *dry cooling*. Several power plants in the world and a few in the United States use dry cooling to conserve water.

Remembering that the area under the process curve on a *T-s* diagram represents the heat transfer for internally reversible processes, we see that the area under process curve 2-3 represents the heat transferred to the water in the boiler and the area under the process curve 4-1 represents the heat rejected in the condenser. The difference between these two (the area enclosed by the cycle) is the net work produced during the cycle.

Energy Analysis of the Ideal Rankine Cycle

All four components associated with the Rankine cycle (the pump, boiler, turbine, and condenser) are steady-flow devices, and thus all four processes that make up the Rankine cycle can be analyzed as steady-flow processes. The kinetic and potential energy changes of the steam are usually small relative to the work and heat transfer terms and are therefore usually neglected. Then the *steady-flow energy equation* per unit mass of steam reduces to

$$(q_{in} - q_{out}) + (w_{in} - w_{out}) = h_e - h_i \qquad \text{(kJ/kg)} \qquad (9\text{-}1)$$

The boiler and the condenser do not involve any work, and the pump and the turbine are assumed to be isentropic. Then the conservation of energy relation for each device can be expressed as follows:

Pump ($q = 0$):
$$w_{pump,in} = h_2 - h_1 \qquad (9\text{-}2)$$

or, from Eq. 6-53,
$$w_{pump,in} = v(P_2 - P_1) \qquad (9\text{-}3)$$

where
$$h_1 = h_{f@P_1} \quad \text{and} \quad v \cong v_1 = v_{f@P_1} \qquad (9\text{-}4)$$

Boiler ($w = 0$):
$$q_{in} = h_3 - h_2 \qquad (9\text{-}5)$$

Turbine ($q = 0$):
$$w_{turb,out} = h_3 - h_4 \qquad (9\text{-}6)$$

Condenser ($w = 0$):
$$q_{out} = h_4 - h_1 \qquad (9\text{-}7)$$

The *thermal efficiency* of the Rankine cycle is determined from

$$\eta_{th} = \frac{w_{net}}{q_{in}} = 1 - \frac{q_{out}}{q_{in}} \qquad (9\text{-}8)$$

where
$$w_{net} = q_{in} - q_{out} = w_{turb,out} - w_{pump,in}$$

The conversion efficiency of power plants in the United States is often expressed in terms of **heat rate,** which is the amount of heat supplied, in Btu's, to generate 1 kWh of electricity. The smaller the heat rate, the greater the efficiency. Considering that 1 kWh = 3412 Btu, the relation between the heat rate and the thermal efficiency can be expressed as

$$\eta_{th} = \frac{3412 \text{ (Btu/kWh)}}{\text{Heat rate (Btu/kWh)}} \qquad (9\text{-}9)$$

For example, a heat rate of 11,363 Btu/kWh is equivalent to 30 percent thermal efficiency.

The thermal efficiency can also be interpreted as the ratio of the area enclosed by the cycle on a *T-s* diagram to the area under the heat-addition process. The use of these relations is illustrated in the following example.

EXAMPLE 9-1 The Simple Ideal Rankine Cycle

Consider a steam power plant operating on the simple ideal Rankine cycle. The steam enters the turbine at 3 MPa and 350°C and is condensed in the condenser at a pressure of 75 kPa. Determine the thermal efficiency of this cycle.

Solution The schematic of the power plant and the *T-s* diagram of the cycle are shown in Fig. 9-3. We note that the power plant involves steady-flow components and operates on the ideal Rankine cycle. Therefore, the pump and the turbine are isentropic, there are no pressure drops in the boiler and condenser, and steam leaves the condenser and enters the pump as saturated liquid at the condenser pressure.

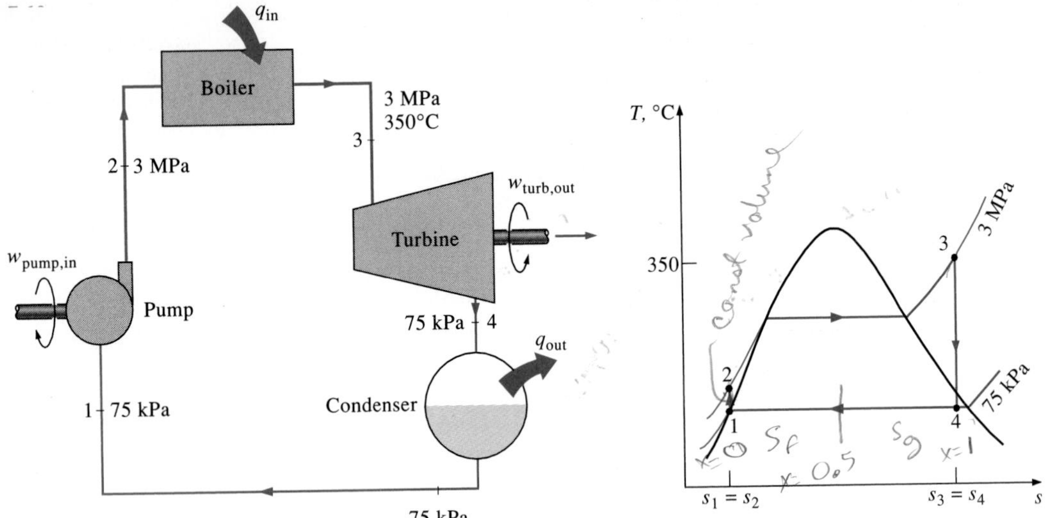

FIGURE 9-3
Schematic and *T-s* diagram for Example 9-1.

Assumptions **1** Steady operating conditions exist. **2** Kinetic and potential energy changes are negligible.

Analysis First we determine the enthalpies at various points in the cycle, using data from steam tables (Tables A-4, A-5, and A-6):

State 1: $\left.\begin{array}{l} P_1 = 75 \text{ kPa} \\ \text{Sat. liquid} \end{array}\right\}$ $\begin{array}{l} h_1 = h_{f\,@\,75\text{ kPa}} = 384.39 \text{ kJ/kg} \\ v_1 = v_{f\,@\,75\text{ kPa}} = 0.001037 \text{ m}^3\text{/kg} \end{array}$

State 2: $P_2 = 3$ MPa

$s_2 = s_1$

$$w_{\text{pump, in}} = v_1(P_2 - P_1) = (0.001037 \text{ m}^3\text{/kg})[(3000 - 75) \text{ kPa}]\left(\frac{1 \text{ kJ}}{1 \text{ kPa}\cdot\text{m}^3}\right)$$
$$= 3.03 \text{ kJ/kg}$$

$$h_2 = h_1 + w_{\text{pump, in}} = (384.39 + 3.03) \text{ kJ/kg} = 387.42 \text{ kJ/kg}$$

State 3: $\left.\begin{array}{l} P_3 = 3 \text{ MPa} \\ T_3 = 350\text{°C} \end{array}\right\}$ $\begin{array}{l} h_3 = 3115.3 \text{ kJ/kg} \\ s_3 = 6.7428 \text{ kJ/(kg}\cdot\text{K)} \end{array}$

State 4: $P_4 = 75$ kPa (sat. mixture)

$s_4 = s_3$

$$x_4 = \frac{s_4 - s_f}{s_{fg}} = \frac{6.7428 - 1.213}{6.2434} = 0.886$$

$$h_4 = h_f + x_4 h_{fg} = 384.39 + 0.886(2278.6) = 2403.2 \text{ kJ/kg}$$

Thus, $q_{in} = h_3 - h_2 = (3115.3 - 387.42) \text{ kJ/kg} = 2727.88 \text{ kJ/kg}$

$q_{out} = h_4 - h_1 = (2403.2 - 384.39) \text{ kJ/kg} = 2018.81 \text{ kJ/kg}$

and $\eta_{th} = 1 - \dfrac{q_{out}}{q_{in}} = 1 - \dfrac{2018.81 \text{ kJ/kg}}{2727.88 \text{ kJ/kg}} = \mathbf{0.260 \text{ or } 26.0\%}$

The thermal efficiency could also be determined from

$$w_{\text{turb, out}} = h_3 - h_4 = (3115.3 - 2403.2) \text{ kJ/kg} = 712.1 \text{ kJ/kg}$$

$$w_{\text{net}} = w_{\text{turb, out}} - w_{\text{pump, in}} = (712.1 - 3.03) \text{ kJ/kg} = 709.07 \text{ kJ/kg}$$

or

$$w_{\text{net}} = q_{\text{in}} - q_{\text{out}} = (2727.88 - 2018.81) \text{ kJ/kg} = 709.07 \text{ kJ/kg}$$

and

$$\eta_{\text{th}} = \frac{w_{\text{net}}}{q_{\text{in}}} = \frac{709.07 \text{ kJ/kg}}{2727.88 \text{ kJ/kg}} = \mathbf{0.260 \text{ or } 26.0\%}$$

That is, this power plant converts 26 percent of the heat it receives in the boiler to net work. An actual power plant operating between the same temperature and pressure limits will have a lower efficiency because of the irreversibilities such as friction.

Discussion Notice that the back work ratio ($r_{\text{bw}} = w_{\text{in}}/w_{\text{out}}$) of this power plant is 0.004, and thus only 0.4 percent of the turbine work output is required to operate the pump. Having such low back work ratios is characteristic of vapor power cycles. This is in contrast to the gas power cycles, which typically have very high back work ratios (about 40 to 80 percent).

It is also interesting to note the thermal efficiency of a Carnot cycle operating between the same temperature limits

$$\eta_{\text{th, Carnot}} = 1 - \frac{T_{\min}}{T_{\max}} = 1 - \frac{(91.78 + 273) \text{ K}}{(350 + 273) \text{ K}} = 0.414$$

The difference between the two efficiencies is due to the large temperature difference between the steam and the combustion gases during the heat-addition process.

9-3 ■ DEVIATION OF ACTUAL VAPOR POWER CYCLES FROM IDEALIZED ONES

The actual vapor power cycle differs from the ideal Rankine cycle, as illustrated in Fig. 9-4*a*, as a result of irreversibilities in various components. Fluid friction and undesired heat loss to the surroundings are the two most common sources of irreversibilities.

Fluid friction causes pressure drops in the boiler, the condenser, and the piping between various components. As a result, steam leaves the boiler at a

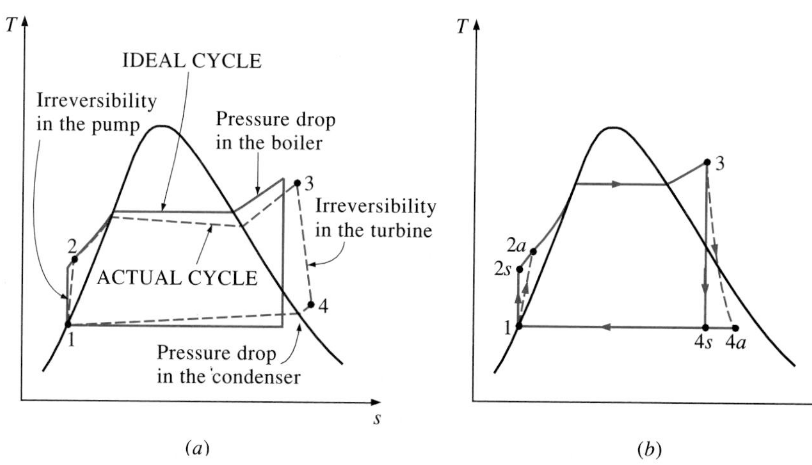

FIGURE 9-4

(*a*) Deviation of actual vapor power cycle from the ideal Rankine cycle. (*b*) The effect of pump and turbine irreversibilities on the ideal Rankine cycle.

somewhat lower pressure. Also, the pressure at the turbine inlet is somewhat lower than that at the boiler exit due to the pressure drop in the connecting pipes. The pressure drop in the condenser is usually very small. To compensate for these pressure drops, the water must be pumped to a sufficiently higher pressure than the ideal cycle calls for. This requires a larger pump and larger work input to the pump.

The other major source of irreversibility is the *heat loss* from the steam to the surroundings as the steam flows through various components. To maintain the same level of net work output, more heat needs to be transferred to the steam in the boiler to compensate for these undesired heat losses. As a result, cycle efficiency decreases.

Of particular importance are the irreversibilities occurring within the pump and the turbine. A pump requires a greater work input, and a turbine produces a smaller work output as a result of irreversibilities. Under ideal conditions, the flow through these devices is isentropic. The deviation of actual pumps and turbines from the isentropic ones can be accurately accounted for, however, by utilizing *adiabatic efficiencies,* defined as

$$\eta_P = \frac{w_s}{w_a} = \frac{h_{2s} - h_1}{h_{2a} - h_1} \tag{9-10}$$

and

$$\eta_T = \frac{w_a}{w_s} = \frac{h_3 - h_{4a}}{h_3 - h_{4s}} \tag{9-11}$$

where states 2a and 4a are the actual exit states of the pump and the turbine, respectively, and 2s and 4s are the corresponding states for the isentropic case (Fig. 9-4b).

Other factors also need to be considered in the analysis of actual vapor power cycles. In actual condensers, for example, the liquid is usually subcooled to prevent the onset of *cavitation,* the rapid vaporization and condensation of the fluid at the low-pressure side of the pump impeller, which may damage it. Additional losses occur at the bearings between the moving parts as a result of friction. Steam that leaks out during the cycle and air that leaks into the condenser represent two other sources of loss. Finally, the power consumed by the auxiliary equipment such as fans that supply air to the furnace should also be considered in evaluating the performance of actual power plants.

The effect of irreversibilities on the thermal efficiency of a steam power cycle is illustrated below with an example.

EXAMPLE 9-2 An Actual Steam Power Cycle

A steam power plant operates on the cycle shown in Fig. 9-5. If the adiabatic efficiency of the turbine is 87 percent and the adiabatic efficiency of the pump is 85 percent, determine (*a*) the thermal efficiency of the cycle and (*b*) the net power output of the plant for a mass flow rate of 15 kg/s.

Solution The schematic of the power plant and the *T-s* diagram of the cycle are shown in Fig. 9-5. The temperatures and pressures of steam at various points are also indicated on the figure. We note that the power plant involves steady-

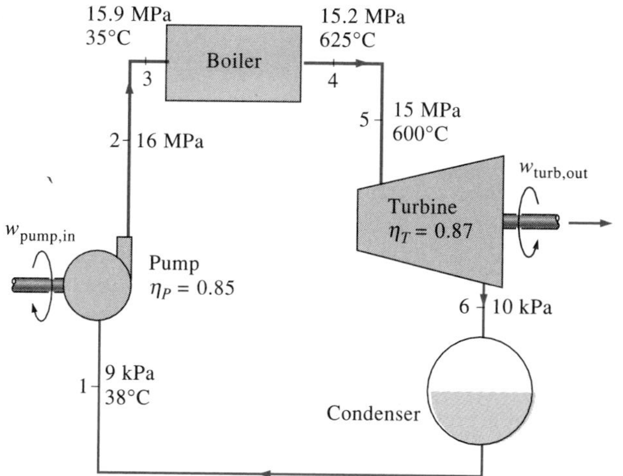

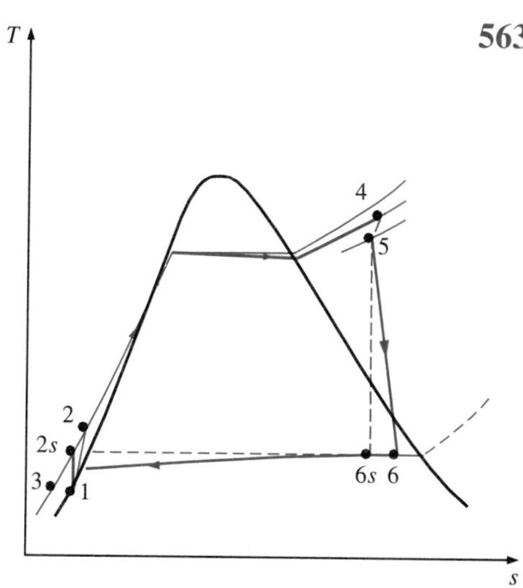

FIGURE 9-5

Schematic and *T-s* diagram for Example 9-2.

flow components and operates on the Rankine cycle, but the imperfections at various components are accounted for.

Assumptions **1** Steady operating conditions exist. **2** Kinetic and potential energy changes are negligible.

Analysis (*a*) The thermal efficiency of a cycle is the ratio of the net work output to the heat input, and it is determined as follows:

Pump work input:

$$w_{\text{pump, in}} = \frac{w_{s,\text{ pump, in}}}{\eta_P} = \frac{v_1(P_2 - P_1)}{\eta_P}$$

$$= \frac{(0.001008\text{ m}^3/\text{kg})[(16{,}000 - 9)\text{ kPa}]}{0.85}\left(\frac{1\text{ kJ}}{1\text{ kPa}\cdot\text{m}^3}\right)$$

$$= 19.0\text{ kJ/kg}$$

Turbine work output:

$$w_{\text{turb, out}} = \eta_T w_{s,\text{ turb, out}}$$

$$= \eta_T(h_5 - h_{6s}) = 0.87(3582.3 - 2115.7)\text{ kJ/kg}$$

$$= 1275.9\text{ kJ/kg}$$

Boiler heat input: $q_{\text{in}} = h_4 - h_3 = (3645.7 - 146.7)\text{ kJ/kg} = 3499.0\text{ kJ/kg}$

Thus, $w_{\text{net}} = w_{\text{turb, out}} - w_{\text{pump, in}} = (1275.9 - 19.0)\text{ kJ/kg} = 1256.9\text{ kJ/kg}$

$$\eta_{\text{th}} = \frac{w_{\text{net}}}{q_{\text{in}}} = \frac{1256.9\text{ kJ/kg}}{3499.0\text{ kJ/kg}} = \mathbf{0.359\text{ or }35.9\%}$$

Without the irreversibilities, the thermal efficiency of this cycle would be 43.0 percent (see Example 9-3*c*).

(*b*) The power produced by this power plant is determined from

$$\dot{W}_{\text{net}} = \dot{m}(w_{\text{net}}) = (15\text{ kg/s})(1256.9\text{ kJ/kg}) = \mathbf{18{,}854\text{ kW}}$$

9-4 ■ HOW CAN WE INCREASE THE EFFICIENCY OF THE RANKINE CYCLE?

Steam power plants are responsible for the production of most electric power in the world, and even small increases in thermal efficiency can mean large savings from the fuel requirements. Therefore, every effort is made to improve the efficiency of the cycle on which steam power plants operate.

The basic idea behind all the modifications to increase the thermal efficiency of a power cycle is the same: *Increase the average temperature at which heat is transferred to the working fluid in the boiler, or decrease the average temperature at which heat is rejected from the working fluid in the condenser.* That is, the average fluid temperature should be as high as possible during heat addition and as low as possible during heat rejection. Next we discuss three ways of accomplishing this for the simple ideal Rankine cycle.

1 Lowering the Condenser Pressure

(*Lowers* $T_{low, av}$)

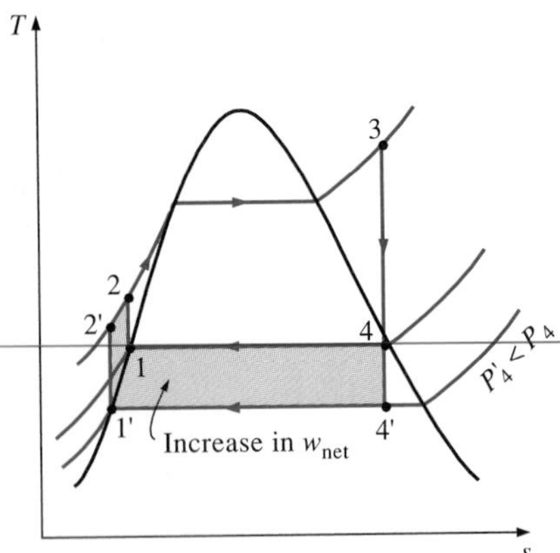

FIGURE 9-6

The effect of lowering the condenser pressure on the ideal Rankine cycle.

Steam exists as a saturated mixture in the condenser at the saturation temperature corresponding to the pressure inside the condenser. Therefore, lowering the operating pressure of the condenser automatically lowers the temperature of the steam, and thus the temperature at which heat is rejected.

The effect of lowering the condenser pressure on the Rankine cycle efficiency is illustrated on a *T-s* diagram in Fig. 9-6. For comparison purposes, the turbine inlet state is maintained the same. The colored area on this diagram represents the increase in net work output as a result of lowering the condenser pressure from P_4 to P'_4. The heat input requirements also increase (represented by the area under curve 2′-2), but this increase is very small. Thus the overall effect of lowering the condenser pressure is an increase in the thermal efficiency of the cycle.

To take advantage of the increased efficiencies at low pressures, the condensers of steam power plants usually operate well below the atmospheric pressure. This does not present a major problem since the vapor power cycles operate in a closed loop. However, there is a lower limit on the condenser pressure that can be used. It cannot be lower than the saturation pressure corresponding to the temperature of the cooling medium. Consider, for example, a condenser that is to be cooled by a nearby river at 15°C. Allowing a temperature difference of 10°C for effective heat transfer, the steam temperature in the condenser must be above 25°C; thus the condenser pressure must be above 3.2 kPa, which is the saturation pressure at 25°C.

Lowering the condenser pressure is not without any side effects, however. For one thing, it creates the possibility of air leakage into the condenser. More importantly, it increases the moisture content of the steam at the final stages of the turbine, as can be seen from Fig. 9-6. The presence of large quantities of moisture is highly undesirable in turbines because it decreases the turbine efficiency and erodes the turbine blades. Fortunately, this problem can be corrected, as discussed later in this chapter.

2 Superheating the Steam to High Temperatures (*Increases* $T_{high, av}$)

The average temperature at which heat is added to the steam can be increased without increasing the boiler pressure by superheating the steam to high temperatures. The effect of superheating on the performance of vapor power cycles is illustrated on a *T-s* diagram in Fig. 9-7. The colored area on this diagram represents the increase in the net work. The total area under the process curve 3-3′ represents the increase in the heat input. Thus both the net work and heat input increase as a result of superheating the steam to a higher temperature. The overall effect is an increase in thermal efficiency, however, since the average temperature at which heat is added increases.

Superheating the steam to higher temperatures has another very desirable effect: It decreases the moisture content of the steam at the turbine exit, as can be seen from the *T-s* diagram (the quality at state 4′ is higher than that at state 4).

The temperature to which steam can be superheated is limited, however, by metallurgical considerations. Presently the highest steam temperature allowed at the turbine inlet is about 620°C (1150°F). Any increase in this value depends on improving the present materials or finding new ones that can withstand higher temperatures. Ceramics are very promising in this regard.

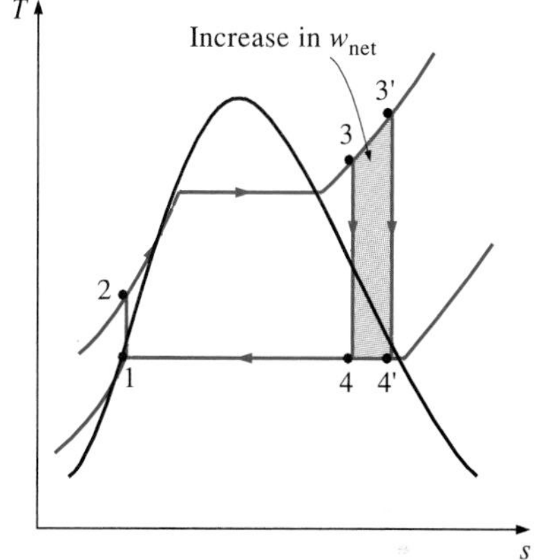

FIGURE 9-7

The effect of superheating the steam to higher temperatures on the ideal Rankine cycle.

3 Increasing the Boiler Pressure (*Increases* $T_{high, av}$)

Another way of increasing the average temperature during the heat-addition process is to increase the operating pressure of the boiler, which automatically raises the temperature at which boiling takes place. This, in turn, raises the average temperature at which heat is added to the steam and thus raises the thermal efficiency of the cycle.

The effect of increasing the boiler pressure on the performance of vapor power cycles is illustrated on a *T-s* diagram in Fig. 9-8. Notice that for a fixed turbine inlet temperature, the cycle shifts to the left and the moisture content of steam at the turbine exit increases. This undesirable side effect can be corrected, however, by reheating the steam, as discussed in the next section.

Operating pressures of boilers have gradually increased over the years from about 2.7 MPa (400 psia) in 1922 to over 30 MPa (4500 psia) today, generating enough steam to produce a net power output of 1000 MW or more in a large power plant. Today many modern steam power plants operate at supercritical pressures ($P > 22.09$ MPa) and have thermal efficiencies of about 40 percent for fossil-fuel plants and 34 percent for nuclear plants. There are about 170 supercritical-pressure steam power plants in operation in the United States. The lower efficiencies of nuclear power plants are due to the lower maximum temperatures used in those plants for safety reasons. The United States has 112 nuclear power plants, which generate about 21 percent of the nation's electricity. (In contrast, 75 percent of the electricity in France comes from nuclear plants.) The *T-s* diagram of a supercritical Rankine cycle is shown in Fig. 9-9.

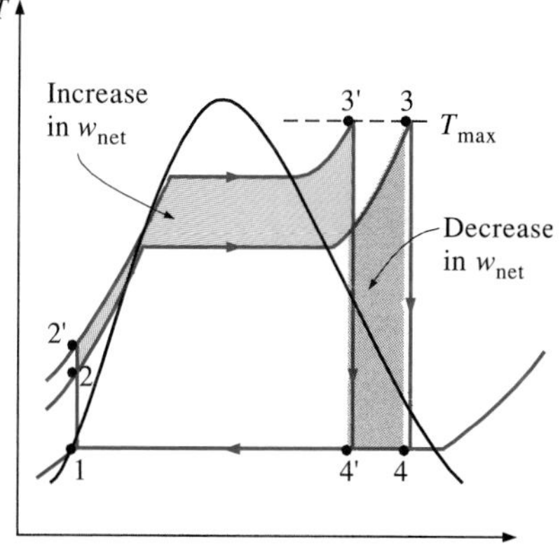

FIGURE 9-8

The effect of increasing the boiler pressure on the ideal Rankine cycle.

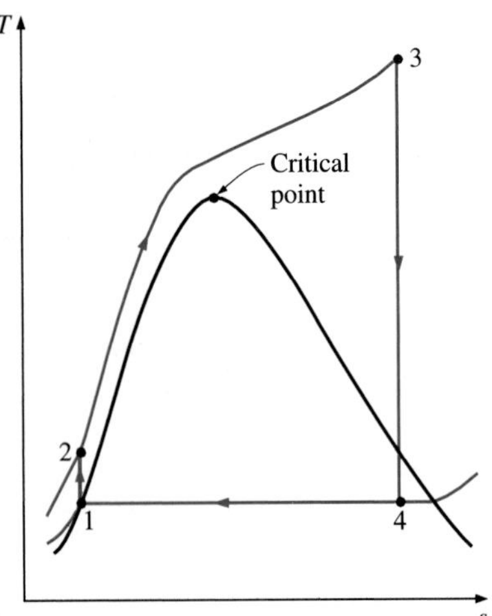

FIGURE 9-9
A supercritical Rankine cycle.

The effects of lowering the condenser pressure, superheating to a higher temperature, and increasing the boiler pressure on the thermal efficiency of the Rankine cycle are illustrated below with an example.

EXAMPLE 9-3 Effect of Boiler Pressure and Temperature on Efficiency

Consider a steam power plant operating on the ideal Rankine cycle. The steam enters the turbine at 3 MPa and 350°C and is condensed in the condenser at a pressure of 10 kPa. Determine (*a*) the thermal efficiency of this power plant, (*b*) the thermal efficiency if steam is superheated to 600°C instead of 350°C, and (*c*) the thermal efficiency if the boiler pressure is raised to 15 MPa while the turbine inlet temperature is maintained at 600°C.

Solution The *T-s* diagrams of the cycle for all three cases are given in Fig. 9-10.

Analysis (*a*) This is the steam power plant discussed in Example 9-1, except that the condenser pressure is lowered to 10 kPa. The thermal efficiency is determined in a similar manner:

State 1: $\left.\begin{array}{l} P_1 = 10\text{ kPa} \\ \text{Sat. liquid} \end{array}\right\}$ $\begin{array}{l} h_1 = h_{f\,@\,10\text{ kPa}} = 191.83\text{ kJ/kg} \\ v_1 = v_{f\,@\,10\text{ kPa}} = 0.001010\text{ m}^3\text{/kg} \end{array}$

State 2: $P_2 = 3\text{ MPa}$
$s_2 = s_1$

$$w_{\text{pump, in}} = v_1(P_2 - P_1) = (0.001010\text{ m}^3\text{/kg})[(3000 - 10)\text{kPa}]\left(\frac{1\text{ kJ}}{1\text{ kPa}\cdot\text{m}^3}\right)$$
$$= 3.01\text{ kJ/kg}$$
$$h_2 = h_1 + w_{\text{pump, in}} = (191.83 + 3.01)\text{ kJ/kg} = 194.84\text{ kJ/kg}$$

State 3: $\left.\begin{array}{l} P_3 = 3\text{ MPa} \\ T_3 = 350°\text{C} \end{array}\right\}$ $\begin{array}{l} h_3 = 3115.3\text{ kJ/kg} \\ s_3 = 6.7428\text{ kJ/(kg}\cdot\text{K)} \end{array}$

State 4: $\begin{array}{l} P_4 = 10\text{ kPa} \\ s_4 = s_3 \end{array}$ (sat. mixture)

$$x_4 = \frac{s_4 - s_f}{s_{fg}} = \frac{6.7428 - 0.6493}{7.5009} = 0.812$$
$$h_4 = h_f + x_4 h_{fg} = 191.83 + 0.812(2392.8) = 2134.8\text{ kJ/kg}$$

Thus, $$q_{\text{in}} = h_3 - h_2 = (3115.3 - 194.84)\text{ kJ/kg} = 2920.46\text{ kJ/kg}$$
$$q_{\text{out}} = h_4 - h_1 = (2134.8 - 191.83)\text{ kJ/kg} = 1942.97\text{ kJ/kg}$$

and $$\eta_{\text{th}} = 1 - \frac{q_{\text{out}}}{q_{\text{in}}} = 1 - \frac{1942.97\text{ kJ/kg}}{2920.46\text{ kJ/kg}} = \mathbf{0.335\text{ or }33.5\%}$$

Therefore, the thermal efficiency increases from 26.0 to 33.5 percent as a result of lowering the condenser pressure from 75 to 10 kPa. At the same time, however, the quality of the steam decreases from 0.886 to 0.812 (in other words, the moisture content increases from 11.4 to 18.8 percent).

(*b*) States 1 and 2 remain the same in this case, and the enthalpies at state 3 (3 MPa and 600°C) and state 4 (10 kPa and $s_4 = s_3$) are determined to be

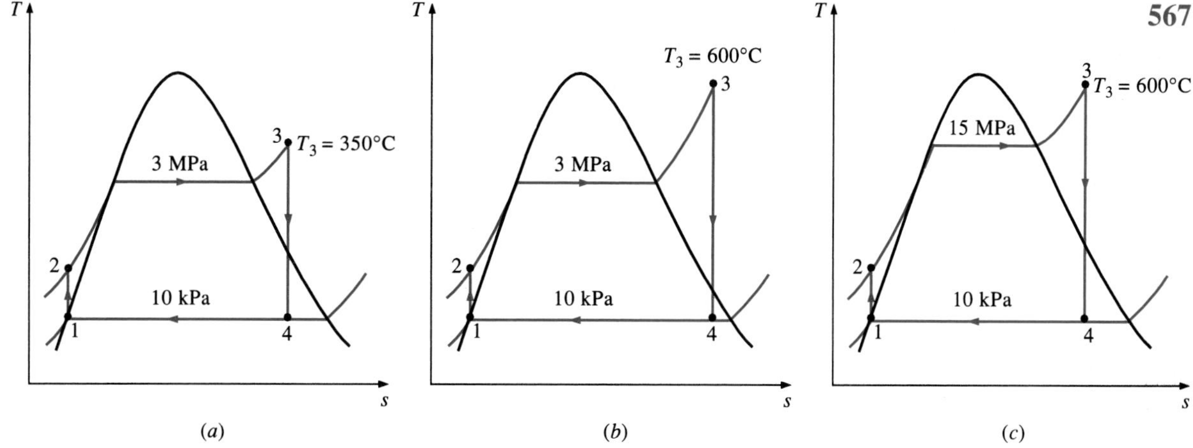

FIGURE 9-10
T-s diagrams of the three cycles discussed in Example 9-3.

$$h_3 = 3682.8 \text{ kJ/kg}$$
$$h_4 = 2378.8 \text{ kJ/kg} \qquad (x_4 = 0.914)$$

Thus,
$$q_{in} = h_3 - h_2 = 3682.3 - 194.84 = 3487.46 \text{ kJ/kg}$$
$$q_{out} = h_4 - h_1 = 2378.8 - 191.83 = 2186.97 \text{ kJ/kg}$$

and
$$\eta_{th} = 1 - \frac{q_{out}}{q_{in}} = 1 - \frac{2186.97 \text{ kJ/kg}}{3487.46 \text{ kJ/kg}} = \mathbf{0.373 \text{ or } 37.3\%}$$

Therefore, the thermal efficiency increases from 33.5 to 37.3 percent as a result of superheating the steam from 350 to 600°C. At the same time, the quality of the steam increases from 0.812 to 0.914 (in other words, the moisture content decreases from 18.8 to 8.6 percent).

(*c*) State 1 remains the same in this case, but the other states change. The enthalpies at state 2 (15 MPa and $s_2 = s_1$), state 3 (15 MPa and 600°C), and state 4 (10 kPa and $s_4 = s_3$) are determined in a similar manner to be

$$h_2 = 206.94 \text{ kJ/kg}$$
$$h_3 = 3582.3 \text{ kJ/kg}$$
$$h_4 = 2115.7 \text{ kJ/kg} \qquad (x_4 = 0.804)$$

Thus,
$$q_{in} = h_3 - h_2 = 3582.3 - 206.94 = 3375.36 \text{ kJ/kg}$$
$$q_{out} = h_4 - h_1 = 2115.7 - 191.83 = 1923.87 \text{ kJ/kg}$$

and
$$\eta_{th} = 1 - \frac{q_{out}}{q_{in}} = 1 - \frac{1923.87 \text{ kJ/kg}}{3375.36 \text{ kJ/kg}} = \mathbf{0.430 \text{ or } 43.0\%}$$

Discussion The thermal efficiency increases from 37.3 to 43.0 percent as a result of raising the boiler pressure from 3 to 15 MPa while maintaining the turbine inlet temperature at 600°C. At the same time, however, the quality of the steam decreases from 0.914 to 0.804 (in other words, the moisture content increases from 8.6 to 19.6 percent).

9-5 ■ THE IDEAL REHEAT RANKINE CYCLE

We noted in the last section that increasing the boiler pressure increases the thermal efficiency of the Rankine cycle, but it also increases the moisture content of the steam to unacceptable levels. Then it is natural to ask the following question:

> *How can we take advantage of the increased efficiencies at higher boiler pressures without facing the problem of excessive moisture at the final stages of the turbine?*

Two possibilities come to mind:

1. Superheat the steam to very high temperatures before it enters the turbine. This would be the desirable solution since the average temperature at which heat is added would also increase, thus increasing the cycle efficiency. This is not a viable solution, however, since it will require raising the steam temperature to metallurgically unsafe levels.

2. Expand the steam in the turbine in two stages, and reheat it in between. In other words, modify the simple ideal Rankine cycle with a **reheat** process. Reheating is a practical solution to the excessive moisture problem in turbines, and it is used frequently in modern steam power plants.

The *T-s* diagram of the ideal reheat Rankine cycle and the schematic of the power plant operating on this cycle are shown in Fig. 9-11. The ideal reheat Rankine cycle differs from the simple ideal Rankine cycle in that the expansion process takes place in two stages. In the first stage (the high-

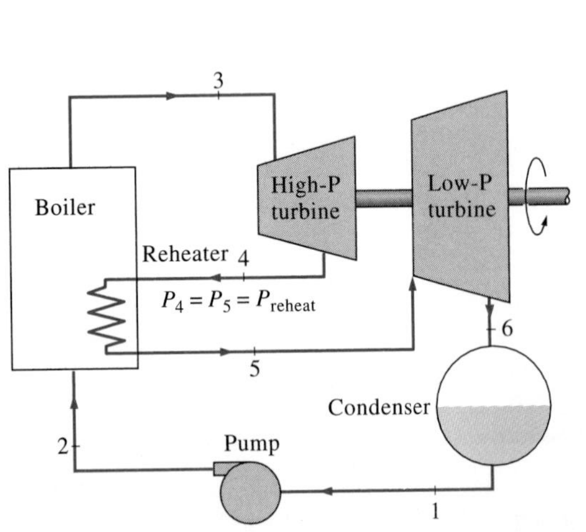

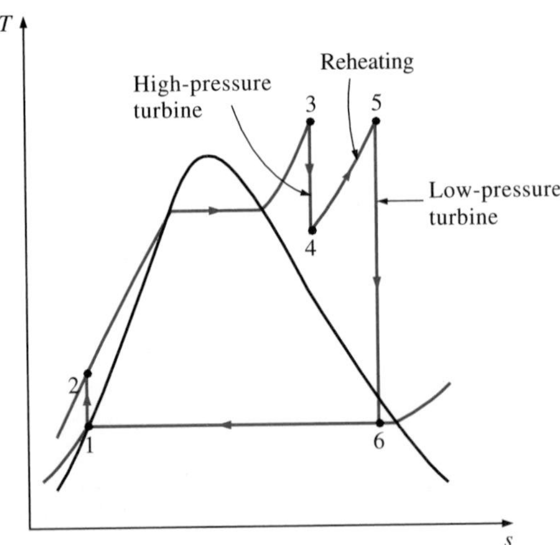

FIGURE 9-11
The ideal reheat Rankine cycle.

pressure turbine), steam is expanded isentropically to an intermediate pressure and sent back to the boiler where it is reheated at constant pressure, usually to the inlet temperature of the first turbine stage. Steam then expands isentropically in the second stage (low-pressure turbine) to the condenser pressure. Thus the total heat input and the total turbine work output for a reheat cycle become

$$q_{in} = q_{primary} + q_{reheat} = (h_3 - h_2) + (h_5 - h_4) \quad (9\text{-}12)$$

and

$$w_{turb,out} = w_{turb,I} + w_{turb,II} = (h_3 - h_4) + (h_5 - h_6) \quad (9\text{-}13)$$

The incorporation of the single reheat in a modern power plant improves the cycle efficiency by 4 to 5 percent by increasing the average temperature at which heat is added to the steam.

The average temperature during the reheat process can be increased by increasing the number of expansion and reheat stages. As the number of stages is increased, the expansion and reheat processes approach an isothermal process at the maximum temperature, as shown in Fig. 9-12. The use of more than two reheat stages, however, is not practical. The theoretical improvement in efficiency from the second reheat is about half of that which results from a single reheat. If the turbine inlet pressure is not high enough, double reheat would result in superheated exhaust. This is undesirable as it would cause the average temperature for heat rejection to increase and thus the cycle efficiency to decrease. Therefore, double reheat is used only on supercritical-pressure ($P > 22.09$ MPa) power plants. A third reheat stage would increase the cycle efficiency by about half of the improvement attained by the second reheat. This gain is too small to justify the added cost and complexity.

The reheat cycle was introduced in the mid 1920s, but it was abandoned in the 1930s because of the operational difficulties. The steady increase in boiler pressures over the years made it necessary to reintroduce single reheat in the late 1940s and double reheat in the early 1950s.

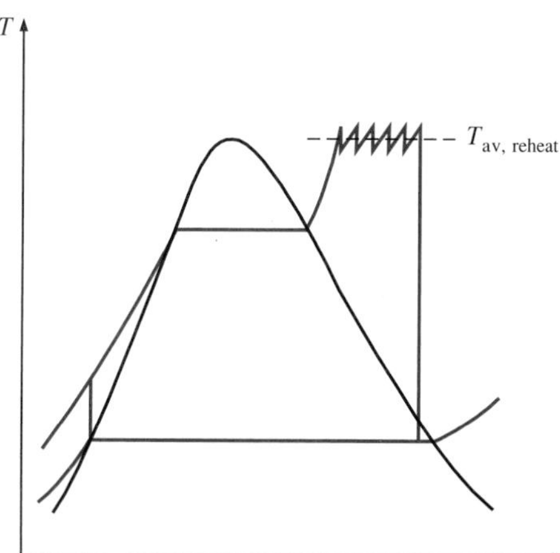

FIGURE 9-12

The average temperature at which heat is added during reheating increases as the number of reheat stages is increased.

The reheat temperatures are very close or equal to the turbine inlet temperature. The optimum reheat pressure is about one-fourth of the maximum cycle pressure. For example, the optimum reheat pressure for a cycle with a boiler pressure of 12 MPa is about 3 MPa.

Remember that the sole purpose of the reheat cycle is to reduce the moisture content of the steam at the final stages of the expansion process. If we had materials that could withstand sufficiently high temperatures, there would be no need for the reheat cycle.

EXAMPLE 9-4 The Ideal Reheat Rankine Cycle

Consider a steam power plant operating on the ideal reheat Rankine cycle. Steam enters the high-pressure turbine at 15 MPa and 600°C and is condensed in the condenser at a pressure of 10 kPa. If the moisture content of the steam at the exit of the low-pressure turbine is not to exceed 10.4 percent, determine (*a*) the pressure at which the steam should be reheated and (*b*) the thermal efficiency of the cycle. Assume the steam is reheated to the inlet temperature of the high-pressure turbine.

Solution The schematic of the power plant and the *T-s* diagram of the cycle are shown in Fig. 9-13. We note that the power plant involves steady-flow components and operates on the ideal reheat Rankine cycle. Therefore, the pump and the turbines are isentropic, there are no pressure drops in the boiler and condenser, and steam leaves the condenser and enters the pump as saturated liquid at the condenser pressure.

Assumptions **1** Steady operating conditions exist. **2** Kinetic and potential energy changes are negligible.

Analysis (*a*) The reheat pressure is determined from the requirement that the entropies at states 5 and 6 be the same:

FIGURE 9-13

Schematic and *T-s* diagram for Example 9-4.

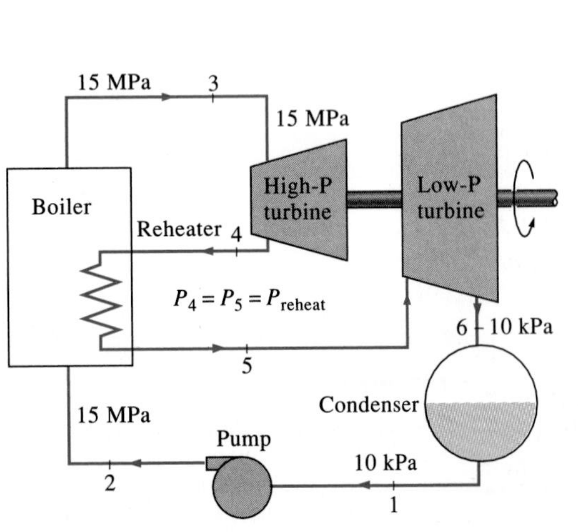

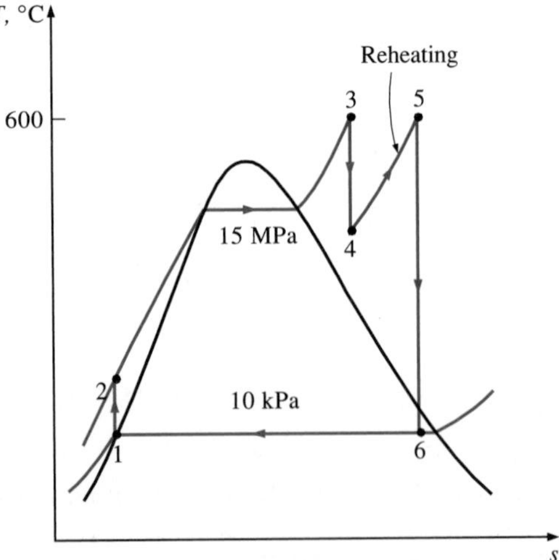

State 6: $P_6 = 10\text{ kPa}$

$x_6 = 0.896$ (sat. mixture)

$s_6 = s_f + x_6 s_{fg} = 0.6493 + 0.896(7.5009) = 7.370\text{ kJ/(kg}\cdot\text{K)}$

Also, $h_6 = h_f + x_6 h_{fg} = 191.83 + 0.896(2392.8) = 2335.8\text{ kJ/kg}$

Thus,

State 5: $\left.\begin{array}{l} T_5 = 600°\text{C} \\ s_5 = s_6 \end{array}\right\}$ $P_5 = \mathbf{4.0\ MPa}$, $h_5 = 3674.4\text{ kJ/kg}$

Therefore, steam should be reheated at a pressure of 4 MPa or lower to prevent a moisture content above 10.4 percent.

(*b*) To determine the thermal efficiency, we need to know the enthalpies at all other states:

State 1: $\left.\begin{array}{l} P_1 = 10\text{ kPa} \\ \text{Sat. liquid} \end{array}\right\}$ $h_1 = h_{f@\,10\text{ kPa}} = 191.83\text{ kJ/kg}$, $v_1 = v_{f@\,10\text{ kPa}} = 0.001010\text{ m}^3\text{/kg}$

State 2: $P_2 = 15\text{ MPa}$

$s_2 = s_1$

$$w_{\text{pump,in}} = v_1(P_2 - P_1) = (0.001010\text{ m}^3\text{/kg})[(15{,}000 - 10)\text{kPa}]\left(\frac{1\text{ kJ}}{1\text{ kPa}\cdot\text{m}^3}\right)$$
$$= 15.11\text{ kJ/kg}$$

$$h_2 = h_1 + w_{\text{pump,in}} = (191.83 + 15.11)\text{ kJ/kg} = 206.94\text{ kJ/kg}$$

State 3: $\left.\begin{array}{l} P_3 = 15\text{ MPa} \\ T_3 = 600°\text{C} \end{array}\right\}$ $h_3 = 3582.3\text{ kJ/kg}$, $s_3 = 6.6776\text{ kJ/(kg}\cdot\text{K)}$

State 4: $\left.\begin{array}{l} P_4 = 4\text{ MPa} \\ s_4 = s_3 \end{array}\right\}$ $h_4 = 3154.3\text{ kJ/kg}$, $(T_4 = 375.5°\text{C})$

Thus
$$q_{\text{in}} = (h_3 - h_2) + (h_5 - h_4)$$
$$= (3582.3 - 206.94)\text{ kJ/kg} + (3674.4 - 3154.3)\text{ kJ/kg}$$
$$= 3895.46\text{ kJ/kg}$$
$$q_{\text{out}} = h_6 - h_1 = (2335.8 - 191.83)\text{ kJ/kg}$$
$$= 2143.97\text{ kJ/kg}$$

and
$$\eta_{\text{th}} = 1 - \frac{q_{\text{out}}}{q_{\text{in}}} = 1 - \frac{2143.97\text{ kJ/kg}}{3895.46\text{ kJ/kg}} = \mathbf{0.450\ or\ 45.0\%}$$

Discussion This problem was worked out in Example 9-3*c* for the same pressure and temperature limits but without the reheat process. A comparison of the two results reveals that reheating reduces the moisture content from 19.6 to 10.4 percent while increasing the thermal efficiency from 43.0 to 45.0 percent.

9-6 ■ THE IDEAL REGENERATIVE RANKINE CYCLE

A careful examination of the *T-s* diagram of the Rankine cycle redrawn in Fig. 9-14 reveals that heat is added to the working fluid during process 2-2′ at a relatively low temperature. This lowers the average temperature at which heat is added and thus the cycle efficiency.

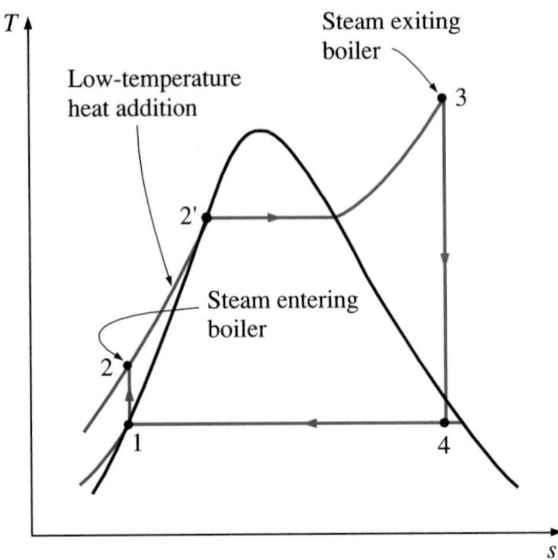

FIGURE 9-14
The first part of the heat-addition process in the boiler takes place at relatively low temperatures.

To remedy this shortcoming, we look for ways to raise the temperature of the liquid leaving the pump (called the *feedwater*) before it enters the boiler. One such possibility is to compress the feedwater isentropically to a high temperature, as in the Carnot cycle. This, however, would involve extremely high pressures and is therefore impractical. Another possibility is to transfer heat to the feedwater from the expanding steam in a counterflow heat exchanger built into the turbine, that is, to use **regeneration.** This solution is also impractical because it is difficult to design such a heat exchanger and because it would increase the moisture content of the steam at the final stages of the turbine.

A practical regeneration process in steam power plants is accomplished by extracting, or "bleeding," steam from the turbine at various points. This steam, which could have produced more work by expanding further in the turbine, is used to heat the feedwater instead. The device where the feedwater is heated by regeneration is called a **regenerator,** or a **feedwater heater.**

Regeneration not only improves cycle efficiency, but also provides a convenient means of deaerating the feedwater (removing the air that leaks in at the condenser) to prevent corrosion in the boiler. It also helps control the large volume flow rate of the steam at the final stages of the turbine (due to the large specific volumes at low pressures). Therefore, regeneration is used in all modern steam power plants since its introduction in the early 1920s.

A feedwater heater is basically a heat exchanger where heat is transferred from the steam to the feedwater either by mixing the two fluid streams (open feedwater heaters) or without mixing them (closed feedwater heaters). Regeneration with both types of feedwater heaters is discussed below.

Open Feedwater Heaters

An **open** (or **direct-contact**) **feedwater heater** is basically a *mixing chamber,* where the steam extracted from the turbine mixes with the feedwater exiting the pump. Ideally, the mixture leaves the heater as a saturated liquid

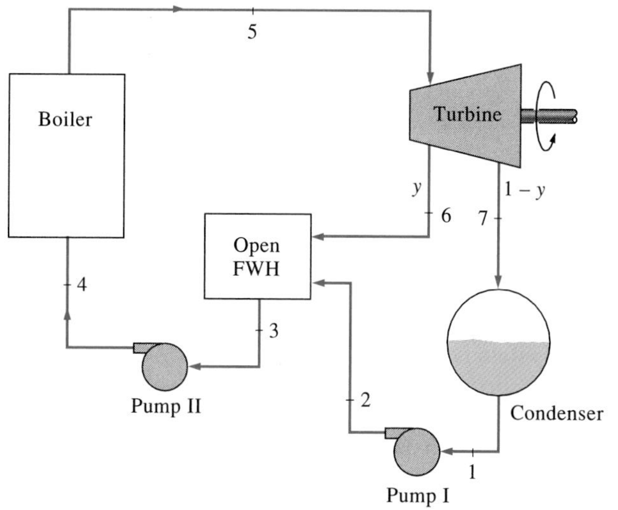

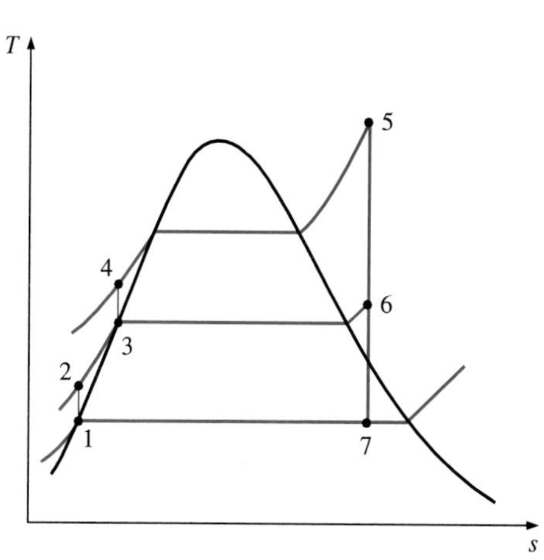

FIGURE 9-15
The ideal regenerative Rankine cycle with an open feedwater heater.

at the heater pressure. The schematic of a steam power plant with one open feedwater heater (also called *single-stage regenerative cycle*) and the *T-s* diagram of the cycle are shown in Fig. 9-15.

In an ideal regenerative Rankine cycle, steam enters the turbine at the boiler pressure (state 5) and expands isentropically to an intermediate pressure (state 6). Some steam is extracted at this state and routed to the feedwater heater, while the remaining steam continues to expand isentropically to the condenser pressure (state 7). This steam leaves the condenser as a saturated liquid at the condenser pressure (state 1). The condensed water, which is also called the *feedwater,* then enters an isentropic pump, where it is compressed to the feedwater heater pressure (state 2) and is routed to the feedwater heater, where it mixes with the steam extracted from the turbine. The fraction of the steam extracted is such that the mixture leaves the heater as a saturated liquid at the heater pressure (state 3). A second pump raises the pressure of the water to the boiler pressure (state 4). The cycle is completed by heating the water in the boiler to the turbine inlet state (state 5).

In the analysis of steam power plants, it is more convenient to work with quantities expressed per unit mass of the steam flowing through the boiler. For each 1 kg of steam leaving the boiler, y kg expands partially in the turbine and is extracted at state 6. The remaining $(1 - y)$ kg expands completely to the condenser pressure. Therefore, the mass flow rates are different in different components. If the mass flow rate through the boiler is $\dot{m}$, for example, it will be $(1 - y)\dot{m}$ through the condenser. This aspect of the regenerative Rankine cycle should be considered in the analysis of the cycle as well as in the interpretation of the areas on the *T-s* diagram. In light of Fig. 9-15, the heat and work interactions of a regenerative Rankine cycle with one feedwater heater can be expressed per unit mass of steam flowing through the boiler as follows:

$$q_{in} = h_5 - h_4 \tag{9-14}$$

$$q_{out} = (1 - y)(h_7 - h_1) \tag{9-15}$$

$$w_{turb,out} = (h_5 - h_6) + (1 - y)(h_6 - h_7) \tag{9-16}$$

$$w_{pump,in} = (1 - y)w_{pumpI,in} + w_{pumpII,in} \tag{9-17}$$

where

$$y = \dot{m}_6/\dot{m}_5 \qquad \text{(fraction of steam extracted)}$$

$$w_{pumpI,in} = v_1(P_2 - P_1)$$

$$w_{pumpII,in} = v_3(P_4 - P_3)$$

The thermal efficiency of the Rankine cycle increases as a result of regeneration. This is because regeneration raises the average temperature at which heat is added to the steam in the boiler by raising the temperature of the water before it enters the boiler. The cycle efficiency increases further as the number of feedwater heaters is increased. Many large plants in operation today use as many as eight feedwater heaters. The optimum number of feedwater heaters is determined from economical considerations. The use of an additional feedwater heater cannot be justified unless it saves more from the fuel costs than its own cost.

Closed Feedwater Heaters

Another type of feedwater heater frequently used in steam power plants is the **closed feedwater heater,** in which heat is transferred from the extracted steam to the feedwater without any mixing taking place. The two streams now can be at different pressures, since they do not mix. The schematic of a steam power plant with one closed feedwater heater and the *T-s* diagram of the cycle are shown in Fig. 9-16. In an ideal closed feedwater heater, the feedwater is heated to the exit temperature of the extracted steam, which ideally leaves the heater as a saturated liquid at the extraction pressure. In actual power plants, the feedwater leaves the heater below the exit temperature of the extracted steam because a temperature difference of at least a few degrees is required for any effective heat transfer to take place.

The condensed steam is then either pumped to the feedwater line or routed to another heater or to the condenser through a device called a **trap.** A trap allows the liquid to be throttled to a lower pressure region but *traps* the vapor. The enthalpy of steam remains constant during this throttling process.

The open and closed feedwater heaters can be compared as follows. Open feedwater heaters are simple and inexpensive and have good heat transfer characteristics. They also bring the feedwater to the saturation state. But for each heater, a pump is required to handle the feedwater. The closed feedwater heaters are more complex because of the internal tubing network, and thus they are more expensive. Heat transfer in closed feedwater heaters is also less effective since the two streams are not allowed to be in direct contact. However, closed feedwater heaters do not require a separate pump for each heater since the extracted steam and the feedwater can be at different pressures. Most steam power plants use a combination of open and closed feedwater heaters, as shown in Fig. 9-17.

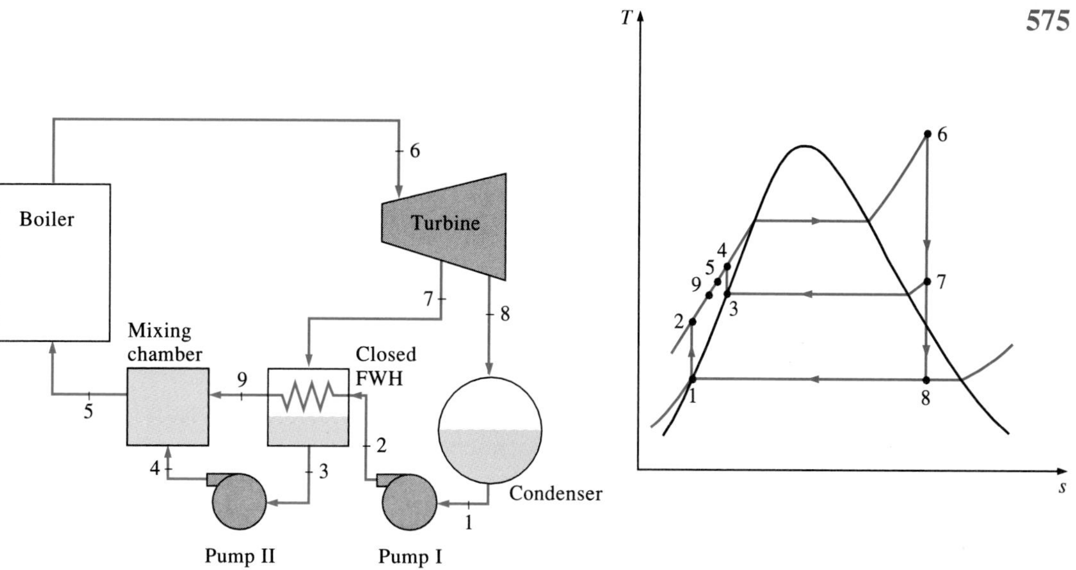

FIGURE 9-16

The ideal regenerative Rankine cycle with a closed feedwater heater.

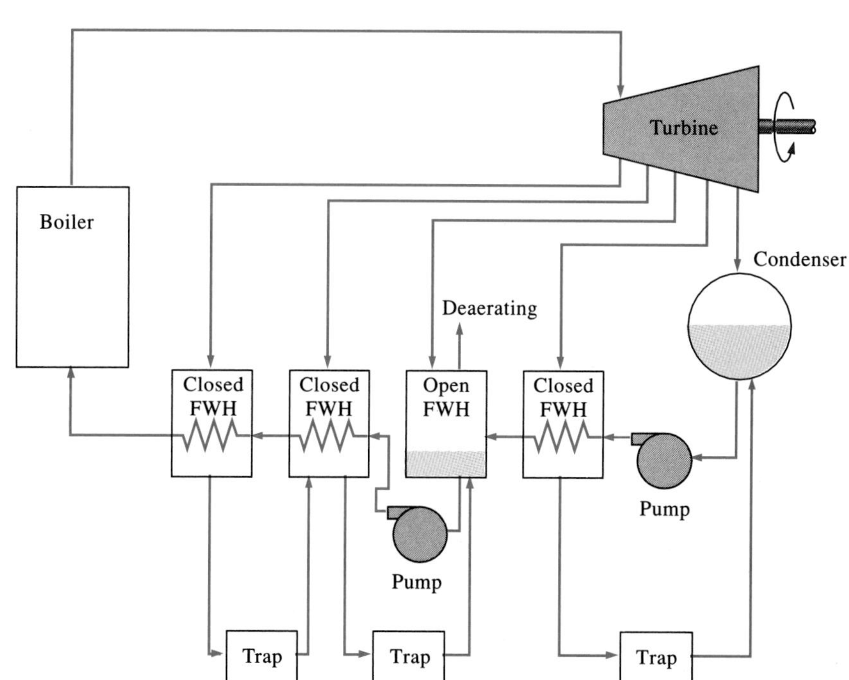

FIGURE 9-17

A steam power plant with one open and three closed feedwater heaters.

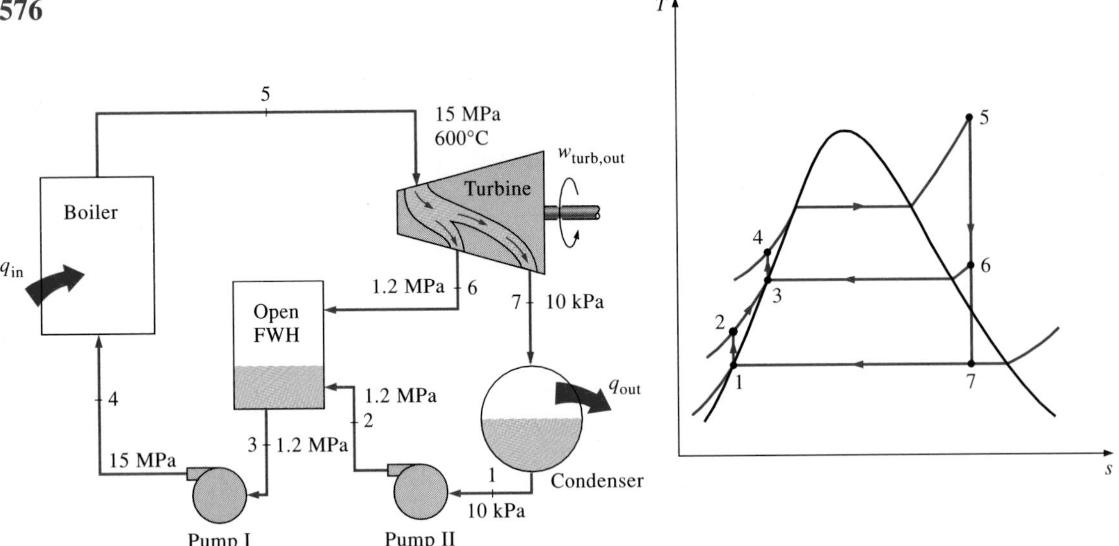

FIGURE 9-18
Schematic and T-s diagram for Example 9-5.

EXAMPLE 9-5 The Ideal Regenerative Rankine Cycle

Consider a steam power plant operating on the ideal regenerative Rankine cycle with one open feedwater heater. Steam enters the turbine at 15 MPa and 600°C and is condensed in the condenser at a pressure of 10 kPa. Some steam leaves the turbine at a pressure of 1.2 MPa and enters the open feedwater heater. Determine the fraction of steam extracted from the turbine and the thermal efficiency of the cycle.

Solution The schematic of the power plant and the T-s diagram of the cycle are shown in Fig. 9-18. We note that the power plant involves steady-flow components and operates on the ideal regenerative Rankine cycle. Therefore, the pumps and the turbines are isentropic; there are no pressure drops in the boiler, condenser, and feedwater heater; and steam leaves the condenser and the feedwater heater as saturated liquid.

Assumptions **1** Steady operating conditions exist. **2** Kinetic and potential energy changes are negligible.

Analysis First, we determine the enthalpies at various states:

State 1: $\left.\begin{array}{l} P_1 = 10 \text{ kPa} \\ \text{Sat. liquid} \end{array}\right\}$ $\begin{array}{l} h_1 = h_{f\,@\,10\text{ kPa}} = 191.83 \text{ kJ/kg} \\ v_1 = v_{f\,@\,10\text{ kPa}} = 0.00101 \text{ m}^3\text{/kg} \end{array}$

State 2: $P_2 = 1.2$ MPa

$s_2 = s_1$

$$w_{\text{pumpI,in}} = v_1(P_2 - P_1) = (0.00101 \text{ m}^3\text{/kg})[(1200 - 10) \text{ kPa}]\left(\frac{1 \text{ kJ}}{1 \text{ kPa}\cdot\text{m}^3}\right)$$

$$= 1.20 \text{ kJ/kg}$$

$$h_2 = h_1 + w_{\text{pumpI,in}} = (191.83 + 1.20) \text{ kJ/kg} = 193.03 \text{ kJ/kg}$$

State 3: $\left.\begin{array}{l} P_3 = 1.2 \text{ MPa} \\ \text{Sat. liquid} \end{array}\right\} \quad h_3 = h_{f\,@\,1.2\text{ MPa}} = 798.65 \text{ kJ/kg}$

State 4: $P_4 = 15 \text{ MPa}$

$s_4 = s_3$

$$w_{\text{pumpII, in}} = v_3(P_4 - P_3)$$

$$= (0.001139 \text{ m}^3\text{/kg})[(15{,}000 - 1200) \text{ kPa}]\left(\frac{1 \text{ kJ}}{1 \text{ kPa} \cdot \text{m}^3}\right)$$

$$= 15.72 \text{ kJ/kg}$$

$$h_4 = h_3 + w_{\text{pumpII, in}} = (798.65 + 15.72) \text{ kJ/kg} = 814.37 \text{ kJ/kg}$$

State 5: $\left.\begin{array}{l} P_5 = 15 \text{ MPa} \\ T_5 = 600°\text{C} \end{array}\right\} \quad \begin{array}{l} h_5 = 3582.3 \text{ kJ/kg} \\ s_5 = 6.6776 \text{ kJ/(kg} \cdot \text{K)} \end{array}$

State 6: $\left.\begin{array}{l} P_6 = 1.2 \text{ MPa} \\ s_6 = s_5 \end{array}\right\} \quad \begin{array}{l} h_6 = 2859.5 \text{ kJ/kg} \\ (T_6 = 218.3°\text{C}) \end{array}$

State 7: $\begin{array}{l} P_7 = 10 \text{ kPa} \\ s_7 = s_5 \end{array} \quad x_7 = \dfrac{s_7 - s_f}{s_{fg}} = \dfrac{6.6776 - 0.6493}{7.5009} = 0.804$

$$h_7 = h_f + x_7 h_{fg} = 191.83 + 0.804(2392.8) = 2115.6 \text{ kJ/kg}$$

The energy analysis of open feedwater heaters is identical to the energy analysis of mixing chambers discussed in Chap. 4. The feedwater heaters are generally well insulated ($\dot{Q} = 0$), and they do not involve any work interactions ($\dot{W} = 0$). By neglecting the kinetic and potential energies of the streams, the steady-flow conservation of energy equation reduces for a feedwater heater to

$$\dot{E}_{\text{in}} = \dot{E}_{\text{out}}$$

$$\sum \dot{m}_i h_i = \sum \dot{m}_e h_e$$

or

$$y h_6 + (1 - y) h_2 = 1(h_3)$$

where y is the fraction of steam extracted from the turbine ($= \dot{m}_6/\dot{m}_5$). Solving for y and substituting the enthalpy values, we find

$$y = \frac{h_3 - h_2}{h_6 - h_2} = \frac{798.65 - 193.03}{2859.5 - 193.03} = \mathbf{0.227}$$

Thus,

$$q_{\text{in}} = h_5 - h_4 = (3582.3 - 814.37) \text{ kJ/kg} = 2767.93 \text{ kJ/kg}$$

$$q_{\text{out}} = (1 - y)(h_7 - h_1) = (1 - 0.227)(2115.6 - 191.83) \text{ kJ/kg}$$

$$= 1487.1 \text{ kJ/kg}$$

and

$$\eta_{\text{th}} = 1 - \frac{q_{\text{out}}}{q_{\text{in}}} = 1 - \frac{1487.1 \text{ kJ/kg}}{2767.93 \text{ kJ/kg}} = \mathbf{0.463 \text{ or } 46.3\%}$$

Discussion This problem was worked out in Example 9-3*c* for the same pressure and temperature limits but without the regeneration process. A comparison of the two results reveals that the thermal efficiency of the cycle has increased from 43.0 to 46.3 percent as a result of regeneration. The net work output decreases by 171 kJ/kg, but the heat input decreases by 607 kJ/kg, which results in a net increase in the thermal efficiency.

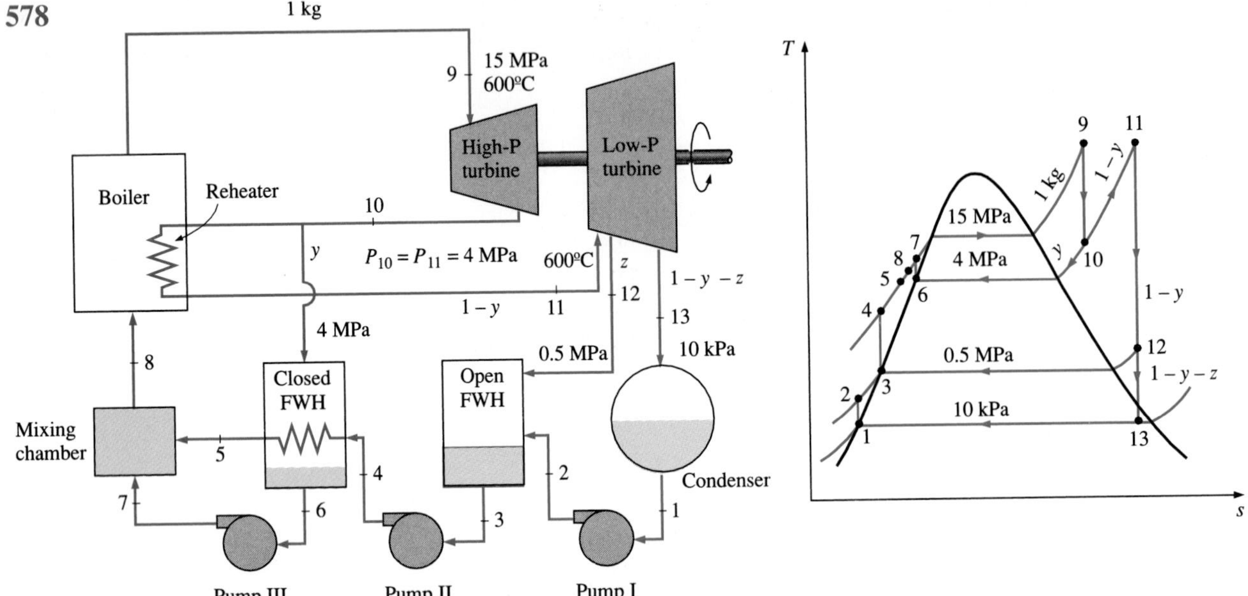

FIGURE 9-19
Schematic and *T-s* diagram for Example 9-6.

EXAMPLE 9-6 The Ideal Reheat-Regenerative Rankine Cycle

Consider a steam power plant that operates on an ideal reheat–regenerative Rankine cycle with one open feedwater heater, one closed feedwater heater, and one reheater. Steam enters the turbine at 15 MPa and 600°C and is condensed in the condenser at a pressure of 10 kPa. Some steam is extracted from the turbine at 4 MPa for the closed feedwater heater, and the remaining steam is reheated at the same pressure to 600°C. The extracted steam is completely condensed in the heater and is pumped to 15 MPa before it mixes with the feedwater at the same pressure. Steam for the open feedwater heater is extracted from the low-pressure turbine at a pressure of 0.5 MPa. Determine the fraction of steam extracted from the turbine each time as well as the thermal efficiency of the cycle.

Solution The schematic of the power plant and the *T-s* diagram of the cycle are shown in Fig. 9-19. We note that the power plant involves steady-flow components and operates on the ideal reheat–regenerative Rankine cycle. Therefore, the pumps and the turbines are isentropic; there are no pressure drops in the boiler, reheater, condenser, and feedwater heaters; and steam leaves the condenser and the feedwater heaters as saturated liquid.

Assumptions **1** Steady operating conditions exist. **2** Kinetic and potential energy changes are negligible.

Analysis The enthalpies at the various states and the pump work per unit mass of fluid flowing through them are

$$
\begin{aligned}
h_1 &= 191.83 \text{ kJ/kg} & h_9 &= 3582.3 \text{ kJ/kg} \\
h_2 &= 192.32 \text{ kJ/kg} & h_{10} &= 3154.3 \text{ kJ/kg} \\
h_3 &= 640.23 \text{ kJ/kg} & h_{11} &= 3674.4 \text{ kJ/kg} \\
h_4 &= 656.08 \text{ kJ/kg} & h_{12} &= 3014.3 \text{ kJ/kg} \\
h_5 &= 1087.31 \text{ kJ/kg} & h_{13} &= 2335.8 \text{ kJ/kg} \\
h_6 &= 1087.31 \text{ kJ/kg} & w_{\text{pumpI, in}} &= 0.49 \text{ kJ/kg} \\
h_7 &= 1101.08 \text{ kJ/kg} & w_{\text{pumpII, in}} &= 15.85 \text{ kJ/kg} \\
h_8 &= 1089.7 \text{ kJ/kg} & w_{\text{pumpIII, in}} &= 13.77 \text{ kJ/kg}
\end{aligned}
$$

The fractions of steam extracted are determined from the mass and energy balances of the feedwater heaters:

Closed feedwater heater:

$$\dot{E}_{\text{in}} = \dot{E}_{\text{out}}$$

$$yh_{10} + (1 - y)h_4 = (1 - y)h_5 + yh_6$$

$$y = \frac{h_5 - h_4}{(h_{10} - h_6) + (h_5 - h_4)} = \frac{1087.31 - 656.08}{(3154.3 - 1087.31) + (1087.31 - 656.08)} = \mathbf{0.173}$$

Open feedwater heater:

$$\dot{E}_{\text{in}} = \dot{E}_{\text{out}}$$

$$zh_{12} + (1 - y - z)h_2 = (1 - y)h_3$$

$$z = \frac{(1 - y)(h_3 - h_2)}{h_{12} - h_2} = \frac{(1 - 0.173)(640.23 - 192.32)}{3014.13 - 192.32} = \mathbf{0.131}$$

The enthalpy at state 8 is determined by applying the conservation of mass and energy equations to the mixing chamber, which is assumed to be insulated:

$$
\begin{aligned}
\dot{E}_{\text{in}} &= \dot{E}_{\text{out}} \\
(1)h_8 &= (1 - y)h_5 + yh_7 \\
h_8 &= (1 - 0.173)(1087.31) \text{ kJ/kg} + 0.173(1101.08) \text{ kJ/kg} \\
&= 1089.7 \text{ kJ/kg}
\end{aligned}
$$

Thus,

$$
\begin{aligned}
q_{\text{in}} &= (h_9 - h_8) + (1 - y)(h_{11} - h_{10}) \\
&= (3582.5 - 1089.7) \text{ kJ/kg} \\
&\quad + (1 - 0.173)(3674.4 - 3154.3) \text{ kJ/kg} \\
&= 2922.7 \text{ kJ/kg} \\
q_{\text{out}} &= (1 - y - z)(h_{13} - h_1) \\
&= (1 - 0.173 - 0.131)(2335.8 - 191.83) \text{ kJ/kg} \\
&= 1492.2 \text{ kJ/kg}
\end{aligned}
$$

and

$$\eta_{\text{th}} = 1 - \frac{q_{\text{out}}}{q_{\text{in}}} = 1 - \frac{1492.2 \text{ kJ/kg}}{2922.7 \text{ kJ/kg}} = \mathbf{0.489 \text{ or } 48.9\%}$$

Discussion This problem was worked out in Example 9-4 for the same pressure and temperature limits with reheat but without the regeneration process. A comparison of the two results reveals that the thermal efficiency of the cycle has increased from 45.0 to 48.9 percent as a result of regeneration.

The thermal efficiency of this cycle could also be determined from

$$\eta_{th} = \frac{w_{net}}{q_{in}} = \frac{w_{turb,out} - w_{pump,in}}{q_{in}}$$

where

$$w_{turb,out} = (h_9 - h_{10}) + (1 - y)(h_{11} - h_{12}) + (1 - y - z)(h_{12} - h_{13})$$
$$w_{pump,in} = (1 - y - z)w_{pumpI,in} + (1 - y)w_{pumpII,in} + (y)w_{pumpIII,in}$$

9-7 ■ SECOND-LAW ANALYSIS OF VAPOR POWER CYCLES

The ideal Carnot cycle is a *totally reversible cycle,* and thus it does not involve any irreversibilities. The ideal Rankine cycles (simple, reheat, or regenerative), however, are only *internally reversible,* and they may involve irreversibilities external to the system, such as heat transfer through a finite temperature difference. A second-law analysis of these cycles will reveal where the largest irreversibilities occur and what their magnitudes are.

Relations for exergy and irreversibility (or exergy destruction) for steady-flow systems are developed in Chap. 7. The irreversibility for a steady-flow system can be expressed, in the rate form, as

$$\dot{I} = T_0\dot{S}_{gen} = T_0(\dot{S}_{out} - \dot{S}_{in}) = T_0\left(\sum \dot{m}_e s_e + \frac{\dot{Q}_{out}}{T_{b,out}} - \sum \dot{m}_i s_i - \frac{\dot{Q}_{in}}{T_{b,in}}\right) \quad \text{(kW)} \quad (9\text{-}18)$$

or on a unit-mass basis for a one-inlet, one-exit, steady-flow device as

$$i = T_0 s_{gen} = T_0\left(s_e - s_i + \frac{q_{out}}{T_{b,out}} - \frac{q_{in}}{T_{b,in}}\right) \quad \text{(kJ/kg)} \quad (9\text{-}19)$$

where $T_{b,in}$ and $T_{b,out}$ are the temperatures of the system boundary where heat is transferred into and out of the system, respectively.

The irreversibility of a cycle depends on the magnitude of the heat transfer with the high- and low-temperature reservoirs involved, and their temperatures, as explained in Chap. 8. It can be expressed on a unit-mass basis as

$$i = T_0\left(\sum \frac{q_{out}}{T_{b,out}} - \sum \frac{q_{in}}{T_{b,in}}\right) \quad \text{(kJ/kg)} \quad (9\text{-}20)$$

For a cycle that involves only heat transfer with a source at T_H and a sink at T_L, the irreversibility becomes

$$i = T_0\left(\frac{q_{out}}{T_L} - \frac{q_{in}}{T_H}\right) \quad \text{(kJ/kg)} \quad (9\text{-}21)$$

The exergy of a fluid stream ψ at any state can be determined from

$$\psi = (h - h_0) - T_0(s - s_0) + \frac{\mathcal{V}^2}{2} + gz \qquad \text{(kJ/kg)} \qquad (9\text{-}22)$$

where the subscript "0" denotes the state of the surroundings. The reversible work for any process can be determined by finding the exergy change of the working fluid during that process.

EXAMPLE 9-7 Irreversibility of an Ideal Rankine Cycle

Determine the irreversibility of the Rankine cycle (all four processes as well as the cycle) discussed in Example 9-1, assuming that heat is transferred to the steam in a furnace at 1600 K and heat is rejected to a cooling medium at 290 K and 100 kPa. Also, determine the exergy of the steam leaving the turbine.

Solution In Example 9-1, the heat input was determined to be 2727.88 kJ/kg, and the heat rejected to be 2018.81 kJ/kg.

Analysis Processes 1-2 and 3-4 are isentropic ($s_1 = s_2$, $s_3 = s_4$) and therefore do not involve any internal or external irreversibilities, that is,

$$i_{12} = \mathbf{0} \qquad \textit{and} \qquad i_{34} = \mathbf{0}$$

Processes 2-3 and 4-1 are constant-pressure heat-addition and rejection processes, respectively, and they are internally reversible. But the heat transfer between the working fluid and the source or the sink takes place through a finite temperature difference, rendering both processes irreversible. The irreversibility associated with each process is determined from Eq. 9-19. The entropy of the steam at each state is determined from the steam tables:

$$s_2 = s_1 = s_{f\,@\,75\text{ kPa}} = 1.213\text{ kJ/(kg}\cdot\text{K)}$$

$$s_4 = s_3 = 6.7428\text{ kJ/(kg}\cdot\text{K)} \qquad \text{(at 3 MPa, 350°C)}$$

Thus,

$$\begin{aligned} i_{23} &= T_0\left(s_3 - s_2 + \frac{q_{\text{in},23}}{T_{\text{source}}}\right) \\ &= (290\text{ K})\left[(6.7428 - 1.213)\text{ kJ/(kg}\cdot\text{K)} - \frac{2727.88\text{ kJ/kg}}{1600\text{ K}}\right] \\ &= \mathbf{1109.2\ kJ/kg} \end{aligned}$$

$$\begin{aligned} i_{41} &= T_0\left(s_1 - s_4 + \frac{q_{\text{out},41}}{T_{\text{sink}}}\right) \\ &= (290\text{ K})\left[(1.213 - 6.7428)\text{ kJ/(kg}\cdot\text{K)} + \frac{2018.81\text{ kJ/kg}}{290\text{ K}}\right] \\ &= \mathbf{415.2\ kJ/kg} \end{aligned}$$

Therefore, the irreversibility of the cycle is

$$\begin{aligned} i_{\text{cycle}} &= i_{12} + i_{23} + i_{34} + i_{41} \\ &= 0 + 1109.2\text{ kJ/kg} + 0 + 415.2\text{ kJ/kg} \\ &= \mathbf{1524.4\ kJ/kg} \end{aligned}$$

The irreversibility of the cycle could also be determined from Eq. 9-21. Notice that the largest irreversibility in the cycle occurs during the heat-addition process. Therefore, any attempt to reduce the irreversibilities should start with this process. Raising the turbine inlet temperature of the steam, for example, would reduce the temperature difference and thus the irreversibility.

The exergy (maximum work potential) of the steam leaving the turbine is determined from Eq. 9-22. Disregarding the kinetic and potential energies of the steam, it reduces to

$$\psi_4 = (h_4 - h_0) - T_0(s_4 - s_0) + \frac{V_4^2}{2}^{\nearrow 0} + gz_4^{\nearrow 0}$$
$$= (h_4 - h_0) - T_0(s_4 - s_0)$$

where

$$h_0 = h_{@\ 290\ K,\ 100\ kPa} \cong h_{f\ @\ 290\ K} = 71.34\ kJ/kg$$
$$s_0 = s_{@\ 290\ K,\ 100\ kPa} \cong s_{f\ @\ 290\ K} = 0.2533\ kJ/(kg \cdot K)$$

Thus, $\psi_4 = (2403.2 - 71.34)\ kJ/kg - (290\ K)[(6.7428 - 0.2533)\ kJ/(kg \cdot K)]$
$= \mathbf{449.9\ kJ/kg}$

Discussion Note that 449.9 kJ/kg of work could be obtained from the steam leaving the turbine if it is brought to the state of the surroundings in a reversible manner.

9-8 ■ COGENERATION

In all the cycles discussed so far, the sole purpose was to convert a portion of the heat transferred to the working fluid to work, which is the most valuable form of energy. The remaining portion of the heat is rejected to rivers, lakes, oceans, or the atmosphere as waste heat, because its quality (or grade) is too low to be of any practical use. Wasting a large amount of heat is a price we have to pay to produce work, because electrical or mechanical work is the only form of energy on which many engineering devices (such as a fan) can operate.

Many systems or devices, however, require energy input in the form of heat, called *process heat*. Some industries that rely heavily on process heat are chemical, pulp and paper, oil production and refining, steel making, food processing, and textile industries. Process heat in these industries is usually supplied by steam at 5 to 7 atm and 150 to 200°C (300 to 400°F). Energy is usually transferred to the steam by burning coal, oil, natural gas, or another fuel in a furnace.

FIGURE 9-20
A simple process-heating plant.

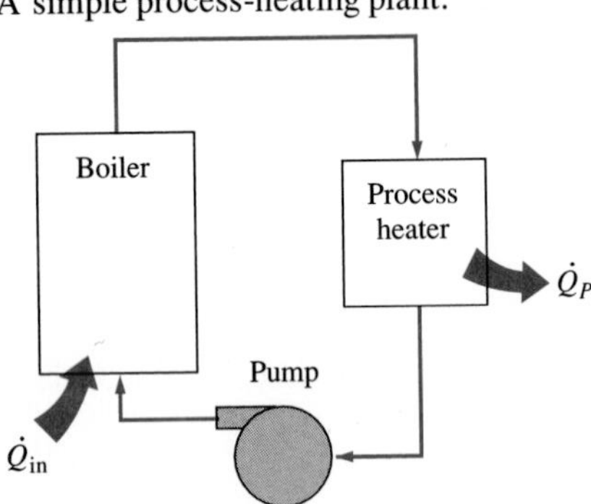

Now let us examine the operation of a process-heating plant closely. Disregarding any heat losses in the piping, all the heat transferred to the steam in the boiler is used in the process-heating units, as shown in Fig. 9-20. Therefore, process heating seems like a perfect operation with practically no waste of energy. From the second-law point of view, however, things do not look so perfect. The temperature in furnaces is typically very high (around 1370°C), and thus the energy in the furnace is of very high quality. This high-quality energy is transferred to water to produce steam at about 200°C or below (a highly irreversible process). Associated with this irreversibility is, of course, a loss in exergy or work potential. It is simply not wise to use high-quality energy to accomplish a task that could be accomplished with low-quality energy.

Industries that use large amounts of process heat also consume a large amount of electric power. Therefore, it makes economical as well as engineering sense to use the already-existing work potential to produce power instead of letting it go to waste. The result is a plant that produces electricity while meeting the process-heat requirements of certain industrial processes. Such a plant is called a *cogeneration plant*. In general, **cogeneration** is *the production of more than one useful form of energy (such as process heat and electric power) from the same energy source.*

Either a steam-turbine (Rankine) cycle or a gas-turbine (Brayton) cycle or even a combined cycle (discussed later) can be used as the power cycle in a cogeneration plant. The schematic of an ideal steam-turbine cogeneration plant is shown in Fig. 9-21. Let us say this plant is to supply process heat $\dot{Q}_p$ at 500 kPa at a rate of 100 kW. To meet this demand, steam is expanded in the turbine to a pressure of 500 kPa, producing power at a rate of, say, 20 kW. The flow rate of the steam can be adjusted such that steam leaves the process-heating section as a saturated liquid at 500 kPa. Steam is then pumped to the boiler pressure and is heated in the boiler to state 3. The pump work is usually very small and can be neglected. Disregarding any heat losses, the rate of heat input in the boiler is determined from an energy balance to be 120 kW.

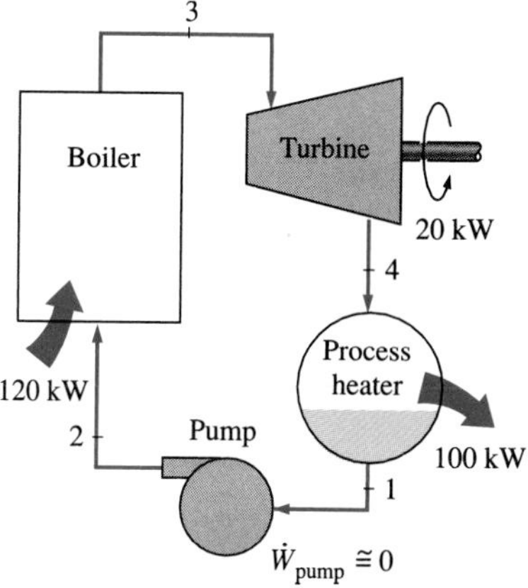

FIGURE 9-21
An ideal cogeneration plant.

Probably the most striking feature of the ideal steam-turbine cogeneration plant shown in Fig. 9-21 is the absence of a condenser. Thus no heat is rejected from this plant as waste heat. In other words, all the energy transferred to the steam in the boiler is utilized as either process heat or electric power. Thus it is appropriate to define a **utilization factor** ε_u for a cogeneration plant as

$$\varepsilon_u = \frac{\text{Net work output} + \text{Process heat delivered}}{\text{Total heat input}} = \frac{\dot{W}_{net} + \dot{Q}_p}{\dot{Q}_{in}} \quad (9\text{-}23)$$

or

$$\varepsilon_u = 1 - \frac{\dot{Q}_{out}}{\dot{Q}_{in}} \quad (9\text{-}24)$$

where $\dot{Q}_{out}$ represents the heat rejected in the condenser. Strictly speaking, $\dot{Q}_{out}$ also includes all the undesirable heat losses from the piping and other components, but they are usually small and thus neglected. It also includes combustion inefficiencies such as incomplete combustion and stack losses when the utilization factor is defined on the basis of the heating value of the fuel. The utilization factor of the ideal steam-turbine cogeneration plant is obviously 100 percent. Actual cogeneration plants have utilization factors as high as 70 percent. Future cogeneration plants are expected to have even higher utilization factors.

Notice that without the turbine, we would need to supply heat to the steam in the boiler at a rate of only 100 kW instead of at 120 kW. The additional 20 kW of heat supplied is converted to work. Therefore, a cogeneration power plant is equivalent to a process-heating plant combined with a power plant that has a thermal efficiency of 100 percent.

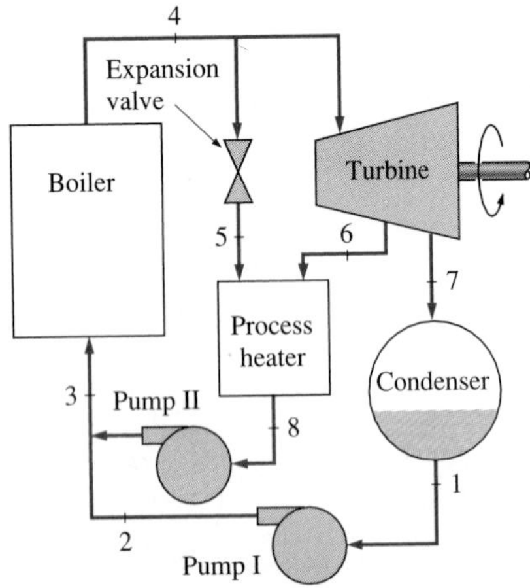

FIGURE 9-22

A cogeneration plant with adjustable loads.

The ideal steam-turbine cogeneration plant described above is not practical because it cannot adjust to the variations in power and process-heat loads. The schematic of a more practical (but more complex) cogeneration plant is shown in Fig. 9-22. Under normal operation, some steam is extracted from the turbine at some predetermined intermediate pressure P_6. The rest of the steam expands to the condenser pressure P_7 and is then cooled at constant pressure. The heat rejected from the condenser represents the waste heat for the cycle.

At times of high demand for process heat, all the steam is routed to the process-heating units and none to the condenser ($\dot{m}_7 = 0$). The waste heat is zero in this mode. If this is not sufficient, some steam leaving the boiler is throttled by an expansion or pressure-reducing valve (PRV) to the extraction pressure P_6 and is directed to the process-heating unit. Maximum process heating is realized when all the steam leaving the boiler passes through the PRV ($\dot{m}_5 = \dot{m}_4$). No power is produced in this mode. When there is no demand for process heat, all the steam passes through the turbine and the condenser ($\dot{m}_5 = \dot{m}_6 = 0$), and the cogeneration plant operates as an ordinary steam power plant. The rates of heat input, heat rejected, and process heat supply as well as the power produced for this cogeneration plant can be expressed as follows:

$$\dot{Q}_{in} = \dot{m}_3(h_4 - h_3) \tag{9-25}$$

$$\dot{Q}_{out} = \dot{m}_7(h_7 - h_1) \tag{9-26}$$

$$\dot{Q}_p = \dot{m}_5 h_5 + \dot{m}_6 h_6 - \dot{m}_8 h_8 \tag{9-27}$$

$$\dot{W}_{turb} = (\dot{m}_4 - \dot{m}_5)(h_4 - h_6) + \dot{m}_7(h_6 - h_7) \tag{9-28}$$

Under optimum conditions, a cogeneration plant simulates the ideal cogeneration plant discussed earlier. That is, all the steam expands in the turbine to the extraction pressure and continues to the process-heating unit. No steam passes through the PRV or the condenser; thus, no waste heat is rejected ($\dot{m}_4 = \dot{m}_6$ and $\dot{m}_5 = \dot{m}_7 = 0$). This condition may be difficult to achieve in practice because of the constant variations in the process-heat and power loads. But the plant should be designed so that the optimum operating conditions are approximated most of the time.

The use of cogeneration dates to the beginning of this century when power plants were integrated to a community to provide district heating, that is, space, hot water, and process heating for residential and commercial buildings. The district heating systems lost their popularity in the 1940s owing to low fuel prices. But the rapid rise in fuel prices in the 1970s brought about renewed interest in district heating.

Cogeneration plants have proved to be economically very attractive. Consequently, more and more such plants have been installed in recent years. It is projected that 15 percent of the nation's electricity will be generated from cogeneration plants by the end of the century.

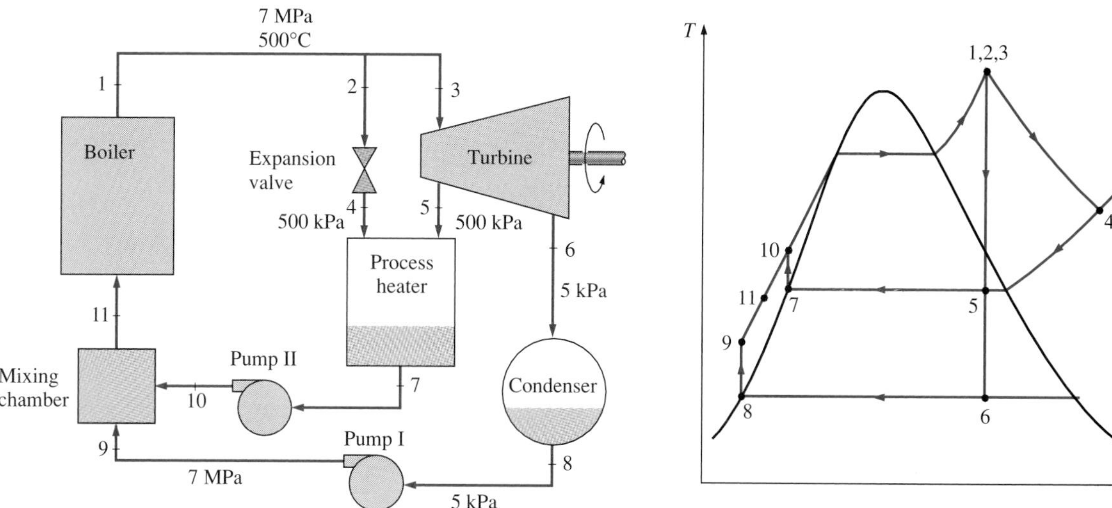

FIGURE 9-23

Schematic and *T-s* diagram for Example 9-8.

EXAMPLE 9-8 An Ideal Cogeneration Plant

Consider the cogeneration plant shown in Fig. 9-23. Steam enters the turbine at 7 MPa and 500°C. Some steam is extracted from the turbine at 500 kPa for process heating. The remaining steam continues to expand to 5 kPa. Steam is then condensed at constant pressure and pumped to the boiler pressure of 7 MPa. At times of high demand for process heat, some steam leaving the boiler is throttled to 500 kPa and is routed to the process heater. The extraction fractions are adjusted so that steam leaves the process heater as a saturated liquid at 500 kPa. It is subsequently pumped to 7 MPa. The mass flow rate of steam through the boiler is 15 kg/s. Disregarding any pressure drops and heat losses in the piping and assuming the turbine and the pump to be isentropic, determine (*a*) the maximum rate at which process heat can be supplied, (*b*) the power produced and the utilization factor when no process heat is supplied, and (*c*) the rate of process heat supply when 10 percent of the steam is extracted before it enters the turbine and 70 percent of the steam is extracted from the turbine at 500 kPa for process heating.

Solution The schematic of the cogeneration plant and the *T-s* diagram of the cycle are shown in Fig. 9-23. We note that the power plant involves steady-flow components and operates on an ideal cycle. Therefore, the pumps and the turbines are isentropic; there are no pressure drops in the boiler, process heater, and condenser; and steam leaves the condenser and the process heater as saturated liquid.

Assumptions **1** Steady operating conditions exist. **2** Pressure drops and heat losses in piping are negligible. **3** Kinetic and potential energy changes are negligible.

Analysis The work inputs to the pumps and the enthalpies at various states are as follows:

$$w_{\text{pumpI, in}} = v_8(P_9 - P_8) = (0.001005 \text{ m}^3/\text{kg})[(7000 - 5) \text{ kPa}]\left(\frac{1 \text{ kJ}}{1 \text{ kPa}\cdot\text{m}^3}\right)$$
$$= 7.03 \text{ kJ/kg}$$
$$w_{\text{pumpII, in}} = v_7(P_{10} - P_7) = (0.001095 \text{ m}^3/\text{kg})[(7000 - 500) \text{ kPa}]\left(\frac{1 \text{ kJ}}{1 \text{ kPa}\cdot\text{m}^3}\right)$$
$$= 7.12 \text{ kJ/kg}$$
$$h_1 = h_2 = h_3 = h_4 = 3410.3 \text{ kJ/kg}$$
$$h_5 = 2738.2 \text{ kJ/kg} \qquad (x_5 = 0.995)$$
$$h_6 = 2071.9 \text{ kJ/kg} \qquad (x_6 = 0.798)$$
$$h_7 = h_{f\,@\,500\text{ kPa}} = 640.23 \text{ kJ/kg}$$
$$h_8 = h_{f\,@\,5\text{ kPa}} = 137.82 \text{ kJ/kg}$$
$$h_9 = h_8 + w_{\text{pumpI, in}} = (137.82 + 7.03) \text{ kJ/kg} = 144.85 \text{ kJ/kg}$$
$$h_{10} = h_7 + w_{\text{pumpII, in}} = (640.23 + 7.12) \text{ kJ/kg} = 647.35 \text{ kJ/kg}$$

(*a*) The maximum rate of process heat is achieved when all the steam leaving the boiler is throttled and sent to the process heater and none is sent to the turbine (that is, $\dot{m}_4 = \dot{m}_7 = \dot{m}_1 = 15$ kg/s and $\dot{m}_3 = \dot{m}_5 = \dot{m}_6 = 0$). Thus,

$$\dot{Q}_{p,\text{max}} = \dot{m}_1(h_4 - h_7) = (15 \text{ kg/s})[(3410.3 - 640.23) \text{ kJ/kg}] = \mathbf{41{,}551 \text{ kW}}$$

The utilization factor is 100 percent in this case since no heat is rejected in the condenser, heat losses from the piping and other components are assumed to be negligible, and combustion losses are not considered.

(*b*) When no process heat is supplied, all the steam leaving the boiler will pass through the turbine and will expand to the condenser pressure of 5 kPa (that is, $\dot{m}_3 = \dot{m}_6 = \dot{m}_1 = 15$ kg/s and $\dot{m}_2 = \dot{m}_5 = 0$). Maximum power will be produced in this mode, which is determined to be

$$\dot{W}_{\text{turb, out}} = \dot{m}(h_3 - h_6) = (15 \text{ kg/s})[(3410.3 - 2071.9) \text{ kJ/kg}] = 20{,}076 \text{ kW}$$
$$\dot{W}_{\text{pump, in}} = (15 \text{ kg/s})(7.03 \text{ kJ/kg}) = 105 \text{ kW}$$
$$\dot{W}_{\text{net, out}} = \dot{W}_{\text{turb, out}} - \dot{W}_{\text{pump, in}} = (20{,}076 - 105) \text{ kW} = \mathbf{19{,}971 \text{ kW}}$$
$$\dot{Q}_{\text{in}} = \dot{m}_1(h_1 - h_{11}) = (15 \text{ kg/s})[(3410.3 - 144.85) \text{ kJ/kg}] = 48{,}982 \text{ kW}$$

Thus,
$$\varepsilon_u = \frac{\dot{W}_{\text{net}} + \dot{Q}_p}{\dot{Q}_{\text{in}}} = \frac{(19{,}971 + 0) \text{ kJ/kg}}{48{,}982 \text{ kJ/kg}} = \mathbf{0.408 \text{ or } 40.8\%}$$

That is, 40.8 percent of the energy is utilized for a useful purpose. Notice that the utilization factor is equivalent to the thermal efficiency in this case.

(*c*) Neglecting any kinetic and potential energy changes, an energy balance on the process heater yields

$$\dot{E}_{\text{in}} = \dot{E}_{\text{out}}$$
$$\dot{m}_4h_4 + \dot{m}_5h_5 = \dot{Q}_{p,\text{out}} + \dot{m}_7h_7$$

or
$$Q_{p,\text{out}} = \dot{m}_4h_4 + \dot{m}_5h_5 - \dot{m}_7h_7$$

where

$\dot{m}_4 = (0.1)(15 \text{ kg/s}) = 1.5 \text{ kg/s}$
$\dot{m}_5 = (0.7)(15 \text{ kg/s}) = 10.5 \text{ kg/s}$
$\dot{m}_7 = \dot{m}_4 + \dot{m}_5 = 1.5 + 10.5 = 12 \text{ kg/s}$

Thus,
$$\dot{Q}_{p,\text{out}} = (1.5 \text{ kg/s})(3410.3 \text{ kJ/kg}) + (10.5 \text{ kg/s})(2738.2 \text{ kJ/kg}) - (12 \text{ kg/s})(640.23 \text{ kJ/kg})$$
or
$$= \mathbf{26{,}184 \text{ kW}}$$

Discussion Note that 26,184 kW of the heat transferred will be utilized in the process heater. We could also show that 10,299 kW of power is produced in this case, and the rate of heat input in the boiler is 42,951 kW. Thus the utilization factor is 84.9 percent.

9-9 ■ BINARY VAPOR CYCLES

With the exception of a few specialized applications, the working fluid predominantly used in vapor power cycles is water. Water is the *best* working fluid presently available, but it is far from being the *ideal* one. The binary cycle is an attempt to overcome some of the shortcomings of water and to approach the *ideal* working fluid by using two fluids. Before we discuss the binary cycle, let us list the characteristics of a working fluid most suitable for vapor power cycles:

1. A high critical temperature and a safe maximum pressure. A critical temperature above the metallurgically allowed maximum temperature (about 620°C) makes it possible to transfer a considerable portion of the heat isothermally at the maximum temperature as the fluid changes phase. This will make the cycle approach the Carnot cycle. Very high pressures at the maximum temperature are undesirable because they create material-strength problems.

2. Low triple-point temperature. A triple-point temperature below the temperature of the cooling medium will prevent any solidification problems.

3. A condenser pressure that is not too low. Condensers usually operate below atmospheric pressure. Pressures well below the atmospheric pressure create air-leakage problems. Therefore, a substance whose saturation pressure at the ambient temperature is too low is not a good candidate.

4. A high enthalpy of vaporization (h_{fg}) so that heat transfer will approach being isothermal and large mass flow rates will not be needed.

5. A saturation dome that resembles an inverted U. This will eliminate the formation of excessive moisture in the turbine and the need for reheating.

6. Good heat transfer characteristics (high thermal conductivity).

7. Other properties such as being inert, inexpensive, readily available, and nontoxic.

Not surprisingly, no fluid possesses all these characteristics. Water comes the closest, although it does not fare well with respect to characteristics 1, 3, and 5. We can cope with its subatmospheric condenser pressure by careful sealing, and with the inverted V-shaped saturation dome by reheating, but there is not much we can do about item 1. Water has a low critical temperature (374°C, well below the 620°C limit) and very high saturation pressures at high temperatures (16.5 MPa at 350°C).

Well, we cannot change the way water behaves during the high-temperature part of the cycle, but we certainly can replace it with a more suitable fluid. The result is a power cycle that is actually a combination of two cycles, one in the high-temperature region and the other in the low-temperature region. Such a cycle is called a **binary vapor cycle.** In binary vapor cycles, the condenser of the high-temperature cycle (also called the *topping cycle*) serves as the boiler of the low-temperature cycle (also called the *bottoming cycle*). That is, the heat output of the high-temperature cycle is used as the heat input to the low-temperature one.

Some working fluids found suitable for the high-temperature cycle are mercury, sodium, potassium, and sodium–potassium mixtures. The schematic and *T-s* diagram for a mercury–water binary vapor cycle are shown in Fig. 9-24. The critical temperature of mercury is 898°C (well above the current 620°C metallurgical limit), and its critical pressure is only about 18 MPa. This makes mercury a very suitable working fluid for the topping cycle. Mercury is not suitable as the sole working fluid for the entire cycle, however, since at a condenser temperature of 32°C its saturation pressure is 0.07 Pa. A power plant cannot operate at this vacuum because of air-leakage problems. At an acceptable condenser pressure of 7 kPa, the saturation temperature of mercury is 237°C, which is too high as the minimum temperature in the cycle. Therefore, the use of mercury as a working fluid is limited to the high-temperature cycles. Other disadvantages of mercury are its toxicity and high cost. The mass flow rate of mercury in binary vapor cycles is several times that of water because of its low enthalpy of vaporization.

It is evident from the *T-s* diagram in Fig. 9-24 that the binary vapor cycle approximates the Carnot cycle more closely than the steam cycle for the same temperature limits. Therefore, the thermal efficiency of a power plant can be increased by switching to binary cycles. The use of mercury–water binary cycles in the United States dates back to 1928. Several such plants have been built since then in the New England area, where fuel costs are typically higher. A small (40-MW) mercury–steam power plant that was in service in New Hampshire in 1950 had a higher thermal efficiency than most of the large modern power plants in use at that time.

Studies show that thermal efficiencies of 50 percent or higher are possible with binary vapor cycles. However, binary vapor cycles are not economically attractive because of their high initial cost and the competition offered by the combined gas–steam power plants.

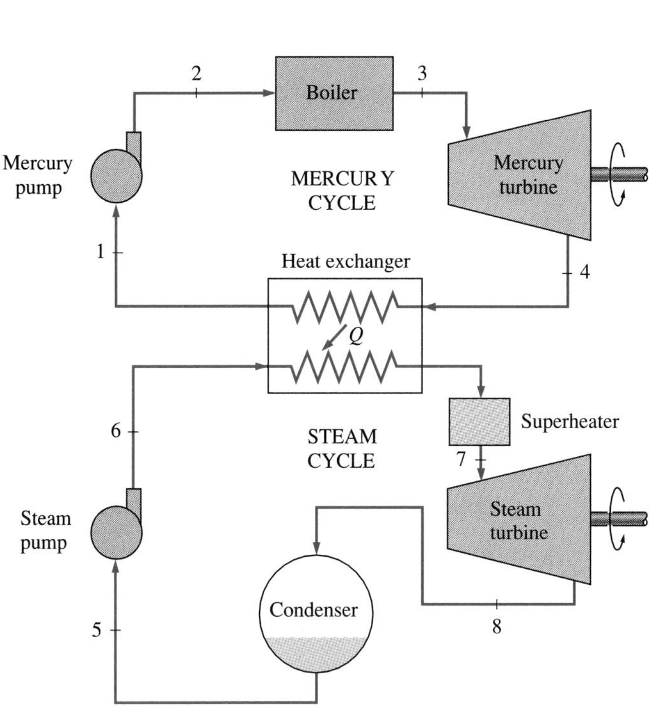

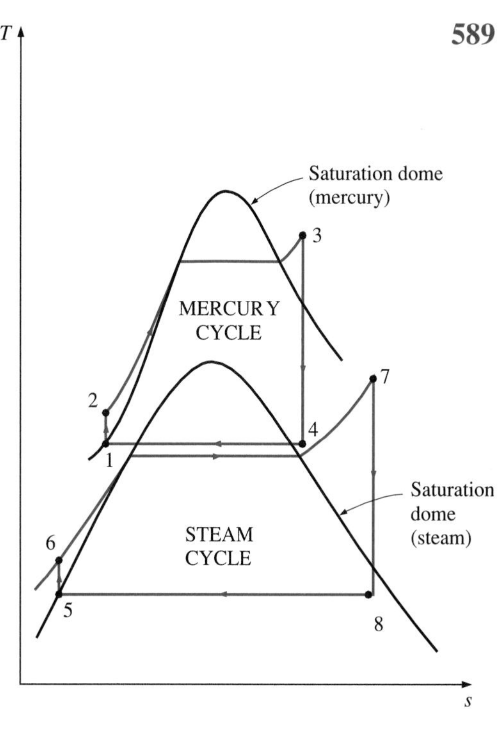

FIGURE 9-24
Mercury–water binary vapor cycle.

9-10 ■ COMBINED GAS–VAPOR POWER CYCLES

The continued quest for higher thermal efficiencies has resulted in rather innovative modifications to conventional power plants. The *binary vapor cycle* discussed above is one such modification. A more popular modification involves a gas power cycle topping a vapor power cycle, which is called the **combined gas–vapor cycle,** or just the **combined cycle.** The combined cycle of greatest interest is the gas-turbine (Brayton) cycle topping a steam-turbine (Rankine) cycle, which has a higher thermal efficiency than either of the cycles executed individually.

Gas-turbine cycles typically operate at considerably higher temperatures than steam cycles. The maximum fluid temperature at the turbine inlet is about 620°C (1150°F) for modern steam power plants, but over 1150°C (2100°F) for gas-turbine power plants. It is over 1500°C at the burner exit of turbojet engines. The use of higher temperatures in gas turbines is made possible by recent developments in cooling the turbine blades and coating the blades with high-temperature-resistant materials such as ceramics. Because of the higher average temperature at which heat is supplied, gas-turbine cycles have a greater potential for higher thermal efficiencies. However, the gas-turbine cycles have one inherent disadvantage: The gas leaves the gas

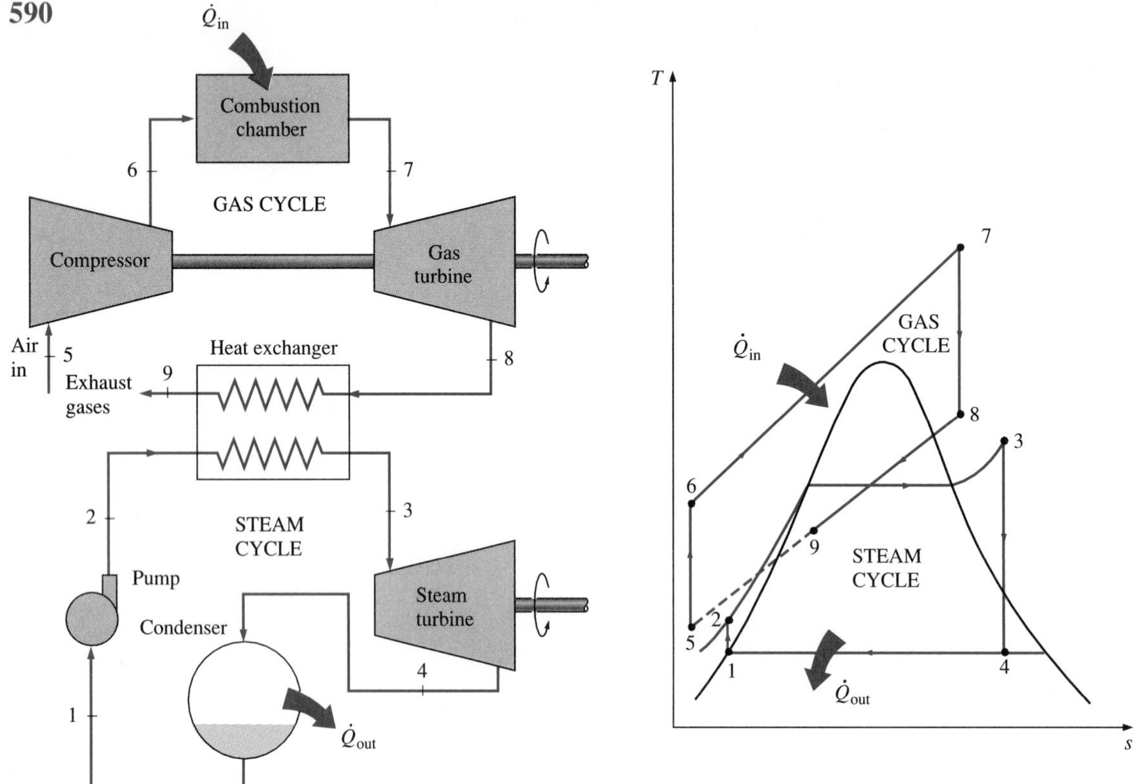

FIGURE 9-25

Combined gas–steam power plant.

turbine at very high temperatures (usually above 500°C), which wipes out any potential gains in the thermal efficiency. The situation can be improved somewhat by using regeneration, but the improvement is limited. Consequently, the thermal efficiency of gas-turbine power plants, in general, is relatively low.

It makes engineering sense to take advantage of the very desirable characteristics of the gas-turbine cycle at high temperatures *and* to use the high-temperature exhaust gases as the energy source for the bottoming cycle such as a steam power cycle. The result is a combined gas–steam cycle, as shown in Fig. 9-25. In this cycle, energy is recovered from the exhaust gases by transferring it to the steam in a heat exchanger that serves as the boiler. In general, more than one gas turbine is needed to supply sufficient heat to the steam. Also, the steam cycle may involve regeneration as well as reheating. Energy for the reheating process can be supplied by burning some additional fuel in the oxygen-rich exhaust gases.

Recent developments in gas-turbine technology have made the combined gas–steam cycle economically very attractive. The combined cycle increases the efficiency without appreciably increasing the initial cost. Consequently, many new power plants operate on combined cycles, and many more existing steam- or gas-turbine plants are being converted to combined-cycle power plants. Thermal efficiencies well over 40 percent are reported as a result of conversion.

A 1090-MW Tohoku combined plant that was put in commercial operation in 1985 in Niigata, Japan, is reported to operate at a thermal efficiency of 44 percent. This plant has two 191-MW steam turbines and six 118-MW gas turbines. Hot combustion gases enter the gas turbines at 1154°C, and steam enters the steam turbines at 500°C. Steam is cooled in the condenser by cooling water at an average temperature of 15°C. The compressors have a pressure ratio of 14, and the mass flow rate of air through the compressors is 443 kg/s.

A 1350-MW combined-cycle power plant built in Ambarli, Turkey, in 1988 by Siemens of Germany is the first commercially operating thermal plant in the world to attain an efficiency level as high as 52.5 percent at design operating conditions. This plant has six 150-MW gas turbines and three 173-MW steam turbines. Some recent combined-cycle power plants have achieved efficiencies above 60 percent.

EXAMPLE 9-9 A Combined Gas–Steam Power Cycle

Consider the combined gas–steam power cycle shown in Fig. 9-26. The topping cycle is a gas-turbine cycle that has a pressure ratio of 8. Air enters the compressor at 300 K and the turbine at 1300 K. The adiabatic efficiency of the compressor is 80 percent, and that of the gas turbine is 85 percent. The bottoming cycle is a simple ideal Rankine cycle operating between the pressure limits of 7 MPa and 5 kPa. Steam is heated in a heat exchanger by the exhaust gases to a temperature of 500°C. The exhaust gases leave the heat exchanger at 450 K. Determine (*a*) the ratio of the mass flow rates of the steam and the combustion gases and (*b*) the thermal efficiency of the combined cycle.

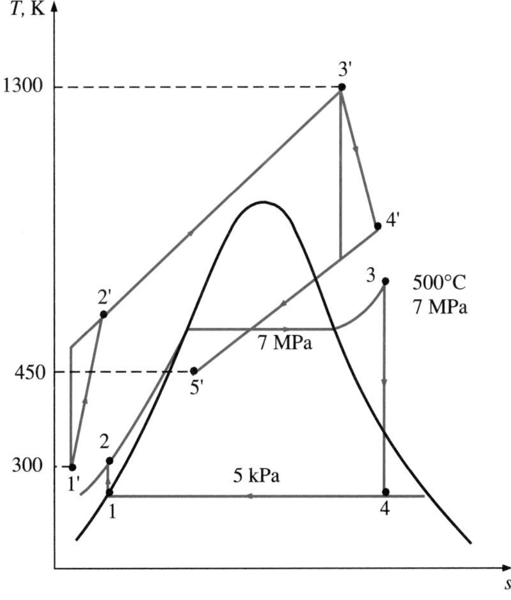

FIGURE 9-26
T-s diagram of the gas–steam combined cycle described in Example 9-9.

Solution The *T-s* diagrams of both cycles are given in Fig. 9-26. The gas-turbine cycle alone was analyzed in Example 8-6, and the steam cycle in Example 9-8*b*, with the following results:

Gas cycle: $h_4' = 880.14$ kJ/kg $\quad (T_4' = 853$ K$)$

$q_{in} = 790.58$ kJ/kg $\quad w_{net} = 210.63$ kJ/kg $\quad \eta_{th} = 26.6\%$

$h_5' = h_{@\,450\,K} = 451.80$ kJ/kg

Steam cycle: $h_2 = 144.85$ kJ/kg $\quad (T_2 = 33°C)$

$h_3 = 3410.3$ kJ/kg $\quad (T_3 = 500°C)$

$w_{net} = 1331.4$ kJ/kg $\quad \eta_{th} = 40.8\%$

Analysis (*a*) The ratio of mass flow rates is determined from an energy balance on the heat exchanger:

$$\dot{E}_{in} = \dot{E}_{out}$$

$$\dot{m}_g h_5' + \dot{m}_s h_3 = \dot{m}_g h_4' + \dot{m}_s h_2$$

$$\dot{m}_s(h_3 - h_2) = \dot{m}_g(h_4' - h_5')$$

$$\dot{m}_s(3410.3 - 144.85) = \dot{m}_g(880.14 - 451.80)$$

Thus,

$$\frac{\dot{m}_s}{\dot{m}_g} = y = \mathbf{0.131}$$

That is, a kg of exhaust gases can heat only 0.131 kg of steam from 33 to 500°C as they are cooled from 853 to 450 K. Then the total net work output per kilogram of combustion gases becomes

$$\begin{aligned} w_{net} &= w_{net,\,gas} + y w_{net,\,steam} \\ &= (210.63 \text{ kJ/kg gas}) + (0.131 \text{ kg steam/kg gas})(1331.4 \text{ kJ/kg steam}) \\ &= 385.04 \text{ kJ/kg gas} \end{aligned}$$

Therefore, for each kg of combustion gases produced, the combined plant will deliver 385.04 kJ of work. The net power output of the plant is determined by multiplying this value by the mass flow rate of the working fluid in the gas-turbine cycle.

(*b*) The thermal efficiency of the combined cycle is determined from

$$\eta_{th} = \frac{w_{net}}{q_{in}} = \frac{385.04 \text{ kJ/kg gas}}{790.58 \text{ kJ/kg gas}} = \mathbf{0.487 \text{ or } 48.7\%}$$

Discussion Note that this combined cycle will convert to useful work 48.7 percent of the energy supplied to the gas in the combustion chamber. This value is considerably higher than the thermal efficiency of the gas-turbine cycle (26.6 percent) or the steam-turbine cycle (40.7 percent) operating alone.

9-11 ■ SUMMARY

The *Carnot cycle* is not a suitable model for vapor power cycles because it cannot be approximated in practice. The model cycle for vapor power cycles is the *Rankine cycle,* which is composed of four internally reversible processes: constant-pressure heat addition in a boiler, isentropic expansion in a turbine, constant-pressure heat rejection in a condenser, and isentropic compression in a pump. Steam leaves the condenser as a saturated liquid at the condenser pressure.

The thermal efficiency of the Rankine cycle can be increased by increasing the average temperature at which heat is added to the working fluid and/or by decreasing the average temperature at which heat is rejected to the cooling medium. The average temperature during heat rejection can be decreased by lowering the turbine exit pressure. Consequently, the condenser pressure of most vapor power plants is well below the atmospheric pressure. The average temperature during heat addition can be increased by raising the boiler pressure or by superheating the fluid to high temperatures. There is a limit to the degree of superheating, however, since the fluid temperature is not allowed to exceed a metallurgically safe value.

Superheating has the added advantage of decreasing the moisture content of the steam at the turbine exit. Lowering the exhaust pressure or raising the boiler pressure, however, increases the moisture content. To take advantage of the improved efficiencies at higher boiler pressures and lower condenser pressures, steam is usually *reheated* after expanding partially in the high-pressure turbine. This is done by extracting the steam after partial extraction in the high-pressure turbine, sending it back to the boiler where it is reheated at constant pressure, and returning it to the low-pressure turbine for complete expansion to the condenser pressure. The average temperature during the reheat process, and thus the thermal efficiency of the cycle, can be increased by increasing the number of expansion and reheat stages. As the number of stages is increased, the expansion and reheat processes approach an isothermal process at maximum temperature. Reheating also decreases the moisture content at the turbine exit.

Another way of increasing the thermal efficiency of the Rankine cycle is by *regeneration*. During a regeneration process, liquid water (feedwater) leaving the pump is heated by some steam bled off the turbine at some intermediate pressure in devices called *feedwater heaters*. The two streams are mixed in open feedwater heaters, and the mixture leaves as a saturated liquid at the heater pressure. In closed feedwater heaters, heat is transferred from the steam to the feedwater without mixing.

The production of more than one useful form of energy (such as process heat and electric power) from the same energy source is called *cogeneration*. Cogeneration plants produce electric power while meeting the process heat requirements of certain industrial processes. This way, more of the energy transferred to the fluid in the boiler is utilized for a useful purpose. The faction of energy that is used for either process heat or power generation is called the *utilization factor* of the cogeneration plant.

The overall thermal efficiency of a power plant can be increased by using *binary cycles* or *combined cycles*. A binary cycle is composed of two separate cycles, one at high temperatures (topping cycle) and the other at relatively low temperatures. The most common combined cycle is the gas–steam combined cycle where a gas-turbine cycle operates at the high-temperature range and a steam-turbine cycle at the low-temperature range.

Steam is heated by the high-temperature exhaust gases leaving the gas turbine. Combined cycles have a higher thermal efficiency than the steam- or gas-turbine cycles operating alone.

REFERENCES AND SUGGESTED READING

1. R. L. Bannister and G. J. Silvestri. "The Evolution of Central Station Steam Turbines." *Mechanical Engineering,* February 1989, pp. 70–78.

2. R. L. Bannister, G. J. Silvestri, A. Hizume, and T. Fujikawa. "High Temperature Supercritical Steam Turbines." *Mechanical Engineering,* February 1987, pp. 60–65.

3. W. Z. Black and J. G. Hartley. *Thermodynamics.* New York: Harper & Row, 1985.

4. M. D. Burghard. *Engineering Thermodynamics with Applications.* 3rd ed. New York: Harper & Row, 1986.

5. M. M. El-Wakil. *Powerplant Technology.* New York: McGraw-Hill, 1984.

6. R. C. Fellinger and W. J. Cook. *Introduction to Engineering Thermodynamics.* Dubuque, IA: William C. Brown, 1985.

7. B. V. Karlekar. *Thermodynamics for Engineers.* Englewood Cliffs, NJ: Prentice Hall, 1983.

8. K. W. Li and A. P. Priddy. *Power Plant System Design.* New York: John Wiley & Sons, 1985.

9. D. C. Look, Jr., and H. J. Sauer, Jr. *Engineering Thermodynamics.* Boston: PWS Engineering, 1986.

10. H. Sorensen. *Energy Conversion Systems.* New York: John Wiley & Sons, 1983.

11. *Steam, Its Generation and Use.* 39th ed. New York: Babcock and Wilcox Co., 1978.

12. *Turbomachinery* 28, no. 2 (March/April 1987) Norwalk, CT: Business Journals, Inc.

13. G. J. Van Wylen and R. E. Sonntag. *Fundamentals of Classical Thermodynamics.* 3rd ed. New York: John Wiley & Sons, 1985.

14. K Wark. *Thermodynamics.* 5th ed. New York: McGraw-Hill, 1988.

15. J. Weisman and R. Eckart. *Modern Power Plant Engineering.* Englewood Cliffs, NJ: Prentice Hall, 1985.

Carnot Vapor Cycle

9-1C Why is excessive moisture in steam undesirable in steam turbines? What is the highest moisture content allowed?

9-2C Why is the Carnot cycle not a realistic model for steam power plants?

9-3E Water enters the boiler of a steady-flow Carnot engine as a saturated liquid at 120 psia and leaves with a quality of 0.95. Steam leaves the turbine at a pressure of 14.7 psia. Show the cycle on a *T-s* diagram relative to the saturation lines, and determine (*a*) the thermal efficiency, (*b*) the quality at the end of the isothermal heat-rejection process, and (*c*) the net work output.
Answers: (*a*) 16.1 percent, (*b*) 0.1245, (*c*) 134.4 Btu/lbm

9-4 A steady-flow Carnot cycle uses water as the working fluid. Water changes from saturated liquid to saturated vapor as heat is transferred to it from a source at 250°C. Heat rejection takes place at a pressure of 20 kPa. Show the cycle on a *T-s* diagram relative to the saturation lines, and determine (*a*) the thermal efficiency; (*b*) the amount of heat rejected, in kJ/kg; and (*c*) the net work output.

9-5 Consider a steady-flow Carnot cycle with water as the working fluid. The maximum and minimum temperatures in the cycle are 350 and 60°C. The quality of water is 0.891 at the beginning of the heat-rejection process and 0.1 at the end. Show the cycle on a *T-s* diagram relative to the saturation lines, and determine (*a*) the thermal efficiency, (*b*) the pressure at the turbine inlet, and (*c*) the net work output.
Answers: (*a*) 0.465, (*b*) 1.40 MPa, (*c*) 1624 kJ/kg

The Simple Rankine Cycle

9-6C What four processes make up the simple ideal Rankine cycle?

9-7C Consider a simple ideal Rankine cycle with fixed turbine inlet conditions. What is the effect of lowering the condenser pressure on

Pump work input:	(*a*) increases, (*b*) decreases, (*c*) remains the same
Turbine work output:	(*a*) increases, (*b*) decreases, (*c*) remains the same
Heat supplied:	(*a*) increases, (*b*) decreases, (*c*) remains the same
Heat rejected:	(*a*) increases, (*b*) decreases, (*c*) remains the same
Cycle efficiency:	(*a*) increases, (*b*) decreases, (*c*) remains the same
Moisture content at turbine exit:	(*a*) increases, (*b*) decreases, (*c*) remains the same

*Students are encouraged to answer *all* the concept "C" questions.

9-8C Consider a simple ideal Rankine cycle with fixed turbine inlet temperature and condenser pressure. What is the effect of increasing the boiler pressure on

Pump work input:	(*a*) increases, (*b*) decreases, (*c*) remains the same
Turbine work output:	(*a*) increases, (*b*) decreases, (*c*) remains the same
Heat supplied:	(*a*) increases, (*b*) decreases, (*c*) remains the same
Heat rejected:	(*a*) increases, (*b*) decreases, (*c*) remains the same
Cycle efficiency:	(*a*) increases, (*b*) decreases, (*c*) remains the same
Moisture content at turbine exit:	(*a*) increases, (*b*) decreases, (*c*) remains the same

9-9C Consider a simple ideal Rankine cycle with fixed boiler and condenser pressures. What is the effect of superheating the steam to a higher temperature on

Pump work input:	(*a*) increases, (*b*) decreases, (*c*) remains the same
Turbine work output:	(*a*) increases, (*b*) decreases, (*c*) remains the same
Heat supplied:	(*a*) increases, (*b*) decreases, (*c*) remains the same
Heat rejected:	(*a*) increases, (*b*) decreases, (*c*) remains the same
Cycle efficiency:	(*a*) increases, (*b*) decreases, (*c*) remains the same
Moisture content at turbine exit:	(*a*) increases, (*b*) decreases, (*c*) remains the same

9-10C How do actual vapor power cycles differ from idealized ones?

9-11C Compare the pressures at the inlet and the exit of the boiler for (*a*) actual and (*b*) ideal cycles.

9-12C The entropy of steam increases in actual steam turbines as a result of irreversibilities. In an effort to control entropy increase, it is proposed to cool the steam in the turbine by running cooling water around the turbine casing. It is argued that this will reduce the entropy and the enthalpy of the steam at the turbine exit and thus increase the work output. How would you evaluate this proposal?

9-13C Is it possible to maintain a pressure of 10 kPa in a condenser that is being cooled by river water entering at 20°C?

9-14 A steam power plant operates on a simple ideal Rankine cycle between the pressure limits of 3 MPa and 50 kPa. The temperature of the steam at the turbine inlet is 400°C, and the mass flow rate of steam through the cycle is 25 kg/s. Show the cycle on a *T-s* diagram with respect to saturation lines, and determine (*a*) the thermal efficiency of the cycle and (*b*) the net power output of the power plant.

9-15 Consider a 300-MW steam power plant that operates on a simple ideal Rankine cycle. Steam enters the turbine at 10 MPa and 500°C and is cooled in the condenser at a pressure of 10 kPa. Show the cycle on a *T-s* diagram with respect to saturation lines, and determine (*a*) the quality of the steam at the turbine exit, (*b*) the thermal efficiency of the cycle, and (*c*) the mass flow rate of the steam. *Answers:* (*a*) 0.793, (*b*) 40.2 percent, (*c*) 235.4 kg/s

9-16 Repeat Prob. 9-15 assuming an adiabatic efficiency of 85 percent for both the turbine and the pump.
Answers: (*a*) 0.874, (*b*) 34.1 percent, (*c*) 277.8 kg/s

9-17E A steam power plant operates on a simple ideal Rankine cycle between the pressure limits of 1250 and 2 psia. The mass flow rate of steam through the cycle is 75 lbm/s. The moisture content of the steam at the turbine exit is not to exceed 10 percent. Show the cycle on a *T-s* diagram with respect to saturation lines, and determine (*a*) the minimum turbine inlet temperature, (*b*) the rate of heat input in the boiler, and (*c*) the thermal efficiency of the cycle.

9-18E Repeat Prob. 9-17E assuming an adiabatic efficiency of 85 percent for both the turbine and the pump.

9-19 Consider a coal-fired steam power plant that produces 300 MW of electric power. The power plant operates on a simple ideal Rankine cycle with turbine inlet conditions of 5 MPa and 450°C and a condenser pressure of 25 kPa. The coal used has a heating value (energy released when the fuel is burned) of 29,300 kJ/kg. Assuming that 75 percent of this energy is transferred to the steam in the boiler and that the electric generator has an efficiency of 96 percent, determine (*a*) the overall plant efficiency (the ratio of net electric power output to the energy input as fuel) and (*b*) the required rate of coal supply. *Answers:* (*a*) 24.6 percent, (*b*) 150.1 t/h

9-20 Consider a solar-pond power plant that operates on a simple ideal Rankine cycle with refrigerant-134a as a the working fluid. The refrigerant enters the turbine as a saturated vapor at 1.6 MPa and leaves at 0.7 MPa. The mass flow rate of the refrigerant is 6 kg/s. Show the cycle on a *T-s* diagram with respect to saturation lines, and determine (*a*) the thermal efficiency of the cycle and (*b*) the power output of this plant.

9-21 Consider a steam power plant that operates on a simple ideal Rankine cycle and has a net power output of 30 MW. Steam enters the turbine at 7 MPa and 500°C and is cooled in the condenser at a pressure of 10 kPa by running cooling water from a lake through the tubes of the condenser at a rate of 2000 kg/s. Show the cycle on a *T-s* diagram with respect to saturation lines, and determine (*a*) the thermal efficiency of the cycle, (*b*) the mass flow rate of the steam, and (*c*) the temperature rise of the cooling water.
Answers: (*a*) 38.9 percent, (*b*) 24.0 kg/s, (*c*) 5.63°C

9-22 Repeat Prob. 9-21 assuming an adiabatic efficiency of 87 percent for both the turbine and the pump.
Answers: (*a*) 33.8 percent, (*b*) 27.65 kg/s, (*c*) 7.03°C

The Reheat Rankine Cycle

9-23C How do the following quantities change when a simple ideal Rankine cycle is modified with reheating? Assume the mass flow rate is maintained the same.

Pump work input:	(*a*) increases, (*b*) decreases, (*c*) remains the same
Turbine work output:	(*a*) increases, (*b*) decreases, (*c*) remains the same
Heat supplied:	(*a*) increases, (*b*) decreases, (*c*) remains the same
Heat rejected:	(*a*) increases, (*b*) decreases, (*c*) remains the same
Moisture content at turbine exit:	(*a*) increases, (*b*) decreases, (*c*) remains the same

9-24C Show the ideal Rankine cycle with three stages of reheating on a *T-s* diagram. Assume the turbine inlet temperature is the same for all stages. How does the cycle efficiency vary with the number of reheat stages?

9-25C Consider a simple Rankine cycle and an ideal Rankine cycle with three reheat sages. Both cycles operate between the same pressure limits. The maximum temperature is 700°C in the simple cycle and 500°C in the reheat cycle. Which cycle do you think will have a higher thermal efficiency?

9-26 A steam power plant operates on the ideal reheat Rankine cycle. Steam enters the high-pressure turbine at 8 MPa and 500°C and leaves at 3 MPa. Steam is then reheated at constant pressure to 500°C before it expands to 20 kPa in the low-pressure turbine. Determine the turbine work output, in kJ/kg, and the thermal efficiency of the cycle. Also, show the cycle on a *T-s* diagram with respect to saturation lines.

9-27 Consider a steam power plant that operates on a reheat Rankine cycle and has a net power output of 150 MW. Steam enters the high-pressure turbine at 10 MPa and 500°C and the low-pressure turbine at 1 MPa and 500°C. Steam leaves the condenser as a saturated liquid at a pressure of 10 kPa. The adiabatic efficiency of the turbine is 80 percent, and that of the pump is 95 percent. Show the cycle on a *T-s* diagram with respect to saturation lines, and determine (*a*) the quality (or temperature, if superheated) of the steam at the turbine exit, (*b*) the thermal efficiency of the cycle, and (*c*) the mass flow rate of the steam. *Answers:* (*a*) 87.5°C, (*b*) 34.1 percent, (*c*) 117.5 kg/s

9-28 Repeat Prob. 9-27 assuming both the pump and the turbine are isentropic. *Answers:* (*a*) 0.948, (*b*) 41.4 percent, (*c*) 93.8 kg/s

9-29E Steam enters the high-pressure turbine of a steam power plant that operates on the ideal reheat Rankine cycle at 800 psia and 900°F and leaves as saturated vapor. Steam is then reheated to 800°F before it expands to a pressure of 1 psia. Heat is transferred to the steam in the boiler at a rate of 6×10^4 Btu/s. Steam is cooled in the condenser by the cooling water from a nearby river, which enters the condenser at 45°F. Show the cycle on a *T-s* diagram with respect to saturation lines, and determine (*a*) the pressure at which reheating takes place, (*b*) the net power output and thermal efficiency, and (*c*) the minimum mass flow rate of the cooling water required.

9-30 A steam power plant operates on an ideal reheat Rankine cycle between the pressure limits of 9 MPa and 10 kPa. The mass flow rate of steam through the cycle is 25 kg/s. Steam enters both stages of the turbine at 500°C. If the moisture content of the steam at the exit of the low-pressure turbine is not to exceed 10 percent, determine (*a*) the pressure at which reheating takes place, (*b*) the total rate of heat input in the boiler, and (*c*) the thermal efficiency of the cycle. Also, show the cycle on a *T-s* diagram with respect to saturation lines.

Regenerative Rankine Cycle

9-31C How do the following quantities change when the simple ideal Rankine cycle is modified with regeneration? Assume the mass flow rate through the boiler is the same.

Turbine work output:	(*a*) increases, (*b*) decreases, (*c*) remains the same
Heat supplied:	(*a*) increases, (*b*) decreases, (*c*) remains the same
Heat rejected:	(*a*) increases, (*b*) decreases, (*c*) remains the same
Moisture content at turbine exit:	(*a*) increases, (*b*) decreases, (*c*) remains the same

9-32C During a regeneration process, some steam is extracted from the turbine and is used to heat the liquid water leaving the pump. This does not seem like a smart thing to do since the extracted steam could produce some more work in the turbine. How do you justify this action?

9-33C How do open feedwater heaters differ from closed feedwater heaters?

9-34C Consider a simple ideal Rankine cycle and an ideal regenerative Rankine cycle with one open feedwater heater. The two cycles are very much alike, except the feedwater in the regenerative cycle is heated by extracting some steam just before it enters the turbine. How would you compare the efficiencies of these two cycles?

9-35C Devise an ideal regenerative Rankine cycle that has the same thermal efficiency as the Carnot cycle. Show the cycle on a *T-s* diagram.

9-36 A steam power plant operates on an ideal regenerative Rankine cycle. Steam enters the turbine at 6 MPa and 450°C and is condensed in the condenser at 20 kPa. Steam is extracted from the turbine at 0.4 MPa to heat the feedwater in an open feedwater heater. Water leaves the feedwater heater as a saturated liquid. Show the cycle on a *T-s* diagram, and determine (*a*) the net work output per kilogram of steam flowing through the boiler and (*b*) the thermal efficiency of the cycle. *Answers:* (*a*) 1016 kJ/kg, (*b*) 37.8 percent

9-37 Repeat Prob. 9-36 by replacing the open feedwater heater with a closed feedwater heater. Assume that the feedwater leaves the heater at the condensation temperature of the extracted steam and that the extracted steam leaves the heater as a saturated liquid and is pumped to the line carrying the feedwater.

9-38 A steam power plant operates on an ideal regenerative Rankine cycle with two open feedwater heaters. Steam enters the turbine at 10 MPa and 600°C and exhausts to the condenser at 5 kPa. Steam is extracted from the turbine at 0.6 and 0.2 MPa. Water leaves both feedwater heaters as a saturated liquid. The mass flow rate of steam through the boiler is 18 kg/s. Show the cycle on a *T-s* diagram, and determine (*a*) the net power output of the power plant and (*b*) the thermal efficiency of the cycle.
Answers: (*a*) 24.5 MW, (*b*) 46.3 percent

9-39 Consider an ideal steam regenerative Rankine cycle with two feedwater heaters, one closed and one open. Steam enters the turbine at 12.5 MPa and 550°C and exhausts to the condenser at 10 kPa. Steam is extracted from the turbine at 0.8 MPa for the closed feedwater heater and at 0.3 MPa for the open one. The feedwater is heated to the condensation temperature of the extracted steam in the closed feedwater heater. The extracted steam leaves the closed feedwater heater as a saturated liquid, which is subsequently throttled to the open feedwater heater. Show the cycle on a *T-s* diagram with respect to saturation lines, and determine (*a*) the mass flow rate of steam through the boiler for a net power output of 250 MW and (*b*) the thermal efficiency of the cycle.

9-40 A steam power plant operates on an ideal reheat–regenerative Rankine cycle and has a net power output of 120 MW. Steam enters the high-pressure turbine at 10 MPa and 550°C and leaves at 0.8 MPa. Some steam is extracted at this pressure to heat the feedwater in an open feedwater heater. The rest of the steam is reheated to 500°C and is expanded in the low-pressure turbine to the condenser pressure of 10 kPa. Show the cycle on a *T-s* diagram with respect to saturation lines, and determine (*a*) the mass flow rate of steam through the boiler and (*b*) the thermal efficiency of the cycle.
Answers: (*a*) 81.9 kg/s, (*b*) 44.4 percent

9-41 Repeat Prob. 9-40, but replace the open feedwater heater with a closed feedwater heater. Assume that the feedwater leaves the heater at the condensation temperature of the extracted steam and that the extracted steam leaves the heater as a saturated liquid and is pumped to the line carrying the feedwater.

9-42E A steam power plant operates on an ideal reheat–regenerative Rankine cycle with one reheater and two open feedwater heaters. Steam enters the high-pressure turbine at 1500 psia and 1100°F and leaves the low-pressure turbine at 1 psia. Steam is extracted from the turbine at 250 and 40 psia, and it is reheated to 1000°F at a pressure of 140 psia. Water leaves both feedwater heaters as a saturated liquid. Heat is transferred to the steam in the boiler at a rate of 6×10^5 Btu/s. Show the cycle on a *T-s* diagram with respect to saturation lines, and determine (*a*) the mass flow rate of steam through the boiler, (*b*) the net power output of the plant, and (*c*) the thermal efficiency of the cycle.

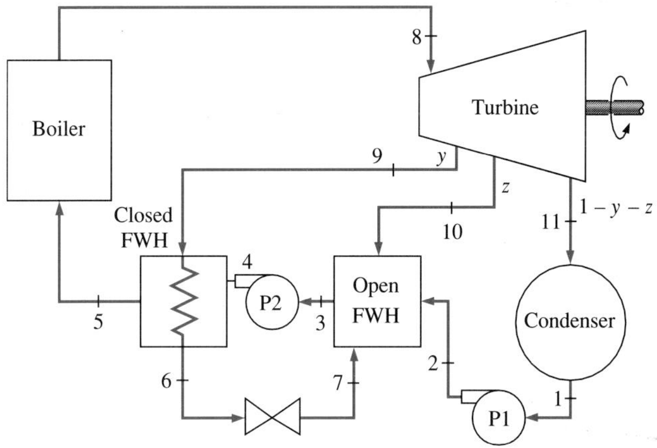

FIGURE P9-39

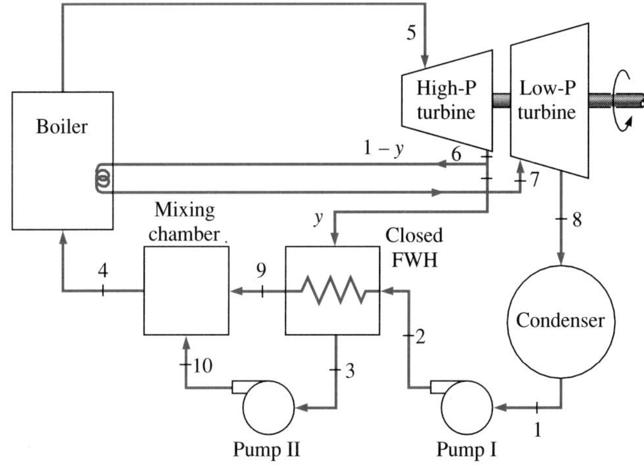

FIGURE P9-41

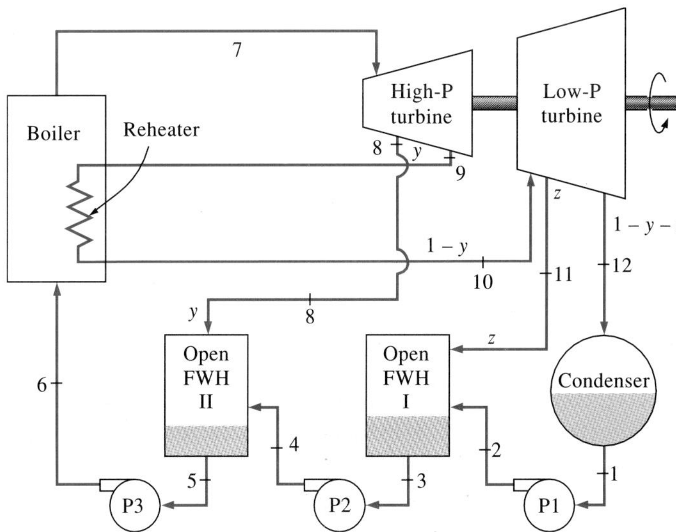

FIGURE P9-42E

Second-Law Analysis of Vapor Power Cycles

9-43 Determine the irreversibilities associated with each of the processes of the Rankine cycle described in Prob. 9-14, assuming a source temperature of 1500 K and a sink temperature of 290 K.

9-44 Determine the irreversibilities associated with each of the processes of the Rankine cycle described in Prob. 9-15, assuming a source temperature of 1500 K and a sink temperature of 290 K.
Answers: 0, 1111.5 kJ/kg, 0, 172.8 kJ/kg

9-45 Determine the irreversibility associated with the heat rejection process in prob. 9-21. Assume a source temperature of 1500 K and a sink temperature of 290 K. Also, determine the exergy of the steam at the boiler exit. Take $P_0 = 100$ kPa.

9-46 Determine the irreversibilities associated with each of the processes of the reheat Rankine cycle described in Prob. 9-26. Assume a source temperature of 1500 K and a sink temperature of 290 K.

9-47 Determine the irreversibilities associated with the heat addition process and the expansion process in Prob. 9-27. Assume a source temperature of 1500 K and a sink temperature of 290 K. Also, determine the exergy of the steam at the boiler exit. Take $P_0 = 100$ kPa.
Answers: 1267 kJ/kg, 249.9 kJ/kg, 1462.8 kJ/kg

9-48 Determine the irreversibility associated with the regenerative cycle described in Prob. 9-36. Assume a source temperature of 1500 K and a sink temperature of 290 K. *Answer:* 1154.7 kJ/kg

9-49 Determine the irreversibilities associated with the reheating and regeneration processes described in Prob. 9-40. Assume a source temperature of 1500 K and a sink temperature of 300 K.

Cogeneration

9-50C How is the utilization factor ε_u for cogeneration plants defined? Could ε_u be unity for a cogeneration plant that does not produce any power?

9-51C Consider a cogeneration cycle for which the utilization factor is 1. Is the irreversibility associated with this cycle necessarily zero? Explain.

9-52C Consider a cogeneration cycle for which the utilization factor is 0.5. Can the irreversibility associated with this cycle be zero? If yes, under what conditions?

9-53C What is the difference between cogeneration and regeneration?

9-54 Steam enters the turbine of a cogeneration plant at 7 MPa and 500°C. One-fourth of the steam is extracted from the turbine at 600-kPa pressure for process heating. The remaining steam continues to expand to 10 kPa. The extracted steam is then condensed and mixed with feedwater at constant pressure and the mixture is pumped to the boiler pressure of 7 MPa. The

FIGURE P9-54

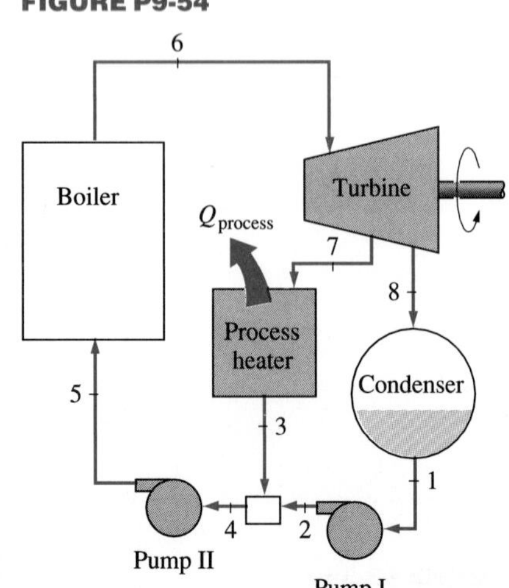

mass flow rate of steam through the boiler is 30 kg/s. Disregarding any pressure drops and heat losses in the piping, and assuming the turbine and the pump to be isentropic, determine the net power produced and the utilization factor of the plant.

9-55E A large food-processing plant requires 8 lbm/s of saturated or slightly superheated steam at 80 psia, which is extracted from the turbine of a cogeneration plant. The boiler generates steam at 1000 psia and 1000°F at a rate of 20 lbm/s, and the condenser pressure is 2 psia. Steam leaves the process heater as a saturated liquid. It is then mixed with the feedwater at the same pressure and this mixture is pumped to the boiler pressure. Assuming both the pumps and the turbine have adiabatic efficiencies of 86 percent, determine (*a*) the rate of heat transfer to the boiler and (*b*) the power output of the cogeneration plant. *Answers:* (*a*) 26,661 Btu/s, (*b*) 8102 kW

9-56 Steam is generated in the boiler of a cogeneration plant at 10 MPa and 450°C at a steady rate of 15 kg/s. In normal operation, steam expands in a turbine to a pressure of 0.5 MPa and is then routed to the process heater, where it supplies the process heat. Steam leaves the process heater as a saturated liquid and is pumped to the boiler pressure. In this mode, no steam passes through the condenser, which operates at 20 kPa.

(*a*) Determine the power produced and the rate at which process heat is supplied in this mode.

(*b*) Determine the power produced and the rate of process heat supplied if only 60 percent of the steam is routed to the process heater and the remainder is expanded to the condenser pressure.

9-57 Consider a cogeneration power plant modified with regeneration. Steam enters the turbine at 6 MPa and 450°C and expands to a pressure of 0.4 MPa. At this pressure, 60 percent of the steam is extracted from the turbine, and the remainder expands to 10 kPa. Part of the extracted steam is used to heat the feedwater in an open feedwater heater. The rest of the

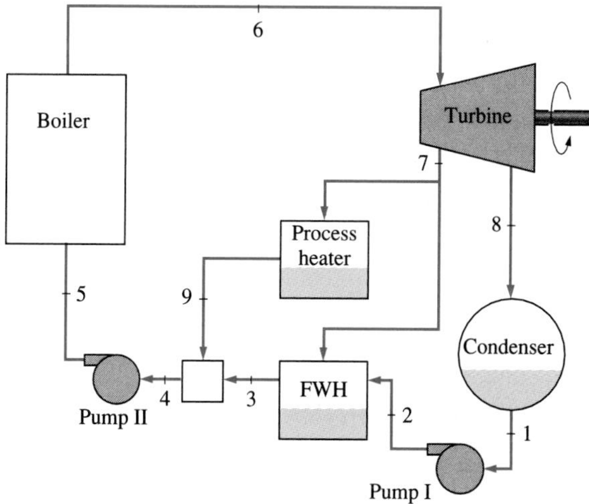

FIGURE P9-57

extracted steam is used for process heating and leaves the process heater as a saturated liquid at 0.4 MPa. It is subsequently mixed with the feedwater leaving the feedwater heater, and the mixture is pumped to the boiler pressure. Assuming the turbines and the pumps to be isentropic, show the cycle on a *T-s* diagram with respect to saturation lines, and determine the mass flow rate of steam through the boiler for a net power output of 15 MW.

Answer: 17.7 kg/s

9-58E Steam is generated in the boiler of a cogeneration plant at 800 psia and 900°F at a rate of 15 lbm/s. The plant is to produce power while meeting the process steam requirements for a certain industrial application. One-third of the steam leaving the boiler is throttled to a pressure of 120 psia and is routed to the process heater. The rest of the steam is expanded in an isentropic turbine to a pressure of 120 psia and is also routed to the process heater. Steam leaves the process heater at 240°F. Neglecting the pump work, determine (*a*) the net power produced, (*b*) the rate of process heat supply, and (*c*) the utilization factor of this plant.

Binary Vapor and Combined Gas–Vapor Power Cycles

9-59C What is a binary power cycle? What is its purpose?

9-60C By writing an energy balance on the heat exchanger of a binary vapor power cycle, obtain a relation for the ratio of mass flow rates of two fluids in terms of their enthalpies.

9-61C Why is steam not an ideal working fluid for vapor power cycles?

9-62C Why is mercury a suitable working fluid for the topping portion of a binary vapor cycle but not for the bottoming cycle?

9-63C What is the difference between the binary vapor power cycle and the combined gas–steam power cycle?

9-64C In combined gas–steam cycles, what is the energy source for the steam?

9-65C Why is the combined gas–steam cycle more efficient than either of the cycles operated alone?

9-66 The gas-turbine portion of a combined gas–steam power plant has a pressure ratio of 16. Air enters the compressor at 300 K at a rate of 14 kg/s and is heated to 1500 K in the combustion chamber. The combustion gases leaving the gas turbine are used to heat the steam to 400°C at 10 MPa in a heat exchanger. The combustion gases leave the heat exchanger at 420 K. The steam leaving the turbine is condensed at 15 kPa. Assuming all the compression and expansion processes to be isentropic, determine (*a*) the mass flow rate of the steam, (*b*) the net power output, and (*c*) the thermal efficiency of the combined cycle. For air, assume constant specific heats at room temperature. *Answers:* (*a*) 1.275 kg/s, (*b*) 7818 kW, (*c*) 66.3 percent

9-67 Consider a combined gas–steam power plant that has a net power output of 450 MW. The pressure ratio of the gas-turbine cycle is 14. Air enters the compressor at 300 K and the turbine at 1400 K. The combustion gases leaving the gas turbine are used to heat the steam at 8 MPa to 400°C in a heat exchanger. The combustion gases leave the heat exchanger at 460 K. An open feedwater heater incorporated with the steam cycle operates at a pressure of 0.6 MPa. The condenser pressure is 20 kPa. Assuming all the compression and expansion processes to be isentropic, determine (*a*) the mass flow rate ratio of air to steam, (*b*) the required rate of heat input in the combustion chamber, and (*c*) the thermal efficiency of the combined cycle.

9-68 Repeat Prob. 9-67 assuming adiabatic efficiencies of 100 percent for the pump, 82 percent for the compressor, and 86 percent for the gas and steam turbines.

Review Problems

9-69 Show that the thermal efficiency of a combined gas–steam power plant η_{cc} can be expressed as

$$\eta_{cc} = \eta_g + \eta_s - \eta_g \eta_s$$

where $\eta_g = W_g/Q_{in}$ and $\eta_s = W_s/Q_{g,out}$ are the thermal efficiencies of the gas and steam cycles, respectively. Using this relation, determine the thermal efficiency of a combined power cycle that consists of a topping gas-turbine cycle with an efficiency of 40 percent and a bottoming steam-turbine cycle with an efficiency of 30 percent.

9-70 It can be shown that the thermal efficiency of a combined gas–steam power plant η_{cc} can be expressed in terms of the thermal efficiencies of the gas- and the steam-turbine cycles as

$$\eta_{cc} = \eta_g + \eta_s - \eta_g \eta_s$$

Prove that the value of η_{cc} is greater than either of η_g or η_s. That is, the combined cycle is more efficient than either of the gas-turbine or steam-turbine cycles alone.

9-71 Consider a steam power plant operating on the ideal Rankine cycle with reheat between the pressure limits of 25 MPa and 10 kPa with a maximum cycle temperature of 600°C and a moisture content of 12 percent at the turbine exit. For a reheat temperature of 600°C, determine the reheat pressures of the cycle for the cases of (*a*) single and (*b*) double reheat.

9-72E The Stillwater geothermal power plant in Nevada, which started full commercial operation in 1986, is designed to operate with seven identical units. Each of these seven units consists of a pair of power cycles, labeled Level I and Level II, operating on the simple Rankine cycle using an organic fluid as the working fluid.

The heat source for the plant is geothermal water (brine) entering the vaporizer (boiler) of Level I of each unit at 325°F at a rate of 384,286 lbm/h and delivering 22.79 MBtu/h ("M" stands for "million"). The organic fluid that enters the vaporizer at 202.2°F at a rate of 157,895 lbm/h leaves it at 282.4°F and 225.8 psia as saturated vapor. This saturated vapor expands in

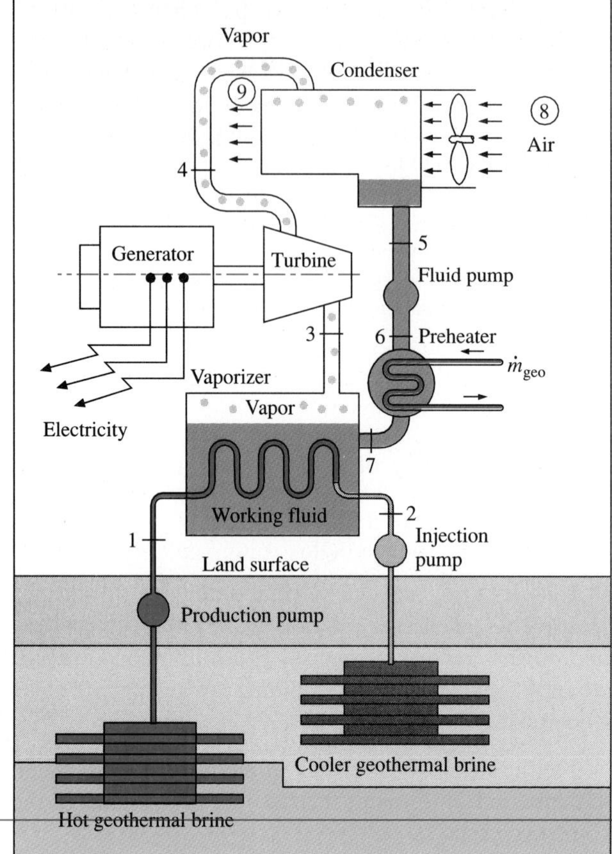

FIGURE P9-72E

Schematic of a binary geothermal power plant. (*Courtesy of ORMAT Energy Systems, Inc.*)

the turbine to 95.8°F and 19.0 psia and produces 1271 kW of electric power. About 200 kW of this power is used by the pumps, the auxiliaries, and the six fans of the condenser. Subsequently, the organic working fluid is condensed in an air-cooled condenser by air that enters the condenser at 55°F at a rate of 4,195,100 lbm/h and leaves at 84.5°F. The working fluid is pumped and then preheated in a preheater to 202.4°F by absorbing 11.14 MBtu/h of heat from the geothermal water (coming from the vaporizer of Level II) entering the preheater at 211.8°F and leaving at 154.0°F.

Taking the average specific heat of the geothermal water to be 1.03 Btu/(lbm · °F), determine (*a*) the exit temperature of the geothermal water from the vaporizer, (*b*) the rate of heat rejection from the working fluid to the air in the condenser, (*c*) the mass flow rate of the geothermal water at the preheater, and (*d*) the thermal efficiency of the Level I cycle of this geothermal power plant. *Answers:* (*a*) 267.4°F, (*b*) 29.7 MBtu/h, (*c*) 187,120 lbm/h, (*d*) 10.8 percent.

9-73 Steam enters the turbine of a steam power plant that operates on a simple ideal Rankine cycle at a pressure of 6 MPa, and it leaves as a saturated vapor at 7.5 kPa. Heat is transferred to the steam in the boiler at a rate of 4×10^4 kJ/s. Steam is cooled in the condenser by the cooling water from a nearby river, which enters the condenser at 18°C. Show the cycle on a *T-s* diagram with respect to saturation lines, and determine (*a*) the turbine inlet temperature, (*b*) the net power output and thermal efficiency, and (*c*) the minimum mass flow rate of the cooling water required.

9-74 A steam power plant operates on an ideal Rankine cycle with two stages of reheat and has a net power output of 300 MW. Steam enters all three stages of the turbine at 500°C. The maximum pressure in the cycle is 15 MPa, and the minimum pressure is 5 kPa. Steam is reheated at 5 MPa the first time and at 1 MPa the second time. Show the cycle on a *T-s* diagram with respect to saturation lines, and determine (*a*) the thermal efficiency of the cycle and (*b*) the mass flow rate of the steam.
Answers: (*a*) 45.5 percent, (*b*) 161.6 kg/s

9-75 Consider a steam power plant that operates on a regenerative Rankine cycle and has a net power output of 150 MW. Steam enters the turbine at 10 MPa and 500°C and the condenser at 10 kPa. The adiabatic efficiency of the turbine is 80 percent, and that of the pumps is 95 percent. Steam is extracted from the turbine at 0.5 MPa to heat the feedwater in an open feedwater heater. Water leaves the feedwater heater as a saturated liquid. Show the cycle on a *T-s* diagram, and determine (*a*) the mass flow rate of steam through the boiler and (*b*) the thermal efficiency of the cycle. Also, determine the irreversibility associated with the regeneration process. Assume a source temperature of 1300 K and a sink temperature of 303 K.

9-76 Repeat Prob. 9-75 assuming both the pump and the turbine are isentropic.

9-77 Consider an ideal reheat–regenerative Rankine cycle with one open feedwater heater. The boiler pressure is 10 MPa, the condenser pressure is

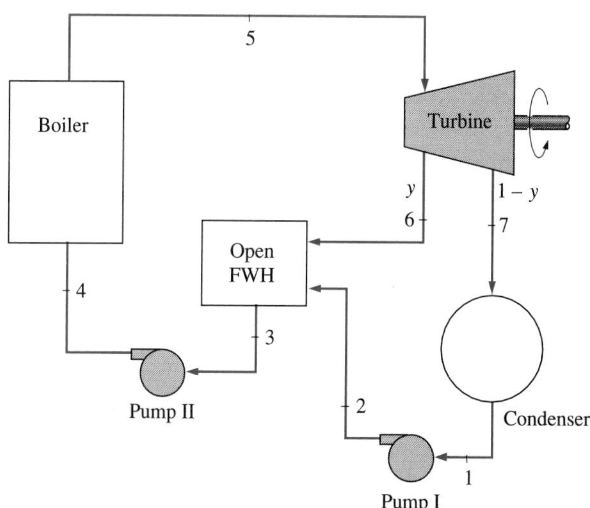

FIGURE P9-75

15 kPa, the reheater pressure is 1 MPa, and the feedwater pressure is 0.6 MPa. Steam enters both the high- and low-pressure turbines at 500°C. Show the cycle on a *T-s* diagram with respect to saturation lines, and determine (*a*) the fraction of steam extracted for regeneration and (*b*) the thermal efficiency of the cycle. *Answers:* (*a*) 0.144, (*b*) 42.1 percent

9-78 Repeat Prob. 9-77 assuming an adiabatic efficiency of 84 percent for the turbines and 100 percent for the pumps.

9-79 A steam power plant operates on an ideal reheat–regenerative Rankine cycle with one reheater and two feedwater heaters, one open and one closed. Steam enters the high-pressure turbine at 15 MPa and 600°C and the low-pressure turbine at 1 MPa and 500°C. The condenser pressure is 5 kPa. Steam is extracted from the turbine at 0.6 MPa for the closed feedwater heater and at 0.2 MPa for the open feedwater heater. In the closed feedwater heater, the feedwater is heated to the condensation temperature of the extracted steam. The extracted steam leaves the closed feedwater heater as a saturated liquid, which is subsequently throttled to the open feedwater heater. Show the cycle on a *T-s* diagram with respect to saturation lines. Determine (*a*) the fraction of steam extracted from the turbine for the open feedwater heater, (*b*) the thermal efficiency of the cycle, and (*c*) the net power output for a mass flow rate of 35 kg/s through the boiler.

9-80 Consider a cogeneration power plant that is modified with reheat and that produces 3 MW of power and supplies 7 MW of process heat. Steam enters the high-pressure turbine at 8 MPa and 500°C and expands to a pressure of 1 MPa. At this pressure, part of the steam is extracted from the turbine and routed to the process heater, while the remainder is reheated to 500°C and expanded in the low-pressure turbine to the condenser pressure of

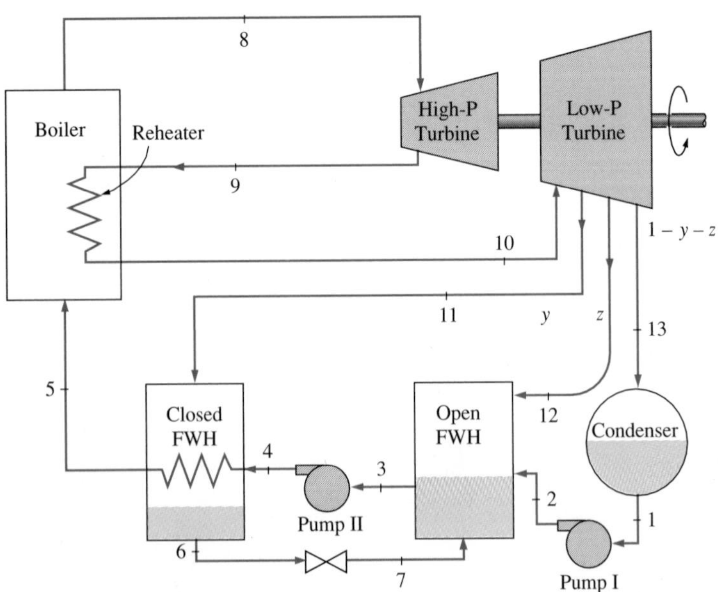

FIGURE P9-79

15 kPa. The condensate from the condenser is pumped to 1 MPa and is mixed with the extracted steam, which leaves the process heater as a compressed liquid at 120°C. The mixture is then pumped to the boiler pressure. Assuming the turbine to be isentropic, show the cycle on a *T-s* diagram with respect to saturation lines, and disregarding pump work, determine (*a*) the rate of heat input in the boiler and (*b*) the fraction of steam extracted for process heating.

9-81 The gas-turbine cycle of a combined gas–steam power plant has a pressure ratio of 8. Air enters the compressor at 290 K and the turbine at 1400 K. The combustion gases leaving the gas turbine are used to heat the steam at 15 MPa to 450°C in a heat exchanger. The combustion gases leave the heat exchanger at 247°C. Steam expands in a high-pressure turbine to a pressure of 3 MPa and is reheated in the combustion chamber to 500°C before it expands in a low-pressure turbine to 10 kPa. The mass flow rate of steam is 20 kg/s. Assuming all the compression and expansion processes to be isentropic, determine (*a*) the mass flow rate of air in the gas-turbine cycle, (*b*) the rate of total heat input, and (*c*) the thermal efficiency of the combined cycle. *Answers:* (*a*) 175.2 kg/s, (*b*) 186,840 kW, (*c*) 55.6 percent

9-82 Repeat Prob. 9-81 assuming adiabatic efficiencies of 100 percent for the pump, 80 percent for the compressor, and 85 percent for the gas and steam turbines.

9-83 Starting with Eq. 9-20, show that the irreversibility associated with a simple ideal Rankine cycle can be expressed as $i = q_{in}(\eta_{th,\,Carnot} - \eta_{th})$, where η_{th} is efficiency of the Rankine cycle and $\eta_{th,\,Carnot}$ is the efficiency of the Carnot cycle operating between the same temperature limits.

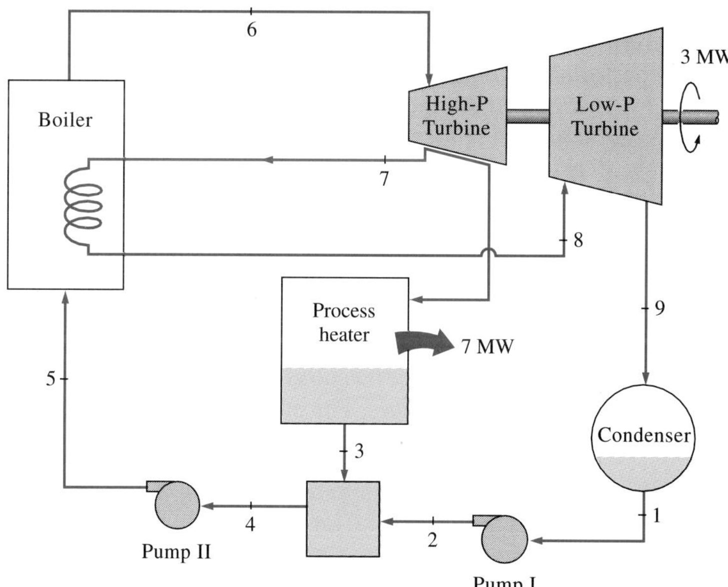

FIGURE P9-80

9-84 Steam is to be supplied from a boiler to a high-pressure turbine whose isentropic efficiency is 75 percent at conditions to be determined. The steam is to leave the high-pressure turbine as a saturated vapor at 1.4 MPa, and the turbine is to produce 1 MW of power. Steam at the turbine exit is extracted at a rate of 1000 kg/min and routed to a process heater while the rest of the steam is supplied to a low-pressure turbine whose isentropic efficiency is 60 percent. The low-pressure turbine allows the steam to expand to 0.2 kPa pressure and produces 0.8 MW of power. Determine the temperature, pressure, and the flow rate of steam at the inlet of the high-pressure turbine.

Computer, Design, and Essay Problems

9-85 Write a computer program to determine the effect of condenser pressure on the performance of a simple ideal Rankine cycle. Assume steam enters the turbine at 5 MPa and 500°C, and neglect the pump work. Determine the thermal efficiency of the cycle, and plot it against condenser pressure for condenser pressures of 100, 75, 50, 25, 10, and 5 kPa.

9-86 Write a computer program to determine the effect of boiler pressure on the performance of a simple ideal Rankine cycle. Assume steam enters the turbine at 500°C and exits at 10 kPa. Neglect the pump work. Determine the thermal efficiency of the cycle, and plot it against boiler pressure for boiler pressures of 0.5, 1, 3, 6, 10, 15, and 20 MPa.

9-87 Write a computer program to determine the effect of superheating the steam on the performance of a simple ideal Rankine cycle. Assume steam enters the turbine at 3 MPa and exits at 10 kPa. Neglect the pump work. Determine the thermal efficiency of the cycle, and plot it against turbine inlet temperature for temperature values of 250, 300, 400, 500, 700, 900, and 1100°C.

9-88 Write a computer program to determine the effect of reheat pressure on the performance of an ideal reheat Rankine cycle. The maximum and minimum pressures in the cycle are 15 MPa and 10 kPa, respectively. Steam enters both stages of the turbine at 500°C. Determine the thermal efficiency of the cycle, and plot it against reheat pressure for pressure values of 12.5, 10, 7, 5, 2, 1, 0.5, and 0.1 MPa.

9-89 Write a computer program to determine the effect of the number of reheat stages on the performance of an ideal reheat Rankine cycle. The maximum and minimum pressures in the cycle are 15 MPa and 10 kPa, respectively. Steam enters all stages of the turbine at 500°C. Determine the thermal efficiency of the cycle, and plot it against the number of reheat stages of one, two, four, and eight reheat stages. For each case, maintain roughly the same pressure ratio across each turbine stage.

9-90 Write a computer program to determine the effect of extraction pressure on the performance of an ideal regenerative Rankine cycle with one open feedwater heater. Steam enters the turbine at 15 MPa and 600°C and the condenser at 10 kPa. Neglecting the pump work, determine the thermal efficiency of the cycle, and plot it against extraction pressure for extraction pressures of 12.5, 10, 7, 5, 2, 1, 0.5, 0.1, and 0.05 MPa.

9-91 Write a computer program to determine the effect of the number of regeneration stages on the performance of an ideal regenerative Rankine cycle. Steam enters the turbine at 15 MPa and 600°C and the condenser at 5 kPa. Neglecting the pump work, determine the thermal efficiency of the cycle, and plot it against the number of regeneration stages for 1, 2, 3, 4, 5, 6, 8, and 10 regeneration stages. For each case, maintain about the same temperature difference between any two regeneration stages.

9-92 Design a steam power cycle that can achieve a cycle thermal efficiency of at least 40 percent under the conditions that all turbines have isentropic efficiencies of 85 percent and all pumps have isentropic efficiencies of 60 percent. Prepare an engineering report describing your design. Your design report must include, but is not limited to, the following:

(*a*) Discussion of various cycles attempted to meet the goal as well as the positive and negative aspects of your design.

(*b*) System figures and *T-s* diagrams with labeled states and temperature, pressure, enthalpy, and entropy information for your design.

(*c*) Sample calculations.

9-93 Contact your power company and obtain information on the thermodynamic aspects of their most recently built power plant. If it is a conventional power plant, find out why it is preferred over a highly efficient combined power plant.

9-94 Several geothermal power plants are in operation in the United States and more are being built since the heat source of a geothermal plant is hot geothermal water, which is "free energy." An 8-MW geothermal power plant is being considered at a location where geothermal water at 160°C is available. Geothermal water is to serve as the heat source for a closed Rankine power cycle with refrigerant-134a as the working fluid (see Fig. P9-72E). Specify suitable temperatures and pressures for the cycle, and determine the thermal efficiency of the cycle. Justify your selections.

9-95 A 10-MW geothermal power plant is being considered at a site where geothermal water at 230°C is available. Geothermal water is to be flashed into a chamber to a lower pressure where part of the water evaporates. The liquid is returned to the ground while the vapor is used to drive the steam turbine. The pressures at the turbine inlet and the turbine exit are to remain above 200 kPa and 8 kPa, respectively. High-pressure flash chambers yield a small amount of steam with high exergy whereas lower-pressure flash chambers yield considerably more steam but at a lower exergy. By trying several pressures, determine the optimum pressure of the flash chamber to maximize the power production per unit mass of geothermal water withdrawn. Also, determine the thermal efficiency for each case assuming 10 percent of the power produced is used to drive the pumps and other auxiliary equipment.

FIGURE P9-95

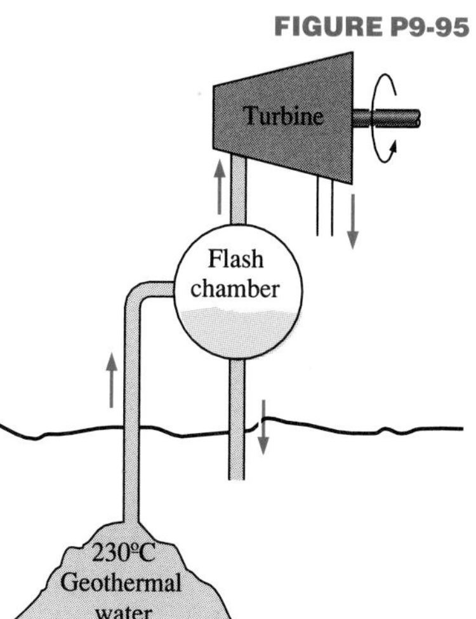

9-96 A natural gas–fired furnace in a textile plant is used to provide steam at 130°C. At times of high demand, the furnace supplies heat to the steam at a rate of 30 MJ/s. The plant also uses up to 6 MW of electrical power purchased from the local power company. The plant management is considering converting the existing process plant into a cogeneration plant to meet both their process-heat and power requirements. Your job is to come up with some designs. Designs based on a gas turbine or a steam turbine are to be considered. First decide whether a system based on a gas turbine or a steam turbine will best serve the purpose, considering the cost and the complexity. Then propose your design for the cogeneration plant complete with pressures and temperatures and the mass flow rates. Show that the proposed design meets the power and process-heat requirements of the plant.

9-97E A photographic equipment manufacturer uses a flow of 64,500 lbm/h of steam in its manufacturing process. Presently the spent steam at 3.8 psig and 224°F is exhausted to the atmosphere. Do the preliminary design of a system to use the energy in the waste steam economically. If electricity is produced, it can be generated about 8000 h/yr and its value is $0.05/kWh. If the energy is used for space heating, the value is also $0.05/kWh, but it can only be used about 3000 h/yr (only during the "heating season"). If the steam is condensed and the liquid H_2O is recycled through the process, its value is $0.50/100 gal. Make all assumptions as realistic as possible. Sketch the system you propose. Make a separate list of required components and their specifications (capacity, efficiency, etc.). The final result will be the calculated annual dollar value of the energy use plan (actually a *saving* because it will replace electricity or heat and/or water that would otherwise have to be purchased).

9-98 Design the condenser of a steam power plant that has a thermal efficiency of 40 percent and generates 10 MW of net electric power. Steam enters the condenser as saturated vapor at 10 kPa, and it is to be condensed outside horizontal tubes through which cooling water from a nearby river flows. The temperature rise of the cooling water is limited to 8°C, and the velocity of the cooling water in the pipes is limited to 6 m/s to keep the pressure drop at an acceptable level. From prior experience, the average heat flux based on the outer surface of the tubes can be taken to be 12,000 W/m^2. Specify the pipe diameter, total pipe length, and the arrangement of the pipes to minimize the condenser volume.

9-99 Water-cooled steam condensers are commonly used in steam power plants. Obtain information about water-cooled steam condensers by doing a literature search on the topic and also by contacting some condenser manufacturers. In a report, describe the various types, the way they are designed, the limitation on each type, and the selection criteria.

9-100 Steam boilers have long been used to provide process heat as well as to generate power. Write an essay on the history of steam boilers and the evolution of modern supercritical steam power plants. What was the role of the American Society of Mechanical Engineers in this development?

9-101 The technology for power generation using geothermal energy is well-established, and numerous geothermal power plants throughout the world are currently generating electricity economically. Binary geothermal plants utilize a volatile secondary fluid such as isobutane, n-pentane, and R-114 in a closed loop. Consider a binary geothermal plant with R-114 as the working fluid that is flowing at a rate of 600 kg/s. The R-114 is vaporized in a boiler at 115°C by the geothermal fluid that enters at 165°C, and is condensed at 30°C outside the tubes by cooling water that enters the tubes at 18°C. Based on prior experience, the average heat flux based on the outer surface of the tubes can be taken to be 4600 W/m^2. The enthalpy of vaporization of R-114 at 30°C is h_{fg} = 121.5 kJ/kg.

Specify (*a*) the length, diameter, and number of tubes and their arrangement in the condenser to minimize overall volume of the condenser; (*b*) the mass flow rate of cooling water; and (*c*) the flow rate of make-up water needed if a cooling tower is used to reject the waste heat from the cooling water. The liquid velocity is to remain under 6 m/s and the length of the tubes is limited to 8 m.

Software Problems

9-102 Repeat Prob. 9-85 using the enclosed software.

9-103 Repeat Prob. 9-86 using the enclosed software.

9-104 Repeat Prob. 9-87 using the enclosed software.

9-105 Repeat Prob. 9-88 using the enclosed software.

9-106 Repeat Prob. 9-90 using the enclosed software.

Refrigeration Cycles

CHAPTER 10

A major application area of thermodynamics is *refrigeration,* which is the transfer of heat from a lower temperature region to a higher temperature one. Devices that produce refrigeration are called *refrigerators* (or *heat pumps*), and the cycles on which they operate are called *refrigeration cycles.* The most frequently used refrigeration cycle is the *vapor-compression refrigeration cycle* in which the refrigerant is vaporized and condensed alternately and is compressed in the vapor phase. Another well-known refrigeration cycle is the *gas refrigeration cycle* in which the refrigerant remains in the gaseous phase throughout. Other refrigeration cycles discussed in this chapter are *cascade refrigeration,* where more than one refrigeration cycle is used; *absorption refrigeration,* where the refrigerant is dissolved in a liquid before it is compressed; and *thermoelectric refrigeration,* where refrigeration is produced by the passage of electric current through two dissimilar materials.

10-1 ■ REFRIGERATORS AND HEAT PUMPS

We all know from experience that heat flows in the direction of decreasing temperature, that is, from high-temperature regions to low-temperature ones. This heat-transfer process occurs in nature without requiring any devices. The reverse process, however, cannot occur by itself. The transfer of heat from a low-temperature region to a high-temperature one requires special devices called **refrigerators.**

Refrigerators are cyclic devices, and the working fluids used in the refrigeration cycles are called **refrigerants.** A refrigerator is shown schematically in Fig. 10-1*a*. Here Q_L is the magnitude of the heat removed from the refrigerated space at temperature T_L, Q_H is the magnitude of the heat rejected to the warm space at temperature T_H, and $W_{net, in}$ is the net work input to the refrigerator. As discussed in Chap. 5, Q_L and Q_H represent magnitudes and thus are positive quantities.

Another device that transfers heat from a low-temperature medium to a high-temperature one is the **heat pump.** Refrigerators and heat pumps are essentially the same devices; they differ in their objectives only. The objective of a refrigerator is to maintain the refrigerated space at a low temperature by removing heat from it. Discharging this heat to a higher-temperature medium is merely a necessary part of the operation, not the purpose. The objective of a heat pump, however, is to maintain a heated space at a high temperature. This is accomplished by absorbing heat from a low-temperature source, such as well water or cold outside air in winter, and supplying this heat to a warmer medium such as a house (Fig. 10-1*b*).

The performance of refrigerators and heat pumps is expressed in terms of the **coefficient of performance** (COP), which was defined in Chap. 5 as

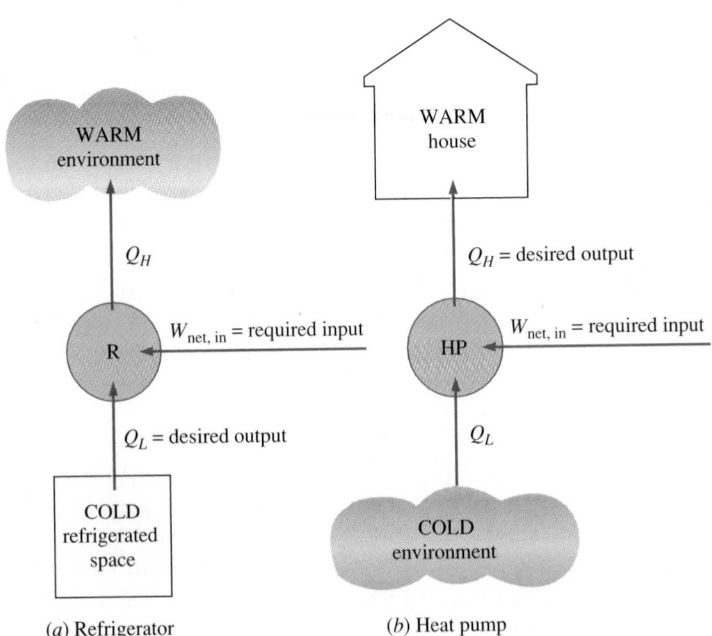

FIGURE 10-1
The objective of a refrigerator is to remove heat (Q_L) from the cold medium; the objective of a heat pump is to supply heat (Q_H) to a warm medium.

$$\text{COP}_{\text{R}} = \frac{\text{Desired output}}{\text{Required input}} = \frac{\text{Cooling effect}}{\text{Work input}} = \frac{Q_L}{W_{\text{net, in}}} \quad (10\text{-}1)$$

$$\text{COP}_{\text{HP}} = \frac{\text{Desired output}}{\text{Required input}} = \frac{\text{Heating effect}}{\text{Work input}} = \frac{Q_H}{W_{\text{net, in}}} \quad (10\text{-}2)$$

These relations can also be expressed in the rate form by replacing the quantities Q_L, Q_H, and $W_{\text{net, in}}$ by $\dot{Q}_L$, $\dot{Q}_H$, and $\dot{W}_{\text{net, in}}$, respectively. Notice that both COP_{R} and COP_{HP} can be greater than 1. A comparison of Eqs. 10-1 and 10-2 reveals that

$$\text{COP}_{\text{HP}} = \text{COP}_{\text{R}} + 1 \quad (10\text{-}3)$$

for fixed values of Q_L and Q_H. This relation implies that $\text{COP}_{\text{HP}} > 1$ since COP_{R} is a positive quantity. That is, a heat pump will function, at worst, as a resistance heater, supplying as much energy to the house as it consumes. In reality, however, part of Q_H is lost to the outside air through piping and other devices, and COP_{HP} may drop below unity when the outside air temperature is too low. When this happens, the system normally switches to the fuel (natural gas, propane, oil, etc.) or resistance-heating mode.

The *cooling capacity* of a refrigeration system—that is, the rate of heat removal from the refrigerated space—is often expressed in terms of **tons of refrigeration.** The capacity of a refrigeration system that can freeze 1 ton (2000 lbm) of liquid water at 0°C (32°F) into ice at 0°C in 24 h is said to be 1 ton. One ton of refrigeration is equivalent to 211 kJ/min or 200 Btu/min. The cooling load of a typical 200-m^2 residence is in the 3-ton (10-kW) range.

10-2 ■ THE REVERSED CARNOT CYCLE

You will recall from the preceding chapters that the Carnot cycle is a totally reversible cycle that consists of two reversible isothermal and two isentropic processes. It has the maximum thermal efficiency for given temperature limits, and it serves as a standard against which actual power cycles can be compared.

Since it is a reversible cycle, all four processes that comprise the Carnot cycle can be reversed. Reversing the cycle will also reverse the directions of any heat and work interactions. The result is a cycle that operates in the counterclockwise direction, which is called the **reversed Carnot cycle.** A refrigerator or heat pump that operates on the reversed Carnot cycle is called a **Carnot refrigerator** or a **Carnot heat pump.**

Consider a reversed Carnot cycle executed within the saturation dome of a refrigerant, as shown in Fig. 10-2. The refrigerant absorbs heat isothermally from a low-temperature source at T_L in the amount of Q_L (process 1-2), is compressed isentropically to state 3 (temperature rises to T_H), rejects heat isothermally to a high-temperature sink at T_H in the amount of Q_H (process 3-4), and expands isentropically to state 1 (temperature drops to T_L). The refrigerant changes from a saturated vapor state to a saturated liquid state in the condenser during process 3-4.

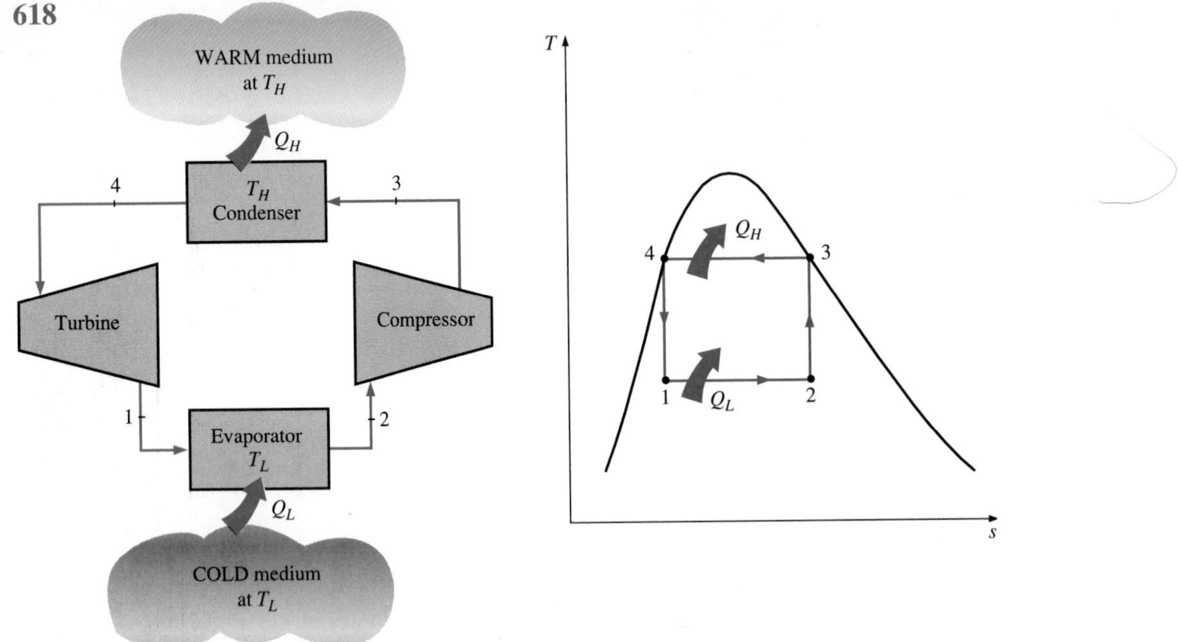

FIGURE 10-2
Schematic of a Carnot refrigerator and *T-s* diagram of the reversed Carnot cycle.

The coefficients of performance of Carnot refrigerators and heat pumps were determined in Sec. 5-11 to be

$$\mathrm{COP}_{\mathrm{R,\,Carnot}} = \frac{1}{T_H/T_L - 1} \tag{10-4}$$

and

$$\mathrm{COP}_{\mathrm{HP,\,Carnot}} = \frac{1}{1 - T_L/T_H} \tag{10-5}$$

Notice that both COPs increase as the difference between the two temperatures decreases, that is, as T_L rises or T_H falls.

The reversed Carnot cycle is the *most efficient* refrigeration cycle operating between two specific temperature levels. Therefore, it is natural to look at it first as a prospective ideal cycle for refrigerators and heat pumps. If we could, we certainly would adapt it as the ideal cycle. But as explained below, the reversed Carnot cycle is not a suitable model for refrigeration cycles.

The two isothermal heat transfer processes are not difficult to achieve in practice since maintaining a constant pressure automatically fixes the temperature of a two-phase mixture at the saturation value. Therefore, processes 1-2 and 3-4 can be approached closely in actual evaporators and condensers. However, processes 2-3 and 4-1 cannot be approximated closely in practice. This is because process 2-3 involves the compression of a liquid–vapor mixture, which requires a compressor that will handle two phases, and process 4-1 involves the expansion of high-moisture-content refrigerant.

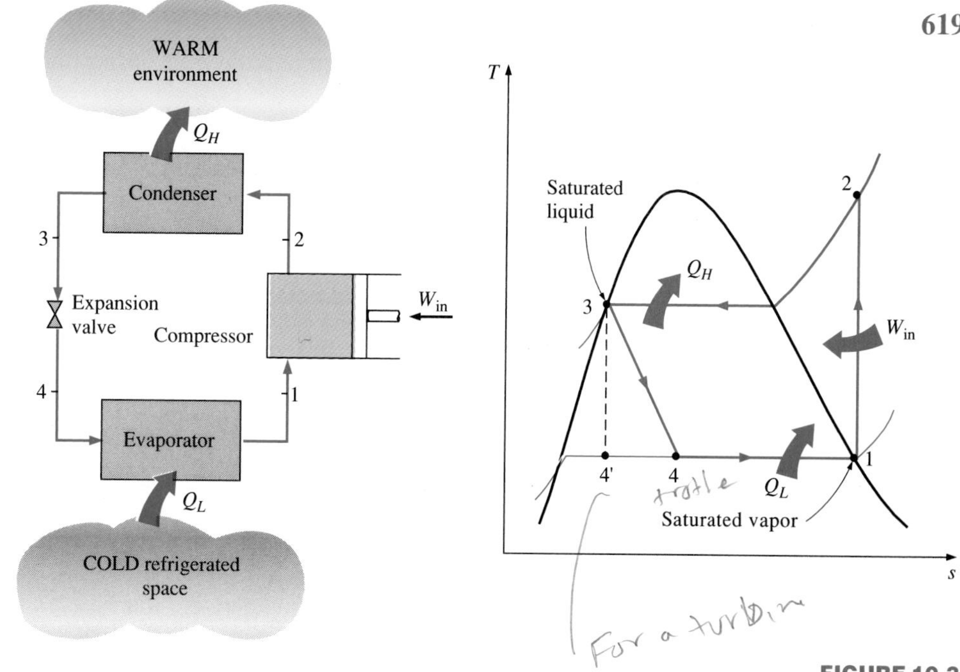

FIGURE 10-3
Schematic and *T-s* diagram for the ideal vapor-compression refrigeration cycle.

It seems as if these problems could be eliminated by executing the reversed Carnot cycle outside the saturation region. But in this case we will have difficulty in maintaining isothermal conditions during the heat-absorption and heat-rejection processes. Therefore, we conclude that the reversed Carnot cycle cannot be approximated in actual devices and is not a realistic model for refrigeration cycles. However, the reversed Carnot cycle can serve as a standard against which actual refrigeration cycles are compared.

10-3 ■ THE IDEAL VAPOR-COMPRESSION REFRIGERATION CYCLE

Many of the impracticalities associated with the reversed Carnot cycle can be eliminated by vaporizing the refrigerant completely before it is compressed and by replacing the turbine with a throttling device, such as an expansion valve or capillary tube. The cycle that results is called the **ideal vapor-compression refrigeration cycle,** and it is shown schematically and on a *T-s* diagram in Fig. 10-3. The vapor-compression refrigeration cycle is the most widely used cycle for refrigerators, air-conditioning systems, and heat pumps. It consists of four processes:

1-2 Isentropic compression in a compressor

2-3 Constant pressure heat rejection in a condenser

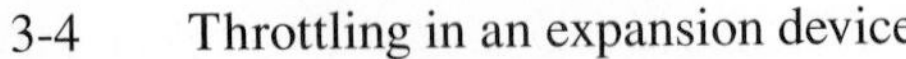

3-4 Throttling in an expansion device

4-1 Constant pressure heat absorption in an evaporator

In an ideal vapor-compression refrigeration cycle, the refrigerant enters the compressor at state 1 as saturated vapor and is compressed isentropically to the condenser pressure. The temperature of the refrigerant increases during this isentropic compression process to well above the temperature of the surrounding medium. The refrigerant then enters the condenser as superheated vapor at state 2 and leaves as saturated liquid at state 3 as a result of heat rejection to the surroundings. The temperature of the refrigerant at this state is still above the temperature of the surroundings.

The saturated liquid refrigerant at state 3 is throttled to the evaporator pressure by passing it through an expansion valve or capillary tube. The temperature of the refrigerant drops below the temperature of the refrigerated space during this process. The refrigerant enters the evaporator at state 4 as a low-quality saturated mixture, and it completely evaporates by absorbing heat from the refrigerated space. The refrigerant leaves the evaporator as saturated vapor and reenters the compressor, completing the cycle.

In a household refrigerator, the freezer compartment where heat is absorbed by the refrigerant serves as the evaporator. The coils behind the refrigerator, where heat is dissipated to the kitchen air, serve as the condenser (Fig. 10-4).

Remember that the area under the process curve on a *T-s* diagram represents the heat transfer for internally reversible processes. The area under the process curve 4-1 represents the heat absorbed by the refrigerant in the evaporator, and the area under the process curve 2-3 represents the heat rejected in the condenser. A rule of thumb is that the *COP improves by 2 to 4 percent for each °C the evaporating temperature is raised or the condensing temperature is lowered.*

Another diagram frequently used in the analysis of vapor-compression refrigeration cycles is the *P-h* diagram, as shown in Fig. 10-5. On this diagram, three of the four processes appear as straight lines, and the heat transfer in the condenser and the evaporator is proportional to the lengths of the corresponding process curves.

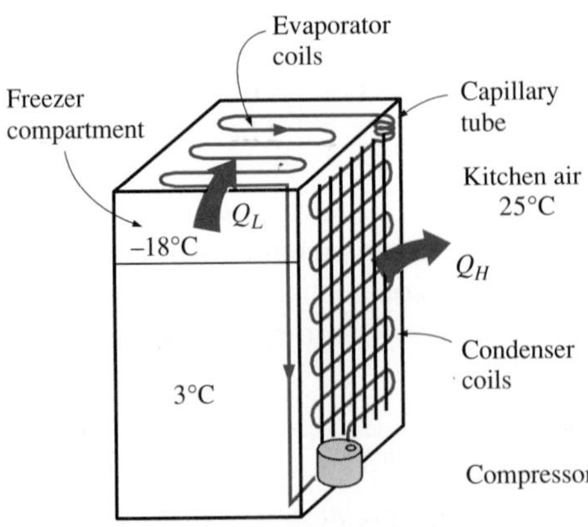

FIGURE 10-4
An ordinary household refrigerator.

Notice that unlike the ideal cycles discussed before, the ideal vapor-compression refrigeration cycle is not an internally reversible cycle since it involves an irreversible (throttling) process. This process is maintained in the cycle to make it a more realistic model for the actual vapor-compression refrigeration cycle. If the throttling device were replaced by an isentropic turbine, the refrigerant would enter the evaporator at state 4′ instead of state 4. As a result, the refrigeration capacity would increase (by the area under process curve 4′-4 in Fig. 10-3) and the net work input would decrease (by the amount of work output of the turbine). Replacing the expansion valve by a turbine is not practical, however, since the added benefits cannot justify the added cost and complexity.

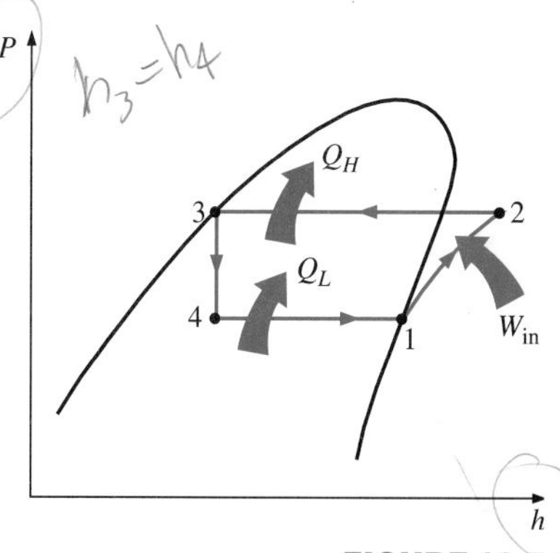

FIGURE 10-5
The *P-h* diagram of an ideal vapor-compression refrigeration cycle.

All four components associated with the vapor-compression refrigeration cycle are steady-flow devices, and thus all four processes that make up the cycle can be analyzed as steady-flow processes. The kinetic and potential energy changes of the refrigerant are usually small relative to the work and heat transfer terms, and therefore they can be neglected. Then the steady-flow energy equation on a unit-mass basis reduces to

$$(q_{in} - q_{out}) + (w_{in} - w_{out}) = h_e - h_i \tag{10-6}$$

The condenser and the evaporator do not involve any work, and the compressor can be approximated as adiabatic. Then the COPs of refrigerators and heat pumps operating on the vapor-compression refrigeration cycle can be expressed as

$$\text{COP}_\text{R} = \frac{q_L}{w_{\text{net,in}}} = \frac{h_1 - h_4}{h_2 - h_1} \tag{10-7}$$

and

$$\text{COP}_\text{HP} = \frac{q_H}{w_{\text{net,in}}} = \frac{h_2 - h_3}{h_2 - h_1} \tag{10-8}$$

where $h_1 = h_{g@P_1}$ and $h_3 = h_{f@P_3}$ for the ideal case.

Vapor-compression refrigeration dates back to 1834 when the Englishman Jacob Perkins received a patent for a closed-cycle ice machine using ether or other volatile fluids as refrigerants. A working model of this machine was built, but it was never produced commercially. In 1850, Alexander Twining began to design and build vapor-compression ice machines using ethyl ether, which is the commercially used refrigerant in vapor-compression systems. Initially, vapor-compression refrigeration systems were large and were mainly used for ice making, brewing, and cold storage. They lacked automatic control and were steam-engine driven. In the 1890s, electric motor-driven smaller machines equipped with automatic control started to replace the older units, and refrigeration systems began to appear in butcher shops and households. By 1930, the continued improvements made it possible to have vapor-compression refrigeration systems that were relatively efficient, reliable, small, and inexpensive.

EXAMPLE 10-1 The Ideal Vapor Compression Refrigeration Cycle

A refrigerator uses refrigerant-134a as the working fluid and operates on an ideal vapor-compression refrigeration cycle between 0.14 and 0.8 MPa. If the mass flow rate of the refrigerant is 0.05 kg/s, determine (*a*) the rate of heat removal from the refrigerated space and the power input to the compressor, (*b*) the rate of heat rejection to the environment, and (*c*) the COP of the refrigerator.

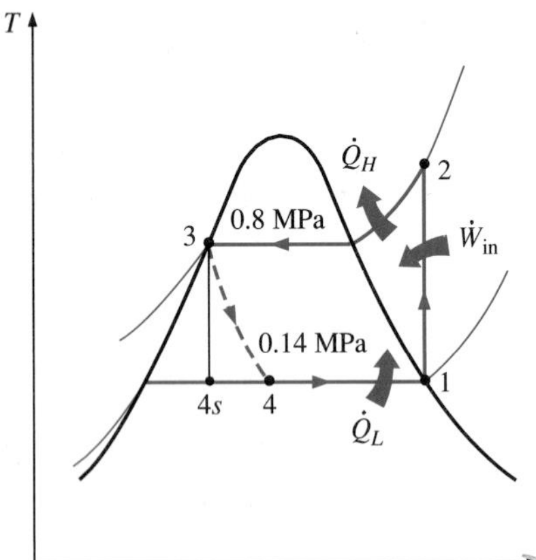

FIGURE 10-6
T-s diagram of the ideal vapor-compression refrigeration cycle described in Example 10.1.

Solution The *T-s* diagram of the refrigeration cycle is shown in Fig. 10-6. We note that this is an ideal vapor-compression refrigeration cycle that involves steady-flow components. Therefore, the compressor is isentropic and the refrigerant leaves the condenser as a saturated liquid and enters the compressor as saturated vapor.

Assumptions **1** Steady operating conditions exist. **2** Kinetic and potential energy changes are negligible.

Analysis From the refrigerant-134a tables, the enthalpies of the refrigerant at all four states are determined as follows:

$$P_1 = 0.14 \text{ MPa} \longrightarrow h_1 = h_{g\,@\,0.14\text{ MPa}} = 236.04\text{kJ/kg}$$
$$s_1 = s_{g\,@\,0.14\text{ MPa}} = 0.9322 \text{ kJ/(kg·K)}$$

$$\left.\begin{array}{l} P_2 = 0.8 \text{ MPa} \\ s_2 = s_1 \end{array}\right\} \quad h_2 = 272.05 \text{ kJ/kg}$$

$$P_3 = 0.8 \text{ MPa} \longrightarrow h_3 = h_{f\,@\,0.8\text{ MPa}} = 93.42 \text{ kJ/kg}$$
$$h_4 \cong h_3 \text{ (throttling)} \longrightarrow h_4 = 93.42 \text{ kJ/kg}$$

(*a*) The rate of heat removal from the refrigerated space and the power input to the compressor are determined from their definitions:

$$\dot{Q}_L = \dot{m}(h_1 - h_4) = (0.05 \text{ kg/s})[(236.04 - 93.42) \text{ kJ/kg}] = \mathbf{7.13 \text{ kW}}$$

and $$\dot{W}_{in} = \dot{m}(h_2 - h_1) = (0.05 \text{ kg/s})[(272.05 - 236.04) \text{ kJ/kg}] = \mathbf{1.80 \text{ kW}}$$

(*b*) The rate of heat rejection from the refrigerant to the environment is determined from

$$\dot{Q}_H = \dot{m}(h_2 - h_3) = (0.05 \text{ kg/s})[(272.05 - 93.42) \text{ kJ/kg}] = \mathbf{8.93 \text{ kW}}$$

It could also be determined from

$$\dot{Q}_H = \dot{Q}_L + \dot{W}_{in} = 7.13 + 1.80 = 8.93 \text{ kW}$$

(*c*) The coefficient of performance of the refrigerator is determined from its definition:

$$\text{COP}_R = \frac{\dot{Q}_L}{\dot{W}_{in}} = \frac{7.13 \text{ kW}}{1.80 \text{ kW}} = \mathbf{3.96}$$

That is, this refrigerator removes about 4 units of energy from the refrigerated space for each unit of electric energy it consumes.

Discussion It would be interesting to see what happens if the throttling valve were replaced by an isentropic turbine. The enthalpy at state 4*s* [the turbine exit with P_{4s} = 0.14 MPa, and $s_{4s} = s_3$ = 0.3459 kJ/(kg · K)] in this case would be 86.92 kJ/kg, and the turbine would produce 0.34 kW of power. This would decrease the power input to the refrigerator from 1.80 to 1.46 kW and increase the rate of heat removal from the refrigerated space from 7.13 to 7.46 kW. As a result, the COP of the refrigerator would increase from 3.96 to 5.11, an increase of 29 percent.

10-4 ■ ACTUAL VAPOR-COMPRESSION REFRIGERATION CYCLES

An actual vapor-compression refrigeration cycle differs from the ideal one in several ways, owing mostly to the irreversibilities that occur in various components. Two common sources of irreversibilities are fluid friction (causes pressure drops) and heat transfer to or from the surroundings. The *T-s* diagram of an actual vapor-compression refrigeration cycle is shown in Fig. 10-7.

In the ideal cycle, the refrigerant leaves the evaporator and enters the compressor as *saturated vapor*. In practice, however, it may not be possible to control the state of the refrigerant so precisely. Instead, it is easier to design the system so that the refrigerant is slightly superheated at the compressor inlet. This slight overdesign ensures that the refrigerant is completely vaporized when it enters the compressor. Also, the line connecting the evaporator to the compressor is usually very long; thus the pressure drop caused by fluid friction and heat transfer from the surroundings to the refrigerant can be very significant. The result of superheating, heat gain in the connecting line, and pressure drops in the evaporator and the connecting line is an increase in the specific volume, thus an increase in the power input requirements to the compressor since steady-flow work is proportional to the specific volume.

The *compression process* in the ideal cycle is internally reversible and adiabatic, and thus isentropic. The actual compression process, however, will involve frictional effects, which increase the entropy, and heat transfer, which

FIGURE 10-7
Schematic and *T-s* diagram for the actual vapor-compression refrigeration cycle.

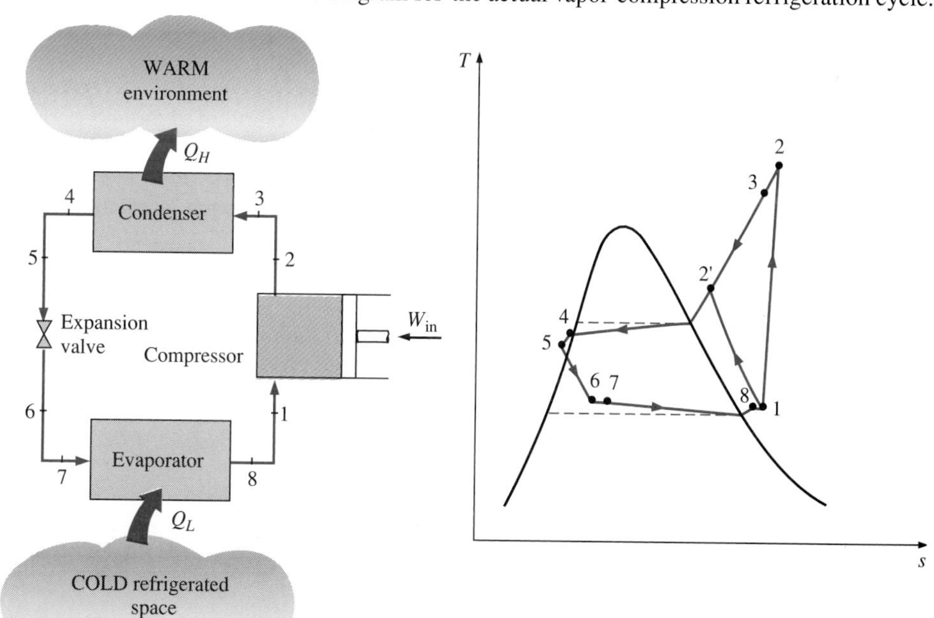

may increase or decrease the entropy, depending on the direction. Therefore, the entropy of the refrigerant may increase (process 1-2) or decrease (process 1-2′) during an actual compression process, depending on which effects dominate. The compression process 1-2′ may be even more desirable than the isentropic compression process since the specific volume of the refrigerant and thus the work input requirement are smaller in this case. Therefore, the refrigerant should be cooled during the compression process whenever it is practical and economical to do so.

In the ideal case, the refrigerant is assumed to leave the condenser as *saturated liquid* at the compressor exit pressure. In actual situations, however, it is unavoidable to have some pressure drop in the condenser as well as in the lines connecting the condenser to the compressor and to the throttling valve. Also, it is not easy to execute the condensation process with such precision that the refrigerant is a saturated liquid at the end, and it is undesirable to route the refrigerant to the throttling valve before the refrigerant is completely condensed. Therefore, the refrigerant is subcooled somewhat before it enters the throttling valve. We do not mind this at all, however, since the refrigerant in this case enters the evaporator with a lower enthalpy and thus can absorb more heat from the refrigerated space. The throttling valve and the evaporator are usually located very close to each other, so the pressure drop in the connecting line is small.

EXAMPLE 10-2 The Actual Vapor-Compression Refrigeration Cycle

Refrigerant-134a enters the compressor of a refrigerator as superheated vapor at 0.14 MPa and −10°C at a rate of 0.05 kg/s and leaves at 0.8 MPa and 50°C. The refrigerant is cooled in the condenser to 26°C and 0.72 MPa and is throttled to 0.15 MPa. Disregarding any heat transfer and pressure drops in the connecting lines between the components, determine (*a*) the rate of heat removal from the refrigerated space and the power input to the compressor, (*b*) the adiabatic efficiency of the compressor, and (*c*) the coefficient of performance of the refrigerator.

Solution The *T-s* diagram of the refrigeration cycle is shown in Fig. 10-8. We note that this is a nonideal vapor-compression refrigeration cycle that involves steady-flow components. The refrigerant leaves the condenser as a compressed liquid and enters the compressor as superheated vapor.

Assumptions **1** Steady operating conditions exist. **2** Kinetic and potential energy changes are negligible.

Analysis The enthalpies of the refrigerant at various states are determined from the refrigerant tables to be

$$\left.\begin{array}{l} P_1 = 0.14 \text{ MPa} \\ T_1 = -10^\circ\text{C} \end{array}\right\} \quad h_1 = 243.40 \text{ kJ/kg}$$

$$\left.\begin{array}{l} P_2 = 0.8 \text{ MPa} \\ T_2 = 50^\circ\text{C} \end{array}\right\} \quad h_2 = 284.39 \text{ kJ/kg}$$

$$\left.\begin{array}{l} P_3 = 0.72 \text{ MPa} \\ T_3 = 26^\circ\text{C} \end{array}\right\} \quad h_3 \cong h_{f\,@\,25^\circ\text{C}} = 85.75 \text{ kJ/kg}$$

$$h_4 \cong h_3(\text{throttling}) \longrightarrow h_4 = 85.75 \text{ kJ/kg}$$

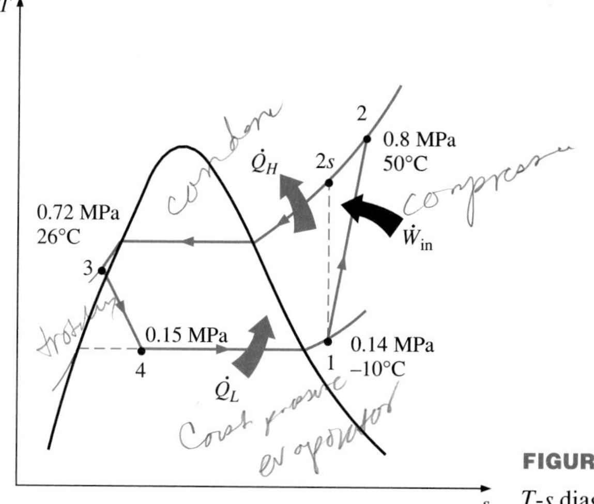

FIGURE 10-8

T-s diagram for Example 10-2.

(*a*) The rate of heat removal from the refrigerated space and the power input to the compressor are determined from their definitions:

$$\dot{Q}_L = \dot{m}(h_1 - h_4) = (0.05\text{ kg/s})[(243.40 - 85.75)\text{ kJ/kg}] = \mathbf{7.88}\text{ kW}$$

and $$\dot{W}_{in} = \dot{m}(h_2 - h_1) = (0.05\text{ kg/s})[(284.39 - 243.40)\text{ kJ/kg}] = \mathbf{2.05\ kW}$$

(*b*) The adiabatic efficiency of the compressor is determined from

$$\eta_C \cong \frac{h_{2s} - h_1}{h_2 - h_1}$$

where the enthalpy at state 2*s* [P_{2s} = 0.8 MPa and $s_{2s} = s_1$ = 0.9606 kJ/(kg · K)] is 281.05 kJ/kg. Thus,

$$\eta_C = \frac{281.05 - 243.40}{284.39 - 243.40} = \mathbf{0.919\ or\ 91.9\%}$$

(*c*) The coefficient of performance of the refrigerator is determined from its definition:

$$\text{COP}_R = \frac{\dot{Q}_L}{\dot{W}_{in}} = \frac{7.88\text{ kW}}{2.05\text{ kW}} = \mathbf{3.84}$$

Discussion This problem is identical to the one worked out in Example 10-1, except that the refrigerant is slightly superheated at the compressor inlet and subcooled at the condenser exit. Also, the compressor is not isentropic. As a result, the heat removal rate from the refrigerated space increases (by 10.5 percent), but the power input to the compressor increases even more (by 13.9 percent). Consequently, the COP of the refrigerator decreases from 3.96 to 3.84.

10-5 ■ SELECTING THE RIGHT REFRIGERANT

When designing a refrigeration system, there are several refrigerants from which to choose, such as chlorofluorocarbons (CFCs), ammonia, hydrocarbons (propane, ethane, ethylene, etc.), carbon dioxide, air (in the air-conditioning of aircraft), and even water (in applications above the freezing point). The right choice of refrigerant depends on the situation at hand. Of these, CFCs such as R-11, R-12, R-22, R-134a, and R-502 account for over 90 percent of the market in the United States.

Ethyl ether was the first commercially used refrigerant in vapor-compression systems in 1850, followed by ammonia, carbon dioxide, methyl chloride, sulphur dioxide, butane, ethane, propane, isobutane, gasoline, and chlorofluorocarbons, among others.

The industrial and heavy-commercial sectors were very satisfied with *ammonia,* and still are, although ammonia is toxic. The advantages of ammonia over other refrigerants are its low cost, higher COPs (and thus lower energy cost), more favorable thermodynamic and transport properties and thus higher heat transfer coefficients (requires smaller and lower-cost heat exchangers), greater detectability in the event of a leak, and no effect on the ozone layer. The major drawback of ammonia is its toxicity, which makes it unsuitable for domestic use. Ammonia is predominantly used in food refrigeration facilities such as the cooling of fresh fruits, vegetables, meat, and fish; refrigeration of beverages and dairy products such as beer, wine, milk, and cheese; freezing of ice cream and other foods; ice production; and low-temperature refrigeration in the pharmaceutical and other process industries.

It is remarkable that the early refrigerants used in the light-commercial and household sectors such as sulfur dioxide, ethyl chloride, and methyl chloride were highly toxic. The widespread publicity of a few instances of leaks that resulted in serious illnesses and death in the 1920s caused a public cry to ban or limit the use of these refrigerants, creating a need for the development of a safe refrigerant for household use. At the request of Frigidaire Corporation, General Motors' research laboratory developed R-21, the first member of the CFC family of refrigerants, within three days in 1928. Of several CFCs developed, the research team settled on R-12 as the refrigerant most suitable for commercial use and gave the CFC family the trade name "Freon." Commercial production of R-11 and R-12 was started in 1931 by a company jointly formed by General Motors and E. I. du Pont de Nemours and Co., Inc. The versatility and low cost of CFCs made them the refrigerants of choice. CFCs were also widely used in aerosols, foam insulations, and the electronic industry as solvents to clean computer chips.

R-11 is used primarily in large-capacity water chillers serving air-conditioning systems in buildings. R-12 is used in domestic refrigerators and freezers, as well as automotive air conditioners. R-22 is used in window air conditioners, heat pumps, air conditioners of commercial buildings, and large industrial refrigeration systems, and offers strong competition to ammonia. R-502 (a blend of R-115 and R-22) is the dominant refrigerant used in commercial refrigeration systems such as those in supermarkets because it allows low temperatures at evaporators while operating at single-stage compression.

The ozone crisis has caused a major stir in the refrigeration and air-conditioning industry and has triggered a critical look at the refrigerants in use. It was realized in the mid-1970s that CFCs allow more ultraviolet radiation into the earth's atmosphere by destroying the protective ozone layer while preventing the infrared radiation from escaping the earth and thus contributing to the greenhouse effect that causes global warming. As a result, the use of some CFCs is banned by international treaties. Fully halogenated CFCs (such as R-11, R-12, and R-115) do the most damage to the ozone layer. The nonfully halogenated refrigerants such as R-22 have about 5 percent of the ozone-depleting capability of R-12. CFCs that are friendly to the ozone layer that protects the earth from harmful ultraviolet rays and at the same time do not contribute to the greenhouse effect have been developed. Currently R-12 is being replaced by the recently developed chlorine-free R-134a.

Two important parameters that need to be considered in the selection of a refrigerant are the temperatures of the two media (the refrigerated space and the environment) with which the refrigerant exchanges heat.

To have heat transfer at a reasonable rate, a temperature difference of 5 to 10°C should be maintained between the refrigerant and the medium with which it is exchanging heat. If a refrigerated space is to be maintained at −10°C, for example, the temperature of the refrigerant should remain at about −20°C while it absorbs heat in the evaporator. The lowest pressure in a refrigeration cycle occurs in the evaporator, and this pressure should be above atmospheric pressure to prevent any air leakage into the refrigeration system. Therefore, a refrigerant should have a saturation pressure of 1 atm or higher at −20°C in this particular case. Ammonia and R-134a are two such substances.

The temperature (and thus the pressure) of the refrigerant on the condenser side depends on the medium to which heat is rejected. Lower temperatures in the condenser (thus higher COPs) can be maintained if the refrigerant is cooled by liquid water instead of air. The use of water cooling cannot be justified economically, however, except in large industrial refrigeration systems. The temperature of the refrigerant in the condenser cannot fall below the temperature of the cooling medium (about 20°C for a household refrigerator), and the saturation pressure of the refrigerant at this temperature should be well below its critical pressure if the heat rejection process is to be approximately isothermal. If no single refrigerant can meet the temperature requirements, then two or more refrigeration cycles with different refrigerants can be used in series. Such a refrigeration system is called a *cascade system*, and is discussed later in this chapter.

Other desirable characteristics of a refrigerant include being nontoxic, noncorrosive, nonflammable, and chemically stable; having a high enthalpy of vaporization (minimizes the mass flow rate); and, of course, being available at low cost.

In the case of heat pumps, the minimum temperature (and pressure) for the refrigerant may be considerably higher since heat is usually extracted from media that are well above the temperatures encountered in refrigeration systems.

10-6 ■ HEAT PUMP SYSTEMS

Heat pumps are generally more expensive to purchase and install than other heating systems, but they save money in the long run in some areas because they lower the heating bills. Despite their relatively higher initial costs, the popularity of heat pumps is increasing. About one-third of all single-family homes built in the United States in 1984 are heated by heat pumps.

The most common energy source for heat pumps is atmospheric air (air-to-air systems), although water and soil are also used. The major problem with air-source systems is *frosting*, which occurs in humid climates when the temperature falls below 2 to 5°C. The frost accumulation on the evaporator coils is highly undesirable since it seriously disrupts the heat transfer. The coils can be defrosted, however, by reversing the heat pump cycle (running it as an air conditioner). This results in a reduction in the efficiency of the system. Water-source systems usually use well water from depths of up to 80 m in the temperature range of 5 to 18°C, and they do not have a frosting problem. They typically have higher COPs but are more complex and require easy access to a large body of water such as underground water. Soil-source systems are also rather involved since they require long tubing placed deep in the ground where the soil temperature is relatively constant. The COP of heat pumps usually ranges between 1.5 and 4, depending on the particular system used and the temperature of the source. A new class of recently developed heat pumps that use variable-speed electric motor drives are at least twice as energy efficient as their predecessors.

Both the capacity and the efficiency of a heat pump fall significantly at low temperatures. Therefore, most air-source heat pumps require a supplementary heating system such as electric resistance heaters or an oil or gas furnace. Since water and soil temperatures do not fluctuate much, supplementary heating may not be required for water-source or soil-source systems. But the heat pump system must be large enough to meet the maximum heating load.

Heat pumps and air conditioners have the same mechanical components. Therefore, it is not economical to have two separate systems to meet the heating and cooling requirements of a building. One system can be used as a heat pump in winter and an air conditioner in summer. This is accomplished by adding a reversing valve to the cycle, as shown in Fig. 10-9. As a result of this modification, the condenser of the heat pump (located indoors) functions as the evaporator of the air conditioner in summer. Also, the evaporator of the heat pump (located outdoors) serves as the condenser of the air conditioner. This feature increases the competitiveness of the heat pump. Such dual-purpose window units are commonly used in motels.

Heat pumps are most competitive in areas that have a large cooling load during the cooling season and a relatively small heating load during the heating season, such as in the southern parts of the United States. In these areas, the heat pump can meet the entire cooling and heating needs of residential or commercial buildings. The heat pump is least competitive in areas where the heating load is significant and the cooling load is small, such as in the northern parts of the United States.

10-7 ■ INNOVATIVE VAPOR-COMPRESSION REFRIGERATION SYSTEMS

The simple vapor-compression refrigeration cycle discussed above is the most widely used refrigeration cycle, and it is adequate for most refrigeration applications. The ordinary vapor-compression refrigeration systems are sim-

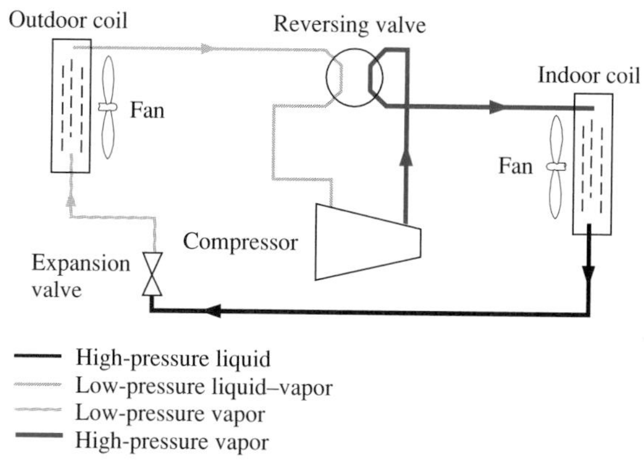

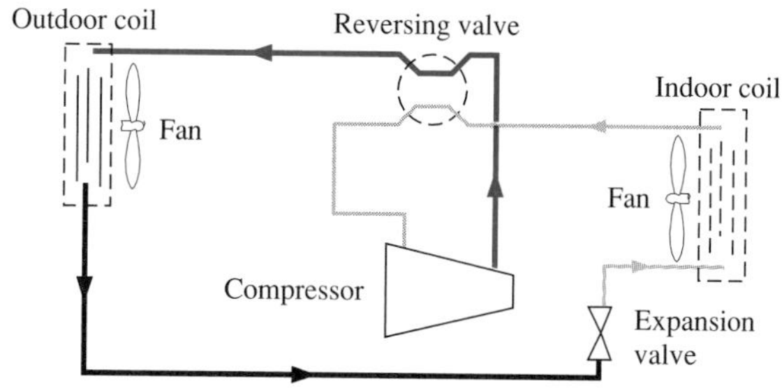

FIGURE 10-9

A heat pump can be used to heat a house in winter and to cool it in summer.

ple, inexpensive, reliable, and practically maintenance-free (when was the last time you serviced your household refrigerator?). However, for large industrial applications *efficiency,* not simplicity, is the major concern. Also, for some applications the simple vapor-compression refrigeration cycle is inadequate and needs to be modified. We shall now discuss a few such modifications and refinements.

Cascade Refrigeration Systems

Some industrial applications require moderately low temperatures, and the temperature range they involve may be too large for a single vapor-compression refrigeration cycle to be practical. A large temperature range also means a large pressure range in the cycle and a poor performance for a reciprocating compressor. One way of dealing with such situations is to perform the refrigeration process in stages, that is, to have two or more refrigeration cycles that operate in series. Such refrigeration cycles are called **cascade refrigeration cycles.**

A two-stage cascade refrigeration cycle is shown in Fig. 10-10. The two cycles are connected through the heat exchanger in the middle, which serves as the evaporator for the topping cycle (cycle A) and the condenser for the bottoming cycle (cycle B). Assuming the heat exchanger is well insulated and the kinetic and potential energies are negligible, the heat transfer from the fluid in the bottoming cycle should be equal to the heat transfer to the fluid

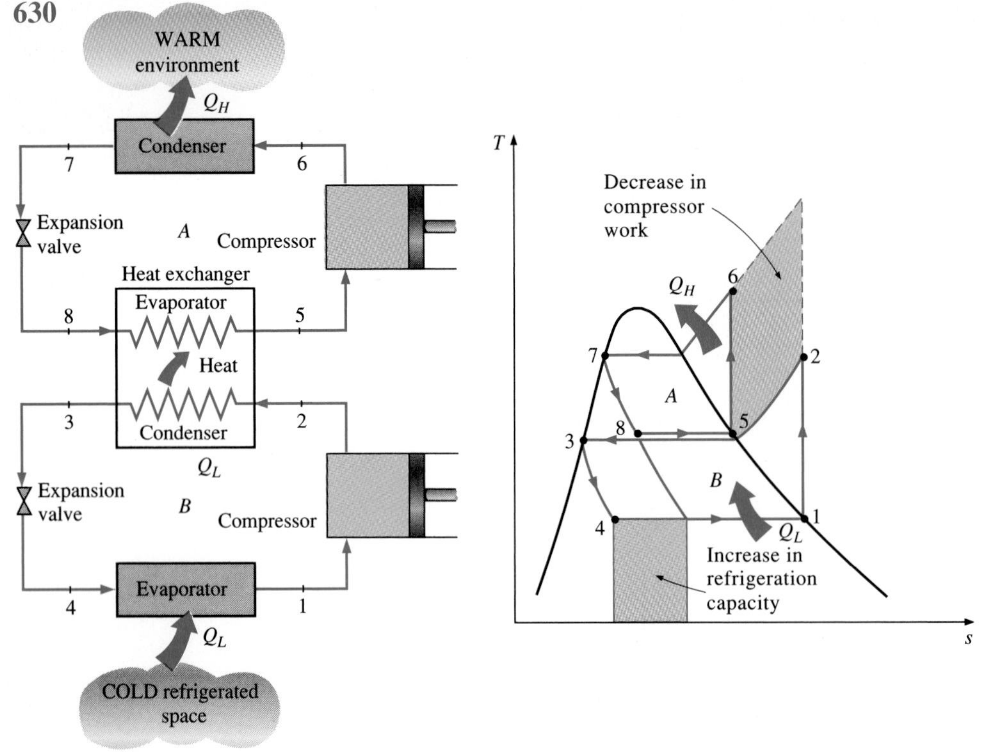

FIGURE 10-10

A two-stage cascade refrigeration system with the same refrigerant in both stages.

in the topping cycle. Thus, the ratio of mass flow rates through each cycle should be

$$\dot{m}_A(h_5 - h_8) = \dot{m}_B(h_2 - h_3) \quad \longrightarrow \quad \frac{\dot{m}_A}{\dot{m}_B} = \frac{h_2 - h_3}{h_5 - h_8} \tag{10-9}$$

Also,

$$\text{COP}_{\text{R, cascade}} = \frac{\dot{Q}_L}{\dot{W}_{\text{net, in}}} = \frac{\dot{m}_B(h_1 - h_4)}{\dot{m}_A(h_6 - h_5) + \dot{m}_B(h_2 - h_1)} \tag{10-10}$$

In the cascade system shown in the figure, the refrigerants in both cycles are assumed to be the same. This is not necessary, however, since there is no mixing taking place in the heat exchanger. Therefore, refrigerants with more desirable characteristics can be used in each cycle. In this case, there would be a separate saturation dome for each fluid, and the *T-s* diagram for one of the cycles would be different. Also, in actual cascade refrigeration systems, the two cycles would overlap somewhat since a temperature difference between the two fluids is needed for any heat transfer to take place.

It is evident from the *T-s* diagram in Fig. 10-10 that the compressor work decreases and the amount of heat absorbed from the refrigerated space increases as a result of cascading. Therefore, cascading improves the COP of a refrigeration system. Some refrigeration systems use three or four stages of cascading.

EXAMPLE 10-3 A Two-Stage Cascade Refrigeration Cycle

Consider a two-stage cascade refrigeration system operating between the pressure limits of 0.8 and 0.14 MPa. Each stage operates on an ideal vapor-compression refrigeration cycle with refrigerant-134a as the working fluid. Heat rejection from the lower cycle to the upper cycle takes place in an adiabatic counterflow heat exchanger where both streams enter at about 0.32 MPa. (In practice, the working fluid of the lower cycle will be at a higher pressure and temperature in the heat exchanger for effective heat transfer.) If the mass flow rate of the refrigerant through the upper cycle is 0.05 kg/s, determine (*a*) the mass flow rate of the refrigerant through the lower cycle, (*b*) the rate of heat removal from the refrigerated space and the power input to the compressor, and (*c*) the coefficient of performance of this cascade refrigerator.

Solution The *T-s* diagram of the refrigeration cycle is shown in Fig. 10-11. The topping cycle is labeled cycle *A* and the bottoming cycle, cycle *B*. We note that this is a two-stage cascade ideal vapor-compression refrigeration cycle that involves steady-flow components. For both cycles, the refrigerant leaves the condenser as a compressed liquid and enters the compressor as superheated vapor.

Assumptions **1** Steady operating conditions exist. **2** Kinetic and potential energy changes are negligible. **3** The heat exchanger is adiabatic.

Analysis The enthalpies of the refrigerant at all eight states are determined from the refrigerant tables and are indicated on the *T-s* diagram.

(*a*) The mass flow rate of the refrigerant through the lower cycle is determined from the steady-flow energy balance on the adiabatic heat exchanger,

$$\sum \dot{m}_e h_e = \sum \dot{m}_i h_i \longrightarrow \dot{m}_A h_5 + \dot{m}_B h_3 = \dot{m}_A h_8 + \dot{m}_B h_2$$

$$\dot{m}_A(h_5 - h_8) = \dot{m}_B(h_2 - h_3)$$

$$(0.05 \text{ kg/s})[(248.66 - 93.42) \text{ kJ/kg}] = \dot{m}_B[(252.71 - 53.31) \text{ kJ/kg}]$$

$$\dot{m}_B = \mathbf{0.039 \text{ kg/s}}$$

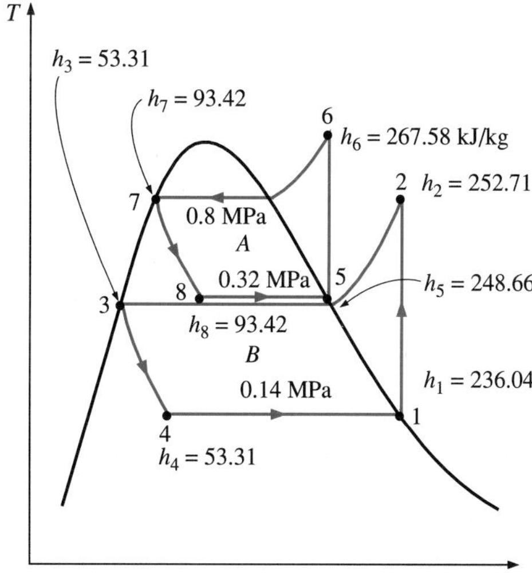

FIGURE 10-11

T-s diagram of the cascade refrigeration cycle described in Example 10-3.

(*b*) The rate of heat removal by a cascade cycle is the rate of heat absorption in the evaporator of the lowest stage. The power input to a cascade cycle is the sum of the power inputs to all of the compressors:

$$\dot{Q}_L = \dot{m}_B(h_1 - h_4) = (0.039 \text{ kg/s})[(236.04 - 53.31) \text{ kJ/kg}] = \mathbf{7.13 \text{ kW}}$$

$$\begin{aligned}\dot{W}_{in} &= \dot{W}_{compI, in} + \dot{W}_{compII, in} = \dot{m}_A(h_6 - h_5) + \dot{m}_B(h_2 - h_1)\\ &= (0.05 \text{ kg/s})[(267.58 - 248.66) \text{ kJ/kg}]\\ &\quad + (0.039 \text{ kg/s})[(252.71 - 236.04) \text{ kJ/kg}]\\ &= \mathbf{1.60 \text{ kW}}\end{aligned}$$

(*c*) The COP of a refrigeration system is the ratio of the refrigeration rate to the net power input:

$$\text{COP}_R = \frac{\dot{Q}_L}{\dot{W}_{net, in}} = \frac{7.13 \text{ kW}}{1.60 \text{ kW}} = \mathbf{4.46}$$

Discussion This problem was worked out in Example 10-1 for a single-stage refrigeration system. Notice that the COP of the refrigeration system increases from 3.96 to 4.46 as a result of cascading. The COP of the system can be increased even more by increasing the number of cascade stages.

2 Multistage Compression Refrigeration Systems

When the fluid used throughout the cascade refrigeration system is the same, the heat exchanger between the stages can be replaced by a mixing chamber (called a *flash chamber*) since it has better heat transfer characteristics. Such systems are called **multistage compression refrigeration systems.** A two-stage compression refrigeration system is shown in Fig. 10-12.

In this system, the liquid refrigerant expands in the first expansion valve to the flash chamber pressure, which is the same as the compressor interstage pressure. Part of the liquid vaporizes during this process. This saturated vapor (state 3) is mixed with the superheated vapor from the low-pressure compressor (state 2), and the mixture enters the high-pressure compressor at state 9. This is, in essence, a regeneration process. The saturated liquid (state 7) expands through the second expansion valve into the evaporator, where it picks up heat from the refrigerated space.

The compression process in this system resembles a two-stage compression with intercooling, and the compressor work decreases. Care should be exercised in the interpretations of the areas on the *T-s* diagram in this case since the mass flow rates are different in different parts of the cycle.

EXAMPLE 10-4 A Two-Stage Refrigeration Cycle with a Flash Chamber

Consider a two-stage compression refrigeration system operating between the pressure limits of 0.8 and 0.14 MPa. The working fluid is refrigerant-134a. The refrigerant leaves the condenser as a saturated liquid and is throttled to a flash chamber operating at 0.32 MPa. Part of the refrigerant evaporates during this flashing process, and this vapor is mixed with the refrigerant leaving the low-pressure compressor. The mixture is then compressed to the condenser pressure by the high-pressure compressor. The liquid in the flash chamber is throttled to

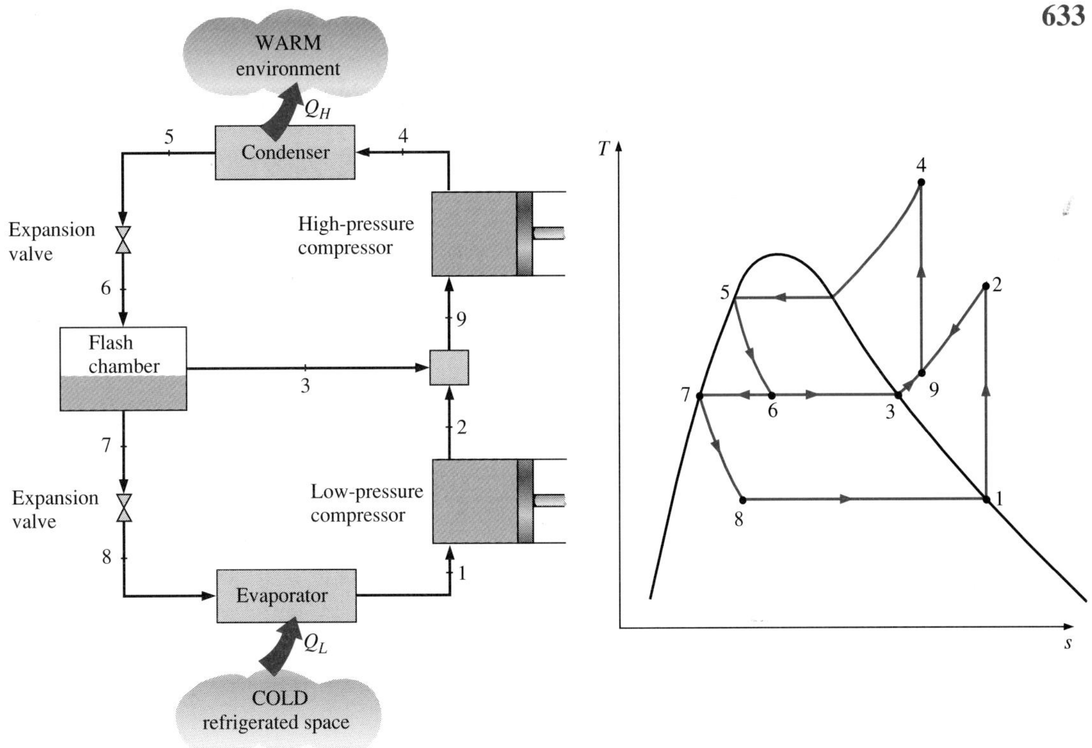

FIGURE 10-12

A two-stage compression refrigeration system with a flash chamber.

the evaporator pressure and cools the refrigerated space as it vaporizes in the evaporator. Assuming the refrigerant leaves the evaporator as a saturated vapor and both compressors are isentropic, determine (*a*) the fraction of the refrigerant that evaporates as it is throttled to the flash chamber, (*b*) the amount of heat removed from the refrigerated space and the compressor work per unit mass of refrigerant flowing through the condenser, and (*c*) the coefficient of performance.

Solution The *T-s* diagram of the refrigeration cycle is shown in Fig. 10-13. We note that this is a two-stage ideal vapor-compression refrigeration cycle that involves steady-flow components. The refrigerant leaves the condenser as compressed liquid and enters the compressor as superheated vapor.

Assumptions **1** Steady operating conditions exist. **2** Kinetic and potential energy changes are negligible. **3** The flash chamber is adiabatic.

Analysis The enthalpies of the refrigerant at various states are determined from the refrigerant tables and are indicated on the *T-s* diagram.

(*a*) The fraction of the refrigerant that evaporates as it is throttled to the flash chamber is simply the quality at state 6, which is

$$x_6 = \frac{h_6 - h_f}{h_{fg}} = \frac{93.42 - 53.31}{195.35} = \mathbf{0.205}$$

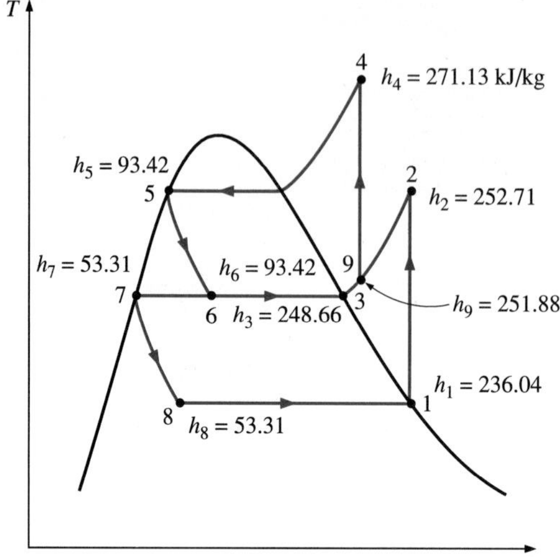

FIGURE 10-13

T-s diagram of the two-stage compression refrigeration cycle described in Example 10-4.

(*b*) The amount of heat removed from the refrigerated space and the compressor work input per unit mass of refrigerant flowing through the condenser are

$$q_L = (1 - x_6)(h_1 - h_8)$$
$$= (1 - 0.205)[(236.04 - 53.31)\text{ kJ/kg}] = \mathbf{145.27\ kJ/kg}$$

and $$w_{in} = w_{compI,in} + w_{compII,in} = (1 - x_6)(h_2 - h_1) + (1)(h_4 - h_9)$$

The enthalpy at state 9 is determined from an energy balance on the mixing chamber,

$$\dot{E}_{out} = \dot{E}_{in}$$
$$\sum \dot{m}_e h_e = \sum \dot{m}_i h_i$$
$$(1)h_9 = x_6 h_3 + (1 - x_6)h_2$$
$$h_9 = (0.205)(248.66) + (1 - 0.205)(252.71) = 251.88\text{ kJ/kg}$$

Also, $s_9 = 0.9292$ kJ/(kg · K). Thus the enthalpy at state 4 (0.8 MPa, $s_4 = s_9$) is $h_4 = 271.13$ kJ/kg. Substituting,

$$w_{in} = (1 - 0.205)[(251.8 - 236.04)\text{ kJ/kg}] + (271.13 - 251.88)\text{ kJ/kg}$$
$$= \mathbf{31.84\ kJ/kg}$$

(*c*) The coefficient of performance is determined from

$$\text{COP}_R = \frac{q_L}{w_{in}} = \frac{145.27\text{ kJ/kg}}{31.84\text{ kJ/kg}} = \mathbf{4.56}$$

Discussion This problem was worked out in Example 10-1 for a single-stage refrigeration system (COP = 3.96) and in Example 10-3 for a two-stage cascade refrigeration system (COP = 4.46). Notice that the COP of the refrigeration system increased considerably relative to the single-stage compression but did not change much relative to the two-stage cascade compression.

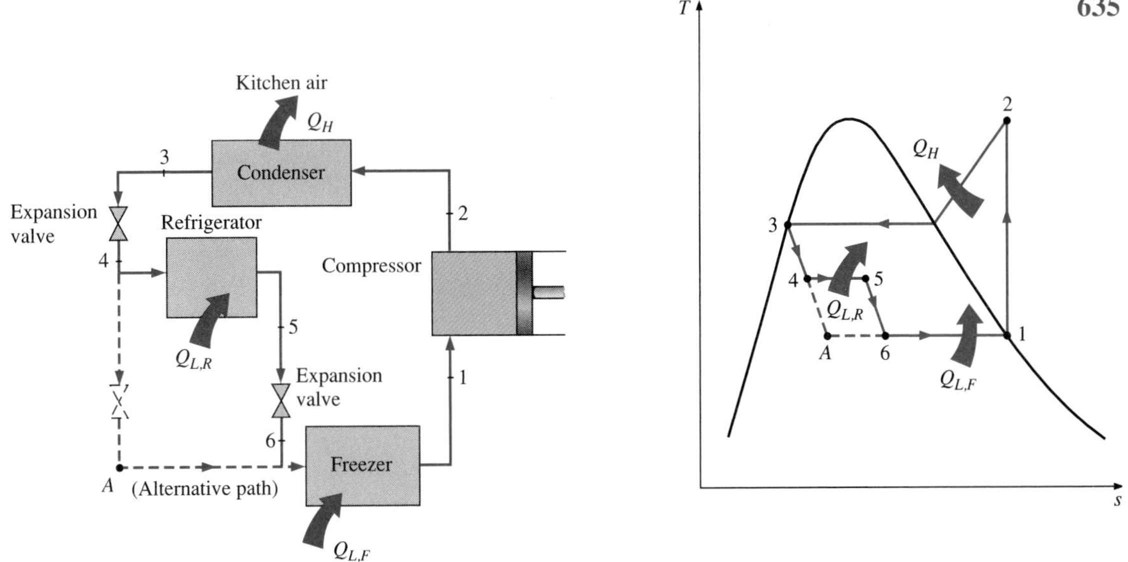

FIGURE 10-14

Schematic and *T-s* diagram for a refrigerator–freezer unit with one compressor.

3 Multipurpose Refrigeration Systems with a Single Compressor

Some applications require refrigeration at more than one temperature. This could be accomplished by using a separate throttling valve and a separate compressor for each evaporator operating at different temperatures. However, such a system will be bulky and probably uneconomical. A more practical and economical approach would be to route all the exit streams from the evaporators to a single compressor and let it handle the compression process for the entire system.

Consider, for example, an ordinary refrigerator–freezer unit. A simplified schematic of the unit and the *T-s* diagram of the cycle are shown in Fig. 10-14. Most refrigerated goods have a high water content, and the refrigerated space must be maintained above the ice point to prevent freezing. The freezer compartment, however, is maintained at about −18°C. Therefore, the refrigerant should enter the freezer at about −25°C to have heat transfer at a reasonable rate in the freezer. If a single expansion valve and evaporator were used, the refrigerant would have to circulate in both compartments at about −25°C, which would cause ice formation in the neighborhood of the evaporator coils and dehydration of the produce. This would not be acceptable to a household. This problem can be eliminated by throttling the refrigerant to a higher pressure (hence temperature) for use in the refrigerated space and then throttling it to the minimum pressure for use in the freezer. The entire refrigerant leaving the freezer compartment is subsequently compressed by a single compressor to the condenser pressure.

4 Liquefaction of Gases

The liquefaction of gases has always been an important area of refrigeration since many important scientific and engineering processes at cryogenic temperatures (temperatures below about −100°C) depend on liquefied gases. Some examples of such processes are the separation of oxygen and nitrogen from air, preparation of liquid propellants for rockets, study of material properties at low temperatures, and study of some exciting phenomena such as superconductivity.

At temperatures above the critical-point value, a substance exists in the gas phase only. The critical temperatures of helium, hydrogen, and nitrogen (three commonly used liquefied gases) are −268, −240, and −147°C, respectively. Therefore, none of these substances will exist in liquid form at atmospheric conditions. Furthermore, low temperatures of this magnitude cannot be obtained by ordinary refrigeration techniques. Then the question that needs to be answered in the liquefaction of gases is this: *How can we lower the temperature of a gas below its critical-point value?*

Several cycles, some complex and others simple, are used successfully for the liquefaction of gases. Below we discuss the Linde-Hampson cycle which is shown schematically and on a *T-s* diagram in Fig. 10-15.

FIGURE 10-15
Linde-Hampson system for liquefying gases.

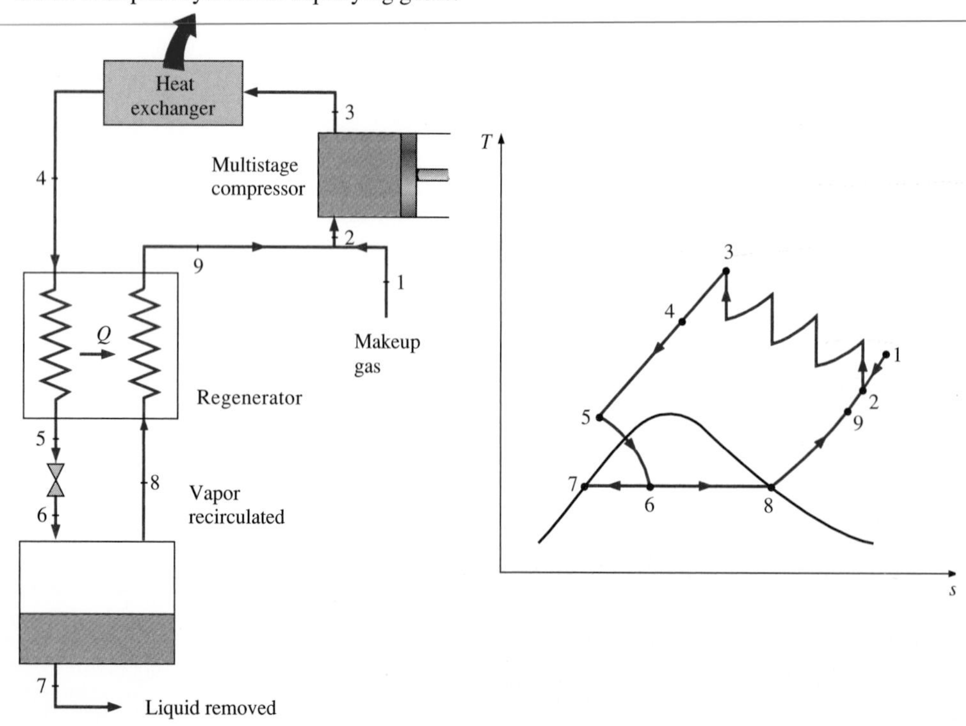

Makeup gas is mixed with the uncondensed portion of the gas from the previous cycle, and the mixture at state 2 is compressed by a multistage compressor to state 3. The compression process approaches an isothermal process due to intercooling. The high-pressure gas is cooled in an after-cooler by a cooling medium or by a separate external refrigeration system to state 4. The gas is further cooled in a regenerative counterflow heat exchanger by the uncondensed portion of gas from the previous cycle to state 5, and it is throttled to state 6, which is a saturated liquid–vapor mixture state. The liquid (state 7) is collected as the desired product, and the vapor (state 8) is routed through the regenerator to cool the high-pressure gas approaching the throttling valve. Finally, the gas is mixed with fresh makeup gas, and the cycle is repeated.

This and other refrigeration cycles used for the liquefaction of gases can also be used for the solidification of gases.

10-8 ■ GAS REFRIGERATION CYCLES

As explained in Sec. 10-2, the Carnot cycle (the standard of comparison for power cycles) and the reversed Carnot cycle (the standard of comparison for refrigeration cycles) are identical, except that the reversed Carnot cycle operates in the reverse direction. This suggests that the power cycles discussed in earlier chapters can be used as refrigeration cycles by simply reversing them. In fact, the vapor-compression refrigeration cycle is essentially a modified Rankine cycle operating in reverse. Another example is the reversed Stirling cycle, which is the cycle on which Stirling refrigerators operate. In this section, we discuss the *reversed Brayton cycle,* better known as the **gas refrigeration cycle.**

Consider the gas refrigeration cycle shown in Fig. 10-16. The surroundings are at T_0, and the refrigerated space is to be maintained at T_L. The gas is compressed during process 1-2. The high-pressure, high-temperature gas at state 2 is then cooled at constant pressure to T_0 by rejecting heat to the surroundings. This is followed by an expansion process in a turbine, during which the gas temperature drops to T_4. (Can we achieve the cooling effect by using a throttling valve instead of a turbine?) Finally, the cool gas absorbs heat from the refrigerated space until its temperature rises to T_1.

All the processes described above are internally reversible, and the cycle executed is the *ideal* gas refrigeration cycle. In actual gas refrigeration cycles, the compression and expansion processes will deviate from the isentropic ones, and T_3 will be higher than T_0 unless the heat exchanger is infinitely large.

On a *T-s* diagram, the area under process curve 4-1 represents the heat removed from the refrigerated space; the enclosed area 1-2-3-4-1 represents the net work input. The ratio of these areas is the COP for the cycle, which may be expressed as

$$\mathrm{COP_R} = \frac{q_L}{w_{\text{net, in}}} = \frac{q_L}{w_{\text{comp, in}} - w_{\text{turb, out}}} \qquad \text{(10-11)}$$

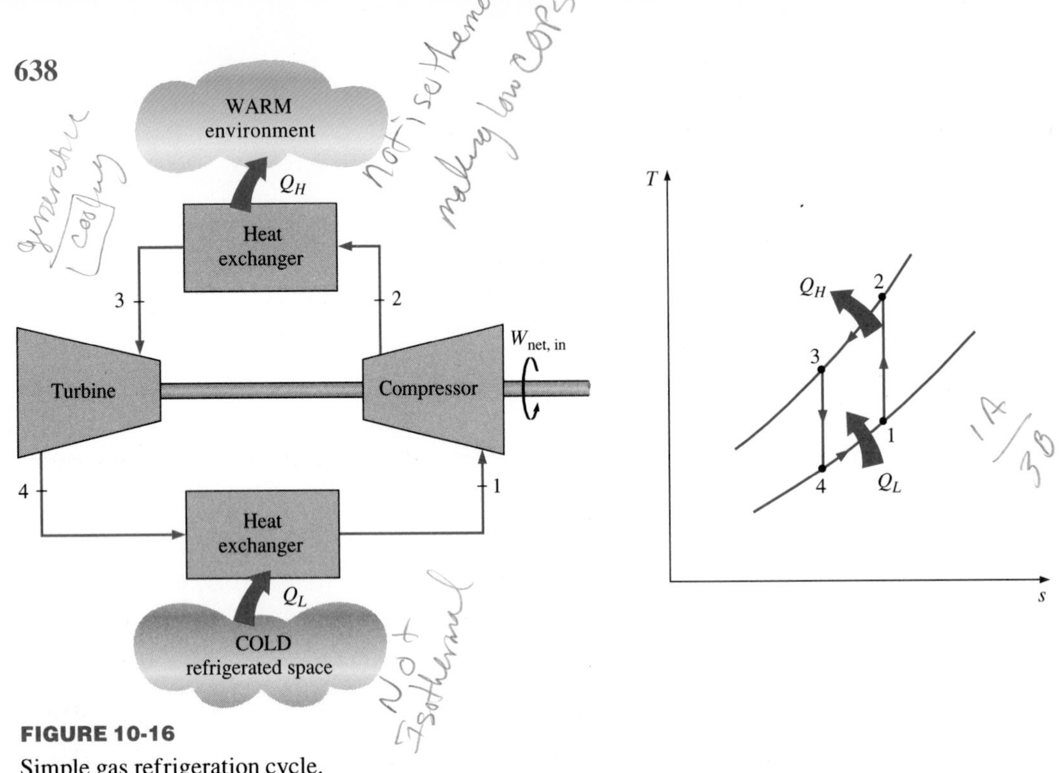

FIGURE 10-16
Simple gas refrigeration cycle.

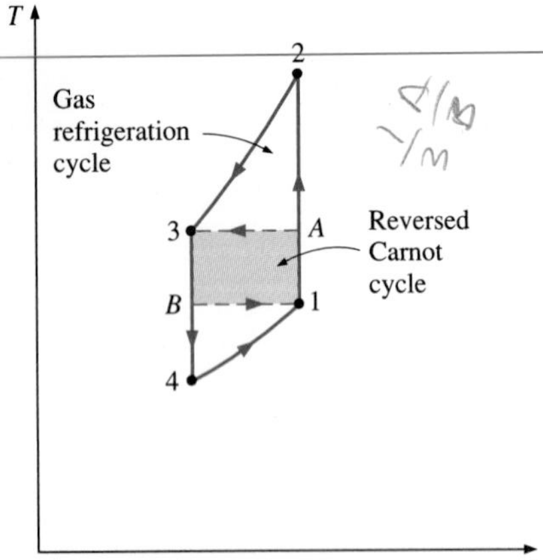

FIGURE 10-17
A reversed Carnot cycle produces more refrigeration (area under $B1$) with less work input (area $1A3B$).

where

$$q_L = h_1 - h_4$$
$$w_{turb,out} = h_3 - h_4$$
$$w_{comp,in} = h_2 - h_1$$

The gas refrigeration cycle deviates from the reversed Carnot cycle because the heat-transfer processes are not isothermal. In fact, the gas temperature varies considerably during heat-transfer processes. Consequently, the gas refrigeration cycles have lower COPs relative to the vapor-compression refrigeration cycles or the reversed Carnot cycle. This is also evident from the *T-s* diagram in Fig. 10-17. The reversed Carnot cycle consumes a fraction of the net work (rectangular area $1A3B$) but produces a greater amount of refrigeration (triangular area under $B1$).

Despite their relatively low COPs, the gas refrigeration cycles have two desirable characteristics: They involve simple, lighter components, which make them suitable for aircraft cooling, and they can incorporate regeneration, which makes them suitable for liquefaction of gases and cryogenic applications. An open-cycle aircraft cooling system is shown in Fig. 10-18. Atmospheric air is compressed by a compressor, cooled by the surrounding air, and expanded in a turbine. The cool air leaving the turbine is then directly routed to the cabin.

The regenerative gas cycle is shown in Fig. 10-19. Regenerative cooling is achieved by inserting a counterflow heat exchanger into the cycle. Without regeneration, the lowest turbine inlet temperature is T_0, the temperature of

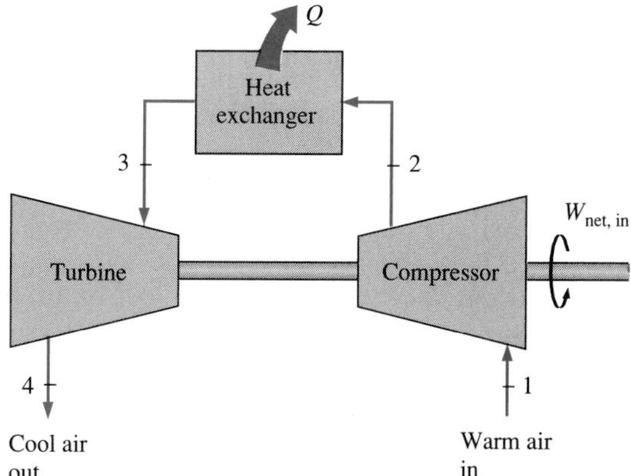

FIGURE 10-18
An open-cycle aircraft cooling system.

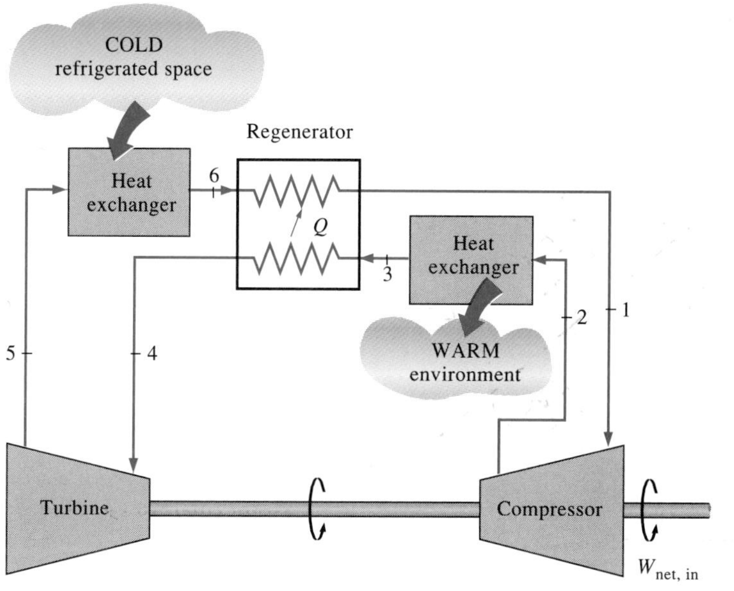

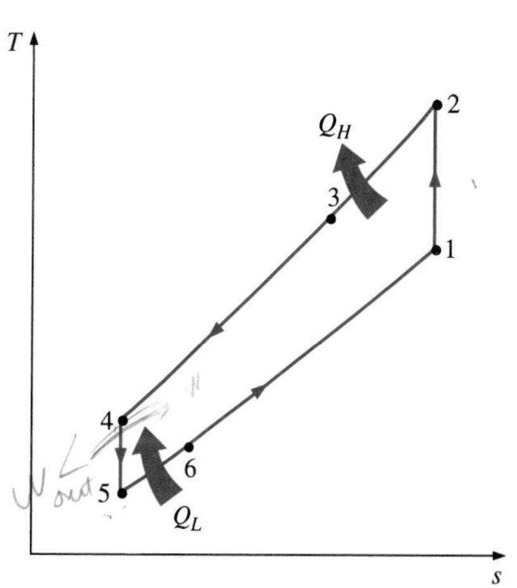

FIGURE 10-19
Gas refrigeration cycle with regeneration.

the surroundings or any other cooling medium. With regeneration, the high-pressure gas is further cooled to T_4 before expanding in the turbine. Lowering the turbine inlet temperature automatically lowers the turbine exit temperature, which is the minimum temperature in the cycle. Extremely low temperatures can be achieved by repeating this process.

EXAMPLE 10-5 The Simple Ideal Gas Refrigeration Cycle

An ideal-gas refrigeration cycle using air as the working medium is to maintain a refrigerated space at 0°F while rejecting heat to the surrounding medium at 80°F. The pressure ratio of the compressor is 4. Determine (*a*) the maximum and minimum temperatures in the cycle, (*b*) the coefficient of performance, and (*c*) the rate of refrigeration for a mass flow rate of 0.1 lbm/s.

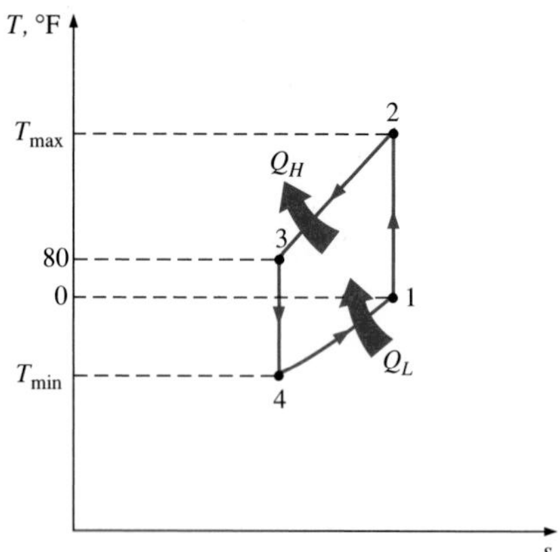

FIGURE 10-20

T-s diagram of the ideal-gas refrigeration cycle described in Example 10-5.

Solution The *T-s* diagram of the gas refrigeration cycle is shown in Fig. 10-20. We note that this is an ideal gas-compression refrigeration cycle that involves steady-flow components. Therefore, both the compressor and the turbine are isentropic, and the air is cooled to the environment temperature before it enters the turbine.

Assumptions **1** Steady operating conditions exist. **2** Air is an ideal gas. **3** Kinetic and potential energy changes are negligible.

Analysis (*a*) The maximum and minimum temperatures in the cycle are determined from the isentropic relations of ideal gases for the compression and expansion processes:

$$T_1 = 460 \text{ R} \longrightarrow h_1 = 109.90 \text{ Btu/lbm} \quad \text{and} \quad P_{r1} = 0.7913$$

$$P_{r2} = \frac{P_2}{P_1} P_{r1} = (4)(0.7913) = 3.165 \longrightarrow \begin{cases} h_2 = 163.5 \text{ Btu/lbm} \\ T_2 = \mathbf{683\ R\ (or\ 223°F)} \end{cases}$$

$$T_3 = 540 \text{ R} \longrightarrow h_3 = 129.06 \text{ Btu/lbm} \quad \text{and} \quad P_{r3} = 1.3860$$

$$P_{r4} = \frac{P_4}{P_3} P_{r3} = (0.25)(1.386) = 0.3465 \longrightarrow \begin{cases} h_4 = 86.7 \text{ Btu/lbm} \\ T_4 = \mathbf{363R\ (or\ -97°F)} \end{cases}$$

Therefore, the highest and the lowest temperatures in the cycle are 223 and −97°F, respectively.

(*b*) The COP of this ideal gas refrigeration cycle is determined from Eq. 10-11:

$$\text{COP}_R = \frac{q_L}{w_{net,\,in}} = \frac{q_L}{w_{comp,\,in} - w_{turb,\,out}}$$

where

$$q_L = h_1 - h_4 = 109.9 - 86.7 = 23.2 \text{ Btu/lbm}$$
$$w_{turb,\,out} = h_3 - h_4 = 129.06 - 86.7 = 42.36 \text{ Btu/lbm}$$
$$w_{comp,\,in} = h_2 - h_1 = 163.5 - 109.9 = 53.6 \text{ Btu/lbm}$$

Thus,

$$\text{COP}_R = \frac{23.2}{53.6 - 42.36} = \mathbf{2.06}$$

It is worth noting that an ideal vapor-compression cycle working under similar conditions would have a COP greater than 3.

(*c*) The rate of refrigeration is

$$\dot{Q}_{refrig} = \dot{m}(q_L) = (0.1 \text{ lbm/s})(23.2 \text{ Btu/lbm}) = \mathbf{2.32\ Btu/s}$$

10-9 ■ ABSORPTION REFRIGERATION SYSTEMS

Another form of refrigeration that becomes economically attractive when there is a source of inexpensive heat energy at a temperature of 100 to 200°C is **absorption refrigeration.** Some examples of inexpensive heat energy sources include geothermal energy, solar energy, and waste heat from cogeneration or process steam plants, and even natural gas when it is available at a relatively low price.

As the name implies, absorption refrigeration systems involve the absorption of a *refrigerant* by a *transport medium*. The most widely used absorption refrigeration system is the ammonia–water system, where ammonia (NH_3) serves as the refrigerant and water (H_2O) as the transport medium. Other absorption refrigeration systems include water–lithium bromide and water–lithium chloride systems, where water serves as the refrigerant. The latter two systems are limited to applications such as air-conditioning where the minimum temperature is above the freezing point of water.

To understand the basic principles involved in absorption refrigeration, we examine the NH_3–H_2O system shown in Fig. 10-21. The ammonia–water refrigeration machine was patented by the Frenchman Ferdinand Carre in 1859. Within a few years, the machines based on this principle were being built in the United States primarily to make ice and store food. You will immediately notice from the figure that this system looks very much like the vapor-compression system, except that the compressor has been replaced by a complex absorption mechanism consisting of an absorber, a pump, a generator, a regenerator, a valve, and a rectifier. Once the pressure of NH_3 is raised by the components in the box (this is the only thing they are set up to do), it is cooled and condensed in the condenser by rejecting heat to the surroundings, is throttled to the evaporator pressure, and picks up heat from the refrigerated space as it flows through the evaporator. So, there is nothing new there. Here is what happens in the box:

Ammonia vapor leaves the evaporator and enters the absorber, where it dissolves and reacts with water to form $NH_3 \cdot H_2O$. This is an exothermic

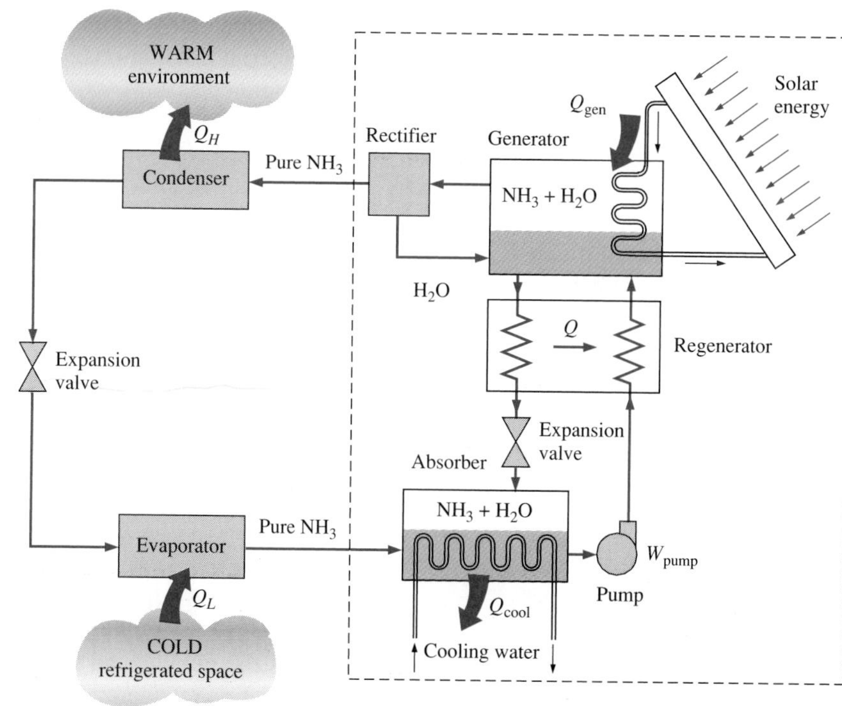

FIGURE 10-21
Ammonia absorption refrigeration cycle.

reaction; thus heat is released during this process. The amount of NH_3 that can be dissolved in H_2O is inversely proportional to the temperature. Therefore, it is necessary to cool the absorber to maintain its temperature as low as possible, hence to maximize the amount of NH_3 dissolved in water. The liquid $NH_3 + H_2O$ solution, which is rich in NH_3, is then pumped to the generator. Heat is transferred to the solution from a source to vaporize some of the solution. The vapor, which is rich in NH_3, passes through a rectifier, which separates the water and returns it to the generator. The high-pressure pure NH_3 vapor then continues its journey through the rest of the cycle. The hot $NH_3 + H_2O$ solution, which is weak in NH_3, then passes through a regenerator, where it transfers some heat to the rich solution leaving the pump, and is throttled to the absorber pressure.

Compared with vapor-compression systems, absorption refrigeration systems have one major advantage: A liquid is compressed instead of a vapor. The steady-flow work is proportional to the specific volume, and thus the work input for absorption refrigeration systems is very small (on the order of one percent of the heat supplied to the generator) and often neglected in the cycle analysis. The operation of these systems is based on heat transfer from an external source. Therefore, absorption refrigeration systems are often classified as *heat-driven systems.*

The absorption refrigeration systems are much more expensive than the vapor-compression refrigeration systems. They are more complex and occupy more space, they are much less efficient thus requiring much larger cooling towers to reject the waste heat, and they are more difficult to service since they are less common. Therefore, absorption refrigeration systems should be considered only when the unit cost of thermal energy is low and is projected to remain low relative to electricity. Absorption refrigeration systems are primarily used in large commercial and industrial installations.

The COP of absorption refrigeration systems is defined as

$$\text{COP}_\text{R} = \frac{\text{Desired output}}{\text{Required input}} = \frac{Q_L}{Q_\text{gen} + W_{pump,\text{in}}} \cong \frac{Q_L}{Q_\text{gen}} \tag{10-12}$$

The maximum COP of an absorption refrigeration system is determined by assuming that the entire cycle is totally reversible (i.e., the cycle involves no irreversibilities and any heat transfer is through a differential temperature difference). The refrigeration system would be reversible if the heat from the source (Q_gen) were transferred to a Carnot heat engine, and the work output of this heat engine ($W = \eta_\text{th, rev}\, Q_\text{gen}$) is supplied to a Carnot refrigerator to remove heat from the refrigerated space. Note that $Q_L = W \times \text{COP}_\text{R, rev} = \eta_\text{th, rev}\, Q_\text{gen}\text{COP}_\text{R, rev}$. Then the overall COP of an absorption refrigeration system under reversible conditions becomes (Fig. 10-22)

$$\text{COP}_\text{rev, absorption} = \frac{Q_L}{Q_\text{gen}} = \eta_\text{th, rev}\, \text{COP}_\text{R, rev} = \left(1 - \frac{T_0}{T_s}\right)\left(\frac{T_L}{T_0 - T_L}\right) \tag{10-13}$$

where T_L, T_0, and T_s are the absolute temperatures of the refrigerated space, environment, and heat source, respectively. Any absorption refrigeration sys-

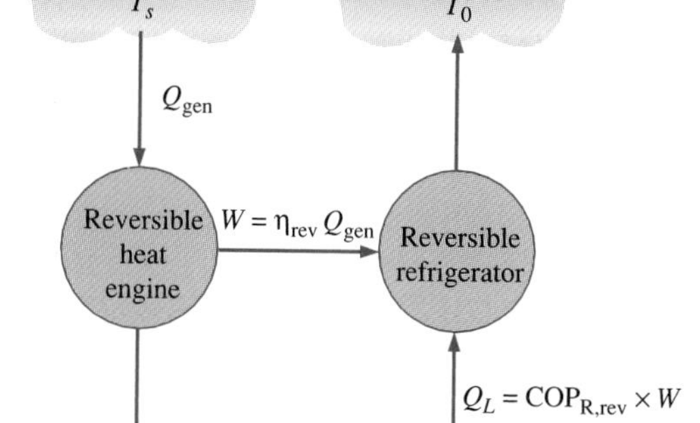

$$W = \eta_{rev} Q_{gen} = \left(1 - \frac{T_0}{T_s}\right) Q_{gen}$$

$$Q_L = COP_{R,rev} W = \left(\frac{T_L}{T_0 - T_s}\right) W$$

$$COP_{rev,\ absorption} = \frac{Q_L}{Q_{gen}} = \left(1 - \frac{T_0}{T_s}\right)\left(\frac{T_L}{T_0 - T_s}\right)$$

FIGURE 10-22
Determining the maximum COP of an absorption refrigeration system.

tem that receives heat from a source at T_s and removes heat from the refrigerated space at T_L while operating in an environment at T_0 will have a lower COP than the one determined from Eq. 10-13. For example, when the source is at 120°C, the refrigerated space is at −10°C, and the environment is at 25°C, the maximum COP that an absorption refrigeration system can have is 1.8. The COP of actual absorption refrigeration systems is usually less than 1.

Air-conditioning systems based on absorption refrigeration, called *absorption chillers,* perform best when the heat source can supply heat at a high temperature with little temperature drop. The absorption chillers are typically rated at an input temperature of 116°C (240°F). The chillers will perform at lower temperatures, but their cooling capacity decreases sharply with decreasing source temperature, about 12.5 percent for each 6°C (10°F) drop in the source temperature. For example, the capacity will go down to 50 percent when the supply water temperature drops to 93°C (200°F). In that case, one needs to double the size (and thus the cost) of the chiller to achieve the same cooling. The COP of the chiller is affected less by the decline of the source temperature. The COP drops by 2.5 percent for each 6°C (10°F) drop in the source temperature. The nominal COP of single stage absorption chillers at 116°C (240°F) is 0.65 to 0.70. Therefore, for each ton of refrigeration, a heat input of (12,000 Btu/h)/0.65 = 18,460 Btu/h will be required. At 88°C (190°F), the COP will drop by 12.5 percent and thus the heat input will increase by 12.5 percent for the same cooling effect. Therefore, the economic aspects must be evaluated carefully before any absorption refrigeration system is considered, especially when the source temperature is below 93°C (200°F).

Another absorption refrigeration system that is quite popular with campers is a propane-fired system invented by two Swedish undergraduate students. In this system, the pump is replaced by a third fluid (hydrogen), which makes it a truly portable unit.

10-10 ■ THERMOELECTRIC POWER GENERATION AND REFRIGERATION SYSTEMS

All the refrigeration systems discussed above involve many moving parts and bulky, complex components. Then this question comes to mind: Is it really necessary for a refrigeration system to be so complex? Can we not achieve the same effect in a more direct way? The answer to this question is *yes*. It is possible to use electric energy more directly to produce cooling without involving any refrigerants and moving parts. Below we discuss one such system, called a *thermoelectric refrigerator.*

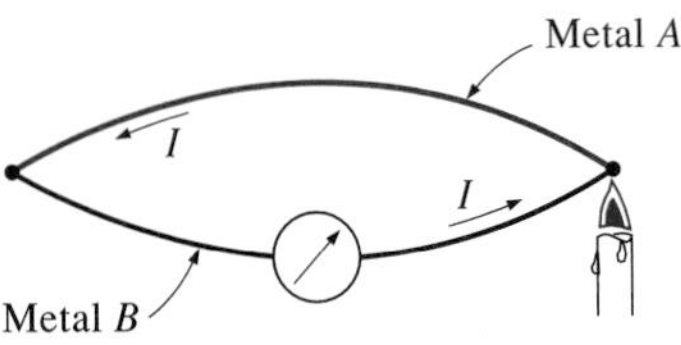

FIGURE 10-23

When one of the junctions of two dissimilar metals is heated, a current I flows through the closed circuit.

Consider two wires made from different metals joined at both ends (junctions), forming a closed circuit. Ordinarily, nothing will happen. But when one of the ends is heated, something interesting happens: A current flows continuously in the circuit, as shown in Fig. 10-23. This is called the **Seebeck effect,** in honor of Thomas Seebeck, who made this discovery in 1821. The circuit that incorporates both thermal and electrical effects is called a **thermoelectric circuit,** and a device that operates on this circuit is called a **thermoelectric device.**

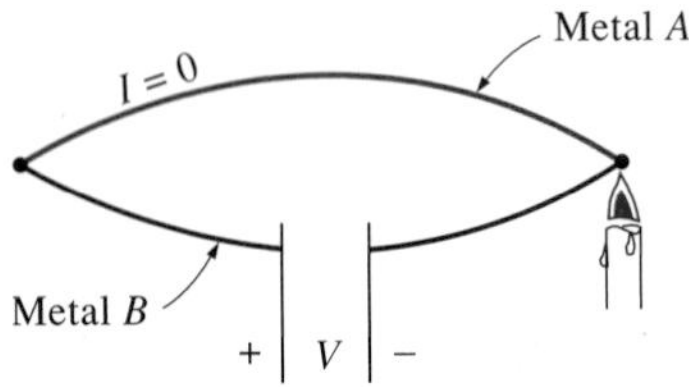

FIGURE 10-24

When a thermoelectric circuit is broken, a potential difference is generated.

The Seebeck effect has two major applications: temperature measurement and power generation. When the thermoelectric circuit is broken, as shown in Fig. 10-24, the current ceases to flow, and we can measure the driving force (the electromotive force) or the voltage generated in the circuit by a voltmeter. The voltage generated is a function of the temperature difference and the materials of the two wires used. Therefore, temperature can be measured by simply measuring voltages. The two wires used to measure the temperature in this manner form a *thermocouple,* which is the most versatile and most widely used temperature measurement device. A common T-type thermocouple, for example, consists of copper and constantan wires, and it produces about 40 μV per °C difference.

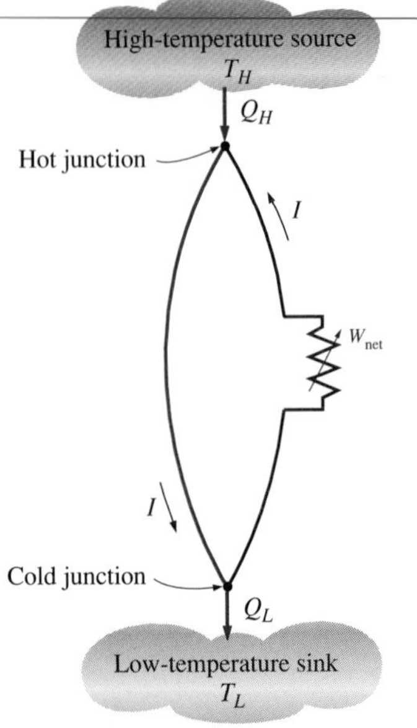

FIGURE 10-25

Schematic of a simple thermoelectric power generator.

The Seebeck effect also forms the basis for thermoelectric power generation. The schematic diagram of a **thermoelectric generator** is shown in Fig. 10-25. Heat is transferred from a high-temperature source to the hot junction in the amount of Q_H, and it is rejected to a low-temperature sink from the cold junction in the amount of Q_L. The difference between these two quantities is the net electrical work produced, that is, $W_e = Q_H - Q_L$. It is evident from Fig. 10-25 that the thermoelectric power cycle closely resembles an ordinary heat engine cycle, with electrons serving as the working fluid. Therefore, the thermal efficiency of a thermoelectric generator operating between the temperature limits of T_H and T_L is limited by the efficiency of a Carnot cycle operating between the same temperature limits. Thus, in the

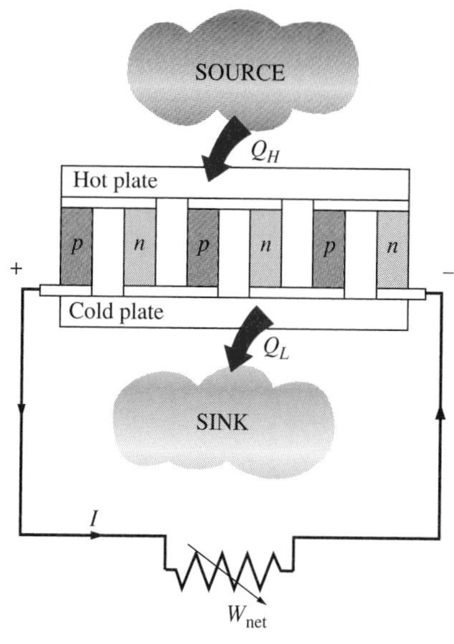

FIGURE 10-26
A thermoelectric power generator.

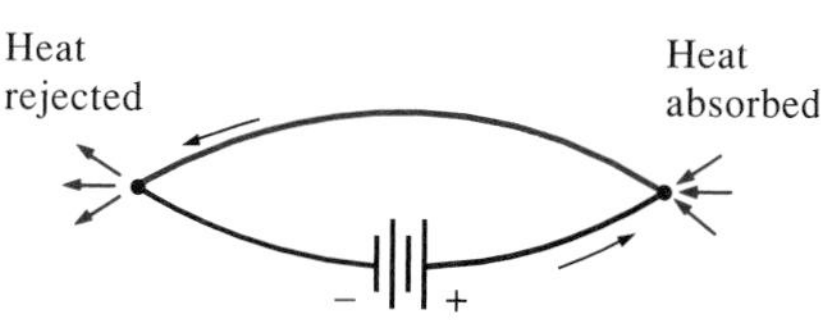

FIGURE 10-27
When a current is passed through the junction of two dissimilar materials, the junction is cooled.

absence of any irreversibilities (such as I^2R heating, where R is the total electrical resistance of the wires), the thermoelectric generator will have the Carnot efficiency.

The major drawback of thermoelectric generators is their low efficiency. The future success of these devices depends on finding materials with more desirable characteristics. For example, the voltage output of thermoelectric devices has been increased several times by switching from metal pairs to semiconductors. A practical thermoelectric generator using n-type (heavily doped to create excess electrons) and p-type (heavily doped to create a deficiency of electrons) materials connected in series is shown in Fig. 10-26. Despite their low efficiencies, thermoelectric generators have definite weight and reliability advantages and are presently used in rural areas and in space applications. For example, silicon–germanium-based thermoelectric generators have been powering *Voyager* spacecraft since 1980 and are expected to continue generating power for many more years.

If Seebeck had been fluent in thermodynamics, he would probably have tried reversing the direction of flow of electrons in the thermoelectric circuit (by externally applying a potential difference in the reverse direction) to create a refrigeration effect. But this honor belongs to Jean Charles Athanase Peltier, who discovered this phenomenon in 1834. He noticed during his experiments that when a small current was passed through the junction of two dissimilar wires, the junction was cooled, as shown in Fig. 10-27. This is called the **Peltier effect,** and it forms the basis for **thermoelectric refrigeration.** A practical thermoelectric refrigeration circuit using semiconductor materials is shown in Fig. 10-28. Heat is absorbed from the refrigerated space in the amount of Q_L and rejected to the warmer environment in the amount of Q_H. The difference between these two quantities is the net electrical work that needs to be supplied; that is, $W_e = Q_H - Q_L$. Thermoelectric refrigerators presently cannot compete with vapor-compression refrigeration systems because of their low coefficient of performance. They are available in the market, however, and are preferred in some applications because of their small size, simplicity, quietness, and reliability.

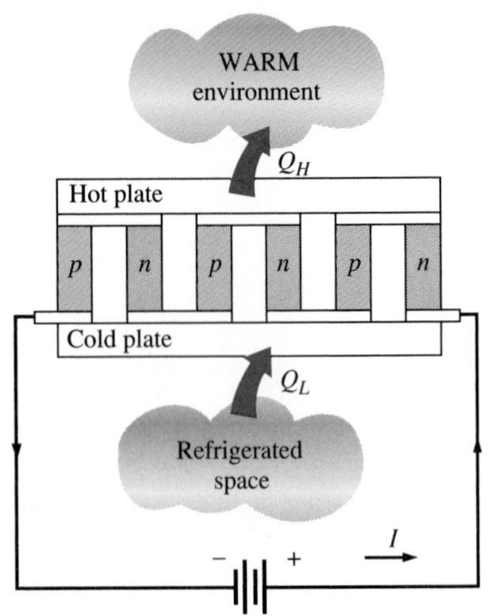

FIGURE 10-28
A thermoelectric refrigerator.

10-11 ■ SUMMARY

The transfer of heat from lower temperature regions to higher temperature ones is called *refrigeration*. Devices that produce refrigeration are called *refrigerators,* and the cycles on which they operate are called *refrigeration cycles.* The working fluids used in refrigerators are called *refrigerants.* Refrigerators used for the purpose of heating a space by transferring heat from a cooler medium are called *heat pumps.*

The performance of refrigerators and heat pumps is expressed in terms of *coefficient of performance* (COP), defined as

$$\text{COP}_\text{R} = \frac{\text{Desired output}}{\text{Required input}} = \frac{\text{Cooling effect}}{\text{Work input}} = \frac{Q_L}{W_{\text{net, in}}}$$

$$\text{COP}_\text{HP} = \frac{\text{Desired ouput}}{\text{Required input}} = \frac{\text{Heating effect}}{\text{Work input}} = \frac{Q_H}{W_{\text{net, in}}}$$

The standard of comparison for refrigeration cycles is the *reversed Carnot cycle.* A refrigerator or heat pump that operates on the reversed Carnot cycle is called a *Carnot refrigerator* or a *Carnot heat pump,* and their COPs are

$$\text{COP}_\text{R, Carnot} = \frac{1}{T_H/T_L - 1}$$

$$\text{COP}_\text{HP, Carnot} = \frac{1}{1 - T_L/T_H}$$

The most widely used refrigeration cycle is the *vapor-compression refrigeration cycle.* In an ideal vapor-compression refrigeration cycle, the refrigerant enters the compressor as a saturated vapor and is cooled to the saturated liquid state in the condenser. It is then throttled to the evaporator pressure and vaporizes as it absorbs heat from the refrigerated space.

Very low temperatures can be achieved by operating two or more vapor-compression systems in series, called *cascading.* The COP of a refrigeration system also increases as a result of cascading. Another way of improving the performance of a vapor-compression refrigeration system is by using *multistage compression with regenerative cooling.* A refrigerator with a single compressor can provide refrigeration at several temperatures by throttling the refrigerant in stages. The vapor-compression refrigeration cycle can also be used to liquefy gases after some modifications.

The power cycles can be used as refrigeration cycles by simply reversing them. Of these, the *reversed Brayton cycle,* which is also known as the *gas refrigeration cycle,* is used to cool aircraft and to obtain very low (cryogenic) temperatures after it is modified with regeneration. The work output of the turbine can be used to reduce the work input requirements to the compressor. Thus the COP of a gas refrigeration cycle is

$$\mathrm{COP_R} = \frac{q_L}{w_{\mathrm{net,in}}} = \frac{q_L}{w_{\mathrm{comp,in}} - w_{\mathrm{turb,out}}}$$

Another form of refrigeration that becomes economically attractive when there is a source of inexpensive heat energy at a temperature of 100 to 200°C is *absorption refrigeration,* where the refrigerant is absorbed by a transport medium and compressed in liquid form. The most widely used absorption refrigeration system is the ammonia–water system, where ammonia serves as the refrigerant and water as the transport medium. The work input to the pump is usually very small, and the COP of absorption refrigeration systems is defined as

$$\mathrm{COP_R} = \frac{\text{Desired output}}{\text{Required input}} = \frac{Q_L}{Q_{\mathrm{gen}} + W_{pump,\mathrm{in}}} \cong \frac{Q_L}{Q_{\mathrm{gen}}}$$

The maximum COP an absorption refrigeration system can have is determined by assuming totally reversible conditions, which yields

$$\mathrm{COP_{rev,\,absorption}} = \eta_{\mathrm{th,rev}}\,\mathrm{COP_{R,rev}} = \left(1 - \frac{T_0}{T_s}\right)\left(\frac{T_L}{T_0 - T_L}\right)$$

where T_0, T_L, and T_s are the absolute temperatures of the environment, refrigerated space, and heat source, respectively.

A refrigeration effect can also be achieved without using any moving parts by simply passing a small current through a closed circuit made up of two dissimilar materials. This effect is called the *Peltier effect,* and a refrigerator that works on this principle is called a *thermoelectric refrigerator.*

REFERENCES AND SUGGESTED READING

1. *ASHRAE, Handbook of Fundamentals.* Atlanta: American Society of Heating, Refrigerating, and Air-Conditioning Engineers, 1985.

2. W. Z. Black and J. G. Hartley. *Thermodynamics.* New York: Harper & Row, 1985.

3. R. C. Fellinger and W. J. Cook. *Introduction to Engineering Thermodynamics.* Dubuque, IA: William C. Brown, 1985.

4. *Heat Pump Systems—A Technology Review.* OECD Report, Paris, 1982.

5. J. B. Jones and G. A. Hawkins. *Engineering Thermodynamics.* 2nd ed. New York: John Wiley & Sons, 1986.

6. B. Nagengast. "A Historical Look at CFC Refrigerants." *ASHRAE Journal* 30, no. 11 (November 1988), pp. 37–39.

7. W. F. Stoecker. "Growing Opportunities for Ammonia Refrigeration." *Proceedings of the Meeting of the International Institute of Ammonia Refrigeration*, Austin, Texas, 1989.

8. W. F. Stoecker and J. W. Jones. *Refrigeration and Air Conditioning*. 2nd ed. New York: McGraw-Hill, 1982.

9. G. J. Van Wylen and R. E. Sonntag. *Fundamentals of Classical Thermodynamics*. 3rd ed. New York: John Wiley & Sons, 1985.

10. K. Wark. *Thermodynamics*. 5th ed. New York: McGraw-Hill, 1988.

PROBLEMS*

The Reversed Carnot Cycle

10-1C Why is the reversed Carnot cycle executed within the saturation dome not a realistic model for refrigeration cycles?

10-2C What is the difference between a refrigerator and a heat pump?

10-3 A steady-flow Carnot refrigeration cycle uses refrigerant-134a as the working fluid. The refrigerant changes from saturated vapor to saturated liquid at 30°C in the condenser as it rejects heat. The evaporator pressure is 120 kPa. Show the cycle on a *T-s* diagram relative to saturation lines, and determine (*a*) the coefficient of performance, (*b*) the amount of heat absorbed from the refrigerated space, and (*c*) the net work input.
Answers: (*a*) 4.44, (*b*) 140.4 kJ/kg, (*c*) 31.61 kJ/kg

10-4E Refrigerant-134a enters the condenser of a steady-flow Carnot refrigerator as a saturated vapor at 90 psia, and it leaves with a quality of 0.05. The heat absorption from the refrigerated space takes place at a pressure of 30 psia. Show the cycle on a *T-s* diagram relative to saturation lines, and determine (*a*) the coefficient of performance, (*b*) the quality at the beginning of the heat-absorption process, and (*c*) the net work input.

Ideal and Actual Vapor-Compression Refrigeration Cycles

10-5C Does the ideal vapor-compression refrigeration cycle involve any internal irreversibilities?

10-6C Why is the throttling valve not replaced by an isentropic turbine in the ideal vapor-compression refrigeration cycle?

10-7C It is proposed to use water instead of refrigerant-134a as the working fluid in air-conditioning applications where the minimum temperature never falls below the freezing point. Would you support this proposal? Explain.

*Students are encouraged to answer *all* the concept "C" questions.

10-8C In a refrigeration system, would you recommend condensing the refrigerant-134a at a pressure of 0.7 or 1.0 MPa if heat is to be rejected to a cooling medium at 15°C? Why?

10-9C Does the area enclosed by the cycle on a *T-s* diagram represent the net work input for the reversed Carnot cycle? How about for the ideal vapor-compression refrigeration cycle?

10-10C Consider two vapor-compression refrigeration cycles. The refrigerant enters the throttling valve as a saturated liquid at 30°C in one cycle and as subcooled liquid at 30°C in the other one. The evaporator pressure for both cycles is the same. Which cycle do you think will have a higher COP?

10-11C The COP of vapor-compression refrigeration cycles improves when the refrigerant is subcooled before it enters the throttling valve. Can the refrigerant be subcooled indefinitely to maximize this effect, or is there a lower limit? Explain.

10-12 A refrigerator uses refrigerant-134a as the working fluid and operates on an ideal vapor-compression refrigeration cycle between 0.12 and 0.7 MPa. The mass flow rate of the refrigerant is 0.05 kg/s. Show the cycle on a *T-s* diagram with respect to saturation lines. Determine (*a*) the rate of heat removal from the refrigerated space and the power input to the compressor, (*b*) the rate of heat rejection to the environment, and (*c*) the coefficient of performance. *Answers:* (*a*) 7.35 kW, 1.82 kW; (*b*) 9.17 kW; (*c*) 4.04

10-13 If the throttling valve in Prob. 10-12 is replaced by an isentropic turbine, determine the percentage increase in the COP and in the rate of heat removal from the refrigerated space. *Answers:* 4.2 percent, 4.2 percent

10-14 Consider a 300 kJ/min refrigeration system that operates on an ideal vapor-compression refrigeration cycle with refrigerant-134a as the working fluid. The refrigerant enters the compressor as saturated vapor at 140 kPa and is compressed to 800 kPa. Show the cycle on a *T-s* diagram with respect to saturation lines, and determine (*a*) the quality of the refrigerant at the end of the throttling process, (*b*) the coefficient of performance, and (*c*) the power input to the compressor.

10-15 Repeat Prob. 10-14 assuming an adiabatic efficiency of 85 percent for the compressor. Also, determine the irreversibility rate associated with the compression process in this case. Take $T_0 = 298$ K.

10-16 Refrigerant-134a enters the compressor of a refrigerator as superheated vapor at 0.14 MPa and −10°C at a rate of 0.04 kg/s, and it leaves at 0.7 MPa and 50°C. The refrigerant is cooled in the condenser to 24°C and 0.65 MPa, and it is throttled to 0.15 MPa. Disregarding any heat transfer and pressure drops in the connecting lines between the components, show the cycle on a *T-s* diagram with respect to saturation lines, and determine (*a*) the rate of heat removal from the refrigerated space and the power input to the compressor, (*b*) the adiabatic efficiency of the compressor, and (*c*) the COP of the refrigerator.

Answers: (*a*) 6.42 kW, 1.72 kW; (*b*) 80.7 percent; (*c*) 3.73

10-17E An ice-making machine operates on the ideal vapor-compression cycle, using refrigerant-134a. The refrigerant enters the compressor as saturated vapor at 20 psia and leaves the condenser as saturated liquid at 100 psia. Water enters the ice machine at 55°F and leaves as ice at 25°F. For an ice production rate of 20 lbm/h, determine the power input to the ice machine (169 Btu of heat needs to be removed from each lbm of water at 55°F to turn it into ice at 25°F).

10-18 Refrigerant-134a enters the compressor of a refrigerator at 140 kPa and −10°C at a rate of 0.2 m^3/min and leaves at 1 MPa. The isentropic efficiency of the compressor is 78 percent. The refrigerant enters the throttling valve at 0.95 MPa and 30°C and leaves the evaporator as saturated vapor at −18.5°C. Show the cycle on a *T-s* diagram with respect to saturation lines, and determine (*a*) the power input to the compressor, (*b*) the rate of heat removal from the refrigerated space, and (*c*) the pressure drop and rate of heat gain in the line between the evaporator and the compressor.
Answers: (*a*) 1.25 kW; (*b*) 3.31 kW; (*c*) 2.9 kPa, 0.164 kW

Selecting the Right Refrigerant

10-19C When selecting a refrigerant for a certain application, what qualities would you look for in the refrigerant?

10-20C Consider a refrigeration system using refrigerant-134a as the working fluid. If this refrigerator is to operate in an environment at 30°C, what is the minimum pressure to which the refrigerant should be compressed? Why?

10-21C A refrigerant-134a refrigerator is to maintain the refrigerated space at −10°C. Would you recommend an evaporator pressure of 0.12 or 0.14 MPa for this system? Why?

10-22 A refrigerator that operates on the ideal vapor-compression cycle with refrigerant-134a is to maintain the refrigerated space at −10°C while rejecting heat to the environment at 25°C. Select reasonable pressures for the evaporator and the condenser, and explain why you chose those values.

10-23 A heat pump that operates on the ideal vapor-compression cycle with refrigerant-134a is used to heat a house and maintain it at 20°C by using underground water at 10°C as the heat source. Select reasonable pressures for the evaporator and the condenser, and explain why you chose those values.

Heat Pump Systems

10-24C What are the advantages and disadvantages of heat pumps? How do they compare to other heating systems?

10-25C Do you think a heat pump system will be more cost-effective in New York or in Miami? Why?

10-26C What is a water-source heat pump? How does the COP of a water-source heat pump system compare to that of an air-source system?

10-27E A heat pump that operates on the ideal vapor-compression cycle with refrigerant-134a is used to heat a house and maintain it at 75°F by using underground water at 50°F as the heat source. The house is losing heat at a rate of 90,000 Btu/h. The evaporator and condenser pressures are 50 and 120 psia, respectively. Determine the power input to the heat pump and the electric power saved by using a heat pump instead of a resistance heater.
Answers: 3.68 hp, 31.69 hp

10-28 A heat pump that operates on the ideal vapor-compression cycle with refrigerant-134a is used to heat water from 15 to 54°C at a rate of 0.18 kg/s. The condenser and evaporator pressures are 1.4 and 0.32 MPa, respectively. Determine the power input to the heat pump.

10-29 A heat pump using refrigerant-134a heats a house by using underground water at 8°C as the heat source. The house is losing heat at a rate of 60,000 kJ/h. The refrigerant enters the compressor at 280 kPa and 0°C, and it leaves at 1 MPa and 60°C. The refrigerant exits the condenser at 30°C. Determine (*a*) the power input to the heat pump, (*b*) the rate of heat absorption from the water, and (*c*) the increase in electric power input if an electric resistance heater is used instead of a heat pump.
Answers: (*a*) 3.65 kW, (*b*) 13.02 kW, (*c*) 13.02 kW

Innovative Refrigeration Systems

10-30C What is cascade refrigeration? What are the advantages and disadvantages of cascade refrigeration?

10-31C How does the COP of a cascade refrigeration system compare to the COP of a simple vapor-compression cycle operating between the same pressure limits?

10-32C A certain application requires maintaining the refrigerated space at −32°C. Would you recommend a simple refrigeration cycle with refrigerant-134a or a two-stage cascade refrigeration cycle with a different refrigerant at the bottoming cycle? Why?

10-33C Consider a two-stage cascade refrigeration cycle and a two-stage compression refrigeration cycle with a flash chamber. Both cycles operate between the same pressure limits and use the same refrigerant. Which system would you favor? Why?

10-34C Can a vapor-compression refrigeration system with a single compressor handle several evaporators operating at different pressures? How?

10-35C Is it possible to have liquid helium at room temperature?

10-36C In the liquefaction process, why are gases compressed to very high pressures?

10-37 Consider a two-stage cascade refrigeration system operating between the pressure limits of 0.8 and 0.14 MPa. Each stage operates on the ideal vapor-compression refrigeration cycle with refrigerant-134a as the working fluid. Heat rejection from the lower cycle to the upper cycle takes place in an adiabatic counterflow heat exchanger where both streams enter at about 0.4 MPa. If the mass flow rate of the refrigerant through the upper cycle is 0.12 kg/s, determine (*a*) the mass flow rate of the refrigerant through the lower cycle, (*b*) the rate of heat removal from the refrigerated space and the power input to the compressor, and (*c*) the coefficient of performance of this cascade refrigerator.

Answers: (*a*) 0.0966 kg/s; (*b*) 16.8 kW, 3.77 kW; (*c*) 4.46

10-38 Repeat Prob. 10-37 for a heat exchanger pressure of 0.5 MPa.

10-39 A two-stage compression refrigeration system operates with refrigerant-134a between the pressure limits of 1 and 0.14 MPa. The refrigerant leaves the condenser as a saturated liquid and is throttled to a flash chamber operating at 0.5 MPa. The refrigerant leaving the low-pressure compressor at 0.5 MPa is also routed to the flash chamber. The vapor in the flash chamber is then compressed to the condenser pressure by the high-pressure compressor, and the liquid is throttled to the evaporator pressure. Assuming the refrigerant leaves the evaporator as saturated vapor and both compressors are isentropic, determine (*a*) the fraction of the refrigerant that evaporates as it is throttled to the flash chamber, (*b*) the amount of heat removed from the refrigerated space for a mass flow rate of 0.25 kg/s through the condenser, and (*c*) the coefficient of performance.

10-40 Repeat Prob. 10-39 for a flash chamber pressure of 0.32 MPa.

Gas Refrigeration Cycle

10-41C How does the ideal-gas refrigeration cycle differ from the Brayton cycle?

10-42C Devise a refrigeration cycle that works on the reversed Stirling cycle. Also, determine the COP for this cycle.

10-43C How does the ideal-gas refrigeration cycle differ from the Carnot refrigeration cycle?

10-44C How is the ideal-gas refrigeration cycle modified for aircraft cooling?

10-45C In gas refrigeration cycles, can we replace the turbine by an expansion valve as we did in vapor-compression refrigeration cycles? Why?

10-46C How do we achieve very low temperatures with gas refrigeration cycles?

10-47 An ideal-gas refrigeration cycle using air as the working fluid is to maintain a refrigerated space at −23°C while rejecting heat to the surrounding medium at 27°C. If the pressure ratio of the compressor is 3, determine (*a*) the maximum and minimum temperatures in the cycle, (*b*) the coefficient of performance, and (*c*) the rate of refrigeration for a mass flow rate of 0.15 kg/s.

10-48 Air enters the compressor of an ideal-gas refrigeration cycle at 12°C and 50 kPa and the turbine at 47°C and 250 kPa. The mass flow rate of air through the cycle is 0.08 kg/s. Assuming variable specific heats for air, determine (*a*) the rate of refrigeration, (*b*) the net power input, and (*c*) the coefficient of performance. *Answers:* (*a*) 6.67 kW, (*b*) 3.88 kW, (*c*) 1.72

10-49E Air enters the compressor of an ideal-gas refrigeration cycle at 40°F and 10 psia and the turbine at 120°F and 30 psia. The mass flow rate of air through the cycle is 0.5 lbm/s. Determine (*a*) the rate of refrigeration, (*b*) the coefficient of performance, and (*c*) the net power input.

10-50 Repeat Prob. 10-48 for a compressor adiabatic efficiency of 80 percent and a turbine adiabatic efficiency of 85 percent.

10-51 A gas refrigeration cycle with a pressure ratio of 3 uses helium as the working fluid. The temperature of the helium is −10°C at the compressor inlet and 50°C at the turbine inlet. Assuming adiabatic efficiencies of 82 percent for both the turbine and the compressor, determine (*a*) the minimum temperature in the cycle, (*b*) the coefficient of performance, and (*c*) the mass flow rate of the helium for a refrigeration rate of 5 kW.

10-52 A gas refrigeration system using air as the working fluid has a pressure ratio of 4. Air enters the compressor at −7°C. The high-pressure air is cooled to 27°C by rejecting heat to the surroundings. It is further cooled to −15°C by regenerative cooling before it enters the turbine. Assuming both the turbine and the compressor to be isentropic and using constant specific heats at room temperature, determine (*a*) the lowest temperature that can be obtained by this cycle, (*b*) the coefficient of performance of the cycle, and (*c*) the mass flow rate of air for a refrigeration rate of 12 kW.
Answers: (*a*) − 99.4°C, (*b*) 1.12, (*c*) 0.234 kg/s

10-53 Repeat Prob. 10-52 assuming adiabatic efficiencies of 75 percent for the compressor and 80 percent for the turbine.

Absorption Refrigeration Systems

10-54C What is absorption refrigeration? How does an absorption refrigeration system differ from a vapor-compression refrigeration system?

10-55C What are the advantages and disadvantages of absorption refrigeration?

10-56C Can water be used as a refrigerant in air-conditioning applications? Explain.

10-57C In absorption refrigeration cycles, why is the fluid in the absorber cooled, and the fluid in the generator heated?

10-58C How is the coefficient of performance of an absorption refrigeration system defined?

10-59C What are the functions of the rectifier and the regenerator in an absorption refrigeration system?

10-60 An absorption refrigeration system that receives heat from a source at 130°C and maintains the refrigerated space at −5°C is claimed to have a COP of 2. If the environment temperature is 27°C, can this claim be valid? Justify your answer.

10-61 An absorption refrigeration system receives heat from a source at 110°C and maintains the refrigerated space at −20°C. If the temperature of the environment is 25°C, what is the maximum COP this absorption refrigeration system can have?

10-62 Heat is supplied to an absorption refrigeration system from a geothermal well at 130°C at a rate of 10^5 kJ/h. The environment is at 25°C, and the refrigerated space is maintained at −30°C. Determine the maximum rate at which this system can remove heat from the refrigerated space.
Answer: 1.15×10^5 kJ/h

10-63E Heat is supplied to an absorption refrigeration system from a geothermal well at 250°F at a rate of 10^5 Btu/h. The environment is at 80°F, and the refrigerated space is maintained at 0°F. If the COP of the system is 0.7, determine the rate at which this system can remove heat from the refrigerated space.

Thermoelectric Power Generation and Refrigeration Systems

10-64C What is a thermoelectric circuit?

10-65C Describe the Seebeck and the Peltier effects.

10-66C Consider a circular copper wire formed by connecting the two ends of a copper wire. The connection point is now heated by a burning candle. Do you expect any current to flow through the wire?

10-67C An iron and a constantan wire are formed into a closed circuit by connecting the ends. Now both junctions are heated and are maintained at the same temperature. Do you expect any electric current to flow through this circuit?

10-68C A copper and a constantan wire are formed into a closed circuit by connecting the ends. Now one junction is heated by a burning candle while the other is maintained at room temperature. Do you expect any electric current to flow through this circuit?

10-69C How does a thermocouple work as a temperature measurement device?

10-70C Why are semiconductor materials preferable to metals in thermoelectric refrigerators?

10-71C Is the efficiency of a thermoelectric generator limited by the Carnot efficiency? Why?

10-72E A thermoelectric generator receives heat from a source at 240°F and rejects the waste heat to the environment at 80°F. What is the maximum thermal efficiency this thermoelectric generator can have?
Answer: 22.9 percent

10-73 A thermoelectric refrigerator removes heat from a refrigerated space at −5°C at a rate of 130 W and rejects it to an environment at 20°C. Determine the maximum coefficient of performance this thermoelectric refrigerator can have and the minimum required power input.
*Answer:*10.72, 12.1 W

10-74 A thermoelectric cooler has a COP of 0.1 and removes heat from a refrigerated space at a rate of 180 W. Determine the required power input to the thermoelectric cooler, in W.

10-75E A thermoelectric cooler has a COP of 0.15 and removes heat from a refrigerated space at a rate of 35 Btu/min. Determine the required power input to the thermoelectric cooler, in hp.

10-76 A thermoelectric refrigerator is powered by a 12-V car battery that draws 3 A of current when running. The refrigerator resembles a small ice chest and is claimed to cool nine canned drinks, 0.350-L each, from 25°C to 3°C in 12 h. Determine the average COP of this refrigerator.

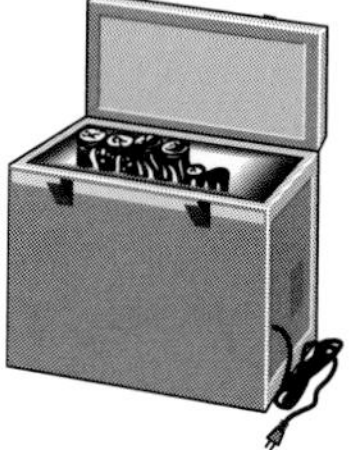

FIGURE P10-76

10-77E Thermoelectric coolers that plug into the cigarette lighter of a car are commonly available. One such cooler is claimed to cool a 12-oz (0.771-lbm) drink from 78°F to 38°F or to heat a cup of coffee from 75°F to 130°F in about 15 min in a well-insulated cup holder. Assuming an average COP of 0.2 in the cooling mode, determine (*a*) the average rate of heat removal from the drink, (*b*) the average rate of heat supply to the coffee, and (*c*) the electric power drawn from the battery of the car, all in W.

Review Problems

10-78 Consider a steady-flow Carnot refrigeration cycle that uses refrigerant-134a as the working fluid. The maximum and minimum temperatures in the cycle are 20 and −20°C, respectively. The quality of the refrigerant is 0.2 at the beginning of the heat absorption process and 0.85 at the end. Show the cycle on a *T*-*s* diagram relative to saturation lines, and determine (*a*) the coefficient of performance, (*b*) the condenser and evaporator pressures, and (*c*) the net work input.

10-79 A large refrigeration plant is to be maintained at −15°C, and it requires refrigeration at a rate of 100 kW. The condenser of the plant is to be cooled by liquid water, which experiences a temperature rise of 8°C as it flows over the coils of the condenser. Assuming the plant operates on the ideal vapor-compression cycle using refrigerant-134a between the pressure limits of 120 and 700 kPa, determine (*a*) the mass flow rate of the refrigerant, (*b*) the power input to the compressor, and (*c*) the mass flow rate of the cooling water.

10-80 Repeat Prob. 10-79 assuming the compressor has an isentropic efficiency of 75 percent. Also, determine the rate of exergy destruction associated with the compression process in this case. Take $T_0 = 25°C$.

10-81 A heat pump that operates on the ideal vapor-compression cycle with refrigerant-134a is used to heat a house. The mass flow rate of the refrigerant is 0.15 kg/s The condenser and evaporator pressures are 900 and 240 kPa, respectively. Show the cycle on a *T-s* diagram with respect to saturation lines, and determine (*a*) the rate of heat supply to the house, (*b*) the volume flow rate of the refrigerant at the compressor inlet, and (*c*) the COP of this heat pump.

10-82 Derive a relation for the COP of the two-stage refrigeration system with a flash chamber as shown in Fig. 10-12 in terms of the enthalpies and the quality at state 6. Consider a unit mass in the condenser.

10-83 Consider a two-stage compression refrigeration system operating between the pressure limits of 0.8 and 0.14 MPa. The working fluid is refrigerant-134a. The refrigerant leaves the condenser as a saturated liquid and is throttled to a flash chamber operating at 0.4 MPa. Part of the refrigerant evaporates during this flashing process, and this vapor is mixed with the refrigerant leaving the low-pressure compressor. The mixture is then compressed to the condenser pressure by the high-pressure compressor. The liquid in the flash chamber is throttled to the evaporator pressure, and it cools the refrigerated space as it vaporizes in the evaporator. Assuming the refrigerant leaves the evaporator as saturated vapor and both compressors are isentropic, determine (*a*) the fraction of the refrigerant that evaporates as it is throttled to the flash chamber, (*b*) the amount of heat removed from the refrigerated space and the compressor work per unit mass of refrigerant flowing through the condenser, and (*c*) the coefficient of performance.
Answers: (*a*) 0.165; (*b*) 145.3 kJ/kg, 32.5 kJ/kg; (*c*) 4.47

10-84 An aircraft on the ground is to be cooled by a gas refrigeration cycle operating with air on an open cycle. Air enters the compressor at 30°C and 100 kPa and is compressed to 250 kPa. Air is cooled to 70°C before it enters the turbine. Assuming both the turbine and the compressor to be isentropic, determine the temperature of the air leaving the turbine and entering the cabin. *Answer:* −9°C

10-85 Consider a regenerative gas refrigeration cycle using helium as the working fluid. Helium enters the compressor at 100 kPa and −10°C and is compressed to 300 kPa. Helium is then cooled to 20°C by water. It then enters the regenerator where it is cooled further before it enters the turbine. Helium leaves the refrigerated space at −25°C and enters the regenerator. Assuming both the turbine and the compressor to be isentropic, determine (*a*) the temperature of the helium at the turbine inlet, (*b*) the coefficient of performance of the cycle, and (*c*) the net power input required for a mass flow rate of 0.3 kg/s.

10-86 An absorption refrigeration system is to remove heat from the refrigerated space at −10°C at a rate of 8 kW while operating in an environment at

25°C. Heat is to be supplied from a solar pond at 85°C. What is the minimum rate of heat supply required? *Answer:* 6.35 kW

10-87 It is proposed to run a thermoelectric generator in conjunction with a solar pond that can supply heat at a rate of 10^6 kJ/h at 80°C. The waste heat is to be rejected to the environment at 30°C. What is the maximum power this thermoelectric generator can produce?

10-88 A typical 200-m^2 house can be cooled adequately by a 3.5-ton air conditioner whose COP is 4.0. Determine the rate of heat gain of the house when the air conditioner is running continuously to maintain a constant temperature in the house.

10-89 Rooms with floor areas of up to 15-m^2 are cooled adequately by window air conditioners whose cooling capacity is 5000 Btu/h. Assuming the COP of the air conditioner to be 3.2, determine the rate of heat gain of the room, in Btu/h, when the air conditioner is running continuously to maintain a constant room temperature.

10-90 A heat pump water heater (HPWH) heats water by absorbing heat from the ambient air and transferring it to water. The heat pump has a COP of 2.2 and consumes 2 kW of electricity when running. Determine if this heat pump can be used to meet the cooling needs of a room most of the time for "free" by absorbing heat from the air in the room. The rate of heat gain of a room is usually less than 5000 kJ/h.

10-91 The vortex tube (also known as a Ranque or Hirsch tube) is a device that produces a refrigeration effect by expanding pressurized gas such as air in a tube (instead of a turbine as in the reversed Brayton cycle). It was invented and patented by Ranque in 1931 and improved by Hirsch in 1945, and is commercially available in various sizes.

The vortex tube is simply a straight circular tube equipped with a nozzle, as shown in the figure. The compressed gas at temperature T_1 and pressure P_1

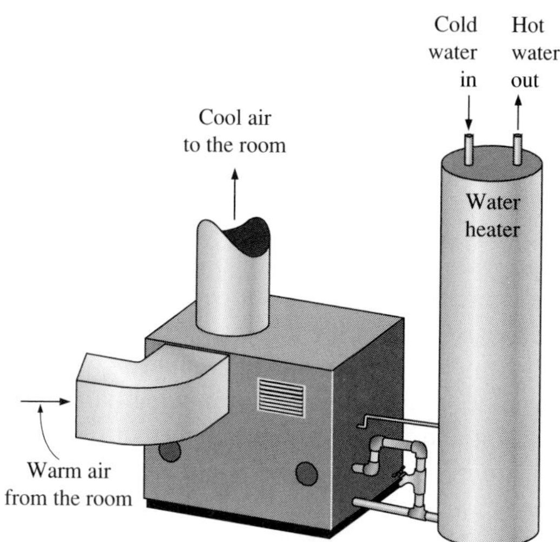

FIGURE P10-90

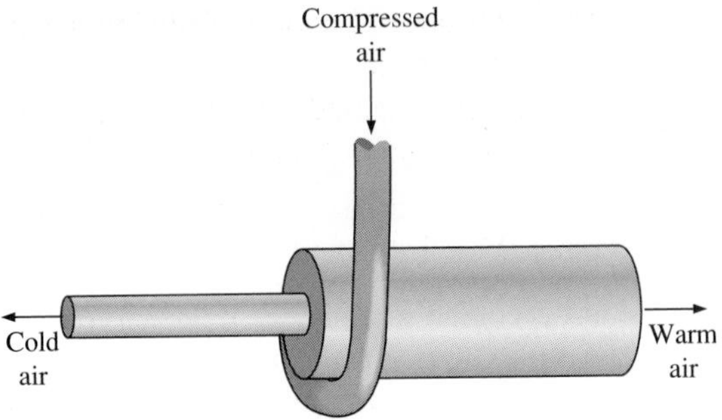

FIGURE P10-91

is accelerated in the nozzle by expanding it to nearly atmospheric pressure and is introduced into the tube tangentially at a very high (typically supersonic) velocity to produce a swirling motion (vortex) within the tube. The rotating gas is allowed to exit through the full-size tube that extends to the right, and the mass flow rate is controlled by a valve located about 30 diameters downstream. A smaller amount of air at the core region is allowed to escape to the left through a small aperture at the center. It is observed that the gas that is in the core region and escapes through the central aperture is cold while the gas that is in the peripheral region and escapes through the full-size tube is hot. If the temperature and the mass flow rate of the cold stream are T_c and $\dot{m}_c$, respectively, the rate of refrigeration in the vortex tube can be expressed as

$$\dot{Q}_{\text{refrig, vortex tube}} = \dot{m}_c(h_1 - h_c) = \dot{m}_c C_p(T_1 - T_c)$$

where C_p is the specific heat of the gas and $T_1 - T_c$ is the temperature drop of the gas in the vortex tube (the cooling effect). Temperature drops as high as 60°C (or 108°F) are obtained at high pressure ratios of about 10. The coefficient of performance of a vortex tube can be defined as the ratio of the refrigeration rate as given above to the power used to compress the gas. It ranges from about 0.1 to 0.15, which is well below the COPs of ordinary vapor compression refrigerators.

This interesting phenomenon can be explained as follows: the centrifugal force creates a radial pressure gradient in the vortex, and thus the gas at the periphery is pressurized and heated by the gas at the core region, which is cooled as a result. Also, energy is transferred from the inner layers toward the outer layers as the outer layers slow down the inner layers because of fluid viscosity that tends to produce a solid vortex. Both of these effects cause the energy and thus the temperature of the gas in the core region to decline. The conservation of energy requires the energy of the fluid at the outer layers to increase by an equivalent amount.

The vortex tube has no moving parts, and thus it is inherently reliable and durable. The ready availability of the compressed air at pressures up to 10 atm in most industrial facilities makes the vortex tube particularly attractive in such settings. Despite its low efficiency, the vortex tube has found application in small-scale industrial spot-cooling operations such as cooling of soldered parts or critical electronic components, cooling drinking water, and cooling the suits of workers in hot environments.

Consider a vortex tube that receives compressed air at 500 kPa and 300 K and supplies 25 percent of it as cold air at 100 kPa and 278 K. The ambient air is at 300 K and 100 kPa, and the compressor has an isentropic efficiency of 80 percent. The air suffers a pressure drop of 35 kPa in the aftercooler and the compressed air lines between the compressor and the vortex tube.

(*a*) Without performing any calculations, explain how the COP of the vortex tube would compare to the COP of an actual air refrigeration system based on the reversed Brayton cycle for the same pressure ratio. Also, compare the minimum temperatures that can be obtained by the two systems for the same inlet temperature and pressure.

(*b*) Assuming the vortex tube to be adiabatic and using specific heats at room temperature, determine the exit temperature of the hot fluid stream.

(*c*) Show, with calculations, that this process does not violate the second law of thermodynamics.

(*d*) Determine the coefficient of performance of this refrigeration system, and compare it to the COP of a Carnot refrigerator.

10-92 Repeat Prob. 10-91 for a pressure of 600 kPa at the vortex tube intake.

Computer, Design, and Essay Problems

10-93 Write a computer program to determine the effect of the evaporator pressure on the COP of an ideal vapor-compression refrigeration cycle. Assume the condenser pressure is kept constant at 1 MPa. Calculate the COP of the refrigeration cycle for the following evaporator pressures: 100, 120, 140, 160, 200, 280, 320, 400, and 500 kPa. Plot the COPs against the evaporator pressure.

10-94 Write a computer program to determine the effect of the condenser pressure on the COP of an ideal vapor-compression refrigeration cycle. Assume the evaporator pressure is maintained constant at 120 kPa. Calculate the COP of the refrigeration cycle for the following condenser pressures: 400, 500, 600, 700, 800, 900, 1000, and 1400 kPa. Plot the COPs against the condenser pressure.

10-95 Design a vapor-compression refrigeration system that will maintain the refrigerated space at $-15°C$ while operating in an environment at 20°C using refrigerant-134a as the working fluid.

10-96 Write an essay on air-, water-, and soil-based heat pumps. Discuss the advantages and the disadvantages of each system. For each system identify the conditions under which that system is preferable over the other two. In what situations would you not recommend a heat pump heating system?

10-97 Consider a solar pond power plant operating on a closed Rankine cycle. Using refrigerant-134a as the working fluid, specify the operating temperatures and pressures in the cycle, and estimate the required mass flow rate of refrigerant-134a for a net power output of 50 kW. Also, estimate the surface area of the pond for this level of continuous power production. Assume that the solar energy is incident on the pond at a rate of 500 W per m^2 of pond area at noontime, and that the pond is capable of storing 15 percent of the incident solar energy in the storage zone.

10-98 Design a thermoelectric refrigerator that is capable of cooling a canned drink in a car. The refrigerator is to be powered by the cigarette lighter of the car. Draw a sketch of your design. Semiconductor components for building thermoelectric power generators or refrigerators are available from several manufacturers. Using data from one of these manufacturers, determine how many of these components you need in your design, and estimate the coefficient of performance of your system. A critical problem in the design of thermoelectric refrigerators is the effective rejection of waste heat. Discuss how you can enhance the rate of heat rejection without using any devices with moving parts such as a fan.

10-99 It is proposed to use a solar-powered thermoelectric system installed on the roof to cool residential buildings. The system consists of a thermoelectric refrigerator that is powered by a thermoelectric power generator whose top surface is a solar collector. Discuss the feasibility and the cost of such a system, and determine if the proposed system installed on one side of the roof can meet a significant portion of the cooling requirements of a typical house in your area.

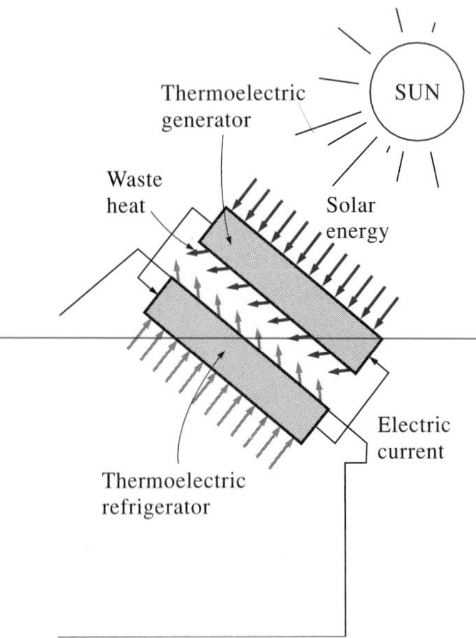

FIGURE P10-99

10-100 A refrigerator using R-12 as the working fluid keeps the refrigerated space at −15°C in an environment at 30°C. You are asked to redesign this refrigerator by replacing R-12 with the ozone-friendly R-134a. What changes in the pressure levels would you suggest in the new system? How do you think the COP of the new system will compare to the COP of the old system?

10-101 In the 1800s, before the development of modern air-conditioning, it was proposed to cool air for buildings with the following procedure using a large piston-cylinder device ["John Gorrie: Pioneer of Cooling and Ice Making," *ASHRAE Journal* 33, no. 1 (Jan. 1991)]:

1. Pull in a charge of outdoor air.
2. Compress it to a high pressure.
3. Cool the charge of air using outdoor air.
4. Expand it back to atmospheric pressure.
5. Discharge the charge of air into the space to be cooled.

Suppose the goal is to cool a room 6 m × 10 m × 2.5 m. Outdoor air is at 30°C, and it has been determined that 10 air changes per hour supplied to the room at 10°C could provide adequate cooling. Do a preliminary design of the system and do calculations to see if it would be feasible. (You may make optimistic assumptions for the analysis.)

(*a*) Sketch the system showing how you will drive it and how step 3 will be accomplished.

(*b*) Determine what pressure will be required (step 2).

(*c*) Estimate (guess) how long step 3 will take and what size will be needed for the piston-cylinder to provide the required air changes and temperature.

(*d*) Determine the work required in step 2 for one cycle and per hour.

(*e*) Discuss any problems you see with the concept of your design. (Include discussion of any changes that may be required to offset optimistic assumptions.)

10-102 Solar or photovoltaic (PV) cells convert sunlight to electricity and are commonly used to power calculators, satellites, remote communication systems, and even pumps. The conversion of light to electricity is called the *photoelectric effect*. It was first discovered in 1839 by Frenchman Edmond Becquerel, and the first PV module, which consisted of several cells connected to each other, was built in 1954 by Bell Laboratories. The PV modules today have conversion efficiencies of about 12 to 15 percent. Noting that the solar energy incident on a normal surface on earth at noontime is about 1000 W/m^2 during a clear day, PV modules on a 1-m^2 surface can provide as much as 150 W of electricity. The annual average daily solar energy incident on a horizontal surface in the United States ranges from about 2 to 6 kWh/m^2.

A PV-powered pump is to be used in Arizona to pump water for wildlife from a depth of 180 m at an average rate of 400 L/day. Assuming a reasonable efficiency for the pumping system, which can be defined as the ratio of the increase in the potential energy of the water to the electrical energy consumed by the pump, and taking the conversion efficiency of the PV cells to be 0.13 to be on the conservative side, determine the size of the PV module that needs to be installed, in m^2.

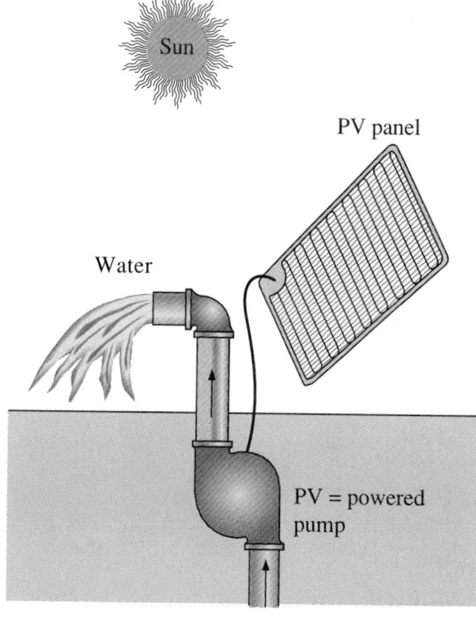

FIGURE P10-102

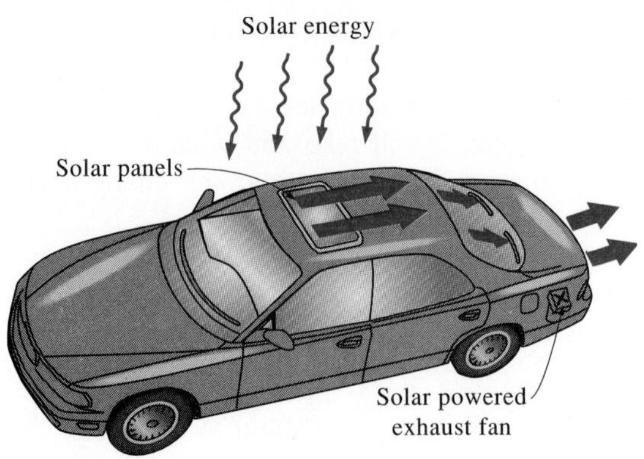

FIGURE P10-103

10-103 The temperature in a car parked in the sun can approach 100°C when the outside air temperature is just 25°C, and it is desirable to ventilate the parked car to avoid such high temperatures. However, the ventilating fans may run down the battery if they are powered by it. To avoid that happening, it is proposed to use the PV cells discussed in Prob. 10-102 to power the fans. It is determined that the air in the car should be replaced once every minute to avoid excessive rise in the interior temperature. Determine if this can be accomplished by installing PV cells on part of the roof of the car. Also, find out if any car is currently ventilated this way.

10-104 A company owns a refrigeration system whose refrigeration capacity is 200 tons (1 ton of refrigeration = 211 kJ/min), and you are to design a forced-air cooling system for fruits whose diameters do not exceed 7 cm under the following conditions: The fruits are to be cooled from 28°C to an average temperature of 8°C. The air temperature is to remain above −2°C and below 10°C at all times, and the velocity of air approaching the fruits must remain under 2 m/s. The cooling section can be as wide as 3.5 m and as high as 2 m.

Assuming reasonable values for the average fruit density, specific heat, and porosity (the fraction of air volume in a box), recommend reasonable values for (*a*) the air velocity approaching the cooling section, (*b*) the product-cooling capacity of the system, in kg · fruit/h, and (*c*) the volume flow rate of air.

Thermodynamic Property Relations

CHAPTER 11

In the preceding chapters we made extensive use of the property tables. We tend to take the property tables for granted, but thermodynamic laws and principles are of little use to engineers without them. In this chapter, we focus our attention on how the property tables are prepared and how some unknown properties can be determined from limited available data.

It will come as no surprise that some properties such as temperature, pressure, volume, and mass can be measured directly. Other properties such as density and specific volume can be determined from these using some simple relations. But properties such as internal energy, enthalpy, and entropy are not so easy to determine because they cannot be measured directly or related to easily measurable properties through some simple relations. Therefore, it is essential that we develop some fundamental relations between commonly encountered thermodynamic properties and express the properties that cannot be measured directly in terms of easily measurable properties.

By the nature of the material, this chapter makes extensive use of partial derivatives. Therefore, we start by reviewing them. Then we develop the Maxwell relations, which form the basis for many thermodynamic relations. Next we discuss the Clapeyron equation, which enables us to determine the enthalpy of vaporization from P, v, and T measurements alone, and we develop general relations for C_v, C_p, du, dh, and ds that are valid for all pure substances under all conditions. Then we discuss the Joule-Thomson coefficient, which is a measure of the temperature change with pressure during a throttling process. Finally, we develop a method of evaluating the Δh, Δu, and Δs of real gases through the use of generalized enthalpy and entropy departure charts.

11-1 ■ A LITTLE MATH—Partial Derivatives and Associated Relations

Many of the expressions developed in this chapter are based on the state postulate, which expresses that the state of a simple, compressible substance is completely specified by any two independent, intensive properties. All other properties at that state can be expressed in terms of those two properties. Mathematically speaking,

$$z = z\,(x, y)$$

where x and y are the two independent properties that fix the state and z represents any other property. Most basic thermodynamic relations involve differentials. Therefore, we start by reviewing the derivatives and various relations among derivatives to the extent necessary in this chapter.

Consider a function f that depends on a single variable x, that is, $f = f(x)$. Figure 11-1 shows such a function that starts out flat but gets rather steep as x increases. The steepness of the curve is a measure of the degree of dependence of f on x. In our case, the function f depends on x more strongly at larger x values. The steepness of a curve at a point is measured by the slope of a line tangent to the curve at that point, and it is equivalent to the **derivative** of the function at that point defined as

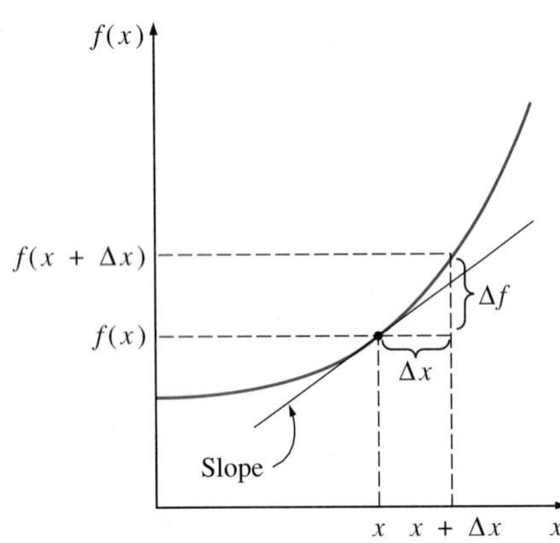

FIGURE 11-1
The derivative of a function at a specified point represents the slope of the function at that point.

$$\frac{df}{dx} = \lim_{\Delta x \to 0} \frac{\Delta f}{\Delta x} = \lim_{\Delta x \to 0} \frac{f(x + \Delta x) - f(x)}{\Delta x} \qquad (11\text{-}1)$$

Therefore, *the derivative of a function $f(x)$ with respect to x represents the rate of change of f with x.*

EXAMPLE 11-1 Approximating Differential Quantities by Differences

The C_p of ideal gases depends on temperature only, and it is expressed as $C_p(T) = dh(T)/dT$. Determine the C_p of air at 300 K, using the enthalpy data from Table A-17, and compare it to the value listed in Table A-2*b*.

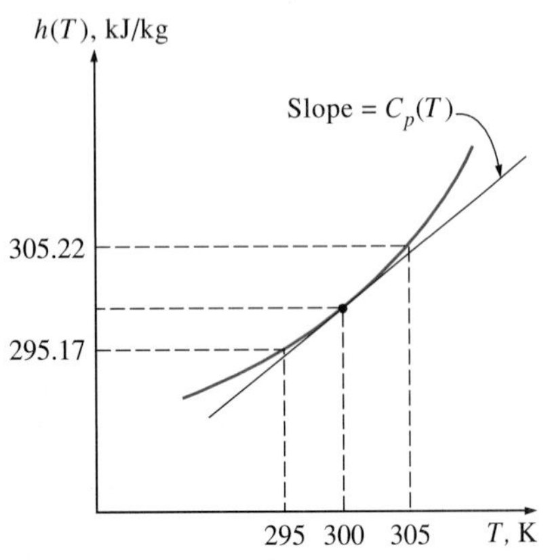

FIGURE 11-2
Schematic for Example 11-1.

Solution The C_p values of air at various temperatures are listed in Table A-2*b*. The listed value at 300 K is 1.005 kJ/(kg · K).

This value could also be determined by differentiating the function $h(T)$ with respect to T and evaluating the result at $T = 300$ K. However, the function $h(T)$ is not available. But we can still determine the C_p value approximately by replacing the differentials in the $C_p(T)$ relation by the corresponding differences in the neighborhood of the specified point (Fig. 11-2):

$$C_p(300\text{ K}) = \left[\frac{dh(T)}{dT}\right]_{T=300\text{ K}} \cong \left[\frac{\Delta h(T)}{\Delta T}\right]_{T\cong 300\text{ K}} = \frac{h(305\text{ K}) - h(295\text{ K})}{(305 - 295)\text{ K}}$$

$$= \frac{(305.22 - 295.17)\text{ kJ/kg}}{(305 - 295)\text{ K}} = \mathbf{1.005\ kJ/(kg \cdot K)}$$

which is identical to the listed value. Therefore, differential quantities can be viewed as differences. They can even be replaced by differences, whenever necessary, to obtain approximate results. The widely used finite difference numerical method is based on this simple principle.

Now consider a function that depends on two (or more) variables, such as $z = z\,(x, y)$. This time the value of z depends on both x and y. It is sometimes desirable to examine the dependence of z on only one of the variables. This is done by allowing one variable to change while holding the others constant and observing the change in the function. The variation of $z(x, y)$ with x when y is held constant is called the **partial derivative** of z with respect to x, and it is expressed as

$$\left(\frac{\partial z}{\partial x}\right)_y = \lim_{\Delta x \to 0}\left(\frac{\Delta z}{\Delta x}\right)_y = \lim_{\Delta x \to 0}\frac{z\,(x + \Delta x, y) - z\,(x, y)}{\Delta x} \qquad (11\text{-}2)$$

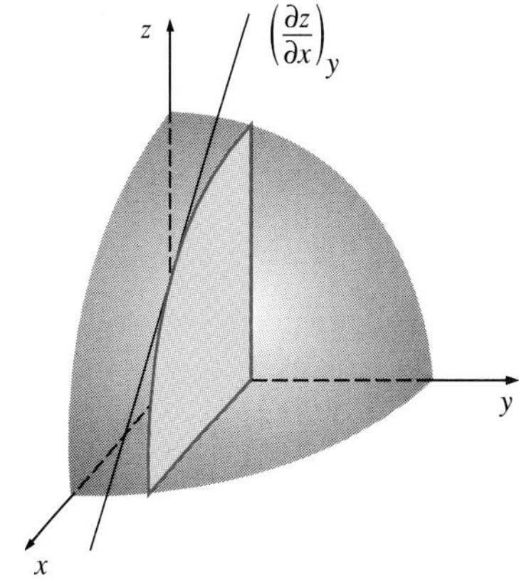

FIGURE 11-3

Geometric representation of partial derivative $(\partial z/\partial x)_y$.

This is illustrated in Fig. 11-3. The symbol ∂ represents differential changes, just like the symbol d. They differ in that the symbol d represents the *total* differential change of a function and reflects the influence of all variables, whereas ∂ represents the *partial* differential change due to the variation of a single variable.

Note that the changes indicated by d and ∂ are identical for independent variables, but not for dependent variables. For example, $(\partial x)_y = dx$ but $(\partial z)_y \neq dz$. [In our case, $dz = (\partial z)_x + (\partial z)_y$.] Also note that the value of the partial derivative $(\partial z/\partial x)_y$, in general, will be different at different y values.

To obtain a relation for the total differential change in $z\,(x, y)$ for simultaneous changes in x and y, consider a small portion of the surface $z\,(x, y)$ shown in Fig. 11-4. When the independent variables x and y change by Δx and Δy, respectively, the dependent variable z changes by Δz, which can be expressed as

$$\Delta z = z\,(x + \Delta x, y + \Delta y) - z\,(x, y)$$

Adding and subtracting $z\,(x, y + \Delta y)$, we get

$$\Delta z = z(x + \Delta x, y + \Delta y) - z(x, y + \Delta y) + z(x, y + \Delta y) - z(x, y)$$

or

$$\Delta z = \frac{z(x + \Delta x, y + \Delta y) - z(x, y + \Delta y)}{\Delta x}\Delta x + \frac{z(x, y + \Delta y) - z(x, y)}{\Delta y}\Delta y$$

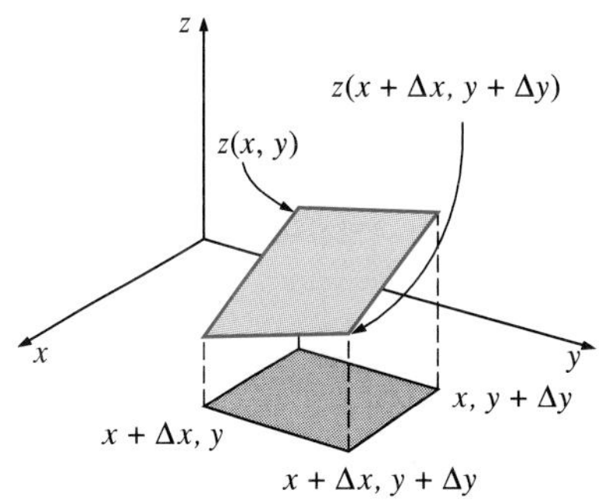

FIGURE 11-4

Geometric representation of total derivative dz for a function $z\,(x, y)$.

Taking the limits as $\Delta x \to 0$ and $\Delta y \to 0$ and using the definitions of partial derivatives, we obtain

$$dz = \left(\frac{\partial z}{\partial x}\right)_y dx + \left(\frac{\partial z}{\partial y}\right)_x dy \qquad (11\text{-}3)$$

Equation 11-3 is the fundamental relation for the **total differential** of a dependent variable in terms of its partial derivatives with respect to the independent variables. This relation can easily be extended to include more independent variables.

EXAMPLE 11-2 Total Differential versus Partial Differential

Consider an ideal gas at 300 K and 0.86 m^3/kg. The state of the gas changes to 302 K and 0.87 m^3/kg as a result of some disturbance. Using Eq. 11-3, estimate the change in the pressure of the gas.

Solution Strictly speaking, Eq. 11-3 is valid for differential changes in variables. But it can also be used with reasonable accuracy if these changes are small. The changes in T and v, respectively, can be expressed as

$$dT \cong \Delta T = (302 - 300)\text{ K} = 2\text{ K}$$

and

$$dv \cong \Delta v = (0.87 - 0.86)\text{ m}^3/\text{kg} = 0.01\text{ m}^3/\text{kg}$$

An ideal gas obeys the relation $Pv = RT$. Solving for P yields

$$P = \frac{RT}{v}$$

Note that R is a constant and $P = P(T, v)$. Applying Eq. 11-3 and using average values for T and v,

$$\begin{aligned} dP &= \left(\frac{\partial P}{\partial T}\right)_v dT + \left(\frac{\partial P}{\partial v}\right)_T dv \\ &= \frac{R\,dT}{v} - \frac{RT\,dv}{v^2} \\ &= [0.287\text{ kPa}\cdot\text{m}^3/(\text{kg}\cdot\text{K})]\left[\frac{2\text{ K}}{0.865\text{ m}^3/\text{kg}} - \frac{(301\text{ K})(0.01\text{ m}^2/\text{kg})}{(0.865\text{ m}^3/\text{kg})^2}\right] \\ &= 0.664\text{ kPa} - 1.155\text{ kPa} \\ &= \mathbf{-0.491\text{ kPa}} \end{aligned}$$

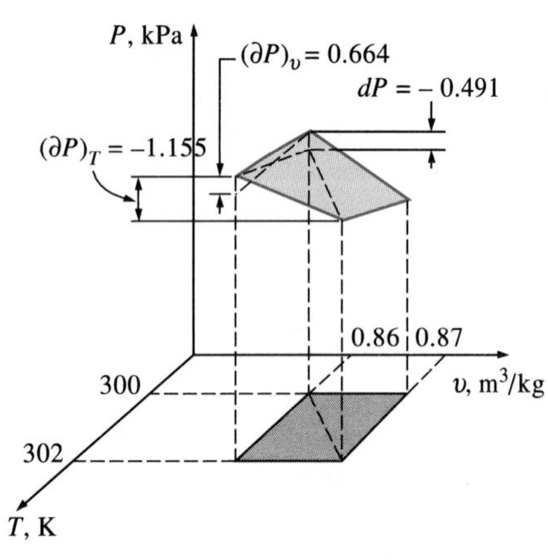

FIGURE 11-5

Geometric representation of the disturbance discussed in Example 11-2.

Therefore, the pressure will decrease by 0.491 kPa as a result of this disturbance. Notice that if the temperature had remained constant ($dT = 0$), the pressure would decrease by 1.155 kPa as a result of the 0.01 m^3/kg increase in specific volume. However, if the specific volume had remained constant ($dv = 0$), the pressure would increase by 0.664 kPa as a result of the 2-K rise in temperature (Fig. 11-5). That is,

$$\left(\frac{\partial P}{\partial T}\right)_v dT = (\partial P)_v = 0.664\text{ kPa}$$

$$\left(\frac{\partial P}{\partial v}\right)_T dv = (\partial P)_T = -1.155\text{ kPa}$$

and

$$dP = (\partial P)_v + (\partial P)_T = 0.664 - 1.155 = -0.491\text{ kPa}$$

Discussion Of course, we could have solved this problem easily (and exactly) by evaluating the pressure from the ideal-gas relation $P = RT/v$ at the final state (302 K and 0.87 m^3/kg) and the initial state (300 K and 0.86 m^3/kg) and taking their difference. This yields −0.491 kPa, which is exactly the value obtained above. Thus the small finite quantities (2 K, 0.01 m^3/kg) can be approximated as differential quantities with reasonable accuracy.

Now let us rewrite Eq. 11-3 as

$$dz = M\,dx + N\,dy \tag{11-4}$$

where

$$M = \left(\frac{\partial z}{\partial x}\right)_y \quad \text{and} \quad N = \left(\frac{\partial z}{\partial y}\right)_x$$

Taking the partial derivative of M with respect to y and of N with respect to x yields

$$\left(\frac{\partial M}{\partial y}\right)_x = \frac{\partial^2 z}{\partial x\,\partial y} \quad \text{and} \quad \left(\frac{\partial N}{\partial x}\right)_y = \frac{\partial^2 z}{\partial y\,\partial x}$$

The order of differentiation is immaterial for properties since they are continuous point functions and have exact differentials. Therefore, the two relations above are identical:

$$\left(\frac{\partial M}{\partial y}\right)_x = \left(\frac{\partial N}{\partial x}\right)_y \tag{11-5}$$

This is an important relation for partial derivatives, and it is used in calculus to test whether a differential dz is exact or inexact. In thermodynamics, this relation forms the basis for the development of the Maxwell relations discussed in the next section.

Finally, we develop two important relations for partial derivatives—the reciprocity and the cyclic relations. The function $z = z(x, y)$ can also be expressed as $x = x(y, z)$ if y and z are taken to be the independent variables. Then the total differential of x becomes, from Eq. 11-3,

$$dx = \left(\frac{\partial x}{\partial y}\right)_z dy + \left(\frac{\partial x}{\partial z}\right)_y dz \tag{11-6}$$

Eliminating dx by combining Eqs. 11-3 and 11-6, we have

$$dz = \left[\left(\frac{\partial z}{\partial x}\right)_y \left(\frac{\partial x}{\partial y}\right)_z + \left(\frac{\partial z}{\partial y}\right)_x\right] dy + \left(\frac{\partial x}{\partial z}\right)_y \left(\frac{\partial z}{\partial x}\right)_y dz$$

Rearranging,

$$\left[\left(\frac{\partial z}{\partial x}\right)_y \left(\frac{\partial x}{\partial y}\right)_z + \left(\frac{\partial z}{\partial y}\right)_x\right] dy = \left[1 - \left(\frac{\partial x}{\partial z}\right)_y \left(\frac{\partial z}{\partial x}\right)_y\right] dz \tag{11-7}$$

The variables y and z are independent of each other and thus can be varied independently. For example, y can be held constant ($dy = 0$), and z can be varied over a range of values ($dz \neq 0$). Therefore, for this equation to be valid at all times, the terms in the brackets must equal zero, regardless of the values of y and z. Setting the terms in each bracket equal to zero gives

$$\left(\frac{\partial x}{\partial z}\right)_y\left(\frac{\partial z}{\partial x}\right)_y = 1 \longrightarrow \left(\frac{\partial x}{\partial z}\right)_y = \frac{1}{(\partial z/\partial x)_y} \quad (11\text{-}8)$$

$$\left(\frac{\partial z}{\partial x}\right)_y\left(\frac{\partial x}{\partial y}\right)_z = -\left(\frac{\partial z}{\partial y}\right)_x \longrightarrow \left(\frac{\partial x}{\partial y}\right)_z\left(\frac{\partial y}{\partial z}\right)_x\left(\frac{\partial z}{\partial x}\right)_y = -1 \quad (11\text{-}9)$$

The first relation is called the **reciprocity relation,** and it shows that the inverse of a partial derivative is equal to its reciprocal (Fig. 11-6). The second relation is called the **cyclic relation,** and it is frequently used in thermodynamics (Fig. 11-7).

Function: $z + 2xy - 3y^2z = 0$

1) $z = \dfrac{2xy}{3y^2-1} \rightarrow \left(\dfrac{\partial z}{\partial x}\right)_y = \dfrac{2y}{3y^2-1}$

2) $x = \dfrac{3y^2z - z}{2y} \rightarrow \left(\dfrac{\partial x}{\partial z}\right)_y = \dfrac{3y^2-1}{2y}$

Thus, $\left(\dfrac{\partial z}{\partial x}\right)_y = \dfrac{1}{\left(\dfrac{\partial x}{\partial z}\right)_y}$

FIGURE 11-6

Demonstration of the reciprocity relation for the function $z + 2xy - 3y^2z = 0$.

FIGURE 11-7

Partial differentials are powerful tools that are supposed to make life easier, not harder.

EXAMPLE 11-3 Verification of Cyclic and Reciprocity Relations

Using the ideal-gas equation of state, verify (*a*) the cyclic relation and (*b*) the reciprocity relation at constant P.

Solution The ideal-gas equation of state $Pv = RT$ involves the three variables P, v, and T. Any two of these can be taken as the independent variables, with the remaining one being the dependent variable.

Analysis (*a*) Replacing x, y, and z in Eq. 11-9 by P, v, and T, respectively, we can express the cyclic relation for an ideal gas as

$$\left(\frac{\partial P}{\partial v}\right)_T\left(\frac{\partial v}{\partial T}\right)_P\left(\frac{\partial T}{\partial P}\right)_v = -1$$

where

$$P = P(v, T) = \frac{RT}{v} \longrightarrow \left(\frac{\partial P}{\partial v}\right)_T = -\frac{RT}{v^2}$$

$$v = v(P, T) = \frac{RT}{P} \longrightarrow \left(\frac{\partial v}{\partial T}\right)_P = \frac{R}{P}$$

$$T = T(P, v) = \frac{Pv}{R} \longrightarrow \left(\frac{\partial T}{\partial P}\right)_v = \frac{v}{R}$$

Substituting yields

$$\left(-\frac{RT}{v^2}\right)\left(\frac{R}{P}\right)\left(\frac{v}{R}\right) = -\frac{RT}{Pv} = -1$$

which is the desired result.

(*b*) The reciprocity rule for an ideal gas at P = constant can be expressed as

$$\left(\frac{\partial v}{\partial T}\right)_P = \frac{1}{(\partial T/\partial v)_P}$$

Performing the differentiations and substituting, we have

$$\frac{R}{P} = \frac{1}{P/R} \longrightarrow \frac{R}{P} = \frac{R}{P}$$

Thus the proof is complete.

The Maxwell Relations

The equations that relate the partial derivatives of properties P, v, T, and s of a simple compressible system to each other are called the *Maxwell relations.* They are obtained from the four Gibbs equations by exploiting the exactness of the differentials of thermodynamic properties.

Two of the Gibbs relations were derived in Chap. 6 and are expressed as

$$du = T\,ds - P\,dv \tag{11-10}$$

$$dh = T\,ds + v\,dP \tag{11-11}$$

The other two Gibbs relations are based on two new combination properties—the **Helmholtz function** a and the **Gibbs function** g, defined as

$$a = u - Ts \tag{11-12}$$

$$g = h - Ts \tag{11-13}$$

Differentiating, we get

$$da = du - T\,ds - s\,dT$$

$$dg = dh - T\,ds - s\,dT$$

Simplifying the above relations by using Eqs. 11-10 and 11-11, we obtain the other two Gibbs relations for simple compressible systems:

$$da = -s\,dT - P\,dv \tag{11-14}$$

$$dg = -s\,dT + v\,dP \tag{11-15}$$

A careful examination of the four Gibbs relations reveals that they are of the form

$$dz = M\,dx + N\,dy \tag{11-4}$$

with

$$\left(\frac{\partial M}{\partial y}\right)_x = \left(\frac{\partial N}{\partial x}\right)_y \tag{11-5}$$

since u, h, a, and g are properties and thus have exact differentials. Applying Eq. 11-5 to each of them, we obtain

$$\left(\frac{\partial T}{\partial v}\right)_s = -\left(\frac{\partial P}{\partial s}\right)_v \tag{11-16}$$

$$\left(\frac{\partial T}{\partial P}\right)_s = \left(\frac{\partial v}{\partial s}\right)_P \tag{11-17}$$

$$\left(\frac{\partial s}{\partial v}\right)_T = \left(\frac{\partial P}{\partial T}\right)_v \tag{11-18}$$

$$\left(\frac{\partial s}{\partial P}\right)_T = -\left(\frac{\partial v}{\partial T}\right)_P \tag{11-19}$$

These are called the **Maxwell relations** (Fig. 11-8). They are extremely valuable in thermodynamics because they provide a means of determining

FIGURE 11-8

Maxwell relations are extremely valuable in thermodynamic analysis.

$$\left(\frac{\partial T}{\partial v}\right)_s = -\left(\frac{\partial P}{\partial s}\right)_v$$

$$\left(\frac{\partial T}{\partial P}\right)_s = \left(\frac{\partial v}{\partial s}\right)_P$$

$$\left(\frac{\partial s}{\partial v}\right)_T = \left(\frac{\partial P}{\partial T}\right)_v$$

$$\left(\frac{\partial s}{\partial P}\right)_T = -\left(\frac{\partial v}{\partial T}\right)_P$$

the change in entropy, which cannot be measured directly, by simply measuring the changes in properties P, v, and T. Note that the Maxwell relations given above are limited to simple compressible systems. However, other similar relations can be written just as easily for nonsimple systems such as those involving electrical, magnetic, and other effects.

EXAMPLE 11-4 Verification of the Maxwell Relations

Verify the validity of the last Maxwell relation (Eq. 11-19) for steam at 250°C and 300 kPa.

Solution The last Maxwell relation states that for a simple compressible substance, the change in entropy with pressure at constant temperature is equal to the negative of the change in specific volume with temperature at constant pressure.

Analysis If we had explicit analytical relations for the entropy and specific volume of steam in terms of other properties, we could easily verify this by performing the indicated derivations. However, all we have for steam are tables of properties listed at certain intervals. Therefore, the only course we can take to solve this problem, without taking a trip to the library, is to replace the differential quantities in Eq. 11-19 with corresponding finite quantities, using property values from the tables (Table A-6 in this case) at or about the specified state.

$$\left(\frac{\partial s}{\partial P}\right)_T \stackrel{?}{=} -\left(\frac{\partial v}{\partial T}\right)_P$$

$$\left(\frac{\Delta s}{\Delta P}\right)_{T=250°C} \stackrel{?}{\cong} -\left(\frac{\Delta v}{\Delta T}\right)_{P=300\text{ kPa}}$$

$$\left[\frac{s_{400\text{ kPa}} - s_{200\text{ kPa}}}{(400-200)\text{ kPa}}\right]_{T=250°C} \stackrel{?}{\cong} -\left[\frac{v_{300°C} - v_{200°C}}{(300-200)°C}\right]_{P=300\text{ kPa}}$$

$$\frac{(7.3789 - 7.7086)\text{ kJ/(kg}\cdot\text{K)}}{(400-200)\text{ kPa}} \stackrel{?}{\cong} -\frac{(0.8753 - 0.7163)\text{ m}^3\text{/kg}}{(300-200)°C}$$

$$-0.00165\text{ m}^3\text{/(kg}\cdot\text{K)} \cong -0.00159\text{ m}^3\text{/(kg}\cdot\text{K)}$$

since kJ = kPa · m^3 and K ≅ °C for temperature differences. The two values are within 4 percent of each other. This difference is due to replacing the differential quantities by relatively large finite quantities. Based on the close agreement between the two values, the steam seems to satisfy Eq. 11-19 at the specified state.

Discussion This example shows that the entropy change of a simple compressible system during an isothermal process can be determined from a knowledge of the easily measurable properties P, v, and T alone.

11-3 ■ THE CLAPEYRON EQUATION

The Maxwell relations have far-reaching implications in thermodynamics and are frequently used to derive useful thermodynamic relations. The Clapeyron equation is one such relation, and it enables us to determine the enthalpy change associated with a phase change (such as the enthalpy of vaporization h_{fg}) from a knowledge of P, v, and T data alone.

Consider the third Maxwell relation, Eq. 11-18:

$$\left(\frac{\partial P}{\partial T}\right)_v = \left(\frac{\partial s}{\partial v}\right)_T$$

During a phase-change process, the pressure is the saturation pressure, which depends on the temperature only and is independent of the specific volume. That is, $P_{sat} = f(T_{sat})$. Therefore, the partial derivative $(\partial P/\partial T)_v$ can be expressed as a total derivative $(dP/dT)_{sat}$, which is the slope of the saturation curve on a *P-T* diagram at a specified saturation state (Fig. 11-9). This slope is independent of the specific volume, and thus it can be treated as a constant during the integration of Eq. 11-18 between two saturation states at the same temperature. For an isothermal liquid–vapor phase-change process, for example, the integration yields

$$s_g - s_f = \left(\frac{dP}{dT}\right)_{sat} (v_g - v_f) \qquad (11\text{-}20)$$

or

$$\left(\frac{dP}{dT}\right)_{sat} = \frac{s_{fg}}{v_{fg}} \qquad (11\text{-}21)$$

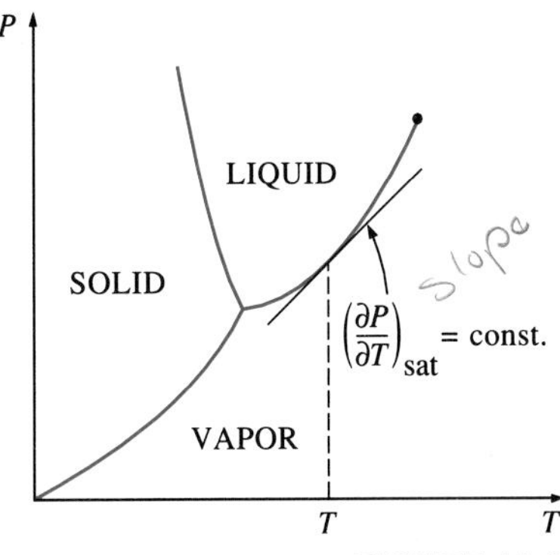

FIGURE 11-9

The slope of the saturation curve on a *P-T* diagram is constant at a constant *T* or *P*.

During this process the pressure also remains constant. Therefore, from Eq. 11-11,

$$dh = T\,ds + v\,dP^{\nearrow 0} \longrightarrow \int_f^g dh = \int_f^g T\,ds \longrightarrow h_{fg} = Ts_{fg}$$

Substituting this result into Eq. 11-21, we obtain

$$\left(\frac{dP}{dT}\right)_{sat} = \frac{h_{fg}}{Tv_{fg}} \qquad (11\text{-}22)$$

which is called the **Clapeyron equation** after the French engineer and physicist E. Clapeyron (1799–1864). This is an important thermodynamic relation since it enables us to determine the enthalpy of vaporization h_{fg} at a given temperature by simply measuring the slope of the saturation curve on a *P-T* diagram and the specific volume of saturated liquid and saturated vapor at the given temperature.

The Clapeyron equation is applicable to any phase-change process that occurs at constant temperature and pressure. It can be expressed in a general form as

$$\left(\frac{dP}{dT}\right)_{sat} = \frac{h_{12}}{Tv_{12}} \qquad (11\text{-}23)$$

where the subscripts 1 and 2 indicate the two phases.

EXAMPLE 11-5 Evaluating the h_{fg} of a Substance from the *P-v-T* Data

Using the Clapeyron equation, estimate the value of the enthalpy of vaporization of refrigerant-134a at 20°C, and compare it with the tabulated value.

Solution From Eq. 11-22,

$$h_{fg} = Tv_{fg}\left(\frac{dP}{dt}\right)_{sat}$$

where, from Table A-11,

$$v_{fg} = (v_g - v_f)_{@\,20°C} = 0.0358 - 0.0008157 = 0.0350 \text{ m}^3/\text{kg}$$

$$\left(\frac{dP}{dT}\right)_{sat,\,20°C} \cong \left(\frac{\Delta P}{\Delta T}\right)_{sat,\,20°C} = \frac{P_{sat\,@\,24°C} - P_{sat\,@\,16°C}}{24°C - 16°C}$$

$$= \frac{645.66 - 504.16 \text{ kPa}}{8°C} = 17.69 \text{ kPa/K}$$

since $\Delta T(°C) \equiv \Delta T(K)$. Substituting, we get

$$h_{fg} = (293.15)(0.0350 \text{ m}^3/\text{kg})(17.69 \text{ kPa/K})\left(\frac{1 \text{ kJ}}{1 \text{ kPa}\cdot\text{m}^3}\right)$$

$$= \mathbf{181.50 \text{ kJ/kg}}$$

The tabulated value of h_{fg} at 20°C is 181.09 kJ/kg. The small difference between the two values is due to the approximation used in determining the slope of the saturation curve at 20°C.

The Clapeyron equation can be simplified for liquid–vapor and solid–vapor phase changes by utilizing some approximations. At low pressures $v_g \gg v_f$, and thus $v_{fg} \cong v_g$. By treating the vapor as an ideal gas, we have $v_g = RT/P$. Substituting these approximations into Eq. 11-22, we find

$$\left(\frac{dP}{dT}\right)_{sat} = \frac{Ph_{fg}}{RT^2}$$

or

$$\left(\frac{dP}{P}\right)_{sat} = \frac{h_{fg}}{R}\left(\frac{dT}{T^2}\right)_{sat}$$

For small temperature intervals h_{fg} can be treated as a constant at some average value. Then integrating this equation between two saturation states yields

$$\ln\left(\frac{P_2}{P_1}\right)_{sat} \cong \frac{h_{fg}}{R}\left(\frac{1}{T_1} - \frac{1}{T_2}\right)_{sat} \qquad (11\text{-}24)$$

This equation is called the **Clapeyron–Clausius equation,** and it can be used to determine the variation of saturation pressure with temperature. It can also be used in the solid–vapor region by replacing h_{fg} by h_{ig} (the enthalpy of sublimation) of the substance.

EXAMPLE 11-6 Extrapolating Tabular Data with the Clapeyron Equation

Estimate the saturation pressure of refrigerant-134a at −50°F, using the data available in the refrigerant tables.

Solution Table A-11E lists saturation data at temperatures −40°F and above. Therefore, we should either resort to other sources or use extrapolation to obtain saturation data at lower temperatures. Equation 11-24 provides an intelligent way to extrapolate:

$$\ln\left(\frac{P_2}{P_1}\right)_{sat} \cong \frac{h_{fg}}{R}\left(\frac{1}{T_1} - \frac{1}{T_2}\right)_{sat}$$

In our case $T_1 = -40°F$ and $T_2 = -50°F$. For refrigerant-134a, $R = 0.01946$ Btu/(lbm · R). Also from Table A-11E at −40°F, we read $h_{fg} = 95.82$ Btu/lbm and $P_1 = P_{sat\,@\,-40°F} = 7.490$ psia. Substituting these values into Eq. 11-24 gives

$$\ln\left(\frac{P_2}{7.490 \text{ psia}}\right) \cong \frac{95.82 \text{ Btu/lbm}}{0.01946 \text{ Btu/(lbm}\cdot\text{R)}}\left(\frac{1}{420 \text{ R}} - \frac{1}{410 \text{ R}}\right)$$

$$P_2 \cong \mathbf{5.63 \text{ psia}}$$

Therefore, according to Eq. 11-24, the saturation pressure of refrigerant-134a at −50°F is 5.63 psia. The actual value, obtained from another source, is 5.505 psia. Thus the value predicted by Eq. 11-24 is in error by about 2 percent, which is quite acceptable for most purposes. (If we had used linear extrapolation instead, we would have obtained 5.06 psia, which is in error by 8 percent.)

11-4 ■ GENERAL RELATIONS FOR du, dh, ds, C_v, AND C_p

The state postulate established that the state of a simple compressible system is completely specified by two independent, intensive properties. Therefore, at least theoretically, we should be able to calculate all the properties of a system at any state once two independent, intensive properties are available. This is certainly good news for properties that cannot be measured directly such as internal energy, enthalpy, and entropy. But the calculation of these properties from measurable ones depends on the availability of simple and accurate relations between the two groups.

In this section we develop general relations for changes in internal energy, enthalpy, and entropy in terms of pressure, specific volume, temperature, and specific heats alone. We also develop some general relations involving specific heats. The relations developed will enable us to determine the *changes* in these properties. The property values at specified states can be determined only after the selection of a reference state, the choice of which is quite arbitrary.

1 Internal Energy Changes

We choose the internal energy to be a function of T and v; that is, $u = u(T, v)$ and take its total differential (Eq. 11-3),

$$du = \left(\frac{\partial u}{\partial T}\right)_v dT + \left(\frac{\partial u}{\partial v}\right)_T dv$$

Using the definition of C_v, we have

$$du = C_v\,dT + \left(\frac{\partial u}{\partial v}\right)_T dv \qquad (11\text{-}25)$$

Now we choose the entropy to be a function of T and v; that is, $s = s(T, v)$ and take its total differential,

$$ds = \left(\frac{\partial s}{\partial T}\right)_v dT + \left(\frac{\partial s}{\partial v}\right)_T dv \qquad (11\text{-}26)$$

Substituting this into the $T\,ds$ relation $du = T\,ds - P\,dv$ yields

$$du = T\left(\frac{\partial s}{\partial T}\right)_v dT + \left[T\left(\frac{\partial s}{\partial v}\right)_T - P\right] dv \qquad (11\text{-}27)$$

Equating the coefficients of dT and dv in Eqs. 11-25 and 11-27 gives

$$\left(\frac{\partial s}{\partial T}\right)_v = \frac{C_v}{T}$$

$$\left(\frac{\partial u}{\partial v}\right)_T = T\left(\frac{\partial s}{\partial v}\right)_T - P \qquad (11\text{-}28)$$

Using the third Maxwell relation (Eq. 11-18), we get

$$\left(\frac{\partial u}{\partial v}\right)_T = T\left(\frac{\partial P}{\partial T}\right)_v - P$$

Substituting this into Eq. 11-25, we obtain the desired relation for du:

$$du = C_v\,dT + \left[T\left(\frac{\partial P}{\partial T}\right)_v - P\right] dv \qquad (11\text{-}29)$$

The change in internal energy of a simple compressible system associated with a change of state from (T_1, v_1) to (T_2, v_2) is determined by integration:

$$u_2 - u_1 = \int_{T_1}^{T_2} C_v\,dT + \int_{v_1}^{v_2} \left[T\left(\frac{\partial P}{\partial T}\right)_v - P\right] dv \qquad (11\text{-}30)$$

2 Enthalpy Changes

The general relation for dh is determined in exactly the same manner. This time we choose the enthalpy to be a function of T and P, that is, $h = h(T, P)$, and take its total differential,

$$dh = \left(\frac{\partial h}{\partial T}\right)_P dT + \left(\frac{\partial h}{\partial P}\right)_T dP$$

Using the definition of C_p, we have

$$dh = C_p\,dT + \left(\frac{\partial h}{\partial P}\right)_T dP \qquad (11\text{-}31)$$

Now we choose the entropy to be a function of T and P; that is, we take $s = s(T, P)$ and take its total differential,

$$ds = \left(\frac{\partial s}{\partial T}\right)_P dT + \left(\frac{\partial s}{\partial P}\right)_T dP \tag{11-32}$$

Substituting this into the $T\,ds$ relation $dh = T\,ds + v\,dP$ gives

$$dh = T\left(\frac{\partial s}{\partial T}\right)_P dT + \left[v + T\left(\frac{\partial s}{\partial P}\right)_T\right] dP \tag{11-33}$$

Equating the coefficients of dT and dP in Eqs. 11-31 and 11-33, we obtain

$$\left(\frac{\partial s}{\partial T}\right)_P = \frac{C_P}{T}$$
$$\left(\frac{\partial h}{\partial P}\right)_T = v + T\left(\frac{\partial s}{\partial P}\right)_T \tag{11-34}$$

Using the fourth Maxwell relation (Eq. 11-19), we have

$$\left(\frac{\partial h}{\partial P}\right)_T = v - T\left(\frac{\partial v}{\partial T}\right)_P$$

Substituting this into Eq. 11-31, we obtain the desired relation for dh:

$$dh = C_p\,dT + \left[v - T\left(\frac{\partial v}{\partial T}\right)_P\right] dP \tag{11-35}$$

The change in enthalpy of a simple compressible system associated with a change of state from (T_1, P_1) to (T_2, P_2) is determined by integration:

$$h_2 - h_1 = \int_{T_1}^{T_2} C_p\,dT + \int_{P_1}^{P_2}\left[v - T\left(\frac{\partial v}{\partial T}\right)_P\right] dP \tag{11-36}$$

In reality, one needs only to determine either $u_2 - u_1$ from Eq. 11-30 or $h_2 - h_1$ from Eq. 11-36, depending on which is more suitable to the data at hand. The other can easily be determined by using the definition of enthalpy $h = u + Pv$:

$$h_2 - h_1 = u_2 - u_1 + (P_2v_2 - P_1v_1) \tag{11-37}$$

3 Entropy Changes

Below we develop two general relations for the entropy change of a simple compressible system.

The first relation is obtained by replacing the first partial derivative in the total differential ds (Eq. 11-26) by Eq. 11-28 and the second partial derivative by the third Maxwell relation (Eq. 11-18), yielding

$$ds = \frac{C_v}{T}\,dT + \left(\frac{\partial P}{\partial T}\right)_v dv \tag{11-38}$$

and

$$s_2 - s_1 = \int_{T_1}^{T_2} \frac{C_v}{T}\, dT + \int_{v_1}^{v_2} \left(\frac{\partial P}{\partial T}\right)_v dv \tag{11-39}$$

The second relation is obtained by replacing the first partial derivative in the total differential of ds (Eq. 11-32) by Eq. 11-34, and the second partial derivative by the fourth Maxwell relation (Eq. 11-19), yielding

$$ds = \frac{C_p}{T}\, dT - \left(\frac{\partial v}{\partial T}\right)_P dP \tag{11-40}$$

and

$$s_2 - s_1 = \int_{T_1}^{T_2} \frac{C_p}{T}\, dT - \int_{P_1}^{P_2} \left(\frac{\partial v}{\partial T}\right)_P dP \tag{11-41}$$

Either relation can be used to determine the entropy change. The proper choice will depend on the available data.

4 Specific Heats C_v and C_p

We mentioned in Chap. 3 that the specific heats of an ideal gas depend on temperature only. For a general pure substance, however, the specific heats depend on specific volume or pressure as well as the temperature. Below we develop some general relations to relate the specific heats of a substance to pressure, specific volume, and temperature.

At low pressures gases behave as ideal gases, and their specific heats essentially depend on temperature only. These specific heats are called *zero pressure,* or *ideal-gas, specific heats* (denoted C_{v0} and C_{p0}), and they are relatively easier to determine. Thus it is desirable to have some general relations that will enable us to calculate the specific heats at higher pressures (or lower specific volumes) from a knowledge of C_{v0} or C_{p0} and the P-v-T behavior of the substance. Such relations are obtained by applying the test of exactness (Eq. 11-5) on Eqs. 11-38 and 11-40, which yields

$$\left(\frac{\partial C_v}{\partial v}\right)_T = T\left(\frac{\partial^2 P}{\partial^2 T}\right)_v \tag{11-42}$$

and

$$\left(\frac{\partial C_p}{\partial P}\right)_T = -T\left(\frac{\partial^2 v}{\partial^2 T^2}\right)_P \tag{11-43}$$

The deviation of C_p from C_{p0} with increasing pressure, for example, is determined by integrating Eq. 11-43 from zero pressure to any pressure P along an isothermal path:

$$(C_p - C_{p0})_T = -T \int_0^P \left(\frac{\partial^2 v}{\partial T^2}\right)_P dP \qquad (11\text{-}44)$$

The integration on the right-hand side requires a knowledge of the P-v-T behavior of the substance alone. The notation indicates that v should be differentiated twice with respect to T while P is held constant. The resulting expression should be integrated with respect to P while T is held constant.

Another desirable general relation involving specific heats is one that relates the two specific heats C_P and C_v. The advantage of such a relation is obvious: We will need to determine only one specific heat (usually C_p) and calculate the other one using that relation and the P-v-T data of the substance. We start the development of such a relation by equating the two ds relations (Eqs. 11-38 and 11-40) and solving for dT:

$$dT = \frac{T(\partial P/\partial T)_v}{C_p - C_v}\,dv + \frac{T(\partial v/\partial T)_P}{C_p - C_v}\,dP$$

Choosing $T = T(v, P)$ and differentiating, we get

$$dT = \left(\frac{\partial T}{\partial v}\right)_P dv + \left(\frac{\partial T}{\partial P}\right)_v dP$$

Equating the coefficient of either dv or dP of the above two equations gives the desired result:

$$C_p - C_v = T\left(\frac{\partial v}{\partial T}\right)_P \left(\frac{\partial P}{\partial T}\right)_v \qquad (11\text{-}45)$$

An alternative form of this relation is obtained by using the cyclic relation:

$$\left(\frac{\partial P}{\partial T}\right)_v \left(\frac{\partial T}{\partial v}\right)_P \left(\frac{\partial v}{\partial P}\right)_T = -1 \longrightarrow \left(\frac{\partial P}{\partial T}\right)_v = -\left(\frac{\partial v}{\partial T}\right)_P \left(\frac{\partial P}{\partial v}\right)_T$$

Substituting the result into Eq. 11-45 gives

$$C_p - C_v = -T\left(\frac{\partial v}{\partial T}\right)_P^2 \left(\frac{\partial P}{\partial v}\right)_T \qquad (11\text{-}46)$$

This relation can be expressed in terms of two other thermodynamic properties called the **volume expansivity** β and the **isothermal compressibility** α, which are defined as (Fig. 11-10)

$$\beta = \frac{1}{v}\left(\frac{\partial v}{\partial T}\right)_P \qquad (11\text{-}47)$$

and

$$\alpha = -\frac{1}{v}\left(\frac{\partial v}{\partial P}\right)_T \qquad (11\text{-}48)$$

Substituting these two relations into Eq. 11-46, we obtain a third general relation for $C_p - C_v$:

FIGURE 11-10
The volume expansivity (also called the *coefficient of volumetric expansion*) is a measure of the change in volume with temperature at constant pressure.

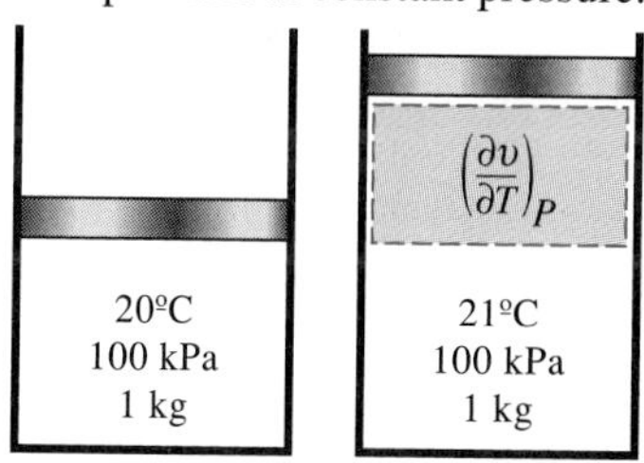

(*a*) A substance with a large β

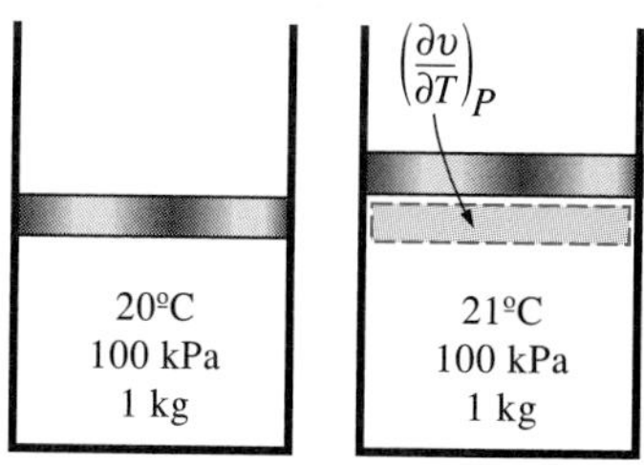

(*b*) A substance with a small β

$$C_p - C_v = \frac{vT\beta^2}{\alpha} \tag{11-49}$$

It is called the *Mayer relation* in honor of the German physician and physicist J. R. Mayer (1814–1878). We can draw several conclusions from this equation:

1. The isothermal compressibility α is a positive quantity for all substances in all phases. The volume expansivity could be negative for some substances (such as liquid water below 4°C), but its square is always positive or zero. The temperature T in this relation is absolute temperature, which is also positive. Therefore we conclude that *the constant-pressure specific heat is always greater than or equal to the constant-volume specific heat*:

$$C_p \geq C_v \tag{11-50}$$

2. The difference between C_p and C_v approaches zero as the absolute temperature approaches zero.

3. The two specific heats are identical for truly incompressible substances since $v =$ constant. The difference between the two specific heats is very small and is usually disregarded for substances that are *almost* incompressible, such as liquids and solids.

EXAMPLE 11-7 Internal Energy Change of a van der Waals Gas

Derive a relation for the internal energy change of a gas that obeys the van der Waals equation of state. Assume that in the range of interest C_v varies according to the relation $C_v = c_1 + c_2 T$, where c_1 and c_2 are constants.

Solution The change in internal energy of any simple compressible system in any phase during any process can be determined from Eq. 11-30:

$$u_2 - u_1 = \int_{T_1}^{T_2} C_v\, dT + \int_{v_1}^{v_2} \left[T\left(\frac{\partial P}{\partial T}\right)_v - P \right] dv$$

The van der Waals equation of state was discussed in Chap. 2. It can be expressed as

$$P = \frac{RT}{v-b} - \frac{a}{v^2}$$

Then

$$\left(\frac{\partial P}{\partial T}\right)_v = \frac{R}{v-b}$$

Thus,

$$T\left(\frac{\partial P}{\partial T}\right)_v - P = \frac{RT}{v-b} - \frac{RT}{v-b} + \frac{a}{v^2} = \frac{a}{v^2}$$

Substituting gives

$$u_2 - u_1 = \int_{T_1}^{T_2} (c_1 + c_2 T)\, dT + \int_{v_1}^{v_2} \frac{a}{v^2}\, dv$$

Integrating yields

$$u_2 - u_1 = c_1 (T_2 - T_1) + \frac{c_2}{2}(T_2^2 - T_1^2) + a\left(\frac{1}{v_1} - \frac{1}{v_2}\right)$$

which is the desired relation.

EXAMPLE 11-8 Internal Energy as a Function of Temperature Alone

Show that the internal energy of (*a*) an ideal gas and (*b*) an incompressible substance is a function of temperature only, $u = u(T)$.

Solution The differential change in the internal energy of a general simple compressible system is given by Eq. 11-29 as

$$du = C_v\, dT + \left[T\left(\frac{\partial P}{\partial T}\right)_v - P\right] dv$$

Analysis (*a*) For an ideal gas $Pv = RT$. Then

$$T\left(\frac{\partial P}{\partial T}\right)_v - P = T\left(\frac{R}{v}\right) - P = P - P = 0$$

Thus, $$du = C_v\, dT$$

To complete the proof, we need to show that C_v is not a function of v either. This is done with the help of Eq. 11-42:

$$\left(\frac{\partial C_v}{\partial v}\right)_T = T\left(\frac{\partial^2 P}{\partial T^2}\right)_v$$

For an ideal gas $P = RT/v$. Then

$$\left(\frac{\partial P}{\partial T}\right)_v = \frac{R}{v} \quad \text{and} \quad \left(\frac{\partial^2 P}{\partial T^2}\right)_v = \left[\frac{\partial (R/v)}{\partial T}\right]_v = 0$$

Thus, $$\left(\frac{\partial C_v}{\partial v}\right)_T = 0$$

which states that C_v does not change with specific volume. That is, C_v is not a function of specific volume either. Therefore we conclude that the internal energy of an ideal gas is a function of temperature only (Fig. 11-11).

(*b*) For an incompressible substance, v = constant and thus $dv = 0$. Also from Eq. 11-49, $C_p = C_v = C$ since $\alpha = \beta = 0$ for incompressible substances. Then Eq. 11-29 reduces to

$$du = C\, dT$$

Again we need to show that the specific heat C depends on temperature only and not on the pressure or the specific volume. This is easily done with the help of Eq. 11-43:

$$\left(\frac{\partial C_p}{\partial P}\right)_T = -T\left(\frac{\partial^2 v}{\partial T^2}\right)_P = 0$$

since v = constant. Therefore we conclude that the internal energy of a truly incompressible substance depends on temperature only.

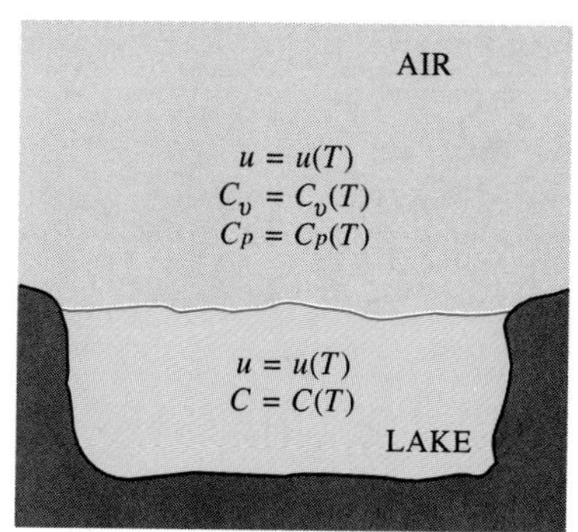

FIGURE 11-11

The internal energies and specific heats of ideal gases and incompressible substances depend on temperature only.

EXAMPLE 11-9 The Specific Heat Difference of an Ideal Gas

Show that $C_p - C_v = R$ for an ideal gas.

Solution This relation is easily proved by showing that the right-hand side of Eq. 11-46 is equivalent to the gas constant R of the ideal gas:

$$C_p - C_v = -T\left(\frac{\partial v}{\partial T}\right)_P^2 \left(\frac{\partial P}{\partial v}\right)_T$$

$$P = \frac{RT}{v} \longrightarrow \left(\frac{\partial P}{\partial v}\right)_T = -\frac{RT}{v^2} = -\frac{P}{v}$$

$$v = \frac{RT}{P} \longrightarrow \left(\frac{\partial v}{\partial T}\right)_P^2 = \left(\frac{R}{P}\right)^2$$

Substituting,

$$-T\left(\frac{\partial v}{\partial T}\right)_P^2 \left(\frac{\partial P}{\partial v}\right)_T = -T\left(\frac{R}{P}\right)^2\left(-\frac{P}{v}\right) = R$$

Therefore,

$$C_p - C_v = R$$

11-5 ■ THE JOULE-THOMSON COEFFICIENT

$T_1 = 20°C$ $T_2\{\geq = \leq\} 20°C$
$P_1 = 800$ kPa $P_2 = 200$ kPa

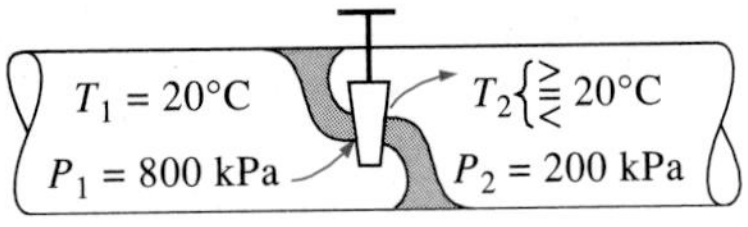

FIGURE 11-12
The temperature of a fluid may increase, decrease, or remain constant during a throttling process.

When a fluid passes through a restriction such as a porous plug, a capillary tube, or an ordinary valve, its pressure decreases. As we have shown in Chap. 4, the enthalpy of the fluid remains approximately constant during such a throttling process. You will remember that a fluid may experience a large drop in its temperature as a result of throttling, which forms the basis of operation for refrigerators and air conditioners. This is not always the case, however. The temperature of the fluid may remain unchanged, or it may even increase during a throttling process (Fig. 11-12).

The temperature behavior of a fluid during a throttling (h = constant) process is described by the **Joule-Thomson coefficient,** defined as

$$\mu = \left(\frac{\partial T}{\partial P}\right)_h \tag{11-51}$$

Thus the Joule-Thomson coefficient is a measure of the change in temperature with pressure during a constant-enthalpy process. Notice that if

$$\mu_{JT} \begin{cases} < 0 & \text{temperature increases} \\ = 0 & \text{temperature remains constant} \\ > 0 & \text{temperature decreases} \end{cases}$$

during a throttling process.

A careful look at its defining equation reveals that the Joule-Thomson coefficient represents the slope of h = constant lines on a T-P diagram. Such diagrams can be easily constructed from temperature and pressure measurements alone during throttling processes. A fluid at a fixed temperature and pressure T_1 and P_1 (thus fixed enthalpy) is forced to flow through a porous plug, and its temperature and pressure downstream (T_2 and P_2) are measured. The experiment is repeated for different sizes of porous plugs, each giving a different set of T_2 and P_2. Plotting the temperatures against the pressures gives us an h = constant line on a T-P diagram, as shown in Fig. 11-13. Repeating the experiment for different sets of inlet pressure and temperature and plotting the results, we can construct a T-P diagram for a substance with several h = constant lines, as shown in Fig. 11-14.

Some constant-enthalpy lines on the T-P diagram pass through a point of zero slope or zero Joule-Thomson coefficient. The line that passes through these points is called the **inversion line,** and the temperature at a point where a constant-enthalpy line intersects the inversion line is called the **inversion temperature.** The temperature at the intersection of the $P = 0$ line (ordinate) and the upper part of the inversion line is called the **maximum inversion temperature.** Notice that the slopes of the h = constant lines are negative ($\mu_{JT} < 0$) at states to the right of the inversion line and positive ($\mu_{JT} > 0$) to the left of the inversion line.

A throttling process proceeds along a constant-enthalpy line in the direction of decreasing pressure, that is, from right to left. Therefore, the temperature of a fluid will increase during a throttling process that takes place on the right-hand side of the inversion line. However, the fluid temperature will decrease during a throttling process that takes place on the left-hand side of the inversion line. It is clear from this diagram that a cooling effect cannot be achieved by throttling unless the fluid is below its maximum inversion temperature. This presents a problem for substances whose maximum inversion temperature is well below room temperature. For hydrogen, for example, the maximum inversion temperature is $-68°C$. Thus hydrogen must be cooled below this temperature if any further cooling is to be achieved by throttling.

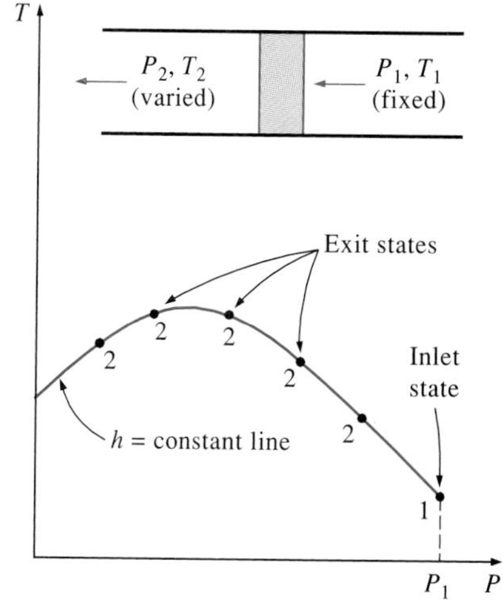

FIGURE 11-13
The development of an h = constant line on a P-T diagram.

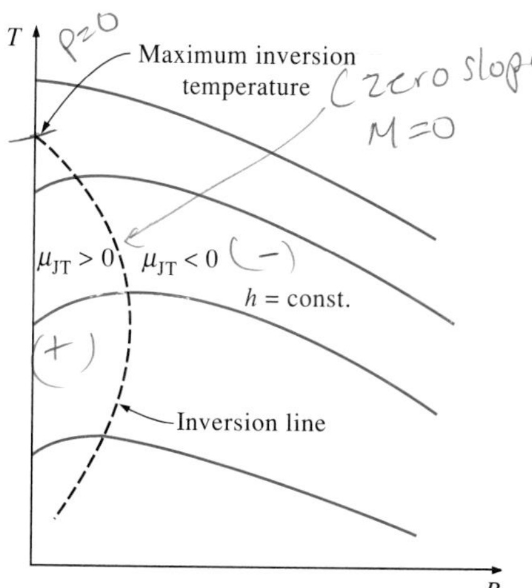

FIGURE 11-14
Constant-enthalpy lines of a substance on a T-P diagram.

Next we would like to develop a general relation for the Joule-Thomson coefficient in terms of the specific heats, pressure, specific volume, and temperature. This is easily accomplished by modifying the generalized relation for enthalpy change (Eq. 11-35)

$$dh = C_p\,dT + \left[v - T\left(\frac{\partial v}{\partial T}\right)_P\right]dP$$

For an h = constant process we have $dh = 0$. Then this equation can be rearranged to give

$$-\frac{1}{C_p}\left[v - T\left(\frac{\partial v}{\partial T}\right)_P\right] = \left(\frac{\partial T}{\partial P}\right)_h = \mu_{JT} \tag{11-52}$$

which is the desired relation. Thus, the Joule-Thomson coefficient can be determined from a knowledge of the constant-pressure specific heat and the P-v-T behavior of the substance. Of course, it is also possible to predict the constant-pressure specific heat of a substance by using the Joule-Thomson coefficient, which is relatively easy to determine, together with the P-v-T data for the substance.

EXAMPLE 11-10 Joule-Thomson Coefficient of an Ideal Gas

Show that the Joule-Thomson coefficient of an ideal gas is zero.

Solution For an ideal gas $v = RT/P$, and thus

$$\left(\frac{\partial v}{\partial T}\right)_P = \frac{R}{P}$$

Substituting this into Eq. 11-52 yields

$$\mu_{JT} = \frac{-1}{C_p}\left[v - T\left(\frac{\partial v}{\partial T}\right)_P\right] = \frac{-1}{C_p}\left[v - T\frac{R}{P}\right] = -\frac{1}{C_p}(v - v) = 0$$

This result is not surprising since the enthalpy of an ideal gas is a function of temperature only, $h = h(T)$, which requires that the temperature remain constant when the enthalpy remains constant. Therefore, a throttling process cannot be used to lower the temperature of an ideal gas (Fig. 11-15).

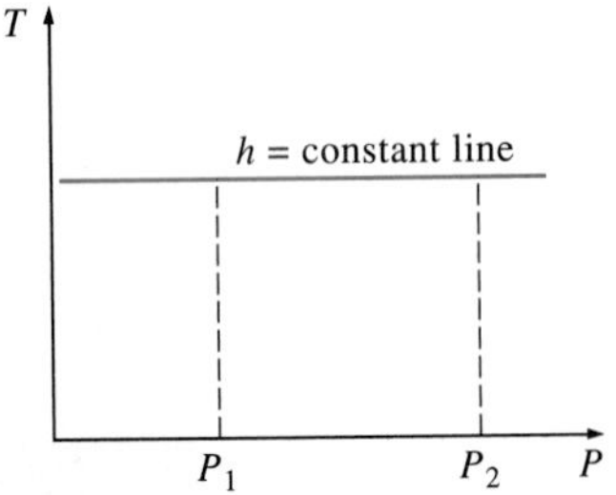

FIGURE 11-15

The temperature of an ideal gas remains constant during a throttling process since h = constant and T = constant lines on a T-P diagram coincide.

11-6 ■ THE Δh, Δu, AND Δs OF REAL GASES

We have mentioned several times that gases at low pressures behave as ideal gases and obey the relation $Pv = RT$. The properties of ideal gases are relatively easy to evaluate since the properties u, h, C_v, and C_p depend on temperature only. At high pressures, however, gases deviate considerably from ideal-gas behavior, and it becomes necessary to account for this deviation. In Chap. 2 we accounted for the deviation in properties P, v, and T by either using more complex equations of state or evaluating the compressibility factor Z from the compressibility charts. Now we extend the analysis to evaluate the changes in the enthalpy, internal energy, and entropy of nonideal (real) gases, using the general relations for du, dh, and ds developed earlier.

1 Enthalpy Changes of Real Gases

The enthalpy of a real gas, in general, depends on the pressure as well as on the temperature. Thus the enthalpy change of a real gas during a process can be evaluated from the general relation for dh (Eq. 11-36)

$$h_2 - h_1 = \int_{T_1}^{T_2} C_p\, dT + \int_{P_1}^{P_2} \left[v - T\left(\frac{\partial v}{\partial T}\right)_P\right] dP$$

where P_1, T_1 and P_2, T_2 are the pressures and temperatures of the gas at the initial and the final states, respectively. For an isothermal process $dT = 0$, and the first term vanishes. For a constant-pressure process, $dP = 0$, and the second term vanishes.

Properties are point functions, and thus the change in a property between two specified states is the same no matter which process path is followed. This fact can be exploited to greatly simplify the integration of Eq. 11-36.

Consider, for example, the process shown on a *T-s* diagram in Fig. 11-16. The enthalpy change during this process $h_2 - h_1$ can be determined by performing the integrations in Eq. 11-36 along a path that consists of two isothermal (T_1 = constant and T_2 = constant) lines and one isobaric (P_0 = constant) line instead of the actual process path, as shown in Fig. 11-16.

Although this approach increases the number of integrations, it also simplifies them since one property remains constant now during each part of the process. The pressure P_0 can be chosen to be very low or zero, so that the gas can be treated as an ideal gas during the P_0 = constant process. Using a superscript asterisk (*) to denote an ideal-gas state, we can express the enthalpy change of a real gas during process 1-2 as

$$h_2 - h_1 = (h_2 - h_2^*) + (h_2^* - h_1^*) + (h_1^* - h_1) \tag{11-53}$$

where, from Eq. 11-36,

$$h_2 - h_2^* = 0 + \int_{P_2^*}^{P_2} \left[v - T\left(\frac{\partial v}{\partial T}\right)_P \right]_{T=T_2} dP$$

$$= \int_0^{P_2} \left[v - T\left(\frac{\partial v}{\partial T}\right)_P \right]_{T=T_2} dP \tag{11-54}$$

$$h_2^* - h_1^* = \int_{T_1}^{T_2} C_p\, dT + 0 = \int_{T_1}^{T_2} C_{p0}\,(T)\, dT \tag{11-55}$$

$$h_1^* - h_1 = 0 + \int_{P_1}^{P_1^*} \left[v - T\left(\frac{\partial v}{\partial T}\right)_P \right]_{T=T_1} dP$$

$$= -\int_0^{P_1} \left[v - T\left(\frac{\partial v}{\partial T}\right)_P \right]_{T=T_1} dP \tag{11-56}$$

The difference between h and h^* is called the **enthalpy departure,** and it represents the variation of the enthalpy of a gas with pressure at a fixed temperature. The calculation of enthalpy departure requires a knowledge of the *P-v-T* behavior of the gas. In the absence of such data, we can use the relation $Pv = ZRT$, where Z is the compressibility factor. Substituting $v = ZRT/P$ and simplifying Eq. 11-56, we can write the enthalpy departure at any temperature T and pressure P as

$$(h^* - h)_T = -RT^2 \int_0^P \left(\frac{\partial Z}{\partial T}\right)_P \frac{dP}{P}$$

The above equation can be generalized by expressing it in terms of the reduced coordinates, using $T = T_{cr}T_R$ and $P = P_{cr}P_R$. After some manipulations, the enthalpy departure can be expressed in a nondimensionalized form as

$$Z_h = \frac{(\bar{h}^* - \bar{h})_T}{R_u T_{cr}} = T_R^2 \int_0^{P_R} \left(\frac{\partial Z}{\partial T_R}\right)_{P_R} d(\ln P_R) \tag{11-57}$$

where Z_h is called the **enthalpy departure factor.** The integral in the above equation can be performed graphically or numerically by employing data from the compressibility charts for various values of P_R and T_R. The values of

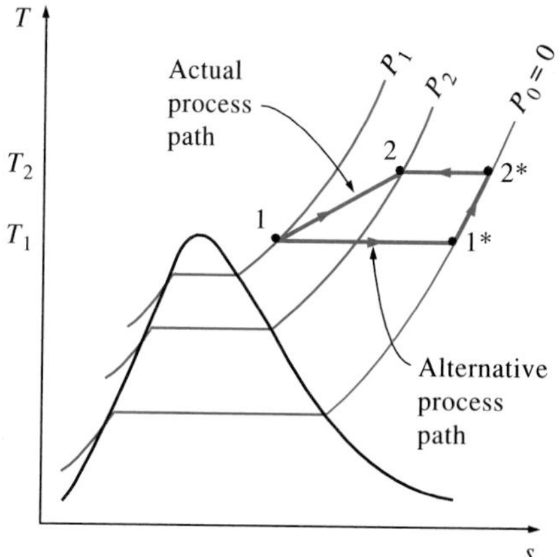

FIGURE 11-16
An alternative process path to evaluate the enthalpy changes of real gases.

Z_h are presented in graphical form as a function of P_R and T_R in Fig. A-31. This graph is called the **generalized enthalpy departure chart,** and it is used to determine the deviation of the enthalpy of a gas at a given P and T from the enthalpy of an ideal gas at the same T. By replacing h^* by h_{ideal} for clarity, Eq. 11-53 for the enthalpy change of a gas during a process 1-2 can be rewritten as

$$\bar{h}_2 - \bar{h}_1 = (\bar{h}_2 - \bar{h}_1)_{\text{ideal}} - R_u T_{\text{cr}}(Z_{h_2} - Z_{h_1}) \quad (11\text{-}58)$$

or

$$h_2 - h_1 = (h_2 - h_1)_{\text{ideal}} - RT_{\text{cr}}(Z_{h_2} - Z_{h_1}) \quad (11\text{-}59)$$

where the values of Z_h are determined from the generalized enthalpy departure chart and $(\bar{h}_2 - \bar{h}_1)_{\text{ideal}}$ is determined from the ideal-gas tables. Notice that the last terms on the right-hand side are zero for an ideal gas.

2 Internal Energy Changes of Real Gases

The internal energy change of a real gas is determined by relating it to the enthalpy change through the definition $\bar{h} = \bar{u} + P\bar{v} = \bar{u} + ZR_uT$:

$$\bar{u}_2 - \bar{u}_1 = (\bar{h}_2 - \bar{h}_1) - R_u(Z_2 T_2 - Z_1 T_1) \quad (11\text{-}60)$$

3 Entropy Changes of Real Gases

The entropy change of a real gas is determined by following an approach similar to that used above for the enthalpy change. There is some difference in derivation, however, owing to the dependence of the ideal-gas entropy on pressure as well as the temperature.

The general relation for ds was expressed as (Eq. 11-41)

$$s_2 - s_1 = \int_{T_1}^{T_2} \frac{c_p}{T}\, dT - \int_{P_1}^{P_2} \left(\frac{\partial v}{\partial T}\right)_P dP$$

where P_1, T_1 and P_2, T_2 are the pressures and temperatures of the gas at the initial and the final states, respectively. The thought that comes to mind at this point is to perform the integrations in the above equation first along a T_1 = constant line to zero pressure, then along the $P = 0$ line to T_2, and finally along the T_2 = constant line to P_2, as we did for the enthalpy. This approach is not suitable for entropy-change calculations, however, since it involves the value of entropy at zero pressure, which is infinity. We can avoid this difficulty by choosing a different (but more complex) path between the two states, as shown in Fig. 11-17. Then the entropy change can be expressed as

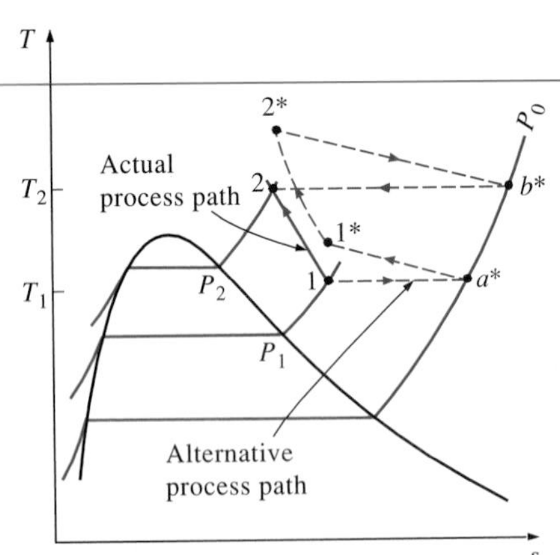

FIGURE 11-17
An alternative process path to evaluate the entropy changes of real gases during process 1-2.

$$s_2 - s_1 = (s_2 - s_b^*) + (s_b^* - s_2^*) + (s_2^* - s_1^*) + (s_1^* - s_a^*) + (s_a^* - s_1) \quad (11\text{-}61)$$

States 1 and 1* are identical ($T_1 = T_1^*$ and $P_1 = P_1^*$) and so are states 2 and 2*. States 1* and 2* exist only in the imagination, and the gas is assumed to behave as an ideal gas at these two states as well as at the states between the two. Therefore, the entropy change during process 1*-2* can be determined from the entropy-change relations for ideal gases. The calculation of entropy change between an actual state and the corresponding imaginary ideal-gas state is more involved, however, and requires the use of generalized entropy departure charts, as explained below.

Consider a gas at a pressure P and temperature T. To determine how much different the entropy of this gas would be if it were an ideal gas at the same temperature and pressure, we consider an isothermal process from the actual state P, T to zero (or close to zero) pressure and back to the imaginary ideal-gas state P^*, T^* (denoted by superscript *), as shown in Fig. 11-17. The entropy change during this isothermal process can be expressed as

$$(s_P - s_P^*)_T = (s_P - s_0^*)_T + (s_0^* - s_P^*)_T$$
$$= -\int_0^P \left(\frac{\partial v}{\partial T}\right)_P dP - \int_P^0 \left(\frac{\partial v^*}{\partial T}\right)_P dP$$

where $v = ZRT/P$ and $v^* = v_{\text{ideal}} = RT/P$. Performing the differentiations and rearranging, we obtain

$$(s_P - s_P^*)_T = \int_0^P \left[\frac{(1-Z)R}{P} - \frac{RT}{P}\left(\frac{\partial Zr}{\partial T}\right)_P\right] dP$$

By substituting $T = T_{\text{cr}}T_R$ and $P = P_{\text{cr}}P_R$ and rearranging, the entropy departure can be expressed in a nondimensionalized form as

$$Z_s = \frac{(\bar{s}^* - \bar{s})_{T,P}}{R_u} = \int_0^{P_R} \left[Z - 1 + T_R\left(\frac{\partial Z}{\partial T_R}\right)_{P_R}\right] d(\ln P_R) \qquad (11\text{-}62)$$

The difference $(\bar{s}^* - \bar{s})_{T,P}$ is called the **entropy departure** and Z_s is called the **entropy departure factor.** The integral in the above equation can be performed by using data from the compressibility charts. The values of Z_s are presented in graphical form as a function of P_R and T_R in Fig. A-32. This graph is called the **generalized entropy departure chart,** and it is used to determine the deviation of the entropy of a gas at a given P and T from the entropy of an ideal gas at the same P and T. Replacing s^* by s_{ideal} for clarity, we can rewrite Eq. 11-61 for the entropy change of a gas during a process 1-2 as

$$\bar{s}_2 - \bar{s}_1 = (\bar{s}_2 - \bar{s}_1)_{\text{ideal}} - R_u(Z_{s_2} - Z_{s_1}) \qquad (11\text{-}63)$$

or

$$s_2 - s_1 = (s_2 - s_1)_{\text{ideal}} - R(Z_{s_2} - Z_{s_1}) \qquad (11\text{-}64)$$

where the values of Z_s are determined from the generalized entropy departure chart and the entropy change $(\bar{s}_2 - \bar{s}_1)_{\text{ideal}}$ is determined from the ideal-gas relations for entropy change. Notice that the last terms on the right-hand side are zero for an ideal gas.

EXAMPLE 11-11 The Δh and Δs of Oxygen at High Pressures

Determine the enthalpy change and the entropy change of oxygen per unit mole as it undergoes a change of state from 220 K and 5 MPa to 300 K and 10 MPa (*a*) by assuming ideal-gas behavior and (*b*) by accounting for the deviation from ideal-gas behavior.

Solution The critical temperature and pressure of oxygen are $T_{cr} = 154.8$ K and $P_{cr} = 5.08$ MPa (Table A-1), respectively. The oxygen remains above its critical temperature; therefore, it is in the gas phase, but its pressure is quite high. Therefore, the oxygen will deviate from ideal-gas behavior and should be treated as a real gas.

Analysis (*a*) If the O_2 is assumed to behave as an ideal gas, its enthalpy will depend on temperature only, and the enthalpy values at the initial and the final temperatures can be determined from the ideal-gas table of O_2 (Table A-19) at the specified temperatures:

$$(\bar{h}_2 - \bar{h}_1)_{\text{ideal}} = \bar{h}_{2,\,ideal} - \bar{h}_{1,\,ideal}$$
$$= (8736 - 6404) \text{ kJ/kmol}$$
$$= \mathbf{2332 \text{ kJ/kmol}}$$

The entropy depends on both temperature and pressure even for ideal gases. Under the ideal-gas assumption, the entropy change of oxygen is determined from

$$(\bar{s}_2 - \bar{s}_1)_{\text{ideal}} = \bar{s}_2^\circ - \bar{s}_1^\circ - R_u \ln \frac{P_2}{P_1}$$
$$= (205.213 - 196.171) \text{ kJ/(kmol·K)}$$
$$- [8.314 \text{ kJ/(kmol·K)}] \times \ln \frac{10 \text{ MPa}}{5 \text{ MPa}}$$
$$= \mathbf{3.28 \text{ kJ/(kmol·K)}}$$

(*b*) The deviation from the ideal-gas behavior can be accounted for by determining the enthalpy and entropy departures from the generalized charts at each state:

$$\left.\begin{aligned} T_{R_1} &= \frac{T_1}{T_{cr}} = \frac{220}{154.8} = 1.42 \\ P_{R_1} &= \frac{P_1}{P_{cr}} = \frac{5}{5.08} = 0.98 \end{aligned}\right\} \quad Z_{h_1} = 0.53,\ Z_{s1} = 0.25$$

and

$$\left.\begin{aligned} T_{R_2} &= \frac{T_2}{T_{cr}} = \frac{300}{154.8} = 1.94 \\ P_{R_2} &= \frac{P_2}{P_{cr}} = \frac{10}{5.08} = 1.97 \end{aligned}\right\} \quad Z_{h_2} = 0.48,\ Z_{s_2} = 0.20$$

Then the enthalpy and entropy changes of oxygen during this process are determined by substituting the values above into Eqs. 11-58 and 11-63,

$$\bar{h}_2 - \bar{h}_1 = (\bar{h}_2 - \bar{h}_1)_{\text{ideal}} - R_u T_{cr}(Z_{h_2} - Z_{h_1})$$
$$= 2332 \text{ kJ/kmol} - [8.314 \text{ kJ/(kmol·K)}][154.8 \text{ K}(0.48 - 0.53)]$$
$$= \mathbf{2396 \text{ kJ/kmol}}$$

and $$\bar{s}_2 - \bar{s}_1 = (\bar{s}_2 - \bar{s}_1)_{\text{ideal}} - R_u(Z_{s_2} - Z_{s_1})$$
$$= 3.28 \text{ kJ/(kmol·K)} - [8.314 \text{ kJ/(kmol·K)}](0.20 - 0.25)$$
$$= \mathbf{3.68 \text{ kJ/(kmol·K)}}$$

Therefore, in this case, the ideal-gas assumption would underestimate the enthalpy change of the oxygen by 2.7 percent and the entropy change by 10.9 percent.

11-7 ■ SUMMARY

Some thermodynamic properties can be measured directly, but many others cannot. Therefore, it is necessary to develop some relations between these two groups so that the properties that cannot be measured directly can be evaluated. The derivations are based on the fact that properties are point functions, and the state of a simple, compressible system is completely specified by any two independent, intensive properties.

The equations that relate the partial derivatives of properties P, v, T, and s of a simple compressible substance to each other are called the *Maxwell relations*. They are obtained from the four Gibbs equations, expressed as

$$du = T\,ds - P\,dv$$
$$dh = T\,ds + v\,dP$$
$$da = -s\,dT - P\,dv$$
$$dg = -s\,dT + v\,dP$$

The *Maxwell relations* are

$$\left(\frac{\partial T}{\partial v}\right)_s = -\left(\frac{\partial P}{\partial s}\right)_v$$
$$\left(\frac{\partial T}{\partial P}\right)_s = \left(\frac{\partial v}{\partial s}\right)_P$$
$$\left(\frac{\partial s}{\partial v}\right)_T = \left(\frac{\partial P}{\partial T}\right)_v$$
$$\left(\frac{\partial s}{\partial P}\right)_T = -\left(\frac{\partial v}{\partial T}\right)_P$$

The *Clapeyron equation* enables us to determine the enthalpy change associated with a phase change from a knowledge of P, v, and T data alone. It is expressed as

$$\left(\frac{dP}{dT}\right)_{\text{sat}} = \frac{h_{fg}}{Tv_{fg}}$$

For liquid–vapor and solid–vapor phase-change processes at low pressures, it can be approximated as

$$\ln\left(\frac{P_2}{P_1}\right)_{\text{sat}} \cong \frac{h_{fg}}{R}\left(\frac{T_2 - T_1}{T_1 T_2}\right)_{\text{sat}}$$

The changes in internal energy, enthalpy, and entropy of a simple, compressible substance can be expressed in terms of pressure, specific volume, temperature, and specific heats alone as

$$du = C_v\,dT + \left[T\left(\frac{\partial P}{\partial T}\right)_v - P\right]dv$$

$$dh = C_p\,dT + \left[v - T\left(\frac{\partial v}{\partial T}\right)_P\right]dP$$

$$ds = \frac{C_v}{T}\,dT + \left(\frac{\partial P}{\partial T}\right)_v dv$$

or

$$ds = \frac{C_p}{T}\,dT - \left(\frac{\partial v}{\partial T}\right)_P dP$$

For specific heats, we have the following general relations:

$$\left(\frac{\partial C_v}{\partial v}\right)_T = T\left(\frac{\partial^2 P}{\partial T^2}\right)_v$$

$$\left(\frac{\partial C_p}{\partial P}\right)_T = -T\left(\frac{\partial^2 v}{\partial T^2}\right)_P$$

$$C_{p,T} - C_{p0,T} = -T\int_0^P \left(\frac{\partial^2 v}{\partial T^2}\right)_P dP$$

$$C_p - C_v = -T\left(\frac{\partial v}{\partial T}\right)_P^2 \left(\frac{\partial P}{\partial v}\right)_T$$

$$C_p - C_v = \frac{vT\beta^2}{\alpha}$$

where β is the *volume expansivity* and α is the *isothermal compressibility*, defined as

$$\beta = \frac{1}{v}\left(\frac{\partial v}{\partial T}\right)_P \quad \text{and} \quad \alpha = -\frac{1}{v}\left(\frac{\partial v}{\partial P}\right)_T$$

The difference $C_p - C_v$ is equal to R for ideal gases and to zero for incompressible substances.

The temperature behavior of a fluid during a throttling (h = constant) process is described by the *Joule-Thomson coefficient*, defined as

$$\mu_{JT} = \left(\frac{\partial T}{\partial P}\right)_h$$

The Joule-Thomson coefficient is a measure of the change in temperature of a substance with pressure during a constant-enthalpy process, and it can also be expressed as

$$\mu_{JT} = -\frac{1}{C_p}\left[v - T\left(\frac{\partial v}{\partial T}\right)_P\right]$$

The enthalpy, internal energy, and entropy changes of real gases can be determined accurately by utilizing *generalized enthalpy* or *entropy departure charts* to account for the deviation from the ideal-gas behavior by using the following relations:

$$\bar{h}_2 - \bar{h}_1 = (\bar{h}_2 - \bar{h}_1)_{\text{ideal}} - R_u T_{\text{cr}}(Z_{h_2} - Z_{h_1})$$

$$\bar{u}_2 - \bar{u}_1 = (\bar{h}_2 - \bar{h}_1) - R_u(Z_2 T_2 - Z_1 T_1)$$

$$\bar{s}_2 - \bar{s}_1 = (\bar{s}_2 - \bar{s}_1)_{\text{ideal}} - R_u(Z_{s_2} - Z_{s_1})$$

where the values of Z_h and Z_s are determined from the generalized charts.

REFERENCES AND SUGGESTED READING

1. W. Z. Black and J. G. Hartley. *Thermodynamics.* New York: Harper & Row, 1985.

2. J. R. Howell and R. O. Buckius. *Fundamentals of Engineering Thermodynamics.* New York: McGraw-Hill, 1987.

3. J. B. Jones and G. A. Hawkins. *Engineering Thermodynamics.* 2nd ed. New York: John Wiley & Sons, 1986.

4. G. J. Van Wylen and R. E. Sonntag. *Fundamentals of Classical Thermodynamics.* 3rd ed. New York: John Wiley & Sons, 1985.

5. K. Wark. *Thermodynamics.* 5th ed. New York: McGraw-Hill, 1988.

PROBLEMS*

Partial Derivatives and Associated Relations

11-1C Consider the function $z(x, y)$. Plot a differential surface on x-y-z coordinates and indicate ∂x, dx, ∂y, dy, $(\partial z)_x$, $(\partial z)_y$, and dz.

11-2C What is the difference between partial differentials and ordinary differentials?

11-3C Consider the function $z(x, y)$, its partial derivatives $(\partial z/\partial x)_y$ and $(\partial z/\partial y)_x$, and the total derivative dz/dx.

(*a*) How do the magnitudes $(\partial x)_y$ and dx compare?
(*b*) How do the magnitudes $(\partial z)_y$ and dz compare?
(*c*) Is there any relation among dz, $(\partial z)_x$, and $(\partial z)_y$?

11-4C Consider a function $z(x, y)$ and its partial derivative $(\partial z/\partial y)_x$. Under what conditions is this partial derivative equal to the total derivative dz/dy?

11-5C Consider a function $z(x, y)$ and its partial derivative $(\partial z/\partial y)_x$. If this partial derivative is equal to zero for all values of x, what does it indicate?

---

*Students are encouraged to answer *all* the concept "C" questions.

11-6C Consider a function $z(x, y)$ and its partial derivative $(\partial z/\partial y)_x$. Can this partial derivative still be a function of x?

11-7C Consider a function $f(x)$ and its derivative df/dx. Can this derivative be determined by evaluating dx/df and taking its inverse?

11-8 Consider air at 350 K and 0.90 m^3/kg. Using Eq. 11-3, determine the change in pressure corresponding to an increase of (*a*) 1 percent in temperature at constant specific volume, (*b*) 1 percent in specific volume at constant temperature, and (*c*) 1 percent in both the temperature and specific volume.

11-9 Prove for an ideal gas that (*a*) the P = constant lines on a T-v diagram are straight lines and (*b*) the high-pressure lines are steeper than the low-pressure lines.

11-10 Derive a relation for the slope of the v = constant lines on a T-P diagram for a gas that obeys the van der Waals equation of state.
Answer: $(v - b)/R$

11-11 Nitrogen gas at 400 K and 100 kPa behaves as an ideal gas. Estimate the C_p and C_v of the nitrogen at this state, using enthalpy and internal energy data from Table A-18, and compare them to the values listed in Table A-2*b*.

11-12E Nitrogen gas at 600 R and 15 psia behaves as an ideal gas. Estimate the C_p and C_v of the nitrogen at this state, using enthalpy and internal energy data from Table A-18E, and compare them to the values listed in Table A-2E*b*. *Answers:* 0.249 Btu/(lbm · R), 0.178 Btu/(lbm · R)

11-13 Consider an ideal gas at 400 K and 100 kPa. As a result of some disturbance, the conditions of the gas change to 404 K and 98 kPa. Estimate the change in the specific volume of the gas using (*a*) Eq. 11-3 and (*b*) the ideal-gas relation at each state.

11-14 Using the equation of state $P(v - a) = RT$, verify (*a*) the cyclic relation and (*b*) the reciprocity relation at constant v.

The Maxwell Relations

11-15 Verify the validity of the last Maxwell relation (Eq. 11-19) for refrigerant-134a at 80°C and 1.2 MPa.

11-16E Verify the validity of the last Maxwell relation (Eq. 11-19) for steam at 900°F and 450 psia.

11-17 Using the Maxwell relations, determine a relation for $(\partial s/\partial P)_T$ for a gas whose equation of state is $P(v - b) = RT$. *Answer:* $-R/P$

11-18 Using the Maxwell relations, determine a relation for $(\partial s/\partial v)_T$ for a gas whose equation of state is $(P - a/v^2)(v - b) = RT$.

11-19 Using the Maxwell relations and the ideal-gas equation of state, determine a relation for $(\partial s/\partial v)_T$ for an ideal gas. *Answer:* R/v

The Clapeyron Equation

11-20C What is the value of the Clapeyron equation in thermodynamics?

11-21C Does the Clapeyron equation involve any approximations, or is it exact?

11-22C What approximations are involved in the Clapeyron-Clausius equation?

11-23 Using the Clapeyron equation, estimate the enthalpy of vaporization of refrigerant-134a at 30°C, and compare it to the tabulated value.

11-24 Using the Clapeyron equation, estimate the enthalpy of vaporization of steam at 200 kPa, and compare it to the tabulated value.

11-25 Calculate the h_{fg} and s_{fg} of steam at 150°C from the Clapeyron equation, and compare them to the tabulated values.

11-26E Determine the h_{fg} of refrigerant-134a at 50°F on the basis of (*a*) the Clapeyron equation and (*b*) the Clapeyron-Clausius equation. Compare your results to the tabulated h_{fg} value.

General Relations for du, dh, ds, C_v, and C_p

11-27C Can the variation of specific heat C_p with pressure at a given temperature be determined from a knowledge of P-v-T data alone?

11-28 Show that the enthalpy of an ideal gas is a function of temperature only and that for an incompressible substance it also depends on pressure.

11-29 Derive expressions for (*a*) Δu, (*b*) Δh, and (*c*) Δs for a gas that obeys the van der Waals equation of state for an isothermal process.

11-30 Derive expressions for (*a*) Δu, (*b*) Δh, and (*c*) Δs for a gas whose equation of state is $P(v - a) = RT$ for an isothermal process.
Answers: (*a*) 0, (*b*) $a(P_2 - P_1)$, (*c*) $-R \ln (P_2/P_1)$

11-31 Derive expressions for $(\partial u/\partial P)_T$ and $(\partial h/\partial v)_T$ in terms of P, v, and T only.

11-32 Derive an expression for the specific-heat difference $C_p - C_v$ for (*a*) an ideal gas, (*b*) a van der Waals gas, and (*c*) an incompressible substance.

11-33 Estimate the specific-heat difference $C_p - C_v$ for liquid water at 10 MPa and 40°C. *Answer:* 0.1094 kJ/(kg · K)

11-34E Estimate the specific-heat difference $C_p - C_v$ for liquid water at 1000 psia and 150°F. *Answer:* 0.0560 Btu/(lbm · R)

11-35 Derive a relation for the volume expansivity β and the isothermal compressibility α (*a*) for an ideal gas and (*b*) for a gas whose equation of state is $P(v - a) = RT$.

11-36 Estimate the volume expansivity β and the isothermal compressibility α of refrigerant-134a at 140 kPa and 20°C.

The Joule-Thomson Coefficient

11-37C What does the Joule-Thomson coefficient represent?

11-38C Describe the inversion line and the maximum inversion temperature.

11-39C The pressure of a fluid always decreases during an adiabatic throttling process. Is this also the case for the temperature?

11-40C Does the Joule-Thomson coefficient of a substance change with temperature at a fixed pressure?

11-41C Will the temperature of helium change if it is throttled adiabatically from 300 K and 500 kPa to 100 kPa?

11-42 Consider a gas whose equation of state is $P(v - a) = RT$, where a is a positive constant. Is it possible to cool this gas by throttling?

11-43 Derive a relation for the Joule-Thomson coefficient and the inversion temperature for a gas whose equation of state is $(P + a/v^2)\,v = RT$.

11-44 Estimate the Joule-Thomson coefficient of nitrogen at (*a*) 2 MPa and 250 K and (*b*) 6 MPa and 275 K.

11-45E Estimate the Joule-Thomson coefficient of nitrogen at (*a*) 200 psia and 500 R and (*b*) 2000 psia and 400 R.

11-46 Estimate the Joule-Thomson coefficient of refrigerant-134a at 0.6 MPa and 100°C.

11-47 Steam is throttled slightly from 1 MPa and 300°C. Will the temperature of the steam increase, decrease, or remain the same during this process?

The Δh, Δu, and Δs of Real Gases

11-48C What is the enthalpy departure?

11-49C On the generalized enthalpy departure chart, the normalized enthalpy departure values seem to approach zero as the reduced pressure P_R approaches zero. How do you explain this behavior?

11-50C Why is the generalized enthalpy departure chart prepared by using P_R and T_R as the parameters instead of P and T?

11-51 Determine the enthalpy of nitrogen, in kJ/kg, at 175 K and 8 MPa using (*a*) data from the ideal-gas nitrogen table and (*b*) the generalized enthalpy departure chart. Compare your results to the actual value of 125.5 kJ/kg. *Answers:* (*a*) 181.48 kJ/kg, (*b*) 121.6 kJ/kg

11-52E Determine the enthalpy of nitrogen, in Btu/lbm, at 400 R and 2000 psia using (*a*) data from the ideal-gas nitrogen table and (*b*) the generalized enthalpy chart. Compare your results to the actual value of 177.8 Btu/lbm.

11-53 What is the error involved in the (*a*) enthalpy and (*b*) internal energy of CO_2 at 350 K and 10 MPa if it is assumed to be an ideal gas?
Answers: (*a*) 50%, (*b*) 49%

11-54 Determine the enthalpy change and the entropy change of nitrogen per unit mole as it undergoes a change of state from 225 K and 6 MPa to 300 K and 10 MPa, (*a*) by assuming ideal-gas behavior and (*b*) by accounting for the deviation from ideal-gas behavior through the use of generalized charts.

11-55 Determine the enthalpy change and the entropy change of CO_2 per unit mass as it undergoes a change of state from 250 K and 7 MPa to 280 K and 12 MPa, (*a*) by assuming ideal-gas behavior and (*b*) by accounting for the deviation from ideal-gas behavior.

11-56 Methane is compressed adiabatically by a steady-flow compressor from 2 MPa and −10°C to 10 MPa and 110°C at a rate of 1.4 kg/s. Using the generalized charts, determine the required power input to the compressor.
Answer: −351 kW

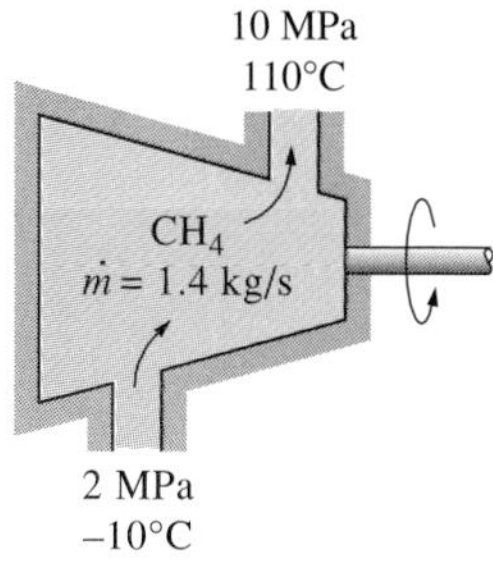

FIGURE P11-56

11-57 Propane is compressed isothermally by a piston-cylinder device from 100°C and 1 MPa to 4 MPa. Using the generalized charts, determine the work done and the heat transfer per unit mass of propane.

11-58E Propane is compressed isothermally by a piston-cylinder device from 200°F and 200 psia to 800 psia. Using the generalized charts, determine the work done and the heat transfer per unit mass of the propane.
Answers: −45.3 Btu/lbm, −132.1 Btu/lbm

11-59 Determine the exergy destruction associated with the process described in Prob. 11-57. Assume T_0 = 25°C.

11-60 Carbon dioxide enters an adiabatic nozzle at 8 MPa and 450 K with a low velocity and leaves at 2 MPa and 350 K. Using the generalized enthalpy departure chart, determine the exit velocity of the carbon dioxide.
Answer: 384 m/s

11-61 A 0.4-m^3 well-insulated rigid tank contains oxygen at 220 K and 10 MPa. A paddle wheel placed in the tank is turned on, and the temperature of the oxygen rises to 250 K. Using the generalized charts, determine (*a*) the final pressure in the tank and (*b*) the paddle-wheel work done during this process. *Answers:* (*a*) 12,190 kPa, (*b* 1881 kJ

Review Problems

11-62 For $\beta \geq 0$, prove that at every point of a single-phase region of an *h-s* diagram, the slope of a constant-pressure (P = constant) line is greater than the slope of a constant-temperature (T = constant) line, but less than the slope of a constant-volume (v = constant) line.

11-63 Using the cyclic relation and the first Maxwell relation, derive the other three Maxwell relations.

11-64 Starting with the relation $dh = T\,ds + v\,dP$, show that the slope of a constant-pressure line on an *h-s* diagram (*a*) is constant in the saturation region and (*b*) increases with temperature in the superheated region.

11-65 Derive relations for (*a*) Δu, (*b*) Δh, and (*c*) Δs of a gas that obeys the equation of state $(P + a/v^2)v = RT$ for an isothermal process.

11-66 Show that

$$C_v = -T\left(\frac{\partial v}{\partial T}\right)_s\left(\frac{\partial P}{\partial T}\right)_v \quad \text{and} \quad C_p = T\left(\frac{\partial P}{\partial T}\right)_s\left(\frac{\partial v}{\partial T}\right)_P$$

11-67 Estimate the C_p of nitrogen at 200 kPa and 400 K, using (*a*) the relation in the above problem and (*b*) its definition. Compare your results to the value listed in Table A-2*b*.

11-68 Steam is throttled from 4.5 MPa and 400°C to 3.5 MPa. Estimate the temperature change of the steam during this process and the average Joule-Thomson coefficient. *Answers:* −7.44°C, 7.44°C/MPa

11-69 A rigid tank contains 2 m^3 of argon at −100°C and 1 MPa. Heat is now transferred to argon until the temperature in the tank rises to 0°C. Using the generalized charts, determine (*a*) the mass of the argon in the tank, (*b*) the final pressure, and (*c*) the heat transfer.
Answers: (*a*) 58.5 kg, (*b*) 1531 kPa, (*c*) 2085 kJ

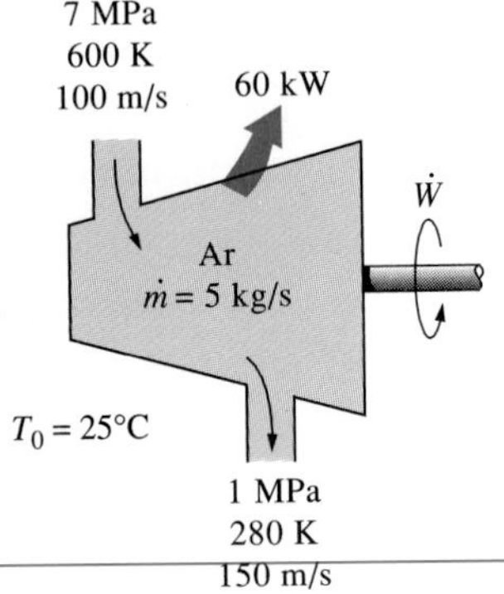

FIGURE P11-70

11-70 Argon gas enters a turbine at 7 MPa and 600 K with a velocity of 100 m/s and leaves at 1 MPa and 280 K with a velocity of 150 m/s at a rate of 5 kg/s. Heat is being lost to the surroundings at 25°C at a rate of 60 kW. Using the generalized charts, determine (*a*) the power output of the turbine and (*b*) the exergy destruction associated with the process.

11-71E Argon gas enters a turbine at 1000 psia and 1000 R with a velocity of 300 ft/s and leaves at 150 psia and 500 R with a velocity of 450 ft/s at a rate of 12 lbm/s. Heat is being lost to the surroundings at 75°F at a rate of 80 Btu/s. Using the generalized charts, determine (*a*) the power output of the turbine and (*b*) the exergy destruction associated with the process.
Answers: (*a*) 937 hp, (*b*) 102.5 Btu/s

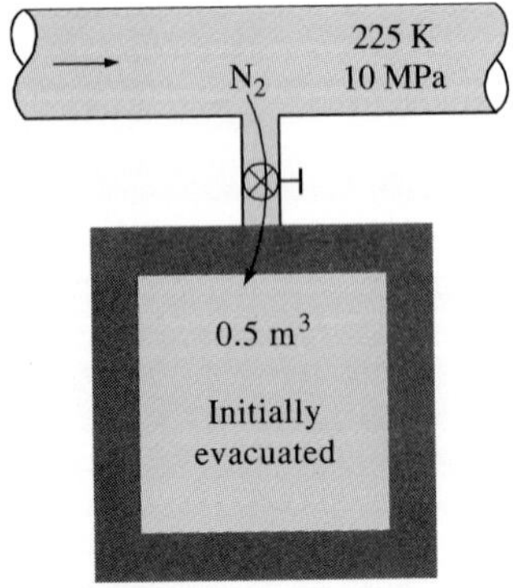

FIGURE P11-72

11-72 An adiabatic 0.5-m^3 storage tank that is initially evacuated is connected to a supply line that carries nitrogen at 225 K and 10 MPa. A valve is opened, and nitrogen flows into the tank from the supply line. The valve is closed when the pressure in the tank reaches 10 MPa. Determine the final temperature in the tank (*a*) treating nitrogen as an ideal gas and (*b*) using generalized charts. Compare your results to the actual value of 293 K.

11-73 For a homogeneous (single-phase) simple pure substance, the pressure and temperature are independent properties, and any property can be expressed as a function of these two properties. Taking $v = v(P, T)$, show that the change in specific volume can be expressed in terms of the volume expansivity β and isothermal compressibility α as

$$\frac{dv}{v} = \beta\, dT - \alpha\, dP$$

Also, assuming constant average values for β and α, obtain a relation for the ratio of the specific volumes v_2/v_1 as a homogeneous system undergoes a process from state 1 to state 2.

11-74 Repeat Prob. 11-73 for an isobaric process.

11-75 The volume expansivity of water at 20°C is $\beta = 0.207 \times 10^{-6}\ \text{K}^{-1}$. Treating this value as a constant, determine the change in volume of 1 m^3 of water as it is heated from 15°C to 25°C at constant pressure.

11-76 The volume expansivity β values of copper at 300 K and 500 K are $49.2 \times 10^{-6}\ \text{K}^{-1}$ and $54.2 \times 10^{-6}\ \text{K}^{-1}$, respectively, and β varies almost linearly in this temperature range. Determine the percent change in the volume of a copper block as it is heated from 300 K to 500 K at atmospheric pressure.

11-77 Starting with $\mu_{JT} = (1/c_p)\,[T(\partial v/\partial T)_p - v]$ and noting that $Pv = ZRT$, where $Z = Z(P, T)$ is the compressibility factor, show that the position of the Joule-Thomson coefficient inversion curve on the *T-P* plane is given by the equation $(\partial Z/\partial T)_P = 0$.

11-78 Consider an infinitesimal reversible adiabatic compression or expansion process. By taking $s = s(P, v)$ and using the Maxwell relations, show that for this process Pv^k = constant, where k is the *isentropic expansion exponent* defined as

$$k = -\frac{v}{P}\left(\frac{\partial P}{\partial v}\right)_s$$

Also, show that the isentropic expansion exponent k reduces to the specific heat ratio C_p/C_v for an ideal gas.

Computer, Design, and Essay Problems

11-79 Consider the function $z = z(x, y)$. Write an essay on the physical interpretation of the ordinary derivative dz/dx and the partial derivative $(\partial z/\partial x)_y$. Explain how these two derivatives are related to each other and when they become equivalent.

11-80 There have been several attempts to represent the thermodynamic relations geometrically, the best known of these being Koenig's thermodynamic square shown in the figure. There is a systematic way of obtaining the four Maxwell relations as well as the four relations for du, dh, dg, and da from this figure. By comparing these relations to Koenig's diagram, come up with the rules to obtain these eight thermodynamic relations from this diagram.

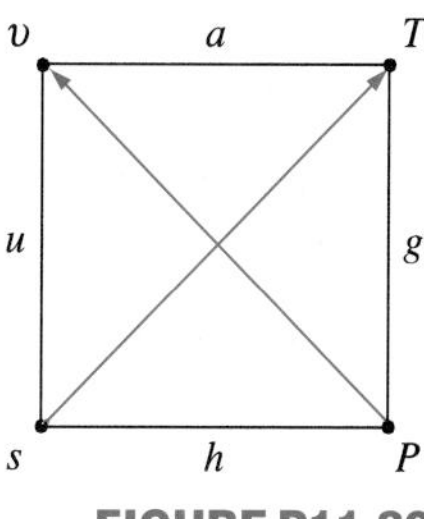

FIGURE P11-80

11-81 Several attempts have been made to express the partial derivatives of the most common thermodynamic properties in a compact and systematic manner in terms of measurable properties. The work of P. W. Bridgman is perhaps the most fruitful of all, and it resulted in the well-known Bridgman's table. The 28 entries in that table are sufficient to express the partial derivatives of the eight common properties P, T, v, s, u, h, f, and g in terms of the six properties P, v, T, C_p, β, and α, which can be measured directly or indirectly with relative ease. Obtain a copy of Bridgman's table and explain, with examples, how it is used.

Gas Mixtures

CHAPTER 12

Up to this point, we have limited our consideration to thermodynamic systems that involve a single pure substance such as water, refrigerant-134a, or nitrogen. Many important thermodynamic applications, however, involve *mixtures* of several pure substances rather than a single pure substance. Therefore, it is important to develop an understanding of mixtures and learn how to handle them.

In this chapter, we deal with nonreacting gas mixtures. A nonreacting gas mixture can be treated as a pure substance since it is usually a homogeneous mixture of different gases. The properties of a gas mixture obviously will depend on the properties of the individual gases (called *components* or *constituents*) as well as on the amount of each gas in the mixture. Therefore, it is possible to prepare tables of properties for mixtures. This has been done for common mixtures such as air. It is not practical to prepare property tables for every conceivable mixture composition, however, since the number of possible compositions is endless. Therefore, we need to develop rules for determining mixture properties from a knowledge of mixture composition and the properties of the individual components. We do this first for ideal-gas mixtures and then for real-gas mixtures. The basic principles involved are also applicable to liquid or solid mixtures, called *solutions*.

12-1 ■ COMPOSITION OF A GAS MIXTURE: Mass and Mole Fractions

To determine the properties of a mixture, we need to know the *composition* of the mixture as well as the properties of the individual components. There are two ways to describe the composition of a mixture: either by specifying the number of moles of each component, called **molar analysis,** or by specifying the mass of each component, called **gravimetric analysis.**

Consider a gas mixture composed of k components. You will agree that the mass of the mixture m_m is the sum of the masses of the individual components, and the mole number of the mixture N_m is the sum of the mole numbers of the individual components* (Figs. 12-1 and 12-2). That is,

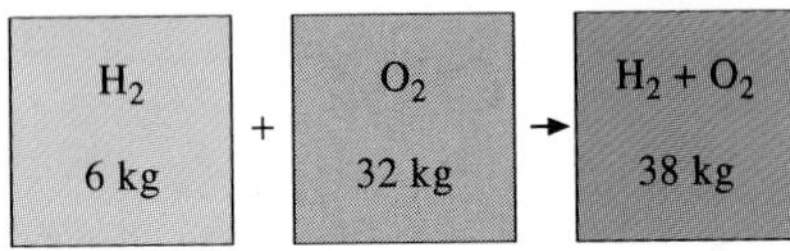

FIGURE 12-1
The mass of a mixture is equal to the sum of the masses of its components.

$$m_m = \sum_{i=1}^{k} m_i \quad \text{and} \quad N_m = \sum_{i=1}^{k} N_i \qquad (12\text{-}1a, b)$$

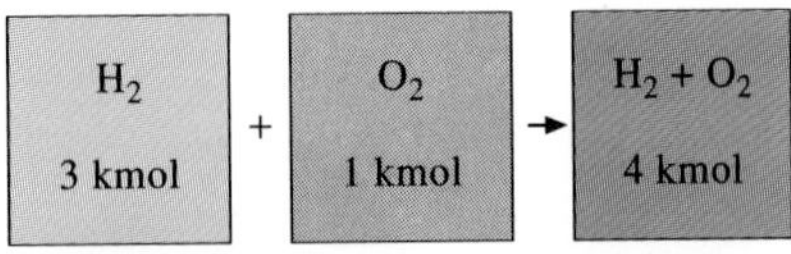

FIGURE 12-2
The number of moles of a nonreacting mixture is equal to the sum of the number of moles of its components.

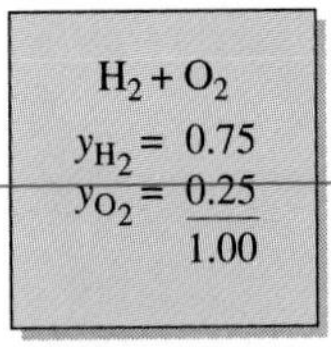

The ratio of the mass of a component to the mass of the mixture is called the **mass fraction** (mf), and the ratio of the mole number of a component to the mole number of the mixture is called the **mole fraction** y:

$$\text{mf}_i = \frac{m_i}{m_m} \quad \text{and} \quad y_i = \frac{N_i}{N_m} \qquad (12\text{-}2a, b)$$

Dividing Eq. 12-1a by m_m or Eq. 12-1b by N_m, we can easily show that the sum of the mass fractions or mole fractions for a mixture is equal to 1 (Fig. 12-3):

$H_2 + O_2$
$y_{H_2} = 0.75$
$y_{O_2} = 0.25$
1.00

FIGURE 12-3
The sum of the mole fractions of a mixture is equal to 1.

$$\sum_{i=1}^{k} \text{mf}_i = 1 \quad \text{and} \quad \sum_{i=1}^{k} y_i = 1 \qquad (12\text{-}3a, b)$$

The mass of a substance can be expressed in terms of the mole number N and molar mass M of the substance as

$$m = NM \qquad \text{(kg)}$$

Then the **apparent** (or **average**) **molar mass** of a mixture can be expressed as

$$M_m = \frac{m_m}{N_m} = \frac{\sum m_i}{N_m} = \frac{\sum N_i M_i}{N_m} = \sum_{i=1}^{k} y_i M_i \qquad \text{(kg/kmol)} \qquad (12\text{-}4)$$

Then the **average** (or **apparent**) **gas constant** of the mixture can be determined from

$$R_m = \frac{R_u}{M_m} \qquad [\text{kJ/(kg}\cdot\text{K)}] \qquad (12\text{-}5)$$

*Throughout this chapter, the subscript m will denote the gas mixture and the subscript i will denote any single component of the mixture.

EXAMPLE 12-1 Mass and Mole Fractions of a Gas Mixture

Consider a gas mixture that consists of 3 kg of O_2, 5 kg of N_2, and 12 kg of CH_4, as shown in Fig. 12-4. Determine (*a*) the mass fraction of each component, (*b*) the mole fraction of each component, and (*c*) the average molar mass and gas constant of the mixture.

3 kg O_2
5 kg N_2
12 kg CH_4

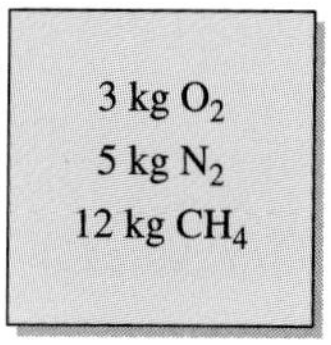

FIGURE 12-4
Schematic for Example 12-1.

Solution The schematic of the gas mixture is given in Fig. 12-4. We note that this is a gas mixture that consists of three gases of known masses.

Analysis (*a*) The total mass of the mixture is

$$m_m = m_{O_2} + m_{N_2} + m_{CH_4} = 3 + 5 + 12 = 20 \text{ kg}$$

Then the mass fraction of each component becomes

$$\text{mf}_{O_2} = \frac{m_{O_2}}{m_m} = \frac{3 \text{ kg}}{20 \text{ kg}} = \mathbf{0.15}$$

$$\text{mf}_{N_2} = \frac{m_{N_2}}{m_m} = \frac{5 \text{ kg}}{20 \text{ kg}} = \mathbf{0.25}$$

$$\text{mf}_{CH_4} = \frac{m_{CH_4}}{m_m} = \frac{12 \text{ kg}}{20 \text{ kg}} = \mathbf{0.60}$$

(*b*) To find the mole fractions, we need to determine the mole numbers of each component first:

$$N_{O_2} = \frac{m_{O_2}}{M_{O_2}} = \frac{3 \text{ kg}}{32 \text{ kg/kmol}} = 0.094 \text{ kmol}$$

$$N_{N_2} = \frac{m_{N_2}}{M_{N_2}} = \frac{5 \text{ kg}}{28 \text{ kg/kmol}} = 0.179 \text{ kmol}$$

$$N_{CH_4} = \frac{m_{CH_4}}{M_{CH_4}} = \frac{12 \text{ kg}}{16 \text{ kg/kmol}} = 0.750 \text{ kmol}$$

Thus, $N_m = N_{O_2} + N_{N_2} + N_{CH_4} = 0.094 + 0.179 + 0.750 = 1.023 \text{ kmol}$

and

$$y_{O_2} = \frac{N_{O_2}}{N_m} = \frac{0.094 \text{ kmol}}{1.023 \text{ kmol}} = \mathbf{0.092}$$

$$y_{N_2} = \frac{N_{N_2}}{N_m} = \frac{0.179 \text{ kmol}}{1.023 \text{ kmol}} = \mathbf{0.175}$$

$$y_{CH_4} = \frac{N_{CH_4}}{N_m} = \frac{0.750 \text{ kmol}}{1.023 \text{ kmol}} = \mathbf{0.733}$$

(*c*) The average molar mass and gas constant of the mixture are determined from their definitions,

$$M_m = \frac{m_m}{N_m} = \frac{20 \text{ kg}}{1.023 \text{ kmol}} = \mathbf{19.6 \text{ kg/kmol}}$$

or

$$\begin{aligned} M_m = \sum y_i M_i &= y_{O_2} M_{O_2} + y_{N_2} M_{N_2} + y_{CH_4} M_{CH_4} \\ &= (0.092)(32) + (0.175)(28) + (0.733)(16) \\ &= 19.6 \text{ kg/kmol} \end{aligned}$$

Also, $R_m = \frac{R_u}{M_m} = \frac{8.314 \text{ kJ/(kmol} \cdot \text{K)}}{19 \text{ kg/kmol}} = \mathbf{0.424 \text{ kJ/(kg} \cdot \text{K)}}$

12-2 ■ *P-v-T* BEHAVIOR OF GAS MIXTURES: Ideal and Real Gases

An ideal gas is defined in Chap. 2 as a gas whose molecules are spaced far apart so that the behavior of a molecule is not influenced by the presence of other molecules—a situation encountered at low densities. We also mentioned that real gases approximate this behavior closely when they are at a low pressure or high temperature relative to their critical-point values. The *P-v-T* behavior of an ideal gas is expressed by the simple relation $Pv = RT$, which is called the *ideal-gas equation of state*. The *P-v-T* behavior of real gases is expressed by more complex equations of state or by $Pv = ZRT$, where Z is the compressibility factor.

When two or more ideal gases are mixed, the behavior of a molecule normally is not influenced by the presence of other similar or dissimilar molecules, and therefore a nonreacting mixture of ideal gases also behaves as an ideal gas. Air, for example, is conveniently treated as an ideal gas in the range where nitrogen and oxygen behave as ideal gases. When a gas mixture consists of real (nonideal) gases, however, the prediction of the *P-v-T* behavior of the mixture becomes rather involved.

The prediction of the *P-v-T* behavior of gas mixtures is usually based on two models: *Dalton's law of additive pressures* and *Amagat's law of additive volumes*. Both models are described and discussed below.

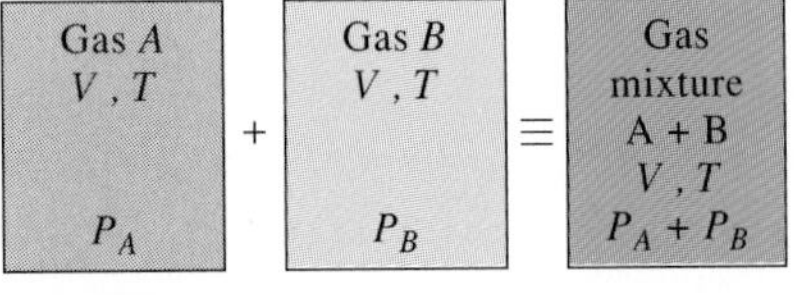

FIGURE 12-5

Dalton's law of additive pressures for a mixture of two ideal gases.

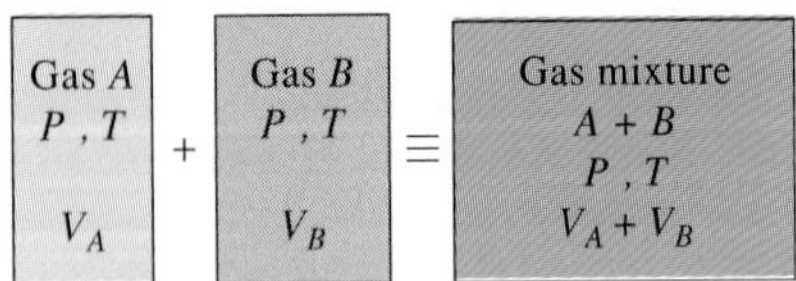

FIGURE 12-6

Amagat's law of additive volumes for a mixture of two ideal gases.

Dalton's law of additive pressures: *The pressure of a gas mixture is equal to the sum of the pressures each gas would exert if it existed alone at the mixture temperature and volume (Fig. 12-5).*

Amagat's law of additive volumes: *The volume of a gas mixture is equal to the sum of the volumes each gas would occupy if it existed alone at the mixture temperature and pressure (Fig. 12-6).*

Dalton's and Amagat's laws hold exactly for ideal-gas mixtures, but only approximately for real-gas mixtures. This is due to intermolecular forces that may be significant for real gases at high densities. For ideal gases, these two laws are identical and give identical results.

Dalton's and Amagat's laws can be expressed as follows:

Dalton's law: $$P_m = \sum_{i=1}^{k} P_i(T_m, V_m) \quad (12\text{-}6)$$

Amagat's law: $$V_m = \sum_{i=1}^{k} V_i(T_m, P_m) \quad (12\text{-}7)$$

exact for ideal gases, approximate for real gases

In these relations, P_i is called the **component pressure** and V_i is called the **component volume** (Fig. 12-7). Note that V_i is the volume a component *would* occupy if it existed alone at T_m and P_m, not the actual volume occupied by the component in the mixture. (In a vessel that holds a gas mixture, each component fills the entire volume of the vessel. Therefore, the volume of each component is equal to the volume of the vessel.) Also, the ratio P_i/P_m is called the **pressure fraction** and the ratio V_i/V_m is called the **volume fraction** of component i.

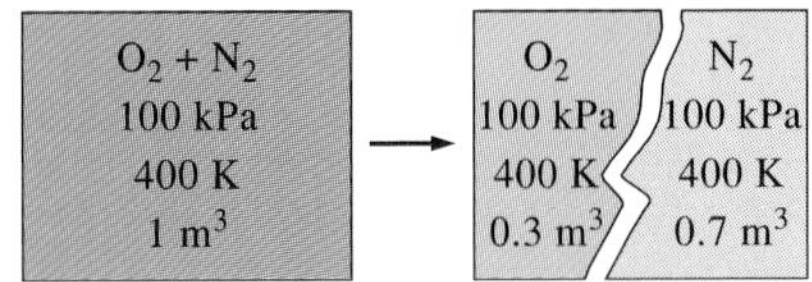

FIGURE 12-7

The volume a component would occupy if it existed alone at the mixture *T* and *P* is called the *component volume* (for ideal gases, it is equal to the partial volume y_iV_m).

1 Ideal-Gas Mixtures

For ideal gases, P_i and V_i can be related to y_i by using the ideal-gas relation for both the components and the gas mixture:

$$\frac{P_i(T_m, V_m)}{P_m} = \frac{N_iR_uT_m/V_m}{N_mR_uT_m/V_m} = \frac{N_i}{N_m} = y_i$$

$$\frac{V_i(T_m, P_m)}{V_m} = \frac{N_iR_uT_m/P_m}{N_mR_uT_m/P_m} = \frac{N_i}{N_m} = y_i$$

Therefore,

$$\frac{P_i}{P_m} = \frac{V_i}{V_m} = \frac{N_i}{N_m} = y_i \qquad (12\text{-}8)$$

Equation 12-8 is strictly valid for ideal-gas mixtures since it is derived by assuming ideal-gas behavior for the gas mixture and each of its components. The quantity y_iP_m is called the **partial pressure** (identical to the *component pressure* for ideal gases), and the quantity y_iV_m is called the **partial volume** (identical to the *component volume* for ideal gases). *Note that for an ideal-gas mixture, the mole fraction, the pressure fraction, and the volume fraction of a component are identical.*

The composition of an ideal-gas mixture (such as the exhaust gases leaving a combustion chamber) is frequently determined by a volumetric analysis and Eq. 12-8. A sample gas at a known volume, pressure, and temperature is passed into a vessel containing reagents that absorb one of the gases. The volume of the remaining gas is then measured at the original pressure and temperature. The ratio of the reduction in volume to the original volume (volume fraction) represents the mole fraction of that particular gas.

2 Real-Gas Mixtures

Dalton's law of additive pressures and Amagat's law of additive volumes can also be used for real gases, often with reasonable accuracy. This time, however, the component pressures or component volumes should be evaluated from relations that take into account the deviation of each component from ideal-gas behavior. One way of doing that is to use more exact equations of state (van der Waals, Beattie–Bridgeman, Benedict–Webb–Rubin, etc.) instead of the ideal-gas equation of state. Another way is to use the compressibility factor (Fig. 12-8) as

FIGURE 12-8

One way of predicting the *P-v-T* behavior of a real-gas mixture is to use compressibility factors.

$$P_mV_m = Z_mN_mR_uT_m$$

$$Z_m = \sum_{i=1}^{k} y_iZ_i$$

$$PV = ZNR_uT \tag{12-9}$$

The compressibility factor of the mixture Z_m can be expressed in terms of the compressibility factors of the individual gases Z_i by applying Eq. 12-9 to both sides of Dalton's law or Amagat's law expression and simplifying. We obtain

$$Z_m = \sum_{i=1}^{k} y_i Z_i \tag{12-10}$$

where Z_i is determined either at T_m and V_m (Dalton's law) or at T_m and P_m (Amagat's law) for each individual gas. It may seem that using either law will give the same result, but it does not.

The compressibility-factor approach, in general, gives more accurate results when the Z_i's in Eq. 12-10 are evaluated by using Amagat's law instead of Dalton's law. This is because Amagat's law involves the use of mixture pressure P_m, which accounts for the influence of intermolecular forces between the molecules of different gases. Dalton's law disregards the influence of dissimilar molecules in a mixture on each other. As a result, it tends to underpredict the pressure of a gas mixture for a given V_m and T_m. Therefore, Dalton's law is more appropriate for gas mixtures at low pressures. Amagat's law is more appropriate at high pressures.

Note that there is a significant difference between using the compressibility factor for a single gas and for a mixture of gases. The compressibility factor predicts the *P-v-T* behavior of single gases rather accurately, as discussed in Chap. 2, but not for mixtures of gases. When we use compressibility factors for the components of a gas mixture, we account for the influence of like molecules on each other; the influence of dissimilar molecules remains largely unaccounted for. Consequently, a property value predicted by this approach may be considerably different from the experimentally determined value.

Pseudopure substance

$$P'_{cr,m} = \sum_{i=1}^{k} y_i P_{cr,i}$$

$$T'_{cr,m} = \sum_{i=1}^{k} y_i T_{cr,i}$$

FIGURE 12-9

Another way of predicting the *P-v-T* behavior of a real-gas mixture is to treat it as a pseudopure substance with critical properties P'_{cr} and T'_{cr}.

Another approach for predicting the *P-v-T* behavior of a gas mixture is to treat the gas mixture as a pseudopure substance (Fig. 12-9). One such method, proposed by W. B. Kay in 1936 and called **Kay's rule,** involves the use of a *pseudocritical pressure* $P'_{cr,m}$ and *pseudocritical* temperature $T'_{cr,m}$ for the mixture, defined in terms of the critical pressures and temperatures of the mixture components as

$$P'_{cr,m} = \sum_{i=1}^{k} y_i P_{cr,i} \quad \text{and} \quad T'_{cr,m} = \sum_{i=1}^{k} y_i T_{cr,i} \tag{12-11a, b}$$

The compressibility factor of the mixture Z_m is then easily determined by using these pseudocritical properties. The result obtained by using Kay's rule is accurate to within about 10 percent over a wide range of temperatures and pressures, which is acceptable for most engineering purposes.

Another way of treating a gas mixture as a pseudopure substance is to use a more accurate equation of state such as the van der Waals, Beattie–

Bridgeman, or Benedict–Webb–Rubin equation for the mixture, and to determine the constant coefficients in terms of the coefficients of the components. In the van der Waals equation, for example, the two constants for the mixture are determined from

$$a_m = \left(\sum_{i=1}^{k} y_i a_i^{1/2}\right)^2 \quad \text{and} \quad b_m = \sum_{i=1}^{k} y_i b_i \qquad (12\text{-}12a, b)$$

where expressions for a_i and b_i are given in Chap. 2.

EXAMPLE 12-2 *P-v-T* **Behavior of Nonideal Gas Mixtures**

A rigid tank contains 2 kmol of N_2 and 6 kmol of CO_2 gases at 300 K and 15 MPa. Estimate the volume of the tank on the basis of (*a*) the ideal-gas equation of state, (*b*) Kay's rule, (*c*) compressibility factors and Amagat's law, and (*d*) compressibility factors and Dalton's law.

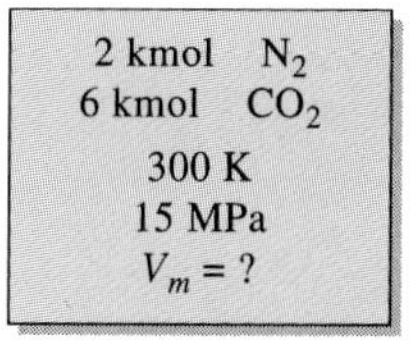

2 kmol N_2
6 kmol CO_2
300 K
15 MPa
V_m = ?

FIGURE 12-10
Schematic for Example 12-2.

Solution A sketch of the tank containing the gas mixture is given in Fig. 12-10. We note that the tank contains a mixture of two gases of known masses at a specified pressure and temperature.

Assumptions Stated in each section.

Analysis (*a*) When the mixture is assumed to behave as an ideal gas, the volume of the mixture is easily determined from the ideal-gas relation for the mixture:

$$V_m = \frac{N_m R_u T_m}{P_m} = \frac{(8\text{ kmol})[8.314\text{ kPa}\cdot\text{m}^3/(\text{kmol}\cdot\text{K})](300\text{ K})}{15{,}000\text{ kPa}} = \mathbf{1.330\ m^3}$$

since $N_m = N_{N_2} + N_{CO_2} = 2 + 6 = 8$ kmol

(*b*) To use Kay's rule, we need to determine the pseudocritical temperature and pseudocritical pressure of the mixture by using the critical-point properties of N_2 and CO_2 from Table A-1. But first we need to determine the mole fraction of each component:

$$y_{N_2} = \frac{N_{N_2}}{N_m} = \frac{2\text{ kmol}}{8\text{ kmol}} = 0.25 \quad \text{and} \quad y_{CO_2} = \frac{N_{CO_2}}{N_m} = \frac{6\text{ kmol}}{8\text{ kmol}} = 0.75$$

$$T'_{cr,m} = \sum y_i T_{cr,i} = y_{N_2} T_{cr,N_2} + y_{CO_2} T_{cr,CO_2}$$
$$= (0.25)(126.2\text{ K}) + (0.75)(304.2\text{ K}) = 259.7\text{ K}$$

$$P'_{cr,m} = \sum y_i P_{cr,i} = y_{N_2} P_{cr,N_2} + y_{CO_2} P_{cr,CO_2}$$
$$= (0.25)(3.39\text{ MPa}) + (0.75)(7.39\text{ MPa}) = 6.39\text{ MPa}$$

Then

$$\left.\begin{aligned} T_R &= \frac{T_m}{T'_{cr,m}} = \frac{300\text{ K}}{259.7\text{ K}} = 1.16 \\ P_R &= \frac{P_m}{P'_{cr,m}} = \frac{15\text{ MPa}}{6.39\text{ MPa}} = 2.35 \end{aligned}\right\} \quad Z_m = 0.49 \qquad \text{(Fig. A-30}b\text{)}$$

Thus,

$$V_m = \frac{Z_m N_m R_u T_m}{P_m} = Z_m V_{ideal} = (0.49)(1.330\text{ m}^3) = \mathbf{0.652\ m^3}$$

(*c*) When Amagat's law is used in conjunction with compressibility factors, Z_m is determined from Eq. 12-10. But first we need to determine the Z of each component on the basis of Amagat's law:

$$\text{N}_2: \quad \left.\begin{aligned} T_{R,\text{N}_2} &= \frac{T_m}{T_{\text{cr},\text{N}_2}} = \frac{300\text{ K}}{126.2\text{ K}} = 2.38 \\ P_{R,\text{N}_2} &= \frac{P_m}{P_{\text{cr},\text{N}_2}} = \frac{15\text{ MPa}}{3.39\text{ MPa}} = 4.42 \end{aligned}\right\} \quad Z_{\text{N}_2} = 1.02 \qquad \text{(Fig. A-30}b\text{)}$$

$$\text{CO}_2: \quad \left.\begin{aligned} T_{R,\text{CO}_2} &= \frac{T_m}{T_{\text{cr},\text{CO}_2}} = \frac{300\text{ K}}{304.2\text{ K}} = 0.99 \\ P_{R,\text{CO}_2} &= \frac{P_m}{P_{\text{cr},\text{CO}_2}} = \frac{15\text{ MPa}}{7.39\text{ MPa}} = 2.03 \end{aligned}\right\} \quad Z_{\text{CO}_2} = 0.30 \qquad \text{(Fig. A-30}b\text{)}$$

Mixture:

$$\begin{aligned} Z_m &= \sum y_i Z_i = y_{\text{N}_2} Z_{\text{N}_2} + y_{\text{CO}_2} Z_{\text{CO}_2} \\ &= (0.25)(1.02) + (0.75)(0.30) = 0.48 \end{aligned}$$

Thus,

$$V_m = \frac{Z_m N_m R_u T_m}{P_m} = Z_m V_{\text{ideal}} = (0.48)(1.330\text{ m}^3) = \mathbf{0.638\ m^3}$$

The compressibility factor in this case turned out to be almost the same as the one determined by using Kay's rule. This is not always the case, however.

(*d*) When Dalton's law is used in conjunction with compressibility factors, Z_m is again determined from Eq. 12-10. But this time the Z of each component is to be determined at the mixture temperature and volume, which is not known. Therefore, an iterative solution is required. We start the calculations by assuming that the volume of the gas mixture is 1.330 m³, the value determined by assuming ideal-gas behavior.

The T_R values in this case are identical to those obtained in part (*c*) and remain constant. The pseudoreduced volume is determined from its definition, Eq. 2-21:

$$\begin{aligned} v_{R,\text{N}_2} &= \frac{\bar{v}_{\text{N}_2}}{R_u T_{\text{cr},\text{N}_2}/P_{\text{cr},\text{N}_2}} = \frac{V_m/N_{\text{N}_2}}{R_u T_{\text{cr},\text{N}_2}/P_{\text{cr},\text{N}_2}} \\ &= \frac{1.33\text{ m}^3/(2\text{ kmol})}{[8.314\text{ kPa}\cdot\text{m}^3/(\text{kmol}\cdot\text{K})](126.2\text{ K})/(3390\text{ kPa})} = 2.15 \end{aligned}$$

Similarly

$$v_{R,\text{CO}_2} = \frac{1.33\text{ m}^3/(6\text{ kmol})}{[8.314\text{ kPa}\cdot\text{m}^3/(\text{kmol}\cdot\text{K})](304.2\text{ K})/(7390\text{ kPa})} = 0.648$$

From Fig. A-30*b*, we read $Z_{\text{N}_2} = 0.99$ and $Z_{\text{CO}_2} = 0.56$. Thus,

$$Z_m = y_{\text{N}_2} Z_{\text{N}_2} + y_{\text{CO}_2} Z_{\text{CO}_2} = (0.25)(0.99) + (0.75)(0.56) = 0.67$$

and

$$V_m = \frac{Z_m N_m R T_m}{P_m} = Z_m V_{\text{ideal}} = (0.67)(1.330\text{ m}^3) = 0.891\text{ m}^3$$

This is 33 percent lower than the assumed value. Therefore, we should repeat the calculations, using the new value of V_m. When the calculations are repeated we obtain 0.738 m³ after the second iteration, 0.678 m³ after the third iteration, and 0.648 m³ after the fourth iteration. This value does not change with more iterations. Therefore,

$$V_m = 0.648 \text{ m}^3$$

Discussion Notice that the results obtained in parts (*b*), (*c*), and (*d*) are very close. But they are very different from the values obtained from the ideal-gas relation. Therefore, treating a mixture of gases as an ideal gas may yield unacceptable errors at high pressures.

12-3 ■ PROPERTIES OF GAS MIXTURES: Ideal and Real Gases

Consider a gas mixture that consists of 2 kg of N_2 and 3 kg of CO_2. The mass (an *extensive property*) of this mixture is, to nobody's surprise, 5 kg. How did we do it? Well, we simply added the mass of each component. This example suggests a simple way of evaluating the **extensive properties** of a nonreacting ideal- or real-gas mixture: *Just add the contributions of each component of the mixture* (Fig. 12-11). Then the total internal energy, enthalpy, and entropy of a gas mixture can be expressed, respectively, as

2 kmol A
6 kmol B
U_A = 1000 kJ
U_B = 1800 kJ
↓
U_m = 2800 kJ

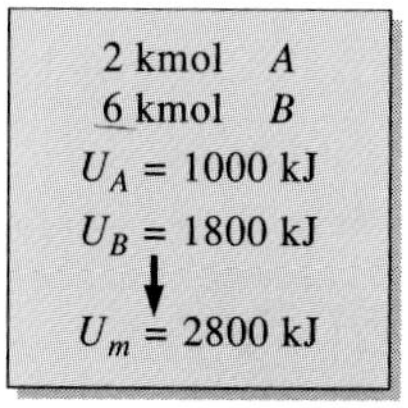

FIGURE 12-11
The extensive properties of a mixture are determined by simply adding the properties of the components.

$$U_m = \sum_{i=1}^{k} U_i = \sum_{i=1}^{k} m_i u_i = \sum_{i=1}^{k} N_i \bar{u}_i \qquad \text{(kJ)} \qquad (12\text{-}13)$$

$$H_m = \sum_{i=1}^{k} H_i = \sum_{i=1}^{k} m_i h_i = \sum_{i=1}^{k} N_i \bar{h}_i \qquad \text{(kJ)} \qquad (12\text{-}14)$$

$$S_m = \sum_{i=1}^{k} S_i = \sum_{i=1}^{k} m_i s_i = \sum_{i=1}^{k} N_i \bar{s}_i \qquad \text{(kJ/K)} \qquad (12\text{-}15)$$

By following a similar logic, the changes in internal energy, enthalpy, and entropy of a gas mixture during a process can be expressed, respectively, as

$$\Delta U_m = \sum_{i=1}^{k} \Delta U_i = \sum_{i=1}^{k} m_i \Delta u_i = \sum_{i=1}^{k} N_i \Delta \bar{u}_i \qquad \text{(kJ)} \qquad (12\text{-}16)$$

$$\Delta H_m = \sum_{i=1}^{k} \Delta H_i = \sum_{i=1}^{k} m_i \Delta h_i = \sum_{i=1}^{k} N_i \Delta \bar{h}_i \qquad \text{(kJ)} \qquad (12\text{-}17)$$

$$\Delta S_m = \sum_{i=1}^{k} \Delta S_i = \sum_{i=1}^{k} m_i \Delta s_i = \sum_{i=1}^{k} N_i \Delta \bar{s}_i \qquad \text{(kJ/K)} \qquad (12\text{-}18)$$

Now reconsider the same mixture, and assume that both N_2 and CO_2 are at 25°C. The temperature (an *intensive* property) of the mixture is, as you would expect, also 25°C. Notice that we did not add the component temperatures to determine the mixture temperature. Instead, we used some kind of averaging scheme, a characteristic approach for determining the **intensive properties** of a gas mixture. The internal energy, enthalpy, and entropy of a gas mixture *per unit mass* or *per unit mole* of the mixture can be determined by dividing the equations above by the mass or the mole number of the mixture (m_m or N_m). We obtain (Fig. 12-12)

FIGURE 12-12
The intensive properties of a mixture are determined by weighted averaging.

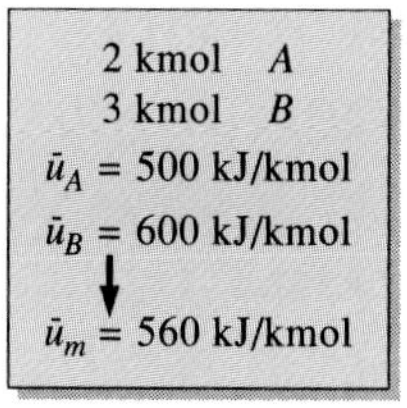

2 kmol A
3 kmol B
$\bar{u}_A$ = 500 kJ/kmol
$\bar{u}_B$ = 600 kJ/kmol
↓
$\bar{u}_m$ = 560 kJ/kmol

$$u_m = \sum_{i=1}^{k} \text{mf}_i u_i \quad \text{and} \quad \bar{u}_m = \sum_{i=1}^{k} y_i \bar{u}_i \qquad \text{(kJ/kg or kJ/kmol)} \tag{12-20}$$

$$h_m = \sum_{i=1}^{k} \text{mf}_i h_i \quad \text{and} \quad \bar{h}_m = \sum_{i=1}^{k} y_i \bar{h}_i \qquad \text{(kJ/kg or kJ/kmol)} \tag{12-20}$$

$$s_m = \sum_{i=1}^{k} \text{mf}_i s_i \quad \text{and} \quad \bar{s}_m = \sum_{i=1}^{k} y_i \bar{s}_i \qquad [\text{(kJ/(kg}\cdot\text{K) or kJ/(kmol}\cdot\text{K)}] \tag{12-21}$$

Similarly, the specific heats of a gas mixture can be expressed as

$$C_{v,m} = \sum_{i=1}^{k} \text{mf}_i C_{v,i} \quad \text{and} \quad \bar{C}_{v,m} = \sum_{i=1}^{k} y_i \bar{C}_{v,i} \qquad [\text{kJ/(kg}\cdot{}^\circ\text{C) or kJ/(kmol}\cdot{}^\circ\text{C)}] \tag{12-22}$$

$$C_{p,m} = \sum_{i=1}^{k} \text{mf}_i C_{p,i} \quad \text{and} \quad \bar{C}_{p,m} = \sum_{i=1}^{k} y_i \bar{C}_{p,i} \qquad [\text{kJ/(kg}\cdot{}^\circ\text{C) or kJ/(kmol}\cdot{}^\circ\text{C)}] \tag{12-23}$$

Notice that *properties per unit mass involve mass fractions* (mf_i) *and properties per unit mole involve mole fractions* (y_i).

The relations given above are generally valid and are applicable to both ideal- and real-gas mixtures. (In fact, they are also applicable to nonreacting liquid and solid solutions.) The only major difficulty associated with these relations is the determination of properties for each individual gas in the mixture. The analysis can be simplified greatly, however, by treating the individual gases as an ideal gas, if doing so does not introduce a significant error.

1 Ideal-Gas Mixtures

The gases that comprise a mixture are often at a high temperature and low pressure relative to the critical-point values of individual gases. In such cases, the gas mixture and its components can be treated as ideal gases with negligible error. Under the ideal-gas approximation, the properties of a gas are not influenced by the presence of other gases, and each gas component in the mixture behaves as if it exists alone at the mixture temperature T_m and mixture volume V_m. This principle is known as the **Gibbs–Dalton law,** which is an extension of Dalton's law of additive pressures. Also, the h, u, C_v, and C_p of an ideal gas depend on temperature only and are independent of the pressure or the volume of the ideal-gas mixture. The partial pressure of a component in an ideal-gas mixture is simply $P_i = y_i P_m$, where P_m is the mixture pressure.

Evaluation of Δu or Δh of the components of an ideal-gas mixture during a process is relatively easy since it requires only a knowledge of the initial and final temperatures. Care should be exercised, however, in evaluating the Δs of the components since the entropy of an ideal gas depends on the pressure or volume of the component as well as on its temperature. The entropy change of individual gases in an ideal-gas mixture during a process can be determined from

$$\Delta s_i = s^\circ_{i,2} - s^\circ_{i,1} - R_i \ln \frac{P_{i,2}}{P_{i,1}} \cong C_{p,i} \ln \frac{T_{i,2}}{T_{i,1}} - R_i \ln \frac{P_{i,2}}{P_{i,1}} \qquad (12\text{-}24)$$

or

$$\Delta \bar{s}_i = \bar{s}^\circ_{i,2} - \bar{s}^\circ_{i,1} - R_u \ln \frac{P_{i,2}}{P_{i,1}} \cong \bar{C}_{p,i} \ln \frac{T_{i,2}}{T_{i,1}} - R_u \ln \frac{P_{i,2}}{P_{i,1}} \qquad (12\text{-}25)$$

where $P_{i,2} = y_{i,2}P_{m,2}$ and $P_{i,1} = y_{i,1}P_{m,1}$. Notice that the partial pressure P_i of each component is used in the evaluation of the entropy change, not the mixture pressure P_m (Fig. 12-13).

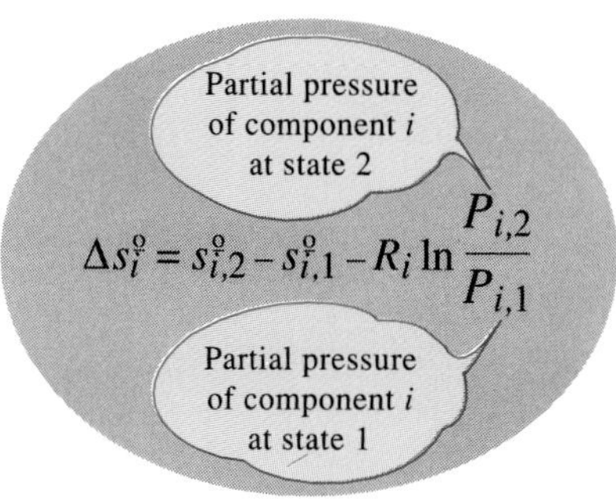

FIGURE 12-13

Partial pressures (not the mixture pressure) are used in the evaluation of entropy changes of ideal-gas mixtures.

EXAMPLE 12-3 Mixing Two Ideal Gases in a Tank

An insulated rigid tank is divided into two compartments by a partition. One compartment contains 7 kg of oxygen gas at 40°C and 100 kPa, and the other compartment contains 4 kg of nitrogen gas at 20°C and 150 kPa. Now the partition is removed, and the two gases are allowed to mix. Determine (*a*) the mixture temperature and (*b*) the mixture pressure after equilibrium has been established.

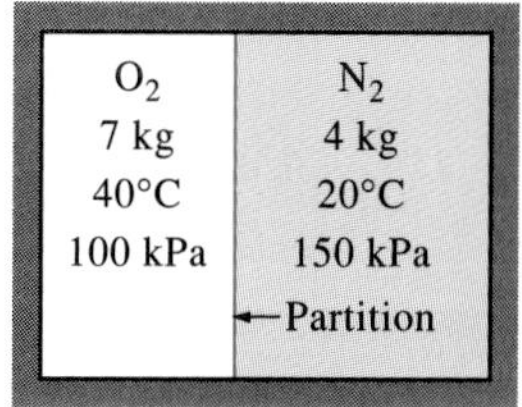

FIGURE 12-14

Schematic for Example 12-3.

Solution We take the entire contents of the tank (both compartments) as the system (Fig. 12-14). This is a *closed system* since no mass crosses the boundary during the process. We note that the volume of a rigid tank is constant and thus $V_2 = V_1$ and there is no boundary work done.

Assumptions **1** We assume both gases to be ideal gases, and their mixture to be an ideal-gas mixture. This assumption is reasonable since both the oxygen and nitrogen are well above their critical temperatures and well below their critical pressures. **2** The tank is insulated and thus there is no heat transfer. **3** There are no other forms of work involved.

Analysis (*a*) Noting that there is no energy transfer to or from the tank, the energy balance for the system can be expressed as

$$E_{in} - E_{out} = \Delta E_{system}$$

$$0 = \Delta U = \Delta U_{N_2} + \Delta U_{O_2}$$

$$[mC_v(T_m - T_1)]_{N_2} + [mC_v(T_m - T_1)]_{O_2} = 0$$

By using C_v values at room temperature (from Table A-2*a*), the final temperature of the mixture is determined to be

$$(4 \text{ kg})[0.743 \text{ kJ/(kg} \cdot {}^\circ\text{C)}](T_m - 20^\circ\text{C}) + (7 \text{ kg})[0.658 \text{ kJ/(kg} \cdot {}^\circ\text{C)}](T_m - 40^\circ\text{C}) = 0$$

$$T_m = \mathbf{32.2^\circ C}$$

(*b*) The final pressure of the mixture is determined from the ideal-gas relation

$$P_mV_m = N_mR_uT_m$$

where

$$N_{O_2} = \frac{m_{O_2}}{M_{O_2}} = \frac{7\text{ kg}}{32\text{ kg/kmol}} = 0.219\text{ kmol}$$

$$N_{N_2} = \frac{m_{N_2}}{M_{N_2}} = \frac{4\text{ kg}}{28\text{ kg/kmol}} = 0.143\text{ kmol}$$

$$N_m = N_{O_2} + N_{N_2} = 0.219 + 0.143 = 0.362\text{ kmol}$$

and

$$V_{O_2} = \left(\frac{NR_uT_1}{P_1}\right)_{O_2} = \frac{(0.219\text{ kmol})[8.314\text{ kPa}\cdot\text{m}^3/(\text{kmol}\cdot\text{K})](313\text{ K})}{100\text{ kPa}} = 5.70\text{ m}^3$$

$$V_{N_2} = \left(\frac{NR_uT_1}{P_1}\right)_{N_2} = \frac{(0.143\text{ kmol})[8.314\text{ kPa}\cdot\text{m}^3/(\text{kmol}\cdot\text{k})]](293\text{ K})}{150\text{ kPa}} = 2.32\text{ m}^3$$

$$V_m = V_{O_2} + V_{N_2} = 5.70 + 2.32 = 8.02\text{ m}^3$$

Thus,

$$P_m = \frac{N_mR_uT_m}{V_m} = \frac{(0.362\text{ kmol})[8.314\text{ kPa}\cdot\text{m}^3/(\text{kmol}\cdot\text{K})](305.2\text{ K})}{8.02\text{ m}^3}$$

$$= \mathbf{114.5\text{ kPa}}$$

Discussion We could also determine the mixture pressure by using $P_mV_m = m_mR_mT_m$, where R_m is the apparent gas constant of the mixture. This would not be easier, however, since the evaluation of R requires the determination of the mole fractions of the components.

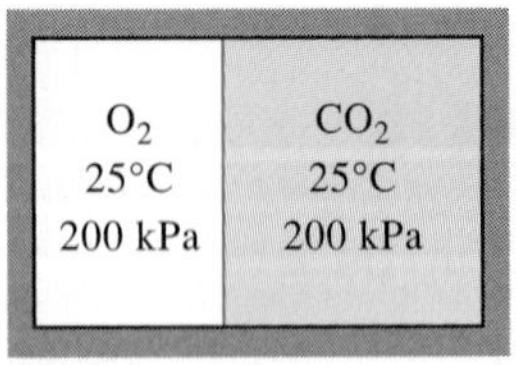

O_2 25°C 200 kPa	CO_2 25°C 200 kPa

FIGURE 12-15
Schematic for Example 12-4.

EXAMPLE 12-4 Exergy Destruction during Mixing of Ideal Gases

An insulated rigid tank is divided into two compartments by a partition, as shown in Fig. 12-15. One compartment contains 3 kmol of O_2, and the other compartment contains 5 kmol of CO_2. Both gases are initially at 25°C and 200 kPa. Now the partition is removed, and the two gases are allowed to mix. Assuming the surroundings are at 25°C and both gases behave as ideal gases, determine the entropy change and exergy destruction associated with this process.

Solution We take the entire contents of the tank (both compartments) as the system (Fig. 12-15). This is a *closed system* since no mass crosses the boundary during the process. We note that the volume of a rigid tank is constant, and there is no energy transfer as heat or work. Also, both gases are initially at the same temperature and pressure.

Assumptions Both gases and their mixture are ideal gases.

Analysis When two ideal gases initially at the same temperature and pressure are mixed by removing a partition between them, the mixture will also be at the same temperature and pressure. (Can you prove it? Will this be true for nonideal gases?) Therefore, the temperature and pressure in the tank will still be 25°C and 200 kPa, respectively, after the mixing. The entropy change of each component gas can be determined from Eqs. 12-18 and 12-25:

$$\Delta S_m = \sum \Delta S_i = \sum N_i \Delta \bar{s}_i = \sum N_i \left(\bar{C}_{p,i} \ln \frac{T_{i,2}}{T_{i,1}} \overset{0}{\nearrow} - R_u \ln \frac{P_{i,2}}{P_{i,1}} \right)$$

$$= -R_u \sum N_i \ln \frac{y_i P_{m,2}}{P_{i,1}} = -R_u \sum N_i \ln y_i$$

since $P_{m,2} = P_{i,1} = 200$ kPa. It is obvious that the entropy change is independent of the composition of the mixture in this case and depends on only the mole fraction of the gases in the mixture. What is not so obvious is that if the same gas in two different chambers is mixed at constant temperature and pressure, the entropy change is zero.

Substituting the known values, we see that the entropy change is

$$N_m = N_{O_2} + N_{CO_2} = (3 + 5) \text{ kmol} = 8 \text{ kmol}$$

$$y_{O_2} = \frac{N_{O_2}}{N_m} = \frac{3 \text{ kmol}}{8 \text{ kmol}} = 0.375$$

$$y_{CO_2} = \frac{N_{CO_2}}{N_m} = \frac{5 \text{ kmol}}{8 \text{ kmol}} = 0.625$$

$$\Delta S_m = -R_u (N_{O_2} \ln y_{O_2} + N_{CO_2} \ln y_{CO_2})$$

$$= -[8.314 \text{ kJ/(kmol} \cdot \text{K)}][(3 \text{ kmol})(\ln 0.375) + (5 \text{ kmol})(\ln 0.625)]$$

$$= \mathbf{44.0 \ kJ/K}$$

The exergy destruction associated with this mixing process is determined from

$$X_{\text{destroyed}} = T_0 S_{\text{gen}} = T_0 (\Delta S_{\text{sys}} + \Delta S_{\text{surr}} \overset{0}{\nearrow}) = T_0 \Delta S_{\text{sys}}$$

$$= (298 \text{ K})(44.0 \text{ kJ/K})$$

$$= \mathbf{13{,}112 \ kJ}$$

This result shows that mixing processes are highly irreversible.

2 Real-Gas Mixtures

When the components of a gas mixture do not behave as ideal gases, the analysis becomes more complex because the properties of real (nonideal) gases such as u, h, C_v, and C_p depend on the pressure (or specific volume) as well as on the temperature. In such cases, the effects of deviation from ideal-gas behavior on the mixture properties should be accounted for.

Consider two nonideal gases contained in two separate compartments of an adiabatic rigid tank at 100 kPa and 25°C. The partition separating the two gases is removed, and the two gases are allowed to mix. What do you think the final pressure in the tank will be? You are probably tempted to say 100 kPa, which would be true for ideal gases. However, this is not true for nonideal gases because of the influence of the molecules of different gases on each other (deviation from Dalton's law, Fig. 12-16).

When real-gas mixtures are involved, it may be necessary to account for the effect of nonideal behavior on the mixture properties such as enthalpy

FIGURE 12-16

It is difficult to predict the behavior of nonideal-gas mixtures because of the influence of dissimilar gas molecules on each other.

Real gas *A*	Real gas *B*
25°C	25°C
0.4 m³	0.6 m³
100 kPa	100 kPa

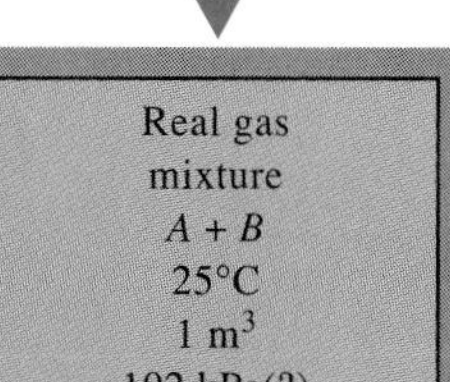

and entropy. One way of doing that is to use compressibility factors in conjunction with generalized equations and charts developed in Chap. 11 for real gases.

Consider the following $T\,ds$ relation for a gas mixture:

$$dh_m = T_m\,ds_m + v_m\,dP_m$$

It can also be expressed as

$$d\left(\sum \mathrm{mf}_i h_i\right) = T_m d\left(\sum \mathrm{mf}_i s_i\right) + \left(\sum \mathrm{mf}_i v_i\right) dP_m$$

or

$$\sum \mathrm{mf}_i (dh_i - T_m\,ds_i - v_i\,dP_m) = 0$$

which yields

$$dh_i = T_m\,ds_i + v_i\,dP_m \qquad \text{(12-26)}$$

This is an important result because Eq. 12-26 is the starting equation in the development of the generalized relations and charts for enthalpy and entropy. It suggests that the generalized property relations and charts for real gases developed in Chap. 11 can also be used for the components of real-gas mixtures. But the reduced temperature T_R and reduced pressure P_R for each component should be evaluated by using the mixture temperature T_m and mixture pressure P_m. This is because Eq. 12-26 involves the mixture pressure P_m, not the component pressure P_i.

The approach described above is somewhat analogous to Amagat's law of additive volumes (evaluating mixture properties at the mixture pressure and temperature), which holds exactly for ideal-gas mixtures and approximately for real-gas mixtures. Therefore, the mixture properties determined with this approach will not be exact, but they will be sufficiently accurate for most purposes.

What if the mixture volume and temperature are specified instead of the mixture pressure and temperature? Well, there is no need to panic. Just evaluate the mixture pressure, using Dalton's law of additive pressures, and then use this value (which is only approximate) as the mixture pressure.

Another way of evaluating the properties of a real-gas mixture is to treat the mixture as a pseudopure substance having pseudocritical properties, determined in terms of the critical properties of the component gases by using Kay's rule. The approach is quite simple, and the accuracy is usually acceptable.

FIGURE 12-17
Schematic for Example 12-5.

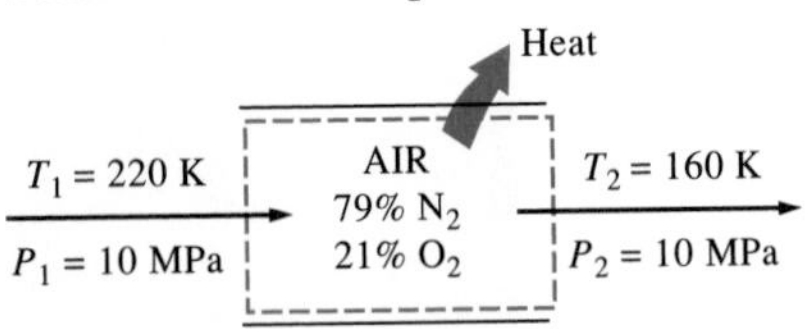

EXAMPLE 12-5 Heating of a Nonideal Gas Mixture

Air is a mixture of N_2, O_2, and small amounts of other gases, and it can be approximated as 79 percent N_2 and 21 percent O_2 on a mole basis. During a steady-flow process, the air is cooled from 220 to 160 K at a constant pressure of 10 MPa (Fig. 12-17). Determine the heat transfer during this process per kmol of air, using (*a*) the ideal-gas approximation, (*b*) Kay's rule, and (*c*) Amagat's law.

Solution We take the *cooling section* as the system (Fig. 12-17). This is a *control volume* since mass crosses the system boundary during the process. We note that heat is transferred out of the system.

Assumptions **1** This is a steady-flow process since there is no change with time at any point and thus $\Delta m_{CV} = 0$ and $\Delta E_{CV} = 0$. **2** The kinetic and potential energy changes are negligible.

Analysis The critical properties are $T_{cr} = 126.2$ K and $P_{cr} = 3.39$ MPa for N_2 and $T_{cr} = 154.8$ K and $P_{cr} = 5.08$ MPa for O_2. Both gases remain above their critical temperatures, but they are also above their critical pressures. Therefore, air will probably deviate from ideal-gas behavior, and thus it should be treated as a real-gas mixture.

The energy balance for this steady-flow system can be expressed on a unit-mole basis as

$$e_{in} - e_{out} = \Delta e_{system}^{\nearrow 0} = 0 \longrightarrow e_{in} = e_{out} \longrightarrow \bar{h}_1 = \bar{h}_2 + \bar{q}_{out}$$

$$\bar{q}_{out} = \bar{h}_1 - \bar{h}_2 = y_{N_2}(\bar{h}_1 - \bar{h}_2)_{N_2} + y_{O_2}(\bar{h}_1 - \bar{h}_2)_{O_2}$$

where the enthalpy change for either component can be determined from the generalized enthalpy departure chart (Fig. A-31) and Eq. 11-58:

$$\bar{h}_1 - \bar{h}_2 = \bar{h}_{1,\text{ideal}} - \bar{h}_{2,\text{ideal}} - R_u T_{cr}(Z_{h1} - Z_{h2})$$

The first two terms on the right-hand side of this equation represent the ideal-gas enthalpy change of the component. The terms in parentheses represent the deviation from the ideal-gas behavior, and their evaluation requires a knowledge of reduced pressure P_R and reduced temperature T_R, which are calculated at the mixture temperature T_m and mixture pressure P_m.

(*a*) If the N_2 and O_2 mixture is assumed to behave as an ideal gas, the enthalpy of the mixture will depend on temperature only, and the enthalpy values at the initial and the final temperatures can be determined from the ideal-gas tables of N_2 and O_2 (Tables A-18 and A-19):

$$T_1 = 220\text{ K} \longrightarrow \bar{h}_{1,\text{ideal},N_2} = 6391\text{ kJ/kmol}$$

$$\bar{h}_{1,\text{ideal},O_2} = 6404\text{ kJ/kmol}$$

$$T_2 = 160\text{ K} \longrightarrow \bar{h}_{2,\text{ideal},N_2} = 4648\text{ kJ/kmol}$$

$$\bar{h}_{2,\text{ideal},O_2} = 4657\text{ kJ/kmol}$$

$$\bar{q}_{out} = y_{N_2}(\bar{h}_1 - \bar{h}_2)_{N_2} + y_{O_2}(\bar{h}_1 - \bar{h}_2)_{O_2}$$
$$= (0.79)(6391 - 4648)\text{ kJ/kmol} + (0.21)(6404 - 4657)\text{ kJ/kmol}$$
$$= \mathbf{1744\ kJ/kmol}$$

(*b*) Kay's rule is based on treating a gas mixture as a pseudopure substance whose critical temperature and pressure are, respectively,

$$T'_{cr,m} = \sum y_i T_{cr,i} = y_{N_2} T_{cr,N_2} + y_{O_2} T_{cr,O_2}$$
$$= (0.79)(126.2\text{ K}) + (0.21)(154.8\text{ K}) = 132.2\text{ K}$$

and

$$P'_{cr,m} = \sum y_i P_{cr,i} = y_{N_2} P_{cr,N_2} + y_{O_2} P_{cr,O_2}$$
$$= (0.79)(3.39\text{ MPa}) + (0.21)(5.08\text{ MPa}) = 3.74\text{ MPa}$$

Then

$$T_{R,1} = \frac{T_{m,1}}{T'_{cr,m}} = \frac{220\text{ K}}{132.2\text{ K}} = 1.66 \quad \Bigg\} \quad Z_{h_1,m} = 1.0$$

$$P_R = \frac{P_m}{P'_{cr,m}} = \frac{10\text{ MPa}}{3.74\text{ MPa}} = 2.67$$

$$T_{R,2} = \frac{T_{m,2}}{T'_{cr,m}} = \frac{160\text{ K}}{132.2\text{ K}} = 1.21 \quad \Bigg\} \quad Z_{h_2,m} = 2.6$$

Also,

$$\begin{aligned}\bar{h}_{m_1,\text{ ideal}} &= y_{N_2}\bar{h}_{1,\text{ ideal},N_2} + y_{O_2}\bar{h}_{1,\text{ ideal},O_2}\\ &= (0.79)(6391\text{ kJ/kmol}) + (0.21)(6404\text{ kJ/kmol})\\ &= 6394\text{ kJ/kmol}\end{aligned}$$

$$\begin{aligned}\bar{h}_{m_2,\text{ ideal}} &= y_{N_2}\bar{h}_{2,\text{ ideal},N_2} + y_{O_2}\bar{h}_{2,\text{ ideal},O_2}\\ &= (0.79)(4648\text{ kJ/kmol}) + (0.21)(4657\text{ kJ/kmol})\\ &= 4650\text{ kJ/kmol}\end{aligned}$$

Therefore,

$$\begin{aligned}\bar{q}_{out} &= (\bar{h}_{m_1,\text{ ideal}} - \bar{h}_{m_2,\text{ ideal}}) - R_uT_{cr}(Z_{h_1} - Z_{h_2})_m\\ &= [(6394 - 4650)\text{ kJ/kmol}] - [8.314\text{ kJ/(kmol}\cdot\text{K)}](132.2\text{ K})(1.0 - 2.6)\\ &= \mathbf{3503\text{ kJ/kmol}}\end{aligned}$$

(*c*) The reduced temperatures and pressures for both N_2 and O_2 at the initial and final states and the corresponding enthalpy departure factors are, from Fig. A-31,

N_2:

$$T_{R_1,N_2} = \frac{T_{m,1}}{T_{cr,N_2}} = \frac{220\text{ K}}{126.2\text{ K}} = 1.74 \quad \Bigg\} \quad Z_{h_1,N_2} = 0.9$$

$$P_{R_1,N_2} = P_{R_2,N_2} = \frac{P_m}{P_{cr,N_2}} = \frac{10\text{ MPa}}{3.39\text{ MPa}} = 2.95$$

$$T_{R_2,N_2} = \frac{T_{m,2}}{T_{cr,N_2}} = \frac{160\text{ K}}{126.2\text{ K}} = 1.27 \quad \Bigg\} \quad Z_{h_2,N_2} = 2.4$$

O_2:

$$T_{R_1,O_2} = \frac{T_{m,1}}{T_{cr,O_2}} = \frac{220\text{ K}}{154.8\text{ K}} = 1.42 \quad \Bigg\} \quad Z_{h_1,O_2} = 1.3$$

$$P_{R_1,O_2} = P_{R_2,O_2} = \frac{P_m}{P_{cr,O_2}} = \frac{10\text{ MPa}}{5.08\text{ MPa}} = 1.97$$

$$T_{R_1,O_2} = \frac{T_{m,2}}{T_{cr,O_2}} = \frac{160\text{ K}}{154.8\text{ K}} = 1.03 \quad \Bigg\} \quad Z_{h_2,O_2} = 4.0$$

From Eq. 11-58,

$$\begin{aligned}(\bar{h}_1 - \bar{h}_2)_{N_2} &= (\bar{h}_{1,\text{ ideal}} - \bar{h}_{2,\text{ ideal}})_{N_2} - R_uT_{cr}(Z_{h_1} - Z_{h_2})_{N_2}\\ &= [(6391 - 4648)\text{ kJ/kmol}] - [8.314\text{ kJ/(kmol}\cdot\text{K)}](126.2\text{ K})(0.9 - 2.4)\\ &= 3317\text{ kJ/kmol}\end{aligned}$$

$$\begin{aligned}(\bar{h}_1 - \bar{h}_2)_{O_2} &= (\bar{h}_{1,\text{ ideal}} - \bar{h}_{2,\text{ ideal}})_{O_2} - R_uT_{cr}(Z_{h1} - Z_{h2})_{O_2}\\ &= [(6404 - 4657)\text{ kJ/kmol}] - [8.314\text{ kJ/(kmol}\cdot\text{K)}](154.8\text{ K})(1.3 - 4.0)\\ &= 5222\text{ kJ/kmol}\end{aligned}$$

Therefore,

$$\begin{aligned}\bar{q}_{out} &= y_{N_2}(\bar{h}_1 - \bar{h}_2)_{N_2} + y_{O_2}(\bar{h}_1 - \bar{h}_2)_{O_2}\\ &= (0.79)(3317\text{ kJ/kmol}) + (0.21)(5222\text{ kJ/kmol})\\ &= \mathbf{3717\text{ kJ/kmol}}\end{aligned}$$

Discussion This result is about 6 percent greater than the result obtained in part (*b*) by using Kay's rule. But it is more than twice the result obtained by assuming the mixture to be an ideal gas.

12-4 ■ SUMMARY

A mixture of two or more gases of fixed chemical composition is called a *nonreacting gas mixture*. The composition of a gas mixture is described by specifying either the *mole fraction* or the *mass fraction* of each component, defined as

$$\mathrm{mf}_i = \frac{m_i}{m_m} \quad \text{and} \quad y_i = \frac{N_i}{N_m}$$

where

$$m_m = \sum_{i=1}^{k} m_i \quad \text{and} \quad N_m = \sum_{i=1}^{k} N_i$$

The *apparent* (or average) *molar mass* and *gas constant* of a mixture are expressed as

$$M_m = \frac{m_m}{N_m} = \sum_{i=1}^{k} y_i M_i \qquad \text{(kg/kmol)}$$

and

$$R_m = \frac{R_u}{M_m} \qquad [\text{kJ/(kg} \cdot \text{K)}]$$

Dalton's law of additive pressures states that the pressure of a gas mixture is equal to the sum of the pressures each gas would exert if it existed alone at the mixture temperature and volume. *Amagat's law of additive volumes* states that the volume of a gas mixture is equal to the sum of the volumes each gas would occupy if it existed alone at the mixture temperature and pressure. Dalton's and Amagat's laws hold exactly for ideal-gas mixtures, but only approximately for real-gas mixtures. They can be expressed as

Dalton's law:

$$P_m = \sum_{i=1}^{k} P_i(T_m, V_m)$$

Amagat's law:

$$V_m = \sum_{i=1}^{k} V_i(T_m, P_m)$$

Here P_i is called the *component pressure* and V_i is called the *component volume*. Also, the ratio P_i/P_m is called the *pressure fraction* and the ratio V_i/V_m is called the *volume fraction* of component i. For *ideal gases*, P_i and V_i can be related to y_i by

$$\frac{P_i}{P_m} = \frac{V_i}{V_m} = \frac{N_i}{N_m} = y_i$$

The quantity $y_i P_m$ is called the *partial pressure* and the quantity $y_i V_m$ is called the *partial volume*. The *P-v-T* behavior of real-gas mixtures can be predicted by using generalized compressibility charts. The compressibility factor of the mixture can be expressed in terms of the compressibility factors of the individual gases as

$$Z_m = \sum_{i=1}^{k} y_i Z_i$$

where Z_i is determined either at T_m and V_m (Dalton's law) or at T_m and P_m (Amagat's law) for each individual gas. The *P-v-T* behavior of a gas mixture can also be predicted approximately by *Kay's rule,* which involves treating a gas mixture as a pure substance with pseudocritical properties determined from

$$P'_{\text{cr},m} = \sum_{i=1}^{k} y_i P_{\text{cr},i} \quad \text{and} \quad T'_{\text{cr},m} = \sum_{i=1}^{k} y_i T_{\text{cr},i}$$

The *extensive properties* of a gas mixture, in general, can be determined by summing the contributions of each component of the mixture. The evaluation of *intensive properties* of a gas mixture, however, involves averaging in terms of mass or mole fractions:

$$U_m = \sum_{i=1}^{k} U_i = \sum_{i=1}^{k} m_i u_i = \sum_{i=1}^{k} N_i \bar{u}_i \qquad \text{(kJ)}$$

$$H_m = \sum_{i=1}^{k} H_i = \sum_{i=1}^{k} m_i h_i = \sum_{i=1}^{k} N_i \bar{h}_i \qquad \text{(kJ)}$$

$$S_m = \sum_{i=1}^{k} S_i = \sum_{i=1}^{k} m_i s_i = \sum_{i=1}^{k} N_i \bar{s}_i \qquad \text{(kJ/K)}$$

and

$$u_m = \sum_{i=1}^{k} \text{mf}_i u_i \quad \text{and} \quad \bar{u}_m = \sum_{i=1}^{k} y_i \bar{u}_i \qquad \text{(kJ/kg or kJ/kmol)}$$

$$h_m = \sum_{i=1}^{k} \text{mf}_i h_i \quad \text{and} \quad \bar{h}_m = \sum_{i=1}^{k} y_i \bar{h}_i \qquad \text{(kJ/kg or kJ/kmol)}$$

$$s_m = \sum_{i=1}^{k} \text{mf}_i s_i \quad \text{and} \quad \bar{s}_m = \sum_{i=1}^{k} y_i \bar{s}_i \qquad [\text{kJ/(kg}\cdot\text{K) or kJ(kmol}\cdot\text{K)}]$$

$$C_{v,m} = \sum_{i=1}^{k} \text{mf}_i C_{v,i} \quad \text{and} \quad \bar{C}_{v,m} = \sum_{i=1}^{k} y_i \bar{C}_{v,i}$$

$$C_{p,m} = \sum_{i=1}^{k} \text{mf}_i C_{p,i} \quad \text{and} \quad \bar{C}_{p,m} = \sum_{i=1}^{k} y_i \bar{C}_{p,i}$$

These relations are applicable to both ideal- and real-gas mixtures. The properties or property changes of individual components can be determined by using ideal-gas or real-gas relations developed in earlier chapters.

REFERENCES AND SUGGESTED READING

1. W. Z. Black and J. G. Hartley. *Thermodynamics*. New York: Harper & Row, 1985.

2. J. P. Holman. *Thermodynamics*. 3rd ed. New York: McGraw-Hill, 1980.

3. J. R. Howell and R. O. Buckius. *Fundamentals of Engineering Thermodynamics*. New York: McGraw-Hill, 1987.

4. J. B. Jones and G. A. Hawkins. *Engineering Thermodynamics*. 2nd ed. New York: John Wiley & Sons, 1986.

5. G. J. Van Wylen and R. E. Sonntag. *Fundamentals of Classical Thermodynamics*. 3rd ed. New York: John Wiley & Sons, 1985.

6. K. Wark. *Thermodynamics*. 5th ed. New York: McGraw-Hill, 1988.

PROBLEMS*

Composition of Gas Mixtures

12-1C What is the *apparent gas constant* for a gas mixture? Can it be larger than the largest gas constant in the mixture?

12-2C Consider a mixture of two gases. Can the apparent molar mass of this mixture be determined by simply taking the arithmetic average of the molar masses of the individual gases? When will this be the case?

12-3C What is the *apparent molar mass* for a gas mixture? Does the mass of every molecule in the mixture equal the apparent molar mass?

12-4C Consider a mixture of several gases of identical masses. Will all the mass fractions be identical? How about the mole fractions?

12-5C The sum of the mole fractions for an ideal-gas mixture is equal to 1. Is this also true for a real-gas mixture?

12-6C What are mass and mole fractions?

12-7C Using the definitions of mass and mole fractions, derive a relation between them.

12-8C Somebody claims that the mass and mole fractions for a mixture of CO_2 and N_2O gases are identical. Is this true? Why?

---

*Students are encouraged to answer *all* the concept "C" questions.

12-9C The sum of the mass fractions for an ideal-gas mixture is equal to 1. Is this also true for a real-gas mixture?

12-10 The composition of moist air is given on a molar basis to be 78 percent N_2, 20 percent O_2, and 2 percent water vapor. Determine the mass fractions of the constituents of air.

12-11 A gas mixture has the following composition on a mole basis: 60 percent N_2 and 40 percent CO_2. Determine the gravimetric analysis of the mixture, its molar mass, and gas constant.

12-12 A gas mixture consists of 5 kg of O_2, 8 kg of N_2, and 10 kg of CO_2. Determine (*a*) the mass fraction of each component, (*b*) the mole fraction of each component, and (*c*) the average molar mass and gas constant of the mixture.

12-13 Determine the mole fractions of a gas mixture that consists of 75 percent CH_4 and 25 percent CO_2 by mass. Also, determine the gas constant of the mixture.

12-14 A gas mixture consists of 8 kmol of H_2 and 2 kmol of N_2. Determine the mass of each gas and the apparent gas constant of the mixture.
Answers: 16 kg, 56 kg, 1.155 kJ/(kg · K)

12-15E A gas mixture consists of 5 lbmol of H_2 and 2 lbmol of N_2. Determine the mass of each gas and the apparent gas constant of the mixture.

P-v-T Behavior of Gas Mixtures

12-16C Is a mixture of ideal gases also an ideal gas? Give an example.

12-17C Express Dalton's law of additive pressures. Does this law hold exactly for ideal-gas mixtures? How about nonideal-gas mixtures?

12-18C Express Amagat's law of additive volumes. Does this law hold exactly for ideal-gas mixtures? How about nonideal-gas mixtures?

12-19C How is the *P-v-T* behavior of a component in an ideal-gas mixture expressed? How is the *P-v-T* behavior of a component in a real-gas mixture expressed?

12-20C What is the difference between the *component pressure* and the *partial pressure*? When are these two equivalent?

12-21C What is the difference between the *component volume* and the *partial volume*? When are these two equivalent?

12-22C In a gas mixture, which component will have the higher partial pressure—the one with the higher mole number or the one with the larger molar mass?

12-23C Consider a rigid tank that contains a mixture of two ideal gases. A valve is opened and some gas escapes. As a result, the pressure in the tank drops. Will the partial pressure of each component change? How about the pressure fraction of each component?

12-24C Consider a rigid tank that contains a mixture of two ideal gases. The gas mixture is heated, and the pressure and temperature in the tank rise. Will the partial pressure of each component change? How about the pressure fraction of each component?

12-25C Is this statement correct? *The volume of an ideal-gas mixture is equal to the sum of the volumes of each individual gas in the mixture.* If not, how would you correct it?

12-26C Is this statement correct? *The temperature of an ideal-gas mixture is equal to the sum of the temperatures of each individual gas in the mixture.* If not, how would you correct it?

12-27C Is this statement correct? *The pressure of an ideal-gas mixture is equal to the sum of the partial pressures of each individual gas in the mixture.* If not, how would you correct it?

12-28C Explain how a real-gas mixture can be treated as a pseudopure substance using Kay's rule.

12-29 A rigid tank contains 4 kmol of O_2 and 5 kmol of CO_2 gases at 290 K and 150 kPa. Estimate the volume of the tank. *Answer:* 144.7 m^3

12-30 A rigid tank contains 0.5 kmol of Ar and 2 kmol of N_2 at 250 kPa and 280 K. The mixture is now heated to 400 K. Determine the volume of the tank and the final pressure of the mixture.

12-31 A gas mixture at 400 K and 150 kPa consists of 1 kg of CO_2 and 3 kg of CH_4. Determine the partial pressure of each gas and the apparent molar mass of the gas mixture.

12-32E A gas mixture at 600 R and 20 psia consists of 1 lbm of CO_2 and 3 lbm of CH_4. Determine the partial pressure of each gas and the apparent molar mass of the gas mixture.

12-33 A 0.3-m^3 rigid tank contains 0.6 kg of N_2, and 0.4 kg of O_2 at 300 K. Determine the partial pressure of each gas and the total pressure of the mixture. *Answers:* 178.1 kPa, 103.9 kPa, 282.0 kPa

12-34 A gas mixture at 290 K and 250 kPa has the following volumetric analysis: 65 percent N_2, 20 percent O_2, and 15 percent CO_2. Determine the mass fraction and partial pressure of each gas.

12-35 A rigid tank that contains 2 kg of N_2 at 25°C and 200 kPa is connected to another rigid tank that contains 3 kg of O_2 at 25°C and 500 kPa. The valve connecting the two tanks is opened, and the two gases are allowed to mix. If the final mixture temperature is 25°C, determine the volume of each tank and the final mixture pressure. *Answers:* 0.884 m^3, 0.465 m^3, 303.4 kPa

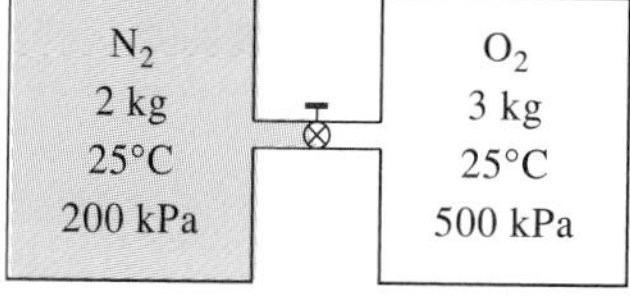

FIGURE P12-35

12-36 A volume of 0.3 m^3 of O_2 at 200 K and 8 MPa is mixed with 0.5 m^3 of N_2 at the same temperature and pressure, forming a mixture at 200 K and 8 MPa. Determine the volume of the mixture, using (*a*) the ideal-gas equation of state, (*b*) Kay's rule, and (*c*) the compressibility chart and Amagat's law.
Answers: (*a*) 0.8 m^3, (*b*) 0.79 m^3, (*c*) 0.80 m^3

12-37 A rigid tank contains 1 kmol of Ar gas at 220 K and 5 MPa. A valve is now opened, and 3 kmol of N_2 gas is allowed to enter the tank at 190 K and 8 MPa. The final mixture temperature is 200 K. Determine the pressure of the mixture, using (*a*) the ideal-gas equation of state and (*b*) the compressibility chart and Dalton's law.

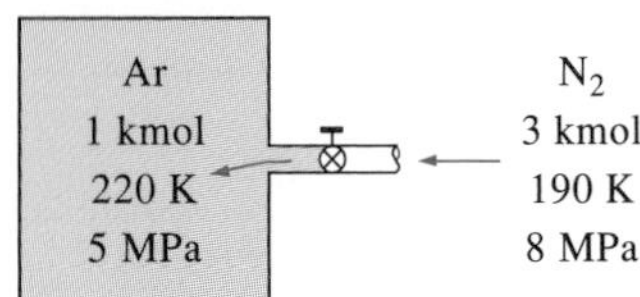

FIGURE P12-37

12-38E A rigid tank contains 1 lbmol of argon gas at 400 R and 750 psia. A valve is now opened, and 3 lbmol of N_2 gas is allowed to enter the tank at 340 R and 1200 psia. The final mixture temperature is 360 R. Determine the pressure of the mixture, using (*a*) the ideal-gas equation of state and (*b*) the compressibility chart and Dalton's law. *Answers:* (*a*) 2700 psia, (*b*) 2472 psia

Properties of Gas Mixtures

12-39C Is the total internal energy of an ideal-gas mixture equal to the sum of the internal energies of each individual gas in the mixture? Answer the same question for a real-gas mixture.

12-40C Is the specific internal energy of a gas mixture equal to the sum of the specific internal energy of each individual gas in the mixture?

12-41C Answer Probs. 12-39C and 12-40C for entropy.

12-42C Is the total internal energy change of an ideal-gas mixture equal to the sum of the internal energy changes of each individual gas in the mixture? Answer the same question for a real-gas mixture.

12-43C When evaluating the entropy change of the components of an ideal-gas mixture, do we have to use the partial pressure of each component or the total pressure of the mixture?

12-44C Suppose we want to determine the enthalpy change of a real-gas mixture undergoing a process. The enthalpy change of each individual gas is determined by using the generalized enthalpy chart, and the enthalpy change of the mixture is determined by summing them. Is this an exact approach? Explain.

12-45 An insulated rigid tank is divided into two compartments by a partition. One compartment contains 0.2 kmol of CO_2 at 27°C and 200 kPa, and the other compartment contains 3 kmol of H_2 gas at 40°C and 400 kPa. Now the partition is removed, and the two gases are allowed to mix. Determine (*a*) the mixture temperature and (*b*) the mixture pressure after equilibrium has been established. Assume constant specific heats at room temperature for both gases.

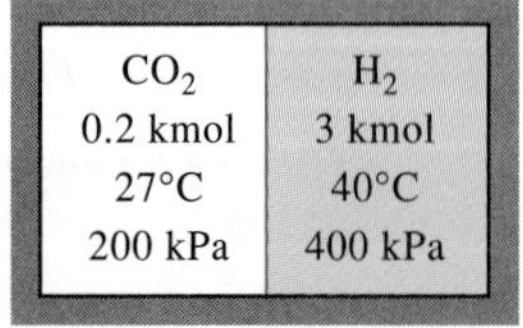

FIGURE P12-45

12-46 A 0.9-m^3 rigid tank is divided into two equal compartments by a partition. One compartment contains Ne at 20°C and 100 kPa, and the other compartment contains Ar at 50°C and 200 kPa. Now the partition is removed, and the two gases are allowed to mix. Heat is lost to the surrounding air at 20°C during this process in the amount of 15 kJ. Determine (*a*) the final mixture temperature and (*b*) the final mixture pressure.
Answers: (*a*) 16.2°C, (*b*) 138.9 kPa

12-47 Ethane (C_2H_6) at 20°C and 200 kPa and methane (CH_4) at 45°C and 200 kPa enter an adiabatic mixing chamber. The mass flow rate of ethane is 9 kg/s, which is twice the mass flow rate of methane. Determine (*a*) the mixture temperature and (*b*) the rate of entropy generation during this process, in kW/K.

12-48 An equimolar mixture of helium and argon gases is to be used as the working fluid in a closed-loop gas-turbine cycle. The mixture enters the turbine at 1.6 MPa and 1500 K and expands isentropically to a pressure of 200 kPa. Determine the work output of the turbine per unit mass of the mixture.

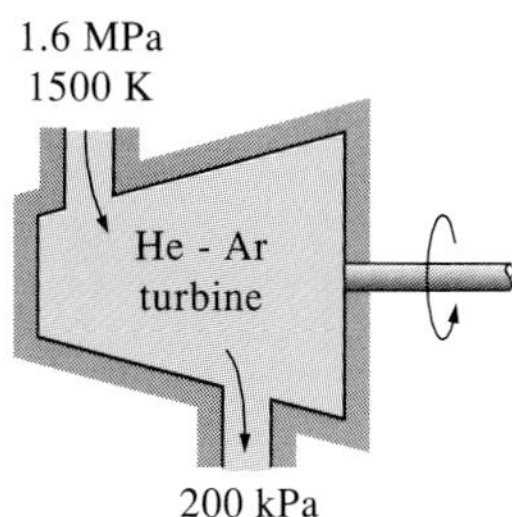

FIGURE P12-48

12-49E A mixture of 80 percent N_2 and 20 percent CO_2 gases (on a mass basis) enters the nozzle of a turbojet engine at 90 psia and 1800 R with a low velocity, and it expands to a pressure of 12 psia. If the isentropic efficiency of the nozzle is 92 percent, determine (*a*) the exit temperature and (*b*) the exit velocity of the mixture. Assume constant specific heats at room temperature.

12-50 A piston-cylinder device contains a mixture of 0.2 kg of H_2 and 1.6 kg of N_2 at 100 kPa and 300 K. Heat is now transferred to the mixture at constant pressure until the volume is doubled. Assuming constant specific heats at the average temperature, determine (*a*) the heat transfer and (*b*) the entropy change of the mixture.

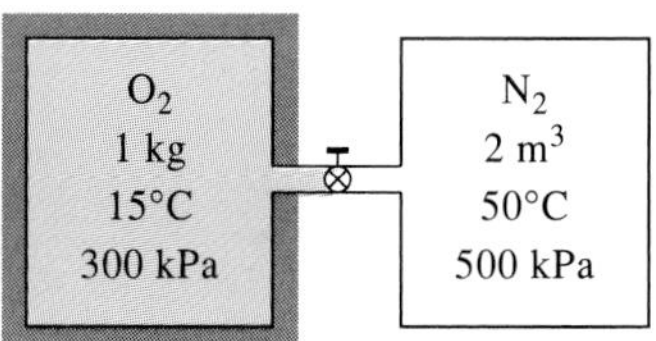

FIGURE P12-51

12-51 An insulated tank that contains 1 kg of O_2 at 15°C and 300 kPa is connected to a 2-m^3 uninsulated tank that contains N_2 at 50°C and 500 kPa. The valve connecting the two tanks is opened, and the two gases form a homogeneous mixture at 25°C. Determine (*a*) the final pressure in the tank, (*b*) the heat transfer, and (*c*) the entropy generated during this process. Assume T_0 = 25°C. *Answers:* (*a*) 444.6 kPa, (*b*) 187.2 kJ loss, (*c*) 0.962 kJ/K

12-52 A piston-cylinder device contains 6 kg of H_2 and 21 kg of N_2 at 160 K and 5 MPa. Heat is now transferred to the device, and the mixture expands at constant pressure until the temperature rises to 200 K. Determine the heat transfer during this process by treating the mixture (*a*) as an ideal gas and (*b*) as a nonideal gas and using Amagat's law.
Answers: (*a*) 3139 kJ, (*b*) 3790 kJ

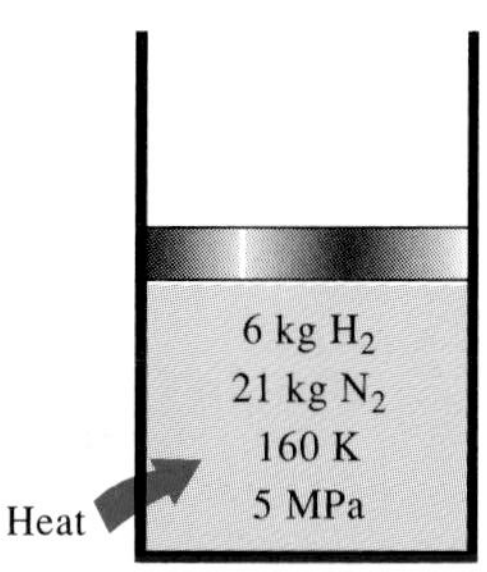

FIGURE P12-52

12-53 Determine the total entropy change and exergy destruction associated with the process described in Prob. 12-52 by treating the mixture (*a*) as an ideal gas and (*b*) as a nonideal gas and using Amagat's law. Assume constant specific heats at room temperature and take T_0 = 25°C.

12-54 Air, which may be considered as a mixture of 79 percent N_2 and 21 percent O_2 by mole numbers, is compressed isothermally at 200 K from 4 to 8 MPa in a steady-flow device. The compression process is internally reversible, and the mass flow rate of air is 2.9 kg/s. Determine the power input to the compressor and the rate of heat rejection by treating the mixture (*a*) as an ideal gas and (*b*) as a nonideal gas and using Amagat's law.
Answers: (*a*) 126.8 kW, 126.8 kW; (*b*) 108.6 kW, 158 kW

FIGURE P12-54
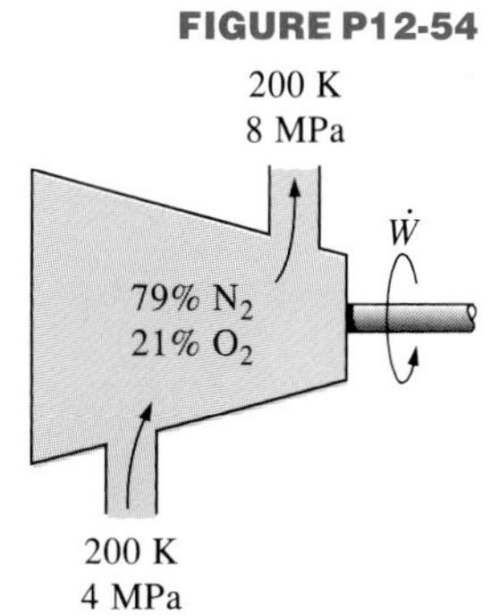

Review Problems

12-55 Air has the following composition on a mole basis: 21 percent O_2, 78 percent N_2, and 1 percent Ar. Determine the gravimetric analysis of air and its molar mass. *Answers:* 23.2 percent O_2, 75.4 percent N_2, 1.4 percent Ar; 28.96 kg/kmol

12-56 Using Amagat's law, show that

$$Z_m = \sum_{i=1}^{k} y_i Z_i$$

for a real-gas mixture of k gases, where Z is the compressibility factor.

12-57 Using Dalton's law, show that

$$Z_m = \sum_{i=1}^{k} y_i Z_i$$

for a real-gas mixture of k gases, where Z is the compressibility factor.

12-58 A rigid tank contains 2 kmol of N_2 and 6 kmol of CH_4 gases at 200 K and 10 MPa. Estimate the volume of the tank, using (*a*) the ideal-gas equation of state, (*b*) Kay's rule, and (*c*) the compressibility chart and Amagat's law.

12-59 A steady stream of equimolar N_2 and CO_2 mixture at 100 kPa and 27°C is to be separated into N_2 and CO_2 gases at 100 kPa and 27°C. Determine the minimum work required per unit mass of mixture to accomplish this separation process. Assume $T_0 = 27$°C.

12-60 A gas mixture consists of O_2 and N_2. The ratio of the mole numbers of N_2 to O_2 is 3:1. This mixture is heated during a steady-flow process from 180 to 210 K at a constant pressure of 8 MPa. Determine the heat transfer during this process per mole of the mixture, using (*a*) the ideal-gas approximation and (*b*) Kay's rule.

12-61 Determine the total entropy change and exergy destruction associated with the process described in Prob. 12-60, using (*a*) the ideal-gas approximation and (*b*) Kay's rule. Assume constant specific heats and $T_0 = 25$°C.

12-62 A rigid tank contains a mixture of 4 kg of He and 8 kg of O_2 at 170 K and 7 MPa. Heat is now transferred to the tank, and the mixture temperature rises to 220 K. Treating the He as an ideal gas and the O_2 as a nonideal gas, determine (*a*) the final pressure of the mixture and (*b*) the heat transfer.

Computer, Design, and Essay Problems

12-63 Write an interactive program to determine the mole fractions of the components of a mixture of n gases when the mass fractions are given, and to determine the mass fractions of the components when the mole fractions are given.

12-64 Write an interactive program to determine the apparent gas constant, specific heat at constant volume, and internal energy of a mixture of n ideal gases when the mass fractions and other properties of the constituent gases are given.

12-65 Write an interactive program to determine the molar mass, specific heat at constant pressure, and enthalpy of a mixture of n ideal gases when the mole fractions and other properties of the constituent gases are given.

12-66 Write an interactive program to determine the entropy change of a mixture of n ideal gases when the mass fractions and other properties of the constituent gases are given.

12-67 Prolonged exposure to mercury even at relatively low but toxic concentrations in the air is known to cause permanent mental disorders, insomnia, and pain and numbness in the hands and the feet, among other things. Therefore, the maximum allowable concentration of mercury vapor in the air at work places is regulated by federal agencies. These regulations require that the average level of mercury concentration in the air does not exceed 0.1 mg/m^3.

Consider a mercury spill that occurs in an airtight storage room at 20°C in San Francisco during an earthquake. Calculate the highest level of mercury concentration in the air that can occur in the storage room, in mg/m^3, and determine if it is within the safe level. The vapor pressure of mercury at 20°C is 0.173 Pa. Propose some guidelines to safeguard against the formation of toxic concentrations of mercury vapor in air in storage rooms and laboratories.

Gas–Vapor Mixtures and Air-Conditioning

CHAPTER 13

At temperatures below the critical temperature, the gas phase of a substance is frequently referred to as a *vapor*. The term *vapor* implies a gaseous state that is close to the saturation region of the substance, raising the possibility of condensation during a process.

In Chap. 12, we discussed mixtures of gases that were usually above their critical temperatures. Therefore, we were not concerned about any of the gases condensing during a process. Not having to deal with two phases greatly simplified the analysis. When we are dealing with a gas–vapor mixture, however, the vapor may condense out of the mixture during a process, forming a two-phase mixture. This may complicate the analysis considerably. Therefore, a gas–vapor mixture needs to be treated differently from an ordinary gas mixture.

Several gas–vapor mixtures are encountered in engineering. In this chapter, we consider the *air–water-vapor mixture,* which is the most commonly used gas–vapor mixture in practice. We also discuss *air-conditioning,* which is the primary application area of air–water-vapor mixtures.

13-1 ■ DRY AND ATMOSPHERIC AIR

Air is a mixture of nitrogen, oxygen, and small amounts of some other gases. The air in the atmosphere normally contains some water vapor (or *moisture*) and is referred to as **atmospheric air.** By contrast, air that contains no water vapor is called **dry air.** It is often convenient to treat air as a mixture of water vapor and dry air since the composition of dry air remains relatively constant, but the amount of water vapor changes as a result of condensation and evaporation from oceans, lakes, rivers, showers, and even the human body. Although the amount of water vapor in the air is small, it plays a major role in human comfort. Therefore, it is an important consideration in air-conditioning applications.

DRY AIR	
T, °C	C_p, kJ/(kg · °C)
−10	1.0038
0	1.0041
10	1.0045
20	1.0049
30	1.0054
40	1.0059
50	1.0065

FIGURE 13-1

The C_p of air can be assumed to be constant at 1.005 kJ/(kg · °C) in the temperature range −10 to 50°C with an error under 0.2 percent.

The temperature of air in air-conditioning applications ranges from about −10 to about 50°C. In this range, the dry air can be treated as an ideal gas with a constant C_p value of 1.005 kJ/(kg · K) [0.240 Btu/(lbm · R)] with negligible error (under 0.2 percent), as illustrated in Fig. 13-1. Taking 0°C as the reference temperature, the enthalpy and enthalpy change of dry air can be determined from

$$h_{\text{dry air}} = C_p T = [1.005 \text{ kJ/(kg} \cdot \text{°C)}]T \qquad \text{(kJ/kg)} \qquad (13\text{-}1a)$$

and

$$\Delta h_{\text{dry air}} = C_p \, \Delta T = [1.005 \text{ kJ/(kg} \cdot \text{°C)}]\Delta T \qquad \text{(kJ/kg)} \qquad (13\text{-}1b)$$

where T is the air temperature in °C and the ΔT is the change in temperature. In air-conditioning processes we are concerned with the *changes* in enthalpy Δh, which is independent of the reference point selected.

Would it not be convenient to also treat the water vapor in the air as an ideal gas? You would probably be willing to sacrifice some accuracy for such convenience. Well, it turns out that we can have the convenience without much sacrifice. At 50°C, the saturation pressure of water is 12.3 kPa. As can be seen from Fig. 2-48, at pressures below this value, water vapor can be treated as an ideal gas with negligible error (under 0.2 percent), even when it is a saturated vapor. Therefore, the water vapor in the air behaves as if it existed alone and obeys the ideal-gas relation $Pv = RT$. Then the atmospheric air can be treated as an ideal-gas mixture whose pressure is the sum of the partial pressure of dry air* P_a and that of the water vapor P_v:

$$P = P_a + P_v \qquad \text{(kPa)} \qquad (13\text{-}2)$$

The partial pressure of water vapor is usually referred to as the **vapor pressure.** It is the pressure the water vapor would exert if it existed alone at the temperature and volume of the mixture.

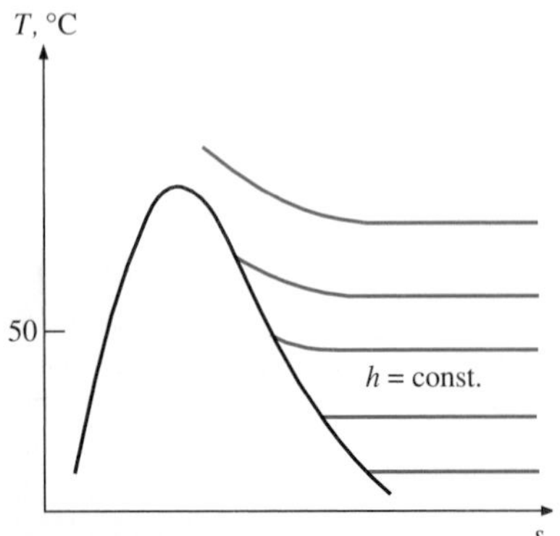

FIGURE 13-2

At temperatures below 50°C, the h = constant lines coincide with the T = constant lines in the superheated vapor region of water.

Since water vapor is an ideal gas, the enthalpy of water vapor is a function of temperature only, that is, $h = h(T)$. This can also be observed from the *T-s* diagram of water given in Fig. A-9 and Fig. 13-2 where the constant-enthalpy lines coincide with constant-temperature lines at temperatures below 50°C. Therefore, *the enthalpy of water vapor in air can be taken to be*

*Throughout this chapter, the subscript a will denote dry air and the subscript v will denote water vapor.

equal to the enthalpy of the saturated vapor at the same temperature. That is,

$$h_v(T, \text{low } P) \cong h_g(T) \qquad (13\text{-}3)$$

The enthalpy of water vapor at 0°C is 2501.3 kJ/kg. The average C_p value of water vapor in the temperature range −10 to 50°C can be taken to be 1.82 kJ/(kg · °C). Then the enthalpy of water vapor can be determined approximately from

$$h_g(T) \cong 2501.3 + 1.82T \qquad \text{(kJ/kg)} \qquad T \text{ in °C} \qquad (13\text{-}4)$$

or

$$h_g(T) \cong 1061.5 + 0.435T \qquad \text{(Btu/lbm)} \qquad T \text{ in °F} \qquad (13\text{-}5)$$

in the temperature range −10 to 50°C (or 15 to 120°F), with negligible error, as shown in Fig. 13-3.

WATER VAPOR

T, °C	h_g, kJ/kg Table A-4	h_g, kJ/kg Eq. 13-4	Difference, kJ/kg
−10	2482.9	2483.1	−0.2
0	2501.3	2501.3	0.0
10	2519.8	2519.5	0.3
20	2538.1	2537.7	0.4
30	2556.3	2555.9	0.4
40	2574.1	2574.1	0.2
50	2592.1	2592.3	−0.2

FIGURE 13-3

In the temperature range −10 to 50°C, the h_g of water can be determined from Eq. 13-4 with negligible error.

13-2 ■ SPECIFIC AND RELATIVE HUMIDITY OF AIR

The amount of water vapor in the air can be specified in various ways. Probably the most logical way is to specify directly the mass of water vapor present in a unit mass of dry air. This is called **absolute** or **specific humidity** (also called *humidity ratio*) and is denoted by ω:

$$\omega = \frac{m_v}{m_a} \qquad \text{(kg water vapor/kg dry air)} \qquad (13\text{-}6)$$

The specific humidity can also be expressed as

$$\omega = \frac{m_v}{m_a} = \frac{P_vV/(R_vT)}{P_aV/(R_aT)} = \frac{P_v/R_v}{P_a/R_a} = 0.622\frac{P_v}{P_a} \qquad (13\text{-}7)$$

or

$$\omega = \frac{0.622\,P_v}{P - P_v} \qquad \text{(kg water vapor/kg dry air)} \qquad (13\text{-}8)$$

where P is the total pressure.

Consider 1 kg of dry air. By definition, dry air contains no water vapor, and thus its specific humidity is zero. Now let us add some water vapor to this dry air. The specific humidity will increase. As more vapor or moisture is added, the specific humidity will keep increasing until the air can hold no more moisture. At this point, the air is said to be saturated with moisture, and it is called **saturated air.** Any moisture introduced into saturated air will condense. The amount of water vapor in saturated air at a specified temperature and pressure can be determined from Eq. 13-8 by replacing P_v by P_g, the saturation pressure of water at that temperature (Fig. 13-4).

The amount of moisture in the air has a definite effect on how comfortable we feel in an environment. However, the comfort level depends more on the amount of moisture the air holds (m_v) relative to the maximum amount of

FIGURE 13-4

For saturated air, the vapor pressure is equal to the saturation pressure of water.

AIR
25°C, 100 kPa

($P_{\text{sat, }H_2O\text{ @ 25°C}}$ = 3.169 kPa)

P_v = 0 → dry air
P_v < 3.169 kPa → unsaturated air
P_v = 3.169 kPa → saturated air

moisture the air can hold at the same temperature (m_g). The ratio of these two quantities is called the **relative humidity** ϕ (Fig. 13-5)

$$\phi = \frac{m_v}{m_g} = \frac{P_vV/(R_vT)}{P_gV/(R_vT)} = \frac{P_v}{P_g} \qquad (13\text{-}9)$$

where $$P_g = P_{\text{sat @ } T} \qquad (13\text{-}10)$$

Combining Eqs. 13-8 and 13-9, we can also express the relative humidity as

$$\phi = \frac{\omega P}{(0.622 + \omega)P_g} \quad \text{and} \quad \omega = \frac{0.622\phi P_g}{P - \phi P_g} \qquad (13\text{-}11a, b)$$

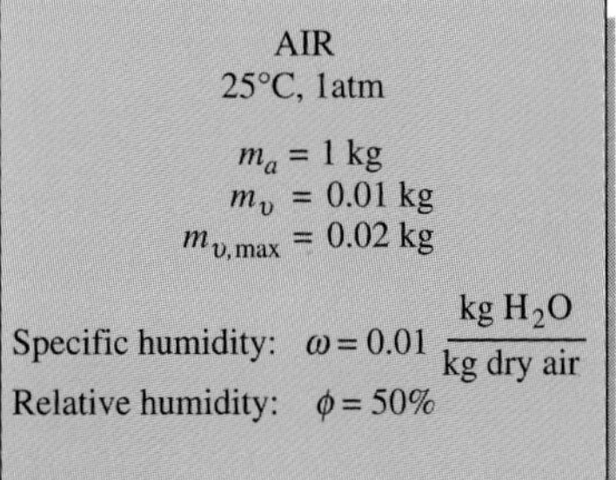

FIGURE 13-5
Specific humidity is the actual amount of water vapor in 1 kg of dry air, whereas relative humidity is the ratio of the actual amount of moisture in the air to the maximum amount of moisture air can hold at that temperature.

The relative humidity ranges from 0 for dry air to 1 for saturated air. Note that the amount of moisture air can hold depends on its temperature. Therefore, the relative humidity of air changes with temperature even when its specific humidity remains constant.

Atmospheric air is a mixture of dry air and water vapor, and thus the enthalpy of air is expressed in terms of the enthalpies of the dry air and the water vapor. In most practical applications, the amount of dry air in the air–water-vapor mixture remains constant, but the amount of water vapor changes. Therefore, the enthalpy of atmospheric air is expressed *per unit mass of dry air* instead of per unit mass of the air–water-vapor mixture.

The total enthalpy (an extensive property) of atmospheric air is the sum of the enthalpies of the dry air and the water vapor:

$$H = H_a + H_v = m_ah_a + m_vh_v$$

Dividing by m_a gives

$$h = \frac{H}{m_a} = h_a + \frac{m_v}{m_a}h_v = h_a + \omega h_v$$

or

$$h = h_a + \omega h_g \qquad \text{(kJ/kg dry air)} \qquad (13\text{-}12)$$

since $h_v \cong h_g$ (Fig. 13-6).

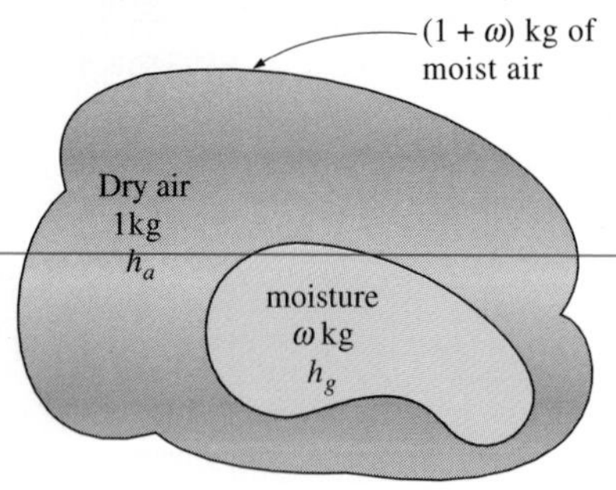

FIGURE 13-6
The enthalpy of moist (atmospheric) air is expressed per unit mass of dry air, not per unit mass of moist air.

Also note that the ordinary temperature of atmospheric air is frequently referred to as the **dry-bulb temperature** to differentiate it from other forms of temperatures that shall be discussed.

EXAMPLE 13-1 The Amount of Water Vapor in Room Air

A 5-m × 5-m × 3-m room shown in Fig. 13-7 contains air at 25°C and 100 kPa at a relative humidity of 75 percent. Determine (*a*) the partial pressure of dry air, (*b*) the specific humidity of the air, (*c*) the enthalpy per unit mass of the dry air, and (*d*) the masses of the dry air and water vapor in the room.

FIGURE 13-7
Schematic for Example 13-1.

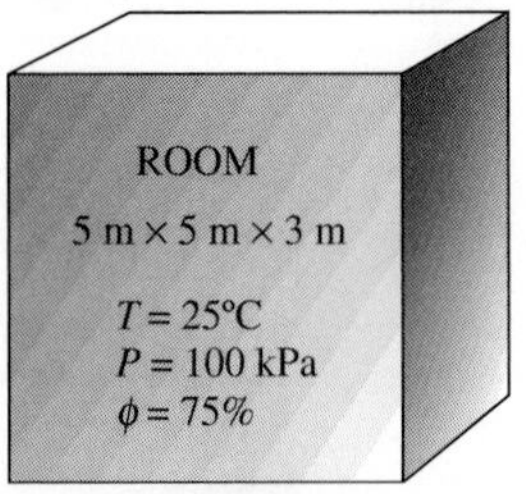

Solution A sketch of the room is given in Fig. 13-7. Both the air and the vapor fill the entire room, and thus the volume of each gas is equal to the volume of the room.

Assumptions The air and the water vapor are ideal gases.

Analysis (*a*) The partial pressure of dry air can be determined from Eq. 13-2:

$$P_a = P - P_v$$

where $P_v = \phi P_g = \phi P_{sat\,@\,25°C} = (0.75)(3.169\text{ kPa}) = 2.38\text{ kPa}$

Thus $P_a = (100 - 2.38)\text{ kPa} = \mathbf{97.62\ kPa}$

(*b*) The specific humidity of air is determined from Eq. 13-8:

$$\omega = \frac{0.622P_v}{P - P_v} = \frac{(0.622)(2.38\text{ kPa})}{(100 - 2.38)\text{ kPa}} = \mathbf{0.0152\ kg\ H_2O/kg\ dry\ air}$$

(*c*) The enthalpy of air per unit mass of dry air is determined from Eq. 13-12, where h_g is taken from Table A-4:

$$h = h_a + \omega h_v \cong C_p T + \omega h_g$$
$$= [1.005\text{ kJ/(kg}\cdot°\text{C)}](25°\text{C}) + (0.0152)(2547.2\text{ kJ/kg})$$
$$= \mathbf{63.8\ kJ/kg\ dry\ air}$$

The enthalpy of water vapor (2547.2 kJ/kg) could also be determined from the approximation given by Eq. 13-4:

$$h_{g\,@\,25°C} \cong 2501.3 + 1.82(25) = 2546.8\text{ kJ/kg}$$

which is very close to the value obtained from Table A-4.

(*d*) Both the dry air and the water vapor fill the entire room completely. Therefore, the volume of each gas is equal to the volume of the room:

$$V_a = V_v = V_{room} = (5)(5)(3) = 75\text{ m}^3$$

The masses of the dry air and the water vapor are determined from the ideal-gas relation applied to each gas separately:

$$m_a = \frac{P_a V_a}{R_a T} = \frac{(97.62\text{ kPa})(75\text{ m}^3)}{[0.287\text{ kPa}\cdot\text{m}^3/(\text{kg}\cdot\text{K})](298\text{ K})} = \mathbf{85.61\ kg}$$

$$m_v = \frac{P_v V_v}{R_v T} = \frac{(2.38\text{ kPa})(75\text{ m}^3)}{[0.4615\text{ kPa}\cdot\text{m}^3/(\text{kg}\cdot\text{K})](298\text{ K})} = \mathbf{1.3\ kg}$$

The mass of the water vapor in the air could also be determined from Eq. 13-6:

$$m_v = \omega m_a = (0.0152)(85.61\text{ kg}) = 1.3\text{ kg}$$

13-3 ■ DEW-POINT TEMPERATURE

If you live in humid weather, you are probably used to waking up most summer mornings and finding the grass wet. You know it did not rain the night before. So what happened? Well, the excess moisture in the air simply condensed on the cool surfaces, forming what we call *dew*. In summer, a considerable amount of water vaporizes during the day. As the temperature falls during the night, so does the "moisture capacity" of air, which is the maximum amount of moisture air can hold. (What happens to the relative humidity during this process?) After a while, the moisture capacity of the air equals the moisture content of the air. At this point, the air is saturated, and

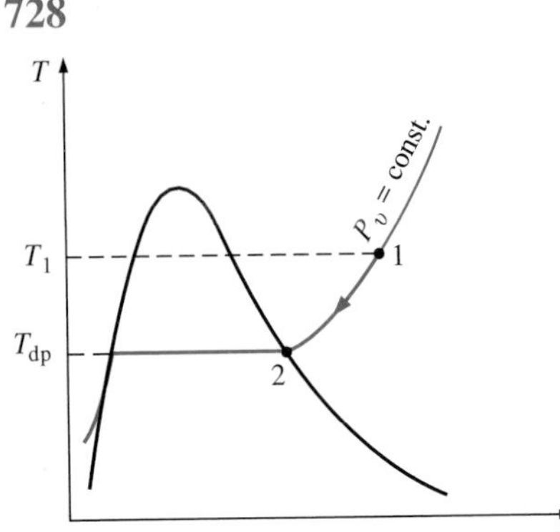

FIGURE 13-8

Constant-pressure cooling of moist air and the dew-point temperature on the *T-s* diagram of water.

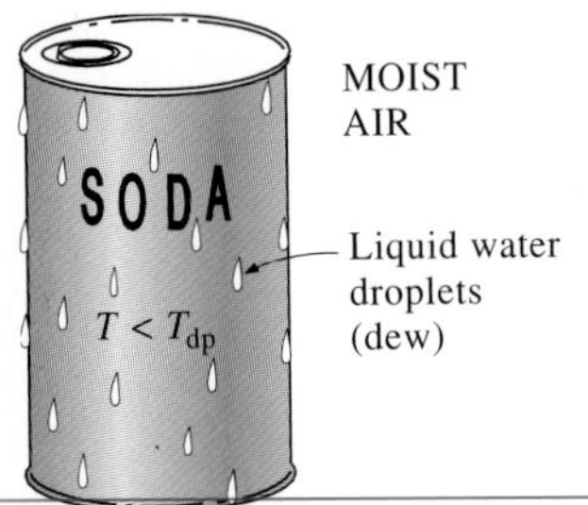

FIGURE 13-9

When the temperature of a cold drink is below the dew-point temperature of the surrounding air, it "sweats."

FIGURE 13-10

Schematic for Example 13-2.

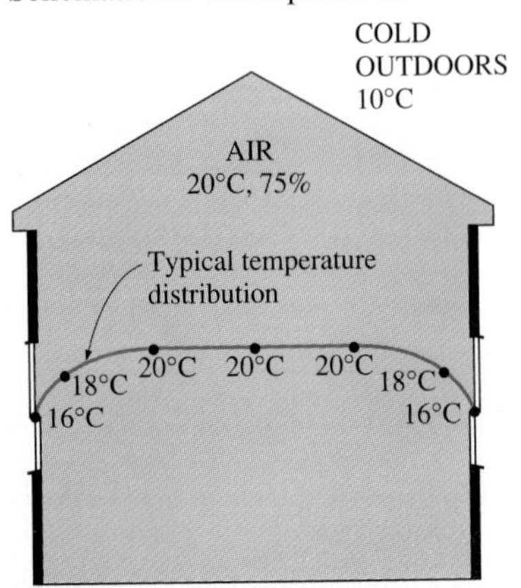

its relative humidity is 100 percent. Any further drop in temperature results in the condensation of some of the moisture, and this is the beginning of dew formation.

The **dew-point temperature** T_{dp} is defined as *the temperature at which condensation begins when the air is cooled at constant pressure.* In other words, T_{dp} is the saturation temperature of water corresponding to the vapor pressure:

$$T_{dp} = T_{sat\,@\,P_v} \tag{13-13}$$

This is also illustrated in Fig. 13-8. As the air cools at constant pressure, the vapor pressure P_v remains constant. Therefore, the vapor in the air (state 1) undergoes a constant-pressure cooling process until it strikes the saturated vapor line (state 2). The temperature at this point is T_{dp}, and if the temperature drops any further, some vapor condenses out. As a result, the amount of vapor in the air decreases, which results in a decrease in P_v. The air remains saturated during the condensation process and thus follows a path of 100 percent relative humidity (the saturated vapor line). The ordinary temperature and the dew-point temperature of saturated air are identical.

You have probably noticed that when you buy a cold canned drink from a vending machine on a hot and humid day, dew forms on the can. The formation of dew on the can indicates that the temperature of the drink is below the dew-point temperature of the surrounding air (Fig. 13-9).

The dew-point temperature of room air can be determined easily by cooling some water in a metal cup by adding small amounts of ice and stirring. The temperature of the outer surface of the cup when dew starts to form on the surface is the dew-point temperature of the air.

EXAMPLE 13-2 Fogging of the Windows in a House

In cold weather, condensation frequently occurs on the inner surfaces of the windows due to the lower air temperatures near the window surface. Consider a house, shown in Fig. 13-10, that contains air at 20°C and 75 percent relative humidity. At what window temperature will the moisture in the air start condensing on the inner surfaces of the windows?

Solution The temperature distribution in a house, in general, is not uniform. When the outdoor temperature drops in winter, so does the indoor temperature near the walls and the windows. Therefore, the air near the walls and the windows remains at a lower temperature than at the inner parts of a house even though the total pressure and the vapor pressure remain constant throughout the house. As a result, the air near the walls and the windows will undergo a P_v = constant cooling process until the moisture in the air starts condensing. This will happen when the air reaches its dew-point temperature T_{dp}. The dew point is determined from Eq. 13-13 to be

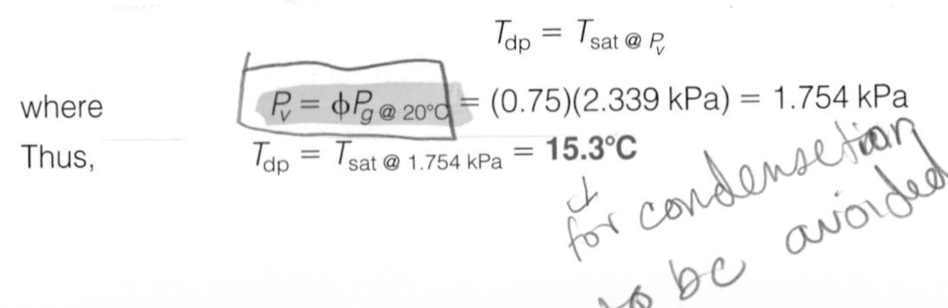

$$T_{dp} = T_{sat\,@\,P_v}$$

where $P_v = \phi P_{g\,@\,20°C} = (0.75)(2.339\text{ kPa}) = 1.754\text{ kPa}$

Thus, $T_{dp} = T_{sat\,@\,1.754\text{ kPa}} = \mathbf{15.3°C}$

for condensation to be avoided

Therefore, the inner surface of the window should be maintained above 15.3°C if condensation on the window surfaces is to be avoided.

13-4 ■ ADIABATIC SATURATION AND WET-BULB TEMPERATURES

Relative humidity and specific humidity are frequently used in engineering and atmospheric sciences, and it is desirable to relate them to easily measurable quantities such as temperature and pressure. One way of determining the relative humidity is to determine the dew-point temperature of air, as discussed in the last section. Knowing the dew-point temperature, we can determine the vapor pressure P_v and thus the relative humidity. This approach is simple, but not quite practical.

Another way of determining the absolute or relative humidity is related to an *adiabatic saturation process,* shown schematically and on a *T-s* diagram in Fig. 13-11. The system consists of a long insulated channel that contains a pool of water. A steady stream of unsaturated air that has a specific humidity of ω_1 (unknown) and a temperature of T_1 is passed through this channel. As the air flows over the water, some water will evaporate and mix with the airstream. The moisture content of air will increase during this process, and its temperature will decrease, since part of the latent heat of vaporization of the water that evaporates will come from the air. If the channel is long enough, the airstream will exit as saturated air (ϕ = 100 percent) at temperature T_2, which is called the **adiabatic saturation temperature.**

If makeup water is supplied to the channel at the rate of evaporation at temperature T_2, the adiabatic saturation process described above can be

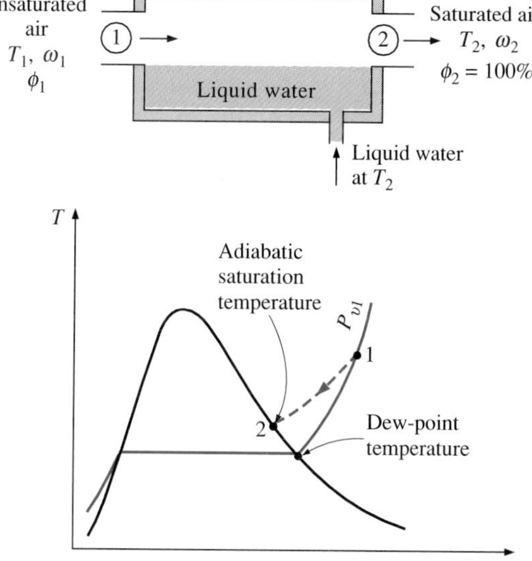

FIGURE 13-11

The adiabatic saturation process and its representation on a *T-s* diagram.

analyzed as a steady-flow process. The process involves no heat or work interactions, and the kinetic and potential energy changes can be neglected. Then the conservation of mass and conservation of energy relations for this two-inlet, one-exit steady-flow system reduces to the following:

Mass balance:

$$\dot{m}_{a_1} = \dot{m}_{a_2} = \dot{m}_a \qquad \text{(The mass flow rate of dry air remains constant)}$$

$$\dot{m}_{w_1} + \dot{m}_f = \dot{m}_{w_2}$$

or

$$\dot{m}_a\omega_1 + \dot{m}_f = \dot{m}_a\omega_2 \qquad \text{(The mass flow rate of vapor in the air increases by an amount equal to the rate of evaporation } \dot{m}_f\text{)}$$

Thus,

$$\dot{m}_f = \dot{m}_a(\omega_2 - \omega_1)$$

Energy balance:

$$\sum \dot{m}_i h_i = \sum \dot{m}_e h_e \qquad (\text{since } \dot{Q} = 0 \text{ and } \dot{W} = 0)$$

$$\dot{m}_{a_1} h_1 + \dot{m}_f h_{f2} = \dot{m}_{a_2} h_2$$

or

$$\dot{m}_a h_1 + \dot{m}_a(\omega_2 - \omega_1)h_{f_2} = \dot{m}_a h_2$$

Dividing by $\dot{m}_a$ gives

$$h_1 + (\omega_2 - \omega_1)h_{f_2} = h_2$$

or

$$(C_pT_1 + \omega_1 h_{g1}) + (\omega_2 - \omega_1)h_{f_2} = (C_pT_2 + \omega_2 h_{g_2})$$

which yields

$$\omega_1 = \frac{C_p(T_2 - T_1) + \omega_2 h_{fg_2}}{h_{g_1} - h_{f_2}} \tag{13-14}$$

where, from Eq. 13-11*b*,

$$\omega_2 = \frac{0.622P_{g_2}}{P_2 - P_{g_2}} \tag{13-15}$$

since $\phi_2 = 100$ percent. Thus we conclude that the specific humidity (and relative humidity) of air can be determined from Eqs. 13-14 and 13-15 by measuring the pressure and temperature of the air at the inlet and the exit of an adiabatic saturator.

If the air entering the channel is already saturated, then the adiabatic saturation temperature T_2 will be identical to the inlet temperature T_1, in which case Eq. 13-14 yields $\omega_1 = \omega_2$. In general, the adiabatic saturation temperature will be between the inlet and dew-point temperatures.

The adiabatic saturation process discussed above provides a means of determining the absolute or relative humidity of air, but it requires a long channel or a spray mechanism to achieve saturation conditions at the exit. A more practical approach is to use a thermometer whose bulb is covered with a cotton wick saturated with water and to blow air over the wick, as shown in Fig. 13-12. The temperature measured in this matter is called the

FIGURE 13-12

A simple arrangement to measure the wet-bulb temperature.

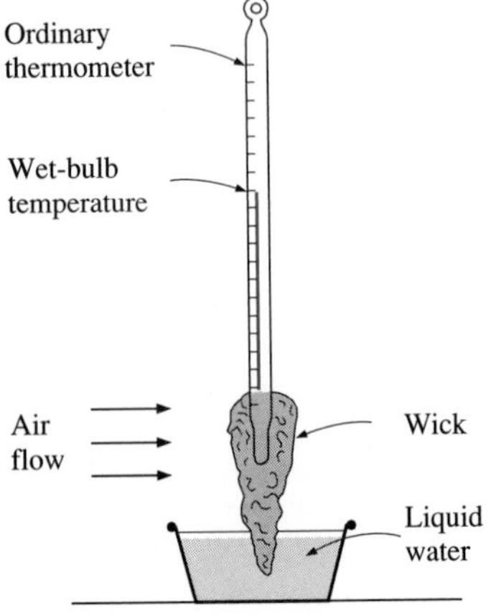

wet-bulb temperature T_{wb}, and it is commonly used in air-conditioning applications.

The basic principle involved is similar to that in adiabatic saturation. When unsaturated air passes over the wet wick, some of the water in the wick evaporates. As a result, the temperature of the water drops, creating a temperature difference (which is the driving force for heat transfer) between the air and the water. After a while, the heat loss from the water by evaporation equals the heat gain from the air, and the water temperature stabilizes. The thermometer reading at this point is the wet-bulb temperature. The wet-bulb temperature can also be measured by placing the wet-wicked thermometer in a holder attached to a handle and rotating the holder rapidly, that is, by moving the thermometer instead of the air. A device that works on this principle is called a *sling psychrometer* and is shown in Fig. 13-13. Usually a dry-bulb thermometer is also mounted on the frame of this device so that both the wet- and dry-bulb temperatures can be read simultaneously.

Advances in electronics made it possible to measure humidity directly in a fast and reliable way. It appears that sling psychrometers and wet-wicked thermometers are about to become things of the past. Today, hand-held electronic humidity measurement devices based on the capacitance change in a thin polymer film as it absorbs water vapor are capable of sensing and digitally displaying the relative humidity within 1 percent accuracy in a matter of seconds.

In general, the adiabatic saturation temperature and the wet-bulb temperature are not the same. But for air–water-vapor mixtures at atmospheric pressure, the wet-bulb temperature happens to be approximately equal to the adiabatic saturation temperature. Therefore, the wet-bulb temperature T_{wb} can be used in Eq 13-14 in place of T_2 to determine the specific humidity of air.

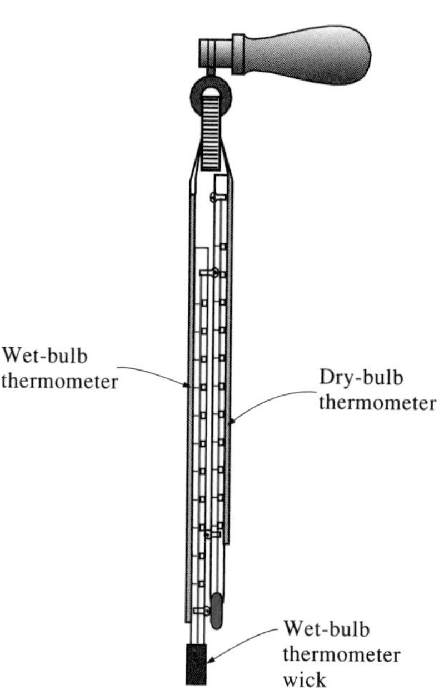

FIGURE 13-13
Sling psychrometer.

EXAMPLE 13-3 The Specific and Relative Humidity of Air

The dry- and the wet-bulb temperatures of atmospheric air at 1 atm (101.325-kPa) pressure are measured with a sling psychrometer and determined to be 25 and 15°C, respectively. Determine (*a*) the specific humidity, (*b*) the relative humidity, and (*c*) the enthalpy of the air.

Solution (*a*) The specific humidity ω_1 is determined from Eq. 13-14,

$$\omega_1 = \frac{C_p(T_2 - T_1) + \omega_2 h_{fg_2}}{h_{g_1} - h_{f_2}}$$

where T_2 is the wet-bulb temperature and ω_2 is determined from Eq. 13-15 to be

$$\omega_2 = \frac{0.622 P_{g_2}}{P_2 - P_{g_2}} = \frac{(0.622)(1.705 \text{ kPa})}{(101.325 - 1.705) \text{ kPa}}$$

$$= 0.01065 \text{ kg } H_2O/\text{kg dry air}$$

Thus,

$$\omega_1 = \frac{[1.005 \text{ kJ/(kg} \cdot {}^\circ\text{C)}][(15 - 25)^\circ\text{C}] + (0.01065)(2465.9 \text{ kJ/kg})}{(2547.2 - 62.99) \text{ kJ/kg}}$$

$$= \mathbf{0.00653 \text{ kg } H_2O/\text{kg dry air}}$$

(*b*) The relative humidity ϕ_1 is determined from Eq. 13-11*a* to be

$$\phi_1 = \frac{\omega_1 P_2}{(0.622 + \omega_1)P_{g_1}} = \frac{(0.00653)(101.325 \text{ kPa})}{(0.622 + 0.00653)(3.169 \text{ kPa})} = \mathbf{0.332 \text{ or } 33.2\%}$$

(*c*) The enthalpy of air per unit mass of dry air is determined from Eq. 13-12:

$$h_1 = h_{a_1} + \omega_1 h_{v_1} \cong C_p T_1 + \omega_1 h_{g_1}$$
$$= [1.005 \text{ kJ/(kg} \cdot \text{°C)}](25\text{°C}) + (0.00653)(2547.2 \text{ kJ/kg})$$
$$= \mathbf{41.8 \text{ kJ/kg dry air}}$$

13-5 ■ THE PSYCHROMETRIC CHART

The state of the atmospheric air at a specified pressure is completely specified by two independent intensive properties. The rest of the properties can be calculated easily from the relations above. The sizing of a typical air-conditioning system involves numerous such calculations, which may eventually get on the nerves of even the most patient engineers. Therefore, there is clear motivation to do these calculations once and to present the data in the form of easily readable charts. Such charts are called **psychrometric charts,** and they are used extensively in air-conditioning applications. A psychrometric chart for a pressure of 1 atm (101.325 kPa or 14.696 psia) is given in Fig. A-33 in SI units and in Fig. A-33E in English units. Psychrometric charts at other pressures (for use at considerably higher elevations than sea level) are also available.

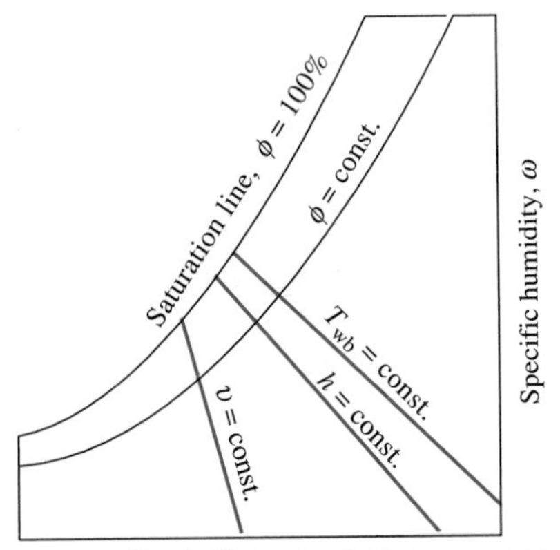

FIGURE 13-14

Schematic for a psychrometric chart.

The basic features of the psychrometric chart are illustrated in Fig. 13-14. The dry-bulb temperatures are shown on the horizontal axis, and the specific humidity is shown on the vertical axis. (Some charts also show the vapor pressure on the vertical axis since at a fixed total pressure P there is a one-to-one correspondence between the specific humidity ω and the vapor pressure P_v, as can be seen from Eq. 13-8.) On the left end of the chart, there is a curve (called the *saturation line*) instead of a straight line. All the saturated air states are located on this curve. Therefore, it is also the curve of 100 percent relative humidity. Other constant relative-humidity curves have the same general shape.

Lines of constant wet-bulb temperature have a downhill appearance to the right. Lines of constant specific volume (in m^3/kg dry air) look similar, except they are steeper. Lines of constant enthalpy (in kJ/kg dry air) lie very nearly parallel to the lines of constant wet-bulb temperature. Therefore, the constant-wet-bulb-temperature lines are used as constant-enthalpy lines in some charts.

FIGURE 13-15

For saturated air, the dry-bulb, wet-bulb, and dew-point temperatures are identical.

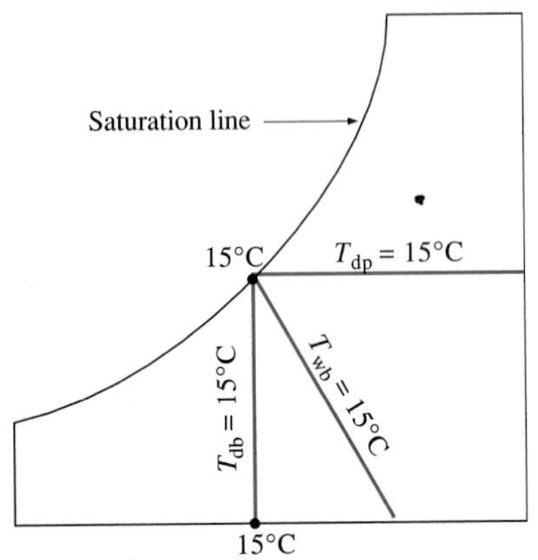

For saturated air, the dry-bulb, wet-bulb, and dew-point temperatures are identical (Fig. 13-15). Therefore, the dew-point temperature of atmospheric air at any point on the chart can be determined by drawing a horizontal line (a line of ω = constant or P_v = constant) from the point to the saturated curve. The temperature value at the intersection point is the dew-point temperature.

The psychrometric chart also serves as a valuable aid in visualizing the air-conditioning processes. An ordinary heating or cooling process, for example, will appear as a horizontal line on this chart if no humidification or dehumidification is involved (that is, ω = constant). Any deviation from a horizontal line indicates that moisture is added or removed from the air during the process.

EXAMPLE 13-4 The Use of the Psychrometric Chart

Consider a room that contains air at 1 atm, 35°C, and 40 percent relative humidity. Using the psychrometric chart, determine (*a*) the specific humidity, (*b*) the enthalpy (in kJ/kg dry air), (*c*) the wet-bulb temperature, (*d*) the dew-point temperature, and (*e*) the specific volume of the air (in m^3/kg dry air).

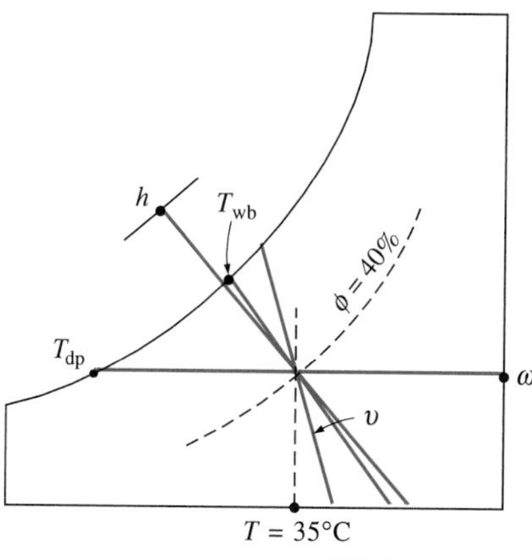

FIGURE 13-16

Schematic for Example 13-4.

Solution At a given total pressure, the state of atmospheric air is completely specified by two independent properties such as the dry-bulb temperature and the relative humidity. Other properties are determined by directly reading their values at the specified state.

Analysis (*a*) The specific humidity is determined by drawing a horizontal line from the specified state to the right until it intersects with the ω axis, as shown in Fig. 13-16. At the intersection point we read

$$\omega = \mathbf{0.0142\ kg\ H_2O/kg\ dry\ air}$$

(*b*) The enthalpy of air per unit mass of dry air is determined by drawing a line parallel to the h = constant lines from the specific state until it intersects the enthalpy scale. At the intersection point we read

$$h = \mathbf{71.5\ kJ/kg\ dry\ air}$$

(*c*) The wet-bulb temperature is determined by drawing a line parallel to the T_{wb} = constant lines from the specified state until it intersects the saturation line. At the intersection point we read

$$T_{wb} = \mathbf{24°C}$$

(*d*) The dew-point temperature is determined by drawing a horizontal line from the specified state to the left until it intersects the saturation line. At the intersection point we read

$$T_{dp} = \mathbf{19.4°C}$$

(*e*) The specific volume per unit mass of dry air is determined by noting the distances between the specified state and the v = constant lines on both sides of the point. The specific volume is determined by visual interpolation to be

$$v = \mathbf{0.893\ m^3/kg\ dry\ air}$$

13-6 ■ HUMAN COMFORT AND AIR-CONDITIONING

Human beings have an inherent weakness—they want to feel comfortable. They want to live in an environment that is neither hot nor cold, neither very humid nor very dry. But comfort does not come easily since the desires of the human body and the weather usually are not quite compatible. Achieving comfort requires a constant struggle against the factors that cause discomfort, such as high or low temperatures and high or low humidity. As engineers, it is our duty to help people feel comfortable. (Besides, it keeps us employed.)

It did not take long for people to realize that they could not change the weather in an area. All they can do is change it in a confined space such as a house or a workplace (Fig. 13-17). In the past, this was partially accomplished by fire and simple indoor heating systems. Today, modern air-conditioning systems can heat, cool, humidify, dehumidify, clean, and even deodorize the air—in other words, *condition* the air to peoples' desires. Air-conditioning systems are designed to *satisfy* the needs of the human body; therefore, it is essential that we understand the thermodynamic aspects of the body.

FIGURE 13-17

We cannot change the weather, but we can change the climate in a confined space by air-conditioning.

The human body can be viewed as a heat engine whose energy input is food. As with any other heat engine, the human body generates waste heat that must be rejected to the environment if the body is to continue operating. The rate of heat generation depends on the level of the activity. For an average adult male, it is about 87 W when sleeping, 115 W when resting or doing office work, 230 W when bowling, and 440 W when doing heavy physical work. The corresponding numbers for an adult female are about 15 percent less. (This difference is due to the body size, not the body temperature. The deep-body temperature of a healthy person is maintained constant at 37°C.) A body will feel comfortable in environments in which it can dissipate this waste heat comfortably (Fig. 13-18).

FIGURE 13-18

A body feels comfortable when it can freely dissipate its waste heat, and no more.

Heat transfer is proportional to the temperature difference. Therefore in cold environments, a body will lose more heat than it normally generates, which results in a feeling of discomfort. The body tries to minimize the energy deficit by cutting down the blood circulation near the skin (causing a pale look). This lowers the skin temperature, which is about 34°C for an average person, and thus the heat transfer rate. A low skin temperature causes discomfort. The hands, for example, feel painfully cold when the skin temperature reaches 10°C (50°F). We can also reduce the heat loss from the body either by putting barriers (additional clothes, blankets, etc.) in the path of heat or by increasing the rate of heat generation within the body by exercising. For example, the comfort level of a resting person dressed in warm winter clothing in a room at 10°C (50°F) is roughly equal to the comfort level of an identical person doing moderate work in a room at about −23°C (−10°F). Or we can just cuddle up and put our hands between our legs to reduce the surface area through which heat flows.

In hot environments, we have the opposite problem—we do not seem to be dissipating enough heat from our bodies, and we feel as if we are going to burst. We dress lightly to make it easier for heat to get away from our bodies, and we reduce the level of activity to minimize the rate of waste heat generation in the body. We also turn on the fan to continuously replace the warmer air layer that forms around our bodies as a result of body heat by the cooler air in other parts of the room. When doing light work or walking slowly, about half of the rejected body heat is dissipated through perspiration as *latent heat* while the other half is dissipated through convection and radiation as *sensible heat*. When resting or doing office work, most of the heat (about

70 percent) is dissipated in the form of sensible heat whereas when doing heavy physical work, most of the heat (about 60 percent) is dissipated in the form of latent heat. The body helps out by perspiring or sweating more. As this sweat evaporates, it absorbs latent heat from the body and cools it. Perspiration is not much help, however, if the relative humidity of the environment is close to 100 percent. Prolonged sweating without any fluid intake will cause dehydration and reduced sweating, which may lead to a rise in body temperature and a heat stroke.

Another important factor that affects human comfort is heat transfer by radiation between the body and the surrounding surfaces such as walls and windows. The sun's rays travel through space by radiation. You warm up in front of a fire even if the air between you and the fire is quite cold. Likewise, in a warm room you will feel chilly if the ceiling or the wall surfaces are at a considerably lower temperature. This is due to direct heat transfer between your body and the surrounding surfaces by radiation. Radiant heaters are commonly used for heating hard-to-heat places such as car repair shops.

The comfort of the human body depends primarily on three factors: the (dry-bulb) temperature, relative humidity, and air motion (Fig. 13-19). The temperature of the environment is the single most important index of comfort. Most people feel comfortable when the environment temperature is between 22 and 27°C (72 and 80°F). The relative humidity also has a considerable effect on comfort since it affects the amount of heat a body can dissipate through evaporation. Relative humidity is a measure of air's ability to absorb more moisture. High relative humidity slows down heat rejection by evaporation, and low relative humidity speeds it up. Most people prefer a relative humidity of 40 to 60 percent.

Air motion also plays an important role in human comfort. It removes the warm, moist air that builds up around the body and replaces it with fresh air. Therefore, air motion improves heat rejection by both convection and evaporation. Air motion should be strong enough to remove heat and moisture from the vicinity of the body, but gentle enough to be unnoticed. Most people feel comfortable at an airspeed of about 15 m/min. Very-high-speed air motion causes discomfort instead of comfort. For example, an environment at 10°C (50°F) with 48 km/h winds feels as cold as an environment at −7°C (20°F) with 3 km/h winds as a result of the body-chilling effect of the air motion (the *wind-chill factor*). Other factors that affect comfort are air cleanliness, odor, noise, and radiation effect.

FIGURE 13-19
A comfortable environment.

13-7 ■ AIR-CONDITIONING PROCESSES

Maintaining a living space or an industrial facility at the desired temperature and humidity requires some processes called air-conditioning processes. These processes include *simple heating* (raising the temperature), *simple cooling* (lowering the temperature), *humidifying* (adding moisture), and *dehumidifying* (removing moisture). Sometimes two or more of these processes are needed to bring the air to a desired temperature and humidity level.

Various air-conditioning processes are illustrated on the psychrometric chart in Fig. 13-20. Notice that simple heating and cooling processes appear

FIGURE 13-20
Various air-conditioning processes.

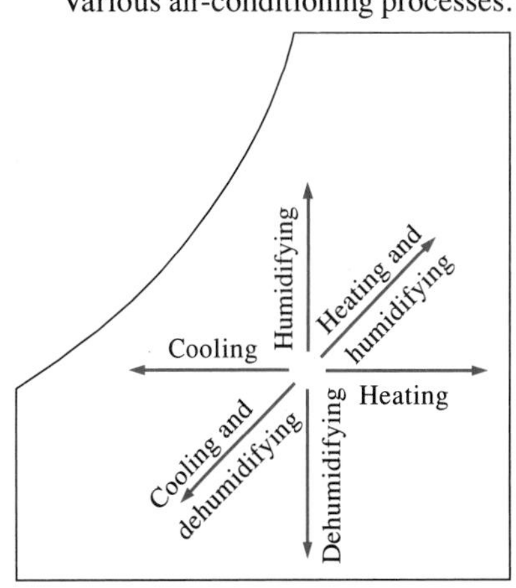

as horizontal lines on this chart since the moisture content of the air remains constant (ω = constant) during these processes. Air is commonly heated and humidified in winter and cooled and dehumidified in summer. Notice how these processes appear on the psychrometric chart.

Most air-conditioning processes can be modeled as steady-flow processes, and thus the *mass balance* relation $\dot{m}_{in} = \dot{m}_{out}$ can be expressed for *dry air* and *water* as

Mass balance for dry air: $$\sum \dot{m}_{a,i} = \sum \dot{m}_{a,e} \qquad \text{(kg/s)} \tag{13-16}$$

Mass balance for water: $$\sum \dot{m}_{w,i} = \sum \dot{m}_{w,e} \quad \text{or} \quad \sum \dot{m}_{a,i}\omega_i = \sum \dot{m}_{a,e}\omega_e \tag{13-17}$$

where the subscripts *i* and *e* denote the inlet and the exit states, respectively. Disregarding the kinetic and potential energy changes, the *steady-flow energy balance* relation $\dot{E}_{in} = \dot{E}_{out}$ can be expressed in this case as

$$\dot{Q}_{in} + \dot{W}_{in} + \sum \dot{m}_i h_i = \dot{Q}_{out} + \dot{W}_{out} + \sum \dot{m}_e h_e \tag{13-18}$$

The work term usually consists of the *fan work input,* which is small relative to the other terms in the energy balance relation. Next we examine some commonly encountered processes in air-conditioning.

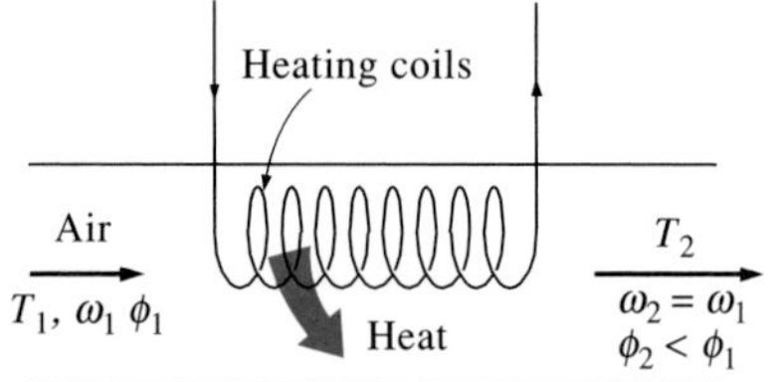

FIGURE 13-21

During simple heating, specific humidity remains constant, but relative humidity decreases.

1 Simple Heating and Cooling (ω = constant)

Many residential heating systems consist of a stove, a heat pump, or an electric resistance heater. The air in these systems is heated by circulating it through a duct that contains the tubing for the hot gases or the electric resistance wires, as shown in Fig. 13-21. The amount of moisture in the air remains constant during this process since no moisture is added to or removed from the air. That is, the specific humidity of the air remains constant (ω = constant) during a heating (or cooling) process with no humidification or dehumidification. Such a heating process will proceed in the direction of increasing dry-bulb temperature following a line of constant specific humidity on the psychrometric chart, which appears as a horizontal line.

Notice that the relative humidity of air decreases during a heating process even if the specific humidity ω remains constant. This is because the relative humidity is the ratio of the moisture content to the moisture capacity of air at the same temperature, and moisture capacity increases with temperature. Therefore, the relative humidity of heated air may be well below comfortable levels, causing dry skin, respiratory difficulties, and an increase in static electricity.

A cooling process at constant specific humidity is similar to the heating process discussed above, except the dry-bulb temperature decreases and the relative humidity increases during such a process, as shown in Fig. 13-22. Cooling can be accomplished by passing the air over some coils through which a refrigerant or cool water flows.

FIGURE 13-22

During simple cooling, specific humidity remains constant, but relative humidity increases.

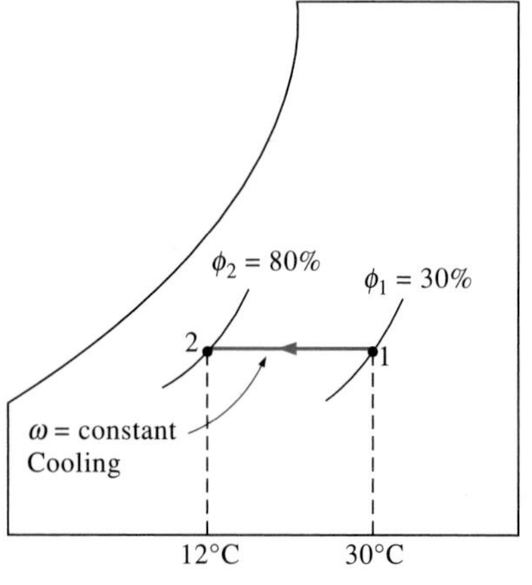

The conservation of mass equations for a heating or cooling process that involves no humidification or dehumidification reduce to $\dot{m}_{a_1} = \dot{m}_{a_2} = \dot{m}_a$

for dry air and $\omega_1 = \omega_2$ for water. Neglecting any fan work that may be present, the conservation of energy equation in this case reduces to

$$\dot{Q} = \dot{m}_a(h_2 - h_1) \qquad \text{or} \qquad q = h_2 - h_1$$

where h_1 and h_2 are enthalpies per unit mass of dry air at the inlet and the exit of the heating or cooling section, respectively.

2 Heating with Humidification

Problems associated with the low relative humidity resulting from simple heating can be eliminated by humidifying the heated air. This is accomplished by passing the air first through a heating section (process 1-2) and then through a humidifying section (process 2-3), as shown in Fig. 13-23.

The location of state 3 depends on how the humidification is accomplished. If steam is introduced in the humidification section, this will result in humidification with additional heating ($T_3 > T_2$). If humidification is accomplished by spraying water into the airstream instead, part of the latent heat of vaporization will come from the air, which will result in the cooling of the heated airstream ($T_3 < T_2$). Air should be heated to a higher temperature in the heating section in this case to make up for the cooling during the humidification process.

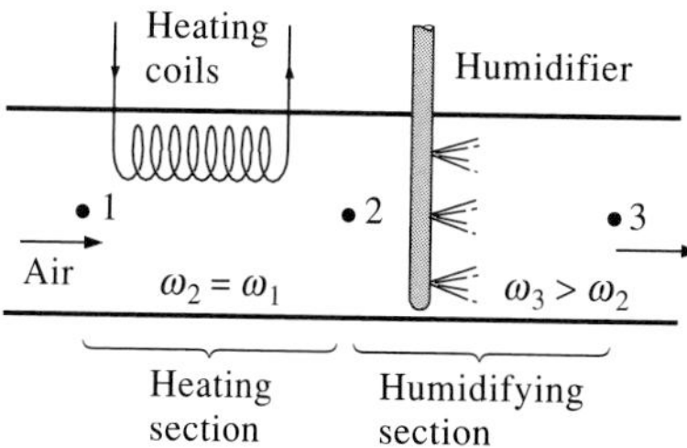

FIGURE 13-23
Heating the humidification.

EXAMPLE 13-5 Heating and Humidification of Air

An air-conditioning system is to take in outdoor air at 10°C and 30 percent relative humidity at a steady rate of 45 m^3/min and to condition it to 25°C and 60 percent relative humidity. The outdoor air is first heated to 22°C in the heating section and then humidified by the injection of hot steam in the humidifying section. Assuming the entire process takes place at a pressure of 100 kPa, determine (*a*) the rate of heat supply in the heating section and (*b*) the mass flow rate of the steam required in the humidifying section.

Solution We will take the system to be the *heating* or the *humidifying section,* as appropriate. The schematic of the system and the psychrometric chart of the process are shown in Fig. 13-24. We note that the amount of water vapor in the air remains constant in the heating section ($\omega_1 = \omega_2$) but increases in the humidifying section ($\omega_3 > \omega_2$).

Assumptions **1** This is a steady-flow process and thus the mass flow rate of dry air remains constant during the entire process. **2** Dry air and water vapor are ideal gases. **3** The kinetic and potential energy changes are negligible.

Analysis (*a*) Applying the mass and energy balances on the heating section gives

Dry air mass balance: $\qquad \dot{m}_{a_1} = \dot{m}_{a_2} = \dot{m}_a$

Water mass balance: $\qquad \dot{m}_{a_1}\omega_1 = \dot{m}_{a_2}\omega_2 \rightarrow \omega_1 = \omega_2$

Energy: $\qquad \dot{Q}_{in} + \dot{m}_a h_1 = \dot{m}_a h_2 \rightarrow \dot{Q}_{in} = \dot{m}_a(h_2 - h_1)$

The psychrometric chart offers great convenience in determining the properties of moist air. However, its use is limited to a specified pressure only, which is 1 atm (101.325 kPa) for the one given in the appendix. At pressures other than 1 atm,

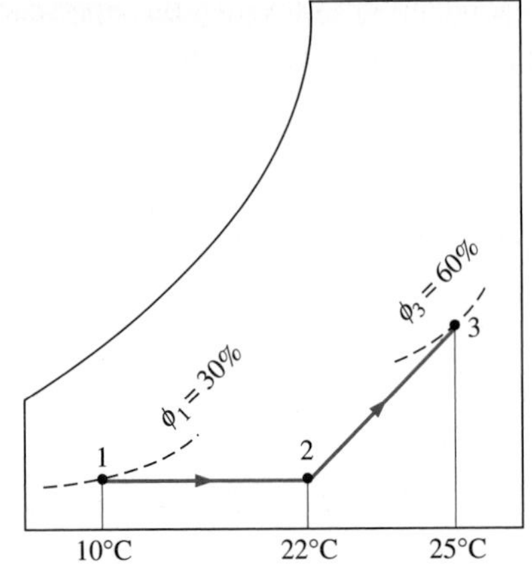

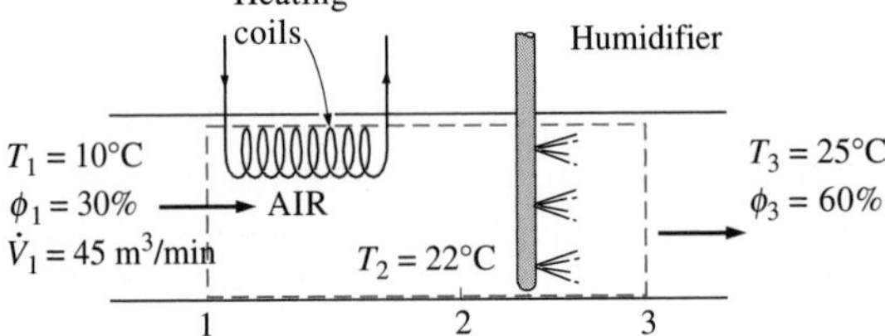

FIGURE 13-24
Schematic and psychrometric chart for Example 13-5.

either other charts for that pressure or the relations developed earlier should be used. In our case, the choice is clear:

$$P_{v_1} = \phi_1 P_{g_1} = \phi_1 P_{\text{sat @ 10°C}} = (0.3)(1.2276\text{ kPa}) = 0.368\text{ kPa}$$

$$P_{a_1} = P_1 - P_{v_1} = (100 - 0.368)\text{ kPa} = 99.632\text{ kPa}$$

$$v_1 = \frac{R_a T_1}{P_{a_1}} = \frac{[0.287\text{ kPa}\cdot\text{m}^3/(\text{kg}\cdot\text{K})](283\text{ K})}{99.632\text{ kPa}} = 0.815\text{ m}^3/\text{kg dry air}$$

$$\dot{m}_a = \frac{\dot{V}_1}{v_1} = \frac{45\text{ m}^3/\text{min}}{0.815\text{ m}^3/\text{kg}} = 55.2\text{ kg/min}$$

$$\omega_1 = \frac{0.622P_{v_1}}{P_1 - P_{v_1}} = \frac{0.622(0.368\text{ kPa})}{(100 - 0.368)\text{ kPa}} = 0.0023\text{ kg H}_2\text{O/kg dry air}$$

$$h_1 = C_pT_1 + \omega_1 h_{g_1} = [1.005\text{ kJ/(kg}\cdot\text{°C)}](10\text{°C}) + (0.0023)(2519.8\text{ kJ/kg})$$
$$= 15.8\text{ kJ/kg dry air}$$

$$h_2 = C_pT_2 + \omega_2 h_{g_2} = [1.005\text{ kJ/(kg}\cdot\text{°C)}](22\text{°C}) + (0.0023)(2541.7\text{ kJ/kg})$$
$$= 28.0\text{ kJ/kg dry air}$$

since $\omega_2 = \omega_1$. Then the rate of heat transfer to the air in the heating section becomes

$$\dot{Q}_{\text{in}} = \dot{m}_a(h_2 - h_1) = (55.2\text{ kg/min})[(28.0 - 15.8)\text{ kJ/kg}]$$
$$= \mathbf{673.4\ kJ/min}$$

(*b*) The mass balance for water in the humidifying section can be expressed as

$$\dot{m}_{a_2}\omega_2 + \dot{m}_w = \dot{m}_{a_3}\omega_3$$

or

$$\dot{m}_w = \dot{m}_a(\omega_3 - \omega_2)$$

where

$$\omega_3 = \frac{0.622\phi_3 P_{g_3}}{P_3 - \phi_3 P_{g_3}} = \frac{0.622(0.60)(3.169 \text{ kPa})}{[100 - (0.60)(3.169)]\text{kPa}}$$

$$= 0.01206 \text{ kg H}_2\text{O/kg dry air}$$

Thus,

$$\dot{m}_w = (55.2 \text{ kg/min})(0.01206 - 0.0023)$$

$$= \mathbf{0.539 \text{ kg/min}}$$

3 Cooling with Dehumidification

The specific humidity of air remains constant during a simple cooling process, but its relative humidity increases. If the relative humidity reaches undesirably high levels, it may be necessary to remove some moisture from the air, that is, to dehumidify it. This requires cooling the air below its dew-point temperature.

The cooling process with dehumidifying is illustrated schematically and on the psychrometric chart in Fig. 13-25 in conjunction with Example 13-6. Hot, moist air enters the cooling section at state 1. As it passes through the cooling coils, its temperature decreases and its relative humidity increases at constant specific humidity. If the cooling section is sufficiently long, air will reach its dew point (state 2, saturated air). Further cooling of air results in the condensation of part of the moisture in the air. Air remains saturated during the entire condensation process, which follows a line of 100 percent relative humidity until the final state (state 3) is reached. The water vapor that condenses out of the air during this process is removed from the cooling section through a separate channel. The condensate is usually assumed to leave the cooling section at T_3.

The cool, saturated air at state 3 is usually routed directly to the room, where it mixes with the room air. In some cases, however, the air at state 3 may be at the right specific humidity but at a very low temperature. In such cases, the air is passed through a heating section where its temperature is raised to a more comfortable level before it is routed to the room.

EXAMPLE 13-6 Cooling and Dehumidification of Air

Air enters a window air conditioner at 1 atm, 30°C, and 80 percent relative humidity at a rate of 10 m^3/min, and it leaves as saturated air at 14°C. Part of the moisture in the air that condenses during the process is also removed at 14°C. Determine the rates of heat and moisture removal from the air.

Solution We take the *cooling section* to be the system. The schematic of the system and the psychrometric chart of the process are shown in Fig. 13-25. We

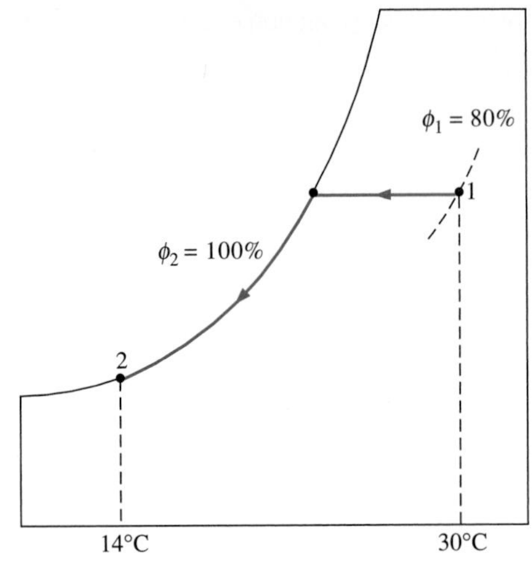

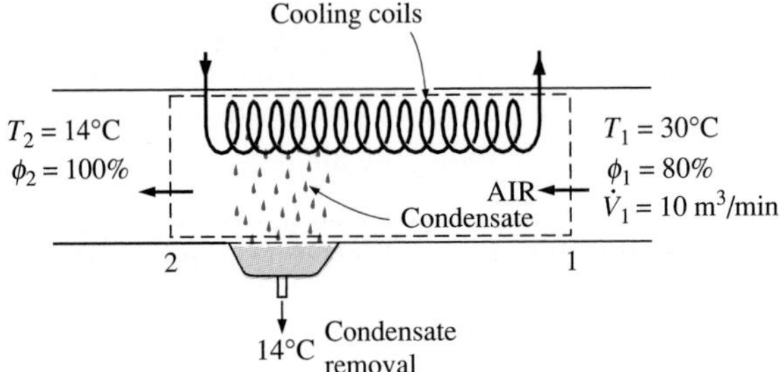

FIGURE 13-25
Schematic and psychrometric chart for Example 13-6.

note that the amount of water vapor in the air decreases during the process ($\omega_2 < \omega_1$) due to dehumidification.

Assumptions **1** This is a steady-flow process and thus the mass flow rate of dry air remains constant during the entire process. **2** Dry air and the water vapor are ideal gases. **3** The kinetic and potential energy changes are negligible.

Analysis Applying the mass and energy balances on the cooling and dehumidification section gives

Dry air mass balance: $\dot{m}_{a_1} = \dot{m}_{a_2} = \dot{m}_a$

Water mass balance: $\dot{m}_{a_1}\omega_1 = \dot{m}_{a_2}\omega_2 + \dot{m}_w \rightarrow \dot{m}_w = \dot{m}_a(\omega_1 - \omega_2)$

Energy balance: $\sum \dot{m}_i h_i = \dot{Q}_{out} + \sum \dot{m}_e h_e \rightarrow \dot{Q}_{out} = \dot{m}_a(h_1 - h_2) - \dot{m}_w h_w$

The inlet and the exit states of the air are completely specified, and the total pressure is 1 atm. Therefore, we can determine the properties of the air at both states from the psychrometric chart to be

$$h_1 = 85.4 \text{ kJ/kg dry air}$$

$$\omega_1 = 0.0216 \text{ kg H}_2\text{O/kg dry air}$$

$$v_1 = 0.889 \text{ m}^3\text{/kg dry air}$$

and $$h_2 = 39.3 \text{ kJ/kg dry air}$$

$$\omega_2 = 0.0100 \text{ kg H}_2\text{O/kg dry air}$$

Also, $$h_w = h_{f @ 14°C} = 58.8 \text{ kJ/kg} \qquad \text{(Table A-4)}$$

Then $$\dot{m}_{a_1} = \frac{\dot{V}_1}{v_1} = \frac{10 \text{ m}^3\text{/min}}{0.889 \text{ m}^3\text{/kg dry air}} = 11.25 \text{ kg/min}$$

$$\dot{m}_w = (11.25 \text{ kg/min})(0.0216 - 0.0100) = \mathbf{0.131 \text{ kg/min}}$$

$$\dot{Q}_{out} = (11.25 \text{ kg/min})[(85.4 - 39.3) \text{ kJ/kg}] - (0.131 \text{ kg/min})(58.8 \text{ kJ/kg})$$

$$= \mathbf{511 \text{ kJ/min}}$$

Therefore, this air-conditioning unit removes moisture and heat from the air at rates of 0.131 kg/min and 511 kJ/min, respectively.

4 Evaporative Cooling

Conventional cooling systems operate on a refrigeration cycle, and they can be used in any part of the world. But they have a high initial and operating cost. In desert (hot and dry) climates, we can avoid the high cost of cooling by using *evaporative coolers,* also known as *swamp coolers.*

Evaporative cooling is based on a simple principle: As water evaporates, the latent heat of vaporization is absorbed from the water body and the surrounding air. As a result, both the water and the air are cooled during the process. This approach has been used for thousands of years to cool water. A porous jug or pitcher filled with water is left in an open, shaded area. A small amount of water leaks out through the porous holes, and the pitcher "sweats." In a dry environment, this water evaporates and cools the remaining water in the pitcher (Fig. 13-26).

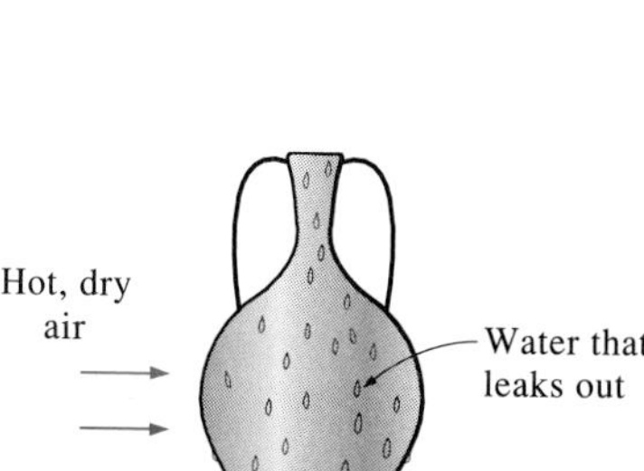

FIGURE 13-26
The water in a porous jug left in an open, breezy area cools as a result of evaporative cooling.

You have probably noticed that on a hot, dry day the air feels a lot cooler when the yard is watered. This is because water absorbs heat from the air as it evaporates. An evaporative cooler works on the same principle. The evaporative cooling process is shown schematically and on a psychrometric chart in Fig. 13-27. Hot, dry air at state 1 enters the evaporative cooler, where it is sprayed with liquid water. Part of the water evaporates during this process by absorbing heat from the airstream. As a result, the temperature of the airstream decreases and its humidity increases (state 2). In the limiting case, the air will leave the cooler saturated at state 2′. This is the lowest temperature that can be achieved by this process.

The evaporative cooling process is essentially identical to the adiabatic saturation process since the heat transfer between the airstream and the surroundings is usually negligible. Therefore, the evaporative cooling process

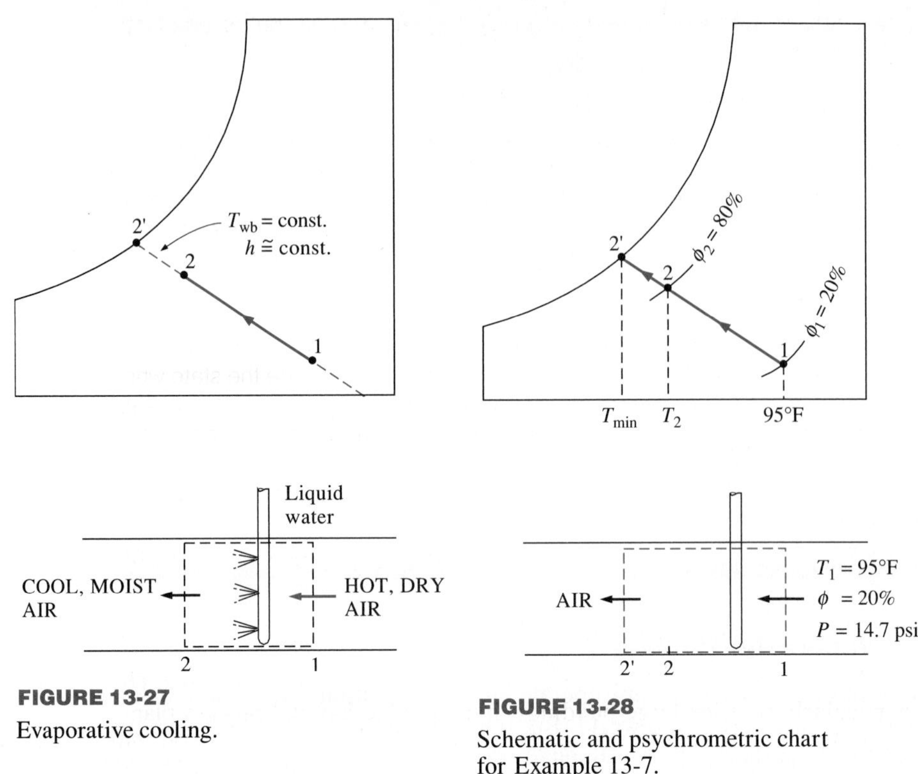

FIGURE 13-27
Evaporative cooling.

FIGURE 13-28
Schematic and psychrometric chart for Example 13-7.

follows a line of constant wet-bulb temperature on the psychrometric chart. (Note that this will not exactly be the case if the liquid water is supplied at a temperature different from the exit temperature of the airstream.) Since the constant-wet-bulb-temperature lines almost coincide with the constant-enthalpy lines, the enthalpy of the airstream can also be assumed to remain constant. That is,

$$T_{wb} \cong \text{constant} \tag{13-19}$$

and

$$h \cong \text{constant} \tag{13-20}$$

during an evaporative cooling process. This is a reasonably accurate approximation, and it is commonly used in air-conditioning calculations.

EXAMPLE 13-7 Evaporative Cooling of Air by a Swamp Cooler

Air enters an evaporative (or swamp) cooler at 14.7 psi, 95°F, and 20 percent relative humidity, and it exits at 80 percent relative humidity. Determine (*a*) the exit temperature of the air and (*b*) the lowest temperature to which the air can be cooled by this evaporative cooler.

Solution The schematic of the evaporative cooler and the psychrometric chart of the process are shown in Fig. 13-28.

Analysis (*a*) If we assume the liquid water is supplied at a temperature not much different from the exit temperature of the airstream, the evaporative cooling process follows a line of constant wet-bulb temperature on the psychrometric chart. That is,

$$T_{wb} \cong \text{constant}$$

The wet-bulb temperature at 95°F and 20 percent relative humidity is determined from the psychrometric chart to be 66.0°F. The intersection point of the $T_{wb} = 66.0$°F and the $\phi = 80$ percent lines is the exit state of the air. The temperature at this point is the exit temperature of the air, and it is determined from the psychrometric chart to be

$$T_2 = \mathbf{70.4°F}$$

(*b*) In the limiting case, the air will leave the evaporative cooler saturated ($\phi = 100$ percent, and the exit state of the air in this case will be the state where the $T_{wb} = 66.0$°F line intersects the saturation line. For saturated air, the dry- and the wet-bulb temperatures are identical. Therefore, the lowest temperature to which the air can be cooled is the wet-bulb temperature, which is

$$T_{min} = T_{2'} = \mathbf{66.0°F}$$

5 Adiabatic Mixing of Airstreams

Many air-conditioning applications require the mixing of two airstreams. This is particularly true for large buildings, most production and process plants, and hospitals, which require that the conditioned air be mixed with a certain fraction of fresh outside air before it is routed into the living space. The mixing is accomplished by simply merging the two airstreams, as shown in Fig. 13-29.

The heat transfer with the surroundings is usually small, and thus the mixing processes can be assumed to be adiabatic. Mixing processes normally involve no work interactions, and the changes in kinetic and potential energies, if any, are negligible. Then the mass and energy balances for the adiabatic mixing of two airstreams reduce to

Mass of dry air: $$\dot{m}_{a_1} + \dot{m}_{a_2} = \dot{m}_{a_3} \tag{13-21}$$

Mass of water vapor: $$\omega_1\dot{m}_{a_1} + \omega_w\dot{m}_{a_2} = \omega_3\dot{m}_{a_3} \tag{13-22}$$

Energy: $$\dot{m}_{a_1}h_1 + \dot{m}_{a_2}h_2 = \dot{m}_{a_3}h_3 \tag{13-23}$$

Eliminating $\dot{m}_{a_3}$ from the relations above, we obtain

$$\frac{\dot{m}_{a_1}}{\dot{m}_{a_2}} = \frac{\omega_2 - \omega_3}{\omega_3 - \omega_1} = \frac{h_2 - h_3}{h_3 - h_1} \tag{13-24}$$

This equation has an instructive geometric interpretation on the psychrometric chart. It shows that the ratio of $\omega_2 - \omega_3$ to $\omega_3 - \omega_1$ is equal to the ratio of $\dot{m}_{a_1}$ to $\dot{m}_{a_2}$. The states that satisfy this condition are indicated by the dashed line *AB*. The ratio of $h_2 - h_3$ to $h_3 - h_1$ is also equal to the ratio of $\dot{m}_{a_1}$ to $\dot{m}_{a_2}$, and the states that satisfy this condition are indicated by the dashed line *CD*. The only state that satisfies both conditions is the intersection

FIGURE 13-29

When two airstreams at states 1 and 2 are mixed adiabatically, the state of the mixture lies on the straight line connecting the two states.

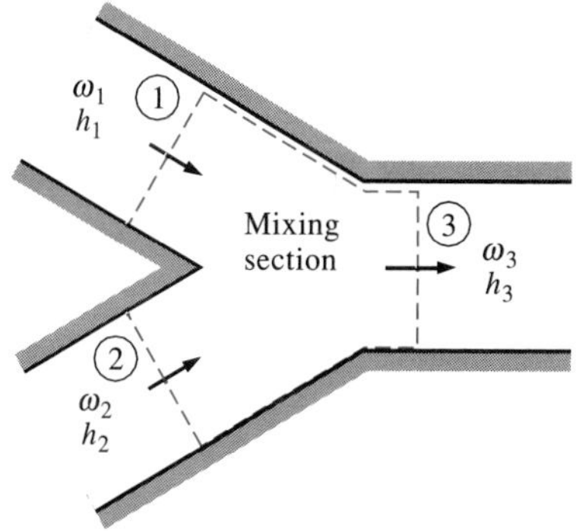

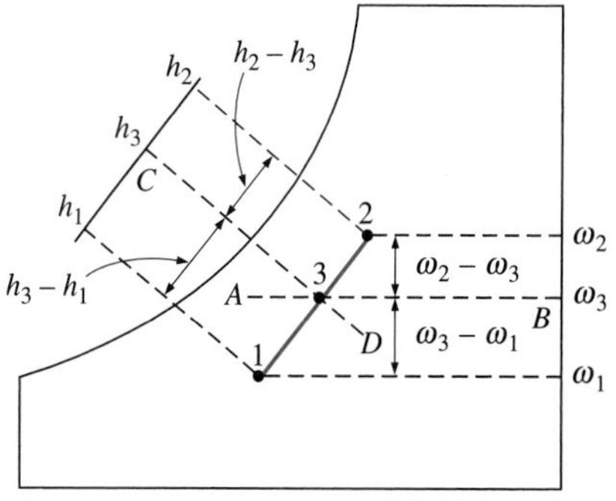

point of these two dashed lines, which is located on the straight line connecting states 1 and 2. Thus we conclude that *when two airstreams at two different states (states 1 and 2) are mixed adiabatically, the state of the mixture (state 3) will lie on the straight line connecting states 1 and 2 on the psychrometric chart, and the ratio of the distances 2-3 and 3-1 is equal to the ratio of mass flow rates $\dot{m}_{a_1}$ and $\dot{m}_{a_2}$.*

The concave nature of the saturation curve and the conclusion above lead to an interesting possibility. When states 1 and 2 are located close to the saturation curve, the straight line connecting the two states will cross the saturation curve, and state 3 may lie to the left of the saturation curve. In this case, some water will inevitably condense during the mixing process.

EXAMPLE 13-8 Mixing of Conditioned Air with Outdoor Air

Saturated air leaving the cooling section of an air-conditioning system at 14°C at a rate of 50 m³/min is mixed adiabatically with the outside air at 32°C and 60 percent relative humidity at a rate of 20 m³/min. Assuming that the mixing process occurs at a pressure of 1 atm, determine the specific humidity, the relative humidity, the dry-bulb temperature, and the volume flow rate of the mixture.

Solution We take the *mixing section* of the streams to be the system. The schematic of the system and the psychrometric chart of the process are shown in Fig. 13-30. We note that this is a steady-flow mixing process.

Assumptions **1** Steady operating conditions exist. **2** Dry air and the water vapor are ideal gases. **3** The kinetic and potential energy changes are negligible. **4** The mixing section is adiabatic.

Analysis The properties of each inlet stream are determined from the psychrometric chart to be

$$h_1 = 39.4 \text{ kJ/kg dry air}$$
$$\omega_1 = 0.010 \text{ kg H}_2\text{O/kg dry air}$$
$$v_1 = 0.826 \text{ m}^3\text{/kg dry air}$$

and

$$h_2 = 79.0 \text{ kJ/kg dry air}$$
$$\omega_2 = 0.0182 \text{ kg H}_2\text{O/kg dry air}$$
$$v_2 = 0.889 \text{ m}^3\text{/kg dry air}$$

Then the mass flow rates of dry air in each stream are

$$\dot{m}_{a_1} = \frac{\dot{V}_1}{v_1} = \frac{50 \text{ m}^3\text{/min}}{0.826 \text{ m}^3\text{/kg dry air}} = 60.5 \text{ kg/min}$$

$$\dot{m}_{a_2} = \frac{\dot{V}_2}{v_2} = \frac{20 \text{ m}^3\text{/min}}{0.889 \text{ m}^3\text{/kg dry air}} = 22.5 \text{ kg/min}$$

From the mass balance of dry air,

$$\dot{m}_{a_3} = \dot{m}_{a_1} + \dot{m}_{a_2} = (60.5 + 22.5) \text{ kg/min} = 83 \text{ kg/min}$$

The specific humidity and the enthalpy of the mixture can be determined from Eq. 13-24, and are obtained by combining the mass and energy balances for the adiabatic mixing of two streams.

FIGURE 13-30
Schematic and psychrometric chart for Example 13-8.

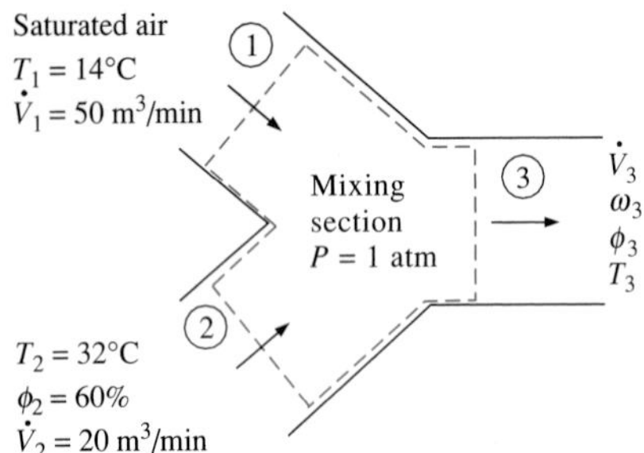

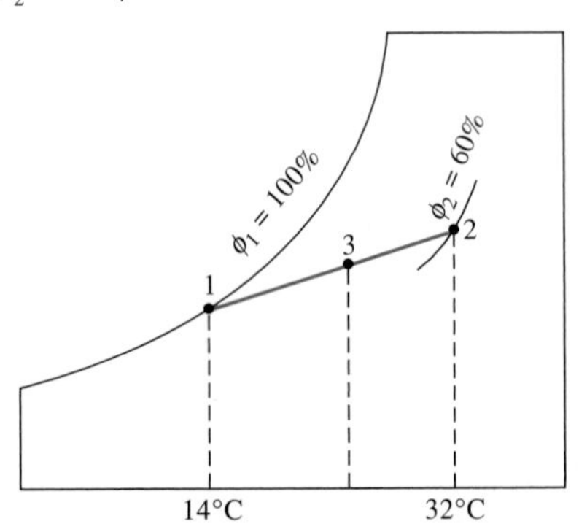

$$\frac{\dot{m}_{a_1}}{\dot{m}_{a_2}} = \frac{\omega_2 - \omega_3}{\omega_3 - \omega_1} = \frac{h_2 - h_3}{h_3 - h_1}$$

$$\frac{60.5}{22.5} = \frac{0.0182 - \omega_3}{\omega_3 - 0.010} = \frac{79.0 - h_3}{h_3 - 39.4}$$

which yield

$$\omega_3 = \mathbf{0.0122\ kg\ H_2O/kg\ dry\ air}$$
$$h_3 = 50.1\ \text{kJ/kg dry air}$$

These two properties fix the state of the mixture. Other properties of the mixture are determined from the psychrometric chart:

$$T_3 = \mathbf{19.0°C}$$
$$\phi_3 = \mathbf{89\%}$$
$$v_3 = 0.844\ \text{m}^3\text{/kg dry air}$$

Finally, the volume flow rate of the mixture is determined from

$$\dot{V}_3 = \dot{m}_{a_3} v_3 = (83\ \text{kg/min})(0.844\ \text{m}^3\text{/kg}) = \mathbf{70.1\ m^3/min}$$

Discussion Notice that the volume flow rate of the mixture is approximately equal to the sum of the volume flow rates of the two incoming streams. This is typical in air-conditioning applications.

6 Wet Cooling Towers

Power plants, large air-conditioning systems, and some industries generate large quantities of waste heat that is often rejected to cooling water from nearby lakes or rivers. In some cases, however, the water supply is limited or thermal pollution is a serious concern. In such cases, the waste heat must be rejected to the atmosphere, with cooling water recirculating and serving as a transport medium for heat between the source and the sink (the atmosphere). One way of achieving this is through the use of wet cooling towers.

A **wet cooling tower** is essentially a semienclosed evaporative cooler. An induced-draft counterflow wet cooling tower is shown schematically in Fig. 13-31. Air is drawn into the tower from the bottom and leaves through the top. Warm water from the condenser is pumped to the top of the tower and is sprayed into this airstream. The purpose of spraying is to expose a large surface area of water to the air. As the water droplets fall under the influence of gravity, a small fraction of water (usually a few percent) evaporates and cools the remaining water. The temperature and the moisture content of the air increase during this process. The cooled water collects at the bottom of the tower and is pumped back to the condenser to pick up additional waste heat. Makeup water must be added to the cycle to replace the water lost by evaporation and air draft. To minimize water carried away by the air, drift eliminators are installed in the wet cooling towers above the spray section.

The air circulation in the cooling tower described above is provided by a fan, and therefore it is classified as a forced-draft cooling tower. Another

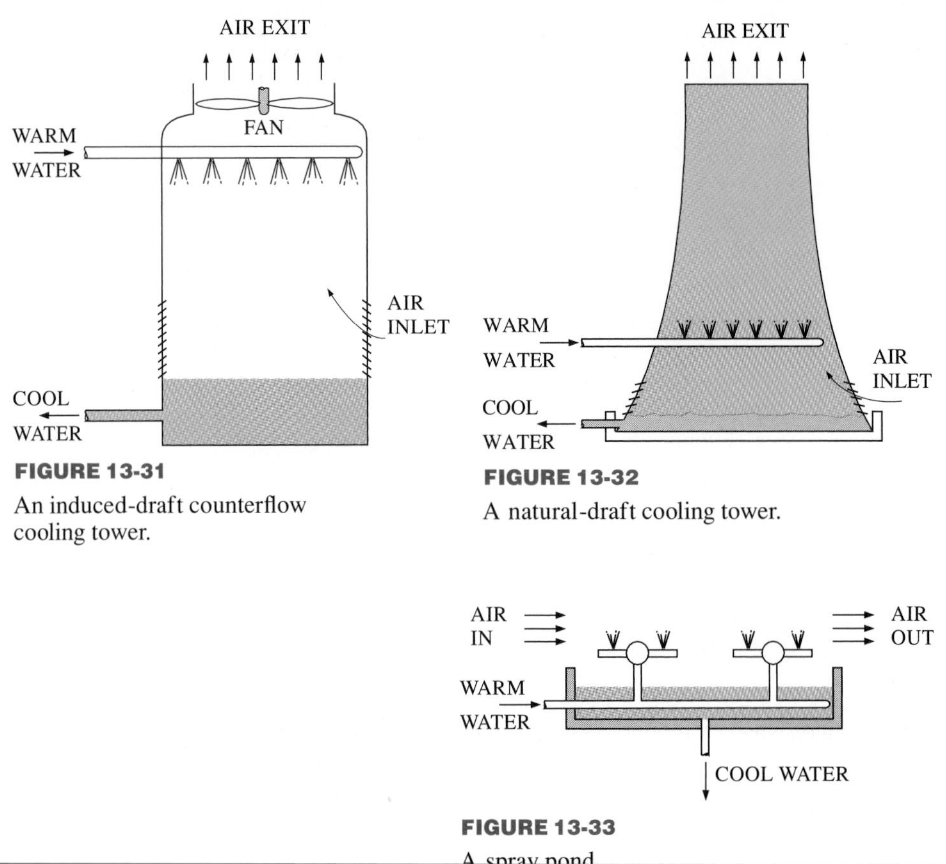

FIGURE 13-31
An induced-draft counterflow cooling tower.

FIGURE 13-32
A natural-draft cooling tower.

FIGURE 13-33
A spray pond.

popular type of cooling tower is the **natural-draft cooling tower,** which looks like a large chimney and works like an ordinary chimney. The air in the tower has a high water-vapor content, and thus it is lighter than the outside air. Consequently, the light air in the tower rises, and the heavier outside air fills the vacant space, creating an airflow from the bottom of the tower to the top. The flow rate of air is controlled by the conditions of the atmospheric air. Natural-draft cooling towers do not require any external power to induce the air, but they cost a lot more to build than forced-draft cooling towers. The natural-draft cooling towers are hyperbolic in profile, as shown in Fig. 13-32, and some are over 100 m high. The hyperbolic profile is for greater structural strength, not for any thermodynamic reason.

The idea of a cooling tower started with the **spray pond,** where the warm water is sprayed into the air and is cooled by the air as it falls into the pond, as shown in Fig. 13-33. Some spray ponds are still in use today. But they require 25 to 50 times the area of a cooling tower, water loss due to air drift is high, and they are unprotected against dust and dirt.

We could also dump the waste heat into a still **cooling pond,** which is basically a large lake open to the atmosphere. But heat transfer from the pond

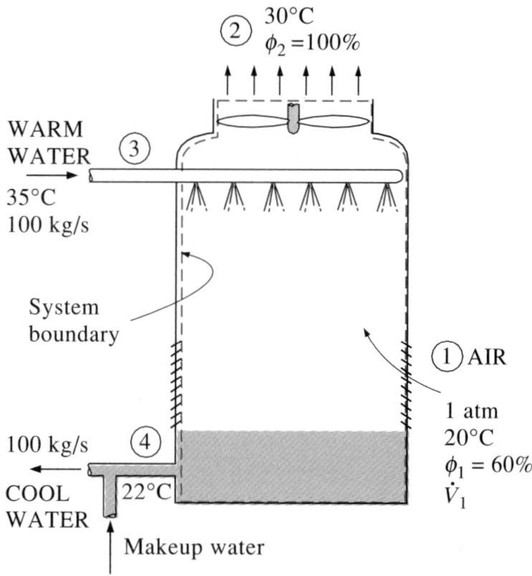

FIGURE 13-34
Schematic for Example 13-9.

surface to the atmosphere is very slow, and we would need about 20 times the area of a spray pond in this case to achieve the same cooling.

EXAMPLE 13-9 Cooling of a Power Plant by a Cooling Tower

Cooling water leaves the condenser of a power plant and enters a wet cooling tower at 35°C at a rate of 100 kg/s. The water is cooled to 22°C in the cooling tower by air that enters the tower at 1 atm, 20°C, and 60 percent relative humidity and leaves saturated at 30°C. Neglecting the power input to the fan, determine (*a*) the volume flow rate of air into the cooling tower and (*b*) the mass flow rate of the required makeup water.

Solution We take the entire *cooling tower* to be the system, which is shown schematically in Fig. 13-34. We note that the mass flow rate of liquid water decreases by an amount equal to the amount of water that vaporizes in the tower during the cooling process. The water lost through evaporation must be made up later in the cycle to maintain steady operation.

Assumptions **1** Steady operating conditions exist and thus the mass flow rate of dry air remains constant during the entire process. **2** Dry air and the water vapor are ideal gases. **3** The kinetic and potential energy changes are negligible. **4** The cooling tower is adiabatic.

Analysis Applying the mass and energy balances on the cooling tower gives

Dry air mass balance:
$$\dot{m}_{a_1} = \dot{m}_{a_2} = \dot{m}_a$$

Water mass balance:
$$\dot{m}_3 + \dot{m}_{a_1}\omega_1 = \dot{m}_4 + \dot{m}_{a_2}\omega_2$$

or
$$\dot{m}_3 - \dot{m}_4 = \dot{m}_a(\omega_2 - \omega_1) = \dot{m}_{\text{makeup}}$$

Energy balance:
$$\sum \dot{m}_i h_i = \sum \dot{m}_e h_e \;\rightarrow\; \dot{m}_{a_1}h_1 + \dot{m}_3 h_3 = \dot{m}_{a_2}h_2 + \dot{m}_4 h_4$$

or
$$\dot{m}_3 h_3 = \dot{m}_a(h_2 - h_1) + (\dot{m}_3 - \dot{m}_{\text{makeup}})h_4$$

Solving for $\dot{m}_a$ gives

$$\dot{m}_a = \frac{\dot{m}_3(h_3 - h_4)}{(h_2 - h_1) - (\omega_2 - \omega_1)h_4}$$

From the psychrometric chart,

$$h_1 = 42.2 \text{ kJ/kg dry air}$$
$$\omega_1 = 0.0087 \text{ kg } H_2O\text{/kg dry air}$$
$$v_1 = 0.842 \text{ m}^3\text{/kg dry air}$$

and

$$h_2 = 100.0 \text{ kJ/kg dry air}$$
$$\omega_2 = 0.0273 \text{ kg } H_2O\text{/kg dry air}$$

From Table A-4,

$$h_3 \cong h_{f\,@\,22°C} = 146.68 \text{ kJ/kg } H_2O$$
$$h_4 \cong h_{f\,@\,22°C} = 92.33 \text{ kJ/kg } H_2O$$

Substituting,

$$\dot{m}_a = \frac{(100 \text{ kg/s})[(146.68 - 92.33) \text{ kJ/kg}]}{[(100.0 - 42.2) \text{ kJ/kg}] - [(0.0273 - 0.0087)(92.33) \text{ kJ/kg}]}$$
$$= 96.9 \text{ kg/s}$$

Then the volume flow rate of air into the cooling tower becomes

$$\dot{V}_1 = \dot{m}_a v_1 = (96.9 \text{ kg/s})(0.842 \text{ m}^3\text{/kg}) = \mathbf{81.6\ m^3/s}$$

(*b*) The mass flow rate of the required makeup water is determined from

$$\dot{m}_{\text{makeup}} = \dot{m}_a(\omega_2 - \omega_1) = (96.9 \text{ kg/s})(0.0273 - 0.0087)$$
$$= \mathbf{1.80\ kg/s}$$

Discussion Note that over 98 percent of the cooling water is saved and recirculated in this case.

13-8 ■ SUMMARY

In this chapter we discussed the air–water-vapor mixture, which is the most commonly encountered gas–vapor mixture in practice. The air in the atmosphere normally contains some water vapor, and it is referred to as *atmospheric air*. By contrast, air that contains no water vapor is called *dry air*. In the temperature range encountered in air-conditioning applications, both the dry air and the water vapor can be treated as ideal gases. The enthalpy change of dry air during a process can be determined from

$$\Delta h_{\text{dry air}} = C_p\, \Delta T = [1.005 \text{ kJ/(kg} \cdot \text{°C)}]\, \Delta T \qquad \text{(kJ/kg)}$$

The atmospheric air can be treated as an ideal-gas mixture whose pressure is the sum of the partial pressure of dry air P_a and that of the water vapor P_v,

$$P = P_a + P_v \qquad \text{(kPa)}$$

The enthalpy of water vapor in the air can be taken to be equal to the enthalpy of the saturated vapor at the same temperature:

$$h_v(T, \text{low } P) \cong h_g(T) \cong 2501.3 + 1.82T \qquad \text{(kJ/kg)} \qquad T \text{ in °C}$$
$$\cong 1061.5 + 0.435T \qquad \text{(Btu/lbm)} \qquad T \text{ in °F}$$

in the temperature range -10 to 50°C (15 to 120°F).

The mass of water vapor present in 1 unit mass of dry air is called the *specific* or *absolute humidity* ω,

$$\omega = \frac{m_v}{m_a} = \frac{0.622P_v}{P - P_v} \qquad \text{(kg H}_2\text{O/kg dry air)}$$

where P is the total pressure of air and P_v is the vapor pressure. There is a limit on the amount of vapor the air can hold at a given temperature. Air that is holding as much moisture as it can is called *saturated air*. The ratio of the amount of moisture air holds (m_v) to the maximum amount of moisture air can hold at the same temperature (m_g) is called the *relative humidity* ϕ,

$$\phi = \frac{m_v}{m_g} = \frac{P_vV/(R_vT)}{P_gV/(R_vT)} = \frac{P_v}{P_g}$$

where $P_g = P_{\text{sat @ }T}$

The relative and specific humidities can also be expressed as

$$\phi = \frac{\omega P}{(0.622 + \omega)P_g} \qquad \text{and} \qquad \omega = \frac{0.622\phi P_g}{P - \phi P_g}$$

Relative humidity ranges from 0 for dry air to 1 for saturated air.

The enthalpy of atmospheric air is expressed *per unit mass of dry air*, instead of per unit mass of the air–water-vapor mixture, as

$$h = h_a + \omega h_g \qquad \text{(kJ/kg dry air)}$$

The ordinary temperature of atmospheric air is referred to as the *dry-bulb temperature* to differentiate it from other forms of temperatures. The temperature at which condensation begins if the air is cooled at constant pressure is called the *dew-point temperature* T_{dp}:

$$T_{\text{dp}} = T_{\text{sat @ }P_v}$$

Relative humidity and specific humidity of air can be determined by measuring the *adiabatic saturation temperature* of air, which is the temperature the air attains after flowing over water in a long channel until it is saturated,

$$\omega_1 = \frac{C_p(T_2 - T_1) + \omega_2 h_{fg_2}}{h_{g_1} - h_{f_2}} \qquad \text{where} \qquad \omega_2 = \frac{0.622P_{g_2}}{P_2 - P_{g_2}}$$

and T_2 is the adiabatic saturation temperature. A more practical approach in air-conditioning applications is to use a thermometer whose bulb is covered with a cotton wick saturated with water and to blow air over the wick. The temperature measured in this manner is called the *wet-bulb temperature* T_{wb}, and it is used in place of the adiabatic saturation temperature. The properties of atmospheric air at a specified total pressure are presented in the form of easily readable charts, called *psychrometric charts*. The lines of constant enthalpy and the lines of constant wet-bulb temperature are very nearly parallel on these charts.

The needs of the human body and the conditions of the environment are not quite compatible. Therefore, it often becomes necessary to change the conditions of a living space to make it more comfortable. Maintaining a living space or an industrial facility at the desired temperature and humidity may require simple heating (raising the temperature), simple cooling (lowering the temperature), humidifying (adding moisture), or dehumidifying (removing moisture). Sometimes two or more of these processes are needed to bring the air to the desired temperature and humidity level.

Most air-conditioning processes can be modeled as steady-flow processes, and therefore they can be analyzed by applying the steady-flow mass (for both dry air and water) and energy balances,

Dry air mass: $$\sum \dot{m}_{a,i} = \sum \dot{m}_{a,e}$$

Water mass: $$\sum \dot{m}_{w,i} = \sum \dot{m}_{w,e} \quad \text{or} \quad \sum \dot{m}_{a,i} w_i = \sum \dot{m}_{a,e} w_e$$

Energy: $$\dot{Q}_{\text{in}} + \dot{W}_{\text{in}} + \sum \dot{m}_i h_i = \dot{Q}_{\text{out}} + \dot{W}_{\text{out}} + \sum \dot{m}_e h_e$$

where subscripts i and e denote inlet and exit states, respectively. The changes in kinetic and potential energies are assumed to be negligible.

During a simple heating or cooling process, the specific humidity remains constant, but the temperature and the relative humidity change. Sometimes air is humidified after it is heated, and some cooling processes include dehumidification. In dry climates, the air can be cooled via evaporative cooling by passing it through a section where it is sprayed with water. In locations with limited water supply, large amounts of waste heat can be rejected to the atmosphere with minimum water loss through the use of cooling towers.

REFERENCES AND SUGGESTED READING

1. ASHRAE. *1981 Handbook of Fundamentals.* Atlanta, GA: American Society of Heating, Refrigerating, and Air-Conditioning Engineers, 1981.

2. W. Z. Black and J. G. Hartley. *Thermodynamics.* New York: Harper & Row, 1985.

3. S. M. Elonka. "Cooling Towers." *Power,* March 1963.

4. J. B. Jones and G. A. Hawkins. *Engineering Thermodynamics.* 2nd ed. New York: John Wiley & Sons, 1986.

5. D. C. Look, Jr., and H. J. Sauer, Jr. *Engineering Thermodynamics.* Boston: PWS Engineering, 1986.

6. W. F. Stoecker and J. W. Jones. *Refrigeration and Air Conditioning.* 2nd ed. New York: McGraw-Hill, 1982.

7. K. Wark. *Thermodynamics.* 5th ed. New York: McGraw-Hill, 1988.

8. L. D. Winiarski and B. A. Tichenor. "Model of Natural Draft Cooling Tower Performance." *Journal of the Sanitary Engineering Division, Proceedings of the American Society of Civil Engineers,* August 1970.

Dry and Atmospheric Air: Specific and Relative Humidity

13-1C Is it possible to obtain saturated air from unsaturated air without adding any moisture? Explain.

13-2C Is the relative humidity of saturated air necessarily 100 percent?

13-3C Moist air is passed through a cooling section where it is cooled and dehumidified. How do (*a*) the specific humidity and (*b*) the relative humidity of air change during this process?

13-4C What is the difference between dry air and atmospheric air?

13-5C Can the water vapor in air be treated as an ideal gas? Explain.

13-6C What is vapor pressure?

13-7C How would you compare the enthalpy of water vapor at 20°C and 2 kPa with the enthalpy of water vapor at 20°C and 0.5 kPa?

13-8C What is the difference between the specific humidity and the relative humidity?

13-9C How will (*a*) the specific humidity and (*b*) the relative humidity of the air contained in a well-sealed room change as it is heated?

13-10C How will (*a*) the specific humidity and (*b*) the relative humidity of the air contained in a well-sealed room change as it is cooled?

13-11C Consider a tank that contains moist air at 3 atm and whose walls are permeable to water vapor. The surrounding air at 1 atm pressure also contains some moisture. Is it possible for the water vapor to flow into the tank from surroundings? Explain.

13-12C Why are the chilled water lines always wrapped with vapor barrier jackets?

13-13C Explain how vapor pressure of the ambient air is determined when the temperature, total pressure, and the relative humidity of the air are given.

13-14 A tank contains 21 kg of dry air and 0.3 kg of water vapor at 30°C and 100 kPa total pressure. Determine (*a*) the specific humidity, (*b*) the relative humidity, and (*c*) the volume of the tank.

13-15 A room contains air at 20°C and 98 kPa at a relative humidity of 85 percent. Determine (*a*) the partial pressure of dry air, (*b*) the specific humidity of the air, and (*c*) the enthalpy per unit mass of dry air.

---

*Students are encouraged to answer *all* the concept "C" questions.

13-16E A room contains air at 70°F and 14.6 psia at a relative humidity of 85 percent. Determine (*a*) the partial pressure of dry air, (*b*) the specific humidity of the air, and (*c*) the enthalpy per unit mass of dry air.

Answers: (*a*) 14.291 psia, (*b*) 0.0134 lbm H_2O/lbm dry air, (*c*) 31.43 Btu/lbm dry air

13-17 Determine the masses of dry air and the water vapor contained in a 120-m^3 room at 98 kPa, 23°C, and 50 percent relative humidity.

Answers: 136.4 kg, 1.25 kg

Dew-Point, Adiabatic Saturation, and Wet-Bulb Temperatures

13-18C What is the dew-point temperature?

13-19C Andy and Wendy both wear glasses. On a cold winter day, Andy comes from the cold outside and enters the warm house while Wendy leaves the house and goes outside. Whose glasses are more likely to be fogged? Explain.

13-20C In summer, the outer surface of a glass filled with iced water frequently "sweats." How can you explain this sweating?

13-21C In some climates, cleaning the ice off the windshield of a car is a common chore on winter mornings. Explain how ice forms on the windshield during some nights even when there is no rain or snow.

13-22C When are the dry-bulb and dew-point temperatures identical?

13-23C When are the adiabatic saturation and wet-bulb temperatures equivalent for atmospheric air?

13-24 A house contains air at 25°C and 65 percent relative humidity. Will any moisture condense on the inner surfaces of the windows when the temperature of the window drops to 12°C?

13-25 After a long walk in the 8°C outdoors, a person wearing glasses enters a room at 25°C and 40 percent relative humidity. Determine whether the glasses will become fogged.

13-26 Repeat Prob. 13-25 for a relative humidity of 70 percent.

13-27E A thirsty woman opens the refrigerator and picks up a cool canned drink at 40°F. Do you think the can will "sweat" as the woman enjoys the drink in a room at 80°F and 40 percent relative humidity?

13-28 The dry- and the wet-bulb temperatures of atmospheric air at 95 kPa are 25 and 20°C, respectively. Determine (*a*) the specific humidity, (*b*) the relative humidity, and (*c*) the enthalpy of the air, in kJ/kg dry air.

13-29 The air in a room has a dry-bulb temperature of 22°C and a wet-bulb temperature of 16°C. Assuming a pressure of 100 kPa, determine (*a*) the specific humidity, (*b*) the relative humidity, and (*c*) the dew-point temperature.

Answers: (*a*) 0.0091 kg H_2O/kg dry air, (*b*) 54.1 percent, (*c*) 12.4°C

13-30E The air in a room has a dry-bulb temperature of 70°F and a wet-bulb temperature of 60°F. Assuming a pressure of 14.7 psia, determine (*a*) the specific humidity, (*b*) the relative humidity, and (*c*) the dew-point temperature.

Answers: (*a*) 0.0087 lbm H_2O/lbm dry air, (*b*) 55.8 percent, (*c*) 53.2°F

Psychrometric Chart

13-31C How do constant-enthalpy and constant-wet-bulb-temperature lines compare on the psychrometric chart?

13-32C At what states on the psychrometric chart are the dry-bulb, wet-bulb, and dew-point temperatures identical?

13-33C How is the dew-point temperature at a specified state determined on the psychrometric chart?

13-34C Can the enthalpy values determined from a psychrometric chart at sea level be used at higher elevations?

13-35 The air in a room is at 1 atm, 32°C, and 60 percent relative humidity. Using the psychrometric chart, determine (*a*) the specific humidity, (*b*) the enthalpy (in kJ/kg dry air), (*c*) the wet-bulb temperature, (*d*) the dew-point temperature, and (*e*) the specific volume of the air (in m^3/kg dry air).

13-36 A room contains air at 1 atm, 26°C, and 70 percent relative humidity. Using the psychrometric chart, determine (*a*) the specific humidity, (*b*) the enthalpy (in kJ/kg dry air), (*c*) the wet-bulb temperature, (*d*) the dew-point temperature, and (*e*) the specific volume of the air (in m^3/kg dry air).

13-37E A room contains air at 1 atm, 82°F, and 70 percent relative humidity. Using the psychrometric chart, determine (*a*) the specific humidity, (*b*) the enthalpy (in Btu/lbm dry air), (*c*) the wet-bulb temperature, (*d*) the dew-point temperature, and (*e*) the specific volume of the air (in ft^3/lbm dry air).

13-38 The air in a room has a pressure of 1 atm, a dry-bulb temperature of 24°C, and a wet-bulb temperature of 17°C. Using the psychrometric chart, determine (*a*) the specific humidity, (*b*) the enthalpy (in kJ/kg dry air), (*c*) the relative humidity, (*d*) the dew-point temperature, and (*e*) the specific volume of the air (in m^3/kg dry air).

Human Comfort and Air-Conditioning

13-39C What does a modern air-conditioning system do besides heating or cooling the air?

13-40C How does the human body respond to (*a*) hot weather, (*b*) cold weather, and (*c*) hot and humid weather?

13-41C What is the radiation effect? How does it affect human comfort?

13-42C How does the air motion in the vicinity of the human body affect human comfort?

13-43C Consider a tennis match in cold weather where both players and spectators wear the same clothes. Which group of people will feel colder? Why?

13-44C Why do you think little babies are more susceptible to cold?

13-45C How does humidity affect human comfort?

13-46C What are humidification and dehumidification?

13-47C What is metabolism? What is the range of metabolic rate for an average man? Why are we interested in the metabolic rate of the occupants of a building when we deal with heating and air-conditioning?

13-48C Why is the metabolic rate of women, in general, lower than that of men? What is the effect of clothing on the environmental temperature that feels comfortable?

13-49C What is sensible heat? How is the sensible heat loss from a human body affected by the (*a*) skin temperature, (*b*) environment temperature, and (*c*) air motion?

13-50C What is latent heat? How is the latent heat loss from the human body affected by the (*a*) skin wettedness and (*b*) relative humidity of the environment? How is the rate of evaporation from the body related to the rate of latent heat loss?

13-51 An average person produces 0.25 kg of moisture while taking a shower and 0.05 kg while bathing in a tub. Consider a family of four who each shower once a day in a bathroom that is not ventilated. Taking the heat of vaporization of water to be 2450 kJ/kg, determine the contribution of showers to the latent heat load of the air conditioner per day in summer.

13-52 An average (1.82 kg or 4.0 lbm) chicken has a basal metabolic rate of 5.47 W and an average metabolic rate of 10.2 W (3.78 W sensible and 6.42 W latent) during normal activity. If there are 100 chickens in a breeding room, determine the rate of total heat generation and the rate of moisture production in the room. Take the heat of vaporization of water to be 2430 kJ/kg.

13-53 A department store expects to have 80 customers and 15 employees at peak times in summer. Determine the contribution of people to the total cooling load of the store.

13-54E In a movie theater in winter, 500 people, each generating heat at a rate of 100 W, are watching a movie. The heat losses through the walls, windows, and the roof are estimated to be 150,000 Btu/h. Determine if the theater needs to be heated or cooled.

13-55 For an infiltration rate of 1.2 air changes per hour (ACH), determine sensible, latent, and total infiltration heat load of a building at sea level, in kW, that is 20 m long, 13 m wide, and 3 m high when the outdoor air is at 32°C and 50 percent relative humidity. The building is maintained at 24°C and 50 percent relative humidity at all times.

Simple Heating and Cooling

13-56C How do relative and specific humidities change during a simple heating process? Answer the same question for a simple cooling process.

13-57C Why does a simple heating or cooling process appear as a horizontal line on the psychrometric chart?

13-58 Air enters a heating section at 95 kPa, 15°C, and 30 percent relative humidity at a rate of 6 m^3/min, and it leaves at 25°C. Determine (*a*) the rate of heat transfer in the heating section and (*b*) the relative humidity of the air at the exit. *Answers:* (*a*) 69.3 kJ/min, (*b*) 16.1 percent

13-59E A heating section consists of a 15-in.-diameter duct that houses a 4-kW electric resistance heater. Air enters the heating section at 14.7 psia, 50°F, and 40 percent relative humidity at a velocity of 25 ft/s. Determine (*a*) the exit temperature, (*b*) the exit relative humidity of the air, and (*c*) the exit velocity. *Answers:* (*a*) 56.8°F, (*b*) 30.8 percent, (*c*) 25.4 ft/s

13-60 Air enters a 40-cm-diameter cooling section at 1 atm, 32°C, and 30 percent relative humidity at 18 m/s. Heat is removed from the air at a rate of 1200 kJ/min. Determine (*a*) the exit temperature, (*b*) the exit relative humidity of the air, and (*c*) the exit velocity.

Answers: (*a*) 24.4°C, (*b*) 46.6 percent, (*c*) 17.6 m/s

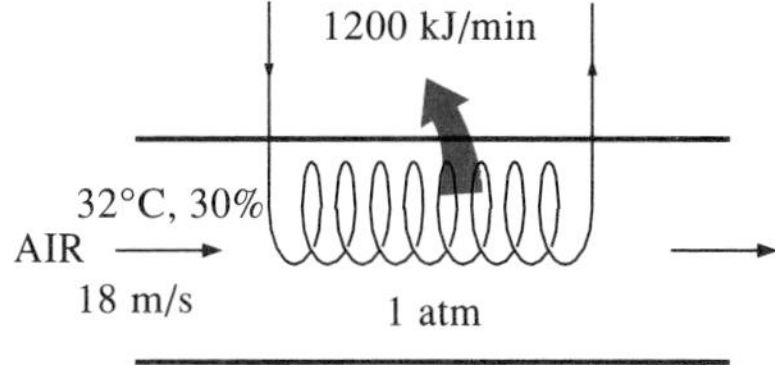

FIGURE P13-60

Heating with Humidification

13-61C Why is heated air sometimes humidified?

13-62 Air at 1 atm, 15°C, and 60 percent relative humidity is first heated to 20°C in a heating section and then humidified by introducing water vapor. The air leaves the humidifying section at 25°C and 65 percent relative humidity. Determine (*a*) the amount of steam added to the air, and (*b*) the amount of heat transfer to the air in the heating section.

Answers: (*a*) 0.0065 kg H_2O/kg dry air, (*b*) 5.1 kJ/kg dry air

13-63E Air at 14.7 psia, 55°F, and 60 percent relative humidity is first heated to 72°F in a heating section and then humidified by introducing water vapor. The air leaves the humidifying section at 75°F and 65 percent relative humidity. Determine (*a*) the amount of steam added to the air, in lbm H_2O/lbm dry air, and (*b*) the amount of heat transfer to the air in the heating section, in Btu/lbm dry air.

13-64 An air-conditioning system operates at a total pressure of 1 atm and consists of a heating section and a humidifier that supplies wet steam (saturated water vapor) at 100°C. Air enters the heating section at 10°C and 70 percent relative humidity at a rate of 70 m^3/min, and it leaves the humidifying section at 20°C and 60 percent relative humidity. Determine (*a*) the temperature and relative humidity of air when it leaves the heating section, (*b*) the rate of heat transfer in the heating section, and (*c*) the rate at which water is added to the air in the humidifying section.

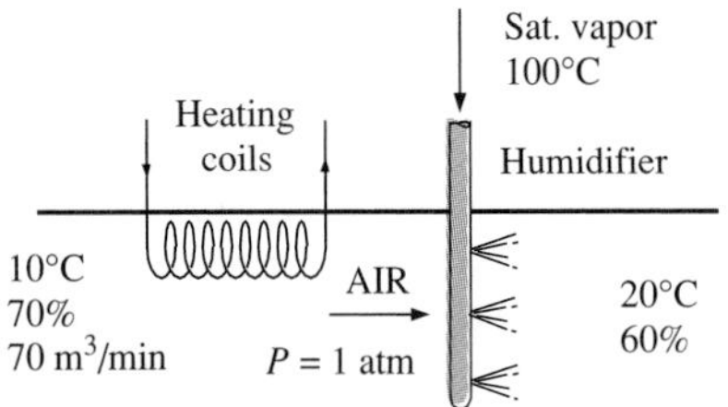

FIGURE P13-64

13-65 Repeat Prob. 13-64 for a total pressure of 95 kPa for the airstream. *Answers:* (*a*) 19.5°C, 37.7 percent; (*b*) 782 kJ/min; (*c*) 0.29 kg/min

Cooling with Dehumidification

13-66 Why is cooled air sometimes reheated in summer before it is discharged to a room?

13-67 Air enters a window air conditioner at 1 atm, 32°C, and 70 percent relative humidity at a rate of 8 m^3/min, and it leaves as saturated air at 12°C. Part of the moisture in the air that condenses during the process is also removed at 12°C. Determine the rates of heat and moisture removal from the air. *Answers:* 462.4 kJ/min, 0.112 kg/min

13-68 An air-conditioning system is to take in air at 1 atm, 34°C, and 70 percent relative humidity and deliver it at 22°C and 50 percent relative humidity. The air flows first over the cooling coils, where it is cooled and dehumidified, and then over the resistance heating wires, where it is heated to the desired temperature. Assuming that the condensate is removed from the cooling section at 10°C, determine (*a*) the temperature of air before it enters the heating section, (*b*) the amount of heat removed in the cooling section, and (*c*) the amount of heat transferred in the heating section, both in kJ/kg dry air.

13-69 Air enters a 30-cm-diameter cooling section at 1 atm, 35°C, and 60 percent relative humidity at 120 m/min. The air is cooled by passing it over a cooling coil through which cold water flows. The water experiences a temperature rise of 8°C. The air leaves the cooling section saturated at 20°C. Determine (*a*) the rate of heat transfer, (*b*) the mass flow rate of the water, and (*c*) the exit velocity of the airstream.

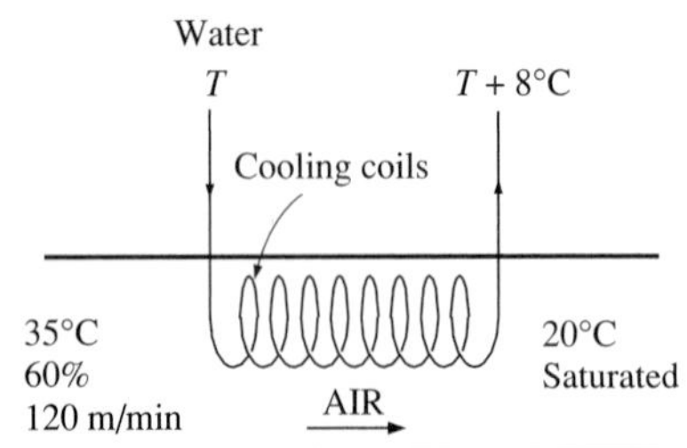

FIGURE P13-69

13-70 Repeat Prob. 13-69 for a total pressure of 95 kPa for air. *Answers:* (*a*) −293.3 kJ/min, (*b*) 8.77 kg/min, (*c*) 113 m/min

13-71E Air enters a 1-ft-diameter cooling section at 14.7 psia, 90°F, and 60 percent relative humidity at 600 ft/min. The air is cooled by passing it over a cooling coil through which cold water flows. The water experiences a temperature rise of 14°F. The air leaves the cooling section saturated at 70°F. Determine (*a*) the rate of heat transfer, (*b*) the mass flow rate of the water, and (*c*) the exit velocity of the airstream.

13-72E Repeat Prob. 13-71E for a total pressure of 14.4 psia for air.

Evaporative Cooling

13-73C Does an evaporation process have to involve heat transfer? Describe a process that involves both heat and mass transfer.

13-74C During evaporation from a water body to air, under what conditions will the latent heat of vaporization be equal to the heat transfer from the air?

13-75C What is evaporative cooling? Will it work in humid climates?

13-76 Air enters an evaporative cooler at 1 atm, 36°C, and 20 percent relative humidity at a rate of 10 m^3/min, and it leaves with a relative humidity of 90 percent. Determine (*a*) the exit temperature of the air and (*b*) the required rate of water supply to the evaporative cooler.

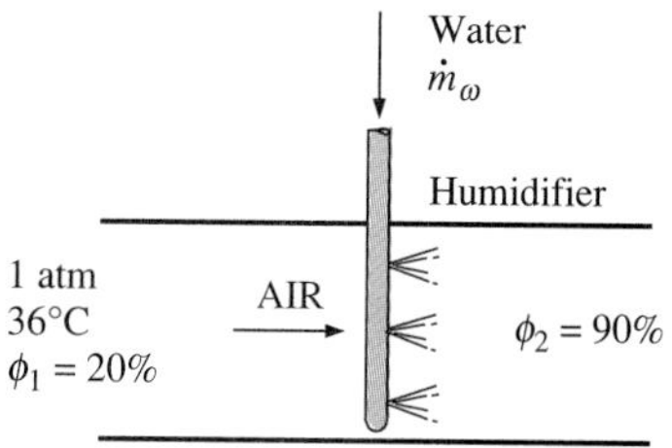

FIGURE P13-76

13-77E Air enters an evaporative cooler at 14.7 psia, 90°F, and 20 percent relative humidity at a rate of 200 ft^3/min, and it leaves with a relative humidity of 90 percent. Determine (*a*) the exit temperature of the air and (*b*) the required rate of water supply to the evaporative cooler.
Answers: (*a*) 64°F, (*b*) 0.08 lbm/min

13-78 Air enters an evaporative cooler at 95 kPa, 35°C, and 30 percent relative humidity and exits saturated. Determine the exit temperature of the air. *Answer:* 21.1°C

13-79E Air enters an evaporative cooler at 14.5 psia, 93°F, and 30 percent relative humidity and exits saturated. Determine the exit temperature of the air.

13-80 Air enters an evaporative cooler at 1 atm, 32°C, and 30 percent relative humidity at a rate of 6 m^3/min and leaves at 22°C. Determine (*a*) the final relative humidity and (*b*) the amount of water added to the air.

13-81 What is the lowest temperature that air can attain in an evaporative cooler if it enters at 1 atm, 29°C, and 40 percent relative humidity?
Answer: 19.3°C

13-82 Air at 1 atm, 15°C, and 60 percent relative humidity is first heated to 30°C in a heating section and then passed through an evaporative cooler where its temperature drops to 25°C. Determine (*a*) the exit relative humidity and (*b*) the amount of water added to the air in kg H_2O/kg dry air.

Adiabatic Mixing of Airstreams

13-83C Two unsaturated airstreams are mixed adiabatically. It is observed that some moisture condenses during the mixing process. Under what conditions will this be the case?

13-84C Consider the adiabatic mixing of two airstreams. Does the state of the mixture on the psychrometric chart have to be on the straight line connecting the two states?

13-85 Two airstreams are mixed steadily and adiabatically. The first stream enters at 32°C and 40 percent relative humidity at a rate of 20 m^3/min, while the second stream enters at 12°C and 90 percent relative humidity at a rate of 25 m^3/min. Assuming that the mixing process occurs at a pressure of 1 atm, determine the specific humidity, the relative humidity, the dry-bulb temperature, and the volume flow rate of the mixture.

Answers: 0.0096 kg H_2O/kg dry air, 63.4 percent, 20.6°C, 45.0 m^3/min

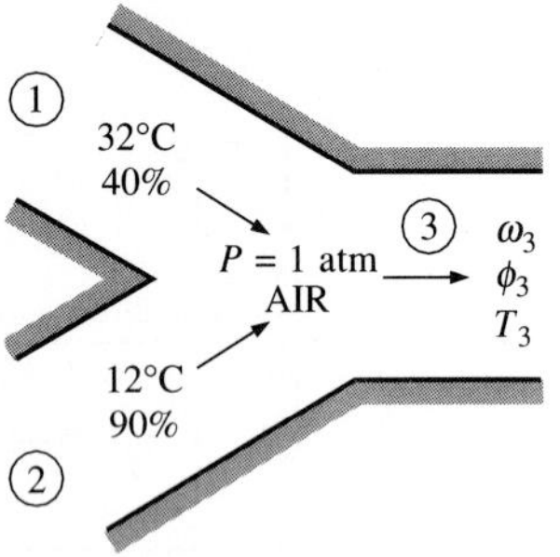

FIGURE P13-85

13-86 Repeat Prob. 13-85 for a total mixing-chamber pressure of 95 kPa.

13-87E During an air-conditioning process, 900 ft^3/min of conditioned air at 65°F and 30 percent relative humidity is mixed adiabatically with 300 ft^3/min of outside air at 80°F and 90 percent relative humidity at a pressure of 1 atm. Determine (*a*) the temperature, (*b*) the specific humidity, and (*c*) the relative humidity of the mixture.

Answers: (*a*) 68.7°F, (*b*) 0.0085 lbm H_2O/lbm dry air, (*c*) 52.1 percent

13-88 A stream of warm air with a dry-bulb temperature of 40°C and a wet-bulb temperature of 32°C is mixed adiabatically with a stream of saturated cool air at 18°C. The dry air mass flow rates of the warm and cool airstreams are 8 and 6 kg/s, respectively. Assuming a total pressure of 1 atm, determine (*a*) the temperature, (*b*) the specific humidity, and (*c*) the relative humidity of the mixture.

Wet Cooling Towers

13-89C How does a natural-draft wet cooling tower work?

13-90C What is a spray pond? How does its performance compare to the performance of a wet cooling tower?

13-91 The cooling water from the condenser of a power plant enters a wet cooling tower at 40°C at a rate of 45 kg/s. The water is cooled to 25°C in the cooling tower by air that enters the tower at 1 atm, 23°C, and 60 percent relative humidity and leaves saturated at 32°C. Neglecting the power input to the fan, determine (*a*) the volume flow rate of air into the cooling tower and (*b*) the mass flow rate of the required makeup water.

13-92E The cooling water from the condenser of a power plant enters a wet cooling tower at 110°F at a rate of 100 lbm/s. The water is cooled to 80°F in the cooling tower by air that enters the tower at 1 atm, 76°F, and 60 percent relative humidity and leaves saturated at 95°F. Neglecting the power input to the fan, determine (*a*) the volume flow rate of air into the cooling tower and (*b*) the mass flow rate of the required makeup water.

Answers: (*a*) 1311 ft^3/s, (*b*) 2.39 lbm/s.

FIGURE 13-93

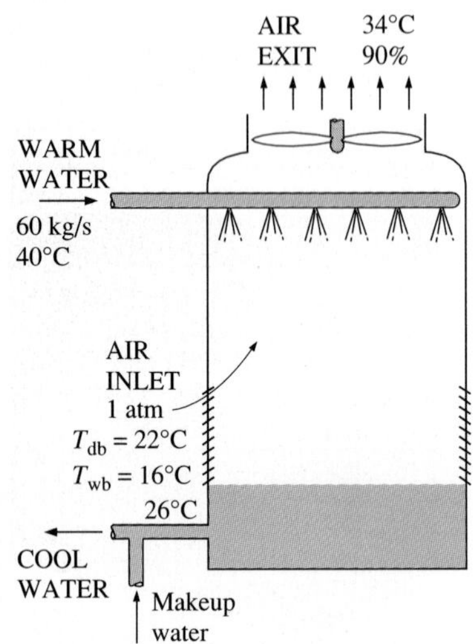

13-93 A wet cooling tower is to cool 60 kg/s of water from 40 to 26°C. Atmospheric air enters the tower at 1 atm with dry- and wet-bulb temperatures of 22 and 16°C, respectively, and leaves at 34°C with a relative humidity of 90 percent. Using the psychrometric chart, determine (*a*) the volume flow rate of air into the cooling tower and (*b*) the mass flow rate of the required makeup water. *Answers:* (*a*) 44.9 m^3/s, (*b*) 1.16kg/s

13-94 A wet cooling tower is to cool 110 kg/s of cooling water from 40 to 25°C at a location where the atmospheric pressure is 96 kPa. Atmospheric air enters the tower at 20°C and 70 percent relative humidity and leaves saturated at 35°C. Neglecting the power input to the fan, determine (*a*) the volume flow rate of air into the cooling tower and (*b*) the mass flow rate of the required makeup water. *Answers:* (*a*) 73.1 m^3/s, (*b*) 2.29 kg/s

Review Problems

13-95 The condensation of the water vapor in compressed-air lines is a major concern in industrial facilities, and the compressed air is often dehumidified to avoid the problems associated with condensation. Consider a compressor that compresses ambient air from the local atmospheric pressure of 92 kPa to a pressure of 800 kPa (absolute). The compressed air is then cooled to the ambient temperature as it flows through the compressed-air lines. Disregarding any pressure losses, determine if there will be any condensation in the compressed-air lines on a day when the ambient air is at 25°C and 40 percent relative humidity.

13-96E The relative humidity of air at 80°F and 14.7 psia is increased from 30 percent to 90 percent during a humidification process at constant temperature and pressure. Determine the percent error involved in assuming the density of air to have remained constant.

13-97 Dry air whose molar analysis is 78.1 percent N_2, 20.9 percent O_2, and 1 percent Ar flows over a water body until it is saturated. If the pressure and temperature of air remain constant at 1 atm and 25°C during the process, determine (*a*) the molar analysis of the saturated air and (*b*) the density of air before and after the process. What do you conclude from your results?

13-98E Determine the mole fraction of the water vapor at the surface of a lake whose surface temperature is 60°F, and compare it to the mole fraction of water in the lake, which is very nearly 1.0. The air at the lake surface is saturated, and the atmospheric pressure at lake level can be taken to be 13.8 psia.

13-99 Determine the mole fraction of dry air at the surface of a lake whose temperature is 15°C. The air at the lake surface is saturated, and the atmospheric pressure at lake level can be taken to be 100 kPa.

13-100 Consider a room that is cooled adequately by an air conditioner whose cooling capacity is 5000 Btu/h. If the room is to be cooled by an evaporative cooler that removes heat at the same rate by evaporation, determine how much water needs to be supplied to the cooler per hour.

13-101E The capacity of evaporative coolers is usually expressed in terms of the flow rate of air in ft^3/min (or cfm), and a practical way of determining the required size of an evaporative cooler for an 8-ft-high house is to multiply

FIGURE P13-101E

the floor area of the house by 4 (by 3 in dry climates and by 5 in humid climates). For example, the capacity of an evaporative cooler for a 30-ft-long, 40-ft-wide house is $1200 \times 4 = 4800$ cfm. Develop an equivalent rule of thumb for the selection of an evaporative cooler in SI units for 2.4-m-high houses whose floor areas are given in m^2.

13-102 A cooling tower with a cooling capacity of 100 tons (440 kW) is claimed to evaporate 15,800 kg of water per day. Is this a reasonable claim?

13-103 The U.S. Department of Energy estimates that 190,000 barrels of oil would be saved per day if every household in the United States raised the thermostat setting in summer by 6°F (3.3°C). Assuming the average cooling season to be 120 days and the cost of oil to be $20/barrel, determine how much money would be saved per year.

13-104E The thermostat setting of a house can be lowered by 2°F by wearing a light long-sleeved sweater, or by 4°F by wearing a heavy long-sleeved sweater for the same level of comfort. If each °F reduction in thermostat setting reduces the heating cost of a house by 4 percent at a particular location, determine how much the heating costs of a house can be reduced by wearing heavy sweaters if the annual heating cost of the house is $600.

13-105 The air-conditioning costs of a house can be reduced by up to 10 percent by installing the outdoor unit (the condenser) of the air conditioner at a location shaded by trees and shrubs. If the air-conditioning costs of a house are $500 a year, determine how much the trees will save the home owner in the 20-year life of the system.

13-106 A 5-m^3 tank contains saturated air at 25°C and 97 kPa. Determine (*a*) the mass of the dry air, (*b*) the specific humidity, and (*c*) the enthalpy of the air per unit mass of the dry air.

Answers: (*a*) 5.49 kg, (*b*) 0.0210 kg H_2O/kg dry air, (*c*) 78.62 kJ/kg dry air

13-107E Air at 15 psia, 60°F, and 50 percent relative humidity flows in a 15-in.-diameter duct at a velocity of 50 ft/s. Determine (*a*) the dew-point temperature, (*b*) the volume flow rate of air, and (*c*) the mass flow rate of dry air.

13-108 Air enters a cooling section at 97 kPa, 35°C, and 20 percent relative humidity at a rate of 12 m^3/min, where it is cooled until the moisture in the air starts condensing. Determine (*a*) the temperature of the air at the exit and (*b*) the rate of heat transfer in the cooling section.

13-109 Outdoor air enters an air-conditioning system at 10°C and 40 percent relative humidity at a steady rate of 14 m^3/min, and it leaves at 25°C and 55 percent relative humidity. The outdoor air is first heated to 22°C in the heating section and then humidified by the injection of hot steam in the humidifying section. Assuming the entire process takes place at a pressure of 1 atm, determine (*a*) the rate of heat supply in the heating section and (*b*) the mass flow rate of the steam required in the humidifying section.

13-110 Air enters an air-conditioning system that uses refrigerant-134a at 30°C and 70 percent relative humidity at a rate of 4 m^3/min. The refrigerant enters the cooling section at 700 kPa with a quality of 20 percent and leaves as saturated vapor. The air is cooled to 20°C at a pressure of 1 atm. Determine (*a*) the rate of dehumidification, (*b*) the rate of heat transfer, and (*c*) the mass flow rate of the refrigerant.

13-111 Repeat Prob. 13-110 for a total pressure of 95 kPa for air.

13-112 An air-conditioning system operates at a total pressure of 1 atm and consists of a heating section and an evaporative cooler. Air enters the heating section at 10°C and 70 percent relative humidity at a rate of 50 m^3/min, and it leaves the evaporative cooler at 20°C and 60 percent relatively humidity. Determine (*a*) the temperature and relative humidity of the air when it leaves the heating section, (*b*) the rate of heat transfer in the heating section, and (*c*) the rate of water added to the air in the evaporative cooler.

Answers: (*a*) 28.3°C, 23.0 percent; (*b*) 1160 kJ/min; (*c*) 0.21 kg/min

13-113 Repeat Prob. 13-112 for a total pressure of 96 kPa.

13-114 Conditioned air at 13°C and 90 percent relative humidity is to be mixed with outside air at 34°C and 40 percent relative humidity at 1 atm. If it is desired that the mixture have a relative humidity of 60 percent, determine (*a*) the ratio of the dry air mass flow rates of the conditioned air to the outside air and (*b*) the temperature of the mixture. Use the psychrometric chart.

13-115 A natural-draft cooling tower is to remove 50 MW of waste heat from the cooling water that enters the tower at 42°C and leaves at 27°C. Atmospheric air enters the tower at 1 atm with dry- and wet-bulb temperatures of 23 and 18°C, respectively, and leaves saturated at 37°C. Determine (*a*) the mass flow rate of the cooling water, (*b*) the volume flow rate of air into the cooling tower, and (*c*) the mass flow rate of the required makeup water.

Computer, Design, and Essay Problems

13-116 The condensation and even freezing of moisture in building walls without effective vapor retarders are of real concern in cold climates as they undermine the effectiveness of the insulation. Investigate how the builders in your area are coping with this problem, whether they are using vapor retarders or vapor barriers in the walls, and where they are located in the walls. Prepare a report on your findings, and explain the reasoning for the current practice.

13-117 Write an interactive computer program to determine the properties of atmospheric air at sea level when the dry-bulb temperature and relative humidity are given.

13-118 Write an interactive computer program to determine the properties of atmospheric air at any altitude when the dry-bulb and wet-bulb temperatures are given.

13-119 The air-conditioning needs of a large building can be met by a single central system or by several individual window units. Considering that both approaches are commonly used in practice, the right choice depends on the situation on hand. Identify the important factors that need to be considered in decision making, and discuss the conditions under which an air-conditioning system that consists of several window units is preferable over a large single central system, and vice versa.

13-120 Identify the major sources of heat gain in your house in summer, and propose ways of minimizing them and thus reducing the cooling load.

13-121 Write an essay on different humidity measurement devices, including electronic ones, and discuss the advantages and disadvantages of each device.

13-122 Design an inexpensive evaporative cooling system suitable for use in your house. Show how you would obtain a water spray, how you would provide airflow, and how you would prevent water droplets from drifting into the living space.

Software Problems

13-123 Repeat Prob. 13-35 using the enclosed software.

13-124 Repeat Prob. 13-36 using the enclosed software.

13-125 Repeat Prob. 13-38 using the enclosed software.

13-126 Repeat Prob. 13-85 using the enclosed software.

13-127E Repeat Prob. 13-87E using the enclosed software.

13-128 Repeat Prob. 13-88 using the enclosed software.

13-129 Repeat Prob. 13-114 using the enclosed software.

Chemical Reactions

CHAPTER 14

In the preceding chapters we limited our consideration to nonreacting systems. That is, the chemical composition of all the systems under consideration remained unchanged during a process. This was the case even with mixing processes during which a homogeneous mixture is formed from two or more fluids without the occurrence of any chemical reactions. In this chapter, we specifically deal with systems whose chemical composition changes during a process, that is, systems that involve *chemical reactions.*

When dealing with nonreacting systems, we need to consider only the *sensible internal energy* (associated with temperature and pressure changes) and the *latent internal energy* (associated with phase changes). When dealing with reacting systems, however, we also need to consider the *chemical internal energy,* which is the energy associated with the destruction and formation of chemical bonds between the atoms. The energy balance relations developed for nonreacting systems are equally applicable to reacting systems, but the energy terms in the latter case should be modified to include the chemical energy of the system.

In this chapter we focus on a particular type of chemical reaction, known as *combustion,* because of its importance in engineering. The reader should keep in mind, however, that the principles developed are equally applicable to other chemical reactions.

We start this chapter with a general discussion of fuels and combustion. Then we apply the mass and energy balances to reacting systems. In this regard we discuss the adiabatic flame temperature, which is the highest temperature a reacting mixture can attain. Finally, we examine the second-law aspects of chemical reactions.

14-1 ■ FUELS AND COMBUSTION

Any material that can be burned to release energy is called a **fuel.** Most familiar fuels consist primarily of hydrogen and carbon. They are called **hydrocarbon fuels** and are denoted by the general formula C_nH_m. Hydrocarbon fuels exist in all phases, some examples being coal, gasoline, and natural gas.

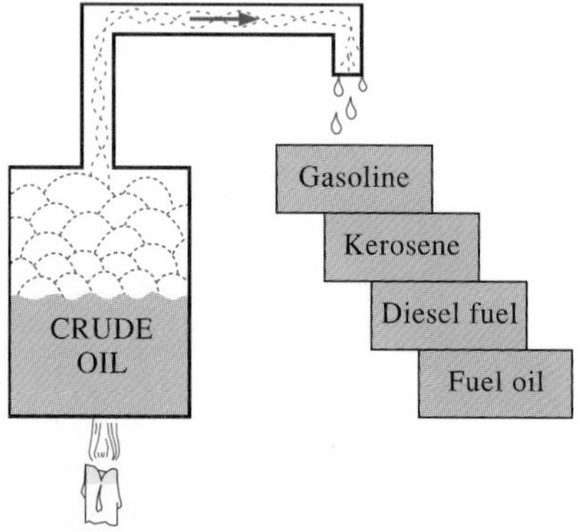

FIGURE 14-1

Most liquid hydrocarbon fuels are obtained from crude oil by distillation.

The main constituent of coal is carbon. Coal also contains varying amounts of oxygen, hydrogen, nitrogen, sulfur, moisture, and ash. It is difficult to give an exact mass analysis for coal since its composition varies considerably from one geographical location to the next and even within the same geographical location. Most liquid hydrocarbon fuels are a mixture of numerous hydrocarbons and are obtained from crude oil by distillation (Fig. 14-1). The most volatile hydrocarbons vaporize first, forming what we know as gasoline. The less volatile fuels obtained during distillation are kerosene, diesel fuel, and fuel oil. The composition of a particular fuel depends on the source of the crude oil as well as on the refinery.

Although liquid hydrocarbon fuels are mixtures of many different hydrocarbons, they are usually considered to be a single hydrocarbon for convenience. For example, gasoline is treated as **octane,** C_8H_{18}, and the diesel fuel as **dodecane,** $C_{12}H_{26}$. Another common liquid hydrocarbon fuel is **methyl alcohol,** CH_3OH, which is also called *methanol* and is used in some gasoline blends. The gaseous hydrocarbon fuel natural gas, which is a mixture of methane and smaller amounts of other gases, is sometimes treated as **methane,** CH_4, for simplicity.

FIGURE 14-2

Combustion is a chemical reaction during which a fuel is oxidized and a large quantity of energy is released.

A chemical reaction during which a fuel is oxidized and a large quantity of energy is released is called **combustion** (Fig. 14-2). The oxidizer most often used in combustion processes is air, for obvious reasons—it is free and readily available. Pure oxygen O_2 is used as an oxidizer only in some specialized applications, such as cutting and welding, where air cannot be used. Therefore, a few words about the composition of air are in order.

On a mole or a volume basis, dry air is composed of 20.9 percent oxygen, 78.1 percent nitrogen, 0.9 percent argon, and small amounts of carbon dioxide, helium, neon, and hydrogen. In the analysis of combustion processes, the argon in the air is treated as nitrogen, and the gases that exist in trace amounts are disregarded. Then dry air can be approximated as 21 percent oxygen and 79 percent nitrogen by mole numbers. Therefore, each mole of oxygen entering a combustion chamber will be accompanied by 0.79/0.21 = 3.76 mol of nitrogen (Fig. 14-3). That is,

$$1 \text{ kmol } O_2 + 3.76 \text{ kmol } N_2 = 4.76 \text{ kmol air} \qquad (14\text{-}1)$$

At ordinary combustion temperatures, nitrogen behaves as an inert gas and does not react with other chemical elements. But even then the presence of nitrogen greatly affects the outcome of a combustion process since nitrogen usually enters a combustion chamber in large quantities at low temperatures and exits at considerably higher temperatures, absorbing a large proportion of the chemical energy released during combustion. Throughout this chapter, nitrogen is assumed to remain perfectly inert. Keep in mind, however, that at very high temperatures, such as those encountered in internal combustion engines, a small fraction of nitrogen reacts with oxygen, forming hazardous gases such as nitric oxide.

The air that enters a combustion chamber normally contains some water vapor (or moisture), which also deserves consideration. For most combustion processes, the moisture in the air can also be treated as an inert gas, like nitrogen. At very high temperatures, however, some water vapor dissociates into H_2 and O_2 as well as into H, O, and OH. When the combustion gases are cooled below the dew-point temperature of the water vapor, some moisture will condense out. It is important to be able to predict the dew-point temperature since the water droplets often combine with the sulfur dioxide that may be present in the combustion gases, forming sulfuric acid, which is highly corrosive.

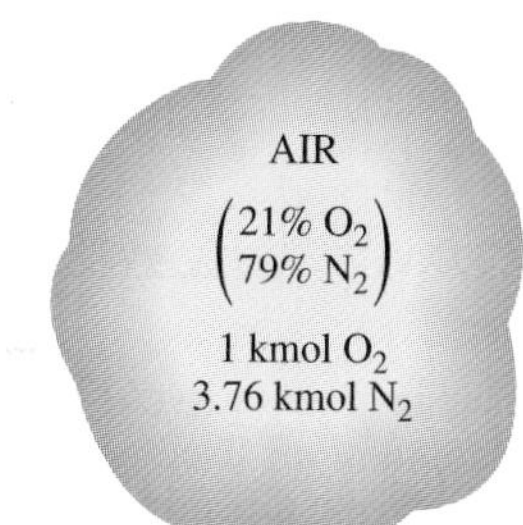

FIGURE 14-3

Each kmol of O_2 in air is accompanied by 3.76 kmol of N_2.

During a combustion process, the components that exist before the reaction are called **reactants** and the components that exist after the reaction are called **products** (Fig. 14-4). Consider, for example, the combustion of 1 kmol of carbon with 1 kmol of pure oxygen, forming carbon dioxide,

$$C + O_2 \longrightarrow CO_2 \tag{14-2}$$

Here C and O_2 are the reactants since they exist before combustion, and CO_2 is the product since it exists after combustion. Note that a reactant does not have to react chemically in the combustion chamber. For example, if carbon is burned with air instead of pure oxygen, both sides of the combustion equation will include N_2. That is, the N_2 will appear both as a reactant and as a product.

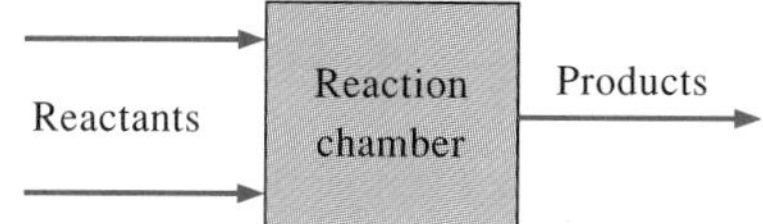

FIGURE 14-4

In a steady-flow combustion process, the components that enter the reaction chamber are called reactants and the components that exit are called products.

We should also mention that bringing a fuel into intimate contact with oxygen is not sufficient to start a combustion process. (Thank goodness it is not. Otherwise, the whole world would be on fire now.) The fuel must be brought above its **ignition temperature** to start the combustion. The minimum ignition temperatures of various substances in atmospheric air are approximately 260°C for gasoline, 400°C for carbon, 580°C for hydrogen, 610°C for carbon monoxide, and 630°C for methane. Moreover, the proportions of the fuel and air must be in the proper range for combustion to begin. For example, natural gas will not burn in air in concentrations less than 5 percent or greater than about 15 percent.

As you will recall from your chemistry courses, chemical equations are balanced on the basis of the **conservation of mass principle** (or the **mass balance**), which can be stated as follows: *The total mass of each element is conserved during a chemical reaction* (Fig. 14-5). That is, the total mass of each element on the right-hand side of the reaction equation (the products) must be equal to the total mass of that element on the left-hand side (the reactants) even though the elements exist in different chemical compounds in the reactants and products. Also, the total number of atoms of each element is conserved during a chemical reaction since the total number of atoms of an element is equal to the total mass of the element divided by its atomic mass.

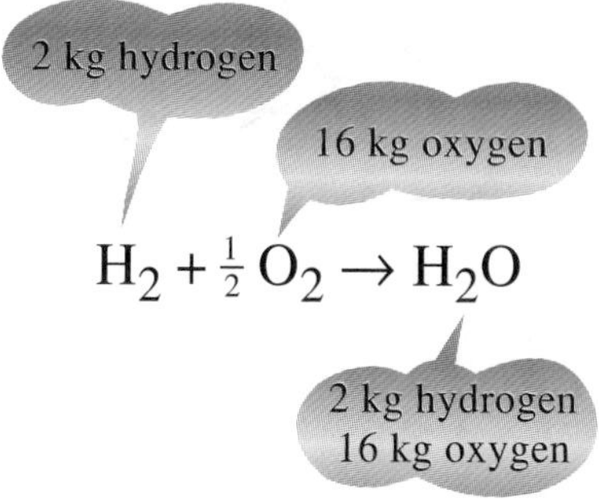

FIGURE 14-5

The mass (and number of atoms) of each element is conserved during a chemical reaction.

For example, both sides of Eq. 14-2 contain 12 kg of carbon and 32 kg of oxygen, even though the carbon and the oxygen exist as elements in the

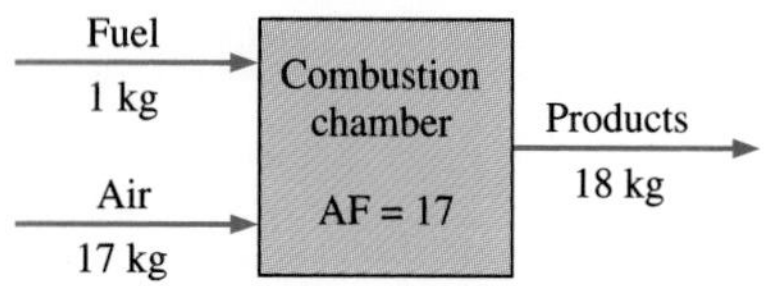

FIGURE 14-6

The air–fuel ratio (AF) represents the amount of air used per unit mass of fuel during a combustion process.

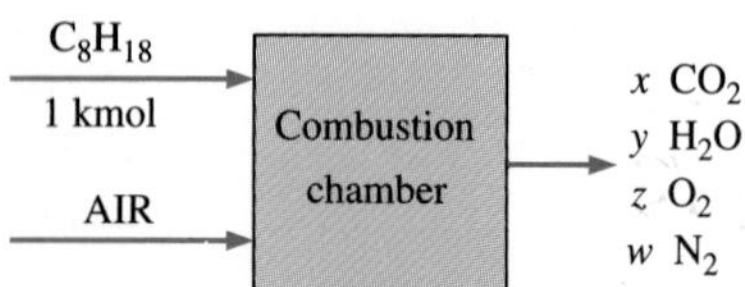

FIGURE 14-7

Schematic for Example 14-1.

reactants and as a compound in the product. Also, the total mass of reactants is equal to the total mass of products, each being 44 kg. (It is common practice to round the molar masses to the nearest integer if great accuracy is not required.) However, notice that the total mole number of the reactants (2 kmol) is not equal to the total mole number of the products (1 kmol). That is, *the total number of moles is not conserved during a chemical reaction.*

A frequently used quantity in the analysis of combustion processes to quantify the amounts of fuel and air is the **air–fuel ratio** AF. It is usually expressed on a mass basis and is defined as *the ratio of the mass of air to the mass of fuel* for a combustion process (Fig. 14-6). That is,

$$AF = \frac{m_{air}}{m_{fuel}} \tag{14-3}$$

The mass m of a substance is related to the number of moles N through the relation $m = NM$, where M is the molar mass.

The air–fuel ratio can also be expressed on a mole basis as the ratio of the mole numbers of air to the mole numbers of fuel. But we will use the former definition. The reciprocal of air–fuel ratio is called the **fuel–air ratio.**

EXAMPLE 14-1 Balancing the Combustion Equation

One kmol of octane (C_8H_{18}) is burned with air that contains 20 kmol of O_2, as shown in Fig. 14-7. Assuming the products contain only CO_2, H_2O, O_2, and N_2, determine the mole number of each gas in the products and the air–fuel ratio for this combustion process.

Solution The amount of fuel and the amount of oxygen in the air are given. The amount of the products and the AF are to be determined.

Assumptions The combustion products contain CO_2, H_2O, O_2, and N_2 only.

Analysis The chemical equation for this combustion process can be written as

$$C_8H_{18} + 20(O_2 + 3.76N_2) \longrightarrow xCO_2 + yH_2O + zO_2 + wN_2$$

where the terms in the parentheses represent the composition of dry air that contains 1 kmol of O_2 and x, y, z, and w represent the unknown mole numbers of the gases in the products. These unknowns are determined by applying the mass balance to each of the elements—that is, by requiring that the total mass or mole number of each element in the reactants be equal to that in the products:

$$\begin{array}{llll} \text{C:} & 8 = x & \longrightarrow & x = \mathbf{8} \\ \text{H:} & 18 = 2y & \longrightarrow & y = \mathbf{9} \\ \text{O} & 40 = 2x + y + 2z & \longrightarrow & z = \mathbf{7.5} \\ \text{N}_2\text{:} & (20)(3.76) = w & \longrightarrow & w = \mathbf{75.2} \end{array}$$

Substituting yields

$$C_8H_{18} + 20(O_2 + 3.76N_2) \longrightarrow 8CO_2 + 9H_2O + 7.5O_2 + 75.2N_2$$

Note that the coefficient 20 in the balanced equation above represents the number of moles of *oxygen,* not the number of moles of air. The latter is obtained by adding $20 \times 3.76 = 75.2$ moles of nitrogen to the 20 moles of oxygen, giving a total of 95.2 moles of air. The air–fuel ratio (AF) is determined from Eq. 14-3 by taking the ratio of the mass of the air and the mass of the fuel,

$$
\begin{aligned}
AF &= \frac{m_{air}}{m_{fuel}} = \frac{(NM)_{air}}{(NM)_C + (NM)_{H_2}} \\
&= \frac{(20 \times 4.76\text{ kmol})(29\text{ kg/kmol})}{(8\text{ kmol})(12\text{ kg/kmol}) + (9\text{ kmol})(2\text{ kg/kmol})} \\
&= \mathbf{24.2\text{ kg air/kg fuel}}
\end{aligned}
$$

That is, 24.2 kg of air is used to burn each kilogram of fuel during this combustion process.

14-2 ■ THEORETICAL AND ACTUAL COMBUSTION PROCESSES

It is often instructive to study the combustion of a fuel by assuming that the combustion is complete. A combustion process is **complete** if all the carbon in the fuel burns to CO_2, all the hydrogen burns to H_2O, and all the sulfur (if any) burns to SO_2. That is, all the combustible components of a fuel are burned to completion during a complete combustion process (Fig. 14-8). Conversely, the combustion process is **incomplete** if the combustion products contain any unburned fuel or components such as C, H_2, CO, or OH.

Insufficient oxygen is an obvious reason for incomplete combustion, but it is not the only one. Incomplete combustion occurs even when more oxygen is present in the combustion chamber than is needed for complete combustion. This may be attributed to insufficient mixing in the combustion chamber during the limited time that the fuel and the oxygen are in contact. Another cause of incomplete combustion is *dissociation,* which becomes important at high temperatures.

Oxygen is more strongly attracted to hydrogen than it is to carbon. Therefore, the hydrogen in the fuel normally burns to completion, forming H_2O, even when there is less oxygen than needed for complete combustion. Some of the carbon, however, ends up as CO or just as plain C particles in the products.

The minimum amount of air needed for the complete combustion of a fuel is called the **stoichiometric** or **theoretical air.** Thus, when a fuel is completely burned with theoretical air, no uncombined oxygen will be present in the product gases. The theoretical air is also referred to as the *chemically correct* amount of air, or 100 percent theoretical air. A combustion process with less than the theoretical air is bound to be incomplete. The ideal combustion process during which a fuel is burned completely with theoretical air is called the **stoichiometric** or **theoretical combustion** of that fuel (Fig. 14-9). For example, the theoretical combustion of methane is

$$CH_4 + 2(O_2 + 3.76N_2) \longrightarrow CO_2 + 2H_2O + 7.52N_2$$

Notice that the products of the theoretical combustion contain no unburned methane and no C, H_2, CO, OH, or free O_2.

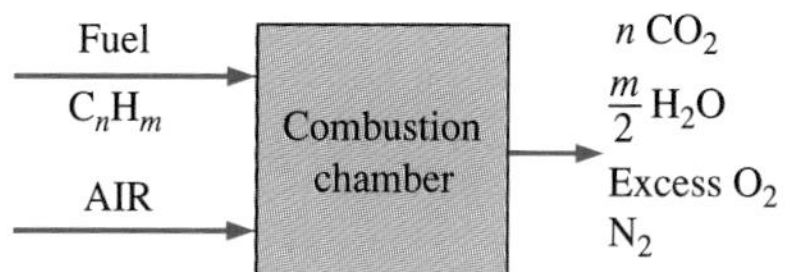

FIGURE 14-8
A combustion process is complete if all the combustible components of the fuel are burned to completion.

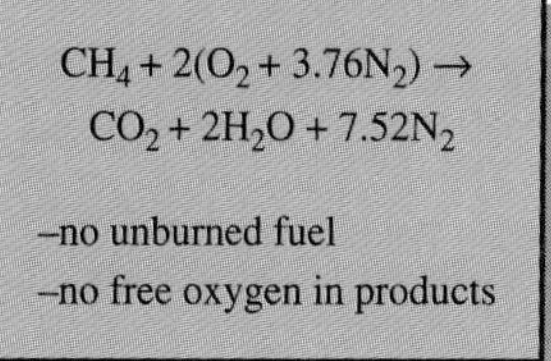

FIGURE 14-9
The complete combustion process with no free oxygen in the products is called theoretical combustion.

In actual combustion processes, it is common practice to use more air than the stoichiometric amount to increase the chances of complete combustion or to control the temperature of the combustion chamber. The amount of air in excess of the stoichiometric amount is called **excess air.** The amount of excess air is usually expressed in terms of the stoichiometric air as **percent excess air** or **percent theoretical air.** For example, 50 percent excess air is equivalent to 150 percent theoretical air, and 200 percent excess air is equivalent to 300 percent theoretical air. Of course, the stoichiometric air can be expressed as 0 percent excess air or 100 percent theoretical air. Amounts of air less than the stoichiometric amount are called **deficiency of air** and are often expressed as **percent deficiency of air.** For example, 90 percent theoretical air is equivalent to 10 percent deficiency of air. The amount of air used in combustion processes is also expressed in terms of the **equivalence ratio,** which is the ratio of the actual fuel–air ratio to the stoichiometric fuel–air ratio.

Predicting the composition of the products is relatively easy when the combustion process is assumed to be complete and the exact amounts of the fuel and air used are known. All one needs to do in this case is simply apply the mass balance to each element that appears in the combustion equation, without needing to take any measurements. Things are not so simple, however, when one is dealing with actual combustion processes. For one thing, actual combustion processes are hardly ever complete, even in the presence of excess air. Therefore, it is impossible to predict the composition of the products on the basis of the mass balance alone. Then the only alternative we have is to measure the amount of each component in the products directly.

A commonly used device to analyze the composition of combustion gases is the **Orsat gas analyzer.** In this device, a sample of the combustion gases is collected and cooled to room temperature and pressure, at which point its volume is measured. The sample is then brought into contact with a chemical that absorbs the CO_2. The remaining gases are returned to the room temperature and pressure, and the new volume they occupy is measured. The ratio of the reduction in volume to the original volume is the volume fraction of the CO_2, which is equivalent to the mole fraction if ideal-gas behavior is assumed (Fig. 14-10). The volume fractions of the other gases are determined by repeating this procedure. In Orsat analysis the gas sample is collected over water and is maintained saturated at all times. Therefore, the vapor pressure of water remains constant during the entire test. For this reason the presence of water vapor in the test chamber is ignored and data are reported on a dry basis. But the amount of H_2O formed during combustion is easily determined by balancing the combustion equation.

BEFORE: 100 kPa, 25°C, Gas sample including CO_2, 1 liter

AFTER: 100 kPa, 25°C, Gas sample without CO_2, 0.9 liter

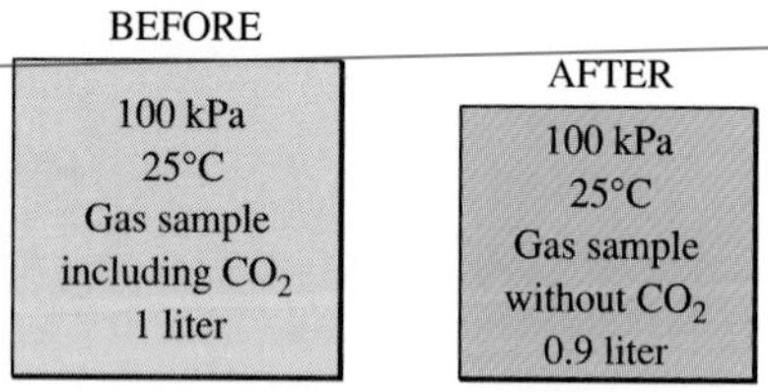

$$y_{CO_2} = \frac{V_{CO_2}}{V} = \frac{0.1}{1} = 0.1$$

FIGURE 14-10

Determining the mole fraction of the CO_2 in combustion gases by using the Orsat gas analyzer.

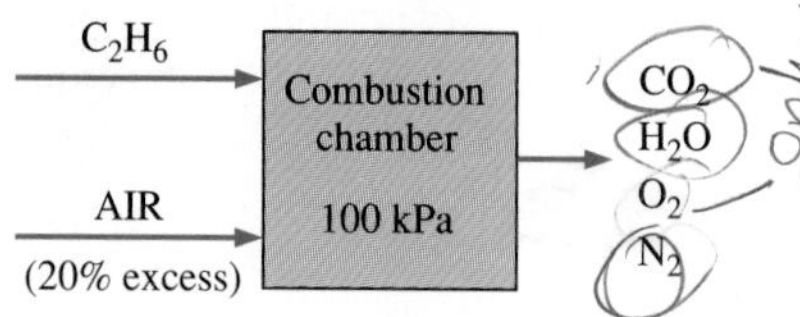

FIGURE 14-11

Schematic for Example 14-2.

EXAMPLE 14-2 Dew-Point Temperature of Combustion Products

Ethane (C_2H_6) is burned with 20 percent excess air during a combustion process, as shown in Fig. 14-11. Assuming complete combustion and a total pressure of 100 kPa, determine (*a*) the air–fuel ratio and (*b*) the dew-point temperature of the products.

Solution The fuel is burned completely with excess air and thus the products will contain CO_2, H_2O, N_2, and some excess O_2 only. The AF and the dew point of the products are to be determined.

Assumptions **1** Combustion is complete. **2** Combustion gases are ideal gases.

Analysis The combustion equation in this case can be written as

$$C_2H_6 + 1.2a_{th}(O_2 + 3.76N_2) \longrightarrow 2CO_2 + 3H_2O + 0.2a_{th}O_2 + (1.2 \times 3.76)a_{th}N_2$$

where a_{th} is the stoichiometric coefficient for air. We have automatically accounted for the 20 percent excess air by using the factor $1.2a_{th}$ instead of a_{th} for air. The stoichiometric amount of oxygen ($a_{th}O_2$) will be used to oxidize the fuel, and the remaining excess amount ($0.2a_{th}O_2$) will appear in the products as unused oxygen. Notice that the coefficient of N_2 is the same on both sides of the equation. You will also notice that we did the C and H_2 balance in our heads as we wrote the combustion equation because it is so obvious. The coefficient a_{th} is determined from the O_2 balance to be

$$O_2: \quad 1.2a_{th} = 2 + 1.5 + 0.2a_{th} \longrightarrow a_{th} = 3.5$$

Substituting gives

$$C_2H_6 + 4.2(O_2 + 3.76N_2) \longrightarrow 2CO_2 + 3H_2O + 0.7O_2 + 15.79N_2$$

(*a*) The air–fuel ratio is determined from Eq. 14-3 by taking the ratio of the mass of the air to the mass of the fuel,

$$AF = \frac{m_{air}}{m_{fuel}} = \frac{(4.2 \times 4.76 \text{ kmol})(29 \text{ kg/kmol})}{(2 \text{ kmol})(12 \text{ kg/kmol}) + (3 \text{ kmol})(2 \text{ kg/kmol})}$$
$$= \mathbf{19.3 \text{ kg air/kg fuel}}$$

That is, 19.3 kg of air is supplied for each kilogram of fuel during this combustion process.

(*b*) The dew-point temperature of the products is the temperature at which the water vapor in the products starts to condense as the products are cooled. You will recall from Chap. 13 that the dew-point temperature of a gas–vapor mixture is the saturation temperature of the water vapor corresponding to its partial pressure. Therefore, we need to determine the partial pressure of the water vapor P_v in the products first. Assuming ideal-gas behavior for the combustion gases, we have

$$P_v = \left(\frac{N_v}{N_{prod}}\right)(P_{prod}) = \left(\frac{3 \text{ kmol}}{21.49 \text{ kmol}}\right)(100 \text{ kPa}) = 13.96 \text{ kPa}$$

Thus, $T_{dp} = T_{sat\,@\,13.96\text{ kPa}} = \mathbf{52.3°C}$ (Table A-5)

EXAMPLE 14-3 Combustion of Gaseous Fuel with Moist Air

A certain natural gas has the following volumetric analysis: 72 percent CH_4, 9 percent H_2, 14 percent N_2, 2 percent O_2, and 3 percent CO_2. This gas is now burned with the stoichiometric amount of air that enters the combustion chamber at 20°C, 1 atm, and 80 percent relative humidity, as shown in Fig. 14-12. Assuming complete combustion and a total pressure of 1 atm, determine the dew-point temperature of the products.

FIGURE 14-12
Schematic for Example 14-3.

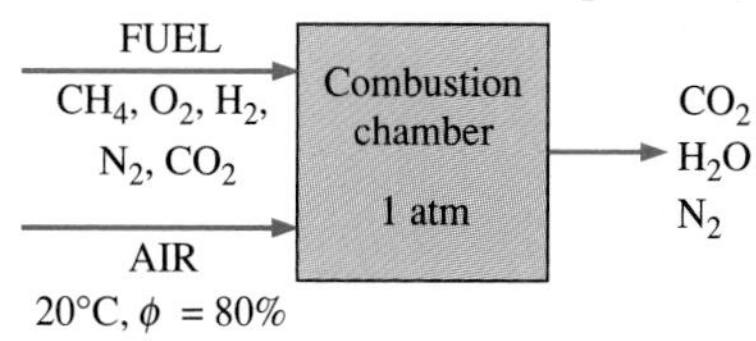

Solution A gaseous fuel is burned with the stoichiometric amount of moist air. The dew point of the products is to be determined. We note that the moisture in the air does not react with anything; it simply shows up as additional H_2O in the products. Therefore, for simplicity, we will balance the combustion equation by using dry air and then add the moisture later to both sides of the equation.

Assumptions **1** The fuel is burned completely and thus all the carbon in the fuel will burn to CO_2 and all the hydrogen to H_2O. **2** The fuel is burned with the stoichiometric amount of air and thus there will be no free O_2 in the product gases. **3** Combustion gases are ideal gases.

Analysis Considering 1 kmol of fuel,

$$\overbrace{(0.72CH_4 + 0.09H_2 + 0.14N_2 + 0.02O_2 + 0.03CO_2)}^{\text{fuel}} + \overbrace{a_{th}(O_2 + 3.76\,N_2)}^{\text{dry air}} \longrightarrow xCO_2 + yH_2O + zN_2$$

The unknown coefficients in the above equation are determined from mass balances on various elements,

$$\text{C:} \quad 0.72 + 0.03 = x \longrightarrow x = 0.75$$

$$\text{H:} \quad 0.72 \times 4 + 0.09 \times 2 = 2y \longrightarrow y = 1.53$$

$$\text{O}_2\text{:} \quad 0.02 + 0.03 + a_{th} = x + \frac{y}{2} \longrightarrow a_{th} = 1.465$$

$$\text{N}_2\text{:} \quad 0.14 + 3.76a_{th} = z \longrightarrow z = 5.648$$

Next we determine the amount of moisture that accompanies $4.76a_{th} = (4.76)(1.465) = 6.97$ kmol of dry air. The partial pressure of the moisture in the air is

$$P_{v,\,air} = \phi_{air} P_{sat\,@\,20°C} = (0.80)(2.339\text{ kPa}) = 1.871\text{ kPa}$$

Assuming ideal-gas behavior, the number of moles of the moisture in the air $N_{v,\,air}$ is

$$N_{v,\,air} = \left(\frac{P_{v,\,air}}{P_{total}}\right) N_{total} = \left(\frac{1.871\text{ kPa}}{101.325\text{ kPa}}\right)(6.97 + N_{v,\,air})$$

which yields $N_{v,\,air} = 0.131$ kmol

The balanced combustion equation is obtained by substituting the coefficients determined earlier and adding 0.131 kmol of H_2O to both sides of the equation:

$$\overbrace{(0.72CH_4 + 0.09H_2 + 0.14N_2 + 0.02O_2 + 0.03CO_2)}^{\text{fuel}} + \overbrace{1.465(O_2 + 3.76N_2)}^{\text{dry air}} + \overbrace{0.131H_2O}^{\text{moisture}} \longrightarrow 0.75CO_2 + \overbrace{1.661H_2O}^{\text{includes moisture}} + 5.648N_2$$

The dew-point temperature of the products is the temperature at which the water vapor in the products starts to condense as the products are cooled. Again, assuming ideal-gas behavior, the partial pressure of the water vapor in the combustion gases is

$$P_{v,\,prod} = \left(\frac{N_{v,\,prod}}{N_{prod}}\right) P_{prod} = \left(\frac{1.661\text{ kmol}}{8.059\text{ kmol}}\right)(101.325\text{ kPa}) = 20.88\text{ kPa}$$

Thus, $T_{dp} = T_{sat\,@\,20.88\text{ kPa}} = \mathbf{60.9°C}$

Discussion If the combustion process were achieved with dry air instead of moist air, the products would contain less moisture, and the dew-point temperature in this case would be 59.5°C.

EXAMPLE 14-4 Reverse Combustion Analysis

Octane (C_8H_{18}) is burned with dry air. The volumetric analysis of the products on a dry basis is (Fig. 14-13)

CO_2:	10.02 percent
O_2:	5.62 percent
CO:	0.88 percent
N_2:	83.48 percent

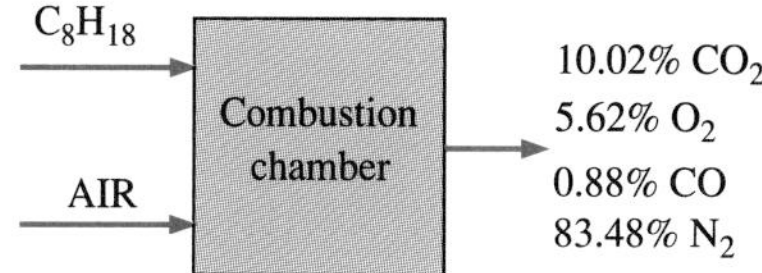

FIGURE 14-13
Schematic for Example 14-4.

Determine (*a*) the air–fuel ratio, (*b*) the percentage of theoretical air used, and (*c*) the fraction of the H_2O that condenses as the products are cooled to 25°C at 100 kPa.

Solution Combustion products whose composition is given are cooled to 25°C. The AF, the percent theoretical air used, and the fraction of water vapor that condenses are to be determined. Note that we know the relative composition of the products, but we do not know how much fuel or air is used during the combustion process. However, they can be determined from mass balances. The H_2O in the combustion gases will start condensing when the temperature drops to dew-point temperature.

Assumptions Combustion gases are ideal gases.

Analysis For ideal gases, the volume fractions are equivalent to the mole fractions. Considering 100 kmol of dry products for convenience, the combustion equation can be written as

$$xC_8H_{18} + a(O_2 + 3.76N_2) \longrightarrow 10.02CO_2 + 0.88CO + 5.62O_2 + 83.48N_2 + bH_2O$$

The unknown coefficients *x*, *a*, and *b* are determined from mass balances,

$$N_2: \quad 3.76a = 83.48 \longrightarrow a = 22.20$$

$$C: \quad 8x = 10.02 + 0.88 \longrightarrow x = 1.36$$

$$H: \quad 18x = 2b \longrightarrow b = 12.24$$

$$O_2: \quad a = 10.02 + 0.44 + 5.62 + \frac{b}{2} \longrightarrow 22.20 = 22.20$$

The O_2 balance is not necessary, but it can be used to check the values obtained from the other mass balances, as we did above. Substituting, we get

$$1.36C_8H_{18} + 22.2(O_2 + 3.76N_2) \longrightarrow 10.02CO_2 + 0.88CO + 5.62O_2 + 83.48N_2 + 12.24H_2O$$

The combustion equation for 1 kmol of fuel is obtained by dividing the above equation by 1.36,

$$C_8H_{18} + 16.32(O_2 + 3.76N_2) \longrightarrow 7.37CO_2 + 0.65CO + 4.13O_2 + 61.38N_2 + 9H_2O$$

(*a*) The air–fuel ratio is determined by taking the ratio of the mass of the air to the mass of the fuel (Eq. 14-3),

$$AF = \frac{m_{air}}{m_{fuel}} = \frac{(16.32 \times 4.76 \text{ kmol})(29 \text{ kg/kmol})}{(8 \text{ kmol})(12 \text{ kg/kmol}) + (9 \text{ kmol})(2 \text{ kg/kmol})}$$
$$= \mathbf{19.76 \text{ kg air/kg fuel}}$$

(*b*) To find the percentage of theoretical air used, the need to know the theoretical amount of air, which is determined from the theoretical combustion equation of the fuel,

$$C_8H_{18} + a_{th}(O_2 + 3.76N_2) \longrightarrow 8CO_2 + 9H_2O + 3.76a_{th}N_2$$

O_2: $$a_{th} = 8 + 4.5 \longrightarrow a_{th} = 12.5$$

Then, $$\text{Percentage of theoretical air} = \frac{m_{air,\,act}}{m_{air,\,th}} = \frac{N_{air,\,act}}{N_{air,\,th}}$$
$$= \frac{(16.32)(4.76) \text{ kmol}}{(12.50)(4.76) \text{ kmol}}$$
$$= \mathbf{131\%}$$

That is, 31 percent excess air was used during this combustion process. Notice that some carbon formed carbon monoxide even though there was considerably more oxygen than needed for complete combustion.

(*c*) For each kmol of fuel burned, 7.37 + 0.65 + 4.13 + 61.38 + 9 = 82.53 kmol of products are formed, including 9 kmol of H_2O. Assuming that the dew-point temperature of the products is above 25°C, some of the water vapor will condense as the products are cooled to 25°C. If N_w kmol of H_2O condenses, there will be $(9 - N_w)$ kmol of water vapor left in the products. The mole number of the products in the gas phase will also decrease to $82.53 - N_w$ as a result. By treating the product gases (including the remaining water vapor) as ideal gases, N_w is determined by equating the mole fraction of the water vapor to its pressure fraction,

$$\frac{N_v}{N_{prod,\,gas}} = \frac{P_v}{P_{prod}}$$
$$\frac{9 - N_w}{82.53 - N_w} = \frac{3.169 \text{ kPa}}{100 \text{ kPa}}$$
$$N_w = \mathbf{6.59 \text{ kmol}}$$

since $P_v = P_{sat\,@\,25°C} = 3.169$ kPa. Therefore, the majority of the water vapor in the products (73 percent of it) will condense as the product gases are cooled to 25°C.

FIGURE 14-14

The microscopic form of energy of a substance consists of sensible, latent, chemical, and nuclear energies.

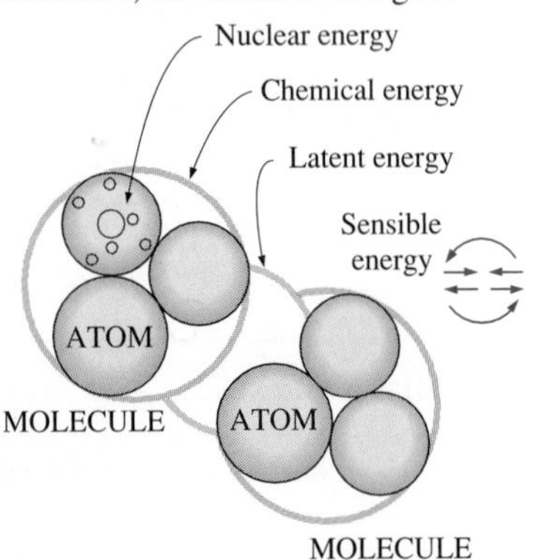

14-3 ■ ENTHALPY OF FORMATION AND ENTHALPY OF COMBUSTION

We mentioned in Chap. 1 that the molecules of a system possess energy in various forms such as *sensible* and *latent energy* (associated with a change of state), *chemical energy* (associated with the molecular structure), and *nuclear energy* (associated with the atomic structure), as illustrated in Fig. 14-14. In this text we do not intend to deal with nuclear energy, so we leave that sleeping giant alone. We also ignored chemical energy until now since the systems considered in previous chapters involved no changes in their chemical structure, and thus no changes in chemical energy. Consequently, all we needed to deal with were the sensible and latent energies.

During a chemical reaction, some chemical bonds that bind the atoms into molecules are broken, and new ones are formed. The chemical energy associated with these bonds, in general, is different for the reactants and the products. Therefore, a process that involves chemical reactions will involve changes in chemical energies, which must be accounted for in an energy balance (Fig. 14-15). Assuming the atoms of each reactant remain intact (no nuclear reactions) and disregarding any changes in kinetic and potential energies, the energy change of a system during a chemical reaction will be due to a change in state and a change in chemical composition. That is,

$$\Delta E_{\text{sys}} = \Delta E_{\text{state}} + \Delta E_{\text{chem}} \tag{14-4}$$

Therefore, when the products formed during a chemical reaction exit the reaction chamber at the inlet state of the reactants, we have $\Delta E_{\text{state}} = 0$ and the energy change of the system in this case is due to the changes in its chemical composition only.

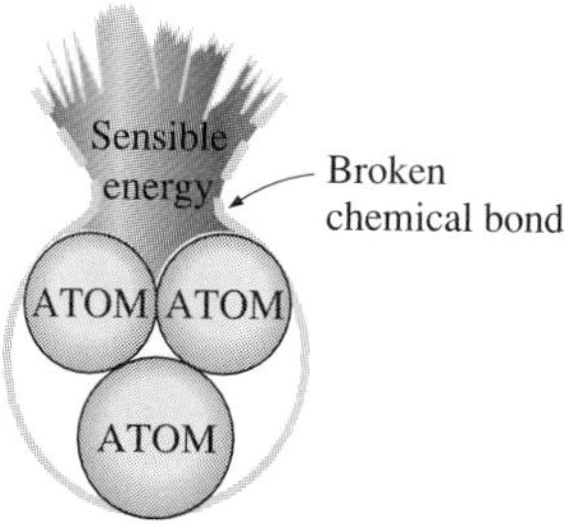

FIGURE 14-15
When the existing chemical bonds are destroyed and new ones are formed during a combustion process, usually a large amount of sensible energy is absorbed or released.

In thermodynamics we are concerned with the *changes* in the energy of a system during a process, and not the energy values at the particular states. Therefore, we can choose any state as the reference state and assign a value of zero to the internal energy or enthalpy of a substance at that state. When a process involves no changes in chemical composition, the reference state chosen has no effect on the results. When the process involves chemical reactions, however, the composition of the system at the end of a process is no longer the same as that at the beginning of the process. In this case it becomes necessary to have a common reference state for all substances. The chosen reference state is 25°C (77°F) and 1 atm, which is known as the **standard reference state.** Property values at the standard reference state are indicated by a superscript "°" (such as $h°$ and $u°$).

When analyzing reacting systems, we must use property values relative to the standard reference state. However, it is not necessary to prepare a new set of property tables for this purpose. We can use the existing tables by subtracting the property values at the standard reference state from the values at the specified state. The ideal-gas enthalpy of N_2 at 500 K relative to the standard reference state, for example, is $\bar{h}_{500\text{ K}} - \bar{h}° = 14{,}581 - 8669 = 5912$ kJ/kmol.

Consider the formation of CO_2 from its elements, carbon and oxygen, during a steady-flow combustion process (Fig. 14-16). Both the carbon and the oxygen enter the combustion chamber at 25°C and 1 atm. The CO_2 formed during this process also leaves the combustion chamber at 25°C and 1 atm. The combustion of carbon is an *exothermic reaction* (a reaction during which chemical energy is released in the form of heat). Therefore, some heat will be transferred from the combustion chamber to the surroundings during this process, which is 393,520 kJ/kmol CO_2 formed. (When one is dealing with chemical reactions, it is more convenient to work with quantities per unit mole than per unit time, even for steady-flow processes.)

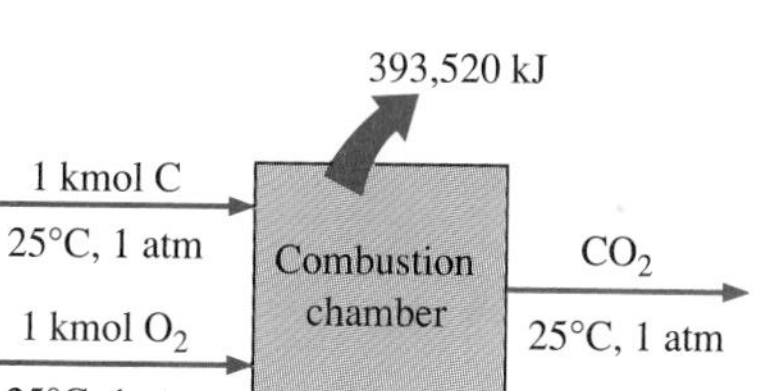

FIGURE 14-16
The formation of CO_2 during a steady-flow combustion process at 25°C and 1 atm.

The process described above involves no work interactions. Therefore, from the steady-flow energy balance relation, the heat transfer during this process must be equal to the difference between the enthalpy of the products and the enthalpy of the reactants. That is,

$$Q = H_{\text{prod}} - H_{\text{react}} = -393{,}520 \text{ kJ/kmol} \tag{14-5}$$

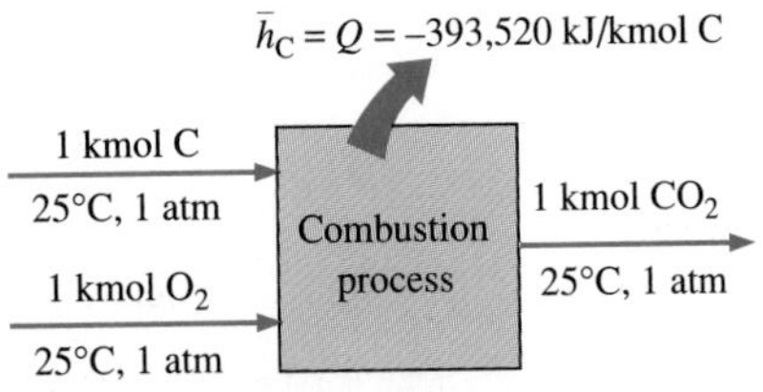

FIGURE 14-17
The enthalpy of combustion represents the amount of energy released as a fuel is burned during a steady-flow process at a specified state.

Since both the reactants and the products are at the same state, the enthalpy change during this process is solely due to the changes in the chemical composition of the system. This enthalpy change will be different for different reactions, and it would be very desirable to have a property to represent the changes in chemical energy during a reaction. This property is the **enthalpy of reaction** h_R, which is defined as *the difference between the enthalpy of the products at a specified state and the enthalpy of the reactants at the same state for a complete reaction.*

For combustion processes, the enthalpy of reaction is usually referred to as the **enthalpy of combustion** h_C, which represents the amount of heat released during a steady-flow combustion process when 1 kmol (or 1 kg) of fuel is burned completely at a specified temperature and pressure (Fig. 14-17). It is expressed as

$$h_C = H_{\text{prod}} - H_{\text{react}} \tag{14-6}$$

which is −393,520 kJ/kmol for C at the standard reference state. The enthalpy of combustion of a particular fuel will be different at different temperatures and pressures. Table A-27 lists h_C values for various fuels at the standard reference state of 25°C and 1 atm.

The enthalpy of combustion is obviously a very useful property for analyzing the combustion processes of fuels. However, there are so many different fuels and fuel mixtures that it is not practical to list h_C values for all possible cases. Besides, the enthalpy of combustion is not of much use when the combustion is incomplete. Therefore a more practical approach would be to have a more fundamental property to represent the chemical energy of an element or a compound at some reference state. This property is the **enthalpy of formation** $\bar{h}_f$, which can be viewed as *the enthalpy of a substance at a specified state due to its chemical composition.*

To establish a starting point, we assign the enthalpy of formation of all stable elements (such as O_2, N_2, H_2, and C) a value of zero at the standard reference state of 25°C and 1 atm. That is, $\bar{h}_f^\circ = 0$ for all stable elements. (This is no different from assigning the internal energy of saturated liquid water a value of zero at 0.01°C.) Perhaps we should clarify what we mean by *stable*. The stable form of an element is simply the chemically stable form of that element at 25°C and 1 atm. Nitrogen, for example, exists in diatomic form (N_2) at 25°C and 1 atm. Therefore, the stable form of nitrogen at the standard reference state is diatomic nitrogen N_2, not monatomic nitrogen N. If an element exists in more than one stable form at 25°C and 1 atm, one of the forms should be specified as the stable form. For carbon, for example, the stable form is assumed to be graphite, not diamond.

Now reconsider the formation of CO_2 (a compound) from its elements C and O_2 at 25°C and 1 atm during a steady-flow process. The enthalpy change during this process was determined to be (Eq. 14-5):

$$H_{prod} - H_{react} = -393{,}520 \text{ kJ/kmol}$$

But $H_{react} = 0$ since both reactants are elements at the standard reference state, and the products consist of 1 kmol of CO_2 at the same state. Therefore, the enthalpy of formation of CO_2 at the standard reference state is −393,520 kJ/kmol (Fig. 14-18). That is,

$$\bar{h}^{\circ}_{f,CO_2} = -393{,}520 \text{ kJ/kmol}$$

The negative sign is due to the fact that the enthalpy of 1 kmol of CO_2 at 25°C and 1 atm is 393,520 kJ less than the enthalpy of 1 kmol of C and 1 kmol of O_2 at the same state. In other words, 393,520 kJ of chemical energy is released (leaving the system as heat) when C and O_2 combine to form 1 kmol of CO_2. Therefore, a negative enthalpy of formation for a compound indicates that heat is released during the formation of that compound from its stable elements. A positive value indicates heat is absorbed.

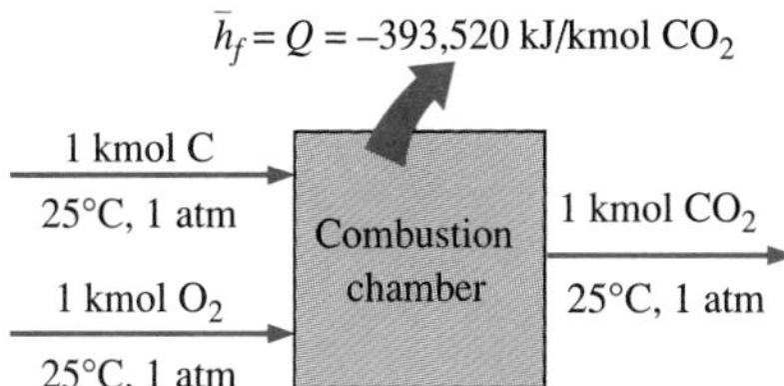

FIGURE 14-18

The enthalpy of formation of a compound represents the amount of energy absorbed or released as the component is formed from its stable elements during a steady-flow process at a specified state.

You will notice that two $\bar{h}^{\circ}_f$ values are given for H_2O in Table A-26, one for liquid water and the other for water vapor. This is because both phases of H_2O are encountered at 25°C, and the effect of pressure on the enthalpy of formation is small. (Note that under equilibrium conditions, water exists only as a liquid at 25°C *and* 1 atm.) The difference between the two enthalpies of formation is equal to the h_{fg} of water at 25°C, which is 2442.3 kJ/kg or 44,000 kJ/kmol.

Another term commonly used in conjunction with the combustion of fuels is the **heating value** of the fuel, which is defined as the amount of heat released when a fuel is burned completely in a steady-flow process and the products are returned to the state of the reactants. In other words, the heating value of a fuel is equal to the absolute value of the enthalpy of combustion of the fuel. That is,

$$\text{Heating value} = |h_C| \qquad \text{(kJ/kg fuel)}$$

The heating value depends on the *phase* of the H_2O in the products. The heating value is called the **higher heating value** (HHV) when the H_2O in the products is in the liquid form, and it is called the **lower heating value** (LHV) when the H_2O in the products is in the vapor form (Fig. 14-19). The two heating values are related by

$$\text{HHV} = \text{LHV} + (N\bar{h}_{fg})_{H_2O} \qquad \text{(kJ/kg fuel)} \qquad (14\text{-}7)$$

where N is the number of moles of H_2O in the products and $\bar{h}_{fg}$ is the enthalpy of vaporization of water at the specified temperature.

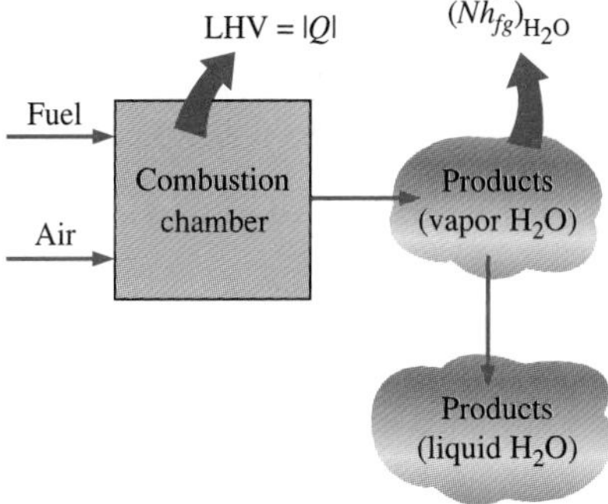

FIGURE 14-19

The higher heating value of a fuel is equal to the sum of the lower heating value of the fuel and the latent heat of vaporization of the H_2O in the products.

The heating value or enthalpy of combustion of a fuel can be determined from a knowledge of the enthalpy of formation for the compounds involved. This is illustrated with the following example.

EXAMPLE 14-5 Evaluation of the Enthalpy of Combustion

Determine the enthalpy of combustion of gaseous octane (C_8H_{18}) at 25°C and 1 atm, using enthalpy-of-formation data from Table A-26. Assume the water in the products is in the liquid form.

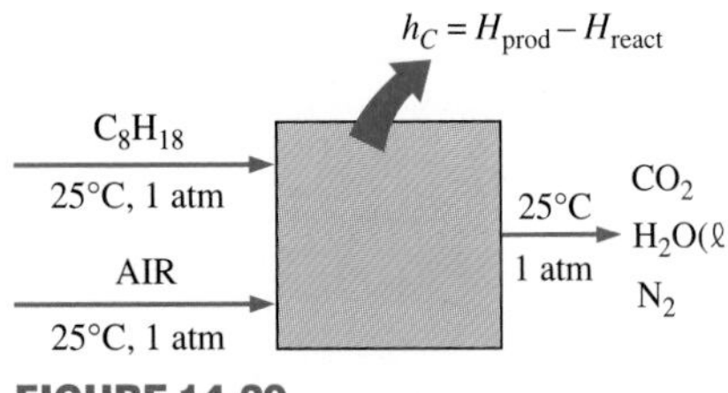

FIGURE 14-20

Schematic for Example 14-5.

Solution The combustion of C_8H_{18} is illustrated in Fig. 14-20. The stoichiometric equation for this reaction is

$$C_8H_{18} + a_{th}(O_2 + 3.76N_2) \longrightarrow 8CO_2 + 9H_2O(\ell) + 3.76a_{th}N_2$$

Both the reactants and the products are at the standard reference state of 25°C and 1 atm. Also, N_2 and O_2 are stable elements, and thus their enthalpy of formation is zero. Then the enthalpy of combustion of C_8H_{18} becomes (Eq. 14-6)

$$\begin{aligned}\bar{h}_C &= H_{prod} - H_{react} \\ &= \sum N_p \bar{h}^\circ_{f,p} - \sum N_r \bar{h}^\circ_{f,r} = (N\bar{h}^\circ_f)_{CO_2} + (N\bar{h}^\circ_f)_{H_2O} - (N\bar{h}^\circ_f)_{C_8H_{18}}\end{aligned}$$

Using $\bar{h}^\circ_f$ values from Table A-26, we get

$$\begin{aligned}\bar{h}_C &= (8\text{ kmol})(-393{,}520\text{ kJ/kmol}) + (9\text{ kmol})(-285{,}830\text{ kJ/kmol}) \\ &\quad - (1\text{ kmol})(-208{,}450\text{ kJ/kmol}) \\ &= \mathbf{-5{,}512{,}180\text{ kJ/kmol } C_8H_{18}}\end{aligned}$$

which is practially identical to the listed value of −5,512,200 kJ in Table A-27. Since the water in the products is assumed to be in the liquid phase, this h_C value corresponds to the HHV of C_8H_{18}.

When the exact composition of the fuel is known, the *enthalpy of combustion* of that fuel can be determined using enthalpy of formation data as shown above. But for fuels that exhibit considerable variations in composition depending on the source, such as coal, natural gas, and fuel oil, it is more practical to determine their enthalpy of combustion experimentally by burning them directly in a bomb calorimeter at constant volume or in a steady-flow device.

14-4 ■ FIRST-LAW ANALYSIS OF REACTING SYSTEMS

The energy balance (or the first-law) relations developed in Chaps. 3 and 4 are applicable to both reacting and nonreacting systems. However, chemically reacting systems involve changes in their chemical energy, and thus it is more convenient to rewrite the energy balance relations so that the changes in chemical energies are explicitly expressed. We do this first for steady-flow systems and then for closed systems.

FIGURE 14-21

The enthalpy of a chemical component at a specified state is the sum of the enthalpy of the component at 25°C, 1 atm ($\bar{h}^\circ_f$), and the sensible enthalpy of the component relative to 25°C, 1 atm.

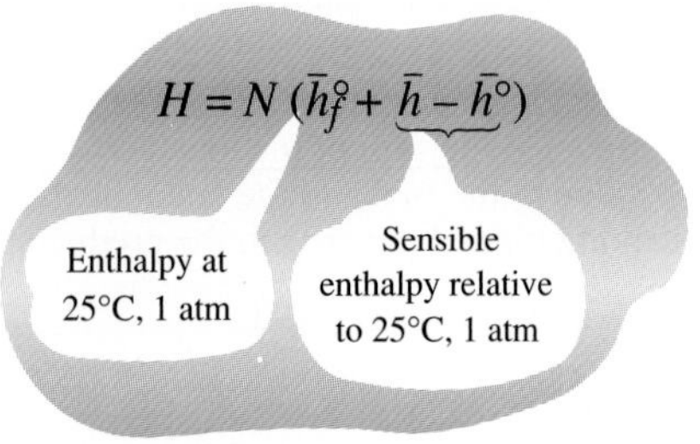

$H = N(\bar{h}^\circ_f + \bar{h} - \bar{h}^\circ)$

Enthalpy at 25°C, 1 atm

Sensible enthalpy relative to 25°C, 1 atm

Steady-Flow Systems

Before writing the energy balance relation, we need to express the enthalpy of a component in a form suitable for use for reacting systems. That is, we need to express the enthalpy such that it is relative to the standard reference state and the chemical energy term appears explicitly. When expressed properly, the enthalpy term should reduce to the enthalpy of formation $\bar{h}^\circ_f$ at the standard reference state. With this in mind, we express the enthalpy of a component on a unit-mole basis as (Fig. 14-21)

$$\text{Enthalpy} = \bar{h}^\circ_f + (\bar{h} - \bar{h}^\circ) \qquad \text{(kJ/kmol)}$$

where the term in the parentheses represents the sensible enthalpy relative to the standard reference state, which is the difference between $\bar{h}$ (the sensible enthalpy at the specified state) and $\bar{h}^\circ$ (the sensible enthalpy at the standard reference state of 25°C and 1 atm). This definition enables us to use enthalpy values from tables regardless of the reference state used in their construction.

When the changes in kinetic and potential energies are negligible, the steady-flow energy balance relation $\dot{E}_{in} = \dot{E}_{out}$ can be expressed for a *chemically reacting steady-flow system* more explicitly as

$$\underbrace{\dot{Q}_{in} + \dot{W}_{in} + \sum \dot{n}_r(\bar{h}_f^\circ + \bar{h} - \bar{h}^\circ)_r}_{\substack{\text{Rate of net energy transfer in} \\ \text{by heat, work, and mass}}} = \underbrace{\dot{Q}_{out} + \dot{W}_{out} + \sum \dot{n}_p(\bar{h}_f^\circ + \bar{h} - \bar{h}^\circ)_p}_{\substack{\text{Rate of net energy transfer out} \\ \text{by heat, work, and mass}}} \qquad (14\text{-}8)$$

where $\dot{n}_p$ and $\dot{n}_r$ represent the molal flow rates of the product p and the reactant r, respectively.

In combustion analysis, it is more convenient to work with quantities expressed *per mole of fuel*. Such a relation is obtained by dividing each term of the equation above by the molal flow rate of the fuel, yielding

$$\underbrace{Q_{in} + W_{in} + \sum N_r(\bar{h}_f^\circ + \bar{h} - \bar{h}^\circ)_r}_{\substack{\text{Energy transfer in per mole of fuel} \\ \text{by heat, work, and mass}}} = \underbrace{Q_{out} + W_{out} + \sum N_p(\bar{h}_f^\circ + \bar{h} - \bar{h}^\circ)_p}_{\substack{\text{Energy transfer out per mole of fuel} \\ \text{by heat, work, and mass}}} \qquad (14\text{-}9)$$

where N_r and N_p represent the number of moles of the reactant r and the product p, respectively, per mole of fuel. Note that $N_r = 1$ for the fuel, and the other N_r and N_p values can be picked directly from the balanced combustion equation. Taking heat transfer *to* the system and work done *by* the system to be *positive* quantities, the energy balance relation above can be expressed more compactly as

$$Q - W = \sum N_p(\bar{h}_f^\circ + \bar{h} - \bar{h}^\circ)_p - \sum N_r(\bar{h}_f^\circ + \bar{h} - \bar{h}^\circ)_r \qquad (14\text{-}10)$$

or as

$$Q - W = H_{prod} - H_{react} \qquad \text{(kJ/kmol fuel)} \qquad (14\text{-}11)$$

where

$$H_{prod} = \sum N_p(\bar{h}_f^\circ + \bar{h} - \bar{h}^\circ)_p \qquad \text{(kJ/kmol fuel)} \qquad (14\text{-}12a)$$

$$H_{react} = \sum N_r(\bar{h}_f^\circ + \bar{h} - \bar{h}^\circ)_r \qquad \text{(kJ/kmol fuel)} \qquad (14\text{-}12b)$$

If the enthalpy of combustion $\bar{h}_C^\circ$ for a particular reaction is available, the steady-flow energy equation per mole of fuel can be expressed as

$$Q - W = \bar{h}_c^\circ + \sum N_p(\bar{h} - \bar{h})_p - \sum N_r(\bar{h} - \bar{h}^\circ)_r \qquad \text{(kJ/kmol)} \qquad (14\text{-}13)$$

The energy balance relations above are sometimes written without the work term since most steady-flow combustion processes do not involve any work interactions.

A combustion chamber normally involves heat output but no heat input. Then the energy balance for a *typical steady-flow combustion process* becomes

$$Q_{out} = \underbrace{\sum N_r(\bar{h}_f^\circ + \bar{h} - \bar{h}^\circ)_r}_{\text{Energy in by mass per mole of fuel}} - \underbrace{\sum N_p(\bar{h}_f^\circ + \bar{h} - \bar{h}^\circ)_p}_{\text{Energy out by mass per mole of fuel}}$$

It expresses that the heat output during a combustion process is simply the difference between the energy of the reactants entering and the energy of the products leaving the combustion chamber.

Closed Systems

The general closed-system energy balance relation $E_{in} - E_{out} = \Delta E_{system}$ can be expressed for a stationary *chemically reacting closed system* as

$$(Q_{in} - Q_{out}) + (W_{in} - W_{out}) = U_{prod} - U_{react} \qquad \text{(kJ/kmol fuel)} \qquad (14\text{-}14)$$

where U_{prod} represents the internal energy of the products and U_{react} represents the internal energy of the reactants. To avoid using another property—*the internal energy of formation* $\bar{u}_f^\circ$—we utilize the definition of enthalpy ($\bar{u} = \bar{h} - P\bar{v}$ or $\bar{u}_f^\circ + \bar{u} - \bar{u}^\circ = \bar{h}_f^\circ + \bar{h} - \bar{h}^\circ - P\bar{v}$) and express the above equation as (Fig. 14-22)

$$U = H - PV$$
$$= N(h_f^\circ + h - h^\circ) - PV$$
$$= N(h_f^\circ + h - h^\circ - P\,v)$$

FIGURE 14-22
An expression for the internal energy of a chemical component in terms of the enthalpy.

$$Q - W = \sum N_p(\bar{h}_f^\circ + \bar{h} - \bar{h}^\circ - P\bar{v})_p - \sum N_r(\bar{h}_f^\circ + \bar{h} - \bar{h}^\circ - P\bar{v})_r \quad (14\text{-}15)$$

where we have taken heat transfer *to* the system and work done *by* the system to be *positive* quantities. The $P\bar{v}$ terms are negligible for solids and liquids, and can be replaced by R_uT for gases that behave as an ideal gas. Also, if desired, the $\bar{h} - P\bar{v}$ terms in Eq. 14-15 can be replaced by $\bar{u}$.

The work term in Eq. 14-15 represents all forms of work, including the boundary work. It was shown in Chap. 3 that $\Delta U + W_b = \Delta H$ for nonreacting closed systems undergoing a quasi-equilibrium P = constant expansion or compression process. This is also the case for chemically reacting systems.

There are several important considerations in the analysis of steady-flow or closed reacting systems. For example, we need to know whether the fuel is a solid, a liquid, or a gas since the enthalpy of formation h_f° of a fuel depends on the phase of the fuel. We also need to know the state of the fuel when it enters the combustion chamber in order to determine its enthalpy. For entropy calculations it is especially important to know if the fuel and air enter the combustion chamber premixed or separately. When the combustion products are cooled to low temperatures, we need to consider the possibility of condensation of some of the water vapor in the product gases.

FIGURE 14-23
Schematic for Example 14-6.

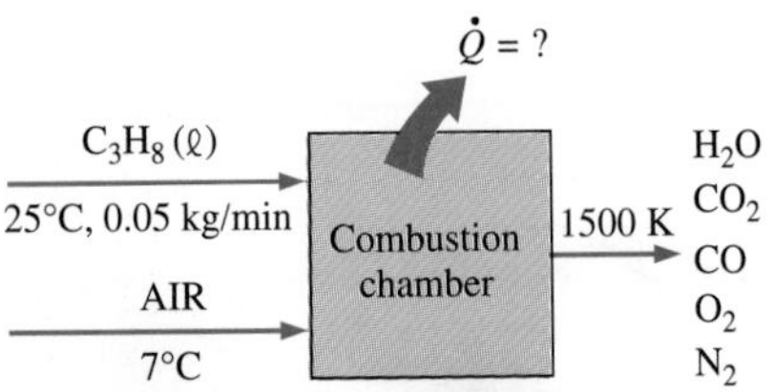

EXAMPLE 14-6 First-Law Analysis of Steady-Flow Combustion

Liquid propane (C_3H_8) enters a combustion chamber at 25°C at a rate of 0.05 kg/min where it is mixed and burned with 50 percent excess air that enters the combustion chamber at 7°C, as shown in Fig. 14-23. An analysis of the combustion gases reveals that all the hydrogen in the fuel burns to H_2O but only 90 percent of the carbon burns to CO_2, with the remaining 10 percent forming CO. If the exit temperature of the combustion gases is 1500 K, determine (*a*) the mass flow rate of air and (*b*) the rate of heat transfer from the combustion chamber.

Solution Liquid propane is burned steadily with excess air. The mass flow rate of air and the rate of heat transfer are to be determined. We note that all the hydrogen in the fuel burns to H_2O but 10 percent of the carbon burns incompletely and forms CO. Also, the fuel is burned with excess air and thus there will be some free O_2 in the product gases.

Assumptions **1** Steady operating conditions exist. **2** Air and the combustion gases are ideal gases. **3** Kinetic and potential energies are negligible.

Analysis The theoretical amount of air is determined from the stoichiometric reaction to be

$$C_3H_8(\ell) + a_{th}(O_2 + 3.76N_2) \longrightarrow 3CO_2 + 4H_2O + 3.76a_{th}N_2$$

O_2 balance: $$a_{th} = 3 + 2 = 5$$

Then the balanced equation for the actual combustion process with 50 percent excess air and some CO in the products becomes

$$C_3H_8(\ell) + 7.5(O_2 + 3.76N_2) \longrightarrow 2.7CO_2 + 0.3CO + 4H_2O + 2.65O_2 + 28.2N_2$$

(*a*) The air–fuel ratio for this combustion process is

$$\text{AF} = \frac{m_{air}}{m_{fuel}} = \frac{(7.5 \times 4.76 \text{ kmol})(29 \text{ kg/kmol})}{(3 \text{ kmol})(12 \text{ kg/kmol}) + (4 \text{ kmol})(2 \text{ kg/kmol})} = 25.53 \text{ kg air/kg fuel}$$

Thus,
$$\dot{m}_{air} = (\text{AF})(\dot{m}_{fuel}) = (23.53 \text{ kg air/kg fuel})(0.05 \text{ kg fuel/min}) = \mathbf{1.18 \text{ kg air/min}}$$

(*b*) The heat transfer for this steady-flow combustion process is determined from the steady-flow energy balance $E_{out} = E_{in}$ applied on the combustion chamber per unit mole of the fuel,

$$Q_{out} + \sum N_p(\bar{h}_f^\circ + \bar{h} - \bar{h}^\circ)_p = \sum N_r(\bar{h}_f^\circ + \bar{h} - \bar{h}^\circ)_r$$

or
$$Q_{out} = \sum N_r(h_f^\circ + \bar{h} - \bar{h}^\circ)_r - \sum N_p(\bar{h}_f^\circ + \bar{h} - \bar{h}^\circ)_p$$

Assuming the air and the combustion products to be ideal gases, we have $h = h(T)$, and using data from the property tables we form the following minitable:

Substance	h_f° **kJ/kmol**	$\bar{h}_{280\text{ K}}$ **kJ/kmol**	$\bar{h}_{298\text{ K}}$ **kJ/kmol**	$\bar{h}_{1500\text{ K}}$ **kJ/kmol**
$C_3H_8(\ell)$	−118.910	—	—	—
O_2	0	8150	8682	49,292
N_2	0	8141	8669	47,073
$H_2O(g)$	−241,820	—	9904	57,999
CO_2	−393,520	—	9364	71.078
CO	−110,530	—	8669	47,517

The $\bar{h}_f^\circ$ of liquid propane is obtained by subtracting the $\bar{h}_{fg}$ of propane at 25°C from the $\bar{h}_f^\circ$ of gas propane. Substituting gives

$$
\begin{aligned}
Q_{out} = &(1 \text{ kmol } C_3H_8)[(-118{,}910 + \bar{h}_{298} - \bar{h}_{298}) \text{ kJ/kmol } C_3H_8] \\
&+ (7.5 \text{ kmol } O_2)[(0 + 8150 - 8682) \text{ kJ/kmol } O_2] \\
&+ (28.2 \text{ kmol } N_2)[(0 + 8141 - 8669) \text{ kJ/kmol } N_2] \\
&- (2.7 \text{ kmol } CO_2)[(-393{,}520 + 71{,}078 - 9364) \text{ kJ/kmol } CO_2] \\
&- (0.3 \text{ kmol } CO)[(-110{,}530 + 47{,}517 - 8669) \text{ kJ/kmol } CO] \\
&- (4 \text{ kmol } H_2O)[(-241{,}820 + 57{,}999 - 9904) \text{ kJ/kmol } H_2O] \\
&- (2.65 \text{ kmol } O_2)[(0 + 49{,}292 - 8682) \text{ kJ/kmol } O_2] \\
&-(28.2 \text{ kmol } N_2)[(0 + 47{,}073 - 8669) \text{ kJ/kmol } N_2] \\
= &\mathbf{363{,}882 \text{ kJ/kmol of } C_3H_8}
\end{aligned}
$$

Thus 363,882 kJ of heat is transferred from the combustion chamber for each kmol (44 kg) of propane. This corresponds to 363,882/44 = 8270.0 kJ of heat loss per kilogram of propane. Then the rate of heat transfer for a mass flow rate of 0.05 kg/min for the propane becomes

$$
\begin{aligned}
\dot{Q}_{out} &= \dot{m}q_{out} = (0.05 \text{ kg/min})(8270.0 \text{ kJ/kg}) = 413.5 \text{ kJ/min} \\
&= \mathbf{6.89 \text{ kW}}
\end{aligned}
$$

EXAMPLE 14-7 First-Law Analysis of Combustion in a Bomb

The constant-volume tank shown in Fig. 14-24 contains 1 lbmol of methane (CH_4) gas and 3 lbmol of O_2 at 77°F and 1 atm. The contents of the tank are ignited, and the methane gas burns completely. If the final temperature is 1800 R, determine (*a*) the final pressure in the tank and (*b*) the heat transfer during this process.

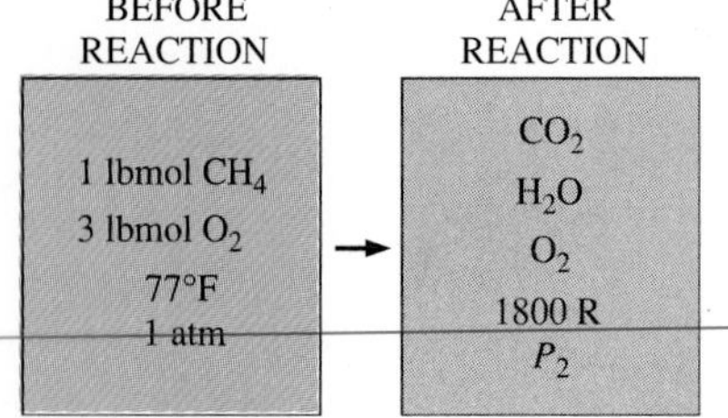

FIGURE 14-24
Schematic for Example 14-7.

Solution Methane is burned in a rigid tank that is a closed system. The pressure in the tank after combustion and the heat transfer are to be determined.

Assumptions **1** The fuel is burned completely and thus all the carbon in the fuel burns to CO_2 and all the hydrogen to H_2O. **2** The fuel, the air, and the combustion gases are ideal gases. **3** Kinetic and potential energies are negligible. **4** There are no work interactions involved.

Analysis The balanced combustion equation is

$$CH_4(g) + 3O_2 \longrightarrow CO_2 + 2H_2O + O_2$$

(*a*) At 1800 R, water exists in the gas phase. Using the ideal gas relation for both the reactants and the products, the final pressure in the tank is determined to be

$$\left.\begin{aligned} P_{react}V &= N_{react}R_uT_{react} \\ P_{prod}V &= N_{prod}R_uT_{prod} \end{aligned}\right\} \qquad P_{prod} = P_{react}\left(\frac{N_{prod}}{N_{react}}\right)\left(\frac{T_{prod}}{T_{react}}\right)$$

Substituting, we get

$$P_{prod} = (1 \text{ atm})\left(\frac{4 \text{ lbmol}}{4 \text{ lbmol}}\right)\left(\frac{1800 \text{ R}}{537 \text{ R}}\right) = \mathbf{3.35 \text{ atm}}$$

(*b*) Noting that the process involves no work interactions, the heat transfer during this constant-volume combustion process can be determined from the energy balance $E_{in} - E_{out} = \Delta E_{system}$ applied on the tank,

$$-Q_{out} = \sum N_p(\bar{h}_f^\circ + \bar{h} - \bar{h}^\circ - P\bar{v})_p - \sum N_r(\bar{h}_f^\circ + \bar{h} - \bar{h}^\circ - P\bar{v})_r$$

Since both the reactants and the products are assumed to be ideal gases, all the internal energy and enthalpies depend on temperature only, and the $P\bar{v}$ terms in this equation can be replaced by R_uT. It yields

$$Q_{out} = \sum N_r(\bar{h}_f^\circ - R_uT)_r - \sum N_p(\bar{h}_f^\circ + \bar{h}_{1800\text{ R}} - \bar{h}_{537\text{ R}} - R_uT)_p$$

since $W = 0$ and the reactants are at the standard reference temperature of 537 R. From $\bar{h}_f^\circ$ and ideal-gas tables in the Appendix,

Substance	$\bar{h}_f^\circ$ Btu/lbmol	$\bar{h}_{537\text{ R}}$ Btu/lbmol	$\bar{h}_{1800\text{ R}}$ Btu/lbmol
CH_4	−32,210	—	—
O_2	0	3725.1	13.485.8
CO_2	−169,300	4027.5	18,391.5
$H_2O(g)$	−104,040	4258.0	15,433.0

Substituting, we have

$$\begin{aligned} Q_{out} = {} & (1\text{ lbmol } CH_4)[(-32{,}210 - 1.986 \times 537)\text{ Btu/lbmol } CH_4] \\ & + (3\text{ lbmol } O_2)[(0 - 1.986 \times 537)\text{ Btu/lbmol } O_2] \\ & - (1\text{ lbmol } CO_2)[(-169{,}300 + 18{,}391.5 - 4027.5 - 1.986 \times 1800)\text{ Btu/lbmol } CO_2] \\ & - (2\text{ lbmol } H_2O)[(-104{,}040 + 15{,}433.0 - 4258.0 - 1.986 \times 1800)\text{ Btu/lbmol } H_2O] \\ & - (1\text{ lbmol } O_2)[(0 + 13{,}485.8 - 3725.1 - 1.986 \times 1800)\text{ Btu/lbmol } O_2] \\ = {} & \mathbf{308{,}729\text{ Btu/lbmol } CH_4} \end{aligned}$$

Discussion On a mass basis, the heat transfer from the tank would be 308,729/16 = 19,296 Btu/lbm.

14-5 ■ ADIABATIC FLAME TEMPERATURE

In the absence of any work interactions and any changes in kinetic or potential energies, the chemical energy released during a combustion process either is lost as heat to the surroundings or is used internally to raise the temperature of the combustion products. The smaller the heat loss, the larger the temperature rise. In the limiting case of no heat loss to the surroundings ($Q = 0$), the temperature of the products will reach a maximum, which is called the **adiabatic flame** or **adiabatic combustion temperature** of the reaction (Fig. 14-25).

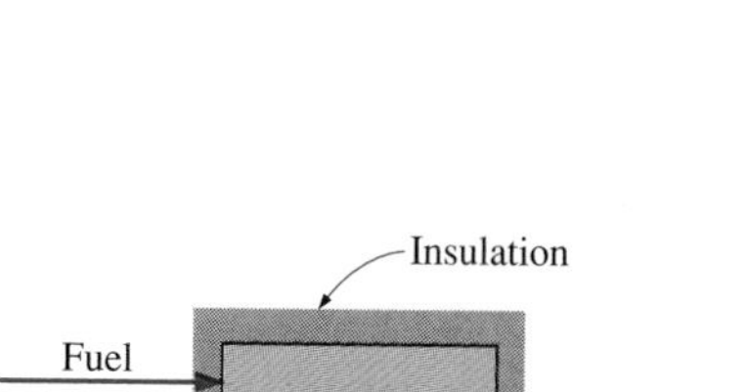

FIGURE 14-25
The temperature of a combustion chamber will be maximum when combustion is complete and no heat is lost to the surroundings ($Q = 0$).

The adiabatic flame temperature of a steady-flow combustion process is determined from Eq. 14-11 by setting $Q = 0$ and $W = 0$. It yields

$$H_{prod} = H_{react} \tag{14-16}$$

or

$$\sum N_p(\bar{h}_f^\circ + \bar{h} - \bar{h}^\circ)_p = \sum N_r(\bar{h}_f^\circ + \bar{h} - \bar{h})_r \tag{14-17}$$

Once the reactants and their states are specified, the enthalpy of the reactants H_{react} can be easily determined. The calculation of the enthalpy of the products H_{prod} is not so straightforward, however, because the temperature of the products is not known prior to the calculations. Therefore, the determination of the adiabatic flame temperature requires the use of an iterative technique unless equations for the sensible enthalpy changes of the combustion products are available. A temperature is assumed for the product gases, and the H_{prod} is determined for this temperature. If it is not equal to H_{react}, calculations are repeated with another temperature. The adiabatic flame temperature is then determined from these two results by interpolation. When the oxidant is air, the product gasses mostly consist of N_2, and a good first guess for the adiabatic flame temperature is obtained by treating the entire product gases as N_2.

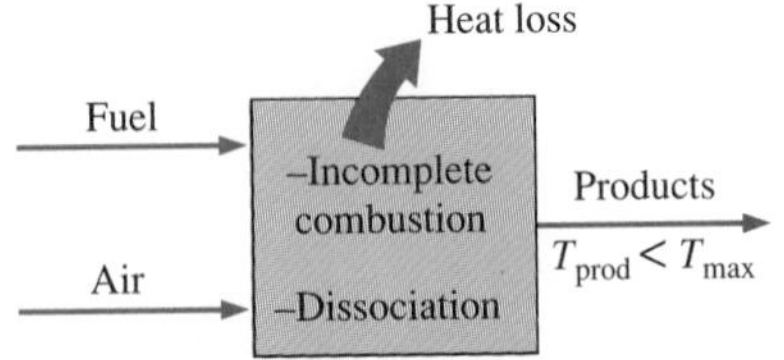

FIGURE 14-26
The maximum temperature encountered in a combustion chamber is lower than the theoretical adiabatic flame temperature.

In combustion chambers, the highest temperature to which a material can be exposed is limited by metallurgical considerations. Therefore, the adiabatic flame temperature is an important consideration in the design of combustion chambers, gas turbines, and nozzles. The maximum temperatures that occur in these devices are considerably lower than the adiabatic flame temperature; however, since the combustion is usually incomplete, some heat loss takes place, and some combustion gases dissociate at high temperatures (Fig. 14-26). The maximum temperature in a combustion chamber can be controlled by adjusting the amount of excess air, which serves as a coolant.

Note that the adiabatic flame temperature of a fuel is not unique. Its value depends on (1) the state of the reactants, (2) the degree of completion of the reaction, and (3) the amount of air used. For a specified fuel at a specified state burned with air at a specified state, *the adiabatic flame temperature attains its maximum value when complete combustion occurs with the theoretical amount of air.*

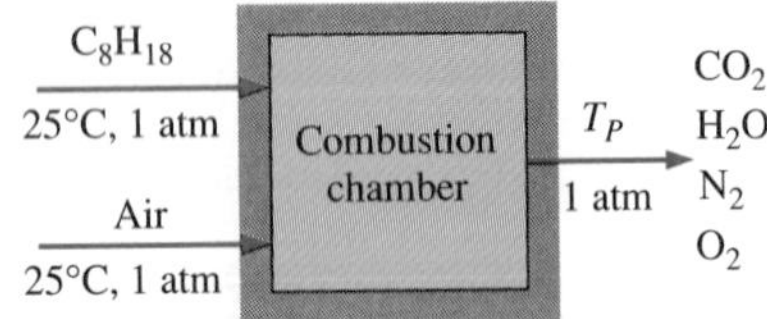

FIGURE 14-27
Schematic for Example 14-8.

EXAMPLE 14-8 Adiabatic Flame Temperature in Steady Combustion

Liquid octane (C_8H_{18}) enters the combustion chamber of a gas turbine steadily at 1 atm and 25°C, and it is burned with air that enters the combustion chamber at the same state, as shown in Fig. 14-27. Disregarding any changes in kinetic and potential energies, determine the adiabatic flame temperature for (*a*) complete combustion with 100 percent theoretical air, (*b*) complete combustion with 400 percent theoretical air, and (*c*) incomplete combustion (some CO in the products) with 90 percent theoretical air.

Solution Liquid octane is burned steadily. The adiabatic flame temperature is to be determined for different cases.

Assumptions **1** This is a steady-flow combustion process. **2** The combustion chamber is adiabatic. **3** There are no work interactions. **4** Air and the combustion gases are ideal gases. **5** Changes in kinetic and potential energies are negligible.

Analysis (*a*) The balanced equation for the combustion process with the theoretical amount of air is

$$C_8H_{18}(\ell) + 12.5(O_2 + 3.76N_2) \longrightarrow 8CO_2 + 9H_2O + 47N_2$$

The adiabatic flame temperature relation $H_{prod} = H_{react}$ in this case reduces to

$$\sum N_p(\bar{h}_f^\circ + \bar{h} - \bar{h}^\circ)_p = \sum N_r \bar{h}_{f,r}^\circ = (N\bar{h}_f^\circ)_{C_8H_{18}}$$

since all the reactants are at the standard reference state and $\bar{h}_f^\circ = 0$ for O_2 and N_2. The $\bar{h}_f^\circ$ and h values of various components at 298 K are

Substance	h_f° kJ/kmol	$\bar{h}_{298\text{ K}}$ kJ/kmol
$C_8H_{18}(\ell)$	−249,950	—
O_2	0	8682
N_2	0	8669
$H_2O(g)$	−241,820	9904
CO_2	−393,520	9364

Substituting, we have

$$\begin{aligned}&(8\text{ kmol } CO_2)[(-393{,}520 + \bar{h}_{CO_2} - 9364)\text{ kJ/kmol } CO_2]\\ &+ (9\text{ kmol } H_2O)[(-241{,}820 + \bar{h}_{H_2O}) - 9904)\text{ kJ/kmol } H_2O]\\ &+ (47\text{ kmol } N_2)[(0 + \bar{h}_{N_2} - 8669)\text{ kJ/kmol } N_2]\\ =\ &(1\text{ kmol } C_8H_{18})(-249{,}950\text{ kJ/kmol } C_8H_{18})\end{aligned}$$

which yields

$$8\bar{h}_{CO_2} + 9\bar{h}_{H_2O} + 47\bar{h}_{N_2} = 5{,}646{,}081\text{ kJ}$$

It appears that we have one equation with three unknowns. But actually we have only one unknown—the temperature of the products T_{prod}—since $h = h(T)$ for ideal gases. Therefore, we will have to use a trial-and-error approach to determine the temperature of the products.

A first guess is obtained by dividing the right-hand side of the equation by the total number of moles, which yields 5,646,081/(8 + 9 + 47) = 88,220 kJ/kmol. This enthalpy value will correspond to about 2650 K for N_2, 2100 K for H_2O, and 1800 K for CO_2. Noting that the majority of the moles are N_2, we see that T_{prod} will be close to 2650 K, but somewhat under it. Therefore, a good first guess is 2400 K. At this temperature,

$$\begin{aligned}8\bar{h}_{CO_2} + 9\bar{h}_{H_2O} + 47\bar{h}_{N_2} &= 8 \times 125{,}152 + 9 \times 103{,}508 + 47 \times 79{,}320\\ &= 5{,}660{,}828\text{ kJ}\end{aligned}$$

This value is higher than 5,646,081 kJ. Therefore, the actual temperature will be slightly under 2400 K. Next we choose 2350 K. It yields

$$8 \times 122{,}091 + 9 \times 100{,}846 + 47 \times 77{,}496 = 5{,}526{,}654$$

which is lower than 5,646,081 kJ. Therefore, the actual temperature of the products is between 2350 and 2400 K. By interpolation, it is found to be T_{prod} = **2394.5 K**

(*b*) The balanced equation for the complete combustion process with 400 percent theoretical air is

$$C_8H_{18}(\ell) + 50(O_2 + 3.76N_2) \longrightarrow 8CO_2 + 9H_2O + 37.5O_2 + 188N_2$$

By following the procedure used in (*a*), the adiabatic flame temperature in this case is determined to be T_{prod} = **962 K**.

Notice that the temperature of the products decreases significantly as a result of using excess air.

(*c*) The balanced equation for the incomplete combustion process with 90 percent theoretical air is

$$C_8H_{18}(\ell) + 11.25(O_2 + 3.76N_2) \longrightarrow 5.5CO_2 + 2.5CO + 9H_2O + 42.3N_2$$

Following the procedure used in (*a*), we find the adiabatic flame temperature in this case to be T_{prod} = **2236 K**.

Notice that the adiabatic flame temperature decreases as a result of incomplete combustion or using excess air. Also, *the maximum adiabatic flame temperature is achieved when complete combustion occurs with the theoretical amount of air.*

14-6 ■ ENTROPY CHANGE OF REACTING SYSTEMS

So far we have analyzed combustion processes from the conservation of mass and the conservation of energy points of view. The thermodynamic analysis of a process is not complete, however, without the examination of the second-law aspects. Of particular interest are the exergy and exergy destruction, both of which are related to entropy.

The entropy balance relations developed in Chap. 6 are equally applicable to both reacting and nonreacting systems provided that the entropies of individual constituents are evaluated properly using a common basis. The **entropy balance** for *any system* (including reacting systems) undergoing *any process* can be expressed as

$$\underbrace{S_{in} - S_{out}}_{\substack{\text{Net entropy transfer}\\ \text{by heat and mass}}} + \underbrace{S_{gen}}_{\substack{\text{Entropy}\\ \text{generation}}} = \underbrace{\Delta S_{system}}_{\substack{\text{Change}\\ \text{in entropy}}} \qquad \text{(kJ/K)} \qquad (14\text{-}18)$$

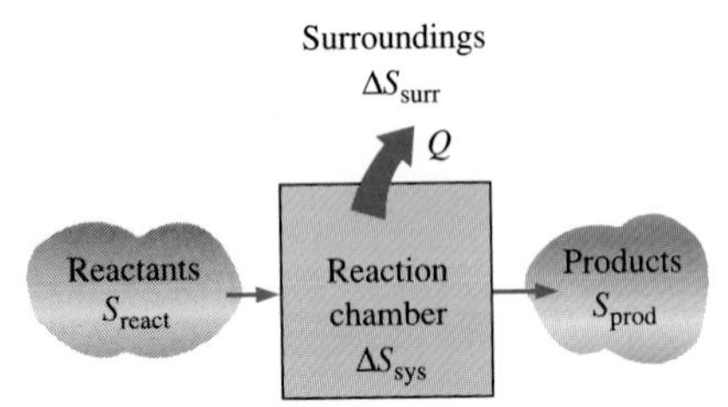

FIGURE 14-28
The entropy change associated with a chemical reaction.

Taking the positive direction of heat transfer to be *to* the system, the entropy balance relation can be expressed more explicitly for a **closed system** as (Fig. 14-28)

$$\sum \frac{Q_k}{T_k} + S_{gen} = S_{prod} - S_{react} \qquad \text{(kJ/K)} \qquad (14\text{-}19)$$

The entropy balance relation above can be stated as: *the entropy change of a closed system during a process is equal to the net entropy transferred through the system boundary by heat transfer and the entropy generated within the system boundaries.* Eq. 14-19 can also be used for a **steady-flow combustion chamber** by using quantities per unit mole of fuel (as we did in the energy balance). For an *adiabatic process* ($Q = 0$), the entropy transfer term in Eq. 14-19 drops out and the entropy balance relation reduces to

$$S_{gen,\,adiabatic} = S_{prod} - S_{react} \geq 0 \qquad (14\text{-}20)$$

The *total* entropy generated during a process can be determined by applying the entropy balance to an *extended system* that includes the system itself and its immediate surroundings where external irreversibilities might be occurring. When evaluating the entropy transfer between an extended system and the surroundings, the boundary temperature of the extended system is simply taken to be the *environment temperature,* as explained in Chap. 6.

The determination of the entropy change associated with a chemical reaction seems to be straightforward, except for one thing: The entropy relations for the reactants and the products (Eq. 14-20) involve the *entropies* of the components, *not entropy changes,* which was the case for nonreacting systems. Thus we are faced with the problem of finding a common base for the entropy of all substances, as we did with enthalpy. The search for such a common base led to the establishment of the **third law of thermodynamics** in the early part of this century. The third law was expressed in Chap. 6 as follows: *The entropy of a pure crystalline substance at absolute zero temperature is zero.*

Therefore, the third law of thermodynamics provides an absolute base for the entropy values for all substances. Entropy values relative to this base are called the **absolute entropy.** The $\bar{s}°$ values listed in Tables A-18 through A-25 for various gases such as N_2, O_2, CO, CO_2, H_2, H_2O, OH, and O are the *ideal-gas absolute entropy values* at the specified temperature and *at a pressure of 1 atm*. The absolute entropy values for various fuels are listed in Table A-26 together with the $\bar{h}_f^\circ$ values at the standard reference state of 25°C and 1 atm.

Equation 14-20 is a general relation for the entropy change of a reacting system. It requires the determination of the entropy of each individual component of the reactants and the products, which in general is not very easy to do. The entropy calculations can be simplified somewhat if the gaseous components of the reactants and the products are approximated as ideal gases. However, entropy calculations are never as easy as enthalpy or internal energy calculations, since entropy is a function of both temperature and pressure even for ideal gases.

When evaluating the entropy of a component of an ideal-gas mixture, we should use the temperature and the partial pressure of the component. Note that the temperature of a component is the same as the temperature of the mixture, and the partial pressure of a component is equal to the mixture pressure multiplied by the mole fraction of the component.

Absolute entropy values at pressures other than $P_0 = 1$ atm for any temperature T can be obtained from the ideal-gas entropy change relation written for an imaginary isothermal process between states (T, P_0) and (T, P), as illustrated in Fig. 14-29:

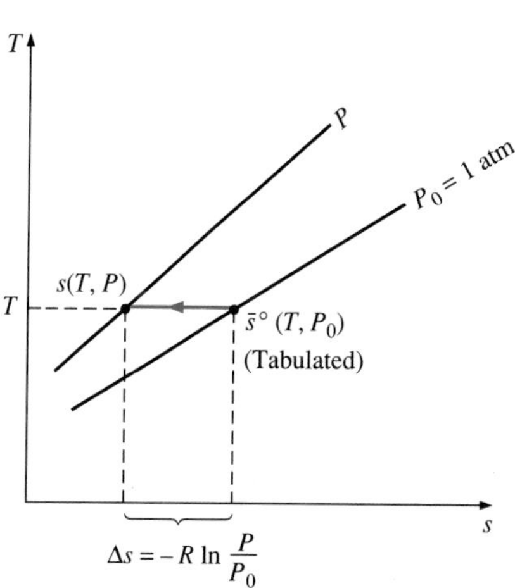

FIGURE 14-29

At a specified temperature, the absolute entropy of an ideal gas at pressures other than $P_0 = 1$ atm can be determined by subtracting $R \ln (P/P_0)$ from the tabulated value at 1 atm.

$$\bar{s}(T, P) = \bar{s}°(T, P_0) - R_u \ln \frac{P}{P_0} \qquad [\text{kJ/(kmol} \cdot \text{K)}] \qquad (14\text{-}21)$$

For the component i of an ideal-gas mixture, this relation can be written as

$$\bar{s}_i(T, P_i) = \bar{s}_i^\circ(T, P_0) - R_u \ln \frac{y_i P_m}{P_0} \qquad [\text{kJ/(kmol} \cdot \text{K)}] \qquad (14\text{-}22)$$

where $P_0 = 1$ atm, P_i is the partial pressure, y_i is the mole fraction of the component, and P_m is the total pressure of the mixture.

If a gas mixture is at a relatively high pressure or low temperature, the deviation from the ideal-gas behavior should be accounted for by incorporating more accurate equations of state or the generalized entropy charts.

14-7 ■ SECOND-LAW ANALYSIS OF REACTING SYSTEMS

Once the total entropy change or the entropy generation is evaluated, the **irreversibility** I or the **exergy destroyed** $X_{\text{destroyed}}$ associated with a chemical reaction can be determined from

$$I = X_{\text{destroyed}} = T_0 S_{\text{gen}} \qquad (\text{kJ}) \qquad (14\text{-}23)$$

where T_0 is the absolute temperature of the surroundings.

When analyzing reacting systems, we are more concerned with the changes in the exergy of reacting systems than with the values of exergy at various states (Fig. 14-30). You will recall from Chap. 7 that the **reversible work** W_{rev} represents the maximum work that can be done during a process. In the absence of any changes in kinetic and potential energies, the reversible work relation for a steady-flow combustion process that involves heat transfer only with the surroundings at T_0 can be obtained by replacing the enthalpy terms in Eqs. 7-23 and 7-52 by $\bar{h}_f^\circ + \bar{h} - \bar{h}^\circ$, yielding

$$W_{\text{rev}} = \sum N_r(\bar{h}_f^\circ + \bar{h} - \bar{h}^\circ - T_0\bar{s})_r - \sum N_p(\bar{h}_f^\circ + \bar{h} - \bar{h}^\circ - T_0\bar{s})_p \quad (14\text{-}24)$$

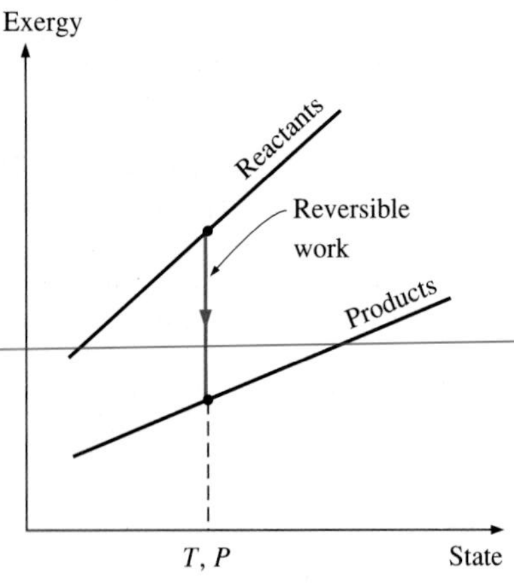

FIGURE 14-30
The difference between the availability of the reactants and of the products during a chemical reaction is the reversible work associated with that reaction.

An interesting situation arises when both the reactants and the products are at the temperature of the surroundings T_0. In that case, $\bar{h} - T_0\bar{s} = (\bar{h} - T_0\bar{s})_{T_0} = \bar{g}_0$, which is, by definition, the **Gibbs function** of a unit mole of a substance at temperature T_0. The W_{rev} relation in this case can be written as

$$W_{\text{rev}} = \sum N_r \bar{g}_{0,r} - \sum N_p \bar{g}_{0,p} \qquad (14\text{-}25)$$

or

$$W_{\text{rev}} = \sum N_r(\bar{g}_f^\circ + \bar{g}_{T_0} - \bar{g}^\circ)_r - \sum N_p(\bar{g}_f^\circ + \bar{g}_{T_0} - \bar{g}^\circ)_p \quad (14\text{-}26)$$

where $\bar{g}_f^\circ$ is the Gibbs function of formation ($\bar{g}_f^\circ = 0$ for stable elements like N_2 and O_2 at the standard reference state of 25°C and 1 atm, just like the enthalpy of formation) and $\bar{g}_{T_0} - \bar{g}^\circ$ represents the value of the sensible Gibbs function of a substance at temperature T_0 relative to the standard reference state.

For the very special case of $T_{\text{react}} = T_{\text{prod}} = T_0 = 25°C$ (i.e., the reactants, the products, and the surroundings are at 25°C) and the partial pressure $P_i = 1$ atm for each component of the reactants and the products, Eq. 14-26 reduces to

$$W_{\text{rev}} = \sum N_r \bar{g}^\circ_{f,r} - \sum n_p \bar{g}^\circ_{f,p} \qquad \text{(kJ)} \qquad (14\text{-}27)$$

We can conclude from the above equation that the $-\bar{g}^\circ_f$ value (the negative of the Gibbs function of formation at 25°C and 1 atm) of a compound represents the *reversible work* associated with the formation of that compound from its stable elements at 25°C and 1 atm in an environment at 25°C and 1 atm (Fig. 14-31). The $\bar{g}^\circ_f$ values of several substances are listed in Table A-26.

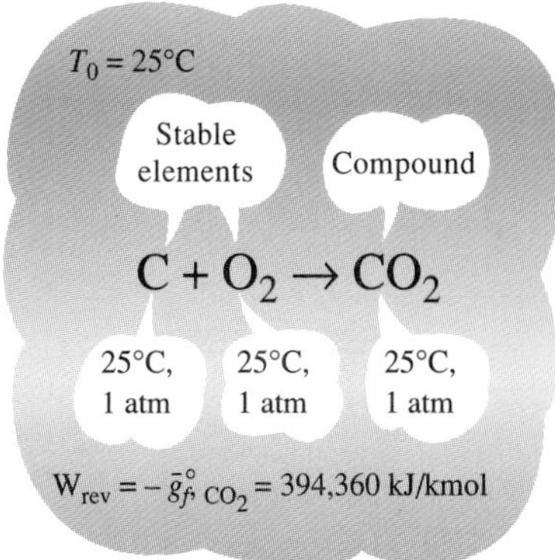

FIGURE 14-31

The negative of the Gibbs function of formation of a compound at 25°C, 1 atm represents the reversible work associated with the formation of that compound from its stable elements at 25°C, 1 atm in an environment that is at 25°C, 1 atm.

EXAMPLE 14-9 Reversible Work Associated with a Combustion Process

One lbmol of carbon at 77°F and 1 atm is burned steadily with 1 lbmol of oxygen at the same state as shown in Fig. 14-32. The CO_2 formed during the process is then brought to 77°F and 1 atm, the conditions of the surroundings. Assuming the combustion is complete, determine the reversible work for this process.

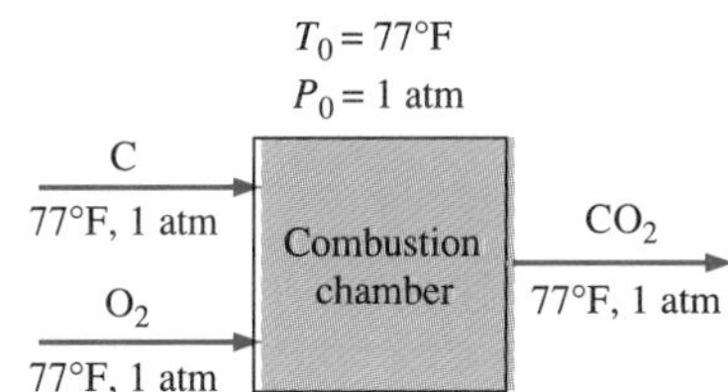

FIGURE 14-32

Schematic for Example 14-9.

Solution Carbon is burned steadily with pure oxygen. The reversible work associated with this process is to be determined. We note that the reactants, products, and the surroundings are at the standard reference state.

Assumptions **1** Combustion is complete. **2** Steady-flow conditions exist during combustion. **3** Air and the combustion gases are ideal gases. **4** Changes in kinetic and potential energies are negligible.

Analysis The combustion equation is

$$C + O_2 \longrightarrow CO_2$$

The C, O_2, and CO_2 are at 77°F and 1 atm, which is the standard reference state and also the state of the surroundings. Therefore, the reversible work in this case is simply the difference between the Gibbs function of formation of the reactants and that of the products (Eq. 14-27):

$$\begin{aligned} W_{\text{rev}} &= \sum N_r \bar{g}^\circ_{f,r} - \sum N_p \bar{g}^\circ_{f,p} \\ &= N_C \bar{g}^{\circ\nearrow 0}_{f,C} + N_{O_2} \bar{g}^{\circ\nearrow 0}_{f,O_2} - N_{CO_2}\bar{g}^\circ_{f,CO_2} = -N_{CO_2}\bar{g}^\circ_{f,CO_2} \\ &= (-1 \text{ lbmol})(-169{,}680 \text{ Btu/lbmol}) \\ &= \mathbf{169{,}680\ Btu} \end{aligned}$$

since the $\bar{g}^\circ_f$ of stable elements at 77°F and 1 atm is zero. Therefore, 169,680 Btu of work could be done as 1 lbmol of C is burned with 1 lbmol of O_2 at 77°F and 1 atm in an environment at the same state. The reversible work in this case represents the exergy of the reactants since the product (the CO_2) is at the state of the surroundings.

We could also determine the reversible work without involving the Gibbs function by using Eq. 14-24:

$$W_{rev} = \sum N_r(\bar{h}_f^\circ + \bar{h} - \bar{h}^\circ - T_0\bar{s})_r - \sum N_p(\bar{h}_f^\circ + \bar{h} - \bar{h}^\circ - T_0\bar{s})_p$$
$$= \sum N_r(\bar{h}_f^\circ - T_0\bar{s})_r - \sum N_p(\bar{h}_f^\circ - T_0\bar{s})_p$$
$$= N_C(\bar{h}_f^\circ - T_0\bar{s}^\circ)_C + N_{O_2}(\bar{h}_f^\circ - T_0\bar{s}^\circ)_{O_2} - N_{CO_2}(\bar{h}_f^\circ - T_0\bar{s}^\circ)_{CO_2}$$

Substituting the enthalpy of formation and absolute entropy values from Table A-26E, we obtain

$$\begin{aligned} W_{rev} = {} & (1\text{ lbmol C})\{0 - (537\text{ R})[1.36\text{ Btu/(lbmol}\cdot\text{R)}]\} \\ & + (1\text{ lbmol O}_2)\{0 - (537\text{ R})[49.00\text{ Btu/(lbmol/R)}]\} \\ & - (1\text{ lbmol CO}_2)\{-169{,}300\text{ Btu/lbmol} - (537\text{ R})[51.07\text{ Btu/(lbmol}\cdot\text{R)}]\} \\ = {} & \mathbf{169{,}681\ Btu} \end{aligned}$$

which is practically identical to the result obtained before.

EXAMPLE 14-10 Second-Law Analysis of Adiabatic Combustion

Methane (CH_4) gas enters a steady-flow adiabatic combustion chamber at 25°C and 1 atm. It is burned with 50 percent excess air that also enters at 25°C and 1 atm, as shown in Fig. 14-33. Assuming complete combustion, determine (*a*) the temperature of the products, (*b*) the entropy generation, and (*c*) the reversible work and exergy destruction. Assume that T_0 = 298 K and the products leave the combustion chamber at 1 atm pressure.

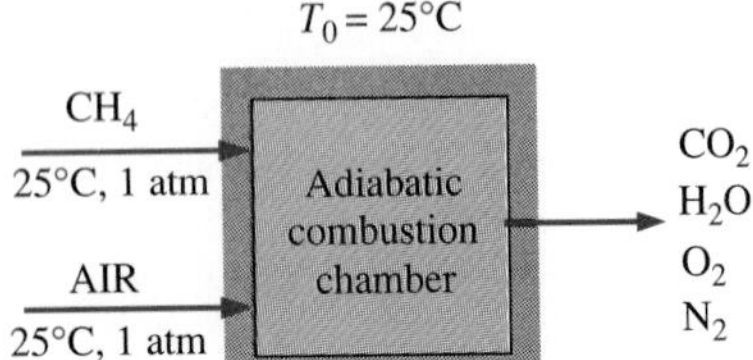

FIGURE 14-33
Schematic for Example 14-10.

Solution Methane is burned with excess air in a steady-flow combustion chamber. The product temperature, entropy generated, reversible work, and exergy destroyed are to be determined.

Assumptions **1** Steady-flow conditions exist during combustion. **2** Air and the combustion gases are ideal gases. **3** Changes in kinetic and potential energies are negligible. **4** The combustion chamber is adiabatic and thus there is not heat transfer. **5** Combustion is complete.

Analysis (*a*) The balanced equation for the complete combustion process with 50 percent excess air is

$$CH_4(g) + 3(O_2 + 3.76N_2) \longrightarrow CO_2 + 2H_2O + O_2 + 11.28N_2$$

Under steady-flow conditions, the adiabatic flame temperature is determined from $H_{prod} = H_{react}$, which reduces to

$$\sum N_p(\bar{h}_f^\circ + \bar{h} - \bar{h}^\circ)_p = \sum N_r\bar{h}_{f,r}^\circ = (N\bar{h}_f^\circ)_{CH_4}$$

since all the reactants are at the standard reference state and $\bar{h}_f^\circ = 0$ for O_2 and N_2. Assuming ideal-gas behavior for air and for the products, the $\bar{h}_f^\circ$ and h values of various components at 298 K can be listed as

Substance	$\bar{h}_f^\circ$ kJ/kmol	$\bar{h}_{298\text{ K}}$ kJ/kmol
$CH_4(g)$	−74,850	—
O_2	0	8682
N_2	0	8669
$H_2O(g)$	−241,820	9904
CO_2	−393,520	9364

Substituting, we have

$$\begin{aligned}
&(1 \text{ kmol } CO_2)[(-393{,}520 + \bar{h}_{CO_2} - 9364) \text{ kJ/kmol } CO_2] \\
&+ (2 \text{ kmol } H_2O)[(-241{,}820 + \bar{h}_{H_2O} - 9904) \text{ kJ/kmol } H_2O] \\
&+ (11.28 \text{ kmol } N_2)[(0 + \bar{h}_{N_2} - 8669) \text{ kJ/kmol } N_2] \\
&+ (1 \text{ kmol } O_2)[(0 + \bar{h}_{O_2} - 8682) \text{ kJ/kmol } O_2] \\
= &(1 \text{ kmol } CH_4)(-74{,}850 \text{ kJ/kmol } CH_4)
\end{aligned}$$

which yields

$$\bar{h}_{CO_2} + 2\bar{h}_{H_2O} + \bar{h}_{O_2} + 11.28\bar{h}_{N_2} = 937{,}950 \text{ kJ}$$

By trial and error, the temperature of the products is found to be

$$T_{prod} = \mathbf{1789.0}$$

(*b*) The entropy generation during this process is determined from Eq. 14-18:

$$S_{gen} = \Delta S_{sys} + \Delta S_{surr} + S_{gen} = \Delta S_{sys}$$

But $(S_{surr} = S_{out}) = 0$ since the process is adiabatic. Thus,

$$S_{gen} = \Delta S_{sys} = S_{prod} - S_{react} = \sum N_p \bar{s}_p - \sum N_r \bar{s}_r$$

The CH_4 is at 25°C and 1 atm, and thus its absolute entropy is $\bar{s}_{CH_4} = 186.16$ kJ/(kmol · K) (Table A-26). The entropy values listed in the ideal-gas tables are for 1 atm pressure. Both the air and the product gases are at a total pressure of 1 atm, but the entropies are to be calculated at the partial pressure of the components, which is equal to $P_i = y_i P_{total}$, where y_i is the mole fraction of component *i*. From Eq. 14-22:

$$S_i = N_i \bar{s}_i(T, P_i) = N_i[\bar{s}_i^\circ(T, P_0) - R_u \ln y_i P_m]$$

The entropy calculations can be represented in tabular form as follows:

	N_i	y_i	$\bar{s}_i^\circ$**(*T*, 1 atm)**	$-R_u$ **ln** $y_i P_m$	$N_i \bar{s}_i$
CH_4	1	1.00	186.16	—	186.16
O_2	3	0.21	205.04	12.98	654.06
N_2	11.28	0.79	191.61	1.96	2183.47
					$S_{react} = 3023.69$
CO_2	1	0.0654	302.517	22.674	325.19
H_2O	2	0.1309	258.957	16.905	551.72
O_2	1	0.0654	264.471	22.674	287.15
N_2	11.28	0.7382	247.977	2.524	2825.65
					$S_{prod} = 3989.71$

Thus,

$$\begin{aligned}
S_{gen} &= S_{prod} - S_{react} = (3989.71 - 3023.69) \text{ kJ/(kmol} \cdot \text{K)}CH_4 \\
&= \mathbf{966.02\ kJ/(kmol \cdot K)CH_4}
\end{aligned}$$

(*c*) The exergy destruction or irreversibility associated with this process is determined from Eq. 14-23,

$$\begin{aligned}
X_{destroyed} = T_0 S_{gen} &= (298 \text{ K})[966.02 \text{ kJ/(kmol} \cdot \text{K)}CH_4] \\
&= \mathbf{287{,}874\ kJ/kmol\ CH_4}
\end{aligned}$$

That is, 287,874 kJ of work potential is wasted during this combustion process for each kmol of methane burned. This example shows that even complete combustion processes are highly irreversible.

This process involves no actual work. Therefore, the reversible work and exergy destroyed are identical:

$$W_{rev} = \mathbf{287{,}874\ kJ/kmol\ CH_4}$$

That is, 287,874 kJ of work could be done during this process, but is not. Instead, the entire work potential is wasted.

EXAMPLE 14-11 Second-Law Analysis of Isothermal Combustion

Methane (CH_4) gas enters a steady-flow combustion chamber at 25°C and 1 atm and is burned with 50 percent excess air, which also enters at 25°C and 1 atm, as shown in Fig. 14-34. After combustion, the products are allowed to cool to 25°C. Assuming complete combustion, determine (*a*) the heat transfer per kmol of CH_4, (*b*) the entropy generation, and (*c*) the reversible work and exergy destruction. Assume that T_0 = 298 K and the products leave the combustion chamber at 1 atm pressure.

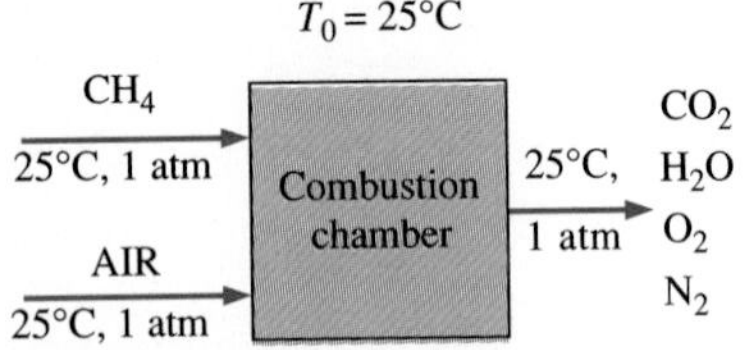

FIGURE 14-34
Schematic for example 14-11.

Solution This is the same combustion process we discussed in Example 14-10, except that the combustion products are brought to the state of the surroundings by transferring heat from them. Thus the combustion equation remains the same:

$$CH_4(g) + 3(O_2 + 3.76N_2) \longrightarrow CO_2 + 2H_2O + O_2 + 11.28N_2$$

At 25°C, part of the water will condense. The amount of water vapor that remains in the products is determined from (see Example 14-3)

$$\frac{N_v}{N_{gas}} = \frac{P_v}{P_{total}} = \frac{3.169\ \text{kPa}}{101.325\ \text{kPa}} = 0.03128$$

and $$N_v = \left(\frac{P_v}{P_{total}}\right) N_{gas} = (0.03128)(13.28 + N_v) \longrightarrow N_v = 0.43\ \text{kmol}$$

Therefore, 1.57 kmol of the H_2O formed will be in the liquid form, which will be removed at 25°C and 1 atm. When one is evaluating the partial pressures of the components in the product gases, the only water molecules that need to be considered are those that are in the vapor phase. As before, all the gaseous reactants and products will be treated as ideal gases.

Heat transfer during this steady-flow combustion process is determined from the steady-flow energy balance $E_{out} = E_{in}$ on the combustion chamber,

$$Q_{out} + \sum N_p \bar{h}^\circ_{f,p} = \sum N_r \bar{h}^\circ_{f,r}$$

since all the reactants and components are at the standard reference of 25°C and the enthalpy of ideal gases depends on temperature only. Solving for Q_{out} and substituting the $\bar{h}^\circ_f$ values, we have

$$\begin{aligned} Q_{out} &= (1 \text{ kmol CH}_4)(-74{,}850 \text{ kJ/kmol CH}_4 \\ &\quad - (1 \text{ kmol CO}_2)(-393{,}520 \text{ kJ/kmol CO}_2) \\ &\quad - [0.43 \text{ kmol H}_2\text{O}(g)][-241{,}820 \text{ kJ/kmol H}_2\text{O}(g)] \\ &\quad - [1.57 \text{ kmol H}_2\text{O}(\ell)][-285.830 \text{ kJ/kmol H}_2\text{O}(\ell)] \\ &= \mathbf{871{,}406 \text{ kJ/kmol CH}_4} \end{aligned}$$

(*b*) The entropy of the reactants was evaluated in Example 14-10 and was determined to be $S_{react} = 3023.69$ kJ/(kmol · K) CH_4. By following a similar approach, the entropy of the products is determined to be

	N_i	y_i	$\bar{s}_i^\circ$**(T, 1 atm)**	$-R_u$ **ln** $y_i P_m$	$N_i \bar{s}_i$
$H_2O(\ell)$	1.57	1.0000	69.92	—	109.77
H_2O	0.43	0.0314	188.83	28.77	93.57
CO_2	1	0.0729	213.80	21.77	235.57
O_2	1	0.0729	205.04	21.77	226.81
N_2	11.28	0.8228	191.61	1.62	2179.63
					$S_{prod} = 2845.35$

Then the total entropy generation during this process is determined from an entropy balance applied on an *extended system* that includes the immediate surroundings of the combustion chamber

$$\begin{aligned} S_{gen} &= S_{prod} - S_{react} + \frac{Q_{out}}{T_{surr}} \\ &= (2845.35 - 3023.69) \text{ kJ/kmol} + \frac{871{,}406 \text{ kJ/kmol}}{298 \text{ K}} \\ &= \mathbf{2745.84 \text{ kJ/(kmol} \cdot \text{K) CH}_4} \end{aligned}$$

(*c*) The exergy destruction and reversible work associated with this process are determined from

$$\begin{aligned} X_{destroyed} = T_0 S_{gen} &= (298 \text{ K})[2745.8 \text{ kJ/(kmol} \cdot \text{K) CH}_4] \\ &= \mathbf{818{,}260 \text{ kJ/kmol CH}_4} \end{aligned}$$

and $$W_{rev} = X_{destroyed} = \mathbf{818{,}260 \text{ kJ/kmol CH}_4}$$

since this process involves no actual work. Therefore, 818,260 kJ of work could be done during this process, but is not. Instead, the entire work potential is wasted. The reversible work in this case represents the exergy of the reactants before the reaction starts since the products are in equilibrium with the surroundings, that is, they are at the dead state.

Discussion Note that, for simplicity, we calculated the entropy of the product gases before they actually entered the atmosphere and mixed with the atmospheric gases. A more complete analysis would consider the composition of the atmosphere and the mixing of the product gases with the gases in the atmosphere, forming a homogeneous mixture. There will be additional entropy generation during this mixing process, and thus additional wasted work potential.

There is an important observation to be made from the preceding two examples. Fuels like methane are commonly burned to provide thermal energy at high temperatures for use in heat engines. However, a comparison of the reversible works obtained in the two examples reveals that the exergy of the reactants (818,260 kJ/kmol CH_4) decreases by 287,874 kJ/kmol as a result of the irreversible adiabatic combustion process alone. That is, the exergy of the hot combustion gases at the end of the adiabatic combustion process is $818{,}260 - 287{,}874 = 530{,}386$ kJ/kmol CH_4. In other words, the work potential of the hot combustion gases is about 65 percent of the work potential of the reactants. It seems that when methane is burned, 35 percent of the work potential is lost before we even start using the thermal energy (Fig. 14-35).

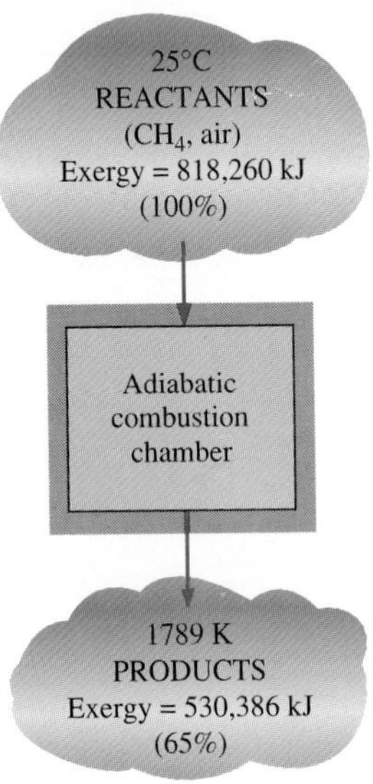

FIGURE 14-35
The availability of methane decreases by 35 percent as a result of irreversible combustion process.

Thus the second law of thermodynamics suggests that there should be a better way of converting the chemical energy to work. The better way is, of course, the less irreversible way, the best being the reversible case. In chemical reactions, the irreversibility is due to uncontrolled electron exchange between the reacting components. The electron exchange can be controlled by replacing the combustion chamber by electrolytic cells, like car batteries. In the electrolytic cells, the electrons are exchanged through conductor wires connected to a load, and the chemical energy is directly converted to electric energy. The energy conversion devices that work on this principle are called **fuel cells.**

The operation of a hydrogen–oxygen fuel cell is illustrated in Fig. 14-36. Hydrogen is ionized at the surface of the anode, and hydrogen ions flow through the electrolyte to the cathode. There is a potential difference between the anode and the cathode, and free electrons flow from the anode to the cathode through an external circuit (such as a generator). Hydrogen ions combine with oxygen and the free electrons at the surface of the cathode, forming water. In steady operation, hydrogen and oxygen continuously enter the fuel cell as reactants, and water leaves as the product.

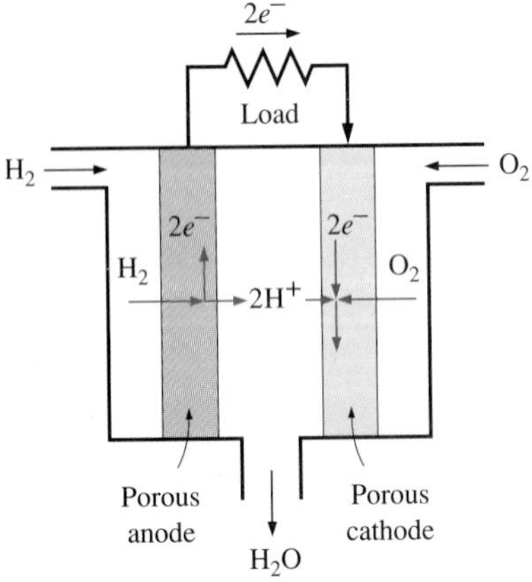

FIGURE 14-36
The operation of a hydrogen–oxygen fuel cell.

Fuel cells are not heat engines, and thus their efficiencies are not limited by the Carnot efficiency. They convert chemical energy to electric energy essentially in an isothermal manner. Despite the irreversible effects such as internal resistance to electron flow, fuel cells have a great potential for higher conversion efficiencies, and they have been used successfully in some small-scale applications. But more research and development are needed before large-scale fuel-cell power plants can be realized.

14-8 ■ SUMMARY

Any material that can be burned to release energy is called a *fuel,* and a chemical reaction during which a fuel is oxidized and a large quantity of energy is released is called *combustion*. The oxidizer most often used in combustion processes is air. The dry air can be approximated as 21 percent oxygen and 79 percent nitrogen by mole numbers. Therefore,

$$1 \text{ kmol } O_2 + 3.76 \text{ kmol } N_2 = 4.76 \text{ kmol air}$$

At ordinary combustion temperatures, nitrogen behaves as an inert gas and does not react with other chemical elements.

During a combustion process, the components that exist before the reaction are called *reactants* and the components that exist after the reaction are called *products*. Chemical equations are balanced on the basis of the *conservation of mass principle,* which states that the total mass of each element is conserved during a chemical reaction. The ratio of the mass of air to the mass of fuel during a combustion process is called the *air–fuel ratio* AF:

$$\text{AF} = \frac{m_{\text{air}}}{m_{\text{fuel}}}$$

where $m_{\text{air}} = (NM)_{\text{air}}$ and $m_{\text{fuel}} = \sum (N_iM_i)_{\text{fuel}}$.

A combustion process is *complete* if all the carbon in the fuel burns to CO_2, all the hydrogen burns to H_2O, and all the sulfur (if any) burns to SO_2. The minimum amount of air needed for the complete combustion of a fuel is called the *stoichiometric* or *theoretical air.* The theoretical air is also referred to as the chemically correct amount of air or 100 percent theoretical air. The ideal combustion process during which a fuel is burned completely with theoretical air is called the *stoichiometric* or *theoretical combustion* of that fuel. The air in excess of the stoichiometric amount is called the *excess air.* The amount of excess air is usually expressed in terms of the stoichiometric air as *percent excess air* or *percent theoretical air.*

During a chemical reaction, some chemical bonds are broken and others are formed. Therefore, a process that involves chemical reactions will involve changes in chemical energies. Because of the changed composition, it is necessary to have a *standard reference state* for all substances, which is chosen to be 25°C (77°F) and 1 atm.

The difference between the enthalpy of the products at a specified state and the enthalpy of the reactants at the same state for a complete reaction is called the *enthalpy of reaction* h_R. For combustion processes, the enthalpy of reaction is usually referred to as the *enthalpy of combustion* h_C, which represents the amount of heat released during a steady-flow combustion process when 1 kmol (or 1 kg) of fuel is burned completely at a specified temperature and pressure. The enthalpy of a substance at a specified state due to its chemical composition is called the *enthalpy of formation* $\overline{h}_f$. The enthalpy of formation of all stable elements is assigned a value of zero at the standard reference state of 25°C and 1 atm. The *heating value* of a fuel is defined as the amount of heat released when a fuel is burned completely in a steady-flow process and the products are returned to the state of the reactants. The heating value of a fuel is equal to the absolute value of the enthalpy of combustion of the fuel,

$$\text{Heating value} = |h_C| \qquad \text{(kJ/kg fuel)}$$

The heating value is called the *higher heating value* (HHV) when the H_2O in the products is in the liquid form, and it is called the *lower heating value* (LHV) when the H_2O in the products is in the vapor form. The two heating values are related by

$$\text{HHV} = \text{LHV} + (N\bar{h}_{fg})_{H_2O} \qquad \text{(kJ/kg fuel)}$$

where N is the number of moles of H_2O in the products and $\bar{h}_{fg}$ is the enthalpy of vaporization of water at 25°C.

Taking heat transfer *to* the system and work done *by* the system to be positive quantities, the conservation of energy relation for chemically reacting steady-flow systems can be expressed per unit mole of fuel as

$$Q - W = \sum N_p(\bar{h}_f^\circ + \bar{h} - \bar{h}^\circ)_p - \sum N_r(\bar{h}_f^\circ + \bar{h} - \bar{h}^\circ)_r$$

where the superscript ° represents properties at the standard reference state of 25°C and 1 atm. For a closed system, it becomes

$$Q - W = \sum N_p(\bar{h}_f^\circ + \bar{h} - \bar{h}^\circ - P\bar{v})_p - \sum N_r(\bar{h}_f^\circ + \bar{h} - \bar{h}^\circ - P\bar{v})_r$$

The $P\bar{v}$ terms are negligible for solids and liquids and can be replaced by R_uT for gases that behave as ideal gases.

In the absence of any heat loss to the surroundings ($Q = 0$), the temperature of the products will reach a maximum, which is called the *adiabatic flame temperature* of the reaction. The adiabatic flame temperature of a steady-flow combustion process is determined from $H_{\text{prod}} = H_{\text{react}}$ or

$$\sum N_p(\bar{h}_f^\circ + \bar{h} - \bar{h}^\circ)_p = \sum N_r(\bar{h}_f^\circ + \bar{h} - \bar{h}^\circ)_r$$

The entropy balance relations developed in Chap. 6 are equally applicable to both reacting and nonreacting systems provided that the entropies of individual constituents are evaluated properly using a common basis.

Taking the positive direction of heat transfer to be *to* the system, the entropy balance relation can be expressed for a *closed system* or *steady-flow combustion chamber* as

$$\sum \frac{Q_k}{T_k} + S_{\text{gen}} = S_{\text{prod}} - S_{\text{react}} \qquad \text{(kJ/K)}$$

For an *adiabatic process* it reduces to

$$S_{\text{gen, adiabatic}} = S_{\text{prod}} - S_{\text{react}} \geq 0$$

The *third law of thermodynamics* states that the entropy of a pure crystalline substance at absolute zero temperature is zero. The third law provides a common base for the entropy of all substances, and the entropy values relative to this base are called the *absolute entropy*. The ideal-gas tables list the absolute entropy values over a wide range of temperatures but at a fixed pressure of $P_0 = 1$ atm. Absolute entropy values at other pressures P for any temperature T are determined from

$$\bar{s}(T, P) = \bar{s}^\circ(T, P_0) - R_u \ln \frac{P}{P_0} \qquad \text{(kJ/(kmol}\cdot\text{K)]}$$

For component i of an ideal-gas mixture, this relation can be written as

$$\bar{s}_i(T, P_i) = \bar{s}_i^\circ(T, P_0) - R_u \ln \frac{y_iP_m}{P_0} \qquad \text{[kJ/(kmol}\cdot\text{K)]}$$

where P_i is the partial pressure, y_i is the mole fraction of the component, and P_m is the total pressure of the mixture in atmospheres.

The *exergy destruction* or *irreversibility* and the *reversible work* associated with a chemical reaction are determined from

$$I = X_{\text{destroyed}} = W_{\text{rev}} - W_{\text{act}} = T_0 S_{\text{gen}} \qquad \text{(kJ)}$$

and

$$W_{\text{rev}} = \sum N_r(\bar{h}_f^\circ + \bar{h} - \bar{h}^\circ - T_0\bar{s})_r - \sum N_p(\bar{h}_f^\circ + \bar{h} - \bar{h}^\circ - T_0\bar{s})_p \qquad \text{(kJ)}$$

When both the reactants and the products are at the temperature of the surroundings T_0, the reversible work can be expressed in terms of the Gibbs functions as

$$W_{\text{rev}} = \sum N_r(\bar{g}_f^\circ + \bar{g}_{T_0} - \bar{g}^\circ)_r - \sum N_p(\bar{g}_f^\circ + \bar{g}_{T_0} - \bar{g}^\circ)_p \qquad \text{(kJ)}$$

REFERENCES AND SUGGESTED READING

1. S. W. Angrist. *Direct Energy Conversion*. 4th ed. Boston: Allyn and Bacon, 1982.

2. W. Z. Black and J. G. Hartley. *Thermodynamics*. New York: Harper & Row, 1985.

3. I. Glassman. *Combustion*. New York: Academic Press, 1977.

4. J. B. Jones and G. A. Hawkins. *Engineering Thermodynamics*. 2nd ed. New York: John Wiley & Sons, 1986.

5. R. Strehlow. *Fundamentals of Combustion*. Scranton, PA: International Textbook Co., 1968.

6. G. J. Van Wylen and R. E. Sonntag. *Fundamentals of Classical Thermodynamics*. 3rd ed. New York: John Wiley & Sons, 1986.

7. K. Wark. *Thermodynamics*. 5th ed. New York: McGraw-Hill, 1988.

PROBLEMS*

Fuels and Combustion

14-1C Is it possible to invent a car that runs on H_2O instead of gasoline?

14-2C What are the approximate chemical compositions of gasoline, diesel fuel, and natural gas?

14-3C How does the presence of N_2 in the air affect the outcome of a combustion process?

14-4C How does the presence of moisture in the air affect the outcome of a combustion process?

---

*Students are encouraged to answer *all* the concept "C" questions.

14-5C What does the dew-point temperature of the product gases represent? How is it determined?

14-6C Will a fuel start burning when it is brought into intimate contact with oxygen?

14-7C Write three different statements that express the conservation of mass principle for a chemical reaction.

14-8C Is the number of atoms of each element conserved during a chemical reaction? How about the total number of moles?

14-9C What is the air–fuel ratio? How is it related to the fuel–air ratio?

14-10C Is the air–fuel ratio expressed on a mole basis identical to the air–fuel ratio expressed on a mass basis?

Theoretical and Actual Combustion Processes

14-11C What are the causes of incomplete combustion?

14-12C Which is more likely to be found in the products of an incomplete combustion of a hydrocarbon fuel, CO or OH? Why?

14-13C What does 100 percent theoretical air represent?

14-14C Are complete combustion and theoretical combustion identical? If not, how do they differ?

14-15C Consider a fuel that is burned with (*a*) 130 percent theoretical air and (*b*) 70 percent excess air. In which case is the fuel burned with more air?

14-16C What is the operation principle of the Orsat gas analyzer?

14-17 Methane (CH_4) is burned with the stoichiometric amount of air during a combustion process. Assuming complete combustion, determine the air–fuel and fuel–air ratios.

14-18 Propane (C_3H_8) is burned with 50 percent excess air during a combustion process. Assuming complete combustion, determine the air–fuel ratio. *Answer:* 23.5 kg air/kg fuel

14-19 Acetylene (C_2H_2) is burned with the stoichiometric amount of air during a combustion process. Assuming complete combustion, determine the air–fuel ratio on a mass and on a mole basis.

14-20 One kmol of ethane (C_2H_6) is burned with an unknown amount of air during a combustion process. An analysis of the combustion products reveals that the combustion is complete, and there are 2 kmol of free O_2 in the products. Determine (*a*) the air–fuel ratio and (*b*) the percentage of theoretical air used during this process.

14-21E Ethylene (C_2H_4) is burned with 200 percent theoretical air during a combustion process. Assuming complete combustion and a total pressure of 14.5 psia, determine (*a*) the air–fuel ratio and (*b*) the dew-point temperature of the products. *Answers:* (*a*) 29.6 lbm air/lbm fuel, (*b*) 100.9°F

14-22 Propylene (C_3H_6) is burned with 50 percent excess air during a combustion process. Assuming complete combustion and a total pressure of 90 kPa, determine (*a*) the air–fuel ratio and (*b*) the temperature at which the water vapor in the products will start condensing.

14-23 Octane (C_8H_{18}) is burned with 250 percent theoretical air, which enters the combustion chamber at 25°C. Assuming complete combustion and a total pressure of 1 atm, determine (*a*) the air–fuel ratio and (*b*) the dew-point temperature of the products.

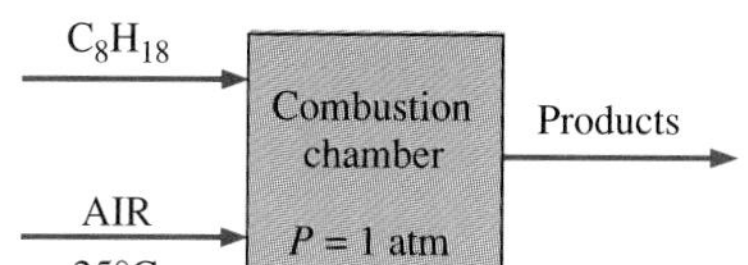

FIGURE P14-23

14-24 Gasoline (assumed C_8H_{18}) is burned steadily with air in a jet engine. If the air–fuel ratio is 24 kg air/kg fuel, determine the percentage of theoretical air used during this process.

14-25 In a combustion chamber, ethane (C_2H_6) is burned at a rate of 6 kg/h with air that enters the combustion chamber at a rate of 132 kg/h. Determine the percentage of excess air used during this process. *Answer:* 37 percent

14-26 One kilogram of butane (C_4H_{10}) is burned with 25 kg of air that is at 35°C and 100 kPa. Assuming that the combustion is complete and the pressure of the products is 100 kPa, determine (*a*) the percentage of theoretical air used and (*b*) the dew-point temperature of the products.

14-27E One lbm of butane (C_4H_{10}) is burned with 25 lbm of air that is at 90°F and 14.7 psia. Assuming that the combustion is complete and the pressure of the products is 14.7 psia, determine (*a*) the percentage of theoretical air used and (*b*) the dew-point temperature of the products.
Answers: (*a*) 161 percent, (*b*) 111.4°F

14-28 A certain natural gas has the following volumetric analysis: 65 percent CH_4, 8 percent H_2, 18 percent N_2, 3 percent O_2, and 6 percent CO_2. This gas is now burned completely with the stoichiometric amount of dry air. What is the air–fuel ratio for this combustion process?

14-29 Repeat Prob. 14-28 by replacing the dry air by moist air that enters the combustion chamber at 25°C, 1 atm, and 85 percent relative humidity.

14-30 A gaseous fuel with a volumetric analysis of 60 percent CH_4, 30 percent H_2, and 10 percent N_2 is burned to completion with 130 percent theoretical air. Determine (*a*) the air–fuel ratio and (*b*) the fraction of water vapor that would condense if the product gases were cooled to 20°C at 1 atm. *Answers:* (*a*) 18.6 kg air/kg fuel, (*b*) 88 percent

14-31 A certain coal has the following analysis on a mass basis: 82 percent C, 5 percent H_2O, 2 percent H_2, 1 percent O_2, and 10 percent ash. The coal is burned with 20 percent excess air. Determine the air–fuel ratio.
Answer: 12.3 kg air/kg coal

14-32 Octane (C_8H_{18}) is burned with dry air. The volumetric analysis of the products on a dry basis is 9.21 percent CO_2, 0.61 percent CO, 7.06 percent O_2, and 83.12 percent N_2. Determine (*a*) the air–fuel ratio and (*b*) the percentage of theoretical air used.

14-33 Carbon is burned with dry air. The volumetric analysis of the products is 10.06 percent CO_2, 0.42 percent CO, 10.69 percent O_2, and 78.83 percent N_2. Determine (*a*) the air–fuel ratio and (*b*) the percentage of theoretical air used.

14-34 Methane (CH_4) is burned with dry air. The volumetric analysis of the products on a dry basis is 5.20 percent CO_2, 0.33 percent CO, 11.24 percent O_2, and 83.23 percent N_2. Determine (*a*) the air–fuel ratio and (*b*) the percentage of theoretical air used.
Answers: (*a*) 34.5 kg air/kg fuel, (*b*) 200 percent

Enthalpy of Formation and Enthalpy of Combustion

14-35C Do nonreacting systems contain any chemical energy? Do they involve any changes in chemical energy during a process?

14-36C What is enthalpy of combustion? How does it differ from the enthalpy of reaction?

14-37C What is enthalpy of formation? How does it differ from the enthalpy of combustion?

14-38C What are the higher and the lower heating values of a fuel? How do they differ? How is the heating value of a fuel related to the enthalpy of combustion of that fuel?

14-39C When are the enthalpy of formation and the enthalpy of combustion identical?

14-40C Does the enthalpy of formation of a substance change with temperature?

14-41C Is it possible to analyze a chemical reaction by using property tables that are prepared using different reference states? How?

14-42C The $\bar{h}_f^\circ$ of N_2 is listed as zero. Does this mean that N_2 contains no chemical energy at the standard reference state?

14-43C Which contains more chemical energy, 1 kmol of H_2 or 1 kmol of H_2O?

14-44 Determine the enthalpy of combustion of methane (CH_4) at 25°C and 1 atm, using the enthalpy of formation data from Table A-26. Assume that the water in the products is in the liquid form. Compare your result to the value listed in Table A-27. *Answer:* −890,330 kJ/kmol

14-45 Repeat Prob. 14-44 for gaseous ethane (C_2H_6).

14-46 Repeat Prob. 14-44 for liquid octane (C_3H_{18}).

First-Law Analysis of Reacting Systems

14-47C Are the energy balance relations different for reacting systems and nonreacting systems?

14-48C Derive an energy balance relation for a reacting closed system undergoing a quasi-equilibrium constant pressure expansion or compression process.

14-49C Consider a complete combustion process during which both the reactants and the products are maintained at the same state. Combustion is achieved with (*a*) 100 percent theoretical air, (*b*) 200 percent theoretical air, and (*c*) the chemically correct amount of pure oxygen. For which case will the amount of heat transfer be the highest? Explain.

14-50C Consider a complete combustion process during which the reactants enter the combustion chamber at 20°C and the products leave at 500°C. Combustion is achieved with (*a*) 100 percent theoretical air, (*b*) 200 percent theoretical air, and (*c*) the chemically correct amount of pure oxygen. For which case will the amount of heat transfer be the lowest? Explain.

14-51 Methane (CH_4) is burned completely with the stoichiometric amount of air during a steady-flow combustion process. If both the reactants and the products are maintained at 25°C and 1 atm and the water in the products exists in the liquid form, determine the heat transfer from the combustion chamber during this process. What would your answer be if combustion were achieved with 50 percent excess air? *Answer:* 890,330 kJ/kmol

14-52 Hydrogen (H_2) is burned completely with the stoichiometric amount of air during a steady-flow combustion process. If both the reactants and the products are maintained at 25°C and 1 atm and the water in the products exists in the liquid form, determine the heat transfer from the combustion chamber during this process. What would your answer be if combustion were achieved with 80 percent excess air?

14-53 Liquid propane (C_3H_8) enters a combustion chamber at 25°C at a rate of 0.4 kg/min where it is mixed and burned with 150 percent excess air that enters the combustion chamber at 12°C. If the combustion is complete and the exit temperature of the combustion gases is 1200 K, determine (*a*) the mass flow rate of air and (*b*) the rate of heat transfer from the combustion chamber. *Answers:* (*a*) 15.7 kg/min, (*b*) 1732 kJ/min

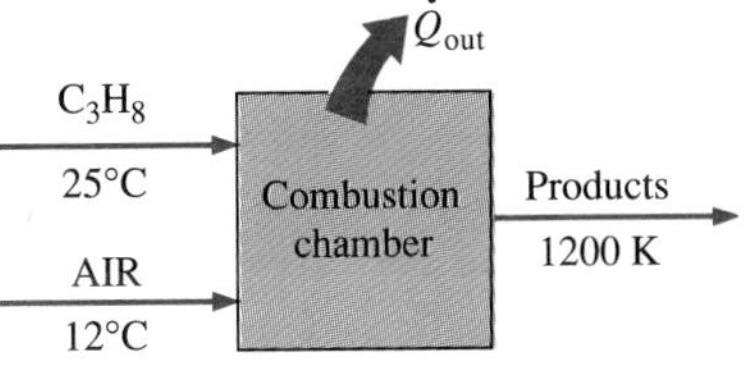

FIGURE P14-53

14-54E Liquid propane (C_3H_8) enters a combustion chamber at 77°F at a rate of 0.3 lbm/min where it is mixed and burned with 150 percent excess air that enters the combustion chamber at 40°F. if the combustion is complete and the exit temperature of the combustion gases is 1800 R, determine (*a*) the mass flow rate of air and (*b*) the rate of heat transfer from the combustion chamber. *Answers:* (*a*) 11.8 lbm/min, (*b*) 1566 Btu/min

14-55 Acetylene gas (C_2H_2) is burned completely with 20 percent excess air during a steady-flow combustion process. The fuel and the air enter the combustion chamber at 25°C, and the products leave at 1500 K. Determine (*a*) the air–fuel ratio and (*b*) the heat transfer for this process.

14-56E Liquid octane (C_3H_{18}) at 77°F is burned completely during a steady-flow combustion process with 180 percent theoretical air that enters the combustion chamber at 77°F. If the products leave at 2500 R, determine (*a*) the air–fuel ratio and (*b*) the heat transfer from the combustion chamber during this process.

14-57 Benzene gas (C_6H_6) at 25°C is burned during a steady-flow combustion process with 95 percent theoretical air that enters the combustion chamber at 25°C. All the hydrogen in the fuel burns to H_2O, but part of the carbon burns to CO. If the products leave at 1000 K, determine (*a*) the mole fraction of the CO in the products and (*b*) the heat transfer from the combustion chamber during this process.
Answers: (*a*) 2.1 percent, (*b*) 2,112,779 kJ/kmol C_6H_6

14-58 Diesel fuel ($C_{12}H_{26}$) at 25°C is burned in a steady-flow combustion chamber with 20 percent excess air that also enters at 25°C. The products leave the combustion chamber at 500 K. Assuming combustion is complete, determine the required mass flow rate of the diesel fuel to supply heat at a rate of 1500 kJ/s. *Answer:* 37.1 g/s

14-59E Diesel fuel ($C_{12}H_{26}$) at 77°F is burned in a steady-flow combustion chamber with 20 percent excess air that also enters at 77°F. The products leave the combustion chamber at 800 R. Assuming combustion is complete, determine the required mass flow rate of the diesel fuel to supply heat at a rate of 1800 Btu/s. *Answer:* 0.1 lbm/s

14-60 Octane gas (C_8H_{18}) at 25°C is burned steadily with 30 percent excess air at 25°C, 1 atm, and 60 percent relative humidity. Assuming combustion is complete and the products leave the combustion chamber at 600 K, determine the heat transfer for this process per unit mass of octane.

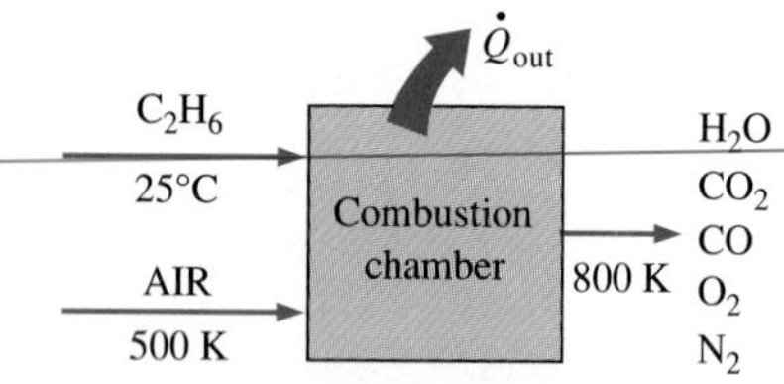

FIGURE P14-61

14-61 Ethane gas (C_2H_6) at 25°C is burned in a steady-flow combustion chamber at a rate of 5 kg/h with the stoichiometric amount of air, which is preheated to 500 K before entering the combustion chamber. An analysis of the combustion gases reveals that all the hydrogen in the fuel burns to H_2O but only 95 percent of the carbon burns to CO_2, the remaining 5 percent forming CO. If the products leave the combustion chamber at 800 K, determine the rate of heat transfer from the combustion chamber.
Answer: 179,420 kJ/h

14-62 A constant-volume tank contains a mixture of 120 g of methane (CH_4) gas and 600 g of O_2 at 25°C and 200 kPa. The contents of the tank are now ignited, and the methane gas burns completely. If the final temperature is 1200 K, determine (*a*) the final pressure in the tank and (*b*) the heat transfer during this process.

14-63 A closed combustion chamber is designed so that it maintains a constant pressure of 150 kPa during a combustion process. The combustion chamber has an initial volume of 0.8 m^3 and contains a stoichiometric mixture of octane (C_8H_{18}) gas and air at 25°C. The mixture is now ignited, and the product gases are observed to be at 1000 K at the end of the combustion process. Assuming complete combustion, and treating both the reactants and the products as ideal gases, determine the heat transfer from the combustion chamber during this process. *Answer:* 2888 kJ

14-64 A constant-volume tank contains a mixture of 1 kmol of benzene (C_6H_6) gas and 30 percent excess air at 25°C and 1 atm. The contents of the tank are now ignited, and all the hydrogen in the fuel burns to H_2O but only 92 percent of the carbon burns to CO_2, the remaining 8 percent forming CO. If the final temperature in the tank is 1000 K, determine the heat transfer from the combustion chamber during this process.

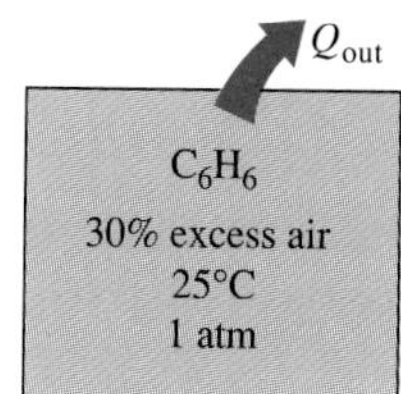

FIGURE P14-64

14-65E A constant-volume tank contains a mixture of 1 lbmol of benzene (C_6H_6) gas and 30 percent excess air at 77°F and 1 atm. The contents of the tank are now ignited, and all the hydrogen in the fuel burns to H_2O but only 29 percent of the carbon burns to CO_2, the remaining 8 percent forming CO. If the final temperature in the tank is 1800 R, determine the heat transfer from the combustion chamber during this process.
Answer: 921,768 Btu

Adiabatic Flame Temperature

14-66C A fuel is completely burned first with the stoichiometric amount of air and then with the stoichiometric amount of pure oxygen. For which case will the adiabatic flame temperature be higher?

14-67C A fuel at 25°C is burned in a well-insulated steady-flow combustion chamber with air that is also at 25°C. Under what conditions will the adiabatic flame temperature of the combustion process be a maximum?

14-68 Hydrogen (H_2) at 7°C is burned with 20 recent excess air that is also at 7°C during an adiabatic steady-flow combustion process. Assuming complete combustion, determine the exit temperature of the product gases.
Answer: 2251.4 K

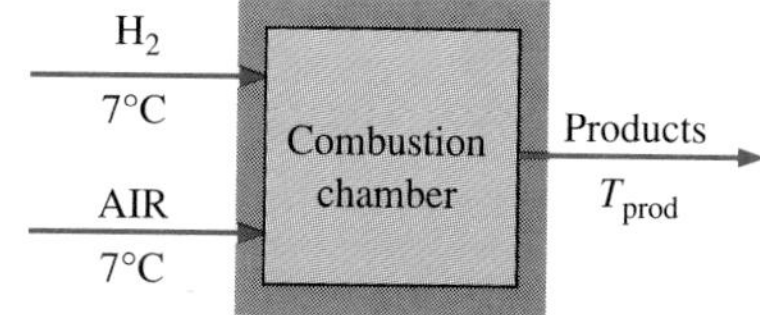

FIGURE P14-68

14-69E Hydrogen (H_2) at 40°F is burned with 20 percent excess air that is also at 40°F during an adiabatic steady-flow combustion process. Assuming complete combustion, find the exit temperature of the product gases.

14-70 Acetylene gas (C_2H_2) at 25°C is burned during a steady-flow combustion process with 30 percent excess air at 27°C. It is observed that 75,000 kJ of heat is being lost from the combustion chamber to the surroundings per kmol of acetylene. Assuming combustion is complete, determine the exit temperature of the product gases. *Answer:* 2303 K

14-71 An adiabatic constant-volume tank contains a mixture of 1 kmol of hydrogen (H_2) gas and the stoichiometric amount of air at 25°C and 1 atm. The contents of the tank are now ignited. Assuming complete combustion, determine the final temperature in the tank.

14-72 Octane gas (C_8H_{18}) at 25°C is burned steadily with 30 percent excess air at 25°C, 1 atm, and 60 percent relative humidity. Assuming combustion is complete and adiabatic, calculate the exit temperature of the product gases.

Entropy Change and Second-Law Analysis of Reacting Systems

14-73C Express the increase of entropy principle for chemically reacting systems.

14-74C What is the importance of the third law of thermodynamics?

14-75C How are the absolute entropy values of ideal gases at pressures different from 1 atm determined?

14-76C Is it a waste of time to calculate the reversible work associated with chemical reactions since most chemical reactions do not involve any work interactions?

14-77C What does the Gibbs function of formation $\bar{g}_f^\circ$ of a compound represent?

14-78 One kmol of H_2 at 25°C and 1 atm is burned steadily with 0.5 kmol of O_2 at the same state. The H_2O formed during the process is then brought to 25°C and 1 atm, the conditions of the surroundings. Assuming combustion is complete, determine the reversible work and irreversibility for this process.

14-79 Ethylene (C_2H_4) gas enters an adiabatic combustion chamber at 25°C and 1 atm and is burned with 20 percent excess air that enters at 25°C and 1 atm. The combustion is complete, and the products leave the combustion chamber at 1 atm pressure. Assuming T_0 = 25°C, determine (*a*) the temperature of the products, (*b*) the entropy generation, and (*c*) the irreversibility.
Answers: (*a*) 2269.6 K, (*b*) 1457.45 kJ/(kmol · K), (*c*) 434,320 kJ/kmol

14-80 Liquid octane (C_8H_{18}) enters a steady-flow combustion chamber at 25°C and 1 atm at a rate of 0.4 kg/min. It is burned with 50 percent excess air that also enters at 25°C and 1 atm. After combustion, the products are allowed to cool to 25°C. Assuming complete combustion and that all the H_2O in the products is in liquid form, determine (*a*) the heat transfer rate from the combustion chamber, (*b*) the entropy generation rate, and (*c*) the reversible work and irreversibility. Assume that T_0 = 298 K and the products leave the combustion chamber at 1 atm pressure.

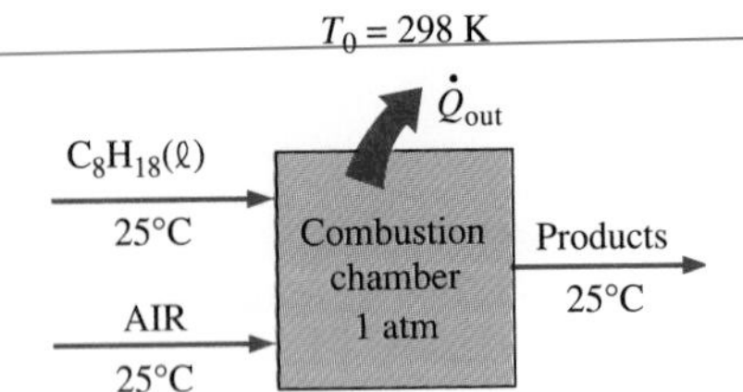

FIGURE P14-80

14-81 Acetylene gas (C_2H_2) is burned completely with 20 percent excess air during a steady-flow combustion process. The fuel and the air enter the combustion chamber separately at 25°C and 1 atm, and heat is being lost from the combustion chamber to the surroundings at 15°C at a rate of 300,000 kJ/kmol C_2H_2. The combustion products leave the combustion chamber at 1 atm pressure. Determine (*a*) the temperature of the products, (*b*) the total entropy change per kmol of C_2H_2, and (*c*) the reversible work and irreversibility during this process.

14-82 A steady-flow combustion chamber is supplied with CO gas at 37°C and 110 kPa at a rate of 0.4 m^3/min and air at 25°C and 110 kPa at a rate of 1.5 kg/min. Heat is transferred to a medium at 800 K, and the combustion products leave the combustion chamber at 900 K. Assuming the combustion is complete and T_0 = 25°C, determine (*a*) the rate of heat transfer from the combustion chamber, (*b*) the reversible work, and (*c*) the rate of exergy destruction.

Answers: (*a*) 3567 kJ/min, (*b*) 1628 kJ/min, (*c*) 1628 kJ/min

14-83E Benzene gas (C_6H_6) at 1 atm and 77°F is burned during a steady-flow combustion process with 95 percent theoretical air that enters the combustion chamber at 77°F and 1 atm. All the hydrogen in the fuel burns to H_2O, but part of the carbon burns to CO. Heat is lost to the surroundings at 77°F, and the products leave the combustion chamber at 1 atm and 1500 R. Determine (*a*) the heat transfer from the combustion chamber and (*b*) the irreversibility.

14-84 Liquid propane (C_3H_8) enters a steady-flow combustion chamber at 25°C and 1 atm at a rate of 0.4 kg/min where it is mixed and burned with 150 percent excess air that enters the combustion chamber at 12°C. If the combustion products leave at 120 K and 1 atm, determine (*a*) the mass flow rate of air, (*b*) the rate of heat transfer from the combustion chamber, and (*c*) the rate of entropy generation during this process. Assume T_0 = 25°C.

Answers: (*a*) 15.7 kg/min, (*b*) 1732 kJ/min, (*c*) 34.2 kJ/(min · K)

Review Problems

14-85 A 1-g sample of a certain fuel is burned in a bomb calorimeter that contains 3 kg of water in the presence of 100 g of air in the reaction chamber. If the water temperature rises by 1.5°C when equilibrium is established, determine the heating value of the fuel, in kJ/kg.

14-86E Hydrogen (H_2) is burned with 100 percent excess air that enters the combustion chamber at 90°F, 14.5 psia, and 60 percent relative humidity. Assuming complete combustion, determine (*a*) the air–fuel ratio and (*b*) the volume flow rate of air required to burn the hydrogen at a rate of 10 lbm/h.

14-87 A gaseous fuel with 80 percent CH_4, 15 percent N_2, and 5 percent O_2 (on a mole basis) is burned to completion with 120 percent theoretical air that enters the combustion chamber at 30°C, 100 kPa, and 60 percent relative humidity. Determine (*a*) the air–fuel ratio and (*b*) the volume flow rate of air required to burn fuel at a rate of 1 kg/min.

14-88 A gaseous fuel with 80 percent CH_4, 15 percent N_2, and 5 percent O_2 (on a mole basis) is burned with dry air that enters the combustion chamber at 25°C and 100 kPa. The volumetric analysis of the products on a dry basis is 3.36 percent CO_2, 0.09 percent CO, 14.91 percent O_2, and 81.64 percent N_2. Determine (*a*) the air–fuel ratio, (*b*) the percent theoretical air used, and (*c*) the volume flow rate of air used to burn fuel at a rate of 1 kg/min.

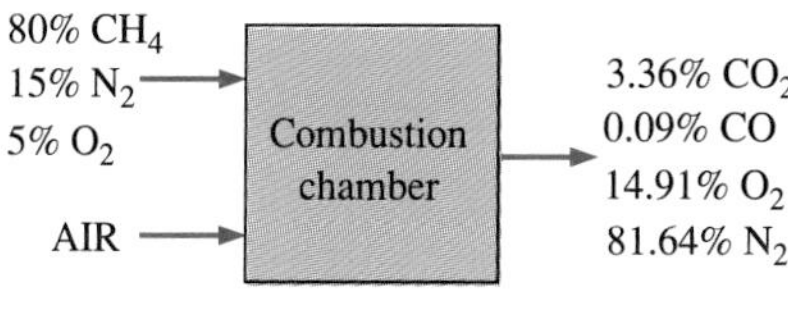

FIGURE P14-88

14-89 A steady-flow combustion chamber is supplied with CO gas at 37°C and 110 kPa at a rate of 0.4 m³/min and air at 25°C and 110 kPa at a rate of 1.5 kg/min. The combustion products leave the combustion chamber at 900 K. Assuming combustion is complete, determine the rate of heat transfer from the combustion chamber.

14-90 Methane gas (CH_4) at 25°C is burned steadily with dry air that enters the combustion chamber at 17°C. The volumetric analysis of the products on a dry basis is 5.20 percent CO_2, 0.33 percent CO, 11.24 percent O_2, and 83.23 percent N_2. Determine (*a*) the percentage of theoretical air used and (*b*) the heat transfer from the combustion chamber per kmol of CH_4 if the combustion products leave at 700 K.

14-91 An 8-m^3 rigid tank initially contains a mixture of 1 kmol of hydrogen (H_2) gas and the stoichiometric amount of air at 25°C. The contents of the tank are ignited, and all the hydrogen in the fuel burns to H_2O. If the combustion products are cooled to 25°C, determine (*a*) the fraction of the H_2O that condenses and (*b*) the heat transfer from the combustion chamber during this process.

14-92 Propane gas (C_3H_8) enters a steady-flow combustion chamber at 1 atm and 25°C and is burned with air that enters the combustion chamber at the same state. Disregarding any changes in kinetic and potential energies, determine the adiabatic flame temperature for (*a*) complete combustion with 100 percent theoretical air, (*b*) complete combustion with 300 percent theoretical air, and (*c*) incomplete combustion (some CO in the products) with 95 percent theoretical air.

14-93 Determine the highest possible temperature that can be obtained when liquid gasoline (assumed C_8H_{18}) at 25°C is burned steadily with air at 25°C and 1 atm. What would your answer be if pure oxygen at 25°C were used to burn the fuel instead of air?

14-94E Determine the work potential of 1 lbmol of diesel fuel ($C_{12}H_{26}$) at 77°F and 1 atm in an environment at the same state.
Answer: 3,315,224 Btu

14-95 Liquid octane (C_8H_{18}) enters a steady-flow combustion chamber at 25°C and 8 atm at a rate of 1.2 kg/min. It is burned with 200 percent excess air that is compressed and preheated to 500 K and 8 atm before entering the combustion chamber. After combustion, the products enter an adiabatic turbine at 1300 K and 8 atm and leave at 950 K and 2 atm. Assuming complete combustion and T_0 = 25°C, determine (*a*) the heat transfer rate from the combustion chamber, (*b*) the power output of the turbine, and (*c*) the reversible work and irreversibility for the entire process.
Answers: (*a*) 1154 kJ/min; (*b*) 486.6 kW; (*c*) 863.7 kW, 377.1 kW

14-96 The combustion of a fuel usually results in an increase in pressure when the volume is held constant, or an increase in volume when the pressure is held constant, because of the increase in the number of moles and the temperature. The increase in pressure or volume will be maximum when the combustion is complete and when it occurs adiabatically with the theoretical amount of air.

Consider the combustion of methyl alcohol vapor $CH_3OH(g)$ with the stoichiometric amount of air in an 0.8-L combustion chamber. Initially, the mixture is at 25°C and 98 kPa. Determine (*a*) the maximum pressure that can occur in the combustion chamber if the combustion takes place at constant volume and (*b*) the maximum volume of the combustion chamber if the combustion occurs at constant pressure.

14-97 Repeat Prob. 14-96 using methane $CH_4(g)$ as the fuel instead of methyl alcohol.

14-98 The furnace of a particular power plant can be considered to consist of two chambers: an adiabatic combustion chamber where the fuel is burned completely and adiabatically, and a heat exchanger where heat is transferred to a Carnot heat engine isothermally. The combustion gases in the heat exchanger are well-mixed so that the heat exchanger is at a uniform temperature at all times that is equal to the temperature of the exiting product gases, T_p. The work output of the Carnot heat engine can be expressed as

$$W = Q\eta_C = Q\left(1 - \frac{T_0}{T_p}\right)$$

where Q is the magnitude of the heat transfer to the heat engine and T_0 is the temperature of the environment. The work output of the Carnot engine will be zero either when $T_p = T_{af}$ (which means the product gases will enter and exit the heat exchanger at the adiabatic flame temperature T_{af}, and thus $Q = 0$) or when $T_p = T_0$ (which means the temperature of the product gases in the heat exchanger will be T_0, and thus $\eta_C = 0$), and will reach a maximum somewhere in between. Treating the combustion products as ideal gases with constant specific heats and assuming no change in their composition in the heat exchanger, show that the work output of the Carnot heat engine will be maximum when

$$T_p = \sqrt{T_{af}T_0}$$

Also, show that the maximum work output of the Carnot engine in this case becomes

$$W_{max} = CT_{af}\left(1 - \sqrt{\frac{T_0}{T_{af}}}\right)^2$$

where C is a constant whose value depends on the composition of the product gases and their specific heats.

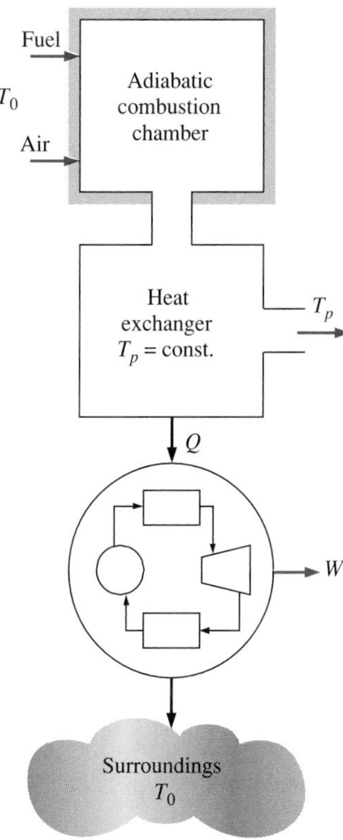

FIGURE P14-98

14-99 The furnace of a particular power plant can be considered to consist of two chambers: an adiabatic combustion chamber where the fuel is burned completely and adiabatically and a counterflow heat exchanger where heat is transferred to a reversible heat engine. The mass flow rate of the working fluid of the heat engine is such that the working fluid is heated from T_0 (the temperature of the environment) to T_{af} (the adiabatic flame temperature) while the combustion products are cooled from T_{af} to T_0. Treating the combustion products as ideal gases with constant specific heats and assuming no change in their composition in the heat exchanger, show that the work output of this reversible heat engine is

$$W = CT_0\left(\frac{T_{af}}{T_0} - 1 - \ln\frac{T_{af}}{T_0}\right)$$

where C is a constant whose value depends on the composition of the product gases and their specific heats.

Also, show that the *effective flame temperature* T_e of this furnace is

$$T_e = \frac{T_{af} - T_0}{\ln (T_{af}/T_0)}$$

That is, the work output of the reversible engine would be the same if the furnace above is considered to be an isothermal furnace at a constant temperature T_e.

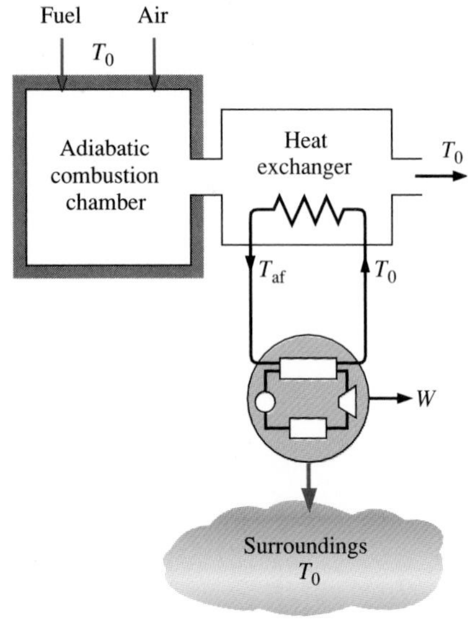

FIGURE P14-99

Computer, Design, and Essay Problems

14-100 Write a computer program to determine the effect of the amount of air on the adiabatic combustion temperature of liquid octane (C_8H_{18}). Assume both the air and the octane are initially at 25°C. Determine the adiabatic combustion temperature for 75, 90, 100, 120, 150, 200, 300, 500, and 800 percent theoretical air. Assume the hydrogen in the fuel always burns to H_2O and the carbon to CO_2, except when there is a deficiency of air. In the latter case, assume that part of the carbon forms CO.

14-101 Write a computer program for determining the heat transfer during the complete combustion of a hydrocarbon fuel C_nH_m in a steady-flow combustion chamber when the percent of excess air and the temperatures of the reactants and the products are specified. Use the functional form of the specific heats in Table A-2*c*, and assume the fuel to enter the combustion chamber at 25°C. How can you use this program to determine the adiabatic flame temperature of a fuel?

14-102 Design a combustion process suitable for use in a gas-turbine engine. Discuss possible fuel selections for the several applications of the engine.

14-103 Constant-volume vessels that contain flammable mixtures of hydrocarbon vapors and air at low pressures are frequently used. Although the ignition of such mixtures is very unlikely as there is no source of ignition in the tank, the Safety and Design Codes require that the tank withstand four times the pressure that may occur should an explosion take place in the tank. For operating gage pressures under 25 kPa, determine the pressure for which these vessels must be designed in order to meet the requirements of the codes for (*a*) acetylene $C_2H_2(g)$, (*b*) propane $C_3H_8(g)$, and (*c*) *n*-octane $C_8H_{18}(g)$. Justify any assumptions that you make.

14-104 The safe disposal of hazardous waste material is a major environmental concern for industrialized societies and creates challenging problems for engineers. The disposal methods commonly used include landfilling, burying in the ground, recycling, and incineration or burning. Incineration is frequently used as a practical means for the disposal of combustible waste such as organic materials. The EPA regulations require that the waste material be burned almost completely above a specified temperature without polluting the environment. Maintaining the temperature above a certain level, typically about 1100°C, necessitates the use of a fuel when the combustion of the waste material alone is not sufficient to obtain the minimum specified temperature.

A certain industrial process generates a liquid solution of ethanol and water as the waste product at a rate of 10 kg/s. The mass fraction of ethanol in the solution is 0.2. This solution is to be burned using methane CH_4 in a steady-flow combustion chamber. Propose a combustion process that will accomplish this task with a minimal amount of methane. State your assumptions.

14-105 Obtain the following information about a power plant that is closest to your town: the net power output; the type and amount of fuel; the power consumed by the pumps, fans, and other auxiliary equipment; stack gas losses; and the rate of heat rejection at the condenser. Using these data, determine the rate of heat loss from the pipes and other components, and calculate the thermal efficiency of the plant.

14-106 What is oxygenated fuel? How would the heating value of oxygenated fuels compare to those of comparable hydrocarbon fuels on a unit-mass basis? Why is the use of oxygenated fuels mandated in some major cities in winter months?

14-107 A promising method of power generation by direct energy conversion is through the use of magnetohydrodynamic (MHD) generators. Write an essay on the current status of MHD generators. Explain their operation principles and how they differ from conventional power plants. Discuss the problems that need to be overcome before MHD generators can become economical.

Software Problems

14-108 Using the enclosed software, determine the maximum adiabatic flame temperature for various fuels listed in the software. Assume the reactants to enter the combustion chamber at 25°C.

14-109 Using the enclosed software, determine the minimum percent of excess air that needs to be used for various fuels listed in the software if the adiabatic flame temperature is not to exceed 1500 K. Assume the reactants enter the combustion chamber at 25°C.

14-110 Repeat Prob. 14-109 for adiabatic flame temperatures of (*a*) 1200 K, (*b*) 1750 K, and (*c*) 2000 K.

14-111 Using the enclosed software, determine the adiabatic flame temperature of CH_4 when both the fuel and the air enter the combustion chamber at 25°C for the cases of 0, 20, 40, 60, 80, 100, 200, 500, and 1000 percent excess air.

14-112 Repeat Prob. 14-111 for (*a*) C_2H_4, (*b*) C_3H_8, and (*c*) C_8H_{18}.

14-113 Using the enclosed software, determine the rate of heat transfer for various fuels listed in the software when they are burned completely in a steady-flow combustion chamber with the theoretical amount of air. Assume the reactants enter the combustion chamber at 298 K and the products leave at 1200 K.

14-114 Repeat Prob. 14-113 for (*a*) 50, (*b*) 100, and (*c*) 200 percent excess air.

14-115 Using the enclosed software, determine the fuel among the listed ones that gives the highest temperature when burned in a constant-volume chamber with the theoretical amount of air. Assume the reactants are at the standard reference state.

Chemical and Phase Equilibrium

CHAPTER 15

In Chap. 14 we analyzed combustion processes under the assumption that combustion is complete when there is sufficient time and oxygen. Often this is not the case, however. A chemical reaction may reach a state of equilibrium before reaching completion even when there is sufficient time and oxygen.

A system is said to be in *equilibrium* if no changes occur within the system when it is isolated from its surroundings. An isolated system is in *mechanical equilibrium* if no changes occur in pressure, in *thermal equilibrium* if no changes occur in temperature, in *phase equilibrium* if no transformations occur from one phase to another, and in *chemical equilibrium* if no changes occur in the chemical composition of the system. The conditions of mechanical and thermal equilibrium are straightforward, but the conditions of chemical and phase equilibrium can be rather involved.

The equilibrium criterion for reacting systems is based on the second law of thermodynamics; more specifically, the increase of entropy principle. For adiabatic systems, chemical equilibrium is established when the entropy of the reacting system reaches a maximum. Most reacting systems encountered in practice are not adiabatic, however. Therefore, we need to develop an equilibrium criterion applicable to any reacting system.

In this chapter, we develop a general criterion for chemical equilibrium and apply it to reacting ideal-gas mixtures. We then extend the analysis to simultaneous reactions. Finally, we discuss phase equilibrium for nonreacting systems.

15-1 ■ CRITERION FOR CHEMICAL EQUILIBRIUM

Consider a reaction chamber that contains a mixture of CO, O_2, and CO_2 at a specified temperature and pressure. Let us try to predict what will happen in this chamber (Fig. 15-1). Probably the first thing that comes to mind is a chemical reaction between CO and O_2 to form more CO_2:

$$CO + \tfrac{1}{2}O_2 \longrightarrow CO_2$$

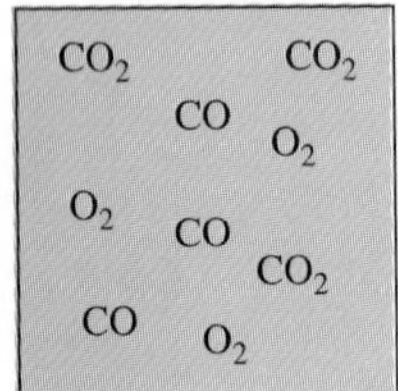

FIGURE 15-1

A reaction chamber that contains a mixture of CO_2, CO, and O_2 at a specified temperature and pressure.

This reaction is certainly a possibility, but it is not the only possibility. It is also possible that some CO_2 in the combustion chamber dissociated into CO and O_2. Yet a third possibility would be to have no reactions among the three components at all, that is, for the system to be in chemical equilibrium. It appears that although we know the temperature, pressure, and composition (thus the state) of the system, we are unable to predict whether the system is in chemical equilibrium. In this chapter we develop the necessary tools to fix that.

Assume that the CO, O_2, and CO_2 mixture mentioned above is in chemical equilibrium at the specified temperature and pressure. The chemical composition of this mixture will not change unless the temperature or the pressure of the mixture is changed. That is, a reacting mixture, in general, will have different equilibrium compositions at different pressures and temperatures. Therefore, when developing a general criterion for chemical equilibrium, we consider a reacting system at a fixed temperature and pressure.

Taking the positive direction of heat transfer to be to the system, the increase of entropy principle for a reacting or nonreacting system was expressed in Chap. 6 as

$$dS_{sys} \geq \frac{\delta Q}{T} \tag{15-1}$$

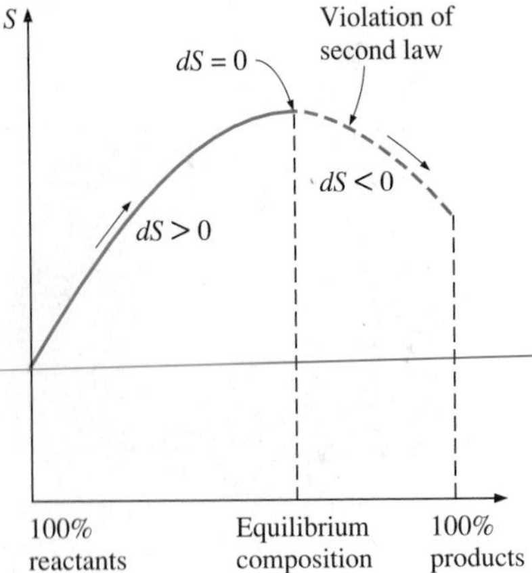

FIGURE 15-2

Equilibrium criteria for a chemical reaction that takes place adiabatically.

For adiabatic systems it reduces to $dS_{sys} \geq 0$. That is, a chemical reaction in an adiabatic chamber proceeds in the direction of increasing entropy. When the entropy reaches a maximum, the reaction stops (Fig. 15-2). Therefore, entropy is a very useful property in the analysis of reacting adiabatic systems.

When a reacting system involves heat transfer, the increase of entropy principle relation (Eq. 15-1) becomes impractical to use, however, since it requires a knowledge of heat transfer between the system and its surroundings. A more practical approach would be to develop a relation for the equilibrium criterion in terms of the properties of the reacting system only. Such a relation is developed below.

Consider a reacting (or nonreacting) simple compressible system of fixed mass with only quasi-equilibrium work modes at a specified temperature T and pressure P (Fig. 15-3). Combining the first- and the second-law relations for this system gives

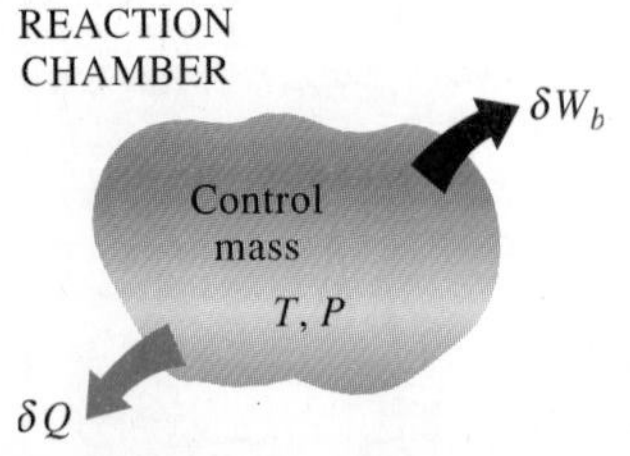

FIGURE 15-3

A control mass undergoing a chemical reaction at a specified temperature and pressure.

$$\left.\begin{aligned} \delta Q - P\,dV &= dU \\ dS &\geq \frac{\delta Q}{T} \end{aligned}\right\} \qquad dU + P\,dV - T\,dS \leq 0 \tag{15-2}$$

Index

D

E

F

G

H

I

Q

R

S

T

U

V

W

Z